EASTERN HORIZON

WESTERN HORIZON

SOUTHERN HORIZON

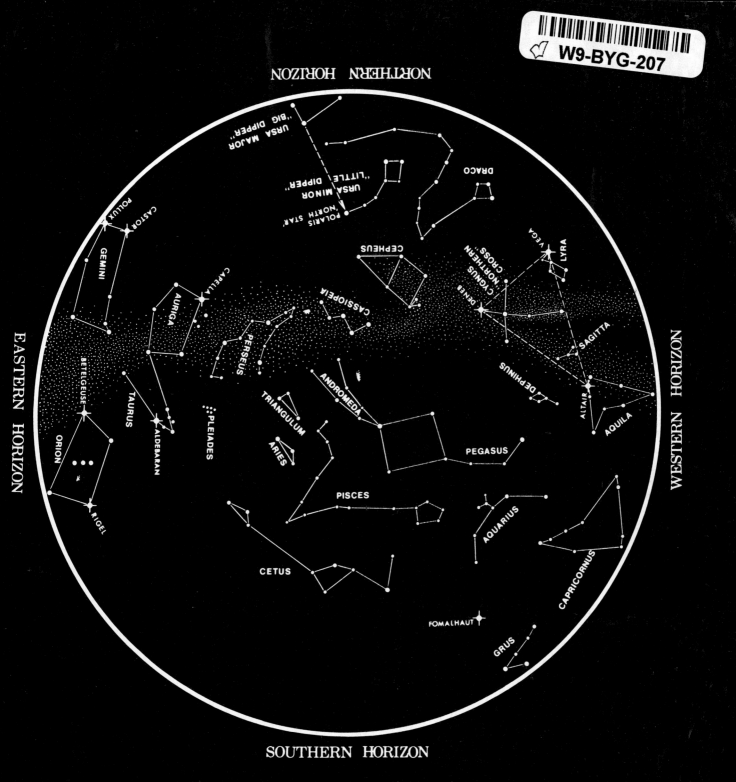

THE NIGHT SKY IN NOVEMBER

Chart time (Local Standard Time):

10 pm . . . First of November
9 pm . . . Middle of November
8 pm . . . Last of November

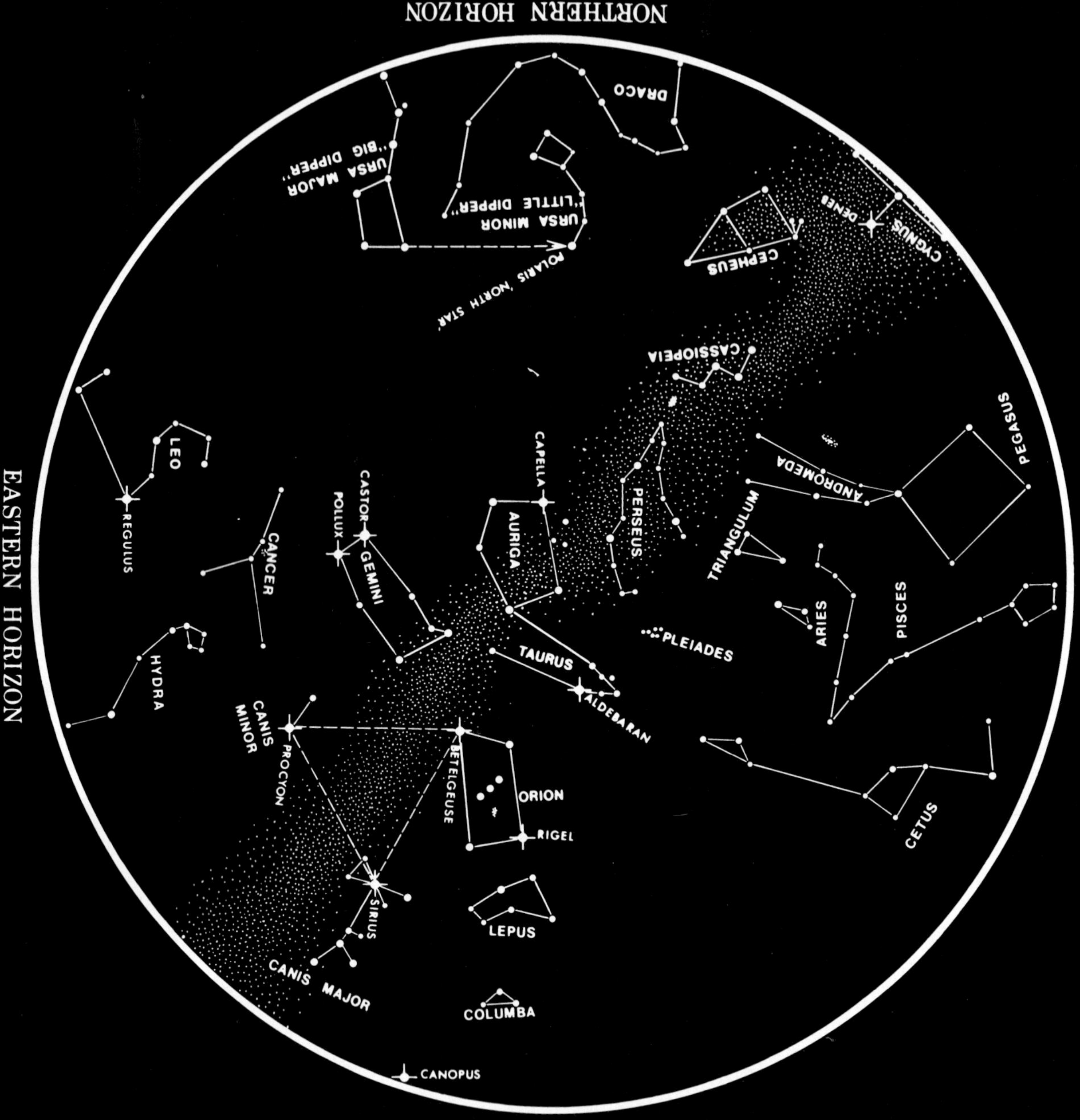

THE NIGHT SKY IN JANUARY

Chart time (Local Standard Time):

10 pm . . . First of January
9 pm . . . Middle of January
8 pm . . . Last of January

DISCOVERING THE UNIVERSE

DISCOVERING THE UNIVERSE

SEVENTH EDITION

Neil F. Comins *University of Maine*

William J. Kaufmann III *San Diego State University*

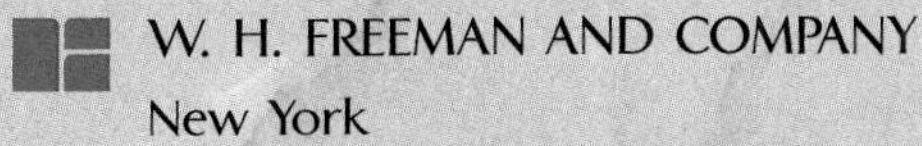

W. H. FREEMAN AND COMPANY
New York

To Horace and Maybelle Rodrigue, caring grandparents and in-laws, and Dorothy Ranta, beloved family matriarch.

Publisher: *Susan Finnemore Brennan*
Acquisitions Editor: *Valerie Raymond*
Marketing Manager: *Scott Guile*
Media Editor: *Victoria Anderson*
Associate Editor: *Amy Shaffer*
Development Editor: *Leslie Carr*
Project Editor: *Bradley Umbaugh*
Cover and Text Designer: *Victoria Tomaselli*
Illustrations: *Fine Line Illustrations*
Illustration Coordinator: *Shawn Churchman*
Photo Editor: *Cecilia Varas*
Photo Researcher: *Tobi Zausner*
Copy Editor and Indexer: *Louise B. Ketz*
Layout: *Marsha Cohen*
Production Manager: *Julia DeRosa*
Composition: *Seven Worldwide*
Manufacturing: *Quebecor World Book Group*

Cover Image: NASA and The Hubble Heritage Team (AURA/STSCI)

ISBN-13: 978-0-7167-6960-6 (ISBN-10: 978-0-7167-6960-3) paperback with CD
ISBN-13: 978-0-7167-6796-1 (ISBN-10: 978-0-7167-6796-1) High School hardback with CD
ISBN-13: 978-0-7167-7584-3 (ISBN-10: 978-0-7167-7584-0) paperback with CD & e-book

Library of Congress Cataloging-in-Publication Control Number: 2004117132

Printed in the United States of America

Third printing

W. H. Freeman and Company
41 Madison Avenue
New York, NY 10010
Houndmills, Basingstoke RG21 6XS, England

www.whfreeman.com

TO CONFINE OUR ATTENTION TO TERRESTRIAL MATTERS WOULD BE TO LIMIT THE HUMAN SPIRIT

—Stephen Hawking

We are in the middle of a major transition in science education. Traditional textbooks and courses in astronomy have focused on supplying the facts that astronomers and astrophysicists have discovered. Although a body of such information is essential to appreciation of the scientific view of the universe, we are also developing and incorporating information that explains *why* scientists accept this information and why scientific discovery is always a work in progress, rarely moving in a straight line. A variety of pedagogic methods incorporated in this text facilitate scientific thinking.

These include

- presenting just enough of the underlying physical *concepts* to make the facts coherent and meaningful
- using both textual and graphical information to present concepts for students who learn in different ways
- helping students to compare their beliefs with the findings of modern science and to understand why the scientific view is correct
- using analogies from everyday life to make cosmic phenomena more concrete
- providing visually rich timelines that connect astronomical discoveries with other events throughout history
- expanding student perspectives by exploring plausible alternative situations (asking "What if ...?" questions)
- linking material presented in the book with more in-depth material offered electronically

I am very pleased to include a wide variety of modern learning techniques and new features in this seventh edition of *Discovering the Universe* while still providing the wide range of factual topics that are the hallmark of this text.

NEW FEATURES BRING THE UNIVERSE INTO CLEARER FOCUS

Everyday Examples Make the Abstract Concrete

- In previous editions, the total mass of meteoritic debris entering the atmosphere per day was just stated. I now show the equivalent mass in terms of an everyday object (an obelisk, in this case).
- To summarize the greenhouse effect, a photo taken inside an automobile shows the temperature gauge with both the exterior and interior temperatures.
- The mass of a primordial black hole is related to the mass of Mount Everest (see the photo below).

FIGURE 14-24 **Mount Everest** Primordial black holes, formed at the beginning of time, may have had masses similar to that of Mount Everest, shown here. Calculations indicate that the universe is old enough for such black holes to have evaporated. Their final particle production rate is so high, it should look as though they are exploding. (© Galen Rowell/Corbis)

• Conservation of angular momentum is illustrated by Olympic skater Sarah Hughes.

• Voids in the universe are illustrated by a slice through a sponge.

• The focusing of jets from the vicinity of black holes by the interstellar medium is shown by analogy to photographs of fluid emerging in air and into water.

• The concept of density is shown in a photograph with different objects of the same mass and different volumes (see the photo below).

R I V U X G

FIGURE 5-12 **The Volumes of Objects with Different Masses** All the objects in this image have the same mass (total number of particles). However, the chemicals from which they form have different densities (number of particles per volume), and so they all take up different amounts of space (volume). (© 2002 Richard Megna/Fundamentals Photographs, NYC)

New Coverage of the Planets and Comparative Planetology

• New information on the discoveries of extrasolar planets in Chapter 5

• New information on Mars in Chapter 7, including the latest images from the Mars rovers

• New information on Saturn from the *Cassini-Huygens* spacecraft in Chapter 8, and on Jupiter and Pluto in Chapter 8

Support for Students

• A new math appendix leads students through the sequence of steps to solve mathematical problems found in the chapters and covers the basics of creating and reading scientific graphs.

• A new appendix provides details about all the constellations.

Information Is Presented "Just in Time"

I have redistributed material from the *Essentials* mini-chapters at the beginning of the four parts of the previous edition back into the chapters where that material is first needed. This change makes the flow of material more seamless.

• Chapter 1 introduces students to the scope of astronomy—the scale of astronomical distances and scientific notation.

• The comprehensibility of the universe and the systematic nature of science are now covered in Chapter 2. Thus readers are introduced early to the idea that the universe is knowable and to the process by which that knowledge evolves, and in this way are prepared for the twists and turns in the history of astronomy and our understanding of the cosmos.

• A new Chapter 5 covers the origins of our solar system (and other systems of planets), preceding discussion of the individual elements of our solar system in the following chapters.

• Apparent magnitude and absolute magnitude are now covered in Chapter 11, where these concepts are used to help draw students into the study of stars.

• The discussion of Cepheid variables and the distances between galaxies is now covered in Chapter 15, as students learn how we know what we know about cosmic distances and the existence of galaxies other than the Milky Way.

SYSTEMATIC PEDAGOGY: FROM WHAT DO YOU THINK? TO WHAT DID YOU THINK?

CHAPTER 5

Matter from Which Planets Form (David Malin/Anglo-Australian Observatory)

WHAT DO YOU THINK?

1 How many stars are there in the solar system?

2 Was the solar system created as a direct result of the formation of the universe?

3 How long has the Earth existed?

4 Is Pluto always the farthest planet from the Sun?

5 What typical shape(s) do moons have?

6 Have any Earthlike planets been discovered orbiting Sunlike stars?

118

WHAT DID YOU THINK?

1 *How many stars are there in the solar system?* The solar system has only one star, the Sun.

2 *Was the solar system created as a direct result of the formation of the universe?* No. All matter and energy were created by the Big Bang. However, much of the material that exists in our solar system was processed inside stars that evolved before the solar system existed. The solar system formed billions of years after the Big Bang occurred.

3 *How long has the* ... along with the ... years ago.

"What Do You Think?" and **"What Did You Think?"** questions in each chapter ask students to consider their present beliefs and actively compare them with the correct science presented in the book. Numbered icons mark the places within the text where each concept is discussed. Encouraging students to think about what they believe is true and then work through the correct science step by step has proved to be an effective teaching technique, especially when time constraints prevent us from working with students individually or in small groups as they try to reconcile incorrect beliefs with proper science. 1

We see visible light from objects even trillions of kilometers (or miles) away with our eyes alone. The discovery that telescopes could make objects in space appear brighter, more detailed, and bigger, coupled with curiosity about the nature of bodies in the sky, led to the realization that we see only the "tip of the cosmic iceberg" with our naked eyes. Telescopes have revolutionized human understanding of the universe. The inventory of the cosmos is far more diverse and interesting than early humans ever imagined, and even today telescopes enable us to discover new things out there. We can think of the energy emitted by bodies in space as the medium of natural cosmic communication, and telescopes as the means by which we gather and read those cosmic messages.

Today, we observe the sky not just from the ground but through telescopes both orbiting the Earth and traveling to other objects in our solar system. Furthermore, we have discovered that visible light is only a tiny fraction of the energy emitted by objects in space.

In this chapter you will discover

- the connection between visible light, radio waves, and other types of electromagnetic radiation
- the debate over what light is and how Einstein resolved it
- how telescopes collect and focus light
- what telescopes can and cannot do
- why different types of telescopes are used for different types of research
- what the new generations of land-based and space-based high-technology telescopes being developed can do
- how astronomers use the entire spectrum of electromagnetic radiation to observe the stars and other astronomical objects and events

THE NATURE OF LIGHT

So far in this text we have used the word "light" in its everyday sense—the stuff to which our eyes are sensitive. This is more properly called "visible light," and it is a form of **electromagnetic radiation**, energy that has properties of both particles, called *photons*, and waves. Detecting electromagnetic radiation with telescopes is the essence of observational astronomy. Although human perception of objects here on Earth and in space comes primarily from the visible light that our eyes detect, this light is only a tiny fraction of all the electromagnetic radiation emitted by objects in the universe. The rest of the electromagnetic radiation is invisible to our eyes, so it has to be observed through specially designed telescopes. We examine telescopes for all types of electromagnetic radiation in this chapter. To understand how telescopes work, we begin studying them by exploring the properties of the electromagnetic radiation they collect.

3-1 Newton discovered that white light is not a fundamental color and debated whether light is composed of particles or waves

From the time of Aristotle until the late seventeenth century, most people believed that white is the fundamental color of light. The colors of the rainbow (or, equivalently, the colors created by light passing through a prism) were believed to be added or created somehow as white light went from one medium through another. Isaac Newton performed experiments during the late 1600s that disproved these beliefs. He started by passing a beam of sunlight through a glass prism, which spread the light out into the colors of the rainbow, as shown in Figure 3-1a. This change in direction as light travels from one medium into another is called **refraction;** and the resulting spread of colors (complete or with colors missing) is called a **spectrum** (plural, **spectra**).

When Newton selected a single color and sent it through a second prism (Figure 3-1b), the light emerging from the first prism was refracted by the second prism, but *it remained the same color.* The fact that colors of refracted light were unchanged by the second prism led Newton to conclude that the colors of a full spectrum (shades of red, orange, yellow, green, blue, and violet) were in fact properties of the light itself, and that white light is a mixture of colors. To prove this last point, he recombined all the spectrum colors, thereby recreating white light. Thus, different colors of the spectrum are different entities. But, Newton wondered, what was the nature of light that it could break apart into distinct colors and then be completely reconstituted?

Back in the mid-1600s, the Dutch scientist Christiaan Huygens proposed that light travels in the form of waves. Newton, on the other hand, performed many experiments in optics that convinced him that light is composed of tiny particles of energy. It turns out that both ideas were right.

In 1801 the English physicist Thomas Young demonstrated that light is indeed composed of waves. Young sent

Chapter opening narratives introduce and launch the chapter's topics and provide students with a context for understanding the material.

Learning objectives underscore the key chapter concepts.

Section headings are written as brief sentences to summarize section content and serve as a quick study guide to the chapter when re-read.

6-6 Visits to the Moon yielded invaluable information about its history

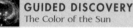

Reaching the Moon became a national obsession for the United States in the early 1960s. A series of successful space missions by the former Soviet Union spurred President John F. Kennedy's goal of landing a person on the Moon by the end of that decade.

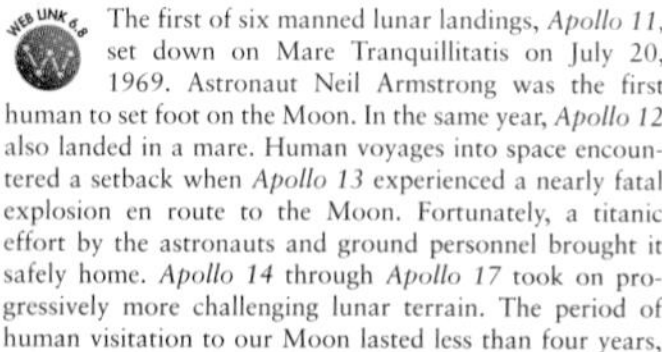

2 The first of six manned lunar landings, *Apollo 11*, set down on Mare Tranquillitatis on July 20, 1969. Astronaut Neil Armstrong was the first human to set foot on the Moon. In the same year, *Apollo 12* also landed in a mare. Human voyages into space encountered a setback when *Apollo 13* experienced a nearly fatal explosion en route to the Moon. Fortunately, a titanic effort by the astronauts and ground personnel brought it safely home. *Apollo 14* through *Apollo 17* took on progressively more challenging lunar terrain. The period of human visitation to our Moon lasted less than four years, culminating in 1972, when *Apollo 17* landed amid rugged mountains just east of Mare Serenitatis (Figure 6-17).

Astronauts discovered that the lunar surface is covered with a layer of fine powder and rock fragments. This layer, ranging in thickness from 1 to 20 m, is called the **regolith**

FIGURE 6-18 The Regolit... a layer of powdered rock that...

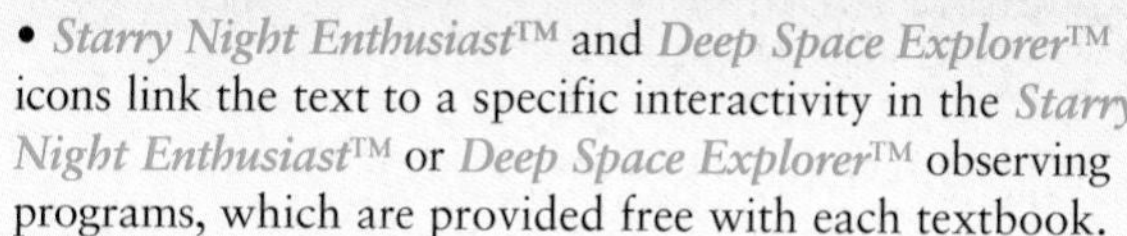

• *Starry Night Enthusiast*™ and *Deep Space Explorer*™ icons link the text to a specific interactivity in the *Starry Night Enthusiast*™ or *Deep Space Explorer*™ observing programs, which are provided free with each textbook.

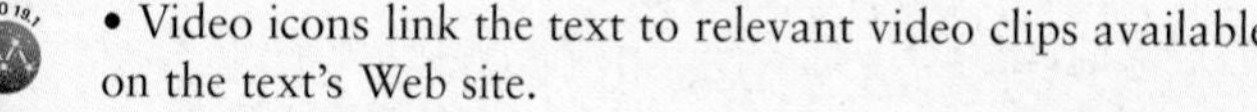

• Video icons link the text to relevant video clips available on the text's Web site.

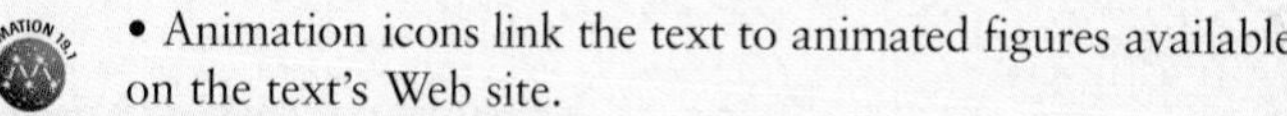

• Animation icons link the text to animated figures available on the text's Web site.

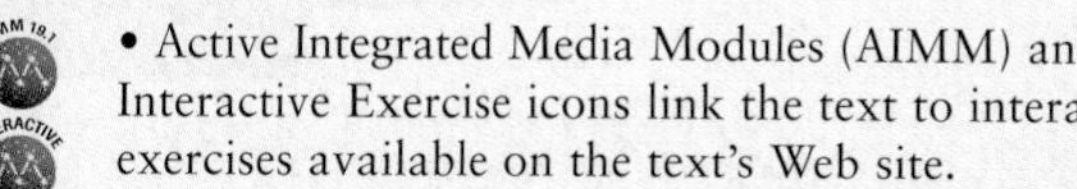

• Active Integrated Media Modules (AIMM) and Interactive Exercise icons link the text to interactive exercises available on the text's Web site.

• Web link icons direct the reader to further information on a particular topic.

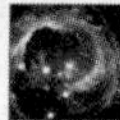

GUIDED DISCOVERY
The Color of the Sun

Different people perceive the Sun to have different colors. To many it appears white, to others yellow. Still others, who notice it at sunset, believe it to be orange or even red. But we have seen that the Sun actually gives off all colors. Moreover, the peak in the Sun's spectrum falls between blue and green. Why doesn't the Sun appear turquoise? Several factors affect our perception of its color.

Before r... through the E... absorbed and... process calle... strongly, follo... low, orange, ... and green pho... than are yello... blue, and gre... Sun's intensity... yellow. (The s... blue light—it ... violet photons...

respond to one of three ranges of colors, which are centered on red, yellow, and blue wavelengths. None of the cones is especially sensitive to blue-green photons. By adding together the color intensities detected by the three types of cones, our brains *re-create* color. After combining all the light it can, the eye is most sensitive to the yellow-green part of the spectrum. We see blue and orange less

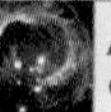

AN ASTRONOMER'S TOOLBOX 1-1
Observational Measurements Using Angles

Ancient mathematicians invented a system of angles and angular measure that is still used today to denote the relative positions and apparent sizes of objects in the sky. To locate stars on the celestial sphere, for example, we do not need to know their distances from Earth. All we need to know is the angle from one star to another in the sky, a property that remains fixed over our lifetimes because the stars are all so far away.

An **arc angle**, often just called an **angle**, is the opening between two lines that meet at a point. Angular measure is a method of describing the size of an angle. The basic unit of angular measure is the **degree**, designated by the symbol °. A full circle is divided into 360°. A right angle measures 90°. As shown in the figure below, the angle between the two "pointer stars" in the Big Dipper is about 5°.

Astronomers also use angular measure to describe the apparent sizes of celestial objects. For example, imagine the full Moon. The angle across the Moon's diameter is

Estimating Angles with the Human Hand Various parts of the adult human hand extended to arm's length can be used to estimate angular distances and sizes in the sky.

• **Guided Discovery** boxes offer hands-on experience with astronomy. Many use the *Starry Night Enthusiast* ™ software.

• Although most of the material in this book is descriptive, some essential equations used in astronomy are introduced. So as not to interrupt the flow of the material, these equations are mostly set off in numbered boxes called **An Astronomer's Toolbox**, which also contain worked examples, additional explanations, and practice doing calculations, with answers at the end of the book. All the equations are summarized in the Appendix.

INSIGHT INTO SCIENCE

Costs and Benefits Science now relies heavily on technology to conduct experiments or make observations. The cost of cutting-edge astronomical observations may run t... or more. The return on... understanding of how t... harness its capabilities, ...

b

R I V U X G

difference. Increasing the exposure time of the smaller diameter telescope **(a)**, will only brighten the image, not improve the resolution. (AURA)

• **Insight into Science** boxes are brief asides within the text that relate topics to the nature of scientific inquiry and encourage critical thinking.

• **Wavelength tabs** show whether the image was made with radio waves (R), infrared radiation (I), visible light (V), ultraviolet light (U), X rays (X), or gamma rays (G). R I V U X G

Summary of Key Ideas

Earth: A Dynamic, Vital World

• The Earth's atmosphere is about four-fifths nitrogen and one-fifth oxygen. This abundance of oxygen is due to the biological processes of life-forms on the planet.

• The Earth's atmosphere is divided into layers named the troposphere, stratosphere, mesosphere, and ionosphere. Ozone molecules in the stratosphere absorb ultraviolet light rays.

• The outermost layer, or crust, of the Earth offers clues to the history of our planet.

• The Earth's surface is divided into huge plates that move over the upper mantle. Movements of these plates, a process called plate tectonics, are caused by convection in the mantle, and upwelling of molten material along cracks in the ocean floor produces seafloor spreading. Plate tectonics is responsible for most of the major features of the Earth's surface, including mountain ranges, volcanoes, and the shapes of the continents and oceans.

• The Moon was molten in its early stages, and the anorthositic crust solidified from low-density magma that floated to the lunar surface. The mare basins were created later by the impact of planetesimals and were then filled with lava from the lunar interior.

• Gravitational and centrifugal interactions between the Earth and the Moon produce tides in the oceans of the Earth and set the Moon in synchronous rotation. The Moon is moving away from the Earth, and consequently, the Earth's rotation rate is decreasing.

WHAT DID YOU THINK?

1 *Can the Earth's ozone layer, which is now being depleted, be naturally replenished?* Yes. Ozone is created continuously from normal oxygen molecules by their interaction with the Sun's ultraviolet radiation.

2 *Who was the first person to walk on the Moon and when did this event occur?* Neil Armstrong was the first person to set foot on the Moon. He and Buzz

For Review

The **Summary of Key Ideas** is a bulleted list of key concepts.

What Did You Think? answers the questions posed at the beginning of the chapter in the What Do You Think? section.

Key Words, Review Questions, and **Advanced Questions** help the reader understand the chapter material.

Key Words

anorthosite, 153
capture theory, 155
cocreation theory, 155
collision-ejection t
continental drift,
convection, 145
core, 144
coronal mass ejec
crust, 141
neap tide, 160
northern lights (aurora borealis), 147

Review Questions

1. Why does the Earth's albedo change daily? seasonally

2. Why is the Earth's surface not riddled with craters as that of the Moon?

3. Describe the structure of the Earth's atmosphere.

4. Describe the process of plate tectonics. Give specific examples of geographic features created by plate tectonics.

Advanced Questions

22. Explain how the outward flow of energy from the Earth's interior drives the process of plate tectonics.

23. Why do some geologists believe that Pangaea was the most recent in a succession of supercontinents?

24. Why are active volcanoes, such as Mount St. Helens, usually located in mountain ranges along the boundaries of tectonic plates?

Discussion Questions

32. If the Earth did not have a magnetic field, do you think auroras would be more common or less common than they are today? Explain.

33. Comment on the idea that without the Moon's presence, life would have developed far more slowly.

34. Identify and compare the advantages and disadvantages of lunar exploration by astronauts as opposed to mobile, unmanned instrument packages and robots.

35. When was the last earthquake near your hometo How far is your hometown from a plate boundary? W kinds of topography (for example, mountains, plains, seashore) dominate the geography of your hometown Does that topography and the frequency of earthquak to be consistent with your hometown's proximity to a boundary?

36. The ice on the Moon is believed to be mixed wit just under the Moon's surface. How might that water economically extracted and purified?

What If . . .

37. The Earth had two moons? Assume one is at our distance and the other is at half that distance. Describ motion of the two moons in the sky and how they mi appear to us. What would be different here on Earth?

38. The Moon orbited the Earth in the opposite direc from the Earth's rotation, rather than in the same direction? What would be different about the Earth and life on it? Assume that today such a counter-revolving Moon is at the same distance as our Moon.

39. Our Moon were one-tenth of its actual distance from the Earth? What would be different about the Earth and life on it? *Hint:* The heights of tides vary as $1/r^3$, where r

Web Questions

43. Use the Web to learn more about plate tectonics. What global changes might accompany the formation and breakup of a supercontinent? How might these changes affect the evolution of life? What life-forms dominated the Earth whe Pangaea existed some 200 million years ago and also wh fragments of the preceding supercontinent were as disp as today's continents?

44. Search the Web for information about "Pangaea Ultima," a supercontinent that may form in the distant future. When is it expected to form? How will it compare to Pangaea of 200 million years ago (see Figure 6-4)?

45. Use the Web to determine the status of the Antarctic and Arctic ozone holes. How has the situation changed over the past few years? Explain why most scientists who study this issue blame chemicals called CFCs for the existence of the

Collaborative Learning

Discussion Questions and **What If ...** questions offer interesting topics to spark lively, insightful debate.

Enrichment

Web Questions take readers to the Web for further study.

Observing Projects

Observing Tips and Tools: You can learn a lot by observing the Moon through binoculars. Note that the Moon will appear right side up through binoculars, but inverted through a telescope. So, if you use a map of the Moon to aid your observations, you will need to take this into account. Inexpensive Moon maps are available, such as "Moon Map" published by *Sky & Telescope* magazine. To determine the lunar phase (when you cannot examine the Moon directly) you can usually find it on most calendars; by checking weather page in your newspaper; by consulting the cu issue of *Sky & Telescope* or *Astronomy* magazine; by usin your *Starry Night Enthusiast*™ program; or on the Web.

50. Use a telescope or binoculars to observe the Moon. Compare the texture of the lunar surface you see on the maria with that of the lunar highlands. How does the visibility of details vary with distance from the terminator (the boundary between day and night on the Moon)? Why?

New Territories

Observing Projects, featuring *Starry Night Enthusiast* ™ and *Deep Space Explorer*™ activities, ask readers to be astronomers themselves.

WHAT IF . . . HUMANS HAD INFRARED-SENSITIVE EYES?

Our eyes are sensitive to less than a trillionth of 1% of the electromagnetic spectrum—what we call "visible light." But this minuscule resource provides an awe-inspiring amount of information about the universe. We interpret visible-light photons as the six colors of the rainbow—red, orange, yellow, green, blue, and violet. These colors combine to form all the others that make our visual world so rich. But the Sun actually emits photons of all wavelengths. So, what would happen if our eyes had evolved to sense another part of the spectrum?

A Darker Vision Gamma rays, X rays, and most ultraviolet radiation do not pass through the Earth's atmosphere. Since the world is illuminated by sunlight, the Earth and sky would look dark, indeed, if our eyes were sensitive only to these wavelengths. Radio waves, in contrast, easily pass through our atmosphere. But to see the same detail from radio waves that we now see from visible-light photons, our eyes would require diameters 10,000 times larger. Each would be the size of a baseball infield!

What about infrared radiation? While not all incoming infrared photons get through our atmosphere, short-wavelength ("near") infrared radiation passes easily through air. Depending on wavelength, the Sun emits between one-half and a ten-billionth as many infrared photons as red-light photons. Fortunately, most of these are in the near infrared.

Heat-Sensitive Vision To see infrared photons, human eyes would need to be only 5 to 10 times larger. Some snakes have evolved infrared vision. Portable infrared "night vision" cameras and goggles are available to us humans. Because everything that emits heat emits infrared photons, infrared sight would be very useful. And not everything we see with infrared sight would be due just to reflected sunlight—hotter objects would be intrinsically brighter than cool ones. For example, seeing infrared would allow us to observe changes in a person's emotional state. Someone who is excited or angry often has more blood near the skin, and thus gives off more infrared radiation (heat) than normal. Conversely, someone who is scared has less blood near the skin and thus emits less heat.

Night Vision The night sky would be a spectacular sight through infrared-sensitive eyes. Gas and dust clouds in the Milky Way absorb visible light, thus preventing the light of distant stars from getting to the Earth. However, because most infrared radiation passes through these clouds, we would be able, unaided, to see distant stars that we cannot see today. On the other hand, the white glow of the Milky Way, which is caused by the scattering of starlight by interstellar clouds, would be dimmer, because the gas and dust clouds do not scatter infrared light as much as they do visible light. (The haze created by the Milky Way would not vanish, however, because when gas and dust clouds are heated by starlight, they emit their own infrared radiation.)

Our concept of stars would be different, too. Many stars, especially young, hot ones, are surrounded by cocoons of gas and dust that emit infrared radiation. This dust is heated by the nearby stars. Instead of appearing as pinpoints, many stars would appear to be surrounded by wild strokes of color, and we would have an Impressionist sky.

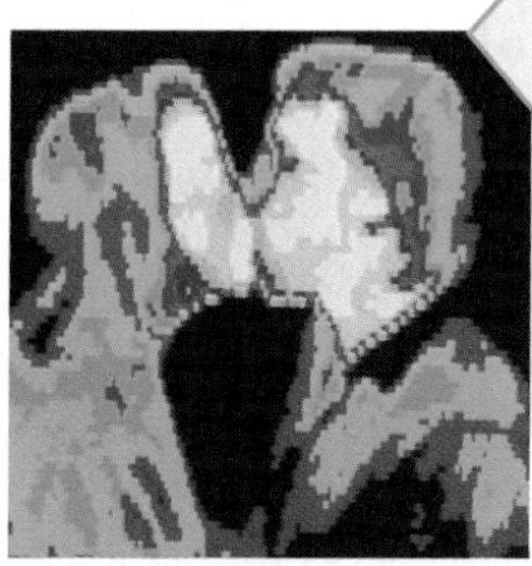

R I V U X G

The infrared (heat) from this kissing couple has been converted into visible light colors so that we can interpret the invisible radiation. The hottest regions are white, with successively cooler areas shown in yellow, orange, red, green, sky blue, dark blue, and violet. (D. Montrose/Custom Medical Stock Photo)

What If . . .

Selected chapters conclude with a **"What If ..."** essay that encourages critical thinking by speculating on how changes in the universe could have profound effects on Earth.

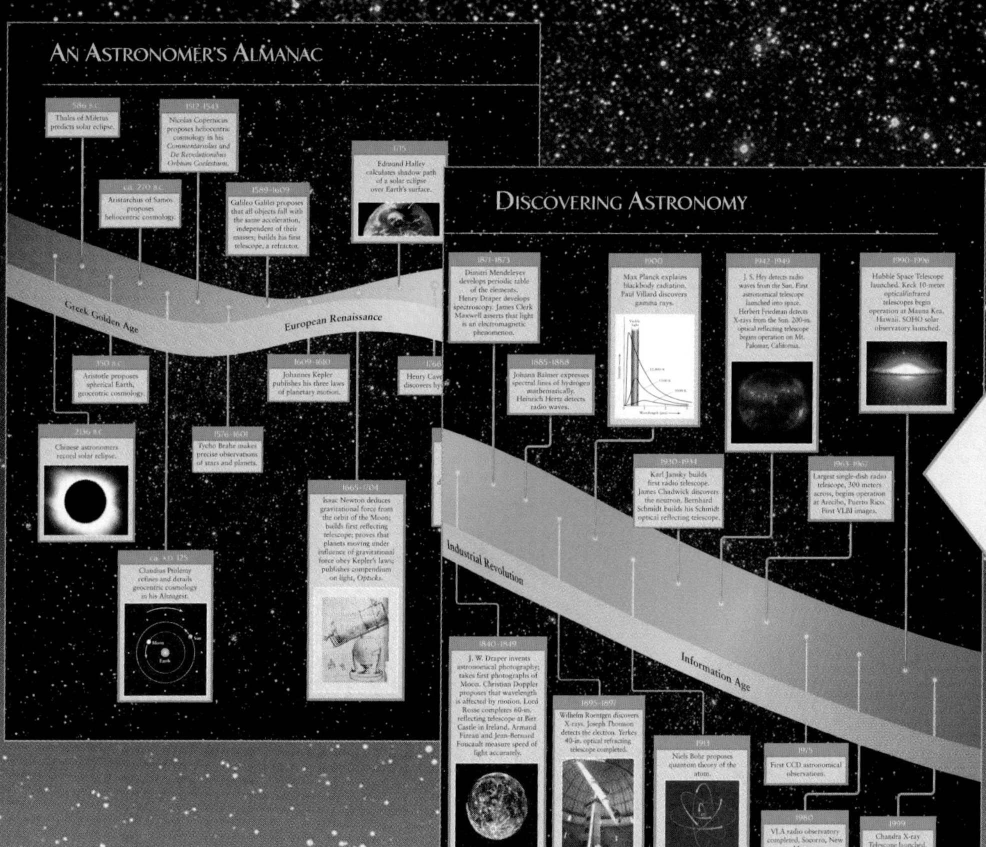

Astronomer's Almanac

Each part opens with **An Astronomer's Almanac,** a dynamic timeline that relates discoveries in astronomy to other historical events, providing strong context for the information presented.

FOR STUDENTS

The NEW! e-book: Affordable, Innovative, and Customizable

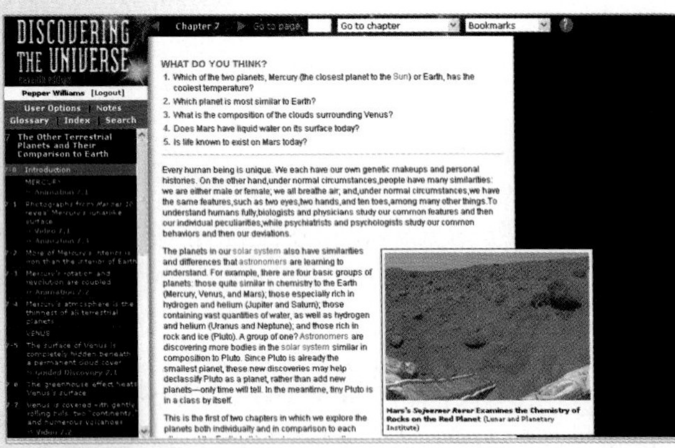

The *Discovering the Universe*, Seventh Edition e-book is a complete online version of the textbook which provides a rich learning experience by taking full advantage of the electronic medium. This online version integrates all the existing media resources and adds features unique to the e-book, such as:

- Easy access from any Internet-connected computer via a standard Web browser
- Quick, intuitive navigation to any section or subsection, as well as any printed book page number
- Integration of all student Web site animated tutorials and activities
- In-text self-quiz questions and links to all glossary entries
- Interactive chapter summary exercises including links to Key Ideas and selected Review, Advanced, and Observing Projects questions
- Text highlighting, down to the level of individual phrases
- A bookmarking feature that allows for quick reference to any page
- A powerful Notes feature that allows students to add notes to any page
- A full glossary and index
- Full-text search, including an option to also search the glossary and index
- Automatic saving of all notes, highlighting, and bookmarks
- The e-book is available FREE with the text (use ISBN: 0-7167-7584-0), or online at www.whfreeman.com/dtu7e

Packaged with this book are two unique software products:

NEW! *Starry Night Enthusiast*™ 4.5 is brilliantly realistic and designed for anyone with an interest in the night sky. See the sky from anywhere on Earth or lift off and visit any solar system body or any location up to 20,000 light-years away. View 2,500,000 stars along with more than 170 deep-space objects like galaxies, star clusters, and nebulae. You can travel 15,000 years in time, check out the view from the International Space Station, and see planets up close from any one of their moons. Included are stunning OpenGL graphics; you can also print handy star charts to explore outside. Learn about the movement of the planets, watch the phases of the moon, and see what the stars looked like on the day you were born!

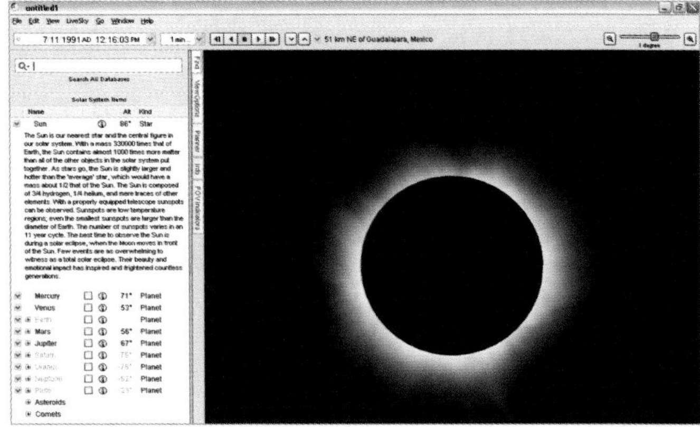

NEW! *Deep Space Explorer*™ developed by Imaginova™ reports the position, type, angular velocity, size, and shape of up to 28,000 of the nearest galaxies. Some 30,000 nearby stars are also yours to explore, including those known to have planets. The program allows you to observe objects from any angle.

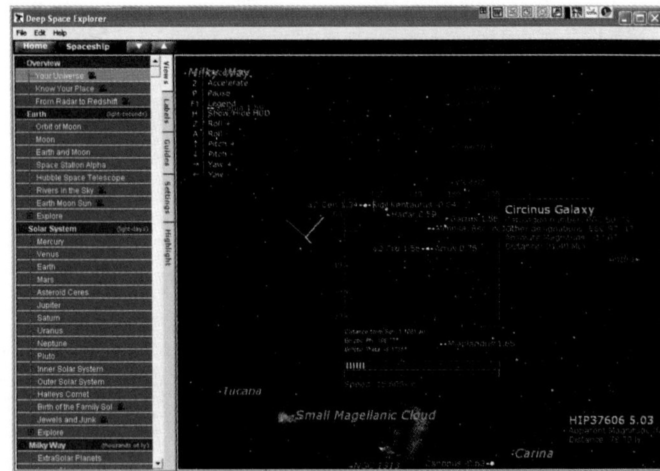

The free Student Companion Web site www.whfreeman.com/dtu7e serves as a study guide and includes these features:

- **Learning Objectives** for each chapter help students formulate study strategies.
- **Self-Quizzes** offer randomized questions and answers with instant feedback referring to specific sections in the text, to help students study, review, and prepare for exams. Instructors can access results through an online database or they can have them e-mailed directly to their accounts.
- **More to Know** expands on key topics in the text.
- **Active Integrated Media Modules (AIMM)** take students deeper into key topics from the text. Topics include:

  Determining Distances to the Stars and Beyond
  Doppler Effect
  Nearest Stars
  Relativistic Redshift

- **Interactive Drag & Drop Exercises** based on text illustrations help students grasp the vocabulary in context.
- **Flashcard** exercises offer help with vocabulary and definitions.
- **Planetary Geology** compares geological features on Earth and other bodies in the solar system.
- **Current Events in Astronomy** (updated every month) keep students up to date with Web links, photos, and information from a variety of resources.
- **Animations and Videos** (updated regularly), both original and NASA-created, are keyed to specific chapters. Updated regularly.
- **Web links** provide a wealth of online resources for the student.
- *Starry Night Enthusiast*™ 4.5 and *Deep Space Explorer*™ Observing Exercises and In-depth lab projects that make use of the student editions of *Starry Night Enthusiast*™ 4.5 and *Deep Space Explorer*™ for an engaging study of the Universe.

FOR INSTRUCTORS

Instructor e-book

The *Discovering the Universe*, Seventh Edition e-book offers instructors flexibility and customization options not previously possible with any printed textbook.

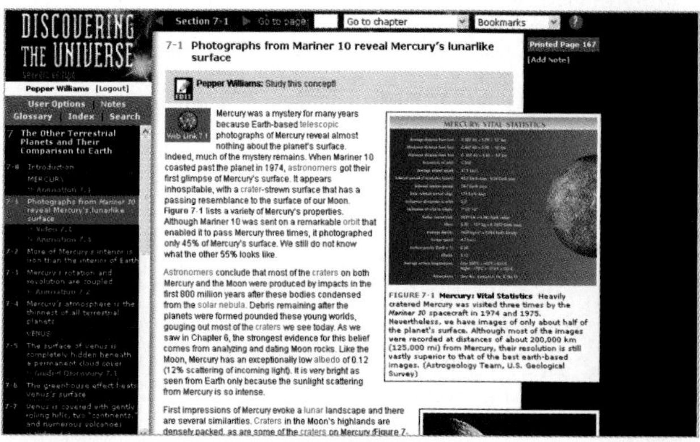

Instructors have access to:

- **E-book Customization:** Instructors can choose the chapters that correspond with their syllabus, and students will get a custom version of the e-book with the selected chapters only.
- **Instructor Notes:** Instructors can choose to create an annotated version of the e-book with their notes on any page. When students in their course log-in, they will see their instructor's notes.
- **Custom Content:** Instructor notes can include text, Web links, and even images, allowing instructors to place any content they choose exactly where they want it.
- **Online Quizzing:** The online quizzes from the student Web site are integrated into the e-book. Students can submit their responses, which are stored in the online grade book.

- **Instructor's Resource CD-ROM, 0-7167-6789-9**
To help instructors create lecture presentations, Web sites, and other resources, this CD-ROM allows instructors to **search** and **export** all these resources by key term or chapter:

  All text images
  Animations, videos, AIMMs, Instructor's Manual, PowerPoint™ and more found on the Web site

- **Instructor's Manual**, by Mark Hollabaugh, *Normandale Community College*. The printable, editable electronic *Instructor's Manual* available in Microsoft Word format includes student learning objectives, answers to end-of-chapter questions, detailed chapter outlines, class discussion and projects, and classroom-tested teaching hints and strategies. The extensive resource guide covers student and instructor reading materials and audiovisual material.
- **PowerPoint™ Lecture Presentation,** by Eric Collins, *University of California, Northridge*. A set of online lecture presentations created in PowerPoint™ allows instructors to tailor their lectures to suit their own needs

using images and notes from the textbook. Available for download from the Web site.

- **Test Bank CD-ROM, 0-7167-6670-1** (Windows, Macintosh, WebCT, and Blackboard formats on disc), by Thomas Krause, *Towson University,* and T. Alan Clark and William J. F. Wilson, *University of Calgary.* More than 3,500 multiple-choice questions are section-referenced. The easy-to-use CD-ROM version includes Windows, Mac, WebCT, and Blackboard versions on a single disc, in a format that lets you add, edit, resequence, and print questions to suit your needs.

- **Diploma Online Testing,** from the Brownstone Research Group (accessible via the test bank CD-ROM). With Diploma, instructors can easily create and administer secure exams over a network and over the Internet, with questions that incorporate multimedia and interactive exercises. The program lets you restrict tests to specific computers or time blocks, and includes an impressive suite of gradebook and result-analysis features.

- **Astronomy Online** booklet with Web site activation code, 0-7167-9669-4, by Timothy F. Slater, *University of Arizona.* Available in HTML and WebCT formats, this site can serve as either the online component for a lecture-based course or a complete online source for a distance learning course. The site combines content from W. H. Freeman and Company astronomy textbooks with powerful features such as online quizzing, animations and simulations, drag-and-drop exercises, active integrated media modules, Web links, and much more. The course can be used as is or tailored to meet the instructor's individual needs. The HTML version can be accessed via: **www.whfreeman.com/aol**. The WebCT version can be ordered via the WebCT content showcase at www.webct.com.

- **Online Course Materials** (WebCT, Blackboard) As a service for adopters, we will provide content files in the appropriate online course format, including the instructor and student resources for this text. The files can be used as is or can be customized to fit specific needs. Course outlines, prebuilt quizzes, links, activities, and a whole array of materials are included.

- **Online Quizzing,** powered by ***Questionmark***™ (accessible via **www.whfreeman.com/dtu7e**) With *Questionmark*™ Perception, instructors can easily and securely quiz students online using prewritten, multiple-choice questions for each text chapter (not from the test bank). Students receive instant feedback and can take the quizzes multiple times. Randomized questioning creates a different set of chapter questions each time a student takes a quiz, with answers reshuffled. Instructors can go into a protected Web site to view results by quiz, student, or question or can get weekly results via e-mail.

- **Overhead Transparencies, 0-7167-6251-X** 100 full-color transparencies of key illustrations, photos, and tables from the text.

- ***The Once and Future Cosmos, Scientific American Special Edition,*** **2002, 0-7167-5708-7** This new *Scientific American Special Edition* takes your students to the frontiers of cosmological inquiry, clarifying the new developments and what they mean.

- ***New Light on the Solar System, Scientific American Special Edition,*** **2003, 0-7167-8603-6** This special edition of *Scientific American* provides the latest developments about our corner of the cosmos, in articles written by the experts who are leading the investigations.

- **Observing Projects using** *Starry Night Enthusiast*™ 4.5, **0-7167-6415-6,** by T. Alan Clark, and William J. F. Wilson, *University of Calgary,* and Marcel Bergman. Available for packaging with the text, and compatible with both Mac and PC, this book contains a variety of comprehensive lab activities for *Starry Night Enthusiast*™4.5. Selected activities will be available online.

ACKNOWLEDGMENTS

I am deeply grateful to the astronomers and teachers who reviewed chapters for this and previous editions.

Gordon Baird, *University of Mississippi*
Henry E. Bass, *University of Mississippi*
J. David Batchelor, *Community College of Southern Nevada*
Jill Bechtold, *University of Arizona*
Peter A. Becker, *George Mason University*
Michael Bennett, *DeAnza College*
John Bieging, *University of Arizona*
Greg Black, *University of Virginia*
John W. Burns, *Mt. San Antonio College*
Gene Byrd, *University of Alabama*
Eugene R. Capriotti, *Michigan State University*
Michael W. Castelaz, *Pisgah Astronomical Research Institute*
Gerald Cecil, *University of North Carolina*
David S. Chandler, *Porterville College*
David Chernoff, *Cornell University*
Tom Christensen, *University of Colorado, Colorado Springs*
Chris Clemens, *University of North Carolina*
Christine Clement, *University of Toronto*
Halden Cohn, *Indiana University*
John Cowan, *University of Oklahoma*
Antoinette Cowie, *University of Hawaii*
Charles Curry, *University of Waterloo*
James J. D'Amario, *Harford Community College*
Purnas Das, *Purdue University*

Peter Dawson, *Trent University*
John M. Dickey, *University of Minnesota, Twin Cities*
John D. Eggert, *Daytona Beach Community College*
Bernd Enders, *College of Marin*
Robert Frostick, *West Virginia State College*
Martin Gaskell, *University of Nebraska*
Bruce Gronich, *University of Texas-El Paso*
Siegbert Hagmann, *Kansas State University*
David Hedin, *Northern Illinois University*
Chuck Higgins, *Penn State University*
James L. Hunt, *University of Guelph*
Nathan Israeloff, *Northeastern College*
Kenneth Janes, *Boston University*
William C. Keel, *University of Alabama*
William Keller, *St. Petersburg Junior College*
Marvin D. Kemple, *Indiana University–Purdue University Indianapolis (IUPUI)*
Pushpa Khare, *University of Illinois at Chicago*
F. W. Kleinhaus, *Indiana University–Purdue University Indianapolis (IUPUI)*
Rob Klinger, *Parkland College*
Patrick M. Len, *Cuesta College*
John Patrick Lestrade, *Mississippi State University*
C. L. Littler, *University of North Texas*
M. A. K. Lohdi, *Texas Tech University*
Michael C. LoPresto, *Henry Ford Community College*
Phyllis Lugger, *Indiana University*
R. M. MacQueen, *Rhodes College*
Robert Manning, *Davidson College*
Paul Mason, *University of Texas-El Paso*
P. L. Matheson, *Salt Lake Community College*
Rahul Mehta, *University of Central Arkansas*
J. Scott Miller, *University of Louisville*
L. D. Mitchell, *Cambria County Area Community College*
J. Ward Moody, *Brigham Young University*
Siobahn M. Morgan, *University of Northern Iowa*
Steven Mutz, *Scottsdale Community College*
Gerald H. Newson, *Ohio State University*
Bob O'Connell, *College of the Redwoods*
William C. Oelfke, *Valencia Community College*
Richard P. Olenick, *University of Dallas*
John P. Oliver, *University of Florida*
Melvyn Jay Oremland, *Pace University*
David Patton, *Trent University*
Jon Pedicino, *College of the Redwoods*
Sidney Perkowitz, *Emory University*
David D. Reid, *Wayne State University*
Adam W. Rengstorf, *Indiana University*
James A. Roberts, *University of North Texas*
Henry Robinson, *Montgomery College*
Dwight P. Russell, *University of Texas in El Paso*
Barbara Ryden, *Ohio State University*
Larry Sessions, *Metropolitan State College*
C. Ian Short, *Florida Atlantic University*
John D. Silva, *University of Massachusetts at Dartmouth*
Michael L. Sitko, *University of Cincinnati*
Earl F. Skelton, *George Washington University*
George F. Smoot, *University of California at Berkeley*
Alex G. Smith, *University of Florida*
David Sturm, *University of Maine, Orono*
Paula Szkody, *University of Washington*
Michael T. Vaughan, *Northeastern University*
Andreas Veh, *Kenai Peninsula College*
John Wallin, *George Mason University*
William F. Welsh, *San Diego State University*
R. M. Williamon, *Emory University*
Edward L. (Ned) Wright, *University of California at Los Angeles*
Jeff S. Wright, *Elon College*
Nicolle E. B. Zellner, *Rensselaer Polytechnic Institute*

I would like to add my special thanks to the wonderfully supportive staff at W. H. Freeman and Company who make the revision process so enjoyable. Among others, these include Valerie Raymond, Leslie Carr, Brad Umbaugh, Vicki Tomaselli, Cecilia Varas, and Erik Gilg. Thanks also go to my copy editor and indexer, Louise B. Ketz, as well as my proofreaders, Michele Kornegay and Barbara Hults. Warm thanks also to David Sturm, University of Maine, Orono, for his help in collecting current data on objects in the solar system; to my UM astronomy colleague David Batusk; and to my wife, Sue, and my sons, James and Josh, for their patience and support while preparing this book.

Neil F. Comins
neil.comins@umit.maine.edu

ABOUT THE AUTHORS

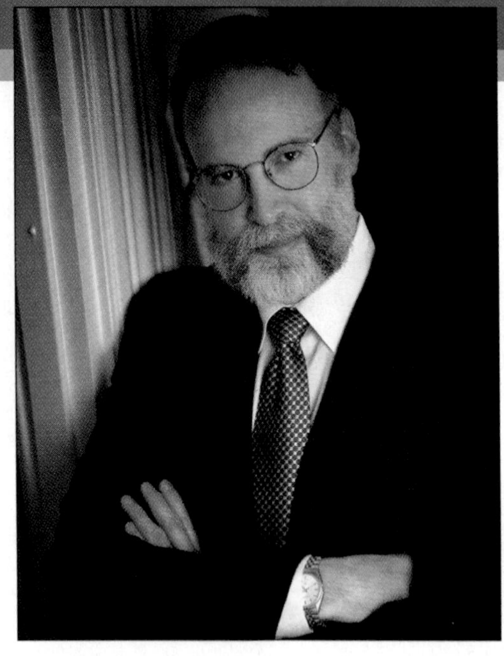

Professor **Neil F. Comins** is on the faculty of the University of Maine. Born in 1951 in New York City, he grew up in New York and New England. He earned a bachelor's degree in engineering physics at Cornell University, a master's degree in physics at the University of Maryland, and a Ph.D. in astrophysics from University College, Cardiff, Wales, under the guidance of Bernard F. Schutz. Dr. Comins's work for his doctorate, on general relativity, was cited in Subramanyan Chandrasekhar's Nobel laureate speech. He has done theoretical and experimental research in general relativity, observational astronomy, computer simulations of galaxy evolution, and science education. The fourth edition of *Discovering the Universe* was the first book in this series that Dr. Comins wrote, having taken over following the death of Bill Kaufmann in 1994. He is also the author of two trade books, *What If the Moon Didn't Exist?* and *Heavenly Errors*. *What If the Moon Didn't Exist?* has been made into planetarium shows, been excerpted for television and radio, and translated into several languages and is the theme for the Mitsubishi Pavilion at the World Expo 2005 in Aichi, Japan. *Heavenly Errors* explores misconceptions people have about astronomy, why such misconceptions are so common, and how to correct them. Dr. Comins has appeared on numerous television and radio shows and gives many public talks. Although he has jumped out of airplanes while in the military, today his activities are a little more sedate: He is a licensed pilot and avid sailor, having once competed against Prince Philip, Duke of Edinburgh.

William J. Kaufmann III was the author of the first three editions of *Discovering the Universe*. Born in New York City on December 27, 1942, he often visited the magnificent Hayden Planetarium as he was growing up. Dr. Kaufmann earned his bachelor's degree magna cum laude in physics from Adelphi University in 1963, a master's degree in physics from Rutgers in 1965, and a Ph.D. in astrophysics from Indiana University in 1968. At 27 he became the youngest director of any major observatory in the United States when he took the helm of the Griffith Observatory in Los Angeles. During his career he also held positions at San Diego State University, UCLA, Caltech, and the University of Illinois. Throughout his professional life as a scientist and educator, Dr. Kaufmann worked to bridge the gap between the scientific community and the general public to help the public share in the advances of astronomy. A prolific author, his many books include *Black Holes and Warped Spacetime, Relativity and Cosmology, The Cosmic Frontiers of General Relativity, Exploration of the Solar System, Planets and Moons, Stars and Nebulas, Galaxies and Quasars,* and *Supercomputing and the Transformation of Science*. Dr. Kaufmann died in 1994.

UNDERSTANDING ASTRONOMY

PART I

Telescopes offer a window on the cosmos.
(Frank Zullo/Photo Researchers, Inc.)

An Astronomer's Almanac

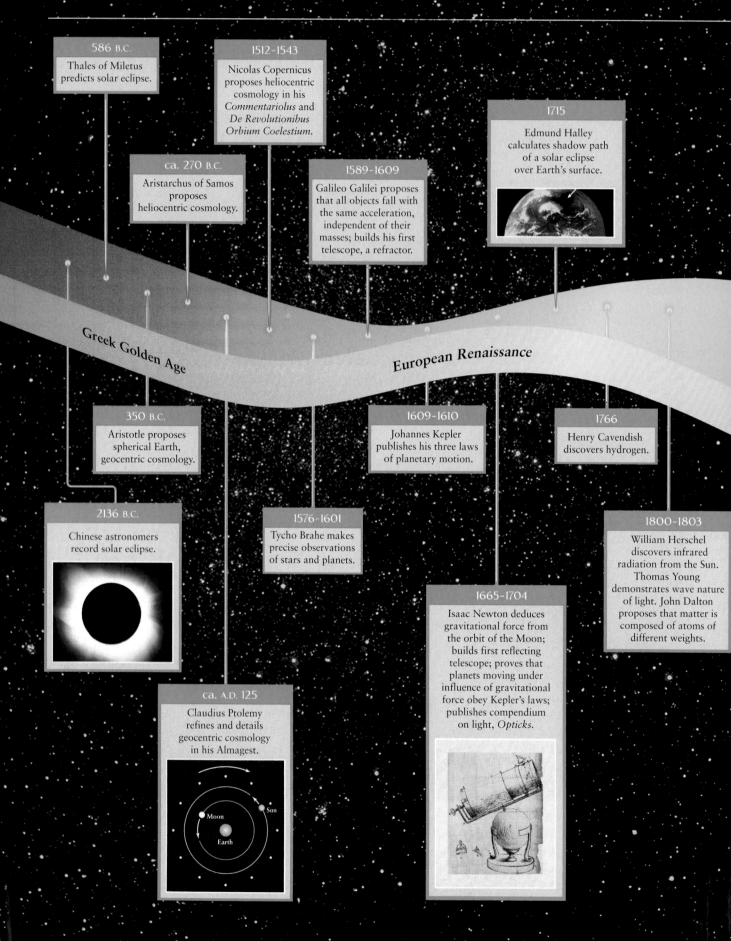

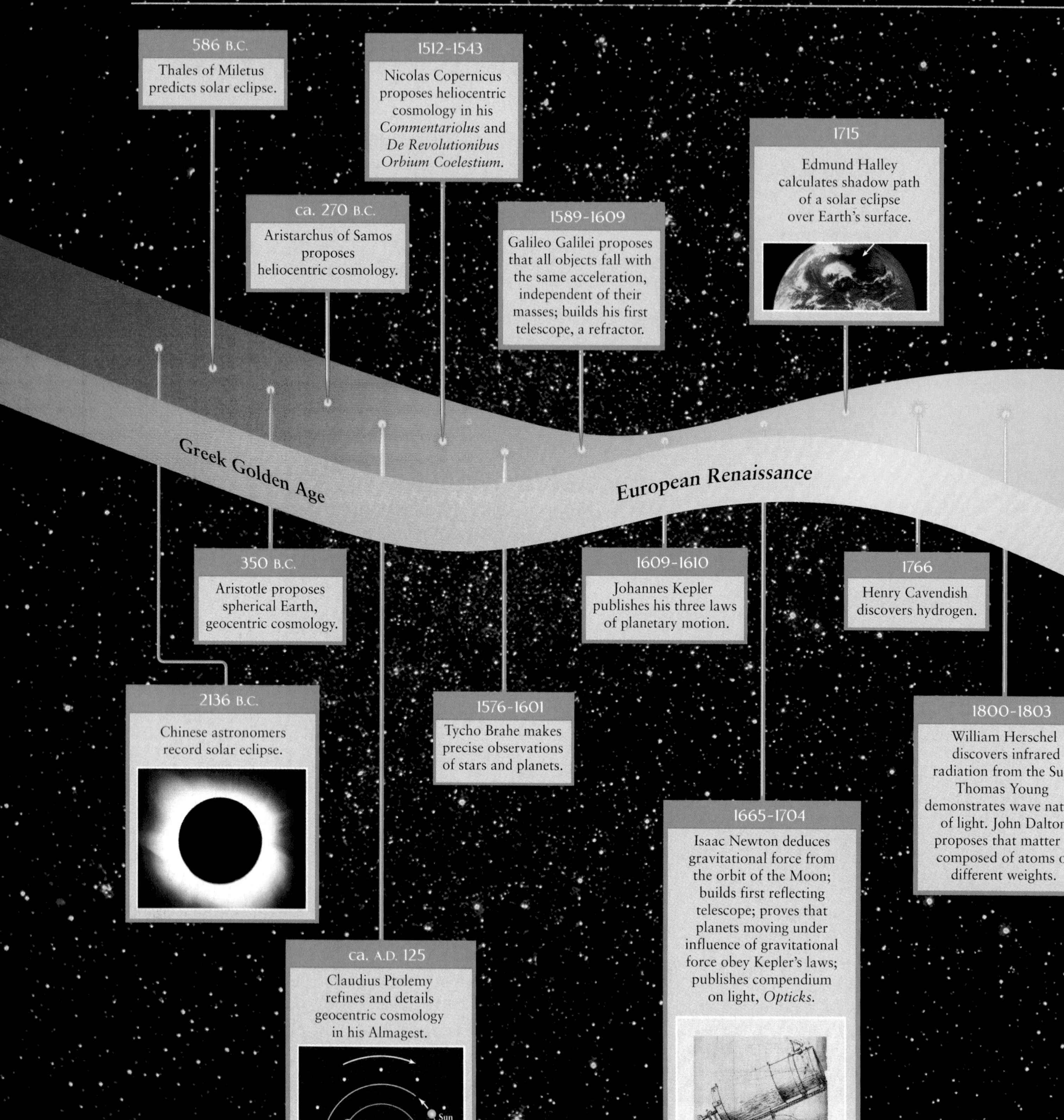

586 B.C.
Thales of Miletus predicts solar eclipse.
1512-1543
Nicolas Copernicus proposes heliocentric cosmology in his *Commentariolus* and *De Revolutionibus Orbium Coelestium*.
1715
Edmund Halley calculates shadow path of a solar eclipse over Earth's surface.
ca. 270 B.C.
Aristarchus of Samos proposes heliocentric cosmology.
1589-1609
Galileo Galilei proposes that all objects fall with the same acceleration, independent of their masses; builds his first telescope, a refractor.
Greek Golden Age
European Renaissance
350 B.C.
Aristotle proposes spherical Earth, geocentric cosmology.
1609-1610
Johannes Kepler publishes his three laws of planetary motion.
1766
Henry Cavendish discovers hydrogen.
2136 B.C.
Chinese astronomers record solar eclipse.
1576-1601
Tycho Brahe makes precise observations of stars and planets.
1800-1803
William Herschel discovers infrared radiation from the Sun. Thomas Young demonstrates wave nature of light. John Dalton proposes that matter is composed of atoms of different weights.
1665-1704
Isaac Newton deduces gravitational force from the orbit of the Moon; builds first reflecting telescope; proves that planets moving under influence of gravitational force obey Kepler's laws; publishes compendium on light, *Opticks*.
ca. A.D. 125
Claudius Ptolemy refines and details geocentric cosmology in his Almagest.
Moon
Sun
Earth

Industrial Revolution

1840-1849

J. W. Draper invents astronomical photography; takes first photographs of Moon. Christian Doppler proposes that wavelength is affected by motion. Lord Rosse completes 60-in. reflecting telescope at Birr Castle in Ireland. Armand Fizeau and Jean-Bernard Foucault measure speed of light accurately.

1871-1873

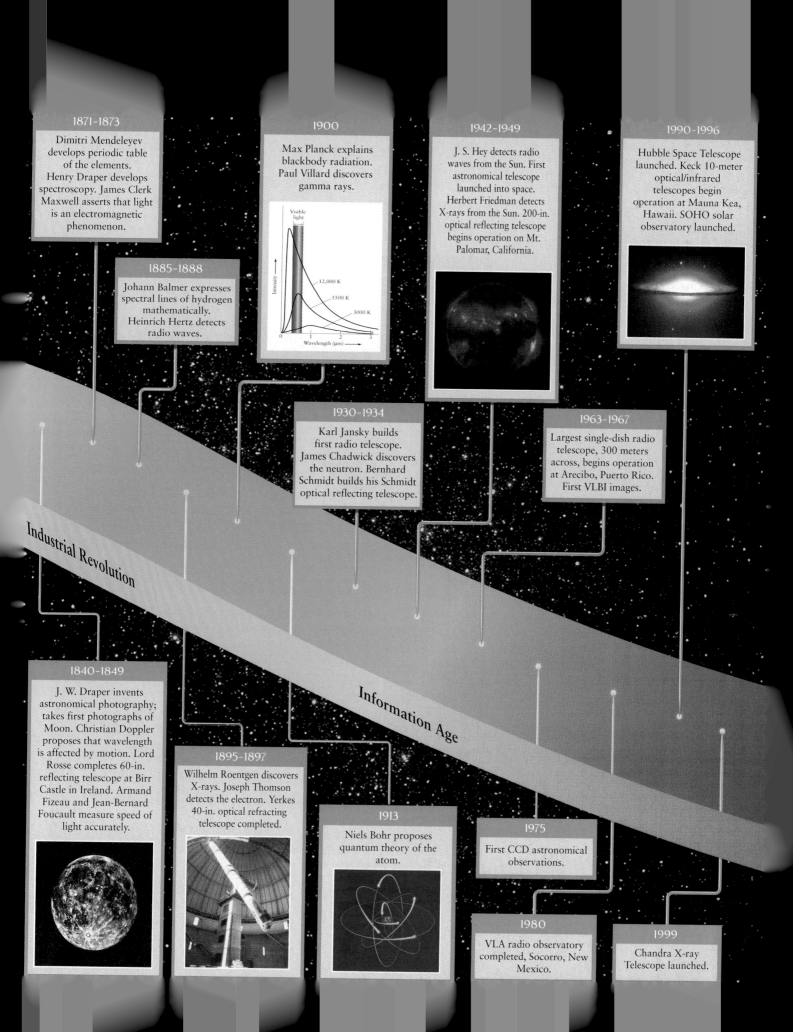

Dimitri Mendeleyev develops periodic table of the elements. Henry Draper develops spectroscopy. James Clerk Maxwell asserts that light is an electromagnetic phenomenon.

1885-1888

Johann Balmer expresses spectral lines of hydrogen mathematically. Heinrich Hertz detects radio waves.

1895-1897

Wilhelm Roentgen discovers X-rays. Joseph Thomson detects the electron. Yerkes 40-in. optical refracting telescope completed.

1900

Max Planck explains blackbody radiation. Paul Villard discovers gamma rays.

Visible light
Intensity
12,000 K
5500 K
3000 K
0 1 2 3
Wavelength (µm)

1913

Niels Bohr proposes quantum theory of the atom.

1930-1934

Karl Jansky builds first radio telescope. James Chadwick discovers the neutron. Bernhard Schmidt builds his Schmidt optical reflecting telescope.

1942-1949

J. S. Hey detects radio waves from the Sun. First astronomical telescope launched into space. Herbert Friedman detects X-rays from the Sun. 200-in. optical reflecting telescope begins operation on Mt. Palomar, California.

Information Age

1963-1967

Largest single-dish radio telescope, 300 meters across, begins operation at Arecibo, Puerto Rico. First VLBI images.

1975

First CCD astronomical observations.

1980

VLA radio observatory completed, Socorro, New Mexico.

1990-1996

Hubble Space Telescope launched. Keck 10-meter optical/infrared telescopes begin operation at Mauna Kea, Hawaii. SOHO solar observatory launched.

1999

Chandra X-ray Telescope launched.

CHAPTER 1

Discovering the Night Sky

Circumpolar Star Trails (Anglo-Australian Observatory/David Malin Images)

WHAT DO YOU THINK?

1 Is the North Star—Polaris—the brightest star in the night sky?

2 Do astronomers regard constellations as the familiar patterns of stars in the sky?

3 What causes the seasons?

4 When is the Earth closest to the Sun?

5 How many zodiac constellations are there?

6 Does the Moon have a dark side that we never see from Earth?

7 Is the Moon ever visible during the daytime?

Answers to these questions appear in the text beside the corresponding numbers in the margins and at the end of the chapter.

You have picked an exciting time to study astronomy. The frontiers of our knowledge about the cosmos are opening as never before. Technology is currently making it possible for astronomers to observe objects that were invisible to them just a few years ago. These new observations have deepened our understanding of virtually every aspect of the universe. We can now watch it expand and see stars explode in distant galaxies; we have discovered planets orbiting nearby stars, as well as newly-born stars enshrouded in clouds of gas and dust, as well as black holes and other remnants of stellar evolution. As we look farther and farther out into the universe (or cosmos)—defined as everything that we can see or that can be seen—we also see farther and farther back into time.

Telescopes are not the only means by which we are deepening our understanding of the skies. We have also begun the process of physically exploring our neighborhood in space. In just the past half century, humans have walked on the Moon and space probes have dug into Martian soil. Other space missions have landed on an asteroid, discovered active volcanoes and barren ice fields on the moons of Jupiter, visited Titan and the shimmering rings of Saturn, and traveled beyond the realm of the planets in our solar system, to mention a few accomplishments. We have even seen the development of a successful commercial spacecraft.

In the best locations, the night sky is truly breathtaking (Figure 1-1). Even if you can't see the thousands of stars visible in clear locations, the *Starry Night Enthusiast*™ software with this book can take you there. The night sky can draw you out of yourself, inviting you to understand what is happening beyond the Earth and inspiring you to think about our place in the universe. Early efforts to comprehend what we see in the sky led our ancestors to make careful observations of the movements of bodies in space. Patterns and relationships began to emerge. They noticed how a few points of light in the night sky move relative to all the others. They came to understand the locations of the Sun and Moon throughout the cycle of lunar phases and at different heights of the tides, along with the variations of the noontime height of the Sun throughout the year.

Hundreds of years ago, the explanations of what was seen in the sky were based on beliefs that had to be accepted on faith—there was no way to test ideas of what the points of light we see are, or whether the Moon really has liquid water oceans, or why those few

R I V U X G

FIGURE 1-1 Sunlight is a curtain that hides virtually everything behind it. Only the Sun and Moon are visible from beyond it during the day. As the Sun sets, we are often treated to beautiful panoramas that can inspire the artist or scientist in many of us. This photograph of saguaro cactus, the night sky, and Comet Halley was taken in northern Mexico. (Courtesy of D. L. Mammana) objects (now known to be planets) move relative to other points of light in the night sky, or why the Sun shines. Times have changed. We are fortunate to be living in an era when science has answers to many of the questions that the sky inspires.

Beautiful, intriguing, and practical, astronomy has something for everyone. This course and this book will help you better understand the universe by sharing what we have learned about some of the questions it poses. As you proceed through this book, I hope that you will gain a new appreciation of the awesome power of the human mind to reach out, to observe, to explore, and to comprehend. One of the great lessons of modern astronomy is that by gaining, sharing, and passing on knowledge, we transcend the limitations of our bodies and the brevity of human life.

In this chapter you will discover

- how astronomers organize the night sky to help them locate objects in it
- that the Earth's spin on its axis causes day and night
- how the tilt of the Earth's axis of rotation and the Earth's motion around the Sun combine to create the seasons
- that the Moon's orbit around the Earth creates the phases of the Moon and lunar and solar eclipses
- how the year is defined and how the calendar was developed

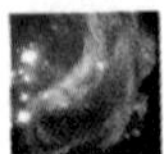

SCALES OF THE UNIVERSE

In learning a new field it is often useful to see the "big picture" before exploring the details. For this reason, we begin by surveying the size, scales of, and major types of objects in the universe.

1-1 Astronomical distances are, well, astronomical

One of the thrills and challenges of studying astronomy is becoming familiar with and comfortable with its vast size. In our everyday lives we typically deal with distances ranging from millimeters to thousands of kilometers. (We use the metric system of units, which is now standard in science. Table E-9 lists some comparisons and conversions between metric and more traditional units of measure.) A hundredth of a meter or a thousand kilometers are numbers that are easy to visualize and write. In astronomy we deal with particles as small as a millionth of a billionth of a meter and systems of stars as large as a thousand billion billion kilometers across. Similarly, the speeds of some things, like light, are so large as to be cumbersome if you have to write them out in words each time. To deal with numbers much larger and much smaller than 1, we use a shorthand throughout this book called **scientific notation** or "powers of ten." (Read Appendix A if you are unfamiliar with scientific notation.)

The size of the universe that we can observe and the range of sizes of the objects in it are truly staggering. Figure 1-2 summarizes the sizes and distances in the universe ranging from atomic particles to the size of the entire universe visible to us. Unlike distance measured on a ruler, moving up 0.5×10^{-2} m (0.5 cm) along the arc brings you to objects 10 times larger. Because of this, going from the size of a proton (roughly 10^{-15} meters) up to the size of an atom (roughly 10^{-10} meters) takes about the same space along the arc as going from the distance between the Earth and the Sun to the distance between the Earth and the nearby stars.

Underlying the fact that astronomy-related objects have such a wide range of sizes is that astronomy is a discipline that synthesizes or brings together information from many other fields of science. We will need to understand what atoms are composed of and how they work; the nature and properties of light; the response of matter and energy to the force of gravity; the creation of energy by fusing particles together in stars; the ability of carbon, and carbon alone, to serve as the foundation of life; and many other things. These concepts will all be introduced as they are needed.

What, then, have astronomers seen of the universe? Figure 1-3 inventories some of the major features we will explore in this text. Planets like Jupiter, rich in hydrogen and helium (Figure 1-3a), have been discovered orbiting a growing list of stars. Whether there are also Earthlike planets orbiting Sunlike stars elsewhere remains to be seen. Much smaller pieces of space debris, some of rock and metal called asteroids and meteoroids (Figure 1-3b), and others of rock and ice called comets (Figure 1-3c), orbit the Sun (Figure 1-3d) and other stars. Vast stores of interstellar gas and dust are found in many galaxies; these are often the incubators of new generations of stars (Figure 1-3e). Stars by the millions, billions, or even trillions are held together by the force of gravity as galaxies (Figure 1-3f). Galaxies

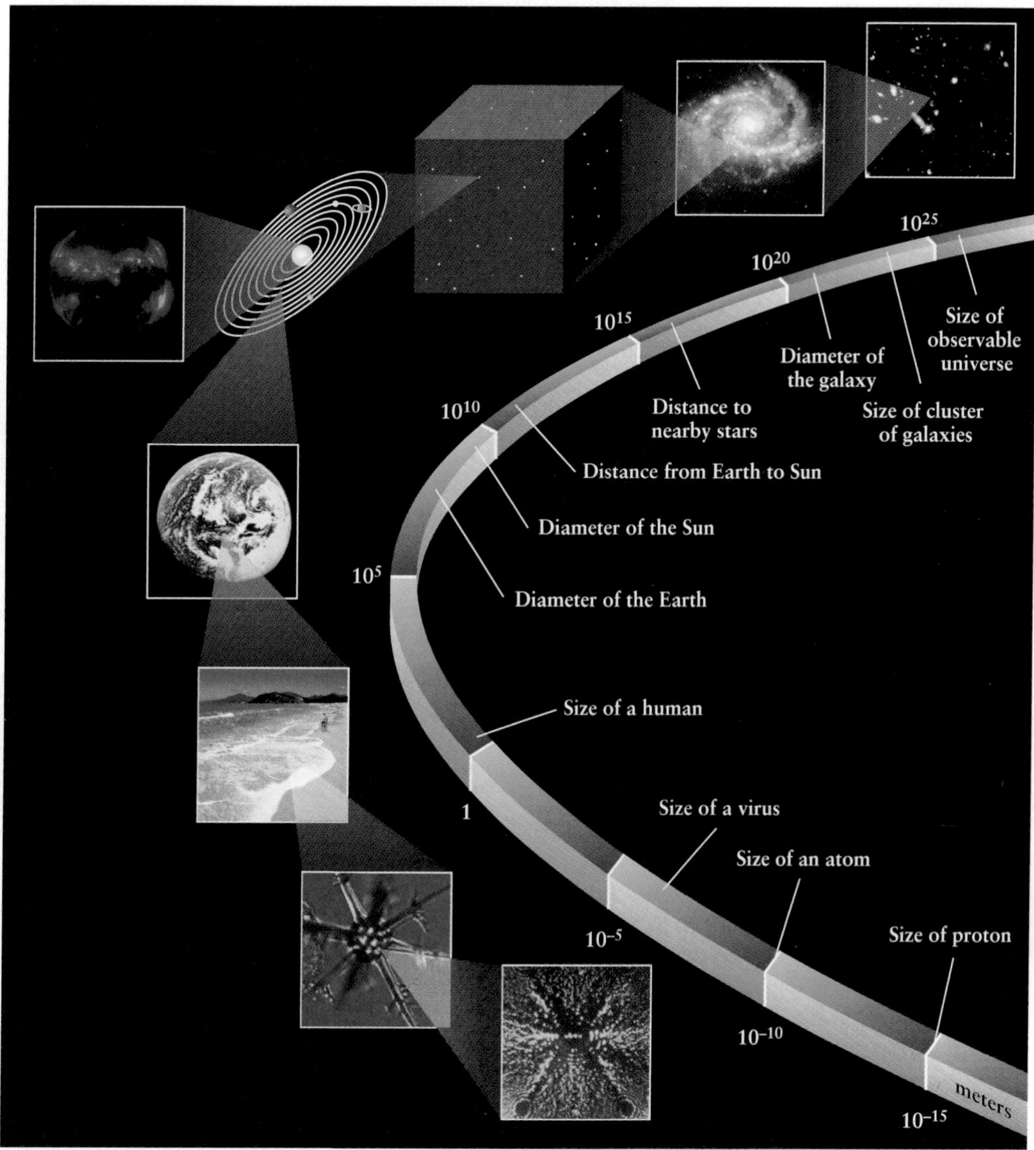

FIGURE 1-2 **The Scales of the Universe** The curve gives the sizes of objects in meters, ranging from subatomic particles at the bottom to the entire observable universe. Every 0.5 cm up along the arc represents a factor of 10 larger. (Top to bottom: R. Williams and the Hubble Deep Field Team [STScI] and NASA; AAT; L. Golub, Naval Observatory, IBM Research, NASA; Richard Bickel/Corbis; Scientific American Books)

like our own Milky Way often contain large amounts of that interstellar gas and dust, as well as regions of space where matter is so dense that it cannot radiate light; these are called black holes (Figure 1-3g). Groups of galaxies are held together by gravity in clusters (Figure 1-3h), and clusters of galaxies are held together by gravity in superclusters. Huge quantities of intergalactic gas are often found between galaxies (Figure 1-3i).

Each object in astronomy is constantly changing; each had an origin, an active period you might consider as its "life"; and each will have an end. We will study these processes along with the important physical concepts upon which they are based. You will also discover that all the matter astronomers see in stars and galaxies is but the tip of the cosmic iceberg—there is much more to the universe, but astronomers do not yet know its nature.

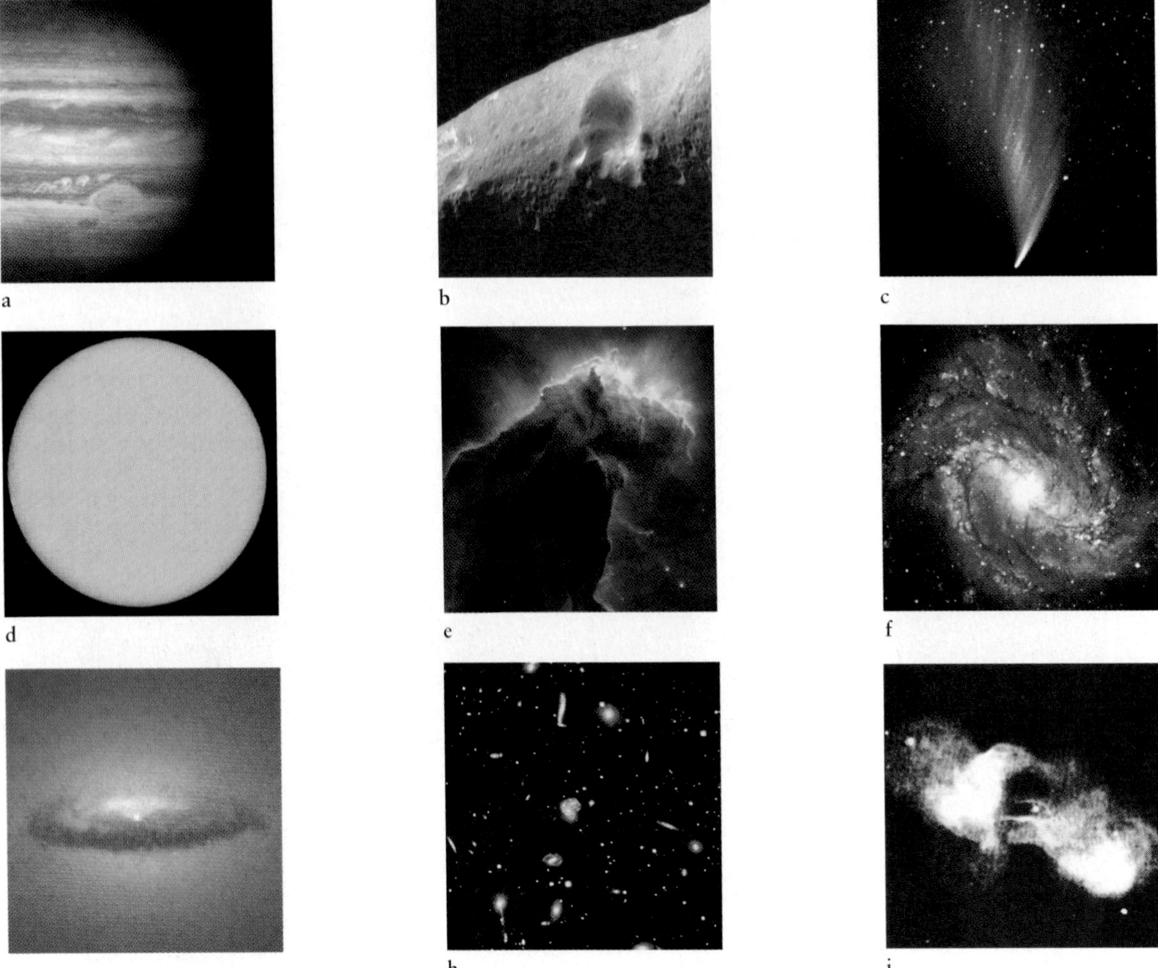

FIGURE 1-3 **Inventory of the Universe** Pictured here are examples of the major categories of objects that have been found throughout the universe. You will discover more about each type in the chapters that follow. **(a)** Planets—in this case, Jupiter; **(b)** rocky and metallic space debris—here, the asteroid Eros; **(c)** rocky and icy space debris—in this case, Comet West; **(d)** stars—such as the Sun; **(e)** interstellar clouds and new stars—in this case, M16; **(f)** galaxies—here, M83; **(g)** black holes—such as one in galaxy NGC 7052 surrounded by a disk of gas and dust; **(h)** clusters of galaxies—in this case, the Hercules cluster; and **(i)** intergalactic gas—here, 3C442. (a: NASA/Hubblesite; b: NASA; c: Peter Stättmayer/European Southern Observatory; d: Big Bear Observatory; e: NASA/Jeff Hester & Paul Scowen; f: Anglo-Australian Observatory; g: NOAO; h: NASA; i: N. F. Comins & F. N. Owen/NRAO)

PATTERNS OF STARS

When you gaze at the sky on a clear, dark night, there seem to be millions of stars twinkling overhead. In reality, the unaided human eye can detect only about 6000 stars over the entire sky. At any one time, you can see roughly 3000 stars in dark skies, because only half of the stars are above the *horizon*, the boundary between the Earth and the sky.

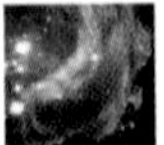

You probably have noticed patterns formed by bright stars and are probably familiar with some common names for these patterns, such as the ladle-shaped Big Dipper and broad-shouldered Orion. These recognizable patterns of stars are called constellations in everyday conversation and they have names derived from ancient legends (Figure 1-4a).

1-2 Constellations make locating stars easy

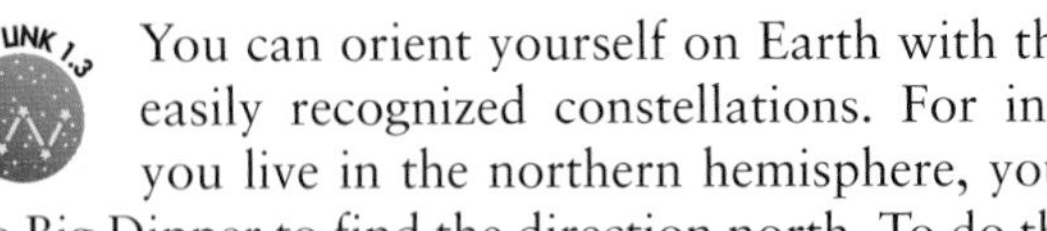

You can orient yourself on Earth with the help of easily recognized constellations. For instance, if you live in the northern hemisphere, you can use the Big Dipper to find the direction north. To do this, locate the Big Dipper and imagine that its bowl is resting on a table (Figure 1-5). If you see the Dipper upside down in the sky, as you frequently will, imagine the dipper resting on an

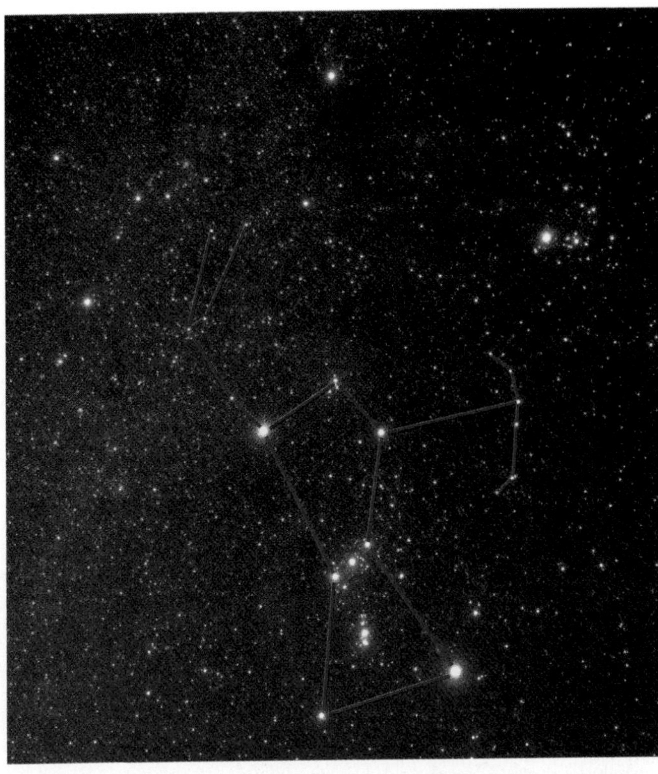

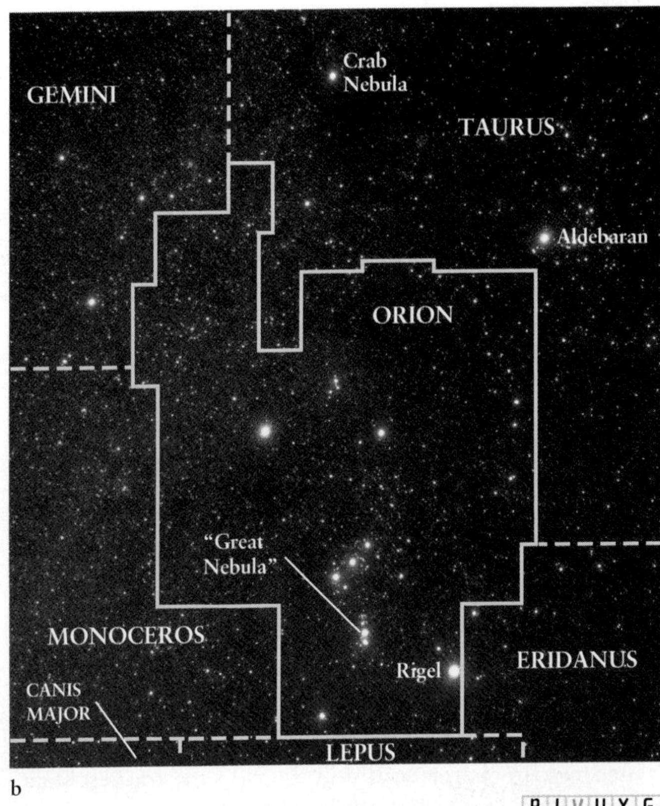

FIGURE 1-4 **The Constellation Orion** **(a)** The pattern of stars called Orion is a prominent winter constellation. From the northern hemisphere, it is easily seen high above the southern horizon from December through March. You can see in this photograph that the various stars have different colors, something to watch for when you observe the night sky. **(b)** The region of the sky called Orion and parts of other nearby constellations are depicted in this photograph. All the stars inside the boundary of Orion are members of that constellation. The celestial sphere is covered by 88 constellations of differing sizes and shapes. (© 2004 Jerry Lodriguss/www.astropix.com)

upside-down table above it. Locate the two stars of the bowl farthest from the Big Dipper's handle. These are called the *pointer stars.* Draw a mental line through these stars leading away from the table, as shown in Figure 1-5. The
1 first *moderately bright* star you then encounter is Polaris, also called the North Star because it is located almost directly over the Earth's North Pole. So, while Polaris is not even among the 20 brightest stars (see Appendix E-5), it is easy to locate. Whenever you face Polaris, you are facing north. East is then on your right, south is behind you, and west is on your left.

The Big Dipper example illustrates the fact that easily recognized constellations make it easy to locate other stars. The most effective way to do this is to use vivid visual connections, especially those of your own devising. For example, imagine gripping the handle of the Big Dipper and slamming its bowl through the imaginary table and onto the head of Leo (the Lion). Leo comprises the first group of bright stars your dipper encounters. As shown in Figure 1-5, the brightest star in this group is Regulus, the dot of the backward question mark that traces the lion's mane. As another example, put the Big Dipper back on its table and then follow the arc of its handle away from its bowl. The first bright star you encounter along that arc beyond the handle is Arcturus in Boötes (the Herdsman). Follow the same arc farther to the prominent bluish star Spica in Virgo (the Virgin). Spotting these stars and remembering their names is easy if you remember the saying "Arc to Arcturus and speed on to Spica."

During the winter months in the northern hemisphere, you can see some of the brightest stars in the sky. Many of them are in the vicinity of the "winter triangle," which connects bright stars in the constellations of Orion (the Hunter), Canis Major (the Larger Dog), and Canis Minor (the Smaller Dog), as shown in Figure 1-6. The winter triangle passes high in the sky at night during the middle of winter. It is easy to find Sirius, the brightest star in the night sky, by locating the belt of Orion and following a straight mental line from it to the left (as you face Orion). The first bright star that you encounter is Sirius.

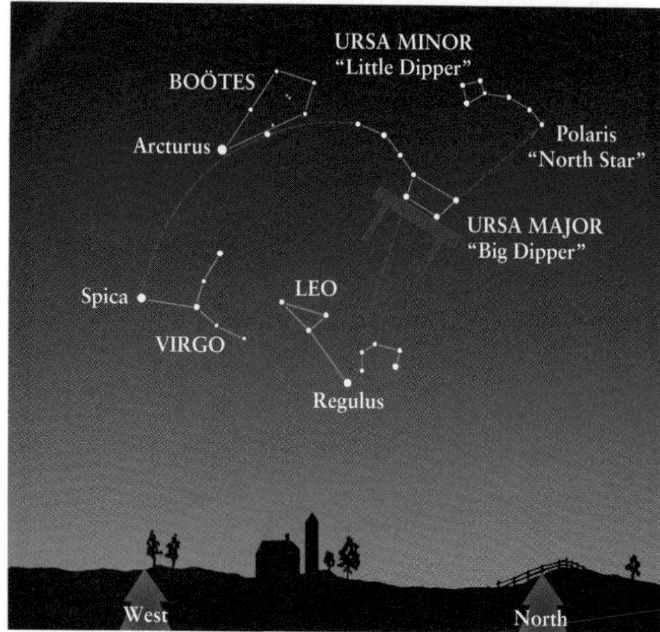

FIGURE 1-5 **The Big Dipper as a Guide** In the northern hemisphere, the Big Dipper is an easily recognized pattern of seven bright stars. This star chart shows how the Big Dipper can be used to point out the North Star as well as the brightest stars in three other constellations. While the Big Dipper appears right side up in this drawing of the sky shortly before sunrise, at other times of the night it appears upside down. Why does this happen?

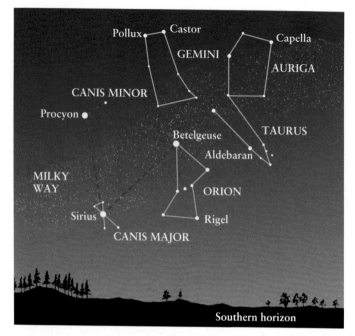

FIGURE 1-6 **The Winter Triangle** This star chart shows the southern sky as it appears during the evening in December. Three of the brightest stars in the sky make up the winter triangle. In addition to the constellations involved in the triangle, Gemini (the Twins), Auriga (the Charioteer), and Taurus (The Bull) are also shown.

INSIGHT INTO SCIENCE

Flexible Thinking Part of learning science is learning to look at things from different perspectives. For example, when learning to identify the prominent constellations, be sure to learn them from different orientations (that is, with the star chart rotated at different angles), so that you can find them at different times of the night and the year.

The "summer triangle," which graces the summer sky as shown in Figure 1-7, connects the bright stars Vega in Lyra (the Lyre), Deneb in Cygnus (the Swan), and Altair in Aquila (the Eagle). A conspicuous portion of the Milky Way forms a beautiful background for these constellations, which are nearly overhead during the middle of summer at midnight. For more on the constellations, see Guided Discovery: The Stars and Constellations.

Astronomers require more accuracy in locating dim objects than is possible simply by moving from constellation to constellation. They have therefore created a celestial map and applied a coordinate system to it analogous to the coordinate system of north-south latitude and east-west

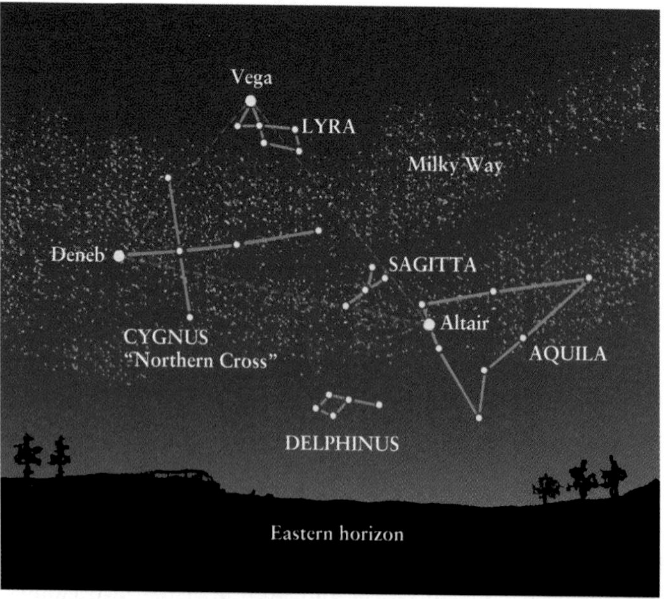

FIGURE 1-7 **The Summer Triangle** This star chart shows the northeastern sky as it appears in the evening in June. In addition to the three constellations involved in the summer triangle, the faint stars in the constellations of Sagitta (the Arrow) and Delphinus (the Dolphin) are also shown.

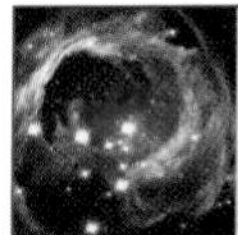

GUIDED DISCOVERY
The Stars and Constellations

WEB LINK 1.4

Many students take an introductory astronomy course expecting to be taught the familiar stars and constellations. Often, teachers just do not have time to cover this material. You can, however, learn them on your own using two steps for memorizing the night sky.

1. Observe easily identified constellations.

2. Use these to find nearby, lesser-known ones.

For example, you may not remember where the star Aldebaran in Taurus is in the sky, but if you remember that Orion is fighting Taurus, you can easily locate Aldebaran by following a line defined by the belt of Orion to the right (away from Sirius). The first bright star you encounter is Aldebaran.

Make your own connections between the constellations. Have fun while you're doing it. Chances are that you will remember the phrase "slam the Big Dipper's bowl downward to hit Leo the Lion on the head" rather than "the first bright group of stars directly below the Big Dipper is Leo."

WEB LINK 1.5

A very efficient way to learn the constellations is to use the board and card game *Stellar 28*. In addition, astronomy computer programs such as *Starry Night Enthusiast*™, included with this text, are available. *Starry Night Enthusiast*™ shows many things besides what constellations are up at night, including the motion, location, and phases of the planets; the location of deep sky objects, such as nebulae; and the sky as seen from any location on Earth on any date.

Another good way to familiarize yourself with the night sky is to use star charts to see which constellations are up each night. You will find a set of star charts from the *Griffith Observer* magazine at the front and back of this book. To use the charts, first select the one that best corresponds to the date and time of your observation. Take the chart outside at night and compare it directly with the sky. Hold the chart vertically and turn it so that the direction you are facing shows at the bottom. Using a flashlight with a red plastic coating over the light will make it easier to read the chart without ruining your night vision. You will find stargazing a truly enjoyable experience.

longitude used to navigate on the Earth. If a star's celestial coordinates are known, it can be quickly located. For such a sky map to be useful in finding stars, the stars must be fixed on it, just as cities are fixed on maps of the Earth.

1-3 The celestial sphere aids in navigating the sky

ANIMATION 1.1

If you look at the night sky year after year, you will see that the stars do indeed appear fixed relative to each other. Furthermore, throughout each night the entire pattern of stars appears to rigidly orbit the Earth. We employ this artificial, Earth-based view of the heavens to make celestial maps by pretending that the stars are attached to the inside of an enormous hollow shell, the **celestial sphere,** with the Earth at its center (Figure 1-8). Visualized another way, you can imagine the half of the celestial sphere that is visible at night as a giant bowl covering the Earth.

Astronomers use the word *constellation* to describe an entire region of the sky and all the objects in that region (see Figure 1-4b), rather than just the recognizable groupings of stars like Orion and the Big Dipper that we referred to earlier.
2 lier. The celestial sphere is divided into 88 unequal regions, and these regions are what astronomers refer to when they use the term **constellations.** (Astronomers call the traditional star patterns *asterisms* rather than constellations

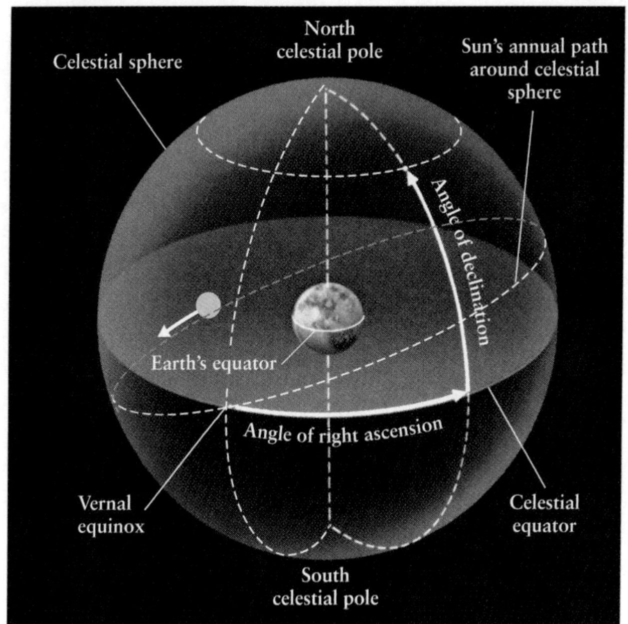

FIGURE 1-8 **The Celestial Sphere** The celestial sphere is the apparent "bowl" or hollow sphere of the sky. The celestial equator and poles are projections of the Earth's equator and axis of rotation onto the celestial sphere. The north celestial pole is therefore located directly over the Earth's North Pole, while the south celestial pole is directly above the Earth's South Pole.

when there is a danger of confusion.) Some constellations, like Ursa Major (the Large Bear), are very large, while others, like Sagitta (the Arrow), are relatively small. To locate stars, we might say "Albireo in the constellation Cygnus (the Swan)," much as we would refer to "Chicago in the state of Illinois," "Cardiff in the county of Glamorgan," or "Ottawa in the province of Ontario."

The stars seem fixed on the celestial sphere only because of their remoteness. In reality, they are at widely varying distances from the Earth, and they do move relative to each other. But we neither see their motion nor perceive their relative distances because the stars are so far from here. You can understand this by imagining a jet plane just one kilometer overhead traveling at 1000 kilometers (620 miles) an hour across the sky. Its motion is unmistakable. But a plane moving at the same speed along the horizon appears to be moving about a hundred times more slowly. And an object at the distance of the Sun traveling at the same speed and moving across the sky would appear to be going nearly a hundred million times more slowly than the plane overhead.

The stars (other than the Sun) are all more than 40 trillion kilometers (25 trillion miles) from us. Therefore, although the patterns of stars in the sky do change, their great distances prevent us from seeing those changes over the course of a human lifetime. Thus, as unrealistic as it is, the celestial sphere is so useful for navigating the heavens that it is used by astronomers even at the most sophisticated observatories around the world.

As shown in Figure 1-8, we can project key geographic features from Earth out into space to establish directions and bearings. If we expand the Earth's equator onto the celestial sphere, we obtain the **celestial equator.** The celestial equator divides the sky into northern and southern hemispheres, just as the Earth's equator divides the Earth into two hemispheres. We can also imagine extending the Earth's North Pole and South Pole out into space along the Earth's axis of rotation. Doing so gives us the **north celestial pole** and the **south celestial pole,** also shown in Figure 1-8. With the celestial equator and poles as reference features, astronomers denote the position of an object in the sky in the same way that latitude and longitude are used to specify a location on Earth.

Just as we need two coordinates (latitude and longitude) to find any location on Earth, two coordinates are needed to locate any object on the celestial sphere. The equivalent to latitude on Earth is **declination** on the celestial sphere. It is measured from 0° to 90° north or south of the celestial equator. The equivalent of longitude on Earth is **right ascension** on the celestial sphere, measured from 0 hrs to 24 hrs around the celestial equator (see Figure 1-8).

In the same way that Greenwich, England, defines the prime meridian or zero of longitude on Earth, there is a zero of right ascension. It is defined by one of the places where the Sun's annual path across the celestial sphere intersects the celestial equator. We will explore later in this chapter why the Sun appears to move in a circle around the celestial sphere during the course of a year. The celestial equator and the Sun's path intersect at two points. The equivalent on the celestial sphere of the Earth's prime meridian is where the Sun crosses the celestial equator moving northward. Angles of right ascension are measured from this point, called the *vernal equinox* (Figure 1-8).

In navigating on the celestial sphere, astronomers measure the distance between objects in terms of angles. If you are not familiar with measuring the separation between objects using this method, read An Astronomer's Toolbox 1-1.

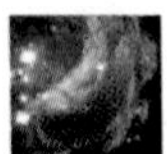

EARTHLY CYCLES

Experience has undoubtedly shown you that the Sun rises and sets at different times each day, with a cycle of rising and setting times that repeats each year. Similarly, the Moon rises and sets at different times each day and repeats its cycle roughly once every 29½ days. Furthermore, as noted in Section 1-2, we do not see the same constellations up in the sky every night of the year, but the cycle of constellations that are up at night repeats each year. The daily and annual rhythms of the sky, the Earth, and all life on it arise from three celestial motions: the Earth's spinning, which causes day and night, as well as causing the apparent motion of the celestial sphere and the constellations fixed on it; the Earth's orbit around the Sun, which creates the seasons, the year, and the change in times at which constellations are up at night; and the Moon's orbit around the Earth, which creates the lunar phases, the cycle of tides, and the spectacular phenomena we call eclipses.

1-4 Earth's rotation creates the day-night cycle and its revolution defines a year

The Earth spins on its axis. Such motion is called **rotation.** We do not feel the Earth's rotation because our planet is so physically large compared to us and its gravitational attraction holds us firmly down. The Earth's rotation causes the constellations—as well as the Sun, Moon, and planets—to appear to rise on the eastern horizon, move across the sky, and set on the western horizon. The Earth's daily rotation, causing the Sun to rise and set, thereby creates day and night. The **diurnal motion,** or daily motion, of the celestial bodies is apparent in time-exposure photographs, such as that shown in Figure 1-9 (on page 14).

Take a friend outside on a clear, warm night to observe the diurnal motion of the stars for yourselves. Soon after dark, find a spot away from bright lights, and note the constellations in the sky relative to some prominent landmarks near you on Earth. A few hours later, check again from the

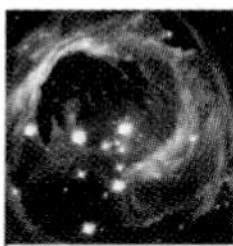

AN ASTRONOMER'S TOOLBOX 1-1
Observational Measurements Using Angles

Ancient mathematicians invented a system of angles and angular measure that is still used today to denote the relative positions and apparent sizes of objects in the sky. To locate stars on the celestial sphere, for example, we do not need to know their distances from Earth. All we need to know is the angle from one star to another in the sky, a property that remains fixed over our lifetimes because the stars are all so far away.

An **arc angle,** often just called an **angle,** is the opening between two lines that meet at a point. Angular measure is a method of describing the size of an angle. The basic unit of angular measure is the **degree,** designated by the symbol °. A full circle is divided into 360°. A right angle measures 90°. As shown in the figure below, the angle between the two "pointer stars" in the Big Dipper is about 5°.

Astronomers also use angular measure to describe the apparent sizes of celestial objects. For example, imagine the full Moon. The angle across the Moon's diameter is nearly ½°. We therefore say that the **angular diameter,** or **angular size,** of the Moon is ½°. Alternatively, astronomers say that the Moon "subtends" an angle of ½°. In this context, subtend means "to extend across."

The adult human hand held up at arm's length provides a means of estimating angles. For example, five fingers at arm's length covers an angle of about 8°, whereas the tip of your finger is about 1½° wide. Various segments of your index finger extended to arm's length can be similarly used to estimate angles a few degrees across, as shown in the figure above on the right.

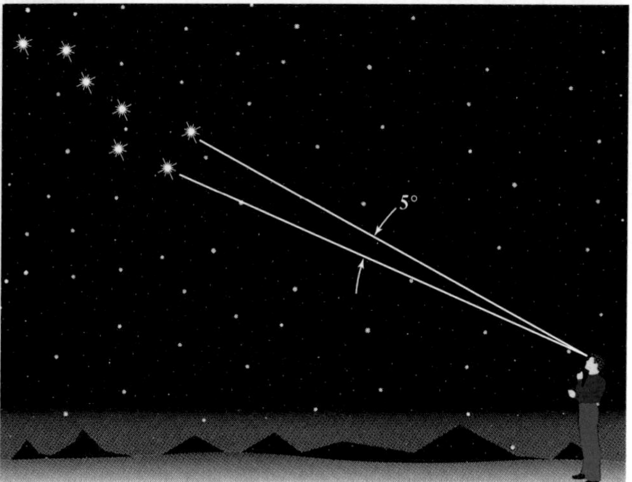

The Big Dipper The angular distance between the two "pointer stars" at the front of the Big Dipper is about 5°. For comparison, the angular diameter of the Moon is about ½°.

Estimating Angles with the Human Hand Various parts of the adult human hand extended to arm's length can be used to estimate angular distances and sizes in the sky.

To talk about smaller angles, we subdivide the degree into 60 arcminutes (abbreviated 60 arcmin or 60′). An arcminute is further subdivided into 60 arcseconds (abbreviated 60 arcsec or 60″). A dime viewed face-on from a distance of 1.6 km (1 mile) has an angular diameter of about 2 arcsec. From everyday experience, we know that an object looks big when it is nearby but small when it is far away. The angular size of an object therefore does not necessarily tell you anything about its actual physical size. For example, the fact that the Moon's angular diameter is ½° does not tell you how big the Moon really is. But if you also happen to know the distance to the Moon, then you can calculate the Moon's physical diameter. In general, the physical diameter of an object can be calculated from the equation:

$$\text{physical diameter} = \text{distance} \times \tan(\text{angular diameter})$$

where tan (angular diameter) means the tangent of the angle denoted "angular diameter." In the Moon's case, using a measured distance (see Appendix E-3) of 384,400 km and an angular diameter of ½°, we find the diameter to be roughly 3350 km. The difference between this and the exact diameter of 3476 km is due primarily to the approximate value of ½° that we have used.

Try these questions: The Sun is 1.5×10^8 km away and has a diameter of 1.4×10^6 km. How large an angle does it make in our sky? How is that relevant to the arc angle of the Moon in our sky? What arc angle would the Moon make in our sky if it were twice as far away? Half as far?

(Answers appear at the end of the book.)

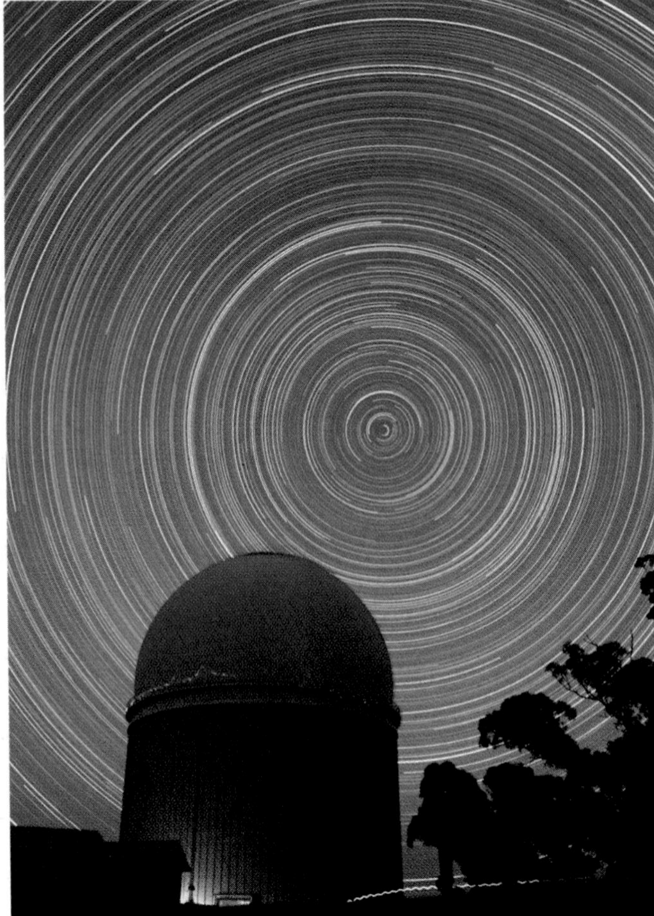

FIGURE 1-9 **Circumpolar Star Trails** This long exposure, taken from Australia's Siding Spring mountain and aimed at the south celestial pole, shows the rotation of the sky. The stars that pass between the pole and the ground are all circumpolar stars. (Anglo-Australian Observatory/David Malin Images)

same place. You will find that the entire pattern of stars (as well as the Moon, if it is visible) has shifted. New constellations will have risen above the eastern horizon, while other constellations will have disappeared below the western horizon. If you check again just before dawn, you will find the stars that were just rising in the east when the night began are now low in the western sky.

Different constellations are visible at night during different times of the year. This occurs because the Earth orbits, or *revolves,* around the Sun. **Revolution** is the motion of any astronomical object around another astronomical object. The Earth takes one year, or about 365¼ days, to go once around the Sun. A year on Earth is measured by the motion of our planet relative to the stars. Draw a straight line from the Sun through the Earth to some star on the opposite side of the Earth from the Sun. As the Earth revolves, that line will inscribe a straight path on the celes-

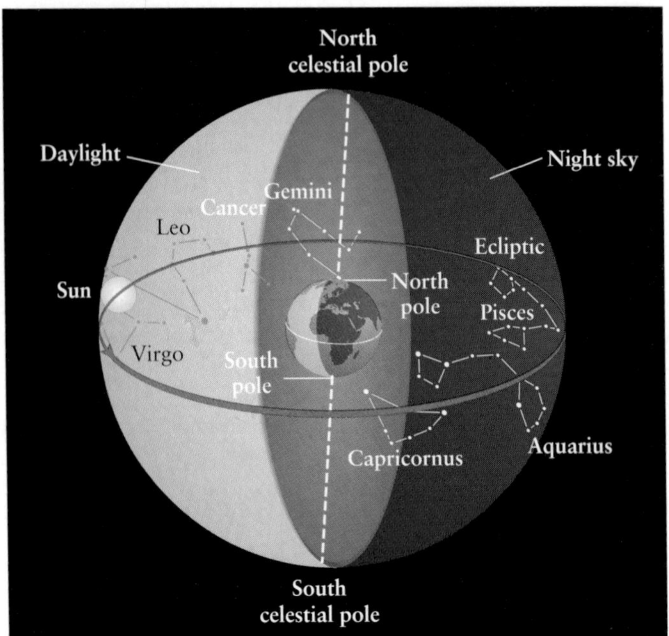

a

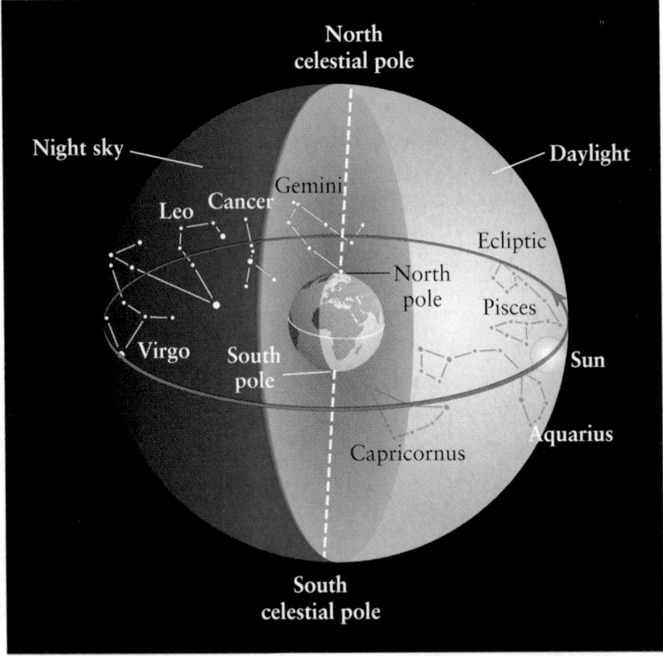

b

FIGURE 1-10 **Why Different Constellations Are Visible at Different Times of the Year** **(a)** On the autumnal equinox each year, the Sun is in the constellation Virgo. As seen from Earth, that part of the sky is in daylight and we see stars only on the other half of the sky, centered around the constellation Pisces. **(b)** Six months later, the Sun is in Pisces. This side of the sky is bright, while the side centered on Virgo is now in darkness.

tial sphere and will return to the original star 365¼ days later. The length of any cycle of motion measured with respect to the stars is called a **sidereal period.**

As a result of the Earth's motion around the Sun, every star rises approximately 4 minutes earlier each day than it did the day (or night) before. This effect accumulates, bringing different constellations up at night throughout the year. Figure 1-10 summarizes this. When the Sun is "in" Virgo (September 18–November 1), for example, the hemisphere containing the Sun and the constellations around Virgo are in daylight (see Figure 1-10a). When the Sun is up, so are Virgo and the surrounding constellations, and so we cannot see them. During that time of year, the constellations on the other side of the celestial sphere, centered on the constellation Pisces, are in darkness. So, when the Sun is "in" Virgo, Pisces and the constellations around it are high in our sky at night.

Six months later, when the Sun is "in" Pisces, that half of the sky is filled with daylight, while Virgo and the constellations around it are high in the night sky (see Figure 1-10b). These arguments apply everywhere on Earth at the same time because the Sun moves through the zodiac constellations very slowly as seen from Earth, taking a year to make one complete circuit.

We spoke earlier of stars rising on the eastern horizon and setting on the western horizon. Depending on your latitude, some of the stars and constellations never disappear below the horizon. Instead, they trace complete circles in the sky over the course of each night see (Figure 1-9). To understand why this happens, imagine that you are standing on the Earth's North Pole at night. Looking straight up, you see Polaris. Because the Earth is spinning around its axis directly under your feet, all the stars appear to move from left to right in horizontal rings around you. The exception is Polaris, which always remains at the North Pole's **zenith.** (Every place has a different zenith, which is the point directly overhead anywhere on Earth.) As seen from the North Pole, no stars rise or set (Figure 1-11). They just seem to revolve around Polaris in horizontal circles. Stars and constellations that never go below the horizon are called **circumpolar.** All stars visible from the North Pole or South Pole are circumpolar.

Now visualize yourself at the equator. All the stars appear to rise straight up in the eastern sky and set straight down in the western sky (Figure 1-12). Polaris is barely visible on the northern horizon. While Polaris never sets, all the other stars do, and therefore none of the stars is circumpolar as seen from the equator.

As you may have concluded from these last two mental exercises, the angle at which the stars rise and set depends on your viewing latitude. Figure 1-13, for example, shows stars setting at 35° north latitude. In Orono, Maine (44°45′ north latitude), where this book was written, stars rise at an angle of nearly 45° to the eastern horizon and set at an angle of 45° to the western horizon. Polaris is fixed at 45° above the horizon in Orono's northern sky, not at the

FIGURE 1-11 **Motion of Stars at the Poles** Because the Earth rotates around its poles, stars seen from these locations appear to move in huge, horizontal circles. This is the same effect you would get by standing up in a room and spinning around; everything would appear to move in circles around you. At the North Pole stars move left to right, while at the South Pole they move right to left.

FIGURE 1-12 **Rising and Setting of Stars at the Equator** Standing on the equator, you are perpendicular to the axis around which the Earth rotates. As seen from there, the stars rise straight up on the eastern horizon and set straight down on the western horizon. This is the same effect you get when driving straight over the crest of a hill; the objects on the other side of the hill appear to move straight upward as you descend.

R I V U X G

FIGURE 1-13 **Rising and Setting of Stars at Middle North Latitudes** Unlike the motion of the stars at the poles (see Figure 1-11), the stars at all other latitudes do change angle above the ground throughout the night. This time-lapse photograph shows stars setting. The latitude determines the angle at which the stars rise and set. (David Miller/DMI) zenith there, as it is at the North Pole or on the horizon as seen from the equator. As another example, except for those stars in the upper corners, the stars whose paths are shown in Figure 1-9 are circumpolar because they are visible all night, every night.

If you live in the northern hemisphere, Polaris is always located above your northern horizon at an angle equal to your latitude. Only the stars and constellations that pass between Polaris and the land directly below it are circumpolar. The farther north you go in the northern hemisphere, the greater the number of stars and constellations that are circumpolar. The southern hemisphere has no Polarislike star at the south celestial pole, but the rest of this argument applies for you; the farther south you go, the greater the number of stars and constellations that are circumpolar.

1-5 Clock times based on the Sun's location created scheduling nightmares

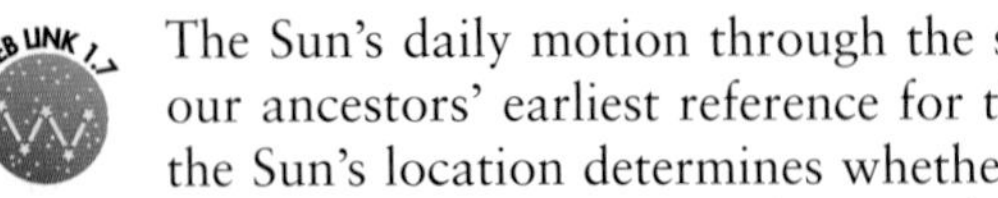

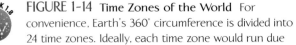

The Sun's daily motion through the sky provided our ancestors' earliest reference for time, because the Sun's location determines whether it is day or night, whether we are awake or asleep, and whether it is time for breakfast or dinner. In other words, the Sun determines the length of the **solar day,** upon which our 24-hour day is based. However, the length of the solar day varies throughout the year. The Sun is not a perfect timekeeper because the Earth's orbit around it is not perfectly circular, as we will study in Chapter 2. Because the Earth moves slightly more rapidly along its orbit when it is near the Sun than when it is farther away, the speed of the Sun across the sky also varies throughout the year. Using the average time

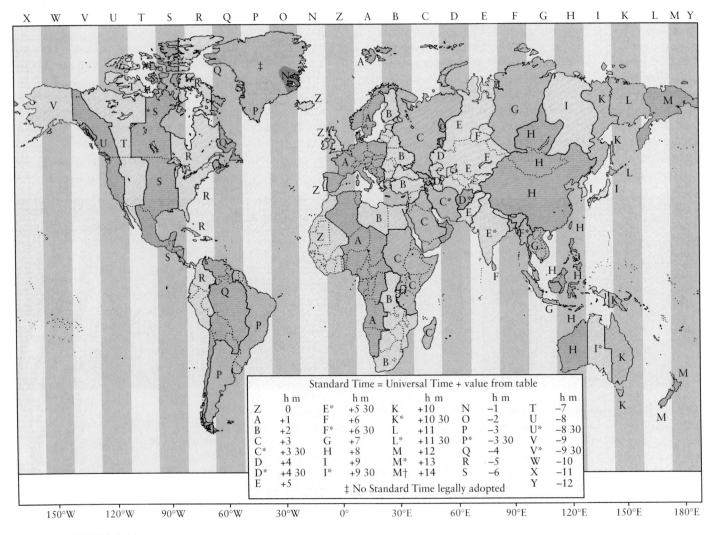

Standard Time = Universal Time + value from table

	h m		h m		h m		h m		h m
Z	0	E*	+5 30	K	+10	N	−1	T	−7
A	+1	F	+6	K*	+10 30	O	−2	U	−8
B	+2	F*	+6 30	L	+11	P	−3	U*	−8 30
C	+3	G	+7	L*	+11 30	P*	−3 30	V	−9
C*	+3 30	H	+8	M	+12	Q	−4	V*	−9 30
D	+4	I	+9	M*	+13	R	−5	W	−10
D*	+4 30	I*	+9 30	M†	+14	S	−6	X	−11
E	+5							Y	−12

‡ No Standard Time legally adopted

FIGURE 1-14 **Time Zones of the World** For convenience, Earth's 360° circumference is divided into 24 time zones. Ideally, each time zone would run due north-south. However, political considerations make many zones irregular. Indeed, there are even a few zones only a half hour wide.

interval between consecutive noontimes to determine the solar day corrects this problem, and this average defines our 24-hour day.

There is also a *sidereal day,* the length of time from when a star is in one place in the sky until it is next in the same place. The solar and sidereal days differ from each other in length because while the Earth rotates, it revolves around the Sun. That change in the Earth's location day by day changes the location of the stars, bringing the stars back to their original positions 4 minutes earlier each day. Therefore, the sidereal day is 23 hours, 56 minutes long, while the solar day is 24 hours long.

Astronomically, noon is defined as the instant when the Sun is highest in the sky. However, because the Earth is rotating eastward, the Sun is highest at different longitudes at different times. For example, astronomical noon in New London, Connecticut, occurs slightly earlier than it does a little farther west, in New Haven, Connecticut. Before the advent of time zones, local time was based on astronomical noon. To travel from New London west to New Haven by train, for example, you had to know the departure time at New London, using New London time, as well as the arrival time in New Haven (say, if someone was going to meet you) in New Haven time. Such time considerations became very confusing and burdensome as society became more complex. Fortunately, in the late nineteenth century, time zones were established to remove this problem. In a **time zone,** everyone agrees to set their clocks alike. Time zones, originally developed for scheduling rail transportation, are based on the time at 0° longitude in Greenwich, England, a location called the *prime meridian,* as mentioned earlier. With some variations due to geopolitical boundaries, every 15° of longitude around the globe begins a new time zone. The resulting 24 time zones are shown in Figure 1-14. Going from one time zone to the next usually requires you to change the time on your wristwatch by exactly 1 hour.

1-6 Calendars based on equal-length years also created scheduling problems

Just as the day originates in the Earth's rotation, the year is the unit of time based on the Earth's revolution about the Sun. The Earth does not take exactly 365 days to orbit the Sun, so the year is not exactly 365 days long. Basing the year on a 365-day cycle led to important events occurring on the wrong day. To resolve this problem, a committee appointed by the Roman statesman Julius Caesar recommended a new calendar that Caesar implemented. Measurement revealed to ancient astronomers that the length of a year is approximately 365 ¼ days. The committee established the system of leap years to accommodate this extra quarter of a day. By adding an extra day to the calendar every four years, Caesar hoped to ensure that seasonal astronomical events, such as the beginning of spring, would occur on the same date year after year.

Caesar's system would have worked if a year were exactly 365 ¼ days long and if the Earth's rotation axis (now pointing toward Polaris, as discussed earlier) never changed direction. Neither assumption is correct. Thus, over time, a discrepancy accumulated between Caesar's "Julian" system and actual time. Annual astronomical events began to fall on different dates each year. To straighten things out, a committee established by Pope Gregory XIII recommended a refinement, thus creating the Gregorian calendar in 1582. Pope Gregory began by dropping ten days (October 5, 1582, was proclaimed to be October 15, 1582), which brought the first day of spring back to March 21. Next, he modified Caesar's system of leap years. Caesar had added February 29 to every calendar year that is evenly divisible by four. For example, 1992, 1996, 2000, and 2004 were all leap years with 366 days. But this system produces an error of about three days every four centuries. To solve the problem, Pope Gregory decreed that century years would be leap years only if evenly divisible by 400. For example, the years 1700, 1800, and 1900 were not leap years under the improved Gregorian system. But the year 2000—which can be divided evenly by 400—was a leap year.

We use the Gregorian system today. It assumes that the year is 365.2425 mean solar days long, which is very close to the length of the *tropical year,* defined as the time interval from one vernal equinox to the next. In fact, the error is only one day in every 3300 years. That won't cause any problems for a long time.

1-7 The seasons result from the tilt of the Earth's rotation axis combined with Earth's revolution around the Sun

Imagine that you could see the stars even during the day, so that you could follow the Sun's apparent motion against the background constellations throughout the year. (The Sun appears to move among the stars, of course, because the Earth orbits around it.) From day to day, the Sun traces a straight path on the celestial sphere. This path is called the **ecliptic.** As you can see in Figure 1-15a, the ecliptic makes a closed circle bisecting the celestial sphere. You can see that the ecliptic is precisely the loop labeled "Sun's annual path around celestial sphere" in Figure 1-8.

The term "ecliptic" has a second use in astronomy. The Earth orbits the Sun in a plane also called the ecliptic. You can see that the two ecliptics exactly coincide. Imagine yourself on the Sun watching the Earth move day by day. The path of the Earth on the celestial sphere as seen from

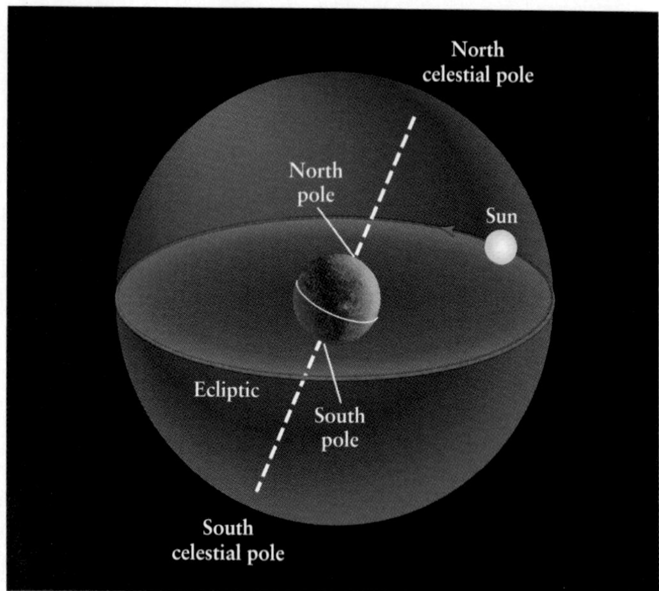

a

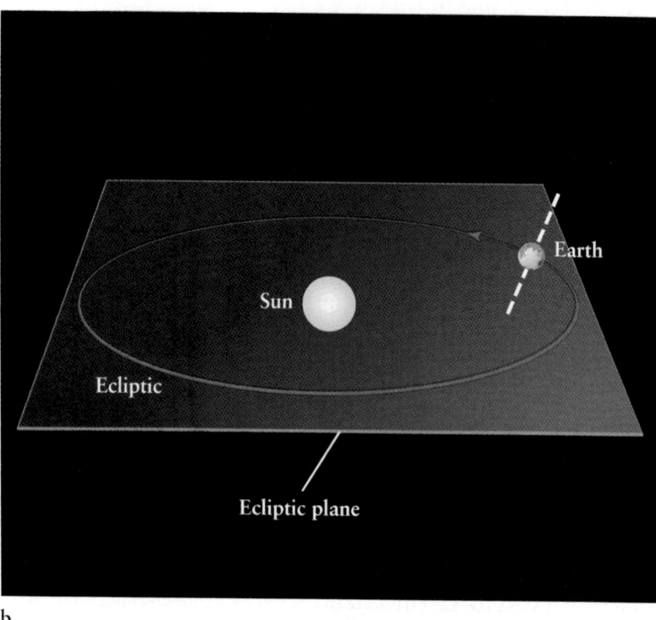

b

FIGURE 1-15 **The Ecliptic** (a) The ecliptic is the apparent annual path of the Sun on the celestial sphere. (b) The ecliptic is also the plane described by the Earth's path around the Sun. The planes created by the two ecliptics exactly coincide. As in (a), the rotation axis of the Earth is shown here tilted $23\frac{1}{2}°$ from being perpendicular to the ecliptic.

the Sun is precisely the same as the path of the Sun as seen from the Earth (Figure 1-15b). Recall also the discussion of the line running from the Sun through the Earth to the celestial sphere in Section 1-4.

INSIGHT INTO SCIENCE

Define Your Terms The words "ecliptic" and "constellation" have more than one scientific meaning. Most often, however, scientists restrict words to one well-defined meaning, because more than one meaning can lead to misunderstandings. Be sure that you understand each word in astronomy and how to use it.

The ecliptic and the celestial equator are different circles tilted $23\frac{1}{2}°$ with respect to each other on the celestial sphere. This occurs because the Earth's rotation axis is tilted $23\frac{1}{2}°$ away from a line perpendicular to the ecliptic, as shown in Figure 1-16. These two circles intersect at only two points, which are exactly opposite each other on the celestial sphere, as shown in Figure 1-17. Each of these two points is called an **equinox** (from the Latin words meaning "equal night"), because when the Sun appears at either point, it is directly over the Earth's equator, and there are 12 hours of daytime and 12 hours of nighttime everywhere on Earth on that day.

Except for tiny changes each year, which are discussed later, the Earth maintains this tilted orientation as it orbits the Sun. Polaris is above the North Pole throughout the year. For half the year, the northern hemisphere is tilted toward the Sun, and, as a result, the Sun rises higher in the northern hemisphere's sky than it does during the other half of the year (Figure 1-18). Equivalently, when the southern hemisphere is tilted toward the Sun, the Sun rises higher in the southern hemisphere's sky.

Consider the location of the Sun as seen from the northern hemisphere throughout the year. (Reverse all the seasonal terminology in the next four paragraphs if you are reading this in the southern hemisphere.) The day that the Sun rises farthest south of east is around December 21 each year (Figure 1-19) and is called the **winter solstice.** The winter solstice is the point on the ecliptic farthest south of the celestial equator (see Figure 1-17). It is also the day when the Sun rises to the lowest height at noon (see Figure 1-19) and it signals the day of the year in the northern hemisphere with the fewest number of daylight hours.

Moving along the ecliptic, day by day the Sun rises earlier, more northerly on the eastern horizon, and goes higher in the sky. Three months later, around March 21, the Sun crosses the celestial equator heading northward. This is called the **vernal equinox** and it is one of the two days on which the Sun rises due east and sets due west (see Figure 1-19). The vernal equinox is the "prime meridian" of the celestial sphere, as discussed earlier. Three months after the vernal equinox, around

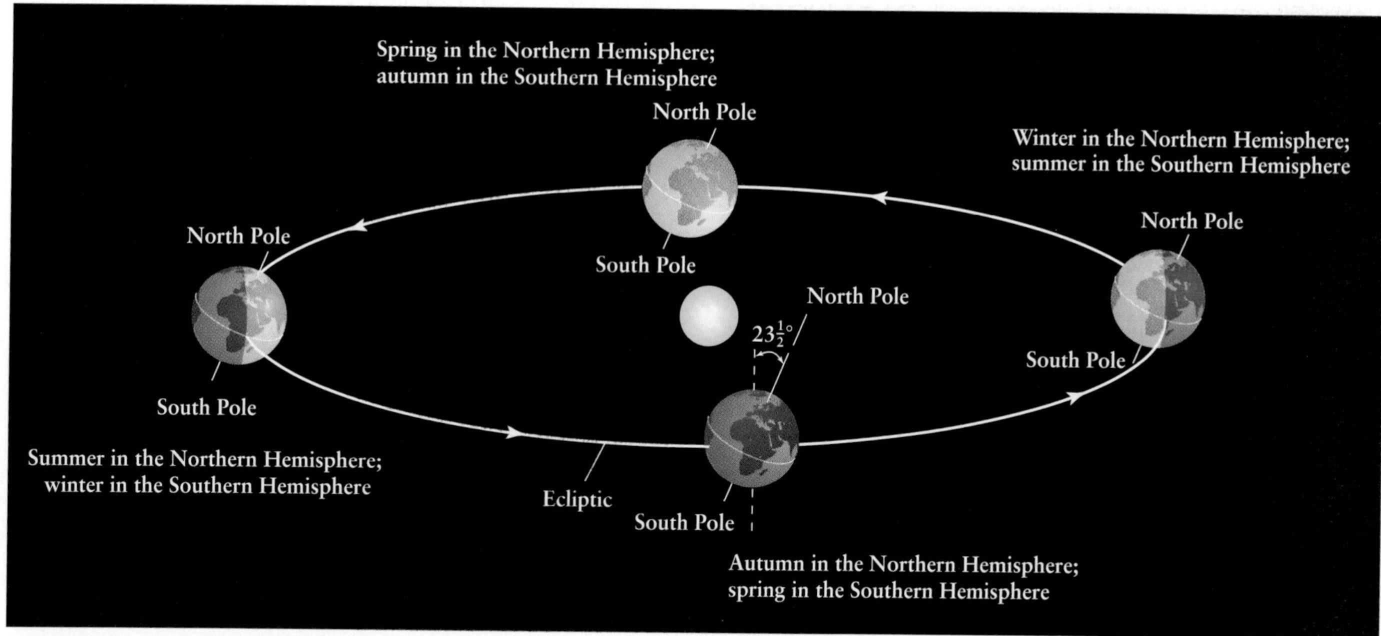

FIGURE 1-16 **The Tilt of the Earth's Axis** The Earth's axis of rotation is tilted $23\frac{1}{2}°$ from being perpendicular to the plane of the Earth's orbit. The Earth maintains this orientation (with its North Pole aimed at the north celestial pole near the star Polaris) throughout the year as it orbits the Sun. Consequently, the amount of solar illumination and the number of daylight hours at any location on Earth vary in a regular fashion throughout the year.

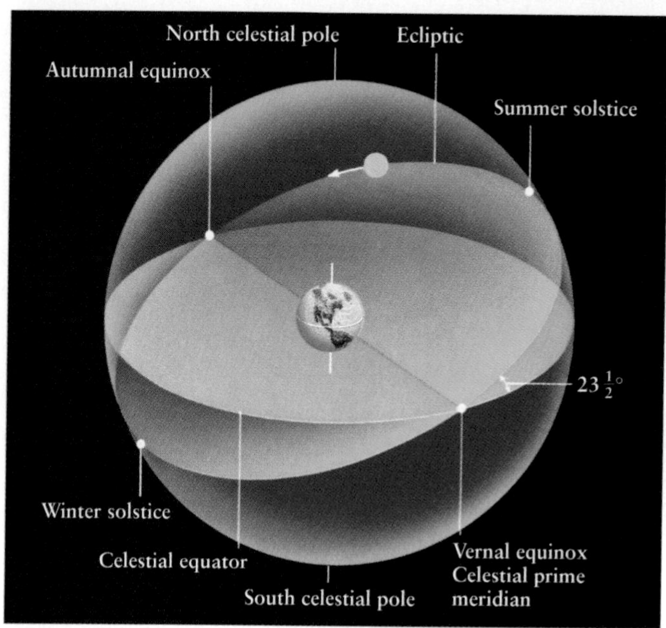

FIGURE 1-17 **The Seasons Are Linked to Equinoxes and Solstices** The ecliptic is inclined to the celestial equator by $23\frac{1}{2}°$ because of the tilt of the Earth's axis of rotation. The ecliptic and the celestial equator intersect at two points called the equinoxes. The northernmost point on the ecliptic is called the summer solstice; the southernmost point is called the winter solstice.

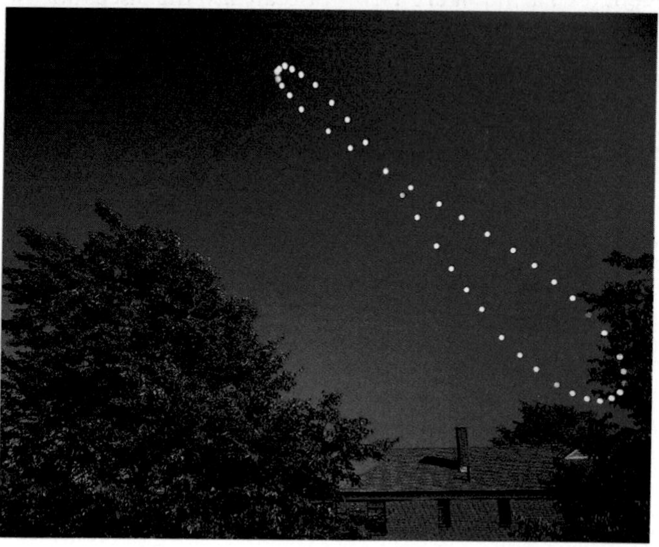

R I V U X G

FIGURE 1-18 **The Height of the Sun** The maximum height of the Sun in the sky varies throughout the year because of the $23\frac{1}{2}°$ tilt of the Earth's axis. This photograph shows the height of the Sun in the sky at the same clock time at 10-day intervals throughout the year, as well as its changing east-west location. The east-west variation in the Sun's location is caused by its rising, reaching its maximum height in the sky, and setting at different places throughout the year (see Figure 1-19). (Dennis Di Cicco)

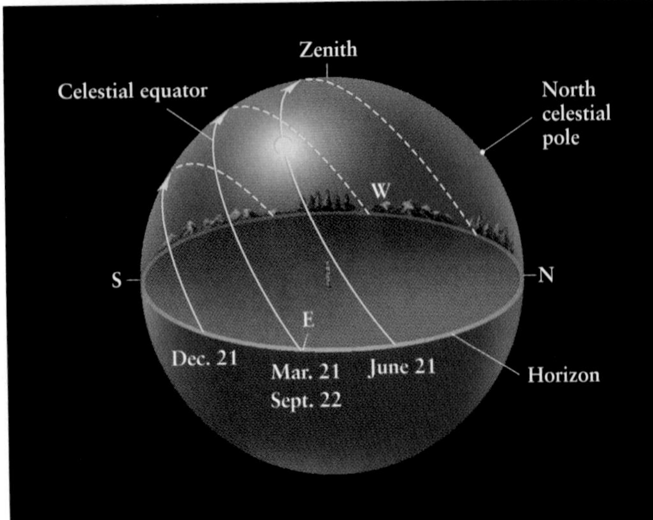
Zenith
Celestial equator
North celestial pole
W
S
N
E
Dec. 21
Mar. 21
Sept. 22
June 21
Horizon

FIGURE 1-19 **The Sun's Daily Path** On the first day of spring and the first day of fall, as seen from middle latitudes in the northern hemisphere, the Sun rises precisely in the east and sets precisely in the west. During summer in the northern hemisphere, the Sun rises in the northeast and sets in the northwest. In winter, the Sun rises in the southeast and sets in the southwest.

June 21, the Sun rises earliest, farthest north of east, and is highest in the sky at noon (see Figure 1-19). This is the **summer solstice** (see Figure 1-17), the day of the year in the northern hemisphere with the most daylight.

From June 21 through December 21, the Sun rises later in the morning and at a location farther south than the preceding day. It is also lower in the sky at noon each succeeding day—the cycle of the previous six months reverses. The **autumnal equinox** occurs around September 22, with the Sun heading southward across the celestial equator, as seen from the Earth.

The higher the Sun rises during the day, the more daylight hours there are. During these days with longer periods of daylight, more light and heat energy from the Sun strike that 3
hemisphere day by day. Furthermore, when the Sun is higher in the sky, its energy is more concentrated on the Earth's surface (Figure 1-20) and so the Earth at such locations gets warmer than at places where the Sun is up for a shorter time and the sunlight is less direct. In other words, the temperature and, hence, the seasons are determined by the duration of daylight there and the height of the Sun in the sky.

To summarize, the Sun is lowest in the northern sky on the winter solstice. This marks the beginning of winter in the northern hemisphere. As the Sun moves northward, the amount of daylight increases daily. The vernal equinox marks a midpoint in the amount of heat deposited by the Sun onto the northern hemisphere and is the beginning of spring. When the Sun reaches the summer solstice, it is highest in the northern sky and is above the horizon for the longest time of the year. This is the beginning of summer. Returning southward, the Sun crosses the celestial equator once again on the autumnal equinox, the beginning of fall.

Contrary to common sense, the changing distance between the Earth and the Sun has only a minor effect on the seasons. If the seasons were caused by the changing distance from the Earth to the Sun, all parts of the Earth should have the same seasons at the same time. In fact, the northern and southern hemispheres have exactly opposite seasons. Furthermore, the Earth is closest to the Sun on or
around January 3 of each year—the dead of winter in the 4
northern hemisphere! The variation in Earth's distance from the Sun would have a greater effect if it were not for

The Sun is high in the midday summer sky...
... so a shaft of sunlight is concentrated onto a small area, which heats the ground effectively and makes the days warm.

a **The Sun in summer**

The Sun is low in the midday winter sky...
... so the same shaft of sunlight is spread out over a larger area and less heating of the ground takes place.

b **The Sun in winter**

FIGURE 1-20 **The Energy Deposited by the Sun** In the northern hemisphere, the angle of the Sun above the southern horizon determines how much heat and light strike each square meter of ground. In the southern hemisphere, the Sun's angle is measured above the northern horizon. **(a)** During the summer at middle latitudes, a shaft of sunlight illuminates a nearly circular patch of ground at noon. **(b)** During the winter, the same shaft of sunlight at noon strikes the ground at a steeper angle, spreading the same amount of sunlight over a larger, oval shape. Because the sunlight's energy is diluted over a larger area and the Sun is above the horizon less during the winter, the ground receives less heat during the winter than during the summer.

the fact that the southern hemisphere has more area covered by oceans than does the northern hemisphere. As a result, when the Earth is closer to the Sun (and the Sun is high in the southern hemisphere's sky), the southern oceans scatter more light and heat directly back into space than occurs when the Sun is over the northern hemisphere during the other half of the year. Had the extra energy sent back into space when we are closer to the Sun been absorbed by our planet, the Earth would, indeed, heat more during this time than when the Sun is over the northern hemisphere.

INSIGHT INTO SCIENCE

Expect the Unexpected Many phenomena in the universe defy commonsense explanations. The process of science requires that we question the obvious, that is, what we think we know. The fact that the changing distance from the Earth to the Sun has a minimal effect on the seasons is an excellent example.

During the northern hemisphere's summer months, when the northern hemisphere is tilted toward the Sun, the Sun rises in the northeast and sets in the northwest. The Sun provides more than 12 hours of daylight in the northern hemisphere and passes high in the sky at noontime. At the summer solstice, the Sun is as far north as it gets, giving the greatest number of daylight hours to the northern hemisphere.

During the northern hemisphere's winter months, when the northern hemisphere is tilted away from the Sun, the Sun rises in the southeast. Daylight lasts for fewer than 12 hours, as the Sun skims low over the southern horizon and sets in the southwest. Night is longest in the northern hemisphere when the Sun is at the winter solstice.

People at different latitudes see the noontime Sun at different angles above the southern horizon each day. The farther from the equator you are, the lower the Sun is in the sky at noon. The farther north or south you go from the equator, less of the Sun's heat and light energy are deposited on each acre (see Figure 1-20) and, therefore, the colder the land is. At latitudes above $66\frac{1}{2}°$ north latitude or below $66\frac{1}{2}°$ south latitude, the Sun does not rise at all during parts of their fall and winter months. During their spring and summer months, those same regions of the Earth have continuous sunlight for weeks or months, hence the name "Land of the Midnight Sun."

The Sun takes one year to complete a trip around the ecliptic. Since there are about $365\frac{1}{4}$ days in a year and 360° in a circle, the Sun appears to move along the ecliptic at a rate of slightly less than 1° per day. The constellations through which the Sun moves throughout the year as it travels along the ecliptic are called **zodiac** constellations. We cannot see the stars of these constellations when the Sun is among them, of course, but we can plot the Sun's path on the celestial sphere to determine through which constellations it moves. One traditionally learns that there are 12 zodiac constellations, which are used by astrologers. These 12 are based on constellation boundaries used in 5
antiquity. Those boundaries have since been redefined by astronomers, and the Sun moves through 13 modern constellations throughout the year. (The thirteenth zodiac constellation is Ophiuchus, the Serpent Holder. The Sun passes through Ophiuchus from December 1 to December 19 each year.) Table 1-1 lists all the zodiac constellations and the dates the Sun passes through them.

TABLE 1-1 The 13 Constellations of the Zodiac

Constellation	Dates of Sun's passage through
Pisces	March 13–April 20
Aries	April 20–May 13
Taurus	May 13–June 21
Gemini	June 21–July 20
Cancer	July 20–August 11
Leo	August 11–September 18
Virgo	September 18–November 1
Libra	November 1–November 22
Scorpius	November 22–December 1
Ophiuchus	December 1–December 19
Sagittarius	December 19–January 19
Capricorn	January 19–February 18
Aquarius	February 18–March 13

1-8 Precession is a slow, circular motion of the Earth's axis of rotation

As noted earlier, the position of the Earth's axis of rotation changes slightly with respect to the celestial sphere (that is, it "points" in a slightly different direction) each year. This change in orientation is caused by gravitational forces from the Moon and Sun pulling on the slight equatorial bulge at the Earth's equator created by the Earth's rotation—our planet's diameter is about 43 km (27 mi) greater at the equator than from pole to pole. **Gravitation** is the universal force of attraction between all matter. The strength of the gravitational force depends on the amount of mass the objects have and the distance between them.

Because of the Earth's tilted axis of rotation, the Sun and Moon are usually not located directly over the Earth's equator. As a result, their gravitational attraction on the

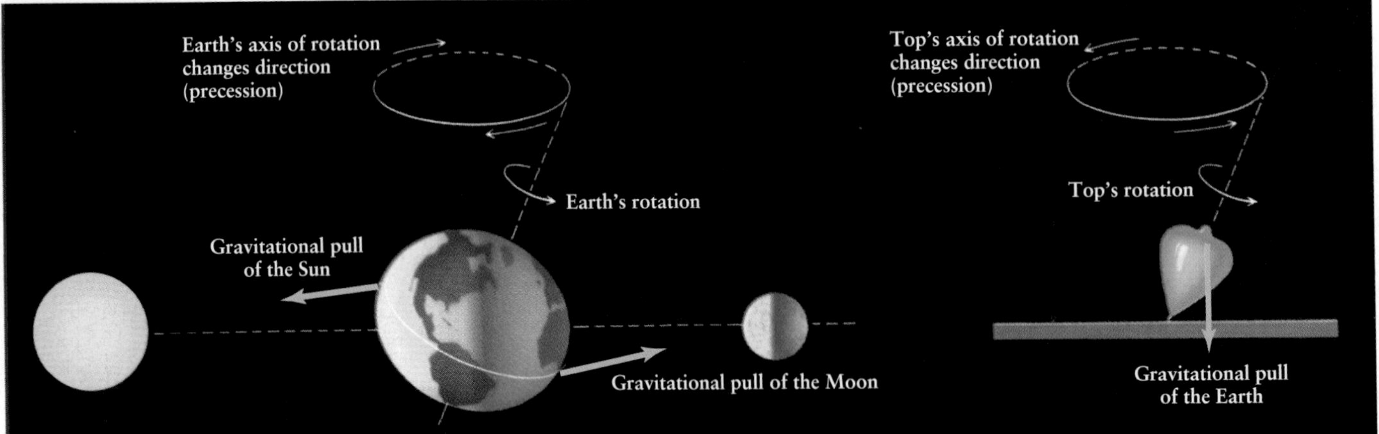

FIGURE 1-21 **Precession** The gravitational pulls of the Moon and the Sun on the Earth's equatorial bulge cause the Earth to precess. As the Earth precesses, its axis of rotation slowly traces out a circle in the sky. The situation is analogous to that of a spinning top. The top of the toy top shows the motion of Earth's North Pole or South Pole, while the point on which the top spins represents the center of the Earth. As the top spins, the Earth's gravitational pull causes the top's axis of rotation to move in a circle—to precess.

Earth tries to force the equatorial bulge to be as close to them as possible. However, the Earth does not respond to these forces from the Sun and Moon by straightening up perpendicular to them on its axis. Instead, it changes the direction in which its axis of rotation points on the celestial sphere—a motion called **precession.** This is exactly the same behavior that is exhibited by a spinning top (Figure 1-21). If the top were not spinning, gravity would pull it over on its side. But when it is spinning, the combined actions of gravity and rotation cause the top's axis of rotation to precess or wobble in a circular path. As with the toy top, the combined actions of gravity from the Sun and Moon and rotation cause the Earth's axis to trace a circle in the sky while remaining tilted about 23½° away from the perpendicular. In the mid-1990s, astronomers simulated the behavior of the Earth and discovered that without a large Moon, the Earth would not keep to a 23½° tilt but rather it would change angle dramatically and sometimes it would wobble wildly!

The Earth's rate of precession is slow compared to human time scales. It takes about 26,000 years for the north celestial pole to trace out a complete circle around the sky, as shown in Figure 1-22. (The south celestial pole executes a similar circle in the southern sky.) At the present time, the Earth's north axis of rotation points within 1° of the star Polaris. In 3000 B.C., it was pointing near the star Thuban in the constellation of Draco (the Dragon). In A.D. 14,000, the pole star will be near Vega in Lyra.

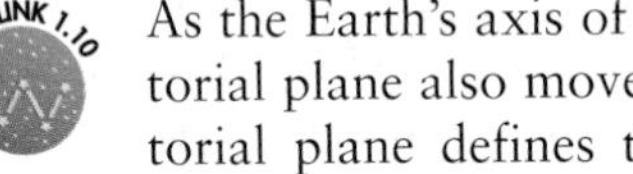

As the Earth's axis of rotation precesses, its equatorial plane also moves. Because the Earth's equatorial plane defines the location of the celestial equator in the sky, the celestial equator also precesses.

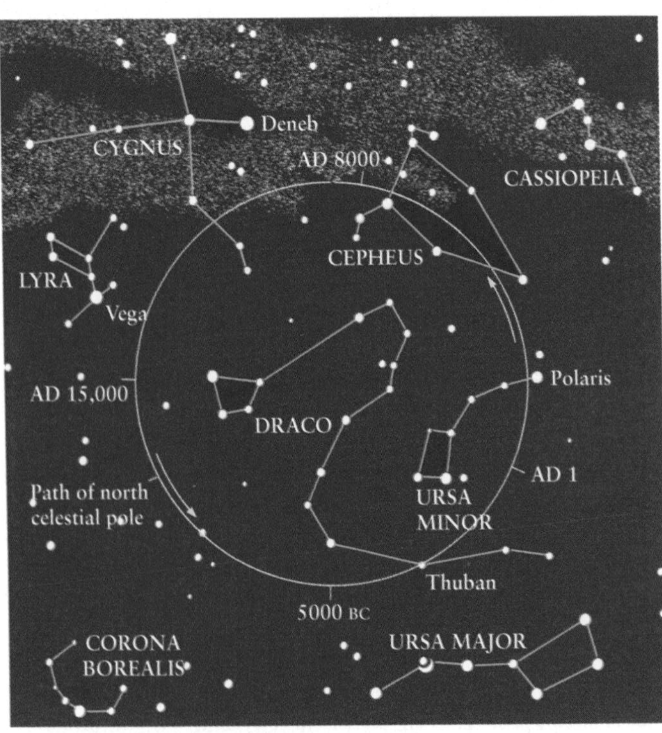

▶ FIGURE 1-22 **The Path of the North Celestial Pole** As the Earth precesses, the north celestial pole slowly traces out a circle among the northern constellations. At the present time, the north celestial pole is near the moderately bright star Polaris, which serves as the pole star. The total precession period is about 26,000 years.

Recall that the intersections of the celestial equator and the ecliptic define the equinoxes and so these key locations in the sky also shift slowly from year to year. This entire phenomenon is often called the **precession of the equinoxes.** This change was discovered by the great Greek astronomer Hipparchus in the second century B.C. Today, the vernal equinox is located in the constellation Pisces (the Fishes). Two thousand years ago, it was located in Aries (the Ram). Around the year A.D. 2600, the vernal equinox will move into Aquarius (the Water Bearer).

1-9 The phases of the Moon originally inspired the concept of the month

The Moon's contribution to the Earth's precession leads to changes in the direction that the Earth's rotation axis points, and these changes take millennia. Other lunar effects are noticeable every day. As the Moon orbits the Earth, it moves from west to east, changing position among the background stars. Its position relative to the Sun also changes, and, as a result, we see different **lunar phases.** As with the Earth and other spherical bodies, the Sun illuminates half of the Moon at all times. The Moon's phase depends on how much of its sunlit hemisphere is facing the Earth. When the Moon is closest to the Sun in the sky, its dark hemisphere faces the Earth. This phase, during which the Moon is at most a tiny crescent, is called the *new* Moon (Figure 1-23).

During the seven days following the new phase, more of the Moon's illuminated hemisphere becomes exposed to our view, resulting in a phase called the *waxing crescent* Moon. At the *first quarter* Moon, we see half of the illuminated hemisphere and half of the dark hemisphere. "Quarter Moon" refers to how far in its cycle the Moon has gone, rather than what fraction of the Moon appears lit by sunlight.

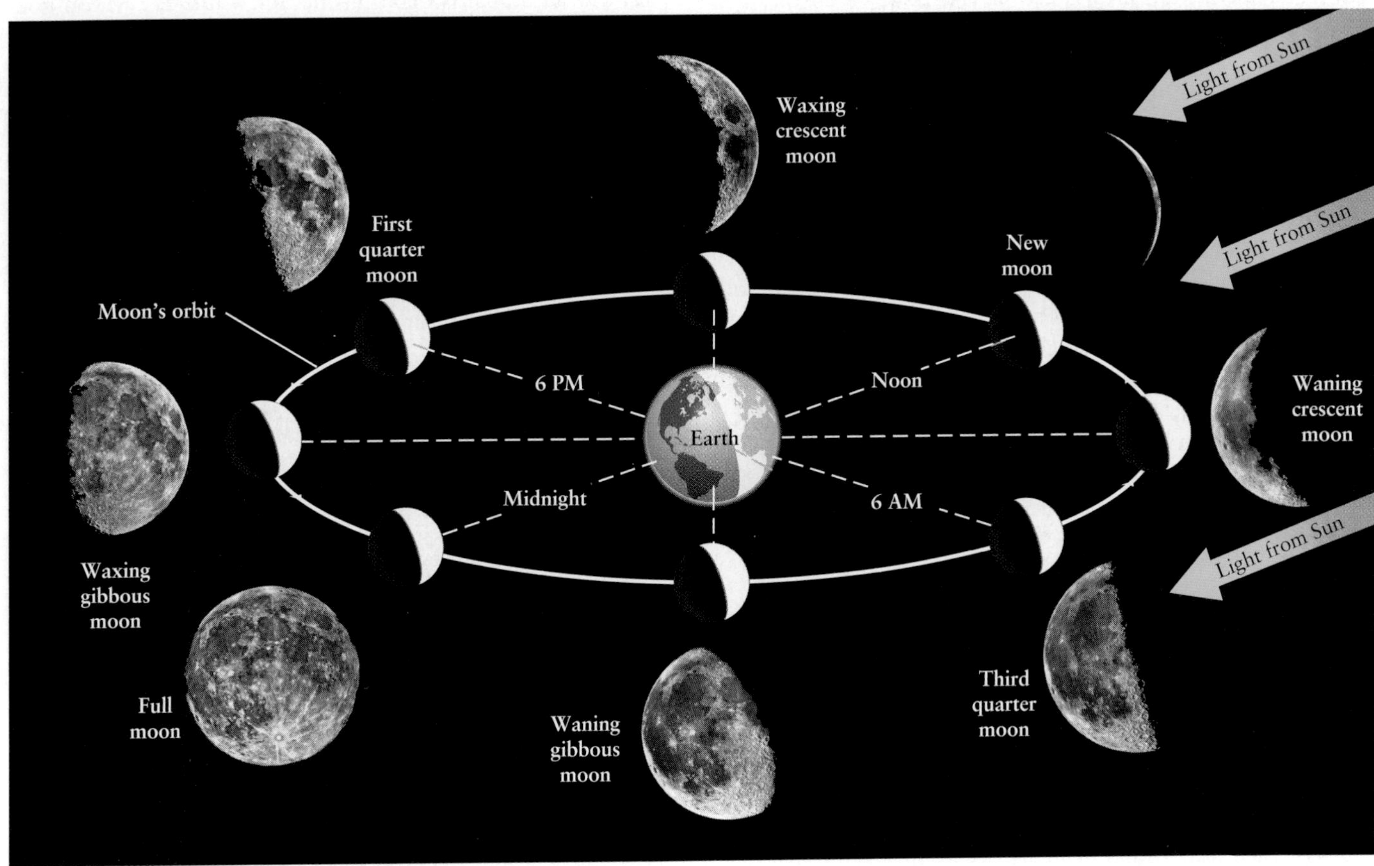

FIGURE 1-23 **The Phases of the Moon** The diagram shows the Moon at eight locations on its orbit as viewed from far above the Earth's North Pole. The corresponding photographs show the resulting lunar phases as seen from Earth. The waning and third quarter photographs look backward, but they are correctly oriented as seen from Earth. Light from the Sun illuminates one-half of the Moon at all times, while the other half is dark. It takes about $29\frac{1}{2}$ days for the Moon to go through all its phases. (Yerkes Observatory and Lick Observatory)

During the next week still more of the illuminated hemisphere can be seen from Earth, giving us the phase called the *waxing gibbous* Moon. Gibbous means "rounded on both sides." When the Moon arrives on the opposite side of the Earth from the Sun, we see virtually all of the fully illuminated hemisphere. This phase is the *full* Moon. Over the following two weeks, we see less and less of the illuminated hemisphere as the Moon continues along its orbit. This movement produces the phases called the *waning gibbous* Moon, *third quarter* Moon, and the *waning crescent* Moon. The Moon completes a full cycle of phases in 29½ days.

6 There is often confusion over the terms "far side" and "dark side" of the Moon. The far side is the side of the Moon facing away from the Earth, while the dark side is the side of the Moon on which the Sun is not shining. By examining the photographs in Figure 1-23, you can see that the same side of the Moon faces the Earth all the time. The half of the Moon that never faces the Earth is the far side. However, the far side is not always the dark side, since we see part of the dark side whenever we see less than a full Moon.

Figure 1-23 shows the Moon at various positions in its orbit. Remember that the bright side of the Moon is on the right (west) side of the waxing Moon, while the bright side is on the left (east) side of the waning Moon. This information can tell you at a glance whether the Moon is waxing or waning. When looking at the Moon through a telescope, the best place to see details is where the shadows are longest. This occurs at the boundary between the bright and dark regions, called the **terminator.**

7 Figure 1-23 also shows local time around the globe, from noon, when the Sun is highest in the sky, to midnight, when it is on the opposite side of the Earth. These time markings roughly indicate when the Moon is highest in the sky. For example, at first quarter, the Moon is 90° east of the Sun in the sky; hence, the Moon is highest at sunset. At full Moon, the Moon is opposite the Sun in the sky; thus, the Moon is highest at midnight. Using this information, you can see that the Moon is visible during the daytime (Figure 1-24) for a part of most days of the year.

Since the dawn of civilization, people have sought accurate timekeeping systems. Ancient Egyptians wanted to know when the Nile would flood, and farmers everywhere needed to know when to plant crops. Migratory tribes wanted to know when the weather would change. Religious leaders scheduled observances in accordance with celestial events. Thus, astronomers have traditionally been responsible for telling time. Indeed, of the four ways in which time cycles are set, three are astronomical in origin: Time is determined by the positions of the Moon, Sun, or stars or, in our own age, by technological means, such as atomic clocks.

The approximately four weeks that the Moon takes to complete one cycle of its phases inspired our ancestors to invent the concept of a month. Astronomers find it useful to define two types of months, depending on whether the Moon's motion is measured relative to the stars or to the Sun. Neither type corresponds exactly to the months of our usual calendar, which have different lengths.

The **sidereal month** is the time it takes the Moon to complete one full orbit of 360° around the Earth (Figure 1-25). As with the sidereal day, the length of the sidereal month is determined by the location of the Moon in its orbit around Earth as measured with respect to the stars. Equivalently, this is the time it takes the Moon to start at one place on the celestial sphere and return to exactly the same place again. The sidereal orbital period of the Moon takes approximately 27.3 days. The **synodic month,** or **lunar month,** is the time it takes the Moon to complete one 29½-day cycle of phases (that is, from new Moon to new Moon or from full Moon to full Moon) and thus is measured with respect to the Sun rather than the stars.

The synodic month is longer than the sidereal month because the Earth is orbiting the Sun while the Moon goes

R I V U X G

FIGURE 1-24 **The Moon during the Day** The Moon is visible at some time during daylight hours virtually every day. The time of day or night it is up in our sky depends on its phase. (Pegasus/Visuals Unlimited)

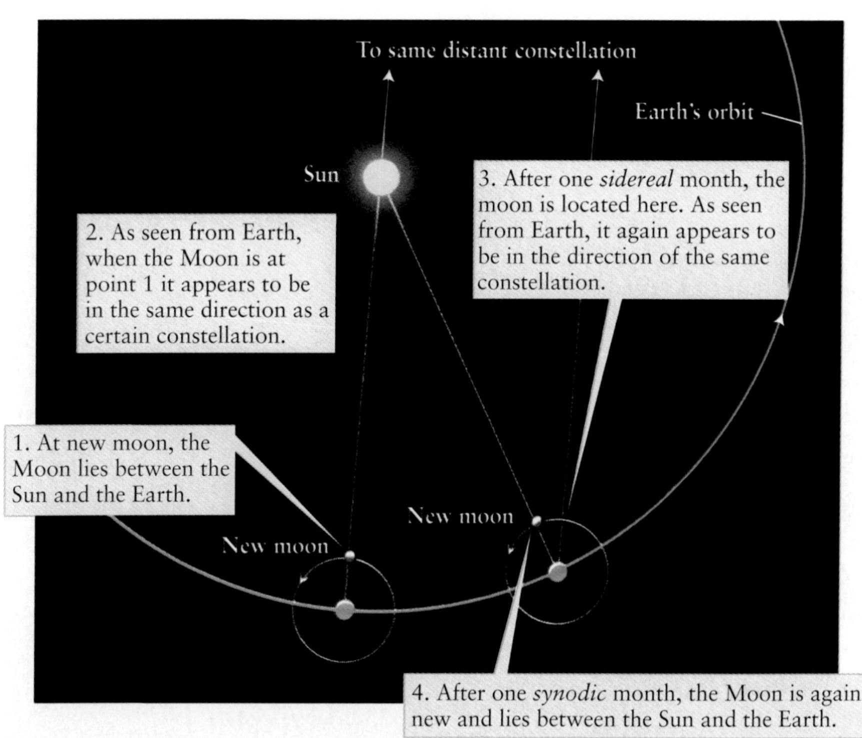

FIGURE 1-25 **The Sidereal and Synodic Months** The sidereal month is the time it takes the Moon to complete one revolution with respect to the background stars. However, because the Earth is constantly moving in its orbit about the Sun, the Moon must travel through more than 360° to get from one new Moon to the next. The synodic month is the time between consecutive new Moons or consecutive full Moons. Thus, the synodic month is slightly longer than the sidereal month.

through its phases. As shown in Figure 1-25, the Moon must travel *more* than 360° along its orbit to complete a cycle of phases (for example, from one new Moon to the next), which takes about 2.2 days longer than the sidereal month.

Both the sidereal month and synodic month vary somewhat, because the gravitational pull of the Sun on the Moon affects the Moon's speed as it orbits the Earth. The sidereal month can vary by as much as 7 hours, while the synodic month can vary by as much as 12 hours.

The terms *synodic* and *sidereal* are also used in discussing the motion of the other bodies in the solar system. The synodic period of a planet is the time between consecutive straight alignments between the Sun, Earth, and that planet (during which time the planet also goes through a cycle of phases, as seen from the Earth). Recall that any orbit measured with respect to the stars is called "sidereal," including orbits of the planets around the Sun, as well as orbits of moons around their planets. The Earth's sidereal year is 365.2564 days. The difference between the sidereal year and the tropical year (the time from one vernal equinox to the next) is due primarily to the Earth's precession (see Section 1-8).

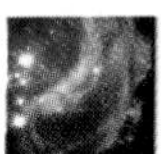

ECLIPSES

Eclipses are among the most spectacular natural phenomena. During a **lunar eclipse,** the brilliant full Moon often darkens to a deep red, while during a **solar eclipse,** broad daylight is transformed into an eerie twilight, as the Sun seems to be blotted from the sky. A lunar eclipse occurs when the Moon passes through the Earth's shadow. This can happen only when the Sun, Earth, and Moon are in a straight line at full Moon. A solar eclipse occurs when the Moon's shadow moves across the Earth's surface. As seen from Earth, the Moon moves in front of the Sun.

1-10 Eclipses occur only when the Moon crosses the ecliptic during the new or full phase

At first glance, it would seem that eclipses should happen at every new and full Moon, but, in fact, they occur much less often. Eclipses occur infrequently because the Moon's orbit is tilted about 5° from the ecliptic, as shown in Figure 1-26. Because of this tilt, the new Moon and full Moon usually occur when the Moon is either above or below the plane of the Earth's orbit. In such positions, a perfect alignment between the Sun, Moon, and Earth is not possible and an eclipse cannot occur.

The Moon crosses the ecliptic at what is called the **line of nodes** (see Figure 1-26). When the Moon crosses the plane of the ecliptic during its new or full phase, an eclipse takes place. By calculating the number of times a new Moon takes place on the line of nodes, we find that at least two and no more than five solar eclipses occur each year. Lunar eclipses occur just about as frequently as solar eclipses, with the maximum number of eclipses (solar plus lunar) possible in a year being seven.

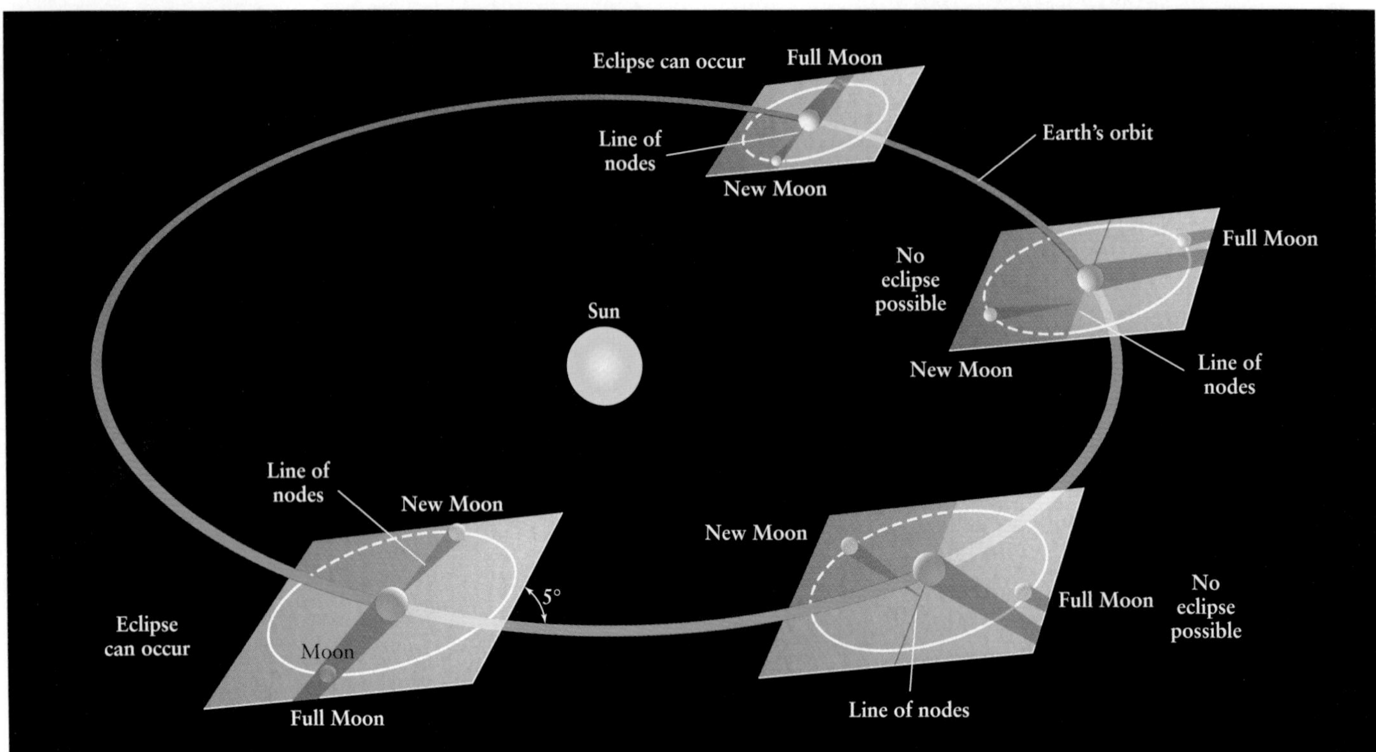

FIGURE 1-26 **Conditions for Eclipses** The Moon must be very nearly on the ecliptic at new Moon for a solar eclipse to occur. A lunar eclipse occurs only if the Moon is very nearly on the ecliptic at full Moon. When new Moon or full Moon phases occur away from the ecliptic, no eclipse is seen, because the Moon and the Earth do not pass through each other's shadow.

1-11 There are three types of lunar eclipse

The Earth's shadow has two distinct parts, as shown in Figure 1-27. The **umbra** is the part of the shadow where all direct sunlight is blocked by the Earth. If you were in the Earth's umbra looking at the Earth, you would not see the Sun behind it at all. The **penumbra** of the shadow is where the Earth blocks only some of the sunlight. If you were in the Earth's penumbra looking at the Earth, you would see a crescent Sun behind it. Depending on how the Moon travels through the Earth's shadow, three kinds of lunar eclipses may occur. A **penumbral eclipse,** when the Moon passes through only the Earth's penumbra, is easy to miss. The Moon still looks full, just a little dimmer than usual and sometimes slightly reddish in color (path 1 in Figure 1-27).

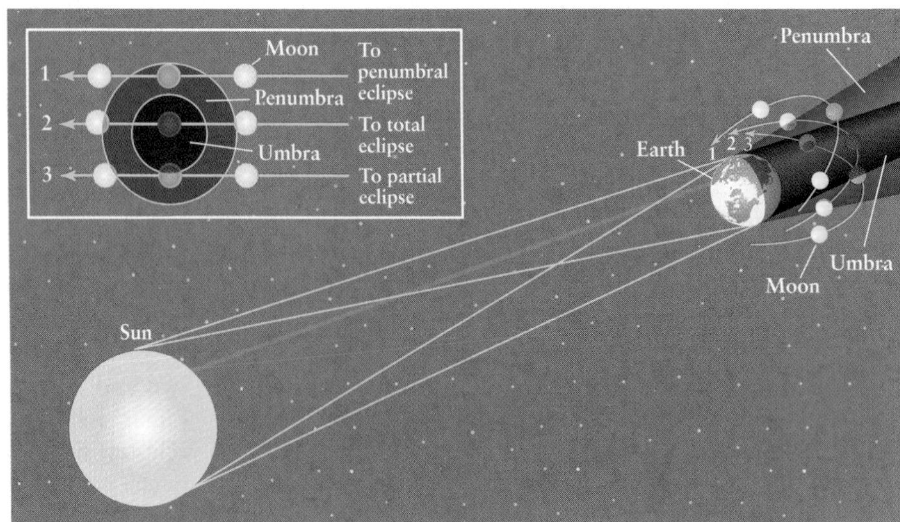

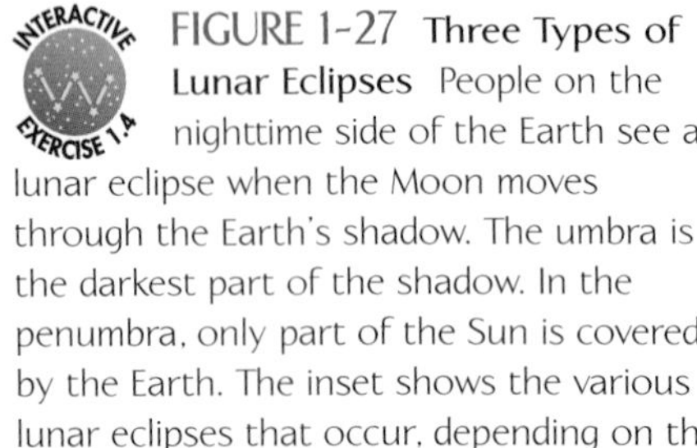
FIGURE 1-27 **Three Types of Lunar Eclipses** People on the nighttime side of the Earth see a lunar eclipse when the Moon moves through the Earth's shadow. The umbra is the darkest part of the shadow. In the penumbra, only part of the Sun is covered by the Earth. The inset shows the various lunar eclipses that occur, depending on the Moon's path through the Earth's shadow.

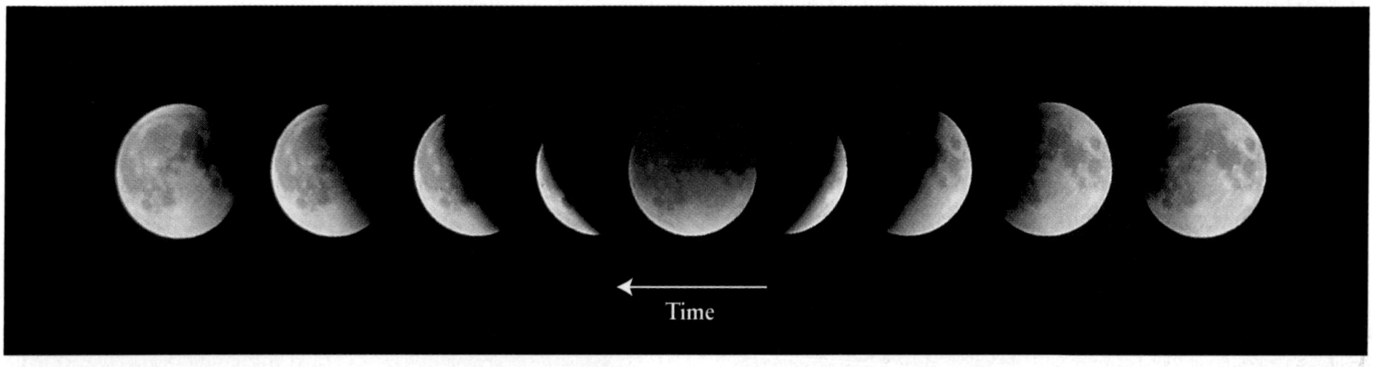

R I V U X G

FIGURE 1-28 **A Total Eclipse of the Moon** This sequence of nine photographs was taken over a 3-hour period during the lunar eclipse of January 20, 2000. During the total phase, the Moon has a distinctly reddish color. (Fred Espenak, NASA/Goddard Space Flight Center; © Fred Espenak, MrEclipse.com)

When just part of the lunar surface passes through the umbra, a bite seems to be taken out of the Moon, and we see a **partial eclipse** (path 3 in Figure 1-27). When the Moon travels completely into the umbra, we see a **total eclipse** of the Moon (path 2 in Figure 1-27). Total lunar eclipses with the maximum duration, lasting for up to 1 hour and 47 minutes, occur when the Moon is closest to the Earth and it travels directly through the center of the umbra. Table 1-2 lists all the total and partial lunar eclipses from mid-2005 through 2008.

Even during a total eclipse, the Moon does not completely disappear. A small amount of sunlight passing through the Earth's atmosphere is bent into the Earth's umbra. The light deflected into the umbra is primarily red and orange, and thus the darkened Moon glows faintly in rust-colored hues, as shown in Figure 1-28. At sunrise and sunset the Sun appears red or orange for the same reason, because at those times red and orange light are deflected from the Sun toward you.

Lunar eclipses are perfectly safe to watch with the naked eye. However, solar eclipses are *never* safe to view without suitable eye protection. ***Viewing the Sun directly for more than a moment without an approved filter causes permanent eye damage.***

1-12 There are also three types of solar eclipse

Because of their different distances from Earth, the Sun and the Moon have nearly the same angular diameter as seen from Earth—about ½°. When the Moon completely covers the Sun, the result is a total solar eclipse. You must be at a location within the Moon's umbra to see a total solar eclipse.

During those few precious moments, hot gases (the **solar corona**) surrounding the Sun can be observed and photographed (Figure 1-29). Astronomers can then learn more about the Sun's temperature, chemistry, and atmospheric activity. You can see in Figure 1-30 that only the tip of the Moon's umbra ever reaches the Earth's surface. As

TABLE 1-2 Lunar Eclipses, 2005–2008

WEB LINK 1.11

Date	Visible from	Type	Duration of totality (h:min)
2005 April 24	Americas, Pacific, eastern Asia	Penumbral	
2005 October 17	Asia, Australia, North America, Pacific	Partial	
2006 March 14	Americas, Europe, Africa, Asia	Penumbral	
2006 September 7	Europe, Africa, Asia, Australia	Partial	
2007 March 3	Americas, Europe, Africa, Asia	Total	1:14
2007 August 28	Americas, eastern Asia, Australia	Total	1:31
2008 February 21	Americas, Europe, Africa	Total	0:51
2008 August 16	South America, Europe, Africa, Asia, Australia	Partial	

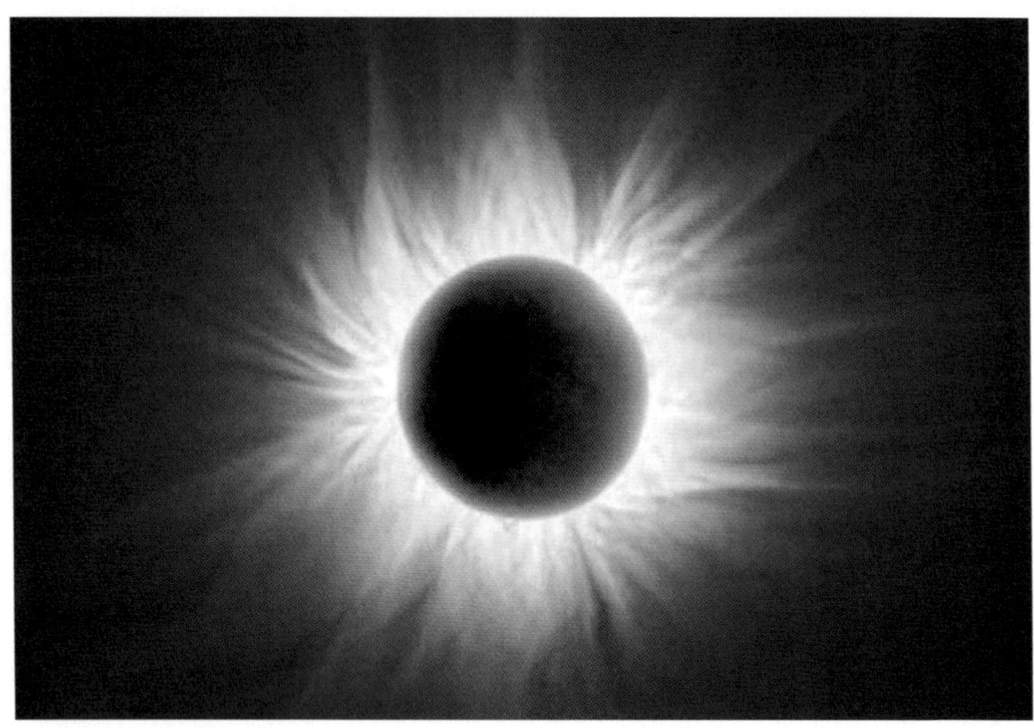

FIGURE 1-29 **A Total Eclipse of the Sun** During a total solar eclipse, the Moon completely covers the Sun's disk, and the solar corona can be photographed. This halo of hot gases extends for millions of kilometers into space. This gorgeous image is a composite of several taken in Chisamba, Zambia, during the June 21, 2001, solar eclipse. (F. Espenak)

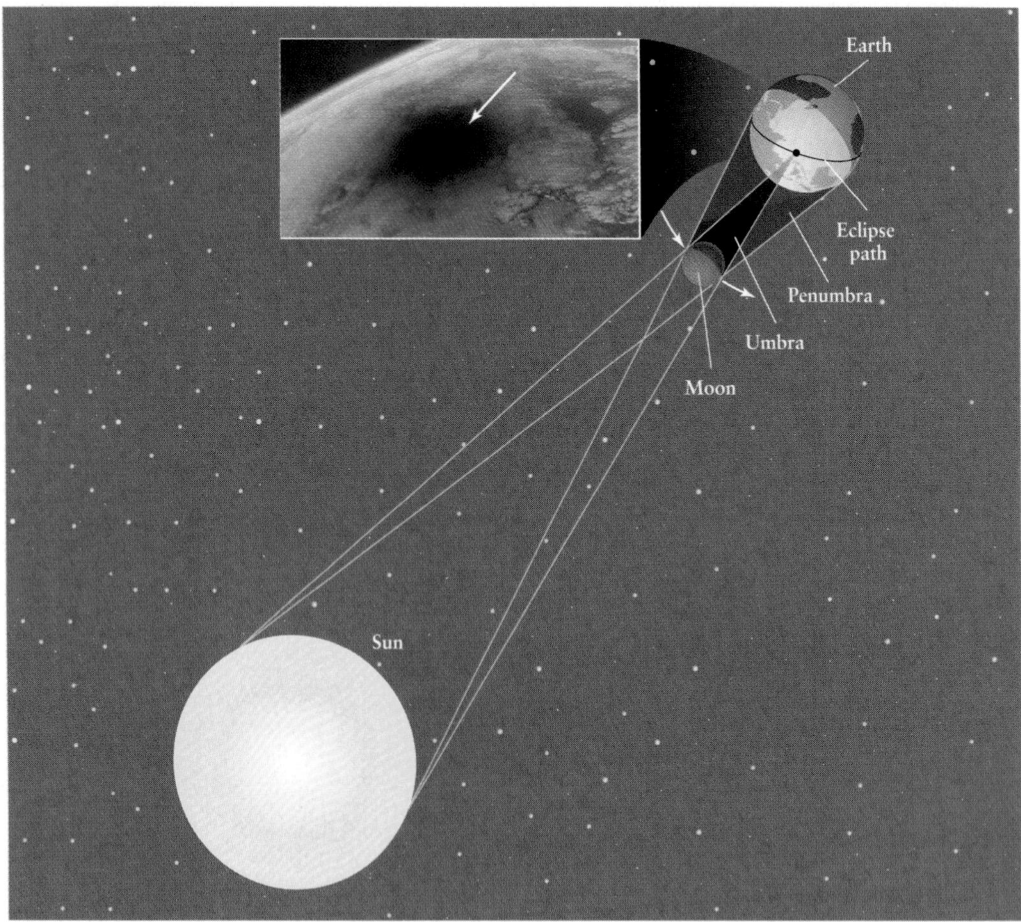

FIGURE 1-30 **The Geometry of a Total Solar Eclipse** During a total solar eclipse, the tip of the Moon's umbra traces an eclipse path across the Earth's surface. People inside the eclipse path see a total solar eclipse, whereas people inside the penumbra see only a partial eclipse. The photograph in this figure shows the Moon's shadow on the Earth. It was taken from the *Mir* space station during the August 11, 1999, total solar eclipse. The Moon's umbra appears as a dark spot on the eastern coast of the United States. (Jean-Pierre Haigneré, Centre National d'Etudes Spatiales, France/GSFS)

TABLE 1-3 Solar Eclipses, 2005–2008

WEB LINK 1.13

Date	Type	Visible from	Total/annular eclipse time (min:s)
2005 April 8	Total	New Zealand, North and South America	0:42
2005 October 3	Annular	Europe, Africa, south Asia	4:32
2006 March 29	Total	Africa, Europe, west Asia	4:07
2006 September 22	Annular	South America, west Africa, Antarctica	7:09
2007 March 19	Partial	Asia, Alaska	
2007 September 11	Partial	South America, Antarctica	
2008 February 7	Annular	Antarctica, east Australia, New Zealand	2:12
2008 August 1	Total	northeast North America, Europe, Asia	2:27

the Earth turns and the Moon orbits, the tip traces an **eclipse path** across the Earth's surface. Only people within this path are treated to the spectacle of a total solar eclipse. The photograph in Figure 1-30 shows the dark spot produced by the Moon's umbra on the Earth's surface during a total solar eclipse.

The Earth's rotation and the orbital motion of the Moon cause the umbra to race along the eclipse path at speeds in excess of 1700 km/h (1050 mph). For this reason, a total eclipse never lasts for more than 7½ minutes at any one location on the eclipse path, and it usually lasts for only a few moments.

The Moon's umbra is also surrounded by a penumbra. During a solar eclipse, the Moon's penumbra extends over a large portion of the Earth's surface. When only the penumbra sweeps across the Earth's surface, the Sun is only partly covered by the Moon. This circumstance results in a *partial eclipse of the Sun.* Similarly, people in the penumbra of a total eclipse see a partial eclipse. In either case, the Sun looks crescent as seen from the Earth.

The Moon's orbit around the Earth is not quite a perfect circle. The distance between the Earth and the Moon, which averages 384,400 km (238,900 mi), varies by a few percent as the Moon goes around the Earth. The width of the eclipse path depends primarily on the Earth-Moon distance during an eclipse. The eclipse path is widest—up to 270 km (170 mi)—when the Moon happens to be at the point in its orbit nearest the Earth. Usually, however, the path is much narrower.

If a solar eclipse occurs when the Moon is farthest from the Earth, then the Moon's umbra falls short of the Earth and no one sees a total eclipse. From the Earth's surface, the Moon then appears too small to cover the Sun completely, and a thin ring or "annulus" of light is seen around the edge of the Moon at mid-eclipse. This type of eclipse is called an **annular eclipse** (Figure 1-31). The length of the Moon's umbra is nearly 5000 km (3100 mi) shorter than the average distance between the Moon and the Earth's surface. Thus, the Moon's shadow often fails to reach the Earth, making annular eclipses occur slightly more often than total eclipses. Table 1-3 lists all the total, partial, and annular solar eclipses from mid-2005 through 2008.

FIGURE 1-31 **An Annular Eclipse of the Sun** This composite of five exposures taken at sunrise in Costa Rica shows the progress of an annular eclipse of the Sun that occurred on December 24, 1974. Note that at mid-eclipse the edge of the Sun is visible around the Moon. (Dennis Di Cicco)

R I V U X G

A total solar eclipse is a dramatic event. The sky begins to darken, the air temperature falls, and the winds increase as the Moon's umbra races toward you. All nature responds: Birds go to roost, flowers close their petals, and crickets begin to chirp as if evening had arrived. As the moment when the Sun becomes totally eclipsed approaches, the landscape is bathed in shimmering bands of light and dark as the last few rays of sunlight peek out from behind the edge of the Moon. Finally, the corona blazes forth in a star-studded daytime sky. It is an awesome sight but should only be viewed through a suitable protective filter.

1-13 Frontiers yet to be discovered

Geologists have discovered a variety of cycles of global temperature change that have occurred over the history of the Earth. Indeed, geologists have discovered that much of the Earth once suffered a global freezing. These changes occur over tens of thousands of years, hundreds of thousands of years, and possibly longer cycles. Some of the causes of these changes have yet to be discovered. Even today the Earth's global climate is changing over the years. All the causes of this change are not yet known. For example, are our activities the only cause of global warming or are there long-term seasonal changes that affect it and, if so, how?

Summary of Key Ideas

Sizes in Astronomy

• Astronomy examines objects that range in size from the parts of an atom ($\sim10^{-15}$ m) to the size of the observable universe ($\sim10^{26}$ m).

Patterns of Stars

• The surface of the celestial sphere is divided into 88 unequal regions called constellations.

Earthly Cycles

• The celestial sphere appears to revolve around the Earth once in each day-night cycle. In fact, it is the Earth's rotation that causes that apparent motion.

• The poles and equator of the celestial sphere are determined by extending the axis of rotation and the equatorial plane of the Earth out onto the celestial sphere.

• Earth's axis of rotation is tilted at an angle of 23½° from the perpendicular to the plane of the Earth's orbit. This tilt causes the seasons.

• Equinoxes and solstices are significant points along the Earth's orbit that are determined by the relationship between the Sun's path on the celestial sphere (the ecliptic) and the celestial equator.

• The Earth's axis of rotation slowly changes direction relative to the stars over thousands of years, a phenomenon called precession. Precession is caused by the gravitational pull of the Sun and Moon on the Earth's equatorial bulge.

• The length of the day is based upon the Earth's rotation rate and the average motion of the Earth around the Sun. These effects combine to produce the 24-hour day upon which our clocks are based.

• The phases of the Moon are caused by the relative positions of the Earth, Moon, and Sun. The Moon completes one cycle of phases in a synodic month, which averages 29½ days.

• The Moon completes one orbit around the Earth with respect to the stars in a sidereal month, which averages 27.3 days.

Eclipses

• The shadow of an object has two parts: the umbra, where direct light from the source is completely blocked; and the penumbra, where the light source is only partially obscured.

• A lunar eclipse occurs when the Moon moves through the Earth's shadow. During a lunar eclipse, the Sun, Earth, and Moon are in alignment, and the Moon is in the plane of the ecliptic.

• A solar eclipse occurs when a strip of the Earth passes through the Moon's shadow. During a solar eclipse, the Sun, Earth, and Moon are in alignment, and the Moon is in the plane of the ecliptic.

• Depending on the relative positions of the Sun, Moon, and Earth, lunar eclipses may be penumbral, partial, or total, and solar eclipses may be annular, partial, or total.

WHAT DID YOU KNOW?

1 *Is the North Star—Polaris—the brightest star in the night sky?* No. Polaris is a star of medium brightness compared with other stars visible to the naked eye.

2 *Do astronomers regard constellations as the familiar patterns of stars in the sky?* Astronomers sometimes use the common definition of a constellation as a pattern of stars. Formally, however, a constellation is an entire region of the celestial sphere and all the stars and other objects in it. Viewed from Earth, the entire sky is covered by 88 different-sized constellations. If there is any room for confusion, astronomers refer to the patterns as asterisms.

3 *What causes the seasons?* The tilt of the Earth's rotation axis with respect to the ecliptic causes the seasons. They are not caused by the changing distance from the Earth to the Sun that results from the shape of Earth's orbit.

4 *When is the Earth closest to the Sun?* On or around January 3 of each year.

5 *How many zodiac constellations are there?* There are 13 zodiac constellations, the lesser-known one being Ophiuchus.

6 *Does the Moon have a dark side that we never see from Earth?* Half of the Moon is always dark. Whenever we see less than a full Moon, we are seeing part of the Moon's dark side. So, the dark side of the Moon is not the same as the far side of the Moon, which we never see from Earth.

7 *Is the Moon ever visible during the daytime?* The Moon is visible at some time during daylight hours almost every day of the year. Different phases are visible during different times of the day.

Key Terms for Review

angle, 13
angular diameter (angular size), 13
annular eclipse, 29
arc angle, 13
autumnal equinox, 20
celestial equator, 12
celestial pole, 12
celestial sphere, 11
circumpolar star, 15
constellation, 11
declination, 12
degree (°), 13
diurnal motion, 12
eclipse path, 29
ecliptic, 17
equinox, 18
gravitation, 21
line of nodes, 25
lunar eclipse, 25
lunar phase, 23
north celestial pole, 12
partial eclipse, 27
penumbra, 26
penumbral eclipse, 26
precession, 22
precession of the equinoxes, 23
revolution, 14
right ascension, 12
rotation, 12
scientific notation, 6
sidereal month, 24
sidereal period, 15
solar corona, 27
solar day, 16
solar eclipse, 25
south celestial pole, 12
summer solstice, 12
synodic month (lunar month), 24
terminator, 24
time zone, 17
total eclipse, 27
umbra, 26
vernal equinox, 18
winter solstice, 18
zenith, 15
zodiac, 21

Review Questions

1. Where is the horizon? **a.** directly overhead, **b.** along the celestial equator, **c.** the boundary between land and sky, **d.** along the path that the Sun follows throughout the day, **e.** the line running from due north, directly overhead, ending due south.

2. How many constellations are there? **a.** 2, **b.** 12, **c.** 48, **d.** 56, **e.** 88.

3. Which of the following lies on the celestial sphere directly over the Earth's equator? **a.** ecliptic, **b.** celestial equator, **c.** north celestial pole, **d.** south celestial pole, **e.** horizon.

4. The length of time it takes the Earth to orbit the Sun is? **a.** an hour, **b.** a day, **c.** a month, **d.** a year, **e.** a century.

5. In Figure 1-8, what is another name for the "Sun's annual path"?

6. How are constellations useful to astronomers?

7. What is the celestial sphere, and why is this ancient concept still useful today?

8. What is the celestial equator, and how is it related to the Earth's equator? How are the north and south celestial poles related to the Earth's axis of rotation?

9. What is the ecliptic, and why is it tilted with respect to the celestial equator?

10. By about how many degrees does the Sun move along the ecliptic each day?

11. Through how many constellations does the Sun move every day?

12. Through how many constellations does the Sun move every year?

13. Why does the tilt of the Earth's axis relative to its orbit cause the seasons as the Earth revolves around the Sun? Draw a diagram to illustrate your answer.

14. What are the vernal and autumnal equinoxes? What are the summer and winter solstices? How are these four points related to the ecliptic and the celestial equator?

15. What is precession, and how does it affect our view of the heavens?

16. How does the daily path of the Sun across the sky change with the seasons?

17. Why is it warmer in the summer than in the winter?

18. Why is it convenient to divide the Earth into time zones?

19. Why does the Moon exhibit phases?

20. What is the difference between a sidereal month and a synodic month? Which is longer? Why?

21. What is the line of nodes, and how is it related to solar and lunar eclipses?

22. What is the difference between the umbra and the penumbra of a shadow?

23. What is a penumbral eclipse of the Moon? Why is it easy to overlook such an eclipse?

24. Which type of eclipse—lunar or solar—have most people seen? Why?

25. How is an annular eclipse of the Sun different from a total eclipse of the Sun? What causes this difference?

26. When is the next leap year?

27. At which phase(s) of the Moon does a solar eclipse occur? A lunar eclipse?

28. Is it safe to watch a solar eclipse without eye protection? A lunar eclipse?

29. During what phase is the Moon "up" least in the daytime?

Advanced Questions

The answers to all computational problems, which are preceded by an asterisk (*), appear at the end of the book.

30. During what phase(s) does the Moon rise after sunrise and before sunset? After sunset and before sunrise? At sunset? At sunrise?

31. Why can't a person in Australia use the Big Dipper to find north?

32. Are there any stars in the sky that are not members of a constellation?

33. At what places on Earth is Polaris seen on the horizon?

34. Where do you have to be on the Earth to see the Sun at your zenith? If you stay at one such location for a full year, on how many days will the Sun pass through the zenith?

35. Where do you have to be on Earth to see the south celestial pole at your zenith? What is the maximum possible elevation (angle) of the Sun above the horizon at that location? On what date is this maximum elevation achieved?

36. Where on the horizon does the Sun rise at the time of the vernal equinox?

37. Consult a star map of the southern hemisphere and determine which, if any, of the bright southern stars could someday become south celestial pole stars.

38. Are there stars in the sky that never set where you live? Are there stars that never rise where you live? Does your answer depend on your location on Earth? Why or why not?

39. Using a diagram, demonstrate that your latitude on Earth is equal to the altitude of the north celestial pole above your northern horizon.

40. Using a star map, determine which bright stars, if any, could someday mark the location of the vernal equinox. Give the approximate years when this should happen.

41. What is the phase of the Moon if it **a.** rises at 3 A.M.? **b.** sets at 9 P.M.? At what time does **c.** the full Moon set? **d.** the first quarter Moon rise?

42. What is the phase of the Moon if, on the first day of spring, the Moon is located at the position of **a.** the vernal equinox, **b.** the summer solstice, **c.** the autumnal equinox, **d.** the winter solstice?

*43. How many more sidereal months than synodic months are there in a year? Why?

44. How do we know that the phases of the Moon are not due to the Moon moving in the Earth's shadow?

45. Do the paths of total solar eclipses fall more frequently on oceans or on land? Explain.

46. Can one ever observe an annular eclipse of the Moon? Why or why not?

47. Which of the five images of the Sun in Figure 1-31 is the first and which is the last in the sequence shown? Justify your answer.

48. During a lunar eclipse, does the Moon enter the Earth's shadow from the east or the west? Explain your answer.

49. Do we see all of the Moon's surface from the Earth? *Hint:* Carefully examine the photographs in Figure 1-23.

50. Explain why the waning gibbous Moon, third quarter Moon, and waning crescent photographs in Figure 1-23 are correctly oriented as seen from Earth.

51. Why is a small crescent of light often observed on the Moon when it is exactly in the new phase?

52. Make a drawing of an annular solar eclipse as seen from space similiar to Figure 1-30. Be sure to make clear how it differs from a total solar eclipse.

*53. Assuming that the Sun makes an angle of ½° in our sky and is at a distance of 1.496×10^{11} m, what is the Sun's diameter? Divide this by 2 to find the Sun's radius and explain why this result is slightly different from the value given in Appendix E-7 at the back of the book.

54. How long was the exposure for the photograph of circumpolar star trails in Figure 1-9?

55. Determine which stars whose paths are shown in Figure 1-9 are *not* circumpolar. An easy way to write the answer is in terms of distance on the photograph from the location of the south celestial pole.

Discussion Questions

56. Examine the list of the 88 constellations in Appendix E-6. Are there any constellations whose names obviously date from modern times? Where are these constellations located? Why do they not have ancient names?

57. In his novel *King Solomon's Mines,* H. Rider Haggard described a total solar eclipse that was seen in both South Africa and the British Isles. Is such an eclipse possible? Why or why not?

58. Describe how a lunar eclipse would look if the Earth had no atmosphere.

59. Examine a listing of total solar eclipses over the next several decades. What are the chances that you might be able to travel to one of the eclipse paths? Do you think you might

go through your entire life without ever seeing a total eclipse of the Sun?

What If . . .

60. The Moon moved about the Earth in an orbit perpendicular to the plane of the Earth's orbit? What would the cycle of lunar phases be? Would solar and lunar eclipses be possible under these circumstances?

61. The Earth's axis of rotation were tilted at a different angle? What would the seasons be like where you are now if the axis of rotation is **a.** 0° and **b.** 45° to its orbital plane? What would be different about the seasons and the day-night cycle if you lived at one of the Earth's poles in these 2 situations?

62. You watched the Earth from the Moon? What would you see for Earth's **a.** daily motion, **b.** motion along the celestial sphere, and **c.** cycle of phases?

63. The Moon didn't rotate? Describe how its surface *features* would appear from the Earth—that is, would we see all sides of it over time? Why or why not? (Ignore the change in phases when discussing its appearance.) *Carefully* study the photographs in Figure 1-23 and state whether the same features are visible at all times or whether we see different features over time. (Again, ignore the phases.) What can you conclude about whether the Moon actually rotates?

Web Questions

AIMM 1.1

*64. Work through the AIMM (Active Integrated Media Module) called "Small-Angle Toolbox" in Chapter 1 of the *Discovering the Universe* Web site. Use it to determine the diameters in kilometers of the Sun, Saturn, and Pluto given the following distances and angular sizes:

Object	Distance (km)	Angular size (")
Sun	1.5×10^8	1800
Saturn	1.5×10^9	16.5
Pluto	6.3×10^9	0.06

65. Search the Web and identify at least four cultures whose stories are used to name modern constellations. Briefly relate the stories of a constellation from each culture.

66. Search the Web for information about the Great Nebula of Orion (see label on Figure 1-4; it is also called the Orion Nebula). Can the Great Nebula be seen with the naked eye? Does it exist alone in space or is it part of a larger system of interstellar material? What has been learned by examining the Great Nebula with telescopes sensitive to infrared light?

67. Search the Web for information about the national flags of Australia, New Zealand, and Brazil, and the state flag of Alaska. What stars are depicted on these flags? Explain any similarities or differences among these flags.

68. Search the Web for the English meaning of the Japanese word *Subaru*. Make a drawing of the Subaru car's emblem and explain it.

69. Use the U.S. Naval Observatory's Web site to determine the times of sunset and sunrise on **a.** your birthday and **b.** the date this assignment is due. Are the times the same? Explain why or why not.

70. Search the Web for information about the next total solar eclipse. Through which major cities, if any, does the path of totality pass? What is the maximum duration of totality? Find a location where this maximum duration is observed. Will the eclipse be visible (even as a partial eclipse) from your present location?

71. Search the Web for information about the next total lunar eclipse. Will the total phase of the eclipse be visible from your present location? If not, will the penumbral phase be visible? Draw a picture of the Sun, Earth, and Moon at totality and indicate your location on the drawing of the Earth.

72. Access the animation "The Moon's Phases" in Chapter 1 of the *Discovering the Universe* Web site. This shows the Earth-Moon system as seen from a vantage point above the Earth's North Pole. **a.** Describe where you would be on the diagram if you are on the equator and the time were 6:00 P.M. **b.** If it were 6:00 P.M. and you were standing on the Earth's equator, would a third-quarter Moon be visible? Why or why not? If it would be visible, describe its appearance.

Observing Projects

If you are going to be using *Starry Night Enthusiast*™, load the program on a suitable computer. You can learn to use the program with the help of the User's Guide that loads with it, with the help of a friend, by experimenting on your own or with on-line help. A brief guide is presented in Appendix G. **Start each Project by resetting the program to the present time by pressing *Ctrl-Shift-H*.**

73. On a clear, cloud-free night, use the star charts within the covers of this book to see how many constellations of the zodiac you can identify. Which ones are easy to find? Which are difficult?

74. Use *Starry Night Enthusiast*™ to see if you can identify the zodiac constellations that are visible in the sky this evening. Set the time to sometime after sunset and turn off the sky motion by clicking on the square to the right of the timestep indicator. Then "grab" the sky and move around until you recognize the zodiac constellations. If you have difficulty finding zodiac constellations, click on *Options* tab, then *Guides: The Ecliptic* to highlight the part of the sky in which 4 the zodiac is located. If you still have trouble, go to *Constellations* and click on *stick figures* to see

the asterisms. If you need even more help, go to *Constellations* again and click on *Auto Identify*. Then, as you center a constellation, its name and classical image will appear. Moving the hand icon over an object will tell you what it is. Make a list of the zodiac constellations up today. Return to home (*Ctrl-Shift-H*). Determine which zodiac constellation the Sun is passing through today using *Favourites/Guides/Atlas*.

75. Use *Starry Night Enthusiast*™ to observe the diurnal motion of the sky. **a.** Using the hand cursor, center your field of view on the northern horizon (if you live in the northern hemisphere) or southern horizon (if you live in the southern hemisphere). Adjust the land to appear roughly flat. Turn off daylight (*Options* tab/*Local View/Daylight*). Close tabs. Set the timestep to 3 minutes (change if this is too slow or fast) and click on the *Forward* button. Do the stars appear to revolve clockwise or counterclockwise? Explain this in terms of the Earth's rotation. Are any of the stars circumpolar? Explain why or why not. **b.** Recenter your field of view on the southern horizon (if you live in the northern hemisphere) or the northern horizon (if you live in the southern hemisphere). Set the program in motion as just described. Explain what you see. Are any of these stars circumpolar? Explain.

76. Use the *Starry Night Enthusiast*™ program to observe the Sun's motion on the celestial sphere. Using the *Options* tab, turn off the daylight and hide the local horizon. Put on the *Ecliptic* and *Celestial Grid.* Note that the Celestial Equator is between the +10 and −10 lines of declination. Center on the Sun by using the *Find* tab. Be sure you see at least 90° of the celestial sphere. **a.** Select *Auto Identify* in the *Options/Constellations* menu to display the constellations at the center of the screen. In which constellation is the Sun located today? Is this the same as the astrological sign for today's date (see Table 1-1)? **b.** Be sure you are locked on the Sun. Set the timestep to 1 day (change if this is too slow or too fast) and click on the ▶ button. Observe the Sun for a full year of simulated time. Note the constellations through which it passes. How many are there? In which constellation does it cross the celestial equator? What path does it follow?

77. Examine the star charts that are published monthly in such popular astronomy magazines as *Sky & Telescope* and *Astronomy.* How do they differ from the star charts within the covers of this book? On a clear, cloud-free night, use one of these star charts to locate the celestial equator and the ecliptic on the night sky. Note the inclination of the Milky Way to the ecliptic and celestial equator. What do your observations tell you about the orientation of the Earth and its orbit around the Sun relative to the rest of the Galaxy?

78. Observe the Moon on each clear night over the course of a month. Note the Moon's location among the constellations and record that location on a star chart that also shows the ecliptic. After a few weeks, your observations will begin to trace the Moon's orbit. Identify the orientation of the line of nodes by marking the points where the Moon's orbit and the ecliptic intersect. On what dates is the Sun near the nodes marked on your star chart? Compare these dates with the dates of the next solar and lunar eclipses.

79. Use *Starry Night Enthusiast*™ to study the Moon's path in the sky and eclipses. Turn off the local horizon and daylight. Turn on the *The Ecliptic* and *Celestial Grid.* Find the Moon on the celestial sphere using the *Find* tab. Change the angle of the sky you see to about 15° using the slide on the upper right of the screen. Note whether the Moon lies on the ecliptic. Set the motion to increment by 6 minutes per timestep (adjust if this is too fast or too slow). Keeping the Moon locked in the center of your screen, set time in motion. **a.** In what direction does the Moon move against the background stars? Does it ever change direction relative to the stars? Why or why not? (Ignore the Moon's rocking motion, which is due to its rising and setting.) **b.** Change the timestep to 1 hour. Determine how many days elapse between successive times the Moon is on the ecliptic. **c.** Move forward in time until the Moon is either full or new *and* it is on the ecliptic. What type of eclipse could occur on that day? **d.** Using the *Live Sky* option, set the program to the date of the next such eclipse. Does the Moon have the same phase as in part (c)? **e.** Moving the time ahead or back by about 5 minutes per timestep, observe the eclipse. Make a series of drawings (with times on each) depicting it.

80. It is quite possible that a lunar eclipse will occur while you are taking this course. If so, look up on the Internet the precise time that it will happen; in the current issue of a reference from the U.S. Naval Observatory, such as the *Astronomical Almanac* or *Astronomical Phenomena;* or in such magazines as *Sky & Telescope* and *Astronomy,* which generally run articles about eclipses the month before they happen. Make arrangements to observe this eclipse. Note the times at which the Moon enters and exits the Earth's umbra.

81. Use *Starry Night Enthusiast*™ to determine when the Sun rises and sets today. Similarly, find out when the Moon rises and sets today. Starting with the time set to last midnight, determine during what hours the Moon is visible during daylight hours today and during what hours it is visible at night.

WHAT IF . . . EARTH'S AXIS LAY ON THE ECLIPTIC?

Imagine the Earth tilted so that its axis of rotation lies in the plane of its orbit about the Sun (see the accompanying figure). We'll call it NeoEarth. NeoEarth still rotates once every 24 hours. We arbitrarily fix its north pole to point toward the star Tau Tauri, a star nearly on the ecliptic, just north of the bright star Aldebaran in the constellation Taurus (see Figure 1-4b or 1-6).

Neo's Seasons Let's see how the seasons unfold on NeoEarth. It is March 21, the date of the spring equinox. The Sun is directly over the equator, as shown in the figure below. For the next three months, the Sun rises higher in the northern hemisphere. Unlike on our Earth, the Sun does not stop moving northward when it is over 23½° north latitude. Rather, the Sun rises farther and farther north day by day until it appears over the north pole of NeoEarth at the summer solstice, around June 22. Three months later, at the autumnal equinox, the Sun again rises over the equator, and day and night have the same length everywhere.

The Sun appears over the south pole of NeoEarth at the winter solstice, around December 22. The Sun's apparent motion through NeoEarth's sky completes the seasonal cycle by moving north, appearing over the equator once again around March 21.

Neo's Climate Day and night take on new meanings for inhabitants of NeoEarth. On our Earth, the regions above the Arctic Circle and below the Antarctic Circle have days or weeks of continual light in summer and days or weeks of constant darkness in winter. But on NeoEarth, *every place* has extended winter periods of constant darkness followed by extended summer periods of constant daylight. Spring and fall on NeoEarth have daily cycles of daylight and darkness, which separate the periods of continual daylight and darkness.

A Neo Day At the latitude of Atlanta, Georgia (33°46′), on NeoEarth, the day-night cycle occurs for only seven and a half months out of the year. During the other four and a half months, there is continuous day or continuous night, coupled with harsh summers and winters. With variations, this sequence of events occurs everywhere on NeoEarth.

The seasonal cycle on NeoEarth prevents the formation of permanent polar ice caps. Polar regions experience the same tropical heating and high temperatures as the equatorial regions of our Earth. The polar regions of NeoEarth in winter are exceptionally cold, so seasonal polar ice caps may form. Because the southern polar cap resting on Antarctica is not permanent, the oceans, and the shorelines on the continents, are higher than those on our Earth.

If seasonal polar ice caps form, the dominant force controlling weather may shift from the jet streams that circle our Earth along lines of latitude to a pole-to-pole flow. Thermal flows created by intense heating at one location and cooling at others may replace our Earth's trade winds and other east-west winds. How else would NeoEarth differ from our world?

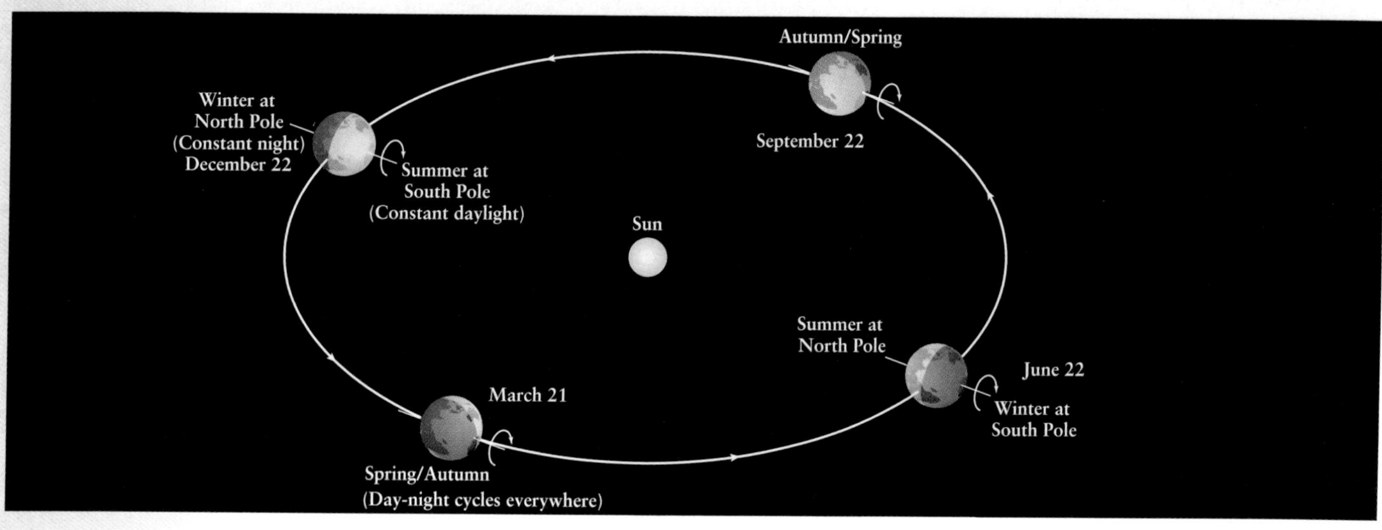

CHAPTER 2

Gravitation and the Waltz of the Planets

WHAT DO YOU THINK?

1 What makes a theory scientific?

2 What is the shape of the Earth's orbit around the Sun?

3 Do the planets orbit the Sun at constant speeds?

4 Do all the planets orbit the Sun at the same speed?

5 How much force does it take to keep an object moving in a straight line at a constant speed?

6 How does an object's mass differ when measured on the Earth and on the Moon?

Answers to these questions appear in the text beside the corresponding numbers in the margins and at the end of the chapter.

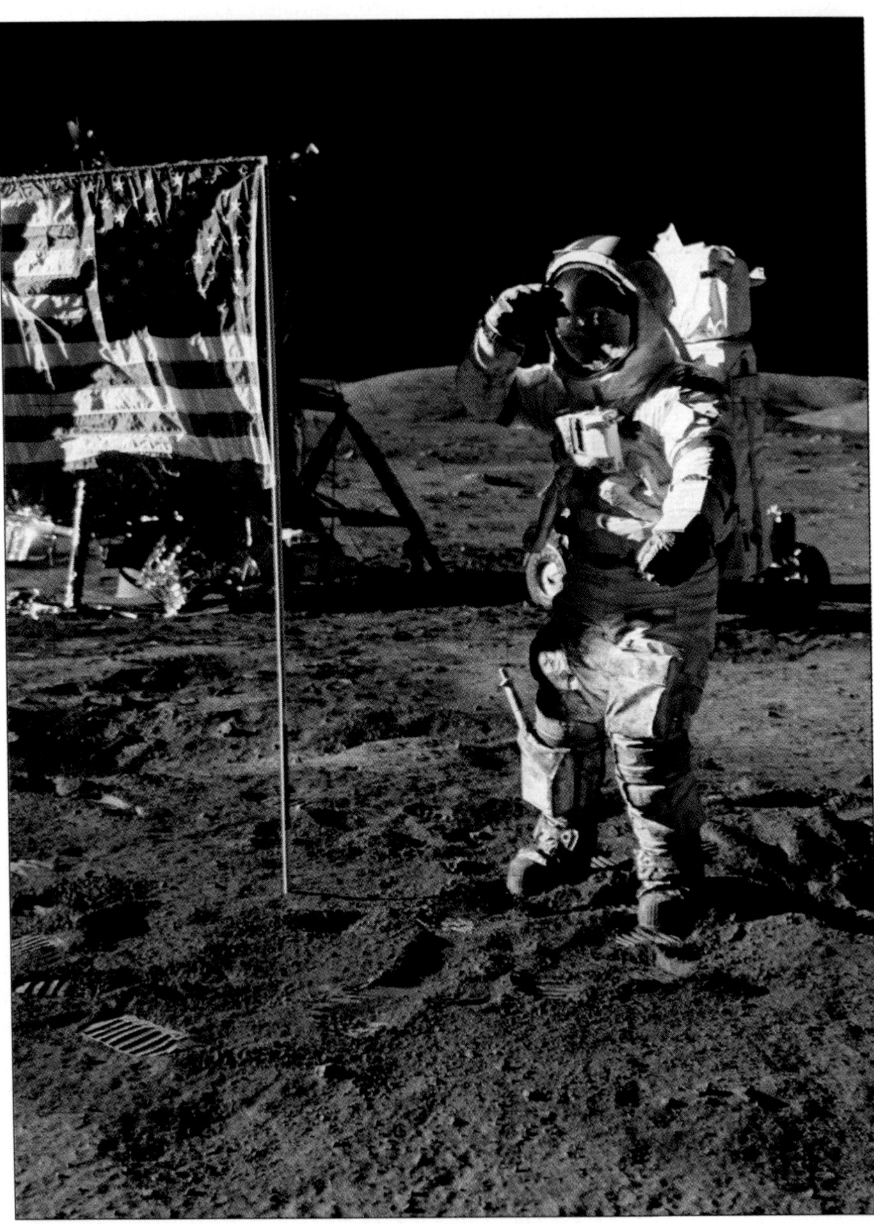

Gravity's Pull (NASA)

R I V U X G

Scientists are a lot like detectives, and the process they follow when trying to explain scientific phenomena has a good deal in common with the activities of sleuths as they try to solve mysteries. This chapter traces how we moved from an Earth-centered view of the universe to a Sun-centered one and how we came to understand the physical principles by which basic motions of the universe operate. The process of this discovery initially involved the efforts of a few determined scientist-sleuths. To unravel the mysteries that puzzled them, such as why planets appear at times to change direction on the celestial sphere or why the explanation that placed the Earth in the center of the universe kept failing to predict the locations of certain bodies, they used careful observations and a willingness to question their own and others' assumptions. Like the detective who suddenly realizes that his prime suspect, the person around whom his whole case has revolved, couldn't have committed the crime because he was actually baking a soufflé at the time, the astronomers and mathematicians who dislodged the Earth-centered view of the universe made some of their biggest leaps forward when they were willing to look beyond their prime suspect, the Earth.

For millennia and across civilizations, understanding of the night sky centered on the actions of demons and heroes, gods and goddesses. The groundwork for modern science was set down by Greek mathematicians and philosophers beginning around 2500 years ago, when Pythagoras and his followers began using mathematics to describe natural phenomena. About 200 years later, Aristotle asserted that the universe is comprehensible: It is governed by regular laws. Just as important, the Greeks were also among the first to leave a written record of their ideas, so that succeeding generations could develop, criticize, and test their conclusions. This concept evolved into writing physical theories quantitatively in mathematical terms so that we can test the theories by observing nature.

The original Greek ideas evolved until a new way to examine, understand, and predict how things work—science—emerged in the seventeenth century. Science provides explanations for activities and events, and it makes predictions about things that have not yet happened or that have not yet been observed. These are incredibly powerful tools that enable us to understand what we see without having to accept events on faith or to fear that things like the force of gravity will change on a whim. Science simplifies and takes some of the uncertainty out of the world. Consider, for example, our understanding of gravity. Until Isaac Newton made the conceptual leap that the force holding the Earth in orbit around the Sun is the same force that holds us onto the Earth, these two effects were considered to be separate and unrelated. Once they were connected, and an equation was written to describe the behavior of the Earth and of falling objects, humans had, for the first time, the ability to predict the future motion of projectiles and other falling objects reliably. Einstein plumbed the depths of gravitational behavior even further, developing equations that made more accurate predictions, whose effects we will examine in Chapter 14. Science gives us confidence that gravity will not change on a whim. Furthermore, science enables us to understand and manipulate an awesome range of nature's properties.

In this chapter you will discover

- what makes a theory scientific
- the clues suggesting that the Earth is not the center of the universe
- the scientific revolution that dethroned Earth from its location at the center of the universe
- Copernicus's argument that the planets orbit the Sun
- why the direction of motion of the planets on the celestial sphere sometimes appears to change
- that Kepler's determination of the shapes of planetary orbits depended on the careful observations of his mentor, Tycho Brahe
- how Isaac Newton formulated an equation to describe the force of gravity
- how Isaac Newton explained why the planets and moons remain in orbit

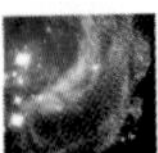

SCIENCE: KEY TO COMPREHENDING THE COSMOS

Understanding how nature works enables us to manipulate the matter and energy that comprise our environment and to thereby create new things to make our lives better. We also use our scientific knowledge to gain even greater insights by creating tools such as telescopes and microscopes to explore more of the universe. Improvements in technology lead to better research equipment, which enables us to make even deeper discoveries about the laws of **physics,** the science that investigates the nature of space, time, matter, and energy, and the relationships between them. In this chapter we will explore the nature of science and use it to see how gravity keeps planets and other objects orbiting the Sun, and moons orbiting their respective planets.

2-1 Science is both a body of knowledge *and* a process of learning about nature

Science is actually two things. First, it is a body of knowledge that we acquire and use to understand and manipulate our environment. The observations of the stars and the motion of the other objects in the sky among the stars described in Chapter 1 are an example of that knowledge. Second, it is a process for gaining more knowledge in a way that assures that the information can be accepted by everyone. Science as a process is also called the **scientific method,** and it describes how scientists go about observing, explaining, and predicting physical reality.

The scientific method (Figure 2-1) can begin in a variety of places, but most often it starts by people making observations or doing experiments. For example, observation that some objects move along the celestial sphere, while others remained fixed, demanded explanation. Observations or experiments then lead scientists to create a **scientific theory** or to compare the results with the predictions of a preexisting theory. A scientific theory is an idea or collection of ideas that proposes to explain the observed phenomenon. Scientific theories are often expressed mathematically as **models.** For example, Newton's *law* (or *theory*, as such ideas are now called) of gravitation is written as an equation that predicts how bodies attract each other. Recall from Chapter 1 that gravitation is the universal attractive force between all matter in the universe. The word "gravity" is often used as shorthand for "gravitation" and both are used in this book.

Scientific theories make testable *predictions* that can be 1
verified using new observations and experiments. Testing predictions is a crucial aspect of the scientific method, which requires that the theory accurately forecast the results of new observations in its realm of validity. Newton's law of gravitation predicts that the Sun's gravitational force makes the planets move in elliptical orbits and it predicts how long it should take each planet to orbit the Sun. As we will see shortly, observations have confirmed most of these predictions.

INSIGHT INTO SCIENCE

Science Is Inclusive Science is truly intended as an inclusive endeavor. In principle, a scientific theory can be created, modified, or tested by anyone inclined to do so. In practice, however, being involved in the scientific enterprise requires that you need to understand the mathematical tools of science. Furthermore, many observations and experiments are very expensive and complicated, requiring the efforts of groups of people. Assuring that people actively involved in science are qualified to do the work is a process intended to prevent the scientific method from being derailed.

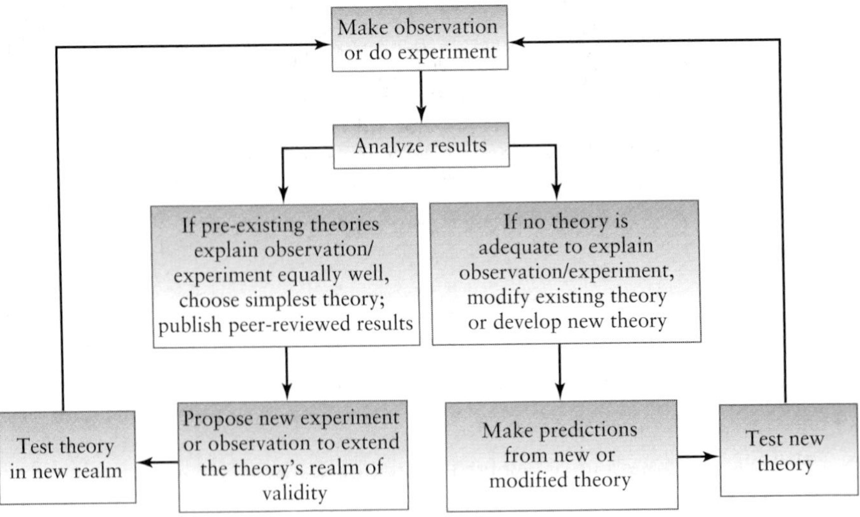

FIGURE 2-1 **The Scientific Method** This flow chart shows the process by which scientists develop new theories. Different scientists start at different places on this chart, including making observations or doing experiments; creating or modifying scientific theories; or making predictions from theories. Anyone interested in some aspect of science and willing to learn the tools of that science can participate in the adventure. ((©) Neil F. Comins)

In order for a theory to be considered scientific, it must also be possible to potentially disprove it. For example, Newton's law of gravitation can be tested and potentially disproved by observations and thus qualifies as a scientific theory. The idea that the Earth was created in six days cannot be tested, much less disproved. It is not a scientific theory but rather a matter of faith.

If the predictions of a theory are inconsistent with observations, the theory is modified, applied only in limited circumstances, or discarded in favor of a more accurate explanation. For example, Newton's law of gravitation is entirely adequate for describing the motion of an apple falling to Earth, of the Space Shuttle orbiting the Earth, or of the Earth orbiting the Sun, but it is inaccurate in the vicinity of a black hole, where matter is especially dense. In this latter realm, Newton's law of gravitation is replaced by Einstein's theory of general relativity, which accurately describes gravitational behavior in a much wider range of conditions than Newton's law, but at the cost of much greater mathematical complexity. When applied to the motion of the Earth around the Sun, general relativity gives the same results as Newton's law of gravitation.

INSIGHT INTO SCIENCE

Theories and Beliefs New theories are personal creations, but science is not a personal belief system. As stated in the previous Insight into Science, scientific theories make predictions that can be tested independently. If everyone who performs tests of the theory's predictions gets results consistent with the theory, the theory is considered valid in some realm. In comparison, belief systems such as which sports team or political system is best are personal matters. People will always hold differing opinions about such issues.

Science also strives to explain as many things as possible with as few theories as possible. We see billions upon billions of objects in the universe. It would be virtually impossible to study all of them separately so that we could come up with a separate theory to explain each one. Fortunately, individual theories explaining each object are not necessary. Scientists overcame this problem by noting that many of the bodies in space appear similar. By categorizing them suitably and then applying the scientific method to these groups of objects, we form a few theories that describe many objects and how they have evolved. These few theories can then be tested and refined as necessary. Such groupings of objects have proven invaluable and they give us insights into the structure and organization of billions of stars and galaxies that are, indeed, very similar to each other.

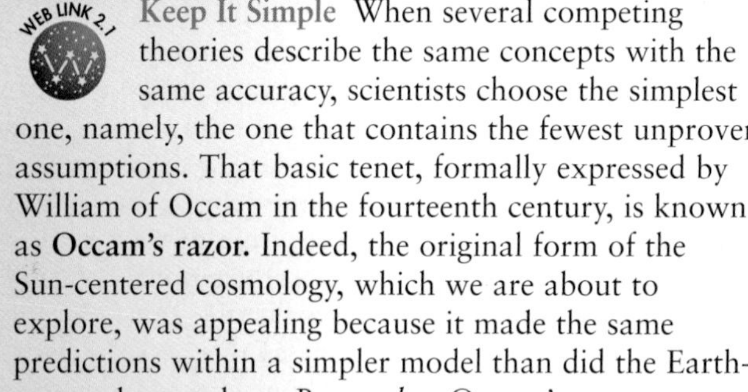

INSIGHT INTO SCIENCE

WEB LINK 2.1

Keep It Simple When several competing theories describe the same concepts with the same accuracy, scientists choose the simplest one, namely, the one that contains the fewest unproven assumptions. That basic tenet, formally expressed by William of Occam in the fourteenth century, is known as **Occam's razor.** Indeed, the original form of the Sun-centered cosmology, which we are about to explore, was appealing because it made the same predictions within a simpler model than did the Earth-centered cosmology. *Remember Occam's razor.*

The scientific method can be summarized in six words: observe, theorize, predict, test, modify, economize. I urge you to watch for applications of the scientific method in action throughout this book. Our first encounter with it is the discovery that the Earth orbits the Sun.

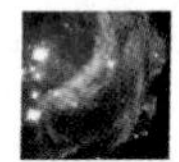

CHANGING OUR EARTH-CENTERED VIEW OF THE UNIVERSE

Early Greek astronomers tried to explain the motion of the five then-known planets: Mercury, Venus, Mars, Jupiter, and Saturn. Most people at that time held a *geocentric* view of the universe: They assumed that the Sun, the Moon, the stars, and the planets revolve about the Earth. A theory of the overall structure and evolution of the universe is called a **cosmology,** so the prevailing Earth-centered cosmology was called *geocentric.* Based on naked eye observations of the motions of heavenly bodies, the geocentric cosmology is so compelling that it held sway for more than 2000 years.

2-2 The belief in a Sun-centered cosmology came slowly

Explaining the motions of the five planets in a geocentric (Earth-centered) universe was one of the main challenges facing the astronomers of antiquity. The Greeks knew that the positions of the planets slowly shift relative to the "fixed" stars in the constellations. In fact, the word *planet* comes from a Greek term meaning "wanderer." They also observed that planets do not move at uniform rates through the constellations. From night to night, as viewed in the northern hemisphere, the planets usually move slowly to the left (eastward) relative to the background stars. This movement is called **direct motion.** Occasionally, however, a planet seems to stop and then back up for

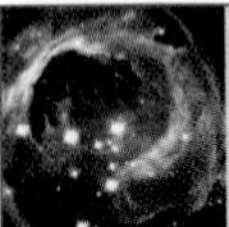

GUIDED DISCOVERY
The Earth-Centered Universe

As we move through the twenty-first century, most of us find it hard to understand why anyone would believe that the Sun, planets, and stars orbit the Earth. After all, we *know* that the Earth spins on its axis. We *know* that the gravitational force from the Sun holds the planets in orbit, just as the Earth's gravitational force holds the Moon in orbit. And we *know* that the stars in the night sky all lie far past the boundaries of our solar system. These facts have become part of our understanding of the motions of the heavenly bodies and we are taught these things from the time we are children.

Psychologists call this, the background information that we use to help explain things, a *conceptual framework*. Any conceptual framework contains all the information we take for granted. For example, when the Sun rises, moves across the sky, and sets today, we take for granted that it is the Earth's rotation that causes the Sun's apparent motion.

Our ancestors possessed a different conceptual framework than the one based on science's understanding of the cosmos. They did not know that the Earth rotates. They did not know that the then-mysterious force that held them to the ground is the same force that attracts the Earth to the Sun and the Moon to the Earth. They did not know that the Sun is a star, just like the fixed points of light in the sky. And they did not know any of the other laws of physics related to motion that we take for granted.

Because they did not feel the Earth move under their feet, nor see any other indication that the Earth is in motion, our forbearers sensed nothing to support the belief that the Earth moves. The obvious conclusion for one who has a prescientific conceptual framework, even today, is that the Earth stays put while objects in the heavens move around it.

This prescientific conceptual framework for understanding the motions of the heavenly bodies was based on the senses and on common sense. That is, people observed motions and drew "obvious," commonsense conclusions. Today, we incorporate the known and tested laws of physics in our understanding of the natural world. Many of these concepts are utterly counterintuitive, and, therefore, the conceptual frameworks we possess are less consistent with common sense than those held in the past. Studying science helps us develop intuition that is consistent with the actual workings of nature.

Geocentric Explanation of the Planets' Retrograde Motion The early Greeks developed many theories to account for the occasional retrograde motion of the planets and the resulting loops that the planets trace out against the background stars. One of the most successful ideas was expounded by the last of the great ancient Greek astronomers, Ptolemy, who lived in Alexandria, Egypt, 1900 years ago. His basic concepts are sketched in the accompanying figure. Each planet is assumed to move in a small circle called an *epicycle*, the center of which moves in a larger circle called a *deferent*, whose center is offset from the Earth. As viewed from Earth, the epicycle moves eastward along the deferent, and both circles rotate in the same direction (counterclockwise).

Most of the time, the motion of the planet on its epicycle adds to the eastward motion of the epicycle on the deferent. Thus, the planet is seen from the Earth to be in direct motion (to the left or eastward) against the background stars throughout most of the year. However, when the planet is on the part of its epicycle nearest the Earth, its motion along the epicycle subtracts from the motion of the epicycle along the deferent. The planet thus appears to slow and then halt its usual movement to the left (eastward motion) among the constellations, and then seems to move to the right (westward) among the stars for a few weeks or months. This concept of epicycles and deferents enabled Greek astronomers to explain the retrograde loops of the planets.

Using the wealth of astronomical data in the library at Alexandria, including records of planetary positions covering

several weeks or months. This reverse movement (to the west relative to the background stars) is called **retrograde motion.**

These planetary motions are much slower than the apparent daily movement of the entire sky caused by the Earth's rotation, and so they are superimposed on it. Therefore, the planets always rise in the east and set in the west, as the stars do. Both direct and retrograde motions are best detected by mapping the nightly position of a planet against the background stars over a long period. For example, Figure 2-2 shows the path of the planet Mars among the stars from September 2009 through July 2010.

The effort to explain retrograde motion resulted in an increasingly contrived and complex cosmological model, especially in explaining retrograde motion (see Guided Discovery: The Earth-Centered Universe). The ancient Greek astronomer Aristarchus proposed a more straightforward explanation of planetary motion, namely, that all the planets, including the Earth, revolve about the Sun. The retrograde motion of Mars in this **heliocentric** (Sun-centered) **cosmology** occurs just because the faster-moving Earth overtakes and passes the red planet, as shown in Figure 2-3. The occasional retrograde movement of a planet is merely the result of our changing viewpoint as we orbit the Sun—

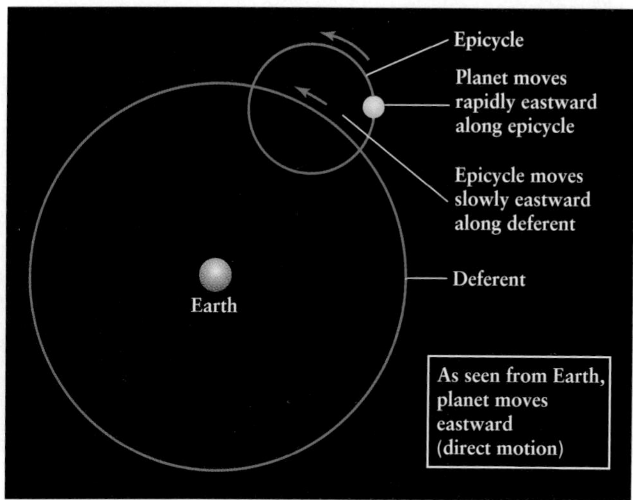

a

b

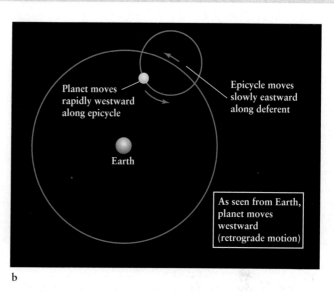

A Geocentric Explanation of Planetary Motion Each planet revolves about an epicycle, which in turn revolves about a deferent centered approximately on the Earth. As seen from Earth, the speed of the planet on the epicycle alternately **(a)** adds to or **(b)** subtracts from the speed of the epicycle on the deferent, thus producing alternating periods of direct and retrograde motion.

hundreds of years, Ptolemy deduced the sizes of the epicycles and deferents and the rates of revolution needed to produce the recorded paths of the planets. After years of arduous work, Ptolemy assembled his calculations in the *Almagest,* in which the positions and paths of the Sun, Moon, and planets were described with unprecedented accuracy. In fact, the *Almagest* was so successful that it became the astronomer's bible. For more than 1000 years, Ptolemy's cosmology endured as a useful description of the workings of the heavens.

Eventually, however, the commonsense explanation of the Earth-centered cosmology began to go awry. Errors and inaccuracies that were unnoticeable in Ptolemy's day compounded and multiplied over the years, especially errors due to precession, the slow change in the direction of the Earth's axis of rotation. Fifteenth-century astronomers made some cosmetic adjustments to the Ptolemaic system. However, the system became less and less satisfactory as more fanciful and arbitrary details were added to keep it consistent with the observed motions of the planets. After Newton's time, scientists knew that orbital motion required a force to be acting on the body. However, nothing in Ptolemy's epicycle theory produced such a force.

an idea that is beautifully simple compared to the geocentric system with all its complex planetary motions.

Because simplicity and accuracy are hallmarks of science, the complex geocentric model eventually gave way to the simpler, more elegant heliocentric cosmology, but dethroning the geocentric model did not occur immediately. In the first place, the Earth just does not seem to move! Combining this with the human desire to be at the center of everything and with geocentric religious teaching kept the belief that the Earth is at the center of everything firmly seated for more than 1300 years after Aristarchus proposed the heliocentric cosmology.

2-3 Copernicus devised the first comprehensive heliocentric cosmology

Over the centuries, more and more accurate observations of the planets' locations led to greater errors in the predictions of geocentric cosmology. In order to reconcile the cosmology with the data, more and more complex motions were attributed to the planets. By the mid-1500s, the geocentric cosmology had become truly unwieldy in its efforts to accurately predict the motions of the planets. It was then that the Polish mathematician, lawyer, physician, economist, cleric, and artist Nicolaus Copernicus resurrected

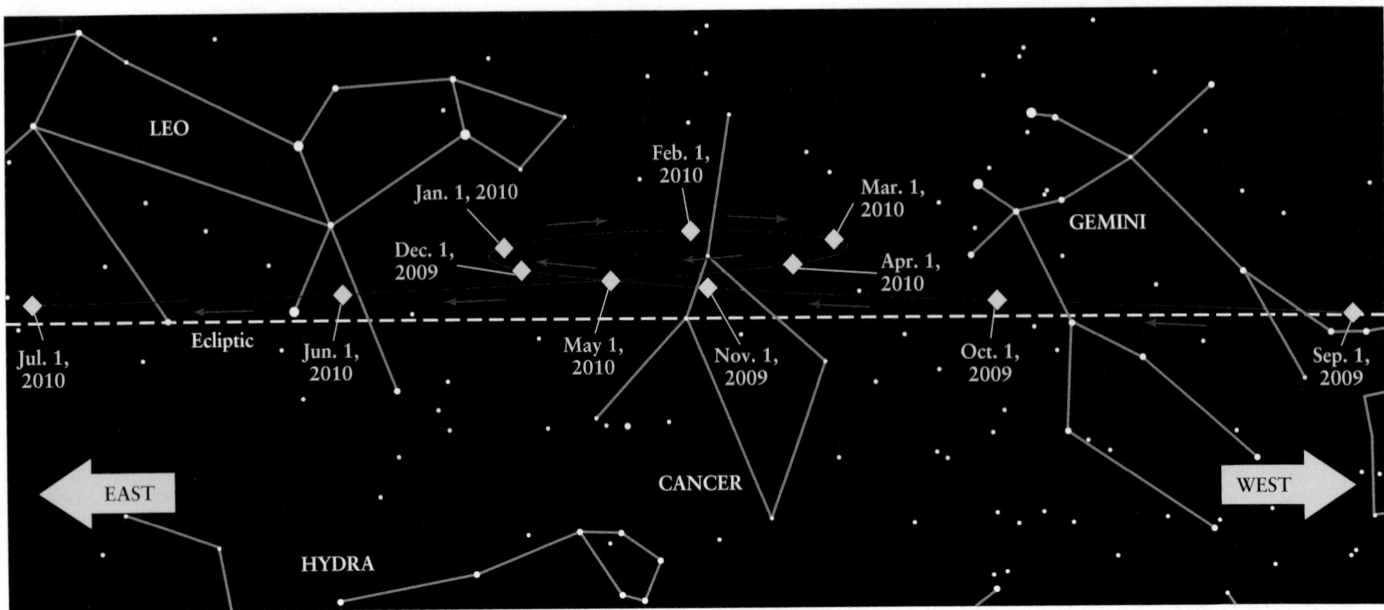

FIGURE 2-2 **The Path of Mars, 2009–2010** From September 2009 through June 2010, Mars moves across the constellations of Gemini, Cancer, and Leo. From December 23 through March 12, 2010, Mars's motion will be retrograde.

Aristarchus's theory. Copernicus (see Guided Discovery: Astronomy's Foundation Builders) was motivated by an effort to simplify the celestial scheme.

After assuming that the planets orbit the Sun rather than the Earth, observations led Copernicus to determine which planets are closer to the Sun than the Earth and which are farther away. Because Mercury and Venus are always observed fairly near the Sun, Copernicus correctly concluded that their orbits must lie inside that of the Earth. The other planets visible to Copernicus—Mars, Jupiter, and Saturn—can sometimes be seen high in the sky in the middle of the night, when the Sun is far below the horizon. This can occur only if the Earth comes between the Sun and a planet. Copernicus therefore concluded (also correctly) that the orbits of Mars, Jupiter, and Saturn lie outside the Earth's orbit.

The geometrical arrangements among the Earth, another planet, and the Sun are called **configurations.** For example, when Mercury or Venus is directly between the Earth and the Sun, as in Figure 2-4, we say the planet is in a

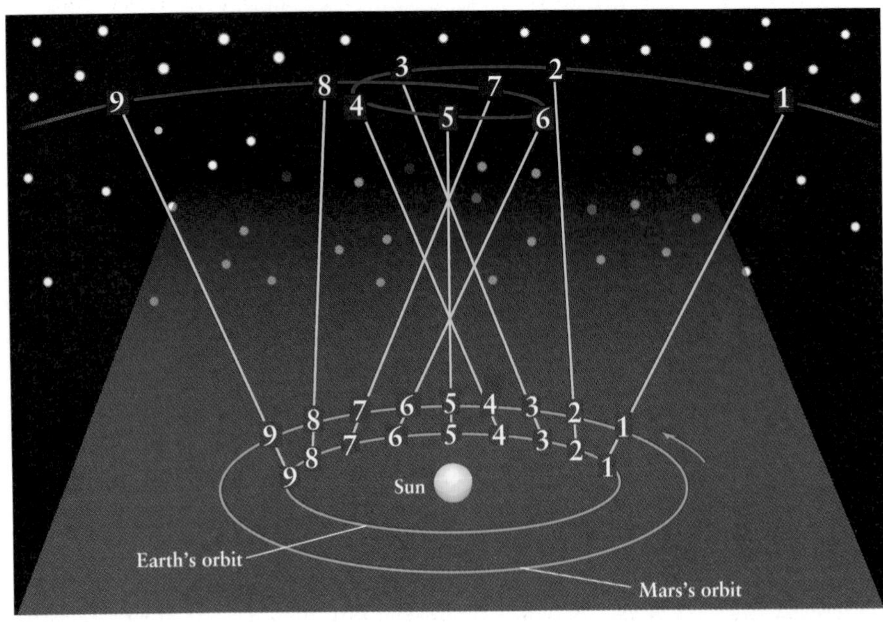

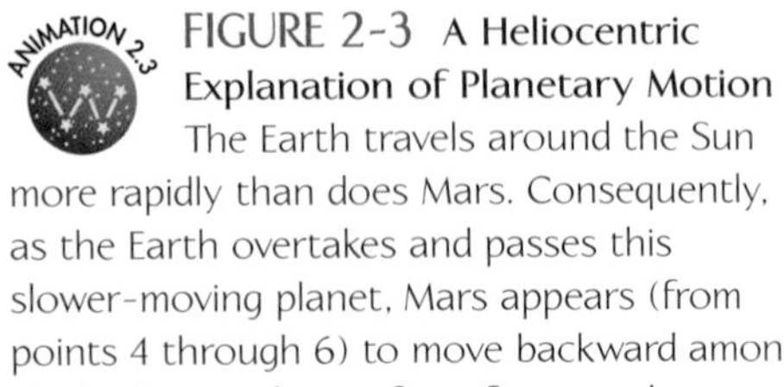

FIGURE 2-3 **A Heliocentric Explanation of Planetary Motion** The Earth travels around the Sun more rapidly than does Mars. Consequently, as the Earth overtakes and passes this slower-moving planet, Mars appears (from points 4 through 6) to move backward among the background stars for a few months.

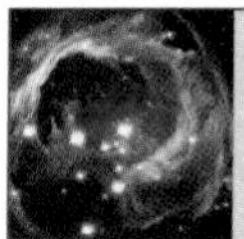

GUIDED DISCOVERY
Astronomy's Foundation Builders

In the two centuries between 1500 and 1700, human understanding of the motion of celestial bodies and the nature of the gravitational force that keeps them in orbit surged forward as never before. Theories related to this subject were developed by brilliant thinkers, whose work established and verified the heliocentric model of the solar system and the role of gravity.

(E. Lessing/Art Resource)

Nicolaus Copernicus (1473–1543) Copernicus was born in Torun, Poland, the youngest of four children. He pursued his higher education in Italy, where he received a doctorate in canon law and studied medicine. Copernicus developed a heliocentric theory of the known universe and published his work in 1543 under the title *De Revolutionibus Orbium Coelestium.* Copernicus became ill and died just after receiving the first copy.

(Painting by Jean-Leon Huens, courtesy of National Geographic Society)

Tycho Brahe (1546–1601) and **Johannes Kepler (1571–1630)** Tycho, depicted within a portrait of Kepler, was born to nobility in the Danish city of Knudstrup, which is now part of Sweden. At age 20 he lost part of his nose in a duel and wore a metal replacement thereafter. In 1576 the Danish king Frederick II built Tycho an astronomical observatory that Tycho named Uraniborg (after Urania, Greek muse of astronomy). Tycho rejected both Copernicus's heliocentric theory and the Ptolemaic geocentric system. He devised a halfway theory called the "Tychonic system." According to Tycho's theory, the Earth is stationary, with the Sun and Moon revolving around it, while all the other planets revolve around the Sun. Tycho died in 1601.

Kepler was educated in Germany, where he spent three years studying mathematics, philosophy, and theology. In 1596, Kepler published a booklet in which he attempted to mathematically predict the planetary orbits. Although his theory was altogether wrong, its boldness and originality attracted the attention of Tycho Brahe, whose staff Kepler joined in 1600. Kepler deduced his three laws from Tycho's observations.

(Art Resource)

Galileo Galilei (1564–1642) Born in Pisa, Italy, Galileo studied medicine and philosophy at the University of Pisa. He abandoned medicine in favor of mathematics. He held the chair of mathematics at the University of Padua and eventually returned to the University of Pisa as a professor of mathematics. There Galileo formulated his famous law of falling bodies: All objects fall with the same acceleration regardless of their weight. In 1609 he constructed a telescope and made a host of discoveries that contradicted the teachings of Aristotle and the Roman Catholic Church. He summed up his life's work on motion, acceleration, and gravity in the book *Dialogues Concerning Two New Sciences,* published in 1632.

(National Portrait Gallery, London)

Isaac Newton (1642–1727) Although he delighted in constructing mechanical devices—sundials, model windmills, a water clock, and a mechanical carriage—Newton showed no exceptional academic ability at Cambridge University, where he received a bachelor's degree in 1665. While pursuing experiments in optics, Newton constructed a reflecting telescope and also discovered that white light is actually a mixture of all colors. His major work on forces and gravitation was the tome *Philosophiae Naturalis Principia Mathematica,* which appeared in 1687. In 1704, Newton published his second great treatise, *Opticks,* in which he described his experiments and theories about light and color. Upon his death in 1727, Newton was buried in Westminster Abbey, the first scientist to be so honored.

configuration called an **inferior conjunction;** when either of these planets is on the opposite side of the Sun from the Earth, its configuration is called a **superior conjunction.**

The angle between the Sun and a planet as viewed from the Earth is called the planet's **elongation.** A planet's elongation varies from zero degrees to a maximum value, depending upon where we see it in its orbit around the Sun. At *greatest eastern* or *greatest western elongation,* Mercury and Venus are as far from the Sun in angle as they can be. This is about 28° for Mercury and about 47° for Venus. When either Mercury or Venus rises before the Sun, it is visible in the eastern sky as a bright "star" and is often called the "morning star." Similarly, when either of these two planets sets after the Sun, it is visible in the western sky and is then called the "evening star." Because these two planets are not always at their greatest elongations, they are often very close in angle to the Sun. This is especially true of Mercury, which is never more than 28° from the Sun and

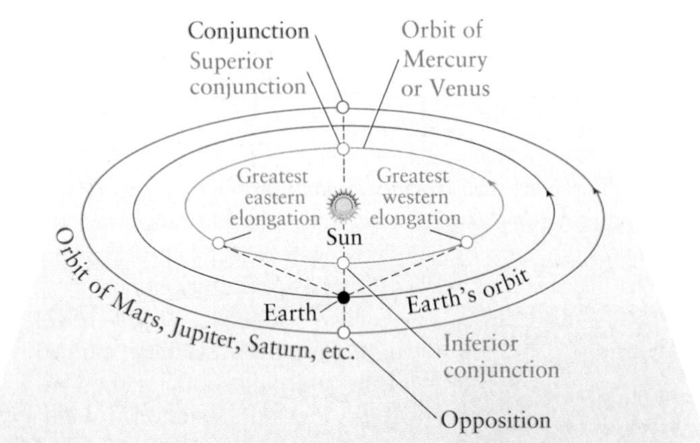

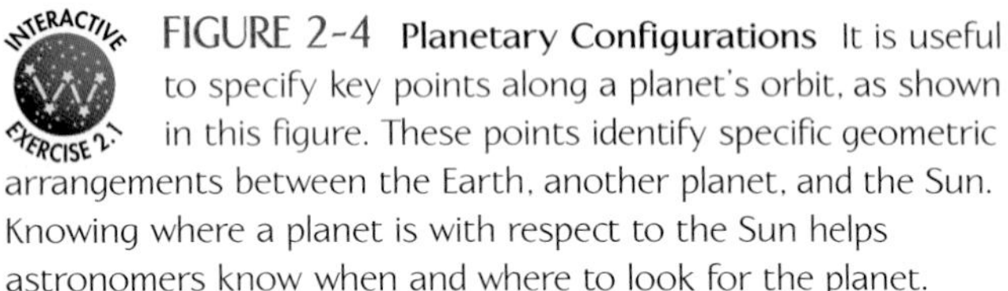

FIGURE 2-4 **Planetary Configurations** It is useful to specify key points along a planet's orbit, as shown in this figure. These points identify specific geometric arrangements between the Earth, another planet, and the Sun. Knowing where a planet is with respect to the Sun helps astronomers know when and where to look for the planet.

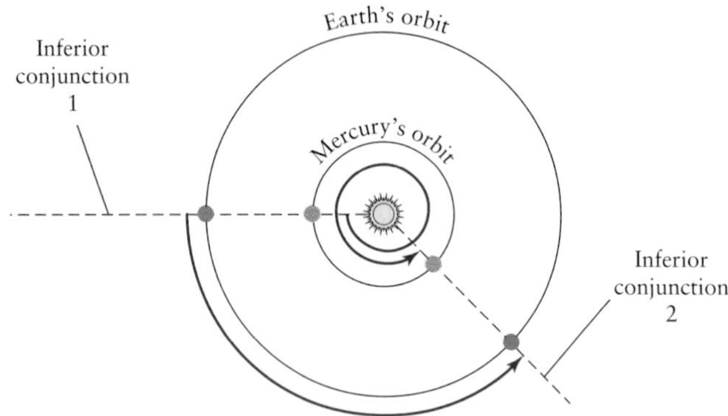

FIGURE 2-5 **Synodic Period** The time between consecutive conjunctions of the Earth and Mercury is 116 days. Typical of synodic periods for all planets, the location of the Earth is different at the beginning and end of the period. You can visualize the synodic periods of the exterior planets by putting the Earth in Mercury's place in this figure and putting one of the outer planets in the Earth's place.

therefore often hard to see. Venus is often nearly halfway up the sky at sunrise or sunset and therefore quite noticeable during much of its orbit. Because they are so bright and sometimes appear to change color due to motion of the Earth's atmosphere, Venus and Mercury are often mistaken for UFOs. (The same motion of the air causes the road in front of your car to shimmer on a hot day.)

Planets whose orbits are larger than Earth's have different configurations. For example, when Mars is located behind the Sun, as seen from Earth, it is said to be in **conjunction.** When it is opposite the Sun in the sky, the planet is at **opposition.** It is not difficult to determine when a planet happens to be located at one of the key positions in Figure 2-4. For example, when Mars is at opposition, it appears high in the sky at midnight.

It is relatively easy to follow a planet as it moves from one configuration to another. However, these observations alone do not tell us the planet's actual orbit, because the Earth, from which we make the observations, is also moving. Copernicus was therefore careful to distinguish between two characteristic time intervals, or *periods,* of each planet.

Recall from your study of the Moon in Chapter 1 that the **sidereal period** of orbit is the true orbital period of any astronomical body. A planet's sidereal period is the time it would take an observer fixed at the Sun's location watching that planet move through the background stars to go from one point on the celestial sphere, around the sphere, and back to that same point again. The sidereal period is the length of a year for each planet.

The other useful time interval Copernicus used is the **synodic period.** The synodic period is the time that elapses between two successive identical configurations as seen from the Earth. It can be from one opposition to the next, for example, or from one conjunction to the next (Figure 2-5). It tells us, for example, when to expect a planet to be closest to the Earth and, therefore, most easily studied.

INSIGHT INTO SCIENCE

Take a Fresh Look When the science is hard to visualize, try another perspective. For example, a planet's sidereal period of orbit is easy to understand from the perspective of the Sun but more complicated from the Earth. The synodic period of each planet, on the other hand, is easily determined from the Earth. As we will see, especially when we study Einstein's theories of relativity, each of these perspectives is called a *frame of reference.*

Thus, nearly 500 years ago, Copernicus was able to obtain the first six entries shown in Table 2-1 (the others are contemporary results included for completeness). Copernicus was then able to devise a straightforward geometric method for determining the distances of the planets from the Sun. His answers turned out to be remarkably close to the modern values, as shown in Table 2-2. From these two tables it is apparent that the farther a planet is from the Sun, the longer it takes to complete its orbit.

Copernicus presented his heliocentric cosmology, including supporting observations and calculations, in a book entitled *De Revolutionibus Orbium Coelestium* (On the Revolutions of the Celestial Spheres), which was published in 1543, the year of his death. Copernicus's great insight was the simplicity of a heliocentric cosmology compared to geocentric views. However, Copernicus incorrectly assumed that the planets travel along circular paths around

TABLE 2-1 Synodic and Sidereal Periods of the Planets (in Earth Years)

	Synodic period	Sidereal period
Mercury	0.318 yr	0.241 yr
Venus	1.599 yr	0.616 yr
Earth	—	1.0 yr
Mars	2.136 yr	1.9 yr
Jupiter	1.092 yr	11.9 yr
Saturn	1.035 yr	29.5 yr
Uranus	1.013 yr	84.0 yr
Neptune	1.008 yr	164.8 yr
Pluto	1.005 yr	248.5 yr

TABLE 2-2 Average Distances of the Planets from the Sun

	Measurement (AU)	
	By Copernicus	Modern
Mercury	0.38	0.39
Venus	0.72	0.72
Earth	1.00	1.00
Mars	1.52	1.52
Jupiter	5.22	5.20
Saturn	9.07	9.54
Uranus	Unknown	19.19
Neptune	Unknown	30.06
Pluto	Unknown	39.53

the Sun. As a result, his predictions were no more accurate than those of a geocentric theory!

2-4 Tycho Brahe made astronomical observations that disproved ancient ideas about the heavens

In November 1572, a bright star suddenly appeared in the constellation Cassiopeia. At first it was even brighter than Venus, but then it began to grow dim. After 18 months it faded from view.

Modern astronomers recognize this event as a supernova explosion, the violent death of a certain type of star (see Chapter 13). In the sixteenth century, however, the prevailing opinion was quite different. Teachings dating back to Aristotle and Plato argued that the heavens were permanent and unalterable. From that perspective, the "new star" of 1572 could not really be a star at all, because the heavens do not change. It must instead be some sort of bright object quite near Earth, many astronomers and theologians of the day argued, perhaps not much farther away than the clouds overhead. A 25-year-old Danish astronomer named Tycho Brahe (see Guided Discovery: Astronomy's Foundation Builders) realized that straightforward observations might reveal the distance to this object.

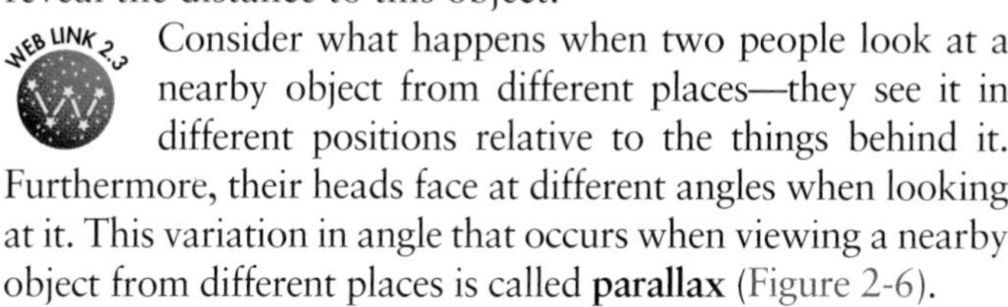

Consider what happens when two people look at a nearby object from different places—they see it in different positions relative to the things behind it. Furthermore, their heads face at different angles when looking at it. This variation in angle that occurs when viewing a nearby object from different places is called **parallax** (Figure 2-6).

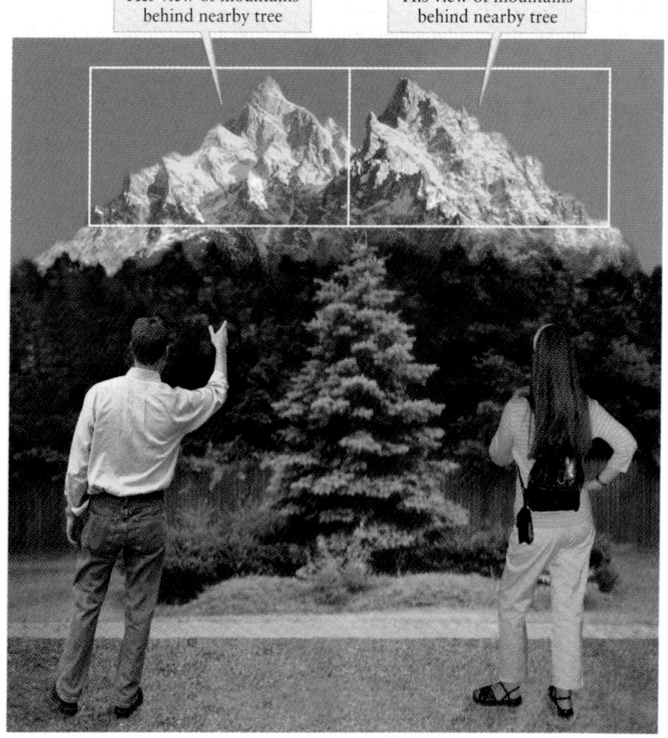

FIGURE 2-6 **Parallax** Nearby objects are viewed at different angles from different places. These objects also appear to be in different places with respect to more distant objects when viewed at the same time by observers located at different positions. Both effects are called parallax, and they are used by astronomers, surveyors, and sailors to determine distances. (Tobi Zausner)

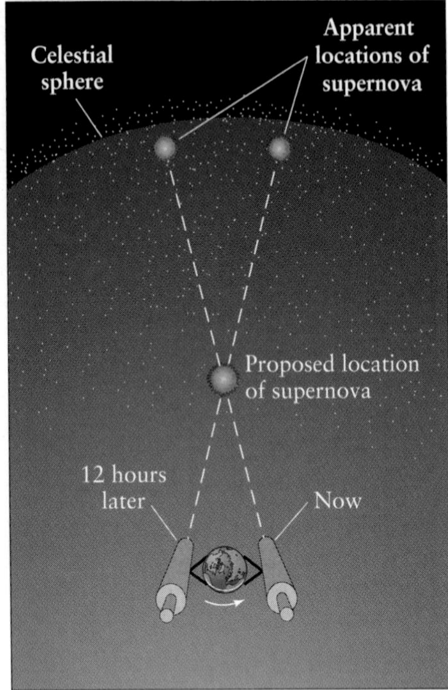

a

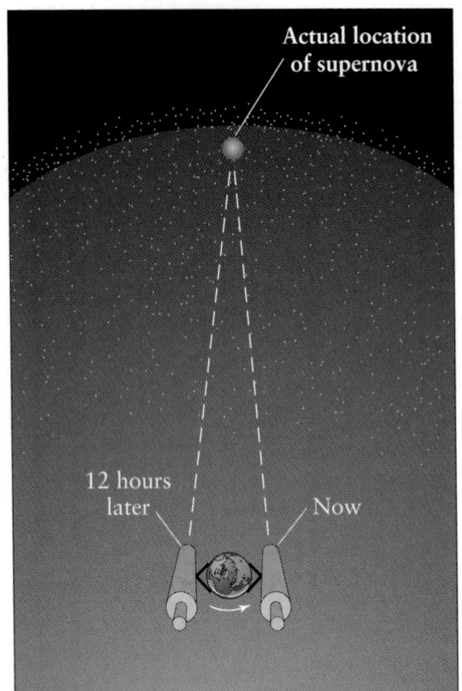

b

FIGURE 2-7 **The Parallax of a Nearby Object in Space** Tycho thought that the Earth does not rotate and that the stars revolve around it. From our modern perspective, the changing position of the supernova would be due to the Earth's rotation as shown in this figure. **(a)** Tycho argued that if an object is near the Earth, its position relative to the background stars should change over the course of a night. **(b)** Tycho failed to measure such changes for the supernova in 1572. This is illustrated in (b) by the two telescopes being parallel to each other. He therefore concluded that the object was far from the Earth.

Tycho reasoned as follows: If the new star is nearby, its position should shift against the background stars over the course of a night, as shown in Figure 2-7a. His careful observations, done in the spirit of the scientific method, failed to disclose any parallax, and so the new star had to be far away, farther from Earth than anyone had imagined (Figure 2-7b). Tycho summarized his findings in a small book, *De Stella Nova* (On the New Star), published in 1573.

Tycho's astronomical records were soon to play an important role in the development of a heliocentric cosmology. From 1576 to 1597, Tycho made comprehensive observations, measuring planetary positions with an accuracy of 1 arcminute, about as precise as is possible with the naked eye. Arcminute (arcmin) is defined in An Astronomer's Toolbox 1-1. Upon his death in 1601, most of these invaluable records were given to his gifted assistant, Johannes Kepler (see Guided Discovery: Astronomy's Foundation Builders).

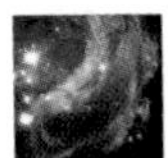

KEPLER'S AND NEWTON'S LAWS

Until Johannes Kepler's time, astronomers had assumed that heavenly objects move in circles. For philosophical and aesthetic reasons, circles were considered the most perfect and most harmonious of all geometric shapes. However, using circular orbits failed to yield accurate predictions for the positions of the planets. For years, Kepler tried to find a shape for orbits that would fit Tycho Brahe's observations of the planets' positions against the background of distant stars. Finally, he began working with a geometric form called an **ellipse.**

2-5 Kepler's laws describe orbital shapes, changing speeds, and the lengths of planetary years

An ellipse can be drawn as shown in Figure 2-8a. Each thumbtack is at a **focus** (plural **foci**). The longest diameter across an ellipse, called the *major axis,* passes through both foci. Half of that distance is called the **semimajor axis,** whose length is usually designated by the letter *a*. In astronomy, the length of the semimajor axis is also the average distance between a planet and the Sun.

To Kepler's delight, the ellipse turned out to be the curve he had been searching for. Predictions of the locations of planets based on elliptical paths were in very close agreement with where the planets actually were. He published this discovery in 1609 in a book known today as *New*

(© 1980 Sidney Harris)

Astronomy. This important discovery is now considered the first of **Kepler's laws:**

Kepler's First Law:
The orbit of a planet about the Sun is an ellipse with the Sun at one focus.

The shapes of ellipses have two extremes. The roundest ellipse is a circle. The most elongated ellipse approaches being a straight line. The shape of a planet's orbit around the Sun is described by its *orbital eccentricity,* designated by the letter e, which ranges from 0 (circular orbit) to just under 1.0 (nearly a straight line). Figure 2-8b shows a sequence of ellipses and their associated eccentricities. Observations have since revealed that there is no object at the second focus of each elliptical planetary orbit.

Tycho's observations also showed Kepler that planets do not move at uniform speeds along their orbits. Rather, a planet moves most rapidly when it is nearest the Sun, a point on its orbit called **perihelion.** Conversely, a planet moves most slowly when it is farthest from the Sun, a point called **aphelion.**

After much trial and error, Kepler discovered a way to describe how fast a planet moves anywhere along its orbit. This discovery, also published in *New Astronomy,* is illustrated in Figure 2-9. Suppose that it takes 30 days for a planet to go from point A to point B. During that time, the line joining the Sun and the planet sweeps out a nearly triangular area (shaded in Figure 2-9). Kepler discovered that the line joining the Sun and the planet sweeps out the same area during any other 30-day interval. In other words, if the planet also takes 30 days to go from point C to point D, then the two shaded segments in Figure 2-9 are equal in area. Kepler's second law, also called the **law of equal areas,** can be stated thus:

Kepler's Second Law:
A line joining a planet and the Sun sweeps out equal areas in equal intervals of time.

The physical content of Kepler's second law is that each planet's speed decreases as it moves from perihelion to aphelion. The speed then increases as the planet moves from aphelion toward perihelion.

Kepler was also able to relate a planet's year to its distance from the Sun. This discovery, published in 1619, stands out because of its impact on future developments in astronomy. Now called *Kepler's third law,*

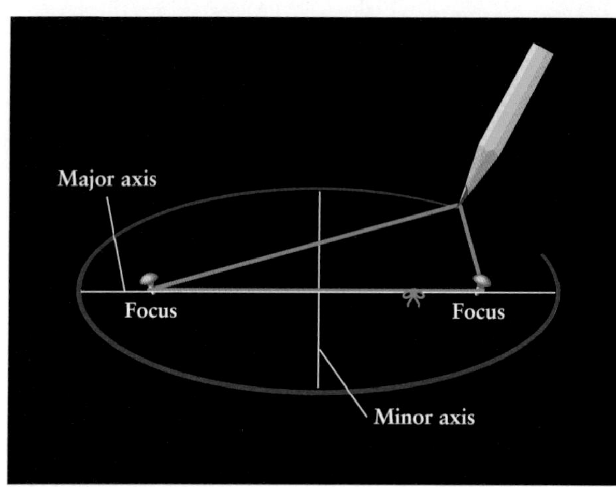

a

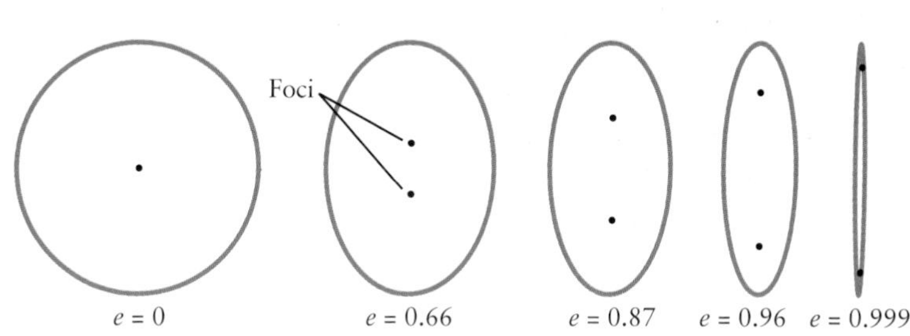

b

FIGURE 2-8 **Ellipse** **(a)** The construction of an ellipse: An ellipse can be drawn with a pencil, a loop of string, and two thumbtacks, as shown in this figure. If the string is kept taut, the pencil traces out an ellipse. The two thumbtacks are located at the two foci of the ellipse. **(b)** A series of ellipses with different eccentricities, e. Eccentricities range between 0 (circle) to just under 1.0 (almost a straight line).

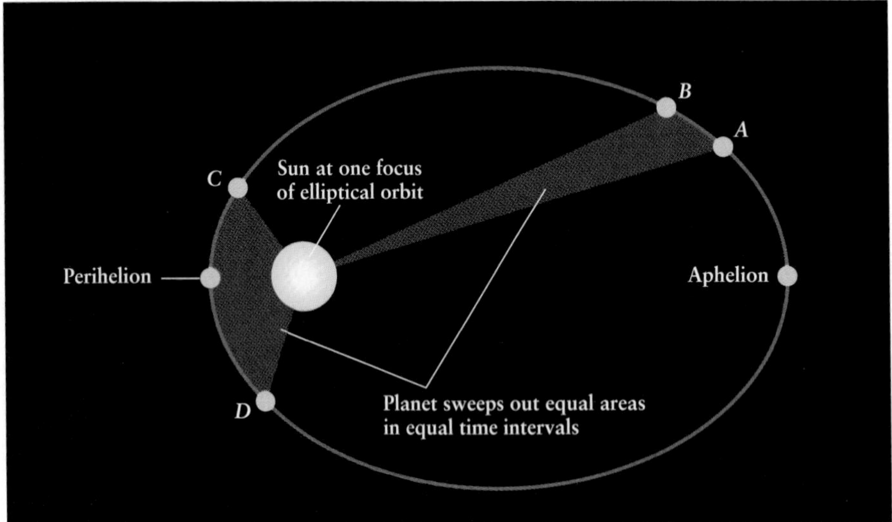

FIGURE 2-9 **Kepler's First and Second Laws** According to Kepler's first law, every planet travels around the Sun along an elliptical orbit with the Sun at one focus. According to his second law, the line joining the planet and the Sun sweeps out equal areas in equal intervals of time. *Note:* This drawing shows a highly elliptical orbit, with e = 0.74. Even though this is a much greater eccentricity than that of any planet in the solar system, the concept still applies to all planets and other orbiting bodies.

it predicts the planet's sidereal period if we know the length of the semimajor axis of the planet's orbit:

Kepler's Third Law:
The square of a planet's sidereal period around the Sun is directly proportional to the cube of the length of its orbit's semimajor axis.

The relationship is easiest to use if we let P represent the sidereal period in Earth years and a represent the length of the semimajor axis measured in **astronomical units.** One astronomical unit is the average distance from the Earth to the Sun. It is commonly used in discussing distances to various objects in the solar system, since no powers of ten need be used with it, as they would if distances between planets were referred to in kilometers or miles. (See An Astronomer's Toolbox 2-1 for more details of astronomical distance units.) Now we can write Kepler's third law as

$$P^2 = a^3$$

TABLE 2-3 A Demonstration of Kepler's Third Law

ANIMATION 2.5

	Sidereal period P (yr)	Semimajor axis a (AU)	P^2	=	a^3
Mercury	0.24	0.39	0.06		0.06
Venus	0.61	0.72	0.37		0.37
Earth	1.00	1.00	1.00		1.00
Mars	1.88	1.52	3.53		3.51
Jupiter	11.86	5.20	140.7		140.6
Saturn	29.46	9.54	867.9		868.3
Uranus	84.01	19.19	7058		7067
Neptune	164.79	30.06	27,160		27,160
Pluto	248.54	39.53	61,770		61,770

4 In other words, a planet closer to the Sun has a shorter year than does a planet farther from the Sun. Combining this with the second law reveals that planets closer to the Sun move more rapidly than those farther away. Using data from Tables 2-1 and 2-2, we can demonstrate Kepler's third law as shown in Table 2-3.

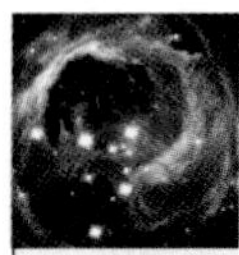

AN ASTRONOMER'S TOOLBOX 2-1
Astronomical Distances

Throughout this book we will find that some of our traditional units of measure become cumbersome. It is fine to use kilometers to measure the diameters of craters on the Moon or the heights of volcanoes on Mars. However, it is as awkward to use kilometers to express distances to planets, stars, or galaxies as it is to talk about the distance from New York City to San Francisco or Paris to London in millimeters. Astronomers have therefore devised new units of measure.

When discussing distances across the solar system, astronomers use a unit of length called the **astronomical unit (AU)**, which is the average distance between the Earth and the Sun:

$$1 \text{ AU} \approx 1.5 \times 10^8 \text{ km} \approx 9.3 \times 10^7 \text{ miles}$$

Jupiter, for example, is an average of 5.2 times farther from the Sun than is the Earth. Thus, the average distance between the Sun and Jupiter can be conveniently stated as 5.2 AU. This can be converted into kilometers or miles using the relationship above.

When talking about distances to the stars, astronomers choose between two different units of length. One is the **light-year (ly)**, which is the distance that light travels in a vacuum (in the absence of air) in one year:

$$1 \text{ ly} \approx 9.46 \times 10^{12} \text{ km} \approx 63{,}000 \text{ AU}$$

One light-year is roughly equal to six trillion miles. Proxima Centauri, the star (other than the Sun) nearest to the Earth, is just over 4.2 ly from Earth.

The second commonly used unit of length is the **parsec (pc)**, the distance at which two objects separated by 1 AU make an angle of 1 arcsecond. Imagine taking a journey far into space, beyond the orbits of the outer planets. Watching the solar system as you move away, the angle between the Sun and the Earth becomes smaller and smaller. When the Sun and Earth are side by side from your perspective and you measure the angle between them as 1/3600° (called 1 arcsecond), you have reached a distance astronomers call 1 parsec, as shown in the figure below. The parsec turns out to be longer than the light-year. Specifically,

$$1 \text{ pc} \approx 3.09 \times 10^{13} \text{ km} \approx 3.26 \text{ ly}$$

Thus, the distance to the nearest star can be stated as 1.3 pc as well as 4.2 ly. Whether one uses light-years or parsecs is a matter of personal taste.

For larger distances, kilolightyears (kly) and megalightyears (Mly) and *kiloparsecs* (kpc) and *megaparsecs* (Mpc) are used. The prefixes "kilo" and "mega" simply mean "thousand" and "million," respectively:

$$1 \text{ kly} = 10^3 \text{ ly}$$

$$1 \text{ Mly} = 10^6 \text{ ly}$$

$$1 \text{ kpc} = 10^3 \text{ pc}$$

$$1 \text{ Mpc} = 10^6 \text{ pc}$$

For example, the distance from Earth to the center of our Milky Way Galaxy is about 8.6 kpc, and the rich cluster of galaxies in the direction of the constellation Virgo is 20 Mpc away.

Try these questions: The nearest star (other than the Sun) is 4.22 ly away. How many miles away is it? How many kilometers? How many parsecs?

(Answers appear at the end of the book.)

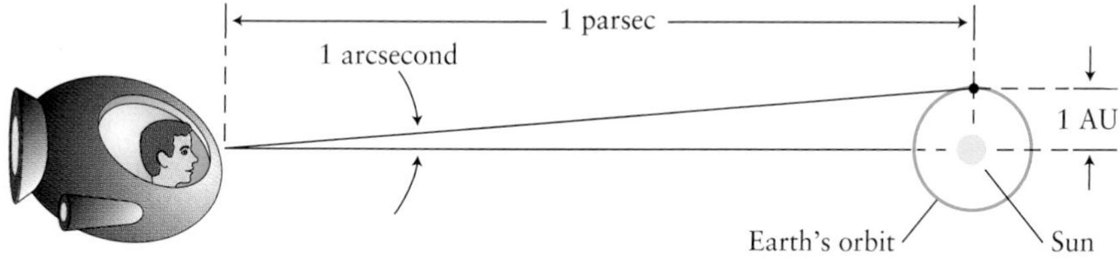

A Parsec The parsec, a unit of length commonly used by astronomers, is equal to 3.26 ly. The parsec is defined as the distance at which 1 AU perpendicular to the observer's line of sight makes an angle of 1 arcsecond.

When Newton derived Kepler's third law using the law of gravitation, discussed later, he discovered that the mass of the planet affects the period of its orbit around the Sun. However, the effect of the planet's mass on the period of its orbit is vanishingly small for all the planets in the solar system, which is why the equation for Kepler's third law, as shown in Table 2-3, gives such good results for the planets' orbits even though it does not take the planet's mass into account.

Kepler's three laws apply not only to the planets orbiting the Sun but also whenever any object orbits another under the influence of their mutual gravitational attraction. Thus, Kepler's laws apply to moons orbiting planets, artificial satellites orbiting the Earth, and even two stars revolving about each other. Throughout this book, we will see that Kepler's three laws have a wide range of practical applications.

INSIGHT INTO SCIENCE

Theories and Explanations Scientific theories (or laws) based on observations can be useful for making predictions even if the reasons that these laws work are unknown. The explanation for Kepler's laws came decades after Kepler deduced them, in 1665, when Newton derived them with his mathematical expression for gravitation, the force that holds the planets in their orbits.

2-6 Galileo's discoveries strongly supported a heliocentric cosmology

While Kepler was making rapid progress in central Europe, an Italian physicist was making equally dramatic observations in southern Europe. Galileo Galilei did not invent the telescope, but he was one of the first people to point the new device toward the sky and publish his observations. He saw things that no one had ever imagined—mountains on the Moon and spots on the Sun. He also discovered that the apparent size of Venus as seen through his telescope was related to the planet's phase (Figure 2-10). Venus appears smallest at gibbous phase and largest at crescent phase. These observations were a big chink in the geocentric cosmology's armor as it could not explain why Venus has phases or changes size, but a heliocentric cosmology explains both. Galileo's observations therefore supported the conclusion that Venus orbits the Sun, not the Earth.

In 1610, Galileo (see Guided Discovery: Astronomy's Foundation Builders) also discovered four moons near Jupiter. In honor of their discoverer, these are today called the **Galilean moons** (or **satellites,** another term for moon). Galileo concluded that the moons are orbiting Jupiter because they move across from one side of the planet to the other. Confirming observations were made in 1620 (Figure 2-11). These observations all provided further proof that the Earth is not at the center of the universe. Like the Earth in orbit around the Sun, Jupiter's four moons obey Kepler's

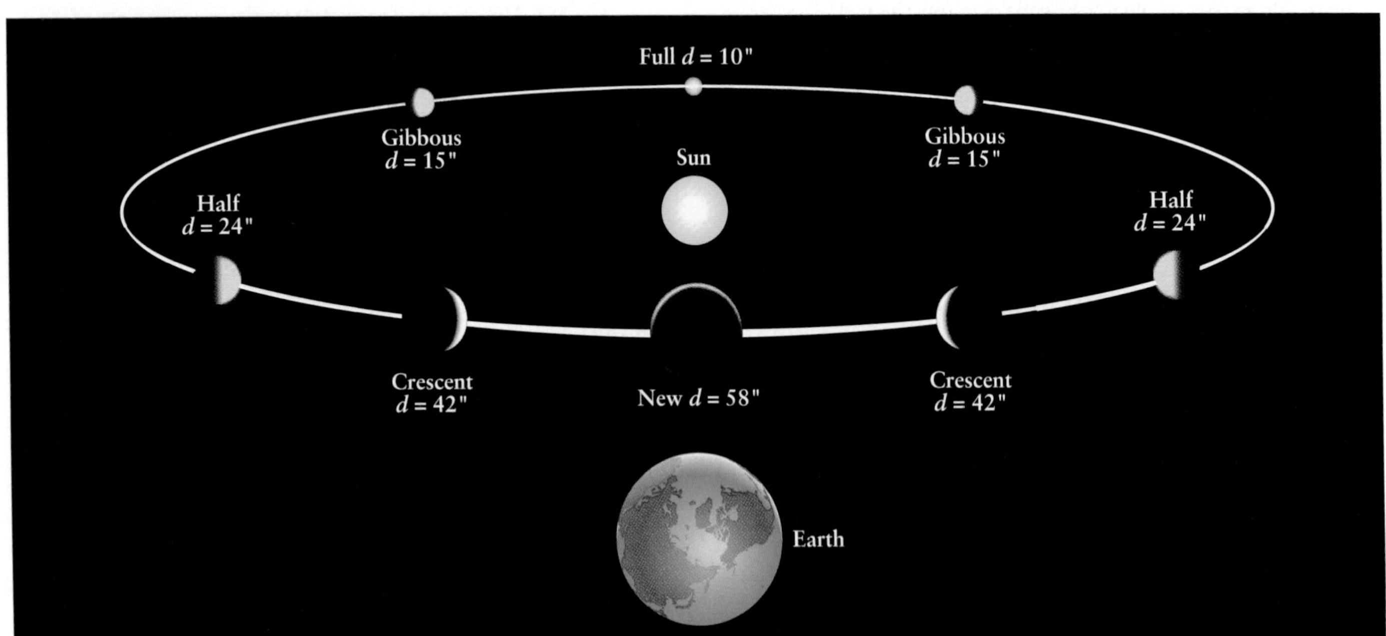

FIGURE 2-10 **The Changing Appearance of Venus** This figure shows how the appearance (phase) of Venus changes as it moves along its orbit. The number below each view is the angular diameter (d) of the planet as seen from Earth, in arcseconds. Note that the phases correlate with the planet's angular size and its angular distance from the Sun, both as seen from Earth. These observations clearly support the idea that Venus orbits the Sun.

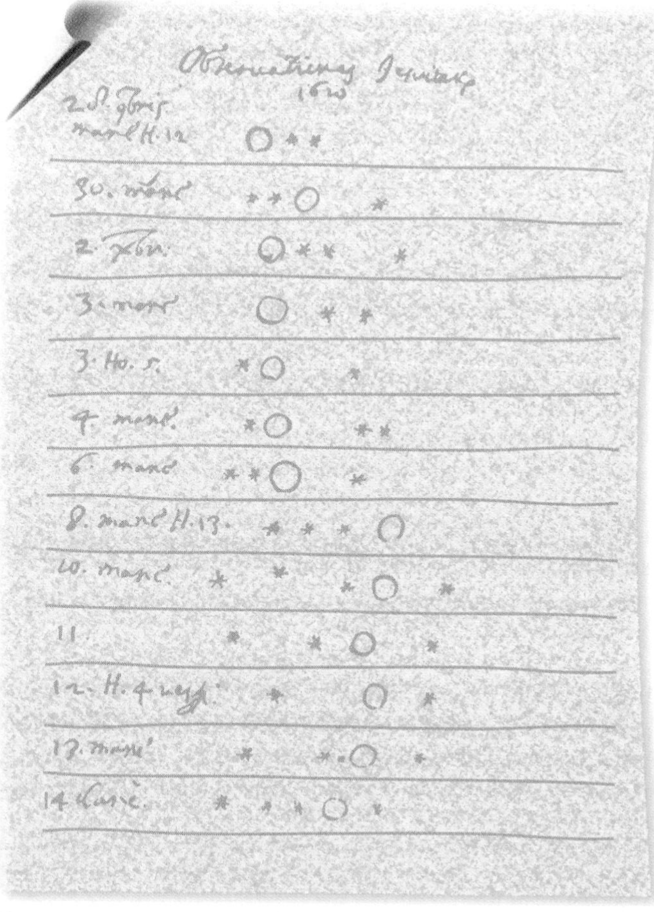

a

b

FIGURE 2-11 **Jupiter and Its Largest Moons** In 1610, Galileo discovered four "stars" that move back and forth across Jupiter. He concluded that they are four moons that orbit Jupiter just as our Moon orbits the Earth. **(a)** This figure shows observations made by Jesuits in 1620. **(b)** This photograph, taken by amateur astronomer C. Holmes, shows the four Galilean satellites alongside an overexposed image of Jupiter. Each satellite would be bright enough to be seen with the unaided eye were it not overwhelmed by the glare of Jupiter. (b: Courtesy of C. Holmes)

third law: The square of a moon's orbital period about Jupiter is directly proportional to the cube of its average distance from the planet.

Galileo's telescopic observations constituted the first fundamentally new astronomical data since humans began recording what they saw in the sky. In contradiction to then-prevailing opinions, these discoveries strongly supported a heliocentric view of the universe. Because Galileo's ideas could not be reconciled with certain passages in the Bible or with the writings of Aristotle and Plato, the Roman Catholic Church condemned him, and he was forced to spend his latter years under house arrest "for vehement suspicion of heresy." He was only cleared of wrongdoing in 1992.

A major stumbling block prevented seventeenth-century thinkers from accepting Kepler's laws and Galileo's conclusions about the heliocentric cosmology. Once anything on Earth is put in motion, it quickly comes to rest. Why didn't the planets orbiting the Sun stop, too?

The scientific method clarified all the mysteries around planets and orbits in the form of equations and laws developed by the brilliant and eccentric (for example, he believed in alchemy) scientist Isaac Newton, who was born on Christmas Day in 1642, less than a year after Galileo died. In the decades that followed, Newton revolutionized science more profoundly than any person before him, and in doing so, he found physical and mathematical proofs of the heliocentric cosmology.

2-7 Newton formulated three laws that describe fundamental properties of physical reality

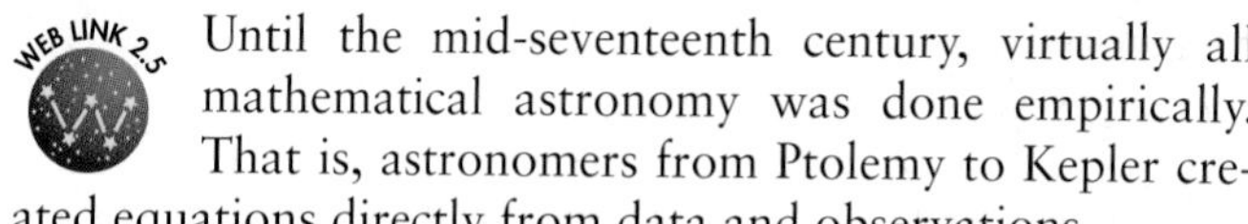

Until the mid-seventeenth century, virtually all mathematical astronomy was done empirically. That is, astronomers from Ptolemy to Kepler created equations directly from data and observations.

Isaac Newton (see Guided Discovery: Astronomy's Foundation Builders) introduced a new approach. He began with three physical assumptions, now called **Newton's laws of motion,** which led to equations that have since been tested and shown to be correct. He also found a formula for the force of **gravity,** the attraction between all objects due to their masses. Putting the assumptions into mathematical form and combining them with the equation for gravity, Newton was able to derive Kepler's three laws and use them to predict the orbits of bodies such as comets and other objects in the solar system. Newton also was able to use these same equations to predict the motions of bodies on and near the Earth, like the path of a projectile or the speed of a falling object.

5 *Newton's first law, the law of inertia:*
A body remains at rest or moves in a straight line at a constant speed unless acted upon by a net outside force.

At first, this law might seem to conflict with your everyday experience. For example, if you shove a chair, it does not move at a constant speed forever but rather comes to rest after sliding only a short distance. From Newton's viewpoint, however, a "net outside force" does indeed act on the moving chair, namely, friction between the chair's legs and the floor. Without friction, the chair would continue in a straight path at a constant speed. An unbalanced outside **force** changes the motion of an object.

Newton's first law tells us that there must be an outside force acting on the planets to continually change their directions and keep them in orbit. If there were no force acting on them, they would move away from the Sun along straight-line paths at constant speeds. Because this does not happen, Newton concluded that some force confines the planets to their elliptical orbits. As we shall see, that force is gravity.

Newton's second assumption describes how a force changes the motion of an object. To appreciate the concepts of force and motion better, we must first understand two quantities that describe motion: velocity and acceleration.

Imagine an object motionless in space. Push on the object and it begins to move. At any moment, you can describe the object's motion by specifying both its speed and direction. Speed and direction of motion together constitute an object's **velocity.** If you continue to push on the object, its speed will increase—it will accelerate.

Acceleration is the rate at which velocity changes with time. Because velocity involves both speed and direction, a slowing down, a speeding up, or a change in direction are all forms of acceleration.

Suppose an object revolved about the Sun in a perfectly circular orbit. As this object moved along its orbit, its speed would remain constant, but its direction of motion would be continuously changing. This body would have acceleration that involved only a change of direction.

Newton's second law, the force law:
The acceleration of an object is proportional to the force acting on it.

In other words, the harder you push on an object that can move, the faster it will accelerate. Newton's law also says that a greater mass accelerates more slowly when acted on by a force than does an object of lesser mass acted on by the same force. That is why by pushing a child's wagon, you can accelerate it faster than you can accelerate a car by pushing on it equally hard. Newton's second law can be succinctly stated as an equation. If a force acts on an object, the object will experience an acceleration such that

$$\text{Force} = \text{mass} \times \text{acceleration}$$

The **mass** of an object is a measure of the total number 6
of particles it contains, which we measure in kilograms. For example, the mass of the Sun is 2×10^{30} kg, the mass of a hydrogen atom is 1.7×10^{-27} kg, and the mass of the author of this book is 83 kg. At rest, the Sun, a hydrogen atom, and I have these same masses regardless of where we happen to be in the universe.

It is important not to confuse the concept of mass with that of weight. Your **weight** is the force with which you push down on a scale due to the gravitational attraction of the body on which you stand or, equivalently, the force with which you push down on a scale when being accelerated (sped up) by a rocket or other device. For example, when you start moving upward in an elevator, the elevator is momentarily accelerating and during that time you feel heavier. If you had a scale under your feet, you would see that while the elevator is accelerating, you actually are heavier than normal until the elevator is moving at a constant speed, at which time you return to your normal weight.

Force is usually expressed in pounds or newtons. For example, the force with which I am pressing down on the ground is 183 pounds. But I weigh 183 pounds only on the Earth. I would weigh 30.5 pounds on the Moon, which has less mass and so it pulls me down with less force. Orbiting in the Space Shuttle, my apparent weight (measured by standing on a scale in the shuttle), would be zero, but my mass would be the same as when I am on Earth. Because I still have inertia in the shuttle, an astronaut there would still have to push me with a force to get me to move. Whenever we describe the properties of planets, stars, or galaxies, we speak of their masses, never of their weights.

Newton's final assumption, called *Newton's third law,* is the *law of action and reaction.*

Newton's third law, the law of action and reaction:
Whenever one body exerts a force on a second body, the second body exerts an equal and opposite force on the first body.

For example, I weigh 183 pounds, and so I press down on the floor with a force of 183 pounds. Newton's third law says that the floor is also pushing up against me with an equal force of 183 pounds. (If it were less, I would fall through the floor, and if it were more, I would be lifted upward.) In the same way, Newton realized that because the Sun is exerting a force on each planet to keep it in orbit, each planet must also be exerting an equal and opposite force on the Sun. As each planet accelerates toward the Sun, the Sun in turn accelerates toward each planet.

The Sun is pulling the planets, so why don't they fall onto it? **Conservation of angular momentum,** a fundamental consequence of Newton's second law of motion, provides the answer. Angular momentum is a measure of how much energy is stored in an object due to its rotation and revolution. The details are presented in An Astronomer's Toolbox 2-2. As the orbiting planets fall Sun-

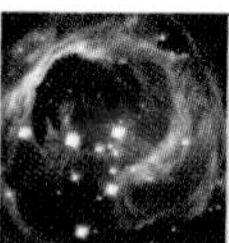

AN ASTRONOMER'S TOOLBOX 2-2

Energy and Momentum

Scientists identify two types of energy that are available to any object. The first, called **kinetic energy,** is associated with the object's motion. For speeds much less than the speed of light, we can write the amount of kinetic energy, KE, in an object as:

$$\mathrm{KE} = \frac{1}{2}mv^2$$

where m is the object's mass (total number of particles) and v is its velocity. Kinetic energy is a measure of how much work the object can do on the outside world or, equivalently, how much work the outside world has done to put the object in motion.

Work is also a rigorously defined concept that often is at odds with our intuition. It is defined as the product of the force, F, acting on an object times the distance, d, over which the object moves in the direction of the force:

$$W = Fd$$

For example, if I exert a horizontal force of 50 newtons (N, a unit of force) and thereby move an object 10 meters in that direction, then I have done 50 N × 10 m = 500 joules of work. (I have used the relationship that 1 newton × 1 meter = 1 joule.)

The second type of energy is called **potential energy.** It represents how much energy is stored in an object as a result of its location in space. For example, if you hold a pencil above the ground, the pencil has potential energy that can be converted into kinetic energy by the Earth's gravitational force. How does that conversion get underway? Just let go of the pencil.

There are various kinds of potential energy, such as the potential energy stored in a battery and the potential energy stored in objects under the influence of gravity. We will focus on *gravitational potential energy.* Far from extremely massive objects, like stars, or extremely dense objects, like black holes, gravitational potential energy can be written as:

$$\mathrm{PE} = \frac{GmM}{r}$$

where $G = 6.668 \times 10^{-11}$ N m^2/kg^2, m is the mass of the object whose gravitational potential energy you are measuring, M is the mass of the object creating the gravitational potential energy, and r is the distance between the centers of masses of the object feeling the gravitational potential energy and the object creating that energy.

Near the surface of the Earth, this equation simplifies to:

$$\mathrm{PE} = mgh$$

where $g = 9.8$ m/s^2 (32 ft/s^2) is the gravitational acceleration at the Earth's surface and h is the height of the object above the Earth's surface.

Potential energy can be converted into kinetic energy and vice versa. By dropping the pencil, its gravitational potential energy begins decreasing while its kinetic energy begins increasing *at the same rate.* The pencil's total energy is conserved. Conversely, if you throw a pencil up in the air, the kinetic energy you give it will immediately begin decreasing, while its potential energy increases at the same rate.

Related to the motion of an object, and hence to its kinetic energy, are the concepts of linear momentum, usually just called **momentum,** and **angular momentum.** Momentum, $\mathbf{p}$, is described by the equation:

$$\mathbf{p} = m\mathbf{v}$$

where $\mathbf{v}$ is the velocity of the object. Both $\mathbf{p}$ and $\mathbf{v}$ are in boldface to indicate that they both represent motion in *some direction* or another, as well as having some numerical value. Simple algebra reveals that kinetic energy and momentum are related by:

$$\mathrm{KE} = \frac{p^2}{2m}$$

Linear momentum, then, indicates how much energy is stored in an object because of its motion in a straight line (its linear motion).

Angular momentum, $\mathbf{L}$, can be expressed mathematically as:

$$\mathbf{L} = I\boldsymbol{\omega}$$

where I is the moment of inertia of an object and $\boldsymbol{\omega}$ (lowercase Greek omega) gives the speed and direction in which it is revolving or rotating. Just as the mass indicates how hard it is to change an object's straight-line motion, the moment of inertia indicates how hard it is to change the rate at which an object rotates or revolves. The moment of inertia depends on an object's mass and shape. Kinetic energy stored in angular motion can be written:

$$\mathrm{KE} = \frac{L^2}{2I}$$

Newton's *first law* can also be expressed in terms of conservation of linear momentum:

A body maintains its linear momentum unless acted upon by a net external force.

(continued on the following page)

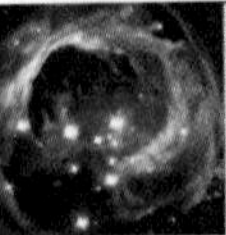

AN ASTRONOMER'S TOOLBOX 2-2 (CONTINUED)
Energy and Momentum

Equivalently, for angular motion we can write the conservation of angular momentum:

A body maintains its angular momentum unless acted upon by a net external torque.

Torques are created when a force acts on an object in some direction other than toward the center of the object's angular motion, as shown in the accompanying figure. The Earth has angular momentum and keeps spinning on its rotational axis and orbiting around the Sun. Likewise, the Moon has angular momentum and keeps spinning on its rotation axis and orbiting the Earth. Virtually all objects in astronomy have angular momentum, and I think it's fair to say that conservation of angular momentum is among the most important laws in the cosmos. After all, it is what keeps the planets in orbit around the Sun, the moons in orbit around the planets, the astronomical bodies rotating at relatively constant rates, and many other rotation-related effects that we will encounter throughout this book.

Try these questions: How does tripling the linear momentum of an object change its kinetic energy? How does halving the angular momentum of an object change its kinetic energy? How much work would you do if you pushed on a desk with a force of 100 N and moved it 20 m? How much work would you do if you pushed on a desk with a force of 500 N and moved it 0 m? What two things can you vary to change the angular momentum of an object?

(Answers appear at the end of the book.)

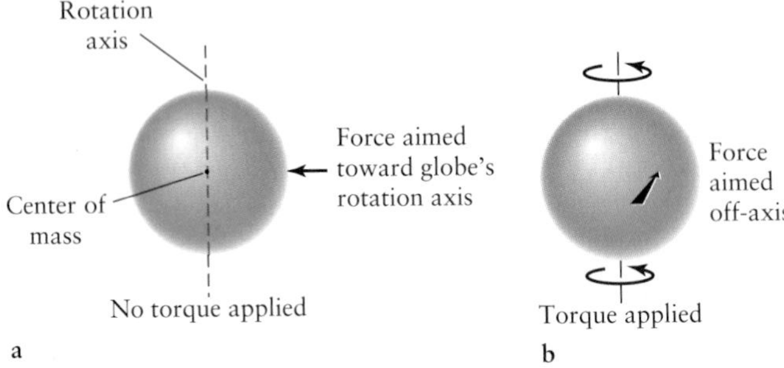

Angular Momentum and Torque **(a)** When a force acts through an object's rotation axis or toward its center of mass, then the force does *not* exert a torque on the object. **(b)** When a force acts in some other direction, then it exerts a torque, causing the body's angular momentum to change. If the object can spin around a fixed axis, like a globe, then the rotation axis is the rod running through it. If the object is not held in place, then the rotation axis is in a line through a point called the object's center of mass. The center of mass of any object is the point that follows an elliptical, parabolic, or hyperbolic path in a gravitational field. All other points in the spinning object wobble as it moves.

a

b

R I V U X G

FIGURE 2-12 **Conservation of Angular Momentum** As this skater brings her arms and outstretched leg in, she must spin faster to conserve her angular momentum. The same effect occurs if you start on the edge of a spinning merry-go-round and then move toward the center—be careful! (Getty Images) ward, their angular momentum provides them with motion perpendicular to that infall, meaning that planets continually fall toward the Sun, but they continually miss it. Because their angular momentum is conserved, planets neither spiral into the Sun nor away from it. Angular momentum remains constant unless acted on by an outside torque.

Angular momentum depends on three things: how fast the body rotates or revolves, how much mass it has, and how spread out that mass is. The greater a body's angular motion or mass, or the more the mass is spread out, the greater its angular momentum. Consider, for example, a twirling ice skater. She rotates with a constant mass, practically free of outside forces. When she wishes to rotate more rapidly, she decreases the spread of her mass distribution by pulling her arms and outstretched leg in closer to her body (Figure 2-12). According to the conservation of angular momentum, as the spread of mass decreases, the rotation rate must increase. In astronomy, we encounter many instances of the same law, as giant objects, such as stars, contract.

We have now reconstructed the central relationships between matter and motion. Scientific belief in the heliocentric cosmology still requires a force to hold the planets in orbit around the Sun and the moons in orbit around the planets. Newton identified that, too.

2-8 Newton's description of gravity accounts for Kepler's laws

Isaac Newton did not invent the idea of gravity. An educated seventeenth-century person would understand that some force pulls things down to the ground. It was Newton, however, who gave us a precise description of the action of gravity, or *gravitation,* as it is more properly called. Using his first law, Newton proved mathematically that the force acting on each of the planets is directed toward the Sun. He expanded this result to the idea that the nature of the force pulling a falling apple straight down to the ground is the same as the nature of the force on the planets from the Sun. More generally, the gravitational force from an object acts to pull every other object directly toward it.

Newton succeeded in formulating a mathematical model describing the behavior of the gravitational force that keeps the planets in their orbits (presented in An Astronomer's Toolbox 2-3).

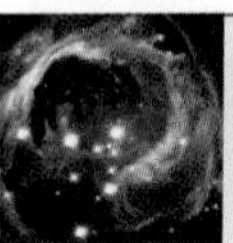

AN ASTRONOMER'S TOOLBOX 2-3
Gravitational Force

From Newton's law of gravitation, if two objects having masses m_1 and m_2 are separated by a distance r, then the gravitational force F between them is

$$F = G\left(\frac{m_1 m_2}{r^2}\right)$$

In this formula, G is the **universal constant of gravitation,** whose value has been determined from laboratory experiments:

$$G = 6.668 \times 10^{-11} \text{ N m}^2 \text{ kg}^{-2}$$

where N is the unit of force, a newton.

The equation $F = G(m_1 m_2/r^2)$ gives, for example, the force from the Sun on the Earth and, equivalently, from the Earth on the Sun. If m_1 is the mass of the Earth (6.0×10^{24} kg), m_2 is the mass of the Sun (2.0×10^{30} kg), and r is the distance from the Earth to the Sun (1.5×10^{11} m),

$$F = 3.6 \times 10^{22} \text{ N}$$

This number can then be used in Newton's second law, $F = ma$, to find the acceleration of the Earth due to the Sun. This yields:

$$a_{\text{Earth}} = \frac{F}{m_1} = 6.0 \times 10^{-3} \text{ m/s}^2$$

Newton's third law says that the Earth exerts the same force on the Sun, so the Sun's acceleration due to the Earth's gravitational force is

$$a_{\text{Sun}} = \frac{F}{m_2} = 1.8 \times 10^{-8} \text{ m/s}^2$$

In other words, the Earth pulls on the Sun, causing the Sun to move toward it. Because of the Sun's greater mass, however, the amount that the Sun accelerates the Earth is more than 300,000 times the amount that the Earth accelerates the Sun.

Try these questions: The Earth's radius is 6.4×10^6 m and 1 kg of mass is equivalent to 2.2 lb of weight on Earth. What is the force that the Earth exerts on you? What is the force that you exert on the Earth? What is the Earth's acceleration on you? What would the Sun's force be on the Earth if our planet were twice as far from the Sun as it is? How does that force compare to the force from the Sun at our present location?

(Answers appear at the end of the book.)

FIGURE 2-13 **Conic Sections** A conic section is any one of a family of curves obtained by slicing a cone with a plane, as shown in this figure. The orbit of one body about another can be an ellipse, a parabola, or a hyperbola. Circular orbits are possible because a circle is just an ellipse for which both foci are at the same point.

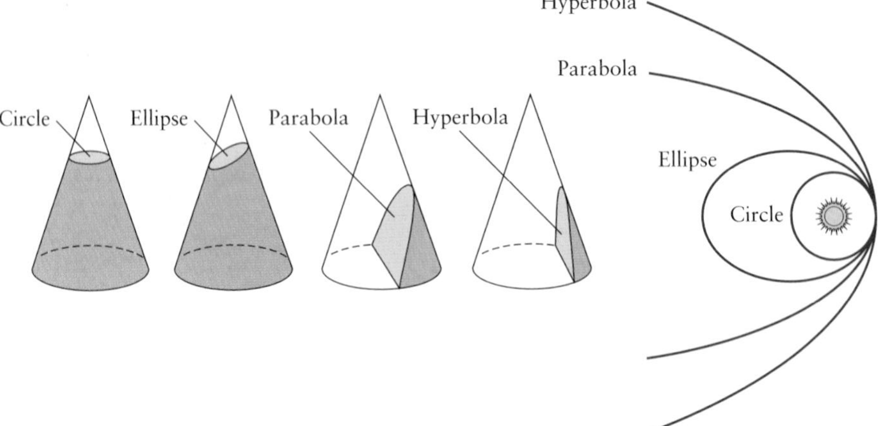

Newton's law of universal gravitation:
Two bodies attract each other with a force that is directly proportional to the product of their masses and inversely proportional to the square of the distance between them.

In other words, gravitational force decreases with distance: Move twice as far away from a body and you feel only one-quarter of the force from it that you felt before. Despite its weakening, the force of gravity from every object extends throughout the universe. Also, an object with twice the mass of another object has twice the gravitational force as the less massive object.

Using his law of gravity, Newton found that he could mathematically explain Kepler's three laws. For example, whereas Kepler discovered by trial and error that the period of orbit, P, and average distance between the Sun and planet, a, are related by $P^2 = a^3$, Newton demonstrated mathematically that this equation (corrected with a tiny contribution due to the mass of the planet, as mentioned earlier) follows from his law of gravitation.

Newton's version of Kepler's third law can be easily recast to predict the orbits of any objects under the influence of a gravitational attraction. Indeed, all three of Kepler's laws apply to all orbiting bodies. Among other examples, this includes moons orbiting planets, artificial satellites orbiting the Earth, and pairs of stars revolving about each other. Throughout this book, we will see that Kepler's three laws have a wide range of practical applications.

Newton also discovered that some objects have nonelliptical orbits around the Sun. His equations led him to conclude that the orbits of some objects are **parabolas** and **hyperbolas** (Figure 2-13). For example, comets hurtling toward the Sun from the depths of space often follow parabolic or hyperbolic orbits.

Newton's ideas turned out to be applicable in an incredibly wide range of situations. The orbits of the planets and their satellites could now be calculated with unprecedented precision. Using Newton's laws, mathematicians proved that the Earth's axis of rotation must precess because of the gravitational pull of the Moon and the Sun on the Earth's equatorial bulge (recall Figure 1-21).

In the spirit of the scientific method, Newton's laws and mathematical techniques were used to predict new phenomena. Newton's friend, Edmund Halley, was intrigued by historical records of a comet that was sighted about every 76 years. Using Newton's methods, Halley worked out the details of the comet's orbit and predicted its return in 1758. It was first sighted on Christmas night of that year, and to this day the comet bears Halley's name (Figure 2-14).

Perhaps the most dramatic early use of the scientific method with Newton's ideas was their role in the discovery of the eighth planet in our solar system. The seventh planet,

R I V U X G

FIGURE 2-14 **Halley's Comet** Halley's Comet orbits the Sun with an average period of about 76 years. During the twentieth century, the comet passed near the Sun twice—once in 1910 and again, shown here, in 1986. The comet will pass close to the Sun again in 2061. While dim in 1986, it nevertheless spread more than 5° across the sky, or 10 times the diameter of the Moon. (Harvard College Observatory/Photo Researchers, Inc.)

a

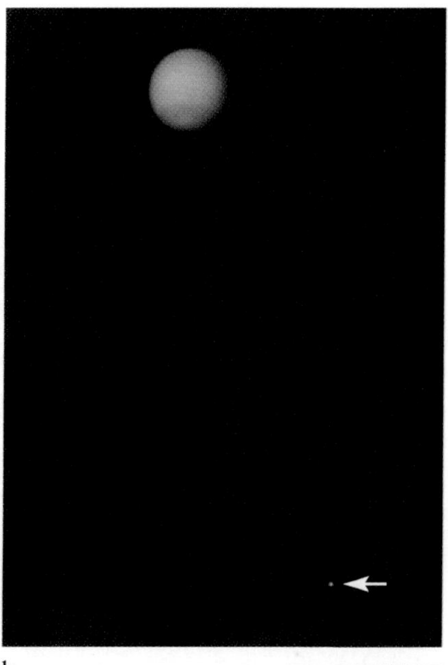
b

R I V U X G

FIGURE 2-15 **Uranus and Neptune** The discovery of Neptune was a major triumph for Newton's laws. In an effort to explain why Uranus (**(a)**, with several of its moons) deviated from its predicted orbit, astronomers predicted the existence of Neptune (**(b)**, with one of its moons indicated by an arrow). Uranus and Neptune are nearly the same size; both have diameters about 4 times that of Earth. (a: Eric Karkoschka (University of Arizona) and NASA; b: NASA)

Uranus, had been discovered accidentally by William Herschel in 1781 during a telescopic survey of the sky. Fifty years later, however, it was clear that Uranus was not following the orbit predicted by Newton's laws. Two mathematicians, John Couch Adams in England and Urbain-Jean-Joseph Leverrier in France, independently calculated that the deviations of Uranus from its predicted orbit could be explained by the gravitational pull of a then-unknown, more distant planet. Each man predicted that the planet would be found at a certain location in the constellation of Aquarius in September 1846. A telescopic search on September 23, 1846, revealed Neptune less than 1° from its calculated position. Although sighted with a telescope, Neptune was really discovered with pencil and paper (Figure 2-15).

INSIGHT INTO SCIENCE

Quantify Predictions Mathematics provides a language that enables science to make quantitative predictions that can be checked by anyone. For example, we have seen in this chapter how Kepler's third law and Newton's universal law of gravitation correctly predict the motion of objects under the influence of the Sun's gravitational attraction.

Over the years, Newton's ideas were successfully used to predict and explain motion here on Earth and throughout the universe. Even today, as we send astronauts into Earth orbit and send probes to the outer planets, Newton's equations are used to calculate the orbits and trajectories of the spacecraft.

It is a testament to Newton's genius that his three laws were precisely the basic ideas needed to understand the motions of the planets. Newton's process of deriving Kepler's laws and the universal law of gravitation helped secure the scientific method as an invaluable tool in our process of understanding the universe.

2-9 Frontiers yet to be discovered

The science related to forces and orbits described in this chapter was well established by the beginning of the nineteenth century. However, questions remained. Careful observation revealed that Newton's law of gravitation gave very slightly inaccurate predictions for the orbital path of Mercury. We will see in Chapter 14 that this problem was resolved by Albert Einstein. Nevertheless, one fundamental question from this chapter's material remains to be answered: What is "mass"? Many physicists studying the building blocks of matter (such as the elementary particles: protons, neutrons, and electrons) believe that mass comes from the presence of a particle permeating the universe called the Higgs boson. When elementary particles interact with it, they gain the property we call mass. Searches for Higgs bosons are underway at high-energy particle accelerators such as CERN, near Geneva, Switzerland, and the Fermi National Accelerator in Illinois. If these experiments are successful, we may know the origin of mass by the end of this decade.

Summary of Key Ideas

Science: Key to Comprehending the Cosmos

• The ancient Greeks laid the groundwork for progress in science. Early Greek astronomers devised a geocentric cosmology, which placed the Earth at the center of the universe.

• The scientific method is a procedure for formulating theories that correctly predict how the universe behaves.

• A scientific theory must be testable, that is, capable of being disproved.

• Theories are tested and verified by observation or experimentation and result in a process that often leads to their refinement or replacement and to the progress of science.

• Observations of the cosmos have led astronomers to discover some fundamental physical laws of the universe.

Origins of a Sun-Centered Universe

• Copernicus's heliocentric (Sun-centered) theory simplified the general explanation of planetary motions compared to the geocentric theory.

• In a heliocentric cosmology, the Earth is but one of several planets that orbit the Sun.

• The sidereal orbital period of a planet is measured with respect to the stars. It determines the length of the planet's year. Its synodic period is measured with respect to the Sun as seen from the moving Earth (for example, from one opposition to the next).

Kepler's and Newton's Laws

• Ellipses describe the paths of the planets around the Sun much more accurately than do circles. Kepler's three laws give important details about elliptical orbits.

• The invention of the telescope led Galileo to new discoveries, such as the phases of Venus and the moons of Jupiter, that supported a heliocentric view of the universe.

• Newton based his explanation of the universe on three assumptions now called Newton's laws of motion. These laws and his law of universal gravitation can be used to deduce Kepler's laws and to describe planetary motions with extreme accuracy.

• The mass of an object is a measure of the amount of matter in the object; its weight is a measure of the force with which the gravity of some other object pulls on the object's mass when the two objects are at rest with respect to each other (or, equivalently, how much the object pushes down on a scale).

• In general, the path of one astronomical object about another, such as that of a comet about the Sun, is an ellipse, a parabola, or a hyperbola.

WHAT DID YOU THINK?

1 *What makes a theory scientific?* A theory is an idea or set of ideas proposed to explain something about the natural world. A theory is scientific if it makes predictions that can be objectively tested and potentially disproved.

2 *What is the shape of the Earth's orbit around the Sun?* All planets have elliptical orbits around the Sun.

3 *Do the planets orbit the Sun at constant speeds?* No. The closer a planet is to the Sun in its orbit, the faster it is moving. It moves fastest at perihelion and slowest at aphelion.

4 *Do all the planets orbit the Sun at the same speed?* No. A planet's speed depends on its average distance from the Sun. The closest planet moves fastest, the most distant planet moves slowest.

5 *How much force does it take to keep an object moving in a straight line at a constant speed?* Unless an object is subject to an outside force, like friction, it takes no force at all to keep it moving in a straight line at a constant speed.

6 *How does an object's mass differ when measured on Earth and on the Moon?* Assuming the object doesn't shed or collect pieces, its mass remains constant whether on the Earth or on the Moon. Its weight, however, is less on the Moon.

Key Words

acceleration, 52
angular momentum, 53
aphelion, 47
astronomical unit, 49
configuration (of a planet), 42
conjunction, 44
conservation of angular momentum, 52
cosmology, 39
direct motion, 39
ellipse, 46
elongation, 43
focus (of an ellipse), 46
force, 52
Galilean moons (satellites), 50
gravity, 51
heliocentric cosmology, 40
hyperbola, 56
inferior conjunction, 43
Kepler's laws, 47
kinetic energy, 53
law of equal areas, 47
law of inertia, 52
law of universal gravitation, 56
light-year, 49
mass, 52
model, 38
momentum, 53
Newton's laws of motion, 51
Occam's razor, 39
opposition, 44
parabola, 56
parallax, 45
parsec, 49
perihelion, 47
physics, 38
potential energy, 53
retrograde motion, 40
scientific method, 38
scientific theory, 38
semimajor axis (of an ellipse), 46
sidereal period, 44
superior conjunction, 43
synodic period, 44
universal constant of gravitation, 55
velocity, 52
weight, 52
work, 53

Review Questions

1. Who wrote down the equation for the law of gravitation? **a.** Copernicus, **b.** Brahe, **c.** Newton, **d.** Galileo, **e.** Kepler.

2. Which of the following most accurately describes the shape of the Earth's orbit around the Sun? **a.** circle, **b.** ellipse, **c.** parabola, **d.** hyperbola, **e.** square.

3. Of the following planets, which takes longest to orbit the Sun? **a.** Earth, **b.** Uranus, **c.** Mercury, **d.** Jupiter, **e.** Venus.

4. What is a Sun-centered model of the solar system called?

5. How long does it take the Earth to complete a sidereal orbit of the Sun?

6. How did Copernicus explain the retrograde motions of the planets?

7. Which planets can never be seen at opposition? Which planets never pass through inferior conjunction?

8. At what configuration (superior conjunction, greatest eastern elongation, etc.) would it be best to observe Mercury or Venus with an Earth-based telescope? At what configuration would it be best to observe Mars, Jupiter, or Saturn? Explain your answers.

9. What are the synodic and sidereal periods of a planet?

10. What are Kepler's three laws? Why are they important?

11. In what ways did the astronomical observations of Galileo support a heliocentric cosmology?

12. How did Newton's approach to understanding planetary motions differ from that of his predecessors?

13. What is the difference between mass and weight?

14. Why was the discovery of Neptune a major confirmation of Newton's universal law of gravitation?

15. Why does an astronaut have to exert a force on a weightless object to move it?

Advanced Questions

The answers to all computational problems, which are preceded by an asterisk (*), appear at the end of the book.

16. Derive the equation KE = $p^2/2m$ in An Astronomer's Toolbox 2-2.

***17.** Convert: **a.** 8.3 pc into light-years; **b.** 6.52 ly into parsecs; **c.** 8450 AU into kilometers; **d.** 2.7×10^3 Mpc into kiloparsecs.

18. Is it possible for an object in the solar system to have a synodic period of exactly one year? Explain your answer.

19. Describe why there is a systematic decrease in the synodic periods of the planets from Mars outward, as shown in Table 2-1.

***20.** A line joining the sun and an asteroid was found to sweep out 5.2 square astronomical units of space in all of 1994. How much area was swept out in 1995? in five years?

***21.** A comet moves in a highly elongated orbit about the Sun within a period of 1000 years. What is the length of the semimajor axis of the comet's orbit? What is the farthest the comet can get from the Sun?

***22.** The orbit of a spacecraft about the Sun has a perihelion distance of 0.5 AU and an aphelion distance of 3.5 AU. What is the spacecraft's orbital period?

***23.** Look up orbital data for the four largest moons of Jupiter on the Internet, in the current issue of a reference from the U.S. Naval Observatory, such as the *Astronomical Almanac* or *Astronomical Phenomena,* or in such magazines as *Sky & Telescope* and *Astronomy.* Demonstrate that these orbits obey Kepler's third law.

24. Make diagrams of Jupiter's phases as seen from Earth and as seen from Saturn.

25. In what direction (left or right, eastward or westward) across the celestial sphere do the planets normally appear to move as seen from Australia? In what direction is retrograde motion as seen from there?

26. The dictionary defines *astrology* as "the study that assumes and attempts to interpret the influence of the heavenly bodies on human affairs." Based on what you know about scientific theory, is astrology a science? Why or why not? Feel free to explore astrology further if you wish before answering this question.

Discussion Questions

27. Which planet would you expect to exhibit the greatest variation in apparent brightness as seen from Earth? Explain your answer.

28. Use two thumbtacks (or pieces of tape), a loop of string, and a pencil to draw several ellipses. Describe how the shapes of the ellipses vary as you change the distance between the thumbtacks.

What If . . .

***29.** The Earth were 2 AU from the Sun? What would the length of the year be? Assuming that such physical properties as rotation rate were as they are today, what else would be different here? What if the Earth was ½ AU from the Sun?

*30. The Earth was moved to a distance of 10 AU from the Sun? How much stronger or weaker would the Sun's gravitational pull be on Earth?

*31. The Earth had twice its present mass? Assume that all other properties of the Earth and its orbit remain the same. What would be the acceleration of the more massive Earth due to the Sun compared to the present acceleration of the Earth from the Sun? *Hint:* Try combining $F = m_1 a$ and the force equation in An Astronomer's Toolbox 2-3, where m_1 is the mass of the Earth in both equations. Since the acceleration determines the period of the planet's orbit, how would the year on the more massive Earth compare to a year today?

32. The Sun suddenly disappeared? What would the Earth's path in space be in response to such an event? Describe how the Earth would change as a result and how humans might survive on a sunless planet.

33. The laws of physics changed from time to time? What coping skills would life require, if it could exist at all? What else besides the laws of physics would scientists need to comprehend in order to understand and make predictions about the universe?

34. Astronomers only collected data (made observations) about the universe and did not create scientific theories? How would our perspective on the cosmos be different from what it is today?

35. The skies of Earth were perpetually cloudy? How might that have changed the history of our understanding of the cosmos and how might humans under such conditions eventually learn what is really "out there"?

36. Scientists remained believers in the first theory of the cosmos that they decided was correct? How might that change the dynamics by which science evolves in the face of new data that conflicts with earlier theories?

Web Questions

37. Search the Web for more information about the scientific method. Then explain in some detail what a theory and a hypothesis are in science and how they are related. What is Occam's razor (sometimes Ockham's razor) and how does it relate to the process of science?

38. Search the Web for information about Galileo. What were his contributions to physics? Which of Galileo's new ideas were later used by Newton to construct his laws of motion? What incorrect beliefs about astronomy did Galileo hold?

39. Search the Web for information about Kepler. Before he realized that the planets move on elliptical paths, what other models of planetary motion did he consider? What was Kepler's idea of the "music of the spheres"?

40. Search the Web for information about Newton. What were some of the contributions that he made to physics other than developing his laws of motion? What contributions did he make to mathematics?

41. **Monitoring the Retrograde Motion of Mars** Access and view the animation "The Path of Mars in 2009–2010" in Chapter 2 of the *Discovering the Universe* Web site. **a.** Through which two constellations does Mars move? **b.** On approximately what date does Mars stop its direct (right to left or eastward) motion among the background stars and begin its retrograde motion? *Hint:* Use the "Stop" and "Start" functions on your animation controls. **c.** Over how many days does Mars move in the retrograde direction?

Observing Projects

42. It is quite probable that within a few weeks of your reading this chapter one of the planets (Saturn or closer to the Sun) will be in opposition or at greatest eastern elongation, making it readily visible in the evening sky. Using the *Starry Night Enthusiast*™ computer program, the Internet, or a reference book, such as the current issue of the *Astronomical Almanac* or the pamphlet *Astronomical Phenomena* (both published by the U.S. Naval Observatory), select a planet that is at or near such a configuration. To use *Starry Night Enthusiast*™, set the time for this evening. Turn planets on using *Options* tab: *Solar System/Planets-Moons*. Also click on the *Labels* box next to planets. Then grab the screen with the mouse and search the sky for planets out to Saturn. If none are up, change the date by, say, three days and search again. Repeat until you have found visible planets up at night. Plan to make your observations at that time. Make a drawing of what you see through the telescope, being careful to include phase, moons, surface features, and nearby stars. If possible, confirm that your observation is a planet by observing a few days later and showing that the object has moved. (This step is vital in confirming discoveries of new bodies in the solar system.)

43. If Jupiter is visible in the evening sky, observe it with a small telescope on five consecutive clear nights. Record the position of the four Galilean satellites by making nightly drawings, just as the Jesuit priests did in 1620 (see Figure 2-11a). From your drawings, can you tell which moon orbits closest to Jupiter and which orbits farthest? Are there nights when you saw fewer than four of the Galilean moons? What happened to the other moons on those nights?

44. Use the *Starry Night Enthusiast*™ software to observe the moons of Jupiter. Set for Atlas mode (*Favourites/Guides/Atlas*). Center and lock on Jupiter (*Find* Tab/*Jupiter*) and set the angular size of the sky you are examining to about 15' using the size button on the

upper right of your screen. Check that you are locked on Jupiter (right click on Jupiter, click on Centre, if necessary). **a.** Note the positions of the moons. Step forward in increments of six hours and draw the positions of the moons at each timestep. **b.** From your drawings, can you tell which moon orbits closest to Jupiter and which orbits farthest away? Explain your reasoning. **c.** Determine the periods of orbits of these moons (change the length of the timestep, if necessary). **d.** Are there timesteps when you see fewer than four Galilean moons? What happened to the other moons at those times?

45. If Venus is visible in the evening sky, observe the planet with a telescope once a week for a month. On each night, make a drawing of the phase that you see. Can you determine from your drawings if the planet is nearer or farther from the Earth than the Sun is? Do your drawings show any changes in the phase from one week to the next? If so, can you deduce if Venus is coming toward us or moving away from us? Explain.

46. Use your *Starry Night Enthusiast*™ software to observe the phases of Venus. Set for Atlas mode (*Favourites/Guides/Atlas*). Center and lock on Venus (*Find* Tab/*Venus*). Set the angular size of the sky to 7'. **a.** Draw the current shape (phase) of Venus. Adjust the timestep to 30 days, make a single timestep, and again draw Venus, to scale. Make a total of 20 timesteps and drawings. **b.** From your drawings, determine when the planet is nearer or farther from the Earth than the Sun is. **c.** Deduce from your drawings when Venus is coming toward us or is moving away from us. **d.** Explain why Venus goes through this particular cycle of phases.

47. Perform observing project 45 using Mars instead of Venus. If you have done project 45, compare your results for the two planets. Why are the cycles of phases as seen from Earth different for the two planets?

48. **Orbital Eccentricities of the Planets** In this exercise, you will observe which planets have the greatest orbital eccentricities. Set *Deep Space Explorer*™ to its "Home" position. Choose **Solar System/Inner Solar System** on the **Views** tab. You will see the orbits of the inner four planets. If you have not done so already, go to **Settings** and uncheck "Use magnitude cutoffs," slide the "Galaxy drawing/Brightness" slide to the middle of the range, and then click on the **Views** tab.

Grab the screen (hold left mouse button on a PC or the mouse button on a Mac) and move the solar system around until the orbits are in the plane of the screen. Using the fact that Venus's orbital eccentricity is .007 (very nearly circular), note which two of the inner four planets have the least circular orbits. Now view the orbits of all the planets by choosing **Solar System/Outer Solar System**. Again, grab the screen and move it until the orbits are in the plane of the screen. which outer planet has the most eccentric orbit? Sketch the orbits of Uranus, Neptune, and Pluto together.

CHAPTER 3

Light and Telescopes

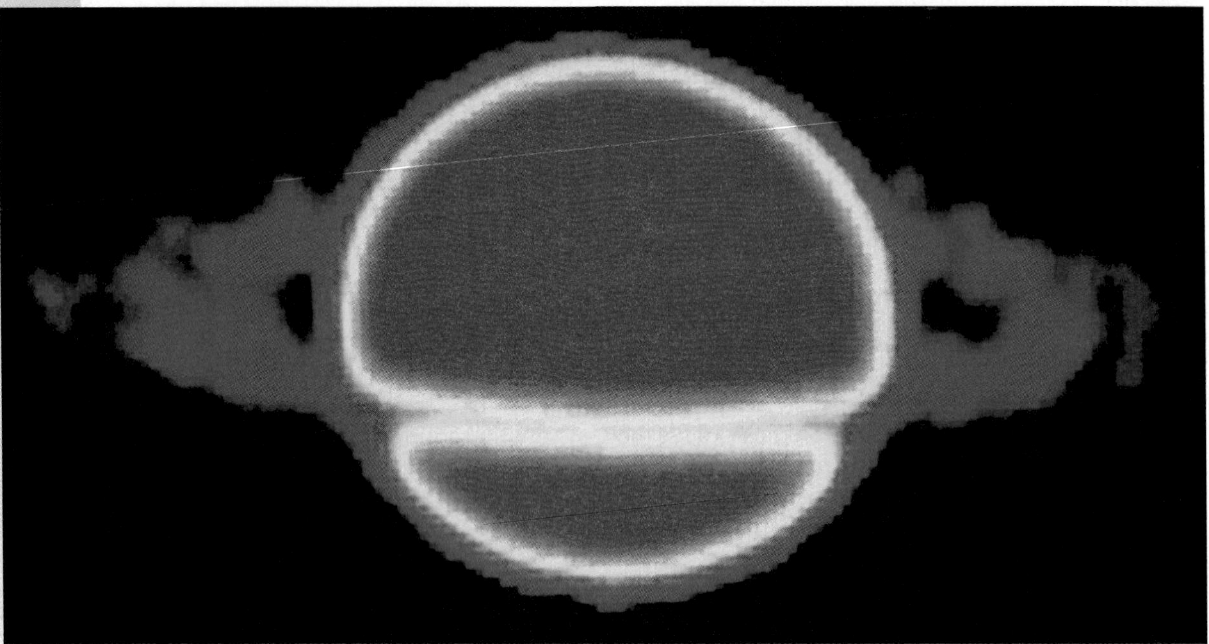

Radio Image of Saturn (Image Courtesy of NRAO/AUI)

R I V U X G

WHAT DO YOU THINK?

1 What is light?

2 Which type of electromagnetic radiation is most dangerous to life?

3 What is the main purpose of a telescope?

4 Why do stars twinkle?

5 What type(s) of electromagnetic radiation can telescopes currently detect?

Answers to these questions appear in the text beside the corresponding question numbers in the margins and at the end of the chapter.

We can see visible light from objects even trillions of kilometers (or miles) away with our eyes alone. The discovery that telescopes could make objects in space appear brighter, more detailed, and bigger, coupled with curiosity about the nature of bodies in the sky, led to the realization that we see only the "tip of the cosmic iceberg" with our naked eyes. Telescopes have revolutionized human understanding of the universe. The inventory of the cosmos is far more diverse and interesting than early humans ever imagined, and even today telescopes enable us to discover new things out there. We can think of the energy emitted by bodies in space as the medium of natural cosmic communication, and telescopes as the means by which we gather and read those cosmic messages.

Today, we observe the sky not just from the ground but through telescopes both orbiting the Earth and traveling to other objects in our solar system. Furthermore, we have discovered that visible light is only a tiny fraction of the energy emitted by objects in space.

In this chapter you will discover

- the connection between visible light, radio waves, and other types of electromagnetic radiation
- the debate over what light is and how Einstein resolved it
- how telescopes collect and focus light
- what telescopes can and cannot do
- why different types of telescopes are used for different types of research
- what the new generations of land-based and space-based high-technology telescopes being developed can do
- how astronomers use the entire spectrum of electromagnetic radiation to observe the stars and other astronomical objects and events

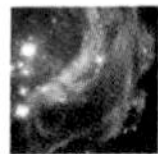

THE NATURE OF LIGHT

So far in this text we have used the word "light" in its everyday sense—the stuff to which our eyes are sensitive. This is more properly called "visible light," and it is a form of **electromagnetic radiation,** energy that has properties of both particles, called *photons*, and waves. Detecting electromagnetic radiation with telescopes is the essence of observational astronomy. Although human perception of objects here on Earth and in space comes primarily from the visible light that our eyes detect, this light is only a tiny fraction of all the electromagnetic radiation emitted by objects in the universe. The rest of the electromagnetic radiation is invisible to our eyes, so it has to be observed through specially designed telescopes. We examine telescopes for all types of electromagnetic radiation in this chapter. To understand how telescopes work, we begin studying them by exploring the properties of the electromagnetic radiation they collect.

3-1 Newton discovered that white light is not a fundamental color and debated whether light is composed of particles or waves

From the time of Aristotle until the late seventeenth century, most people believed that white is the fundamental color of light. The colors of the rainbow (or, equivalently, the colors created by light passing through a prism) were believed to be added or created somehow as white light went from one medium through another. Isaac Newton performed experiments during the late 1600s that disproved these beliefs. He started by passing a beam of sunlight through a glass prism, which spread the light out into the colors of the rainbow, as shown in Figure 3-1a. This change in direction as light travels from one medium into another is called **refraction;** and the resulting spread of colors (complete or with colors missing) is called a **spectrum** (plural, **spectra**).

When Newton selected a single color and sent it through a second prism (Figure 3-1b), the light emerging from the first prism was refracted by the second prism, but *it remained the same color.* The fact that colors of refracted light were unchanged by the second prism led Newton to conclude that the colors of a full spectrum (shades of red, orange, yellow, green, blue, and violet) were in fact properties of the light itself, and that white light is a mixture of colors. To prove this last point, he recombined all the spectrum colors, thereby recreating white light. Thus, different colors of the spectrum are different entities. But, Newton wondered, what was the nature of light that it could break apart into distinct colors and then be completely reconstituted?

Back in the mid-1600s, the Dutch scientist Christiaan Huygens proposed that light travels in the form of waves. Newton, on the other hand, performed many experiments in optics that convinced him that light is composed of tiny particles of energy. It turns out that both ideas were right.

In 1801 the English physicist Thomas Young demonstrated that light is indeed composed of waves. Young sent

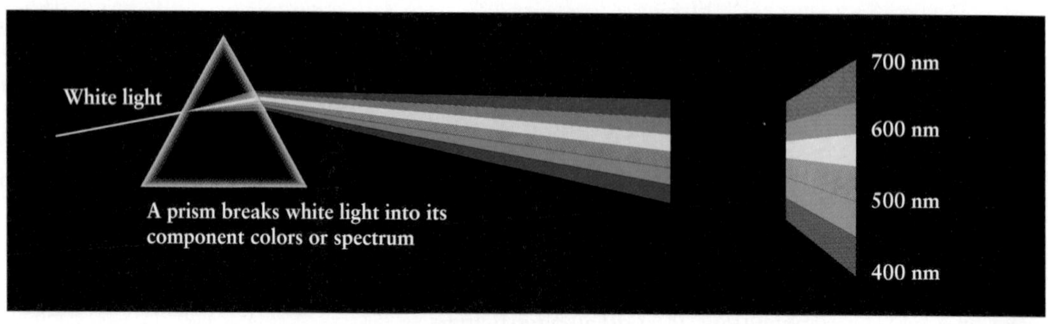

a

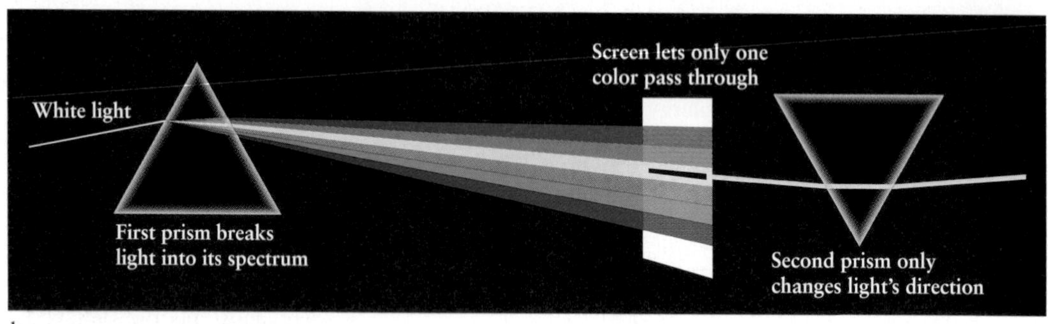

b

FIGURE 3-1 **Prisms and a Spectrum** **(a)** When a beam of white light passes through a glass prism, the light is separated or refracted into a rainbow-colored band called a spectrum. The numbers on the right side of the spectrum indicate wavelengths in nanometers (10^{-9} m). **(b)** Drawing of Newton's experiment showing that glass does not add to the color of light, only changes its direction. Since color is not added, this experiment shows that color is an intrinsic property of light.

light of a single color through two parallel slits (Figure 3-2). He reasoned that if the light were waves, then these waves would behave the way waves on the surface of water behave, flowing through similar gaps. In particular, the light waves from each slit would interact with light waves from the other. For example, when two light waves meet with one wave going up and the other going down, they would interfere with each other and partially or totally cancel each other out, leaving dark regions on the screen. When two waves meet that are both moving up or both moving down, they would reinforce each other and create bright regions. As you can see in Figure 3-2, this is precisely what happened. (If light were just random particles going through the slits, they would not interfere with each other in this way, and the pattern on the screen would just yield two bright regions, one behind each slit.) The analogy with water waves ends here, as light waves were discovered not to travel through a medium, like water waves do on water.

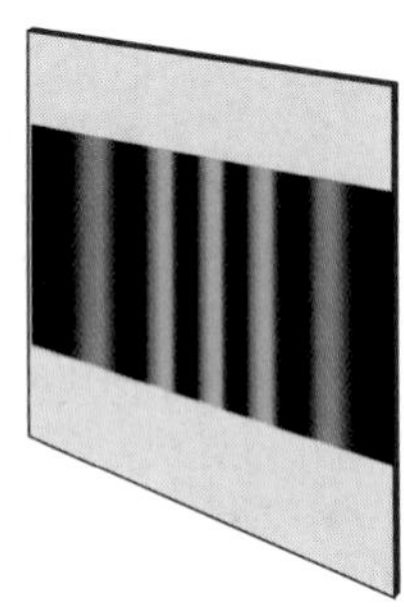

FIGURE 3-2 **Electromagnetic Radiation Travels as Waves** Thomas Young's interference experiment shows that light of a single color passing through a barrier with two slits behaves as waves that create alternating light and dark patterns on a screen. (Univerisity of Colorado, Center for Intergrated Plasma Studies, Boulder, CO)

Further insight into the wave character of light came from calculations by the Scottish physicist James Clerk Maxwell in the 1860s. Maxwell unified the basic properties of electricity and magnetism, which includes light, into four equations. By combining these equations, Maxwell demonstrated that electrical and magnetic effects should travel through space together in the form of waves. Maxwell's suggestion that some of these waves are observed as light was soon confirmed by a variety of experiments. Because of its electric and magnetic properties, visible light is the form of electromagnetic radiation to which our eyes are sensitive. Despite the name "electromagnetic radiation," visible light is not electrically charged.

Newton showed that sunlight is composed of all the colors of the rainbow. Young and others showed that light travels as waves. What makes the colors of the rainbow distinct from each other? The answer is surprisingly simple: Different colors are waves with different wavelengths, the distances between two successive wave crests (Figure 3-3). **Wavelength** is usually designated by λ, the lowercase Greek letter lambda.

The wavelengths of all colors are extremely small, less than a thousandth of a millimeter. To express these tiny dis-

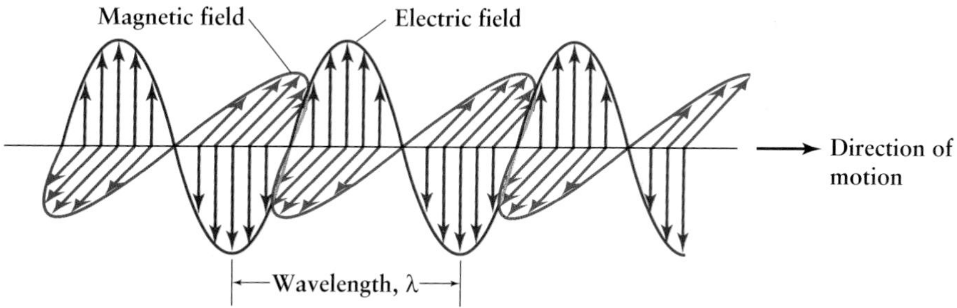

FIGURE 3-3 **Electromagnetic Radiation** All forms of electromagnetic radiation (radio waves, infrared radiation, visible light, ultraviolet radiation, X rays, and gamma rays) consist of oscillating electric and magnetic fields perpendicular to each other that move through empty space at a speed of 3×10^5 km/s. These fields are the mathematical description of the electric and magnetic effects. The distance between two successive crests, denoted λ, is called the wavelength of the light.

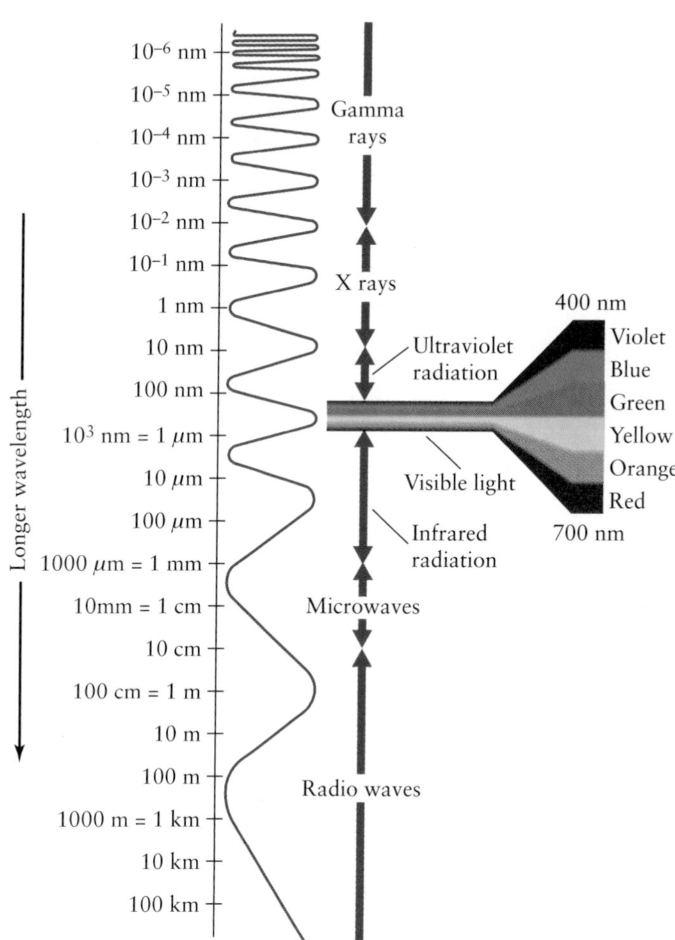

FIGURE 3-4 **The Electromagnetic Spectrum** The full array of all types of electromagnetic radiation is called the electromagnetic spectrum. It extends from the shortest-wavelength gamma rays to the longest-wavelength radio waves. Visible light forms only a tiny portion of the full electromagnetic spectrum. Note that 1 μm (micrometer) is 10^{-6} meters and 1 nm (nanometer) is 10^{-9} meters.

tances conveniently, scientists use a unit of length called the *nanometer* (nm), where 1 nm = 10^{-9} m. Another unit you might encounter when talking to astronomers is the Angstrom, Å, where 1 Å = .1 nm = 10^{-10} m. Experiments demonstrated that visible light has wavelengths covering the range from about 400 nm for the shortest wavelength of violet light to about 700 nm for the longest wavelength of red light. Intermediate colors of the rainbow fall between these wavelengths (Figure 3-4). The complete spectrum of colors from the longest wavelength to the shortest is red, orange, yellow, green, blue, and violet. Referring back to Figure 3-1, you can see that the amount of refraction different colors undergo depends on their wavelengths: The shorter the wavelength, the more the light is refracted.

3-2 Light travels at a finite, but incredibly fast, speed

The fact that we see lightning before we hear the accompanying thunderclap tells us that light travels faster than sound. But does that mean that light travels instantaneously from one place to another, or does it move with a measurable speed?

The first evidence for the finite speed of light came in 1675, when Ole Rømer, a Danish astronomer, carefully timed eclipses of Jupiter's moons (Figure 3-5). Rømer discovered that the moment at which a moon enters Jupiter's shadow depends on the distance between the Earth and Jupiter. When Jupiter is in opposition, that is, when Jupiter and the Earth are on the same side of the Sun, the Earth-Jupiter distance is relatively short compared to when Jupiter is near conjunction. At opposition, Rømer found that eclipses occur slightly earlier than predicted, while they occur slightly later than predicted by Kepler's laws when Jupiter is near conjunction.

Rømer correctly concluded that light travels at a finite speed and so it takes light more time to travel longer distances across space. The greater the distance to Jupiter, the longer the image of an eclipse takes to reach our eyes. From

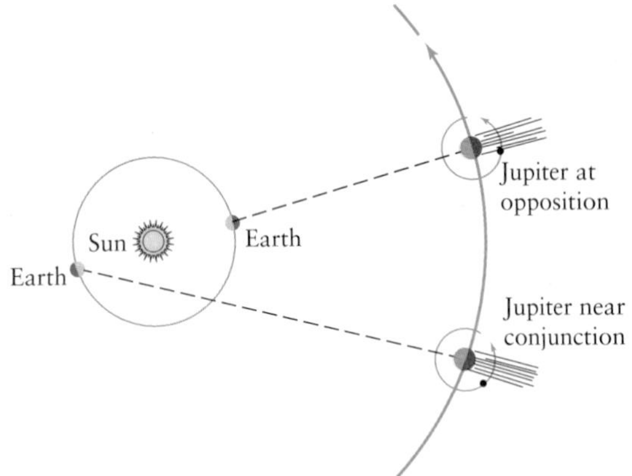

FIGURE 3-5 **Measuring the Speed of Light** An eclipse of one of Jupiter's moons seen at opposition (Earth and Jupiter on the same side of the Sun) appears to occur earlier than an eclipse seen near conjunction (Earth and Jupiter on opposite sides of the Sun). The differences in apparent times of the eclipses are due to the extra time it takes light to travel the additional distance when the planets are near conjunction. The actual time of the eclipse is determined using Kepler's laws.

his timing measurements, Rømer concluded that it takes 16½ minutes for visible light to traverse the diameter of the Earth's orbit (2 AU). Incidentally, Rømer's interpretation of the data requires a heliocentric cosmology—that both the Earth and Jupiter orbit the Sun.

Rømer's subsequent calculation of the speed of light was off by 25% because the value for the astronomical unit (the average distance from the Earth to the Sun) existing at that time was highly inaccurate. Nevertheless, he proved his main point—light travels at a finite speed. The first accurate laboratory measurements of the speed of visible light were performed in the mid-1800s.

Maxwell's equations also reveal that light waves of all wavelengths travel at the same speed in a vacuum (a region that contains no matter), and despite a few atoms per cubic meter, the space between planets and stars is a very good vacuum. The constant speed of light in a vacuum, usually designated by the letter *c*, has been measured to be about 299,792.458 km/s, which we generally round to

$$c = 3.0 \times 10^5 \text{ km/s} = 1.86 \times 10^5 \text{ mi/s}$$

(Standard abbreviations for units of speed, such as km/s for kilometers per second and mi/s for miles per second, will be used throughout the rest of this book.) Light traveling through air, water, glass, or any other substance always moves more slowly than it does in a vacuum. Indeed, in 1999, scientists were able to get light slowed to 61 km/hr (38 mph), and in 2001 they were able to momentarily stop it altogether.

The value *c* is a fundamental property of the universe. The speed of light appears in equations that describe, among other things, atoms, gravity, electricity, magnetism, distance, and time. It has extraordinary universal properties. For example, if you were traveling in space at 99% the speed of light, .99*c*, you would still measure the speed of any light beam moving toward you as *c*, which is also the speed you would measure of any light beam moving away from you!

As objects move toward or away from you, they change color in analogy to how the pitch of sirens change as they pass you. In 1842, Christian Doppler, a professor of mathematics in Prague, pointed out that wavelength, and hence color, is affected by motion. As shown for the observer on the left in Figure 3-6, the wavelengths of electromagnetic radiation from an approaching source are compressed. The circles represent waves emitted from consecutive wave peaks (S1 through S4) as the source moves along. Because each successive wave is emitted from a position slightly closer to her, she sees a shorter wavelength than she would if the source were stationary. All the colors in the spectrum of an approaching source, such as a star, are therefore shifted toward the short-wavelength (blue) end of the spectrum, regardless of its distance. This phenomenon is called a **blueshift.**

Conversely, electromagnetic waves from a receding source are stretched out. The observer on the right in Figure 3-6 sees a longer wavelength than he would if the source were stationary. All the colors in the spectrum of a receding

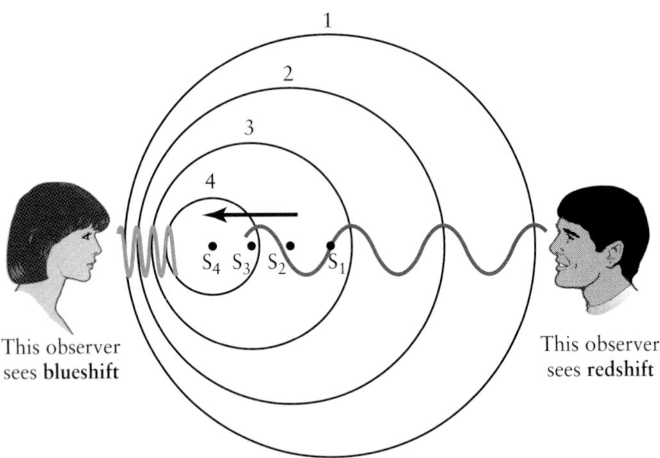

FIGURE 3-6 **The Doppler Effect** Wavelength is affected by motion between the light source and an observer. A source of light is moving toward the left. The four circles (numbered 1 through 4) indicate the location of light waves that were emitted by the moving source when it was at points S_1 through S_4, respectively. Note that the waves are compressed in front of the source but stretched out behind it. Consequently, wavelengths appear shortened (blueshifted) if the source is moving toward the observer and lengthened (redshifted) if the source is moving away from the observer. Motion perpendicular to an observer's line of sight does not affect wavelength.

source, regardless of its distance, are shifted toward the longer-wavelength (red) end of the spectrum, producing a **redshift.** A blueshift or a redshift is also called a **Doppler shift.** The amount of Doppler shift varies directly with approaching or receding speed: When the speeds are small compared to the speed of light, an object approaching twice as fast as another has all its colors (or wavelengths) blueshifted twice as much as does the slower-moving object. Conversely, an object moving away twice as fast as another object has its colors redshifted twice as much as the slower moving object.

3-3 Einstein showed that light sometimes behaves as particles that carry energy

1 WEB LINK 3.2 In 1905, just as scientists were becoming comfortable with the wave nature of light, Albert Einstein proposed that light sometimes acts as particles. Einstein used this idea to explain the photoelectric effect: Shorter wavelengths of light can knock some electrons off the surfaces of metals while longer wavelengths of light cannot, no matter how intense the beam of long-wavelength light. Einstein knew that electrons are bound onto a metal's surface by electric forces and that it takes energy to overcome those forces. Because some colors (or, equivalently, wavelengths) can remove the electrons and others cannot, the electrons must receive different amounts of energy from different colors of light. But how? Einstein proposed that light travels as waves enclosed in discrete packets called **photons,** and that photons with different wavelengths have different amounts of energy. *Specifically, the shorter the wavelength, the higher a photon's energy.*

$$\text{Photon energy} = \frac{\text{Planck's constant} \times \text{speed of light}}{\text{wavelength}}$$

where Planck's constant (named for the German physicist Max Planck) has the value 6.67×10^{-34} J s, where J is the unit of energy called a joule. Einstein's concept of light means that it can act both as waves (as when going through slits) and as particles (as when striking matter).

All photons with the same wavelength are identical to each other, and, therefore, every photon of a given wavelength carries the same amount of energy as every other photon of that wavelength. The energy delivered by a photon is either enough to rip an electron off the surface of the metal or it is not. There is no middle ground. Extensive testing in the twentieth century confirmed both the wave and particle properties of light.

3-4 Light is only one type of electromagnetic radiation

Around 1800, the British astronomer William Herschel discovered **infrared radiation** in an experiment with a prism. When he held a thermometer just beyond the red end of the visible spectrum, the thermometer registered a temperature increase, indicating that it was being heated by an invisible form of energy. We have said that visible light has a narrow range of wavelengths, from about 400 to 700 nm, but Maxwell's equations place no length restrictions on the wavelengths of electromagnetic radiation. Infrared radiation, discovered before Maxwell's equations were formulated, was later identified as electromagnetic radiation with wavelengths slightly longer than red light.

In experiments with electric sparks in 1888, the German physicist Heinrich Hertz succeeded in producing electromagnetic radiation a few centimeters in wavelength, now known as **radio waves.** In 1895, Wilhelm Roentgen invented a machine that produces electromagnetic radiation with wavelengths shorter than 10 nm, now called **X rays.** Modern versions of Roentgen's machine are found in medical and dental offices and airport security checkpoints.

We now know that visible light occupies only a tiny fraction of the full range of possible wavelengths, collectively called the **electromagnetic spectrum.** As shown in Figure 3-4, the electromagnetic spectrum stretches from the longest-wavelength radio waves, through infrared radiation, visible light, ultraviolet radiation, and X rays, to the shortest-wavelength photons, gamma rays. On the long-wavelength side of the visible spectrum, infrared radiation covers the range from about 700 nm to 1 mm. Astronomers interested in infrared radiation often express wavelength in *micrometers* or *microns* (abbreviated μm), where 1 μm = 1000 nm = 10^{-6} m. From roughly 1 mm to 10 cm is the range of microwaves, which are sometimes considered to be infrared radiation and sometimes radio waves. Radio waves are all electromagnetic waves longer than 10 cm.

At wavelengths shorter than those of visible light, **ultraviolet (UV) radiation** extends from about 400 nm down to 10 nm. Next are **X rays,** with wavelengths between about 10 and 0.01 nm, and beyond them are **gamma rays.** These boundaries are all arbitrary and are primarily used as convenient divisions in the electromagnetic spectrum, which is actually continuous.

These various types of electromagnetic radiation share many basic properties. For example, they are all photons, they all travel at the same speed, and they all sometimes behave as particles and sometimes as waves. But because of their different wavelengths (and therefore different energies), they interact very differently with matter. X rays penetrate deep into your body tissues, while visible light is mostly stopped and scattered by the surface layer of skin; your eyes respond to visible light but not to infrared radiation; your radio detects radio waves but not ultraviolet radiation.

The Earth's atmosphere is relatively transparent to visible light, radio waves, short-wavelength infrared, and long-wavelength ultraviolet, meaning that these radiations pass through it to the surface and, therefore, to ground-based

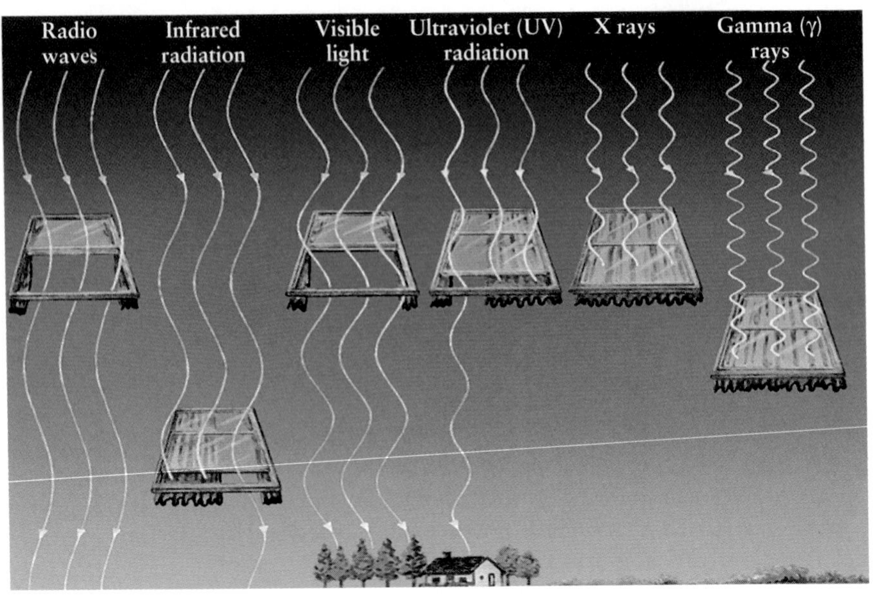

FIGURE 3-7 Windows through the Atmosphere The Earth's atmosphere allows different types of electromagnetic radiation to penetrate into it in varying amounts. Visible light, radio waves, short wavelength infrared, and long wavelength ultraviolet reach all the way to the Earth's surface. The other types of radiation are absorbed or reflected by the gases in the air at different characteristic altitudes (indicated by heights of windows). While the atmosphere does not have actual "windows," astronomers use the term to characterize the passage of radiation through it.

telescopes sensitive to them. Astronomers say that the atmosphere has *windows* for these parts of the electromagnetic spectrum (Figure 3-7). Our bodies detect infrared radiation as heat.

The longest-wavelength ultraviolet radiation, called UV-A, causes tanning and sunburns. Ozone (O_3) in the Earth's atmosphere normally screens out intermediate-wavelength ultraviolet radiation, or UV-B. The ozone in the *ozone layer* high in the atmosphere is being depleted by human-made chemicals, such as chlorofluorocarbons (CFCs) and bromine-rich gases. As a result, more intermediate-UV-B is reaching the Earth's surface, and these highly energetic photons damage living tissue, causing skin cancer and glaucoma, among other diseases.

2 The Earth's atmosphere is completely opaque to the other types of electromagnetic radiation, meaning that they do not reach the Earth's surface. (This is a good thing, because short-wavelength ultraviolet radiation, called UV-C, and X rays and gamma rays are devastating to living tissue. Gamma rays, packing the highest energies, are the deadliest.) Direct observations of these wavelengths must be performed high in the atmosphere or, ideally, from space.

As noted earlier, photons with different energies interact with matter in different ways. Higher energy photons will pass through or rip apart material that lower energy photons will bounce off. Telescope designs for collecting and focusing radiation differ, depending on the energy or wavelength of interest. Knowing the energies of photons and their effects on matter enables astronomers to design both the telescopes to collect them and the devices used to record their presence. In the next section we will consider the lengths to which astronomers have gone to capture visible and invisible (or nonoptical) electromagnetic radiation.

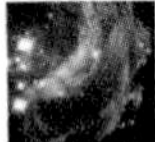

OPTICS AND TELESCOPES

Since the time of Galileo, astronomers have been designing instruments to collect more light than the human eye can collect on its own. As noted earlier, collecting more light enables us to see things more brightly, in more detail, and at a greater distance. There are two basic types of telescopes—those that collect light through lenses, or **refracting telescopes,** and those that collect it from mirrors, or **reflecting telescopes.** The earliest telescopes used lenses to collect light. Lenses have a variety of shortcomings as light-gathering devices and so today, all modern research telescopes use mirrors to collect the light. Lenses are still used in the eyepieces of telescopes to straighten the collected light so that our brains can accurately interpret what we see. Lenses are also used to collect light in many small telescopes you can buy for home use, as well as in binoculars, cameras, and eyeglasses. Reflecting and refracting telescopes have the
same main purpose—to collect as many photons as possible 3
in order to better observe the sky (see Section 3-6). We begin exploring telescopes by discussing how reflecting telescopes work. Then we consider how lenses collect light and finally how astronomers have developed telescopes to see nonvisible electromagnetic radiation.

3-5 Reflecting telescopes use mirrors to concentrate incoming starlight

The first reflecting telescope was built in the eighteenth century by Isaac Newton (Figure 3-8). To understand how these telescopes work, consider first a flow of photons, more commonly a called a *light ray,* moving toward a flat

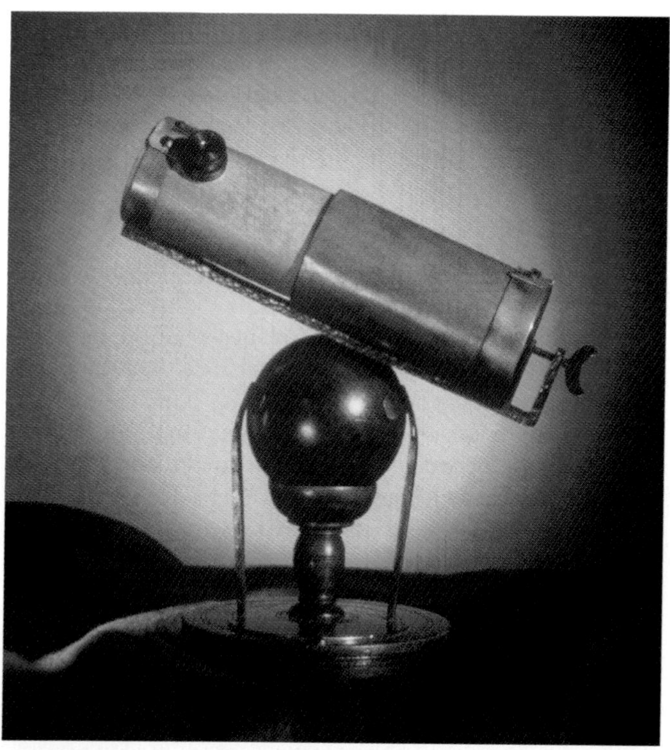

◀ FIGURE 3-8 **Newton's Original Reflecting Telescope** Built in 1672, this reflecting telescope has a spherical primary mirror 3 cm (1.3″) in diameter. Its magnification was 40x. (Royal Greenwich Observatory/Science Photo Library)

mirror. In Figure 3-9a, a light ray strikes the mirror, and we imagine a perpendicular line coming out of the mirror at that point. According to the principle of **reflection**, which follows from Maxwell's equations, the angle between the incoming light ray and the perpendicular (dashed line) is always equal to the angle between the outgoing, reflected light ray and the perpendicular. This rule also applies if the mirror is curved (Figure 3-9b).

Using this principle, Newton determined that a concave (hollowed-out) mirror ground in the shape of a parabola causes all parallel incoming light rays striking the mirror to converge to a **focal point,** as shown in Figure 3-9b. The distance between this **primary mirror** and the focal point, where the image of the distant object is formed, is the **focal length** of the mirror. Focal points exist for light from sources that

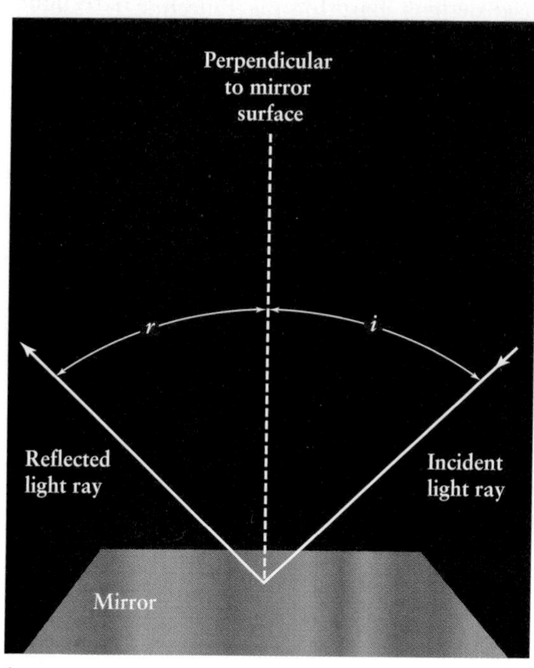

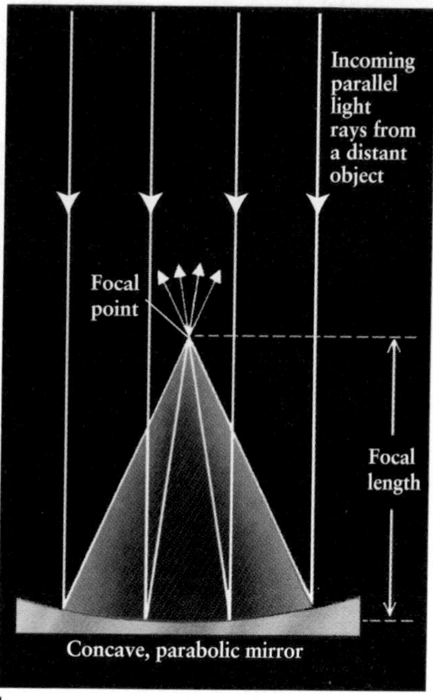

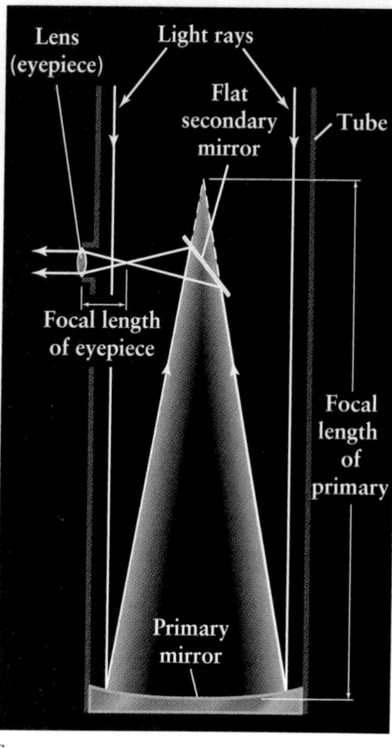

FIGURE 3-9 **Reflection** **(a)** The angle at which a beam of light strikes a mirror (the angle of incidence *i*) is always equal to the angle at which the beam is reflected from the mirror (the angle of reflection *r*). **(b)** A concave, parabolic mirror causes parallel light rays to converge and meet at the focal point. The distance between the mirror and focal point is the focal length. **(c)** A Newtonian telescope uses a flat mirror, called the secondary, to send light toward the side of the telescope. The light rays are made parallel again by passing through a lens, called the eyepiece. The dashed line shows where the focal point of this primary mirror would be if the secondary mirror were not in the way.

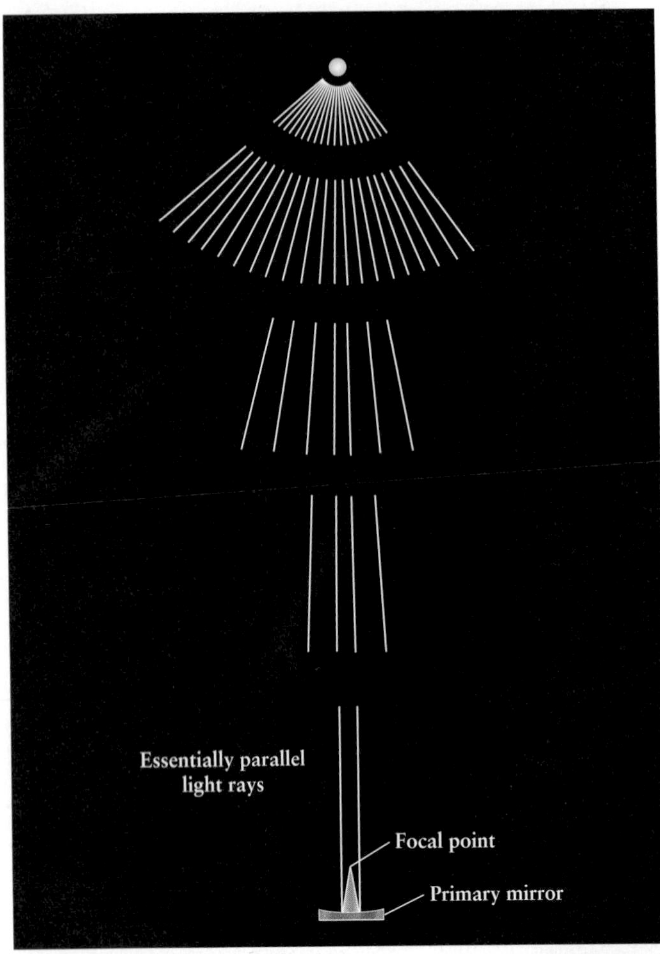

◀ **FIGURE 3-10 Parallel Light Rays from Distant Objects** As light travels away from any object, the light rays, all moving in straight lines, separate. By the time light has traveled trillions of kilometers or miles, only the light rays moving in parallel tracks are still near each other.

are extremely far away, like the stars (Figure 3-10 shows why stars can be considered "far away"). If the object is larger than a point, like the Moon or a planet, then the light will converge to a plane at the focal length called the **focal plane,** rather than to a focal point.

To view the image, Newton placed a small, flat mirror at a 45° angle between the primary mirror and the focal point, as sketched in Figure 3-9c and Figure 3-11a. This **secondary mirror** reflects the light rays to one side of the telescope, and the astronomer views the image through an eyepiece lens. We will discuss how this lens works in Section 3-8. A telescope with this optical design is still called a **Newtonian reflector.**

Newtonian telescopes are very popular with amateur astronomers because they are convenient to use while the observer is standing up. However, they are not used in research observatories because they are lopsided. If astronomers attach their often heavy and bulky research equipment onto its side, the telescope sags and distorts the image in unpredictable ways.

Three basic designs exist for the reflecting telescopes used in research. In the first, a hole is drilled directly through the center of the primary mirror. A convex (outwardly

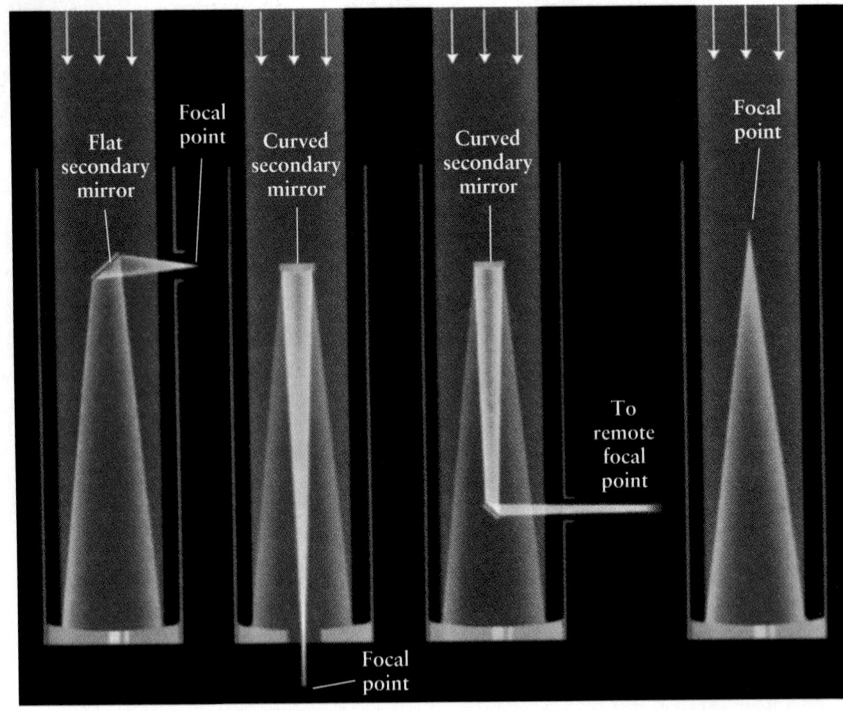

▶ **FIGURE 3-11 Reflecting Telescopes** Four of the most common optical designs for reflecting telescopes: **(a)** Newtonian focus (popular among amateur astronomers) and the three major designs used by researchers—**(b)** Cassegrain focus, **(c)** coudé focus, and **(d)** prime focus.

curved) secondary mirror placed between the primary mirror and its focal point reflects the light rays back through the hole (Figure 3-11b). This is called the **Cassegrain focus.** This secondary mirror extends the telescope's focal length. Light-gathering equipment is bolted to the bottom of the telescope, and the light is brought into focus in it. This design has an advantage over Newtonian telescopes, in that the attached equipment is balanced and does not distort the telescope frame and, hence, the image.

The second design is handy for long and bulky optical equipment that cannot be mounted directly on the telescope or for observations that benefit from extremely long focal lengths. In this design (Figure 3-11c), a series of curved mirrors channels the light rays away from the telescope to a remote focal point called the **coudé focus** (named after a French word meaning "bent like an elbow"). Examples of equipment that require the use of the coudé focus are a variety of *spectrographs,* instruments that separate light from objects into its individual colors in order to determine the objects' chemistry, surface temperature, and motion toward or away from us.

In the third design, an observing device is located at the undeflected focal point, directly in front of the primary mirror. This arrangement is called a **prime focus** (Figure 3-11d). It is used when the image requires the telescope to have the shortest focal length.

3-6 Telescopes brighten, resolve, and magnify

As mentioned earlier, a telescope's most important function is to provide the astronomer with as bright an image as possible—the brighter an object is, the more information about it the astronomer can extract.

The observed brightness of any object depends on the total number of photons collected from it, which in turn depends on the area of the telescope's primary mirror. A larger primary mirror intercepts and collects more light than does a smaller one. For exposures of equal times, a telescope with a large primary mirror produces brighter images and detects fainter objects than a telescope with a smaller primary mirror (Figure 3-12). No wonder telescopes are sometimes called "photon buckets."

R I V U X G

FIGURE 3-12 **Light-Gathering Power** Because a large mirror intercepts more starlight than a small mirror, a large mirror produces a brighter image. The same principle applies to telescopes that collect light using an objective lens rather than a primary mirror. The two photographs of the Andromeda Galaxy were taken through telescopes with different diameters and were exposed for equal lengths of time at equal magnification. (AURA)

INSIGHT INTO SCIENCE

Costs and Benefits Science now relies heavily on technology to conduct experiments or make observations. The cost of cutting-edge astronomical observations may run to hundreds of millions of dollars or more. The return on such investments is a better understanding of how the universe works, how we can harness its capabilities, and our place in it.

The **light-gathering power** of a telescope is directly related to the area of the telescope's primary mirror. Recall that the area and diameter of a circle are related by the formula

$$\text{Area} = \frac{\pi d^2}{4}$$

where d is the diameter of the mirror and π (pi) is about 3.14.

Consequently, *a mirror with twice the diameter of another mirror has 4 times the area of the smaller mirror and therefore collects 4 times as much light as the smaller one in the same amount of time.* For example, a 36-cm-diameter mirror has 4 times the area of an 18-cm-diameter mirror. Therefore, the 36-cm telescope has 4 times the light-gathering power of a telescope half its size.

The second most important function of any telescope is to reveal greater detail of objects that are more than just points of light, often called extended objects. A large telescope increases the sharpness of the image and the degree of detail that can be seen. **Angular resolution** (often called just "resolution") measures the clarity of images (Figure 3-13). The angular resolution of a telescope is measured as the arc angle between two adjacent stars whose images can just barely be distinguished by the telescope. The smaller this angle, the sharper the image. Large, modern telescopes, like the Keck telescopes in Hawaii, which we will discuss later in this chapter, have angular resolutions better than 0.1 arcsec. As a general rule, *a telescope with a primary mirror twice the diameter of another telescope's primary will be able to see twice as much detail as the smaller telescope.*

The final function of a telescope is to make objects appear larger. This property is called **magnification.** Magnification is associated with resolution, because the larger the image, the more detail of it you can potentially see. The magnification of a reflecting telescope is equal to the focal length of the primary mirror divided by the focal length of the eyepiece lens:

$$\text{Magnification} = \frac{\text{focal length of the primary}}{\text{focal length of the eyepiece}}$$

For example, if the primary mirror of a telescope has a focal length of 100 cm and the eyepiece has a focal length of 0.5 cm, then the magnifying power of the telescope is

$$\text{Magnification} = \frac{100 \text{ cm}}{0.5 \text{ cm}} = 200$$

This property is usually expressed as 200×.

There is a limit to the magnification of any telescope. Try to magnify beyond that limit and the image becomes distorted. As a rule, *a telescope with a primary mirror twice the diameter of another telescope's primary will have twice the maximum magnification of the smaller telescope.*

a

b

R I V U X G

FIGURE 3-13 **Resolution** The larger the diameter of a telescope's primary mirror, the greater the detail the telescope can resolve. These two images of the Andromeda Galaxy, taken through telescopes with different diameters, show this difference. Increasing the exposure time of the smaller diameter telescope **(a)**, will only brighten the image, not improve the resolution. (AURA)

3-7 Storing and analyzing light from space is key to understanding the cosmos

The invention of photography during the nineteenth century was a boon for astronomy. By taking a long exposure with a camera mounted at the focal plane of a telescope, an astronomer could record extremely faint features that could not be seen just by looking through the telescope. The reason is that our eyes clear the images in them several times a second, whereas film adds up the intensity of all the photons affecting its emulsion. By keeping telescopes aimed precisely, so that images don't blur, exposures of an hour or more are now quite routine. Astrophotographs make all objects appear brighter and reveal greater detail in extended objects, such as galaxies, star clusters, and planets, than we could otherwise see.

Astronomers have long known, however, that a photographic plate is an inefficient detector of light because it depends on a chemical reaction to produce an image. Typically, only 2% of the light striking film triggers a reaction in the photosensitive material. Thus, roughly 98% of the light falling onto a photographic plate is wasted.

Technology has changed all that. We have now replaced photographic film with highly efficient electronic light detectors called **charge-coupled devices (CCDs)**. CCDs commonly respond to 70% of the light falling on them; their resolution is better than that of film, and they respond more uniformly to light of different colors. Each CCD is a square or rectangle a few centimeters on a side. Several of them are often used together to create a larger image (Figure 3-14). Each CCD is divided into an array of small, light-sensitive squares called picture elements or, more commonly, **pixels.** The largest grouping of CCDs used on a telescope has 378 megapixels (million pixels), compared to most digital cameras, which typically have between 2 and 8 megapixels.

When an image from a telescope is focused on the CCD, an electric charge builds up in each pixel in direct proportion to the intensity of the light falling on that pixel. When the exposure is finished, the charge on each pixel is read into a computer. Figure 3-15 shows one photograph and two CCD images of the same region of the sky, all taken with the same telescope. Notice that many details visible in the CCD images are absent in the ordinary photograph.

3-8 Eyepieces, refracting telescopes, binoculars, and eyeglasses use lenses to change the direction of incoming light

Although light travels at about 300,000 km/s in a vacuum, it moves more slowly through a dense substance such as glass. As light enters the glass, it slows abruptly, much like a person walking from a boardwalk onto a sandy beach: Her pace suddenly slows as she steps from the smooth, hard pavement onto the sand. And just as the person stepping back onto the boardwalk easily resumes her original pace, light exiting a piece of glass resumes its original speed.

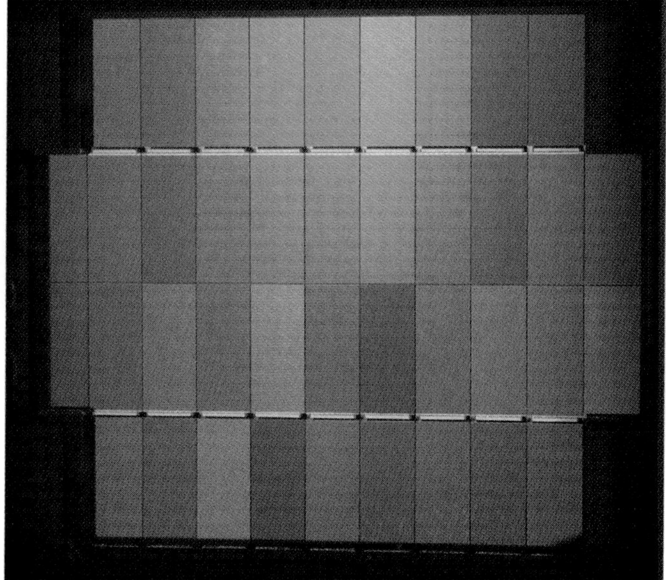

a R I V U X G

b R I V U X G

FIGURE 3-14 **Mosaic of Charge-Coupled Devices (CCDs)** (a) These 40 CCDs combine to provide up to 378 million light-sensitive pixels that store images collected by the *CFHT* telescope on the dormant volcano Mauna Kea, Hawaii. Electronic circuits transfer the data to a waiting computer. (b) This image of the Rosette Nebula, a region of star formation 5000 ly away in the constellation Monoceros (the Unicorn), was taken with this CCD. It shows the incredible detail that can be recorded by large telescopes and high resolution CCDs. (a & b: J. C. Cuillandre/ Canada-France-Hawaii Telescope)

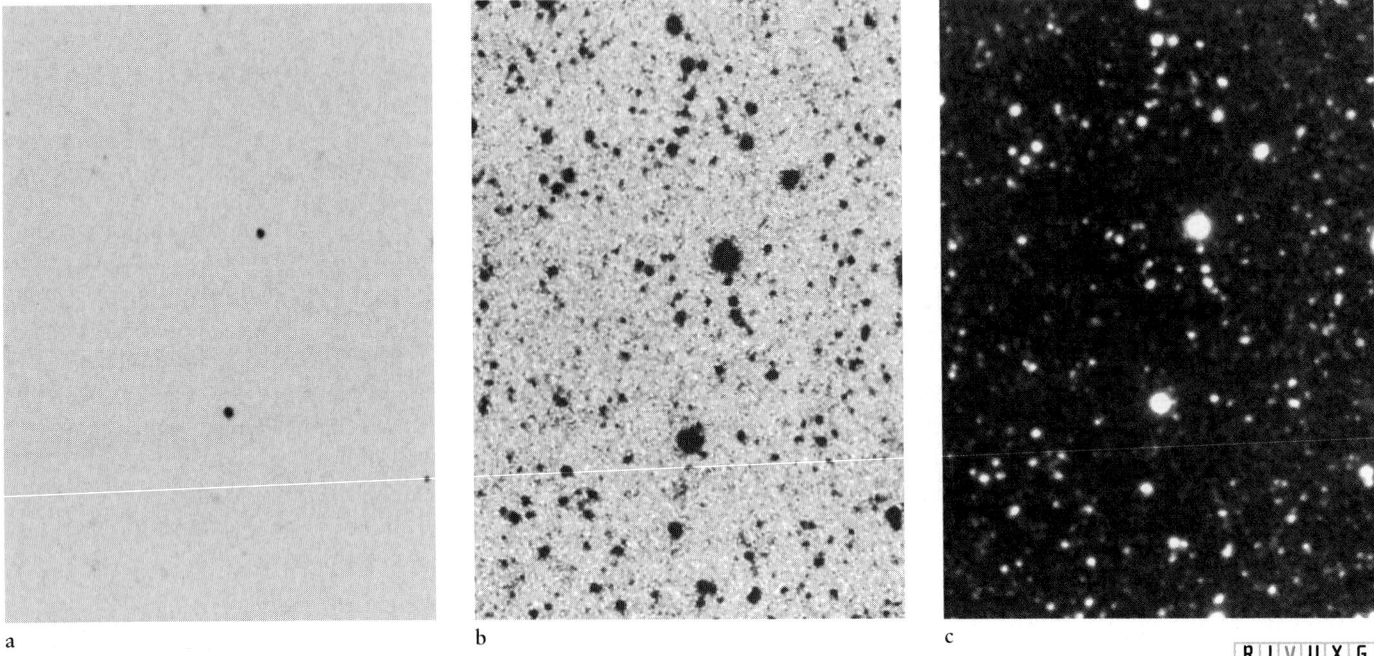

FIGURE 3-15 **Ordinary Photography Versus CCD Images** These three views of the same part of the sky, each taken with the same 4-m telescope, compare CCDs to photographic plates. **(a)** A negative print (black stars and white sky) of a photographic image. **(b)** A negative CCD image. Notice that many faint stars and galaxies virtually invisible in the ordinary photograph can be seen clearly in this CCD image. **(c)** This (positive) color view was produced by combining a series of CCD images taken through colored filters. (Patrick Seitzer, NOAO)

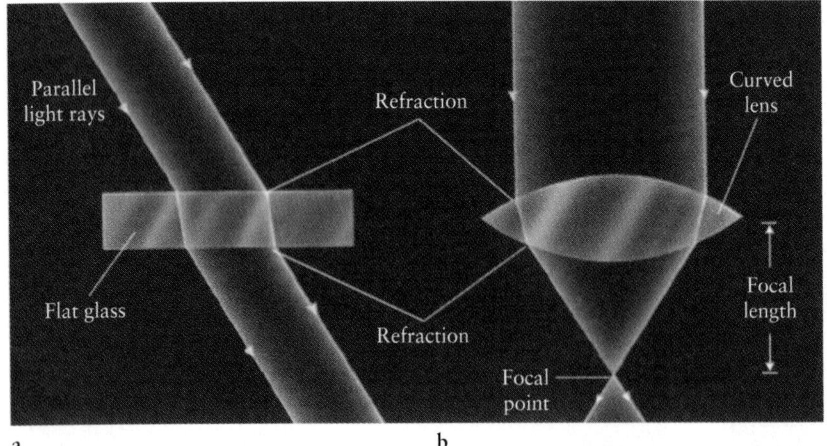

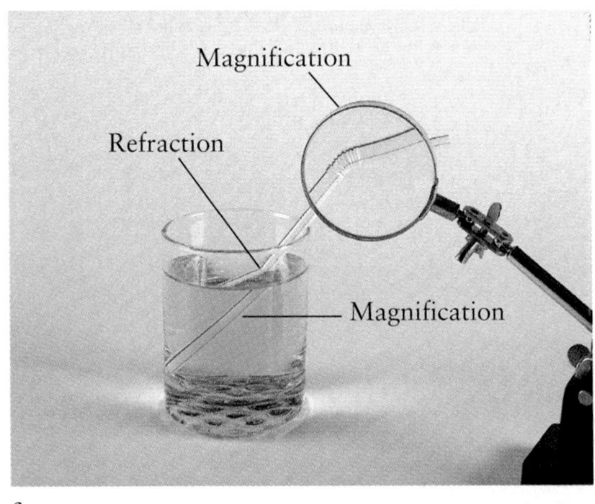

FIGURE 3-16 **Refraction through Uniform and Variable Thickness Glasses** **(a)** Refraction is the change in direction of a light ray when it passes into or out of a transparent medium such as glass. A light ray entering a denser medium, like going from air into water or glass, is bent or refracted to an angle more perpendicular to the surface than the angle at which it was originally traveling. If the glass is flat, then the light leaving the glass is refracted back to the direction it had before entering the glass. There is no overall change in the direction in which the light travels. **(b)** If the glass is in the shape of a suitable convex lens, parallel light rays converge to a focus at the focal point. As with parabolic mirrors, the distance from the lens to the focal point is called the focal length of the lens. **(c)** The straw as seen through the side of the liquid is magnified and offset from the straw above the liquid because the liquid is given a curved shape by the side of the glass. The straw as seen through the top of the liquid is refracted, but does not appear magnified because the top is flat. (c: Ray Moller/Dorling Kindersley)

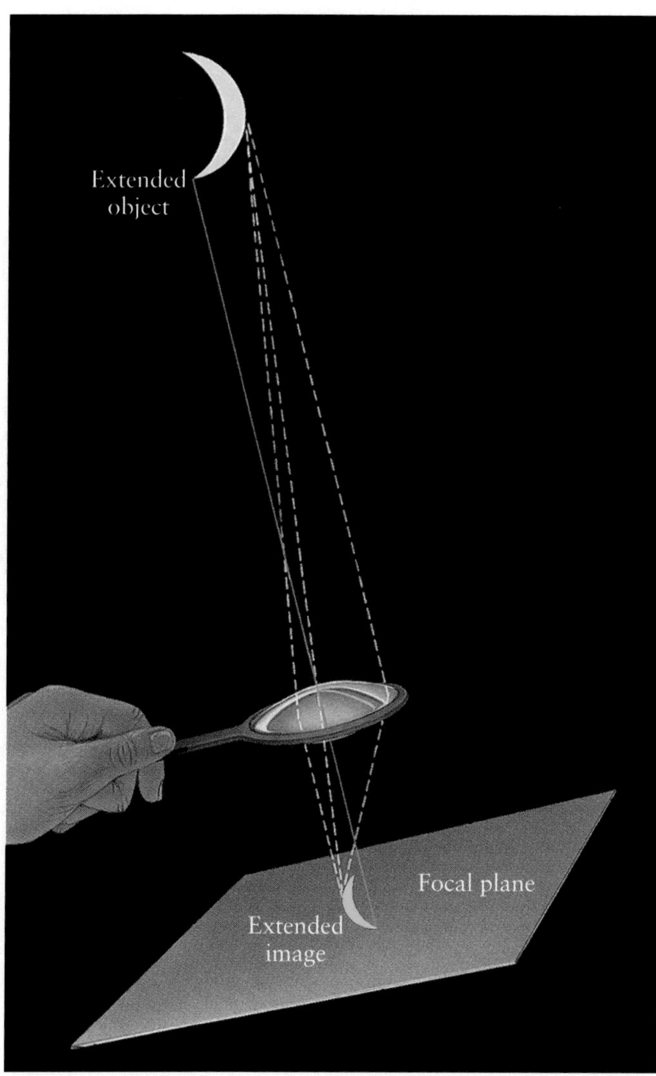

FIGURE 3-17 **Extended Objects Create a Focal Plane** Light from objects larger than points in the sky does not all converge to the focal point of a lens. Rather, the object creates an image at the focal length in what is called the focal plane.

As a result of changing speed, light can also change direction as it passes from one transparent medium into another. As noted earlier, this is called refraction. You see refraction every day when looking through windows. Imagine a stream of photons from a star entering a window, as shown in Figure 3-16a. As a light ray goes from the air into the glass, the light ray's direction changes so that it is more perpendicular to the surface of the glass than it was before entering. Once inside the glass, the light ray travels in a straight line. Upon emerging from the other side, the light ray bends once again, resuming its original direction and speed. The net effect is only a slight, uniform displacement of the objects beyond the glass.

Unlike a window, lenses have surfaces of varying thickness (Figure 3-16b). These curved surfaces force the light rays to emerge from the lens in different directions than they had before entering the lens. Different rays striking the first surface of the lens at different places are refracted by different amounts: For lenses used on telescopes, binoculars, and many eyeglasses, light rays start converging once they enter the glass. As the light emerges from the glass, it converges further. In astronomy, large lenses called **objective lenses** were used instead of primary mirrors (and still are for many telescopes built for home use) and small lenses are used as eyepieces to bring the light rays back to being parallel so that our eyes can make sense of what we see.

Parallel light rays entering an objective lens converge and meet at the focal point of the lens. Its distance from the lens is called the focal length (Figure 3-16b). If the object is close enough or large enough to be more than just a dot as seen through the telescope, all the light from it does not converge at the focal point but rather focuses along a surface, the focal plane (Figure 3-17). These concepts of focal length and focal plane for lenses are the same as those for reflecting telescopes, as discussed earlier.

A **refracting telescope,** or **refractor** (Figure 3-18), is an arrangement of two lenses used to gather light. The

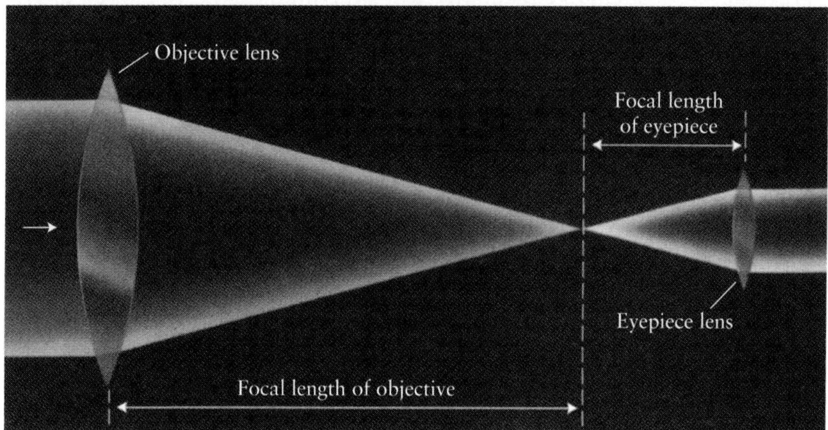

FIGURE 3-18 **Essentials of a Refracting Telescope** A refracting telescope consists of a large, long-focal-length objective lens that collects and focuses light rays and a small, short-focal-length eyepiece lens that restraightens the light rays. The lenses work together to brighten, resolve, and magnify the image formed at the focal plane of the objective lens.

FIGURE 3-19 **The Largest Refracting Telescope** This giant refracting telescope, built in the late 1800s, is housed at Yerkes Observatory near Chicago. The objective lens is 102 cm (40 in.) in diameter, and the telescope tube is 19⅓ m (63½ ft) long. Note the the floor moves down when the telescope is aimed higher. (Yerkes Observatory)

objective lens at the top of the telescope has a large diameter and long focal length. Like a primary mirror, its purpose is to collect as much light as possible. The **eyepiece lens,** at the bottom of the telescope, is smaller and has a short focal length. Because the light striking the focal plane or leaving the eyepiece is now more concentrated, objects are brighter as seen through a telescope. The mathematics of magnification for a refracting telescope is the same as for a reflecting telescope, with the focal length of the objective lens substituting for the focal length of the primary mirror. Likewise, all the rules for the limits on telescopes are the same for refractors as for reflectors.

The largest refracting telescope in the world, completed in 1897 and located at the Yerkes Observatory near Chicago (Figure 3-19), has an objective lens 102 cm (40 in.) in diameter with a focal length of 19⅓ m (63.5 ft). The second largest refracting telescope, located at Lick Observatory near San Jose, California, has an objective lens of 91 cm (36 in.) in diameter. No major refracting telescopes were constructed in the twentieth century or are planned for this century.

3-9 Refractors have more limitations than reflectors

Refracting telescopes suffer from a variety of problems that have limited their use as research instruments. These problems include:

- Different colors of light are refracted by different amounts, and so different colors have different focal lengths. As a result, objects look blurred. This is not a problem for reflectors because the reflecting surface is on the tops of the mirrors so that light never enters the glass.
- It is hard to grind a lens to the very complicated shape necessary to have parallel light rays passing through it all come to the same focal point. This is less of a problem with reflectors since the surface is a simple parabola.
- The weight of the lens can cause it to sag and thereby distort the image. This is not a problem with reflectors since the entire underside of the mirror can be supported, as necessary.
- Air bubbles in the glass cause unwanted refractions and hence distorted images.
- Glass is opaque to certain ranges of wavelength, meaning that they do not get through the glass. These last two points are not problems with reflectors because the light never enters the glass.

The combination of these problems makes images from refractors less accurate than those obtained from reflecting telescopes, which explains why modern research telescopes are all reflectors. The mirrored surfaces of research telescopes are polished so smoothly that the highest bumps are less than 1/500 the thickness of a human hair.

While they are better light collectors than refracting telescopes, reflecting telescopes are not perfect; there are several prices to pay for the advantages they offer over refractors. Two of the most important are *blocked light* and *spherical aberration.* Let's consider each problem briefly.

You have probably noticed that the secondary mirror of a reflector blocks some incoming light, one unavoidable price that astronomers must pay. Typically, a secondary mirror prevents about 10% of the incoming light from reaching the primary mirror. This problem is addressed by constructing primary mirrors with sufficiently large surface areas to compensate for the loss of light. You might also think that because light is missing from the center of the telescope due to blockage by the secondary mirror, a corresponding central "hole" appears in the images. However, this problem does not occur, because light from all parts of each object being observed enters all parts of the telescope (Figure 3-20). Indeed, covering part of the light-collecting mirror or lens darkens the image, but does not limit which parts of the object you can see through the telescope.

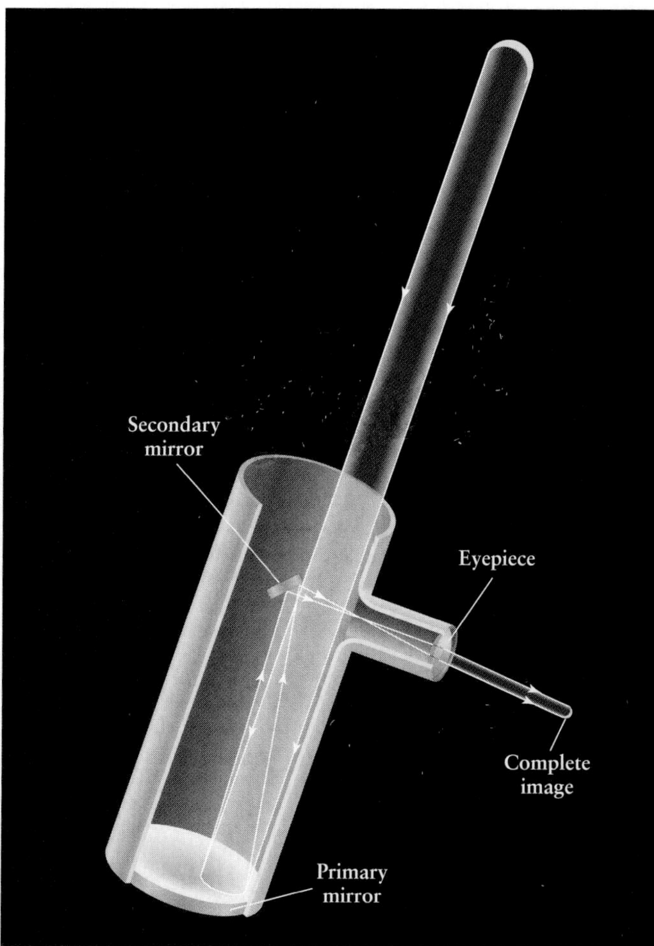

FIGURE 3-20 **The Secondary Mirror Does Not Create a Hole in the Image** Because the light rays from distant objects are parallel, light from the entire object reflects off all parts of the mirror. Therefore, every part of the object sends photons to the eyepiece. This figure shows the reconstruction of the entire Moon from light passing through just part of this telescope. The same drawing applies everywhere on the primary mirror that is not blocked by the secondary mirror.

To make a reflector, an optician traditionally grinds and polishes a large slab of glass into a concave spherical surface—before computer control, grinding a spherical surface was much easier than grinding the ideal parabolic surface. However, light entering a spherical telescope mirror at different distances from the mirror's center comes into focus at different focal lengths, and images taken with all such telescopes appear blurry (Figure 3-21a).

Spherical aberration in reflecting telescopes is avoided by making the mirror parabolic (Figure 3-21b) or by using a thin correcting lens, called a **Schmidt corrector plate,** with a spherical mirror. The corrector plate is located at the top of the telescope (Figure 3-21c). The light coming into the telescope is refracted by the plate just enough to compensate for spherical aberration and to bring all the light into focus at the same focal length. These correctors have the added benefit of focusing light from a larger angle in the sky than would be in focus without the plate. A Schmidt corrector plate enables astronomers to map large areas of the sky with relatively few photographs at moderately high magnification. In other words, the Schmidt corrector plate acts like a wide-angle lens on a camera. Schmidt corrector plates are often used in conjunction with Cassegrain telescopes to create *Schmidt-Cassegrain* telescopes, which are also popular for home use. However, the plate does not allow for as much magnification as a telescope with a parabolic mirror. (If you are interested in buying a telescope, you might find the Guided Discovery: Buying a Telescope helpful.)

While some parabolic primary mirrors have been meticulously ground over the past century, the advent

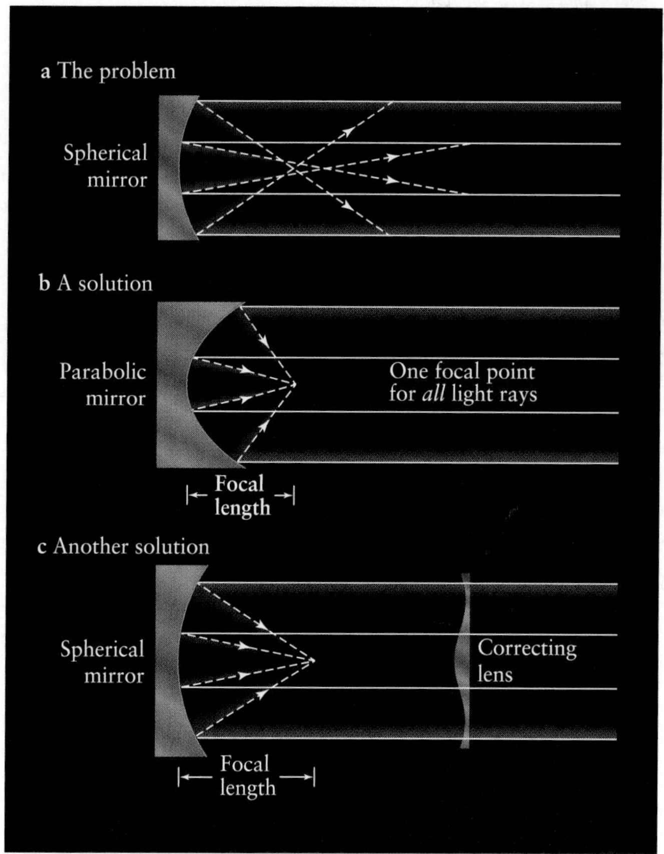

FIGURE 3-21 **Spherical Aberration** (a) Different parts of a spherically concave mirror reflect light to slightly different focal points. This effect, spherical aberration, causes image blurring. This problem can be overcome by (b) using a parabolic mirror or (c) using a Schmidt corrector plate (lens) in front of the telescope.

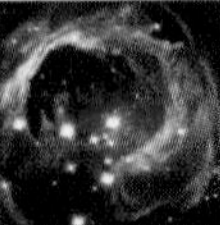

GUIDED DISCOVERY
Buying a Telescope

If you feel elated discovering treasures, you may want to consider viewing the night sky through powerful binoculars or a telescope. Both types of instruments enable you to find many beautiful, exciting objects in space. Binoculars are nice because you can quickly scan the heavens to see craters on the Moon, star clusters, the Great Nebula of Orion, and even the Andromeda Galaxy. Telescopes open new vistas and even enable amateur astronomers to discover new comets, among other things, adding to our scientific database and immortalizing the discoverers' names.

There are four basic types of telescope mounts that steer the telescopes around the sky: The Fork Equatorial Mount, the German Equatorial Mount, the Altitude-Azimuth (Alt-Azimuth) Mount, and the Dobsonian Mount (see figures). The table on the next page presents the advantages and disadvantages of each mount.

Reflecting and refracting telescopes both have pros and cons, as discussed in the text. Perhaps the most significant difference is in their lengths. Refractors and Newtonian reflectors tend to be longer and more bulky than any other types. Newtonians are the least expensive and simplest telescopes to build. As a result, they are often used with the inexpensive Dobsonian mounts. If you plan to take photographs, a Cassegrain telescope is the best instrument because it remains balanced as the telescope tracks objects across the sky.

If you want to see especially large areas of the sky, you may want to purchase a telescope with a *Schmidt corrector plate*. The most common wide-field telescopes are of the Schmidt-Cassegrain design. These telescopes have a Schmidt corrector plate on top and a Cassegrain mirror on the bottom with their eyepieces located underneath.

If you build or buy a telescope with the eyepiece positioned so that you have to crawl under it or climb a ladder to see through it, you will find the experience less satisfying than if you can observe in a comfortable position. If the eyepiece is not easily accessible, you can buy a *diagonal mirror* (a right-angle mirror that goes between the telescope body and the eyepiece) to help correct this problem.

You will also want to get a few different *eyepieces* (three is a good number to start with), so that you can look at large areas of the sky under low magnification and details of small areas under high magnification. As dis-

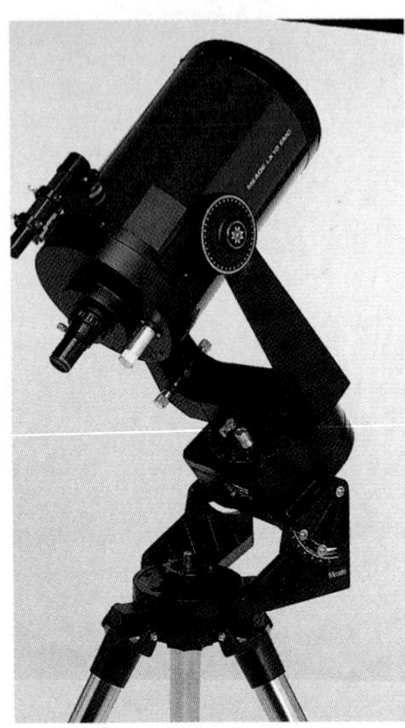

a Fork Equatorial Mount

b German Equatorial Mount

c Alt-Azimuth Mount

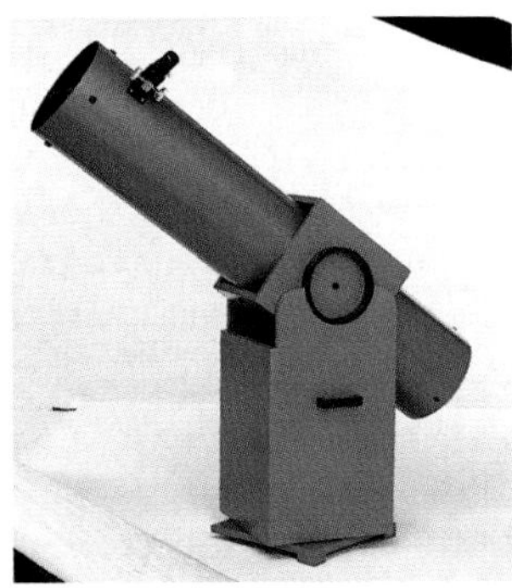

d Dobsonian Mount

Buying a Telescope (a, b, & d: Andy Crawford/Dorling Kindersley; c: Celestron Images)

of computer-controlled grinding and rotating furnaces (Figure 3-22), in which the liquid glass actually spins into a parabolic shape, has now made it economical to cast parabolic mirrors with diameters of several meters.

3-10 Earth's atmosphere hinders astronomical research

Earth's atmosphere affects the light from objects in space before it reaches our telescopes. The air is turbulent and

Buying a Telescope (continued)

Telescope Mounts

Type of Telescope	Pros	Cons
Fork Equatorial	Can track objects in the sky with a clock drive Useful for taking CCD or film photographs Not too heavy or cumbersome	Eyepiece is in an uncomfortable position near celestial poles
German Equatorial	Can track objects in the sky with a clock drive Useful for taking CCD or film photographs Easy to change direction telescope points	Hard at first to learn to set up Takes a long time to set up Very heavy compared to other types of mounts
Alt-Azimuth	Compact—easy to set up, store, and carry Eyepiece is convenient for viewing	Must be computerized in order to track objects in the sky
Dobsonian	Least expensive for a given diameter primary mirror Easy to set up Easy to use	Eyepiece is often in inconvenient position Cannot track objects in the sky or take photographs without specialized equipment

cussed in the text, the magnification of a telescope depends on the focal lengths of the objective lens and the eyepiece. The objective lens or primary mirror is fixed in the telescope. However, all telescopes come with removable eyepieces. Change the eyepiece to another of different focal length, and you change the telescope's magnification. Keep in mind, however, that increasing the magnification decreases the area of the sky that you see.

Unless your telescope is computerized, it is also essential to have a *finder scope,* which is a small, low-magnification, very-wide-field telescope, with crosshairs, attached directly onto your main telescope. If the finder and your main scope are well aligned, you can quickly zero in on an object.

If you plan to take photographs with a CCD camera or other instrument on your telescope or to show the cosmos to large groups of people, you will need a *tracking motor* to enable the telescope to follow the stars automatically. Tracking motors require their own power from batteries, an adaptor for a car cigarette lighter, or a 110-V outlet. The motor should also be able to run at high speed so you can slew (turn the telescope rapidly) from one object to the next. For photography, you will also need a *camera adapter.*

Viewing the Sun is very exciting, especially when you can observe sunspots. If you want to observe the Sun, it is essential that you buy a good *Sun filter.* **Never, ever look at the Sun directly, either through a telescope or with your naked eye. Doing so for even a second can lead to partial blindness!**

Finally, you will need a flashlight with a red plastic film over the lens or one with a red LED light. Because red light does not cause the pupils of your eyes to contract, they will remain dilated (wide open) while you use the red flashlight to inspect your equipment and your star charts.

filled with impurities. You have probably seen turbulence while driving in a car on a hot day, when the road ahead appears to shimmer. Blobs of air, heated by the Earth, move upward to create this effect. Light passing through such a blob is refracted, because each hot air mass has a different density than the cooler air around it. Hence, because each blob behaves like a lens, images of objects beyond it appear distorted.

The atmosphere over our heads is similarly moving and changing density, and the starlight passing through it is

a

b

FIGURE 3-22 **Rotating Furnace for Making Parabolic Telescope Mirrors** **(a)** To make each 8.4-meter primary mirror for the Large Binocular Telescope II on Mount Graham in Arizona, 40,000 pounds of glass are loaded into a rotating furnace and heated to 1450 K (2150°F). This image shows glass fragments loaded into the cylindrical furnace. **(b)** After melting, spinning, and cooling, the mirror's parabolic surface is ready for final smoothing and coating with a highly reflective material. (a: Roger Ressmeyer/Corbis; b: The University of Arizona, Steward Observatory)

similarly refracted. Because the air density changes rapidly, the resulting changes in refraction make the stars appear to change brightness and position rapidly, an effect we see as **twinkling.** When photographed from earth for more than a few seconds, twinkling smears out a star's image, causing it to look like a disk rather than a pinpoint of light (Figure 3-23a). Astronomers use the expression "seeing" to describe how steady the atmosphere is; when the seeing is bad, much twinkling is occurring and, therefore, telescopic images are spread out and blurry.

The angular diameter of a star's smeared-out image, called the **seeing disk,** is a realistic measure of the best possible resolution. The size of the seeing disk varies from one observatory site to another. At Mount Palomar in California, the seeing disk is roughly 1 arcsec (1″).The best conditions on Earth, with a seeing disk of 0.2″, have been

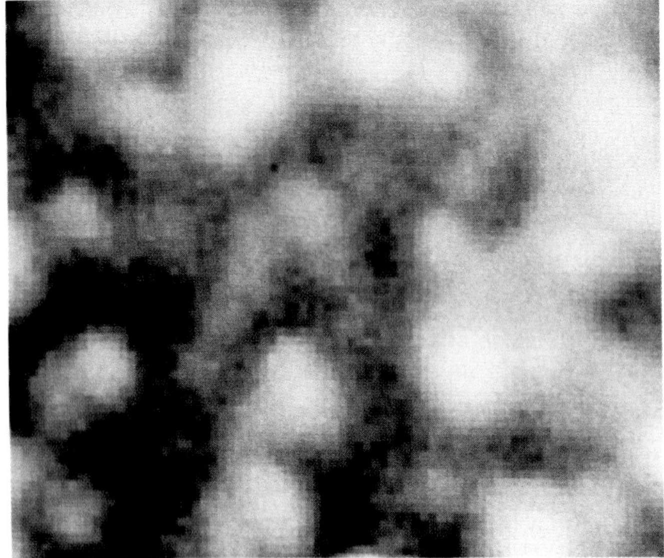

a

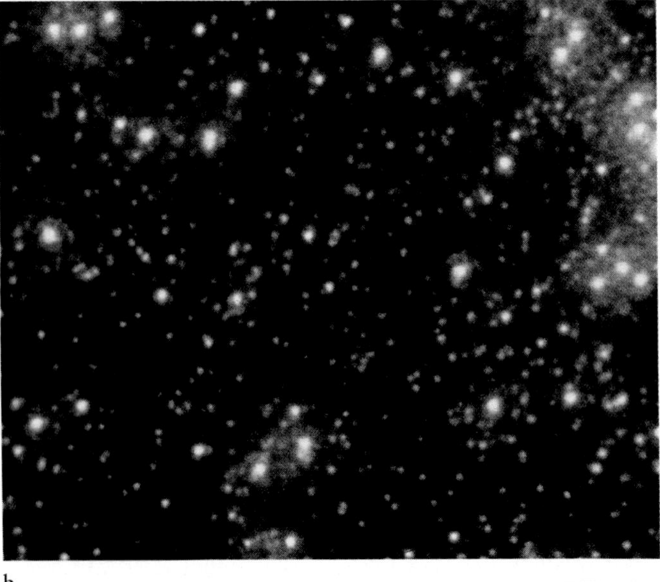

b

R I V U X G

FIGURE 3-23 **Effects of Twinkling** The same star field photographed with **(a)** a ground-based telescope, which is subject to poor seeing conditions that result in stars twinkling, and **(b)** the Hubble Space Telescope, which is free from the effects of twinkling. (NASA/ESA) recorded at the observatory on the 14,000-ft summit of Mauna Kea, the tallest volcano on the island of Hawaii.

Without the effects of the Earth's atmosphere, stars do not twinkle. As a result, photographs taken from telescopes in space reveal stars as much finer points (Figure 3-23b) and more detail for extended objects, such as planets and galaxies. The Hubble Space Telescope, as we will see often in this book, achieves magnificent resolution.

INSIGHT INTO SCIENCE

Research Requires Patience Seeing conditions—indeed, most observing situations in science—are rarely ideal. Besides such natural phenomena, which are beyond their control, scientists must also contend with equipment failures, late deliveries of parts, and design flaws. Patience is not a virtue in cutting-edge science; it is a necessity. Next time you visit an observatory, ask astronomers how the seeing has been lately, but be prepared for a series of expletives!

Light pollution from cities poses another problem for Earth-based telescopes (Figure 3-24). Keep in mind that the larger the primary mirror, the more light it gathers and therefore the more information astronomers can obtain from its images or spectra. The 5-m (200-in.) telescope at the Palomar Observatory between San Diego and Los Angeles, California, was the first truly great large telescope, providing astronomers with invaluable insights into the universe for decades. However, light pollution from the two cities now fills the night sky, seriously reducing the ability of that telescope to collect light from objects in space. Not surprisingly, the best observing sites in the world are high on mountaintops—above smog, water vapor, and clouds—and far from city lights. An even better location, astronomers have discovered, is to observe space from space, eliminating the interference of both the lights of civilization and the Earth's atmosphere.

3-11 The Hubble Space Telescope provides stunning details about the universe

For decades, astronomers dreamed of observatories in space. Such facilities would eliminate the image distortion created by twinkling and by poor atmospheric transparency due to pollution, volcanic debris, and water vapor. They could operate 24 hours a day and over a wide range of wavelengths—from the infrared through the visible range and far out into the gamma ray part of the spectrum. NASA and other space agencies have launched a variety of space telescopes, including four of what NASA calls its Great Observatories. The first Great Observatory to go up was the Hubble Space Telescope (HST).

Soon after HST was placed in orbit from the Space Shuttle in 1990, astronomers discovered that the telescope's 2.4-m primary mirror lacked the proper curvature, which

FIGURE 3-24 **Light Pollution** These two images of Tucson, Arizona, were taken from the Kitt Peak National Observatory, which is 38 linear miles away. They show the dramatic growth in ground light output between 1959 (top) and 1989 (bottom). Since 1972, light pollution, a problem for many observatories around the world, has been at least partially controlled by a series of local ordinances. (NOAO/AURA/NSF/Galaxy)

FIGURE 3-25 **The Hubble Space Telescope (HST)** This photograph of HST hovering above the Space Shuttle's cargo bay was taken in 1993, at completion of the first servicing mission. HST has studied the heavens at infrared, visible light, and ultraviolet wavelengths. (NASA)

caused its star images to be surrounded by a hazy glow. During a repair mission in December 1993, astronauts installed corrective optics that eliminated the problem (Figure 3-25). It was further upgraded in 2002. Now HST has a resolution of better than 0.1″, which is better than can be obtained by telescopes on the Earth's surface without the use of advanced technology (see Section 3-12).

WEB LINK 3.5 The observations taken by HST continue to stagger the imagination. It has made new observations related to the planets in our solar system, planetary systems forming around other stars, the distances to other galaxies, black holes, quasars, the formation of the earliest galaxies, and the age of the universe, among many other things. The universe is like a library, with books that we haven't yet opened. Hubble is one of the best resources astronomers have for reading those books. This success, coupled with rapidly improving technology, has spurred scientists and engineers to begin developing the 6.5-m diameter James Webb Space Telescope to replace Hubble. This new telescope is scheduled for launch in 2011.

3-12 Advanced technology is spawning a new generation of superb ground-based telescopes

The clarity of images taken by the Hubble Space Telescope may suggest that ground-based observational astronomy is a dying practice. However, two exciting techniques, called active optics and adaptive optics, enable telescopes on the ground to match the quality of Hubble—or better it!

Active optics finds the best orientation for the primary mirror in response to changes in temperature and the shape of the telescope mount. It adjusts the mirror every few seconds to help keep the telescope aimed at its target. With active optics, the New Technology Telescope in Chile and the Keck telescopes in Hawaii routinely achieve resolutions as fine as 0.3″ when the resolution is much worse for telescopes without active optics at the same sites.

WEB LINK 3.6 **Adaptive optics** uses sensors to determine the amount of twinkling created by atmospheric turbulence. The stellar motion is neutralized by computer-activated, motorized supports that actually reshape either the primary mirror or a smaller mirror installed farther down the optical path of the telescope. Adaptive optics effectively eliminates atmospheric distortion and produces remarkably sharp images (Figure 3-26). Many large ground-based telescopes now use adaptive optics on at least some observations, resulting in images comparable to those from HST.

WEB LINK 3.7 Until the 1980s, telescopes with primary mirrors of between 2 and 6 m were the largest and most powerful in the world. Now, new technologies in mirror building and computer control allow us to construct much larger telescopes. There are at least 55 reflectors around the world today with primary mirrors measuring 2 m or more in diameter. This number is up by 6 in only 3 years. Among these are 10 with mirrors between 8 and 10 m in diameter, with more than a dozen other very large telescopes under construction.

Because the cost of building very large mirrors is enormous, astronomers have devised less expensive ways to collect the same amount of light. One approach is to make a large mirror out of smaller pieces, fitted together like floor tiles. The largest examples of this segmented-mirror technique are the 10-m (400-in.) Keck I and Keck II telescopes on the summit of Mauna Kea in Hawaii. Thirty-six hexagonal mirrors are mounted side by side in each telescope to collect the same amount of light as a single primary mirror 10 m in diameter (Figure 3-27). Another method, called **interferometry,** combines images from different telescopes. For example, used together to observe the same object, the Keck telescopes have the resolving power of a single 85-meter telescope.

a

b

c

R I V U X G

FIGURE 3-26 **Images from Earth and Space** (a) Image of Saturn from an Earth-based telescope without adaptive optics. (b) Image of Saturn from an Earth-based telescope with adaptive optics. (c) Image of Saturn from the Hubble Space Telescope, which does not incorporate adaptive optics technology. (a & b: Air Force Research Laboratory, Starfire Optical Range, Kirtland AFB, NM; c: Reta Beebe/New Mexico State University, D. Gilmore, L. Bergeron/STScI, and NASA)

FIGURE 3-27 **The 10-m Keck Telescopes** Located on the (hopefully) dormant Mauna Kea volcano in Hawaii, these huge twin telescopes each consist of 36 hexagonal mirrors measuring 1.8 m (5.9 ft) across. Each Keck telescope has the light-gathering, resolving, and magnifying ability of a single mirror 10 m in diameter. **Inset:** View down the Keck I telescope. The hexagonal apparatus near the top of the photograph shows the housing for the 1.4 m secondary mirror. (W. M. Keck Observatory, Courtesy of Richard J. Wainscoat)

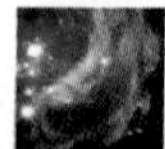

NONOPTICAL ASTRONOMY

Looking back at Figure 3-4 you will see that visible light represents a very tiny fraction of the electromagnetic spectrum. As late as the 1940s, astronomers had no idea how much nonvisible radiation is emitted by objects in space. We now know that many objects in space emitting nonvisible radiation are invisible to our eyes even with the most powerful visible-light telescopes. Newer telescopes gather electromagnetic energy in the nonvisible parts of the spectrum. From the radio waves and infrared radiation being emitted by vast interstellar gas clouds to the X rays from the remnants of stars to bursts of gamma rays of extraordinary power from merging black holes and other sources, our growing ability to see the entire electromagnetic spectrum is revealing astonishing secrets. To get a sense of how much we do not see in the visible part of the spectrum, look at Figure 3-28, which shows ultraviolet, infrared, and visible images of the familiar constellation Orion.

3-13 A radio telescope uses a large concave dish to reflect radio waves

The first evidence of nonvisible radiation from outer space came from the work of a young radio engineer, Karl Jansky, working at Bell Laboratories. Using radio antennas, Jansky was investigating the sources of static that affect short-wavelength radiotelephone communication. In 1932 he realized that a certain kind of radio noise is strongest when the constellation Sagittarius is high in the sky. Because the center of our Galaxy is located in the direction of Sagittarius, Jansky correctly concluded that he was detecting radio waves from elsewhere in the Galaxy. Jansky's accidental

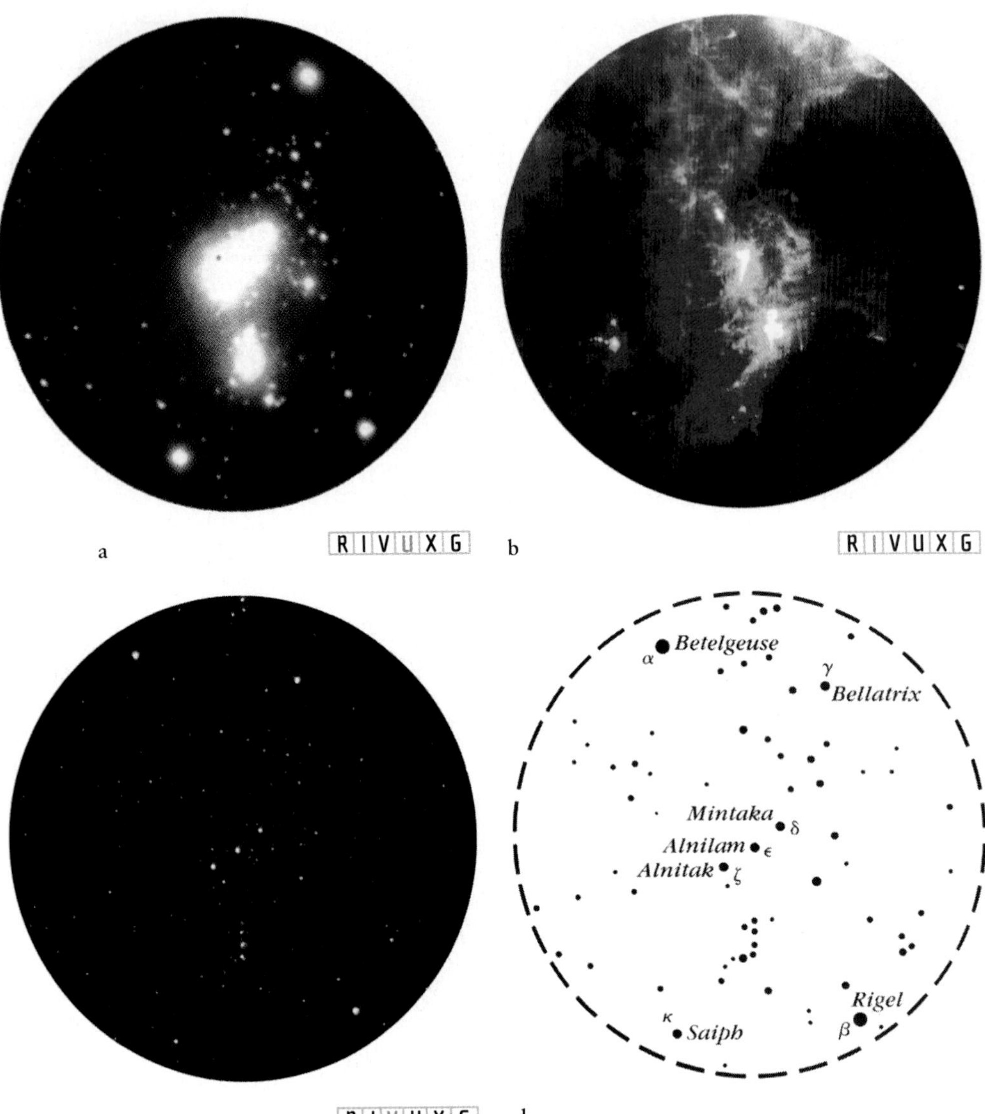

FIGURE 3-28 **Orion as Seen in Ultraviolet, Infrared, and Visible Wavelengths** **(a)** An ultraviolet view of the constellation Orion obtained during a brief rocket flight on December 5, 1975. The 100-s exposure captured wavelengths ranging from 125 to 200 nm. **(b)** A false-color view from the Infrared Astronomical Satellite uses color to display specific ranges of infrared wavelengths: red indicates long-wavelength radiation; green, intermediate-wavelength radiation; and blue, short-wavelength radiation. For comparison, **(c)** an ordinary optical photograph and **(d)** a star chart are included. (a: G. R. Carruthers, NRL; b: NASA; c: R. C. Mitchell, Central Washington University)

discovery of radio signals from space led to a worldwide effort to "see" what nonvisible wavelengths could teach us about the cosmos.

INSIGHT INTO SCIENCE

Think "Outside the Box" Observations and experiments often require scientists to make connections between seemingly unrelated concepts. Jansky's proposal that some radio waves originate in space is an example of a scientist connecting apparently disparate scientific fields—astronomy and radio engineering.

Radio telescopes record radio signals from the sky. Radio waves have the longest wavelengths of all electromagnetic radiation. The angular resolution of any telescope decreases as the wavelength increases. In other words, the longer the wavelength, the fuzzier the picture. Because radio radiation has such long wavelengths, the first, small radio telescopes produced blurry, indistinct images. But by the 1970s, groups of radio telescopes observing the same object were providing radio images with resolution even better than that of visible light telescopes.

Radio telescopes each have a large, reflecting, concave dish (Figure 3-29) that acts exactly like a mirror in an optical telescope. A small antenna tuned to the desired wavelength is located at the prime focus or the Cassegrain focus. The incoming signal is relayed to amplifiers and recording instruments. Very large radio telescopes create sharper radio images because, as with optical telescopes, the bigger the dish, the better the angular resolution. For this reason, most modern radio telescopes have dishes more than 25 m in diameter. Nevertheless, even the

FIGURE 3-29 **A Radio Telescope** Recall that the secondary mirror or prime focus on most telescopes blocks incoming light or other radiation. This new radio telescope at the National Radio Astronomy Observatory in Green Bank, West Virginia, has its prime focus hardware located off-center from the telescope's 100-m reflector. By using this new design, there is no such loss of signal. Such configurations are also common on microwave dishes used to receive satellite transmissions for home televisions. (NRAO/AUI/NSF)

largest radio dish in existence (305-m diameter in Arecibo, Puerto Rico) cannot come close to the resolution of the best optical telescopes. For example, a 6-m optical telescope has 2000 times better resolution than a 6-m radio telescope detecting radio waves of 1-mm wavelength.

To overcome the limitation on resolution set by telescope diameter, radio astronomers often use interferometry to produce high-resolution radio images. Recall from Section 3-12 that interferometry combines the data received simultaneously by two or more telescopes. The telescopes can be kilometers, continents, or even worlds apart. The radio signals received by all the dishes are made to "interfere," or blend together, and with suitable computer-aided processing, the combined image of the source is sharp and clear. The results are impressive: The resolution of such a system is equivalent to that of one gigantic dish with a diameter equal to the distance between the farthest telescopes in the array.

WEB LINK 3.10 Interferometry, exploited for the first time in the late 1940s, gave astronomers their first detailed views of "radio objects" in the sky. More recently, radio telescopes separated by thousands of kilometers have been linked together in **very-long-baseline interferometry (VLBI)**, producing images that are much sharper and clearer than even those from optical telescopes. One of the most complex systems of radio telescopes began operating in 1980 on the Plains of San Agustin near Socorro, New Mexico. Called the Very Large Array (VLA), it consists of 27 concave dishes, each 26 m (85 ft) in diameter. The 27 telescopes are positioned along the three arms of a gigantic Y that can span a distance of 36 km (22 mi). Working together, they can create radio images with 0.1″ resolution. Figure 3-30 shows the VLA. This system produces radio views of the sky with resolutions comparable to that of the very best optical telescopes. The Very Long Baseline Array (VLBA) consists of ten 25-m radio telescopes located across the United States, from Hawaii to New Hampshire. With a maximum baseline of 8000 km, VLBA has a resolving power of 0.001″.

The best angular resolution on Earth is obtained by combining radio data from telescopes on opposite sides of our planet. In that case, features as small as 0.00001″ can be distinguished at radio wavelengths—10,000 times sharper than the best views obtainable from single optical

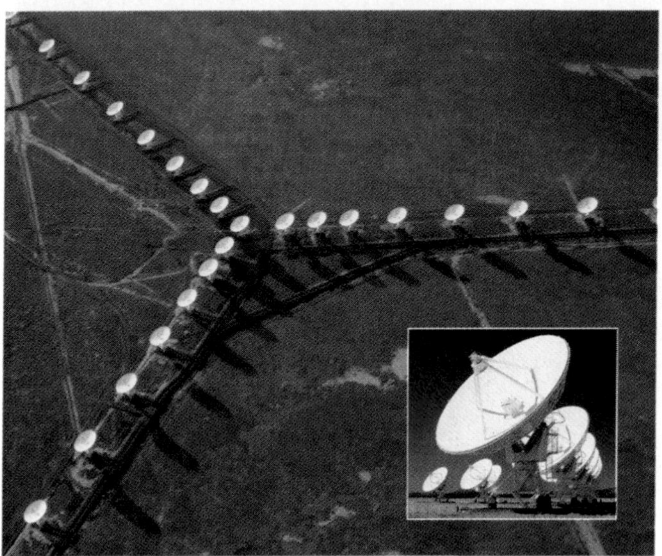

FIGURE 3-30 **The Very Large Array (VLA)** The 27 radio telescopes of the VLA system are arranged along the arms of a Y in central New Mexico. The telescopes can be moved so that the array can detect either wide areas of the sky (when they are close together, as in this photograph) or small areas with higher resolution (when they are farther apart). The inset shows the traditional secondary mirror assembly in the center of each of these antennas. (Jim Sugar/Corbis; inset: David Nunuk/Science Photo Library/Photo Researchers)

telescopes. Radio telescopes are also being put into space and used in even longer-baseline interferometers. The success that radio astronomers had with interferometry is what prompted it to be applied to optical and infrared telescopes, as discussed earlier.

To make radio images more comprehensible, radio astronomers often use false colors or gray scales to display their radio views of astronomical objects. An example of the use of false colors is shown in Figure 3-31. The most intense radio emission is shown in red, the least intense in blue. Intermediate colors of the rainbow represent intermediate levels of radio intensity. Black indicates that there is no detectable radio radiation. Astronomers working at other nonvisible-wavelength ranges also frequently use false-color techniques to display views obtained from their instruments.

a

R I V U X G

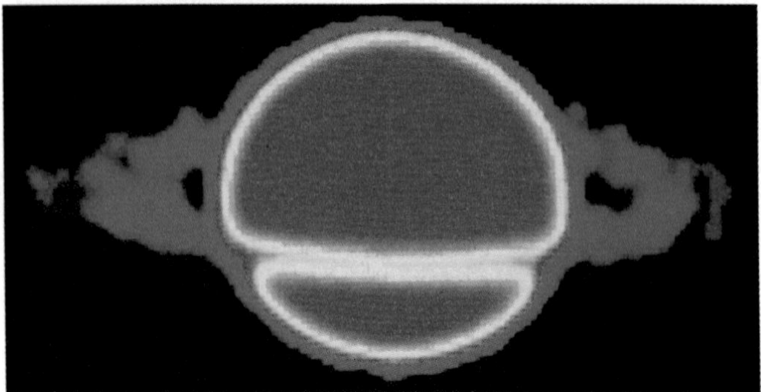

b

R I V U X G

FIGURE 3-31 **Visible and Radio Views of Saturn** **(a)** This picture was taken by a camera on board a spacecraft as it approached Saturn. The view was produced by sunlight scattered from the planet's cloudtops and rings. **(b)** This false-color picture, taken by the VLA, shows radio emission from Saturn at a wavelength of 2 cm. (a: NASA; b: Image courtesy of NRAO/AUI)

Consider an example of how nonoptical astronomy, such as is done with these radio telescopes, can overcome limitations of optical telescopes: Late in the eighteenth century, the astronomer Sir William Herschel observed regions of our Milky Way from which no visible light was emitted. "Surely, there is a hole in the heavens," he reported on seeing the first such region. (He was not referring to a modern "black hole," but rather just to an area that was especially dark to his eyes.) In the twentieth century, astronomers discovered that these regions are actually filled with interstellar gas and dust that prevents visible light from stars beyond the clouds to pass through to us, just as thick clouds of water vapor in our sky obscure the Sun and darken the sky. Radio waves, among other wavelengths, pass through both our clouds and interstellar clouds, and it was only with the advent of radio telescopes that astronomers got their first glimpses of the variety of objects that lie beyond the latter clouds.

3-14 Infrared and ultraviolet telescopes also use reflectors to collect their electromagnetic radiation

As the success of radio astronomy mounted, astronomers started making observations at other nonvisible wavelengths. The next two parts of the spectrum to be explored were infrared and ultraviolet. As with radio telescopes, infrared and ultraviolet telescopes are all reflectors. Because water vapor is the main absorber of infrared radiation from space, locating infrared observatories at sites of low humidity can overcome much of the atmosphere's hindrance. For example, the summit of Mauna Kea on Hawaii is exceptionally dry (most of the moisture in the air is below the height of this volcano), and infrared observations are the primary function of NASA's 3-m Infrared Telescope Facility (IRTF) there.

WEB LINK 3.11

The best way to avoid the obscuring effects of water vapor is to place a telescope in orbit around the Earth. The 1983 Infrared Astronomical Satellite (IRAS), Hubble Space Telescope's Near Infrared Camera and Multi-Object Spectrometer (NICMOS), the Infrared Space Observatory (ISO) launched in 1995, and the Spitzer Space Telescope (Figure 3-32) launched in 2003 have done much to reveal the full richness and variety of the infrared sky. Spitzer is the infrared equivalent to HST, one of NASA's Great Observatories.

Infrared telescopes detect infrared radiation from sources including small bodies in the solar system, the bands of dust in our Galaxy, newly formed stars, the dust disks around stars at various stages of their evolution, and the most distant galaxies whose radiation arrives here primarily at infrared wavelengths. Infrared telescopes have located more than a quarter million infrared sources in the sky, most of which are invisible to optical telescopes. As with radio waves, some infrared radiation passes through

a

b

FIGURE 3-32 **Spitzer Space Telescope** **(a)** The mirror assembly for the Spitzer Space Telescope showing the 85-cm objective mirror. **(b)** Launched in 2003, this Great Observatory is scheduled to observe the infrared cosmos for five years. It is taking images and spectra of planets, comets, gas and dust around other stars and in interstellar space, galaxies, and the large-scale distribution of matter in the universe. **Inset:** An infrared image of a region of star formation invisible to optical telescopes. (a: Balz/SIRTF Science Center; b: NASA/JPL-Caltech; inset: NASA/JPL)

dust and gas allowing astronomers to see, for example, the otherwise invisible surface features on Saturn's cloud-enshrouded moon Titan and, using Earth-based infrared telescopes with adaptive optics, even to photograph individual stars at the center of our Galaxy some 245 trillion km (153 trillion mi) away.

During the early 1970s, ultraviolet astronomy got off the ground—literally and figuratively. Both Apollo and Skylab astronauts used small telescopes above the Earth's atmosphere to give us some of our first views of the ultraviolet sources in space. Small rockets have also been used to place ultraviolet cameras briefly above the Earth's atmosphere. A typical ultraviolet view is shown in Figure 3-28a.

Some of the finest early ultraviolet astronomy was accomplished by the International Ultraviolet Explorer (IUE), which was launched in 1978 and functioned until 1996. The Space Shuttle was transformed into an orbiting observatory twice in the 1990s, carrying aloft and then returning to Earth three ultraviolet telescopes. In 1992 the Extreme Ultraviolet Explorer (EUVE) was launched. As with infrared observations, ultraviolet images reveal sights previously invisible and often unexpected. Many objects in space emit ultraviolet radiation that astronomers can use to study the chemistries of these cosmic bodies. Therefore, astronomers launched the Far Ultraviolet Spectroscopic Explorer (FUSE). It is providing us with information about such things as how much deuterium (hydrogen nuclei with one neutron) was created when the universe formed and the chemical evolution of galaxies.

3-15 X-ray and gamma-ray telescopes cannot use normal reflectors to gather information

Neither X rays nor gamma rays penetrate the Earth's atmosphere without interacting with the gases in the air, so direct observations at these extremely short wavelengths must be made from space. Astronomers got their first look at objects emitting X rays in space during brief rocket flights in the late 1940s. Several small satellites launched during the early 1970s viewed the sources of X rays and gamma rays in space, revealing hundreds of previously unknown short-wavelength objects, including several black holes (see Chapter 14). X-ray telescopes have also been carried on Space Shuttle missions. The insets in Figure 3-33 show how different our Sun appears when seen through X-ray and visible light "eyes."

X-ray photons are tricky to collect. Because of their high energies, X-rays penetrate even highly polished surfaces that they meet nearly head on. Therefore, normal reflecting mirrors cannot be used to focus them. Instead, X-ray telescopes are designed to collect these photons at a fairly shallow angle because X-rays that are barely skimming or grazing a surface can be reflected and thereby focused. This is analogous to skipping a flat rock on water. Throw it at a steep angle and it immediately sinks. Throw it at a shallow angle and it will skip off the surface. Figure 3-34 shows the design of "grazing incidence" X-ray telescopes.

The Chandra X-ray Observatory, a NASA Great Observatory named after the Nobel laureate astronomer Subrahmanyan Chandrasekhar, and the European Space Agency's XMM-Newton were both launched in 1999 and provide astronomers with the highest resolution X-ray

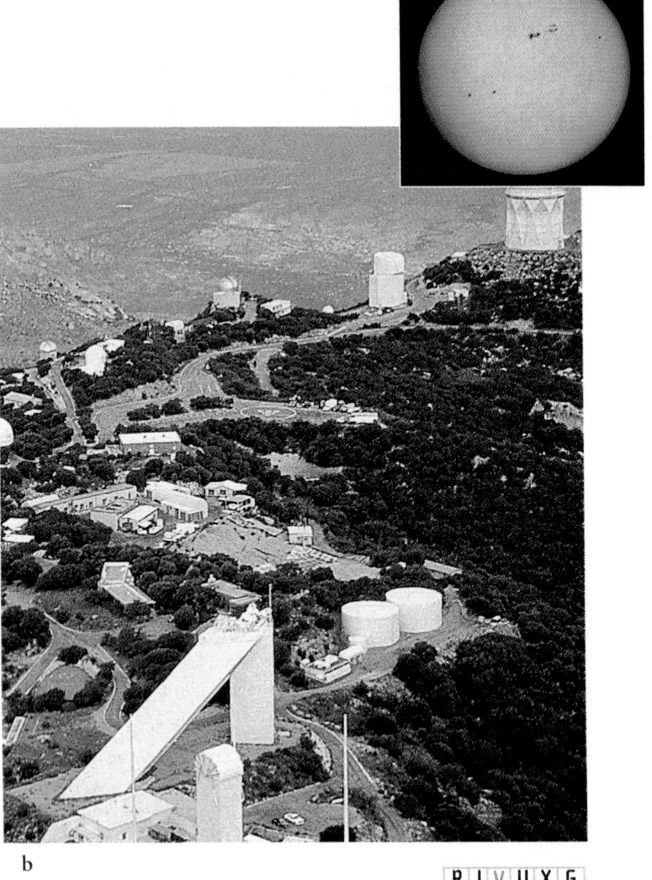

FIGURE 3-33 **Nonvisible and Visible Radiation (a)** This X-ray telescope was carried aloft in 1994 by the Space shuttle. The inset shows an X-ray image of the Sun. **(b)** This solar observatory in Arizona (the inverted V-shaped structure) takes visible-light photographs of the Sun, such as the one shown in the inset. Comparing the images in the two insets reveals how important observing nonvisible radiation from astronomical phenomena is to furthering our understanding of how the universe operates. (Top left and right, L. Golub, Naval Observatory, IBM Research, NASA; bottom left, NASA; bottom right, NOAO)

images available. Thousands of X-ray sources have been discovered all across the sky. Among these are planetary atmospheres, stars (see the Sun in Figure 3-33), stellar remnants, vast clouds of intergalactic gas, jets of gas emitted by galaxies, black holes, quasars, clusters of galaxies, and a diffuse X-ray glow that fills the universe.

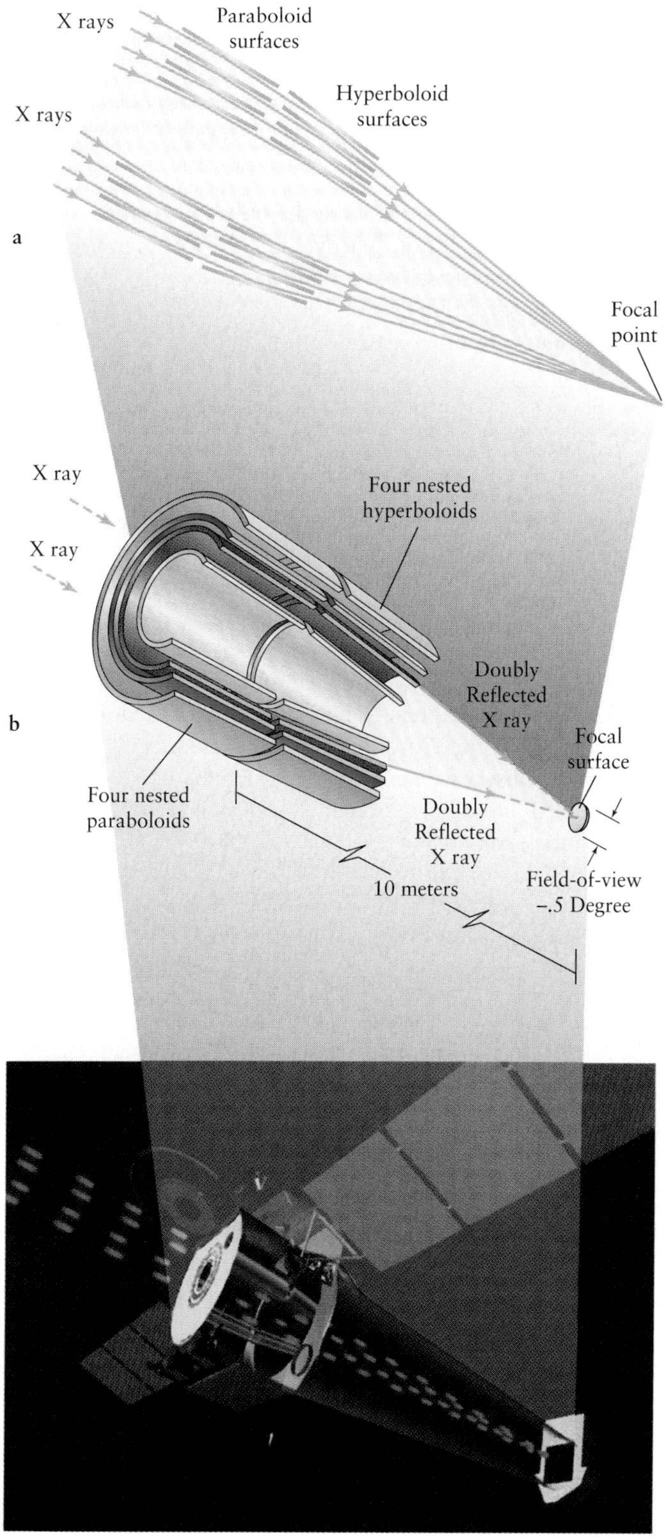

WEB LINK 3.12

The electromagnetic radiation with the shortest wavelengths and the most energy are gamma rays. In 1991 the Compton Gamma Ray Observatory, also a NASA Great Observatory, was carried aloft by the Space Shuttle. Named in honor of Arthur Holly Compton, an American physicist who made important discoveries about gamma rays, this orbiting observatory carried four instruments that performed a variety of observations, giving us tantalizing views of the gamma-ray sky.

Gamma rays are too powerful even for grazing incidence telescopes. Therefore, astronomers have devised other methods of detecting them and determining where they came from. These include absorbing them in crystals; allowing them to pass through tiny holes called collimators whose directions are well determined; and using chambers in which the gamma rays transform into electrons and positrons (positively charged electrons), leaving a track whose direction can be determined. These techniques are nowhere near as precise as those used in other parts of the spectrum, and the best resolution gamma-ray instruments are only accurate to about 5″.

Astronomers are now taking advantage of the interaction between very high energy gamma rays and the Earth's atmosphere to build telescopes on the ground that indirectly detect the gamma rays. When these photons strike the atmosphere, they often cause particles in the air to move faster than the speed of light in air (but still slower than the speed of light in empty space). These particles quickly slow down, emitting short bursts of blue light that travel in the same direction as the incoming gamma rays; by detecting this light, it is possible to determine the direction from which the gamma rays came.

We now have telescopes with which we can see the 5
energy in the universe from virtually all parts of the electromagnetic spectrum (Figure 3-35). Telescopes provide us with more than just stunning images of objects in space. They also provide information about the chemistry of stars and interstellar gas and dust, the motion of objects toward or away from us, whether stars are rotating, and whether they are alone in space or orbiting a companion, among other things. We will explore how we get this information when we study the various objects in space.

◀ FIGURE 3-34 **Grazing Incidence X-ray Telescopes** X rays penetrate objects they strike head on. To focus them, they have to be gently nudged by skimming off cylindrical "mirrors." The shapes of the mirrors optimize the focus. The bottom diagram shows how X rays are focused in the Chandra X-ray Telescope. (a & b: NASA/CXC/SAO; c: NASA/Chandra X-ray Observatory Center/Smithsonian Astrophysical Observatory)

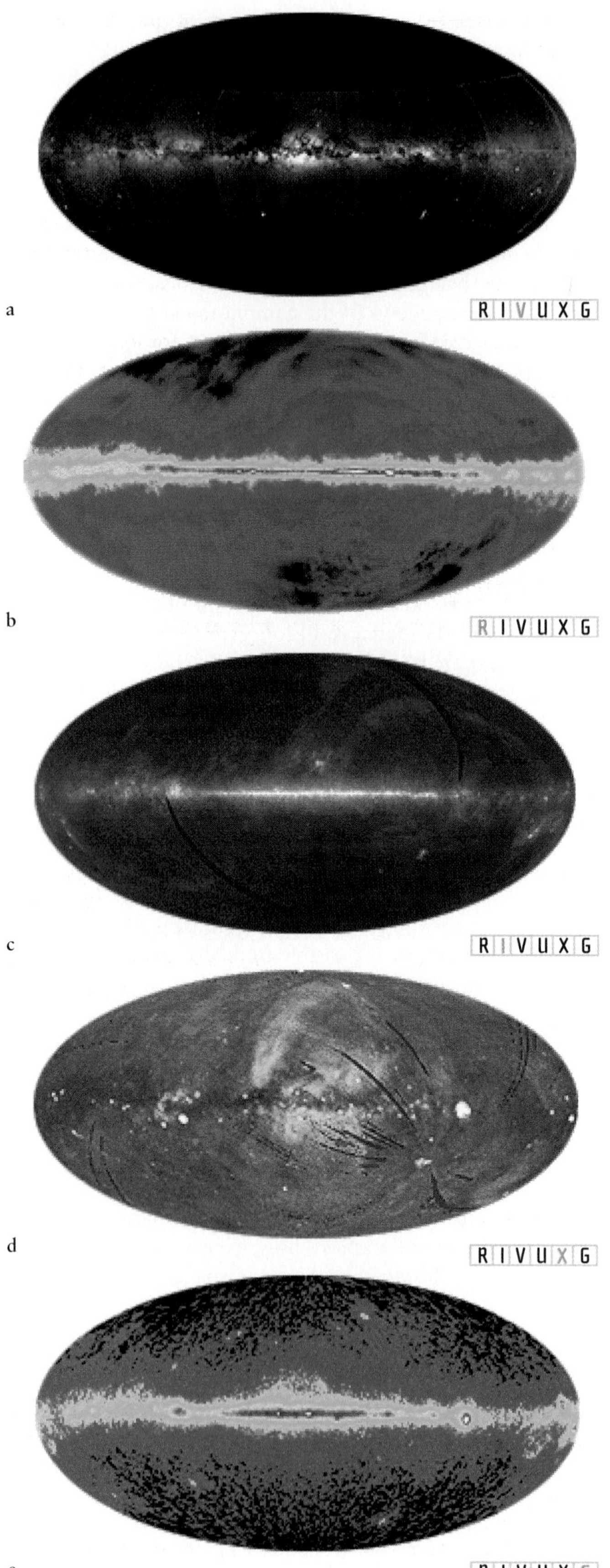

◀ FIGURE 3-35 **Survey of the Universe in Various Parts of the Electromagnetic Spectrum** By mapping the celestial sphere onto a flat surface (like making a map of the Earth), astronomers can see the overall distribution of strong or nearby energy sources in space. The center of our Galaxy's disk cuts these images horizontally in half. Since most of the emissions shown in these diagrams fall in this region, we know that most of the strong sources of various electromagnetic radiation as seen from Earth (except X rays) are in our Galaxy: **(a)** visible light, **(b)** radio waves, **(c)** infrared radiation, **(d)** X rays, **(e)** gamma rays. (GFSC/NASA)

Until recently, photons were the only sources of detailed astronomical information that we had. However, in the past four decades, physicists have built detectors for waves and particles whose nature is not electromagnetic energies, but whose origins are often astronomical and therefore of direct interest to astronomers. Chapters 10 and 14 discuss two of these devices—neutrino detectors and gravity wave detectors.

3-16 Frontiers yet to be discovered

Before the twentieth century, astronomers were like the blind men in the story about trying to describe the elephant. Our perceptions were piecemeal; we could describe parts of the universe, but not the whole. Our ancestors did not have the technology that could enable them to see the big picture of the universe. However, we are beginning to see it, and as you will learn in the chapters that follow, our understanding of the cosmos is therefore increasing dramatically. A vast amount of observational information remains to be gathered. Indeed, literally every planet, moon, piece of interplanetary debris, star, stellar remnant, gas cloud, galaxy, quasar, cluster of galaxies, and supercluster of galaxies has a story to tell. Observational astronomy is so new an activity that we are still making new and often unexpected discoveries almost daily. There is still a lot to discover about the elephant.

Summary of Key Ideas

The Nature of Light

• Photons, compact units of vibrating electric and magnetic fields, all carry energy through space at the same speed, "the speed of light" (300,000 km/s in a vacuum, slower in any medium).

• Radio waves, infrared radiation, visible light, ultraviolet radiation, X rays, and gamma rays are the forms of electromagnetic radiation. They travel as photons, sometimes behaving as particles, sometimes as waves.

• Visible light occupies only a small portion of the electromagnetic spectrum.

• The wavelength of a visible light photon is associated with its color. Wavelengths of visible light range from about 400 nm for violet light to 700 nm for red light.

• Infrared radiation and radio waves have wavelengths longer than those of visible light. Ultraviolet radiation, X rays, and gamma rays have wavelengths that are shorter.

• The motion of an object toward or away from an observer causes the observer to see all the colors from the object to blueshift or redshift, respectively. This is generically called a Doppler shift.

Optics and Telescopes

• A telescope's most important function is to gather as much light as possible. Its second function is to reveal the observed object in as much detail as possible. Often the least important function of a telescope is to magnify objects.

• Reflecting telescopes, or reflectors, produce images by reflecting light rays from concave mirrors to a focal point or focal plane.

• Refracting telescopes, or refractors, produce images by bending light rays as they pass through glass lenses. Glass impurity, opacity to certain wavelengths, and structural difficulties make it inadvisable to build extremely large refractors. Reflectors are not subject to many of the problems that limit the usefulness of refractors.

• Charge-coupled devices (CCDs) are used to record images.

• Earth-based telescopes are being built with active and adaptive optics. These advanced technologies yield resolving power comparable to the Hubble Space Telescope.

Nonoptical Astronomy

• Radio telescopes have large reflecting antennas (dishes) that are used to focus radio waves.

• Very sharp radio images are produced with arrays of radio telescopes linked together in a technique called interferometry.

• The Earth's atmosphere is fairly transparent to most visible light and radio waves, along with some infrared and ultraviolet radiation, arriving from space, but it absorbs much of the electromagnetic radiation at other wavelengths.

• For observations at other wavelengths, astronomers mostly depend upon telescopes carried above the atmosphere by rockets. Satellite-based observatories are giving us a wealth of new information about the universe and permitting coordinated observation of the sky at all wavelengths.

WHAT DID YOU THINK?

1 *What is light?* Light, more properly "visible light," is one form of electromagnetic radiation. All electromagnetic radiation (radio waves, infrared radiation, visible light, ultraviolet radiation, X rays, and gamma rays) has both wave and particle properties.

2 *What type of electromagnetic radiation is most dangerous to life?* Gamma rays have the highest energies of all photons, so they are the most dangerous to life. However, ultraviolet radiation from the Sun is the most common everyday form of dangerous electromagnetic radiation we encounter.

3 *What is the main purpose of a telescope?* A telescope is designed primarily to collect as much light as possible.

4 *Why do stars twinkle?* Rapid changes in the density of the Earth's atmosphere cause passing starlight to change direction, making stars appear to twinkle.

5 *What type(s) of electromagnetic radiation can telescopes currently detect?* Telescopes have been built that can observe the entire electromagnetic spectrum.

Key Words

active optics, 82
adaptive optics, 82
angular resolution (resolution), 72
blocked light, 76
blueshift, 66
Cassegrain focus, 71
charge-coupled device (CCD), 73
coudé focus, 71
Doppler shift, 67
electromagnetic radiation, 63
electromagnetic spectrum, 67
eyepiece lens, 76
focal length, 69
focal plane, 70
focal point, 69
gamma ray, 67
infrared radiation, 67
interferometry, 82
light-gathering power, 72
magnification, 72
Newtonian reflector, 70
objective lens, 75
photon, 67
pixel, 73
primary mirror, 69
prime focus, 71
radio telescope, 84
radio wave, 67
redshift, 67
reflecting telescope (reflector), 68
reflection, 69
refracting telescope (refractor), 68
refraction, 63
Schmidt corrector plate, 77
secondary mirror, 70
seeing disk, 80
spectrum (plural spectra), 63
spherical aberration, 76
twinkling, 80
ultraviolet (UV) radiation, 67
very-long-baseline interferometry (VLBI), 85
wavelength (λ),64
X ray, 67

Review Questions

The answers to all computational problems, which are preceded by an asterisk (*), appear at the end of the book.

1. Describe reflection and refraction. How do these processes enable astronomers to build telescopes?

2. Give everyday examples of refraction and reflection.

3. Describe a reflecting telescope by doing Interactive Exercise 3-1 and transcribe the drawing and correct labels to paper, if requested.

*4. How much more light does a 3-m-diameter telescope collect than a 1-m-diameter telescope?

5. Explain some of the advantages of reflecting telescopes over refracting telescopes.

6. What are the three major functions of a telescope?

7. What is meant by the angular resolution of a telescope?

8. What limits the ability of the 5-m telescope on Mount Palomar to collect starlight? There are several correct answers to this question.

9. What is the Doppler shift, and why is it important to astronomers?

10. Why will many of the very large telescopes of the future make use of multiple mirrors or ultrathin mirrors?

11. What is meant by adaptive optics? What problem does adaptive optics overcome?

12. Compare an optical reflecting telescope and a radio telescope. What do they have in common? How are they different?

13. Why can radio astronomers observe at any time during the day, whereas optical astronomers are mostly limited to observing at night?

14. Why must astronomers use satellites and Earth-orbiting observatories to study the heavens at X-ray wavelengths?

15. What are NASA's four Great Observatories and in what parts of the electromagnetic spectrum do they observe?

16. Why did Rømer's observations of the eclipses of Jupiter's moons support the heliocentric, but not the geocentric, cosmogony?

Advanced Questions

17. Advertisements for telescopes frequently give a magnification for the instrument. Is this a good criterion for evaluating telescopes? Explain your answer.

*18. The observing cage in which an astronomer sits at the prime focus of the 5-m telescope on Mount Palomar is about 1 m in diameter. What fraction of the incoming starlight is blocked by the cage? *Hint:* The area of a circle of diameter d is $\pi d^2/4$, where $\pi \approx 3.14$.

*19. Compare the light-gathering power of the Palomar 5-m telescope to that of the fully dark-adapted human eye, which has a pupil diameter of about 5 mm.

20. Show by means of a diagram why the image formed by a simple refracting telescope is "upside down."

*21. Suppose your Newtonian reflector has a mirror with a diameter of 20 cm and a focal length of 2 m. What magnification do you get with eyepieces whose focal lengths are **a.** 9 mm, **b.** 20 mm, and **c.** 55 mm?

22. Why does no major observatory have a Newtonian reflector as its primary instrument, whereas Newtonian reflectors are extremely popular among amateur astronomers?

23. How can astronomers on the ground detect gamma ray sources in space?

Discussion Questions

24. Discuss the advantages and disadvantages of using a small visible-light telescope in Earth orbit versus a large visible-light telescope on a mountaintop.

25. If you were in charge of selecting a site for a new observatory, what factors would you consider?

What If . . .

26. Telescopes were first invented today? What objects or areas of the sky would you recommend that astronomers explore first? Why?

27. An observatory were established on the Moon? List the advantages and disadvantages for astronomy.

28. We had eyes sensitive to radio waves? How would we be different and how would our visual perceptions of the world be different?

29. Humans were unable to detect any electromagnetic radiation? How would that change our lives and what alternatives might evolve (indeed have, for some species) to provide information about distant objects?

Web Questions

30. Several telescope manufacturers build Schmidt-Cassegrain telescopes. These use a correcting lens in an arrangement like that shown in Figure 3-21c. Consult advertisements on the Web and list the dimensions, weights, and costs of some of these telescopes. Why are they popular among amateur astronomers?

31. Projects are underway to build large optical reflectors in South Africa (SALT, the Southern African Large Telescope) and in the Canary Islands (GTC, the Gran Telescopio Canarias). Search for current information about these telescopes on the Web. What will be the sizes of the primary mirrors? Are the telescopes designed for imaging, for spectroscopy, or both? Will they observe only at visible wavelengths? In what ways do they complement or surpass existing telescopes?

32. Access the Active Integrated Media Module "Telescope Magnification" in Chapter 3 of the *Discovering the Universe* Web site. A common telescope found in department stores is a 3-in. (76-mm) diameter refractor with a f_{obj} = 750 mm that boasts a magnification of 300 times. Use the magnification calculator to determine the magnifications that are achieved by using each of the following commonly found eyepieces on that telescope:

Eyepiece A with focal length 40 mm; Eyepiece B with focal length 25 mm; Eyepiece C with focal length 12 mm; and Eyepiece D with focal length 2.5 mm. Which eyepiece was used in the advertisement and why was that one chosen?

Observing Projects

33. During the daytime, obtain a telescope and several eyepieces of differing focal lengths. If you can determine the telescope's focal length (often printed on it), calculate and record the magnifying power of each eyepiece. Focus the telescope on some familiar object, such as a distant lamppost or tree. **DO NOT FOCUS ON THE SUN! Looking directly at the Sun through a telescope will cause blindness.** Describe the image you see through the telescope. Is it upside down? How does the image move as you slowly and gently shift the telescope left and right or up and down? Examine the distant objects under different magnifications. How do the field of view and the quality of the image change as you go from low magnification to high magnification?

34. On a dark, clear, moonless night, can you see the Milky Way from where you live? If so, briefly describe its appearance. If not, what is interfering with your ability to see it?

35. Determine what fraction of stars visible to the naked eye you can see from your location. On the next dark, clear night, observe the night sky and then run *Starry Night Enthusiast*™. First choose *Hide Daylight* under the *View* button on the top. In the *Options* tab on the left, put the cursor over the words *Local Light Pollution*. Click on the *Local Light Pollution Options* button that appears. Click on *Local Light Pollution* and slide the bar until it matches the sky outside. You can also use the *City Lights* options (hold mouse over each to adjust parameters). If you have used any sky pollution setting, the goal of this project is to estimate what fraction of the visible stars you are not seeing. If you are not using a light pollution setting, the goal is to see what fraction of the stars inner-city sky viewers are missing. Print the screen, if possible. Otherwise, carefully do the following on the screen. Using a roughly 8 cm (3 in.)-square section of the sky on the screen (a square about half the length of a typical pen on each side), count and record the number of stars with your present setting. Now set the sky to either no light pollution in the first case or to local light pollution on with the slide bar to the right of its range in the second and count the stars. What fraction of the stars were you seeing in the first case? What fraction of the stars are large city viewers missing in the second case?

36. Observe the stars when the Moon is either full, new, or in a quarter phase. Record the phase of the Moon and note, qualitatively, whether you see many more stars than those just visible in the well-known asterisms such as the Big Dipper or Orion. You are taking this information so you can compare it with observations on a later date. Repeat these observations on a clear night about a week later, again noting the phase of the Moon and the numbers of visible stars. Compare the numbers of stars on the two nights. On which night did you see more stars? Why? During what three phases of the Moon did you expect that astronomers most like to make their observations?

37. On a clear night, view the Moon, a planet, and a star through a telescope using eyepieces of various focal lengths. (Use your *Starry Night Enthusiast*™ program or consult a source on the Internet or such magazines as *Sky & Telescope* or *Astronomy* to determine the phase of the Moon and the locations of the planets.) How do the images change as you view with increasing magnification? Do they become distorted at any level of magnification?

38. ***Observing the Andromeda Galaxy*** In this exercise, you will explore properties of the Andromeda galaxy. Set *Deep Space Explorer*™ to its "Home" position. If you have not done so already, go to **Settings** and uncheck "Use magnitude cutoffs," slide the "Galaxy drawing/Brightness" slide to the middle of the range, and then click on the View tab.

Grab the screen (hold the left mouse button for a PC, the mouse button for a Mac) and move space around until you see another galaxy beside the Milky Way. Release the mouse and move it to that galaxy. If the galaxy name is "Andromeda," right click (ctrl-click for a Mac) and chose "Centre Andromeda." If you have found another galaxy or nebula, repeat until you have found Andromeda. Make a sketch of Andromeda as seen from the "Home" position. Then zoom toward Andromeda until you are about 1 MLY (million light years) from it. Make another sketch. Repeat this process at about 2.5 MLY. Has the magnification of Andromeda changed? How do you see this effect? Has the resolution of the image of Andromeda improved? If so, give some examples of the improvements, as portrayed on your sketches.

39. ***Extra solar Planets*** In this exercise we will use *Deep Space Explorer*™ to observe the distribution of known stars with planets orbiting them. If you have not done so already, go to **Settings** and uncheck "Use magnitude cutoffs," slide the "Galaxy drawing/Brightness" slide to the middle of the range, and then click on the **Views** tab.

Now find the stars with extra solar planets by clicking on **Milky Way/ExtraSolar Planets.** This will take you to a distance of 158.8 LY from the solar system so that you can see the stars with extra solar planets (marked so by circles) in our vicinity. List 10 stars that have extra solar planets. Now put the cursor over any star for which a name pops up, right click (ctrl-click on Mac) and choose "Center Sun." Now grab the screen (hold left mouse button on a PC or the mouse button on a Mac) and move the mouse around to see the stars with extra solar planets from different perspectives. Describe the distribution of these stars around the solar system. Are they clustered on one side of it? Are they located in a few clumps? Are they located in a thin plane? Explain why they are in the distribution you see, keeping in mind that the Milky Way is a flattened disk of stars, gas, and dust.

WHAT IF . . . HUMANS HAD INFRARED-SENSITIVE EYES?

Our eyes are sensitive to less than a trillionth of 1% of the electromagnetic spectrum—what we call "visible light." But this minuscule resource provides an awe-inspiring amount of information about the universe. We interpret visible-light photons as the six colors of the rainbow—red, orange, yellow, green, blue, and violet. These colors combine to form all the others that make our visual world so rich. But the Sun actually emits photons of all wavelengths. So, what would happen if our eyes had evolved to sense another part of the spectrum?

A Darker Vision Gamma rays, X rays, and most ultraviolet radiation do not pass through the Earth's atmosphere. Since the world is illuminated by sunlight, the Earth and sky would look dark, indeed, if our eyes were sensitive only to these wavelengths. Radio waves, in contrast, easily pass through our atmosphere. But to see the same detail from radio waves that we now see from visible-light photons, our eyes would require diameters 10,000 times larger. Each would be the size of a baseball infield!

What about infrared radiation? While not all incoming infrared photons get through our atmosphere, short-wavelength ("near") infrared radiation passes easily through air. Depending on wavelength, the Sun emits between one-half and a ten-billionth as many infrared photons as red-light photons. Fortunately, most of these are in the near infrared.

Heat-Sensitive Vision To see infrared photons, human eyes would need to be only 5 to 10 times larger. Some snakes have evolved infrared vision. Portable infrared "night vision" cameras and goggles are available to us humans. Because everything that emits heat emits infrared photons, infrared sight would be very useful. And not everything we see with infrared sight would be due just to reflected sunlight—hotter objects would be intrinsically brighter than cool ones. For example, seeing infrared would allow us to observe changes in a person's emotional state. Someone who is excited or angry often has more blood near the skin, and thus gives off more infrared radiation (heat) than normal. Conversely, someone who is scared has less blood near the skin and thus emits less heat.

Night Vision The night sky would be a spectacular sight through infrared-sensitive eyes. Gas and dust clouds in the Milky Way absorb visible light, thus preventing the light of distant stars from getting to the Earth. However, because most infrared radiation passes through these clouds, we would be able, unaided, to see distant stars that we cannot see today. On the other hand, the white glow of the Milky Way, which is caused by the scattering of starlight by interstellar clouds, would be dimmer, because the gas and dust clouds do not scatter infrared light as much as they do visible light. (The haze created by the Milky Way would not vanish, however, because when gas and dust clouds are heated by starlight, they emit their own infrared radiation.)

Our concept of stars would be different, too. Many stars, especially young, hot ones, are surrounded by cocoons of gas and dust that emit infrared radiation. This dust is heated by the nearby stars. Instead of appearing as pinpoints, many stars would appear to be surrounded by wild strokes of color, and we would have an Impressionist sky.

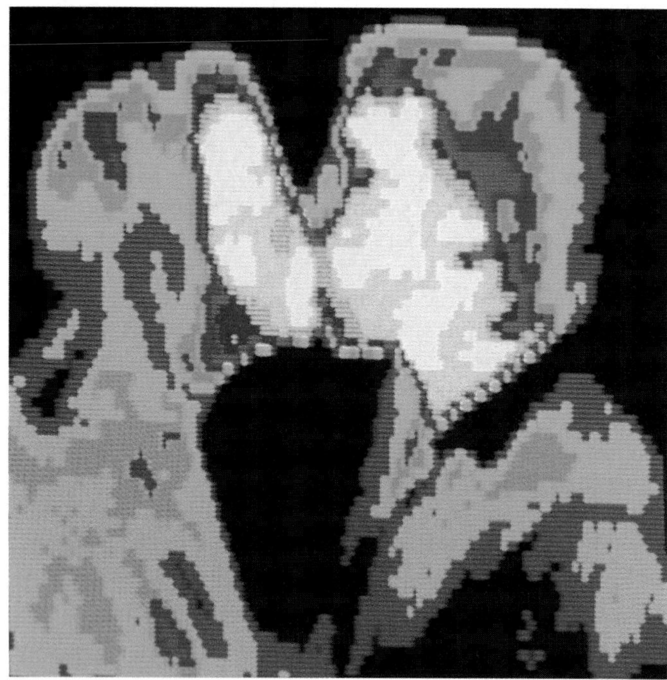

R I V U X G

The infrared (heat) from this kissing couple has been converted into visible light colors so that we can interpret the invisible radiation. The hottest regions are white, with successively cooler areas shown in yellow, orange, red, green, sky blue, dark blue, and violet. (D. Montrose/Custom Medical Stock Photo)

Visible Light and Other Electromagnetic Radiation

CHAPTER 4

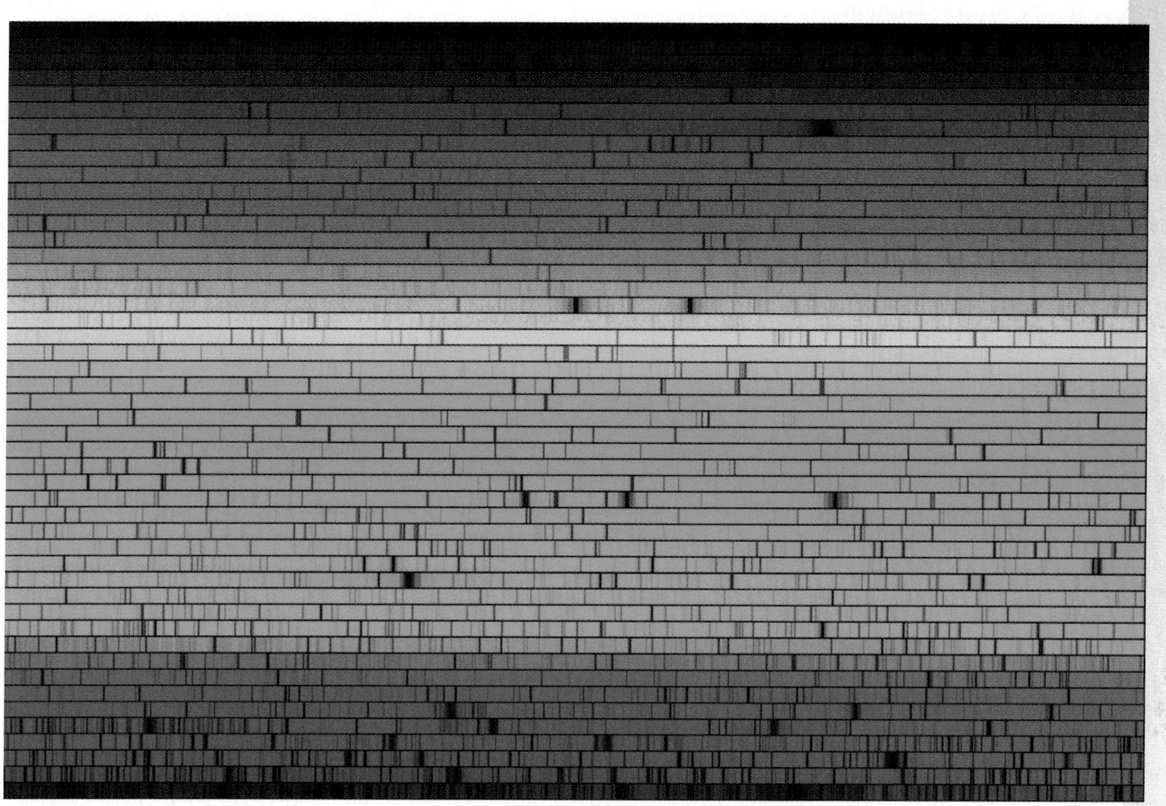

Spectrum of the Sun (N. A. Sharp, NOAO/AURA/NSF)

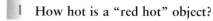

WHAT DO YOU THINK?

1 How hot is a "red hot" object?

2 What color is the Sun?

3 How can we determine the age of space debris found on Earth?

Answers to these questions appear in the text beside the corresponding numbers in the margins and at the end of the chapter.

Nearly everything we know about the cosmos comes from electromagnetic radiation, as introduced in Chapter 3. Visible light provided the first information. We will see in this chapter how scientists have discovered that most objects in space actually emit all types of electromagnetic radiation, and the amounts of each type of radiation they emit tell us the objects' surface temperatures. Human nature carried our ancestors to the next step in understanding when they asked, "Why do we see what we see?" This ongoing adventure of discovery began by understanding the basic way that electromagnetic radiation and matter interact, also covered in this chapter. From this knowledge we are able to determine the physical properties of objects, such as their chemical compositions, masses, sizes, rotation, and motion through space. We do not stop there. In later chapters we will explore related issues such as why stars shine (that is, emit electromagnetic radiation), where different elements on Earth come from, and why objects have the motions they do, among many other things.

In this chapter you will discover

- the origins of electromagnetic radiation
- the structure of atoms
- that stars with different surface temperatures emit different intensities of electromagnetic radiation
- that astronomers can determine the chemical compositions of stars and interstellar clouds by studying the wavelengths of electromagnetic radiation that they absorb or emit
- how to tell whether an object in space is moving toward or away from Earth

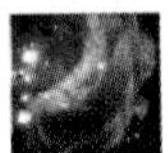

BLACKBODY RADIATION

We begin our study of why we see what we see with a perfectly familiar property of matter—temperature. We will discover that an object's temperature determines the relative numbers of photons that it emits at each wavelength.

4-1 An object's peak color shifts to shorter wavelengths as it is heated

Imagine that you have mounted an iron rod in a vise in a completely darkened room. You cannot see the rod because it emits too few visible photons and you cannot feel its warmth without touching it because it is at the same temperature as the rest of the objects in the room. Now imagine that there is also a propane torch in the room. You light it and heat the iron rod for several seconds, until it begins to glow. At the same time, you can feel the rod warm up, meaning that it is also emitting more infrared radiation.

The rod's first visible color is red—it glows "red hot" 1
(Figure 4-1a). Heated a little more, the rod appears orange and brighter, and feels hotter (Figure 4-1b). Heated more, it appears yellow and brighter, and feels hotter still (Figure 4-1c). After even more heating, the rod appears white-hot and brighter yet. If it does not melt, further heating will make the rod appear blue and even brighter, and feel even hotter. As confirmed by the photographs in Figure 4-1, this experiment shows how the amount of electromagnetic radiation any object emits changes with its temperature.

If you were to put the light emitted by a heated object like the iron rod through a prism, you would discover that all wavelengths are present and that one is brightest. This peak wavelength is the color the object appears to have. The experiment shows two things about the electromagnetic radiation emitted by objects as their temperatures change:

1. As an object heats up, it gets brighter, emitting more electromagnetic radiation at all wavelengths.

2. The brightest color (most intense wavelength) of the emitted radiation changes with temperature.

Most objects big enough to be seen with the naked eye continuously emit electromagnetic radiation as just described. The peak wavelength of any object's emitted radiation is denoted λ_{max}. When the object is relatively cool, like a rock or animal, λ_{max} is a radio or infrared wavelength. When it is hot enough, like a fire or the Sun, λ_{max} is in the range of visible light, giving a hot object its characteristic color. Exceptionally hot stars emit λ_{max} in the ultraviolet part of the electromagnetic spectrum.

Iron bars and stars are approximations of an important class of objects that astrophysicists call **blackbodies.** An ideal blackbody absorbs all the electromagnetic radiation that strikes it. The incoming radiation heats up the blackbody, which then reemits the energy it has absorbed, but with different intensities at each wavelength than it received.

Intensity is a measure of how much energy a blackbody emits per second per square meter on its surface. A **blackbody curve** (Figure 4-2) depicts in detail the electromagnetic radiation a blackbody emits. The

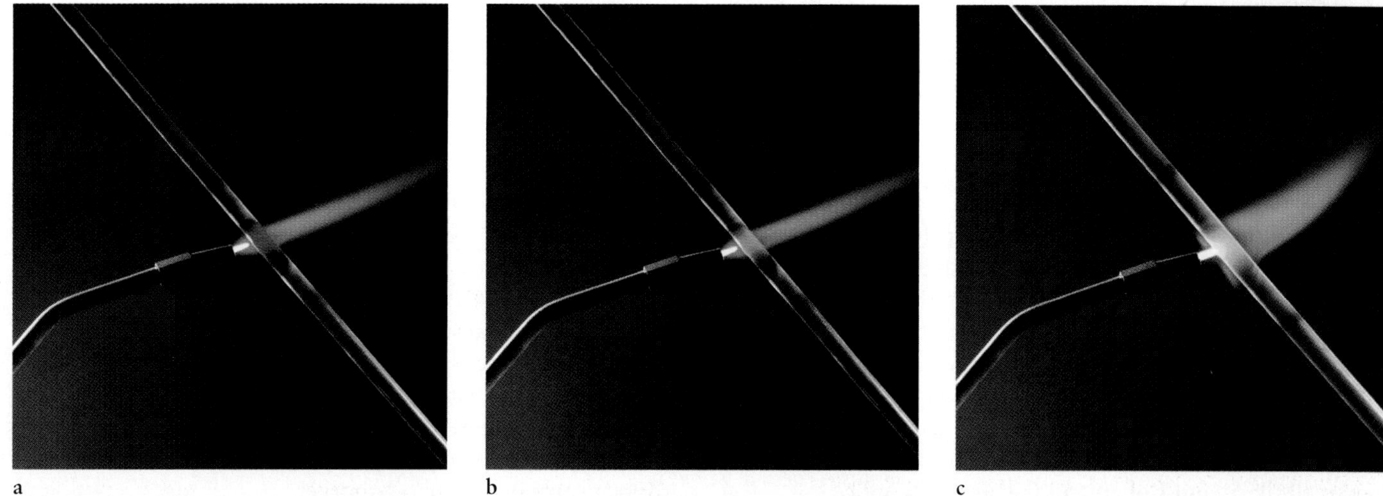
a
b
c

FIGURE 4-1 **Heating a Bar of Iron** This sequence of photographs shows the changing appearance of a piece of iron as it is heated. As the temperature increases, the amount of energy radiated by the bar increases, and so it appears brighter. The apparent color of the bar also changes because, as the temperature goes up, the dominant wavelength of light emitted by the bar decreases. (© 1984 Richard Megna Fundamental Photographs)

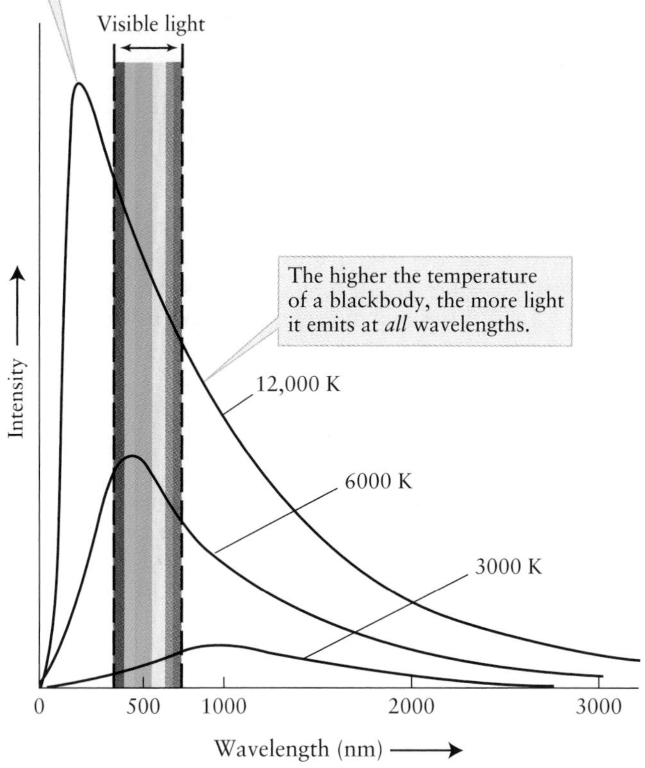
The higher the temperature of a blackbody, the shorter the wavelength of maximum emission (the wavelength at which the curve peaks).
Visible light
The higher the temperature of a blackbody, the more light it emits at *all* wavelengths.
12,000 K
6000 K
3000 K
Intensity
0
500
1000
2000
3000
Wavelength (nm)

FIGURE 4-2 **Blackbody Curves** Three representative blackbody curves are shown here. Each curve shows the intensity of radiation over a wide range of wavelengths emitted by a blackbody at a particular temperature.

emitted intensities from a blackbody only depend on its surface temperature, and so by examining a star's blackbody curve, we are able to determine that temperature. In other words, like reading a thermometer, reading a blackbody curve tells the temperature of an object. (Temperatures throughout this book are expressed in Kelvins. If you are not familiar with the Kelvin temperature scale, review Appendix D "Temperature Scales" at the back of the book.)

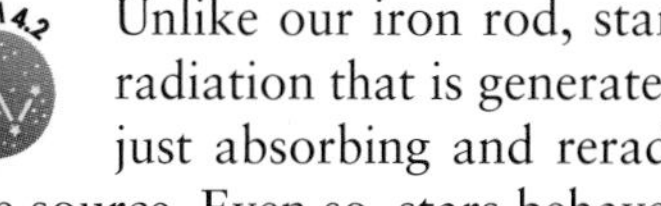
AIMM 4.2

Unlike our iron rod, stars radiate electromagnetic radiation that is generated inside them, rather than just absorbing and reradiating light from an outside source. Even so, stars behave like blackbodies, and the self-generated radiation they emit closely follows the idealized blackbody curves.

4-2 The intensities of different emitted colors reveal a star's temperature

In 1893 the German physicist Wilhelm Wien found that

The dominant wavelength of radiation emitted by a blackbody is inversely proportional to its temperature.

In other words, the hotter any object becomes, the shorter its λ_{max}, and vice versa. **Wien's law** is the mathematical relationship between the location of the peak for each curve in Figure 4-2 and that blackbody's temperature.

Wien's law proves very useful in computing the surface temperature of a star, because all we need to know is the dominant wavelength of the star's electromagnetic radiation—we do not need to know the star's size, distance, or any other physical property.

In 1879 the Austrian physicist Josef Stefan observed that

An object emits energy per unit area at a rate proportional to the fourth power of its temperature in Kelvins.

Think of a toaster. As it becomes hotter, its heating elements get brighter—it emits more energy. Stefan's result says that if you double the temperature of an object (for example, from 500 to 1000 K), the energy emitted from each square meter of the object's surface each second increases by a factor of 2^4, or 16 times. If you triple the temperature (for example, from 500 to 1500 K), the rate at which energy is emitted increases by a factor of 3^4, or 81 times. Stefan's experimental results were put on firm theoretical ground by Ludwig Boltzmann in 1885. The intensity-temperature relationship for blackbodies is named the **Stefan-Boltzmann law** in their honor.

The Stefan-Boltzmann law and Wien's law are powerful tools for understanding the properties of stars. If you would like to learn how to use these two radiation laws to make predictions, see An Astronomer's Toolbox 4-1. These laws describe two basic properties of blackbody radiation, namely, the relative rate at which photons of different wavelengths are emitted and the wavelength that is emitted most intensely. Blackbody curves, such as those shown in Figure 4-2, provide the same information as Wien's law and Stefan-Boltzmann's law, as well as showing the intensity of each blackbody at all wavelengths. None of the blackbody curves for objects with different temperatures overlap. Therefore, when astronomers measure the intensity of radiation from a star at a few wavelengths, they can determine the object's temperature by fitting these measurements to the appropriate blackbody curve.

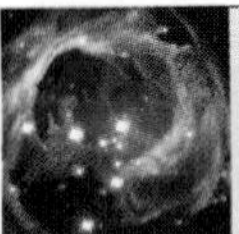

AN ASTRONOMER'S TOOLBOX 4-1
The Radiation Laws

Wien's law can be stated as a simple equation. If λ_{max}, the wavelength of maximum intensity, is measured in meters and T is the temperature of the blackbody measured in Kelvins, then

$$\lambda_{max} = \frac{2.9 \times 10^{-3}}{T}$$

Example: The Sun's maximum intensity is at a wavelength of about 500 nm, or 5×10^{-7} m. From Wien's law, we can calculate the Sun's surface temperature as

$$T = \frac{2.9 \times 10^{-3}}{5 \times 10^{-7}} = 5800 \text{ K}$$

The total energy emitted from each square meter of an object's surface each second is called the **energy flux,** F. In this context, flux means "rate of flow." Using this concept, with T as the temperature of the object in Kelvins and σ (lowercase Greek sigma) a number called the Stefan-Boltzmann constant, we can write the Stefan-Boltzmann law as

$$F = \sigma T^4$$

and

$$\sigma = 5.67 \times 10^{-8} \text{J/(m}^2 \cdot \text{K}^4 \cdot \text{s)}$$

where J is the energy unit joules, m is the unit meters, K is degrees Kelvin, and s is the unit seconds.

Because F is the energy emitted per second from each square meter of an object, multiplying F by the surface area, $4\pi r^2$, where r is the object's radius, yields the total energy emitted by a spherical object each second. Denoted L, and called **luminosity**, we have

$$L = F \cdot 4\pi r^2 = \sigma T^4 4\pi r^2$$

Luminosity is the total energy emitted per second by an entire object.

Compare! We consider one application of the Stefan-Boltzmann law. Suppose you observe two stars of equal size, one with a surface temperature of 10,000 K and the other with the same surface temperature as the Sun (5800 K). You can use the Stefan-Boltzmann law to determine how much brighter the hotter star is:

$$\frac{F_{hotter}}{F_{colder}} = \frac{\sigma \times 10{,}000^4}{\sigma \times 5800^4} \approx 8.8$$

So, the hotter star emits about 8.8 times as much energy from each square meter of its surface than does the cooler star. Because the stars are the same size, the hotter one is therefore 8.8 times brighter than the cooler one.

Try these questions: The color yellow is centered around 550 nm. If the Sun were actually yellow, what would be its surface temperature? To find out why it is not yellow, see Guided Discovery: The Color of the Sun. What is the peak wavelength in nanometers given off by a blackbody at room temperature of 300 K? Referring to Figure 3-4, what part of the electromagnetic spectrum is that in?

(Answers appear at the end of the book.)

GUIDED DISCOVERY
The Color of the Sun

Different people perceive the Sun to have different colors. To many it appears white, to others yellow. Still others, who notice it at sunset, believe it to be orange or even red. But we have seen that the Sun actually gives off all colors. Moreover, the peak in the Sun's spectrum falls between blue and green. Why doesn't the Sun appear turquoise? Several factors affect our perception of its color.

Before reaching our eyes, visible sunlight passes through the Earth's atmosphere. Certain wavelengths are absorbed and reemitted by the molecules in the air, a process called *scattering*. Violet light is scattered most strongly, followed in decreasing order by blue, green, yellow, orange, and red. That means that more violet, blue, and green photons are scattered by the Earth's atmosphere than are yellow photons. The intense scattering of violet, blue, and green has the effect of shifting the peak of the Sun's intensity entering our eyes from blue-green toward yellow. (The sky is blue because of the strong scattering of blue light—it isn't violet because the Sun emits many fewer violet photons than blue ones.)

The perception of a yellow Sun is further enhanced by our eyes themselves. Our eyes do not see all colors equally well. Rather, the light-sensitive cones in our eyes each respond to one of three ranges of colors, which are centered on red, yellow, and blue wavelengths. None of the cones is especially sensitive to blue-green photons. By adding together the color intensities detected by the three types of cones, our brains *re-create* color. After combining all the light it can, the eye is most sensitive to the yellow-green part of the spectrum. We see blue and orange less well, and violet and red most poorly. Although the eye sees yellow and green light about equally well, the dominant color from the Sun is yellow because the air scatters green light, and our eyes are relatively insensitive to blue-green. Therefore, a casual glance at the Sun leaves the impression of a yellow object.

A longer look is extremely dangerous. **Don't try it!** Hypothetically, such a glance would leave the impression of a white Sun. The Sun's light is so intense that it can saturate the color-sensitive cones in our eyes; our brains interpret such saturation of the cones as white.

At sunrise and sunset we see an orange or red Sun. This occurs because close to the horizon the Sun's violet, blue, green, and even yellow photons are all strongly scattered by the thick layer of atmosphere through which they travel, leaving the Sun looking redder and redder as it sets.

While intensity curves of stars "closely follow" blackbody curves, they are not ideal blackbodies. The *differences* between an ideal blackbody curve and the curves seen from actual stars reveal stellar chemistries, the presence of companion stars too dim to see, and the motions of stars toward or away from us, among other things.

2 Figure 4-3 shows how the intensity of sunlight varies with wavelength. The blackbody curve for a body with a temperature of 5800 K is also plotted in Figure 4-3. Note how the observed intensity curve for the Sun nearly reproduces the ideal blackbody curve at most wavelengths. Because the observed intensity curves for most stars and the idealized blackbody curves are so closely correlated, the laws of blackbody radiation can be applied to starlight. The peak of the intensity curve for the Sun is at a wavelength of about 500 nm, which is in the blue-green part of the visible spectrum.

The shapes of the blackbody curves were first derived mathematically in 1900 by the German physicist Max Planck. To do this, he had to assume that electromagnetic radiation is emitted as packets of energy. Light, a wave, behaves like a beam of particles. It was a remarkable result verified in 1905 by Albert Einstein, who called these particles of light photons, as discussed in Section 3-3.

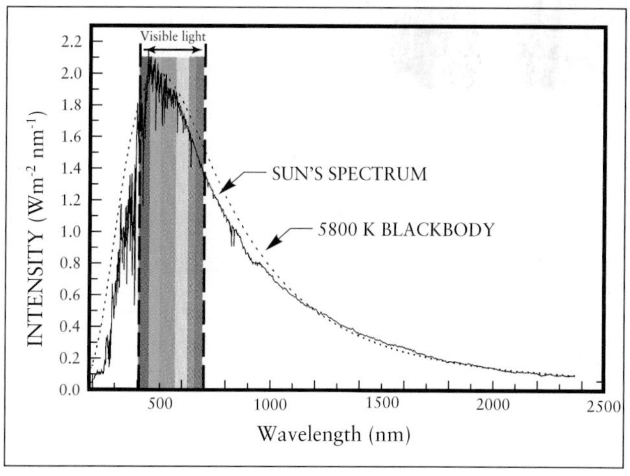

FIGURE 4-3 **The Sun as a Blackbody** This graph compares the intensity of sunlight over a wide range of wavelengths with the intensity of radiation from a blackbody at a temperature of 5800 K. The Earth's atmosphere scatters shorter-wavelength photons more than it scatters longer-wavelength ones. As seen from Earth, this scattering lowers the intensity of sunlight and shifts the blackbody peak to the right from where it actually occurs. To avoid these effects, the measurements of the Sun's intensity in this figure were made in space.

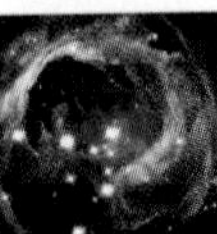

AN ASTRONOMER'S TOOLBOX 4-2
Photon Energies

All photons with the same energy are identical: Their energy depends solely on the photon's wavelength. To find any photon's energy we use the equation introduced in Section 3-3.

$$E = \frac{hc}{\lambda}$$

where Planck's constant h is 6.67×10^{-34} J s, the speed of light c is 300,000 km/s, and the photon's wavelength is λ.

Example: A photon of red light has a wavelength 700 nm. What is its energy? *Note:* All distances in the following equation must be converted to the same units, such as kilometers.

$$E_{red} = \frac{(6.67 \times 10^{-34} \text{ J s})(300{,}000 \text{ km/s})}{700 \text{ nm}}$$

$$E_{red} = 2.86 \times 10^{-19} \text{ J}$$

(A joule, abbreviated J, is a unit of energy.) Each photon of red light with wavelength 700 nm has an energy of 2.86×10^{-19} J.

Compare! In 1 second a 25-watt light bulb emits 25 J.

Try these questions: A photon has energy 4.90×10^{-19} J. Calculate its wavelength in nanometers. Referring to Figure 3-1, what is this photon's color? What is the wavelength of a photon with twice this energy? What is the energy of a green photon? (*Clue:* See Figure 3-1.)

(Answers appear at the end of the book.)

We can now explain our experiment with the iron bar at the beginning of this chapter in terms of the energy and intensity of photons. Recall from Section 3-3 that the energy carried by a photon is inversely proportional to its wavelength. In other words, long-wavelength photons, such as radio waves, carry little energy while short-wavelength photons, like X rays and gamma rays, carry much more energy. When the rod was cool, it mostly emitted lower energy infrared and radio photons; when it was hot enough to glow, it emitted mostly visible photons and more of all types of photons, which is why it felt hotter. The relationship between the energy of a photon and its wavelength is called **Planck's law** as shown in An Astronomer's Toolbox 4-2.

Together, Planck's law and Wien's law relate the temperature of an object to the energy of the photons it emits (Table 4-1). A cool object emits primarily long-wavelength photons that carry little energy, while a hot object gives off

TABLE 4-1 Some Properties of Electromagnetic Radiation

	Wavelength (nm)	Photon energy (eV)*	Blackbody temperature
Radio	$>10^7$	$<10^{-4}$	<0.03
Microwave**	10^7 to 4×10^5	10^{-4} to 3×10^{-3}	0.03 to 30
Infrared	4×10^5 to 7×10^2	3×10^{-3} to 2	30 to 4100
Visible	7×10^2 to 4×10^2	2 to 3	4100 to 7300
Ultraviolet	4×10^2 to 10^1	3 to 10^3	7300 to 3×10^6
X ray	10^1 to 10^{-2}	10^3 to 10^5	3×10^6 to 3×10^8
Gamma ray	$<10^{-2}$	$>10^5$	$>3 \times 10^8$

Note: > means greater than, < means less than.

*1 eV = 1.6×10^{-19} J.

**Microwaves, listed here separately, are often classified as radio waves or infrared radiation.

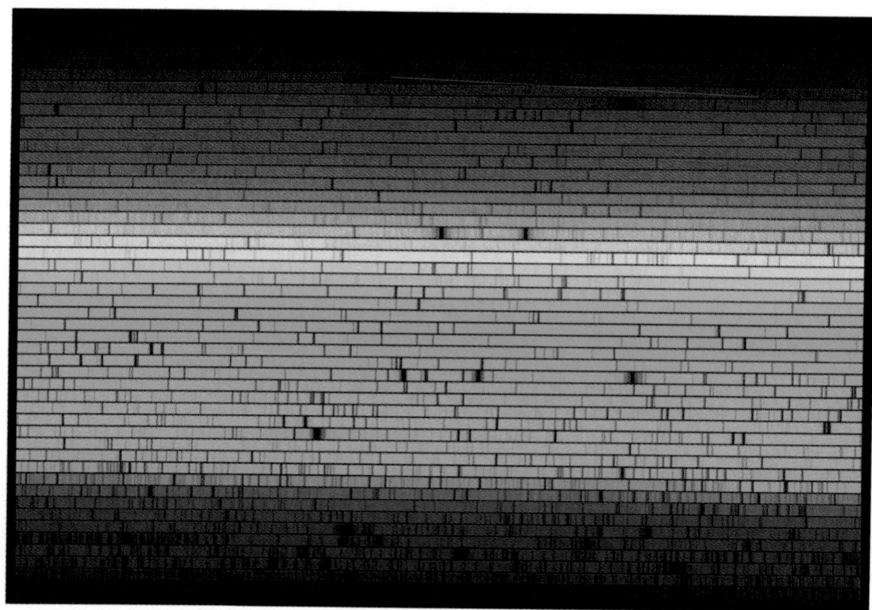

FIGURE 4-4 **Solar Spectra** Starlight passing through a prism or diffraction grating spreads into its component colors. From such spectra we can learn an incredible amount about stars, including their masses, surface temperatures, diameters, chemical compositions, rotation rates, and motions toward or away from us. This image, called a spectrogram, shows the spectrum of the Sun sliced and stacked to fit on this page. In it you can see thousands of absorption lines. As you study spectra, also note the distinct differences between the intensities of the various colors emitted by each star and by different stars. (N. A. Sharp, NOAO/AURA/NSF)

R I V U X G

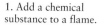

mostly short-wavelength photons that carry much more energy. In later chapters, we will find these ideas invaluable for understanding how stars of various temperatures interact with gas and dust in space.

IDENTIFYING THE ELEMENTS BY ANALYZING THEIR UNIQUE SPECTRA

In 1814, the German optician Joseph von Fraunhofer repeated Newton's classic experiment of shining a beam of sunlight through a prism (recall Figure 3-1). But Fraunhofer magnified the resulting rainbow-colored spectrum. He discovered that the solar spectrum contains hundreds of fine dark lines, which became known as **absorption lines** because the light of these colors has been absorbed by gases between the Sun and the viewer on Earth. Fraunhofer counted more than 600 such lines, and today physicists have detected more than 30,000 of them. Thousands of spectral lines are visible in the photograph of the Sun's spectrum shown in Figure 4-4.

By the mid-1800s, chemists discovered that they could produce spectral lines in the laboratory. Around 1857, the German chemist Robert Bunsen invented a special gas burner that produces a clean, constant flame. Certain chemicals are easy to identify by the distinctive colors emitted when bits of the chemical are sprinkled into the flame of a Bunsen burner.

Bunsen's colleague, Gustav Kirchhoff, suggested that light from the colored flames could best be studied by passing them through a prism. The process of collecting the spectra of an object is called *spectroscopy.* Bunsen and Kirchhoff collaborated in designing and constructing the first **spectroscope.** This device consists of a narrow slit, a prism, and several lenses that straighten the light rays and magnify the spectrum so that it can be closely examined (Figure 4-5).

The chemists discovered that the spectrum from a flame consists of a pattern of thin, bright spectral lines called **emission lines,** against a dark background (not a blackbody spectrum!). They next found that *the number of*

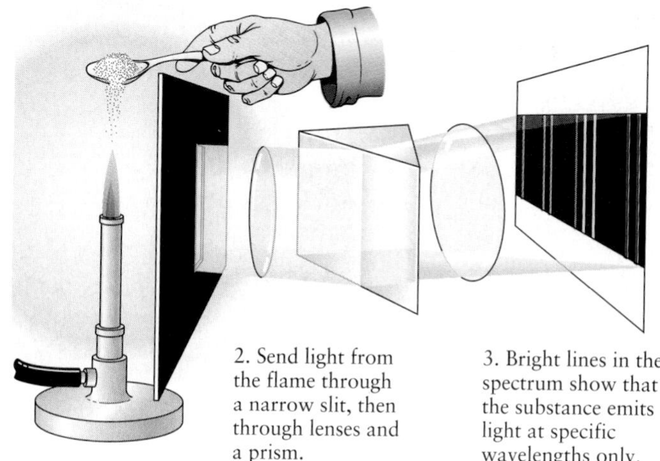
1. Add a chemical substance to a flame.
2. Send light from the flame through a narrow slit, then through lenses and a prism.
3. Bright lines in the spectrum show that the substance emits light at specific wavelengths only.

FIGURE 4-5 **Early Spectroscope** In the mid-1850s, Kirchhoff and Bunsen discovered that when a chemical substance is heated and vaporized, the resulting spectrum exhibits a series of bright spectral lines when passed through a slit, lenses, and a prism. This device is called a spectroscope. In addition, they found that each chemical element produces its own characteristic pattern of spectral lines. The lenses focus and magnify the spectrum.

lines produced and their colors are unique to the element or compound that produces them. Thus was born in 1859 the technique of **spectral analysis,** the identification of chemical substances by their spectral lines.

4-3 Each chemical element produces its own unique set of spectral lines

A chemical **element** is a fundamental substance that cannot be broken down into more basic units and still retain its properties. By the mid-1800s, chemists had already identified such familiar elements as hydrogen, oxygen, carbon, iron, gold, and silver. Spectral analysis promptly led to the discovery of additional elements, many of which are quite rare.

After Bunsen and Kirchhoff recorded the prominent spectral lines of all the known elements, they began to discover other spectral lines in mineral samples. In 1860, for example, they found a new line in the blue portion of the spectrum of mineral water. After chemically isolating the previously unknown element responsible for the line, they named it cesium (from the Latin *caesius,* meaning gray-blue). The next year a new spectral line in the red portion of the spectrum of a mineral sample led to the discovery of the element rubidium (from *rubidus,* for red).

During a solar eclipse in 1868, astronomers found a new spectral line in the light coming from the upper atmosphere of the Sun when the main body of the Sun was hidden by the Moon. This line was attributed to a new element, which was named helium (from the Greek *helios,* meaning Sun). Helium was not actually discovered on Earth until 1895, when it was identified in gases obtained from a uranium compound.

A list of the chemical elements is most conveniently displayed in the form of a **periodic table** (Figure 4-6). Each element has a unique atomic number (explained in Section 4-5), and the elements are arranged in the periodic table by their atomic numbers. With a few exceptions, this sequence also corresponds to increasing average mass of the atoms of the elements. Thus, hydrogen (symbol H) with atomic number 1 is the lightest element. Iron (Fe) has atomic number 26 and is a moderately heavy element.

All the elements in a single vertical column of the periodic table have similar chemical properties. For example, the elements in the far right column are all gases at room

Periodic Table of the Elements

1 H Hydrogen																	2 He Helium
3 Li Lithium	4 Be Beryllium											5 B Boron	6 C Carbon	7 N Nitrogen	8 O Oxygen	9 F Fluorine	10 Ne Neon
11 Na Sodium	12 Mg Magnesium											13 Al Aluminum	14 Si Silicon	15 P Phosphorus	16 S Sulfur	17 Cl Chlorine	18 Ar Argon
19 K Potassium	20 Ca Calcium	21 Sc Scandium	22 Ti Titanium	23 V Vanadium	24 Cr Chromium	25 Mn Manganese	26 Fe Iron	27 Co Cobalt	28 Ni Nickel	29 Cu Copper	30 Zn Zinc	31 Ga Gallium	32 Ge Germanium	33 As Arsenic	34 Se Selenium	35 Br Bromine	36 Kr Kryton
37 Rb Rubidium	38 Sr Strontium	39 Y Yttrium	40 Zr Zirconium	41 Nb Niobium	42 Mo Molybdenum	43 Tc Technetium	44 Ru Ruthenium	45 Rh Rhodium	46 Pd Palladium	47 Ag Silver	48 Cd Cadmium	49 In Indium	50 Sn Tin	51 Sb Antimony	52 Te Tellurium	53 I Iodine	54 Xe Xenon
55 Cs Cesium	56 Ba Barium	57 La Lanthanum	72 Hf Hafnium	73 Ta Tantaium	74 W Tungsten	75 Re Rhenium	76 Os Osmium	77 Ir Iridium	78 Pt Platinum	79 Au Gold	80 Hg Mercury	81 Tl Thallium	82 Pb Lead	83 Bi Bismuth	84 Po Polonium	85 At Astatine	86 Rn Radon
87 Fr Francium	88 Ra Radium	89 Ac Actinium	104 Rf Rutherfordium	105 Db Dubnium	106 Sg Seaborgium	107 Bh Bohrium	108 Hs Hassium	109 Mt Meitnerium	110 Ds Darmstadium	111 Rg Roentgenium	112 Uub Ununbium	113 Uut Ununtium	114 Uug Ununguadium	115 Uup Ununpentium	116 Uuh Ununhexium		

58 Ce Cerium	59 Pr Praseodymium	60 Nd Neodymium	61 Pm Promethium	62 Sm Samarium	63 Eu Europium	64 Gd Gadolinium	65 Tb Terbium	66 Dy Dysprosium	67 Ho Holmium	68 Er Erbium	69 Tm Thulium	70 Yb Ytterbium	71 Lu Lutetium
90 Th Thorium	91 Pa Protactinium	92 U Uranium	93 Np Neptunium	94 Pu Plutonium	95 Am Americium	96 Cm Curium	97 Bk Berkelium	98 Cf Californium	99 Es Einsteinium	100 Fm Fermium	101 Md Mendelevium	102 No Nobelium	103 Lr Lawrencium

FIGURE 4-6 **Periodic Table of the Elements** The periodic table is a convenient listing of the elements arranged according to their atomic numbers and chemical properties.

temperature and normal air pressure, and they rarely react chemically with other elements.

In addition to the 92 naturally occurring elements, Figure 4-6 lists the artificially produced elements. All the human-made elements are heavier than uranium (U), and all are highly radioactive, meaning that they spontaneously decay into lighter elements shortly after being created in the laboratory.

INSIGHT INTO SCIENCE

Seek Relationships Finding how properties relate often provides invaluable insights into how new things work. The periodic table, for example, enables scientists to determine the properties of similar chemical elements. Keep an eye out for other such relationships throughout this book.

During their experiments with spectra of hot sources passing through cooler gas, Bunsen and Kirchhoff saw dark *absorption lines* among the colors of the rainbow created by the hot source (as Fraunhofer had seen earlier). In other experiments, when looking just at a gas without a hot source behind it, they saw bright *emission lines* against an otherwise dark background. They discovered that *the emission lines of a particular gas occur at exactly the same wavelengths as the absorption lines of that gas.*

Because each chemical element produces its own unique pattern of spectral lines, scientists can determine the chemical composition of a remote astronomical object by identifying the lines in its spectrum. For example, Figure 4-7 shows a portion of the Sun's absorption spectrum along with the emission spectrum of an iron sample taken here on Earth. The spectral lines of iron also appear in the Sun's spectrum, so we can reasonably conclude that the Sun's atmosphere contains some vaporized iron. This gas absorbs certain wavelengths from the continuum emitted below it.

After photography was invented, scientists preferred to do spectroscopy by making a permanent photographic record of spectra. A device for photographing a spectrum is called a **spectrograph,** and this instrument is now among the astronomer's most important tools.

In its earliest form, a spectrograph attached to a telescope consisted of a slit, two lenses, and a prism arranged to focus the spectrum of an object on film, similar to the device sketched in Figure 4-5. While conceptually straightforward, this early type of spectrograph has severe drawbacks. A prism does not spread colors evenly: The blue and violet portions of the spectrum are spread out more than the red portion. In addition, because the blue and violet wavelengths must pass through more glass than the red wavelengths (see Figure 3-1), light is absorbed unevenly across the spectrum. Indeed, a glass prism is opaque to ultraviolet wavelengths.

Practical spectrographs used in research today (Figure 4-8a) separate light from objects in space into the colors of the rainbow using a **diffraction grating,** a piece of glass on which thousands of closely spaced parallel grooves are cut. Some of the finest diffraction gratings have more than 10,000 grooves per centimeter. The spacing of the grooves must be very regular. Light rays reflected or transmitted from different parts of the diffraction grating interfere with each other to produce a spectrum. Figure 4-8a shows the design of a modern diffraction grating spectrograph with a CCD (charged coupled device; see Figure 3-14) that has replaced film to record the spectra. This optical device typically mounts at the focal point of a telescope. The image of the object to be examined is focused on the slit. After the spectrum of a star, galaxy, or other object has been recorded, the CCD collects an emission spectrum from a known source, such as a gas composed of helium, neon, and argon,

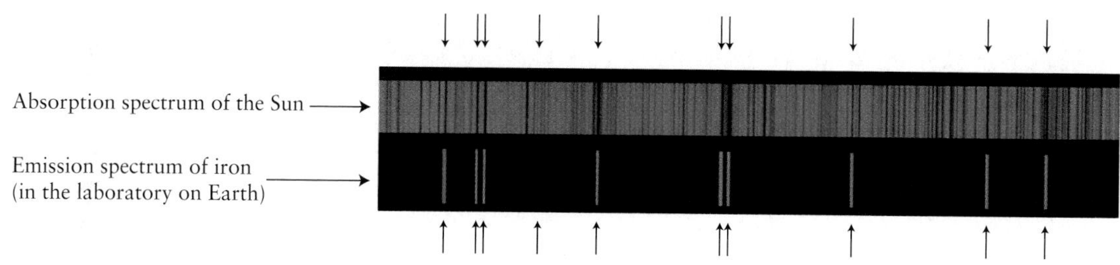
Absorption spectrum of the Sun
Emission spectrum of iron (in the laboratory on Earth)
For each emission line of iron, there is a corresponding absorption line in the solar spectrum; hence there must be iron in the Sun's atmosphere
R I V U X G

FIGURE 4-7 **Iron in the Sun's Atmosphere** The upper spectrum is a portion of the Sun's spectrum from 425 to 430 nm. Numerous dark spectral lines are visible. The lower spectrum is a corresponding portion of the spectrum of vaporized iron. Several bright spectral lines can be seen against the black background. The fact that the iron lines coincide with some of the solar lines proves that there is some iron (albeit a very tiny amount) in the Sun's atmosphere. (Carnegie Observatories)

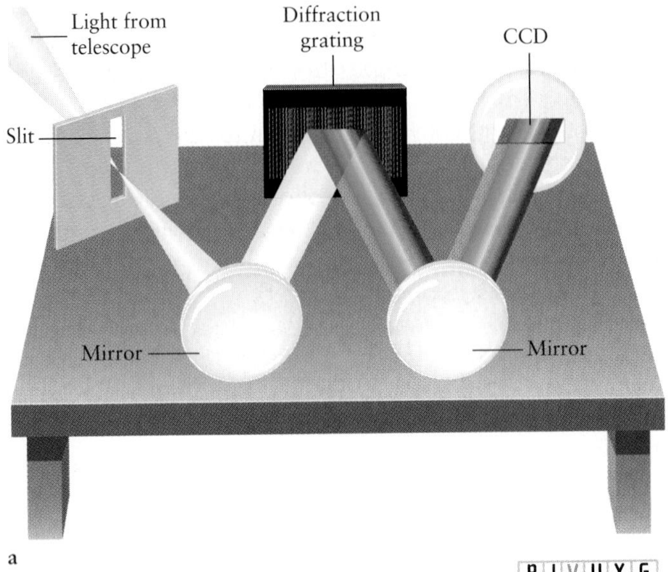

FIGURE 4-8 **A Grating Spectrograph** (a) A grating spectrograph separates light from a telescope into individual colors in order for the object's spectrum to be analyzed. (b) This peacock feather contains numerous natural diffraction gratings. The role of the parallel lines etched in a human-made diffraction grating is played by parallel rods of the protein melanin in the feathers. (b: © Paul Silverman/Fundamental Photos)

that is also focused on the slit. This "comparison spectrum" is placed next to the spectrum of the object, as in Figure 4-7. Because the wavelengths of the spectral lines in the comparison spectrum are already known from laboratory experiments, these lines can be used to identify and measure the wavelengths of the lines in the spectrum of the star or galaxy under study. Figure 4-8b shows natural diffraction gratings.

The spectral data are converted by computer to a graph that plots light intensity against wavelength. Dark lines in the rainbow-colored spectrum appear as depressions or valleys on the graph, while bright lines in the spectrum appear as peaks. For example, Figure 4-9 shows both a picture and a plot of a spectrum for hydrogen in which five dark spectral lines appear.

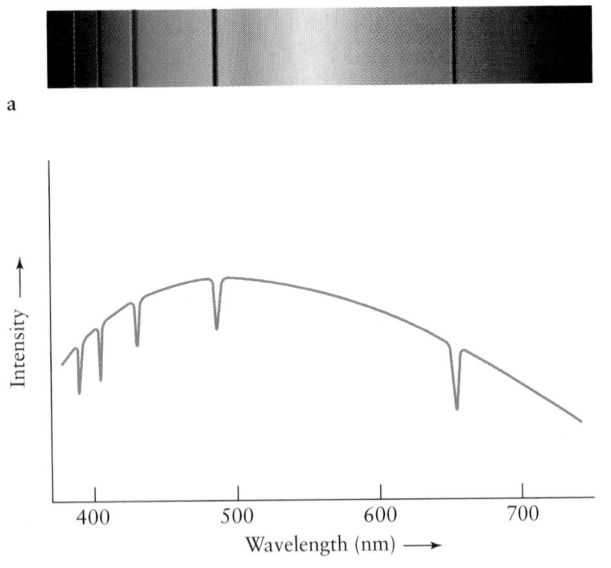

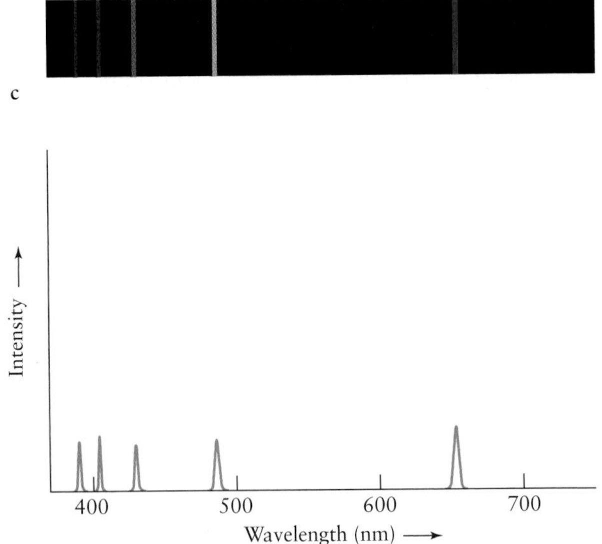

FIGURE 4-9 **Spectrum of Hydrogen Gas** (a) When a CCD is placed at the focus of a spectrograph, a rainbow-colored spectrum is recorded. (b) The spectrum is converted by computer into a graph of intensity versus wavelength. Note that the absorption lines appear as dips in the intensity-versus-wavelength curve. Conversely, when a gas emits only a few wavelengths, (c), its emission spectrum appears as a series of bright lines, which are converted into peaks (d) on a graph of intensity versus wavelength.

4-4 The brightnesses of spectral lines depend on conditions in the spectrum's source

By the early 1860s, Kirchhoff had discovered the conditions under which these different types of spectra are observed. His description is summarized today as **Kirchhoff's laws:**

Law 1 A solid, liquid, or dense gas produces a ***continuous spectrum*** *(also called a* ***continuum****)—a complete rainbow of colors without any spectral lines. This is a blackbody spectrum. The light given off by the iron rod at the beginning of this chapter and by an incandescent lightbulb are examples of continuous spectra that are bright enough for us to see.*

Law 2 A rarefied (opposite of dense) gas produces an ***emission line spectrum****—a series of bright spectral lines against a dark background. The light given off by neon lights and low pressure sodium vapor lights are examples of emission line spectra. The neon has bright red emission lines, while the low pressure sodium has bright yellow lines.*

Law 3 The light from an object with a continuous spectrum that passes through a cool gas produces an ***absorption line spectrum****—a series of dark spectral lines among the colors of the rainbow. Sunlight and the light from other stars pass through several cooler gases on their way to us. Hence, light from stars are absorption line spectra.*

In summary (Figure 4-10), a continuum is seen if there is no gas between an object and the observer, absorption lines are seen if the background object is hotter than gas between us and it, and emission lines are seen if the background is cooler than such a gas.

Consider, for example, the spectrum of the Sun. We know that the Sun's surface emits a continuous blackbody spectrum, but here on Earth many absorption lines are seen in it (see Figure 4-4). Kirchhoff's third law explains why. There must be a cooler gas between the surface of the Sun and the Earth. In fact, there are two: the Sun's lower atmosphere and the Earth's entire atmosphere. Using the catalogs of spectra of different elements, the absorption lines in the Sun's spectrum tell us the chemical composition of the Sun's lower atmosphere and of the Earth's atmosphere.

Cataloging the spectra of different elements is one thing. Understanding how these spectra arise is something else altogether, and it takes us deep into the realm of the building blocks of matter: atoms.

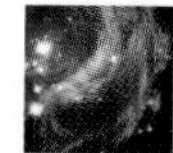

ATOMS AND SPECTRA

An atom is the smallest particle of a chemical element that still has the properties of that element. At the time of Kirchhoff's discoveries, scientists knew that all matter is composed of atoms, but they did not know how atoms were structured. Furthermore, scientists saw that atoms of a gas somehow extract light of specific wavelengths from white light that passes through the gas, leaving dark absorption lines, and they perceived that the atoms then radiate light of precisely the same wavelengths—the bright emission lines (see Figure 4-7). But traditional theories of electromagnetism could not explain this phenomenon. The answer came early in the twentieth century with the development of nuclear physics and quantum mechanics.

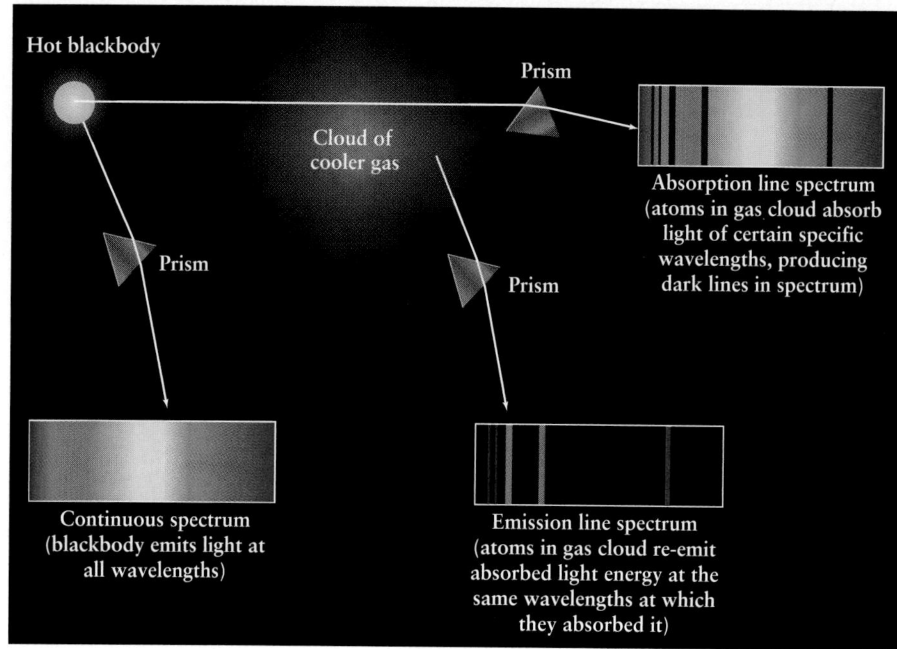

FIGURE 4-10 Continuous, Absorption Line, and Emission Line Spectra This schematic diagram summarizes how different types of spectra are produced. The prisms are added for conceptual clarity, but in real telescopes diffraction gratings are used to separate the colors. A hot, glowing object emits a continuous spectrum. If this source of light is viewed through a cool gas, dark absorption lines appear in the resulting spectrum. When the same gas is viewed against a cold, dark background, its spectrum consists of just bright emission lines. The animation link shows the details of the activity drawn here.

4-5 An atom consists of a small, dense nucleus surrounded by electrons

The internal structure of atoms first came into focus in 1908 when Nobel laureate Ernest Rutherford and his colleagues at the University of Manchester in England were investigating the recently discovered phenomenon of radioactivity. Over time, a **radioactive** element naturally and spontaneously transforms into another element by emitting particles. Certain radioactive elements, such as uranium and radium, were known to emit such particles with considerable speed. It seemed plausible that a beam of these high-speed particles would penetrate a thin sheet of gold. Rutherford and his associates found that almost all the particles did pass through the gold sheet with little or no deflection. To their surprise, however, an occasional particle bounced right back. It must have struck something very dense indeed.

Within a few decades, the nature of atoms fell into place. That dense "something" is now called the **nucleus** of the atom, and it consists of elementary particles called *protons* and *neutrons*. Surrounding the nucleus, one or more electrons orbit. The protons and electrons have a property called *electric charge*. All protons have the exact same positive charge, while all electrons have a negative charge equal in strength to the proton's charge. The terms "positive" and "negative" are arbitrary and just indicate that they are opposite to each other. Particles with opposite charges attract each other. Therefore, protons attract electrons and it is this attraction that keeps electrons in orbit around nuclei.

Particles with the same type of charge, such as a pair of protons or a pair of electrons, repel each other. For atoms that have more than one proton in their nuclei (and many do), the protons are pushing away from each other. In order for nuclei with more than one proton to exist, there must be an attractive force stronger than the repulsion of protons to keep them glued together. The electrically neutral neutrons in the nucleus help provide that attractive force, called the **strong nuclear force.**

For completeness, it is worth knowing that there are four fundamental forces in nature: gravitation, electromagnetism, the strong nuclear force, and the weak nuclear force. Their properties are summarized in Table 4-2. The weak nuclear force is involved in some radioactive decays, such as when a neutron transforms into a proton. We will explore it more in our study of the evolution of the whole universe, in Chapter 18.

The number of protons in an atom's nucleus determines what *element* that atom is. The number of protons the nucleus contains is the element's **atomic number.** All hydrogen nuclei have one proton, all helium nuclei have two, and so forth. There are 92 different types of elements that form naturally. Uranium is the most massive, with 92 protons in its nucleus (see Figure 4-6).

In contrast, the nuclei of most elements can have different numbers of neutrons. For example, hydrogen always has one proton, but it can have either zero, one, or two neutrons; oxygen, with an atomic number of 8, always has eight protons, but it may have eight, nine, or ten neutrons. Each different combination of protons and neutrons is called an **isotope.** There are three isotopes of hydrogen and three isotopes of oxygen. Hydrogen with no neutrons is the most common hydrogen isotope, while oxygen with eight neutrons is by far the most abundant isotope of oxygen.

Some isotopes are stable, meaning that the number of protons and neutrons in their nuclei do not change. However, many elements have isotopes that are unstable and come apart spontaneously. These are the radioactive isotopes. For example, carbon with six neutrons, ^{12}C, is stable, while carbon with eight neutrons, ^{14}C, is unstable. ^{14}C decays into nitrogen with seven neutrons, ^{14}N. To learn how radioactive decay is used to determine the ages of different objects, see An Astronomer's Toolbox 4-3. Using radioactive age dating techniques on rocks from the Moon, astronomers are able to determine roughly when it formed (4.5 billion years ago). Applying the same technique to space debris found on Earth (meteorites), we can determine that the solar system formed some 4.62 billion years ago.

TABLE 4-2 The Four Fundamental Forces of Nature

Name	Strength (compared to the strong force)	Range of Effect (from each object)
Strong force	1	Inside atomic nuclei
Electromagnetic force	$1/137$	Throughout the universe
Weak force	10^{-5}	Inside atomic nuclei
Gravitational force	6×10^{-39}	Throughout the universe

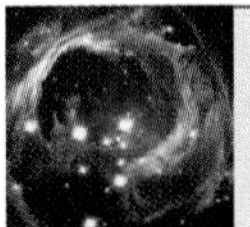

AN ASTRONOMER'S TOOLBOX 4-3
Radioactivity and the Ages of Objects

The isotopes of many elements are radioactive, meaning that the elements spontaneously transform into other elements. Each radioactive isotope has a distinctive *half-life,* the time it takes half of the initial concentration of the isotope to transform into another element. After two half-lives, a radioactive isotope is reduced to ½ × ½ or ¼ of its initial concentration (see the accompanying figure). Among the most important radioactive elements for determining the age of objects in astronomy is the isotope of uranium with 146 neutrons, ^{238}U. The half-life of ^{238}U decaying into lead is 4.5 billion years.

To determine the time since an object, such as a piece of space debris discovered on Earth, solidified, scientists estimate how much lead it had when it formed. Then they measure the amounts of uranium and lead it contains now. Subtracting the amount of original lead, they use the amount of uranium and lead with the graph to determine the object's age.

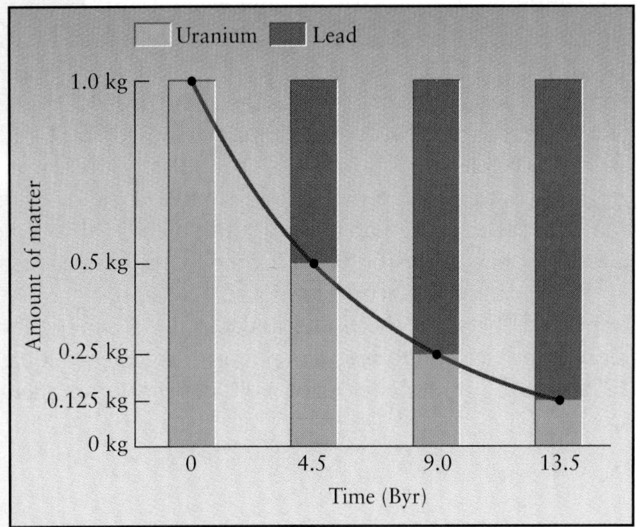

The Transformation of Uranium into Lead This figure shows the rate that 1 kg of uranium decays into lead, as described in the text. It contains ⅛ kg of uranium after 13.5 billion years.

Example: Suppose a piece of space debris discovered on Earth was determined to have equal amounts of lead and uranium. How long ago did this debris form? Assuming that it originally had no lead, we see from the chart that a 1-to-1 mix of lead to uranium occurs 4.5 billion years after the object formed. This is one half-life of uranium.

Compare! This process works with any radioactive isotope. However, some isotopes have such short half-lives that they are not useful in astronomy. For example, the well-known carbon dating used to determine the ages of ancient artifacts on Earth is of little use in astronomy because ^{14}C has a half-life of only 5730 years. Because so much of it has decayed away by then, carbon dating is useful only for time intervals shorter than 100,000 years, usually a period over which little of astronomical importance occurs.

Try these questions: What fraction of a kilogram of radioactive material remains after 3 half-lives have passed? Using the accompanying figure, estimate how much uranium will remain after 6¾ billion years. Approximately how many half-lives of ^{14}C pass in 100,000 years?

(Answers appear at the end of the book.)

Normally, the number of electrons orbiting an atom is equal to the number of protons in the nucleus, thus making the atom electrically neutral. Astronomers denote neutral atoms by writing the atomic symbol followed by the Roman numeral I. For example, neutral hydrogen is written as H I and neutral iron is Fe I.

When an atom contains a different number of electrons than protons, the atom is called an **ion.** The process of creating an ion is called **ionization.** Ions are denoted by the atomic symbol followed by a roman numeral that is one greater than the number of missing electrons. Positively ionized hydrogen (missing its one electron) is denoted H II, while positively ionized iron with seven electrons missing is denoted Fe VIII. It is worth noting that negative ions also exist, where nuclei have more electrons orbiting than they have protons.

Atoms can share electrons and, by doing so, become bound together. Such groups are called **molecules.** They are the essential building blocks of all complex structures, including life.

4-6 Spectra occur because electrons absorb and emit photons with only certain wavelengths

Protons and neutrons have nearly the same mass, which is about 2000 times greater than the mass of an electron. Therefore, at least 99.95% of the mass of an atom

3

is concentrated in the nucleus, the rest residing in its electrons. The electron orbits are far from the nucleus, typically 10,000 times farther away than the radius of the nucleus. This is why you may have heard the statement that matter is mostly empty space.

Be careful not to imagine electrons as miniature planets orbiting a miniature "Sun." Elementary particles are not solid little miniature planets or marbles. Rather, like photons, they all have both wave and particle properties. The science that accurately describes their complex behavior is called **quantum mechanics**.

Quantum mechanics explains that the wave properties of electrons only permit them to exist in certain *allowed orbits* around their nuclei, except when they are making a **transition** from one allowed orbit to another. These orbital conditions are completely unlike planets, which can exist at any distance (in any orbit) around the Sun. Each allowed electron orbit has a well-defined energy associated with it, and every different type of atom and molecule has a unique set of allowed orbits. These orbits and the transitions between them are the key to understanding the spectra we have been discussing.

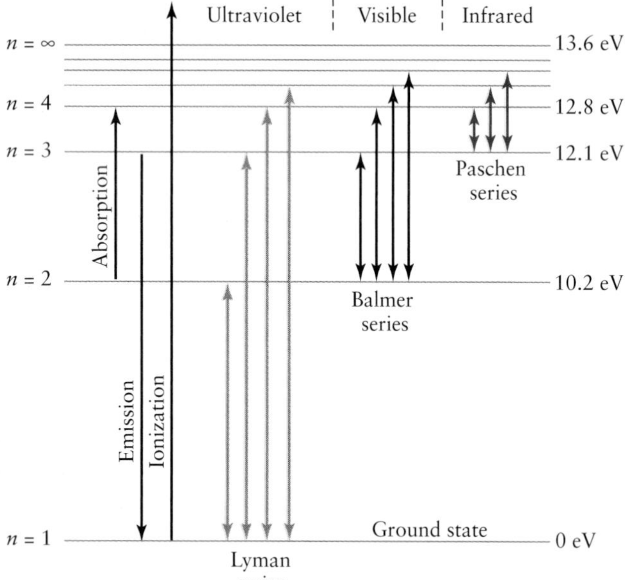

FIGURE 4-11 **Energy Level Diagram of Hydrogen** The activity of a hydrogen atom's electron is conveniently displayed in a diagram showing some of the energy levels at which it can exist. A variety of electron jumps, or transitions, are also shown, including those that produce the most prominent lines in the hydrogen spectrum. For another perspective on the same information, you can try the linked Interactive Exercise.

Consider an atom of the simplest hydrogen isotope, which contains just a single proton in its nucleus orbited by one electron. (This discussion generalizes directly to all the other elements and isotopes, made more complex only because they have more than one electron in orbit.) Figure 4-11 shows this hydrogen atom's lowest energy levels.

Normally, the electron is in the lowest energy allowed orbit or energy level, commonly called the **ground state**; this is labeled $n = 1$ in Figure 4-11. Each allowed orbit with successively higher energy is labeled $n = 2, 3, 4$, and so on. When an electron is in an orbit with more energy than the lowest energy state available to it, it is said to be in an **excited state**.

Electrons change orbits by absorbing or emitting photons. However, electrons cannot absorb just any photon that they encounter. Electrons can only absorb those photons with energies exactly enough to boost them up to a higher-energy allowed orbit. That is, the photons that are absorbed are those with energies equal to the difference between the energies of two allowed orbits. All photons that do not satisfy this condition pass straight through the atom. If the electron starts in its ground state, then it must get exactly the energy necessary to move it to an excited state. If it already is in an excited state, then by absorbing a photon, it must transition to a higher-energy excited state.

For example, referring to Figure 4-11 and Figure 4-12, transitions between the $n = 2$ and the $n = 3$ energy levels require the electron to absorb a photon with energy equal to 12.1 − 10.2 = 1.9 eV (electron volt, a measure of energy).

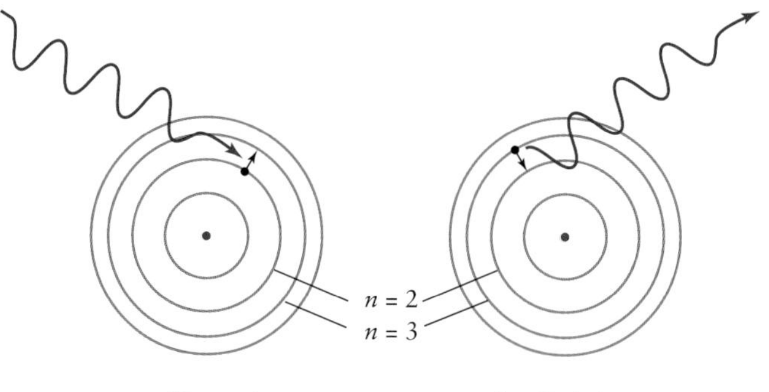

FIGURE 4-12 **The Absorption and Emission of an H_α Photon** This schematic diagram of hydrogen's four lowest allowed orbits shows what happens when a hydrogen atom absorbs or emits an H_α photon, which is red and has a wavelength of 656.28 nm. (a) A photon is absorbed by the electron, causing the electron to transition from orbit $n = 2$ up to orbit $n = 3$. (b) A photon is emitted as the electron makes a transition from orbit $n = 3$ down to orbit $n = 2$.

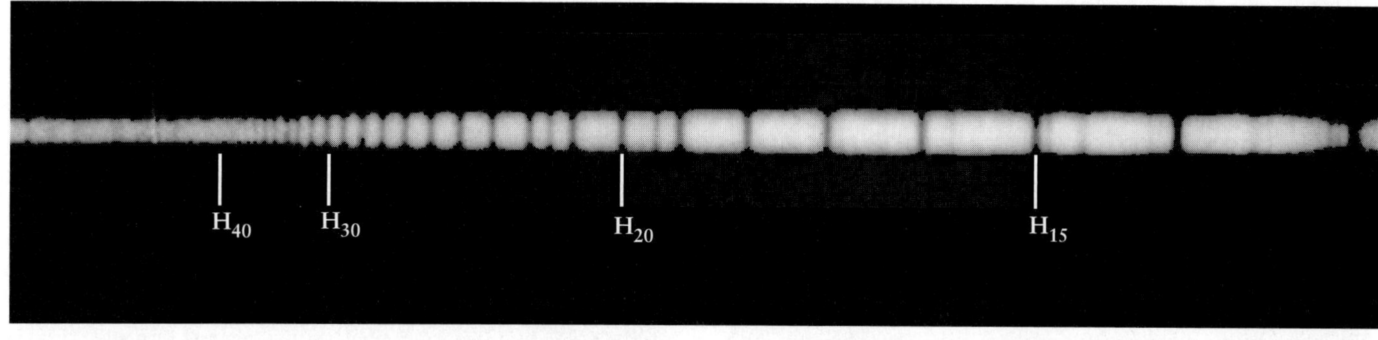

FIGURE 4-13 **Balmer Lines in the Spectrum of a Star** This portion of the spectrum of a star called HD 193182 shows more than two dozen Balmer lines. The series converges at 364.56 nm, just to the left of H_{40}. This star's spectrum also contains the first 12 Balmer lines (H_α through H_{12}), but they are not visible in this particular spectrogram. (Carnegie Observatories)

Recall from An Astronomer's Toolbox 4-2 that a photon's energy corresponds to a certain wavelength. In this case, the photon absorbed by the electron has a wavelength of 656.3 nm (Figure 4-12a); it is a red photon (Figure 3-4).

While some absorption occurs at optical wavelengths, it also happens in many other parts of the electromagnetic spectrum. Indeed, most of the transitions in hydrogen, shown in Figure 4-11 as Lyman series (up from and down to $n = 1$) and Paschen series (up from and down to $n = 3$), are nonvisible photons. Only a few of the Balmer series (up from and down to $n = 2$) transitions are visible. We will refer to Balmer transitions when we study stars. They are named after Johann Balmer, who first calculated their wavelengths. The longest wavelength Balmer line is called H_α, the second H_β, the third H_γ, and so forth, ending with the shortest wavelength Balmer line, H_∞. (The first dozen lines of the series have Greek-letter subscripts; the remainder are identified by numerical subscripts.) Part of the spectrum of a star having Balmer absorption lines H_{13} through H_{40} is shown in Figure 4-13.

Absorption lines are therefore caused by photons taken out of the stream of light by electrons, which move into higher energy allowed orbits. An observer on the right in Figure 4-10 looking at the hot blackbody through the cloud of cooler gas sees dark absorption lines where photons have been absorbed by the cooler gas.

Electrons in excited states are unstable. Sometimes they lose energy by bumping into other particles and suddenly having either too much or too little energy than is necessary to be in that state. More often they are nudged out of orbit by collisions with **virtual particles**. This is where things get really weird. The laws of nature allow pairs of particles to spontaneously appear provided that they are identical, except for having opposite electric charges, and that the pairs of particles annihilate each other and disappear within a very short period of time, typically 10^{-21} seconds. We know that virtual particles actually flash into existence because some have been made into real particles by permanently separating them as a result of collisions with high-speed particles in particle accelerators. Billions and billions of these pairs of virtual particles are seething in and out of existence in your body every second.

As a hydrogen atom's electron orbits in an excited state, it is moving so fast that it encounters virtual particles before they disappear. These push the electron out of its allowed orbit. Such events typically occur within about 10^{-8} seconds after the electron arrives in the excited state. When this happens, the electron is forced to descend to a lower-energy allowed state, either the ground state or some allowed state between where it starts and the ground state. In either case, to make this transition down, the electron must *emit* a photon with energy once again equal to the difference between the energies of the starting and final allowed orbits (see Figure 4-12b).

The emitted photons have the same set of wavelengths as the absorbed photons. Furthermore, the emitted photons are sent out in all directions, causing the gas to glow. The color of the gas depends on the atoms and molecules in it. Hydrogen-rich gas clouds glow red because of the H_α emission (Figure 4-14a), while oxygen emits many green photons, so oxygen-rich gas clouds glow green (Figure 4-14b).

If any electron in a hydrogen atom encounters a photon with more than 13.6 eV, that photon will be absorbed and will knock the electron completely out of orbit and away from the atom. This process is called *photoionization*. Each type of atom and molecule has a different photoionization energy above which all electrons are kicked out of orbit. Photoionization occurs in stars and interstellar nebulas (Figure 4-14b). Indeed, many spectral lines from stars and nebulas are those of ionized atoms that retain at least one electron.

a

R I V U X G

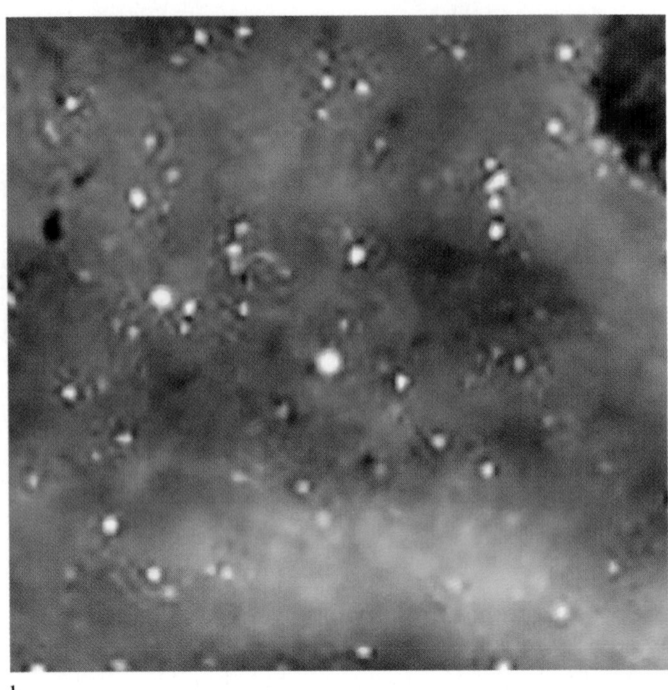

b

R I V U X G

FIGURE 4-14 **Emission Spectra from Interstellar Gas Clouds** **(a)** Stars in this interstellar gas cloud NGC 2363 in the constellation Camelopardus (the Giraffe) emit continuous spectra. Electrons in the cloud's hydrogen gas absorb and reemit the red light from these stars. NGC 2363 is located some 10 million light-years away. **(b)** Part of the Rosette Nebula (NGC 2237), an interstellar gas cloud in the constellation Monoceros (the Unicorn). The green glow is generated by doubly ionized oxygen atoms (oxygen atoms missing two electrons) in the cloud that are emitting 501 nm photons. The Rosette is 3000 ly away. (a: L. Drissen, J.-R. Roy, and C. Robert/Département de Physique and Observatoire du Mont Mégantic, Université Laval; and NASA; b: T. A. Rector, B. Wolpa, M. Hanna, AURA/NOAO/NSF)

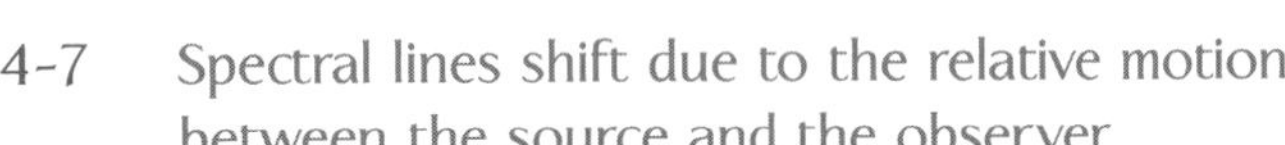

4-7 Spectral lines shift due to the relative motion between the source and the observer

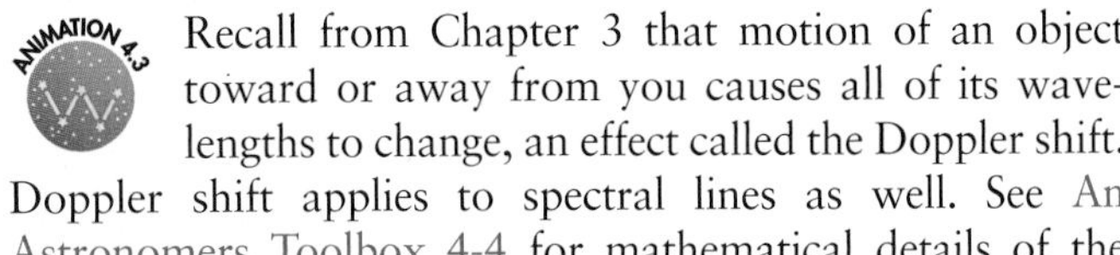

ANIMATION 4.3

Recall from Chapter 3 that motion of an object toward or away from you causes all of its wavelengths to change, an effect called the Doppler shift. Doppler shift applies to spectral lines as well. See An Astronomers Toolbox 4-4 for mathematical details of the Doppler shift.

The speed of an object toward or away from you is called the **radial velocity,** because the motion is along our line of sight, or along the "radius" drawn from Earth to the star. Of course, the star or other object you are observing may also have motion perpendicular to our line of sight, across the celestial sphere. This **proper motion** (Figure 4-15) does not affect the perceived wavelength and cannot be determined by Doppler shift. Proper motion is determined by measuring a star's motion relative to background stars. The star with the greatest proper motion as seen from Earth is Barnard's star (Figure 4-16).

Proper motions are so small that they can be measured only for relatively nearby stars in our Galaxy. However, the radial velocity of virtually every object in space can be

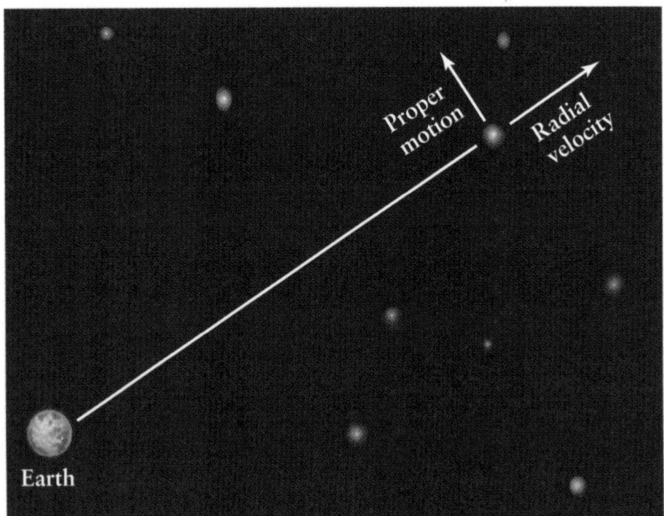

ANIM 4.3

FIGURE 4-15 **Radial and Proper Motions of a Star** The speed of a star toward or away from the Earth is its radial velocity. This motion creates a Doppler shift in the star's spectrum. The motion of the star across the sky, perpendicular to our line of sight, is called proper motion. Proper motion does not affect the star's spectrum.

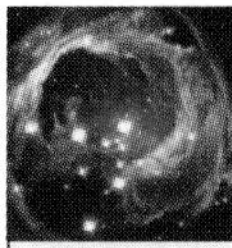

AN ASTRONOMER'S TOOLBOX 4-4
The Doppler Shift

Suppose that λ_0 is the wavelength of a spectral line from a stationary source. This is the wavelength that a reference book would list or a laboratory experiment would yield. If the source is moving, this particular spectral line is shifted to a different wavelength, λ. The size of the wavelength shift is usually written as $\Delta\lambda$ (delta lambda), where $\Delta\lambda = \lambda - \lambda_0$. This $\Delta\lambda$ is the difference between the wavelength that you actually observe in the spectrum of a star or galaxy and the wavelength listed in reference books.

Doppler proved that the wavelength shift is governed by the simple equation

$$\frac{\Delta\lambda}{\lambda_0} = \frac{v}{c}$$

where v is the speed of the source measured along the line of sight between the source and the observer. As usual, c is the speed of light (3×10^5 km/s).

Example: The spectral lines of hydrogen appear in the spectrum of the bright star Vega as shown in the figure below. The hydrogen line H_α has a normal wavelength of 656.285 nm, but in Vega's spectrum the line is located at 656.255 nm. The wavelength shift is –0.030 nm, so the star is approaching us with a speed of –14 km/s. The minus sign indicates that the star is moving toward us.

Try these questions: How fast would a star be moving toward us if all its wavelengths shifted to nine-tenths of their rest wavelengths? If a star is moving away from us at 100 km/s, what is the change, $\Delta\lambda/\lambda_0$? If a star is moving toward us at 0.1 c, what is the change, $\Delta\lambda/\lambda_0$, in its wavelengths?

(Answers appear at the end of the book.)

The Spectrum of Vega Several Balmer lines are seen in this photograph of the spectrum of Vega, the brightest star in the constellation Lyra (the Lyre). All the spectral lines are shifted equally and very slightly toward the blue side of the spectrum, indicating that Vega is approaching us. (NOAO)

determined. For example, Doppler shift measurements of the spectra of hot gases on the Sun's surface reveal that they rise and fall. Doppler shift measurements of stars in double star systems give crucial data about the speeds of the stars orbiting around each other. Doppler measurements of planets, the Sun, and other stars reveal that many of these bodies are rotating. This is determined when the Doppler shift shows that half of a surface is heading toward us, while the other half is simultaneously moving away from us. This can only happen if the bodies are rotating. Furthermore, Doppler measurements reveal that all of the very distant galaxies are moving away from us, meaning that the universe is expanding. The spectra of distant galaxies enable us

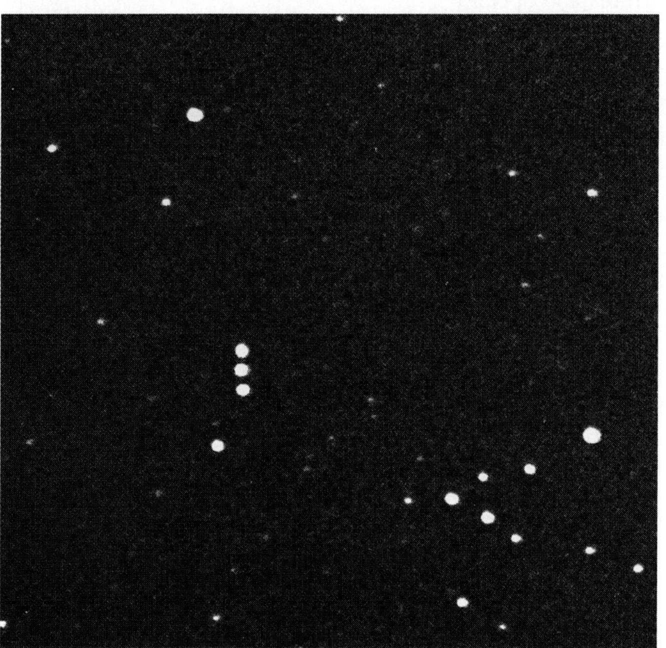

R I V U X G

WEB LINK 4.3 ◀ **FIGURE 4-16 Barnard's Star** This is a series of three superimposed photographs, taken over a four-year period. The line of three dots shows the proper motion of Barnard's star during that time. In addition to having the largest known proper motion (10.3″ per year), Barnard's star is one of the closest stars to Earth. (John Sanford/Science Photo Library/Photo Researchers) to determine the rate of that expansion. In later chapters, we will refer to the Doppler shift whenever we need to convert an observed wavelength shift into a speed toward or away from us.

INSIGHT INTO SCIENCE

Remote Science Astronomical objects are so remote and their activities are often so complex that astrophysicists must use a tremendous amount of physics to interpret observations. For example, a single spectrum can contain information about stars, extrasolar planets, interstellar gas, the Earth's motion and atmosphere, and the performance of the observing telescope and the equipment attached to it. All of these factors must be understood theoretically and accounted for.

With the knowledge of spectroscopy, atomic and nuclear physics, and the Doppler shift, we can now list many of the vital properties of matter in space that are available to us. These include the chemical compositions of stars, interstellar gas clouds, and other objects; the temperatures of these objects; the rotation of bodies in space; the motion of objects toward or away from the Earth; the presence of planets and dim companion stars; and the rate at which the universe is expanding. Knowing these properties, astrophysicists have developed models that predict the ages, masses, distances, internal activity, and companion objects of stars, among other things.

Modern physics was born when Newton set out to understand the motions of the planets. Two and a half centuries later, scientists, including Maxwell, Planck, Einstein, Rutherford, and Doppler, among many others, discovered the basic properties of electromagnetic radiation and the structures of atoms. As we will see in the following chapters, the fruits of their labors have important implications for astronomy even today.

4-8 Frontiers yet to be discovered

Spectra provide much of what we know about the cosmos. Spectra of objects in space continue to yield new insights into the chemical composition of the planets, moons, local space debris, stars, interstellar gas and dust, and galaxies, among other things. Furthermore, as we obtain more accurate Doppler measurements of objects in our Galaxy, in other nearby galaxies, and, indeed, of entire galaxies, we are better able to determine the rate at which the objects in the universe are moving relative to each other. The more we understand of the chemistries and motions of objects in space, the more accurately we will be able to explain the evolution of the solar system, stars, galaxies, and the universe.

Summary of Key Ideas

• By studying the wavelengths of electromagnetic radiation emitted and absorbed by an astronomical object, astronomers can learn about the object's temperature, chemical composition, companion objects, and movement through space.

Blackbody Radiation

• A blackbody is a hypothetical object that perfectly absorbs electromagnetic radiation at all wavelengths. The radiation that it emits depends only on its temperature. Stars closely approximate blackbodies.

• Wien's law states that the dominant wavelength of radiation emitted by a blackbody is inversely proportional to its temperature. The intensities of radiation emitted at various wavelengths by a blackbody at a given temperature are shown as a blackbody curve.

• The Stefan-Boltzmann law relates the temperature of a blackbody to the rate at which it radiates energy.

Discovering Spectra

• Spectroscopy—the study of electromagnetic spectra—provides important information about the chemical composition of remote astronomical objects.

• Kirchhoff's three laws of spectral analysis describe the conditions under which absorption lines, emission lines, and a continuous spectrum can be observed. Spectral lines serve as distinctive "fingerprints" for the chemical elements and chemical compounds comprising a light source.

Atoms and Spectra

• An atom consists of a small, dense nucleus (composed of protons and neutrons) surrounded by electrons. Different elements have different numbers of protons, different isotopes have different numbers of neutrons.

• Quantum mechanics describes the behavior of particles and shows that electrons can only be in certain allowed orbits around the nucleus.

• The spectral lines of a particular element correspond to the various electron transitions between allowed orbits of that element with different energy levels. When an electron shifts from one energy level to another, a photon of the appropriate energy (and hence a specific wavelength) is absorbed or emitted by the atom.

• The spectrum of hydrogen at visible wavelengths consists of the Balmer series, which arises from electron transitions between the second energy level of the hydrogen atom and higher levels.

• Every different element, isotope, and molecule has a different set of spectral lines.

• When an atom loses or gains one or more electrons it is said to be charged. One way for it to lose an electron is for the electron to absorb an energetic photon and thereby fly free from its atom.

• The equation describing the Doppler effect states that the size of a wavelength shift is proportional to the radial velocity between the light source and the observer.

WHAT DID YOU THINK?

1 *How hot is a "red hot" object?* Of all objects that glow visibly from heat stored or generated inside them, those that glow red are the coolest. Those that glow violet are the hottest.

2 *What color is the Sun?* The Sun emits all wavelengths of electromagnetic radiation. The colors it emits most intensely are in the blue-green part of the spectrum. Since the human eye is less sensitive to blue-green than to yellow and the Earth's atmosphere scatters blue-green wavelengths more readily than longer wavelengths, we see the Sun as yellow.

3 *How can we determine the age of space debris found on Earth?* We measure how much long-lived radioactive elements, such as ^{238}U, have decayed in the object. Carbon dating is only reliable for things that formed within the past 100,000 years. It cannot be used for determining the age of debris found on Earth, which was all formed more than 4.5 billion years ago.

Key Words

absorption line, 101
absorption line spectrum, 105
atomic number, 106
blackbody, 96
blackbody curve, 96
continuous spectrum (continuum), 105
diffraction grating, 103
element, 102
emission line, 101
emission line spectrum, 105
energy flux, 98
excited state, 108
ground state, 108
ion, 107
ionization, 107
isotope, 106
Kirchhoff's laws, 105
luminosity, 98
molecule, 107
nucleus (of an atom), 106
periodic table, 102
Planck's law, 100
proper motion, 110
quantum mechanics, 108
radial velocity, 110
radioactive, 106
spectral analysis, 102
spectrograph, 103
spectroscope, 101
Stefan-Boltzmann law, 98
strong nuclear force, 106
transition (of an electron), 108
virtual particles, 109
Wien's law, 97

Review Questions

1. What is a blackbody? What does it mean to say that a star appears almost like a blackbody? If stars appear to be like blackbodies, why are they not black?

2. What is Wien's law? How could you use it to determine the temperature of a star's surface?

3. What is the Stefan-Boltzmann law? How do astronomers use it?

4. Using Wien's law and the Stefan-Boltzmann law, explain the changes in color and intensity that are observed as the temperature of a hot, glowing object increases.

5. What color will an interstellar gas cloud composed of hydrogen glow and why?

6. What is an element? List the names of five different elements and briefly explain what makes them different from each other.

7. How are the three isotopes of hydrogen different from each other?

8. Explain how the spectrum of hydrogen is related to the structure of the hydrogen atom.

9. Why do different elements have different patterns of lines in their spectra?

10. Explain why the Doppler shift tells us only about the motion directly along the line of sight between a light source and an observer and not about motion across the celestial sphere.

Advanced Questions

The answers to all computational problems, which are preceded by an asterisk (*), appear at the end of the book.

***11.** Approximately how many times around the world could a beam of light travel in 1 second?

***12.** The bright star Regulus in the constellation of Leo (the Lion) has a surface temperature of 12,200 K. Approximately what is the dominant wavelength (λ_{max}) of the light it emits?

***13.** The bright star Procyon in the constellation of Canis Minor (the Little Dog) emits the greatest intensity of radiation at a wavelength λ_{max} = 445 nm. Approximately what is the surface temperature of the star in Kelvins?

***14.** The wavelength of H_β in the spectrum of the star Megrez in the Big Dipper is 486.112 nm. Laboratory measurements demonstrate that the normal wavelength of this spectral line is 486.133 nm. Is the star coming toward us or moving away from us? At what speed?

*15. In the spectrum of the bright star Rigel, H_α has a wavelength of 656.331 nm. Is the star coming toward us or moving away from us? How fast?

*16. Imagine driving down a street toward a traffic light. How fast would you have to go so that the red light (700 nm) would appear green (500 nm)?

Discussion Questions

17. Compare the technique of identifying chemicals by their spectral line patterns with that of identifying people by their fingerprints.

18. Suppose you look up at the night sky and observe some of the brightest stars with your naked eye. Is there any way of telling which stars are hotter and which are cooler? Explain.

19. How can we exclude the Earth's atmosphere as the source of the iron absorption lines in Figure 4-7?

What If . . .

*20. The Sun were twice its actual diameter, but still had the same surface temperature? At what wavelength would that new Sun emit its radiation most intensely? How many times brighter than the present Sun would the new Sun be? How might things on Earth be different under the bigger Sun?

21. All the nearby stars were observed to have redshifted spectra? What conclusions could we draw? What other measurement would be useful to have for each star and why?

22. No stars had any Doppler shift? What would that say about stellar motions relative to the solar system? Is that possible? *Clue:* Consider Newton's law of gravitation from Chapter 3.

23. The spectrum of the Sun (Figure 4-4) had several absorption lines missing when taken from above the Earth's atmosphere compared to its spectrum taken at the Earth's surface?

Web Questions

*24. **Measuring Stellar Temperatures** Access the Active Integrated Media Module "Blackbody Curves" in Chapter 4 of the *Discovering the Universe* Web site. (a) Use the module to determine the range of temperatures over which a star's peak wavelength is in the visible spectrum. (b) Determine if any of the following stars have a peak wavelength in the visible spectrum: Rigel, T = 14,000 K; Deneb, T = 9500 K; Arcturus, T = 4500 K; Vega, T = 11,500 K; Betelgeuse, T = 3100 K.

Observing Projects

25. As soon as weather conditions permit, observe a rainbow. Confirm that the colors are in the same order as shown in Figure 3-1a.

26. Obtain a glass prism or a diffraction grating, available from science museum stores, catalogs, or your physics department. Look through the prism or grating at various light sources, such as an ordinary incandescent lightbulb, a neon sign, and a mercury vapor street lamp. *Do not look at the Sun! Looking directly at the Sun causes blindness.* Do you have any trouble seeing spectra? What do you have to do to see a spectrum? Describe the differences in the spectra of the various light sources you observed.

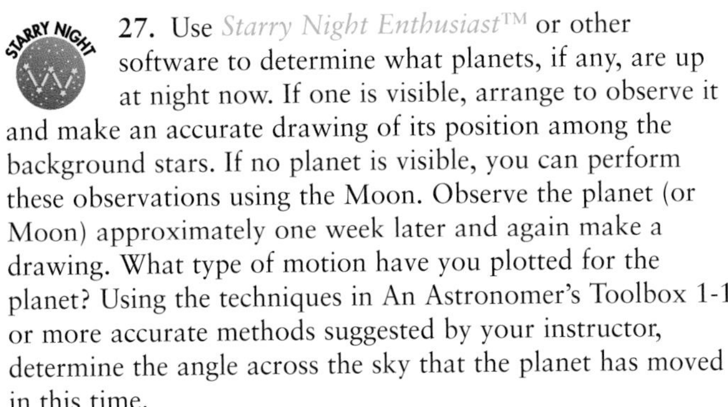

27. Use *Starry Night Enthusiast*™ or other software to determine what planets, if any, are up at night now. If one is visible, arrange to observe it and make an accurate drawing of its position among the background stars. If no planet is visible, you can perform these observations using the Moon. Observe the planet (or Moon) approximately one week later and again make a drawing. What type of motion have you plotted for the planet? Using the techniques in An Astronomer's Toolbox 1-1 or more accurate methods suggested by your instructor, determine the angle across the sky that the planet has moved in this time.

28. Use *Starry Night Enthusiast*™ to examine three distant celestial objects. First view the entire sky *Favourites/Guides/Atlas*. Then search for objects (i), (ii), (iii) listed below (*Find/object name*). Put the cursor on the object and magnify it. Based on its appearance, state whether it is likely to have a continuum spectrum, an absorption line spectrum, or an emission line spectrum, and explain your reasoning. (i) Trifid Nebula, (ii) Ring nebula, (iii) Lagoon Nebula. Your teacher may assign additional objects to find.

UNDERSTANDING THE SOLAR SYSTEM

PART II

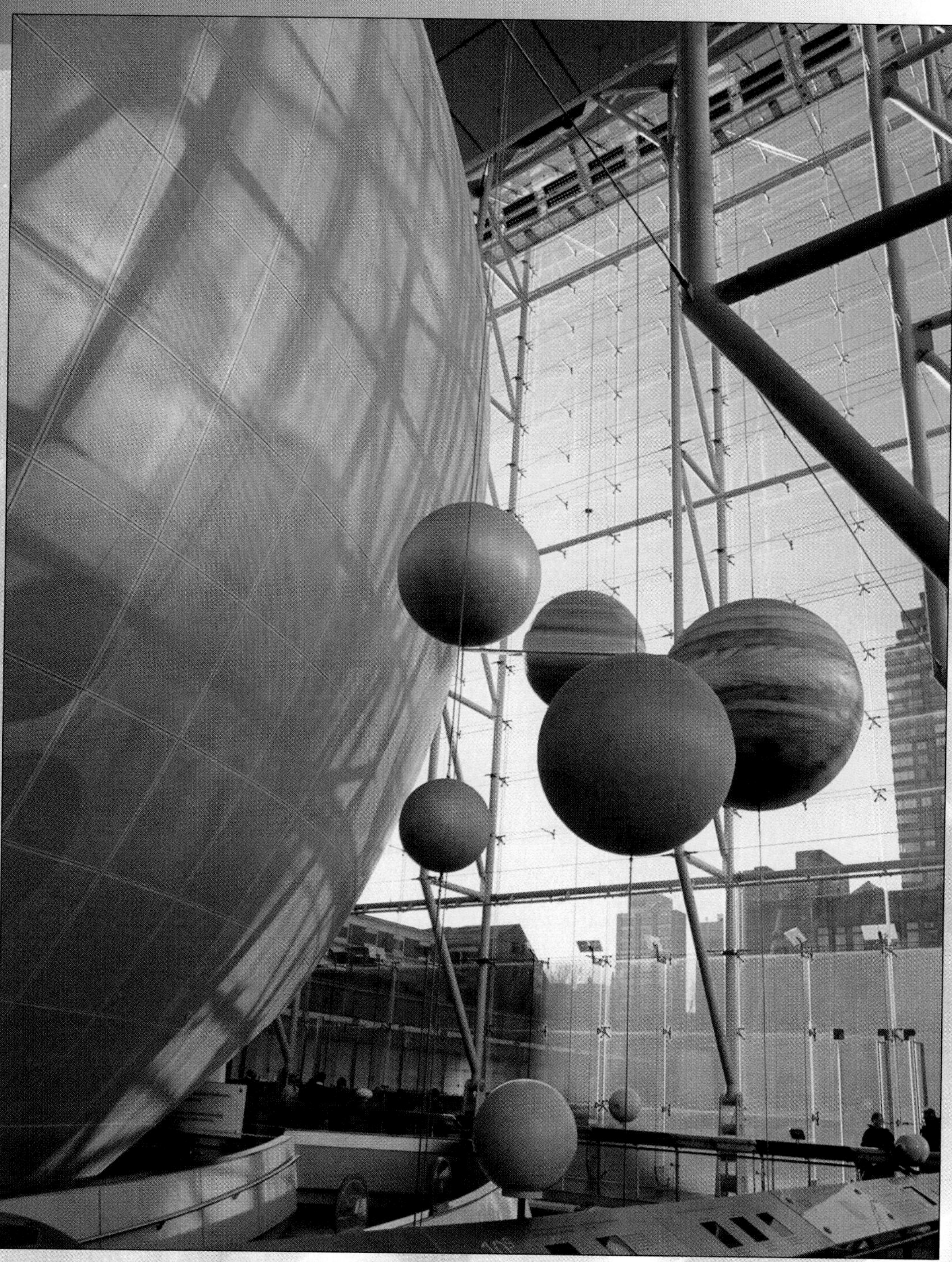

Scale models of the planets and Sun (large sphere) in the Rose Space Center at the American Museum of Natural History in New York City.
(Geray Sweeney/Corbis)

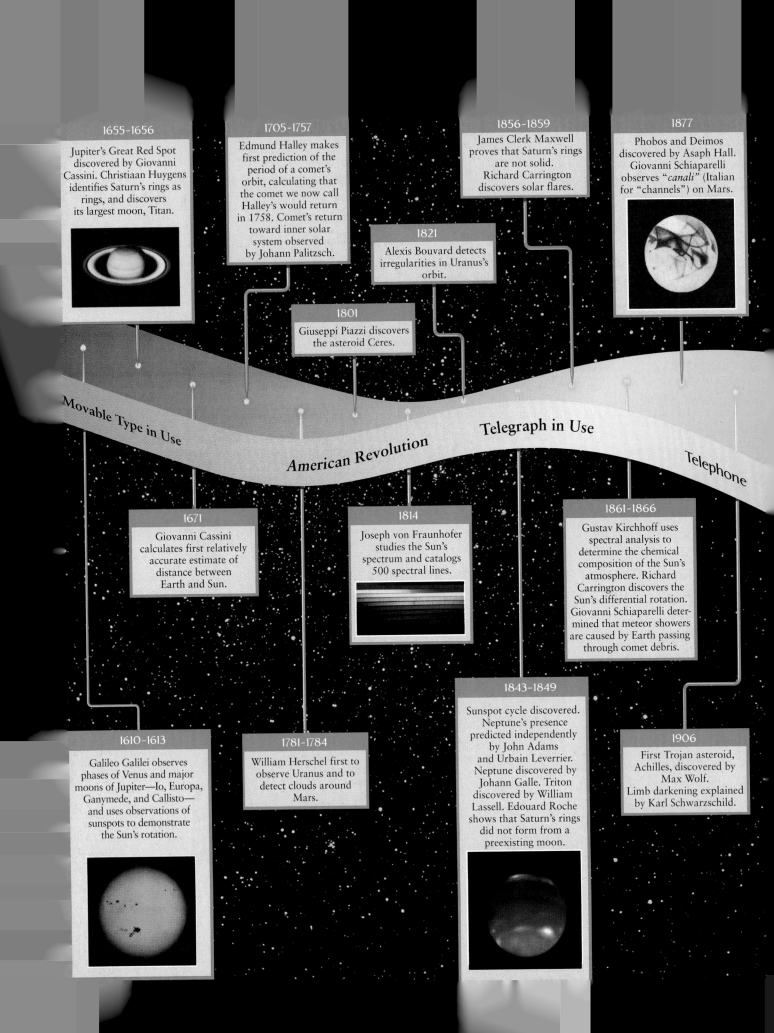

1655-1656

Jupiter's Great Red Spot discovered by Giovanni Cassini. Christiaan Huygens identifies Saturn's rings as rings, and discovers its largest moon, Titan.

1705-1757

Edmund Halley makes first prediction of the period of a comet's orbit, calculating that the comet we now call Halley's would return in 1758. Comet's return toward inner solar system observed by Johann Palitzsch.

1856-1859

James Clerk Maxwell proves that Saturn's rings are not solid. Richard Carrington discovers solar flares.

1877

Phobos and Deimos discovered by Asaph Hall. Giovanni Schiaparelli observes "*canali*" (Italian for "channels") on Mars.

1821

Alexis Bouvard detects irregularities in Uranus's orbit.

1801

Giuseppi Piazzi discovers the asteroid Ceres.

Movable Type in Use

American Revolution

Telegraph in Use

Telephone

1671

Giovanni Cassini calculates first relatively accurate estimate of distance between Earth and Sun.

1814

Joseph von Fraunhofer studies the Sun's spectrum and catalogs 500 spectral lines.

1861-1866

Gustav Kirchhoff uses spectral analysis to determine the chemical composition of the Sun's atmosphere. Richard Carrington discovers the Sun's differential rotation. Giovanni Schiaparelli determined that meteor showers are caused by Earth passing through comet debris.

1610-1613

Galileo Galilei observes phases of Venus and major moons of Jupiter—Io, Europa, Ganymede, and Callisto—and uses observations of sunspots to demonstrate the Sun's rotation.

1781-1784

William Herschel first to observe Uranus and to detect clouds around Mars.

1843-1849

Sunspot cycle discovered. Neptune's presence predicted independently by John Adams and Urbain Leverrier. Neptune discovered by Johann Galle. Triton discovered by William Lassell. Edouard Roche shows that Saturn's rings did not form from a preexisting moon.

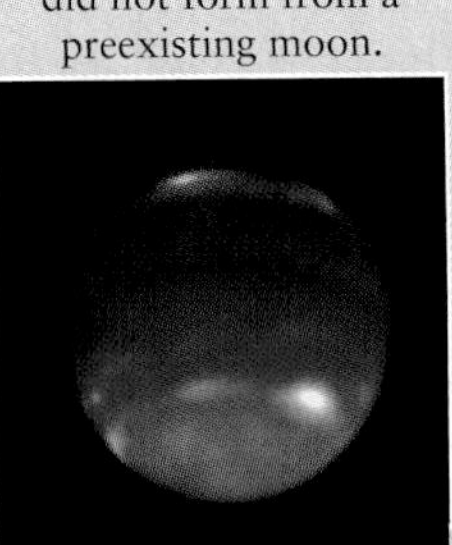

1906

First Trojan asteroid, Achilles, discovered by Max Wolf. Limb darkening explained by Karl Schwarzschild.

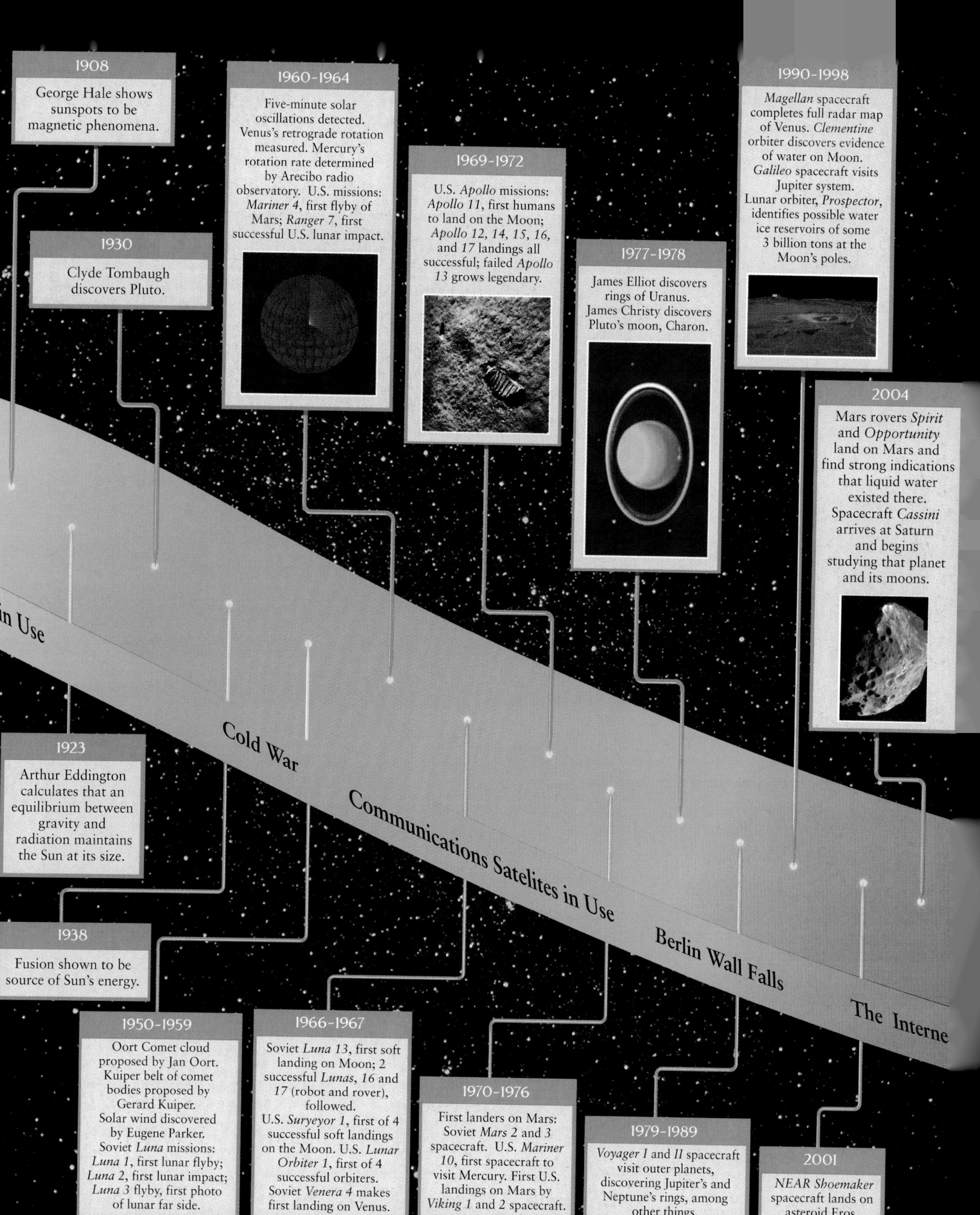
1908
George Hale shows sunspots to be magnetic phenomena.
1930
Clyde Tombaugh discovers Pluto.
1960-1964
Five-minute solar oscillations detected. Venus's retrograde rotation measured. Mercury's rotation rate determined by Arecibo radio observatory. U.S. missions: *Mariner 4*, first flyby of Mars; *Ranger 7*, first successful U.S. lunar impact.
1969-1972
U.S. *Apollo* missions: *Apollo 11*, first humans to land on the Moon; *Apollo 12, 14, 15, 16,* and *17* landings all successful; failed *Apollo 13* grows legendary.
1977-1978
James Elliot discovers rings of Uranus. James Christy discovers Pluto's moon, Charon.
1990-1998
Magellan spacecraft completes full radar map of Venus. *Clementine* orbiter discovers evidence of water on Moon. *Galileo* spacecraft visits Jupiter system. Lunar orbiter, *Prospector*, identifies possible water ice reservoirs of some 3 billion tons at the Moon's poles.
2004
Mars rovers *Spirit* and *Opportunity* land on Mars and find strong indications that liquid water existed there. Spacecraft *Cassini* arrives at Saturn and begins studying that planet and its moons.
in Use
Cold War
Communications Satelites in Use
Berlin Wall Falls
The Interne
1923
Arthur Eddington calculates that an equilibrium between gravity and radiation maintains the Sun at its size.
1938
Fusion shown to be source of Sun's energy.
1950-1959
Oort Comet cloud proposed by Jan Oort. Kuiper belt of comet bodies proposed by Gerard Kuiper. Solar wind discovered by Eugene Parker. Soviet *Luna* missions: *Luna 1*, first lunar flyby; *Luna 2*, first lunar impact; *Luna 3* flyby, first photo of lunar far side.
1966-1967
Soviet *Luna 13*, first soft landing on Moon; 2 successful *Lunas*, *16* and *17* (robot and rover), followed. U.S. *Suryeyor 1*, first of 4 successful soft landings on the Moon. U.S. *Lunar Orbiter 1*, first of 4 successful orbiters. Soviet *Venera 4* makes first landing on Venus.
1970-1976
First landers on Mars: Soviet *Mars 2* and *3* spacecraft. U.S. *Mariner 10*, first spacecraft to visit Mercury. First U.S. landings on Mars by *Viking 1* and 2 spacecraft.
1979-1989
Voyager I and *II* spacecraft visit outer planets, discovering Jupiter's and Neptune's rings, among other things.
2001
NEAR Shoemaker spacecraft lands on asteroid Eros.

CHAPTER 5

Formation of the Solar System and Other Planetary Systems

Matter from Which Planets Form (David Malin/Anglo-Australian Observatory)

WHAT DO YOU THINK?

1. How many stars are there in the solar system?
2. Was the solar system created as a direct result of the formation of the universe?
3. How long has the Earth existed?
4. Is Pluto always the farthest planet from the Sun?
5. What typical shape(s) do moons have?
6. Have any Earthlike planets been discovered orbiting Sunlike stars?

The formation of the solar system would make a good action movie. It began quietly, with lots of small encounters and interactions. Then things started to heat up and become progressively more violent. There were impacts occurring every second. Particles clumped together and sometimes big clumps smashed each other to bits. Eventually, the heavyweights (in this case, bodies destined to become planets and large moons) managed to wipe out most of the smaller players. The drama eventually led to the evolution of life, a subplot featuring the third rock from the Sun. Yes, "The Making of the Solar System" would be a great film—except for one thing: You would have to watch it for at least a hundred million years.

1 The Sun and all the bodies that orbit it make up our **solar system.** The solar system has only one star, the Sun. The stars you see in the night sky are members of a larger system called the *Milky Way Galaxy.* Our solar system is also part of the Milky Way, a topic we discuss in detail in Chapter 15.

In this chapter you will discover

- how the solar system formed
- why the environment of the early solar system was much more violent than it is today
- how the planets are grouped
- how astronomers characterize each planet's "personality"
- how the moons throughout the solar system formed
- what the debris of the solar system is made of
- that planets have been observed around a growing number of stars
- that newly forming star and planet systems are being observed

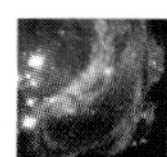

FORMATION OF THE SOLAR SYSTEM

How did the solar system form? How were its varied building blocks of rock, metal, ice, and gas created? How has the solar system changed since its formation, and what does its history tell us about the planets we see today? Within the past few decades, telescopes and space probes, along with the theories of modern science, have finally provided answers to these age-old questions. Our new wealth of information is giving us a rapidly growing understanding of our nearest neighbors in space.

5-1 The solar system formed from a cloud of cold gas and dust

The solar system formed from a fragment of a vast cloud of interstellar gas and dust. You can see a similar cloud just by looking at the middle "star" in Orion's sword. To your naked eye, this "star," called the Orion Nebula (also called the Great Nebula) in Figure 1-4, looks like a fuzzy blob. That is because it is not a single star, but is part of an interstellar cloud region in which new stars and planets are forming today. Recalling from Chapter 3 the limited information that visible light provides, you can see in Figure 3-28 that ultraviolet and infrared images reveal even more of this region.

Our study of the formation of the solar system begins with an inconsistency, namely that the Earth, Moon, and many other bodies orbiting the Sun are composed primarily of heavy elements, such as oxygen, silicon, aluminum, iron, carbon, and calcium, among others, and they contain extremely little hydrogen and helium. Observations of the spectra of the Sun, other stars, and interstellar clouds reveal that hydrogen and helium are by far the most abundant elements in the universe. These two elements account for 98% of the mass of all the material in existence—all the other elements combined account for only 2%. How is it that the Earth, Moon, Mars, Venus, Mercury, and many smaller bodies in the solar system typically contain less than 0.15% hydrogen and helium? Somehow these familiar planets formed from matter that had been enriched with the heavier elements and depleted in hydrogen and helium.

There is a good reason for the overwhelming abundance of hydrogen and helium throughout the universe. Astronomers believe that the universe formed between 13 and 14 billion years ago with a violent event called the *Big Bang.* Only the lightest elements—hydrogen, helium, and a tiny amount of lithium—emerged when the cosmos was formed. The first stars, composed only of these three elements, condensed out of this primeval matter, probably within a few hundred million years after the Big Bang.

The difference in chemical composition between the 2
early universe and the solar system shows that the solar system did not form as a *direct* result of the Big Bang. The solar system came billions of years later, and the heavier elements came to the solar nebula from stars that existed and exploded before the solar system formed.

Stars use various elements, such as hydrogen, helium, and carbon, to create energy. In this process, called *fusion,* lighter elements are transformed into heavier ones, such as hydrogen becoming helium or helium becoming carbon. We call fusion a *thermonuclear reaction* because the tremendous heat generated deep inside stars is what enables nuclei of lighter elements there to combine, or fuse, into heavier elements. We explore details of the creation of this heavier matter in Chapters 12 and 13.

The heavy elements found in the solar system were formed by fusion in stars and then ejected into space. Some of these heavier elements are made deep inside stars while the stars are still shining. A fraction of these newly formed elements travel upward with other gases. Near the ends of their lives, most stars cast their outer layers into space to form clouds of interstellar gas and dust. The denser elements delivered to these outer layers are thereby sent out into space along with lots of the star's hydrogen and helium.

The outer layers of most stars are gradually expelled (Figure 5-1a and b), but some stars end their lives with spectacular detonations called *supernovae* (Figure 5-1c), which blow the stars apart. The heaviest elements found in the universe are created by fusion during supernovas as the outer layers of these exploding stars begin their journeys into space.

As mentioned earlier, the elements other than hydrogen and helium represent only a small fraction of the mass ejected by stars. However, under the right circumstances, the lighter hydrogen and helium are separated from these other elements, leaving dense elements in sufficient concentration to form Earthlike planets. Let us consider how this happens.

Small particles of dust form when stars eject matter, like the soot formed in a fire or in diesel exhaust. Over time, gas and dust are emitted by enough stars to form interstellar clouds, as you see in Orion. Sometimes, the explosive force of a supernova compresses fragments of preexisting interstellar clouds. Other times, pairs of clouds collide, compressing both. When a cloud fragment gets sufficiently dense, the mutual gravitational attraction of gas and dust particles in it causes the fragment to collapse, thereby creating a new star and planetary system (Figure 5-2). The collapsing matter is primarily hydrogen and helium, but, as discussed earlier, the stars that ejected it had converted some of the hydrogen and helium to the other 90 naturally formed elements. Such a collapse is believed to be how the solar system formed—the

a

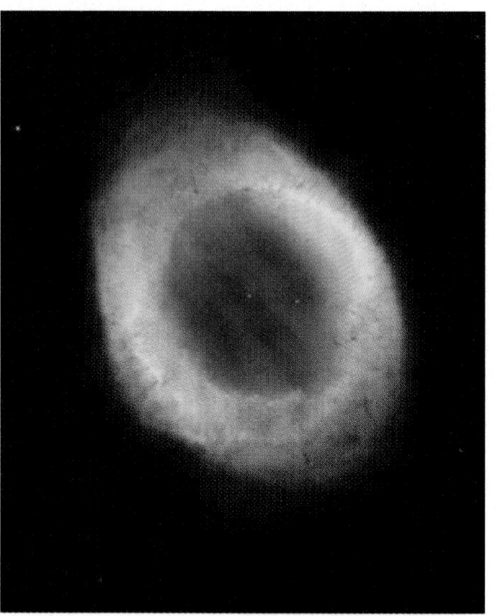

b

c

R I V U X G

FIGURE 5-1 How Stars Lose Mass (a) The brightest star in Scorpius, Antares, is nearing the end of its existence. Strong winds from its surface are expelling large quantities of gas and dust, creating this nebula reminiscent of an Impressionist painting. The scattering of starlight off this material makes it appear especially bright, even at a distance of 604 ly. (b) The Ring Nebula is just over 2000 ly from Earth. The central star shed its outer layers of gas and dust in an expanding spherical shell now about $\frac{1}{2}$ ly across. The relatively gentle emission of this matter is called a planetary nebula, discussed in Chapter 13. (c) A supernova is the most powerful known mechanism for a star to shed mass. This is the Crab Nebula. Although it is about 6000 ly from Earth, it was visible during the day for three weeks during A.D. 1054. (a: David Malin/Anglo-Australian Observatory; b: NASA; c: Malin/Pasachoff/Caltech)

FIGURE 5-2 A Dusty Region of Star Formation These young stars in the constellation Orion are still surrounded by much of the gas and dust from which they formed. The bluish, wispy nebulosity is caused by starlight scattering off abundant interstellar dust grains. These grains are made of heavy elements (such as carbon, silicon, and iron) produced by earlier generations of stars. Astronomers hypothesize that the solar system formed from a fragment of such a cloud of gas and dust. ((©)1980, 1984 Anglo-Australian Observatory)

debris left over from a previous generation of stars. *We are quite literally made of star dust*!

5-2 Gravity, rotation, and heat shaped the young solar system

The cloud fragment from which the solar system formed is called the **solar nebula.** Initially, it was very cold inside the solar nebula—well below the freezing point of water. The nebula began with a diameter of at least 100 AU and a total mass about 2 to 3 times the mass of the Sun. Ice and ice-coated dust grains composed of heavy elements were scattered abundantly across this vast volume. Deep inside the nebula the gravitational attraction caused the gas and dust there to fall rapidly toward its center. (In terms of the physics presented in Chapter 2, the cloud's gravitational potential energy was being converted into kinetic energy—energy of motion). As a result, the density and pressure at the center of the nebula began to increase, producing a concentration of matter called the **protosun.**

As the protosun continued to increase in mass and to contract, atoms at its center collided with one another with increasing speed and frequency. Such collisions create heat, causing the temperature deep inside the protosun to soar. Therefore, the first heat in the solar system came from colliding gas, not from nuclear fusion, which began later.

Rotation also played a key role in the formation of the solar system. When it first formed, the solar nebula was a very slowly rotating ensemble of particles. This rotation occurred because the formation of nebulae is turbulent, meaning that they are created with many slowly swirling regions, like so much smoke rising from a fire. As a result of this rotation, the nebula had angular momentum (see An Astronomer's Toolbox 2-2), which prevented the entire nebula from collapsing into the protosun. A non-swirling collapse would have created a star without any planets or other orbiting matter. The conservation of angular momentum (see Section 2-7) tells us that, like a skater pulling in her arms as she spins (see Figure 2-12), rotating matter falling inward in the solar nebula revolved faster and faster. Even the debris forming the protosun was orbiting as it fell inward, like water spiraling down a drain. Therefore, the protosun was spinning and the Sun that formed from it should be spinning. It is, as we will see in Chapter 10.

The outer parts of the solar nebula collapsed inward more slowly than the central matter that formed the protosun. This outer region was probably a very ragged nebula, but mathematical studies show that the combined effects of gravity and rotation transform even an irregular cloud fragment into a rotating disk with a warm center and cold edges, as shown in Figure 5-3.

While we cannot see our solar system as it was before planets formed, astronomers have found what we believe are similar disks of gas and dust surrounding other young stars, like those shown in Figure 5-4. Called **protoplanetary disks** or **proplyds,** these systems are undergoing the same initial stage of evolution that we are describing here for our solar system.

The protosun's temperature increased as it became denser and more and more atoms in it collided more often. Radiating more and more heat, the temperature around it began to climb. The rising temperature vaporized all the common icy substances in the inner region of the solar nebula and pushed light gases like hydrogen and helium outward. In this inner region, where the Earth orbits today, mostly heavier elements remained. These heavy elements were eventually to come together and form the four inner planets: Mercury, Venus, Earth, and Mars. The removal from their immediate neighborhood of the lighter elements by the protosun explains why the inner planets are composed mostly of heavy elements.

FIGURE 5-3 The Formation of the Solar System This sequence of drawings shows six stages in the formation of the solar system. (a) A slowly rotating cloud of interstellar gas and dust begins to contract because of its own gravity. (b) A central condensation, the protosun, forms as the cloud flattens and rotates faster. (c) A flattened disk of gas and dust surrounds the protosun, which has begun to shine. The next three drawings show the inner region of planet formation. (d) The Sun's rising temperature removes the gas from the inner regions, leaving dust and larger debris revolving in place. (e) The planets have established dominance in their regions of the solar system. (f) The solar system as it appears today.

FIGURE 5-4 Young Circumstellar Disks of Matter These Hubble Space Telescope images show circumstellar disks surrounding stars. (a) A 5-million-year-old star (blocked out) and its huge disk of gas and dust. The spirals of light in the disk are probably made by the gravitational tugs of nearby stars. The disk will contract dramatically before planets form. (b, c) These two stars are 1500 light-years away from Earth in the Orion Nebula. They are both less than 10 million years old and are surrounded by protoplanetary gas and dust. (a: NASA, M. Clampin (StScI), H. Ford (JHU), G. Illingworth (UCO/Lick), J. Krist (StScI), D. Ardila (JHU), D. Golimowski (JHU), the ACS Science Team and ESA; b and c: STScI, M. J. McCaughrean/MPIA, C. R. O'Dell/Rice University, NASA)

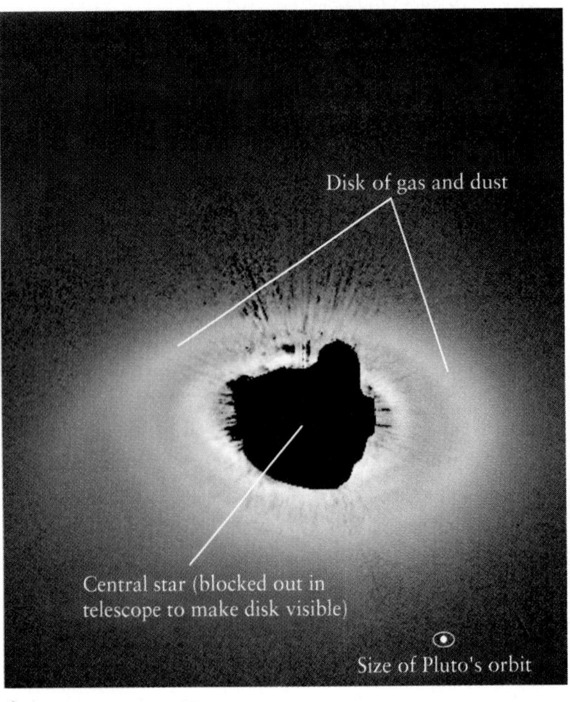

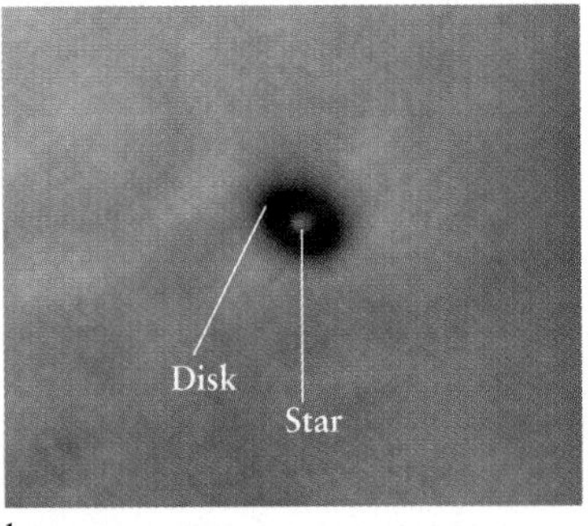

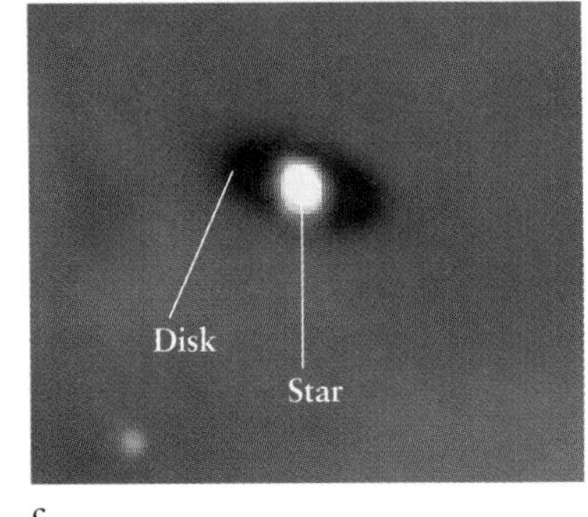

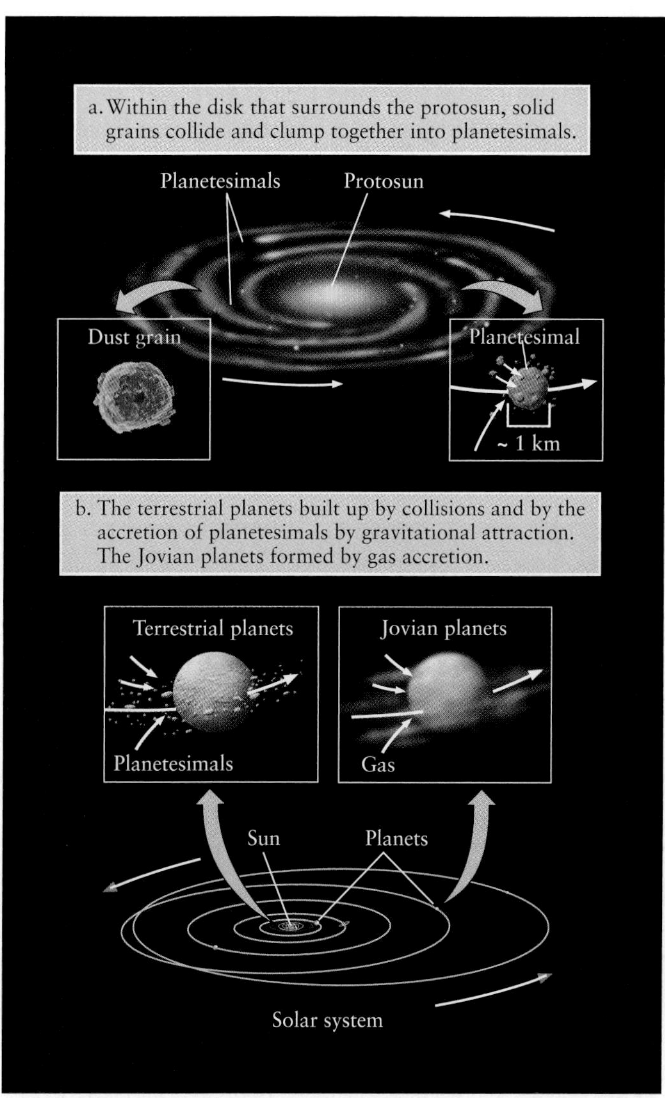
a. Within the disk that surrounds the protosun, solid grains collide and clump together into planetesimals.
Planetesimals
Protosun
Dust grain
Planetesimal
~ 1 km
b. The terrestrial planets built up by collisions and by the accretion of planetesimals by gravitational attraction. The Jovian planets formed by gas accretion.
Terrestrial planets
Planetesimals
Jovian planets
Gas
Sun
Planets
Solar system
R I V U X G

◀ FIGURE 5-5 **Formation of the Planets** (a) Planetesimals about a kilometer in size formed in the solar nebula from small dust grains (inset) sticking together. **Inset:** Collected by a high-flying aircraft, this piece of interplanetary dust is believed to be typical of the state of matter in the protoplanetary disk. It is about 20 μ (0.02 mm) across—about $^1/_5$ the diameter of a typical human hair. (b) Planetesimals in the inner solar system grouped together to form the terrestrial planets, while the giant planets began as terrestrial planets and accumulated massive envelopes with different amounts of hydrogen, helium, water, and other light materials. (inset: Scott Messenger/University of Washington, St. Louis/Science Photo Library/Photo Researchers)

5-3 Collisions in the early solar system led to the formation of planets

The formation of the inner planets was the result of innumerable impacts of small, rocky particles in the solar system's protoplanetary disk. Initially, neighboring dust grains (Figure 5-5a) and pebbles in the solar nebula collided and stuck together. Then, over a period of a few million years, these accumulations of dust and pebbles coalesced into larger objects called **planetesimals**, with diameters of a few kilometers or miles (Figure 5-5b).

During the next stage, the planetesimals collided. Many were pulverized, many fell into the Sun, but eventually enough of them survived, so that their mutual gravitational attraction could bring them together to form larger objects called **protoplanets.** Computer simulations of the inner solar system based on Newton's laws (see Section 2-7) suggest that **accretion,** the coming together of smaller pieces of matter to form larger ones, continued for over 100 million years (Figure 5-5b) and should lead to the formation of fewer than half a dozen planets (Figure 5-6). The agreement between predictions of the number of planets and the number in our inner solar system is encouraging.

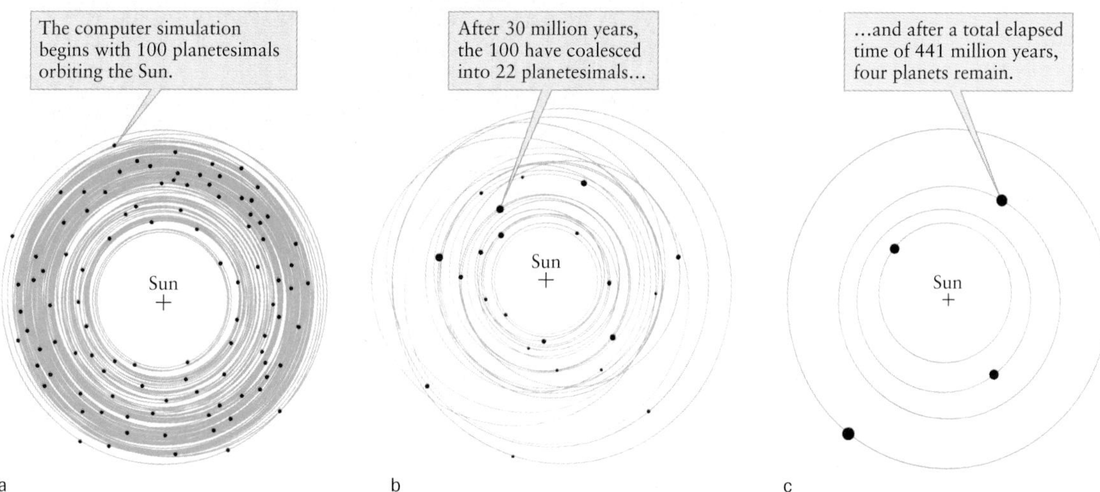
The computer simulation begins with 100 planetesimals orbiting the Sun.
After 30 million years, the 100 have coalesced into 22 planetesimals...
...and after a total elapsed time of 441 million years, four planets remain.
Sun
Sun
Sun
a
b
c

FIGURE 5-6 **Accretion of the Inner Planets** These three drawings show the results of a computer simulation of the formation of the inner planets. In this simulation the formation of the inner planets is nearing completion after only 100 million years.

Observations made in 2004 of star systems undergoing all stages of planet formation indicate that this entire process, often delayed by collisions between large protoplanets, typically takes several hundred million years to complete.

INSIGHT INTO SCIENCE

Computer-Aided Analysis Many equations are so complex that they require computer analysis to enable us to understand their implications. An example is the physics of the formation of the planets and the Sun. Observations then lead to refined computer models.

While the inner regions of the solar system were heating up, temperatures in the outer regions of the solar nebula remained quite cool. Ice and ice-coated dust, along with hydrogen and helium gas, were able to survive in these cooler regions. The large outer planets—Jupiter, Saturn, Uranus, and Neptune—probably began to form from the accretion of planetesimals. For each, a rocky core even bigger than the Earth apparently served as a "seed," and for about a million years the core became coated with additional rock, gas, and ice. When the gas-rich envelope became as massive as the rocky core, the gas and ice began to accumulate at a runaway pace as the growing planet's gravitational attraction pulled more of it onto the surface. Thereafter, the envelope pulled in most of the gas and ice in its vicinity, creating a shell of water

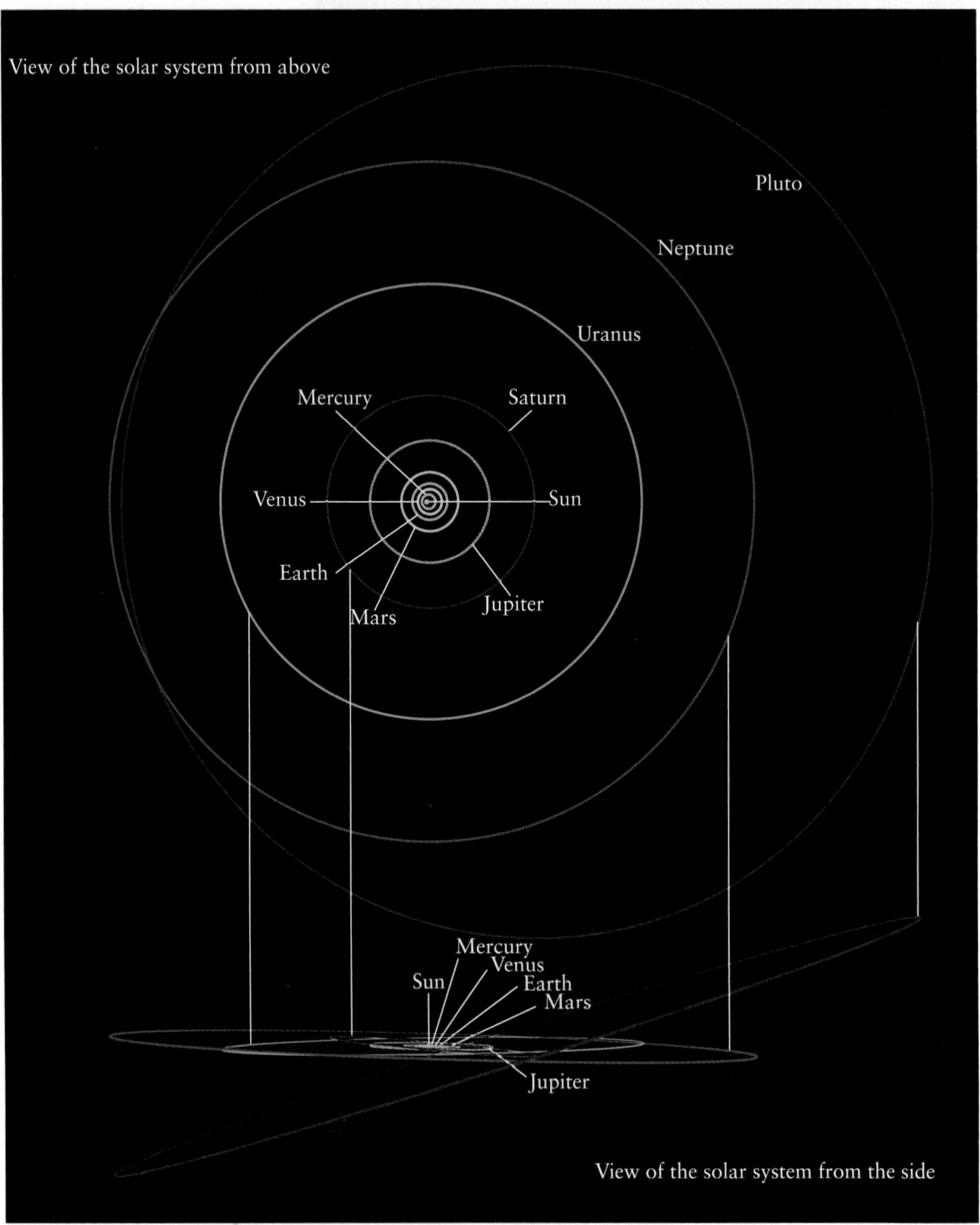

FIGURE 5-7 **The Solar System** This scale drawing shows the distribution of planetary orbits around the Sun. All orbits are counterclockwise because the view is from above the Earth's North Pole. The four terrestrial planets are located close to the Sun; the four giant planets and Pluto orbit at much greater distances from the Sun. Seen from above the disk of the solar system, most of the orbits appear nearly circular. Pluto has the most elliptical orbit of any planet. Indeed, it spends part of the time closer to the Sun than Neptune. It last moved inside Neptune's orbit in 1979 and moved back outside in 1999. Note that Pluto and Neptune will never collide because their orbits are synchronized by their gravitational interaction to keep them apart.

surrounded by a huge volume of hydrogen and helium (see Figure 5-5b).

The outer planets Jupiter, Saturn, Uranus, and Neptune now include different amounts of hydrogen, helium, methane, ammonia, and water. Many of the moons of the outer planets are also partially composed of these low-density substances.

During the millions of years that the planets were forming, so, too, was the Sun. Throughout this time the temperature and pressure at the center of the contracting protosun continued to climb. Finally, the center of the protosun became hot enough to ignite thermonuclear reactions. Hydrogen began fusing into helium in its core, a process that creates and releases huge amounts of energy in the form of electromagnetic radiation, and the Sun was born. This fusion continues today and provides the Earth with the energy necessary for life.

Sunlike stars take approximately 100 million years to form from a nebula and settle down, which means that the Sun probably became a full-fledged star before the accretion of the inner planets was completed. Radiation from the Sun heated and thereby disbursed the remaining gases in the solar system, thereby limiting the sizes of the planets and preventing new planetesimal formation.

If this collapsing, swirling scenario accurately represents the history of the solar system, then the conservation of angular momentum predicts that the planets should all be orbiting in more or less the same plane as the Earth (that is, nearly in the plane of the ecliptic). The angles of the orbital planes of the other planets with respect to the ecliptic are called their **orbital inclinations.** All the planets except Pluto have orbital inclinations of 7° or less. Even Pluto, which is much smaller than any other planet and which may actually be a large version of another type of body orbiting the Sun, only has an orbital inclination of 17° (Figure 5-7).

Conservation of angular momentum further predicts that all the planets should orbit the Sun in the same "sense" as does the Earth. Traveling way above the Earth's North Pole and looking down at the Earth's orbit, we see that our planet is going counterclockwise around the Sun. All the other planets also orbit counterclockwise as observed from the same vantage point, including little Pluto.

3 To find out more about the solar system's origins, we study interplanetary debris: asteroids, meteoroids, and comets. Some of these bodies are believed to be virtually unchanged remnants of the formation of the solar system.

According to radioactive dating of the oldest debris ever discovered from space (see An Astronomer's Toolbox 4-3), the solar system was roughly in the form we know today just under 4.6 billion years ago. By the time the planets were in place, only a few of the dozens of moons were in orbit. Most of the moons in the solar system are just planetesimals captured by the gravitational attractions of the planets.

One moon, ours, was created as the result of a collision between the Earth and a large body in a highly elliptical orbit (see Chapter 6). Once formed, our Moon's surface was scarred by numerous impacts of remnant debris left over after the formation of the major bodies in the solar system (Figure 5-8). These scars, called **craters,** are also found on those planets and moons that do not have appreciable atmospheres or geological activity that would otherwise erase these features. Indeed, the Moon's virtually airless environment has preserved important information about the early history of the solar system. (Note that we capitalize the word "moon" only when we are referring to the Earth's Moon.)

Radioactive dating (see An Astronomer's Toolbox 4-3) of Moon rocks brought back by the Apollo astronauts indicates that the rate of impacts declined dramatically about 3.8 billion years ago. Since that time, impact cratering has proceeded at a very low rate. Thus, most of the craters on the Moon and planets were formed during the first 800 million years of the solar system's history, as the young planets pulled down rocky debris left over from the formative years of the solar nebula.

R I V U X G

FIGURE 5-8 **Our Moon** This photograph, taken by astronauts in 1972, shows thousands of the craters produced by impacts of rocky debris left over from the formation of the solar system. Age-dating of lunar rocks brought back by the astronauts indicates that the Moon is about 4.5 billion years old. Most of the lunar craters were formed during the Moon's first 700 million years of existence, when the rate of bombardment was much greater than it is now. (NASA)

Conditions developed on the young Earth, especially the presence of liquid water, so that biological evolution could begin to occur here. There is strong evidence that liquid water existed on Mars and may still exist inside it and inside Jupiter's moons Europa and Ganymede, raising the hotly debated question of whether simple life evolved on those worlds, too.

5-4 Minor debris from the formation of the solar system still exists

In addition to the nine planets and their moons, many other objects created when the solar system formed still orbit the Sun. Between the orbits of Mars and Jupiter are believed to be millions of such bodies, called **asteroids,** most of them smaller than a kilometer across. This region is called the **asteroid belt.** Asteroids are composed primarily of metal and rock. The largest asteroid, Ceres, has a diameter of about 900 km. The next largest, Pallas and Vesta, are each about 500 km in diameter. Still smaller ones are increasingly numerous. There are many thousands of kilometer-sized asteroids. A close-up picture of the asteroid Gaspra is shown in Figure 5-9. In addition to the asteroids between Mars and Jupiter, other asteroids have highly elliptical orbits that take them across the paths of some of the planets. Some asteroids have their own moons, such as asteroid Ida, which is orbited by heavily cratered Dactyl.

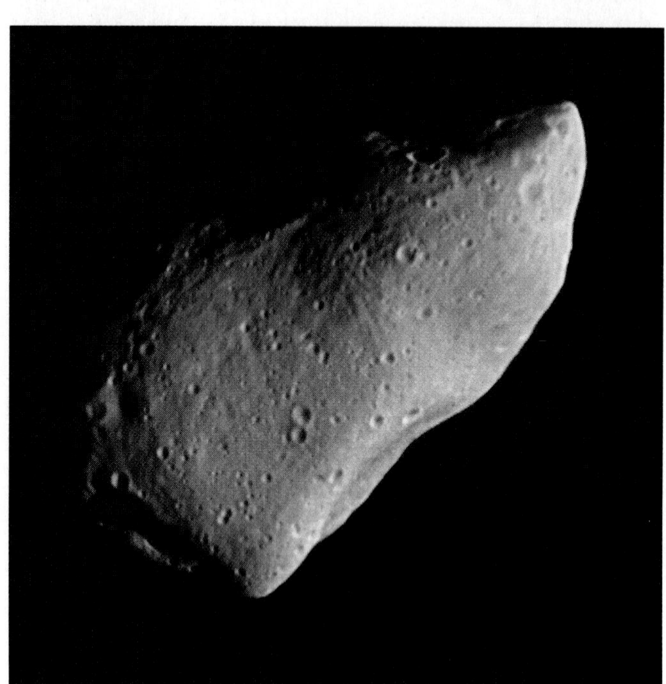

R I V U X G

WEB LINK 5.2 FIGURE 5-9 **An Asteroid** This picture of the asteroid Gaspra was taken in 1991 by the *Galileo* spacecraft on its way to Jupiter. The asteroid measures 12 × 20 × 11 km. Millions of similar chunks of rock orbit the Sun between the orbits of Mars and Jupiter. (NASA)

R I V U X G

FIGURE 5-10 **A Comet** The comet nucleus, the solid part of a comet, is an irregularly shaped chunk of ice and rocky debris typically 10 km in diameter. When a comet passes near the Sun, solar radiation vaporizes some of the comet's ices, and the resulting gases and dust form one or two tails millions of kilometers long. This photograph shows Comet Kohoutek in 1974. (NASA)

Pieces of rocky and metallic debris even smaller than asteroids are called **meteoroids.** Typical meteoroids are boulder-sized or smaller and apparently exist throughout the disk of the solar system, although most are in the asteroid belt. Chemical studies show that many meteoroids are small fragments of asteroids that broke off when larger bodies collided.

Beyond the orbit of Pluto are hundreds of billions of chunks of rock and ice called **comets,** or, more accurately, *comet nuclei.* Some of these comet nuclei orbit in a doughnut-shaped region beyond Neptune's orbit called the *Kuiper belt,* which is centered on the ecliptic. Others are believed to be distributed even further out from the Sun in a spherical distribution called the *Oort comet cloud.* Many comet nuclei have highly elongated orbits that occasionally bring them close to the Sun. When this happens, the Sun's radiation vaporizes some of their ices, producing the long, flowing tails we associate with comets (Figure 5-10).

As you are likely to have heard in the media, objects similar to Pluto have been discovered in the Kuiper belt. Like Pluto, the large bodies out beyond Neptune have more elliptical orbits than the planets. It is likely that the bodies out there are composed of rock and ice, as are Pluto and comet nuclei. Collisions probably break up many of the larger bodies in the Kuiper belt and Oort cloud, thereby creating the smaller debris we call comet nuclei. These collisions also cause some comet nuclei to fall into the inner solar system to become comets.

Examples of the distant objects that have been observed are Quaoar (pronounced kwa-o-ar) and Sedna. Because they are smaller than planets, but larger than comet nuclei, they are sometimes called *planetoids*. Whether Pluto will be reclassified as a planetoid or not remains to be seen. In this book we will maintain the traditional assertion that Pluto is the smallest, and usually most distant, planet from the Sun in the solar system.

TABLE 5-1 Orbital Characteristics of the Planets

	Average distance from Sun (AU)	Average distance from Sun (10^6 km)	Orbital period (yr)
Mercury	0.39	58	0.24
Venus	0.72	108	0.62
Earth	1.00	150	1.00
Mars	1.52	228	1.88
Jupiter	5.20	778	11.86
Saturn	9.54	1427	29.46
Uranus	19.19	2871	84.01
Neptune	30.06	4497	164.79
Pluto	39.53	5914	248.54

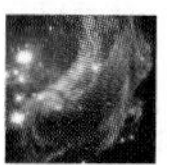

COMPARATIVE PLANETOLOGY

There are two ways of studying objects in the solar system. One is to compare and contrast one feature of all similar objects, such as all the atmospheres or all the surfaces of planets, and then move on to the next feature. This has the advantage of showing the "big picture" about these features. The other approach is to take an object and explore all its properties before moving on to the next object. This has the advantage of connecting all the properties of each object directly to each other, so that you do not, say, confuse the atmosphere of Venus with the surface of Mercury. Since both approaches are valuable, we will do both.

5-5 Comparisons among the nine planets show distinct similarities and significant differences

Before exploring the properties of the individual planets in the upcoming chapters, it is instructive to compare a variety of their orbital and physical properties. We will also provide further comparative planetology information in the form of tables in the chapters on the various planets. The planets that emerged from the accretion of planetesimals were Mercury, Venus, Earth, Mars, Jupiter, Saturn, Uranus, Neptune, and Pluto.

Orbits Table 5-1 lists some orbital characteristics of the nine planets. As shown in Figure 5-7, the orbits of the four inner planets—Mercury, Venus, Earth, and Mars—are crowded close to the Sun. In contrast, the orbits of the four large outer planets—Jupiter, Saturn, Uranus, and Neptune—are widely spaced at greater distances from the Sun. Pluto usually orbits at the far fringes of the part of the solar system inhabited by planets.

Kepler's laws (see Chapter 2) showed us that all the 4
planets have elliptical orbits. Even so, most of their orbits are nearly circular. The exceptions are Mercury and Pluto, whose orbits are noticeably elliptical. In fact, Pluto's orbit sometimes takes it nearer to the Sun than its neighbor, Neptune. In 1979, Pluto passed inside Neptune's orbit. Neptune was the most distant planet from the Sun until February 1999, when Pluto's elliptical orbit once again took it farther out than Neptune (see Figure 5-7). Pluto will be farther from the Sun than Neptune for about the next 230 years.

Size Size is the most obvious physical difference between the planets. The smallest of the four giant outer planets, Neptune, is nearly 4 times larger in diameter than the Earth, the largest of the four inner planets. First place among the giant planets goes to Jupiter, whose diameter is about 11 times bigger than that of Earth. Pluto is the smallest of all the planets, and is even smaller than the seven largest moons, including our own. Figure 5-11 shows the Sun and the planets drawn to the same scale. The diameters of the planets are given in Table 5-2.

Mass Mass, a measure of the total number of particles an object contains, is another characteristic that distinguishes the inner and outer planets. The four inner planets have low masses compared with the giant planets. Again, first place goes to Jupiter, whose mass is 318 times greater than Earth's (see Table 5-2).

Density Size and mass can be combined in a useful way to provide information about the chemical composition of a planet (or any other object). Matter composed of heavy elements, like iron or lead, has more particles (protons and neutrons) packed into the same volume than matter composed of light elements, like hydrogen, helium, or carbon. Therefore,

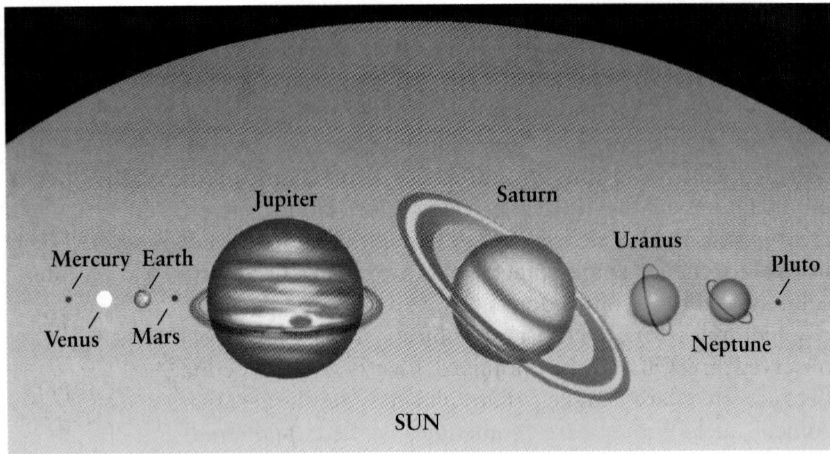

FIGURE 5-11 **The Sun and the Planets** This drawing shows the nine planets in front of the disk of the Sun, with all ten bodies drawn to the same scale. The four planets orbiting nearest the Sun (Mercury, Venus, Earth, and Mars) are small and made of rock and metal. The next two planets (Jupiter and Saturn) are large and composed primarily of hydrogen and helium. Uranus and Neptune are also large and contain much water, as well. Pluto has roughly equal amounts of rock and ice.

objects composed primarily of heavier elements have a greater **average density** than objects composed primarily of lighter elements. Average density is given by the equation

$$\text{Average density} = \frac{\text{total mass}}{\text{total volume}}$$

The chemical composition (kinds of elements) of any object therefore determines the object's average density—how much mass the object has in a unit of volume. For example, a kilogram of copper takes up less space (is denser) than a kilogram of water, even though both have the same mass (Figure 5-12). Average density is expressed in kilograms per cubic meter. To help us grasp this concept, we often compare the average densities of the planets to the density of something familiar, namely liquid water, which has a density of 1000 kg/m^3.

The four inner planets have high average densities compared to water (see Table 5-2). In particular, the average density of the Earth is 5520 kg/m^3. Because the density of typical surface rock is only about 3000 kg/m^3, the Earth must contain a large amount of material inside it that is denser than surface rock. The fact that the inner four planets have densities similar to that of the Earth (in Latin, *terra*) means that they are composed of similar chemicals. Consequently, these are called **terrestrial planets.** We will explore them in Chapters 6 and 7.

In sharp contrast, Jupiter, Saturn, Uranus, and Neptune have relatively low average densities (see Table 5-2). Indeed, Saturn's average density is less than that of water. Their low densities support the belief that the giant outer planets contain significant amounts of the lightest elements, hydrogen and helium. We will see that Jupiter and Saturn

TABLE 5-2 Physical Characteristics of the Planets

	Diameter		Mass		Average density
	(km)	(Earth = 1)	(kg)	(Earth = 1)	(kg/m^3)
Mercury	4,878	0.38	3.3×10^{23}	0.06	5430
Venus	12,100	0.95	4.9×10^{24}	0.81	5250
Earth	12,756	1.00	6.0×10^{24}	1.00	5520
Mars	6,786	0.53	6.4×10^{23}	0.11	3950
Jupiter	142,984	11.21	1.9×10^{27}	317.94	1330
Saturn	120,536	9.45	5.7×10^{26}	95.18	690
Uranus	51,118	4.01	8.7×10^{25}	14.53	1290
Neptune	49,528	3.88	1.0×10^{26}	17.14	1640
Pluto	2,300	0.18	1.3×10^{22}	0.002	2030

R I V U X G

FIGURE 5-12 **The Volumes of Objects with Different Densities** All the objects in this image have the same mass (total number of particles). However, the chemicals from which they form have different densities (number of particles per volume), so they all take up different amounts of space (volume). (© 2002 Richard Megna/Fundamentals Photographs, NYC)

have very similar overall compositions that are primarily hydrogen and helium, while Uranus and Neptune contain vast quantities of water as well as large amounts of hydrogen and helium. These four are sometimes called *Jovian planets* (the Roman god Jupiter was also called Jove), because they all have relatively low densities compared to the Earth's. However, the differences in the chemistries of Jupiter and Saturn compared to those of Uranus and Neptune make the name *giant planets* (giant compared to Earth) more appropriate in describing the four of them. We will explore them as two groups in Chapter 8.

Pluto again is an oddity. Although it is smaller than any of the inner planets, its average density falls between those of the terrestrial and the giant planets. Pluto is probably composed of a mixture of rock and ice, because its average density is between that of rock and ice. Pluto is classified as neither a terrestrial planet nor a giant planet, and it may eventually be considered to the largest member of a group of ice and rock objects that include the comet nuclei.

Spectra Further information about the chemical composition of bodies in the solar system is obtained from their spectra. For solar system objects, this radiation is primarily sunlight scattered off the surface or clouds surrounding each object. As we saw in Chapter 4, the spectra provide us with details of an object's surface or atmospheric chemical composition. It is from spectra that we confirm that the outer layers of the giant planets are primarily composed of hydrogen and helium, and that the surface of Mars is rich in iron oxides, among many other things.

INSIGHT INTO SCIENCE

Astronomical Measurement Astronomers use the laws of physics to infer things we cannot measure directly. For example, we can get an overall idea of the chemical compositions of planets from their masses and volumes: Kepler's laws enable us to determine the mass of each planet from the periods of their moons' orbits. The measured diameters of the planets (determined from their distances from Earth and angular sizes in the sky) yield their volumes. As shown in the equation on page 128, dividing total mass by total volume yields the average density. Comparing this density with the densities of known substances gives us information about the planets' chemistries.

Albedo The surfaces or the upper cloud layers of the planets scatter (send in many directions) different amounts of light. The fraction of incoming light returning directly into space is called a body's **albedo.** An object that scatters no light has an albedo of 0.0; for example, powdered charcoal has an albedo of nearly 0.0. An object that scatters all the light that strikes it (a high-quality mirror comes close) has an albedo of 1.0. The albedo multiplied by 100 gives the percentage of light directly scattered off that body. Three planets (Mercury, Earth, and Mars) have albedos of 0.37 or less. Such low albedos imply relatively dark, solid surfaces exposed to space. Venus, Jupiter, Saturn, Uranus, Neptune, and Pluto have albedos of 0.47 or more. High albedos imply bright layers exposed to space. The albedos of Venus, Jupiter, Saturn, Uranus, and Neptune are high because they are all completely enshrouded by clouds. Pluto's albedo of 0.5 is relatively high, but we do not yet know much about its surface because it is so far away and so small that our telescopes can barely resolve it.

WEB LINK 5.5

Moons Every planet except Mercury and Venus has moons (sometimes called *natural satellites*). There are at least 153 moons in the solar system (up from 99 known in 2001), and more are still being discovered. Unlike our Moon, most moons are irregularly shaped, more like potatoes than spheres. We will discover that there is as much variety among moons as there is among planets.

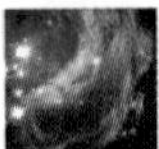

PLANETS OUTSIDE OUR SOLAR SYSTEM

Understanding the process by which the solar system was formed is aided by seeing such formation occurring elsewhere in our Milky Way Galaxy. We will now explore the planets that astronomers have discovered orbiting other stars. Such discoveries may one day help tell us whether we are alone in the universe.

5-6 Planets orbiting other stars have been discovered

WEB LINK 5.6

As indicated earlier in this chapter, astronomers are discovering other stars in early stages of formation that are surrounded by disks of gas and dust called proplyds. These disks are similar to what the solar nebula was believed to be like. Indirect observational evidence for the existence of planets outside the solar system, called *extrasolar planets,* first came from the distortion of proplyd disks.

If one or more planets orbit in a young star and planet-forming disk, then the planet's gravitational pull will affect the disk of gas and dust around it, causing the disk to warp. This is seen in the edge-on disk surrounding the star Beta Pictoris (Figure 5-13). This star and the material orbiting it formed only 20 million years ago. In 2001, astronomers also detected evidence that millions of comet nuclei have formed in this system. By studying such systems as Beta Pictoris, we can also begin addressing such questions as of whether the Earth acquired its water from comets in the young solar system.

Observing extrasolar planets directly as they orbit other stars is extremely challenging, because even high-albedo planets like Jupiter scatter back into space less than a millionth as much light as the stars they orbit emit. In 2004, the first planet orbiting another star was observed directly. This was possible

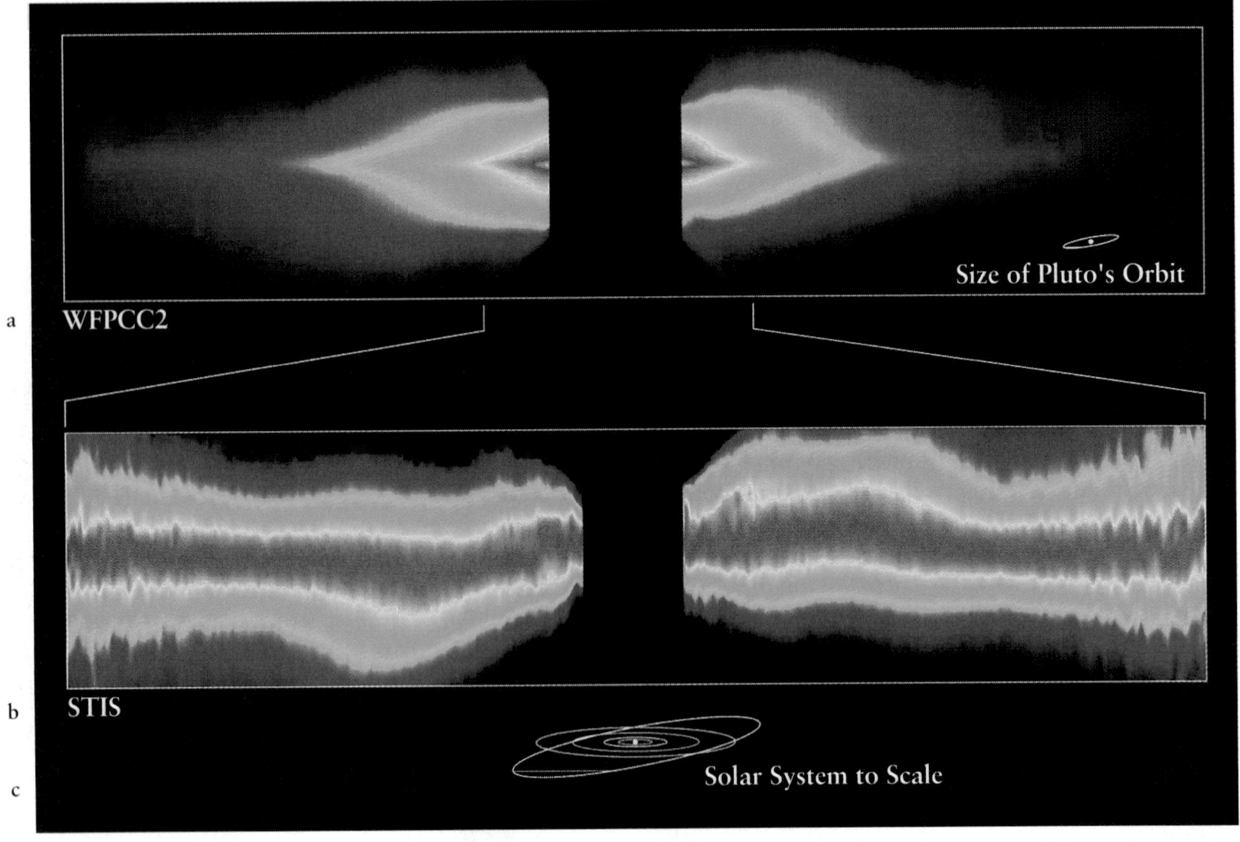

FIGURE 5-13 A Circumstellar Disk of Matter (a) An edge-on disk of material 225 billion km (140 billion miles) across orbiting the star Beta Pictoris (blocked out in this image) 50 ly from Earth. Twenty million years old, this disk is believed to be composed primarily of iceberglike bodies orbiting the star. (b) The central region of the disk is clearly warped in this image. This distortion is believed to be due to the gravitational tug of at least one planet orbiting Beta Pictoris. (c) The solar system drawn to the same scale as the image in (b). (Al Schultz (CSC/STScI), Sally Heap (GSFC/NASA) and NASA)

because the star is especially dim, while the planet is relatively bright. Since 1995, most extrasolar planets have been discovered by their effects on the stars they orbit (Figure 5-14a).

Some planets are discovered by measuring the radial velocity (defined in Section 4-7) of their stars, as measured by the Doppler shifts of the stars' spectra (Figure 5-14b). This motion toward or away from Earth is created by the gravitational pull of the planet. The Doppler shift changes cyclically, and the length of time of one cycle is the period of the planet's orbit.

Other planets are discovered from variations in a star's proper motion, or motion among the background stars (also defined in Section 4-7). A planet's attraction causes its star to deviate from motion in a straight line. This *astrometric* method of discovery looks for just such a wobble (Figure 5-14c).

Yet other planets are located by changes in the brightness of their stars. As a planet passes or transits between us and its star, it partially eclipses the star (Figure 5-14d). In 2003, astronomers used this *transit method* to detect a planet only one-sixteenth as far from its star as Mercury is to the Sun. That newly discovered planet orbits once every 28 hours, 33 minutes. Using this method in 2001, astronomers were also able to study the atmosphere of a planet. As a gas giant passed in front of the star HD 209458, hydrogen and sodium in the planet's atmosphere absorbed certain wavelengths of starlight. The Hubble Space Telescope further revealed that the outer layers of this planet are being heated so much that at least 10,000 tons of hydrogen are evaporating into space from it every second. At this lower limit of mass loss, the planet has lost about 0.1% of its mass over its lifetime of 5 billion years.

The astrometric method also tells us the mass of a planet. The two lowest-mass planets known with confidence have 3 or 4 times the mass of the Earth, and one may have been observed with a mass only slightly greater than that of our Moon. The vast majority, however, range between a half and a few times the mass of Jupiter, and many orbit surprisingly close to their stars (Figure 5-15). From our model of how our solar system formed, these hydrogen- and helium-rich planets should have formed much farther from their stars than they are at present. They are apparently spiraling inward. Astronomers are still working to explain why so many giant planets are so close to their stars.

In 2004, astronomers began discovering extrasolar planets using a property of space discovered by Albert Einstein. His theory of general relativity, explored in detail in Chapter 14, shows that matter warps the surrounding space, causing, among other things, passing light to change direction. This is quite analogous to how light passing through a lens changes direction and is focused—light can also be focused by gravity, an effect called **microlensing.** Figure 5-16 shows how a star with a planet passing between the Earth and a distant star will focus the light from the distant star, causing it to appear to change brightness. This occurs twice, once as the distant star's light is focused toward us by the closer star and again when the light is focused toward us by the planet.

In 1999, astronomers began discovering star systems with more than one planet. This is done by observing some

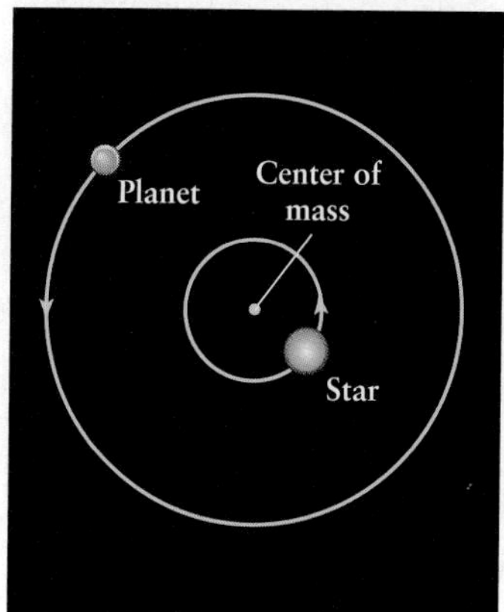

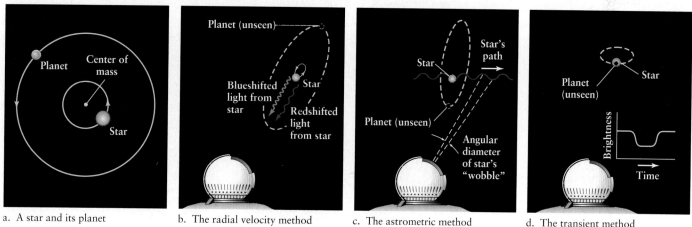

a. A star and its planet

b. The radial velocity method

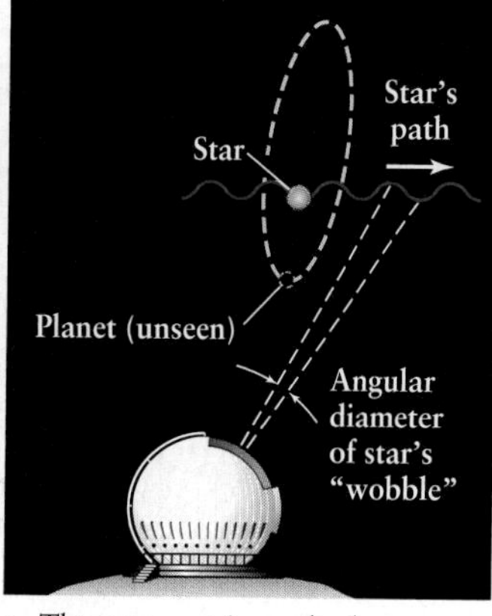

c. The astrometric method

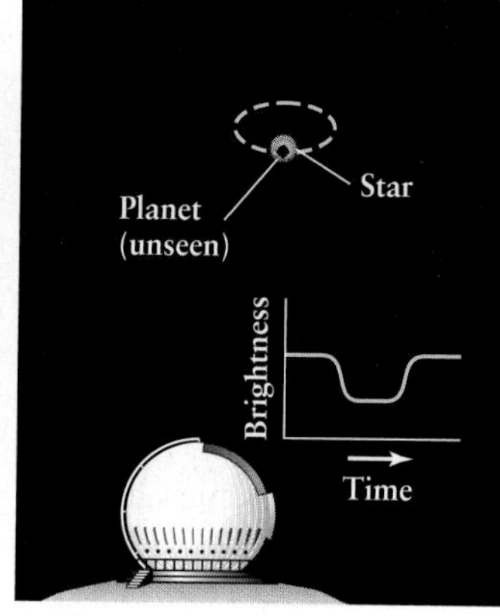

d. The transient method

FIGURE 5-14 **Detecting Planets Orbiting Other Stars**
(a) A planet and its star both orbit around their common center of mass, always staying on opposite sides of that point. The planet's motion around the center of mass often provides astronomers with the information that a planet is present. (b) As a planet moves toward or away from us, its star moves in the opposite direction. Using spectroscopy, we can measure the Doppler shift of the star's spectrum, which reveals the effects of the unseen planet or planets. (c) If a star and its planet are moving across the sky, the motion of the planet causes the star to orbit its center of mass. This motion appears as a wobbling of the star across the celestial sphere. (d) If a planet happens to move in a plane that takes it across its star (that is, it transits the star), as seen from Earth, then the planet will hide some of the starlight, causing the star to dim. This change in brightness will occur periodically and can reveal the presence of a planet.

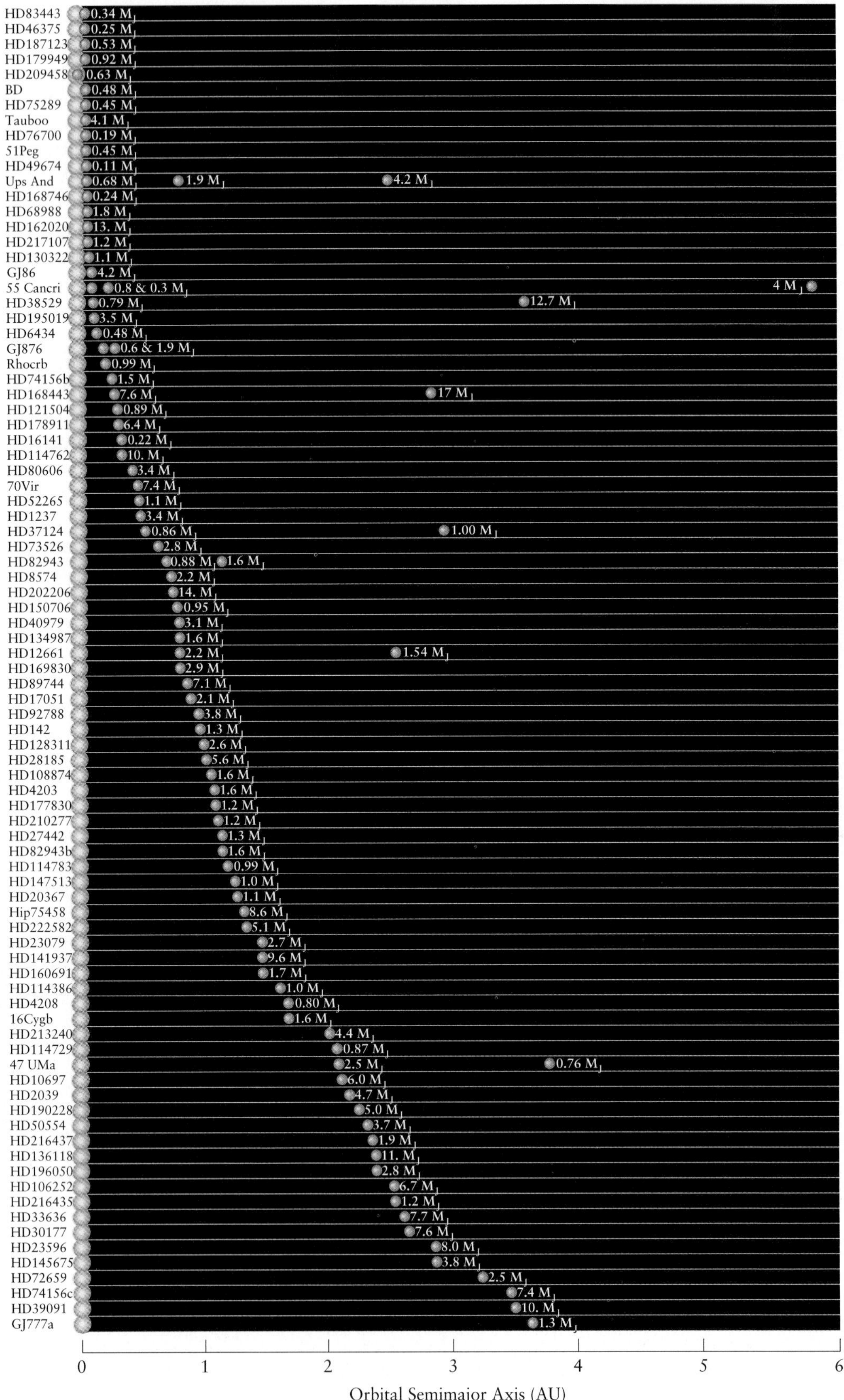

FIGURE 5-15 Planets and Their Stars This figure shows the separation between extrasolar planets and their stars. The corresponding star names are given on the left of each line. Note that many systems have giant planets orbiting much closer than 1 AU from their stars. (California and Carnegie Planet Search)

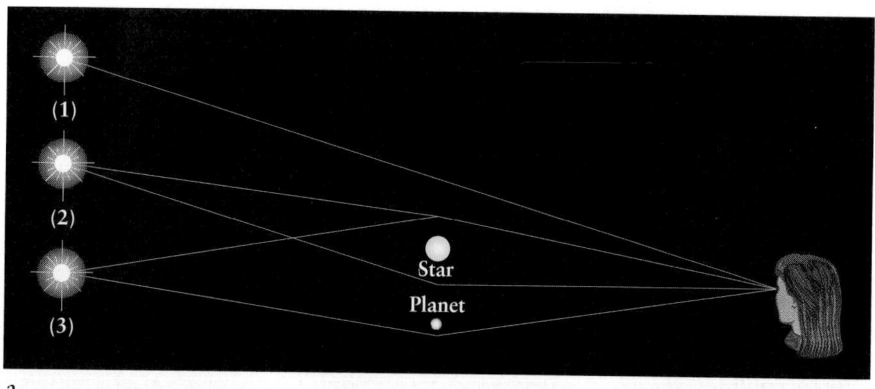

a

(1) No microlensing

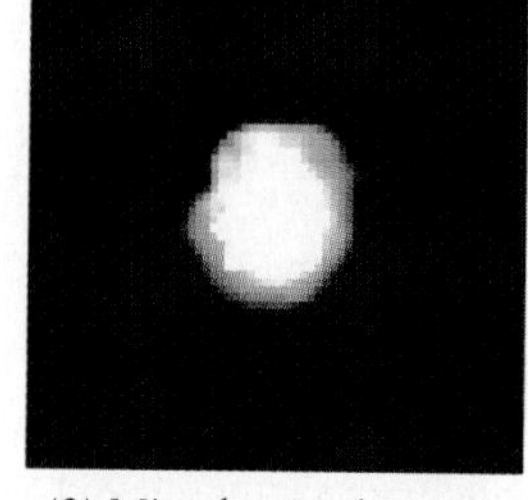

(2) Microlensing by star

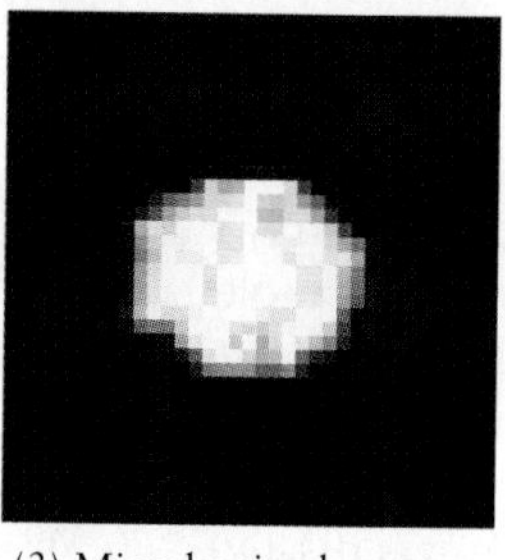

(3) Microlensing by star and planet

b

FIGURE 5-16 **Microlensing Reveals an Extrasolar Planet** (a) Gravitational fields cause light to change direction. As a star with a planet passes between Earth and a more distant star, the light from the distant star is focused toward us, making the distant star appear brighter. (b) The focusing of the distant star's light occurs twice, once by the closer star and once by its planet, making the distant star change brightness. The closer star and planet are 17,000 ly away, while the distant star is 24,000 ly away.

stars wobbling in ways too complicated to be caused by a single planet. In a multiple planet system, each planet contributes a tug to the star. By combining the effects of two or more planets, the observed pattern of the star's motion (Doppler or astrometric) can be recovered. The first star discovered with multiple planets was Upsilon Andromedae (Figure 5-17). So far, 13 stars have been observed with two or more planets orbiting them. Several more of them are depicted in Figure 5-15.

A dramatic consequence of the inward spiraling of planets was discovered in 2001, when the remnants of at

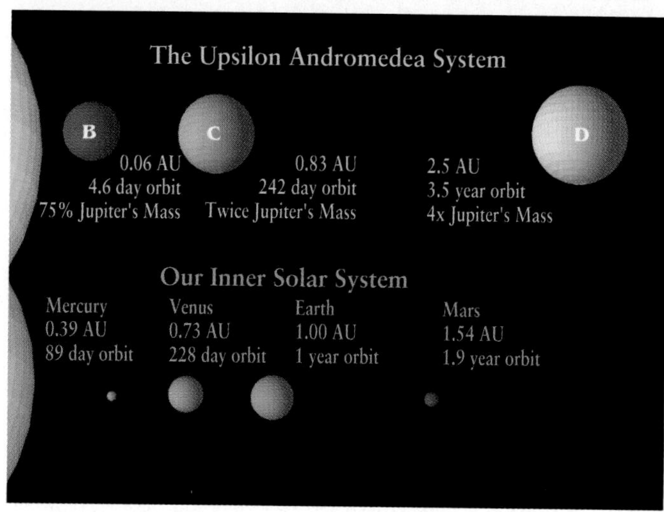

a

b

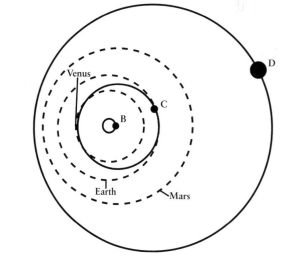

FIGURE 5-17 **A Star with Three Planets** (a) The star Upsilon Andromedae has at least three planets, discovered by measuring the complex Doppler shift of the star. This star system is located 44 ly from Earth and the planets all have masses similar to that of Jupiter. (b) The orbital paths of the planets, labeled B, C, and D, along with the orbits of Venus, Earth, and Mars, are drawn in for comparison. (Adam Contos, Harvard-Smithsonian Center for Astrophysics)

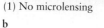

least one planet were discovered in the atmosphere of a star that contains at least two other planets. The star's atmosphere contained a rare form of lithium that is found in planets, but which is destroyed in stars within 30 million years after they form. The presence of this isotope, ^{6}Li, means that at least one planet spiraled so close to the star that it was vaporized.

5-7 Extrasolar planets orbit a breathtaking variety of stars

While some planets are orbiting stars similar to the Sun, many are orbiting burned-out remnants of stars and even orbiting pairs of stars. Most of the planets we have found are around stars in the disk of the Milky Way Galaxy (which is where our solar system resides), but a few have been found in large clumps of stars outside our Galaxy's disk. These clumps, called *globular clusters,* are very old stars that formed shortly after the universe began. Because it is in a globular cluster, the oldest known planet dates back to the cluster's formation, 13 billion years ago.

The vast majority of the 120 extrasolar planets discovered so far have highly elliptical orbits, with eccentricities up to at least to $e = 0.71$ (see Section 2-5). Star systems with massive planets on highly eccentric orbits are unlikely to have life-sustaining Earthlike planets. This is because the changing gravitational force from the massive planet is likely to prevent smaller planets from staying in stable orbits.

Given the growing number of extrasolar planets that are being discovered orbiting a wide range of stars, astronomers are beginning to make estimates for the number of Jupiterlike planets orbiting Sunlike stars in our Milky Way Galaxy. The estimates range between a billion and 30 billion planets. Of these, up to 2 billion are believed to be in orbit around Sunlike stars. All these numbers will be refined as better estimates of the percentage of star systems with planets are determined.

6 None of the *known* extrasolar planets supports life as we know it. In the first place, like our giant planets, most of them lack solid surfaces and they are composed primarily of hydrogen and helium. The terrestrial-type planets among them are orbiting only tiny stellar remnants. These stars exploded long ago with such titanic force that any life on their planets would have been annihilated. Furthermore, most known extrasolar planets are so close to their stars as to be too hot for organic compounds to remain stable.

We are most likely to discover extraterrestrial life on extrasolar planets similar to Earth orbiting at about 1 AU from stars similar to the Sun. This is because stars supporting life have to last long enough for life to evolve and the planets must be at distances at which water can exist as a liquid on their surfaces.

5-8 Frontiers yet to be discovered

Despite their relative closeness to us, there are still countless objects to be discovered and studied in the solar system. While the most poorly understood bodies are the Kuiper belt objects, every other object in the solar system has secrets to be uncovered. Our understanding of the formation of the solar system also has the potential of being greatly advanced by further study of proplyds and planets orbiting other stars.

We begin a detailed exploration of the solar system by examining the two bodies we know best—the Earth and its Moon. By understanding them, we will be better able to make some sense of the remarkably alien neighboring worlds we encounter thereafter.

Summary of Key Ideas

Formation of the Solar System

• Hydrogen, helium, and traces of lithium, the three lightest elements, were formed shortly after the creation of the universe. The heavier elements were produced much later by stars and cast into space when the stars died. By mass, 98% of the matter in the universe is hydrogen and helium.

• The solar system formed 4.6 billion years ago from a swirling, disk-shaped cloud of gas, ice, and dust called the solar nebula.

• The four inner planets formed through the accretion of dust particles into planetesimals and then into larger protoplanets. The four large outer planets probably formed through the runaway accretion of gas and ice onto rocky protoplanetary cores.

• The Sun formed at the center of the solar nebula. After about 100 million years, the temperature at the protosun's center was high enough to ignite thermonuclear reactions. For 800 million years after the Sun formed, impacts of asteroidlike objects on the young planets dominated the history of the solar system.

Comparative Planetology

• The four inner planets of the solar system share many characteristics and are distinctly different from the four giant outer planets and from Pluto.

• The four inner, terrestrial planets are relatively small, have high average densities, and are composed primarily of rock and metal.

• Jupiter and Saturn have large diameters and low densities and are composed primarily of hydrogen and helium. Uranus and Neptune have large quantities of water as well as much hydrogen and helium. Pluto, the smallest of the nine planets, is of intermediate density owing to its rock-ice composition.

• Asteroids are rocky and metallic debris in the solar system larger than about a kilometer in diameter. Meteoroids are smaller pieces of such debris. Comets are debris that contain both ice and rock.

Planets Outside Our Solar System

• Astronomers have observed disks of gas and dust orbiting young stars.

• At least 120 extrasolar planets have been discovered orbiting other stars.

• Most of the extrasolar planets that have been discovered have masses around that of Jupiter.

• Extrasolar planets are discovered indirectly, by their effects on the stars they orbit.

WHAT DID YOU THINK?

1 *How many stars are there in the solar system?* The solar system has only one star, the Sun.

2 *Was the solar system created as a direct result of the formation of the universe?* No. All matter and energy were created by the Big Bang. However, much of the material that exists in our solar system was processed inside stars that evolved before the solar system existed. The solar system formed billions of years after the Big Bang occurred.

3 *How long has the Earth existed?* The Earth formed along with the rest of the solar system 4.6 billion years ago.

4 *Is Pluto always the farthest planet from the Sun?* No. From 1979 to 1999, Neptune was the farthest planet from the Sun. This was because Pluto's orbit is highly eccentric, bringing that planet inside Neptune's orbit for about 20 years once every 250 years.

5 *What typical shape(s) do moons have?* While some moons are spherical, most look roughly like potatoes.

6 *Have any Earthlike planets been discovered orbiting Sunlike stars?* No, nearly all the planets discovered orbiting Sunlike stars are Jupiter-like gas giants.

Key Words

accretion, 123
albedo, 129
asteroid, 126
asteroid belt, 126
average density, 128
comet, 126
crater, 125
meteoroid, 126
microlensing, 131
orbital inclination, 125
planetesimal, 123
protoplanet, 123
protoplanetary disks (proplyds), 121
protosun, 121
solar nebula, 121
solar system, 119
terrestrial planet, 128

Review Questions

1. Why was Pluto recently closer to the Sun than Neptune?

2. Why do astronomers believe that the solar nebula was rotating?

3. To test your understanding of the formation of the solar system, do Interactive Exercise 5-1 on the Web site. Explain what is happening in each figure. You can print out your results, if required.

4. Describe four methods that exist for discovering extrasolar planets.

Advanced Questions

5. What two properties of a planet must be known in order for its average density to be determined? How are these properties determined?

6. How can Neptune have more mass than Uranus but a smaller diameter? *Hint:* See Table 5-2.

What If . . .

7. The Earth had formed around one of the first-generation stars created in the universe? What chemical elements would the Earth then be composed of? Could we humans exist on that version of Earth? Why or why not?

8. The Earth had a highly elliptical orbit, like Pluto? What would be different about the Earth and life on it?

9. The accretion process of planet formation was still going on? How would life in general on Earth be different and how might we be different?

10. The solar system passed through a cloud of gas and dust that was beginning to collapse to form a new star and planet system? What might happen to the Earth and how would the passage affect the appearance of the sky?

Web Questions

11. Search the Web for information about recent observations of protoplanetary disks. What insights about the formation of the solar system have astronomers gained from these observations? Explain the evidence astronomers have, from these observations, that planets are forming in these disks.

12. In 2000, extrasolar planets with masses comparable to that of Saturn were first detected around the stars HD 16141 (also called 79 Ceti) and HD 46375. Search the Web for information about these planets. How do their masses compare to previously discovered extrasolar planets? Do these planets move around their stars in the same kind of orbit as Saturn follows around the Sun? If not, explain the differences.

13. Search the Web for information about the planet orbiting the star HD 209458. This planet has been detected using two methods. What are they? What have astronomers learned about this planet?

14. **Determining Terrestrial Planet Orbital Periods.** Access the animation "Planetary Orbits" in Chapter 5 of the *Discovering the Universe* Web site. Focus on the motions of the inner planets during the last half of the animation. Using the stop and start buttons, determine how many days it takes Mars, Venus, and Mercury to orbit the Sun once if it takes Earth approximately 365 days.

Observing Projects

15. Planetarium software can help you get an overview of our solar system. We will learn more about it in the next five chapters. For now, however, open your *Starry Night Enthusiast*™ program and put the program in Atlas mode (*Favourites/Guides/Atlas*). Stop the daily motion by pressing the ■ button. Grab the sky and move along the celestial equator until the planets (with their names) appear. Describe and comment on the brightnesses of the various planets relative to this evening and the stars. Now go to home setting (ctrl+shift+h) and change the time this evening to determine which, if any, planets are now visible in the night sky where you live. List them.

16. There are many young stars still embedded in the clouds of gas and dust from which they formed. In the winter evening sky in the northern hemisphere, for example, is the famous Orion Nebula (Figure 5-2). In the summer night sky in the northern hemisphere, the Lagoon, Omega, and Trifid nebulae can be observed. Examine some of these nebulae with a telescope. Describe their appearances. Try to determine which stars in your field of view are actually associated with the nebulosity. You can use the *Starry Night Enthusiast*™ program to help you find these nebulae using the *Find* tab. Also, the table below gives their coordinates (right ascension and declination; see Section 1-3).

Nebula	Right ascension	Declination
Lagoon	$18^h\ 03.8^m$	–24° 23'
Omega	18 20.8	–16 11
Trifid	18 02.3	–23 02
Orion	5 35.4	–5 27

17. Use the *Starry Night Enthusiast*™ program to observe magnified images of at least four of the planets. Set atlas mode (*Favourites/Guides/Atlas*). Use *Find* tab, press on the ▼, and select *magnify*. Describe the planet's appearance. From what you observe, state whether you are looking at the planet's surface or a cloud cover and justify your assertion.

Earth and Moon

Aurora Borealis in Alaska (J. Finch/Photo Researchers, Inc.) R I V U X G

WHAT DO YOU THINK?

1 Can the Earth's ozone layer, which is now being depleted, be replenished?

2 Who was the first person to walk on the Moon and when did this event occur?

3 Do we see all parts of the Moon's surface at some time throughout the lunar cycle?

4 Does the Moon rotate and, if so, how fast?

5 What causes the ocean tides?

6 When does the spring tide occur?

Imagine you are entering our solar system from a distant star system, looking for habitable worlds. Your spacecraft approaches the inner solar system from the opposite side of the Sun from the Earth. You encounter Mars, with its thin, chilled, unbreathable atmosphere and barren desert landscapes. Not very promising. Moving much closer to the Sun, the next planet you happen to encounter, Venus, is enshrouded in corrosive clouds hiding a menacingly hot surface. Finally, moving around the Sun and heading back outward you spy Earth, orbiting the Sun between these two relatively forbidding worlds—one too cold, the other too hot. Ever-changing white clouds pirouette above the browns and blues of Earth's continents and oceans. Its close companion, the Moon, pockmarked by countless craters, causes those oceans to move up and down in a ceaseless rhythm. You begin to think that Earth may be what you are looking for.

In this chapter you will discover

- why the Earth is such an ideal environment for life
- that the Earth is constantly in motion inside and out
- how Earth's magnetic field helps protect us
- what made the craters on the Moon
- how the Sun and Moon cause the Earth's tides
- that both the Earth and Moon have two (different) major types of surface features
- that water ice may have been found at the Moon's poles

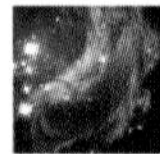

EARTH: A DYNAMIC, VITAL WORLD

Looking at the Earth from outer space, the first thing one notices is that it scatters about 37% of the sunlight it receives right back into space. That scattered light (an albedo of 0.37) would tell an interstellar traveler that the Earth is indeed an appealing world to investigate, because it indicates large quantities of liquid water as well as dry land. Water, the nearly universal solvent in which life on Earth first formed, covers about 71% of our planet's surface.

This is a geologically active, ever-changing world. Earthquakes shake many regions of it. Volcanoes pour huge quantities of molten rock from just under the crust (the outer layer) onto the surface, and gas from within it vents into the atmosphere. Some mountains are still rising, while others are wearing away. Water flow erodes topsoil and carves river valleys. Rain and snow help rid the atmosphere of dust particles. And life teems virtually everywhere, making the Earth unique in the solar system. Figure 6-1 details Earth's important physical and orbital properties. The symbol ⊕ in Figure 6-1 and hereafter is astronomers' shorthand for "Earth."

6-1 The Earth's atmosphere has evolved over billions of years

The Earth's atmosphere is a key part of our planet's dynamical activity and its ability to sustain life. Essential to the existence of complex life is an atmosphere containing oxygen molecules. The Earth's atmosphere is unique among the planets. The air we breathe is predominantly a 4-to-1 mixture of nitrogen to oxygen, two gases that are found only in small amounts in the atmospheres of other planets. This is, however, the third atmosphere to envelop the Earth. The Earth's early atmosphere contained no oxygen.

Composed of trace remnants of hydrogen and helium left over from the formation of the solar system, the first atmosphere did not last very long. These gases are too light to stay near the Earth. Heated by sunlight (meaning that they moved faster and faster), they gained enough kinetic energy to break the Earth's gravitational bond and fly away into space.

The gases of the second atmosphere came from inside the Earth, vented through volcanoes and cracks in the Earth's surface. That atmosphere was composed primarily of carbon dioxide and water along with some nitrogen. There was roughly 100 times as much gas in that second atmosphere as there is in the air today. Carbon dioxide and water store a lot of heat, helping to create what is called the *greenhouse effect,* about which we will say more shortly. The high concentration of these gases early in the Earth's life stored more heat than the atmosphere does today. The Earth back then would have been a serious hothouse had the Sun reached its full intensity at that time. Evidence indicates it did not, and so the high concentrations of carbon dioxide and water in the air prevented the young Earth from freezing.

Oceans began forming when water precipitated out of this atmosphere, as well as from impacts of water-rich space debris such as the comet nuclei introduced in Chapter 5. The rain absorbed and carried down lots of carbon dioxide and the oceans eventually absorbed about half the carbon dioxide in that second atmosphere. (Water holds a lot

EARTH: VITAL STATISTICS

Average distance from the Sun:	1.000 AU = 1.496×10^8 km
Maximum distance from the Sun:	1.017 AU = 1.521×10^8 km
Minimum distance from the Sun:	0.983 AU = 1.471×10^8 km
Orbital eccentricity:	0.017
Average orbital speed:	29.79 km/s
Sidereal period of revolution (yr):	365.26 days (1.00 yr)
Sidereal rotation period:	0.997 days
Solar rotation period (day):	1.00 day = 24.0 hr
Inclination of equator to orbit:	23.5°
Mass:	5.974×10^{24} kg (1 $M_{\oplus}$)
Average density:	5520 kg/m^3
Albedo (average):	0.37
Escape speed:	11.2 km/s
Surface temperature range:	Maximum 60°C = 140°F = 333 K Mean: 14°C = 57°F = 287 K Minimum: −90°C = −130°F = 183 K
Atmospheric composition (by number of molecules):	78.08% nitrogen (N_2), 20.95% oxygen (O_2), 0.035% carbon dioxide (CO_2), about 1% water vapor

R I V U X G

FIGURE 6-1 **A View of Earth's Surface** An oasis in the forbidding void of space, the Earth is a world of unsurpassed beauty and variety. Ever-changing cloud patterns drift through its skies. More than two-thirds of its surface is covered with oceans. This liquid water, in combination with a huge variety of chemicals from its lands, led to the formation and evolution of life over most of the planet's surface. (NASA)

of carbon dioxide, as you can see by the amount of carbon dioxide fizz in a can of soda.) As life evolved in the oceans, much of that dissolved carbon dioxide was transformed into the shells of many creatures, and then, as the animals died, their carbon-rich shells sank to the ocean bottoms. Shells piled up and compressed the shells underneath them into rock such as limestone.

Early plant life in the oceans and then on the shores removed most of the remaining carbon dioxide in the air by converting it into oxygen plus the nutrients that the plants needed to survive. The most efficient of such conversion mechanisms is photosynthesis, which today helps maintain the balance between carbon dioxide and oxygen in the air. What remained in the air was nitrogen and water vapor.

Oxygen that entered the air early in the life of the Earth came from oxygen-rich molecules, like carbon dioxide, that were broken apart by the Sun's ultraviolet radiation. Then came oxygen released as a by-product of ocean plant life activity. In both cases, this oxygen did not stay in the air for long. Oxygen is highly reactive, and early atmospheric oxygen thus combined quickly with many elements on the Earth's surface, notably iron, to form new compounds such as iron oxides (otherwise known as rust).

About two billion years ago, after all the minerals that could combine with oxygen had done so, the atmosphere began to fill with this gas. With the carbon dioxide gone, the air became the nitrogen-oxygen mixture that we breathe today. While nitrogen was only a tiny fraction of that earlier atmosphere, it now dominates the thinner air of today.

Because it has mass, the air is pulled down toward the Earth by gravity, thereby creating a pressure on the surface. We define pressure as a force acting over some area:

$$\text{Pressure} = \frac{\text{force}}{\text{area}}$$

The weight of the air pushing down creates a pressure at sea level of 14.7 pounds per square inch, which is commonly denoted as 1 atmosphere (atm). The atmospheric pressure decreases with increasing altitude, falling by roughly half with every gain of 5.5 km of altitude. Seventy-five percent of the mass of the atmosphere lies within 11 km (approximately 7 mi, or 36,000 ft) of the Earth's surface.

Layers of the Atmosphere

All Earth's weather—clouds, rain, sleet, and snow—occurs in this lowest layer of the atmosphere, called the **troposphere.** Commercial jets generally fly at the top of the troposphere to avoid having to plow through the denser air below. If the Earth were the size of a typical classroom globe, the troposphere would only be as thick as a sheet of paper wrapped around its surface.

Starting at the Earth's surface, atmospheric temperature initially decreases upward, reaching a minimum of 218 K (–67°F) at the top of the troposphere. Above this level is the region called the **stratosphere,** which extends from 11 to 50 km (approximately 7 to 31 mi) above the Earth's surface. This is the realm of the **ozone layer.**

INSIGHT INTO SCIENCE

What's in a Word The word "layer" in the term "ozone layer" suggests a narrow region, perhaps a few meters thick. Not so. The ozone layer extends in altitude over some 40 kilometers of the atmosphere!

The Ozone Layer

The ozone layer has been much in the news. What is it and how threatened is it? Ozone molecules (three oxygen atoms, O_3, bound together) are created in the stratosphere when short-wavelength ultraviolet radiation from the Sun occasionally splits normal oxygen molecules, O_2, up there into two separate oxygen atoms. Each resulting oxygen atom then combines with different O_2 molecules to create ozone, O_3. Once created, ozone efficiently absorbs intermediate wavelength solar ultraviolet rays, thereby heating the air in this layer while preventing much of this lethal radiation from reaching the Earth's surface.

The troposphere (lowest part of the atmosphere) is heated primarily by the Earth and so the temperature in this part of the atmosphere decreases with altitude—the farther away from the heat source of the Earth, the cooler the air gets. Because the ozone in the stratosphere stores significant heat from the Sun above it, the temperature in the stratosphere actually rises with altitude to about 285 K (50°F) at its top (Figure 6-2).

Not much ozone actually exists in the ozone layer. If all of it was compressed to the density of the air we breathe, it would be a layer only a few millimeters (about one-eighth of an inch) thick. It is enough, however, to protect us from some of the Sun's lethal ultraviolet radiation. Ultraviolet radiation has been a double-edged sword in the history of the Earth. Because the Earth's early atmosphere lacked ozone, ultraviolet radiation penetrated to the planet's surface, where it provided much of the energy necessary for life to evolve in the oceans. After life developed, however, the formation of the ozone layer by the same ultraviolet radiation was essential in lessening the amount of ultraviolet radiation that reached the surface, thereby allowing lifeforms to leave the oceans and survive on land.

Today, the ozone layer is still vital, because the energy it absorbs would otherwise cause skin cancer and various eye diseases. In sufficiently high doses, ultraviolet radiation even damages the human immune system by destroying some of the molecules that normally repair cells in our bodies. The ozone layer is being depleted today by human-made chemicals, such as chlorofluorocarbons (CFCs). At present, ozone levels in the mid-latitudes of the northern hemisphere are decreasing by about 4% per decade. Much

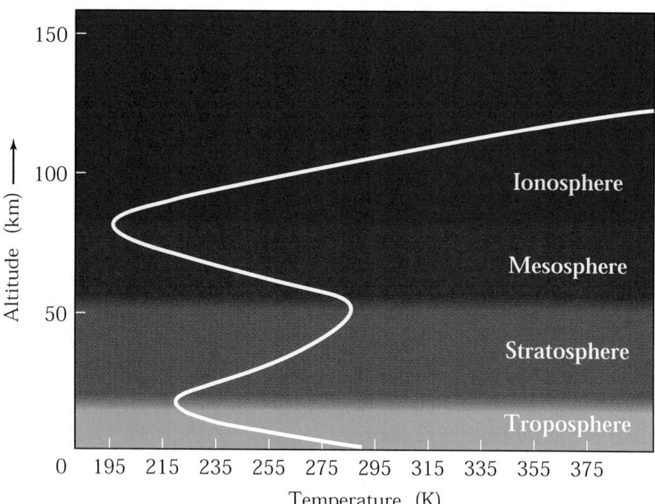

FIGURE 6-2 **Temperature Profile of Earth's Atmosphere** The atmospheric temperature changes with altitude because of the way sunlight interacts with various gases at different heights.

more rapid and dramatic drops in the ozone layer over the Earth's poles, especially over the South Pole, have been observed for more than two decades. These seasonal *ozone holes* typically extend over 24 million square kilometers, or 4.6% of the Earth's surface area.

1 Fortunately, as explained earlier, sunlight creates ozone. If current international efforts aimed at slowing the depletion rate succeed, the ozone layer will naturally replenish itself in a century or so, and the ozone holes will probably stop occurring.

Note in Figure 6-2 that above the stratosphere lies the **mesosphere.** Atmospheric temperature again declines as one moves up through the mesosphere, reaching a minimum of about 200 K (–103°F) at an altitude of about 80 km (50 mi). This minimum marks the bottom of the **ionosphere,** above which the Sun's ultraviolet light heats and ionizes atoms, producing charged particles that reflect radio waves. You can often tune in distant AM radio stations because their transmissions bounce off this region and return to Earth. On the other hand, FM radio stations use much higher frequencies that pass through this layer without bouncing back. That is why you must be quite near an FM station to receive it.

6-2 Plate tectonics produce major changes on the Earth's surface

Land masses protrude through Earth's oceans. These are regions of the Earth's outermost layer, or **crust,** and are composed of relatively low-density rock that floats on denser material below it. By studying the various kinds of rocks at a particular location, geologists can deduce the history of that site. For example, whether an area was once covered by an ancient sea or flooded by lava from volcanoes is readily apparent from the kinds of rocks that are present.

The crust has distinct boundaries (see, for example, Figure 6-3), many occurring beneath the Earth's oceans. These divisions indicate that the continents are separate bodies. Anyone who carefully examines the shapes of Earth's continents (see Figure 6-5a) might conclude that landmasses move. Eastern South America, for example, looks like it would fit snugly against western Africa. Between 1912 and 1915, the German scientist Alfred Wegener published his observations on this remarkable fit between landmasses on either side of the Atlantic Ocean, an idea proposed by Isaac Newton 2 centuries earlier. Wegener was inspired to advocate the theory of **continental drift**—that the continents on either side of the Atlantic Ocean have moved apart.

Wegener argued that a single, gigantic supercontinent called Pangaea (meaning "all lands") began to break up and drift apart some 200 million years ago. Initially, most geologists ridiculed Wegener's ideas. Although it was generally accepted that the continents do float on denser rock beneath them, few geologists could accept that entire continents move across the Earth's face at speeds as great as several centimeters per year. Wegener and other "continental drifters" could not explain what forces would shove the massive continents around or how less-dense continental rock had gotten through the dense rock of the ocean floor.

WEB LINK 6.3 Then, in the mid-1950s, long, underwater mountain ranges were discovered. The Mid-Atlantic Ridge, for example, stretches all the way from Iceland to Antarctica (see Figure 6-3). Careful examination revealed that molten rock from the Earth's interior is being forced upward there, causing the ocean floor to slide apart on either side of the ridge. This **seafloor spreading** is pushing the Americas away from Europe and Africa at a speed of roughly 3 cm per year.

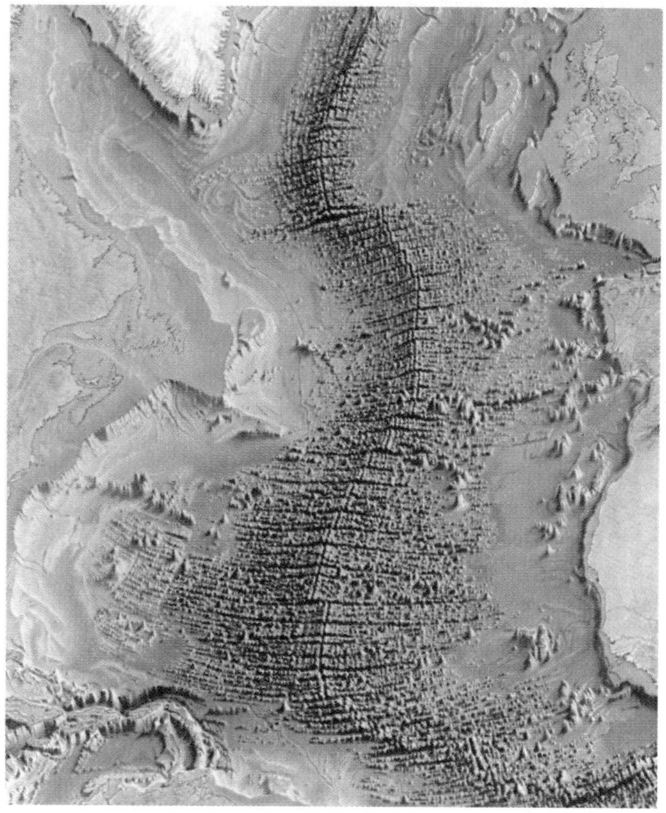

FIGURE 6-3 **The Mid-Atlantic Ridge** This artist's rendition of the bottom of the North Atlantic Ocean shows an unusual mountain range in the middle of the ocean floor. Called the Mid-Atlantic Ridge, these mountains are created by lava seeping up from the Earth's interior along a rift that extends from Iceland to Antarctica. (Courtesy of M. Tharp and B. C. Heezen)

a 237 million years ago: the supercontinent Pangaea

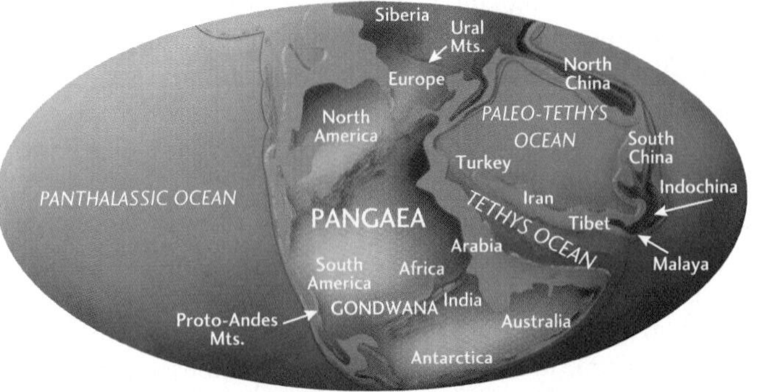

b 152 million years ago: the breakup of Pangaea

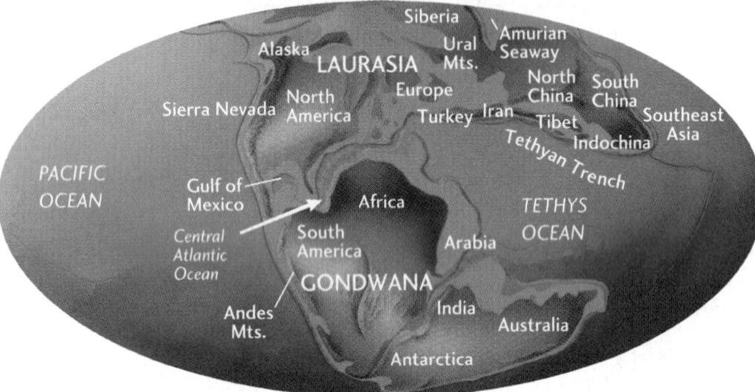

c The continents today

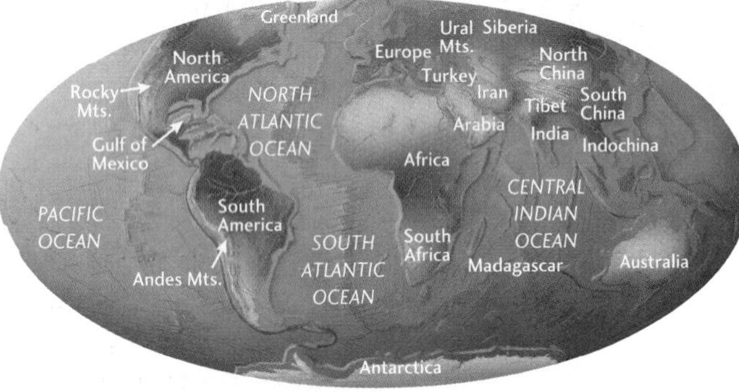

◀ **FIGURE 6-4 The Supercontinent Pangaea** (a) The continents of today are pieces of what was once a bigger, united body called Pangaea. (b) Geologists believe that Pangaea must have first split into two smaller supercontinents, which they call Laurasia and Gondwanaland. (c) These bodies later separated into the continents of today. Gondwanaland split into Africa, South America, Australia, and Antarctica, while Laurasia divided to become North America and Eurasia.

INSIGHT INTO SCIENCE

From Proof to Acceptance Scientific theories may take years or decades of testing and replication before they are accepted. Newton speculated about continental drift in the late seventeenth century. Wegener had proposed plate tectonics by 1912. His theory was not verified until the 1950s and was not accepted until the late 1960s.

Seafloor spreading provided just the physical mechanism that had been missing from the theory of continental drift. It established Wegener's theory of crustal motion, which was developed into the modern theory of **plate tectonics.**

The Earth's surface is made up of about a dozen major thick rafts of rock called tectonic plates that move relative to each other. These plates are floating on the slowly moving, denser matter below them. The plates are not always separate, as they are today. Sometimes they are merged into a single mass called a *supercontinent.* In recent years, geologists have uncovered evidence that points to a whole succession of supercontinents that broke apart and reassembled. Pangaea is only the most recent supercontinent in this cycle, which repeats on average every 500 million years (Figure 6-4).

The boundaries between plates are the sites of some of the most impressive geological activity on our planet. Earthquakes tend to occur at the boundaries of the Earth's crustal plates, where the plates are colliding (convergent boundaries), separating (divergent boundaries), or grinding against each other (transform fault boundaries). The boundaries between tectonic plates (Figure 6-5a) stand out clearly when the epicenters of earthquakes are plotted on a map. The San Andreas Fault, running along the West Coast of the United States, is an example of a fault where the North American and Pacific plates are rubbing against each other (Figure 6-5b). The Red Sea and Gulf of Suez are expanding as Africa and the Middle East move apart (Figure 6-5c). Great mountain ranges, such as the Himalayas, are thrust up by ongoing collisions between the Indian-Australian and Eurasian plates (Figure 6-5d). During such collisions, one plate moves downward (subducts) into the Earth under the other. Where old crust subducts, we find deep trenches, such as the ones off the coasts of Japan and Chile. Tectonic plate motion helps explain why the Earth's oceans have few craters today: Countless impacts occurred on the oceans, which cover most of our planet. Some of the impacting bodies were slowed or obliterated before cratering the ocean floor, while others were big enough or fast enough to hit the floors. The resulting craters in the oceans have been drawn underground and thereby obliterated.

But what powers all this activity? Some enormous energy source has caused shifting among landmasses for billions of years. The answer to that question leads us to a scientific model of the Earth's interior.

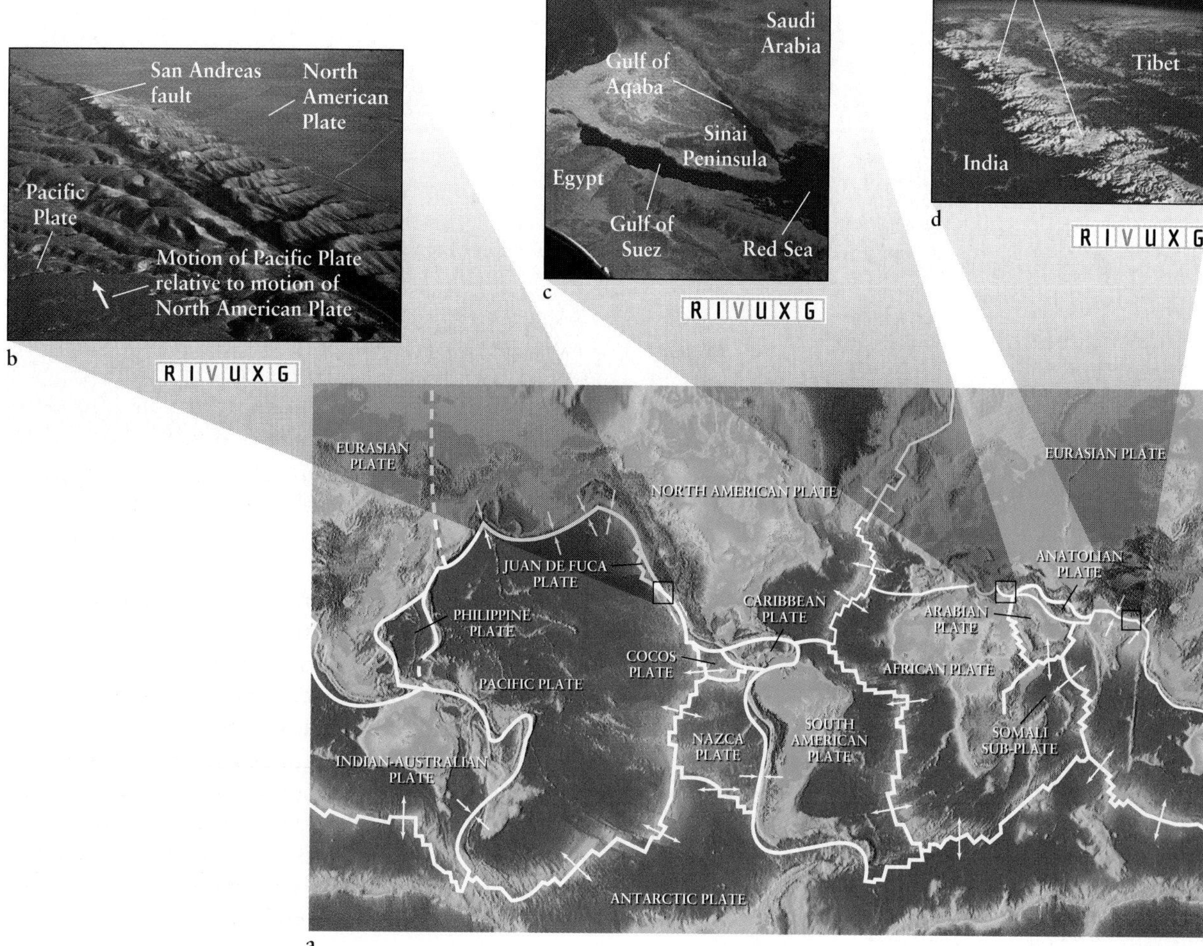

FIGURE 6-5 **The Earth's Major Tectonic Plates** (a) The Earth's surface is divided into a dozen or so rigid plates that move relative to each other. The boundaries of the plates are the scenes of violent seismic and geologic activity, such as earthquakes, volcanoes, rising mountain ranges, and sinking seafloors. The arrows indicate whether plates are moving apart (← →), together (→ ←), or sliding past one another ⇆).

ANIMATION 6.1 (b) Rubbing of Two Plates. The San Andreas fault, running up the west coast of North America, exists because the Pacific Plate is moving northwest along the North American Plate.

ANIMATION 6.2 (c) The Separation of Two Plates. The plates that carry Egypt and Saudi Arabia are moving apart, leaving the trench that contains the Red Sea. This view, taken by astronauts in 1966, shows the northern Red Sea on the right. (d) The Collision of Two Plates. The plates that carry India and China are colliding. As a result, the Himalayas are being thrust upward. In this photograph, taken by astronauts in 1968; Mount Everest is one of the snow-covered peaks near the center. The power of tectonic plate motion was demonstrated on December 26, 2004 in the form of a devastating tsunami (ocean wave) that caused death and destruction in the Indian Ocean. (a: Digital image by Peter W. Sloss, NOAA-NESDIS-NGDC; b: © Craig Aurness/Corbis; c: *Gemini 12*, NASA; d: *Apollo 7*, NASA)

6-3 Earth's interior consists of a rocky mantle and an iron-rich core

Geologists have calculated that the Earth was entirely molten soon after its formation, about 4.6 billion years ago. The violent impact of space debris, along with energy released by the breakup of radioactive elements, heated, melted, and kept the young Earth molten. The process of radioactive atoms breaking apart, called *nuclear fission,* also creates the heat used to generate electricity in nuclear power plants. This differs from the energy generating nuclear fusion in the Sun, wherein lighter atoms are bonded

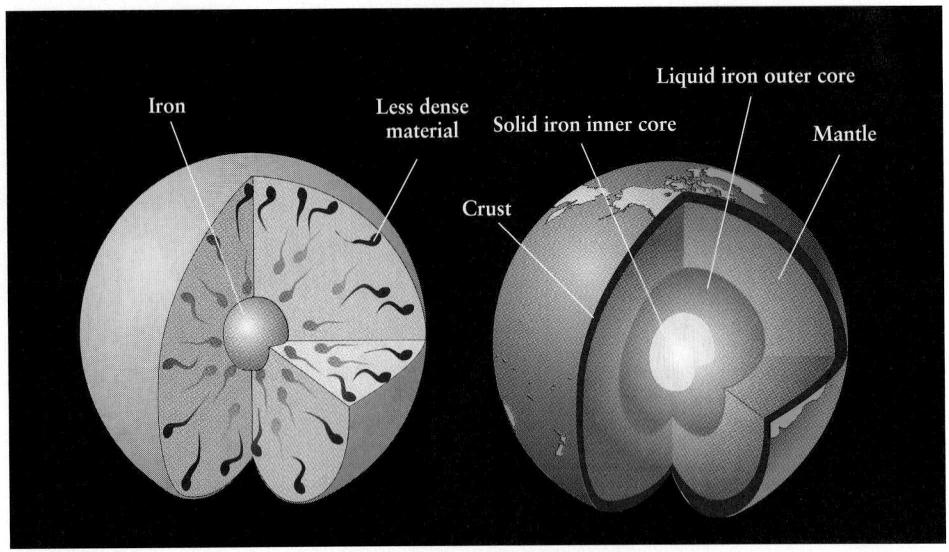

a During differentiation, iron sank to the center and less dense material floated upward

b As a result of differentiation, the Earth has the layered structure that we see today

FIGURE 6-6 **Cutaway Model of the Earth** The early Earth was initially a homogeneous mixture of elements with no continents or oceans. (a) While molten, iron sank to the center and light material floated upward to form a crust. (b) As a result, Earth has a dense iron core and a crust of light rock, with a mantle of intermediate density between them.

together, as mentioned in Chapter 5 and discussed in detail in Chapter 10.

Most of the iron and the other dense elements sank toward the center of the young, liquid Earth, just as a rock sinks in a pond. At the same time, most of the less dense materials were forced upward toward the surface (Figure 6-6a). This process, called **planetary differentiation,** produced a layered structure within the Earth: a very dense central **core** surrounded by a **mantle** of less dense minerals, which in turn was surrounded by a thin crust of relatively light minerals (Figure 6-6b).

Differentiation explains why most of the rocks you find on the ground are composed of lower-density elements like silicon and aluminum. Much of the iron, gold, lead, and other denser elements found on the Earth today actually had to return to the surface via volcanoes and other lava flows. Despite this return of heavier elements to the surface, the average density of crustal rocks is 3000 kg/m^3, considerably less than the average density of the Earth as a whole (5520 kg/m^3).

To understand the Earth's interior, geologists have created a mathematical model that includes the effects of gravity, radioactivity, chemistry, heat transfer, rotation, and other factors. The model predicts that the temperature in the Earth rises steadily from about 290 K on the surface to nearly 5000 K at the center. Radioactivity in the Earth's interior helps replenish heat lost through the planet's surface. In addition to temperature, pressure also increases with increasing depth below the Earth's surface. The deeper you go, the more mass presses down on you. These factors shape the Earth's inner layers.

You might think that temperatures of 5000 K (9000°F) would melt virtually any substance. The mantle, in particular, is composed largely of minerals rich in iron and magnesium, both of which have melting points of only slightly more than 1250 K at the Earth's surface. However, the melting point of anything also depends on the pressure to which it is subjected: The higher the pressure, the higher a material's melting point. The high pressures within the Earth (1.4 million atm at the bottom of the mantle) keep the mantle solid to a depth of about 2900 km and also allow for a solid iron core in the center of the planet. In other words, the inner core's pressure is so high that it forces the very hot atoms there into a solid.

Geologists have tested their models of the Earth's interior by studying the response of the planet to earthquakes. Earthquakes produce a variety of **seismic waves,** vibrations that travel through the Earth either as ripples like ocean waves or by compressing matter like sound waves. Geologists use sensitive **seismographs** to detect and record these vibratory motions. The varying density and composition of the Earth's interior affects the direction of seismic waves traveling through the Earth. By studying the deflection of these waves, geologists have been able to determine the structure of the Earth's interior.

Analysis of seismic recordings led to the discovery that the Earth's solid core has a radius of about 1250 km (775 mi), surrounded by a molten iron core with a thickness of about 2250 km (1375 mi). For comparison, the overall radius of our planet is 6378 km (3963 mi). The cores are composed of roughly 80% iron and 20% lighter elements such as silicon. The interior of our planet therefore has a curious structure: a liquid region sandwiched between a solid inner core and a solid mantle (see Figure 6-6b).

A sophisticated computer model of the Earth's interior predicts that the solid inner core is rotating faster than the

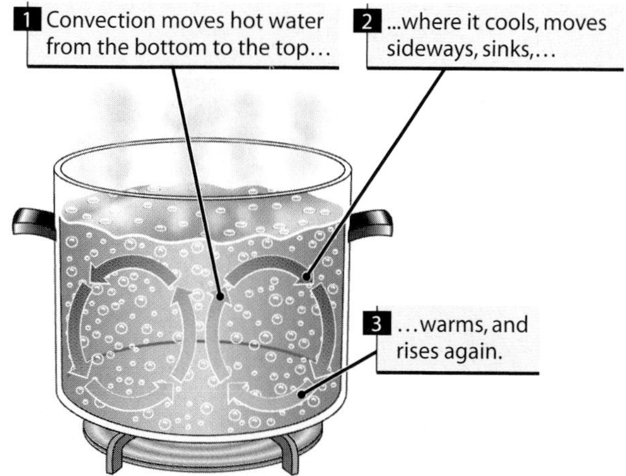

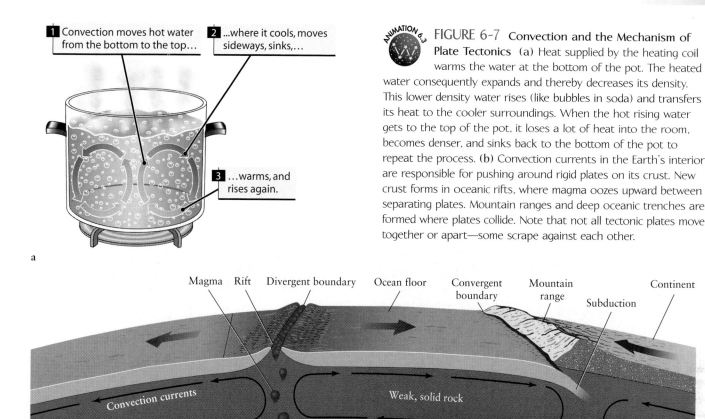

FIGURE 6-7 Convection and the Mechanism of Plate Tectonics (a) Heat supplied by the heating coil warms the water at the bottom of the pot. The heated water consequently expands and thereby decreases its density. This lower density water rises (like bubbles in soda) and transfers its heat to the cooler surroundings. When the hot rising water gets to the top of the pot, it loses a lot of heat into the room, becomes denser, and sinks back to the bottom of the pot to repeat the process. (b) Convection currents in the Earth's interior are responsible for pushing around rigid plates on its crust. New crust forms in oceanic rifts, where magma oozes upward between separating plates. Mountain ranges and deep oceanic trenches are formed where plates collide. Note that not all tectonic plates move together or apart—some scrape against each other.

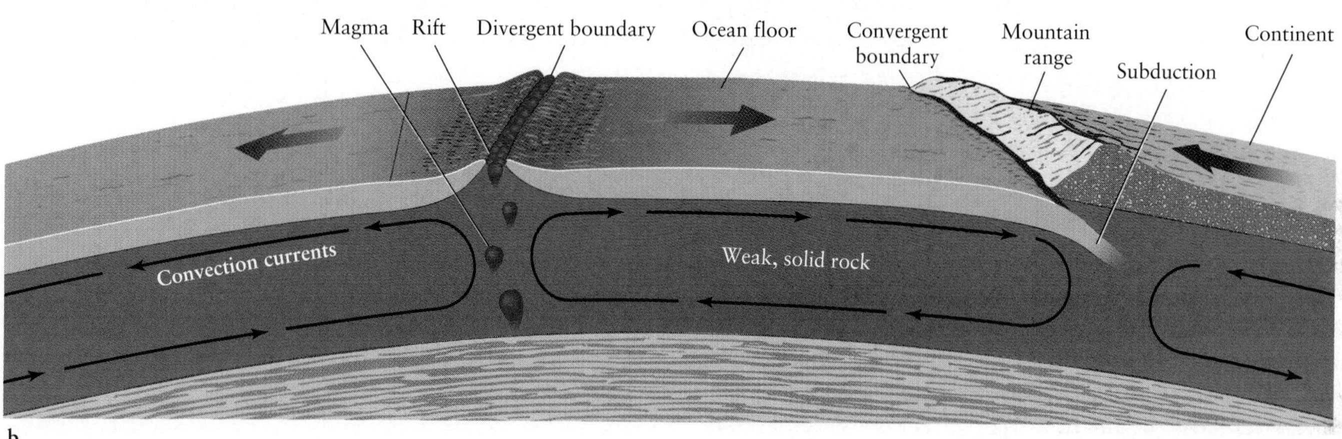

rest of the planet. Seismic evidence supports the model that the core goes around about 2 degrees more per year than the surface. Refined models and observations like these are revealing more of the secrets of the Earth's interior.

The tremendous heat and molten rock inside the Earth provide the energy that drives the plate motion. That heat, from the pressure of the overlying rock and from the decay of radioactive elements, affects the Earth's surface today. This heat transport is accomplished by **convection.** In this process a liquid or gas is heated from below and as a result, it expands, becomes buoyant, and therefore rises, carrying the heat it received with it. Eventually, it gives off the heat, cools, condenses, and sinks back down. You witness convection every time you see simmering water. Heated from below, blobs of hot water move upward. The rising water gives off its heat at its surface (Figure 6-7a).

Heated from below, the rocks of the Earth's upper mantle, although solid, are weak and hot enough to have an oozing, plastic flow (like stretching Silly Putty). As sketched in Figure 6-7b, molten rock, or *magma,* seeps upward along cracks or rifts in the ocean floor, where convection currents of mantle rock are carrying plates apart. Crustal rock sinks back down into the Earth where plates collide. The continents ride on top of the plates as they are pushed around by the convection currents circulating beneath them.

Supercontinents like Pangaea apparently sow the seeds of their own destruction by blocking the flow of heat from the Earth's interior. As soon as a supercontinent forms, temperatures beneath it rise. As heat accumulates, the supercontinent domes upward and cracks. Overheated molten rock wells up to fill the resulting rifts, which continue to widen as pieces of the fragmenting supercontinent are pushed apart.

The Earth's molten, iron-rich interior affects the Earth's exterior in another important way, one that has fewer obvious signs than tectonic plate motion. It is the creation of a magnetic field that extends through the Earth's surface and out into space. As we will see next, that magnetic field results from the convection of molten metal inside the Earth combined with the Earth's rotation.

6-4 The Earth's magnetic field shields us from the solar wind

If you have ever used a compass, you have seen the effect of the Earth's magnetic field. This planetary field is quite similar

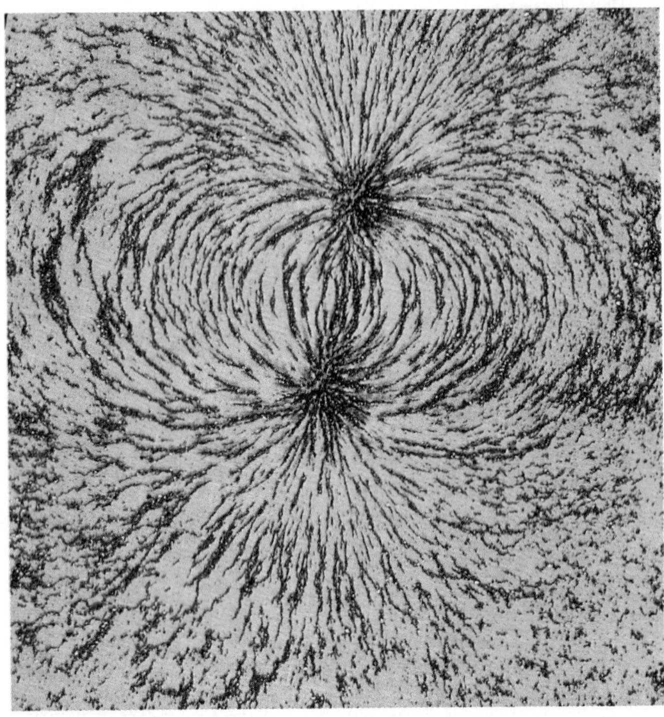

a

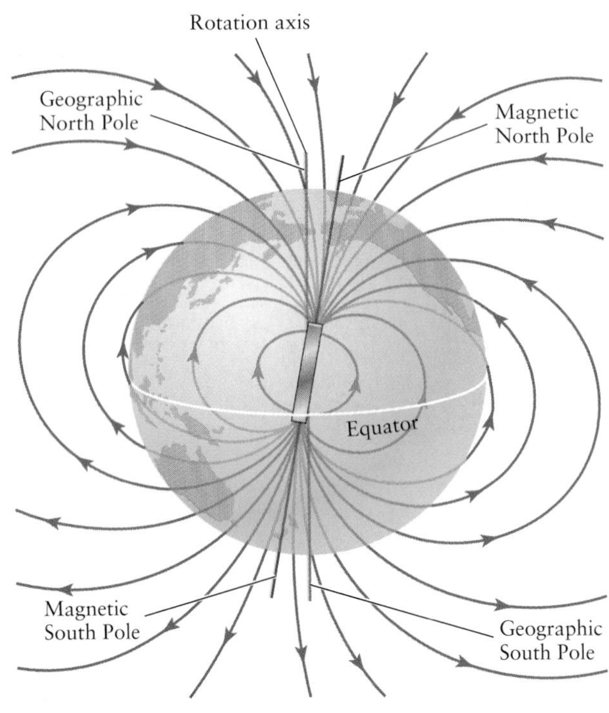

b

FIGURE 6-8 **The Earth's Magnetic Field** (a) The magnetic field of a bar magnet is revealed by the alignment of iron filings on paper. (b) Generated in the Earth's molten, metallic core, the Earth's magnetic field extends far into space. Note that the field is not aligned with the Earth's rotation axis. We will see similar misalignments and flipped magnetic fields when we study other planets. By convention, the magnetic pole near the Earth's north rotation axis is called the magnetic north pole even though it is actually the south pole on a magnet! (a: Jules Bucher/Photo Researchers)

to the field that surrounds a bar magnet (Figure 6-8a). Magnetic fields are created by electrical charges in motion. Thus, the motion of your compass is caused by electrical charges moving inside the Earth. The wires in your house also carry moving charges called electric currents that, in turn, create magnetic fields of their own.

Many geologists believe that convection of molten iron in the Earth's outer core combined with our planet's rotation creates electric currents, which in turn create the Earth's magnetic field. The details of this so-called **dynamo theory** of Earth's magnetic field are still being developed.

Magnetic fields nearly always form complete loops (see Figure 6-8b). The magnetic fields near the Earth's south rotation pole loop out tens of thousands of kilometers into space and then return near the Earth's north rotation pole. The places where the magnetic fields pierce the Earth's surface most intensely are called the north and south magnetic poles. These poles move daily. Evidence in solidified lava also reveals that the Earth's magnetic field completely reverses or flips (the north magnetic pole becomes the south magnetic pole and vice versa) on an irregular schedule ranging from tens of thousands to hundreds of thousands of years. The last reversal was about 600,000 years ago. Today, the magnetic and rotation poles of the Earth are about 11.3° apart; some planets have much larger angles between their magnetic and rotation poles. Figure 6-9 is a scale drawing of the magnetic fields around the Earth, which comprise the Earth's **magnetosphere.**

Our planet's magnetic field protects the Earth's surface from bombardment by energetic particles from space. Most of these particles originate in the Sun (see Chapter 10) and are called the **solar wind.** The solar wind is an erratic flow of electrically charged particles away from the Sun's upper atmosphere. Near the Earth, the particles in the solar wind move at speeds of approximately 400 km/s, or about a million miles per hour.

Magnetic fields can change the direction of moving charged particles. The solar wind particles heading toward the Earth are deflected by our planet's magnetic field. Many particles therefore pass around the Earth. However, the field also traps some of these charged particles in two huge, doughnut-shaped rings called the **Van Allen radiation belts,** which surround the Earth (see Figure 6-9). These belts, discovered in 1958 during the flight of the first successful U.S. Earth-orbiting satellite, were named after the physicist James Van Allen, who insisted that the satellite carry a Geiger counter to detect charged particles.

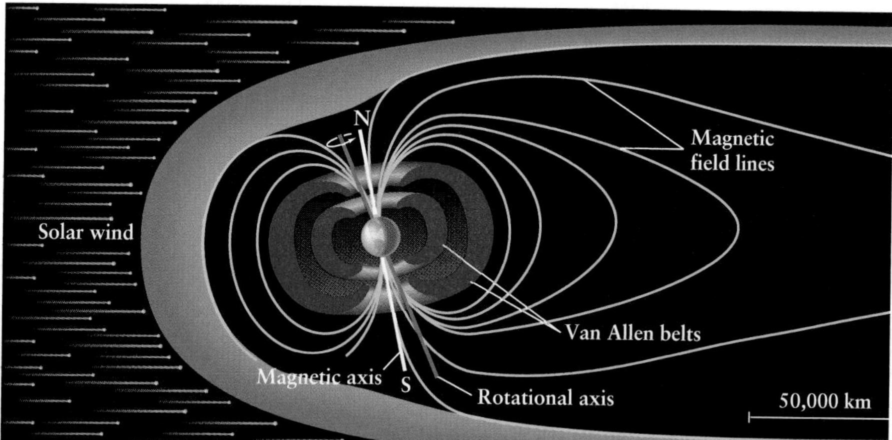

FIGURE 6-9 Earth's Magnetosphere A slice through the Earth's magnetic field, which surrounds the entire planet, carves out a cavity in space that excludes charged particles ejected from the Sun, called the solar wind. The Earth is located in this cavity. Most of the particles of the solar wind are deflected around the Earth by the fields in a turbulent region colored blue in this drawing. Because of the strength of the Earth's magnetic field, our planet traps some charged particles in two huge, doughnut-shaped rings called the Van Allen belts (in red).

INSIGHT INTO SCIENCE

Pace of Discovery Discoveries such as the Van Allen belts remind us of how much remains to be learned, even about our own planet. Scientists continue to make such fundamental discoveries for many reasons. The technology necessary to make certain discoveries, for example, may just have been developed. Or our equations describing various phenomena may be incomplete. Sometimes these equations are so complex that their implications are not yet clear. Progress in science requires unquenchable curiosity and unfailing open-mindedness.

Sometimes the Van Allen belts overload with particles. The particles then leak through the magnetic fields near the poles and cascade down into the Earth's upper atmosphere, usually in a ring-shaped pattern. As these high-speed, charged particles collide with gases in the upper atmosphere, the gases fluoresce (give off light) like the gases in a neon sign. The result is a beautiful, shimmering display called the **northern lights (aurora borealis)** or the **southern lights (aurora Australis)**, depending on the hemisphere in which the phenomenon is observed (Figure 6-10).

Occasionally a violent event on the Sun's surface called a **coronal mass ejection** sends a burst of protons and electrons

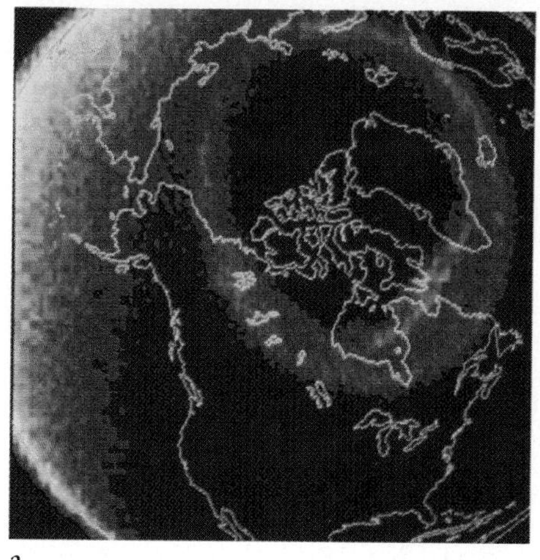

a

R I V U X G

b

R I V U X G

c

R I V U X G

WEB LINK 6.4

FIGURE 6-10 The Northern Lights (Aurora Borealis) A deluge of charged particles from the Sun can overload the Van Allen belts and cascade toward the Earth, producing auroras that can be seen over a wide range of latitudes. (a) View of aurora from the deep space satellite *Dynamics Explorer 1*. The green lines show the locations of land masses under the aurora. Auroras typically occur 100 to 400 km above the Earth's surface. (b) View of aurora from space. Part of a Space Shuttle is visible at the bottom left of the picture. (c) Aurora Borealis in Alaska. The gorgeous aurora seen here is mostly glowing green due to emission by oxygen atoms in our atmosphere. Some auroras remain stationary for hours, while others shimmer, like curtains blowing in the wind. (a: L. A. Frank and J. D. Crave, University of Iowa; b: MTU/Geological & Mining Engineering & Sciences; c: J. Finch/Photo Researchers, Inc.)

straight through the Van Allen belts and into the atmosphere. These events deplete the ozone layer and cause spectacular auroral displays that can often be seen over a wide range of latitudes and longitudes. Such events also disturb radio transmissions and can damage communications satellites and electrical transmission lines.

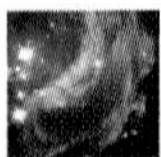

THE MOON AND TIDES

Earth's only natural satellite provides one of the most dramatic sights in the nighttime sky. The Moon is so large and so nearby that some of its surface features are readily visible to the naked eye. Even without a telescope, you can easily see dark gray and light gray areas that cover vast expanses of the Moon (Figure 6-11). People have dreamed of visiting our nearest neighbor for millennia. One of the first movies ever made was *A Trip to the Moon,* based on Jules Verne's book *From the Earth to the Moon.* The wonder of actually sending people to the Moon for the first time was one of the few things that have ever caused humanity to stop its quarreling long enough to participate, even vicariously, in a great adventure. To date 12 people have walked on the Moon. Our exploration of it begins with a study of its surface.

6-5 The Moon's surface is covered with craters, plains, and mountains

Perhaps the most familiar and characteristic features on the Moon are its craters (Figure 6-12a). As discussed in Chapter 5 on the formation of the solar system, all lunar craters

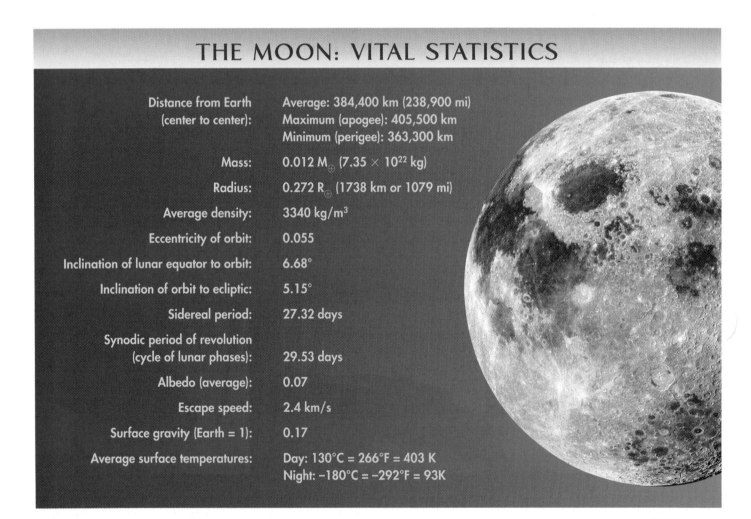

THE MOON: VITAL STATISTICS

Distance from Earth (center to center):	Average: 384,400 km (238,900 mi) Maximum (apogee): 405,500 km Minimum (perigee): 363,300 km
Mass:	0.012 $M_\oplus$ (7.35×10^{22} kg)
Radius:	0.272 $R_\oplus$ (1738 km or 1079 mi)
Average density:	3340 kg/m^3
Eccentricity of orbit:	0.055
Inclination of lunar equator to orbit:	6.68°
Inclination of orbit to ecliptic:	5.15°
Sidereal period:	27.32 days
Synodic period of revolution (cycle of lunar phases):	29.53 days
Albedo (average):	0.07
Escape speed:	2.4 km/s
Surface gravity (Earth = 1):	0.17
Average surface temperatures:	Day: 130°C = 266°F = 403 K Night: −180°C = −292°F = 93K

R I V U X G

WEB LINK 6.5

FIGURE 6-11 **The Moon** Our Moon is one of seven large satellites in the solar system. The Moon's diameter of 3476 km (2160 mi) is slightly less than the distance from New York to San Francisco. This photograph is a composite of first quarter and last quarter views, in which long shadows enhance the surface features. (Lick Observatory, University of California)

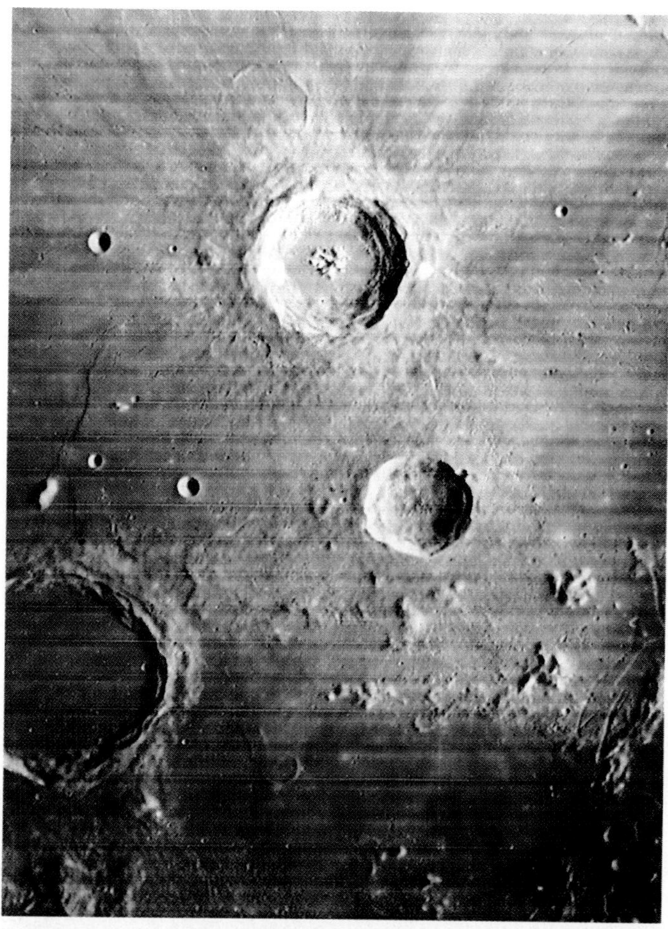

a

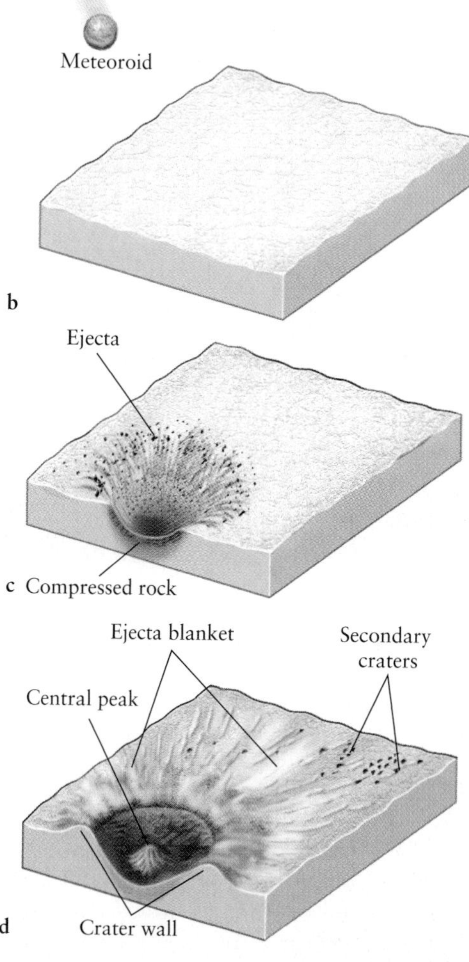

FIGURE 6-12 **Lunar Craters** (a) This photograph, taken from lunar orbit by astronauts, shows the crater Aristillus. Note the crater's central peak; collapsed, terraced crater wall; and ejecta blanket. Numerous tiny craters resulting from the impact pockmark the surrounding lunar surface. The following three drawings show the crater formation process. (b) An incoming meteoroid, (c) upon impact, is pulverized and the surface explodes outward and downward. (d) After the impact, the ground rebounds, creating the central peak and causing the crater walls to collapse. The lighter region is the ejecta blanket. (a: 2004 Lunar and Planetary Institute/Universities Space Research Association)

we have examined are the result of bombardment by meteoritic material. Nearly all these craters are circular, indicating that they were not merely gouged out by moderate-speed rocks, which would typically create oval craters. Instead, the high-speed (upward of 180,000 km/hr or 112,000 mi/hr) collisions with the Moon's surface vaporized the rapidly moving debris, often with the power of a large nuclear bomb. The resulting explosions of hot gas and debris produced the round-rimmed craters we observe today (Figures 6-12 b, c, and d).

Meteoritic impact causes material from the crater site to be ejected onto the surrounding surface. This pulverized rock is called an **ejecta blanket.** You can see some of the more recent ejecta blankets, which are lighter-colored than the older ones (see Figures 6-11 and 6-12). The ejecta blankets darken with time as their surfaces roughen from impacts by the solar wind. The lightness of its ejecta is one clue astronomers use to determine how long ago a crater formed.

Craters larger than about 20 km often form *central peaks* (see Figure 6-12). These occur because the impact compresses the crater floor so much that afterward the crater rebounds and pushes the peak upward. As the peak goes up, the crater walls collapse and form *terraces*.

Through an Earth-based telescope, some 30,000 craters, with diameters ranging from 1 km to more than 100 km, are visible. Following a tradition established in the seventeenth century, the most prominent craters are named after astronomers, physicists, mathematicians, and philosophers, such as Kepler, Copernicus, Pythagoras, Plato, and Aristotle. Close-up photographs from lunar orbit reveal

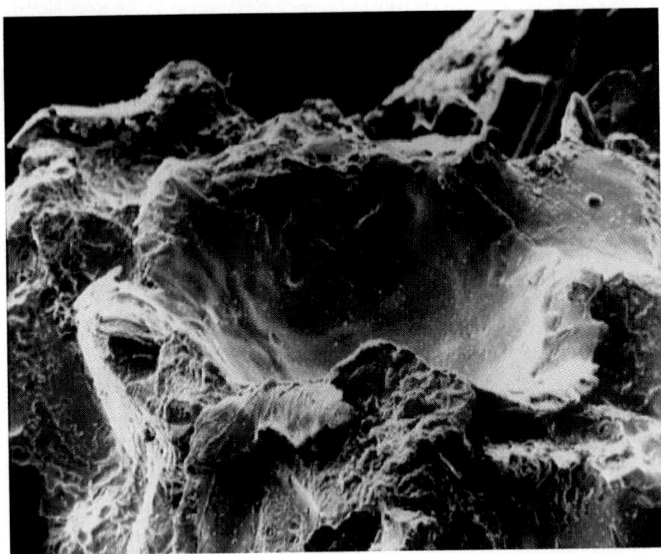

R I V U X G

FIGURE 6-13 **A Microscopic Lunar Crater** This photograph made with a microscope shows tiny microcraters less than 1 mm across on a piece of moon rock. (NASA)

millions of craters too small to be seen with Earth-based telescopes. Indeed, extreme close-up photos of the Moon's surface reveal countless microscopic craters (Figure 6-13). Earth, by comparison, has only about 200 known craters caused by impacts. Many craters on Earth have been drawn inside our planet by plate tectonic motion and obliterated. Many other craters of a few kilometers in diameter and smaller have been worn away (eroded) by weathering effects of wind and water. The Earth's atmosphere vaporized countless space rocks that would otherwise have created small craters here. Moreover, those that do hit are slowed down by the atmosphere so that they produce smaller craters than they would have otherwise. We will discuss Earth's craters further in Chapter 9.

WEB LINK 6.6

Besides its craters, the most obvious characteristic of the Moon visible from Earth is that its surface is various shades of gray (see Figure 6-11). Most prominent are the large, dark gray plains called **maria** (pronounced "mar-ee-uh"). The singular form of this term, **mare** (pronounced "mar-ay"), means "sea" in Latin and was introduced in the seventeenth century when observers using early telescopes thought that these features were large bodies of water. We know now that no water exists in liquid form on our satellite. Nevertheless, we retain these poetic, fanciful names, from Mare Tranquillitatis (Sea of Tranquillity) and Mare Nubium (Sea of Clouds) to Mare Nectaris (Sea of Nectar) and Mare Serenitatis (Sea of Serenity). As we will discuss shortly, maria are basins on the Moon that were filled in with lava after most cratering ended.

The largest of the maria is Mare Imbrium (Sea of Showers). It is roughly circular and measures 1100 km (700 mi) in diameter (Figure 6-14). Although the maria seem quite smooth in telescopic views from the Earth, close-up photographs from lunar orbit and from the surface reveal that they contain small craters and occasional cracks called **rilles** (Figure 6-15). Rilles are the collapsed roofs of underground lava rivers or the beds of long-gone surface lava rivers. When flowing underground lava rivers solidified, the rock it formed shrank and the roofs fell in on the spaces created.

As we will discuss further in Section 6-8, the same side of the Moon always faces the Earth. One of the surprises stemming from the earliest lunar exploration is that only one small mare, Mare Orientale, graces the Moon's far side, the side we never see from Earth (Figure 6-16). Except for that lone mare, craters and mountains cover the entire far side of the Moon.

Detailed measurements by astronauts in lunar orbit demonstrated that the maria on the Moon's Earth-facing side are 2 to 5 km below the average lunar elevation and the one on the far side is 4 to 5 km above the average lunar elevation. The flat, low-lying, dark gray maria cover only

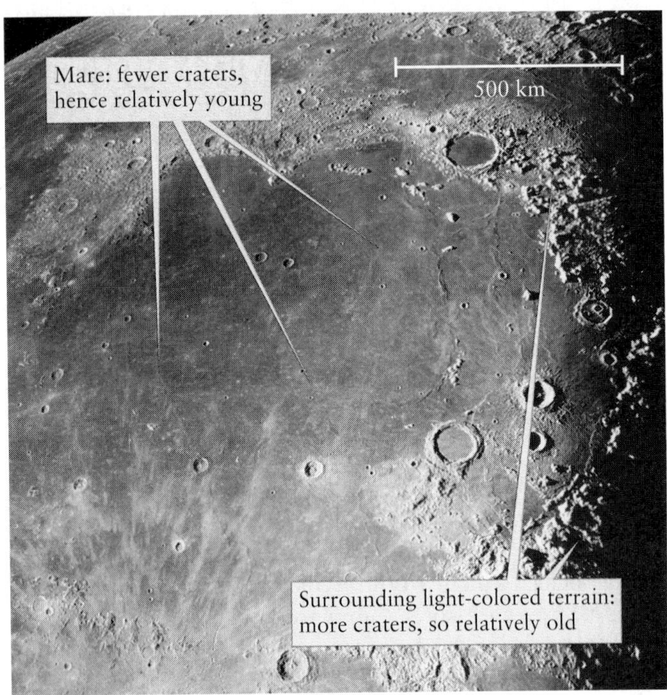

R I V U X G

FIGURE 6-14 **Mare Imbrium and the Surrounding Highlands** Mare Imbrium, the largest of the 14 dark plains that dominate the Earth-facing side of the Moon, is ringed by lighter-colored highlands strewn with craters and towering mountains. The highlands were created by asteroid impacts pushing land together. (NASA)

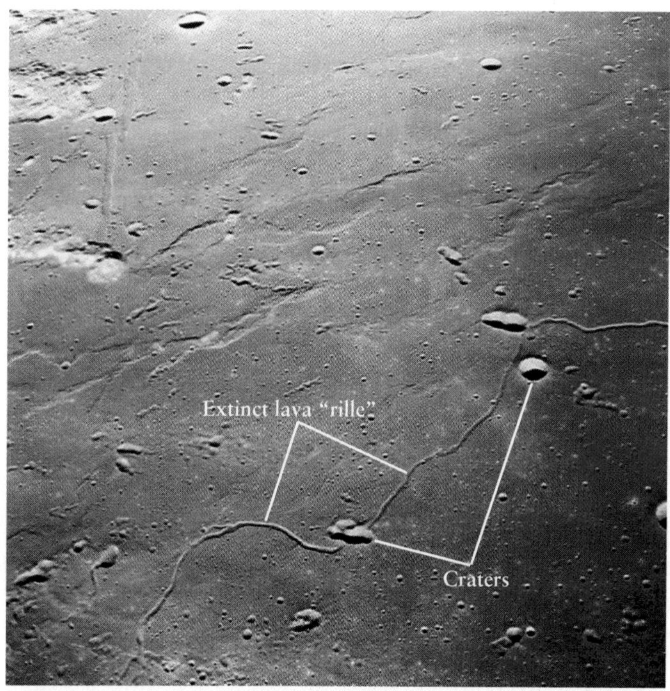

a

R I V U X G

b

R I V U X G

FIGURE 6-15 **Details of a Lunar Mare** **(a)** Close-up views of the lunar surface reveal numerous tiny craters and rilles on the maria. Astronauts in lunar orbit took this photograph of Mare Tranquillitatis (Sea of Tranquility) in 1969 while searching for potential landing sites for the first manned landing. At 1100 km (700 mi) across, it is the same size as the distance from London to Rome or Chicago to Philadelphia. **(b)** Astronaut David Scott on Hadley's rille during the *Apollo 15* mission to the Moon. (a: NASA; b: Kennedy Space Center/NASA)

17% of the lunar surface. The remaining 83% is composed of light gray, heavily cratered, mountainous regions called **highlands** (see Figure 6-11).

Lunar mountain ranges that rim the vast maria basins suggest ancient impacts far more violent than those that produced typical craters. One such mountain range can be seen around the edges of Mare Imbrium (see Figure 6-14). Although they take their names from famous terrestrial ranges, such as the Alps, the Apennines, and the Carpathians, the lunar mountains were not formed through plate tectonics as mountains on Earth were. Analyses of Moon rocks and observations from lunar orbit indicate that violent impacts of huge asteroids thrust up the lunar highlands.

R I V U X G

◀ FIGURE 6-16 **Comparing the Near and Far Sides of the Moon** This is a composite image from the *Galileo* spacecraft. Mare Orientale lies on the boundary between the near and far sides. The broad dark region at lower left is the South Pole–Aitken Basin, an impact scar some 2100 km (1300 mi) across and 12 km (7 mi) deep. (NASA)

6-6 Visits to the Moon yielded invaluable information about its history

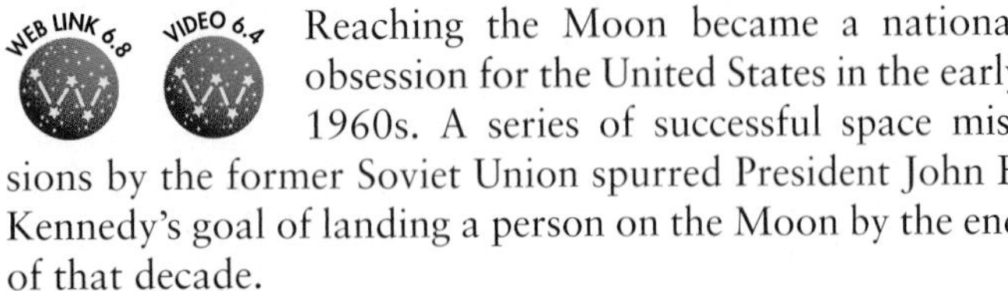

Reaching the Moon became a national obsession for the United States in the early 1960s. A series of successful space missions by the former Soviet Union spurred President John F. Kennedy's goal of landing a person on the Moon by the end of that decade.

2 WEB LINK 6.9 The first of six manned lunar landings, *Apollo 11*, set down on Mare Tranquillitatis on July 20, 1969. Astronaut Neil Armstrong was the first human to set foot on the Moon. In the same year, *Apollo 12* also landed in a mare. Human voyages into space encountered a setback when *Apollo 13* experienced a nearly fatal explosion en route to the Moon. Fortunately, a titanic effort by the astronauts and ground personnel brought it safely home. *Apollo 14* through *Apollo 17* took on progressively more challenging lunar terrain. The period of human visitation to our Moon lasted less than four years, culminating in 1972, when *Apollo 17* landed amid rugged mountains just east of Mare Serenitatis (Figure 6-17).

Astronauts discovered that the lunar surface is covered with a layer of fine powder and rock fragments. This layer, ranging in thickness from 1 to 20 m, is called the **regolith** (from the Greek, meaning "blanket of stone"; Figure 6-18). It formed during 4.5 billion years of relentless micro-meteorite bombardment that pulverized the surface rock. We do not refer to this layer as soil, because the term "soil" suggests the presence of decayed biological matter, which is not found on the Moon's surface. The rough, powdered regolith absorbs most of the light incident on it. Therefore, the Moon's albedo is only 0.07, meaning that it scatters only 7% of the light striking it.

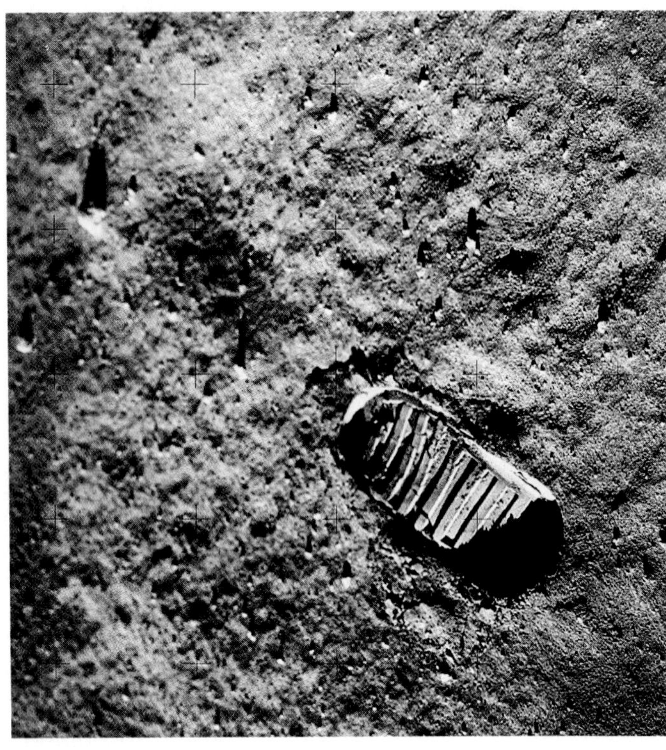

R I V U X G

FIGURE 6-18 **The Regolith** The Moon's surface is covered by a layer of powdered rock that was created over billions of years as a result of bombardment by space debris. Called the regolith, this surface material sticks together like wet sand, as illustrated by this *Apollo 11* astronaut's bootprint. (NASA)

R I V U X G

VIDEO 6.5 FIGURE 6-17 **An Apollo Astronaut on the Moon** *Apollo 17* astronaut Harrison Schmitt enters the Taurus-Littow Valley on the Moon. Here an enormous boulder once slid down a mountain, fracturing on the way. This final Apollo mission landed in the most rugged terrain of any Apollo flight. (NASA)

A major goal of the Apollo program was to study the geology of the Moon and help determine its history. The astronauts brought back a total of 382 kg (842 lb) of lunar rocks, which have provided important information about the early history of the Moon. By carefully measuring trace amounts of radioactive elements, geologists determined that lunar rock from the maria is solidified lava that dates back only 3.1 to 3.8 billion years. This indicates that the maria developed about a billion years after the Moon formed and that the maria were formed from molten rock. Indeed, these Moon rocks are composed mostly of the same minerals that are found in volcanic rocks on Hawaii and Iceland. The rock of these low-lying lunar plains is called **mare basalt** (Figure 6-19). About 17% of the Moon's surface rock is basalt.

FIGURE 6-19 **Mare Basalt** This 1.53-kg (3.38-lb) specimen of mare basalt was brought back by *Apollo 15* astronauts in 1971. Small holes that cover about a third of its surface suggest that gas was dissolved in the lava from which this rock solidified. When the lava reached the airless lunar surface, bubbles formed as the pressure dropped and the gas expanded. Some of the bubbles were frozen in place as the rock cooled. (NASA)

FIGURE 6-20 **Anorthosite** The lunar highlands are covered with this ancient type of rock, which is believed to be the material of the original lunar crust. This sample's dimensions are 18 × 16 × 7 cm. While this sample is medium gray, other anorthosites retrieved from the Moon have been white, while others are darker gray than this one. It was brought back by *Apollo 16* astronauts. (NASA)

In contrast to the dark mare basalt, the lunar highlands are covered with a light-colored rock called **anorthosite** (Figure 6-20). On Earth, anorthositic rock is found only in very old mountain ranges, such as the Adirondacks in the eastern United States. Compared to the mare basalts, which have more of the heavier elements like iron, manganese, and titanium, anorthosite is rich in calcium and aluminum. Anorthosite is therefore less dense than basalt. Many highland rocks brought back to Earth are **impact breccias** (Figure 6-21), which are composites of different rock fused together as a result of meteorite impacts.

In contrast to rocks from maria, typical anorthositic specimens from the highlands are between 4.0 and 4.3 billion years old. These ancient specimens of highland rock probably represent samples of the Moon's original crust (see Figure 6-16).

Since spacecraft began orbiting the Moon, scientists have noted that their orbits do not follow the expected elliptical paths. Every once in a while, usually over a circular mare, the spacecraft momentarily dip Moonward, as they are pulled by a higher than normal concentration of mass near the Moon's surface. These *mass concentrations,* or **mascons**, are local regions of relatively dense rock and metal near the Moon's surface. The mascons found in maria are believed to be material that filled the bottom of the mare after the basins were formed by impacts of very large bodies. New mascons continue to be discovered, including seven found by the *Lunar Prospector* spacecraft in 1998.

WEB LINK 6.10

In 1994, the lunar-orbiting *Clementine* spacecraft sent radio signals to the Moon's surface. These signals scattered off the materials there, and then they were received and analyzed on Earth. The results were intriguing. They suggested that frozen water exists in craters near the Moon's south pole. If so, this ice is not in

FIGURE 6-21 **Impact Breccias** These rocks are created from shattered debris fused together under high temperature and pressure. Such conditions prevail immediately following impacts of space debris on the Moon's surface. (NASA)

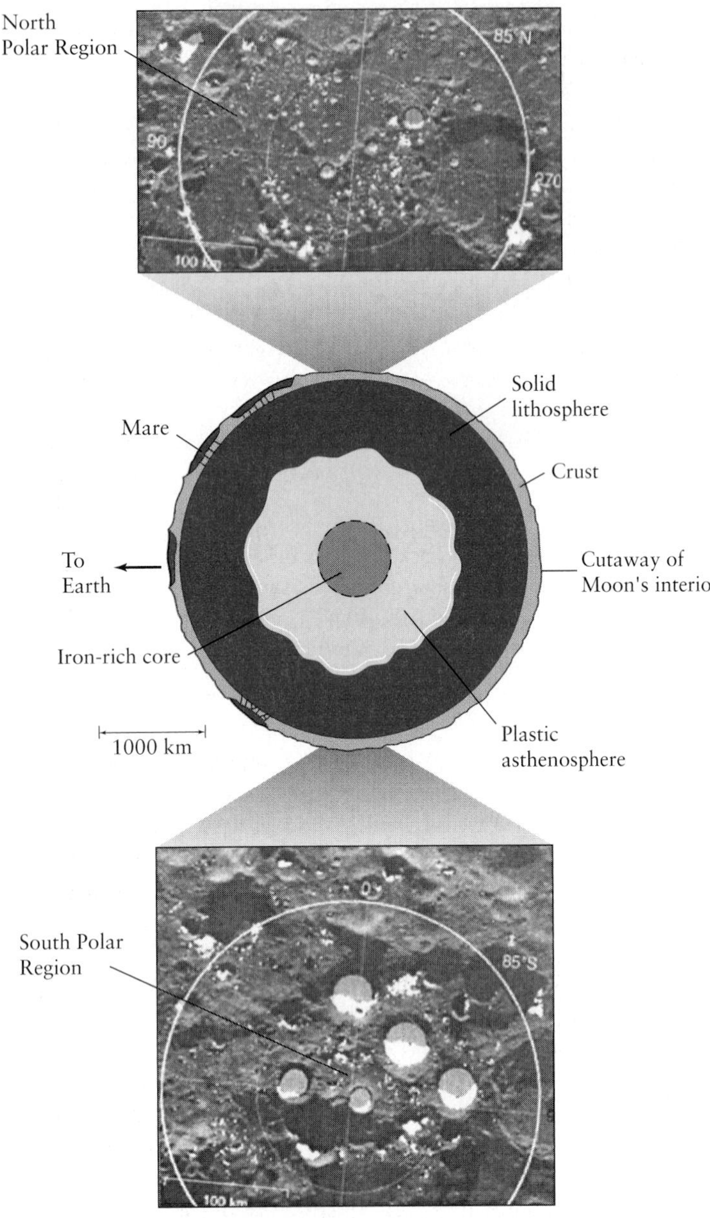

◀ FIGURE 6-22 **The Moon's Interior** Based on seismic experiments left on the Moon by Apollo astronauts and the studies done by *Lunar Prospector*, we know that the Moon has a crust, mantle, and core. The lunar crust has an average thickness of about 60 km on the Earth-facing side but about 100 km on the far side. The crust and solid upper mantle extend to about 800 km, where the nonrigid inner mantle begins. The Moon's core has a radius of between 300 and 425 km. Although the main features of the Moon's interior are analogous to those of the Earth, the proportions and details are quite different. The *Clementine* spacecraft revealed that the south pole has a significant basin. The crust was apparently stripped away by an impact. *Top inset:* A radar image of the Moon's north polar region. The areas computer-colored in white and light gray are regions where the Sun never shines and that may contain water. *Bottom inset:* A radar image of the Moon's south polar region, also showing regions of permanent or near-permanent darkness. (insets: NASA/Galaxy)

danger of being evaporated by sunlight, because, just as is the case with the Earth's poles, the Sun deposits very little energy at the lunar poles (Figure 6-22). In 1998, the *Lunar Prospector*, using different technology, found evidence for ice at the Moon's north pole. However, in 2003, radio experiments from Earth similar to those run on *Clementine* did not see indications of water. Clearly the issue of whether there are large quantities of water on the Moon is still up in the air. If it exists there, the ice probably was deposited on the Moon during comet and asteroid impacts; estimates based on the first observations are that over 3 billion tons of water may be encased there either mixed in with the regolith or in layers below it.

The presence of water ice would greatly simplify the prospect of human settlement of the Moon, because its ice can provide us not only with liquid water but also with oxygen and rocket fuel. The Space Shuttle's primary rocket engine is fueled by liquid oxygen and liquid hydrogen. Separating the Moon's water into liquid oxygen and liquid hydrogen would create the same fuel. An existing water reservoir would make establishing a lunar colony much less expensive and more feasible than lugging all that water from Earth or dragging a water-rich comet to the Moon.

Astronauts found no traces of life on the Moon. Life as we know it requires liquid water, of which the Moon has none. The Moon has an atmosphere of sodium gas so thin as to be considered an excellent vacuum compared to the density of the air we breathe. An atmosphere is held around a planet or moon by that body's gravitational force. If the atmospheric gases are too hot (moving too fast) for the gravitational force to hold them, they will leak into space. Our Moon has so little gravitational attraction (about one-sixth what we feel at the Earth's surface) that it is unable to retain its tenuous atmosphere. The atmosphere is renewed by sunlight striking the Moon, releasing more gas from its rocks. The Earth, in contrast, has enough mass to retain such gases as nitrogen, oxygen, carbon dioxide, water, and argon.

The Moon's slow rotation rate and small iron core, whose presence was confirmed by *Lunar Prospector*, suggest that unlike the Earth, the Moon should not have a strong magnetic field. Spacecraft orbiting the Moon have confirmed this. However, the *Lunar Prospector* has discovered local areas where the magnetic fields are strong enough to keep the solar wind away from the Moon's surface. These anomalous regions have yet to be explained.

6-7 The Moon probably formed from debris cast into space when a huge asteroid struck the young Earth

Before the Apollo program, scientists debated three different theories about the origin of the Moon. The **fission theory** holds that the Moon was pulled out from a rapidly rotating proto-Earth. This theory does not explain why the Moon has virtually no water incorporated in its rocks, as revealed by the bone-dry samples brought back by Apollo astronauts. The **capture theory** posits that the Moon was formed elsewhere in the solar system and then drawn into orbit about the Earth by gravitational forces. However, it is physically very difficult for a planet to capture a large moon. Furthermore, because bodies formed in different places have different overall chemical compositions, this theory does not explain the similar geochemistries of the Moon and the Earth's surface. The **cocreation theory** proposes that the Earth and the Moon were formed near each other at the same time. However, this theory fails to explain why, compared to the Earth, the Moon has less of the denser elements, such as iron, and why certain types of rocks found on the Moon are not found on Earth.

A fourth theory, now held to be correct by most astronomers, was first proposed in the 1980s. It applies the established fact of age-old collisions in the solar system to the formation of the Moon. Called the **collision-ejection theory,** or large impact theory, it proposes that the newly formed Earth was struck at an angle by a Mars-sized asteroid that literally splashed some of the Earth's surface layers into orbit around the young planet. Evidence from Moon rocks places this event at around 4.5 billion years ago, within the first 100 million years of the Earth's existence. A computer simulation of this cataclysm is shown in Figure 6-23. This orbiting matter became a short-lived disk that eventually condensed into the Moon, just as planetesimals condensed to form the Earth.

The collision-ejection theory is consistent with many of the known facts about the Moon. For example, rock vaporized by the impact would have been depleted of *volatile* (easily evaporated) elements and water, leaving the Moon rocks parched, as we now know they are. Also, the Moon has little heavy iron-rich matter because that material had sunk deep into the Earth before the asteroid struck. The material from the Earth that splashed into orbit and formed the Moon was mostly the lighter rock floating on the Earth's surface. Furthermore, most of the debris from the collision would orbit near the plane of the ecliptic as long as the orbit of the impacting asteroid had been near that plane. Also, the impact of an object large enough to create the Moon could have tipped the Earth's axis of rotation and so inaugurated the seasons.

The surface of the newborn Moon probably remained molten for millennia, due to heat released during the impact of rock fragments falling onto the young satellite and the decay of short-lived radioactive isotopes in it. As the Moon gradually cooled, low-density lava floating on its surface began to solidify into the anorthositic crust that exists today in the highlands.

By correlating the ages of Moon rocks with the density of craters at the sites where they were collected, geologists have found that the rate of impacts on the Moon changed over the ages. The ancient, heavily cratered lunar highlands are evidence of intense bombardment that dominated the Moon's early history. This barrage extended from 4.5 billion years ago, when the Moon's surface first solidified, until about 3.8 billion years ago. This period of impacts was punctuated by a severe pounding 4 billion years ago, part of a solar-system-wide reign of terror.

The frequency of impacts gradually tapered off as meteoroids and planetesimals were swept up by the newly formed planets, as discussed in Chapter 5. Recorded among the final scars at the end of this crater-making era are the impacts of more than a dozen objects, each measuring at least 100 km across. As these huge rocks pounded the young Moon, they blasted gigantic craters in regions that would later become the maria. Meanwhile, heat from the decay of long-lived radioactive elements like uranium and thorium began to melt the inside of the Moon again. Then, from 3.8 to 3.1 billion years ago, great floods of molten rock gushed up from the lunar interior through the weak spots in the crust created by these large impacts. Lava filled the impact basins and created the maria we see today. Because most cratering was over by then, maria have few craters. The reason that the far side of the Moon has only one mare apparently stems from the fact that the Moon's crust on the side facing Earth is thinner than on the far side. It was therefore easier for the molten rock inside the Moon to escape through the large craters on our side than those on the far side.

Relatively little large-scale activity has happened on the Moon since those ancient times. Although only a few fresh large craters have been formed, the small-sized debris still plentiful in the solar system has formed many small ones. Back in 1953, an impact was observed that has been connected with the formation of a 1.5-km (0.93-mi) diameter crater. Smaller impacts are observed nearly every year when large groups of space debris are known to be crossing the Moon's path. By and large, however, the astronauts visited a world that has remained largely unchanged for more than 3 billion years. Unlike the Earth, which is still quite geologically active, the Moon is considered to be geologically dead.

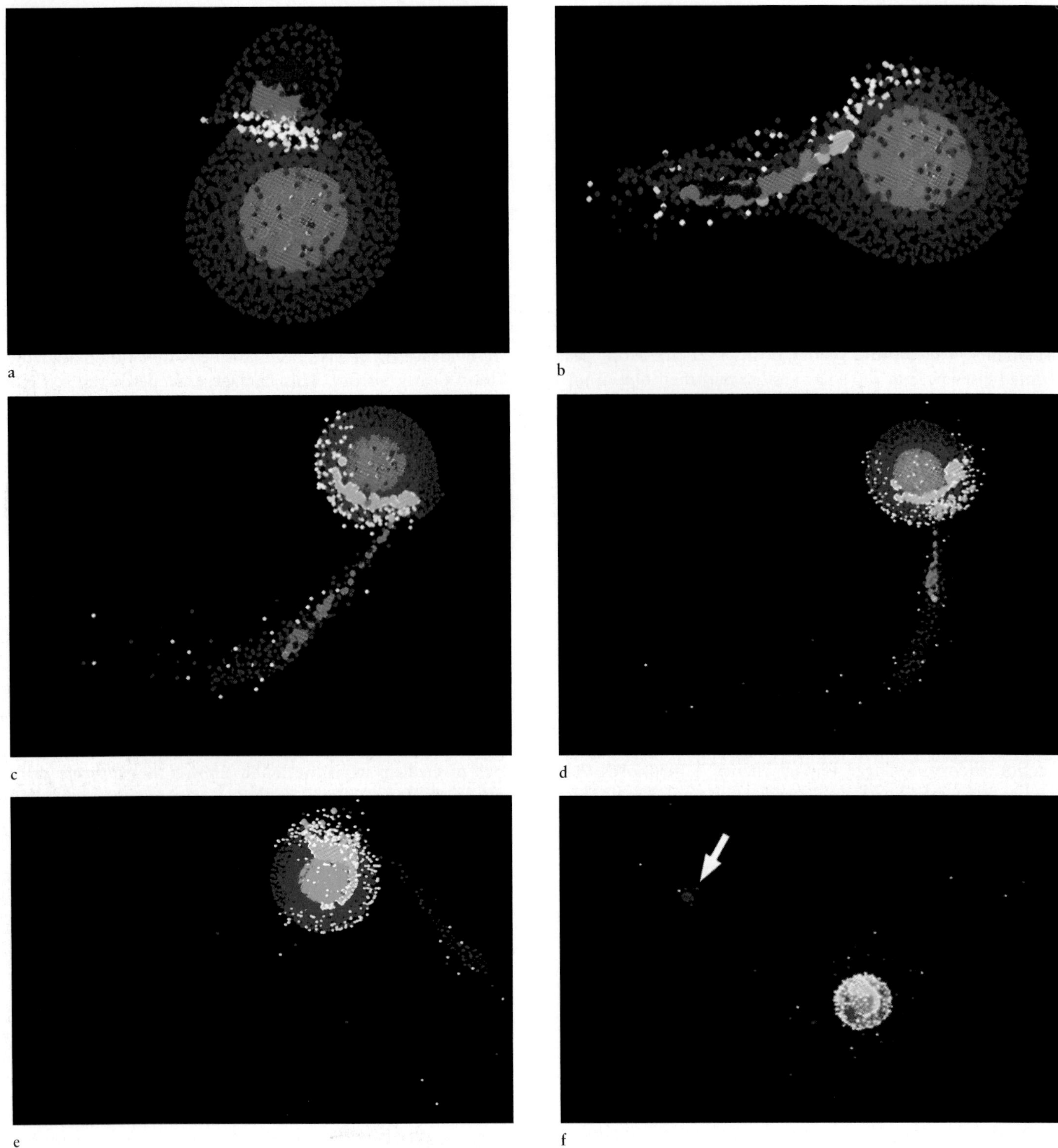

ANIMATION 6.4

FIGURE 6-23 **The Moon's Creation** This computer simulation models the creation of the Moon from material ejected by the impact of a large planetlike body with the young Earth. To follow the ejected material as it moves away from the Earth, successive views (a–f) pan out to show increasingly larger volumes of space. Blue and green indicate iron from the cores of the Earth and the asteroid; red, orange, and yellow indicate rocky mantle material. In this simulation, the impact ejects both mantle and core material, but most of the dense iron falls back onto the Earth. The surviving ejected rocky matter coalesces to form the Moon (indicated by arrow). (W. Benz)

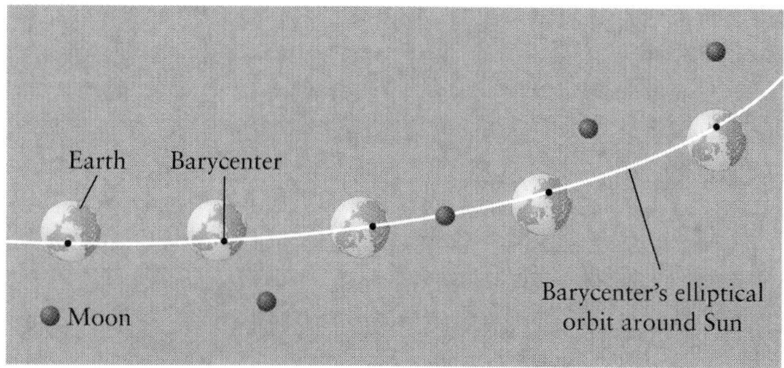

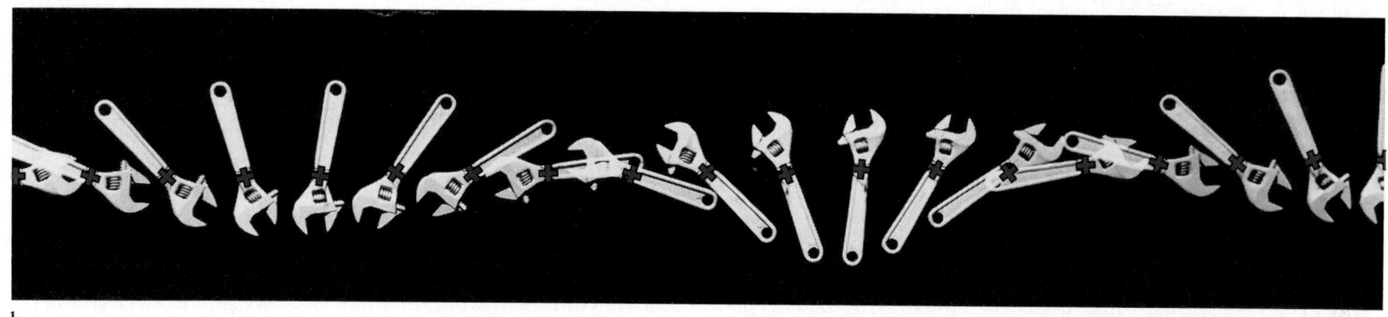

FIGURE 6-24 **Motion of the Earth-Moon System** **(a)** The paths of the Earth and Moon as their barycenter follows an elliptical orbit around the Sun. **(b)** Analogously, this time-lapse image shows a spinning wrench sliding across a table. Note that its center of mass (the red plus) moves in a straight line, while its other parts follow curved paths. (b: Berenice Abbott/Photo Researchers)

6-8 Gravitational force and the orbits of the Moon and Earth produce the tides on Earth and also force the Moon to rotate at the same rate that it revolves around the Earth

Gravitational attraction keeps the Earth and Moon together, while the angular momentum (see An Astronomer's Toolbox 2-2) given to the Moon when it was formed keeps the Moon from being pulled back onto the Earth. Contrary to appearances, the Moon does not orbit the Earth. Rather, both bodies are in motion around their center of mass called their *barycenter* (Figure 6-24a), just as a spinning wrench sliding across a table has its parts move around a similar common point (Figure 6-24b).

3 4 The same surface features on the Moon always face the Earth, regardless of what phase the Moon is in (see Figure 1-23). This occurs because the Moon is in **synchronous rotation** around the Earth: The Moon rotates on its axis at the same rate that it and the Earth orbit their barycenter (Figure 6-25). To see why the Moon must be rotating if it keeps the same side facing us, hold your arm out with your palm facing you. Your palm represents the side of the Moon facing the Earth. Standing still, swing your arm always keeping the palm facing you. The only way you can do this is if your hand turns (rotates) at exactly the same rate that it swings (revolves).

We have seen the Moon's "far side" only from spacecraft (see Figure 6-16). The far side of the Moon is often, and incorrectly, called its "dark side." The dark side of the

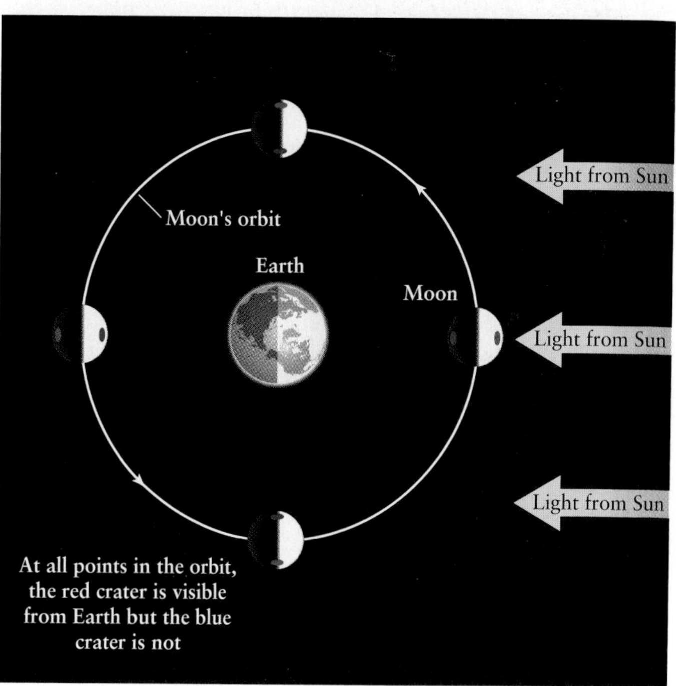

FIGURE 6-25 **Synchronous Rotation of the Moon** This figure shows the motion of the Moon around the Earth as seen from above the Earth's north polar region (ignoring the Earth's orbit around the Earth-Moon barycenter). In order for the Moon to keep the same side facing the Earth as it orbits our planet, the Moon must rotate on its axis at precisely the same rate that it revolves around the Earth.

Moon is exactly like the dark side of the Earth—it is wherever the Sun is below the horizon. Whenever we view less than a full Moon, we see part of its dark side and some sunlight is then striking part of its far side.

The origin of the Moon's synchronous rotation is in the **tidal force** that acts between the Earth and Moon. This tidal force is also what creates ocean tides on Earth. To understand how the Moon came to have synchronous rotation, we investigate the forces acting between the Earth and the Moon. To "kill two birds with one stone," the following discussion focuses on how the Moon creates the tides on Earth. We will then turn this information around and see how the Earth once created tides on the Moon that changed its original rotation rate to synchronous rotation.

Most of our tides are a result of the gravitational tug-of-war between the Earth and Moon as the two bodies

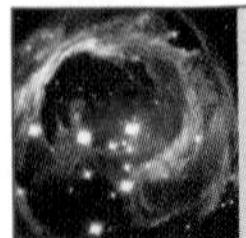

GUIDED DISCOVERY
Tides

Ignoring the Earth's rotation for the moment, let us consider the motion of the Earth and Moon around their barycenter (see Figure 6-24a). The barycenter is located 1068 miles below the Earth's surface on the line between the centers of the Earth and Moon. As a result of its motion around the barycenter, every point on Earth feels an equally strong force away from the Moon. This is unlike the centrifugal force you feel on a merry-go-round, which increases the farther you are from the pivot point. The force away from the Moon is represented at three locations by the force labeled F_{out} in Figure 6-26a.

Recall from Chapter 2 (An Astronomer's Toolbox 2-2) that the gravitational force from one body on another decreases with distance. Therefore, the Moon's gravitational attraction on Earth is greatest at the point on the Earth's surface closest to the Moon. This attraction is greater than the force felt at, say, the center of the Earth, which in turn is greater than the force felt on the opposite side of the Earth from the Moon. The gravitational force in these places is labeled $F_{grav.}$ in Figure 6-26a.

Because forces in the same or opposite directions simply add or subtract, respectively, we can subtract the two forces F_{out} and $F_{grav.}$ at any point on the Earth, due to the presence of the Moon. The resulting net force is labeled F_{tide} in Figure 6-26a. This is the tidal force acting at these places. For example, at the point on the Earth directly below the Moon, the tidal force is very strong and points toward the Moon. At the center of the Earth the two forces cancel each other completely. At the point on the opposite side of the Earth from the Moon, the tidal force is equal in strength to the tidal force directly under the Moon, but directed away from the Moon. The strength of the tidal force at various places is summarized in Figure 6-26b.

Consider the point on the Earth closest to the Moon. The strength of the tidal force, F_{tide}, seems to imply that the Moon's gravitational force lifts the water there to create the high tide. However, the force of gravitation from the Moon is not strong enough to raise that water more than a few inches. Rather, the high tide closest to the Moon occurs because ocean waters from nearly halfway around to the opposite side of the Earth are pulled Moonward by the gravitational force acting on them. This water slides toward the Moon and fills the ocean directly under it, creating a tidal bulge or high tide. Likewise, the net outward force acting on the opposite side of the Earth from the Moon pushes the waters that are just over halfway around from the

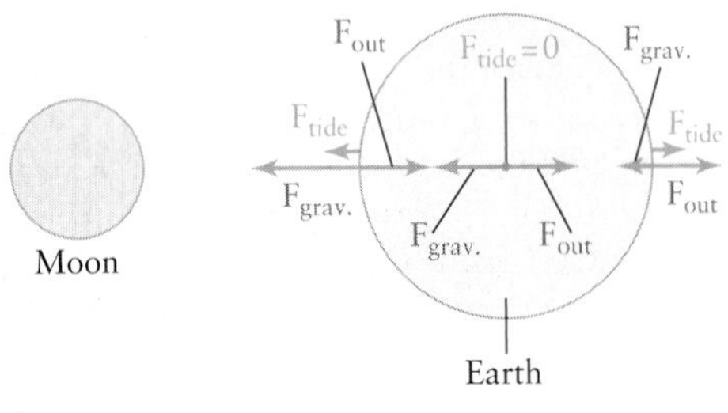

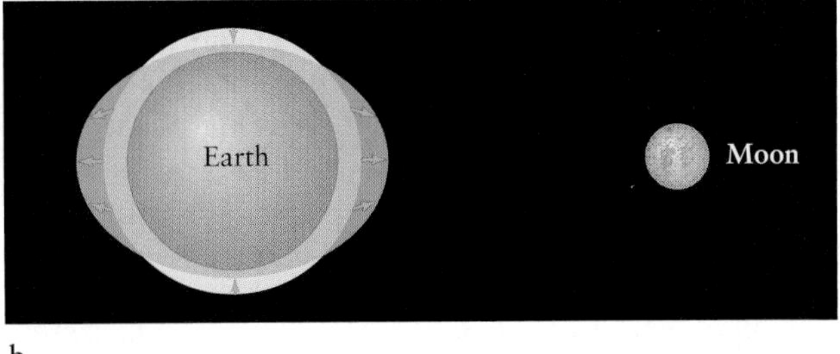

FIGURE 6-26 **Tidal Forces** (a) The Moon induces tidal forces, F_{tide}, on the Earth. At each point, this force is the difference between the force, F_{out}, created by the orbital motion of the two bodies around their barycenter, and the Moon's gravitational force, F_{grav}, at that point. The magnitude and direction of each arrow represents the strength and direction of each force. (b) Water slides along the Earth to create the tides. Ignoring the Earth's rotation and the effects of the continents until Figure 6-27, this figure shows how two high tides are created on the Earth by the Moon's gravitational pull. The Sun has a weaker, but otherwise identical, effect.

orbit their barycenter. As explained in Guided Discovery: Tides along with other details of the tidal process, there are always two high tides on opposite sides of the Earth (see Figure 6-26). Detailed calculations show that the Sun's tidal effect is just less than half the tidal effect of the Moon. In other words, the Moon generates about two-thirds of the tidal activity on Earth, while the Sun generates most of the rest.

When the Sun, Moon, and Earth are aligned, the Sun and Moon create pairs of high tides in the same directions, and the resulting combined tides are the highest high tides of the lunar cycle. These **spring tides** (so-called not because they only occur in the Spring—they occur in all seasons—but, from the German, *springen,* meaning to "spring up") occur at every new and full Moon (see Figure 6-27a). Note that it does not matter whether the Sun and Moon are on

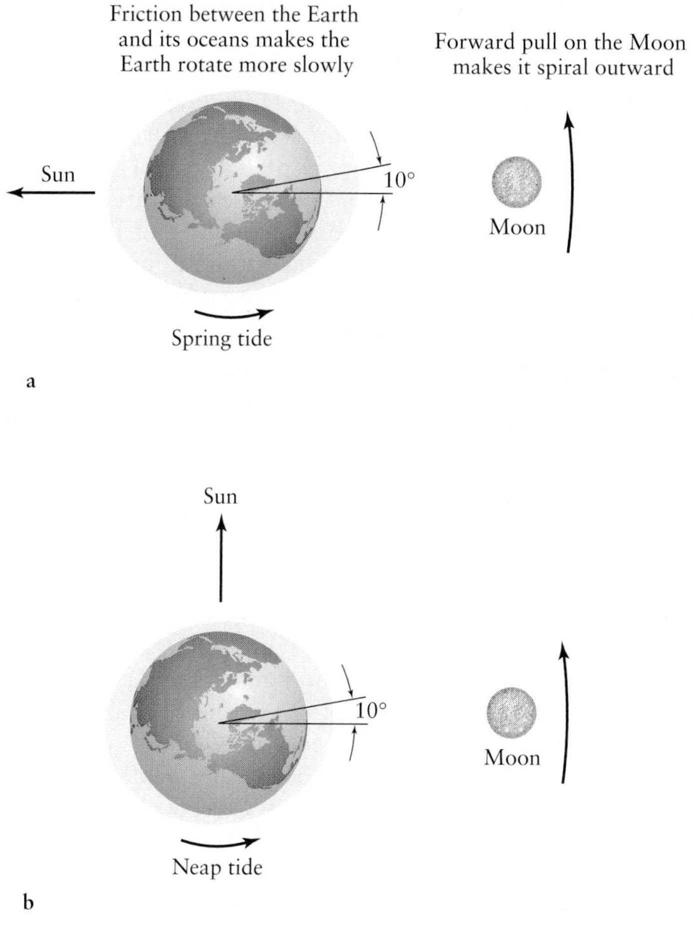

FIGURE 6-27 **Tides on the Earth** The gravitational and centrifugal forces of the Moon and Sun deform the oceans. Due to Earth's rapid rotation, the high tide closest to the Moon leads it around the Earth by about 10°. **(a)** The greatest deformation (spring tides) occurs when the Sun, Earth, and Moon are aligned with the Sun and Moon either on the same or opposite sides of the Earth. **(b)** The least deformation (neap tides) occurs when the Sun, Earth, and Moon form a right angle.

Moon to the side directly away from it, creating a simultaneous, second high tide of equal magnitude on the opposite side of the Earth from the Moon. Where the water has been drained away in this process, low tides occur.

Let's spin the Earth back up. As it rotates, the geographical locations of the high and low tides continually change. Figure 6-26b indicates that high tides occur when the Moon is high in the sky or below our feet, while low tides occur when the Moon is near either horizon. This means that there are two cycles of high and low tides at most places on Earth each day. These cycles do not span exactly 24 hours, however, because the Moon is moving along the celestial sphere, changing the time at which high or low tide occurs at any location from day to day. As a result, the time between consecutive high tides is 12 h 25 min.

Tides are complicated by the fact that the oceans are not uniformly deep and that the shores around continents and islands have a variety of shapes. As a result, there are some places where the oceans have only a single cycle of tides each day. Some places have tides that change negligibly while some channels, such as the Bay of Fundy between the United States and Canada and the Bristol Channel between Wales and England, have two tides that are each much higher than normal.

The Earth's rotation also has the effect of moving the high tides away from the line between the centers of the Earth and Moon. The Earth's eastward rotation is nearly 30 times faster than the Moon's eastward revolution around the Earth. Therefore, the rotating Earth drags the high tides eastward with it. As a result, the high tide that should be directly between the centers of the Earth and Moon is actually 10° ahead of the Moon in its orbit around the Earth (Figure 6-27). This high tide stays at roughly this position because the Moon's gravitational force pulls it westward while it is being dragged eastward by the Earth's rotation, and the two effects balance each other out. A similar argument applies to the tide on the opposite side of the Earth.

We can repeat this analysis for the Earth-Sun system and get exactly the same qualitative results.

the same or opposite sides of the Earth when generating the spring tides since the Sun and Moon both create high tides on opposite sides of the Earth. At first quarter and third quarter, the Sun and Moon form a right angle as seen from the Earth. The gravitational forces they exert compete with each other, so the tidal distortion is the least pronounced. The especially small tidal shifts on these days are called **neap tides** (Figure 6-27b).

Finally, we can explain the Moon's synchronous rotation: When the Moon was young, it was molten and Earth's gravity created huge tides of molten rock up to 60 ft high there. The Earth's tidal force acted to keep the Moon's high tide directly between the Moon's center and the center of the Earth. Friction between this moving, molten rock and the rest of the Moon changed the Moon's rotation rate. As a result, the Moon's rotation rate became the same as its revolution rate. When it solidified, the Moon was locked into this synchronous rotation.

6-9 The Moon is moving away from the Earth

A century ago, Sir George Darwin (son of the evolutionist Charles Darwin) proposed that the Moon is moving away from the Earth. To understand why, recall from the Guided Discovery: Tides and Figure 6-27 that the high tide nearest the Moon is actually 10° ahead of the Moon in its orbit around the Earth. This offset occurs because friction between the ocean and the ocean floor causes the rapidly spinning Earth to drag the tidal bulge ahead of the line between the center of the Earth and center of the Moon. The gravitational force from the high ocean tide nearest the Moon acts back on the Moon, pulling it ahead in its orbit, giving it energy and thereby forcing it to spiral outward.

R I V U X G

FIGURE 6-28 **Lunar Ranging** Beams of laser light are fired through three telescopes at the Observatoire de la Côte d'Azur, France. The light is then reflected back by the corner reflectors placed on the Moon by Apollo astronauts. From the time it takes the light to reach the Moon and return to Earth, astronomers can determine the distance to the Moon to within a few millimeters. (Observatoire de la Côte d'Azur)

To test this concept, *Apollo 11, 14,* and *15* astronauts placed on the Moon sets of reflectors, similar to the orange and red ones found on cars, to help measure the Moon-Earth distance. Pulses of laser light were fired at the Moon from the Earth (Figure 6-28), and the time it took the light to reach the Moon, bounce off a reflector array, and return to the Earth was carefully recorded. Knowing the speed of light and the time it takes for the light pulse to make the round trip, astronomers can calculate the distance to the Moon with an error of only a few millimeters. From such measurements over time, astronomers have established that the Moon is spiraling away from the Earth at a rate of 3.8 cm per year.

This means, of course, that the Moon used to be closer to the Earth. Although the initial distance of the Moon when it coalesced is not known, its present rate of recession tells us that it was at least half its present distance; more accurate calculations suggest one-tenth. At that distance, the tides on the Earth were a thousand times higher than they are today. (What effects would those tides have had on the geology of the young Earth?)

Where does the Moon get the energy necessary to spiral away from the Earth? The answer, of course, is from the Earth itself. As the tides move westward to try to stay under the Moon, they encounter the eastward-moving landmasses, which disrupt the water's motion. Because the oceans move westward, while the planet rotates eastward, the oceans actually push on the continents opposite to the Earth's direction of rotation. The result is that the tides are slowing the Earth's rotation. The energy lost by the Earth as it spins down is gained by the Moon. At present, the day is slowing down by about one-thousandth of a second each century. Geologists calculate that when the Earth first formed, the day was only between 5 and 6 hours long.

Will the Moon ever leave the Earth completely as a result of these dynamics? In fact, it will not. Rather, the Earth's rotation will slow down until it is in synchronous rotation with respect to the Moon. Thereafter, the Moon will remain at a fixed distance over one side of the Earth. However, billions of years before that happens, the Sun is going to self-destruct!

6-10 Frontiers yet to be discovered

The Earth and Moon still hold many mysteries. We have yet to understand the different rotation rate of the Earth's core from its surface, as well as the details of the dynamo effect that generates the Earth's magnetic field. We do not know precisely how fast the Earth was rotating initially. We also need to better understand the impact of humans on the Earth and its atmosphere.

The Moon still inspires many questions and hides many secrets. The six American manned missions and three Soviet soft, robotic lunar landings have barely scratched the Moon's surface, bringing back samples from only nine locations. We still know very little about the Moon's far side. Also, is there really water buried at the poles? Is the Moon's interior molten? Does the Moon really have an iron-rich core? How old are the youngest lunar rocks? Did lava flows occur over western Oceanus Procellarum only two billion years ago, as crater densities there suggest? Is the Moon really geologically dead, or does it just appear dead because our examination of the lunar surface has been so cursory? Such questions can be answered only by returning to the Moon.

Summary of Key Ideas

Earth: A Dynamic, Vital World

• The Earth's atmosphere is about four-fifths nitrogen and one-fifth oxygen. This abundance of oxygen is due to the biological processes of life-forms on the planet.

• The Earth's atmosphere is divided into layers named the troposphere, stratosphere, mesosphere, and ionosphere. Ozone molecules in the stratosphere absorb ultraviolet light rays.

• The outermost layer, or crust, of the Earth offers clues to the history of our planet.

• The Earth's surface is divided into huge plates that move over the upper mantle. Movements of these plates, a process called plate tectonics, are caused by convection in the mantle, and upwelling of molten material along cracks in the ocean floor produces seafloor spreading. Plate tectonics is responsible for most of the major features of the Earth's surface, including mountain ranges, volcanoes, and the shapes of the continents and oceans.

• Study of seismic waves (vibrations produced by earthquakes) shows that the Earth has a small, solid inner core surrounded by a liquid outer core. The outer core is surrounded by the dense mantle, which in turn is surrounded by the thin, low-density crust. The Earth's inner and outer cores are composed primarily of iron with some nickel mixed in. The mantle is composed of iron-rich minerals.

• The Earth's magnetic field produces a magnetosphere that surrounds the planet and blocks the solar wind.

• Some charged particles from the solar wind are trapped in two huge, doughnut-shaped rings called the Van Allen radiation belts. A deluge of particles from a coronal mass ejection by the Sun can initiate an auroral display.

The Moon and Tides

• The Moon has light-colored, heavily cratered highlands and dark-colored, smooth-surfaced maria.

• Many lunar rock samples are solidified lava formed largely of minerals also found in Earth rocks.

• Anorthositic rock in the lunar highlands was formed between 4.0 and 4.3 billion years ago, whereas the mare basalts solidified between 3.1 and 3.8 billion years ago. The Moon's surface has undergone very little geological change over the past 3 billion years.

• Impacts have been the only significant "weathering" agent on the Moon; the Moon's regolith (pulverized rock layer) was formed by meteoritic action. Lunar rocks brought back to Earth contain no water and are depleted of volatile elements.

• Frozen water may have been discovered at the Moon's poles.

• The collision-ejection theory of the Moon's origin, accepted by most astronomers, holds that the young Earth was struck by a huge asteroid, and debris from this collision coalesced to form the Moon.

• The Moon was molten in its early stages, and the anorthositic crust solidified from low-density magma that floated to the lunar surface. The mare basins were created later by the impact of planetesimals and were then filled with lava from the lunar interior.

• Gravitational and centrifugal interactions between the Earth and the Moon produce tides in the oceans of the Earth and set the Moon in synchronous rotation. The Moon is moving away from the Earth, and consequently, the Earth's rotation rate is decreasing.

WHAT DID YOU THINK?

1 *Can the Earth's ozone layer, which is now being depleted, be naturally replenished?* Yes. Ozone is created continuously from normal oxygen molecules by their interaction with the Sun's ultraviolet radiation.

2 *Who was the first person to walk on the Moon and when did this event occur?* Neil Armstrong was the first person to set foot on the Moon. He and Buzz Aldrin flew on the *Apollo 11* spacecraft piloted by Michael Collins. Armstrong and Aldrin set down the *Eagle* lander on the Moon on July 20, 1969.

3 *Do we see all parts of the Moon's surface at some time throughout the lunar cycle of phases?* No. Because the Moon's rotation around Earth is synchronous, we always see the same side. The far side of the Moon has been seen only from spacecraft that pass or orbit it.

4 *Does the Moon rotate and, if so, how fast?* The Moon rotates at the same rate that it revolves around the Earth. If the Moon did not rotate, then, as it

revolved, we would see its entire surface from the Earth, which we do not.

5 *What causes the ocean tides?* The tides are created by orbital and gravitational forces, primarily from the Moon and, to a lesser extent, from the Sun.

6 *When does the spring tide occur?* Spring tides occur twice monthly, during each full and new Moon.

Key Words

anorthosite, 153
capture theory, 155
cocreation theory, 155
collision-ejection theory, 155
continental drift, 141
convection, 145
core, 144
coronal mass ejection, 147
crust, 141
dynamo theory, 146
ejecta blanket, 149
fission theory, 155
highlands, 151
impact breccias, 153
ionosphere, 141
magnetosphere, 146
mantle, 144
mare (plural maria), 150
mare basalt, 152
mascons, 153
mesosphere, 141
neap tide, 160
northern lights (aurora borealis), 147
ozone layer, 140
planetary differentiation, 144
plate tectonics, 142
regolith, 152
rille, 150
seafloor spreading, 141
seismic waves, 144
seismograph, 144
solar wind, 146
southern lights (aurora Australis), 147
spring tide, 159
stratosphere, 140
synchronous rotation, 157
tidal force, 158
troposphere, 140
Van Allen radiation belts, 146

Review Questions

1. Why does the Earth's albedo change daily? seasonally?

2. Why is the Earth's surface not riddled with craters as is that of the Moon?

3. Describe the structure of the Earth's atmosphere.

4. Describe the process of plate tectonics. Give specific examples of geographic features created by plate tectonics.

5. To review the Earth's tectonic plates, do Interactive Exercise 6-1 and print out or sketch the completed diagram.

6. To review the mechanism of plate tectonics, do Interactive Exercise 6-2 and print out or sketch the completed diagram.

7. How do we know about the Earth's interior, given that the deepest wells and mines extend only a few kilometers into its crust?

8. Describe the interior structure of the Earth.

9. Why is the center of the Earth not molten?

10. Describe the Earth's magnetosphere.

11. To review the Earth's magnetosphere, do Interactive Exercise 6-3 and print out or sketch the completed diagram.

12. What are the Van Allen belts?

13. What kind of features can you see on the Moon with a small telescope?

14. On the basis of lunar rocks brought back by the astronauts, explain why the maria are dark-colored, but the lunar highlands are light-colored.

15. Why are there so few craters on the maria?

16. To review the crater formation history in the solar system, do Interactive Exercise 6-4 and print out or sketch the completed diagram.

17. Briefly describe the main differences and similarities between Moon rocks and Earth rocks.

18. How do we know that the maria were formed after the lunar highlands?

19. What is a tidal force? How do tidal forces produce tides in the Earth's oceans?

20. What is the difference between spring tides and neap tides?

21. Why do most scientists support the collision-ejection theory for the Moon's formation?

Advanced Questions

22. Explain how the outward flow of energy from the Earth's interior drives the process of plate tectonics.

23. Why do some geologists believe that Pangaea was the most recent in a succession of supercontinents?

24. Why are active volcanoes, such as Mount St. Helens, usually located in mountain ranges along the boundaries of tectonic plates?

25. Why is more lunar detail visible through a telescope when the Moon is near quarter phase than when it is at full phase?

26. In *The Tragedy of Pudd'nhead Wilson*, Mark Twain wrote, "Everyone is a Moon, and has a dark side which he never shows to anybody." What, if anything, is wrong with Twain's astronomy?

27. Why are the Moon rocks retrieved by astronauts so much older than typical Earth rocks, even though both worlds formed at nearly the same time?

28. Some people who supported the fission theory proposed that the Pacific Ocean basin is the scar left when the Moon pulled away from the Earth. Explain why this idea is wrong.

29. Apollo astronauts left seismometers on the Moon that radioed seismic data back to Earth. The data showed that moonquakes occur more frequently when the Moon is at perigee (closest to the Earth) than at other locations along its orbit. Give an explanation for this finding.

30. Why do you think that no Apollo missions landed on the far side of the Moon?

31. How might studying albedo help astronomers locate habitable worlds orbiting other stars?

Discussion Questions

32. If the Earth did not have a magnetic field, do you think auroras would be more common or less common than they are today? Explain.

33. Comment on the idea that without the Moon's presence, life would have developed far more slowly.

34. Identify and compare the advantages and disadvantages of lunar exploration by astronauts as opposed to mobile, unmanned instrument packages and robots.

35. When was the last earthquake near your hometown? How far is your hometown from a plate boundary? What kinds of topography (for example, mountains, plains, seashore) dominate the geography of your hometown area? Does that topography and the frequency of earthquakes seem to be consistent with your hometown's proximity to a plate boundary?

36. The ice on the Moon is believed to be mixed with rock just under the Moon's surface. How might that water be economically extracted and purified?

What If . . .

37. The Earth had two moons? Assume one is at our Moon's distance and the other is at half that distance. Describe the motion of the two moons in the sky and how they might appear to us. What would be different here on Earth?

38. The Moon orbited the Earth in the opposite direction from the Earth's rotation, rather than in the same direction? What would be different about the Earth and life on it? Assume that today such a counter-revolving Moon is at the same distance as our Moon.

39. Our Moon were one-tenth of its actual distance from the Earth? What would be different about the Earth and life on it? *Hint:* The heights of tides vary as $1/r^3$, where r is the distance between the centers of the Earth and the Moon.

40. The Moon, located where it actually is, were as massive as the Earth? What would be different about the Earth and life on it and about the Moon?

41. The Earth's interior were entirely solid, rather than partly molten? What would be different about the Earth and life on it?

42. The Earth were now in synchronous rotation with the Moon? That is, suppose that the Earth rotated at the same rate that the Moon orbits the Earth. What would be different about the Earth and life on it?

Web Questions

43. Use the Web to learn more about plate tectonics. What global changes might accompany the formation and breakup of a supercontinent? How might these changes affect the evolution of life? What life-forms dominated the Earth when Pangaea existed some 200 million years ago and also when fragments of the preceding supercontinent were as dispersed as today's continents?

44. Search the Web for information about "Pangaea Ultima," a supercontinent that may form in the distant future. When is it expected to form? How will it compare to Pangaea of 200 million years ago (see Figure 6-4)?

45. Use the Web to determine the status of the Antarctic and Arctic ozone holes. How has the situation changed over the past few years? Explain why most scientists who study this issue blame chemicals called CFCs for the existence of the ozone holes.

46. In 1989, representatives of many countries signed a treaty called the Montreal Protocols to protect the ozone layer. Use the Web to learn of the current status of the treaty. How many nations have signed it? Has the treaty been amended? If so, how? List when various substances that destroy the ozone layer are scheduled to be phased out.

47. Several unmanned missions to the Moon were under development as of this writing. These include LUNAR-A and SELENE (Institute of Space and Astronautical Science, Japan) and SMART-1 (European Space Agency). Search the Web for current information about these and any other upcoming lunar missions you can find. When is each scheduled to be launched? What new investigations are planned for each mission? What existing scientific issues may these missions resolve?

48. Refinements to the mass and impact angle of the planetesimal that struck the Earth and created the Moon have recently been made. Search the Web for this information and explain the justification for these new parameters.

49. Some astronomers have observed changes in brightness and color on the Moon, called Lunar Transient Phenomena. Search the Web and explain what these events are and what their origins are believed to be.

Observing Projects

Observing Tips and Tools: You can learn a lot by observing the Moon through binoculars. Note that the Moon will appear right side up through binoculars, but inverted through a telescope. So, if you use a map of the Moon to aid your observations, you will need to take this into account. Inexpensive Moon maps are available, such as "Moon Map" published by *Sky & Telescope* magazine. To determine the lunar phase (when you cannot examine the Moon directly), you can usually find it on most calendars; by checking the weather page in your newspaper; by consulting the current issue of *Sky & Telescope* or *Astronomy* magazine; by using your *Starry Night Enthusiast*™ program; or on the Web.

50. Use a telescope or binoculars to observe the Moon. Compare the texture of the lunar surface you see on the maria with that of the lunar highlands. How does the visibility of details vary with distance from the terminator (the boundary between day and night on the Moon)? Why?

51. If you live near the ocean, observe the tides to see how the times of high and low tides are correlated with the position of the Moon in the sky.

52. Observe the Moon through a telescope or binoculars every few nights over a period of two weeks between new Moon and full Moon. Make sketches of various surface features, such as craters, mountain ranges, and maria. How does the appearance of these features change with the Moon's phase? Which features are most easily seen at a low angle of illumination (near the terminator)? Which features show up best with the Sun nearly overhead as seen from the Moon (far from the terminator)?

53. Use the *Starry Night Enthusiast*™ program to view the Earth as it appears from the Moon (*Options/Viewing Location . . ./View From the Surface of the Moon*). Choose the *Map* tab, click on any large maria, and press *Set Location*. It will take a moment to go there. Grab the sky and move until you find the blue Earth. Slide angle of view to 60 degrees). Set the timestep to between 15 minutes and 1 hour, depending on your computer and click the ► (flow time) button. (a) Describe the phases of the Earth. (b) Observe the surface features of the Earth that you see. Do they change? If so, explain why, considering that there is no significant change in the face of the Moon as seen from Earth.

54. Use *Starry Night Enthusiast*™ to observe the changing appearance of the Moon. Set for today, turn on Atlas mode (*Favourites/Guides/Atlas*) and then find the Moon (Tab:*Find/Moon*). Set the view angle slide (upper right on screen) to about 5 degrees. What phase is the Moon in? Set the timestep to between 1 and 6 hours, depending on your computer's speed, and click the ► (flow time) button. (a) Describe how the phase of the Moon changes. (b) Look carefully at the features near the left-hand and right-hand edges (limbs) of the Moon. Are these features always exactly in the same places relative to the limb? They should be if the Moon's synchronous rotation were the only motion it had relative to the Earth. If not, explain. (*Hint:* A Web search on the word "libration" [not libation] may help.) (c) Does the apparent size of the Moon remain constant? If not, explain what this tells us about the shape of the Moon's orbit around the Earth.

55. ***Exploring Earth*** In this exercise, we use the *Deep Space Explorer*™ program to examine the Earth as seen from space. If you have not done so already, go to **Settings** and uncheck "Use magnitude cutoff," slide the "Galaxy drawing/Brightness" slide to the middle of the range, and then click on the **Views** tab.

Go to the Earth by clicking **Solar System/Earth** on the left side of the star field. You can zoom in and zoom out using the buttons at the upper left of the window (an upward-pointing triangle and a downward-pointing triangle). You can also rotate the Earth by putting the mouse cursor over the image, holding down the mouse button, and moving the mouse. (Hold down the left PC mouse button or the mouse button on a Mac.)

(a) Move the Earth around until you see the Sun almost behind it. From what you now see of the Earth's phase explain how you can easily find the Sun in searches for it at other planet. Go to Mars, **Solar System/Mars,** and using what you have just learned, put the sun behind the planet, thereby confirming what you have just proposed. Explain.

(b) Return to Earth, **Solar System/Earth.** Move the Earth around until South America is visible. Eastern Brazil, at the eastern edge of South America, is near the terminator (the dividing line between the day and night sides of the Earth). As seen from eastern Brazil, is the sun rising or setting? Explain.

56. ***Exploring the Moon*** In this exercise, we will use *Deep Space Explorer*™ to examine the Moon. If you have not done so already, go to Settings and uncheck "Use magnitude cutoffs," slide the "Galaxy drawing/Brightness" slide to the middle of the range, and then click on the **Views** tab.

Go to the Moon by clicking **Earth/Moon** on the left side of the star field. You can zoom in and zoom out using the buttons at the upper left of the window (an upward-pointing triangle and a downward-pointing triangle). You can also rotate the Moon by putting the mouse cursor over the image, holding down the mouse button, and moving the mouse. (Hold down the left PC mouse button or button on a Mac.)

(a) Based *only* on what you can see in the image, what evidence can you find that the Moon is geologically inactive? Explain.

(b) Spreading outward from some of the largest craters on the Moon are straight lines of light-colored material, called *rays,* that were caused by material ejected outward by the impact that caused the crater. Make a map of the lit side of the Moon and indicate craters that have rays. If a labeled map of the Moon is available, use it to identify the craters with rays.

WHAT IF . . . THE MOON DIDN'T EXIST?

Throughout history, people have woven myths about the Moon and its effects on everything from childbirth to stock market activity. Countless romances have begun under a full Moon, and entire nations have dedicated themselves to reaching it.

What if the Moon never existed? Would life even have developed on Earth? If so, how would it be different? Would a self-aware species like ourselves have evolved?

Delayed Origins It would have been much more difficult for life to start evolving on a Moonless Earth. In the first place, the minerals from which life developed in our Earth's young oceans were swept down there in large part by gigantic tides, a thousand times higher than the tides of today, created by the Moon when it was some 10 times closer to Earth than at present. The newly formed Earth, spinning more than 4 times faster than at present, caused these tides to move miles inland and back out to sea every 1½ hours or so.

A Moonless Earth would still have tides, created by the Sun. At most, however, these tides would never be more than one-third as high as Earth's *present* tides. Minerals would wash into the oceans incredibly slowly from the flow of rivers. As a result, it is likely that it would have taken much longer, hundreds of millions or even billions of years longer, for enough minerals to be dissolved for life to firmly establish itself.

Harsh Conditions Animal life's transition from oceans to land would also be much harder because of continuous winds, between 50 and 150 miles per hour, created by the planet's rapid rotation. The resulting waves (wind causes waves) would be enormous and perpetual. Fish with legs near shores would almost certainly be pounded to a pulp rather than being able to sedately walk onto land and then back into the water, as apparently happened on Earth.

But we see on Earth how life develops in the most incredible forms and places. It seems plausible, then, that sea life would find a way to make the transfer to dry land and continue to evolve there.

Allowing then for diverse terrestrial life on the Moonless Earth, what would be different about that life compared to life on our Earth? For one thing, creatures evolving there would have to withstand the perpetual pummeling from winds and the debris they carry. Turtlelike shells are one solution.

Obviously, there would be no eclipses or moonlight—all clear nights would be equally dark and star-filled. Therefore, Earthlike nocturnal animals would be less successful at hunting, foraging, and traveling. Instead, these animals might evolve more enhanced senses to compensate for their inability to see visible light well at night.

Rush Hour Clearly, naked apes do not seem a likely bet on a Moonless Earth, nor do birds battling the ever-present winds. But given enough time, complex, even self-aware life likely would evolve. After all, we evolved because of the challenges faced by our ancestors, and Earth without a Moon would clearly provide challenges of its own.

The physiology of life on Earth evolved based on a 24-hour day. This is most evident in our biological clocks, or circadian rhythms, which are the internal mechanisms that regulate sleeping and waking, eating, and other cyclic activities. Faced with a 6-hour day, these circadian rhythms would be hopelessly out of sync with the natural world. To function on such a world, all its creatures would have to evolve biological clocks based on 6 hours, which certainly could have occurred. Considering all we have to do now, imagine what life would be like with only 6 hours in each day!

Tumbling Earth Finally, rapidly rotating terrestrial planets without large moons are unstable, and calculations show that they periodically tumble—the north pole becomes the south pole and vice versa! If the Earth had no Moon, this process would create enormous stresses on the planet's surface leading to catastrophic earthquakes, volcanic activity, tsunamis, as well as severe atmosphere and magnetic field changes. Our Moon stabilizes the Earth's rotation, preventing such flips. It would be incomparably harder for complex life to persist on a tumbling world than it is on ours.

CHAPTER 7

The Other Terrestrial Planets and Their Comparison to Earth

R I V U X G

Mars's *Sojourner Rover* Examines the Chemistry of Rocks on the Red Planet
(NASA/Lunar and Planetary Institute)

WHAT DO YOU THINK?

1 Which of the two planets, Mercury (the closest planet to the Sun) or Earth, has the coolest temperature?

2 Which planet is most similar to Earth?

3 What is the composition of the clouds surrounding Venus?

4 Does Mars have liquid water on its surface today?

5 Is life known to exist on Mars today?

Every human being is unique. We each have our own genetic makeups and personal histories. On the other hand, people have many similarities: we are either male or female; we all breathe air; and, under normal circumstances, we have the same features, such as two eyes, two hands, and ten toes, among many other things. To understand humans fully, biologists and physicians study our common features and then our individual peculiarities, while psychiatrists and psychologists study our common behaviors and then our deviations.

The planets in our solar system also have similarities and differences that astronomers are learning to understand. For example, there are four basic groups of planets: those quite similar in chemistry to the Earth (Mercury, Venus, and Mars); those especially rich in hydrogen and helium (Jupiter and Saturn); those containing vast quantities of water, as well as hydrogen and helium (Uranus and Neptune); and those rich in rock and ice (Pluto). A group of one? Astronomers are discovering more bodies in the solar system similar in composition to Pluto. Since Pluto is already the smallest planet, these new discoveries may help declassify Pluto as a planet, rather than add new planets—only time will tell. In the meantime, tiny Pluto is in a class by itself.

This is the first of two chapters in which we explore the planets both individually and in comparison to each other and the Earth. In this chapter we examine the three planets similar to the Earth in composition and in their proximity to the Sun. The remaining five outer planets are presented in the next chapter.

In this chapter you will discover

- Mercury, a Sun-scorched planet with a heavily cratered surface and a substantial iron core
- Venus, perpetually shrouded in thick, poisonous clouds and mostly covered by gently rolling hills
- Mars, the inspiration for scientific and popular speculation about extraterrestrial life

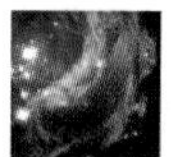

MERCURY

The closest planet to the Sun, Mercury is a truly inhospitable world of temperature extremes, with an atmosphere so incredibly thin that it is left unprotected from countless impacts and a continuous bath of deadly solar radiation. We can only see Mercury from Earth for a few hours before sunrise or a few hours after sunset because its angle from the Sun (its elongation) is always less than 28°. Mercury sometimes appears as one of the brightest objects in the sky. This brightness is misleading, however, since Mercury is less than half the diameter of the Earth and its rough, solid surface is nearly as dark as coal.

7-1 Photographs from *Mariner 10* reveal Mercury's lunarlike surface

Mercury was a mystery for many years because Earth-based telescopic photographs of Mercury reveal almost nothing about the planet's surface. Indeed, much of the mystery remains. When *Mariner 10* coasted past the planet in 1974, astronomers got their first glimpse of Mercury's surface. It appears inhospitable, with a crater-strewn surface that has a passing resemblance to the surface of our Moon. Figure 7-1 lists a variety of Mercury's properties. Although *Mariner 10* was sent on a remarkable orbit that enabled it to pass Mercury three times, it photographed only 45% of Mercury's surface. We still do not know what the other 55% looks like.

Astronomers conclude that most of the craters on both Mercury and the Moon were produced by impacts in the first 800 million years after these bodies condensed from the solar nebula. Debris remaining after the planets were formed pounded these young worlds, gouging out most of the craters we see today. As we saw in Chapter 6, the strongest evidence for this belief comes from analyzing and dating Moon rocks. Like the Moon, Mercury has an exceptionally low albedo of 0.12 (12% scattering of incoming light). It is very bright as seen from Earth only because the sunlight scattering from Mercury is so intense.

First impressions of Mercury evoke a lunar landscape; and there are several similarities. Craters in the Moon's highlands are densely packed, as are some of the craters on Mercury (Figure 7-2). The craters on both worlds have similar features, as discussed for the Moon in Section 6-5.

The most impressive feature discovered by *Mariner 10* was a huge circular region called the Caloris Basin, which measures 1300 km (810 mi) in diameter and lies along the *terminator* (the border between day and night) in Figure 7-3a. It is surrounded by a 2-km-high ring of mountains, beyond which are relatively smooth plains. Like the lunar maria, the Caloris Basin was probably gouged out by the impact of a large meteorite that penetrated the planet's crust. Because

MERCURY: VITAL STATISTICS

Average distance from Sun:	0.387 AU = 5.79×10^7 km
Maximum distance from Sun:	0.467 AU = 6.98×10^7 km
Minimum distance from Sun:	0.307 AU = 4.60×10^7 km
Eccentricity of orbit:	0.21
Average orbital speed:	47.9 km/s
Sidereal period of revolution:	88.0 Earth days = 0.24 Earth year
Sidereal rotation period:	58.7 Earth days
Solar rotation period (day):	176 Earth days
Inclination of equator to orbit:	0.5°
Inclination of orbit to ecliptic:	7° 00' 16"
Radius (equatorial):	2439 km = 0.382 Earth radius
Mass:	3.30×10^{23} kg = 0.0553 Earth mass
Average density:	5430 kg/m^3 = 0.984 Earth density
Escape speed:	4.3 km/s
Surface gravity (Earth = 1):	0.38
Albedo:	0.12
Average surface temperatures:	Day: 350°C = 662°F = 623 K Night: −170°C = −274°F = 103 K
Atmosphere:	Very thin, transient H, He, K, Na, O

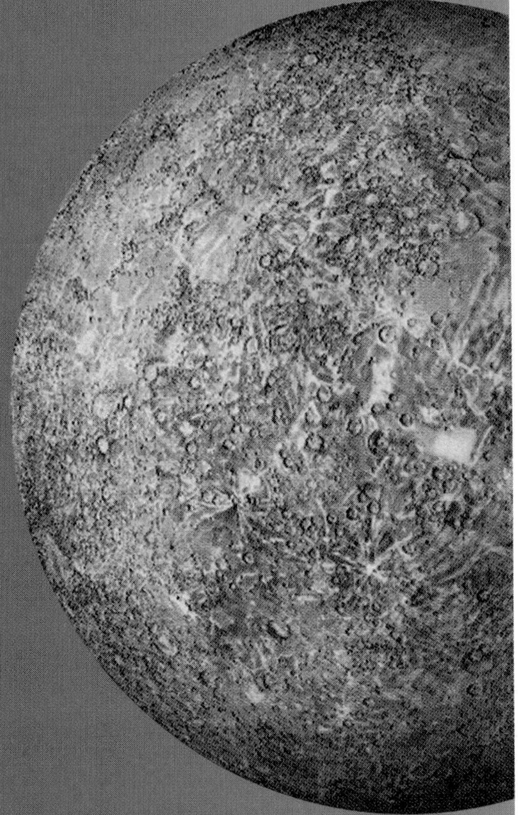

R I V U X G

FIGURE 7-1 **Mercury: Vital Statistics** Heavily cratered Mercury was visited three times by the *Mariner 10* spacecraft in 1974 and 1975. Nevertheless, we have images of only about half of the planet's surface. Although most of the images were recorded at distances of about 200,000 km (125,000 mi) from Mercury, their resolution is still vastly superior to that of the best Earth-based images. (Astrogeology Team, U.S. Geological Survey)

relatively few craters pockmark the lava flows that filled the basin, the Caloris impact must have occurred toward the end of the major crater-making period.

The Caloris impact was a tumultuous event that shook the entire planet. Indeed, the collision affected the side of Mercury directly opposite the Caloris Basin. That area (Figure 7-3b) has a jumbled, hilly region covering nearly half a million square kilometers, about twice the size of Wyoming. The hills, which appear as tiny wrinkles that cover most of the photograph, are about 5 to 10 km wide and between 100 and 1800 m high. Geologists believe that energy from the Caloris impact traveled through the planet and became focused, like light through a lens, as it passed through Mercury. As this concentrated energy reached the far surface of the planet, jumbled hills were pushed up. A similar pair of phenomena to Caloris and the jumbled terrain on the other side of Mercury can be seen in our Moon's Mare Orientale (Figure 7-3c) and chaotic hills on the opposite side. Finding such a similar pair supports the theory that the impact and the jumbled surface are connected.

Looking at Mercury in more detail, one starts to see a variety of differences from the features on the Moon. Unlike the Moon and its maria, Mercury lacks extensive

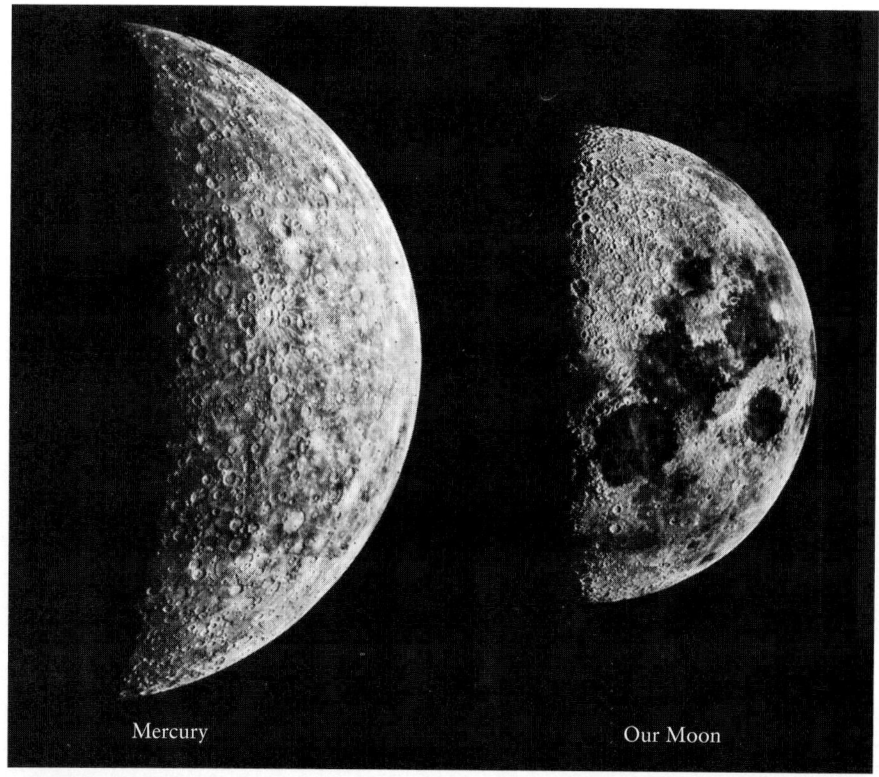

FIGURE 7-2 Mercury and Our Moon Mercury and our Moon are shown here to the same scale. Mercury's radius is 2439 km and the Moon's is 1738 km. For comparison, the distance from New York to Los Angeles is 3944 km (2451 mi). Mercury's surface is more uniformly cratered than that of the Moon. Daytime temperatures at the equator on Mercury reach 700 K (800°F), hot enough to melt lead or tin. (NASA, UCO/Lick Observatory)

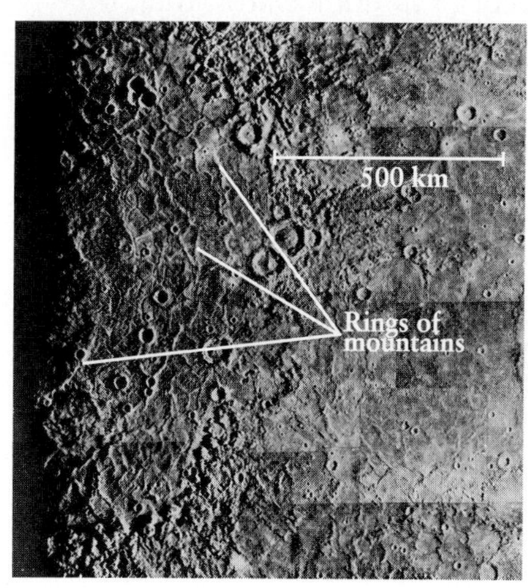

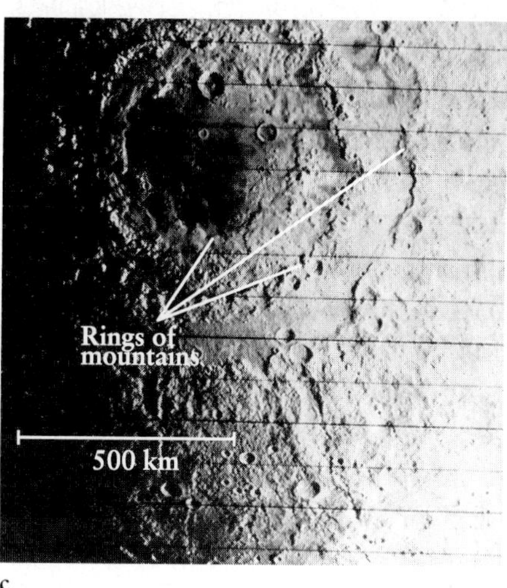

FIGURE 7-3 Major Impacts on Mercury and on Our Moon (a) The Caloris Basin *Mariner 10* sent back this view of a huge impact basin on Mercury's equator. Only about half of the Caloris Basin appears because it happened to lie on the terminator when the spacecraft sped past the planet. Although the center of the impact basin is hidden in the shadows (just beyond the left side of the picture), several semicircular rings of mountains reveal its extent. **(b) Unusual, Hilly Terrain** What look like tiny, fine-grained wrinkles on this picture are actually closely spaced hills, part of a jumbled terrain that covers nearly 500,000 square kilometers on the opposite side of Mercury from the Caloris Basin. The large, smooth-floored crater, Petrarch, has a diameter of 170 km (106 mi). This impact crater was produced more recently than the Caloris Basin. **(c) Mare Orientiale on the Moon** This giant impact feature is 900 km (560 mi) in diameter and is very similar to the Caloris Basin on Mercury. There is also terrain similar to that shown in (b) on the opposite side of the Moon from Mare Orientale. (NASA)

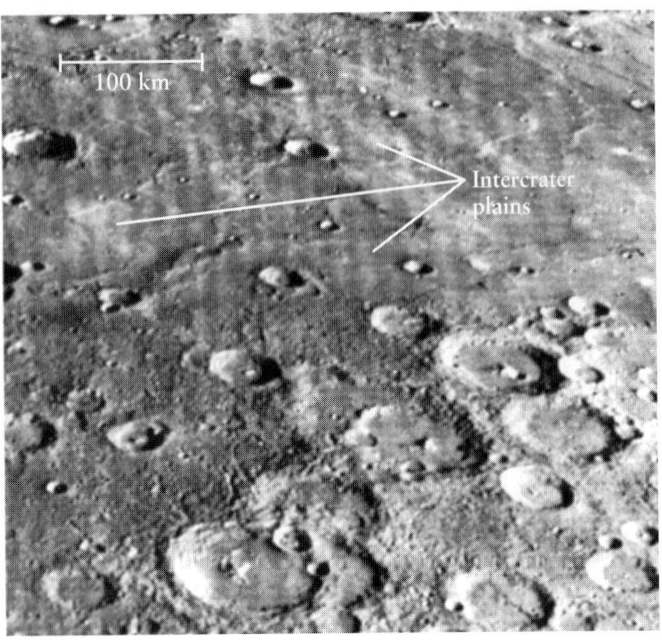
100 km
Intercrater plains
R I V U X G

VIDEO 7.1 FIGURE 7-4 **Mercury's Craters and Plains** This view of Mercury's northern hemisphere was taken by *Mariner 10* as it sped past the planet in 1974. Numerous craters on the bottom half of the image and broad inter-crater plains on the top half cover an area 480 km (300 mi) wide. (NASA)

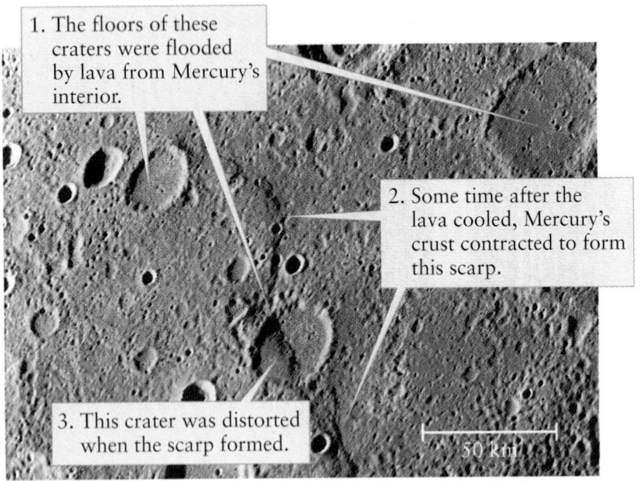
1. The floors of these craters were flooded by lava from Mercury's interior.
2. Some time after the lava cooled, Mercury's crust contracted to form this scarp.
3. This crater was distorted when the scarp formed.
50 km
R I V U X G

ANIMATION 7.2 FIGURE 7-5 **Scarps on Mercury** A long, meandering cliff called Santa Maria Rupes runs from north to south across this *Mariner 10* image of a region near Mercury's equator. This cliff, called a scarp by geologists, is more than 1 km high and runs for several hundred kilometers. Note how the crater in the center of the image was distorted vertically when the scarp formed. (NASA)

craterless regions. Instead of maria, Mercury has plains sparsely filled with relatively small craters. Such plains are not found on the Moon. Figure 7-4 shows a typical close-up view of Mercury. As we learned in Section 6-7, the lunar maria were produced by extensive lava flows that occurred between 3.1 and 3.8 billion years ago. Ancient lava flows also probably formed the Mercurian plains. As large meteorites punctured the planet's thin, newly formed crust, lava welled up from the molten interior to flood low-lying areas. The existence of craters pitting Mercury's plains suggests that these plains formed just over 3.8 billion years ago, near the end of the era of heavy bombardment. Mercury's plains are therefore older than most of the lunar maria, leaving more time for cratering to eradicate marialike features.

Mariner 10 also revealed gently rolling plains and numerous, long cliffs, called **scarps,** meandering across Mercury's surface (Figure 7-5). The scarps are believed to have developed as the planet cooled. Almost everything that cools contracts. Therefore, as Mercury's mantle and molten iron core cooled and contracted, its surface moved inward. Because it was solid, Mercury's crust could not collapse uniformly. Instead, it wrinkled as it contracted, forming the scarps. These features and the lack of recent volcanic activity suggest that the planet's interior is solid to a significant depth. Otherwise, lava would have leaked out as the scarps formed.

7-2 More of Mercury's interior is iron than the interior of Earth

Mercury's average density of 5430 kg/m^3 is quite similar to Earth's (5520 kg/m^3). As we saw in Section 6-3, typical rocks on Earth's surface have a density of only about 3000 kg/m^3

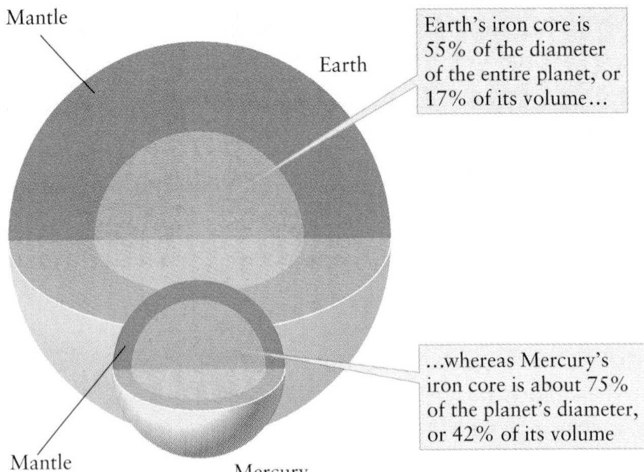
Mantle
Earth
Earth's iron core is 55% of the diameter of the entire planet, or 17% of its volume...
...whereas Mercury's iron core is about 75% of the planet's diameter, or 42% of its volume
Mantle
Mercury

FIGURE 7-6 **The Interiors of Earth and Mercury** Mercury has the highest percentage of iron of any planet in the solar system. Its iron core occupies an exceptionally large fraction of its interior.

because they are composed primarily of lightweight elements. The high average densities of both Mercury and our planet are caused by their dense interiors.

Because Mercury is less dense than the Earth, you might conclude that Mercury has a lower percentage of iron, a heavy element, than our planet. Mercury is actually the most iron-rich planet in the solar system. It is only the Earth's greater mass pressing inward and thereby compressing our planet's inner parts that makes Earth denser than Mercury.

Figure 7-6 shows a scale drawing of Mercury's interior, where an iron core fills 42% of the planet's volume. Surrounding the core is a 600-km-thick rocky mantle. For comparison, the Earth's iron core occupies only 17% of the planet's volume. How much of Mercury's core is molten is still unknown.

Events early in Mercury's history must somehow account for its high iron content. We know that the inner regions of the primordial solar nebula were incredibly hot. Perhaps only iron-rich minerals were able to withstand the solar heat there, and these subsequently formed iron-rich Mercury. According to another theory, an especially intense outflow of particles from the young Sun stripped Mercury of its low-density mantle shortly after the Sun formed. A third possibility is that, during the final stages of planet formation, Mercury was struck by a large planetesimal, just as the Earth was struck by a Mars-sized body that led to the formation of our Moon. Computer simulations show that this cataclysmic collision would have ejected much of the lighter mantle (Figure 7-7).

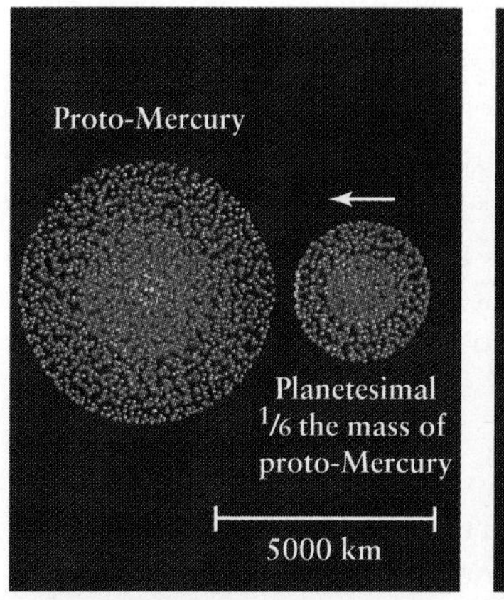

Just before contact

3 minutes after contact

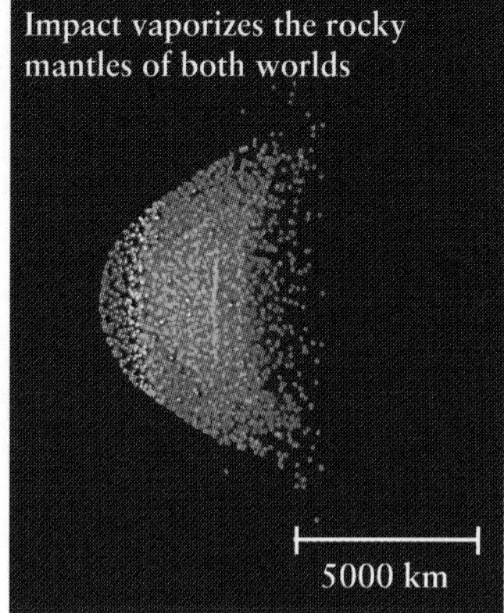

6 minutes after contact

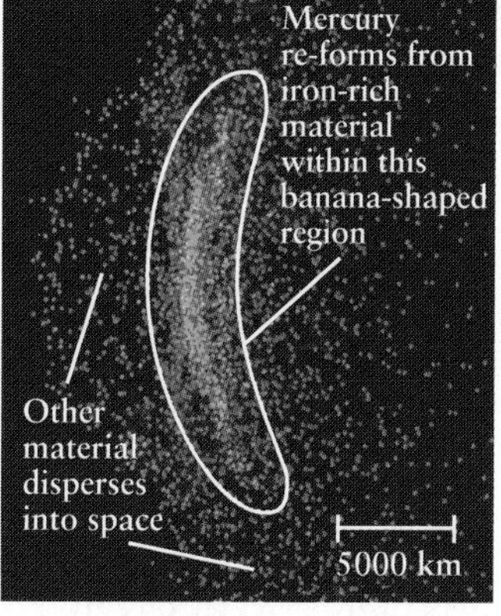

30 minutes after contact

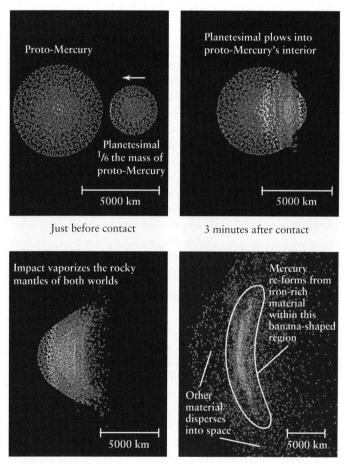

FIGURE 7-7 **The Stripping of Mercury's Mantle** To account for Mercury's high iron content, one theory proposes that a collision with a massive asteroid stripped Mercury of most of its rocky mantle. These four images show a computer simulation of a head-on collision between proto-Mercury and a body one-sixth its mass. Both worlds are shattered by the impact, which vaporizes much of their rocky mantles. Mercury eventually re-forms from the iron-rich debris left behind (material inside the banana-shaped region). The rest of the material (outside the banana) left the vicinity of the planet. (Courtesy of W. Benz, A. G. W. Cameron, and W. Slattery)

7-3 Mercury's rotation and revolution are coupled

Mercury has one of the most unique orbits in our solar system. Recall from Section 6-8 that when the Earth was young, its gravitational force created tides on the Moon, thereby forcing the Moon into synchronous rotation. Something similar (but not identical) happened with the Sun playing the role of the Earth and Mercury in the place of our Moon. To begin with, the Sun created a significant tidal bulge on Mercury that locked into place when the planet solidified. Mercury is so close to the Sun (average separation: 0.387 AU) that the Sun's gravitational force on Mercury's tidal bulges changed the planet's rotation rate. However, Mercury's highly eccentric orbit ($e = 0.21$) prevented the planet from being locked into synchronous orbit like our Moon. Instead, Mercury developed what is called a **3-to-2 spin-orbit coupling.** This means that Mercury undergoes three sidereal rotations (rotations measured with respect to the distant stars, not the Sun), while undergoing two revolutions around the Sun (Figure 7-8).

The major consequence of spin-orbit coupling is that one or the other of Mercury's regions of high tide is facing the Sun whenever the planet is at perihelion (see Figure 7-8). There is some slight variation in each orbit so that the exact high tide doesn't always point toward the Sun.

A Day on Mercury Is Two Years Long

The motion of the Sun across Mercury's sky is unique in the solar system. First, a solar day there (noon to noon) is 176 Earth days long, twice the length of a year! Furthermore, if you were to set up a camera at the location of high tide on Mercury to take a series of pictures when the planet is passing through perihelion, you would see the Sun start to rise in the east, stop high in the sky, move back toward the east, stop again, and then resume its westward

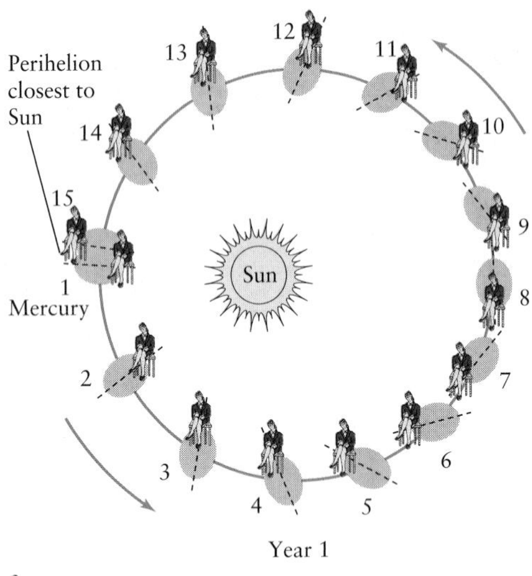

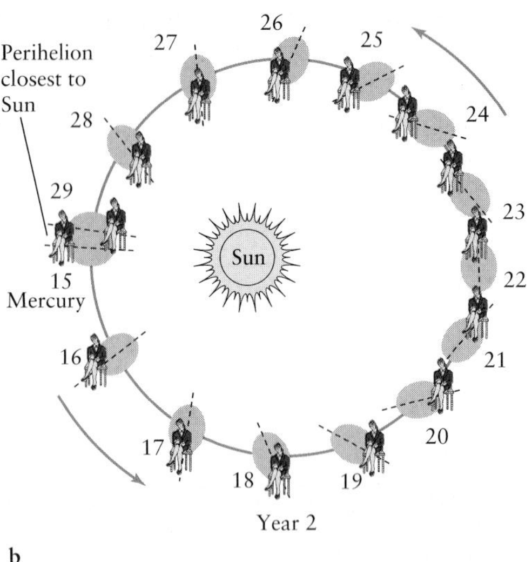

FIGURE 7-8 **3-to-2 Spin-Orbit Coupling** Mercury undergoes three sidereal rotations every two years. You can see this by following the observer on the planet from time 1 to time 15 in (a) and then from 15 to 29 in (b). The observer points to the right 4 times during this interval (at time 1, between times 10 and 11, 19 and 20, and finally at 29). This means that Mercury has revolved three times in exactly two sidereal Mercurian years. Since a sidereal year is 88 Earth days long, a sidereal day on Mercury is 58.7 Earth days long. During the same interval, however, the Sun is at noon as seen from the observer's location only twice: at time 1 and time 29. Therefore, a solar day is 176 Earth days long.

journey. This is analogous to the retrograde motion of the planets that we observe from Earth (see Section 2-2).

Mariner 10 discovered that Mercury has a weak magnetic field. Explaining it presents a challenge to astronomers because the dynamo model used to explain Earth's magnetic field doesn't work here. As we saw in Section 6-4, electric currents flowing in a planet's liquid iron core create a planetwide magnetic field. The Earth creates its electric current by rotating once a day. Mercury rotates 59 times more slowly (see Figure 7-1). This is hardly fast enough to generate the magnetic field observed here.

Besides Pluto, Mercury is the planet from which we have the least observational data. In August 2004, the *Messenger* spacecraft was launched toward Mercury. It will loop past Venus twice and Mercury three times before settling into orbit around Mercury in 2011. During all of its encounters, *Messenger* will provide astronomers with vital information about the innermost planets.

7-4 Mercury's atmosphere is the thinnest of all terrestrial planets

Mercury's mass (the total number of particles that comprise it) is only 5.5% that of the Earth. Like our Moon, the force of gravity is too weak on Mercury to hold a permanent atmosphere, but trace amounts of five different gases have been detected around Mercury. Scientists believe the Sun is the source of the hydrogen and helium gas near Mercury, while sodium and potassium gas escape from rocks inside the planet (a process called *outgassing*, which also occurs on the Earth). Oxygen observed in Mercury's atmosphere may come from polar ice that is slowly evaporating. All these gases drift into space and are continually being replenished in the atmosphere from their respective resources. The average density of this atmosphere is at least 10^{17} times less dense than the air we breathe.

Mercury's Temperature Range Is the Most Extreme in the Solar System

The Earth's thick atmosphere stores heat, which helps 1
explain why it feels warm on a cloudy day or at night. Furthermore, as the Earth rotates it loses heat into space. Because of the Earth's rapid daily rotation, this lost heat is quickly replaced by the Sun during daylight hours and so the average temperature change between day and night on Earth is only about 11 K (20°F). Because of Mercury's slow rotation and minimal atmosphere, the differences in temperature between day and night there are far more noticeable than on the Earth. At noon on Mercury, the surface temperature is 700 K (800°F). At the terminator, where day meets night, the temperature is about 425 K (305°F), and on the night side, the temperature falls as low as 100 K (–280°F)! The resulting daily range of temperature is therefore 600 K (1080° F) on Mercury.

Just as astronomers use scattered radio waves to search for water on the Moon (see Section 6-6), they have also sent radio waves to Mercury. In 1992 they made an extraordinary discovery—evidence for ice near Mercury's poles in craters that are permanently in shadow. Dozens of circular regions, presumably craters, send back signals with characteristics distinct to ice. Confirmation of the presence of this ice will have to wait until the *Messenger* spacecraft visits Mercury late in this decade. If it is there, the origin of this ice, whether from comet impacts, from gases rising from inside the planet and then freezing, or from both sources, also remains to be determined.

INSIGHT INTO SCIENCE

Model-Building In modeling real situations, scientists consider several different effects simultaneously. Omit any crucial property and you get inaccurate results. For example, think of how scientists calculate how long ice can remain at Mercury's poles. They must take into account the planet's distance from the Sun, the tilt of its axis of rotation, its rotation rate, surface features, whether the ice is exposed or mixed with other material, and the chemical composition of Mercury.

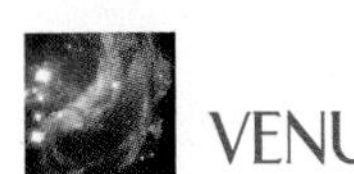

VENUS

2 One of the few things that Mercury and Venus have in common is that neither has a moon. Venus and Earth, on the other hand, have much in common. They have almost the same mass, the same diameter, and the same average density. Indeed, if Venus were located at the same distance from the Sun as is the Earth, then it, too, might well have evolved life. However, Venus is 30% closer to the Sun than the Earth is, and this one difference between the two planets leads to a host of others, making Venus inhospitable to life.

7-5 The surface of Venus is completely hidden beneath a permanent cloud cover

At nearly twice the distance from the Sun as Mercury, Venus is often easy to view without interference from the Sun's glare. At its greatest elongation, Venus is seen high above the western horizon after sunset, where, like Mercury, it is often called the "evening star." High in the eastern sky before sunrise, it is called the "morning star."

Venus is easy to identify because it is often one of the brightest objects in the night sky. Only the Sun and the Moon outshine Venus at its greatest brilliance. Venus is often mistaken for a UFO (Figure 7-9), because when it appears low on the horizon, its bright light is strongly refracted by the Earth's atmosphere, making it appear to rapidly change color and position. You can see the appearance of the inner planets using *Starry Night Enthusiast*™ (see Guided Discovery: The Inner Solar System).

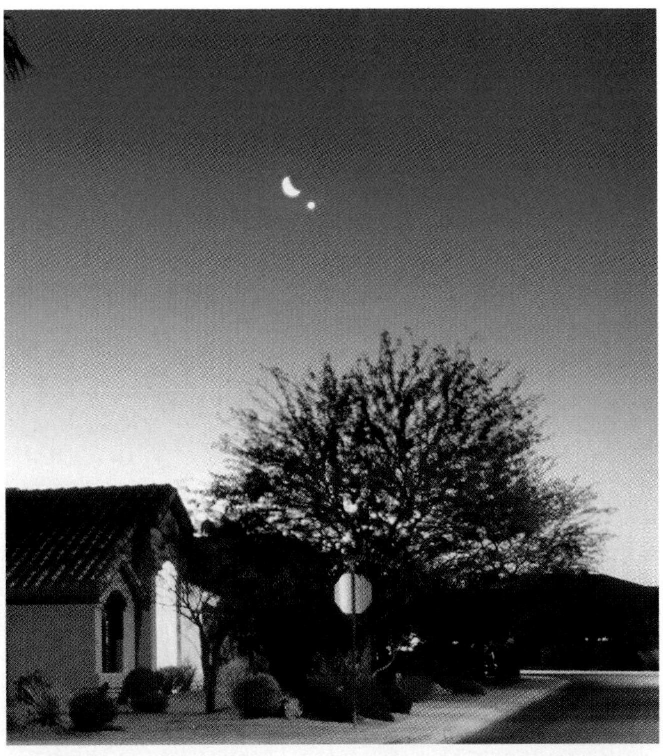

R I V U X G

FIGURE 7-9 **UFO? Just Venus** Venus is bright and is often seen near the horizon where rising and sinking gases in the Earth's atmosphere make it appear to move and change color, like alleged UFOs. (Joe Orman)

Unlike Mercury, Venus is intrinsically bright, because it is completely surrounded by light-colored, highly reflective clouds (Figure 7-10). Since visible light telescopes cannot penetrate this thick, unbroken layer of clouds, we did not even know how fast Venus rotates until 1962. In the 1960s, however, both the United States and the former Soviet Union began sending probes to Venus. The Americans sent fragile, lightweight spacecraft into orbit near the planet. The Soviets, who had more powerful rockets, sent massive vehicles directly into the Venusian atmosphere.

Building spacecraft that could survive the descent into Venus's atmosphere proved to be more frustrating than anyone had expected. Finally, in 1970, a Soviet probe managed to transmit data for a few seconds directly from the Venusian surface. Soviet missions, which continued until 1983, measured a surface temperature of 750 K (900°F) and a pressure of 90 atm. This is the same pressure you would feel if you were swimming 0.82 km (2700 ft) underwater on Earth.

In contrast to Earth's present nitrogen- and oxygen-rich atmosphere (see Section 6-1), Venus's thick atmosphere is 96% carbon dioxide, with the remaining 4% being mostly nitrogen. This atmosphere is remarkably similar to the Earth's early carbon dioxide–rich atmosphere. These gases were vented from

VENUS: VITAL STATISTICS

Average distance from Sun:	0.723 AU = 1.082 × 10^8 km
Maximum distance from Sun:	0.728 AU = 1.089 × 10^8 km
Minimum distance from Sun:	0.718 AU = 1.075 × 10^8 km
Eccentricity of orbit:	0.007
Average orbital speed:	35.0 km/s
Sidereal period of revolution:	224.7 days = 0.615 Earth year
Sidereal rotation period:	243.0 days (retrograde)
Solar rotation period (day):	116.8 Earth days
Inclination of equator to orbit:	177.4°
Inclination of orbit to ecliptic:	3.39°
Radius (equatorial):	6051 km = 0.949 Earth radius
Mass:	4.87 × 10^{24} kg = 0.815 Earth mass
Average density:	5240 kg/m^3 = 0.949 Earth density
Escape speed:	10.4 km/s
Surface gravity (Earth = 1):	0.91
Albedo:	0.59
Average surface temperature:	460°C = 860°F = 733 K
Atmospheric composition (by number of molecules):	96.5% carbon dioxide (CO_2), 3.5% nitrogen (N_2), 0.003% water vapor (H_2O)

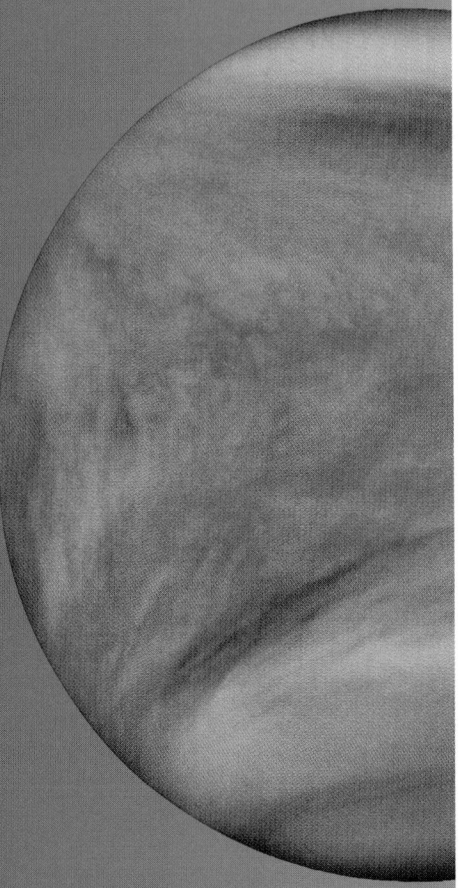

R I V U X G

FIGURE 7-10 **Venus: Vital Statistics** Venus's thick cloud cover efficiently traps heat from the Sun, resulting in a surface temperature even hotter than that on Mercury. Unlike Earth's clouds, which are made of water droplets, Venus's clouds are very dry and contain droplets of concentrated sulfuric acid. This ultraviolet image was taken by the *Pioneer Venus Orbiter* in 1979. (NASA)

inside Venus through volcanoes and other openings. Unlike the Earth, however, Venus has no liquid oceans to dissolve the carbon dioxide or life to convert it into oxygen and carbon compounds, so it remains in the atmosphere there today.

Soviet spacecraft also discovered that Venus's clouds are confined to a 20-km-thick layer located 48 to 68 km above the planet's surface. Above and below the clouds are 20-km-thick layers of haze. Beneath the lower level of haze, the Venusian atmosphere is clear all the way down to the surface. Standing on the surface of Venus, you would experience a perpetually cloudy day.

3 Unlike the clouds on Earth, which appear white from above, the cloudtops of Venus appear yellowish or yellow-orange to the human eye. These colors are typical of sulfur and its compounds. Indeed, spacecraft found substantial amounts of sulfur dust in Venus's upper atmosphere and sulfur dioxide and hydrogen sulfide at lower elevations. The clouds are composed of droplets of concentrated sulfuric acid! Because of the tremendous atmospheric pressure at Venus's surface, the droplets remain suspended as a thick mist, rather than falling as rain.

All of the major chemical compounds spewed into our air by Earth's volcanoes have also been detected in Venus's atmosphere. Among these molecules are large quantities of sulfur compounds. Because many of these substances are very short-lived, they must be constantly replenished by

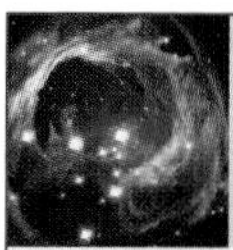

GUIDED DISCOVERY
The Inner Solar System

In this section, we will learn more about the appearances of the inner planets as seen from Earth using your *Starry Night Enthusiast*™ program.

Mercury Find Mercury (tab*Find/Mercury*). The first step is to see that Mercury is always close to the horizon when it is visible in the night sky.

1. *If Mercury is below the horizon:* Choose the "Best Time" option when the "not currently visible" warning comes up. If Mercury is to the west (to the right) of the Sun go to 2a. If Mercury is to the east (to the left) of the Sun, go to 2b.

2a. *If Mercury is already up in the sky:* If Mercury is west (to the right) of the Sun, we want to set up the screen with Mercury as the "morning star." To do this, reset the time to 5 A.M., relocate Mercury, and run time forward or backward until Mercury is just above the eastern horizon and the Sun is below the horizon. This shows Mercury as the "morning star." Now, run time forward at 1 minute per timestep, and you will see the white dot of Mercury fade. (If it is too close to the Sun, it won't be visible at all. Change the date and try again.) Now go to section 3.

2b. If Mercury is east (to the left) of the Sun in the sky, then set the time to 6 P.M., relocate Mercury, and run the time forward or backward until the Sun is about to set. Step forward in time by 1-minute intervals and watch Mercury first appear as the "evening star" and then set on the western horizon. (If it is too close to the Sun, it won't be visible at all. Change the date and try again.)

3. *Now we explore Mercury's orbit.* Go to Atlas mode (*Favourites/Guides/Atlas*) and again find Mercury. Set the sky to about 70°. Make sure you are centered and locked on Mercury (grab Mercury, right click, and then left click on *Centre* if it is not already checked). Set the timestep to 1 day (more or less, depending on your computer speed) and press the right pointing triangle. Describe and explain the motion of the Sun relative to Mercury. You can learn more about Mercury as seen from Earth by trying Observing Project 48 at the end of this chapter.

Venus Repeat the same procedures for Venus as we have described for Mercury, starting at step 1 above. What similarities and differences do you see?

Mars Return to Atlas mode, set the date to 9/12/2003, the time to midnight, and lock on Mars. Zoom in on Mars with the slide until you see its surface features. What is Mars's phase? Zoom away until you have the Sun on the screen or you have gone the maximum distance (about 100° angle on the screen). Determine how far (in angle) the Sun is away from Mars. If the Sun is not visible, you can make this measurement by "dragging" the celestial sphere around and measuring angles from the grid. Does the Sun ever get this far in angle from Mercury or Venus, as seen from Earth? Explain.

new eruptions. In fact, the abundance of sulfur compounds in Venus's atmosphere does vary. These data suggest the possibility that the sulfurous compounds in Venus's atmosphere come from active volcanoes. This question is very much open, however, because of the question of lightning on Venus. Lightning is a frequent event around active volcanoes here on Earth. Astronomers have searched for lightning on Venus's atmosphere, but whether it has been detected there remains controversial.

The results of the early probing of Venus's atmosphere are summarized in Figure 7-11. Both pressure and temperature

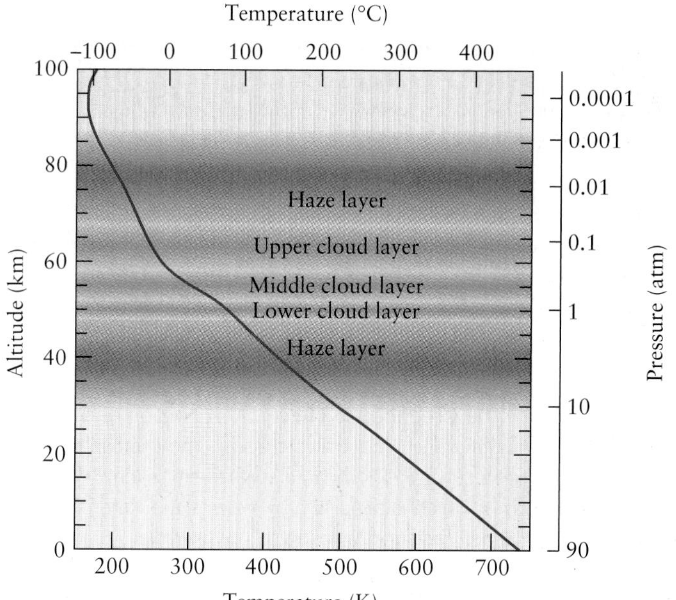

▶ FIGURE 7-11 **Temperature and Pressure in the Venusian Atmosphere** The pressure at the Venusian surface is a crushing 90 atm (1296 lb/in.^{2}). Above the surface, atmospheric pressure decreases smoothly with increasing altitude. The temperature of Venus's atmosphere increases smoothly from a minimum of about 173 K (−150°F) at an altitude of 100 km to a maximum of nearly 750 K (900°F) on the ground.

decrease smoothly with increasing altitude. From the changes in temperature and pressure above a planet's surface, we begin to understand the structure of its atmosphere.

7-6 The greenhouse effect heats Venus's surface

At first, no one could believe reports that the surface temperature on Venus was higher than the surface temperature on Mercury, which, after all, is closer to the Sun. Setting aside this initial skepticism, astronomers found a straightforward explanation—the **greenhouse effect** (Figure 7-12a).

Perhaps you have had the experience of parking your car in the sunshine on a warm summer day. You roll up the windows, lock the doors, and go on an errand. After a few hours, you return to discover that the interior of your automobile has become hotter than the outside air temperature (Figure 7-12b).

What happened to make your car so warm? First, sunlight entered your car through the windows. This radiation was absorbed by the dashboard and the upholstery, raising their temperatures. Because they become warm, your dashboard and upholstery emit infrared radiation, which you detect as heat and which your car windows do not permit to escape. This energy is therefore trapped inside your car and absorbed by the air and interior surfaces. As more sunlight comes through the windows and is trapped, the temperature continues to rise. The same trapping of sunlight warms an actual greenhouse.

Carbon dioxide is responsible for a similar warming of Venus's atmosphere. Like your car windows, carbon dioxide is transparent to visible light but it absorbs infrared radiation. Although most of the visible sunlight striking the Venusian cloudtops is reflected back into space, enough light reaches the Venusian surface to heat it. The warmed surface in turn emits infrared radiation, which cannot escape through Venus's carbon dioxide–rich atmosphere. This trapped radiation produces the high temperatures found on Venus.

Without the greenhouse effect, the surface of Venus would have a noontime temperature of 465 K. Because of it, that temperature is actually a sweltering 750 K, hotter than the hottest spot on Mercury! Furthermore, the thick atmosphere keeps the night side of Venus at nearly the same temperature, unlike the night side of Mercury, where the temperature drops precipitously.

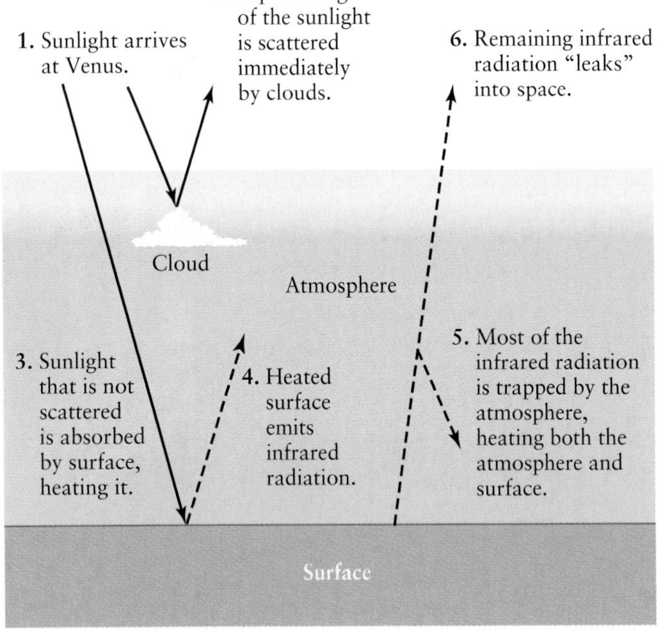
1. Sunlight arrives at Venus.
2. A percentage of the sunlight is scattered immediately by clouds.
6. Remaining infrared radiation "leaks" into space.
Cloud
Atmosphere
3. Sunlight that is not scattered is absorbed by surface, heating it.
4. Heated surface emits infrared radiation.
5. Most of the infrared radiation is trapped by the atmosphere, heating both the atmosphere and surface.
Surface

a

°F/°C
78 EXT
AUTO
89 °F
REAR

b

R I V U X G

FIGURE 7-12 **The Greenhouse Effect** (a) A portion of the sunlight reaching a planet penetrates through the atmosphere and heats the planet's surface. The heated surface emits infrared radiation, much of which is absorbed by water vapor and carbon dioxide. The trapped radiation helps raise the average temperatures of the surface and atmosphere. Some infrared radiation does penetrate the atmosphere and leaks into space. In a state of equilibrium, the rate at which the planet loses energy to space in this way is equal to the rate at which it absorbs energy from the Sun. (b) The glass windows in this car allow visible light to enter, but prevent the infrared released by the car's interior from escaping. The infrared therefore heats the air in the car much higher than the outside air. This also occurs in greenhouses and is called the greenhouse effect. (b: © Tobi Zausner)

This high temperature prevents liquid water from existing on Venus. Just like the Earth, Venus was likely to have initially had water both in the debris from which it formed and from the impacts of ice-bearing bodies, such as comets. The high atmospheric temperature has evaporated any water that came to Venus's surface, which is now bone dry. Furthermore, the atmospheric heat has caused most of the water that evaporated into the air to drift into space.

On Earth, atmospheric water and carbon dioxide both contribute to the greenhouse effect in our air. Warmed by sunlight, the Earth emits infrared radiation, some of which is absorbed by water vapor and carbon dioxide gas, causing the air temperature to rise. The small amounts of these gases in the Earth's atmosphere produce a comparatively gentle greenhouse effect. The heating due to the greenhouse effect is, however, increasing on Earth as our atmosphere acquires more carbon dioxide from burning fossil fuels and other sources. The air retains this gas due to large-scale clear-cutting of forests, which decreases the number of plants that can convert this gas into other carbon compounds and oxygen.

7-7 Venus is covered with gently rolling hills, two "continents," and numerous volcanoes

Soviet spacecraft that landed on Venus provided us with intriguing close-up images of the planet's arid surface. Figure 7-13 is a view of Venus's regolith taken in 1981. Russian scientists believe that this region was covered with a thin layer of lava that contracted and fractured upon cooling to create the rounded, interlocking shapes seen in the photograph. Indeed, measurements by several spacecraft indicate that Venusian rock is quite similar to lava rocks called basalt, which are common on Earth and the Moon.

WEB LINK 7.4 By far the best images of the Venusian surface came from the highly successful *Magellan* spacecraft that arrived at Venus in 1990. In orbit about the planet, *Magellan* sent radar signals through the clouds surrounding Venus. It mapped Venus using a radar altimeter that bounced microwaves off the ground directly below the spacecraft. By measuring the time delay of the radar echo, scientists determined the heights and depths of Venus's hills and valleys. As a result, astronomers have been able to construct a three-dimensional map of the planet.

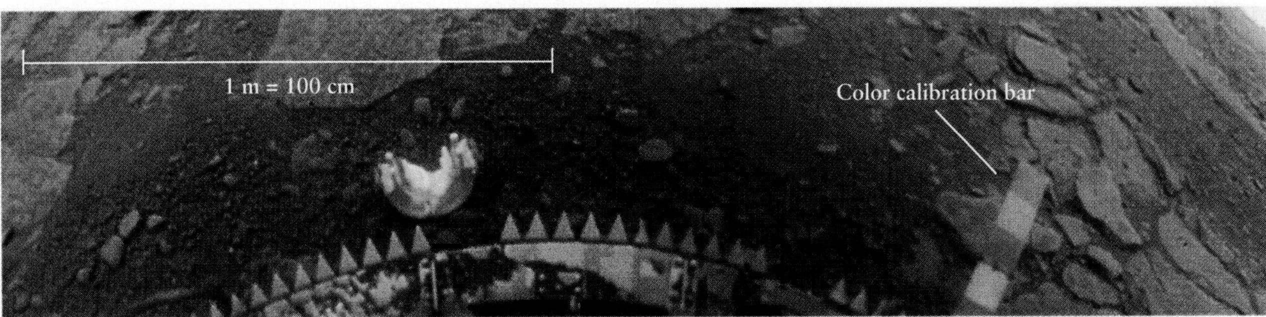

Image from *Venera 13*

Color-corrected image

R I V U X G

FIGURE 7-13 **The Venusian Surface** (a) This color photograph, taken by a Soviet spacecraft, shows rocks that appear orange because the light was filtered through the thick, sulfur-rich clouds. (b) By comparing the apparent color of the spacecraft to the color it was known to be, computers can correct for the sulfurous light. The actual color of the rocks is gray. In this view, the rocky plates covering the ground may be fractured segments of a thin layer of lava. The toothed wheel in each image is part of the landing mechanism that keeps the spherical spacecraft from rolling. (Courtesy of C. M. Pieters and the USSR Academy of Sciences)

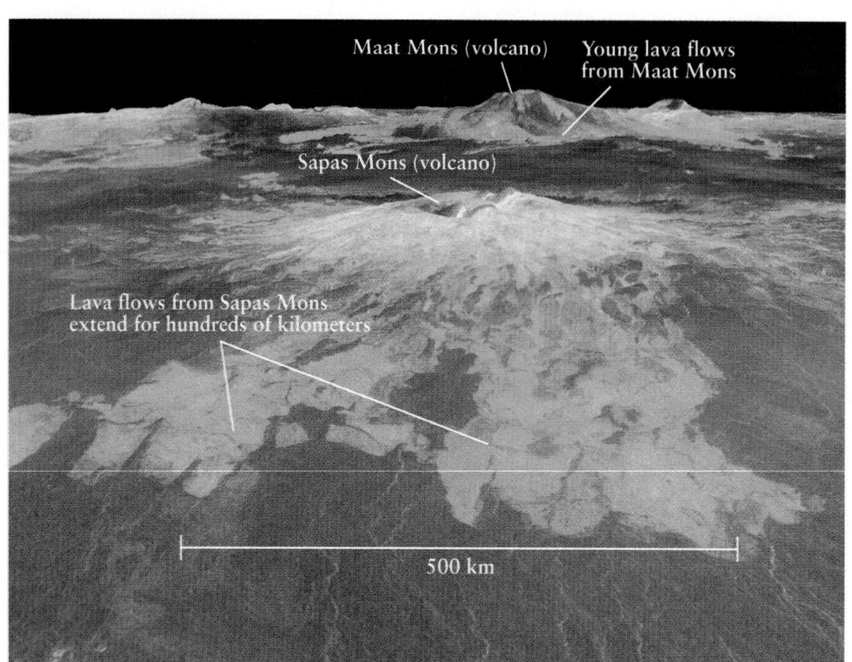

FIGURE 7-14 A Venusian Landscape A computer combined radio images to yield this perspective view of Venus as you would see it from an altitude of 4 km (2.5 mi). The color recreates the color you would see through Venus's thick clouds. The brighter color of the extensive lava flows indicates that they reflect radio waves more strongly. The vertical scale has been exaggerated 10 times to show the gentle slopes of Sapas Mons and Maat Mons, volcanoes named for ancient Phoenician and Egyptian goddesses, respectively. (NASA, JPL Multimission Image Processing Laboratory)

R I V U X G

The detail visible on this map (also called *resolution*) is about 75 m. You could see a football stadium on Venus if there were any. (None were detected.)

Venus is remarkably flat compared to the Earth. More than 80% of Venus's surface is covered with volcanic plains and gently rolling hills created by numerous lava flows. Figure 7-14 is an image using the *Magellan* radar data to create a view of a Venusian landscape.

Global radar images revealed two large highlands, or "continents," rising well above the generally level surface of the planet (Figure 7-15). The continent in the northern hemisphere is Ishtar Terra, named after the Babylonian goddess of love and approximately the same size as Australia. Ishtar Terra is dominated by a high plateau ringed by towering mountains. The highest mountain is Maxwell Montes, whose summit rises to an altitude of 11 km above the average surface. For comparison, Mount Everest on Earth rises 9 km above sea level.

The larger Venusian continent, Aphrodite Terra (named after the Greek goddess called Venus by the Romans), is a

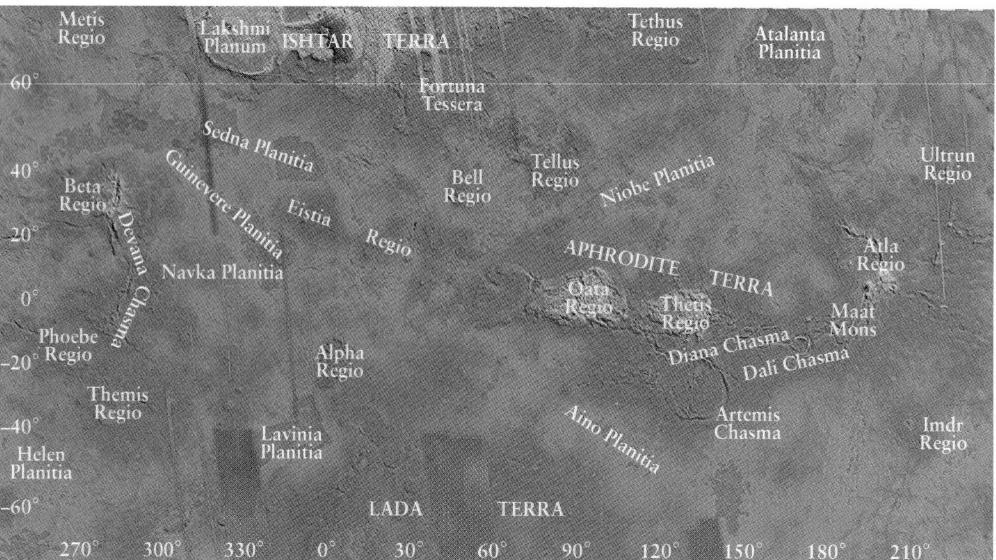

FIGURE 7-15 A Map of Venus This false-color radar map of Venus, analogous to a topographical map of Earth, shows the large-scale surface features of the planet. The equator extends across the middle of the map. Color indicates elevation—red for highest, followed by orange, yellow, green, and blue for lowest. The planet's highest mountain is Maxwell Montes on Ishtar Terra. Scorpion-shaped Aphrodite Terra, a continentlike highland, contains several spectacular volcanoes. Do not confuse the blue and green for oceans and land. (Peter Ford/MIT, NASA/JPL)

R I V U X G

R I V U X G

FIGURE 7-16 **A "Global" View of Venus** A computer using map-generated numerous *Magellan* images creates a simulated globe. Color is used to enhance small-scale structures. Extensive lava flows and lava plains cover about 80% of Venus's relatively flat surface. The bright band running almost east-west is the continent-like highland region Aphrodite Terra. (NASA)

vast belt of highlands just south of the equator. Aphrodite is 16,000 km (10,000 mi) in length and 2000 km (1200 mi) wide, giving it an area about one-half that of Africa. The global view of the surface of Venus in Figure 7-16 shows that most of Aphrodite Terra is covered by vast networks of faults and fractures.

Venus has more than 1600 major volcanoes and volcanic features and fewer than a thousand impact craters (Figure 7-17), as compared to the hundreds of thousands seen on the Moon and Mercury. Today, of course, Venus's thick atmosphere heats, and thereby vaporizes, much of the infalling debris that would otherwise create craters. Because its atmosphere is thicker than Earth's, Venus's atmosphere vaporizes such space junk falling toward it more efficiently than Earth's atmosphere does. However, there should have been a period shortly after Venus formed and before its atmosphere developed (from gases escaping from the planet's interior) when impacts were common.

The low number and random distribution of craters on Venus led to the belief that the planet's surface is periodically erased and replaced. Because geologists have found no large-scale tectonic plate motion on Venus to refresh the surface, the current theory explaining the lack of extensive cratering is that the entire surface of the planet melts! This would occur if the crust is very thick compared to the Earth's crust. A crust 300 km thick (Earth's crust is everywhere less than 65 km thick) would insulate the interior so much that, heated by radioactive elements in it, the mantle could become hot enough to melt the crust on occasion. It has been proposed that every 700 million years or so, the entire surface of Venus liquefies until the pent-up heat escapes, a new solid crust forms, and the process begins anew.

The sulfur content of the air and the traces of active volcanoes on the surface are compelling indicators that, like the Earth, Venus has a molten interior. Because the average density of Venus is similar to that of Earth, its core is predominantly iron. Currents in the molten iron should generate a magnetic field. However, none of the spacecraft sent there has detected one. Rotation of currents creates magnetic fields (see Section 6-4). Therefore, Venus's apparent lack of a magnetic field is plausible only if it rotates exceptionally slowly. In fact, the planet takes 116.8 Earth days to get from one sunrise to the next if you could see the Sun from Venus's surface, (see Figure 7-10).

Unlike the Earth, Venus has **retrograde rotation.** This means that the direction of Venus's orbit around the Sun (counterclockwise as seen from

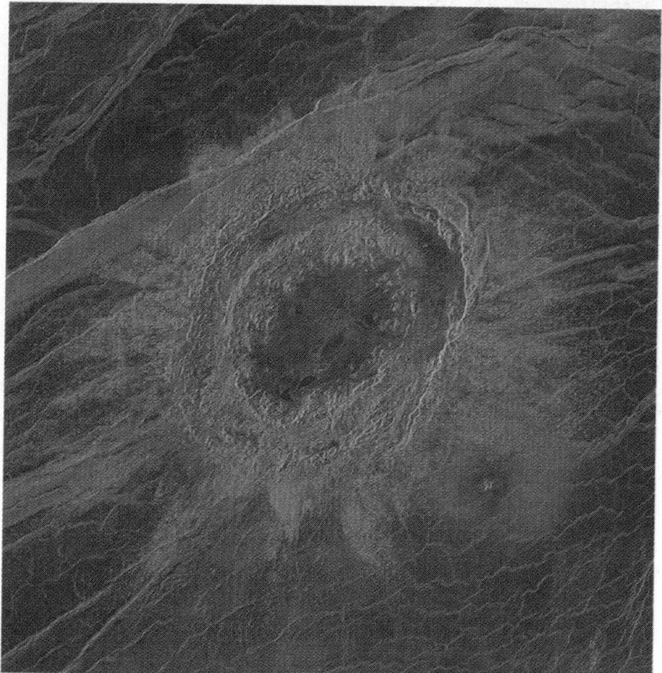

R I V U X G

FIGURE 7-17 **Craters on Venus** Impact craters on Venus tend to occur in clusters, which suggests that they are formed from large, single pieces of infalling debris that are broken up in the atmosphere. Shown here is the fourth largest crater, Crater Klenova. It is 142 km across. (Lunar and Planetary Institute/NASA)

space far above the Earth's North Pole) is *opposite* the direction of its rotation (clockwise as seen from the same vantage point). In other words, sunrise on Venus occurs in the west. Venus's rotation axis is tilted more than 177°, compared to the Earth's 23½° tilt. Because Venus's axis is within 3° of being perpendicular to the plane of its orbit around the Sun, the planet has no seasons. Although we do not know the cause of Venus's retrograde rotation, one likely explanation is that a monumental impact flipped the rotation axis early in the planet's existence.

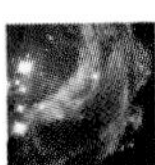

MARS

Mars's distinctive rust-colored hue (Figure 7-18) makes it stand out in the night sky. For centuries Mars has generated more excitement as a possible home for alien life than any other world in our solar system. The idea began following the first well-documented telescopic observations of Mars in 1659 by the Dutch physicist Christiaan Huygens. Huygens identified a prominent, dark surface feature that reemerged about every 24 hours, suggesting a rate of rotation similar to that of Earth.

MARS: VITAL STATISTICS

Average distance from Sun:	1.52 AU = 2.28×10^8 km
Maximum distance from Sun:	1.67 AU = 2.49×10^8 km
Minimum distance from Sun:	1.38 AU = 2.10×10^8 km
Eccentricity of orbit:	0.093
Average orbital speed:	24.1 km/s
Sidereal period of revolution:	687 Earth days = 1.88 Earth years
Sidereal rotation period:	$24^h\ 37^m\ 22^s$
Solar rotation period (day):	$24^h\ 39^m\ 35^s$
Inclination of equator to orbit:	25.19°
Inclination of orbit to ecliptic:	1.85°
Radius (equatorial):	3393 km = 0.53 Earth radius
Mass:	6.42×10^{23} kg = 0.107 Earth mass
Average density:	3950 kg/m^3 = 0.716 Earth density
Escape speed:	5.0 km/s
Surface gravity (Earth = 1):	0.38
Albedo:	0.16
Surface temperatures:	Maximum: 20°C = 70°F = 293 K Mean: −53°C = −63°F = 220 K Minimum: −140°C = −220°F = 133 K
Atmospheric composition (by number of molecules):	95.3% carbon dioxide (CO_2) 2.7% nitrogen (N_2) 0.03% water vapor (H_2O) 2% other gases

FIGURE 7-18 **Mars: Vital Statistics** This photograph, taken from space, shows part of the enormous Valles Marineris, a canyon system on Mars similar to the Grand Canyon in North America. (NASA, USGS)

Belief that advanced life exists on Mars skyrocketed at the end of the nineteenth century after Giovanni Virginio Schiaparelli, an Italian astronomer, reported seeing 40 lines crisscrossing the Martian surface in 1877 (Figure 7-19). He called these dark features *canali*, an Italian term meaning "channels." It was soon mistranslated into English as *canals*, implying the existence on Mars of intelligent creatures capable of engineering feats. This speculation led Percival Lowell, who came from a wealthy Boston family, to finance a major new observatory near Flagstaff, Arizona. By the end of the nineteenth century, Lowell had allegedly observed 160 Martian "canals."

By the beginning of the twentieth century, it was fashionable to speculate that the Martian canals formed an enormous, planetwide irrigation network to transport water from melting polar caps to vegetation near the equator. In view of the planet's reddish, desertlike appearance, Mars was thought to be a dying planet whose inhabitants must go to great lengths to irrigate their farmlands. No doubt the Martians would readily abandon their arid ancestral homeland and invade the Earth for its abundant resources.

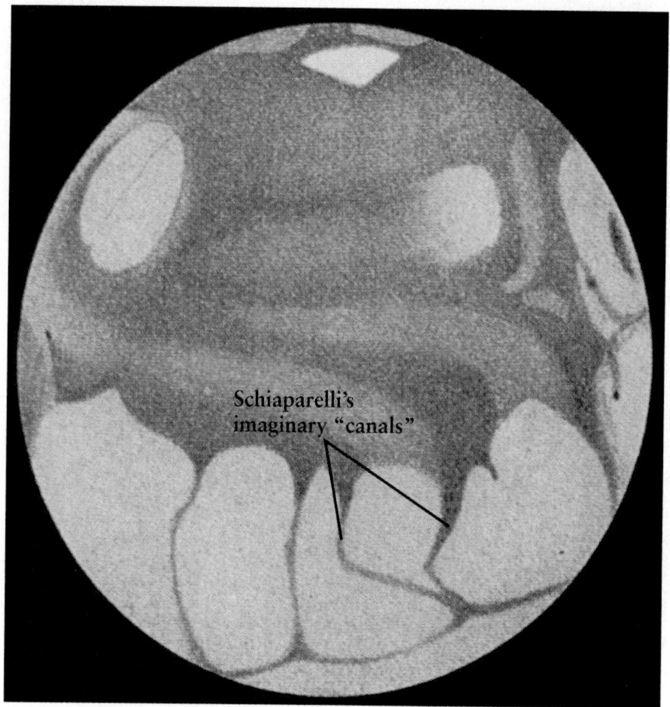

FIGURE 7-19 **The Illusion of Martian Canals** Giovanni Schiaparelli examined Mars through a 20-cm (8-in.) diameter telescope, the same size used by many amateur astronomers today. He drew images of the red planet that showed features perceived by Percival Lowell and others as irrigation canals. Higher resolution images from Earth and spacecraft visiting Mars failed to show the same features. (Michael Hoskin, ed., *The Cambridge Illustrated History of Astronomy*, Cambridge University Press, 1997, p. 286. Illustration by G. V. Schiaparelli. Courtesy Institute of Astronomy, University of Cambridge, UK)

R I V U X G

FIGURE 7-20 ***War of the Worlds*** This image from the 1953 movie *The War of the Worlds* shows alien ships zapping Earth. (Paramount/The Kobal Collection)

INSIGHT INTO SCIENCE

Perception and Reality Our tendency to see patterns, even where they do not exist, makes it easy for us to leap to unjustified conclusions. Scientific images and photographs require objective analysis.

In particular, on October 30, 1938, Orson Welles presented a radio program announcing the invasion of Earth by Martians. His broadcast, live from New York City, was inspired by the 1898 book *War of the Worlds* by H. G. Wells. So realistic was the performance that it caused panic throughout the New York–New Jersey area. Let's begin our exploration of Mars by seeing whether H. G. Wells and the dozens of other authors, radio storytellers, and filmmakers (Figure 7-20) who capitalized on the idea of advanced life on Mars were close to the mark.

7-8 Mars's global features include plains, canyons, craters, and volcanoes

Spacecraft journeying to Mars have sent back pictures since the 1970s. They show that Mars's surface is completely dry and that it has broad plains, shallow craters, enormous volcanoes, and vast canyons (Figure 7-21). The broad northern plain is called Vastitas

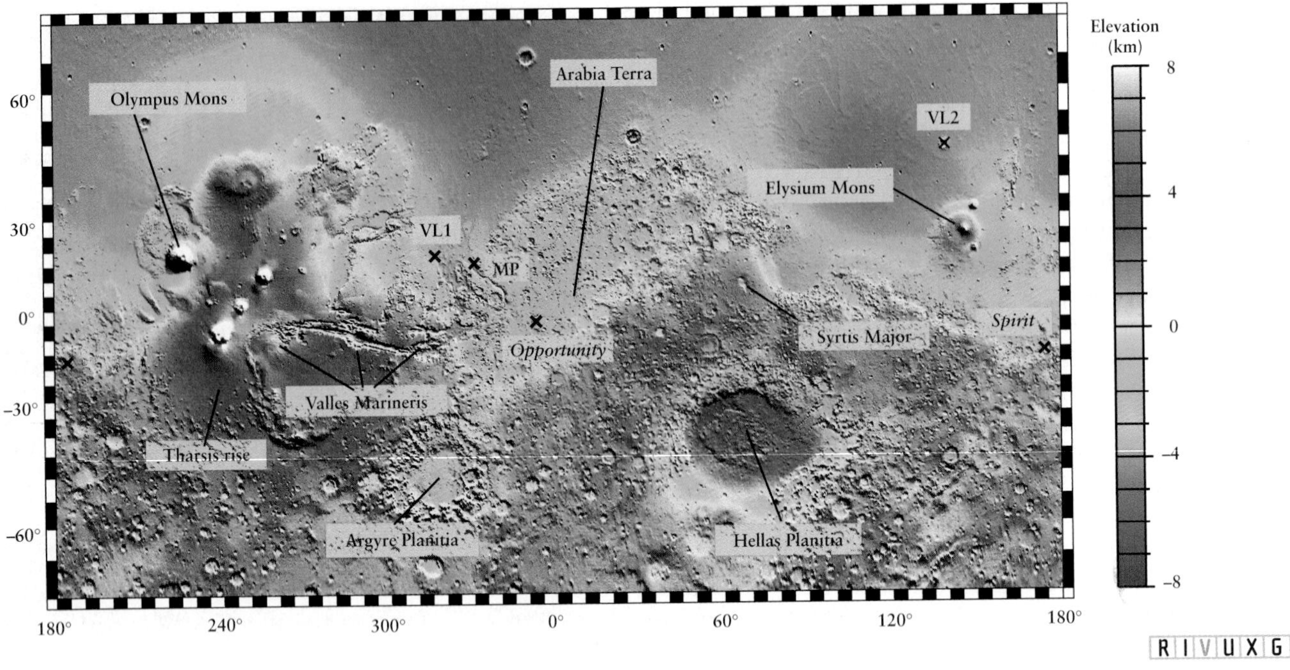

FIGURE 7-21 **The Topography of Mars** The color coding on this map of Mars shows elevations above (positive numbers) or below (negative numbers) the planet's average radius. To produce this map, an instrument on board *Mars Global Surveyor* fired pulses of laser light at the planet's surface, then measured how long it took each reflected pulse to return to the spacecraft. The *Viking Lander 1* (VL1), *Viking Lander 2* (VL2), *Mars Pathfinder* (MP), *Opportunity*, and *Spirit* landing sites are each marked with an ×. (MOLA Science Team, NASA/GSFC)

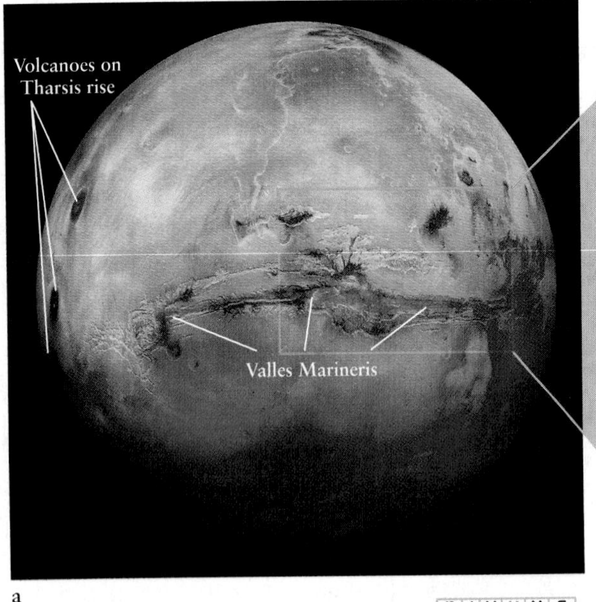

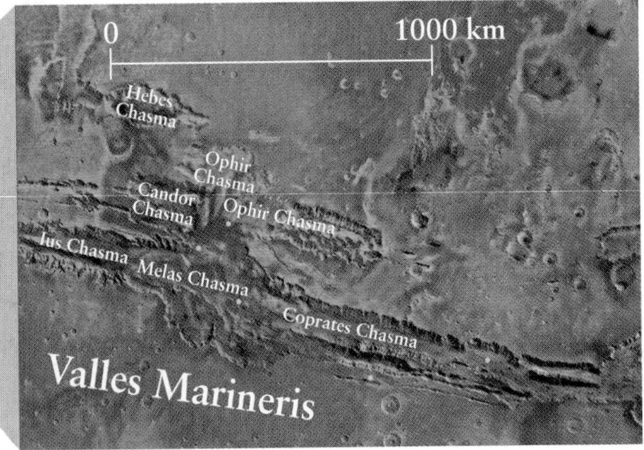

FIGURE 7-22 **Martian Terrain** This high-altitude photograph shows a variety of the features on Mars, including broad, towering volcanoes (left) on the highland called Tharsis Bulge; impact craters (upper right); and vast, windswept plains. The enormous Valles Marineris canyon system crosses horizontally just below the center of the image. *Inset:* Details of the Valles Marineris, which is about 100 km (60 mi) wide. The canyon floor has two major levels. The northern (upper) canyon floor is 8 km (5 mi) beneath the surrounding plateau, whereas the southern canyon floor is only 5 km (3 mi) below the plateau. (USGS/NASA; insert: NASA/GSFP/LTP)

Borealis (the **northern vastness** or **northern lowlands**) and is shown in blue on Figure 7-21. It is some 5 km (3 mi) below the much more heavily cratered and hilly plains in the south, called the **southern highlands.** The cause of the significant difference in heights of the two hemispheres is not known, but it may stem from either early tectonic activity (see Section 6-2 for a discussion of tectonics) or from a powerful early impact.

Between the two hemispheres, *Mariner 9* discovered a vast canyon now called Valles Marineris, which runs roughly parallel to the Martian equator (Figure 7-22). Valles Marineris stretches over 4000 km, about one-fifth the circumference of Mars. The Martian canyons are up to 6 km (4 mi) deep and 190 km (120 mi) wide. Valles Marineris begins with heavily fractured terrain in the west and ends with ancient cratered terrain in the east. If this canyon were located on Earth, it would stretch from New York to Los Angeles. Geologists believe that Valles Marineris is a large crack that formed as the planet cooled. It was enhanced by nearby rising crust to its west and widened further by erosion. Some of the eastern parts of this system appear to have been formed almost entirely by water flow, like Earth's Grand Canyon.

Most of the impact craters are located on Mars's southern hemisphere (Figure 7-23). This suggests that the northern vastness has been resurfaced by some process that eradicated ancient lowland craters, consistent with either of the earlier explanations of the differences in hemispheres.

Most of the volcanoes on Mars are located in the northern hemisphere. The largest volcano, Olympus Mons (Figure 7-24), covers an area as big as the state of Missouri and rises 26 km (16 mi) above the surrounding plains—nearly 3 times the height of Mount Everest. The highest volcano on Earth, Mauna Loa in the Hawaiian Islands, has a summit only 17 km above the ocean floor.

Planetwide, high-resolution photographs of Mars over the past 40 years have failed to show one canal of a size consistent with those allegedly seen from Earth (see Figure 7-19). We now know for certain that Schiaparelli's *canali* were optical illusions, and the science fiction writers were completely off the mark.

INSIGHT INTO SCIENCE

Fact and Fiction Science fiction may foreshadow real scientific discoveries. (Can you think of some examples?) But science also reins in the fantasies of the science fiction writer. A century ago, millions of people, including top scientists, believed that technologically advanced life existed on Mars.

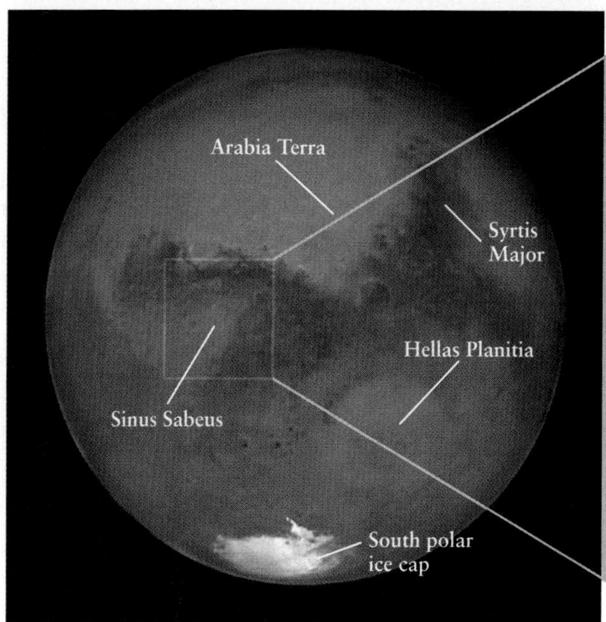

Mars from the Hubble Space Telescope

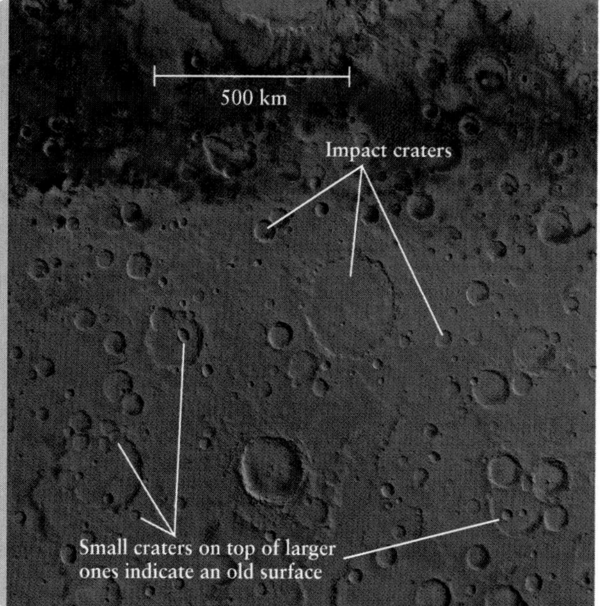

Closeup of Sinus Sabeus region

R I V U X G

FIGURE 7-23 **Craters on Mars** This image was made during the opposition of 2003. Arabia Terra is dotted with numerous flat-bottom craters. Syrtis Major was first identified by Christiaan Huygens in 1659. A single impact carved out Hellas Planitia, which is 5 times the size of Texas. *Inset:* This mosaic of images from the *Viking Orbiter 1* and *2* spacecraft shows an extensively cratered region located south of the Martian equator. Note how worn down these craters are compared to those on Mercury and our Moon. (NASA; J. Bell, Cornell University; and M. Wolff, SSI; inset: USGS)

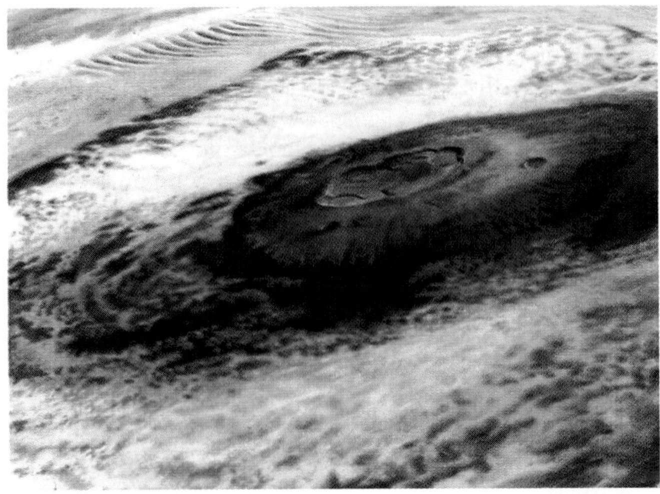

R I V U X G

WEB LINK 7.8 **FIGURE 7-24 The Olympus Caldera** This view of the summit of Olympus Mons is based on a mosaic of six pictures taken by one of the *Viking* orbiters. The caldera consists of overlapping, volcanic craters and measures about 70 km across. The volcano is wreathed in mid-morning clouds brought upslope by cool air currents. The cloudtops are about 8 km below the volcano's peak. (NASA)

7-9 While no canals exist on Mars, it does have some bizarre features

Mars not only lacks canals, but it has no cities, roads, or other signs of civilization. However, astronomers have photographed several surface features that at first glance could have been crafted by intelligent life-forms. In 1976, a hundred years after Schiaparelli's alleged discovery of canali, the *Viking Orbiter 1* spacecraft photographed a feature that appeared to be a humanlike face (Figure 7-25a). However, when the *Mars Orbiter* photographed the surface in greater detail and from a different angle in 1998 (Figure 7-25b), it found no facial features. If anything, the same spot looked more like a giant heel print.

Other photographs show a collection of pyramids and a "happy face." Were any of these things relics of an advanced civilization? Scientists universally believe that the two faces and the pyramids are naturally formed features. The first face is a region of harder rock around which softer debris is being worn away. Similarly, the pyramids are consistent with the winds eroding softer rock around harder rock pushed upward from inside Mars billions of years ago. The same thing happens on Earth. The happy face is made from a large impact crater and several small ones inside it.

INSIGHT INTO SCIENCE

Keep It Simple Was the Martian "face" made by intelligent life or by wind-sculpted hills? When two or more alternative interpretations present themselves, scientists apply the principle of Occam's razor and always choose the one requiring the fewest unproven assumptions.

7-10 Mars's interior is less molten than the inside of the Earth

Part of the Valles Marineris was apparently formed when large regions of Mars's surface moved apart due to tectonic plate motion. This activity is generated by the flow of rock under the planet's surface (see Section 6-2 for more on plate tectonics). However, there is no evidence that plate tectonic activity occurs on Mars today. This indicates that more of

FIGURE 7-25 In the Eye of the Beholder These two images of the same site on Mars, taken 22 years apart, show how the apparent face in **(a)** changed to a more "natural-looking" feature in **(b)**. This transformation was due to weathering of the site, improved camera technology, and the change in angle at which the photograph was taken. (a: NSSDC/NASA; b: NASA/JPL/Malin Space Science Systems)

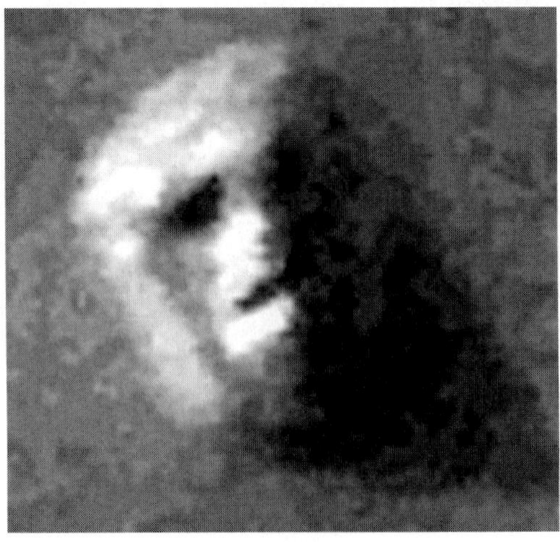

a

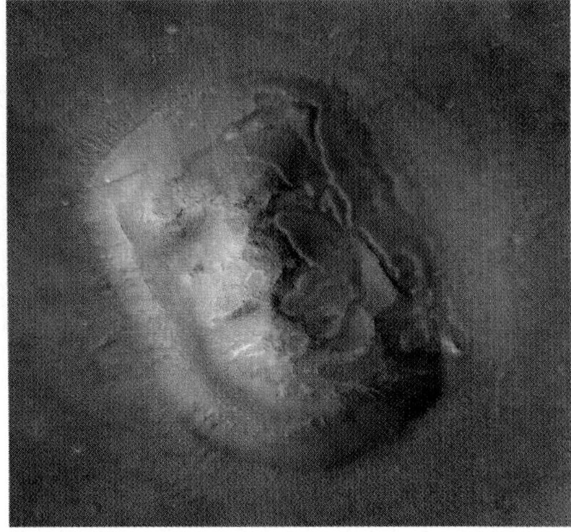

b

R I V U X G

Mars's interior is cooled, solid rock than exists inside the Earth. This belief is supported by the observation that Mars lacks a global magnetic field (see Section 6-4 for discussion of the field-generating dynamo effect). However, observations made in 2003 revealed that the Sun creates tiny tides (less than a centimeter) on the solid body of Mars. As the land rises and sinks ever so slightly through the day on Mars, friction created by this motion in the planet's interior provides enough heat to keep some of the interior molten. Rub your hands vigorously together to see how friction creates heat.

While a global magnetic field is absent today, there are remnants of such a field. In 1997 the *Mars Global Surveyor* discovered patterns of local surface magnetic fields in nine places on Mars like those found where Earth's tectonic plates separate and lock the Earth's magnetic field in the rock (see Section 6-4). Geologists propose that when Mars was young, its internal "dynamo" created a strong magnetic field and that tectonic plates helped craft its surface, while locking traces of its changing magnetic field in the molten rock. However, the planet quickly cooled, its global magnetic field vanished, and tectonic-plate activity apparently ceased nearly 4 billion years ago.

A more molten interior early in its existence is also consistent with the enormous volcanoes that exist on Mars. Volcanoes are places where molten rock oozes or bursts from a body's interior. The volcanoes on Mars are the type where molten rock oozed out, like the volcanoes that created the Hawaiian Islands, but very different from Mount Saint Helens in Washington State, or Mount Etna in Sicily, where the material from inside the Earth is ejected violently. On Earth, the Hawaiian Islands are only the most recent additions to a long chain of volcanoes. They resulted from **hot-spot volcanism,** a process by which molten rock rises to the surface from a fixed hot region far below. The Pacific tectonic plate is slowly moving northwest at a rate of several centimeters per year. As a result, new volcanoes are created above the hot spot, while older ones move off and become extinct, eventually disappearing beneath the ocean.

Because Mars's surface is apparently frozen in place due to the lack of tectonic plate motion, one hot spot can keep pumping lava upward through the same vent for millions of years. One result is Olympus Mons—a single giant volcano rather than a long chain of smaller ones (see Figure 7-24). This volcano's summit has collapsed to form a volcanic crater, called a **caldera,** large enough to contain the state of Rhode Island. Because Mars's solid crust and mantle (the region directly below the surface) are now so thick, they insulate the surface, preventing any more molten rock from escaping. Therefore, we do not expect to see any further volcanic activity there.

7-11 Martian air is thin and often filled with dust

To understand more about the activity on Mars's surface, we must first examine its atmosphere, which has only 0.6% the pressure of the Earth's atmosphere. As on Venus, some 95% of Mars's thin atmosphere is composed of carbon dioxide. The remaining 5% consists of nitrogen, argon, and some traces of oxygen. Unlike the gravitational attraction of tiny Mercury, which is too weak to hold any gases as a permanent atmosphere, Mars's gravitational force is just strong enough to prevent carbon dioxide, nitrogen, argon, and oxygen from escaping into space, but not strong enough to hold down water vapor.

If, as many astronomers expect, there were once vast oceans of water on the surface of Mars, then much of that water has evaporated and drifted into space, never to return. Today, the concentration of water vapor in Mars's atmosphere is 30 times lower than the concentration in the Earth's air. If all the water vapor could somehow be squeezed out of the Martian atmosphere, it would not fill even one of the five Great Lakes of North America. The remainder of the water that may have been on Mars's surface has turned to ice, much of which remains there today, especially at the poles. In 2002, astronomers discovered enough water ice in the upper 1 meter of Mars's south pole to fill two Great Lakes, and it is certain that much more water remains to be discovered inside Mars.

Despite the low density of Mars's air, the red planet experiences Earthlike seasons because of a striking coincidence, first noted in the late 1700s by the German-born English astronomer William Herschel. Just as the Earth's equatorial plane is tilted 23½° from the plane of its orbit, Mars's equator makes an angle of about 25° with its orbit. The Martian seasons last nearly twice as long as Earth's, because Mars takes nearly two Earth years to orbit the Sun. When Mars is near opposition, even telescopes for home use today reveal its daily changes. Dark seasonal markings on the Martian surface can be seen to vary and prominent polar caps shrink noticeably during the spring and summer months (Figure 7-26).

While sometimes blue (Figure 7-27), Mars's atmosphere is often pastel red, sometimes turning shades of pink and russet (Figure 7-28). All these latter colors are due to the fine dust blown from the planet's desertlike surface during windstorms. The dust is iron oxide, familiar here on Earth as rust. Mars's sky also changes color because the amount of dust in the air varies with the season. During the winter, carbon dioxide ice adheres to the dust particles and drags them to the ground. This helps clear and lighten the air. In the summer months, the carbon dioxide is not frozen, and the dust blown by surface winds remains aloft longer.

The sky color also changes over periods of many years. In 1995, for example, the amount of dust was observed to have dropped dramatically compared to that observed in the 1970s. The reason for such long-term changes is still under investigation.

As on Earth, the temperature changes with the seasons on Mars. For example, the temperature at the landing site of the rover *Spirit* near the equator ranged from 188 K to 243 K one day, and from 200 K to 263 K 100 days later (see Figure 7-21 for the site location). During the afternoons, heat from

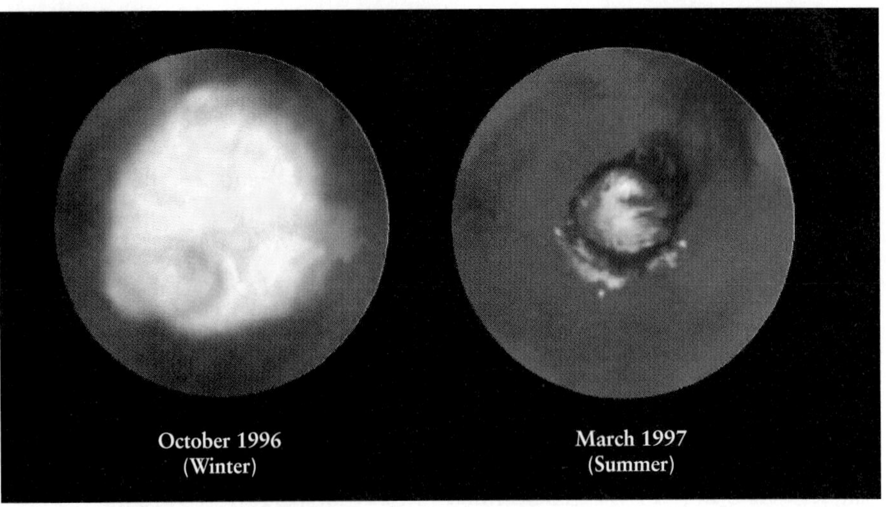

FIGURE 7-26 **Changing Seasons on Mars** During the Martian winter, the temperature drops so low that carbon dioxide freezes out of the Martian atmosphere. A thin coating of carbon dioxide frost covers a broad region around Mars's north pole. During summer in the northern hemisphere, the range of this north polar carbon dioxide cap decreases dramatically. In the summer a ring of dark sand dunes is exposed around Mars's north pole. (S. Lee/J. Bell/M. Wolff/Space Science Institute/NASA)

the planet's surface warms the air and sometimes creates whirlwinds called **dust devils.** A similar phenomenon occurs in dry or desert terrain on Earth. Martian dust devils reach altitudes of 6 km (20,000 ft). As on Earth, spacecraft on Mars detect drops in air pressure as dust devils sweep past. Some dust devils are large enough to be seen by orbiting spacecraft (Figure 7-29).

The winds on Mars help explain why its surface is eroding. Over the past two centuries, astronomers have seen faint surface markings disappear under a reddish-orange haze as thin Martian winds have stirred up finely powdered dust from the Martian regolith. Some storms obscure the entire planet, as happened in 2001. Over the ages, winds have worn down crater walls and deposits of dust have filled in the crater bottoms (Figure 7-30). But the Martian atmosphere is so thin that 3 billion years of sporadic storms have not carried enough of Mars's extremely fine-grained powder to eradicate them completely.

FIGURE 7-27 **Atmospheric Dust on Mars** When Mars's sky is relatively free of dust, it appears similar in color to our sky, as shown in this sunset photo taken by the rover *Opportunity*. The darker, brown color in which the Sun is immersed is due to lingering dust in the sky. When less dust is present, the Sun looks almost white during Martian sunsets. Most of the images we have of Mars's sky show colors like that seen in Figure 7-28. (NASA/JPL)

FIGURE 7-28 **Mars's Rust-Colored Sky** Taken by the *Mars Pathfinder*, this stunning photograph of another world shows the *Sojourner* rover snuggled against a rock named Moe on the Ares Vallis in order to run tests on the rock. At the top of the image, the pink color of the Martian sky is evident. (NASA/Lunar and Planetary Institute)

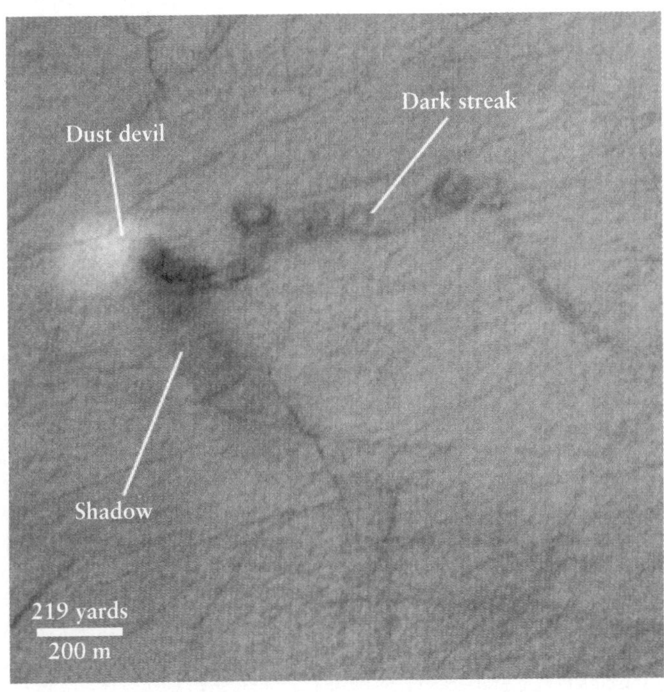

a

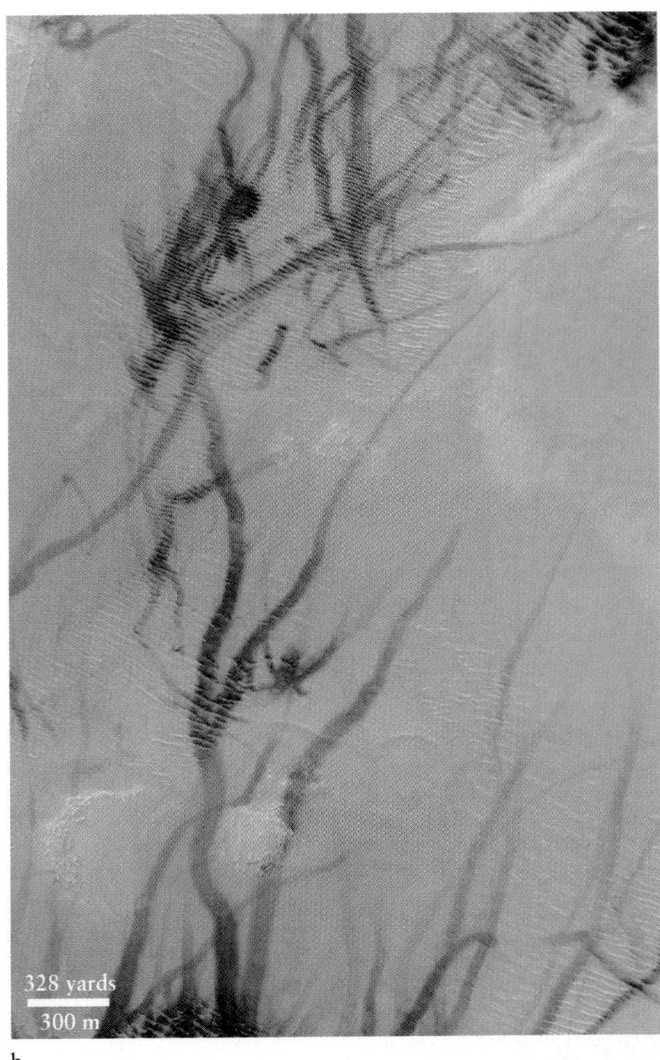

b

R I V U X G

FIGURE 7-29 **Martian Dust Devil** **(a)** This *Mars Global Surveyor* image shows a dust devil as seen from almost directly above. This tower of swirling air and dust casts a long shadow in the afternoon Sun. The dust devil had been moving from right to left before the picture was taken, leaving a dark curlicue-shaped trail in its wake. The area shown is about 1.5 × 1.7 km (about 1 mile on a side). **(b)** These dark streaks are the paths of dust devils on the Argyre Planitia of Mars. The tracks cross hills, sand dunes, and boulder fields, among other features on the planet's surface. (a: NASA/JPL/Malin Space Science Systems; b: NASA/JPL/Malin Space Science Systems)

R I V U X G

FIGURE 7-30 **Crater Endurance** Photographed by the rover *Opportunity*, this crater on Mars is about 130 m (430 ft) across. Rocks are visible in the crater and in vertical cliffs along its walls. By studying such rocks, astronomers hope to understand more of the history of water on Mars. (JPL/NASA)

7-12 Surface features indicate that water once flowed on Mars

4 Despite disproving the theory that Mars has broad canals or any other liquid surface water, Mars-orbiting spacecraft did reveal many dried-up riverbeds (Figure 7-31a), lakes (Figure 7-32a), river deltas (Figures 7-32a and b), where water and debris emptied into lakes and oceans, sedimentation laid down by water flow (Figure 7-33), and other water-related features. Some of the riverbeds include intricate branched patterns and delicate channels meandering among flat-bottomed craters. Rivers on Earth invariably follow similarly winding courses (Figure 7-31b).

Surface rovers, including *Spirit* and *Opportunity*, have found strong evidence that water did indeed create many of these features (Figure 7-34), including the fact that saltwater once flowed on the red planet. Surveys from space show that water persisted on Mars's surface for millions of years or longer. The origin of at least some of this water is believed to be ice-rich bodies that struck the surface, releasing their water into the atmosphere, then to rain down on the surface.

In 2001, astronomers reported evidence that volcanic activity may still be causing episodic flows of both water and lava on Mars's surface. *Mars Global Surveyor* images show channels apparently carved by water flowing down the walls of pits or craters (Figure 7-35). What makes these observations especially intriguing is that they appear to be geologically young, indicating that liquid water may still exist under the Martian surface.

While the flow of water that created the features shown in Figure 7-35 is relatively small, lake or ocean-volume quantities of water are believed to still be trapped under the Martian surface. If so, volcanic heating or the heat generated by impacts from space debris could open fissures enabling water to melt, escape, and carve new riverbeds today. This scenario would require vast amounts of water for several reasons. First, water is liquid over a limited range of temperature and pressure. At low temperatures, water becomes ice; at high temperatures, it becomes steam. Second, the atmospheric pressure also affects the state of water. If the pressure is very low, molecules easily escape from the liquid's surface, causing the water to vaporize. Because the pressure of Mars's atmosphere is only 0.6% that of the Earth's atmosphere, any liquid water on Mars today would quickly transform into ice or furiously boil and evaporate into the thin Martian air, and then be lost into space.

The total amount of water that existed on Mars's surface is not known. By studying flood channels

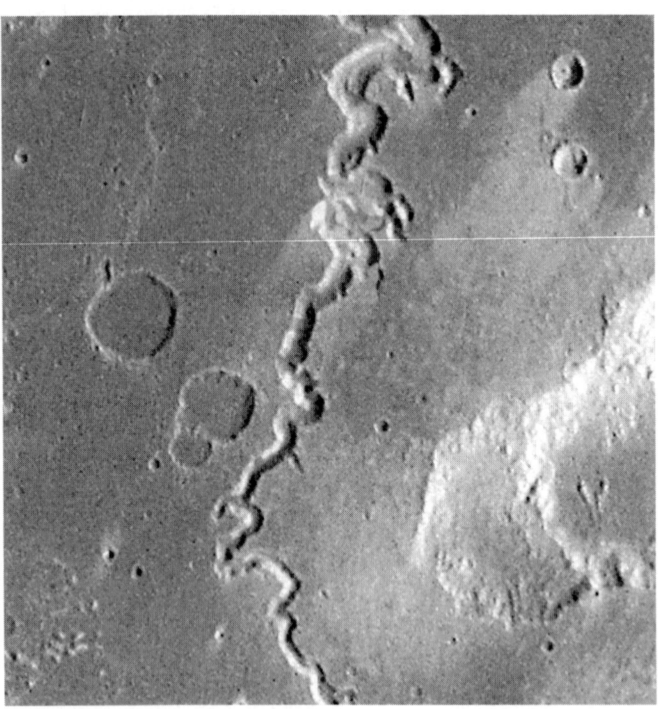

a

R I V U X G

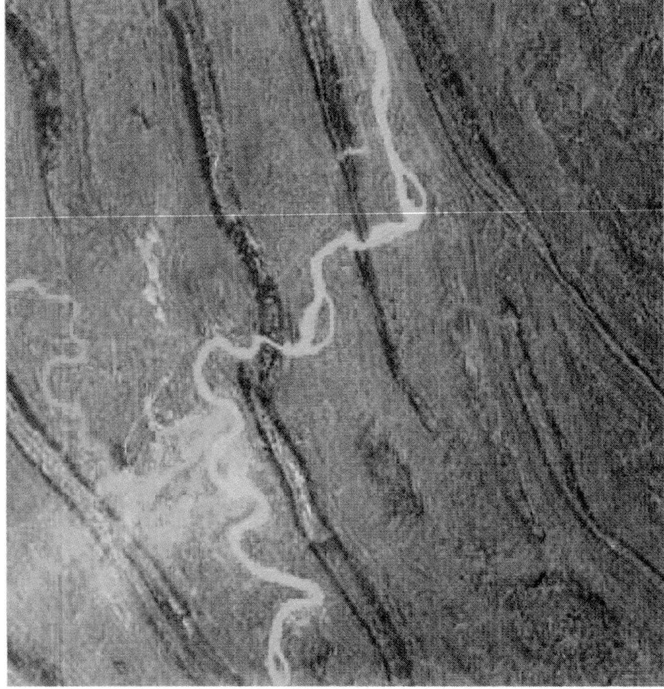

b

R I V U X G

FIGURE 7-31 **Rivers on Mars and Earth** (a) Winding canyons on Mars, such as the one in this *Viking Orbiter I* image, appear to be due to sustained water flow. This belief is supported by the terraces seen on the canyon walls in high resolution *Mars Orbiter* images. Long periods of water flow require that the planet's atmosphere was once thicker and its climate more Earthlike. (b) This image is of the Yangtze River near Chongqing, China. Typical of rivers on Earth, it shows the same snakelike curve as the river channels on Mars. (a: NASA; b: TMSC/NASA)

a

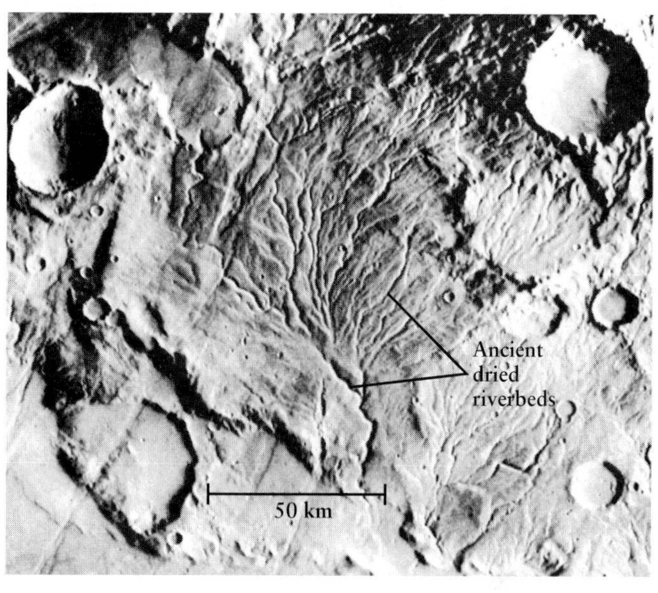

b

R I V U X G

R I V U X G

FIGURE 7-32 **Evidence of Water on Mars** (a) This dry Martian lake, photographed by the *Mars Global Surveyor* with a resolution of 1.5 m, is an excellent example of how geology and astronomy overlap. The features of this dry lake are consistent with those found on lakebeds on Earth. (b) This network of dry riverbeds is located on Mars's cratered southern hemisphere. (a: NASA/JPL; b: NASA)

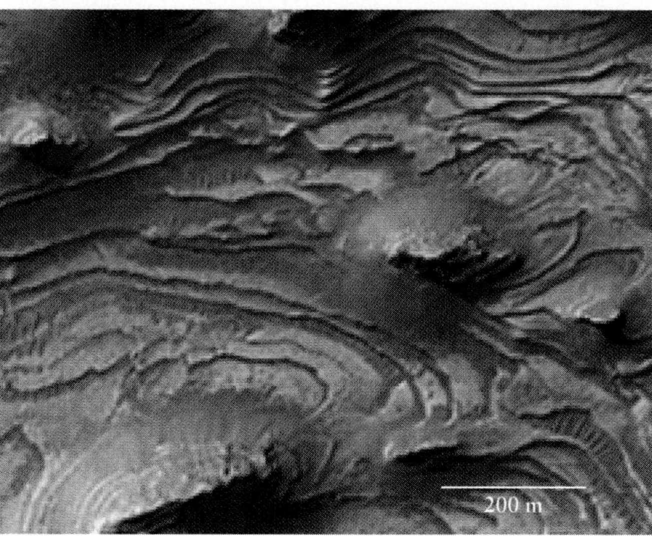

R I V U X G

VIDEO 7.5

FIGURE 7-33 **Ancient Oceans and Lakes on Mars** This *Mars Global Surveyor* image of a portion of *Valles Marineris* reveals terrain with "stair step layers." Such terrain is likely to have been created by sedimentation at the bottom of an ancient body of water. (Malin Space Science Systems/JPL/NASA)

that exist there, geologists estimate that there was enough water to cover the planet to a depth of 500 meters (1500 ft). For comparison, Earth has enough water to cover our planet to a depth of 2700 meters (8900 ft), assuming that the Earth's surface was everywhere a uniform height.

Further evidence that water once flowed on Mars comes from so-called *SNC meteorites* found on Earth (Figure 7-36a). These space rocks are believed to have once been pieces of Mars because their chemistries are consistent with those of rocks studied on Mars's surface and because they contain trace gases in amounts found only in the current Martian atmosphere. The meteorites were ejected into space during especially powerful impacts on that planet's surface. They also contain water-soaked clay, which is not expected to be found on any objects in the solar system besides Mars, Earth, and perhaps Europa. At least 34 meteorites from Mars have been identified.

Some of the water that was on Mars is still near the surface today, frozen at the red planet's poles. Recent measurements indicate that as much as 90% of the ice at the poles is water ice, the remainder being dry ice (carbon dioxide ice). The temperature at Mars's poles, typically 160 K (–170°F), keeps the water there permanently frozen, like the permafrost found in northern Asia and on Antarctica. Seasonal

a

R I V U X G

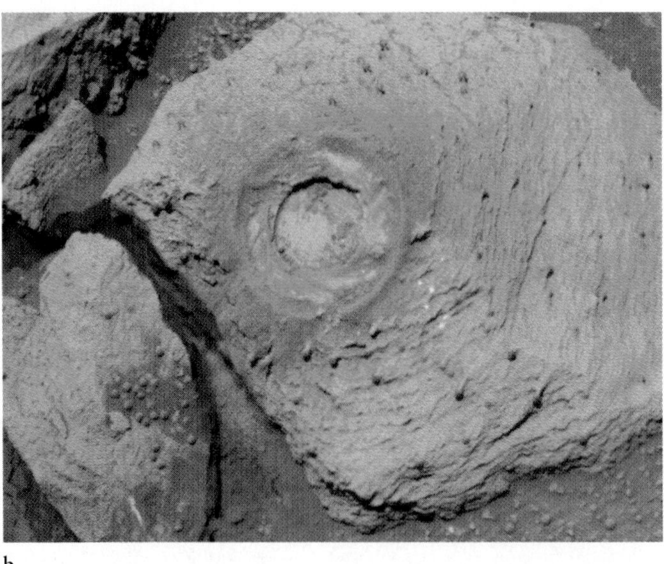

b

R I V U X G

FIGURE 7-34 **Layers of Rock Laid Down by Water** **(a)** This close-up image taken by the rover *Opportunity* shows a small section of rock layers in a location called The Dells. The angled and curved layering seen here is only created on Earth by water flow, strongly suggesting that this sediment was also laid down by water. The nearly spherical rocks, called "blueberries" because they are dark, have been chemically identified as hematite, an iron-rich mineral that is usually formed in water. The rovers have found blueberries strewn in a wide variety of locations. **(b)** Located in a small crater named Fram is a rock called Pilbara that has been dug into by the rover *Opportunity*. The chemistry of the rock, and the blueberries sliced in it and located all around it, are consistent with Pilbara having been formed when this part of Mars was wet. The cut is 4.5 cm (1.8 in.) wide and 7.2 mm (0.28 in.) deep. (a: NASA/JPL/USGS; b: NASA/JPL/Cornell)

variations at Mars's poles (see Figure 7-26) are created by the freezing and evaporating of dry ice, which reaches thickness of 2 meters. Indeed, during the winter, about one-third of the carbon dioxide in Mars's atmosphere comes down as dry ice.

Evidence that water existed as recently as a few million years ago (and possibly still today) just meters below Mars's equator also comes in the form of clusters of cones on the planet's surface. These cones are created when lava flows over water-rich terrain. The lava vaporizes the water, which eventually bursts through the lava, creating the cones. Similar clusters of cones on Earth created by this mechanism are found in Iceland.

The 1997 visit by *Pathfinder* and its little rover (see Figure 7-28) to the Ares Vallis revealed evidence of several floods at the mouth of a dried flood plain. The evidence includes layers of sediment, clumps of rock and sand stuck together like similar groupings created by water on Earth, rounded rocks worn down as they were dragged by the water, and rocks aligned by water flow. The rover performed

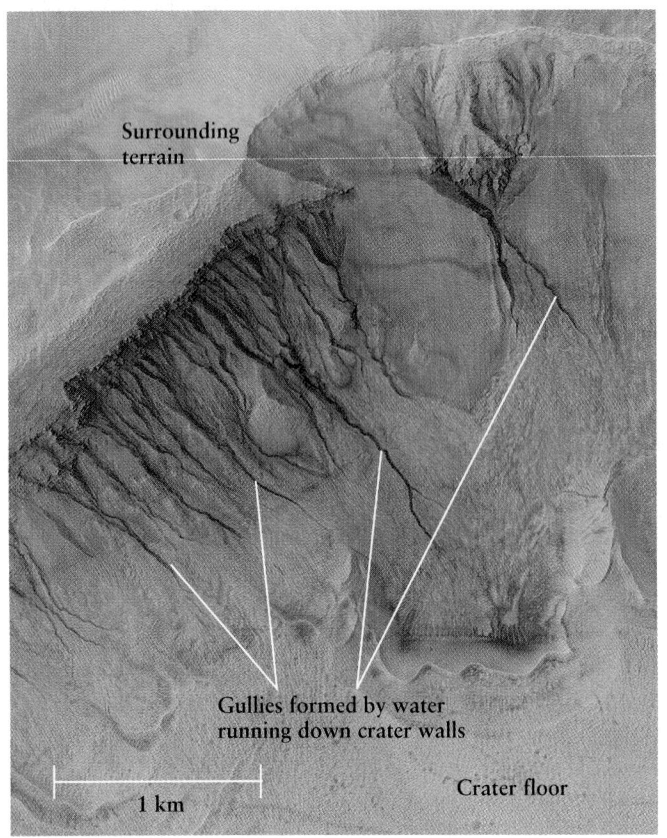

R I V U X G

◀ FIGURE 7-35 **Martian Gullies** This *Mars Global Surveyor* image of a crater wall in Mars's southern hemisphere shows a series of gullies along the crater wall. These gullies are similar to water-flow gullies formed on Earth. The gullies on Mars may have formed when underground water seeped out or when melted snow fell onto the crater walls. (NASA/JPL/Malin Space Science Systems)

a

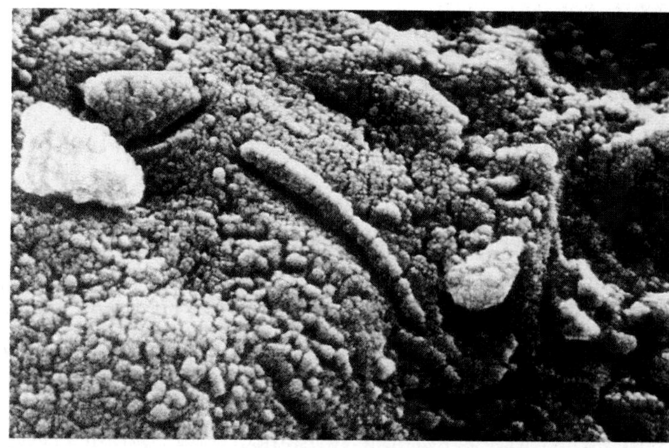

b

R I V U X G

FIGURE 7-36 **A Piece of Mars on Earth** Mars's thin atmosphere does little to protect it from impacts. Some of the debris ejected from impact craters there apparently traveled to Earth. **(a)** This SNC meteorite was recovered in Antarctica. It shows strong evidence of having been exposed to liquid water on Mars, perhaps for hundreds of years. **(b)** These may be the fossil remains of primitive bacterial life on Mars, although nonbiological origins for it have been presented. (a: NASA; b: NASA/JPL)

some 20 chemical analyses of rocks in the landing area. The soil is a rusty color, laden with sulfur. The source of this sulfur is a mystery, as is the large amount of silicon-rich silica discovered on rocks such as Barnacle Bill.

There is observational evidence of ancient shoreline encircling the northern lowlands of Mars. The belief that the northern lowlands were an ocean is supported by the analysis of a 1.2 billion-year-old meteorite from Mars. It has remnants of salt believed to have been deposited there when the Martian ocean dried up.

7-13 Search for microscopic life on Mars continues

The discovery in the 1960s that water once existed on the surface of Mars rekindled speculation about Martian life. Although it was clear that Mars has neither civilizations nor fields of plants, microbial life-forms still seemed possible. Searching for signs of organic matter was one of the main objectives of the ambitious and highly successful *Viking* missions.

WEB LINK 7.12

The two *Viking* spacecraft were launched during the summer of 1975. Each spacecraft consisted of two modules—an orbiter and a lander. Almost a year later, both *Viking* landers set down on rocky plains north of the Martian equator (Figure 7-37).

The landers confirmed the long-held suspicion that the red color of the planet is due to large quantities of iron in its soil. Despite the high iron content of its crust, Mars has a lower average density (3950 kg/m^3) than that of other terrestrial planets (more than 5000 kg/m^3 for Mercury, Venus,

R I V U X G

FIGURE 7-37 **Digging in the Martian Regolith** *Viking*'s mechanical arm with its small scoop protrudes from the right side of this view of the *Viking 1* landing site. Several small trenches dug by the scoop in the Martian regolith appear near the left side of the picture. (NASA)

and Earth). Mars must therefore contain overall a lower percentage of iron than these other planets.

Each *Viking* lander was able to dig into the Martian regolith and retrieve rock samples for analysis (see Figure 7-37). Analysis showed the rocks at both sites to be rich in iron, silicon, and sulfur. The Martian regolith can best be described as an iron-rich clay. The *Viking* landers each carried a compact biological laboratory designed to test for microorganisms in the Martian soil. Three biological experiments were conducted, each based on the idea that living things alter their environment: They eat, they breathe, and they give off waste products. In each experiment, a sample of the Martian regolith was placed in a closed container, with or without a nutrient substance. The container was then examined for any changes in its contents.

The first data returned by the *Viking* biological experiments caused great excitement. In almost every case, rapid and extensive changes were detected inside the sealed containers. However, further analysis showed that these changes were due solely to nonbiological chemical processes. Apparently, the Martian regolith is rich in chemicals that effervesce (fizz) when moistened. A large amount of oxygen is apparently tied up in the regolith in the form of unstable chemicals called *peroxides* and *superoxides,* which break down in the presence of water to release oxygen gas.

The chemical reactivity of the Martian regolith probably comes from ultraviolet radiation that beats down on the planet's surface. Ultraviolet photons easily break apart molecules of carbon dioxide (CO_2) and water vapor (H_2O) by knocking off oxygen atoms, which then become loosely attached to chemicals in the regolith. Ultraviolet photons also produce ozone (O_3) and hydrogen peroxide (H_2O_2), which become incorporated in the regolith. In all these cases, the loosely attached oxygen atoms make the regolith extremely reactive.

Here on Earth, hydrogen peroxide is commonly used as an antiseptic. When you pour this liquid on a wound, it fizzes and froths as the loosely attached oxygen atoms chemically combine with organic material and destroy germs. The *Viking* landers *may* have failed to detect any organic compounds on Mars because the superoxides and peroxides in the Martian regolith make it antiseptic today.

5 In 2003, astronomers detected methane in Mars's atmosphere. This is an intriguing discovery, since sunlight breaks down methane within a few centuries. Therefore, the gas needs to be continuously replenished. This simple compound, CH_4, is often a by-product of biological activity (cows come to mind in this regard). Methane is also trapped inside planets like the Earth and possibly Mars during their formation, and thereafter leaks out. While the methane discovered on Mars is suggestive that simple life might be active under the planet's surface today, it is by no means proof positive that such life exists.

"IT MAY BE JUST A TECHNICALITY, BUT NOW AREN'T WE ALSO EXTRATERRESTRIALS?"

(© Sidney Harris)

Possible signs of ancient life on Mars have been discovered here on Earth. When cut open, several of the Martian meteorites have shown microscopic features that could be fossils of Martian bacterial life and their excretions (see Figure 7-36b). These features, no larger than 500 nm (1⁄100 the diameter of a human hair), are 30 times smaller than bacteria found on Earth. Detailed analysis of the meteorites in 2001 revealed the presence of several organically created features, including tiny spheres found with some of Earth's bacteria, and magnetite crystals, which are also used by some bacteria on Earth as compasses to find food. The debate as to whether the features in these meteorites are fossil evidence of life on Mars has raged unabated for several years, and it is not yet resolved. (It is worth noting that similar claims of fossils from Mars in the 1960s have been disproved.)

The search for microscopic life on Mars is not over. If there is liquid water under the Martian surface, then it is entirely possible that life has evolved in it and may still exist there. This belief is based on the incredibly wide range of places life has developed in water on Earth, from volcanic vents on the bottoms of oceans, to geysers, to a lake under the Antarctic polar ice cap.

7-14 Mars has a crust of varying thickness

By observing the orbit of *Mars Global Surveyor,* astronomers have concluded that the Martian crust is about 40 km thick under the northern lowlands and about 70 km thick under the southern highlands. However, they find that the boundary between the thin and thick crusts does not line up with the boundary between high and low terrain. Current models of Mars suggest that it has a core about 3400 km in diameter and that at least some of that core is molten. However, a detailed understanding of Mars's interior waits for the placing of seismic detectors on the planet's surface.

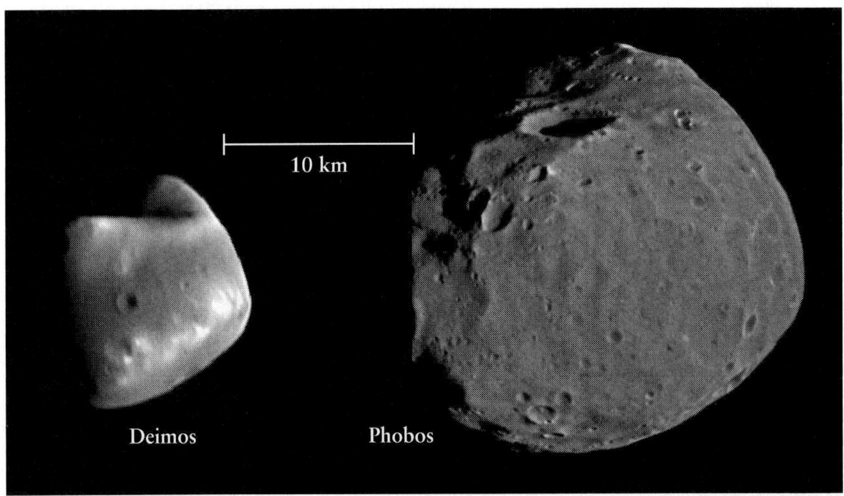

WEB LINK 7.13 FIGURE 7-38 **Phobos and Deimos** Phobos, the larger of Mars's two moons, is potato-shaped and measures approximately 28 × 23 × 20 km. It is dominated by crater Stickney, seen on the top. Deimos is less cratered than Phobos and measures roughly 16 × 12 × 10 km. (JPL/NASA)

7-15 Mars's two moons look more like potatoes than spheres

Two tiny moons orbit close to Mars's surface. Phobos (meaning "fear") and Deimos ("panic") are so small that they were not discovered until 1877. Potato-shaped Phobos is the inner and larger of the two (Figure 7-38). It rises in the west and races across the sky in only 5½ hours, as seen from Mars's equator. It is heavily cratered, and its surface has been transformed into dust at least 1 meter thick by countless tiny impacts over the eons. Football-shaped Deimos is less cratered than Phobos (Figure 7-38). As seen from Mars, Deimos rises in the east and takes about 3 Earth days to creep from one horizon to the other.

Phobos and Deimos were not formed like our Moon, by splashing off Mars. Rather, they are captured planetesimals. Both moons are in synchronous rotation as they orbit the red planet.

INSIGHT INTO SCIENCE

Imagine the Moon The definitions we use for words based on our everyday experience often fail us in astronomy. For example, the word "moon" usually creates an image of a spherical body, like our Moon. In reality, most of the moons in the solar system are unsymmetrical, like Phobos and Deimos.

7-16 Comparisons of planetary features provide new insights

Now that you have explored the terrestrial planets individually, it can be useful to see how their various features compare to each other and to the earth. Earth is the largest and most massive of all four terrestrial planets. In this regard, Venus is almost the sister planet to Earth, with nearly 95% the Earth's diameter and 82% of the Earth's mass. While Mars is most similar to Earth in other ways, such as its history of surface water, it is only half (53%) Earth's diameter and 11% Earth's mass. Mercury, with 38% Earth's diameter and a scant 5.5% Earth's mass, is much more similar to Earth's Moon in dimensions. Mercury is only 1.4 times bigger than the Moon.

The chemistries of the four terrestrial planets are similar, with cores consisting primarily of iron surrounded by rock. Venus, Earth, and Mars are known to have partially molten cores, and it remains to be seen if Mercury's core is also molten. Although Mercury is the smallest terrestrial planet, it has the highest density, meaning that it has the highest percentage of iron of these (and, in fact, of any) planets. This is true probably because Mercury lost more of its outer, rocky layer as a result of impacts than did any other terrestrial planet.

Earth contains by far the highest percentage of water of all the terrestrial planets. Mars contains water frozen on or near its surface, and we have yet to determine whether it has any liquid water inside. Venus contains very little water compared to either Earth or Mars because Venus is so hot that it has evaporated surface water into its atmosphere, and water in its interior has probably been mostly ejected through volcanoes or when the surface periodically melts. Mercury apparently has some water (very little compared to Earth or Mars) frozen at its poles, the result of collisions with water-rich comets. Because its interior is so iron rich, the water-bearing layers were probably blasted into space by impacts early in Mercury's existence.

Venus has by far the densest atmosphere of the terrestrial planets, with about 90 times as much gas as the air we breathe. Furthermore, Venus's atmosphere is composed primarily of carbon dioxide, with a minor component of

THE INNER PLANETS: A COMPARISON

	Interior	Surface	Temperature	Atmosphere	Magnetic Field
Mercury	Not known whether liquid or solid	Heavy cratering, scarps	700 K by day, 100 K by night	H, He, K, Na, O; transient and tenuous	0.1 times Earth's field
Venus	At least partially molten	Light cratering, mostly volcanic plains, gently rolling hills, some volcanoes	750 K	Mostly CO_2, N_2; 90 times denser than Earth's	None detected
Earth	Solid inner core, molten outer core, mantle	Very little cratering, continents and land at ocean floors, weathering, volcanoes, global tectonic plates	200–315 K	Mostly N_2, O_2	Strong global field
Mars	Probably solid core	Moderate cratering, weathering, dormant volcanoes, huge canyons	160–280 K	Mostly CO_2, N_2; 0.006 times as dense as Earth's	Weak, local fields

R I V U X G

(NASA/USGS)

nitrogen. The thick atmosphere has protected Venus's surface from all but the most massive infalling space debris. This, and the fact that Venus's surface may periodically melt, explains why Venus has few impact craters. Venus's atmosphere is most similar in composition to the air around Mars, although the density of Mars's air is only 0.6% as dense as the air we breathe. While Mars's thin atmosphere has enabled many pieces of space debris to strike the planet and form craters, these craters are eroded primarily by wind. Earth's atmosphere was once very similar to that of Venus, but was transformed by water and life into the nitrogen-oxygen atmosphere we have today. As with Venus, Earth's atmosphere protects the surface from most impacts. Many more have been removed by plate tectonic motion, which apparently does not occur today on any other planet. Mercury's gravity is too low to hold any gases as a permanent atmosphere. It is surrounded by a very thin atmosphere of transient gases from the Sun and from the planet's interior. As these drift into space, they are replaced by fresh gas. Because its atmosphere is so thin and its surface unchanged for billions of years, Mercury is the most heavily cratered of all the terrestrial planets.

Some of the temperatures on the terrestrial worlds are surprising at first glance. While Mercury, closest to the Sun, has a hot daytime surface of about 700 K (800°), its lack of atmosphere allows a lot of this heat to escape at night, bringing its nighttime temperature down to a frigid 100 K (–280°), much colder than on any other terrestrial planet. Venus's thick atmosphere creates a greenhouse effect that keeps that planet at 750 K (890°), even hotter than Mercury. Earth's surface temperature ranges from about 330 K (140°) to 180 K (–130°), and Mars is only slightly colder, with temperatures ranging from 280 K (45°) down to 160 K (–170°).

All the terrestrial planets rotate, with Earth's solar day being shortest at 24 hours. This motion combined with its liquid iron core creates a strong magnetic field surrounding our planet. Mars has virtually the same solar day of 24 hrs 39 min, but its molten iron core is much smaller than ours and it has no global magnetic field, only local magnetic fields. The solar days of Venus (117 Earth days) and Mercury (176 Earth days) are both extremely long by Earth standards, and, indeed, a solar day on Mercury is two Mercurian years long! No magnetic field has been detected around Venus. Mercury has a weak global field that may result from the extremely high amount of iron it contains. The table "The Inner Planets: A Comparison" summarizes much of this material.

7-17 Frontiers yet to be discovered

All three terrestrial planets other than Earth have much to reveal. We have yet to see all of Mercury's surface. What is the chemical composition of its surface rocks? What is Mercury's cooling history and how much of its core is molten? Is there really ice at its poles? If so, how did it get there? What will its internal structure reveal?

Does Venus have active volcanoes? If not, what supplies its atmosphere with sulfur compounds? Can we find observational evidence of what caused Venus's rotation axis to flip over? Likewise, can we find further evidence that its surface periodically undergoes significant re-covering?

Some of the most significant questions about our solar system that could be answered are whether life ever existed on Mars, whether there actually is liquid water under the red planet's surface, and whether there is still life there. If so, how far did that life evolve? What are the similarities and differences between such life and life on Earth? Significant similarities might imply a common origin. Furthermore, what is the surface water history of that world? What causes its local magnetic fields? What does its interior look like? How did its axis get tilted? Where did its moons come from? It is likely that we will have answers to many of these questions in the coming few decades.

Summary of Key Ideas

All four inner planets are composed primarily of rock and metal, and thus they are classified as terrestrial.

Mercury

• Even at its greatest orbital elongations, Mercury can be seen from Earth only briefly after sunset or before sunrise.

• The *Mariner 10* spacecraft passed near Mercury in the mid-1970s, providing pictures of its surface. The Mercurian surface is pocked with craters like the Moon's, but extensive, smooth plains lie between these craters. Long cliffs meander across the surface of Mercury. These scarps probably formed as the planet cooled, solidified, and shrank.

• The long-ago impact of a large object formed the huge Caloris Basin on Mercury and shoved up jumbled hills on the opposite side of the planet.

• Mercury has an iron core much like that of the Earth.

Venus

• Venus is similar to the Earth in size, mass, and average density, but it is covered by unbroken, highly reflective clouds that conceal its other features from Earth-based observers.

• While most of Venus's atmosphere is carbon dioxide, its dense clouds contain droplets of concentrated sulfuric acid mixed with yellowish sulfur dust. Active volcanoes on Venus may be a constant source of this sulfurous veil.

• Venus's exceptionally high temperature is caused by the greenhouse effect, as the dense carbon dioxide atmosphere traps and retains heat emitted by the planet. The surface pressure on Venus is 90 atm, and the surface temperature is 750 K. Both temperature and pressure decrease as altitude increases.

• The surface of Venus is surprisingly flat, mostly covered with gently rolling hills. There are two major "continents" and several large volcanoes. The surface of Venus shows evidence of local tectonic activity but not the large-scale motions that play a major role in continually reshaping the Earth's surface.

Mars

• Earth-based observers found that the Martian solar day is nearly the same as that of the Earth, that Mars has polar caps that expand and shrink with the seasons, and that the Martian surface undergoes seasonal color changes.

• A century ago observers reported networks of linear features that many perceived as canals. These observations led to speculation about self-aware life on Mars.

• The Martian surface has many flat-bottomed craters, several huge volcanoes, a vast equatorial canyon, and dried-up riverbeds—but no canals formed by intelligent life. River deltas and dry riverbeds on the Martian surface indicate that large amounts of water once flowed there.

• Liquid water would quickly boil away in Mars's thin present-day atmosphere, but the planet's polar caps contain some frozen water, and a layer of permafrost may exist beneath the regolith.

• The Martian atmosphere is composed mostly of carbon dioxide. The surface pressure is less than 0.01 atm.

• Chemical reactions in the regolith together with ultraviolet radiation from the Sun apparently act to sterilize the Martian surface.

• Mars has no global magnetic fields, but local fields pierce its surface in at least nine places.

• Features that may be fossil remains of bacteria have been found in several meteorites that are believed to have come from Mars.

• Mars has two potato-shaped moons, the captured planetesimals Phobos and Deimos. Both are in synchronous rotation with Mars.

WHAT DID YOU THINK?

1 *Which of the two planets, Mercury (the closest planet to the Sun) or Earth, has the cooler temperature?* The temperature on the daytime side of Mercury is much higher than on Earth, but the temperature on the nighttime side of Mercury is much lower than on Earth, because Mercury rotates so slowly and has little atmosphere to retain heat.

2 *Which planet is most similar to Earth?* Venus is most similar to Earth in size, chemistry, and distance from the Sun. Mars is more similar to Earth (than Venus) in its length of day, seasons, erosion, and in having water ice.

3 *What is the composition of the clouds surrounding Venus?* The clouds are made primarily of sulfuric acid.

4 *Does Mars have liquid water on its surface today?* No, but there are strong indications that it had liquid water on its surface in the past.

5 *Is life known to exist on Mars today?* No current life has yet been discovered on Mars.

Key Words

Review Questions

1. Why is Mercury so difficult to observe? When is the best time to see the planet? *Hint:* Guided Discovery: The Inner Solar System on page 175 can help.

2. Compare the surfaces of Mercury and our Moon. How are they similar? How are they different?

3. Compare the interiors of Mercury and Earth. How are they similar? How are they different?

4. To better understand the interiors of Mercury and Earth, do Interactive Exercise 7-1 on the Web. You can print out the result, if requested.

5. What are the longest features found on Mercury? Why are the examples of this feature probably much older than tectonic features on the Earth?

6. Briefly describe a scientific theory explaining why Mercury has such a large iron core.

7. Astronomers often refer to Venus as the Earth's twin. What physical properties do the two planets have in common? In what ways are the two planets dissimilar?

8. Why is it hotter on Venus than on Mercury?

9. What is the greenhouse effect? What role does it play in the atmospheres of Venus and the Earth?

10. What evidence exists for active volcanoes on Venus?

11. Describe the Venusian surface. What kinds of geological features would you see if you could travel around the planet?

12. Why do astronomers believe that Venus's surface was not molded by the kind of tectonic activity that shaped the Earth's surface?

13. Why is Mars red?

14. When is the best time to observe Mars from Earth? *Hint:* Guided Discovery: The Inner Solar System on page 175 can help.

15. Compare the cratered regions of Mercury, the Moon, and Mars. Assuming that the craters on all three worlds originally had equally sharp rims, what can you conclude about the environmental histories of these worlds?

16. How would you tell which craters on Mars were formed by meteoritic impacts and which by volcanic activity?

17. Compare the volcanoes of Venus, Earth, and Mars. Do you think hot-spot volcanism is or was active on all three worlds? Explain.

18. What geologic features indicate plate tectonic activity once occurred on Mars? What features created by tectonic activity on Earth are not found on Mars?

19. To better understand the surface features of Mars, do Interactive Exercise 7-2 on the Web. You can print out your results, if requested.

20. What is the current knowledge concerning life on Mars? Do you think that today Mars is as barren and sterile as the Moon? Why or why not?

21. To compare the surfaces of Mercury, Venus, and Mars, do Interactive Exercise 7.3 on the Web. You can print out your results, if requested.

22. What evidence have astronomers accumulated that liquid water once existed in large quantities on Mars's surface? What evidence is there that water is still there, under the surface?

Advanced Questions

23. What evidence do we have that the surface features on Mercury were not formed during recent geologic history?

24. Venus takes 440 days to move from greatest western elongation to greatest eastern elongation, but it needs only 144 days to go from greatest eastern elongation to greatest western elongation. With the aid of a diagram, explain why.

25. As seen from Earth, the brightness of Venus changes as it moves along its orbit. Describe the main factors that determine Venus's variations in brightness as seen from Earth. *Hint:* See the discussion of Venus in Chapter 2.

26. How might Venus's cloud cover change if all of Venus's volcanic activity suddenly stopped? How might these changes affect the overall Venusian environment?

27. Compare Venus's continents with those on the Earth. What do they have in common? How are they different?

28. Explain why Mars has the longest synodic period of all the planets, although its sidereal period is only 687 days.

29. With carbon dioxide accounting for about 95% of the atmospheres of both Mars and Venus, why is there little greenhouse effect on Mars today?

30. Could the polar regions of Mars reasonably be expected to harbor life-forms, even though the Martian regolith is sterile at the *Viking* lander sites?

Discussion Questions

31. If you were planning a return mission to Mercury, what features and observations would be of particular interest to you and why?

32. If you were designing a space vehicle to land on Venus, what special features would be necessary? In what ways would this mission and landing craft differ from a spacecraft designed for a similar mission to Mercury?

33. Suppose someone told you that the *Viking* mission failed to detect life on Mars simply because the tests were designed to detect terrestrial life-forms, not Martian life-forms. How would you respond?

34. Compare the scientific opportunities for long-term exploration offered by the Moon and Mars. What difficulties would there be in establishing a permanent base or colony on each of these two worlds?

35. Imagine you are an astronaut living at a base on Mars. Describe your day's activities, what you see, the weather, the spacesuit you are wearing, and so on.

What If . . .

36. Mercury had synchronous rotation? How would the temperatures on such a planet be different than they are today? Where would humans set up camp on such a world?

37. Venus could support self-aware life, but it still had permanent cloud-cover? In what ways would life there be different than it is here? How would their perceptions of the cosmos be different from ours?

38. Mars had the same mass, surface features, and atmosphere as the Earth? In what ways would life there be different than it is here?

39. Mars rotated once every 20 days rather than once every 1.026 days? What would be different?

40. The carbon dioxide content of the Earth's atmosphere increased? What would happen to the Earth? This is not a completely hypothetical question, because carbon dioxide levels are increasing today.

Web Questions

41. Search the Web for the latest information about ongoing and upcoming space missions to Mercury. When are they scheduled to be launched and when are they scheduled to arrive at Mercury? What scientific experiments will they carry? What scientific issues are these instruments intended to resolve?

42. **Elongations of Mercury** Access the animation "Elongations of Mercury" in Chapter 7 of the *Discovering the Universe* Web site. **a.** Note the dates of the greatest eastern and western elongations in the animation. Which time interval is greater: from a greatest eastern elongation to a greatest western elongation or vice versa? **b.** Based on what you observe in the animation, draw a diagram to explain your answer to the question in a.

43. Search the Web for the latest information about proposed future missions to Venus. What scientific experiments will they carry? What scientific issues are these instruments intended to resolve?

44. **Surface Temperature of Venus** Access the Active Integrated Media Module "Wien's Law" in Chapter 4 of the *Discovering the Universe* Web site. **a.** Using the Wien's law calculator, determine Venus's approximate temperature if it emits blackbody radiation with a peak wavelength of 3866 nm. **b.** By trial and error, find the wavelength of maximum emission for a surface temperature of 750 K (for present-day Venus) and a surface temperature of 850 K (as it might become if its greenhouse gas density increases). In what part of the electromagnetic spectrum do these wavelengths lie?

45. In 1999, two NASA spacecraft—*Mars Climate Orbiter* and *Mars Polar Orbiter*—failed to reach their destinations. Search the Web for information on these missions. What were their scientific goals? How and why did the missions fail? Which current and future missions, if any, are intended to replace these missions?

46. Search the Web for information about possible manned missions to Mars. How long would such a mission take? How expensive would they be? What are some advantages and disadvantages of a manned mission compared to an unmanned one?

47. **Conjunctions of Mars** Access and view the animation "The Orbits of Earth and Mars" in Chapter 7 of the *Discovering the Universe* Web site. **a.** The animation highlights three dates when Mars is in opposition, so that the Earth lies directly between Mars and the Sun. By using the "Stop" and "Play" buttons in the animation, find two times during the animation when Mars is in *conjunction*, so that the Sun lies directly between Mars and the Earth (see Figure 2-4). For each conjunction, make a drawing showing the positions of the Sun, the Earth, and Mars, and record the month and year when the conjunction occurs. **b.** When Mars is in conjunction, at approximately what time of day does it rise as seen from Earth? At what time of day does it set? Is Mars suitably placed for telescopic observations when it is in conjunction?

Observing Projects

48. Refer to the following table to determine the dates of the next two or three greatest elongations of Mercury. Consult such magazines as *Sky & Telescope* and *Astronomy*, or use your *Starry Night Enthusiast*™ planetarium software to see if any of these greatest elongations is going to be especially favorable for viewing the planet. If so, make plans to see the innermost planet of the solar system, a rare sight. Set aside several evenings (or mornings) around the date of the favorable elongation to reduce the chances of being "clouded out." Select an observing site that has a clear, unobstructed view of the horizon where the Sun sets (or rises). Make arrangements to have a telescope at your disposal. Search for the planet on the dates you have selected and make a drawing of its appearance through your telescope.

Greatest Elongations of Mercury,* 2005–2008

Year	Greatest western elongations	Greatest eastern elongations
2005	April 26, August 23, December 12	March 12, July 9, November 3
2006	April 8, August 7, November 25	February 24, June 20, October 17
2007	March 22, July 20, November 8	February 7, June 2, September 29
2008	March 3, July 1, October 22	January 22, May 14, September 11

*Mercury can be seen in the eastern sky at dawn for a few days around greatest western elongation. It can be seen in the western sky at dusk for a few days around greatest eastern elongation.

49. Refer to the following table to see if Venus is currently near a greatest elongation. If so, view the planet through a telescope. Make a sketch of the planet's appearance. From your sketch, can you determine whether Venus is closer to us or farther from us than the Sun?

Greatest Elongations of Venus,* 2005–2008

Year	Greatest western elongations	Greatest eastern elongations
2005	(none)	November 3
2006	March 25	(none)
2007	October 28	June 9
2008	(none)	(none)

*Venus can be seen in the eastern sky in the hours before dawn for several months around greatest western elongation. It can be seen in the western sky in the hours after dusk for several months around greatest eastern elongation.

50. Observe Venus through a telescope once a week for a month and make a sketch of the planet's appearance on each occasion. From your sketches, can you determine whether Venus is approaching us or moving away from us?

51. Consult such magazines as *Sky & Telescope* and *Astronomy*, or use your *Starry Night Enthusiast*™ planetarium software to determine Mars's location among the constellations. If Mars is suitably placed for observation, arrange to view the planet through a telescope. Draw a picture of what you see. What magnifying power seems to give you the best image? Can you distinguish any surface features? Can you see a polar cap or dark markings? If not, can you offer an explanation for Mars's bland appearance?

52. Using your *Starry Night Enthusiast*™ software, try the activities in this chapter's Guided Discovery: The Inner Solar System to follow Mercury's orbit as it nears the Sun. With the timestep still set at 1 sidereal day, describe the motion of Mercury relative to the equatorial coordinate grid (that is, the celestial sphere). What can you say about the maximum and minimum angles on the celestial sphere between Mercury and the Sun? Periodically stop the motion, zoom in on Mercury, and determine its phase. Describe the pattern of phases.

53. The Guided Discovery: The Inner Solar System on page 175 will introduce you to Mars's motion. In *Starry Night Enthusiast*™ go into Atlas mode (*Favourites/Guides/Atlas*), find Mars (tab*Find/Mars*), and with the slider set the sky at 60°. Run the program with a timestep of 1 sidereal day. Notice that the equatorial coordinate system usually flows smoothly past Mars. When the stars behind Mars begin to slow down, stop the motion by pressing the ■ icon. Unlock Mars by grabbing it and moving it slightly. Then run the program again. Describe how Mars now moves relative to the coordinate system. Stop the program, run it backward through this entire period, and stop it with Mars back in the center of the screen. Rerun it forward in time, this time using the single-step feature. Make a day-by-day drawing of Mars's motion during this time. What is the name of this motion?

54. **The Sun as Seen from the Various Terrestrial Planets** In this exercise, we will use *Deep Space Explorer*™ to examine how big the Sun appears in the sky. If you have not done so already, go to **Settings** and uncheck "Use magnitude cutoffs," slide the "Galaxy drawing/Brightness" slide to the middle of the range, and then click on the **Views** tab.

You will need a ruler, ideally with millimeters. Be careful not to harm your computer screen when you make the following measurements. First, go to Mercury by clicking **Solar System/Mercury** on the left side of the star field. Looking at Mercury, write down where you expect the Sun to be found (e.g., upper left, directly to the right, etc.) Now grab the screen (hold down left PC mouse button or the mouse button on a Mac) and move the mouse until you see the Sun at one edge of Mercury. Measure the yellow part of the Sun with a ruler and record the Sun's apparent diameter. Repeat this for Venus, Earth, and Mars. How does the Sun vary in diameter from Mercury to Mars? Allowing that these are approximate measurements, divide the Sun's diameter at Earth by its diameter at Mercury. Divide the Sun's diameter at Earth by the Sun's diameter at Venus. Finally, divide the Sun's diameter at Earth by the Sun's diameter at Mars. How do these three numbers compare with the semimajor axes of the planets in AU (see Table E-1). Explain.

55. **The Orbits of Venus and Earth** In this exercise we use *Deep Space Explorer*™ to compare the orbits of Venus and the Earth. If you have not done so already, go to **Settings** and uncheck "Use magnitude cutoffs," slide the "Galaxy drawing/Brightness" slide to the middle of the range, and then click on the **Views** tab.

Now, go to the inner solar system by clicking **Solar System/Inner Solar System** on the left side of the star field. You can zoom in and zoom out using the buttons at the upper left of the window (an upward-pointing triangle and a downward-pointing triangle). You can also rotate the solar system by putting the mouse cursor over the image, holding down the mouse button, and moving the mouse. (Hold down the left PC mouse button or the mouse button on a Mac.) Use these techniques to determine whether the orbits of Venus and of the Earth are in the same plane. If not, use a protractor (if available) to measure the angle between the orbits. Otherwise, qualitatively describe by how much these planes differ.

CHAPTER

The Outer Planets

R I V U X G

Saturn's Rings Are Made of Numerous Thin Ringlets (NASA/JPL)

WHAT DO YOU THINK?

1 Is Jupiter a "failed star?"

2 What is Jupiter's Great Red Spot?

3 Does Jupiter have continents and oceans?

4 Is Saturn the only planet with rings?

5 Are the rings of Saturn solid?

6 Do all moons rise and set as seen from their respective planets?

We can pretty easily imagine a trip to Mars, exploring its vast canyons and icy polar regions. Even Venus and Mercury lend themselves to being compared to Earth and the Moon. However, we find few similarities between the Earth and the outer planets: Jupiter, Saturn, Uranus, Neptune, and Pluto.

Jupiter, Saturn, Uranus, and Neptune lack solid surfaces and are so much larger, rotate so much faster, and have such different chemical compositions from our world that upon seeing them one knows, to paraphrase Dorothy, "You're not on Earth anymore." There is nothing on Earth remotely similar to the swirling red and brown clouds of Jupiter, the ever-changing ring system of Saturn, or the blue-green clouds of Uranus and Neptune. Pluto, even smaller than our Moon and composed of rock and ice, also resist any close comparison to the Earth. We begin our exploration of the outer planets with magnificent Jupiter.

In this chapter you will discover

- Jupiter, an active, vibrant, multicolored world more massive than all the other planets combined
- Jupiter's diverse system of moons
- Saturn, with its spectacular system of thin, flat rings and numerous moons
- what Uranus and Neptune have in common and how they differ from Jupiter and Saturn
- that tiny Pluto and its moon, Charon, orbit each other in synchronous rotation

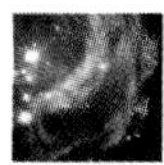

JUPITER

Jupiter's multicolored bands, dotted with ovals of white and brown (Figure 8-1), give the appearance of a world unlike any of the terrestrial planets. Viewed even through a small telescope (see Figure 2-11), you can see up to four of its moons. As the high-resolution images in this chapter reveal, Jupiter is a world of breath-taking beauty. Figure 8-1 shows Jupiter and lists its vital statistics.

Jupiter is the largest planet in the solar system; more 1
than 1300 Earths could be packed into its volume. Using the orbital periods of its moons in Kepler's laws, astronomers have determined that Jupiter is 318 times more massive than Earth. Indeed, Jupiter has more than 2½ times as much mass as all the other planets combined. This huge mass has created the urban myth that Jupiter is a failed star, meaning that it almost has enough matter to shine on its own, like the Sun, which we will study in Chapter 10. The fact is that Jupiter would have to be 75 times more massive than it is in order to generate energy as the Sun does and therefore be classified as a star.

8-1 Jupiter's outer layer is a dynamic area of storms and turbulent gases

Jupiter is permanently covered with clouds (see Figure 8-1). Because it rotates about once every 10 hours—the fastest of any planet—Jupiter's clouds are in perpetual motion and are confined to narrow bands of latitude. Even through a small telescope, you can see on Jupiter dark, reddish bands called **belts**, alternating with light-colored bands called **zones.** These belts and zones are gases flowing east or west, with very little north-south motion. In contrast, winds on slower-rotating Earth wander over vast ranges of latitude (compare Figure 6-1).

Jupiter's belts and zones provide a backdrop for turbulent swirling cloud patterns, as well as rotating storms similar in structure to hurricanes or cyclones on Earth. These storms are known as *white ovals* and *brown ovals* (Figure 8-2). The white ovals are observed to be cool clouds higher than the average clouds in Jupiter's atmosphere. The brown ovals are warmer and lower clouds. They are holes in the normal cloud layer. The various oval features last from hours to centuries. In 1998, two 50-year-old storms on Jupiter were observed to merge into a single storm as large across as the Earth's diameter. Computers show us how the cloud features on Jupiter would look if the planet's atmosphere were unwrapped like a piece of paper (Figure 8-3). You can see changing ripples, plumes, and light-colored wisps in these images, which were taken 4 months apart.

Jupiter's most striking feature is its **Great Red Spot** 2
(Figure 8-4), which is so large that it can be seen through a small telescope. It changes dimensions, and at present it is about 25,000 km long by 12,000 km wide, large enough so that two Earths could easily fit side by side inside it. The Great Red Spot was first observed around 1656, either by the English scientist Robert Hooke or the Italian astronomer Giovanni Cassini. Because earlier telescopes were unlikely to have been able to see it, the Great Red Spot could well have formed long before that time. The Great Red Spot is a hurricane or typhoonlike storm of swirling

JUPITER: VITAL STATISTICS

Average distance from Sun:	5.20 AU = 7.78×10^8 km
Maximum distance from Sun:	5.46 AU = 8.16×10^8 km
Minimum distance from Sun:	4.95 AU = 7.41×10^8 km
Eccentricity of orbit:	0.048
Average orbital speed:	13.1 km/s
Orbital period:	11.86 years
Rotation period:	$9^h\ 50^m\ 28^s$ (equatorial) $9^h\ 55^m\ 30^s$ (internal)
Inclination of equator to orbit:	3.12°
Inclination of orbit to ecliptic:	1.30°
Radius:	71,490 km = 11.21 Earth radii (equatorial) 66,850 km = 10.48 Earth radii (polar)
Mass:	1.90×10^{27} kg = 318 Earth masses
Average density:	1330 kg/m^3 = 0.241 Earth density
Escape speed:	60.2 km/s
Surface gravity (Earth = 1):	2.36
Albedo:	0.52
Average temperature at cloudtops:	−108°C = −162°F = 165 K
Atmospheric composition (by number of molecules):	86.2% hydrogen (H_2), 13.6% helium (He), 0.2% methane (CH_4), ammonia (NH_3), water vapor (H_2O), and other gases

R I V U X G

FIGURE 8-1 **Jupiter as Seen from a Spacecraft** This view was sent back from *Voyager 1* in 1979. Features as small as 600 km across can be seen in the turbulent cloudtops of this giant planet. The complex cloud motions surrounding the Great Red Spot are clearly visible. Note that the clouds at its equator rotate faster than does the interior of the planet, which is why two rotation periods are given. Also, clouds at different latitudes have different rotation rates. The inset image of Earth shows its size relative to Jupiter. (NASA/JPL; inset: NASA)

gases. Heat welling upward from inside Jupiter has maintained this storm for more than three centuries. Consider what life would be like for us if the Earth sustained storms for such long periods.

In 1690, Cassini noticed that the speeds of Jupiter's clouds vary with latitude, an effect called **differential rotation.** Near the poles, the rotation period of Jupiter's atmosphere, 9 h 55 min 30 s, is 5 minutes longer than at the equator. Furthermore, clouds at different latitudes circulate

▶ FIGURE 8-3 **Jupiter's Northern and Southern Hemispheres** Computer processing shows the entire Jovian atmosphere from (a) *Voyager 1* and (b) *Voyager 2* "unwrapped". Notice the dark ovals in the northern hemisphere and the white ovals in the southern hemisphere. The banded structure is absent near the poles. The Great Red Spot moved westward while the white ovals moved eastward during the 4 months between the two *Voyager* flybys. (NASA)

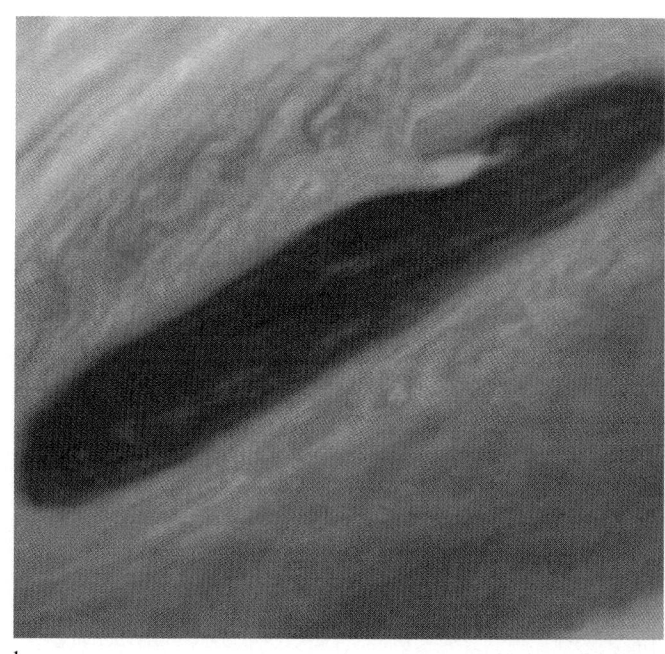

FIGURE 8-2 **Close-ups of Jupiter's Atmosphere** The dynamic winds, rapid rotation, internal heating, and complex chemical composition of Jupiter's atmosphere create its beautiful and complex banded pattern. **(a)** A *Voyager 2* southern hemisphere image showing a white oval that has existed for over 40 years. **(b)** A *Voyager 2* northern hemisphere image showing a brown oval. The white feature overlapping the oval is a high cloud. (NASA)

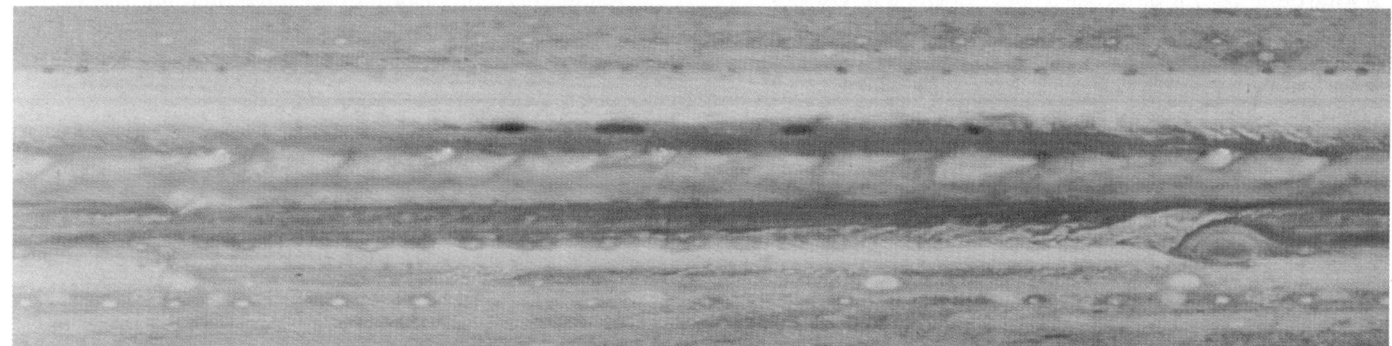

a *Voyager 1* view

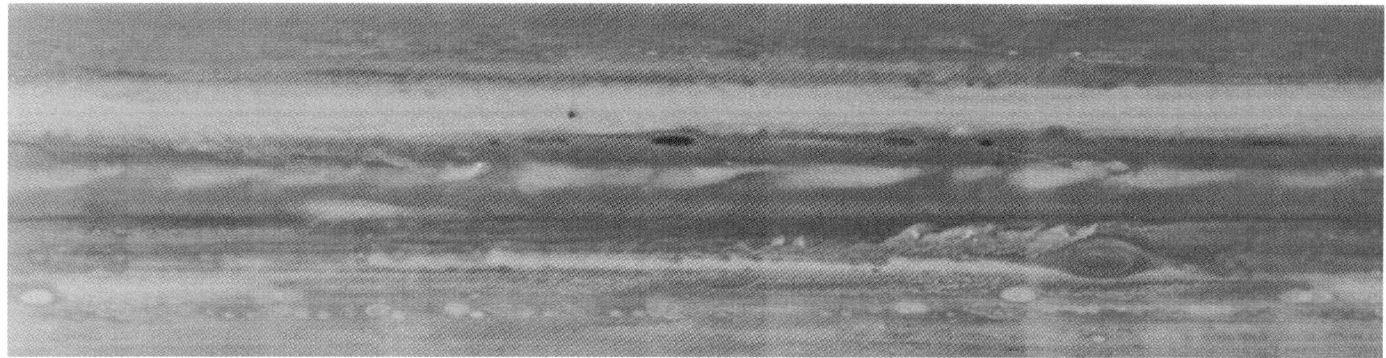

b *Voyager 2* view

R I V U X G

FIGURE 8-4 **The Great Red Spot** This image of the Great Red Spot, taken by the *Galileo* spacecraft in 1996, has been greatly color-enhanced to show more details of the dynamic activity taking place in and around this giant storm. The counterclockwise circulation of gas in the Great Red Spot takes about 6 days to make one rotation. The clouds that encounter it are forced to pass around it, and when other oval features are near it, the entire system becomes particularly turbulent, like the batter in a two-bladed blender. (NASA/JPL)

in opposite directions—some eastward, some westward. At their boundaries the clouds rub against each other, creating beautiful swirling patterns (see Figure 8-2). The interactions of clouds at different latitudes also help provide stability for storms like the Great Red Spot.

Astronomers first determined Jupiter's overall chemical composition from its average density—only 1330 kg/m^3. Recall from Section 5-5 that average density is mass/volume. We determine Jupiter's mass from Kepler's third law, using the orbital periods of its moons, while the trigonometry of Jupiter's distance from Earth and angular size in our sky reveal its diameter and hence its volume. This low density implies that Jupiter is composed primarily of lightweight elements such as hydrogen and helium surrounding
3 a relatively small core of metal and rock. It has no solid continents, islands, or water oceans on its surface.

Spectra from Earth-based telescopes and from the *Galileo* probe sent into Jupiter's upper atmosphere in 1995 give more detail about the chemistry of Jupiter's atmosphere. About 86% of its atoms are hydrogen and 13% are helium. The remainder consist of molecular compounds such as methane (CH_4), ammonia (NH_3), and water vapor (H_2O). Keeping in mind that different elements have different masses, we can convert these percentages of atoms into the masses of various substances in Jupiter's atmosphere: 75% hydrogen, 24% helium, and 1% other substances. Because the interior contains more heavy elements than the surface, the overall mass distribution in Jupiter is calculated to be 71% hydrogen, 24% helium, and 5% all heavier elements.

The descent of the probe from the *Galileo* spacecraft into Jupiter's atmosphere revealed wind speeds of up to 600 km/h (375 mph), higher-than-expected air density and temperature, and lower-than-expected concentrations of water, helium, neon, carbon, oxygen, and sulfur. This probably occurred because the probe descended into a particularly arid region of the atmosphere called a *hot spot,* akin to the air over a desert on Earth.

Observations from spacecraft visiting Jupiter and its moons, combined with the scientific model of Jupiter's atmosphere developed to explain the observations of Jupiter's clouds and chemistry, indicate it has three major cloud layers (Figure 8-5).

The uppermost Jovian cloud layer is composed of crystals of frozen ammonia. Since these crystals and the frozen water in Jupiter's clouds are white, what chemicals create the subtle tones of brown, red, and orange? The answer is as yet unknown. Some scientists think that sulfur compounds, which can assume many different colors depending on their temperature, play an important role. Others think that phosphorus might be involved, especially in the Great Red Spot. The middle cloud layer is primarily ammonium hydrosulfide and the bottom cloud layer is mostly composed of water vapor.

Below its cloud layer, Jupiter's mantle is entirely liquid. Here on Earth, the distinction between the gaseous air and the liquid oceans is very clear—jump off a diving board and you know when you hit the water. However, the conditions on Jupiter under which hydrogen liquefies are different from anything we normally experience. As a result, on Jupiter there is no definite boundary between the planet's gaseous atmosphere and its liquid mantle. The hydrogen gradually gets denser until 1000 km below the cloudtops, the pressure is high enough for the hydrogen to be what we would consider a liquid.

In introducing the solar system, we also noted that a young planet heats up as it coalesces. After it forms, radioactive elements continue to heat its interior. On Earth, this heat leaks out of the surface through volcanoes and other vents. Jupiter loses heat everywhere on its surface, because, unlike the Earth, it has no landmasses to block the heat loss.

As the heat from within Jupiter warms its liquid mantle, blobs of heated hydrogen and helium move upward. When these blobs reach the cloudtops, they give off their heat and descend back into the interior. (The same process, *convection,* drives the motion of the Earth's mantle and its tectonic plates, as well as water simmering on a stove; see Figure 6-7.) Jupiter's rapid, differential rotation draws the convecting gases into bands of winds moving eastward and westward at different speeds around the planet.

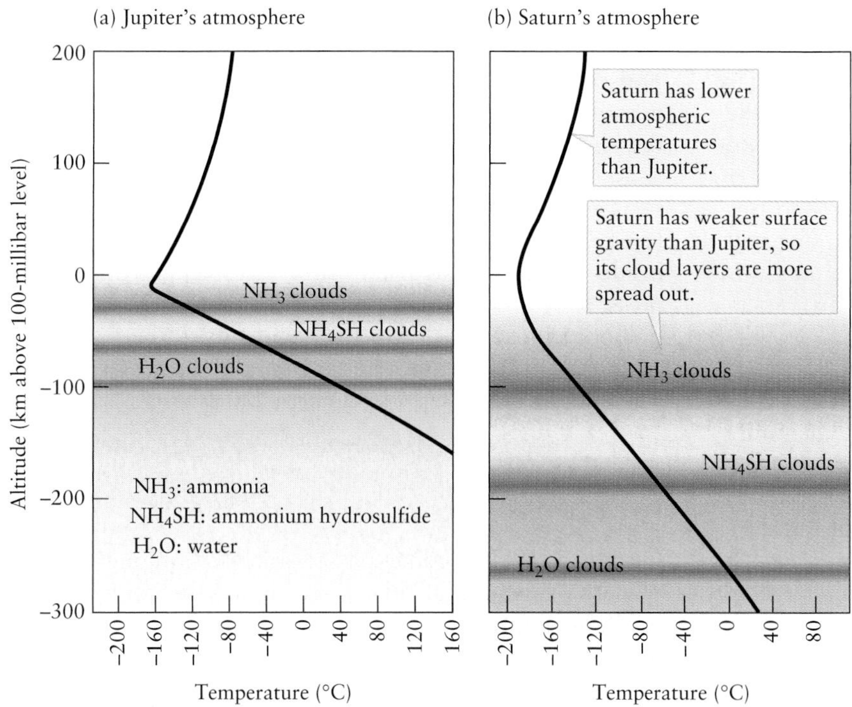

FIGURE 8-5 Jupiter's and Saturn's Upper Layers These graphs display temperature profiles of Jupiter's and Saturn's upper regions, as deduced from measurements at radio and infrared wavelengths. Three major cloud layers are shown in each, along with the colors that predominate at various depths. Data from the *Galileo* spacecraft indicate that Jupiter's cloud layers are not found at all locations around the planet; there are some relatively clear, cloud-free areas.

Astronomers believe that Jupiter's belts and zones result from the combined actions of the planet's convection and rapid differential rotation. Until recently they believed that the light-colored zones are regions of hotter, rising gas, while the dark belts are regions of cooler, descending gas (Figure 8-6). However, high resolution observations by the passing *Cassini* spacecraft on its way to Saturn revealed that there are numerous white clouds that are too small to be seen from Earth rising in the dark belts of Jupiter, while the zones are sinking gas—just the opposite of the original model! The correct explanation for the behavior of gases in the belts and zones may be provided by *Cassini*, when it studies the analogous belts and zones found on Saturn (Section 8-9) in greater detail in the next few years.

8-2 Jupiter's interior has four distinct regions

Descending below Jupiter's clouds, we first encounter liquid, molecular hydrogen and helium. As in Earth's oceans, the pressure in this liquid increases with depth. The gravitational force created by

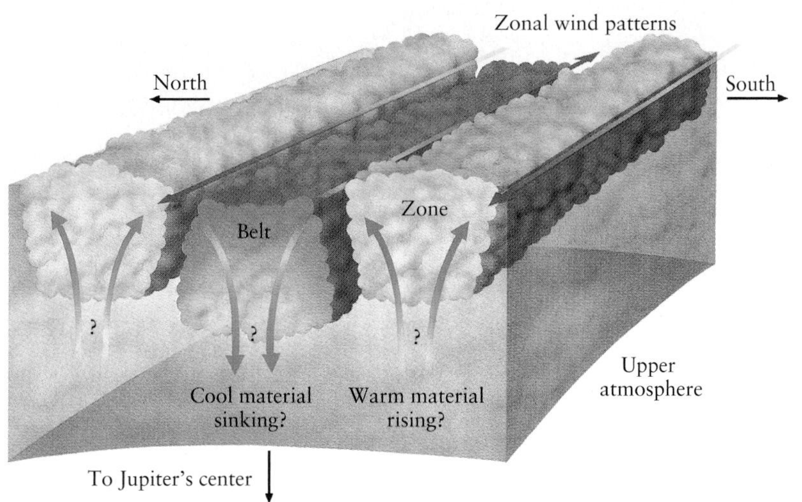

FIGURE 8-6 Original Model of Jupiter's Belts and Zones The light-colored zones and dark-colored belts in Jupiter's atmosphere were believed until recently to be regions of rising and descending gases, respectively. In the zones, gases warmed by heat from Jupiter's interior were thought to rise upward and cool, forming high-altitude clouds. In the belts, cooled gases were thought to descend and undergo an increase in temperature; the cloud layers seen there are at lower altitudes than in the zones. Recent observations suggest that just the opposite may be correct (stay tuned)! In either case, Jupiter's rapid differential rotation shapes the rising and descending gas into bands of winds parallel to the planet's equator. Differential rotation also causes the wind velocities at the boundaries between belts and zones to be predominantly to the east or west.

Jupiter's enormous mass compresses and heats its interior so much that 20,000 km below the cloudtops the pressure is 3 million atmospheres. Below this depth, the pressure is high enough to transform hydrogen into **liquid metallic hydrogen.** Under such conditions hydrogen acts like a metal in its ability to conduct electricity and heat, like the copper wiring in a house. Electric currents running through this rotating, metallic region of Jupiter generate a powerful planetary magnetic field. At the cloud level of Jupiter, this field is 14 times stronger per square meter than Earth's field is at our planet's surface.

For Earth-based astronomers, the only evidence of Jupiter's magnetosphere is a hiss of radio static, which varies cyclically over a period of 9 h 55 min 30 s, the planet's internal rotation rate. The two *Pioneer* and two *Voyager* spacecraft that journeyed past Jupiter in the 1970s revealed the awesome dimensions of Jupiter's magnetosphere: It is nearly 30 million kilometers across. It envelops the orbits of many of its moons. If Jupiter's magnetosphere were visible from Earth, it would cover an area of the sky 16 times larger than the full Moon.

Although the sequence of events in Jupiter's formation is still being worked out, it appears that a terrestrial (rock and metal) protoplanet formed at Jupiter's orbit and attracted hydrogen and helium to form its outer layers. This terrestrial matter is now Jupiter's core. This core is only 4% of Jupiter's mass, which still amounts to nearly 13 times the mass of the entire Earth. Water, carbon dioxide, methane, and ammonia are likely to have existed as ice in the terrestrial protoplanet, and they were also attracted onto the young and growing Jupiter. When astronomers talk about "ice," they generically refer to any or all of these latter four compounds.

The tremendous crushing weight of the bulk of Jupiter above the core—equal to the mass of 305 Earths—compresses the terrestrial core down to a sphere only 10,000 km in radius. By comparison, Earth's diameter is 12,756 km. At the same time, the pressure forced the lighter ices out of the rock and metal, thereby forming a shell of these "ices" between the solid core and the liquid metallic hydrogen layer. Calculations reveal that the temperature and pressure inside Jupiter should make these "ices" liquid there! The pressure at Jupiter's very center is calculated to be about 70 million atmospheres, and the temperature there is about 25,000 K, nearly 4 times hotter than the surface of the Sun. The interior of Jupiter is summarized in Figure 8-20.

8-3 Cometary fragments were observed to strike Jupiter

On July 7, 1992, a comet passed so close to Jupiter that the planet's gravitational tidal force ripped the comet into at least 21 pieces. The debris from this comet was first observed in March 1993 by comet hunters Gene and Carolyn Shoemaker and David Levy. (Because it was the ninth comet they had found together, it was named Shoemaker-Levy 9 in their honor.) Shoemaker-Levy 9 was an unusual comet that actually orbited Jupiter, rather than just orbiting the Sun. Calculations of the comet's orbit showed that the pieces would return to strike Jupiter between July 16 and July 22, 1994 (Figure 8-7).

Recall from Section 5-3 that impacts were extremely common in the first 800 million years of the solar system's existence. From more recent times, two chains of impact craters have been discovered on Earth, consistent with pieces of comets having hit our planet within the past

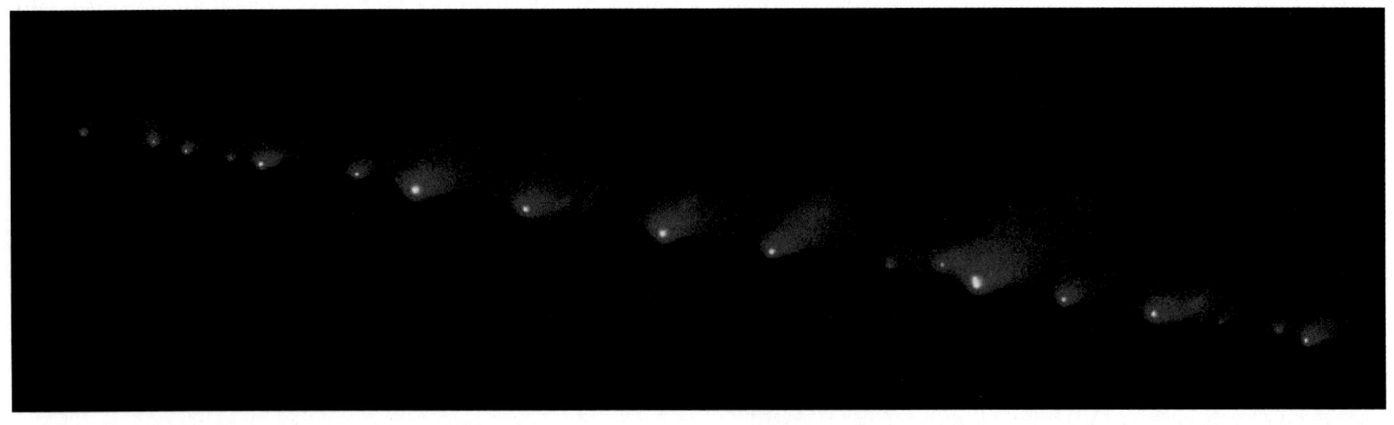

R I V U X G

FIGURE 8-7 **Comet Shoemaker-Levy 9 Debris Approaching Jupiter** The comet was torn apart by Jupiter's gravitational force on July 7, 1992, fracturing into at least 21 pieces. This comet originally orbited Jupiter and its returning debris, shown here in May 1994, struck the planet between July 16 and July 22, 1994. (H. A. Weaver, T. E. Smith, STScI and NASA)

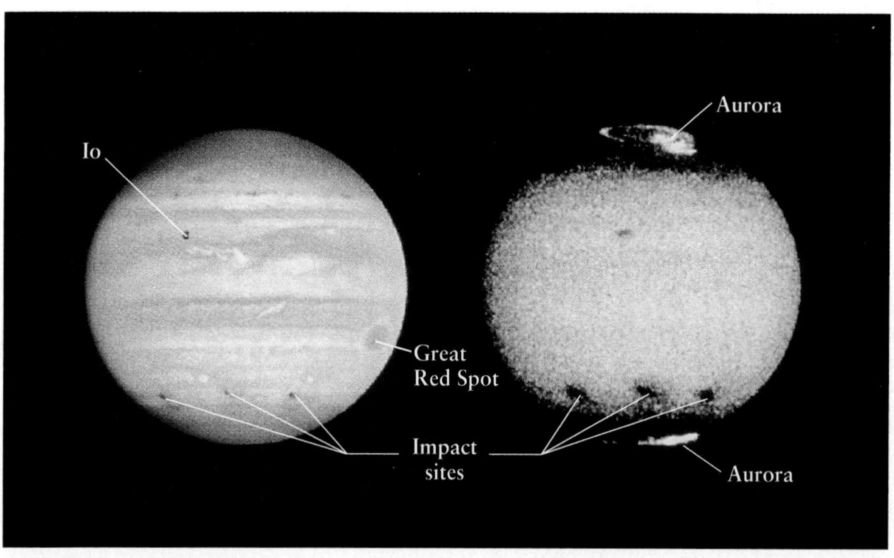

FIGURE 8-8 **Three Impact Sites of Comet Shoemaker-Levy 9 on Jupiter** Shown here are visible (left) and ultraviolet (right) images of Jupiter taken by the Hubble Space Telescope after three pieces of Comet Shoemaker-Levy 9 struck the planet. Astronomers had expected white remnants (the color of condensing ammonia or water vapor); the darkness of the impact sites may have come from carbon compounds in the comet debris. Note the auroras in the ultraviolet image. Auroras and lightning are common on Jupiter, due in part to the planet's strong magnetic fields and dynamic cloud motions. (NASA)

300 million years. However, it is very uncommon for pieces of space debris as large as several kilometers in diameter to collide with planets today. Therefore, the discovery that Shoemaker-Levy 9 would hit Jupiter created great excitement in the astronomical community. Seeing how a planet and a comet respond to such an impact would allow astronomers to deduce information about the planet's atmosphere and interior and also about the striking body's properties.

The impacts occurred as predicted, with most of Earth's major telescopes—as well as those on several spacecraft—watching closely (Figure 8-8). At least 20 fragments from Shoemaker-Levy 9 struck Jupiter, and 15 of them had detectable impact sites. The impacts resulted in fireballs some 10 km in diameter with temperatures of 7500 K, which is hotter than the surface of the Sun. Indeed, the largest fragment gave off as much energy as 600 million megatons of TNT, far more than the energy that could be released by all the nuclear weapons remaining on Earth, combined. Impacts were followed by crescent-shaped ejecta containing a variety of chemical compounds. Ripples or waves spread out from the impact sites through Jupiter's clouds in splotches that lasted for months.

The observations suggest that the pieces of comet did not penetrate very far into Jupiter's upper cloud layer. This fact, in turn, suggests that the pieces were not much larger than a kilometer in diameter. The ejecta from each impact included a dark plume that rose high into Jupiter's atmosphere. The darkness was apparently due to carbon compounds vaporized from the comet bodies. Also detected from the comet were water, sulfur compounds, silicon, magnesium, and iron.

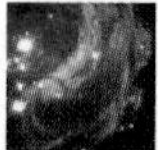

JUPITER'S MOONS AND RINGS

Jupiter hosts at least 63 moons. The four largest—Io, Europa, Ganymede, and Callisto—may have formed at the same time as Jupiter from debris orbiting the planet, just as the planets formed from debris orbiting the Sun, as discussed in Section 5-3. Jupiter's other moons, as we will see, are probably captured planetesimals and smaller pieces of space debris.

Galileo was the first person to observe the four largest moons, in 1610, seen through his meager telescope as pinpoints of light. He called them the "Medicean stars" to attract the attention of the Medicis, rulers of Florence and wealthy patrons of the arts and sciences. To Galileo, the moons provided evidence supporting the then-controversial Copernican cosmology; at that time, Western theologians asserted that all cosmic bodies orbited the Earth. The fact that the Medicean stars orbited Jupiter raised grave concerns in some circles.

To the modern astronomer, these moons are four extraordinary worlds, different both from the rocky terrestrial planets and from hydrogen-rich Jupiter. Now called collectively the **Galilean moons** or **Galilean satellites,** they are named after the mythical lovers and companions of the Greek god Zeus (called Jupiter by the early Romans). From the closest moon outward, they are Io, Europa, Ganymede, and Callisto.

These four worlds were photographed extensively by the *Voyager 1* and *Voyager 2* flybys and by the *Galileo* spacecraft (Figure 8-9). The two inner Galilean satellites, Io and Europa, are about the same size as our Moon. The two outer satellites, Ganymede and Callisto, are roughly the size

THE GALILEAN MOONS, MERCURY, AND THE MOON: VITAL STATISTICS

	Mean distance from Jupiter (km)	Sidereal period (d)	Diameter (km)	Mass (kg)	Mass (Moon = 1)	Mean density (kg/m³)
Io	421,600	1.77	3630	8.94×10^{22}	1.22	3570
Europa	670,900	3.55	3138	4.80×10^{22}	0.65	2970
Ganymede	1,070,000	7.16	5262	1.48×10^{23}	2.01	1940
Callisto	1,883,000	16.69	4800	1.08×10^{23}	1.47	1860
Mercury	–	–	4878	3.30×10^{23}	4.49	5430
Moon	–	–	3476	7.35×10^{22}	1.00	3340

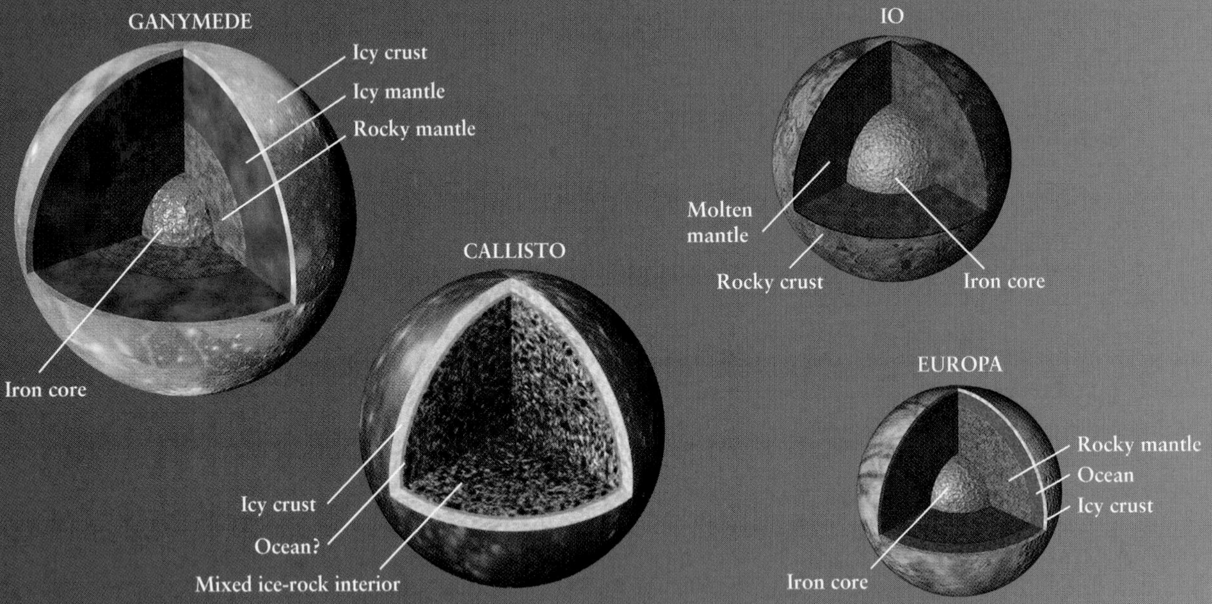

WEB LINK 8.9

FIGURE 8-9 The Galilean Satellites The four Galilean satellites are shown here to the same scale. Io and Europa have diameters and densities comparable to our Moon and are composed primarily of rocky material. Ganymede and Callisto are roughly as big as Mercury, but their low average densities indicate that each contains a thick layer of water and ice. The cross-sectional diagrams of the interiors of the four Galilean moons show the probable internal structures of the moons based on their average densities and on information from the *Galileo* mission. (NASA and NASA/JPL) of Mercury. Figure 8-9 presents comparative information about these six bodies.

8-4 Io's surface is sculpted by volcanic activity

Sulfury Io is among the most exotic moons in our solar system (Figure 8-10). With a density of 3570 kg/m^3, it is neither terrestrial nor "Jovian" (that is, like Jupiter) in chemical composition. Its density is slightly less than that of Earth's surface, suggesting that most of Io is rock and iron, rather than lighter hydrogen, helium, or water. It zooms through its orbit of Jupiter once every 1.8 days. Like our Moon, Io is in synchronous rotation with its planet. Like the Earth, Io has an iron core reaching from its center halfway to its surface.

Images of Io reveal giant plumes rich in sulfur dioxide emitted by geysers, similar to Old Faithful on Earth. Most of this ejected material falls back onto Io's surface; the rest is moving fast enough to escape into space. These geysers are usually associated with Io's 300 or so active volcanoes. Observations indicate that these volcanoes emit 10 trillion tons of matter each year in plumes up to 500 km high. That is enough material to resurface Io to a depth of 1 meter each century (Figure 8-10b). One eruption in 2002 occurred over an area the size of London, some 1600 square kilometers (620 sq mi). The volcanoes also emit basaltic lava flows rich in magnesium and iron. Io's volcanoes are named after gods and goddesses associated with fire in Greek, Norse, Hawaiian, and other mythologies. Io also has numerous black "dots" on its surface, which apparently are dormant volcanic vents. Old lava flows radiate from many of these locations, which are typically 10 to 50 km in diameter and cover 5% of Io's surface.

INSIGHT INTO SCIENCE

What's in a Name? Following up on the preconceptions that common words create (see Insight into Science: Imagine the Moon in Chapter 7), objects with familiar names often have different characteristics than we expect. We tend to envision moons as inert, airless, lifeless, dry places similar to our Moon. As we will see throughout this chapter, starting with Io, some moons have active volcanoes, atmospheres, and probably vast, underground, liquid water oceans.

Just before their discovery, the existence of active volcanoes on Io was predicted from analysis of the gravitational forces to which that moon is subjected. As it rapidly orbits Jupiter, Io repeatedly passes between Jupiter and one or another of the other Galilean satellites. These moons pull on Io, causing it to slightly change its distance from Jupiter. This change in distance creates a change in the tidal forces

a

R I V U X G

b

R I V U X G

FIGURE 8-10 Io (a) This true-color view was taken by the *Galileo* spacecraft in 1999. The range of colors results from surface deposits of sulfur ejected from Io's numerous volcanoes. Plumes from the volcano Prometheus rise up 100 km. Prometheus has been active in every image taken of Io since the *Voyager* flybys of 1979. (b) Photographed in 1999 and then 2000 (shown here), the ongoing lava flow from this volcanic eruption at Tvashtar Catena has considerably altered this region of Io's surface. (a: Moses Milazzo, PIRL, LPL, NASA; b: University of Arizona/JPL/NASA)

(see Section 6-8) acting on it from the planet. As the distance between Io and Jupiter varies, the resulting tidal stresses alternately squeeze and flex the moon. This ongoing motion of Io's interior generates heat through friction, creating as much energy inside Io as the detonation of 2400 tons of TNT every second. Gas and molten rock from inside Io eventually make their way to the moon's surface, where they are ejected through the volcanoes.

Satellite instruments have identified sulfur and sulfur dioxide in the material erupting from Io's volcanoes. Sulfur is normally bright yellow. If heated and suddenly cooled, however, it forms molecules that assume a range of colors, from orange and red to black, which accounts for Io's tremendous range of colors (see Figure 8-10). Sulfur dioxide (SO_2) is an acrid gas commonly discharged from volcanic vents here on Earth and, apparently, on Venus. When eruptions on Io release this gas into the cold vacuum of space, it crystallizes into white flakes, which fall onto the surface and account for the moon's whitish deposits.

The *Galileo* spacecraft detected an atmosphere around Io. Composed of oxygen, sulfur, and sulfur dioxide, it is only one-billionth as dense as the air we breathe. Io's atmosphere can sometimes be seen to glow blue, red, or green, depending on the gases involved. Gases ejected from Io's volcanoes have also been observed extending out into space and forming a doughnut-shaped region around Jupiter called the *Io torus*, about which we will say more shortly.

8-5 Europa apparently harbors liquid water below its surface

Images of Europa's ice and rock surface from *Voyager 2* and the *Galileo* spacecraft suggest that Jupiter's second-closest Galilean moon contains liquid water (Figure 8-11). Europa orbits Jupiter every 3½ days and, like Io, is in synchronous rotation. The changing gravitational tug of Io creates stress inside Europa similar to the magma-generating distortion that Io undergoes. The stress may create enough heat inside Europa to keep the water just a few kilometers below the moon's ice and rock surface in a liquid state.

The ice floes seen in Figure 8-11 support the belief that Europa has liquid water inside. The surface features are similar to those seen in the Arctic region on Earth. The movement of the surface features creating the floes, along with swirls, strips, and ridges, appears to be driven by circulating water underneath the moon's surface, creating tectonic plate motion and tidal flexing. Also indicative of liquid water is the moon's reddish color. The coloring may be due to salt deposits left after liquid water rose to the surface and evaporated. What makes the underground ocean of Europa especially interesting to scientists is that virtually all liquid water locations on Earth support life, and life may well have evolved in Europa's oceans.

Galileo spacecraft images suggest that some of Europa's features have moved within the past few million years, and perhaps are still in motion. Indeed, the chaotic surface revealed by *Galileo* is interpreted as having formed as a result of volcanism on Europa, strengthening the belief that this moon still has a liquid water layer. Replenishment of the surface by tectonic plate motion would explain why only a few small impact craters, such as the crater Pwyll (see Figure 8-11), have survived.

Galileo also photographed red and white domes called *lenticulae* on Europa (Figure 8-12a) that astronomers believe are rising warmed ice mixed with other material, possibly including organic matter. These lenticulae are typically a hundred meters high and 10 km (6 mi) wide. The rising material behaves like the blobs in lava lamps (Figure 8-12b), although the lenticulae are calculated to take 100,000 years to reach the surface from the liquid ocean that is believed to exist inside the moon.

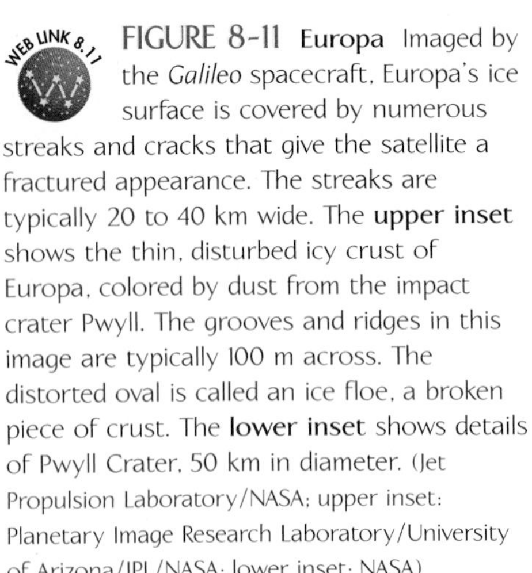

FIGURE 8-11 Europa Imaged by the *Galileo* spacecraft. Europa's ice surface is covered by numerous streaks and cracks that give the satellite a fractured appearance. The streaks are typically 20 to 40 km wide. The **upper inset** shows the thin, disturbed icy crust of Europa, colored by dust from the impact crater Pwyll. The grooves and ridges in this image are typically 100 m across. The distorted oval is called an ice floe, a broken piece of crust. The **lower inset** shows details of Pwyll Crater, 50 km in diameter. (Jet Propulsion Laboratory/NASA; upper inset: Planetary Image Research Laboratory/University of Arizona/JPL/NASA; lower inset: NASA)

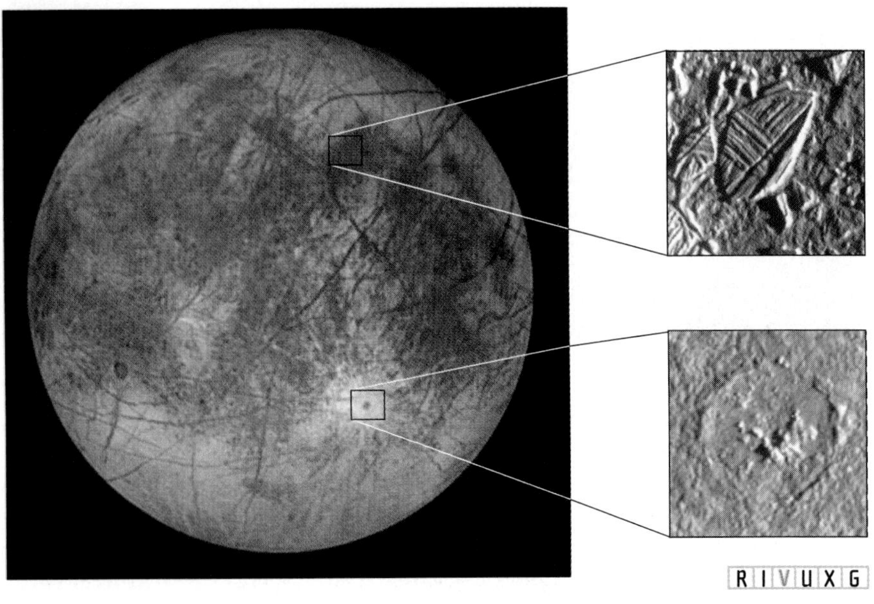

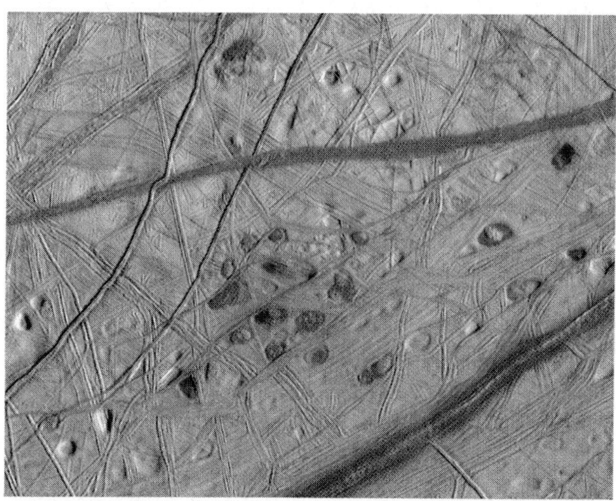

FIGURE 8-12 **Rising Blobs on Europa** (a) This *Galileo* image of Europa shows surface scars attributed to rising warmed ice and debris convecting up from the moon's interior, arriving at and then leaking out at the surface. The white domes are likely to be rising material that has not yet reached the surface. (b) A lava lamp in which warmed material rises through cooler liquid. The rising material in Europa is analogous to the rising motion of the blobs in a lava lamp, except that on Europa, the motion is through ice. (a: NASA/JPL/University of Arizona/University of Colorado; b: C. Varas/W. H. Freeman & Worth Publishers)

Europa's average density of 2970 kg/m^3 is slightly less than Io's. A quarter of its mass may be water. It also has a metallic core of much higher density and a weak magnetic field. In 1995, astronomers discovered an extremely thin atmosphere containing molecular oxygen surrounding Europa. The density of this gas is about 10^{-11} times the density of the air we breathe. The oxygen may come from water molecules broken up on the moon's surface by ultraviolet radiation from the Sun.

8-6 Ganymede is larger than Mercury

Ganymede is the largest satellite in the solar system (Figure 8-13). Its diameter is greater than Mercury's, but its density of 1940 kg/m^3 is much less than that of Mercury. It also has a permanent magnetic field that is twice as strong as Mercury's field. Ganymede orbits Jupiter in synchronous rotation once every 7.2 days. Like its neighbor Europa, Ganymede has an iron-rich core, a rocky mantle, a liquid water ocean, a thin atmosphere, and a covering of dirty ice.

The existence of the ocean is implied by the discovery of a second, continuously changing magnetic field around Ganymede that is generated by Jupiter's magnetic field. As Ganymede orbits Jupiter, the planet's powerful magnetic field creates an electrical current inside the moon, which in turn creates Ganymede's varying magnetic field. (The same process is used to create electrical currents in electrical power stations here on Earth. In the latter situation, magnets are whirled around in the power station. Their magnetic fields cut across wires, thereby pushing electrons in the wire. These moving electrons are the electric current we use.) The best explanation of why current flows inside Ganymede is that liquid saltwater exists there, and saltwater is a good conductor of electricity. This implies, of course, the presence of a liquid ocean. Furthermore, salts have been observed on Ganymede's surface. They were apparently carried upward and deposited there as water leaked out and froze. As in Europa, liquid water may imply the presence of life.

Like our Moon, Ganymede has two very different kinds of terrain. Dark, polygon-shaped regions are its oldest surface features, as judged by their numerous craters. Light-colored, heavily grooved terrain is found between the dark, angular islands. These lighter regions are much less cratered and therefore younger. Ganymede's grooved terrain consists

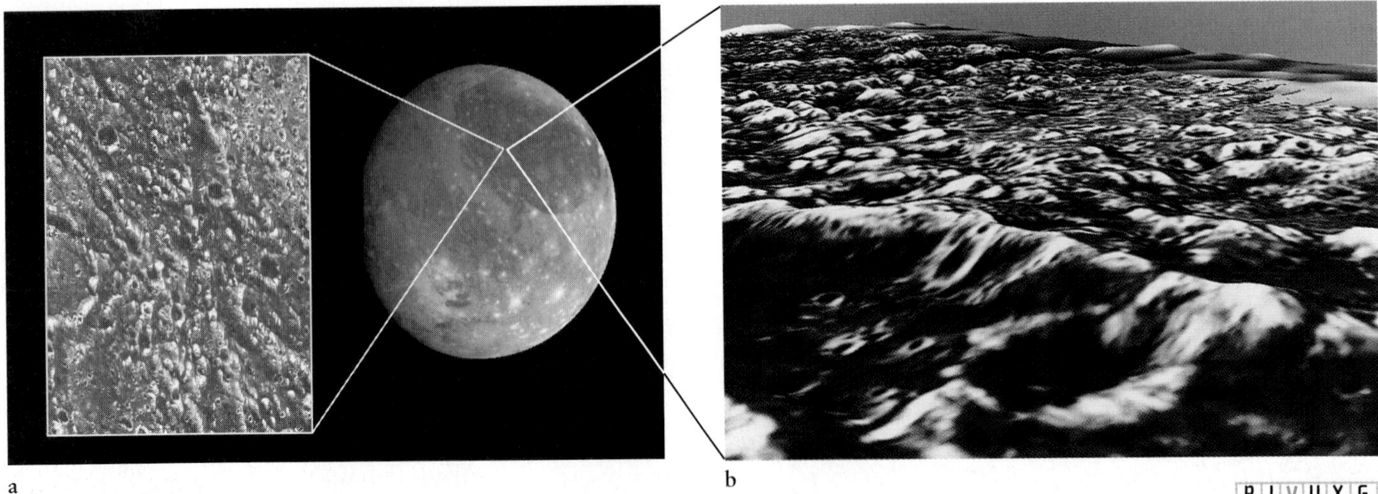

FIGURE 8-13 **Ganymede** This side of Ganymede is dominated by the huge, dark, circular region called Galileo Regio, which is the largest remnant of Ganymede's ancient crust. Darker areas of the moon are older; lighter areas are younger, tectonically deformed regions. The light white areas in and around some craters indicate the presence of water ice. **Inset a** is about 46 × 64 km (29 × 38 mi). Note in this *Galileo* spacecraft image the deep furrows in the moon's icy crust that probably resulted from crustal movement due to impacts and tectonic plate motion. The perspective in **inset b** was created by combining information from two *Galileo* images. (NASA/JPL)

of parallel mountain ridges up to 1 km high and spaced 10 to 15 km apart. These features suggest that the process of plate tectonics may have dominated Ganymede's early history. But unlike Europa, where tectonic activity still occurs today, tectonics on Ganymede bogged down 3 billion years ago as the satellite's crust froze solid.

Another mechanism that may have created Ganymede's large-scale features is a bizarre property of water. Unlike most liquids, which shrink upon solidifying, water expands when it freezes. Seeping up through cracks in Ganymede's original crust, water thus forced apart fragments of that crust. This process could have produced jagged, dark islands of old crust separated by bands of younger, light-colored, heavily grooved ice. Topping off Ganymede's varied features is the discovery that auroras occur there.

8-7 Callisto bears the scars of a huge asteroid impact

Callisto is Jupiter's outermost Galilean moon. It orbits Jupiter in 16.7 days, and, like the other Galilean moons, Callisto's rotation is synchronous. Callisto is 91% as big and 96% as dense as Ganymede. It has a thin atmosphere of hydrogen and carbon dioxide. Like Ganymede, Callisto apparently harbors a substantial liquid water ocean. Its presence is again inferred by Callisto's changing magnetic field. The heat that keeps the ocean liquid apparently comes from energy released by radioactive decay inside the moon.

While numerous large impact craters are scattered over Callisto's dark, ancient, icy crust, it has very few craters smaller than 100 m across (Figure 8-14). Astronomers speculate that the smaller craters have disintegrated. Unlike Ganymede and Europa, Callisto has no younger, grooved terrain. The absence of grooved terrain suggests that tectonic activity never began there: The satellite simply froze too rapidly. It is bitterly cold on Callisto's surface. *Voyager* instruments measured a noontime temperature of 155 K (–180°F), and the nighttime temperature plunges to 80 K (–315°F).

Callisto carries the cold, hard evidence of what happens when one astronomical body strikes another. *Voyager 1* photographed the huge impact basin, Valhalla, on Callisto (see Figure 8-14a). An asteroid-sized object produced Valhalla Basin, which is located on Callisto's Jupiter-facing hemisphere. Like throwing a rock into a calm lake, ripples ran out from the impact site along Callisto's surface, cracking the surface and freezing into place for eternity. The largest remnant rings surrounding the impact crater have diameters of 3000 km. In 2001, the *Galileo* spacecraft revealed spires 80–100 m high on Callisto (Figure 8.14b). These are also believed to have been created by an impact, perhaps the same one. One thing that is missing from Callisto is the jumbled terrain on the side opposite Valhalla Basin, as is seen opposite large impact sites on Mercury and our Moon. The smoothness of Callisto is explained by a model that shows that a liquid water interior would dampen the impact shock and thereby prevent the opposite side from becoming disturbed. This supports the idea that Callisto has liquid water inside. The probable interiors of the Galilean moons are shown in Figure 8-9.

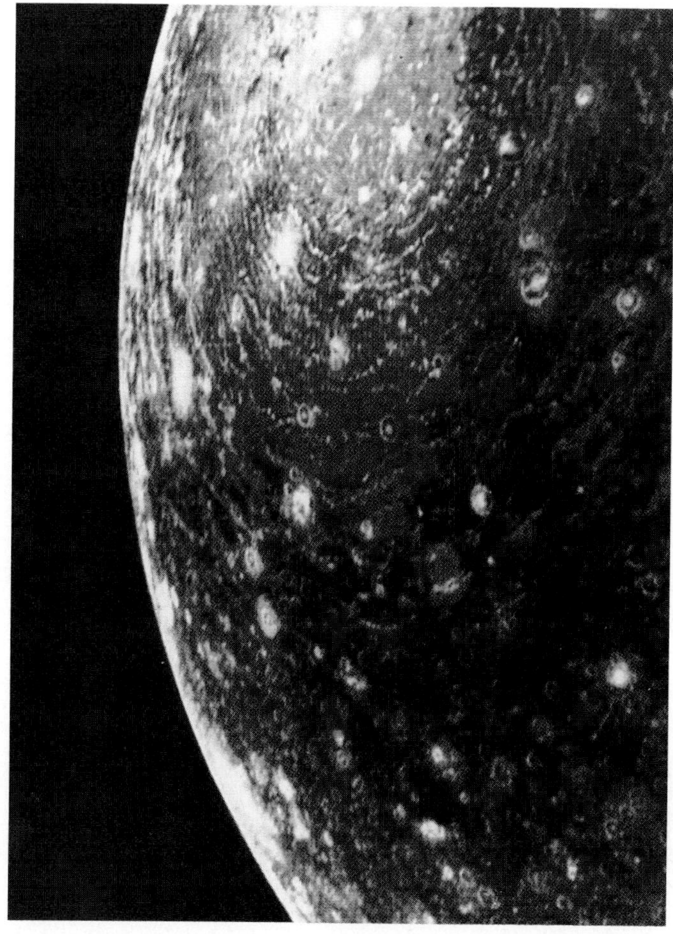

a

FIGURE 8-14 **Callisto** The outermost Galilean satellite is almost exactly the same size as Mercury. Numerous craters pockmark Callisto's icy surface. **(a)** The series of faint, concentric rings that cover much of this image is the result of a huge impact that created the impact basin named Valhalla. Valhalla dominates the Jupiter-facing hemisphere of this frozen, geologically inactive world. **(b)** The two insets in this *Galileo* mission image show spires containing both ice and some dark material. The spires were probably thrown upward as the result of an impact. The spires erode as dark material in them absorb heat from the Sun (to the left of these images). (a: Courtesy of NASA/JPL; b: NASA/JPL/Arizona State University)

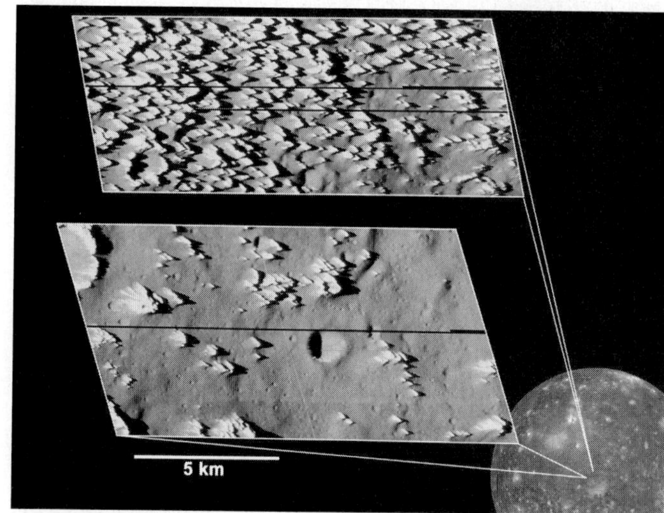

b

R I V U X G

8-8 Other debris orbits Jupiter as smaller moons and ringlets

Besides the four Galilean moons, Jupiter has at least 59 other moons, a set of tenuous ringlets, and two doughnut-shaped toruses of electrically charged gas particles. The non-Galilean moons are all irregular in shape and less than 275 km in diameter. Four of these moons are inside Io's orbit (Figure 8-15); all the other known moons are outside Callisto's orbit. The inner moon Amalthea (Figure 8-15) is red-colored, has about the same density as water, and is apparently made of pieces of rock and ice barely held together by the moon's

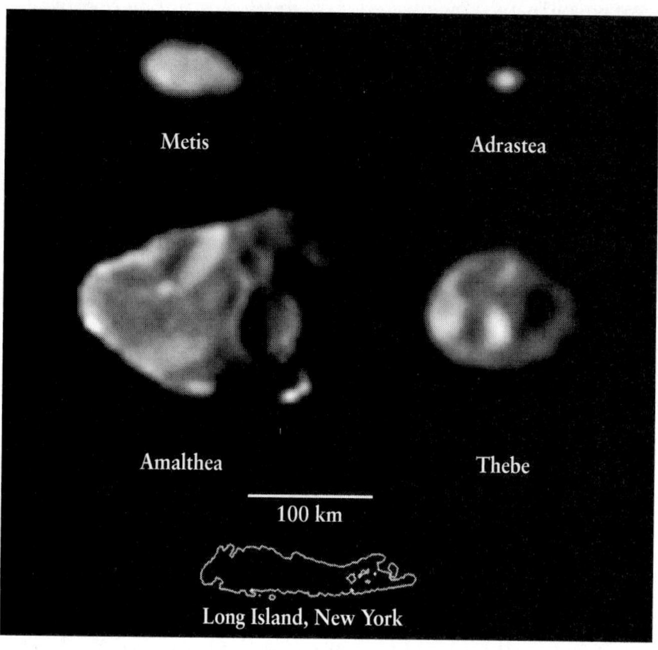

▶ FIGURE 8-15 **Irregularly-Shaped Inner Moons** The four known inner moons of Jupiter are significantly different than the Galilean satellites. They are roughly oval-shaped bodies. While craters have not yet been resolved on Adrastea and Metis, their irregular shapes strongly suggest that they are cratered. All four moons are named for characters in mythology relating to Jupiter (in Greek mythology, Zeus). (NASA/JPL, Cornell University)

R I V U X G

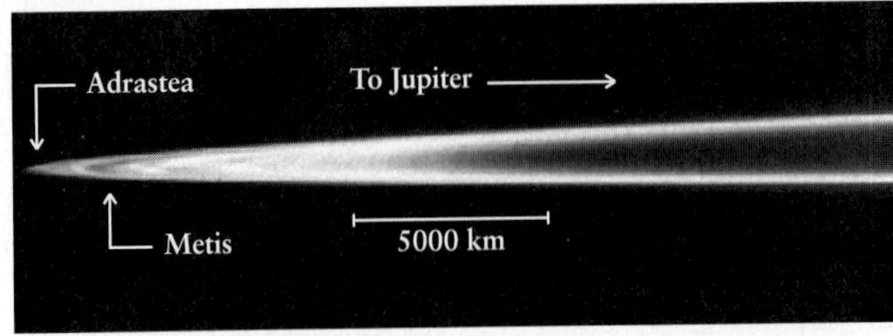

a

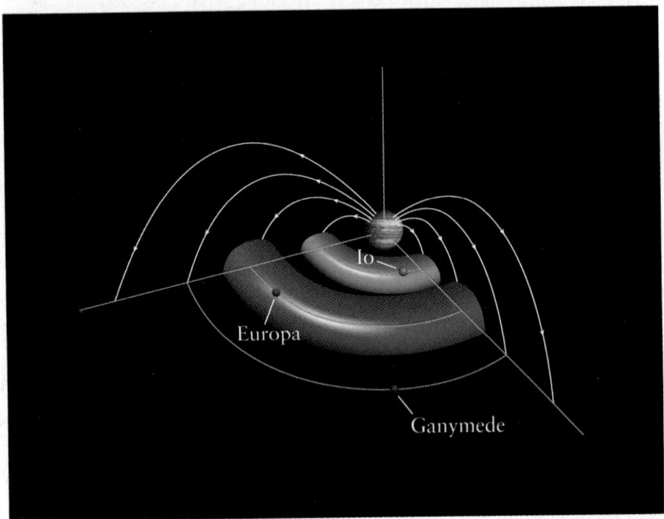

b

R I V U X G

FIGURE 8-16 **Jupiter's Ring and Torus** (a) A portion of Jupiter's faint ring system, photographed by *Voyager 2*. The ring is probably composed of tiny rock fragments. The brightest portion of the ring is about 6000 km wide. The outer edge of the ring is sharply defined, but the inner edge is somewhat fuzzy. A tenuous sheet of material extends from the ring's inner edge all the way down to the planet's cloudtops. (b) Quarter images of Io's and Europa's toruses (also called plasma toruses because the gas particles in them are charged—plasmas). Io is visible in its torus (colored green), while Europa is visible in its torus (colored blue). Some of Jupiter's magnetic field lines are also drawn in. Plasma from toruses flow inward along these field lines toward Jupiter. (a: NASA/JPL, Cornell University; b: NASA/JPL/Johns Hopkins University Applied Physics Laboratory)

own gravity. The Galilean moons along with the smaller moons closer to Jupiter and six of the outer moons orbit in the same direction that Jupiter rotates (**prograde orbits**). The remaining outer moons revolve in the opposite direction (**retrograde orbits**). The outer ones appear to be individual, captured planetesimals, while the inner ones are probably smaller pieces broken off a single larger body.

4 As we will see, Jupiter is only the first of four planets with rings. As astronomers predicted, the cameras aboard *Voyager 1* discovered three **ringlets** around Jupiter. These are bands of particles, the brightest of which is seen in Figure 8-16a, a *Voyager 2* image. They are the darkest, simplest set of rings in the solar system. They consist of very fine dust particles that are continuously pushed out of orbit by the impact of radiation from Jupiter and the Sun. Therefore, these rings are being replenished by material from Io and the other moons that has been knocked free by impacts from tiny pieces of interplanetary space debris.

At least two doughnut-shaped regions of electrically charged gas particles, *plasmas,* orbit Jupiter. One is in the same orbit as Io (Figure 8-16b). The Io torus consists of sulfur and oxygen ions (charged atoms) along with free electrons. These particles were ejected by Io's geysers and they are held in orbit by Jupiter's strong magnetic field. Guided by the field, some of this matter spirals toward Jupiter, thereby creating the aurora seen there (see Figure 8-8).

The other torus is in Europa's orbit (Figure 8-16b). It consists of hydrogen and oxygen ions and electrons created from water molecules kicked off Europa's surface by radiation from Jupiter. The mass of this ring of gas is calculated to be about 6×10^4 tons.

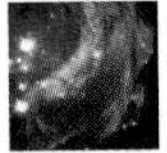

SATURN

Saturn, with its ethereal rings, presents the most spectacular image of all the planets (Figure 8-17). Giant Saturn has 95 times as much mass as the Earth, making it second in mass and size to Jupiter. Like Jupiter, Saturn has a thick, active atmosphere composed predominantly of hydrogen. It also has a strong magnetic field. Ultraviolet images from the Hubble Space Telescope reveal auroras around Saturn like those seen around Jupiter (see Figure 8-8). Saturn's vital statistics are listed in Figure 8-17.

8-9 Saturn's surface and interior are similar to those of Jupiter

Partly obscured by the thick, hazy atmosphere above them, Saturn's clouds lack the colorful contrast visible on Jupiter. Nevertheless, photographs do show faint stripes in Saturn's atmosphere similar to Jupiter's belts and zones (Figure 8-18). Their existence indicates that Saturn also has internal heat that convects the cloud gases. Changing features there show that Saturn's atmosphere, too, has differential rotation—ranging from about 10 hours and 14 minutes at the equator to 10 hours and 40 minutes at high latitudes. As on Jupiter, some of the belts and zones move eastward, while others move westward. Although Saturn lacks a long-lived spot like Jupiter's Great Red Spot, it does have storms, including a major new one discovered by the Hubble Space Telescope in 1994 and a

SATURN: VITAL STATISTICS

Average distance from Sun:	9.57 AU = 1.43×10^9 km
Maximum distance from Sun:	10.1 AU = 1.51×10^9 km
Minimum distance from Sun:	9.06 AU = 1.36×10^9 km
Eccentricity of orbit:	0.056
Average orbital speed:	9.64 km/s
Orbital period:	29.4 years
Rotation period:	$10^h\ 13^m\ 59^s$ (equatorial) $10^h\ 39^m\ 25^s$ (internal)
Inclination of equator to orbit:	26.7°
Inclination of orbit to ecliptic:	2.48°
Radius:	60,270 km = 9.45 Earth radii (equatorial) 54,360 km = 8.52 Earth radii (polar)
Mass:	5.69×10^{26} kg = 95.2 Earth masses
Average density:	687 kg/m^3
Escape speed:	35.5 km/s
Surface gravity (Earth = 1):	0.92
Albedo:	0.46
Average temperature at cloudtops:	−180°C = −292°F = 93 K
Atmospheric composition (by number of molecules):	96.3% hydrogen (H_2), 3.3% helium (He), 0.4% methane (CH_4), ammonia (NH_3), water vapor (H_2O), and other gases

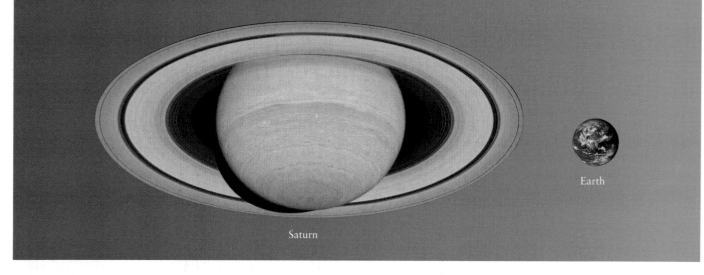

R I V U X G

WEB LINK 8.16

FIGURE 8-17 **Saturn** *Voyager 2* sent back this image when the spacecraft was 34 million kilometers from Saturn. The image of Earth shows its size relative to Saturn. (NASA/ESA/E. Karkoschka; inset: NASA)

FIGURE 8-18 **Belts and Zones on Saturn** *Voyager 1* took this view of Saturn's cloudtops at a distance of 1.8 million kilometers. The color has been enhanced to better show the belts and zones. Note that there is little swirling structure and substantially less contrast between belts and zones on Saturn than there is on Jupiter. (NASA)

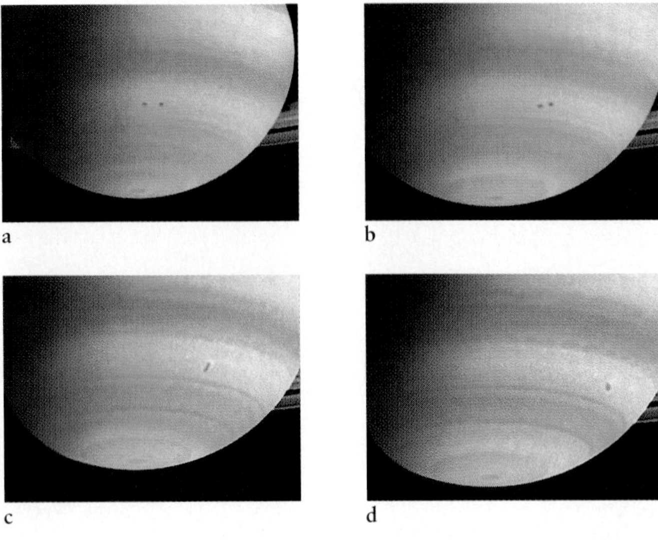

FIGURE 8-19 **Merging Storms on Saturn** This sequence of *Cassini* images shows two hurricanelike storms merging into one on Saturn in 2004. Each storm is about 1000 km (600 mi) across. (NASA/JPL/Space Science Institute)

pair of storms that merged in 2004 (Figure 8-19). In 2004, the *Cassini* spacecraft detected lightning in Saturn's atmosphere.

Saturn's atmosphere is composed of the same basic gases as Jupiter. However, because its mass is lower than Jupiter's, Saturn's gravitational force on its atmosphere is less. Therefore, Saturn's atmosphere is more spread out than that of its larger neighbor (see Figure 8-5). Saturn has among the highest winds of any planet. At its equator, storms have been clocked moving at speeds of 1600 km/h (1000 mph), which is some 10 times faster than the Earth's jet streams and 3 times faster than the fastest winds on Jupiter. The reason for Saturn's extremely high wind speeds is still under investigation.

Astronomers infer that Saturn's interior structure resembles Jupiter's. A layer of molecular hydrogen just below the clouds surrounds a layer of liquid metallic hydrogen, liquid "ices," and a rock and metal core (Figure 8-20).

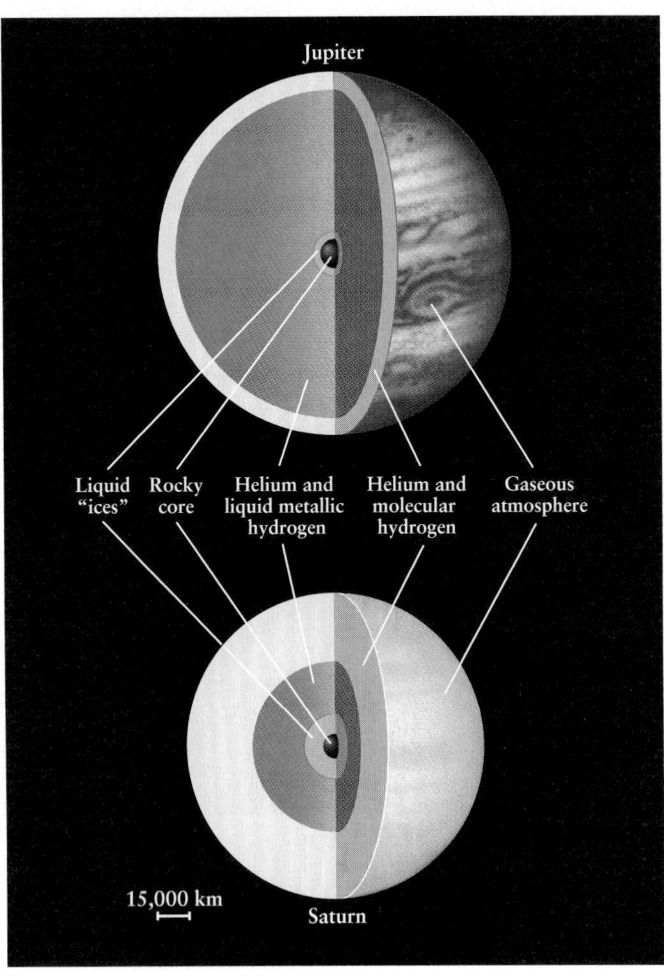

▶ FIGURE 8-20 **Cutaways of Jupiter and Saturn** The interiors of both Jupiter and Saturn are believed to have four regions: a terrestrial core, a liquid "ice" shell, a metallic hydrogen shell, and a normal liquid hydrogen mantle. Their atmospheres are thin layers above the normal hydrogen, which boils upward, creating the belts and zones.

Because of its smaller mass, Saturn's interior is also less compressed than Jupiter's, and the pressure inside Saturn is insufficient to convert as much hydrogen into a liquid metal. Saturn's rocky core is about 7400 km in radius, while its layer of liquid metallic hydrogen is 12,000 km thick (see Figure 8-20). To alchemists, Saturn was associated with the extremely dense element lead. This is wonderfully ironic in that at 687 kg/m^3, Saturn is the least dense body in the entire solar system.

8-10 Saturn's spectacular rings are composed of fragments of ice and ice-coated rock

WEB LINK 8.17

Even when viewed through a telescope from the vicinity of the Earth, Saturn's magnificent rings are among the most interesting objects in the solar system. They are tilted 27° from the perpendicular to Saturn's plane of orbit, so sometimes they are nearly edge-on to us, making them virtually impossible to see (Figure 8-21). Even when the rings are at their maximum tilt from our perspective, Saturn is so far away that our best Earth-mounted telescopes can reveal only their largest features. But even these have turned out to hold surprises. In 1675, Giovanni Cassini discovered an intriguing feature—a dark division in the rings. This 5000-km-wide gap, called the **Cassini division**, separates the dimmer **A ring** from the brighter **B ring**, which lies closer to the planet. By the mid-1800s, astronomers using improved telescopes detected a faint C ring just inside the B ring (Figure 8-22a).

The Cassini division exists because the gravitational force from Saturn's moon Mimas combines with the gravitational force from the planet to keep the region clear of debris. Whenever matter drifts into the Cassini division, Mimas (orbiting at a different rate than the matter in the Cassini division) periodically exerts a force on this matter, thereby pulling it out of the division. This effect is called a **resonance** and is similar to what happens when you push someone on a swing at the right time and so enable them to go higher and higher.

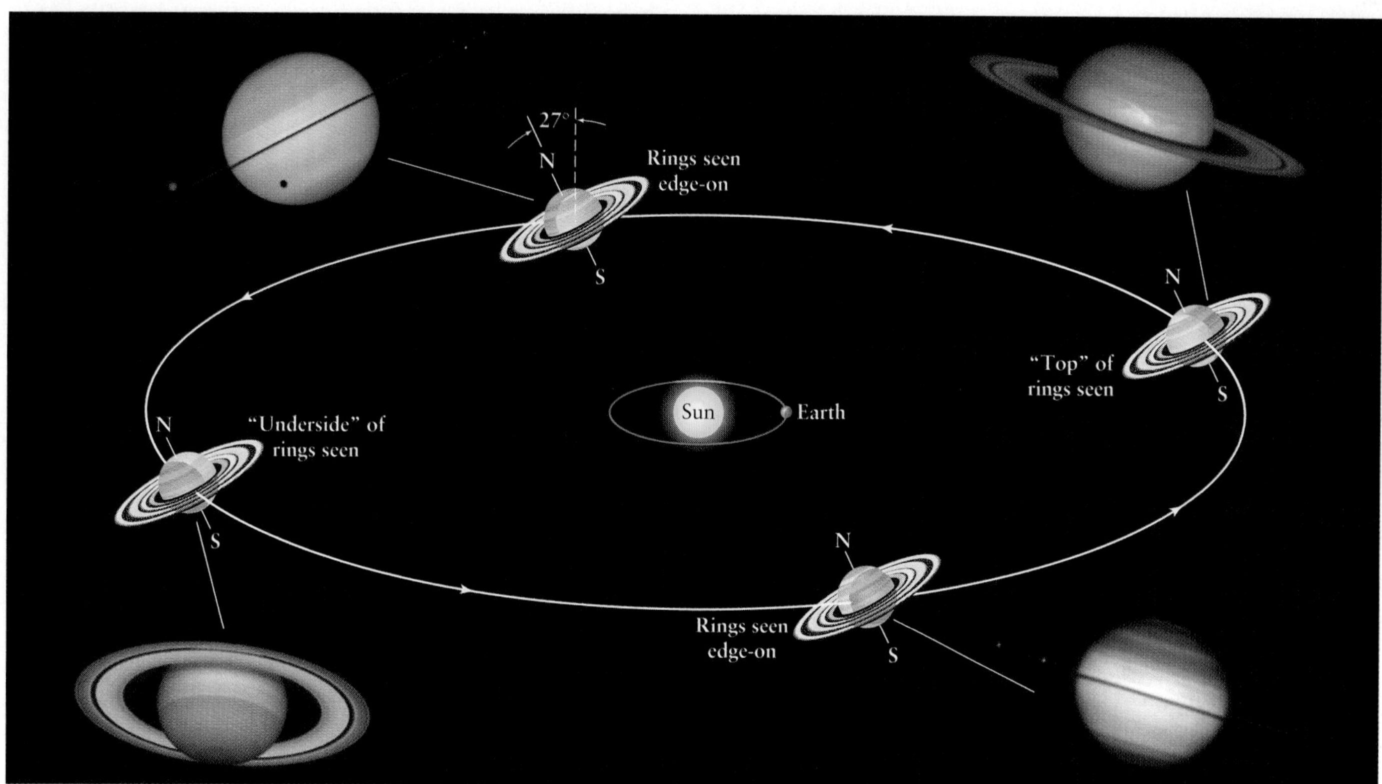

FIGURE 8-21 **Saturn as Seen from Earth** Saturn's rings are aligned with its equator, which is tilted 27° from the plane of Saturn's orbit around the Sun. Therefore, Earth-based observers see the rings at various angles as Saturn moves around its orbit. The plane of Saturn's rings and equator keeps the same orientation in space as the planet goes around its orbit, just as the Earth keeps its 23½° tilt as it orbits the Sun. The accompanying Earth-based photographs show how the rings seem to disappear entirely about every 15 years. (Top left: E. Karkoschka/U. of Arizona Lunar and Planetary Lab and NASA; bottom left: AURA/STScI/NASA; bottom right and top right: A. Bosh/Lowell Obs. and NASA)

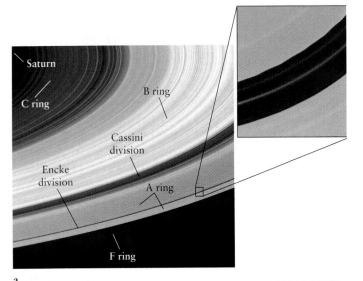

a

R I V U X G

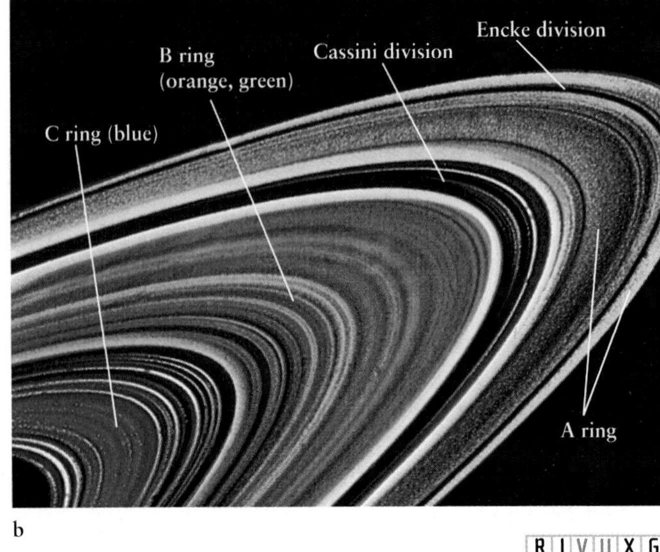

b

R I V U X G

VIDEO 8.5 VIDEO 8.6

FIGURE 8-22 **Numerous Thin Ringlets Constitute Saturn's Rings** **(a)** This *Cassini* image shows that Saturn's rings contain numerous ringlets. **Inset:** As moons orbit near or between rings, they cause the ring ices to develop spiral ripples, often like the grooves in an old-fashioned record. **(b)** Seen from a different perspective, other details of Saturn's rings are visible in this photograph sent back by *Voyager 2*. The rings are actually composed of thousands of closely spaced ringlets. The colors are exaggerated by computer processing to show the different ringlets more clearly. (a: NASA/JPL/Space Science Institute; inset: Courtesy of NASA/JPL; b: NASA/JPL)

A second gap exists in the outer portion of the A ring, named the **Encke division** (Figure 8-22a), after the German astronomer Johann Franz Encke, who allegedly saw it in 1838. (Many astronomers have argued that Encke's report was erroneous, because his telescope was inadequate to resolve such a narrow gap.) The first undisputed observation of the 270-km-wide division was made by the American astronomer James Keeler in the late 1880s, with the newly constructed 36-in. refractor at the Lick Observatory in California. Unlike the Cassini division, the Encke division is kept clear because a small moon, Pan, orbits within it. The gravitational tugs of Pan and other moons cause the rings to ripple, as shown in the inset of Figure 8-22a.

Pictures from the *Cassini* spacecraft and the Hubble Space Telescope show that Saturn's rings are remarkably thin—about 10 meters thick according to recent estimates! This is amazing when you consider that the total ring system has a width (from inner edge to outer edge) of more than 89,000 km.

Because Saturn's rings are very bright (albedo = 0.80), the particles that form them must be highly reflective. Astronomers had long suspected that the rings consist primarily of water ice, along with ice-coated rocks. Spectra taken by Earth-based observatories and by the spacecraft visiting Saturn have confirmed this suspicion. Because the temperature of the rings ranges from 93 K (–290°F) in the sunshine to less than 73 K (–330°F) in Saturn's shadow, frozen water is in no danger of melting or evaporating from the rings.

Saturn's rings are slightly salmon-colored. This coloring suggests that they contain traces of organic molecules, which often have similar hues. It appears that the rings have gained this material by being bombarded with debris from the outer solar system.

To determine the size of the particles in Saturn's rings, 5
scientists measured the brightness of different rings from many angles as the *Voyager* and *Cassini* spacecraft flew past the planet. Astronomers also measured changes in radio signals received from the spacecraft as they passed behind the rings. The largest particles in Saturn's rings are roughly 10 m across, although snowball-sized particles about 10 cm in diameter are more abundant than larger pieces. High-resolution images reveal that the ring structure seen from Earth actually consists of thousands of closely spaced ringlets (Figure 8-22). Furthermore, myriad dust-sized particles fill the Cassini and Encke divisions.

The *Voyager* cameras also sent back the first high-quality pictures of the F ring, a thin set of ringlets just beyond the outer edge of the A ring. Two tiny satellites following orbits on either side of the F ring serve to keep the ring intact (Figure 8-23). The outer of the two satellites orbits Saturn at a slower speed than do the ice particles in the ring, as governed by Kepler's third law. As the ring

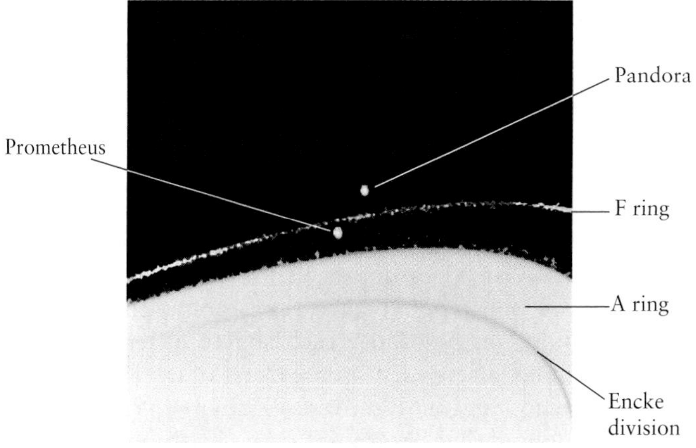

FIGURE 8-23 **The F Ring and Its Two Shepherds** Two tiny satellites, Prometheus and Pandora, each measuring about 50 km across, orbit Saturn on either side of the F ring. The gravitational effects of these two shepherd satellites confine the particles in the F ring to a band about 100 km wide. (NASA)

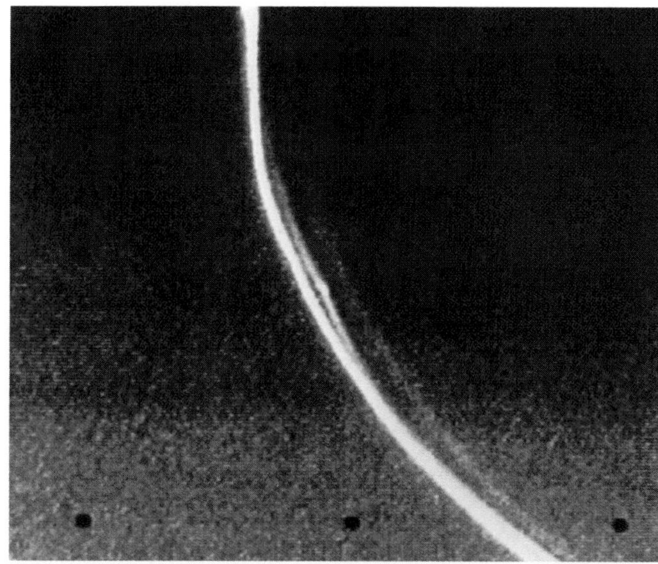

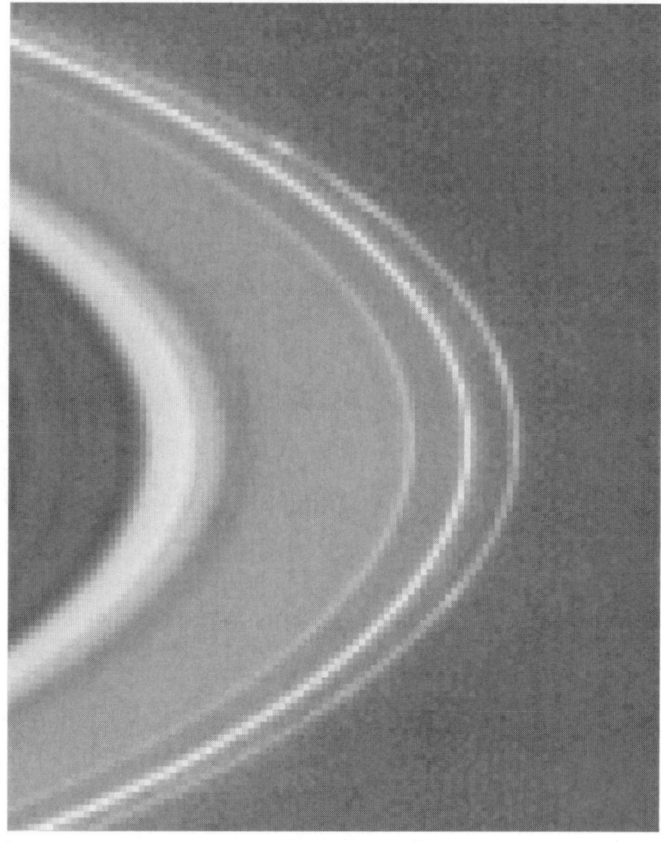

FIGURE 8-24 **Detail of the F Ring** (a) This photograph from *Voyager 1* shows several strands, each measuring roughly 10 km across, that comprise the F ring. The total width of the F ring is about 100 km. (b) The F ring has clumps of material that each orbit in it for a few months or less. The origins of these clumps and how they disperse are not yet known. (a: NASA; b: NASA/JPL/Space Science Institute)

particles pass near it, they receive a tiny, backward gravitational tug, which slows them down, causing them to fall into orbits a bit closer to Saturn. Meanwhile, the inner satellite orbits the planet faster than the F ring particles. Its gravitational force pulls them forward and nudges the particles back into a higher orbit. The combined effect of these two satellites is to focus the icy particles into a well-defined, narrow band about 100 km wide.

Because of their confining influence, these two moons, Prometheus and Pandora, are called **shepherd satellites** or **shepherd moons.** Among the most curious features of the F ring is that the ringlets are sometimes braided or intertwined (Figure 8-24a) and sometimes separate. Although stabilized by the shepherd moons, the F ring does show clumps (Figure 8-24b), whose origins are not yet known.

Saturn's shell of liquid metallic hydrogen produces a planetwide magnetic field that apparently affects its rings. Saturn's much smaller volume of liquid metallic hydrogen produces a magnetic field at Saturn's surface that is only about two-thirds as strong as the magnetic field here at the Earth's surface. Data from spacecraft show that Saturn's magnetosphere contains radiation belts similar to Earth's Van Allen belts. Furthermore, dark **spokes** move around Saturn's rings (Figure 8-25); these are believed to be created by electric charges on the ring material interacting with the planet's magnetic field. The electric charge lifts charged particles out of the plane in which the rings orbit. Spreading the particles out decreases the light scattered from them and therefore makes the rings appear darker. The magnetic field causes the regions of spread particles to change, making the spokes appear to revolve around the planet.

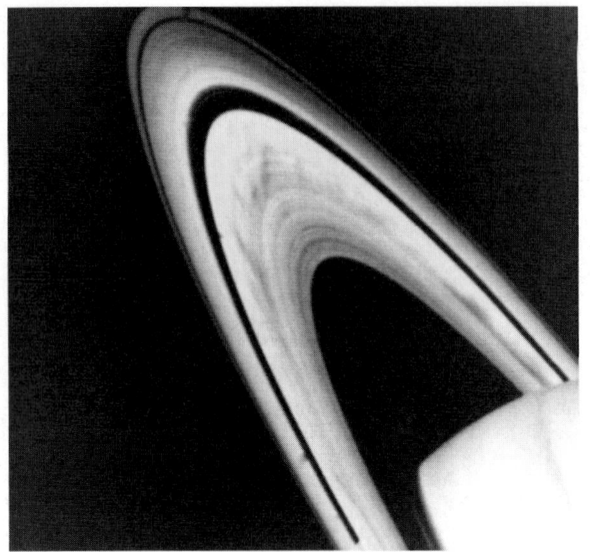

R I V U X G

FIGURE 8-25 **Spokes in Saturn's Rings** Believed to be caused by Saturn's magnetic field moving electrically charged particles that are lifted out of the ring plane, these dark regions move around the rings like the spokes on a rotating wheel. (NASA)

8-11 Titan has a thick, opaque atmosphere rich in nitrogen, methane, and other hydrocarbons

Only 7 of Saturn's 47 known moons are spherical (Figure 8-26). The rest are oblong, suggesting that they are captured asteroids. The 12 moons discovered in 2001 move in clumps, indicating that they may be pieces of a larger moon that was broken up by impacts. Saturn's largest moon, Titan, is second in size to Ganymede among the moons of the solar system and is the only moon to have a dense atmosphere. About 10 times more gas lies above each square meter of Titan's surface than lies above each square meter of the Earth.

Christiaan Huygens discovered Titan in 1655, the same year he proposed that Saturn has rings. By the early 1900s, several scientists had begun to suspect that Titan might have an atmosphere. This moon is larger than (although less massive than) Mercury, and far colder than that planet. Calculations reveal that Titan is cool enough and massive enough to retain heavy gases.

Because of its atmosphere, Titan was a primary target for the *Voyager* missions. Unexpectedly, Titan's thick haze completely blocked any view of its surface. Observations

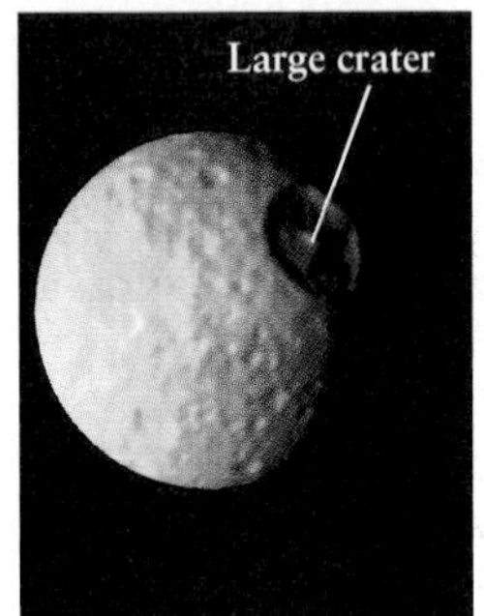

a Mimas (diameter 392 km)

b Enceladus (diameter 500 km)

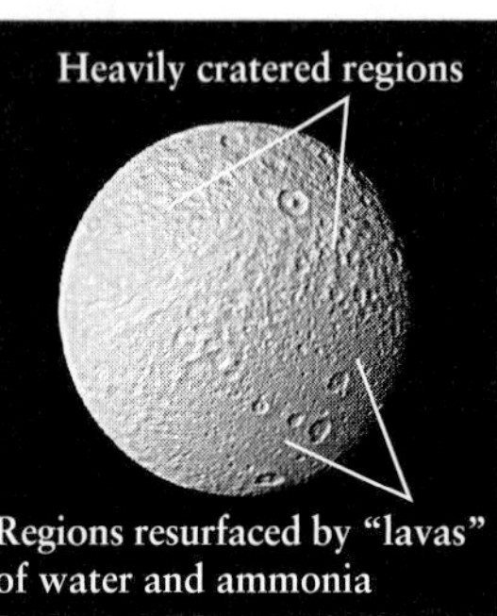

c Tehys (diameter 1060 km)

d Dione (diameter 1120 km)

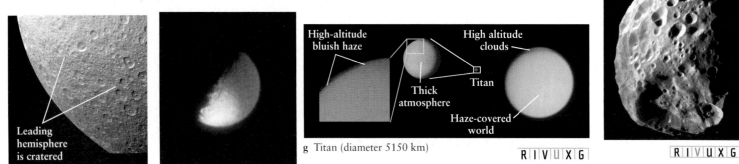

e Rhea (diameter 1530 km)

f Iapetus (diameter 1460 km)

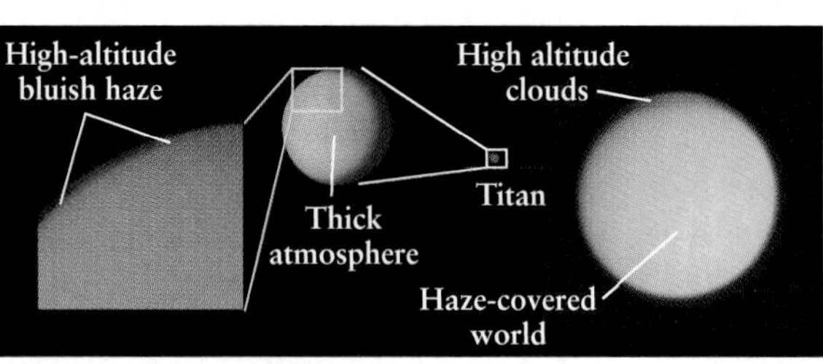

g Titan (diameter 5150 km)

R I V U X G

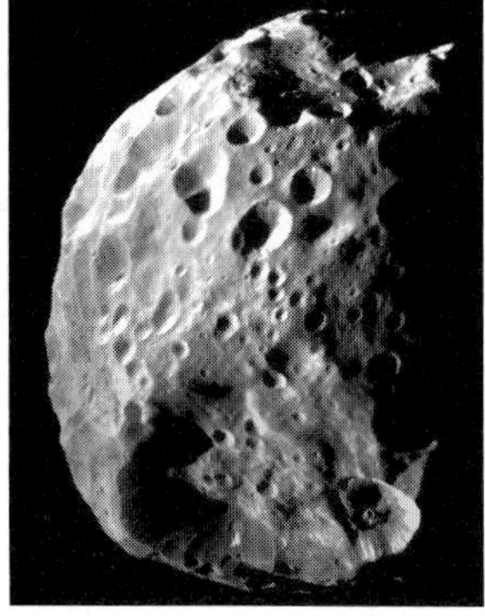

R I V U X G

h Phoebe (diameter 220 km)

FIGURE 8-26 **Saturn's Diverse Moons** (a-g) These *Voyager 1* and *Voyager 2* images show the variety of surface features seen on Saturn's seven spherical moons. They are not shown to scale (refer to the diameters given below each image). Very few features are visible in the thick, unbroken haze that surrounds Saturn's largest satellite, Titan, which has a diameter of 5150 km. The main haze layer is located nearly 300 km above Titan's surface. (h) This *Cassini* image of Phoebe, an irregular moon of Saturn almost as dark as coal, shows how the smaller moons have nonspherical shapes, along with many craters, landslides, grooves, and ridges. Phoebe is barely held in orbit by Saturn. (a-e: JPL/NASA; f: Courtesy of NASA/JPL; g: NASA/JPL/Space Science Institute; h: Courtesy of NASA/JPL/Space Science Institute) also reveal that clouds of methane form and dissipate over periods of hours to days near Titan's south pole.

Voyager data indicated that roughly 90% of Titan's atmosphere is nitrogen. Most of this nitrogen probably formed from the breakdown of ammonia (NH_3) by the Sun's ultraviolet radiation into hydrogen and nitrogen atoms. Because Titan's gravity is too weak to retain hydrogen, this gas has escaped into space, leaving behind ample nitrogen. The unbreathable atmosphere is about 4 times as dense as that of Earth.

The second most abundant gas on Titan is methane, a major component of natural gas. Sunlight interacting with methane induces chemical reactions that produce a variety of other carbon-hydrogen compounds, or **hydrocarbons.** For example, spacecraft have detected small amounts of ethane (C_2H_6), acetylene (C_2H_2), ethylene (C_2H_4), and propane (C_3H_8) in Titan's atmosphere.

Ethane, the most abundant of these compounds, condenses into droplets as it is produced and falls to Titan's surface to form a liquid. Nitrogen combines with various hydrocarbons to produce other compounds that can exist as liquids on Titan. Although one of these compounds, hydrogen cyanide (HCN), is a poison, some of the others are the building blocks of life's organic molecules. Other compounds can form intriguing liquids on Titan: Some molecules can join together in long, repeating molecular chains to form substances called **polymers,** of which rubber, cellulose, and plastics are the best known examples. Many of the hydrocarbons and carbon-nitrogen compounds in Titan's atmosphere can form such polymers. Scientists hypothesize that droplets of lighter polymers remain suspended in Titan's atmosphere to form a mist, while heavier polymer particles settle down onto Titan's surface.

Using the Keck I telescope in 1999, astronomers observed infrared emissions from Titan's surface consistent with a partially liquid surface. Images of the surface taken by the *Cassini* spacecraft (Figure 8-27), which arrived in the Saturn system in 2004, show a lack of craters, indicating that Titan's surface undergoes dynamic change. Radio and optical evidence indicates that the surface is covered with hydrocarbons and the dark regions may be the long-sought lakes or oceans of hydrocarbons. A probe, named Huygens, recently landed on Titan and information about what it has discovered is available on the Web.

We have little reason to suspect that life exists on Titan, because its surface temperature of 95 K (–288°F) is prohibitively cold. Nevertheless, the material on Titan's surface may be similar to what was on the young Earth and so a more detailed study of the chemistry of Titan may shed light on the origins of life on Earth.

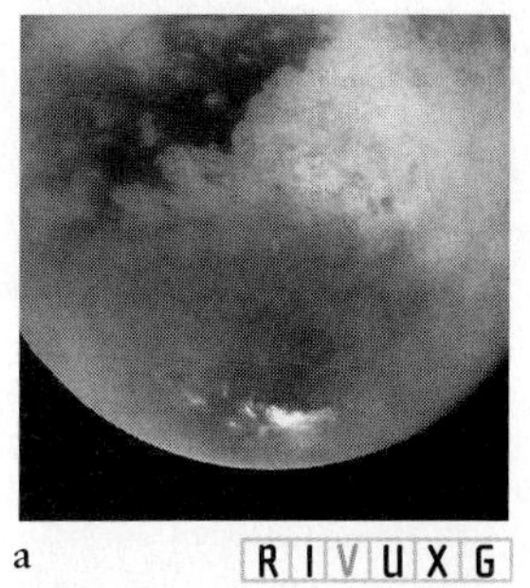

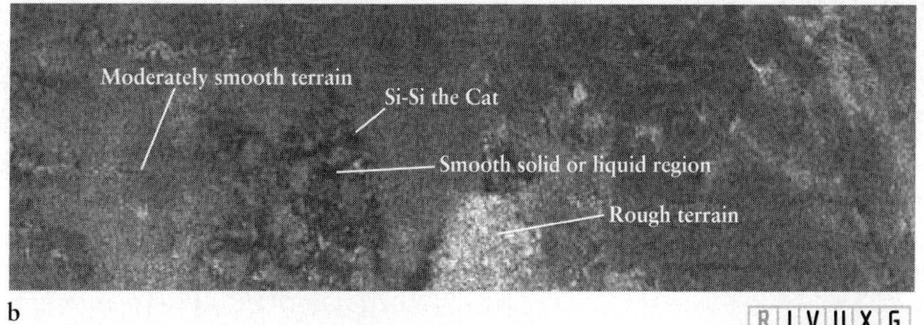

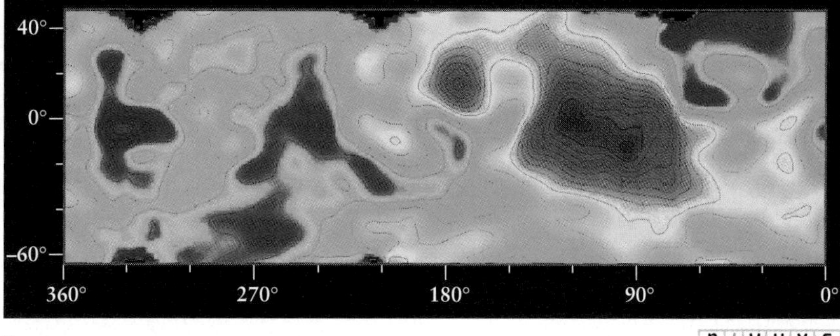

FIGURE 8-27 **Surface Features on Titan** (a) This *Cassini* image shows lighter highlands called Xanadu and dark, flat, lowlands that may be hydrocarbon seas. Resolution is 4.2 km (2.6 mi). (b) This radio image shows a region 250 km by 478 km (155 m × 297 mi) with resolution of 300 m (984 ft). That cat-shaped image may be a hydrocarbon lake peppered with islands. (c) This map shows how well various features on Titan's surface reflect infrared radiation. The darkest features reflect deep blue and the increasingly lighter features are light blue, green, yellow, red, and dark red, respectively. The red feature is Xanadu. (d) The Huygens lander took this image at Titan's surface on Jan 14., 2005. Blocks of ice are strewn around the landscape. (a-c: NASA/JPL/Space Science Institute; d: ESA/NASA/University of Arizona)

URANUS: VITAL STATISTICS

Average distance from Sun:	19.2 AU = 2.87 × 10^9 km
Maximum distance from Sun:	20.0 AU = 3.00 × 10^9 km
Minimum distance from Sun:	18.4 AU = 2.75 × 10^9 km
Eccentricity of orbit:	0.047
Average orbital speed:	6.83 km/s
Orbital period:	84.1 years
Rotation period (internal):	17.24 hours
Inclination of equator to orbit:	97.9°
Inclination of orbit to ecliptic:	0.77°
Radius:	25,560 km = 4.01 Earth radii (equatorial)
Mass:	8.68 × 10^{25} kg = 14.5 Earth masses
Average density:	1318 kg/m^3
Escape speed:	21.3 km/s
Surface gravity (Earth = 1):	0.90
Albedo:	0.56
Average temperature at cloudtops:	−218°C = −360°F = 55 K
Atmospheric composition (by number of molecules):	82.5% hydrogen (H_2), 15.2% helium (He), 2.3% methane (CH_4)

NEPTUNE: VITAL STATISTICS

Average distance from Sun:	30.1 AU = 4.50 × 10^9 km
Maximum distance from Sun:	30.4 AU = 4.54 × 10^9 km
Minimum distance from Sun:	29.8 AU = 4.45 × 10^9 km
Eccentricity of orbit:	0.009
Average orbital speed:	5.5 km/s
Orbital period:	164.8 years
Rotation period (internal):	16.11 hours
Inclination of equator to orbit:	29.6°
Inclination of orbit to ecliptic:	1.77°
Radius:	24,764 km = 3.88 Earth radii (equatorial)
Mass:	1.024 × 10^{26} kg = 17.1 Earth masses
Average density:	1638 kg/m^3
Escape speed:	23.5 km/s
Surface gravity (Earth = 1):	1.1
Albedo:	0.51
Average temperature at cloudtops:	−218°C = −360°F = 55 K
Atmospheric composition (by number of molecules):	79% hydrogen (H_2), 18% helium (He), 3% methane (CH_4)

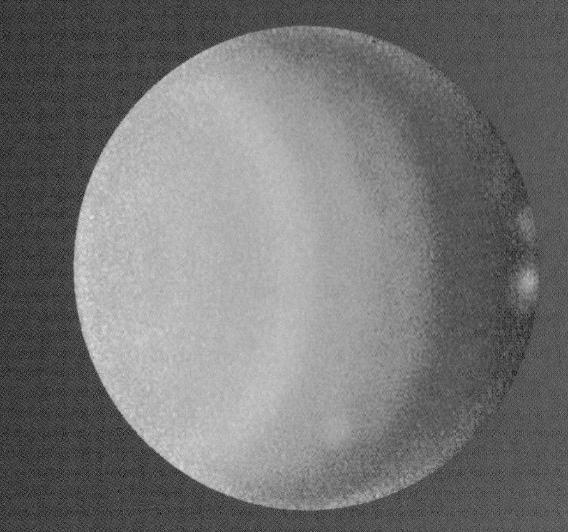

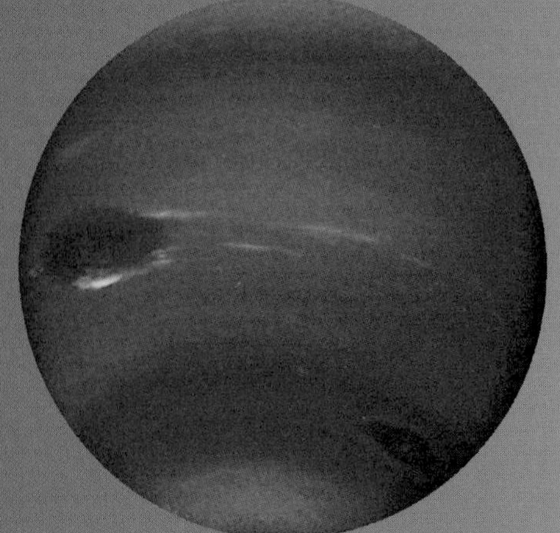

FIGURE 8-28 R I V U X G R I V U X G R I V U X G

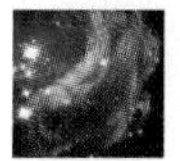

◀ **FIGURE 8-28 Uranus, Earth, and Neptune** These images of Uranus, the Earth, and Neptune are to the same scale. Uranus and Neptune are quite similar in mass, size, and chemical composition. Both planets are surrounded by thin, dark rings, quite unlike Saturn's, which are broad and bright. The clouds on the right of Uranus are each the size of Europe. (NASA)

URANUS

Uranus (pronounced YOUR-uh-nus), so far away that it is virtually invisible to the naked eye, was the first planet system discovered by a telescopic search of the heavens. It is 19.2 AU from the Sun, and it takes Uranus 84 Earth years to orbit the Sun. Since its discovery in 1781, Uranus has orbited the Sun fewer than three times.

8-12 Uranus sports a hazy atmosphere and clouds

Until early 1996, observations of Uranus, the fourth most massive planet in our solar system, revealed few notable features in the visible part of the spectrum. Even the 1986 visit of *Voyager 2* to Uranus showed a remarkably featureless world. It took the Hubble Space Telescope's infrared camera to find what *Voyager*'s visible light camera could not: Uranus has a system of belts and zones. Its hydrogen atmosphere has traces of methane along with a high-altitude haze under which are clear air and ever-changing methane clouds that dwarf the typical cumulus clouds we see on Earth. Uranus's clouds are towering and huge, each typically as large as Europe (Figure 8-28). Along with the rest of the atmosphere, the clouds go around the planet once every 16½ hours.

Uranus contains 14½ times as much mass as the Earth and is 4 times bigger in diameter (see Figure 8-28). Its outer layers are composed predominantly of gaseous hydrogen and helium. The temperature in the upper atmosphere of the planet is so low (about 73 K, or –330°F) that the methane and water there condense to form clouds of ice crystals. Because methane freezes at a lower temperature than water, methane forms higher clouds over Uranus. Methane efficiently absorbs red light, giving Uranus its blue-green color.

Earth-based observations show that Uranus rotates once every 17 hours 14 minutes. Its rotation axis lies very nearly in the plane of its orbit around the Sun (Figure 8-29). Put another way, recall that Earth's rotation axis is tilted

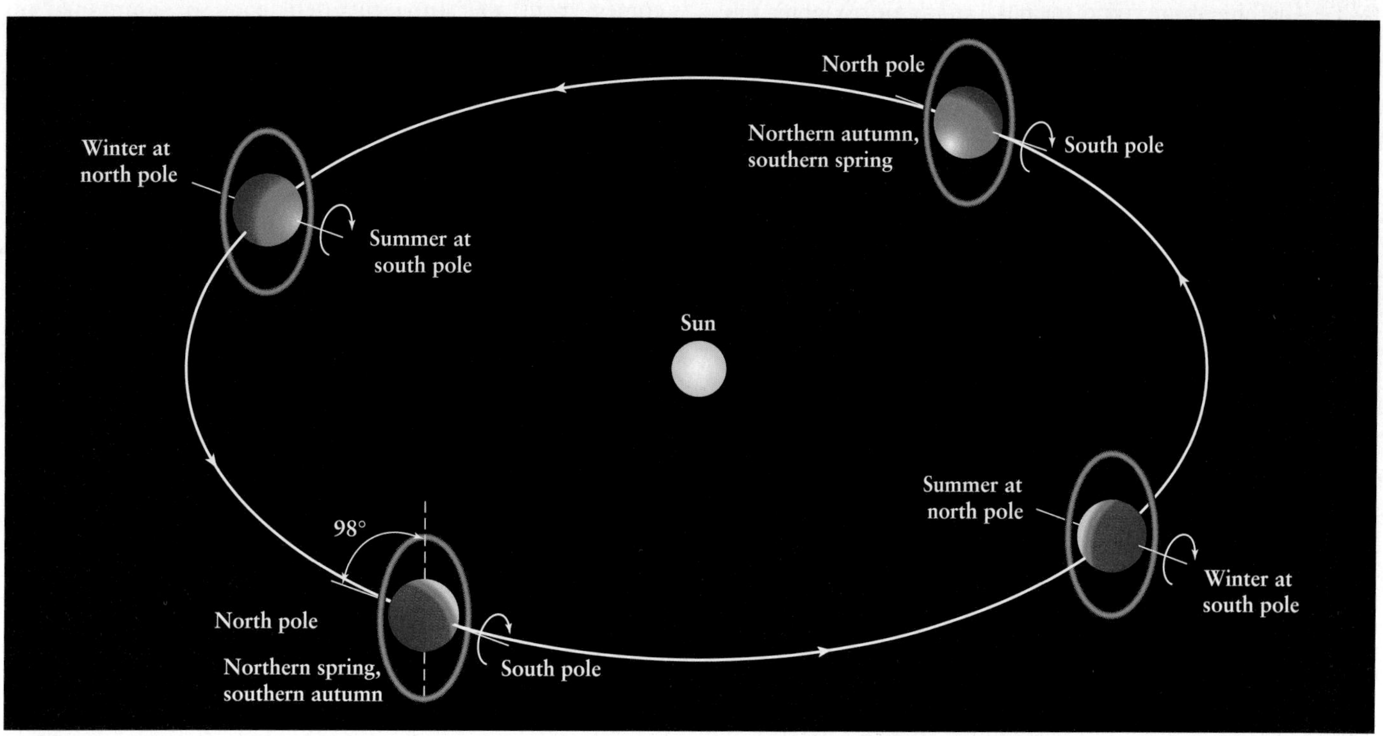

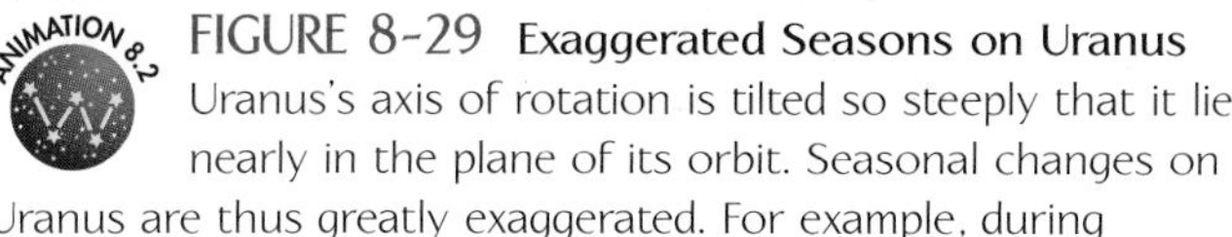

FIGURE 8-29 Exaggerated Seasons on Uranus Uranus's axis of rotation is tilted so steeply that it lies nearly in the plane of its orbit. Seasonal changes on Uranus are thus greatly exaggerated. For example, during midsummer at Uranus's south pole, the Sun appears nearly overhead for many Earth years, while the planet's northern regions are subjected to a long, continuous winter night. Half an orbit later, the seasons are reversed.

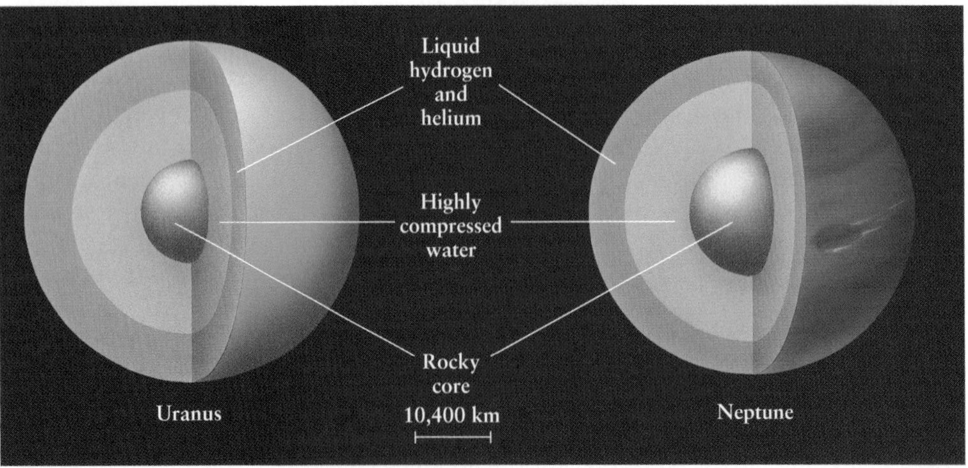

FIGURE 8-30 Cutaways of the Interiors of Uranus and Neptune The interiors of both Uranus and Neptune are believed to have three regions: a terrestrial core surrounded by a liquid water mantle, which is surrounded in turn by liquid hydrogen and helium. Their atmospheres are thin layers at the top of their hydrogen and helium layers.

about 23½° from being perpendicular to the ecliptic. Uranus's rotation axis is inclined 98° from a line perpendicular to its plane of orbit (see lower left drawing in Figure 8-29). Therefore, its rotation axis lies within 8° of the plane of orbit.

Because its north pole points below the plane of its orbit around the Sun, Uranus is one of only three planets with retrograde rotation (Venus and Pluto are the other two). As Uranus orbits the Sun, its north and south poles alternately point almost directly toward or directly away from the Sun, producing exaggerated seasons (see Figure 8-29). In the summertime, near Uranus's north pole, the Sun is almost directly overhead for many Earth years, at which time southern latitudes are subjected to a continuous, frigid winter night. Forty-two Earth years later, the situation is reversed. For more on this, see the essay "What If . . . Earth's Axis Lay on the Ecliptic?" on page 35.

From *Voyager* photographs, planetary scientists have concluded that each of the five largest Uranian moons has probably had at least one shattering impact. A catastrophic collision with an Earth-sized object may also have knocked Uranus on its side, as we see it today.

From its mass and density (1318 kg/m^3), astronomers conclude that Uranus's interior has three layers. The outer 30% of the planet is liquid hydrogen and helium, the next 40% inward is highly compressed liquid water (with some methane and ammonia), and the inner 30% is a rocky core (Figure 8-30). Indirect evidence for the water layer comes from the apparent deficiency of ammonia on Uranus. This gas dissolves easily in water, so an ocean would explain the scarcity of ammonia in the planet's atmosphere.

Uranus's surface magnetic field is about three-quarters that of the Earth. That strength is reasonable, considering

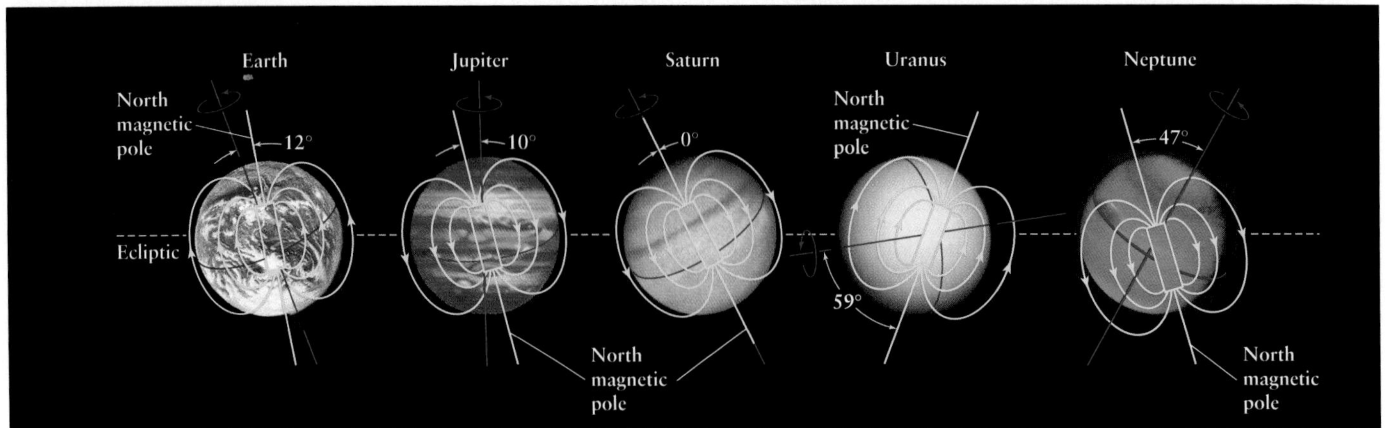

FIGURE 8-31 The Magnetic Fields of Five Planets This drawing shows how the magnetic fields of Earth, Jupiter, Saturn, Uranus, and Neptune are tilted relative to their rotation axes. Note that the magnetic fields of Uranus and Neptune are offset from the centers of the planets and steeply inclined to their rotation axes. Jupiter, Saturn, and Neptune have north magnetic poles on the hemisphere where Earth has its south magnetic pole.

the planet's mass and rotation rate, but everything else about the magnetic field is extraordinary. The field is remarkably tilted—59° from its axis of rotation—and its axis does not even pass through the center of the planet (Figure 8-31).

Because of the large angle between the magnetic field of Uranus and its rotation axis, the magnetosphere of Uranus wobbles considerably as the planet rotates. Such a rapidly changing magnetic field will help us in explaining pulsars, a type of star we will study in Chapter 13.

8-13 A system of rings and satellites revolves around Uranus

Uranus has 10 known rings, each divided into ringlets. The first nine of the thin, dark rings were discovered accidentally in 1977, when Uranus passed in front of a star. The star's light was momentarily blocked by each ring, thereby revealing their existence to astronomers (Figure 8-32). Blocking the light of a more distant object, such as the star here, by something between it and us, such as Uranus's rings, is called an **occultation.** A picture taken while *Voyager* was in Uranus's shadow revealed the tenth ring (Figure 8-33).

Most of Uranus's moons, like its rings, orbit in the plane of the planet's equator. Five of these satellites, ranging in diameter from 480 to nearly 1600 km, were known before the *Voyager* mission. However, *Voyager*'s cameras discovered 10 additional satellites, each smaller than 50 km across. Still others have since been observed from Earth. Several of these tiny, irregularly shaped moons are shepherd satellites whose gravitational pull confines the particles within the thin rings that circle Uranus.

The smallest of Uranus's five main satellites, Miranda, is the most fascinating and bizarre of its 27 known moons. Unusual wrinkled and banded features cover Miranda's surface (Figure 8-34). Its highly varied terrain suggests that it was once seriously disturbed. Perhaps a shattering impact temporarily broke it into several pieces that then recoalesced, or perhaps severe tidal heating, as we saw on Io, moved large pieces of its surface.

Miranda's core originally consisted of dense rock, while its outer layers were mostly ice. If a powerful impact did occur, blocks of debris broken off from Miranda drifted back together through mutual gravitational attraction. Recolliding with that moon, they formed a chaotic mix of rock and ice. In this scenario, the landscape we see today on Miranda is the result of huge, dense rocks trying to settle toward the satellite's center, forcing blocks of less dense ice upward toward the surface.

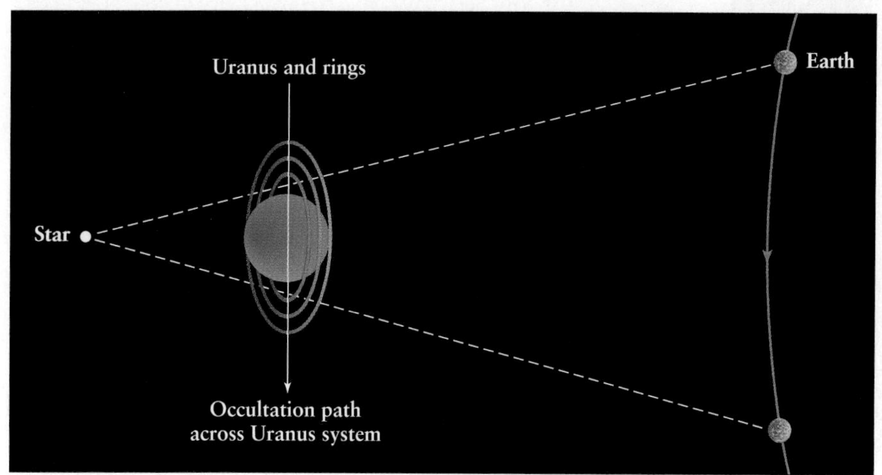

a

FIGURE 8-32 **Discovery of the Rings of Uranus** (a) Light from a star is reduced as the rings move in front of it. (b) With sensitive light detectors, astronomers can detect the variation in light intensity. Such dimming led to the discovery of Uranus's rings. Of course, the star vanishes completely when Uranus occults it.

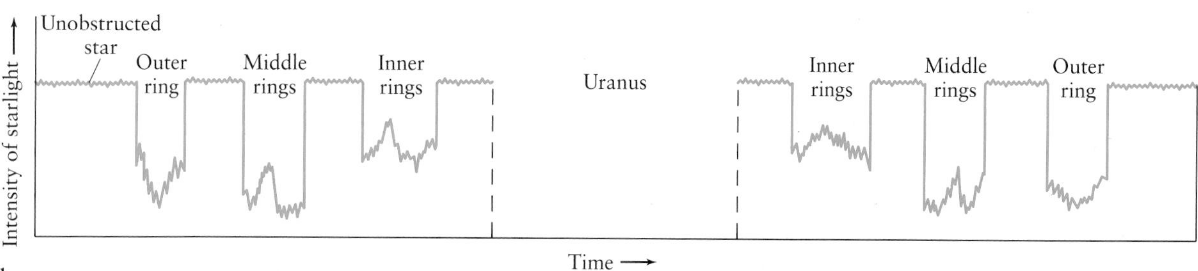

b

FIGURE 8-33 **The Rings and Moons of Uranus** This image of Uranus, its rings, and eight of its moons was taken by the Hubble Space Telescope. **Inset:** Close-up of part of the ring system taken by *Voyager 2* when the spacecraft was in Uranus's shadow looking back toward the Sun. Numerous fine dust particles between the main rings gleam in the sunlight. Uranus's rings are much darker than Saturn's, and this long exposure revealed many very thin rings and dust lanes. The short streaks are star images blurred because of the spacecraft's motion during the exposure. (Inset: NASA)

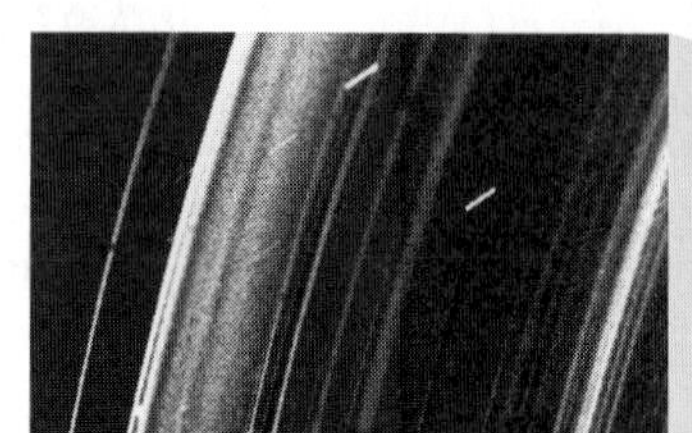

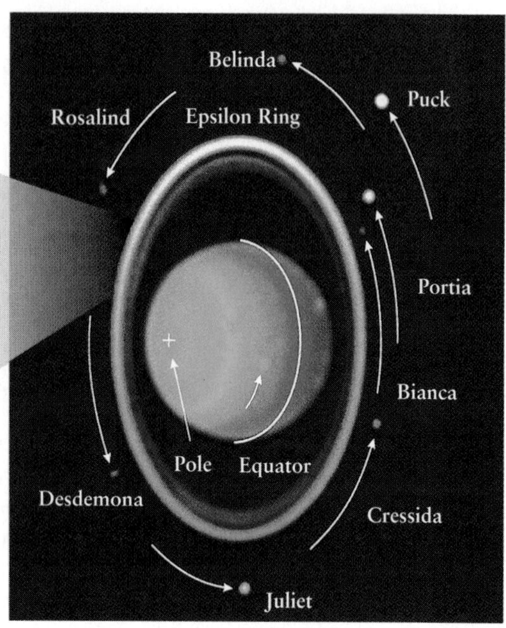

R I V U X G

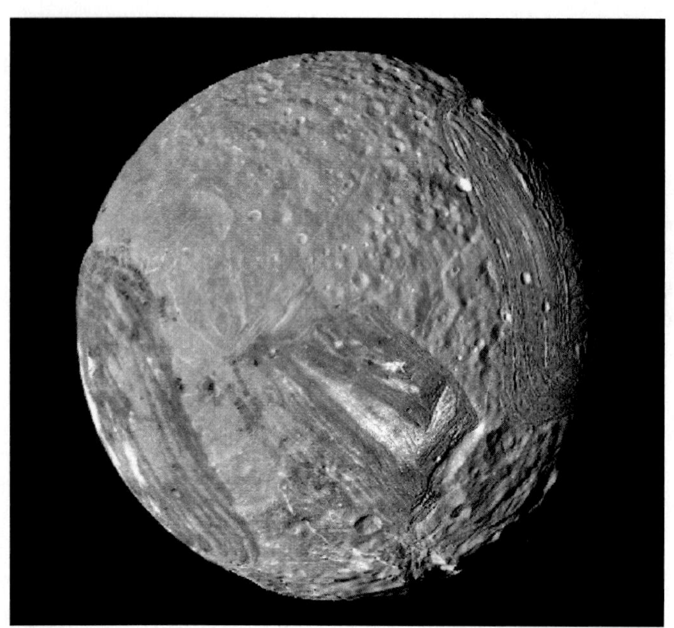

R I V U X G

FIGURE 8-34 **Miranda** The patchwork appearance of Miranda in this mosaic of *Voyager 2* images suggests that this satellite consists of huge chunks of rock and ice that came back together after an ancient, shattering impact by an asteroid or a neighboring Uranian moon. The curious banded features that cover much of Miranda are parallel valleys and ridges that may have formed as dense, rocky material sank toward the satellite's core. At the very bottom of the image—where a "bite" seems to have been taken out of the satellite—is a range of enormous cliffs that jut upward as high as 20 km, twice the height of Mount Everest. (NASA)

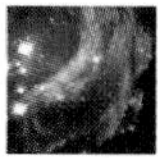

NEPTUNE

Neptune is physically similar to Uranus (review Figures 8-28 and 8-30). Neptune has 17.1 times the Earth's mass, 3.88 times Earth's diameter, and a density of 1640 kg/m^3. Its rapid rotation draws its clouds into belts and zones. Unlike Uranus, however, cloud features can readily be discerned on Neptune. Its whitish, cirruslike clouds consist of methane ice crystals. As on Uranus, the methane absorbs red light, leaving the planet's belts and zones with a banded, bluish appearance (Figure 8-35). Like the other giant planets, the atmosphere of Neptune experiences differential rotation. The winds on Neptune blow at speeds up to 2000 km/h—among the fastest known in the solar system. Neptune's rotation axis is tilted nearly 30 degrees from the plane of its orbit around the Sun. Recall that the tilt of Earth's axis creates our seasons. Hubble Space Telescope observations of Neptune in 2003 showed seasonal changes near its poles.

8-14 Neptune was discovered because it had to be there

Neptune's discovery is storied because it illustrates a scientific prediction leading to an expected discovery. In 1781, the British astronomer William Herschel discovered Uranus. Its position was carefully plotted, and by the 1840s, it was clear that even considering the gravitational effects of all the known bodies in the solar system, Uranus

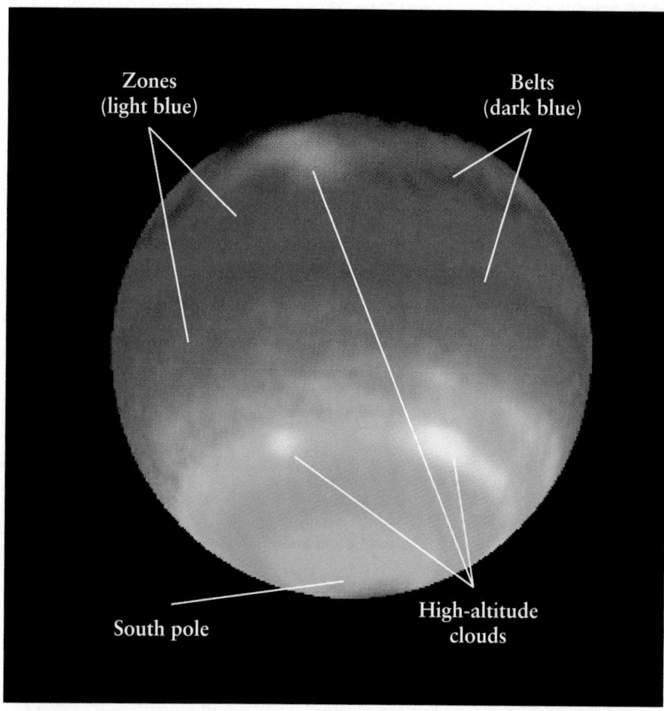

R I V U X G

WEB LINK 8.23

FIGURE 8-35 Neptune's Banded Structure Several Hubble Space Telescope images at different wavelengths were combined to create this enhanced-color view of Neptune. The dark blue and light blue areas are the belts and zones, respectively. The dark belt running across the middle of the image lies just south of Neptune's equator. White areas are high-altitude clouds, presumably of methane ice. The very highest clouds are shown in yellow-red, as seen at the very top of the image. The green belt near the south pole is a region where the atmosphere absorbs blue light, probably indicating some differences in chemical composition. (Lawrence Sromovsky, University of Wisconsin-Madison and STScI/NASA)

was not following the path predicted by Newton's and Kepler's laws. Either the theories behind these laws were wrong, or there had to be another, yet-to-be-discovered body in the solar system pulling on Uranus.

Independent, nearly simultaneous calculations by an English mathematician, John Adams, and a French astronomer, Urbain Leverrier, predicted the same location for the alleged planet. That planet, Neptune, was located in 1846 by the German astronomer Johann Galle, within a degree or two of where it had to be in order to have the observed influence on Uranus.

In August 1989, nearly 150 years after its discovery, *Voyager 2* arrived at Neptune to cap one of NASA's most ambitious and successful space missions. Scientists were overjoyed at the detailed, close-up pictures and wealth of data about Neptune sent back to Earth by the spacecraft.

INSIGHT INTO SCIENCE

Process and Progress Scientific theories enable scientists to make testable predictions (see Chapter 1). Based on the details of Uranus's orbit around the Sun, Newton's law of gravitation predicted that Uranus's orbit was being affected by the gravitational attraction of another planet. The law also predicted where that planet was located, leading to the discovery of Neptune.

At the time *Voyager 2* passed it, a giant storm raged in Neptune's atmosphere. Called the **Great Dark Spot,** it was about half as large as Jupiter's Great Red Spot. The Great Dark Spot (Figure 8-36) was located at about the same latitude on Neptune and occupied a similar proportion of Neptune's surface as the Great Red Spot does on Jupiter. Although these similarities suggested that similar mechanisms created the spots, the Hubble Space Telescope in 1994 showed that the Great Dark Spot had disappeared.

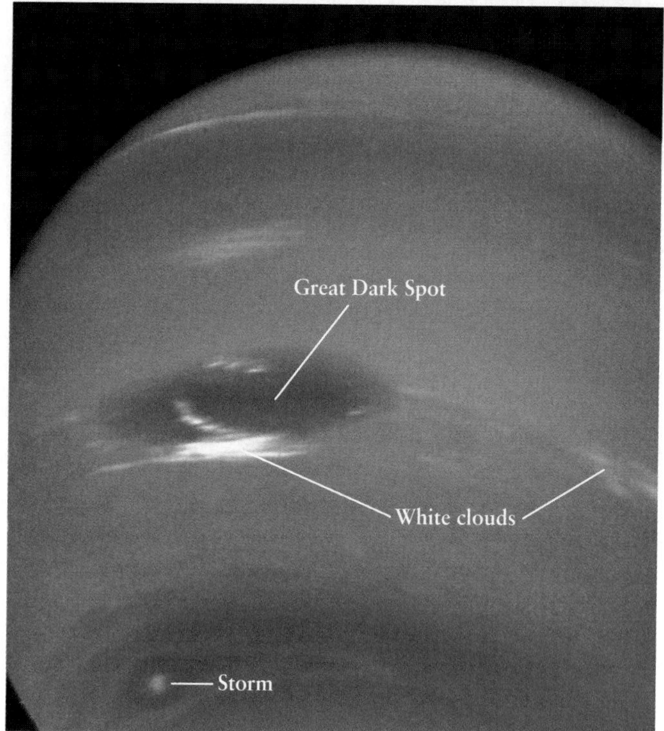

R I V U X G

VIDEO 8.8

FIGURE 8-36 Neptune This view from *Voyager 2* looks down on the southern hemisphere of Neptune. The Great Dark Spot's longer dimension at the time was about the same size as the Earth's diameter. It has since vanished. Note the white, wispy methane clouds. (NASA/JPL)

Then, in April 1995, another spot developed in the opposite hemisphere.

Neptune's interior is believed to be very similar in composition and structure to that of Uranus: a rocky core surrounded by ammonia- and methane-laden water (see Figure 8-30). Neptune's surface magnetic field is about 40 percent that of the Earth. Also, as with Uranus, Neptune's magnetic axis, the line connecting its north and south magnetic poles, is tilted sharply from its rotation axis. In this case, the tilt is 47%. Like Uranus, Neptune's magnetic axis does not pass through the center of the planet (see Figure 8-31).

We saw in Section 8-2 that the magnetic fields of Jupiter and Saturn are believed to be generated by the motions of their liquid metallic hydrogen. However, Uranus and Neptune lack this material, and their magnetic fields have a different origin. These fields are believed to exist because molecules such as ammonia dissolved in their water layers lose electrons (become ionized). These ions, moving with the planets' rotating, fluid interiors, create the same dynamo effect that produces the magnetic field.

FIGURE 8-37 **Neptune's Rings** Two main rings are easily seen in this view alongside overexposed edges of Neptune. In taking this image, the bright planet was hidden so that the dim rings would be visible, hence the black rectangle running down the center of the figure. Careful examination also reveals a faint inner ring. A fainter-still sheet of particles, whose outer edge is located between the two main rings, extends inward toward the planet. (NASA)

8-15 Neptune has rings and has captured most of its moons

Like Uranus, Neptune is surrounded by a system of thin, dark rings (Figure 8-37). It is so cold at these distances from the Sun that both planets' ring particles retain methane ice. Scientists speculate that eons of radiation damage have converted this methane ice into darkish carbon compounds, thus accounting for the low reflectivity of the rings.

Neptune has 13 known moons. Twelve have irregular shapes and highly elliptical orbits, which suggest that Neptune captured them. Triton, discovered in 1846, is spherical and was quickly observed to have a nearly circular, *retrograde* orbit around Neptune. It is difficult to imagine how a planet rotating one way and a satellite revolving the other way could form together. Indeed, only the small outer satellites of Jupiter and Saturn have retrograde orbits, and these bodies are probably captured asteroids. Some scientists have therefore suggested that Triton may have been captured 3 or 4 billion years ago by Neptune's gravity. Upon being captured, Triton was most likely in a highly elliptical orbit. However, the tides that the moon creates on Neptune's liquid surface would in turn have made Triton's orbit more circular.

Triton's average density of about 2100 kg/m^3 indicates that it is an equal mix of rock and ice. It has a thin nitrogen atmosphere. When Triton was younger, the tidal force created on it by Neptune due to the moon's initially elliptical orbit would have caused Triton to stretch and flex, providing enough energy to melt much of its interior and obliterate its original surface features, including craters. Triton's south polar region is shown in Figure 8-38. Note that very few craters are visible. Calculations based on these observations indicate that Triton's present surface is about 100 million years old.

Triton does exhibit some surface features seen on other icy worlds, such as long cracks resembling those on Europa and Ganymede. Other features unique to Triton are quite puzzling. For example, the top part of Figure 8-38 reveals a wrinkled terrain that resembles the skin of a cantaloupe. Triton also has a few frozen lakes like the one shown in Figure 8-39. Some scientists have speculated that these lakelike features are the calderas of extinct ice volcanoes. A mixture of methane, ammonia, and water, which can have a melting point far below that of pure water, could have formed a kind of cold lava on Triton.

Voyager instruments measured a surface temperature of 36 K (–395°F), making Triton the coldest world that our probes have ever visited. Nevertheless, *Voyager* cameras did glimpse two towering plumes of gas extending up to 8 km above the satellite's surface. These are apparently plumes of nitrogen gas warmed by interior radioactive decay and escaping through vents or fissures.

Triton continues to create tides on Neptune. Whereas the tides on Earth cause our Moon to spiral outward, the tides on Neptune cause Triton (in its retrograde orbit) to spiral inward. Within the next quarter of a billion years, Triton will reach the **Roche limit,** the distance at which a planet creates tides on its moon's solid surface high enough to pull its moon apart. Pieces of Triton will then

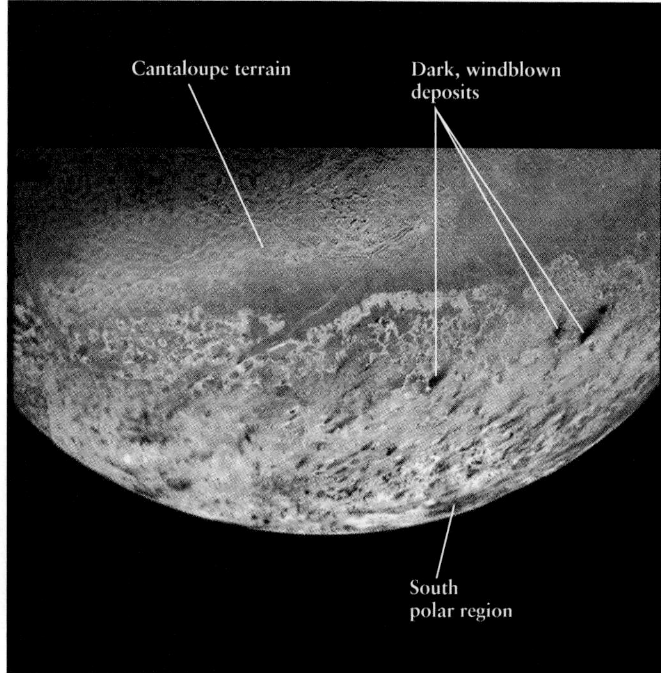

FIGURE 8-38 **Triton's South Polar Cap** Approximately a dozen high-resolution *Voyager 2* images were combined to produce this view of Triton's southern hemisphere. The pinkish polar cap is probably made of nitrogen frost. A notable scarcity of craters suggests that Triton's surface was either melted or flooded by icy lava after the era of bombardment that characterized the early history of the solar system. (NASA/JPL)

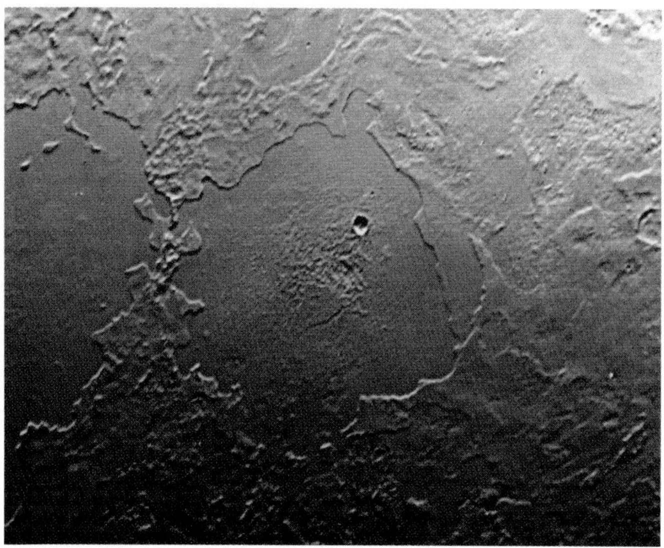

R I V U X G

FIGURE 8-39 **A Frozen Lake on Triton** Scientists believe that the feature in the center of this image is a basin filled with water ice. The flooded basin is about 200 km across. (NASA)

R I V U X G

FIGURE 8-40 **Discovery of Pluto** Pluto was discovered in 1930 by searching for a dim, starlike object that slowly moves against the background stars. These two photographs were taken one day apart. (UC Regents/Lick Observatory)

literally float into space until the entire moon is demolished! By destroying Triton, Neptune will create a new ring system that will be much more substantial than its present one.

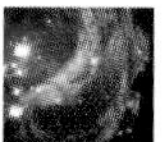

PLUTO AND BEYOND

In 1930, the astronomer Clyde W. Tombaugh was searching for a giant planet that was allegedly affecting Neptune's orbit. What he found was Pluto (Figure 8-40), a planet far too small to have a noticeable gravitational effect on Neptune. Figure 8-41 lists Pluto's vital statistics. Refined observations show that Neptune's orbit is not being affected by

PLUTO: VITAL STATISTICS

Average distance from Sun:	39.5 AU = 5.91×10^9 km
Maximum distance from Sun:	49.4 AU = 7.39×10^9 km
Minimum distance from Sun:	29.7 AU = 4.44×10^9 km
Eccentricity of orbit:	0.250
Average orbital speed:	4.7 km/s
Orbital period:	248.60 years
Rotation period:	6.39 days (retrograde)
Inclination of equator to orbit:	118°
Inclination of orbit to ecliptic:	17.1°
Radius:	about 1190 km = 0.18 Earth radius
Mass:	1.3×10^{22} kg = 0.0021 Earth mass
Average density:	1750 kg/m^3
Escape speed:	1.2 km/s
Surface gravity (Earth = 1):	0.07
Albedo:	0.50
Average temperature at cloudtops:	−233°C = −387°F = 40 K

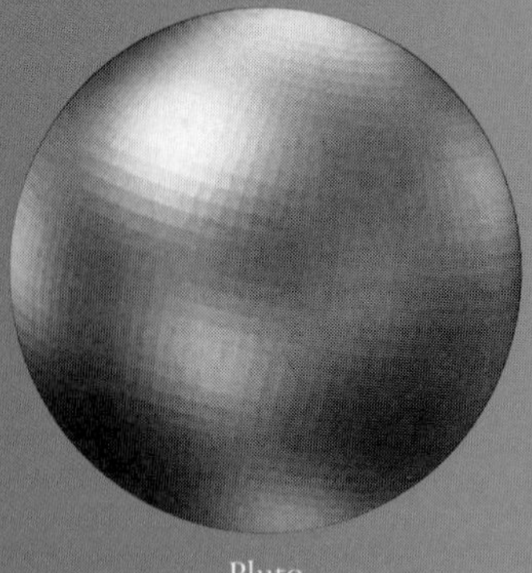

Pluto

R I V U X G

FIGURE 8-41 **Pluto and Its Vital Statistics** This Hubble Space Telescope image of Pluto shows little detail but indicates that the major features of Pluto's surface each cover large amounts of its surface area. (Alan Stern, Southwest Research Institute; Marc Buie, Lowell Observatory; NASA; and ESA)

another giant planet. No additional giant planet has ever been found. Recall from Section 5-4 that bodies beyond Pluto have been found, including Sedna, which is nearly Pluto's size. However, these outer bodies in the solar system are so small and have such low masses that they are not considered to be planets. (Indeed, Pluto may someday be declassified as a planet, but for now it is the farthest planet in the solar system.)

Tombaugh recognized Pluto as a planet because it moved among the background stars from night to night, but he had no idea how strange its orbit is compared with those of the other planets. As shown in Figure 5-7, Pluto's orbit is so elliptical that it is sometimes closer to the Sun than Neptune, as it was from 1979 to 1999. It is now farther away from the Sun than Neptune and will continue to be so for about the next 230 years. Pluto's orbit is tilted with respect to the plane of the ecliptic more than any other planet.

8-16 Pluto and its moon, Charon, are about the same size

Pluto was little understood for half a century until astronomers noticed that its image sometimes appears oblong (Figure 8-42). This observation led to the discovery in 1978 of Pluto's only known moon, Charon (pronounced KAR-en, after the mythical boatman who ferried souls across the River Styx to Hades, the domain ruled by Pluto). From 1985 through 1990, the orbit of Charon was oriented so that Earth-based observers could watch it eclipse Pluto. Astronomers used observations of these eclipses to determine that Pluto is only twice as broad as its satellite: Its diameter is 2380 km and Charon's is 1190 km.

The average distance between Charon and Pluto is less 6
than 1/20 the distance between the Earth and our Moon. Furthermore, Pluto always keeps the same side facing Charon.

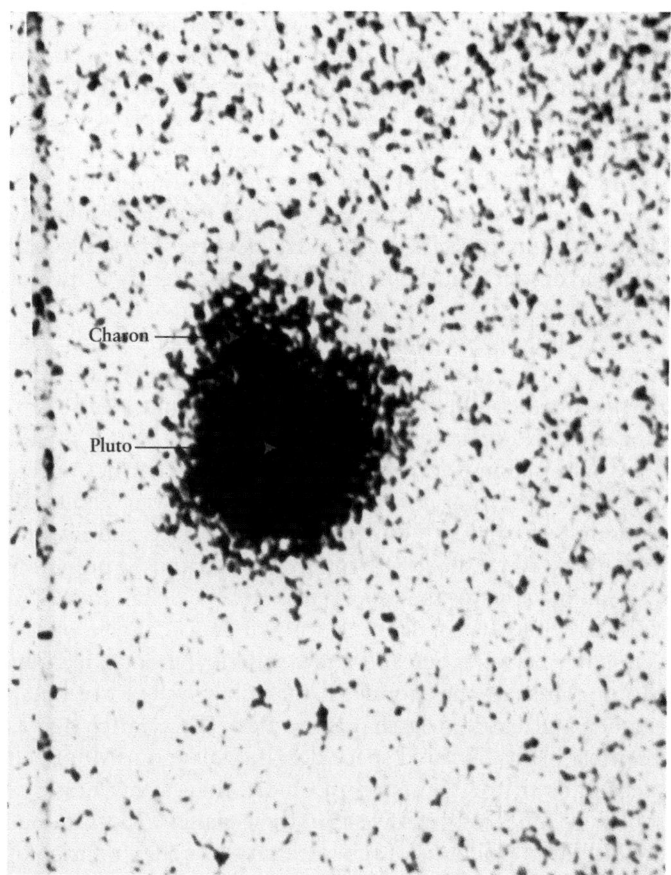

FIGURE 8-42 **Discovery of Charon** Long ignored as just a defect in the photographic emulsion, the bump on the upper left side of this image of Pluto led astronomer James Christy to discover the moon Charon. (U.S. Naval Observatory)

VIDEO 8.12

This is the only case in which a moon and a planet both have synchronous rotation with respect to each other. As seen from the satellite-facing side of Pluto, Charon neither rises nor sets, but instead hovers in the sky, perpetually suspended above the horizon.

The exceptional similarities between Pluto and Charon suggest that this binary system may have formed when Pluto collided with a body of similar size. Perhaps chunks of matter were stripped from this second body, leaving behind a mass, now called Charon, that was vulnerable to capture by Pluto's gravity. Alternatively, perhaps Pluto's gravity captured Charon into orbit during a close encounter between the two worlds.

Both of these are unlikely scenarios. For either to be feasible, many Pluto-sized objects must have existed in the outer regions of the young solar system. One astronomer estimates that there must have been at least a thousand Pluto-sized bodies in order for a collision or close encounter between two of them to have occurred at least once since the solar system formed 4.6 billion years ago.

The best pictures of Pluto and Charon were taken by the Hubble Space Telescope (Figure 8-43). Like Neptune's moon Triton, these worlds are probably composed of nearly equal amounts of rock and ice. The surface of Pluto shows more large-scale features than any object in the solar system other than the Earth. Such features on Earth include oceans and continents, while on Pluto they may include regions dominated by rock, regions dominated by ice, major impact regions, or mountains. We just do not know

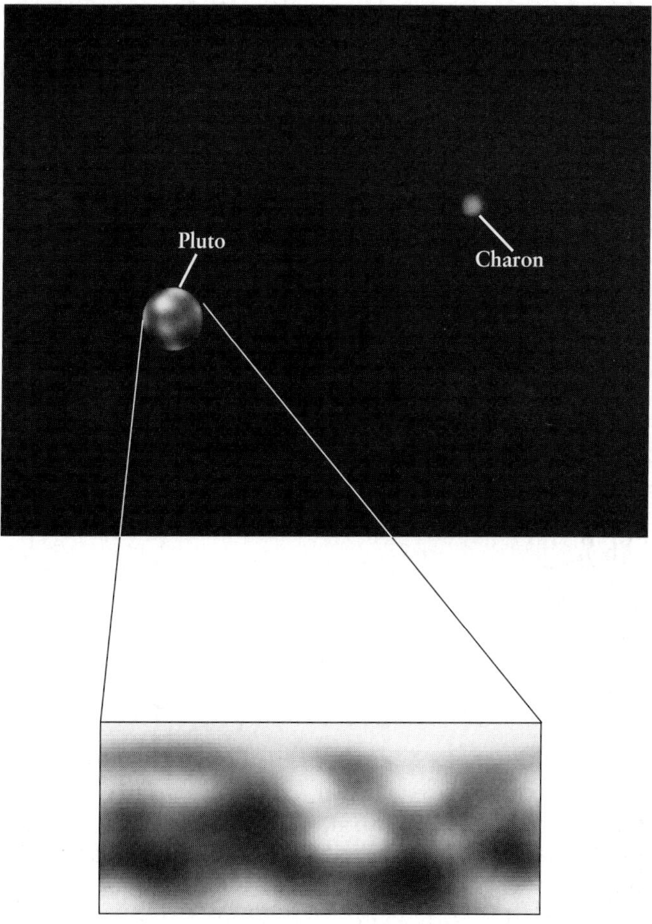

FIGURE 8-43 **Pluto and Charon** This picture of Pluto and its moon, Charon, is a composite of Hubble Space Telescope images. It shows the greatest detail we have of Pluto's surface. Pluto and Charon are separated here by only 19,700 km. **Inset:** A map covering 85% of Pluto's surface. The contrast between dark and light strongly suggests regions covered with ice and regions covered with rocky material. (Pluto: Alan Stern/SwRI, Mar Cuie/Lowell Observatory, NASA, & ESA; Charon: R. Albrecht, ESA/ESO Space Telescope European Coordinating Facility, and NASA; inset: A. Stern/SwRI, M. Buie/Lowell Observatories, NASA, and ESA)

WEB LINK 8.13

yet. Examination of Hubble Space Telescope images has revealed some 12 distinct regions, including polar ice caps. Studies of Pluto's surface temperature show that different regions vary by up to 25 K from each other.

Pluto's spectrum shows that its surface contains frozen nitrogen, methane, and carbon monoxide. The planet is also observed to have a very thin atmosphere of nitrogen and carbon monoxide. In contrast, Charon's surface appears to be covered predominantly with water ice.

Ever since Pluto was discovered, astronomers have searched for other objects in our solar system. Neptune's captured moons support the idea of other bodies at the outskirts of the solar system, and astronomers have begun discovering them by the hundreds. These distant masses of ice and rock, which we will discuss further in Chapter 9, lie in a region now called the Kuiper belt, which is believed to extend out 500 AU from the Sun.

INSIGHT INTO SCIENCE

Paradigm Shift Usually it takes overwhelming evidence contradicting established beliefs before scientists begin looking at old ideas in completely new ways. Such changes in fundamental beliefs are sometimes called *paradigm shifts*. The acceptance of tectonic plate motion is one such change. Until the past few years, astronomers took it for granted that Pluto is a planet. Perhaps another paradigm shift in the near future will reclassify tiny Pluto as the largest object in the Kuiper belt.

8-17 Comparative planetology of the outer planets

Now that we have examined the individual outer planets, it is instructive to compare their various properties. The four giant planets, Jupiter, Saturn, Uranus, and Neptune, are all much larger (roughly 4 to 11 times the Earth's diameter) and more massive (roughly 14 to 318 times the Earth's mass) than the Earth. Pluto, on the other hand, has only 0.2 times the Earth's diameter and 0.002 times the Earth's mass.

The four giant planets have thick hydrogen and helium-rich atmospheres that are permanently and completely covered with clouds, while Pluto has very little atmosphere compared to the Earth and no cloud cover. All the giant planets rotate rapidly compared with the Earth. Their sidereal rotation rates range from about 10 to about 17¼ hours. This rapid rotation draws their clouds into parallel bands called belts and zones. The Earth's slower rotation allows the clouds and winds here to roam over much greater range of latitudes than on the giant planets. Pluto rotates more than 6 times slower than the Earth.

All four giant planets have terrestrial bodies as their cores. That is where their internal similarities to the Earth end. Jupiter and Saturn have small amounts of "ice" surrounding their cores. These ices are surrounded by thick liquid metallic hydrogen layers, which are surrounded in turn by layers of normal hydrogen and helium. In contrast, Uranus's and Neptune's terrestrial cores are surrounded by large amounts of water, which are in turn surrounded by liquid hydrogen and helium. Pluto is apparently a mix of rock and ice.

The four giant planets all have magnetic fields that store particles like the Earth's Van Allen belts. Like the Earth's magnetic field, those of Jupiter, Uranus, and Neptune are tilted relative to their rotation axes. Saturn's magnetic field is along its rotation axis. It is unlikely, but uncertain, that Pluto has a magnetic field. The four giants all have rings. Saturn's rings are the most massive and distinctive, followed by those of Uranus, Jupiter, and Neptune.

Among them, the giant planets have at least 150 moons, while Pluto has one known moon. Each of the giants has between one and a few spherical moons, like our Moon. Most of the moons in the solar system are much smaller and irregularly shaped bodies. These latter bodies are most likely captured space debris. Unlike our Moon, all the moons of the giants are no more than a few hundred thousand times less massive than their planets. Most moons are millions or billions of times less massive than their planets. Recall that our Moon is 81 times less massive than the Earth. Charon, Pluto's known moon, is 7 times less massive than Pluto. Charon's revolution around Pluto takes the same length of time as Pluto's rotation. Therefore, Pluto and Charon are in synchronous rotation with respect to each other, the only planet–moon system in the solar system with this property. The Table, The Outer Planets: A Comparison, summarizes much of this material.

8-18 Frontiers yet to be discovered

The outer planets hold countless new insights into the formation and evolution of the solar system. Considering that most of the known extrasolar planets are Jupiterlike gas giants, our outer planets all have a lot to tell us about planets throughout our Galaxy. Why did the *Galileo* probe fail to detect the atmospheric structures believed to exist on Jupiter? How has the Great Red Spot persisted for so long? Is there life in the oceans of the outer Galilean moons? What, exactly, do the ring particles around each of the planets look like? How long ago did Saturn's ring system form? How long will it last? Are the moving spokes in Saturn's rings really caused by its magnetic field, and, if so, why do the spokes appear as they do? What caused Miranda's surface to become so profoundly disturbed? How did the Pluto–Charon system form? These are but a few of the questions about the outer planets that will be answered during this century.

THE OUTER PLANETS: A COMPARISON

	Interior	Surface	Rings	Atmosphere	Magnetic Field*
Jupiter	Terrestrial core, liquid metallic hydrogen shell, liquid hydrogen mantle	No solid surface, atmosphere gradually thickens to liquid state, belt and zone structure, hurricanelike features	Yes	Primarily H, He	19,000× Earth's total field; at its cloud layer, 14× stronger than Earth's surface field
Saturn	Similar to Jupiter, with bigger terrestrial core and less metallic hydrogen	No solid surface, less distinct belt and zone structure than Jupiter	Yes	Primarily H, He	570× Earth's total field; at its cloud layer, ⅔× Earth's surface field
Uranus	Terrestrial core, liquid water shell, liquid hydrogen and helium mantle	No solid surface, weak belt and zone system, hurricanelike features, color from methane absorption of red, orange, yellow	Yes	Primarily H, He, some CH_4	50× Earth's total field; at its cloud layer, 0.7× Earth's surface field
Neptune	Similar to Uranus	Like Uranus	Yes	Primarily H, He, some CH_4	35× Earth's total field; at its cloud layer, 0.4× Earth's surface field
Pluto	Unknown	Apparently rock and ice	No	Unknown	Unknown

For detailed numerical comparisons between planets, see Appendix Tables E-1 and E-2.

*To see the orientations of these magnetic fields relative to the rotation axes of the planets, see Figure 8-31.

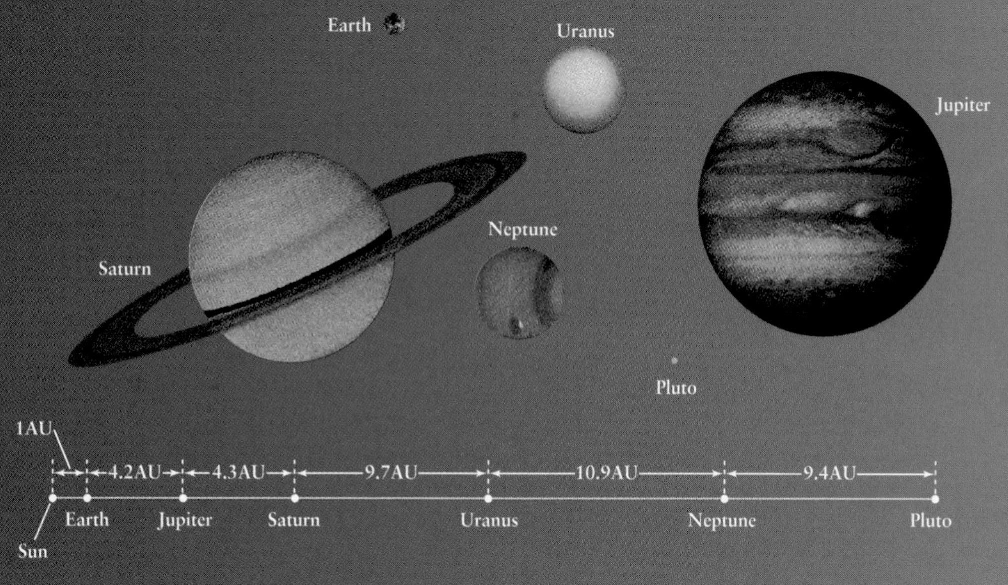

(Stephen P. Meszaros/NASA, M. Buie, K. Horne, and D. Tholen)

Summary of Key Ideas

Jupiter and Saturn

• Jupiter is by far the largest and most massive planet in the solar system.

• Jupiter and Saturn probably have rocky cores surrounded by a thick layer of liquid metallic hydrogen and an outer layer of ordinary liquid hydrogen. Both planets have an overall chemical composition very similar to that of the Sun.

• The visible features of Jupiter exist in the outermost 100 km of its atmosphere. Saturn has similar features, but they are much fainter. Three cloud layers exist in the upper atmospheres of both Jupiter and Saturn. Because Saturn's cloud layers extend through a greater range of altitudes, the colors of the Saturnian atmosphere appear muted.

• The colored ovals visible in the Jovian atmosphere are gigantic storms, some of which (such as the Great Red Spot) are stable and persist for years or even centuries.

• Jupiter and Saturn have strong magnetic fields created by electric currents in their metallic hydrogen layers.

• Four large satellites orbit Jupiter. The two inner Galilean moons, Io and Europa, are roughly the same size as our Moon. The two outer moons, Ganymede and Callisto, are approximately the size of Mercury.

• Io is covered with a colorful layer of sulfur compounds deposited by frequent explosive eruptions from volcanic vents. Europa is covered with a smooth layer of frozen water crisscrossed by an intricate pattern of long cracks.

• The heavily cratered surface of Ganymede is composed of frozen water with large polygons of dark, ancient crust separated by regions of heavily grooved, lighter-colored, younger terrain. Callisto has a heavily cratered ancient crust of frozen water.

• Saturn is circled by a system of thin, broad rings lying in the plane of the planet's equator. Each major ring is composed of a great many narrow ringlets consisting of numerous fragments of ice and ice-coated rock. Jupiter has a much less substantial ring system.

Uranus and Neptune

• Uranus and Neptune are quite similar in appearance, mass, size, and chemical composition. Each has a rocky core surrounded by a thick, watery mantle; the axes of their magnetic fields are steeply inclined to their axes of rotation; and both planets are surrounded by systems of thin, dark rings.

• Uranus is unique in that its axis of rotation lies nearly in the plane of its orbit, producing greatly exaggerated seasons on the planet.

• Uranus has five moderate-sized satellites, the most bizarre of which is Miranda.

• The largest satellite of Neptune, Triton, is an icy world with a tenuous nitrogen atmosphere. Triton moves in a retrograde orbit that suggests it was captured into orbit by Neptune's gravity. It is spiraling down toward Neptune around which it will eventually form a ring system.

Pluto and Beyond

• Pluto, the smallest planet in the solar system, and its satellite, Charon, are icy worlds that may well resemble Triton.

• Other objects orbit the Sun beyond the orbit of Pluto. Hundreds of these Kuiper belt objects have been observed.

WHAT DID YOU THINK?

1 *Is Jupiter a "failed star" or almost a star?* No. Jupiter has 75 times too little mass to shine as a star.

2 *What is Jupiter's Great Red Spot?* The Great Red Spot is a long-lived, oval cloud circulation similar to a hurricane on Earth.

3 *Does Jupiter have continents and oceans?* No. Jupiter is surrounded by a thick atmosphere primarily of hydrogen and helium that becomes liquid as one moves inward. The only solid matter in Jupiter is its core.

4 *Is Saturn the only planet with rings?* No. Four planets (Jupiter, Saturn, Uranus, and Neptune) have rings.

5 *Are the rings of Saturn solid ribbons?* Saturn's rings are all composed of thin, closely spaced ringlets consisting of particles of ice and ice-coated rocks. If they were solid ribbons, Saturn's gravitational tidal force would tear them apart.

6 *Do all moons rise and set as seen from their respective planets?* No. The one exception is Pluto's moon Charon, which remains over the same place on Pluto at all times.

Key Words

A ring, 217
B ring, 217
belt, 201
Cassini division, 217
differential rotation, 202
Encke division, 218
Galilean moon (satellite), 207
Great Dark Spot, 227
Great Red Spot, 201
hydrocarbon, 221
liquid metallic hydrogen, 206
occultation, 225
polymer, 221
prograde orbit, 214
resonance, 217
retrograde orbit, 214
ringlet, 214
Roche limit, 228
shepherd satellite (moon), 219
spoke, 219
zone, 201

Review Questions

1. Which is the most massive planet in the solar system? a. Earth b. Neptune c. Saturn d. Jupiter e. Pluto.

2. Which of the following planets does *not* have rings? Chose only one. a. Mars b. Uranus c. Neptune d. Saturn e. Jupiter.

3. Which is the least massive planet in the solar system? a. Mercury b. Mars c. Pluto d. Jupiter e. Venus.

4. Which planet is presently known to have the most moons? a. Mars b. Saturn c. Uranus d. Jupiter e. Neptune.

5. Describe the appearance of Jupiter's atmosphere. Which features are long-lived and which are relatively fleeting?

6. To test your knowledge of Jupiter's belt and zone structure, do Interactive Exercise 8-1 on the Web. You can print out your results, if required.

7. What causes the belts and zones in Jupiter's atmosphere?

8. To test your knowledge of Jupiter's internal structure, do Interactive Exercise 8-2 on the Web. You can print out your results, if required.

9. What is liquid metallic hydrogen? Which planets contain this substance? What produces this form of hydrogen?

10. Compare and contrast the surface features of the four Galilean satellites, discussing their geologic activity and their evolution.

11. To test your knowledge of the Galilean Moons, do Interactive Exercise 8-3 on the Web. You can print out your results, if required.

12. What energy source powers Io's volcanoes?

13. Why are numerous impact craters found on Ganymede and Callisto but not on Io or Europa?

14. Describe the structure of Saturn's rings. What are they made of?

15. To test your knowledge of Saturn's rings, do Interactive Exercise 8-4 on the Web. You can print out your results, if required.

16. Why do features in Saturn's atmosphere appear to be much fainter and "washed out" compared with features in Jupiter's atmosphere?

17. Explain how shepherd satellites affect some planetary rings. Is "shepherd satellite" an appropriate term for these objects? Explain.

18. Describe Titan's atmosphere. What effect has sunlight had on Titan's atmosphere?

19. Describe the seasons on Uranus. Why are the Uranian seasons different from those on any other planet?

20. Briefly describe the evidence supporting the idea that Uranus was struck by a large planetlike object several billion years ago.

21. Why are Uranus and Neptune distinctly bluer than Jupiter and Saturn?

22. Compare the ring systems of Saturn and Uranus. Why were Uranus's rings unnoticed until the 1970s?

23. How do the orientations of Uranus's and Neptune's magnetic axes differ from those of the other planets?

24. To test your knowledge of planetary magnetic fields, do Interactive Exercise 8-5 on the Web. You can print out your results, if required.

25. Suppose you were standing on Pluto. Describe the motions of Charon relative to the horizon. Under what circumstances would you never see Charon?

26. Describe the circumstantial evidence supporting the idea that Pluto is one of a thousand similar icy worlds that occupy the outer regions of the solar system.

27. Explain why Triton will never collide with Neptune, even though Triton is spiraling toward that planet.

28. What role did Charon play in enabling astronomers to determine Pluto's mass?

Advanced Questions

29. Consult the Internet or such magazines as *Sky & Telescope* or *Astronomy* to determine what space missions are now under way. What data and pictures have they sent back that update information presented in this chapter?

30. Long before the *Voyager* flybys, Earth-based astronomers reported that Io appeared brighter than usual for a few hours after emerging from Jupiter's shadow. Explain this brief brightening of Io.

31. Compare and contrast Valhalla on Callisto with the Caloris Basin on Mercury.

32. As seen by Earth-based observers, the intervals between successive edge-on presentations of Saturn's rings alternate between 13 years 9 months and 15 years 9 months. Why are these two intervals not equal?

33. Compare and contrast the internal structures of Jupiter and Saturn with the internal structures of Uranus and Neptune. Can you propose an explanation for why the differences between these two pairs of planets occurred?

34. Neptune has the third largest mass of all the planets, but Uranus has the third largest diameter. Reconcile these two facts.

Discussion Questions

35. Suppose that you were planning a mission to Jupiter employing an airplanelike vehicle that would spend many

days, even months, flying through the Jovian clouds. What observations, measurements, and analyses should this aircraft make? What dangers might it encounter, and what design problems would you have to overcome?

36. Discuss why astronomers believe that Europa, Ganymede, and Callisto may harbor some sort of marine life. Why do they not expect any life on the surfaces of these worlds?

37. Suppose you were planning separate missions to each of Jupiter's Galilean moons. What questions would you want these missions to answer, and what kinds of data would you want your spacecraft to send back? Given the different environments on the four satellites, how would the designs of the four spacecraft differ?

38. NASA and the Jet Propulsion Laboratory have tentative plans to place spacecraft in orbit about Uranus and Neptune in this century. What kinds of data should be collected and what questions would you like to see answered by these missions?

39. Would you expect the surfaces of Pluto and Charon to be heavily cratered? Explain.

What If . . .

40. Jupiter, at its present location, were a star? What would Earth be like? *Hint:* Recall that to be a star, Jupiter would have to have 75 times more mass than it has today.

41. Jupiter had formed at one-third its present distance of 5.2 AU from the Sun? What would Earth be like?

42. Io were struck by another object of similar size? *Hint:* You can create a variety of different scenarios by imagining the impacting body striking from different directions and with different speeds and at different angles.

43. Jupiter were orbiting in the opposite direction that it actually is? What effects might this have on the other planets? Would this change affect Earth? If so, how?

Web Questions

44. **Moving Weather Systems on Jupiter** Access and view the video "The Great Red Spot" in Chapter 8 of the *Discovering the Universe* Web site. **a.** Near the bottom of the video window you will see a white oval moving from left to right. By stepping through the video one frame at a time, estimate how long it takes this oval to move a distance equal to its horizontal dimension. (*Hint:* You can keep track of time by noticing how many frames it takes a feature in the Great Red Spot, at the center of the video window, to move in a complete circle around the center of the spot. The actual time for this feature to complete a circle is about 6 days.) **b.** The horizontal dimension of the white oval is about 4000 km. At what approximate speed (in km/hr) does the white oval move? (Speed = distance/time.)

45. Search the Web, especially the Web sites at NASA's Jet Propulsion Laboratory and at the European Space Agency, for information about the current status of the *Cassini* mission. When did *Cassini* arrive at Saturn? Where in the Saturn system is it presently? What are the current plans for its tour of Saturn's satellites? What ideas are being considered for the *Cassini* extended mission, to begin in 2008?

46. In 2004, astronomers reported the discovery of two new moons of Saturn. Search the Web for information about these. How did astronomers discover them? Have the observations been confirmed? How large are these moons? What sort of orbits do they follow?

47. **The Rotation Rate of Saturn** Access and view the video "Saturn from the Hubble Space Telescope" in Chapter 8 of the *Discovering the Universe* Web site. The total time that actually elapses in this video is 42.6 hours. Using this information, identify and follow an atmospheric feature and determine the rotation period of Saturn.

48. The discovery of Charon (see Figure 8-42) was made by an astronomer at the U.S. Naval Observatory. Search the Web to find out why the U.S. Navy carries out work in astronomy.

Observing Projects

49. Consult such magazines as *Sky & Telescope* and *Astronomy* or your *Starry Night Enthusiast*™ software to determine whether Jupiter is currently visible in the night sky. If so, make arrangements to view the planet through a telescope. What magnifying power seems to give you the best view? Draw a picture of what you see. Can you see any belts and zones? How many? Can you see the Great Red Spot?

50. Make arrangements to view Jupiter's Great Red Spot through a telescope. Consult the *Sky & Telescope* Web site, which lists the times when the center of the Great Red Spot passes across Jupiter's central region, as seen from Earth. The Great Red Spot is well placed for viewing for 50 minutes before and after this time. You will need a refractor with an objective lens of at least 15 cm (6 in.) diameter or a reflector with an objective of at least 20 cm (8 in.) diameter. Using a pale blue or green filter can increase the color contrast and make the spot more visible. For other useful hints, see the article "Tracking Jupiter's Great Red Spot" by Alan MacRobert (*Sky & Telescope*, September 1997). Sketch Jupiter based on your observations.

51. If it is visible at night, observe Jupiter through a pair of binoculars or telescope. Can you see all four Galilean moons? Make a drawing of what you observe. To identify the moons, use *Starry Night*

Enthusiast™. Start by going to Atlas mode (*Favourites/Guides/Atlas*). Then locate Jupiter (Tab *Find/Jupiter*). Zoom out to 30′ (30 arcsecond) angle. You should be able to see all the moons labeled in their present positions (be sure that the time is *not* changing). Compare the locations of the moons on the screen with your observations to determine which ones you are seeing.

52. Use *Starry Night Enthusiast*™ to observe the motion of the Galilean moons of Jupiter. First go to Atlas mode (*Favourites/Guides/Atlas*). Lock on Jupiter (Tab *Find/Jupiter*) and zoom in to 30′ (30 arcsecond) angle. Set the timestep to 30 minutes and click the ▶ button. You will see the four Galilean moons orbiting Jupiter. **a.** Are all four moons ever on the same side of Jupiter? **b.** Observe the moons passing in front of and behind Jupiter (zoom in and slow the timestep as needed). Explain how your observations tell you that all four satellites orbit Jupiter in the same direction.

53. Consult such magazines as *Sky & Telescope* and *Astronomy* or your *Starry Night Enthusiast*™ software to determine whether Saturn is currently visible in the night sky. If so, view Saturn through a small telescope. Make a sketch of what you see. Estimate the angle at which the rings are tilted to your line of sight. Can you see the Cassini division? Can you see any belts or zones in Saturn's clouds? Do you observe a faint, starlike object near Saturn that might be Titan? What observations could you perform to test whether the starlike object is a Saturnian satellite?

54. Use the *Starry Night Enthusiast*™ program to observe the changing appearance of Saturn. First go to Atlas mode (*Favourites/Guides/Atlas*) and then find Saturn (Tab *Find/Saturn*). Zoom in until Saturn and its rings are clearly visible. Set the timestep to 1 year. Use the single-step time control buttons to observe the changing aspect or orientation of the rings. During which of the next 30 years will we see the rings edge-on?

55. Determine whether Uranus or Neptune is currently visible in the night sky. If so, make arrangements to view them through a telescope. To help you find these planets, use your *Starry Night Enthusiast*™ software or the star chart published each January in *Sky & Telescope* showing the paths of Uranus and Neptune against the background stars. For more detailed images than you can get through a telescope, locate these planets with your *Starry Night Enthusiast*™ software and zoom to high resolution. Also, locate Triton with the software.

56. Use the *Starry Night Enthusiast*™ program to observe Pluto and Charon. First go to Atlas mode (*Favourites/Guides/Atlas*) and find Charon (Tab *Find/Charon*). Zoom in as close as possible. Set the timestep to 3 hours. Use the single-step time button to step through enough time to determine the period of Charon's orbit. How does your answer compare with Pluto's sidereal rotation period given in Figure 8-41? Explain.

57. **Saturn as Seen from Earth** In this exercise we will use *Deep Space Explorer*™ to predict how Saturn will appear as seen from Earth. If you have not done so already, go to **Settings** and uncheck "Use magnitude cutoffs," slide the "Galaxy drawing/Brightness" slide to the middle of the range, and then click on the **Views** tab.

Now, go to Saturn by clicking **Views/Solar System/Saturn** on the left side of the star field. From the image of Saturn that appears, write down where you expect the Sun is located (e.g., above, down to the right, etc.). Grab the screen by holding down the left PC mouse button (the mouse button on a Mac) and move the mouse until the Sun almost disappears behind Saturn's rings. Is the Sun in the same plane as the rings of Saturn? Now go to Earth and locate Saturn by moving around the sky and putting the cursor on bright objects. "Centre Saturn." Zoom in so that Saturn grows until you can see its rings. Which of the locations of Saturn on Figure 8-21 most closely matches the configuration of Saturn and the Sun on *Deep Space Explorer*™? Are the image from Earth of Saturn and the image of the Sun from Saturn consistent with each other?

58. **Pluto, the Sun, and Asterisms** In this exercise, we will use *Deep Space Explorer*™ to locate the Sun as seen from Pluto. If you have not done so already, go to **Settings** and uncheck "Use magnitude cutoffs," slide the "Galaxy drawing/Brightness" slide to the middle of the range, and then click on the **Views** tab.

Now, go to Pluto by clicking **Solar System/Pluto** on the left side of the star field. Grab the screen by holding down the left PC mouse button (the mouse button on Mac) and move the mouse until you see the object you believe is the Sun. Put the cursor over that object and see if you are correct. If not, repeat until you find the Sun. Make a simple drawing of Pluto, the Sun, and a few asterisms showing the Sun's location. Locate the asterism of Orion. Does it look the same from Pluto as it does from the Earth? Explain why or why not.

WHAT IF . . . WE LIVED ON A METAL-POOR EARTH?

Earth provides a wealth of building blocks necessary for the development and evolution of complex life-forms. More than 80 elements on or near Earth's surface combine in countless ways essential for its diversity of flora and fauna. Especially important for life on Earth are metals such as iron. The human body typically contains more than 3 grams of iron, mostly in the form of hemoglobin that helps transport oxygen through the bloodstream. A slight iron imbalance leads to anemia or toxicity. And without abundant metals, most of our technologies would never have developed.

But what if the solar system formed from an interstellar gas cloud containing fewer metals and high-mass elements, that is, a solar system with less iron, nickel, copper, and other metals crucial for life as we know it?

Metal-Rich Versus Metal-Poor Earth formed with almost 6×10^{21} metric tons of star matter, almost one-third of it iron. If the solar system formed from interstellar gas containing relatively few heavy elements, Earth would contain a much lower fraction of such elements as uranium, lead, iron, and nickel. The Earth would be comprised of a correspondingly higher fraction of lower-mass elements, such as carbon, nitrogen, oxygen, silicon, and aluminum. That version of Earth—let's call it Lithia—would be profoundly different from our planet.

The Environment of Lithia Without heavy elements like iron, Lithia's density, gravity, and magnetic fields are much lower than Earth's. Let's assume that the density of Lithia is similar to that of the Moon. (We'll assume that the Moon doesn't change in its characteristics or its distance from Lithia.)

A Barren Landscape With a thin atmosphere and an abundance of lighter elements such as silicon (silicon dioxide is sand), Lithia might resemble a barren desert here on Earth. (Photri)

If Lithia has the same radius as Earth, the force of gravity on Lithia's surface is only 60% that on Earth. That is, you would weigh 40% less on Lithia. Mobile life-forms that evolve on the metal-poor planet require much less muscle strength to get about and less bone density to resist the planet's gravity.

With fewer radioactive elements to provide heat, the core of Lithia cooled off quickly compared with the core of Earth. As a result, Lithia has much less heat flowing upward through its mantle and crust, so there most likely is no plate tectonic activity. Lithia also has less volcanic activity. But without plate motion to spread the lava flows, volcanoes will grow to higher elevations than on Earth. The lack of crustal motions also means that pockets of high-density material will not rise from below, forming the metal ores mined on Earth.

The Moon's orbit and the tides it induces are different, too. With Lithia having 60% of Earth's mass, the Moon at its present distance orbits once every 36 days instead of once every 29.5 days. Thus, the cycle of lunar phases is longer, but the cycle of lunar-induced tides is slightly shorter. Surprisingly, the tides have roughly the same height, because the tidal effects depend only on the Moon's gravity and not Earth's, and the distance between Earth and Lithia and their satellite is the same.

Life on Lithia The low abundance of iron and the resulting lack of a planetary magnetic field might preclude the development of complex life-forms on Lithia. And yet, the incredible diversity of life we find on Earth suggests that evolution on Lithia might well occur. If advanced life-forms evolve on Lithia before the Sun uses up its nuclear fuel and becomes a red giant, the chemical differences between Lithia and Earth dictate that they will differ profoundly from the complex life we find on Earth.

Without the heat of radioactive elements to keep Lithia's core molten for billions of years and with fewer metals, Lithia will have a much weaker magnetic field—if any at all—than Earth. Inhabitants can expect auroras to grace the night sky continuously. On the other hand, high-energy particles and the radiation they create in the atmosphere will bombard any life-forms on the surface continuously and will also adversely affect the ozone layer. A larger amount of the Sun's harmful ultraviolet radiation will reach the surface. Without more protection, life as we know it could not survive.

Vagabonds of the Solar System

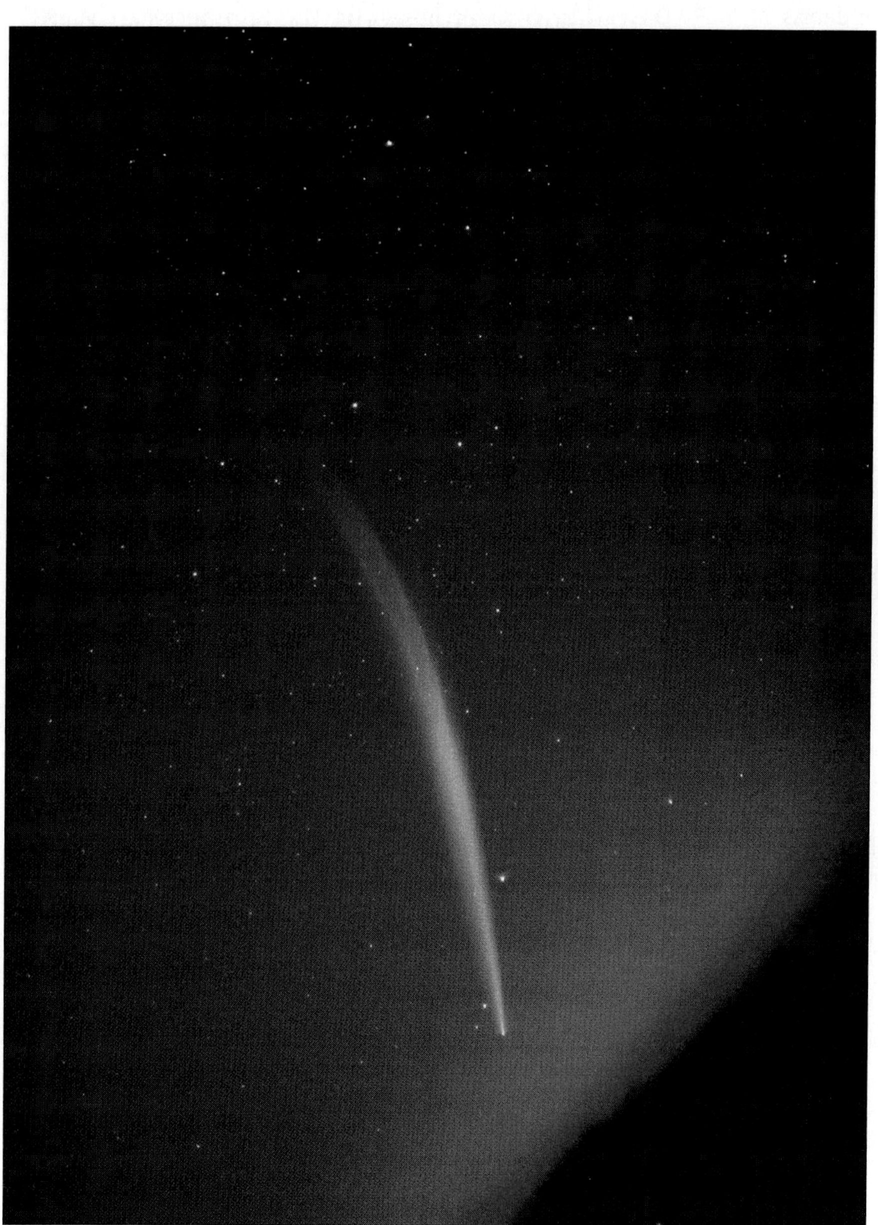

RIVUXG

The Tail of Comet Ikeya-Seki (Roger Lynds/NOAO/AURA/NSF)

WHAT DO YOU THINK?

1 Are the asteroids a planet that was somehow destroyed?

2 How far apart are the asteroids on average?

3 Why do comets have tails?

4 In which direction does a comet's tail point?

5 What is a shooting star?

The formation of the solar system was a messy affair. It left a lot of space junk flying around. Planetesimals collided by the billions to form the planets we see today and their moons. For nearly a billion years, pieces of smaller debris often struck the young planets and moons, scarring and shaping their surfaces. This era of frequent collisions ended some 3.8 billion years ago, but innumerable pieces of debris still orbit the Sun, and dramatic impacts continue today. Large, rocky bodies occasionally pass startlingly close to the Earth—closer even than the Moon. Myriad small bodies penetrate the atmosphere daily. Comets weave glorious trails across the sky. Icy comet bodies may once have provided Earth with water and other material essential to the evolution of life.

These leftovers, the asteroids, meteoroids, and comets, are the vagabonds of the solar system. We have already seen how pieces of Comet Shoemaker-Levy 9 plunged into Jupiter. You may have spotted Comet Hale-Bopp during the summer of 1997. What do the characteristics of these interplanetary travelers reveal about the composition of the solar system and about how life began here?

In this chapter you will discover

- asteroids and meteoroids, pieces of interplanetary rock and metal
- comets, bodies containing large amounts of ice and rocky debris
- space debris that falls through the Earth's atmosphere
- that impacts from space 250 million and 65 million years ago caused mass extinctions of life on Earth
- that a wayward asteroid could again threaten life on Earth

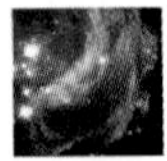

ASTEROIDS: THE MINOR PLANETS

We know that the solar system formed from a rotating disk of gas and dust. The matter that had too much angular momentum to fall onto the protosun coalesced at varying distances into planetesimals. Many of these chunks of rock and metal eventually collided, forming the planets and larger moons of the solar system. Others were captured whole by various planets as small, irregularly shaped moons, like Phobos and Deimos in orbit around Mars. However, many planetesimals still orbit the Sun today in splendid isolation. These are the **asteroids,** sometimes called *minor planets.*

9-1 Most asteroids orbit the Sun between Mars and Jupiter

On New Year's Day 1801, the Sicilian astronomer Giuseppe Piazzi was carefully mapping faint stars in the constellation of Taurus. He noticed a dim, previously uncharted "star" that shifted its position slightly over the next several nights. Uranus had been discovered that way by William Herschel just 20 years before. Was Piazzi's object another planet? There is no official size limit for a planet, as the discussion about Pluto and Sedna earlier in the book has revealed. However, in 1801, all the known planets (Mercury, Venus, Earth, Mars, Jupiter, Saturn, and Uranus) were so much larger than Piazzi's discovery that his lucky sighting was deemed too small to qualify as a full-fledged planet. Rather, he was the first to discover an asteroid.

Later that year, the orbit of this object was determined to lie between Mars and Jupiter. At Piazzi's request, the object was named Ceres (pronounced see-reez), after the patron goddess of Sicily. Ceres is spherical like the planets (Figure 9-1), but its diameter is a scant 940 km (585 mi), only one-quarter the diameter of our Moon.

In 1802, the German astronomer Heinrich Olbers discovered another faint, starlike object that moved against the background stars. He called it Pallas, after the Greek goddess of wisdom. Like Ceres, Pallas orbits the Sun in a nearly circular orbit between the orbits of Mars and Jupiter. Pallas is even dimmer and smaller than Ceres, with a diameter of only 600 km (375 mi).

Only two more of these minor planets—Juno and Vesta—were found until the mid-1800s, when telescopes improved. Astronomers then began to stumble across many more asteroids orbiting the Sun at distances from 2 and 3½ AU, between the orbits of Mars and Jupiter. This region of the solar system is now called the **asteroid belt** (Figure 9-2). Asteroids whose orbits lie entirely within this region are called **belt asteroids.**

The next real breakthrough came in 1891, when the German astronomer Max Wolf applied photographic techniques to the search for asteroids. A total of 300 asteroids had been found up to that time, each painstakingly discovered by scrutinizing the skies for faint, uncharted "stars" whose positions shifted slowly from

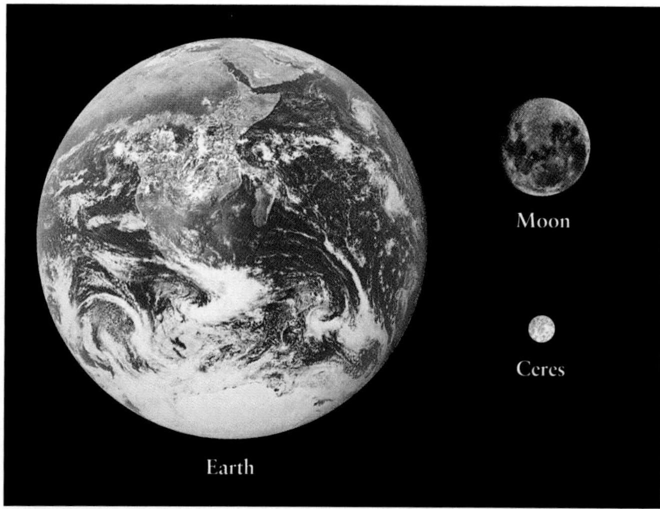

FIGURE 9-1 **Comparison of Ceres with the Moon and the Earth** Ceres, the Moon, and the Earth are shown here to scale. Ceres, shown in this infrared photo (Earth and Moon appear in visible light), is the largest asteroid but is so small that it is not considered a planet. Because it does not orbit a body other than the Sun, it is also not classified as a moon. (NASA)

one night to the next. With the advent of astrophotography, however, the floodgates were opened. Astronomers could simply aim a camera-equipped telescope at the stars and take long exposures. If an asteroid happened to be in the field of view, it left a distinctive trail on the photographic plate (Figure 9-3). Using this technique, Wolf alone discovered 228 asteroids.

INSIGHT INTO SCIENCE

Confirming Observations New findings must be confirmed or replicated by other competent scientists before discoveries and observations are accepted by the scientific community. Therefore, asteroid observations need to be repeated and repeated in order to determine exact orbits and to eliminate the possibility that a sighting is of a known asteroid or other object.

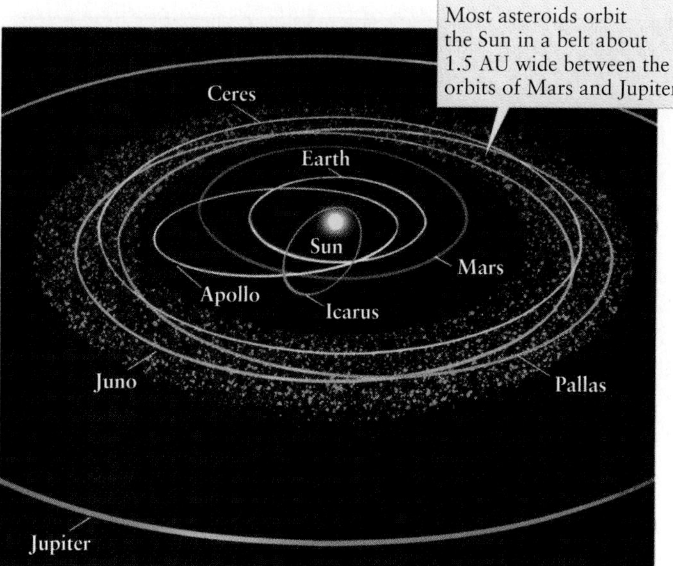

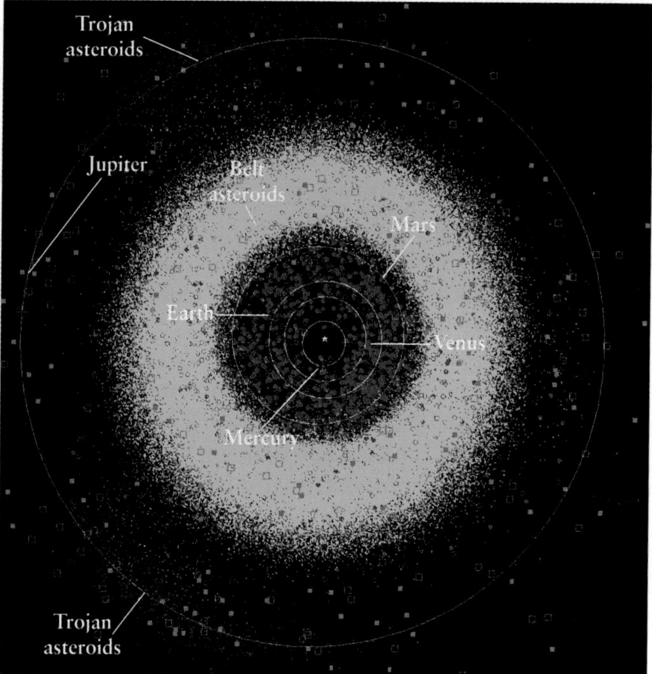

FIGURE 9-2 **Asteroid Orbits** **(a)** An artist's rendition of some asteroids and their orbits. Most asteroids orbit the Sun in a $1^1/_2$-AU-wide belt between the orbits of Mars and Jupiter. The orbits of belt asteroids Ceres, Pallas, and Juno are indicated to scale. Some asteroids, such as Apollo and Icarus, have highly eccentric paths that cross Earth's orbit. Others, called the Trojan asteroids, follow the same orbit as Jupiter. **(b)** Actual positions of all known asteroids at Jupiter's orbit or closer. The locations of the belt asteroids are indicated by green dots. Objects passing closer than 1.3 AU to the Sun are shown by red circles. Objects observed at more than one opposition are indicated by filled circles, objects seen at only one opposition are indicated by outline circles. Jupiter's Trojan asteroids are deep blue squares. Comets are filled and unfilled light-blue squares. (b: Minor Planet Center)

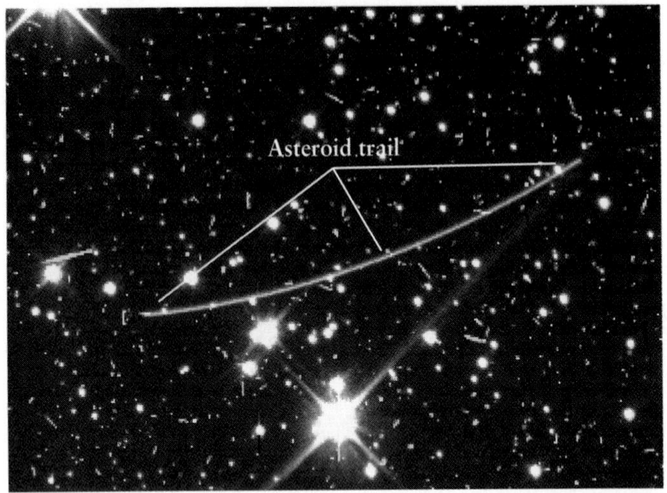

FIGURE 9-3 **Discovering Asteroids** In 1998, the Hubble Space Telescope found this asteroid while observing in the constellation Centaurus. The exposure, tracking stars, shows the asteroid as a 19 arcsecond streak. This asteroid is about 2 km in diameter and was located about 140 million km (87 million miles) from Earth. (R. Evans and K. Stapelfeldt, Jet Propulsion Laboratory and NASA)

Ceres, the largest asteroid, alone accounts for about 30% of the mass of all the known asteroids combined. Only three asteroids—Ceres, Pallas, and Vesta—have diameters greater than 300 km. Records of all objects smaller than planets in our solar system are kept by the Minor Planet Center in Cambridge, Massachusetts. The center reports that as of November 2004, 96,154 asteroids had been confirmed to exist, with more than 24,227,286 other observations awaiting confirmation as new asteroids. The number of asteroids increases dramatically with *decreasing size:* Only 30 asteroids have diameters between 200 and 300 km; 200 more are bigger than 100 km across; there are believed to be tens of millions that are less than 1 km across.

1 You may have heard the common belief that the asteroids were once a single planet that was somehow destroyed. Indeed, it is plausible that the asteroid belt region once had much more mass, possibly several Earth masses, worth of debris. However, Jupiter's gravitational force pulled many planetesimals out of this region before large numbers of them could meet, coalesce, and form a single planet-sized object.

If all the present asteroids had once been part of a single body, it would have had a diameter of only 1500 km, or 12% of the Earth's diameter. This is less than two-thirds the diameter of Pluto and half the diameter of our Moon. Because Pluto just barely makes it into the category of a planet by today's standards, a single body containing the mass of all the asteroids would not qualify as a planet. (We consider Jupiter's gravitational effect in more detail in the next section.)

9-2 Jupiter's gravity creates gaps in the asteroid belt

In 1867, the American astronomer Daniel Kirkwood called attention to gaps in the asteroid belt. These features, called **Kirkwood gaps,** show the influence of Jupiter's gravitational attraction, as best seen in a graph of asteroid orbital periods, like the one in Figure 9-4. Note the gaps at simple fractions (1/3, 2/5, 3/7, and 1/2) of Jupiter's orbital period. The gravitational effect of Jupiter's creating gaps in the asteroid belt resembles the gravitational resonance effect of Mimas creating the Cassini division in Saturn's rings (see Section 8-10).

It is likely that the asteroid belt contains tens of mil- 2
lions of asteroids, yet their typical separation is a staggering 10 million kilometers. This is quite unlike the image that has been created by innumerable popular movies and television shows of asteroids so close together that you must dodge them as you fly past.

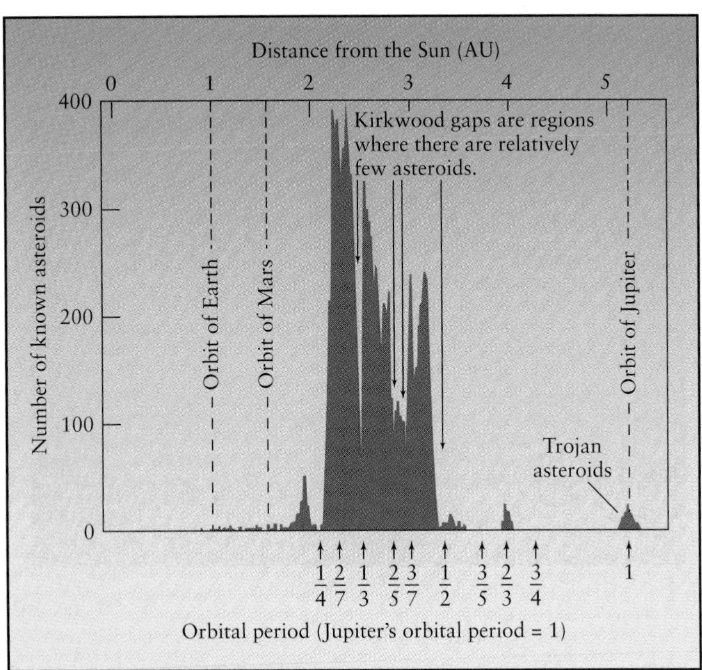

FIGURE 9-4 **The Kirkwood Gaps** This graph displays the number of asteroids at various distances from the Sun. Notice that few orbital periods of asteroids correspond to such simple fractions as 1/3, 2/5, 3/7, and 1/2 of Jupiter's orbital period. Repeated alignments with Jupiter have deflected asteroids away from these orbits. The Trojan asteroids accompany Jupiter as it orbits the Sun.

INSIGHT INTO SCIENCE

Get Real Scientists continually apply the "laws of nature" to new situations, problems, observations, and experiments, even to popular culture. For example, applying Newton's law of gravity to the asteroids reveals that they could never swarm, as science fiction movies suggest. At those close quarters, their gravity would cause them either to collide or to pass so close together that they would subsequently fly rapidly and permanently apart.

Despite the large average separation between asteroids, the gravitational influences of Mars and Jupiter have sent some asteroids caroming into each other at various times over the 4.6 billion years that the solar system has existed. A collision between kilometer-sized asteroids must be an awe-inspiring event. Typical collision velocities are estimated to be 3600 to 18,000 km/h (2000 to 11,000 mph), which is more than sufficient to shatter rock. In some collisions, the resulting fragments may not have enough speed to escape from each other's gravitational attraction, and they reassemble. The asteroid Toutatis appears to be composed of two comparably sized pieces connected to each other, as does the asteroid Castalia. These bodies are therefore likely to have been broken apart and re-formed.

Alternatively, several large fragments may end up orbiting each other. In 1918, the Japanese astronomer Kiyotsugu Hirayama drew attention to groups of asteroids that share nearly identical orbits. These are fragments of parent asteroids. Pursuing this line of research, astronomers in 2002 discovered that two large asteroids collided only a few million years ago (very recently in solar system history) and created about 20 separate families or clusters of smaller asteroids. Asteroids in each of these families orbit together. The largest remnant cluster of this impact is named Karin, after its largest member, an asteroid some 20 km (12 mi) across.

In the early 1990s, the Jupiter-bound *Galileo* spacecraft passed near two asteroids—Gaspra (see Figure 5-9) and Ida (Figure 9-5)—and sent back close-up views. Both asteroids are probably fragments of larger parent bodies that were broken apart by catastrophic collisions. Because Ida's surface is more heavily cratered than Gaspra's, Ida is much older.

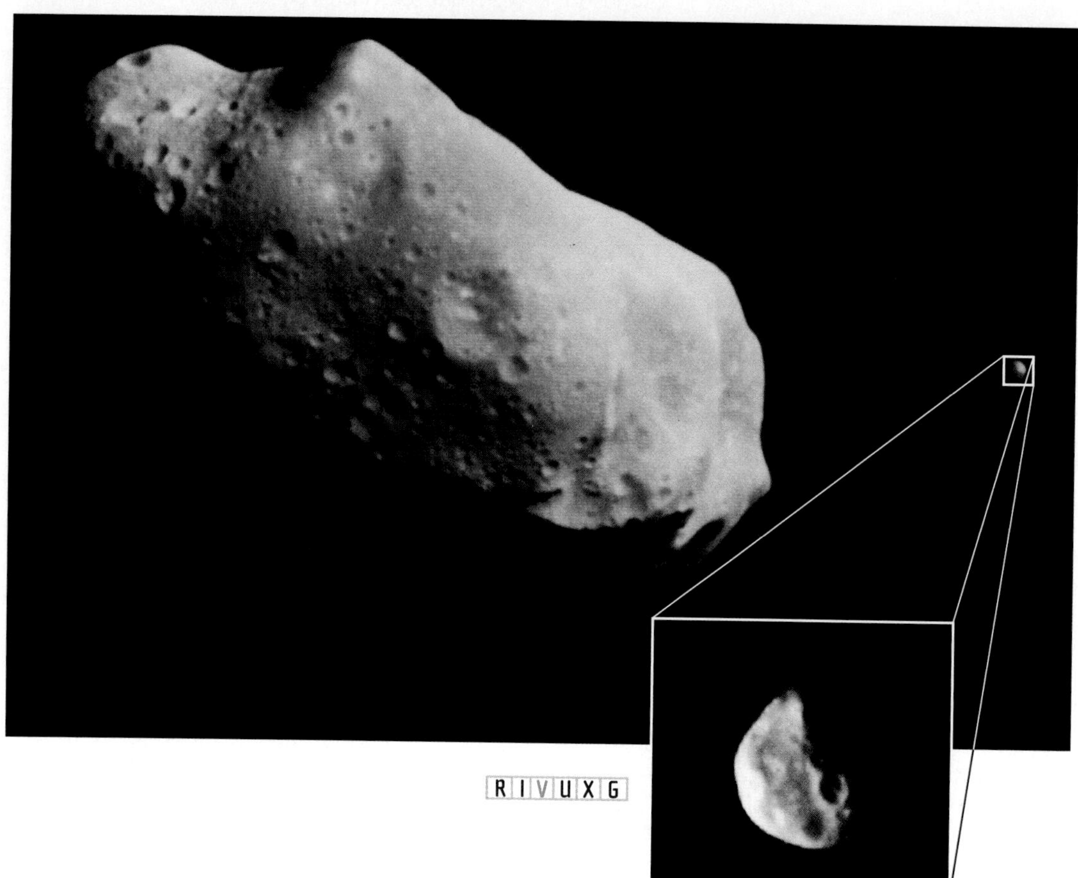

FIGURE 9-5 **Ida and Its Satellite** The 55-km-long rocky asteroid Ida, shown here with its satellite Dactyl, is about twice the size of the younger asteroid Gaspra (see Figure 5-9). **Inset:** Dactyl is also heavily cratered. (NASA)

Many asteroids have smaller satellite asteroids of their own. For example, Dactyl is the satellite of Ida, and Petit-Prince is the satellite of Eugenia. Dactyl is a pockmarked asteroid some 1.5 km in diameter that orbits Ida at a distance of 100 km. Petit-Prince is 13 km across, orbiting 1200 kilometers from the larger body. To date, 57 asteroid/satellite systems have been observed.

9-3 Asteroids exist outside the asteroid belt

While Jupiter's gravitational pull clears out certain orbits within the asteroid belt, it actually captures asteroids at two locations in the path of its own orbit. The gravitational forces of the Sun and Jupiter work together to hold asteroids in orbit at these locations, called **stable Lagrange points,** in honor of the French mathematician Joseph Lagrange, whose calculations explained them. One Lagrange point is located 60° ahead of Jupiter, and the other is 60° behind, as shown in Figure 9-6.

The asteroids trapped at Jupiter's Lagrange points are called **Trojan asteroids,** each named after a hero of the Trojan War. As of October 2004, 1691 Trojan asteroids orbiting with Jupiter have been catalogued (see Figure 9-2b). Neptune has one known Trojan asteroid and astronomers continue to search for Trojans orbiting with other planets.

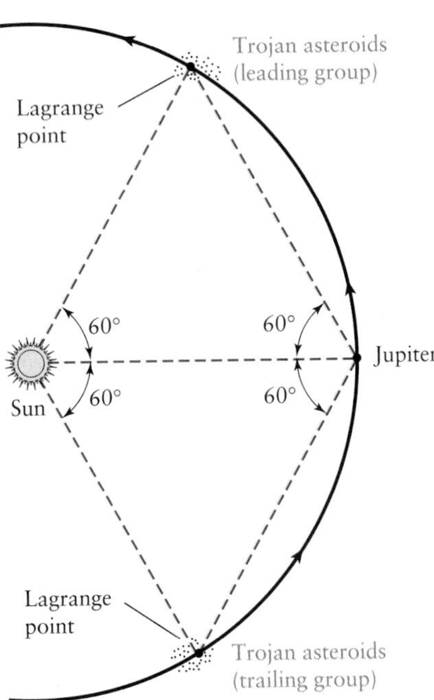

FIGURE 9-6 **Jupiter's Trojan Asteroids** Groups of asteroids orbit at the two stable Lagrange points along Jupiter's orbit, trapped by the combined gravitational forces of Jupiter and the Sun. Asteroids at these locations are named after Homeric heroes of the Trojan War. For more details of the location of Jupiter's Trojan asteroids, see Figure 9-2b.

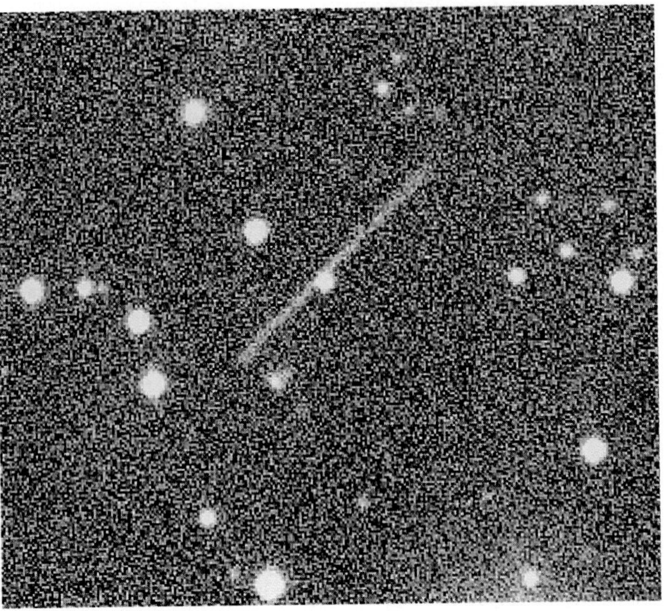

R I V U X G

FIGURE 9-7 **Asteroid 1994 XM1** This image was obtained on December 9, 1994, shortly before the asteroid arrived in the Earth's vicinity. When it passed by the Earth just 12 hours later, asteroid 1994 XM1 was less than half the distance from the Earth to the Moon. (Jim Scotti, Spacematch on Kitt Peak)

WEB LINK 9.4

Some asteroids have highly elliptical orbits that bring them into the inner regions of the solar system (see Figure 9-2). Others have similarly elliptical orbits that extend from the asteroid belt out beyond the farthest reaches of Pluto's orbit. The Amor asteroids cross Mars's orbit, while the **Apollo asteroids** even cross Earth's orbit. At least 1690 *Earth-crossing* asteroids are known, several of which are pairs of asteroids orbiting each other. Not all of these Earth-crossing asteroids pose threats to the Earth, but at least 628 of them have the potential of someday striking the Earth.

There were several close calls in recent years. In 1972, space debris was observed to skip off Earth's atmosphere and retreat back into space. On December 9, 1994, asteroid 1994 XM1 (10 m across—the size of a small bus) (Figure 9-7) passed within 105,000 km (65,000 mi) of the Earth. Two near misses occurred in 2002: one on January 8, when an asteroid passed within 375,000 km (233,000 mi), and the other on June 14, when an asteroid passed only 119,000 km (75,000 mi), which is less than a third the distance to the Moon.

During these close encounters, astronomers can examine the details of asteroids. For example, an asteroid's brightness often varies as it rotates because different surface features scatter different amounts of light. Such data show that typical rotation periods for asteroids are between 5 and 20 hours, although one asteroid,

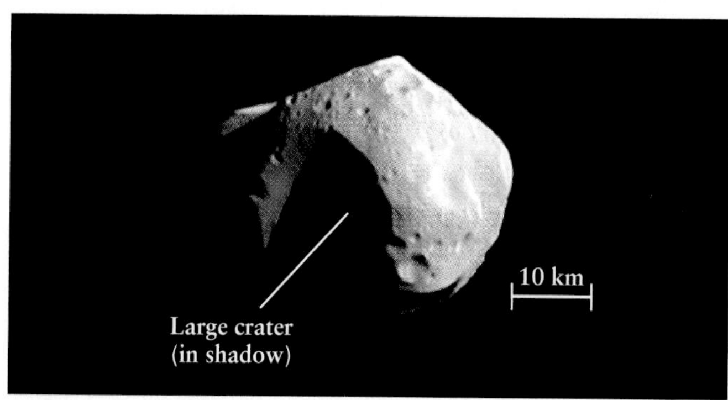

R I V U X G

FIGURE 9-8 **Asteroid Mathilde** Reflecting only half as much light as a charcoal briquette, Mathilde is half as dense as typical stony asteroids. Slightly larger than Ida, irregularly shaped Mathilde measures 66 × 48 × 46 km, rotates once every 17.4 d, and has a mass equivalent to 110 trillion tons. The part of the asteroid shown is about 59 × 47 km. The large crater in shadow is about 20 km across. (Johns Hopkins University, Applied Physics Laboratory)

labeled 1998 KY26, rotates about once every 10.7 minutes. This is the fastest-rotating object known in the solar system.

The proximity of asteroids and the promise of learning more about these ancient and extremely varied members of our solar system prompted NASA to send the *Near Earth Asteroid Rendezvous (NEAR) Shoemaker* spacecraft to visit asteroids Mathilde (Figure 9-8) and Eros. *NEAR Shoemaker* revealed that Mathilde is only 1.3 times denser than water and has an albedo of 0.04, making it darker than charcoal. This heavily cratered, carbon-rich body therefore has only half the density of the other asteroids astronomers have studied, such as Ida. Eros is about 3 times as dense as water and rotates once every 5¼ hours. In February 2000, *NEAR Shoemaker* went into orbit around Eros, and for a year the spacecraft's cameras and other sensors sent back a wealth of information about the asteroid.

NEAR Shoemaker showed that Eros (Figure 9-9) is a solid chunk of rock and metal. Analyzing Eros's spectra reveals that it is probably much the same as it was when it coalesced 4.6 billion years ago. This means that it was never hot enough to differentiate (separate rock from metal). Infrared observations reveal that like our Moon, Eros has a regolith. It also has several substantial craters and is strewn with boulders (Figure 9-9c). On February 12, 2001, NASA engineers landed *NEAR Shoemaker* on Eros so gently that the spacecraft continued to transmit data after landing. Details of Eros as small as 1.4 meters across were imaged by *NEAR Shoemaker.*

The Sun is not the only star with an asteroid belt. In 2001, astronomers discovered that the star Zeta Leporis, 70 light-years from Earth in the constellation Lepus (the

a

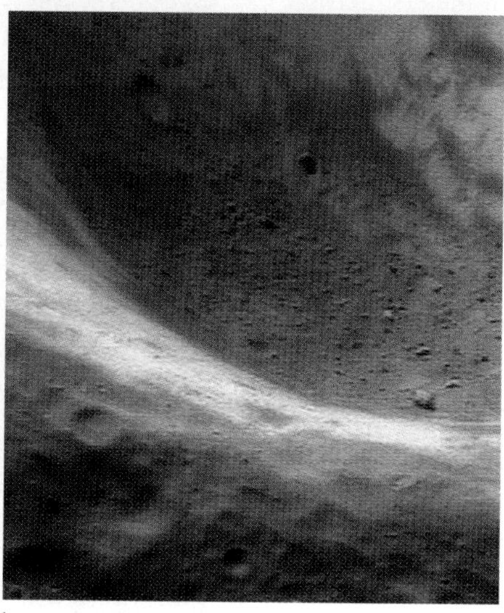
b

c

R I V U X G

FIGURE 9-9 **Asteroid Eros** (a) The *Near Earth Asteroid Rendezvous (NEAR) Shoemaker* spacecraft took this image of asteroid Eros in February 1999. The top of the figure is the asteroid's north polar region. Eros's dimensions are 33 × 13 × 13 km (21 × 8 × 8 mi) and it rotates every 5¼ hours. Its density is 2700 kg/m^3, close to the average density of the Earth's crust and twice as dense as asteroid Mathilde. (b) Looking into the large crater near the top of (a), which is 5.3 km (3.3 miles) across. (c) The penultimate image taken by *NEAR Shoemaker* before it was gently landed on Eros. Taken from an altitude of 250 m (820 ft), the image is only 12 meters across. You can see rocks and boulders buried to different depths in the regolith. (a, b, c: Johns Hopkins Applied Physics Laboratory)

Hare), has a disk of debris that appears to contain asteroids. This star system is less than 0.5 billion years old; astronomers hope it will provide insights into the early evolution of the asteroids and other objects in the disk of gas and dust surrounding the early Sun.

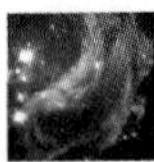

COMETS

While asteroids consist primarily of rock and metal, other pieces of space debris are composed of frozen water, along with rock, metal, and ices of other compounds. We have seen in earlier chapters that water ice, along with carbon dioxide, methane, and ammonia ices, was locked up in planets and moons. Ices in the young solar system also condensed with roughly equal amounts of small rocky and metallic debris into bodies that still remain in orbit around the Sun. These dirty icebergs in space are the **comets.**

9-4 Comets come from far out in the solar system

In the first few hundred million years of the solar system's existence, comets formed in its outer reaches at roughly the distances of Saturn, Uranus, and Neptune. In that region, water was plentiful and the temperature low enough for the ices to condense into chunks several kilometers across. After they formed, gravitational tugs from Uranus and Neptune flung the comets in every direction. In 2001, astronomers were able to measure the temperature at which ammonia ice formed in Comet Linear. This temperature determines the ammonia's solid structure, and so by determining the structure of the ammonia in Comet Linear, we are able to calculate how far the comet was from the Sun when it formed. That comet formed between the orbits of Saturn and Uranus.

Today, the solar system is believed to contain two reservoirs of comets. Most comets that eventually enter the inner solar system are believed to come from a doughnut-shaped region beyond the orbit of Pluto. Comets head inward as a result of collisions or near misses with other comet bodies. This **Kuiper belt,** named after the American astronomer Gerard Kuiper, who first proposed its existence in 1951, is centered on the plane of the ecliptic and extends out some 500 AU from the Sun (Figure 9-10).

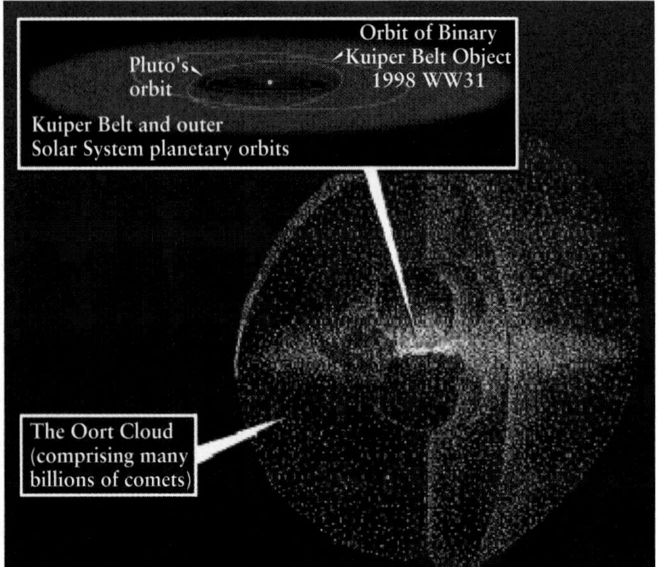

a

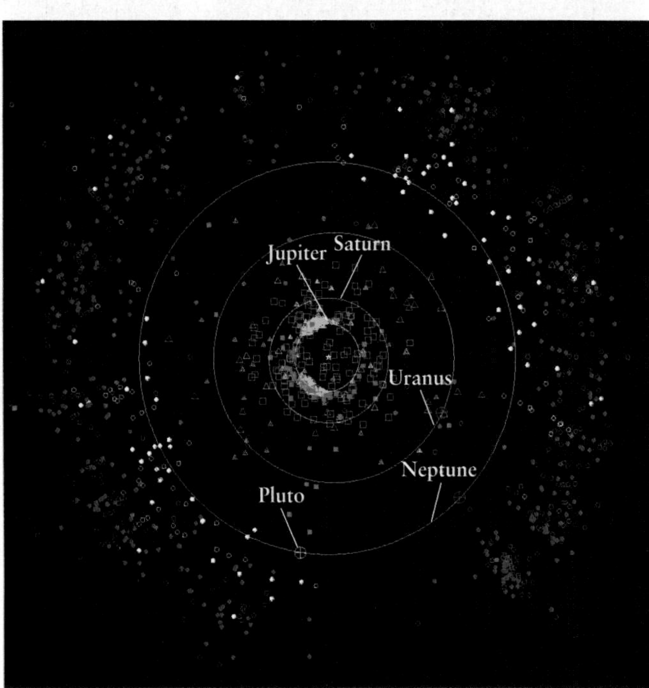

b

R I V U X G

▶ FIGURE 9-10 **The Kuiper Belt and Oort Cloud** (a) The Kuiper belt of comets spreads from Pluto out 500 AU from the Sun, as depicted in this artist's conception. Most of the estimated 200 million belt comets are believed to orbit in or near the plane of the ecliptic. The largest is 1600 km across. Some astronomers believe that Pluto and Charon are members of the Kuiper belt. The spherical Oort comet cloud extends from beyond the Kuiper belt. (b) The current positions of the minor bodies in the outer solar system are shown in this diagram. Unusual high eccentricity objects are shown as cyan triangles. Objects roaming among the outer planets, called Centaur objects, are orange triangles. Plutinos are white circles. Miscellaneous objects are magenta circles, and "classical" or "main-belt" objects are red circles. Objects observed at only one opposition are denoted by open symbols; objects with multiple-opposition orbits are denoted by filled symbols. Comets are filled and unfilled light-blue squares. (a: NASA and A. Field/Space Telescope Science Institute; b: Minor Planet Center)

At least 95 of the Kuiper belt objects orbit the Sun in the same region as Pluto. Because these latter bodies are smaller than Pluto, they are called *Plutinos*. Given the number and locations of all the Kuiper belt objects discovered so far, astronomers estimate that the belt contains at least 200 million comets.

The Kuiper belt consists of two groups of objects. Those with roughly circular orbits are called *classical Kuiper belt objects* and are found between 30 AU (Neptune's orbit) and 50 AU from the Sun. The second group, called *scattered Kuiper belt objects*, have more elliptical orbits that range from 35 AU to at least 200 AU from the Sun. More than 938 Kuiper belt objects have been observed (Figure 9-10 and Figures 9-11a and b). The Kuiper belt object 2002 LM 60, called Quaoar (Kwa-whar), is 1250 km across, some 400 km larger than Ceres or about half the size of Pluto. At least 1% of the Kuiper belt objects are orbiting pairs (Figure 9-11c), including Quaoar.

The vast majority of the several billion comets estimated to exist are believed to lie even farther from the Sun. Unlike the Kuiper belt comets and the rest of the solar system, these comets are believed to have a spherical distribution around the Sun called the **Oort cloud** (Figure 9-10a), named after the Dutch astronomer Jan Oort, who first proposed its existence in the 1950s.

In 2003, astronomers discovered an object larger than Quaoar and farther from the Sun than another known object. Called Sedna (see Section 5-4), this body is less than 1800 km (1100 mi) across and is presently 86 AU from the Sun. It has a highly elliptical orbit (Figure 9-12) that takes it from the outer realm of the main Kuiper belt and possibly into the Oort cloud. Indeed, since the distance from the Sun to the inner edge of the Oort cloud is not known, astronomers are still debating whether it is another Kuiper belt object or the first known Oort cloud object. Sedna has a sidereal rotation period of 40 Earth days, one of the slowest rotations in the solar system. Just as the Moon has slowed down Earth's rotation, astronomers were expecting that Sedna has a moon to slow its rotation down. Observations failed to reveal such a moon; the cause of Sedna's slow rotation is still a mystery.

Astronomers calculate that the Oort cloud extends out at least 50,000 AU, one-fifth the distance to the nearest stars. Most of these comets have orbits so nearly circular that they never even get as close as Pluto is to the Sun. However, occasionally a passing star's gravitational force nudges a distant comet toward the inner solar system. As a result of its inward plunge, comets from the Oort comet cloud, like Hale-Bopp and Hyakutake, have highly elliptical, parabolic, or even hyperbolic orbits (see Section 2-8). Often comets also have orbits that are highly tilted, even perpendicular, to the plane of the ecliptic.

Because the comets in the Kuiper belt and Oort cloud are far from the Sun, they are completely frozen. Solid comet bodies, called **nuclei** (*singular* **nucleus**), are typically between 1 and 10 kilometers across, although some, like that of Comet Halley, are 15 km or more in diameter. The first pictures of a comet's nucleus were obtained when a fleet of

a

b

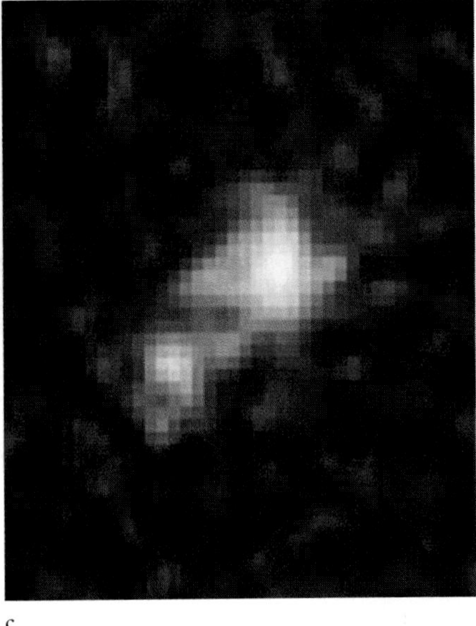
c

R I V U X G

FIGURE 9-11 **Kuiper Belt Objects** (a) and (b) These 1993 images show the discovery (white lines) of one of more than 938 known Kuiper belt objects. These two images of Kuiper belt object 1993 SC were taken 4.6 hours apart, during which time the object moved against the background stars. (c) The Kuiper belt object 1998 WW31 and its moon (lower left). (a & b: Alan Fitzsimmons, Queen's University of Belfast; c: C. Veillet/CFHT)

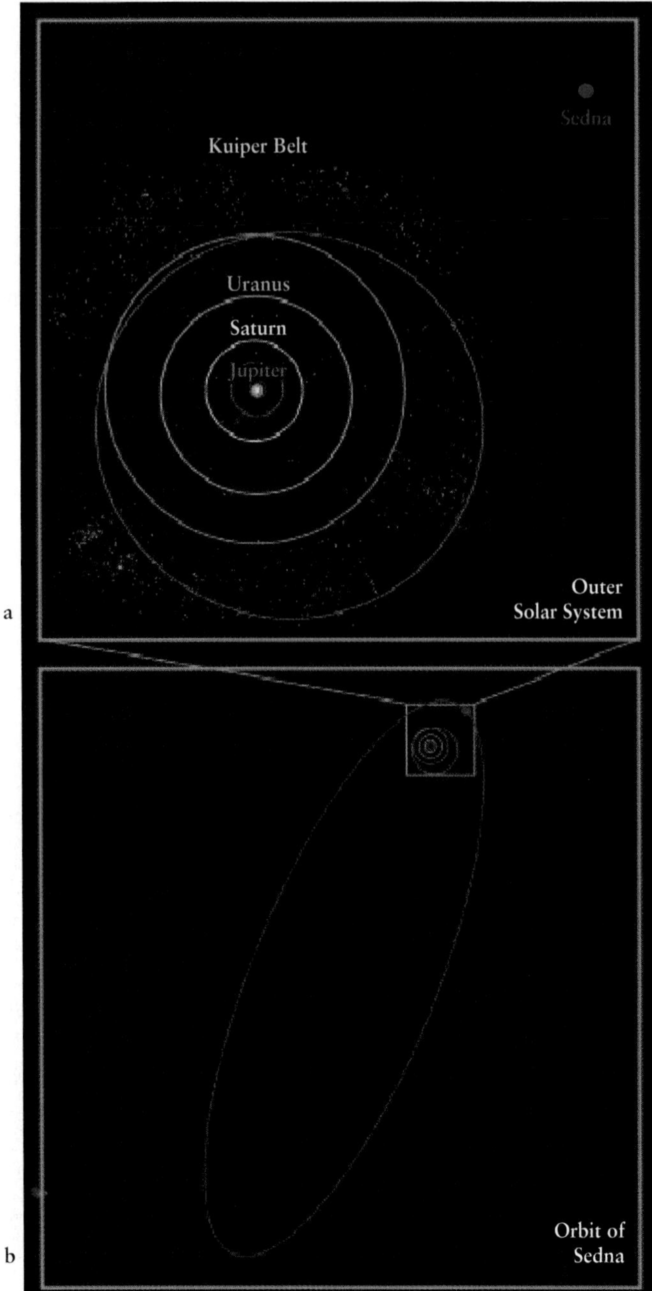

FIGURE 9-12 **Sedna's Orbit** **(a)** The farthest known body beyond Pluto is in a highly elliptical orbit **(b)** that ranges from the outer reaches of the Kuiper belt and possibly extends to the inner Oort comet cloud. (NASA/Caltech)

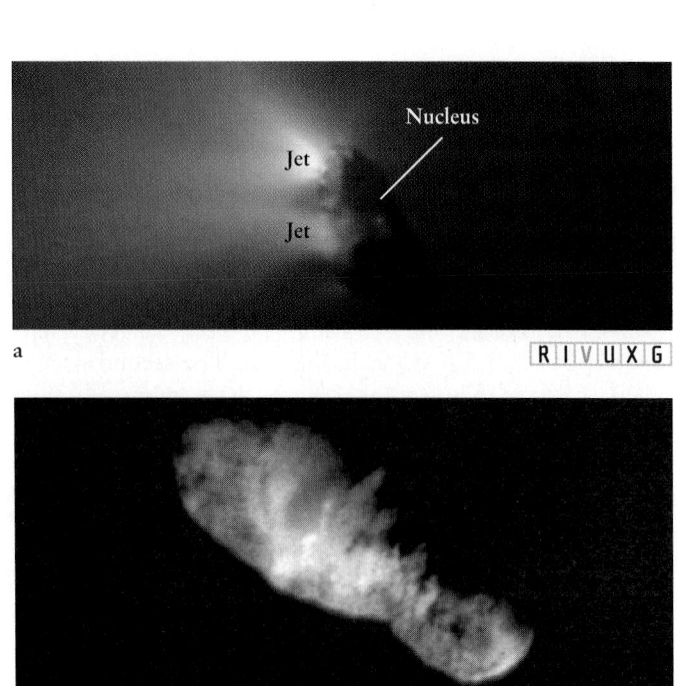

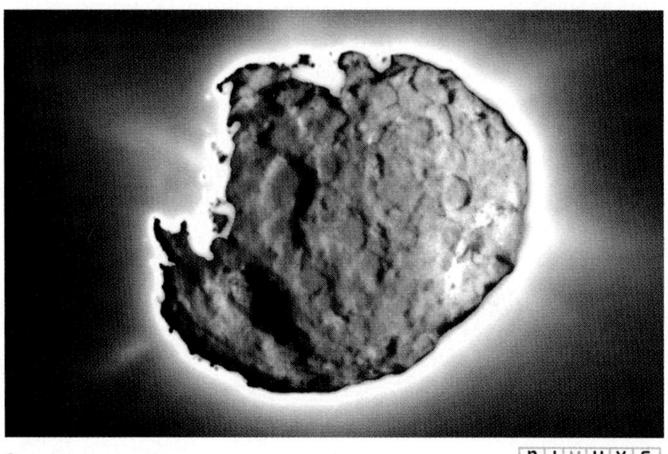

FIGURE 9-13 **Comet Nuclei** **(a)** The nucleus of Comet Halley. This image, taken by the *Giotto* spacecraft, shows the potato-shaped nucleus of the comet. Its dark nucleus measures 15 km in its longest dimension and about 8 km in its shortest. The Sun illuminates the comet from the left. The numerous bright areas on the nucleus are icy outcroppings that reflect more sunlight than surrounding areas of the comet. Two jets of gas can be seen emanating from the left side of the nucleus. **(b)** The nucleus of Comet Borrelly. Taken by *Deep Space 1*, this image has resolution of 45 m (150 ft). The nucleus is 8 km (5 mi) long, and the image was taken from 3417 km (about 2000 mi) away. **(c)** Comet Wild 2. This is two images combined. One is a high-resolution photograph showing the surprisingly heavily cratered comet. The other image is a longer photograph showing gas and dust jetting from the comet. Its tails are millions of kilometers long. (a: Max-Planck-Institut für Aeronomie; b: Deep Space 1 Team, JPL, NASA; c: NASA/JPL)

spacecraft flew past Comet Halley in 1986 (Figure 9-13a). Halley's potato-shaped nucleus is darker than coal, probably because of carbon-rich compounds left behind after its ice evaporated. In 2001, the *Deep Space 1* spacecraft took high-resolution images of Comet Borrelly's nucleus (Figure 9-13b). The ends of the comet are very rugged, while the center

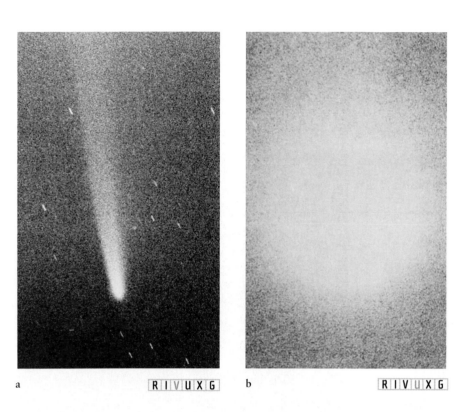

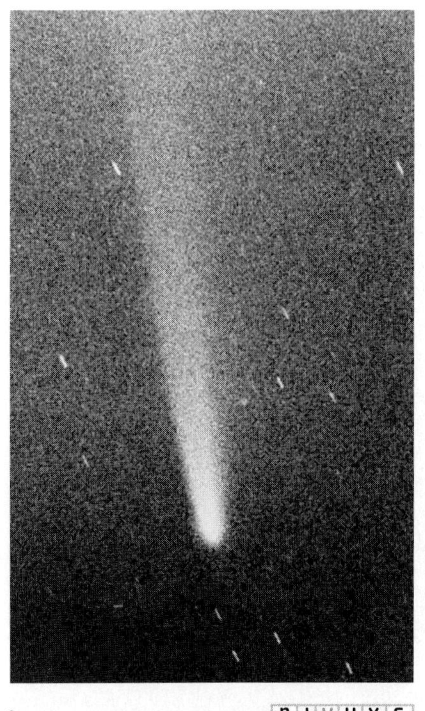

FIGURE 9-14 **Comet Kohoutek and Its Hydrogen Envelope** Comet Kohoutek as seen in visible light (a) and to the same scale at ultraviolet wavelengths (b). The ultraviolet picture reveals a huge hydrogen cloud surrounding the comet's nucleus. (Johns Hopkins University and Naval Research Laboratory)

has long, rolling plains, from which jets of gas appear to originate. To better determine the composition of the gas emitted by comets, the *Stardust* probe has visited Comet Wild 2 (Figure 9-13c), where it collected samples and is scheduled to return with them to Earth in 2006. Wild 2 showed many unexpected features, including towering spires, many craters and cliffs, and more than two dozen jets, rather than the two or three jets seen on other comets.

As a comet nucleus comes within 20 AU of the Sun, solar radiation begins to vaporize the ices on its surface. The liberated gases form an atmosphere, or **coma**, around the nucleus. Because the coma scatters sunlight, it appears as a fuzzy, luminous ball. The largest coma ever measured was more than a million kilometers across—nearly as large as the Sun. Not visible to the human eye is the **hydrogen envelope**, a sphere of tenuous gas surrounding the comet's nucleus and measuring as much as 20 million kilometers in diameter (Figure 9-14).

9-5 Comet tails develop from gases and dust pushed outward by the Sun

3 4 Of course, the most visible and inspiring features of comets are their long, flowing, diaphanous **tails** (Figure 9-15). Comet tails develop from coma gases and dust pushed outward from the Sun. This means that comet tails do not trail behind the nucleus, as the exhaust from a jet plane does in the Earth's atmosphere. Rather, *at the*

FIGURE 9-15 **Comet West** Astronomer Richard M. West first noticed this comet on a photograph taken with a telescope in 1975. After passing near the Sun, Comet West became one of the brightest comets of the 1970s. This photograph shows the comet in the predawn sky in March 1976. (Hans Vehrenberg)

comet's nucleus, the tails always point away from the Sun (Figure 9-16), regardless of the direction of the comet's motion. The implication that something from the Sun was "blowing" the comet's gases radially outward led Ludwig Biermann to predict the existence of the solar wind (see Sections 6-4 and 10-3). This stream of particles from the Sun was actually discovered in 1962, a full decade after it was predicted, by instruments on the spacecraft *Mariner 2*. For two and a half years, NASA's *Genesis* spacecraft collected pristine solar wind particles, returning them to Earth where it crash-landed in September 2004. Despite the disastrous landing, some of the particles that were collected were salvaged and are now being studied to learn more about them.

INSIGHT INTO SCIENCE

Starting with Observations Science often advances by working from observations and experiments that require scientific explanations. For example, seeing comet tails and how they behave led Biermann to wonder what causes them. The simplest physical explanation is that matter from each comet body is being evaporated and pushed upon by something from the Sun, the solar wind. Remember Occam's razor.

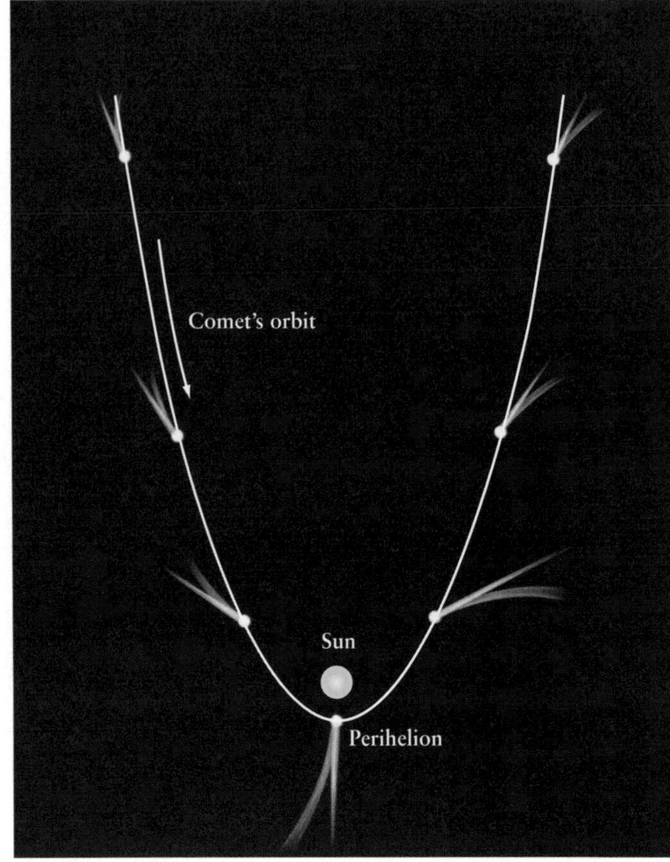

FIGURE 9-16 **The Orbit and Tails of a Comet** The solar wind and sunlight blow a comet's dust particles and ionized atoms away from the Sun. Consequently, comets' tails always point away from the Sun.

The Sun's radiation usually produces two comet tails: a **gas** (or **ion**) **tail** and a **dust tail** (Figure 9-17). Positively charged ions (atoms missing one or more electrons) from the coma are swept from the comet directly away from the

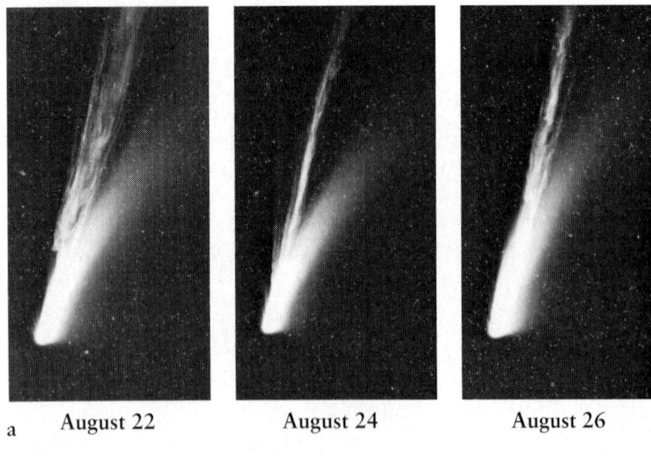

FIGURE 9-17 **The Two Tails of Comet Mrkos** (a) Comet Mrkos dominated the evening sky in August 1957. These three views, taken at two-day intervals, show dramatic changes in the comet's gas tail. In contrast, the slightly curved dust tail remained fuzzy and featureless. (b) Wind blowing smoke from this forest fire causes the smoke column to change shape and direction, just like the solar wind and sunlight cause the tails of comets to change shape and direction. (a: Palomar Observatory; b: Paul Von Baich/The Image Works)

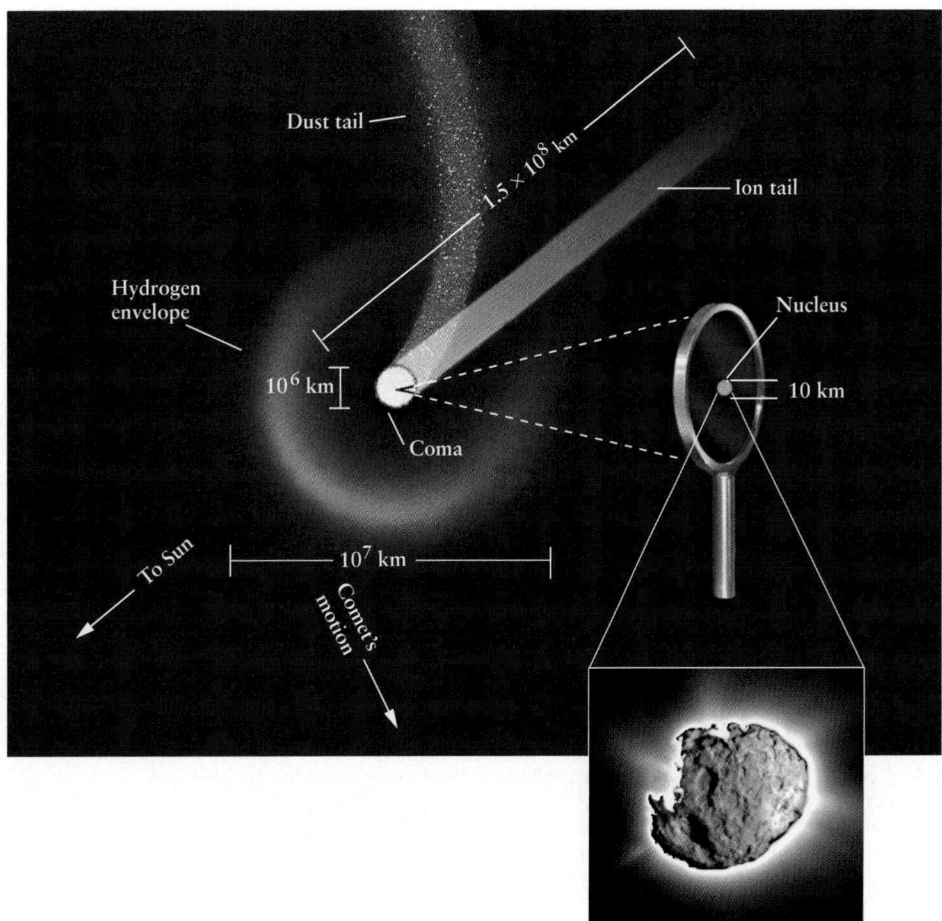

FIGURE 9-18 **The Structure of a Comet** The solid part of a typical comet (the nucleus) is roughly 10 km in diameter. The coma can be as large as 10^5 to 10^6 km across, and the hydrogen envelope is typically 10^7 km in diameter. A comet's tail can be enormous—as long as 1 AU. (This drawing is not to scale.) (Inset: NASA/JPL)

Sun by the solar wind to form the gas tail. This tail often appears blue, because, like the gas in our air, the tail's gas strongly scatters blue light. Ions typically leave the coma at speeds of 1.4 million kilometers per hour. The relatively straight gas tail can change dramatically from night to night (see Figure 9-17). This occurs because the solar wind that pushes the gases changes rapidly and the gases themselves are very light and easily moved about, like the smoke from a fire (Figure 9-17b).

The dust tail is formed when photons strike dust particles that have been freed from the comet's evaporating nucleus. Light exerts pressure on any object that absorbs or scatters it. While this pressure, called **radiation pressure** or **photon pressure**, is quite weak, fine-grained dust particles in a comet's coma are sufficiently light to be pushed from the vicinity of the comet, thus producing a dust tail. The dust tail often is the color of sunlight. The dust particles are massive enough not to flow straight away from the Sun; rather, the dust tail arches in a path that lies between the gas tail and the direction from which the comet came (Figure 9-16 and Figure 9-18).

Figure 9-18 outlines the elements of comets, although many have bigger or smaller features. Some comets, like the one shown in Figure 9-19, have a large, bright coma but short, stubby tails. Others have an inconspicuous coma but one or more tails of astonishing length. The gas tail in Figure 9-20 stretches more than 150 million kilometers (1 AU) in length.

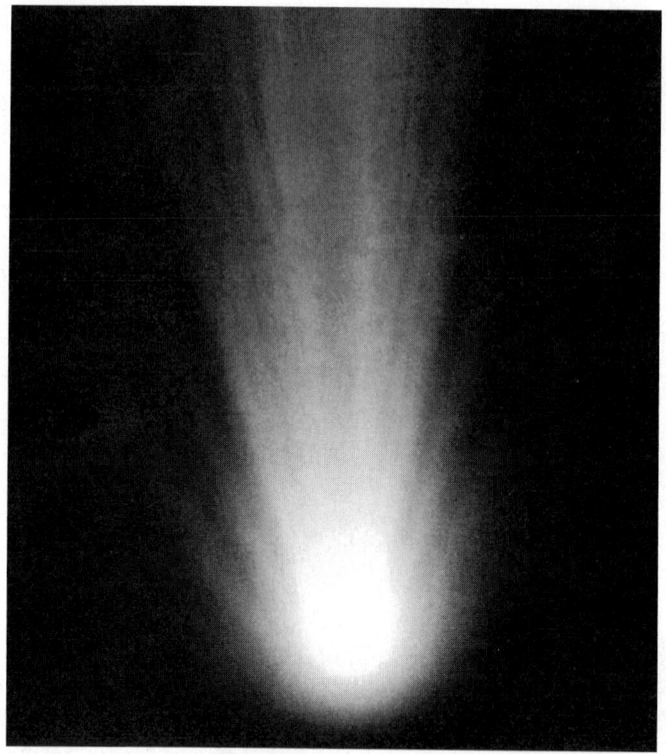
R I V U X G

FIGURE 9-19 **The Head of Comet Brooks** This comet, named after its discoverer, had an exceptionally large, bright coma. It dominated the night skies in October 1911. (UCO/Lick Observatory)

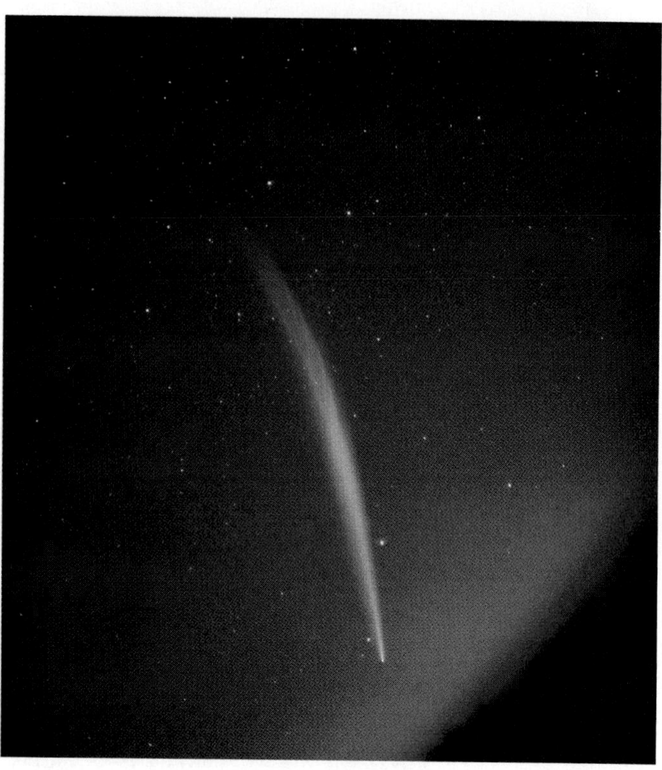
R I V U X G

FIGURE 9-20 **The Tail of Comet Ikeya-Seki** Named after its codiscoverers in Japan, this comet dominated the predawn skies in late October 1965. Although its coma was tiny, its tail spanned over 1 AU. (Roger Lynds/NOAO/AURA/NSF)

9-6 Comets do not last forever

Astronomers, many of them amateurs, typically discover at least a dozen new comets each year. Falling sunward from the Kuiper belt or Oort cloud, most are **long-period comets,** which have such eccentric orbits that they leave the inner solar system after one pass by the Sun and typically take 1 million to 30 million years to return, if ever.

However, sometimes a comet passes so close to a giant planet that the planet's gravitational force changes the comet's orbit, slowing it down and trapping it in the inner solar system (Figure 9-21). The comet then becomes a **short-period comet,** orbiting the Sun in fewer than 200 years. Like Comet Halley, short-period comets appear again and again at predictable intervals. Your next chance to see Comet Halley is in 2061.

Comets cannot survive an infinite number of passages near the Sun. A typical comet is estimated to lose between 1/60th and 1/100th of its mass at each perihelion (closest approach to the Sun). Therefore, a typical comet survives at

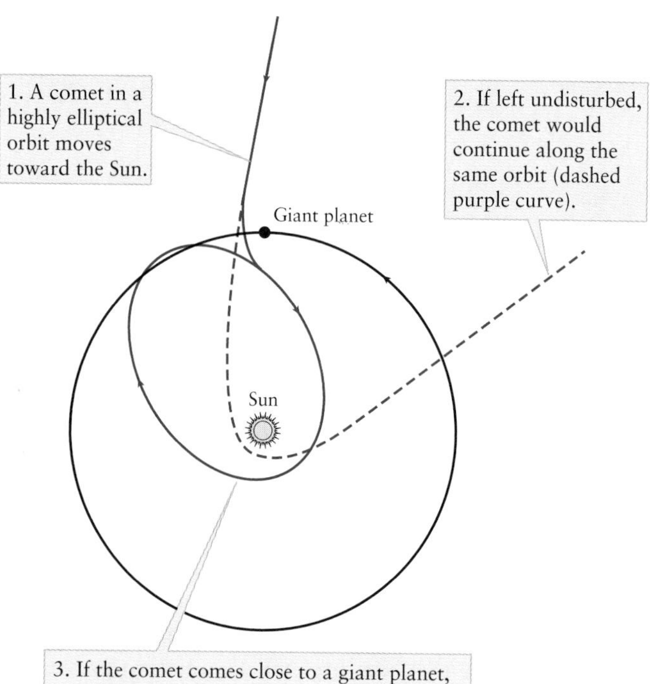
1. A comet in a highly elliptical orbit moves toward the Sun.
2. If left undisturbed, the comet would continue along the same orbit (dashed purple curve).
Giant planet
Sun
3. If the comet comes close to a giant planet, the planet's gravity can deflect the comet into a smaller orbit (solid purple curve).

▶ FIGURE 9-21 **Transforming a Long-Period Comet into a Short-Period Comet** The gravitational force of a giant planet can change a comet's orbit. Initially, on highly elliptical orbits, comets are sometimes deflected into more circular paths that keep them in the inner solar system.

most 100 close passes to the Sun before its ices evaporate completely. Comet Halley has been seen at least 27 times, meaning that it has no more than 73 more close passes to the Sun before it dissipates. Since it returns every 76 years or so, it will survive for fewer than 73 × 76 = 5548 more years. Figure 9-22 shows the nucleus of a comet breaking up shortly after it passed perihelion. Soon thereafter, its remaining dust and rock fragments spread out in a loose collection of debris that continues to circle the Sun along the comet's orbital path.

A comet can also be torn apart when it comes too close to a planet, or it can be destroyed completely by striking a planet, a moon, or the Sun. A spectacular example was Comet Shoemaker-Levy 9. As discussed in Section 8-3, that comet fragmented under the tidal force from Jupiter in 1992. Two years later, with the world's astronomers watching carefully, the pieces returned and struck the planet (see Figures 8-7 and 8-8).

The comet's disintegration provided an important clue to its structure. For Shoemaker-Levy 9 to break up in the first place tells astronomers that its nucleus must have been held together very weakly. It may actually have been composed of separate pieces that had stuck together until Jupiter's tidal force pulled them apart. Since then, astronomers have observed at least four other comets break up, in 1999, 2000 (Comet LINEAR), 2001, and 2002 (Comet Toit-Neujmin-Delporte). These observations suggest that this structure may be typical of other comets as well.

R I V U X G

FIGURE 9-22 The Fragmentation of Comet LINEAR On July 26, 2000, Comet LINEAR passed its perihelion 0.74 AU from the Sun. Within days, the comet was observed to have fragmented, like Comet Shoemaker-Levy 9 and Comet West (among many others) before it. (European Southern Observatory)

Comets occasionally lose mass quickly by ejecting it in bursts. Several times, starting in September 1995, astronomers observed Comet Hale-Bopp eject 7 to 10 times more mass than usual (Figure 9-23). Astronomers believe that this event resulted from surface ice and dust being

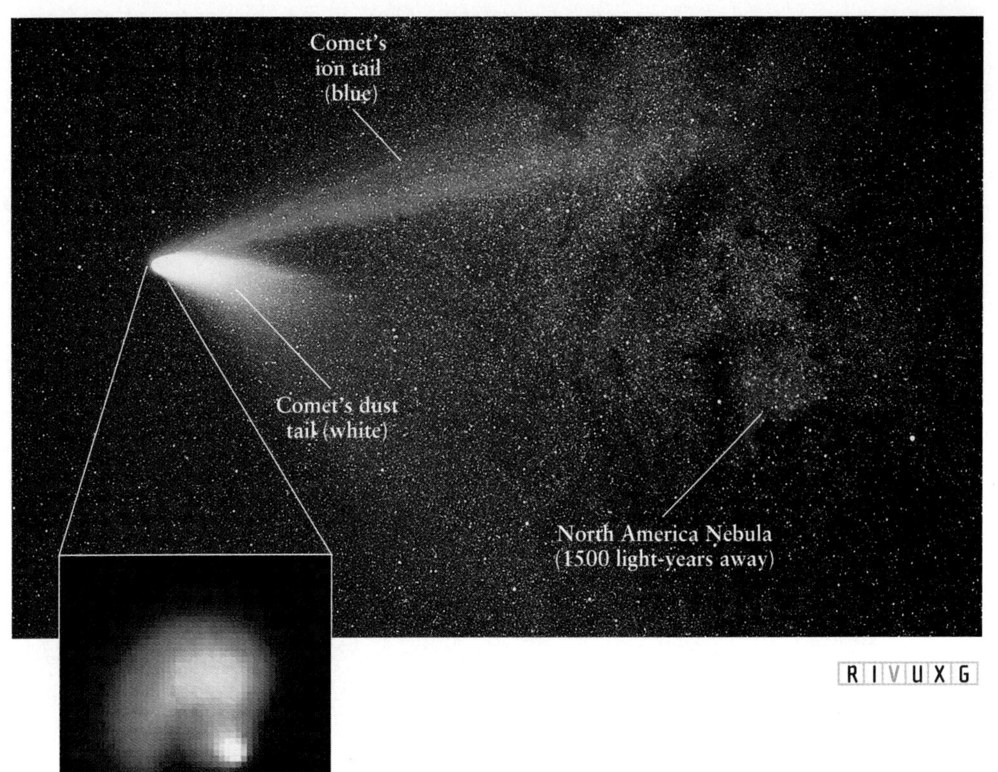

FIGURE 9-23 Comet Hale-Bopp Discovered on July 23, 1995, this comet was at its breathtaking best in mid-1997. **Inset:** Jets of gas and debris were observed shooting out from Comet Hale-Bopp several times. This image shows the comet nucleus (lower bright region), an ejected piece of the comet's surface (upper bright region), and a spiral tail. The ejected piece eventually disintegrated, following the same spiral pattern as the tail. (Tony and Daphne Hallas, Astrophotos)

heated by the Sun and then being rapidly ejected from a local region on the comet's surface called a *vent*. Hale-Bopp rotates with a period of about 12 hours. The ejected matter therefore formed a pinwheel-shaped distribution spiraling from the comet's nucleus at a speed of 109 km/h (68 mph). While the ejected matter did not represent a significant fraction of the comet's mass, its light-reflecting dust made it look like a huge comet fragment.

The Sun is not the only star around which comets orbit. A star labeled CW Leonis, about 500 light-years from Earth, is presently enlarging in size and vaporizing billions of comets around it. The spectra of the water vapor from these bodies have been observed.

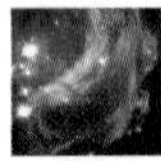

METEOROIDS, METEORS, AND METEORITES

As noted earlier, asteroids occasionally collide with each other, sending fragments into interplanetary space. These smaller pieces, along with rocky and metallic debris from evaporating comets, and material that never coalesced with larger bodies, are still strewn throughout the solar system. **Meteoroids** are rocky and metallic debris smaller than asteroids scattered throughout the solar system. Although no official size standard distinguishes the two, meteoroids are no more than about a hundred meters across and the vast majority of them are smaller than a millimeter.

By studying the chemistry of various asteroids, astronomers are now beginning to identify which asteroids were the origins for other asteroids and for meteoroids. For example, the asteroid Braille (named in honor of Louis Braille, developer of the alphabet for the blind) is a piece of debris 2 km long blasted off the 500-km-diameter asteroid Vesta.

9-7 Small rocky debris peppers the solar system

Passing meteoroids are often pulled by gravity toward 5
Earth's atmosphere. As they move through the atmosphere, they compress and heat the air in front of them so much that the air glows. The glowing gas creates the trail we see, and the meteoroid becomes a **meteor** (Figure 9-24). Common names for these dramatic streaks of light flashing across the sky include *shooting stars, fireballs,* and *bolides.* Fireballs are meteors at least as bright as Venus; bolides are bright meteors that explode in the air. Therefore, "shooting stars" are not stars of any kind, nor are they dying stars.

9-8 Impact craters and meteor showers mark remnants of space debris on Earth, while meteorite impacts are seen on the Moon

The hot air in front of meteoroids causes them to begin vaporizing. Most meteors vaporize completely before they can strike the Earth. Their dust settles to the ground, often carried by raindrops. (This is not the source of acid rain,

FIGURE 9-24 **A Meteor** This brilliant meteor is seen lighting up the dark desert skies of the California desert area of Joshua Tree National Park. To the right of it are the Pleiades. (Wally Pacholka/Astropics.com)

R I V U X G

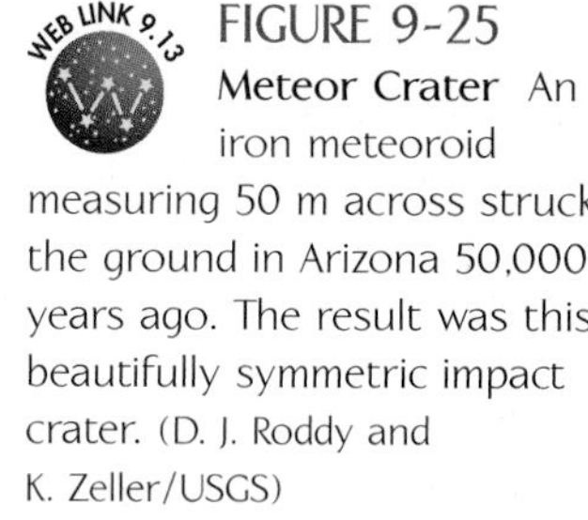

FIGURE 9-25 **Meteor Crater** An iron meteoroid measuring 50 m across struck the ground in Arizona 50,000 years ago. The result was this beautifully symmetric impact crater. (D. J. Roddy and K. Zeller/USGS)

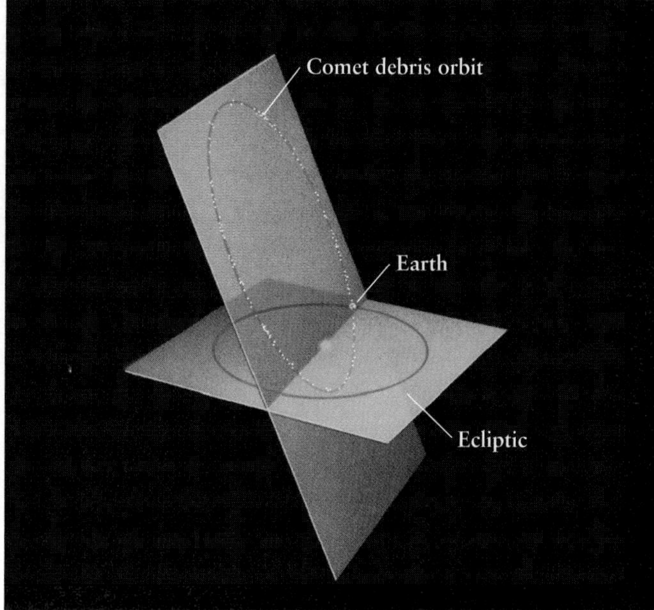

FIGURE 9-26 **The Origin of Meteor Showers** As comets dissipate, they leave debris behind that spreads out along their orbit. When the Earth plows through such material, many meteors can be seen emanating from the same place within a very short time—a meteor shower. As shown in this diagram, many comets have high orbital inclinations.

however, which comes from natural and human-made gases ejected into the atmosphere.) Any part of a meteor that survives its fiery descent to Earth may leave an **impact crater.** Weather and water erosion are wearing away all of the 200 or so impact craters now known on the Earth, and thousands more have long since been drawn into the Earth by the motion of its tectonic plates. Indeed, those known today are all less than 500 million years old because of forces reshaping the Earth's surface.

One of Earth's best-preserved impact craters is Meteor (or Barringer) Crater near Winslow, Arizona (Figure 9-25). Measuring 1.2 km across and 200 m deep, it formed approximately 50,000 years ago when an iron-rich meteoroid some 50 m across (about half the length of a football field) struck the ground at 40,000 km/h (25,000 mph). The blast was like the detonation of a 20-megaton hydrogen bomb.

On a typical clear night, you can expect to see a meteor about every 10 minutes. However, at predictable times throughout each year, the Earth is inundated with them. These **meteor showers** occur when the Earth moves through the orbit of debris left behind by a comet (Figure 9-26 and Figure 9-27).

Some 30 meteor showers can be seen each year. Because the meteors in each shower appear to come from a fixed region of the sky, they are named after the constellation

PROMINENT YEARLY METEOR SHOWERS

Shower	Date of maximum intensity	Typical hourly rate	Constellation
Quadrantids	January 3	40	Boötes
Lyrids	April 22	15	Lyra
Eta Aquarids	May 4	20	Aquarius
Delta Aquarids	July 30	20	Aquarius
Perseids	August 12	80	Perseus
Orionids	October 21	20	Orion
Taurids	November 4	15	Taurus
Leonids	November 16	15	Leo Major
Geminids	December 13	50	Gemini
Ursids	December 22	15	Ursa Minor

R I V U X G

FIGURE 9-27 **Meteor Streaks Seen during a Meteor Shower** This table lists highly active, annual meteor showers, which last for several days. The time exposure, taken in 1998, shows meteors streaking away from the constellation Leo Major. They are part of the Leonid meteor shower. This shower occurs because the Earth is moving through debris left by comet Temple-Tuttle. (Tony and Daphne Hallas/Photo Researchers, Inc.)

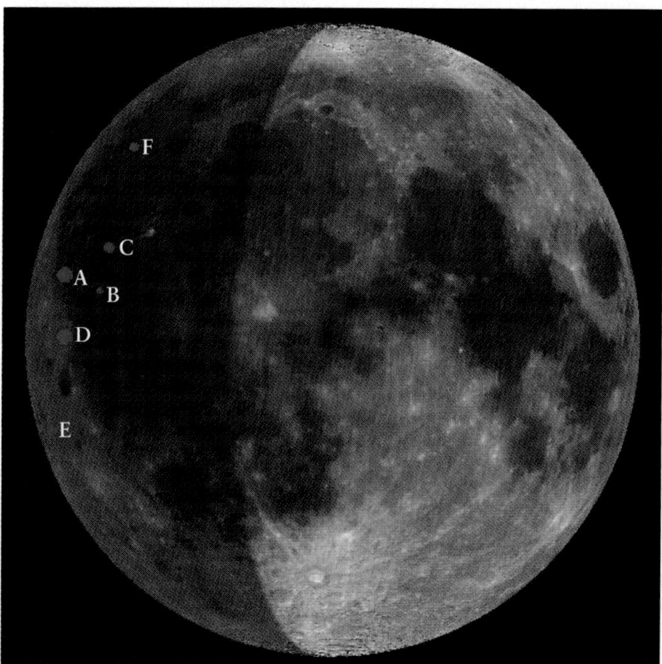

FIGURE 9-28 **Recent Impacts on the Moon** The locations A–F are places on the Moon where impacts were observed from Earth in 1999 during the Leonid meteor shower. The impacting bodies hit the Moon at around 72 km/s (160,000 mph) and had masses of between 1 and 10 kg. Each impact created a short-lived cloud that momentarily heated to between 5×10^4 and 10×10^4 K, much hotter than the surface of the Sun. (NASA)

from which the meteors appear to radiate (Figure 9-27). For example, meteors in the Perseid shower appear to originate in the constellation Perseus. More than one meteor can be seen each minute at the peaks of such prodigious meteor showers as the Perseids, which take place in the summertime and are among the most exciting light shows in astronomy. Except for the Lyrids, meteor showers are best seen after midnight.

The Moon also moves through the debris that creates meteor showers on the Earth. Whereas our atmosphere vaporizes most of the meteors before they strike the Earth, the Moon's thin atmosphere provides it no protection and so the Moon is struck by numerous meteoroids during each meteor shower. The Leonid meteor shower of 1999 provided astronomers with an excellent opportunity to observe contemporary impacts on the Moon. Some of these impacts were even bright enough to see with the naked eye (Figure 9-28). When a 10-kg meteorite strikes the Moon, the impact is as powerful as 10^4 pounds of TNT and the resulting cloud is momentarily between 5×10^4 and 10×10^4 K, which is hotter than the surface of the Sun. Subsequent Leonids have provided further impact sightings on the Moon.

9-9 Meteorites are space debris that land intact

Although most meteors entering Earth's atmosphere completely vaporize, some reach the ground before totally disintegrating. Such pieces of debris are called **meteorites,** and they come from a variety of sources: Many are believed to have been broken off from asteroids by collisions; some are debris that were never part of larger bodies; still others come from the Moon, Mars, and comets. We saw in Section 7-12, for example, how SNC meteorites offer clues to the chemistry of Mars.

Meteorites tell us the age of the solar system. Measuring the various radioactive elements in meteorites allows astronomers to determine how long ago they formed (see An Astronomer's Toolbox 4-3). The oldest known meteorites became solid bodies about 4.57 billion years ago and impacts on Earth as early as 4 billion years ago have been identified, based on how they disturbed preexisting rock here. Therefore, the solar system is at least 4.57 billion years old, and we will use 4.6 billion years as its age throughout the book.

People have been picking up debris from space for thousands of years. The descriptions of meteorites in historical Chinese, Indian, Islamic, Greek, and Roman literature show that early peoples placed special significance on these "rocks from heaven." They also have a practical "impact": Infalling space debris is increasing the Earth's mass by nearly 300 tons per day on average (Figure 9-29).

FIGURE 9-29 **The Mass of Impacts on Earth** The Vatican Obelisk is about 300 tons, the amount of mass that strikes the Earth every day. As a result, the Earth's mass increases by the equivalent of 300 tons every day. (John and Dallas Heaton/Corbis)

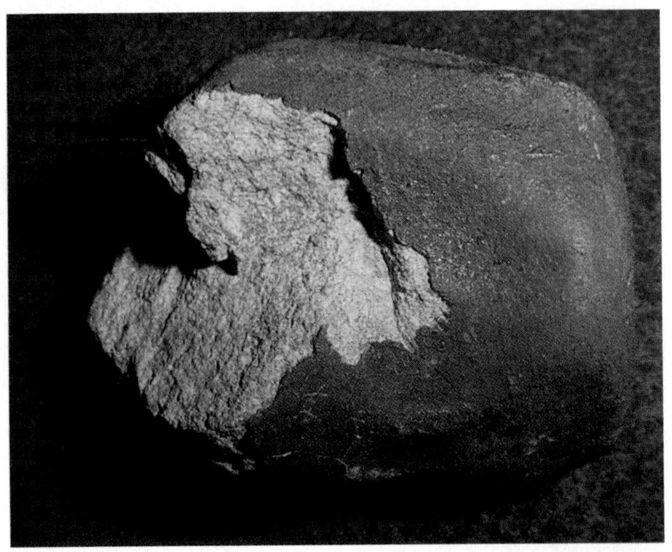

FIGURE 9-30 **Stony Meteorites** (a) Most meteorites that fall to Earth are stones. Many freshly discovered specimens, like the one shown here, are coated with thin, dark crusts. This stony meteorite fell in Morocco. (b) Some stony meteorites contain tiny specks of iron, which can be seen when the stones are cut and polished. This specimen was discovered in Ohio. (The R. N. Hartman Collection)

Stony Meteorites Often Look like Ordinary Rocks

Meteorites are classified as stones (or stony), stony-irons, and irons. Most **stony meteorites** look much like ordinary rocks, although some are covered with a dark *fusion crust* (Figure 9-30a), created when the meteorite's outer layer melts during its fiery descent through the atmosphere (see Figure 9-27). When a stony meteorite is cut in two and polished, tiny flecks of iron are sometimes found in the rock (Figure 9-30b).

Stony meteorites account for about 95% of all meteoritic material that falls on the Earth. Meteorites with a high iron content can be located with a metal detector. They also look unusual and hence are more likely to be noticed. Consequently, the easily found iron and stony-iron meteorites dominate most museum collections.

Iron Meteorites Are Very Dense and Look Noticeably Different Than Rock

Iron meteorites (Figure 9-31a) may also contain from 10% to 20% nickel by weight. Iron is moderately abundant in the universe as well as being one of the most common rock-forming elements, so it is not surprising that iron is an important constituent of asteroids and meteoroids. Another element, iridium, is common in the iron-rich minerals of meteorites but rare in ordinary rocks on the Earth's surface because most of the iridium that collected here when the Earth formed settled deep into the Earth eons ago. Measurements of iridium in the Earth's crust can thus tell us the rate at which meteoritic material has been deposited on the Earth over the ages.

In 1808, Count Alois von Widmanstätten, director of the Imperial Porcelain Works in Vienna, discovered a conclusive test for the most common type of iron meteorite. Most irons have a unique structure of long nickel-iron crystals called **Widmanstätten patterns,** which become visible when the meteorites are cut, polished, and briefly dipped into a dilute solution of acid (Figure 9-31b). Because nickel-iron crystals can grow to lengths of several centimeters only if the molten metal cools slowly over many millions of years, Widmanstätten patterns are never found in counterfeit meteorites, or "meteorwrongs."

Stony-Iron Meteorites Are the Most Exotic of All Space Debris on Earth

The final category of meteorites are the **stony-iron meteorites** which consist of roughly equal amounts of rock and iron. Figure 9-32, for example, shows the greenish mineral olivine suspended in a matrix of iron.

To understand why different types of meteorites exist, we consider their formation. Most meteorites were once pieces of asteroids. Heat from the rapid decay of radioactive isotopes melted newly formed asteroid interiors. Over the next few million years, differentiation occurred, just as in the young Earth. Iron sank toward the asteroid's center, while lighter rock floated up to the asteroid's surface. Iron meteorites are fragments of asteroid cores, and stones are samples of their crusts. Stony-irons are believed to come from the boundary regions between the iron cores and stony crusts.

A class of rare stony meteorites, **carbonaceous chondrites,** shows no evidence of ever having been melted as parts of asteroids. These rarities may therefore be primordial material from which our solar system was created.

FIGURE 9-31 **Iron Meteorites** (a) Irons are composed almost entirely of iron-nickel minerals. The surface of a typical iron is covered with thumbprintlike depressions created as the meteorite's outer layers vaporized during its high-speed descent through the atmosphere. This specimen was found in Argentina. (b) When cut, polished, and etched with a weak acid solution, most iron meteorites exhibit interlocking crystals in designs called Widmanstätten patterns. This meteorite was found in Australia. (a: The R. N. Hartman Collection; b: R. A. Oriti)

Carbonaceous chondrites are meteorites that contain small glass-rich beads called *chondrules*. The meteorities also contain complex carbon compounds, including simple sugars and glycerin, among others. They also have as much as 20% water bound into their minerals. The organic compounds would have been broken down and the water driven out if these meteorites had been significantly heated. Asteroid Mathilde, shown in Figure 9-8, has a very dark gray color and virtually the same spectrum as a carbonaceous chondrite meteorite, so it is likely composed of primordial material.

FIGURE 9-32 **A Stony-Iron Meteorite** Stony-irons account for about 1% of all meteorites that fall to Earth. This specimen, a variety of stony-iron called a pallasite, was found in Antarctica. This specimen is thinly cut and appears to glow because of a light located behind it. (James C. Hartman)

Amino acids, the building blocks of proteins upon which terrestrial life is based, are among the organic compounds occasionally found inside carbonaceous chondrites. These amino acids may be contaminants acquired after the meteoroids entered the Earth's atmosphere. However, space scientists have actually created them in the laboratory under the conditions found in deep space, showing that amino acids may exist out there. Indeed, some scientists suspect that amino acid–rich carbonaceous chondrites may have played a role in the origin of life on Earth.

There Is a Difference between the Percentages of Meteorite Impacts and the Percentages of Meteorite "Finds"

Most stony meteorites are not identified as meteorites because, well, they look like stones. Indeed, the percentages of the stony, iron, and stony-iron meteorites that are discovered are quite different from the percentages that actually land. Nevertheless, astronomers and geologists have a good idea about how many of each type strike land. They get the correct percentages of impacts by carefully surveying areas in which only meteorites land, namely snow and ice-covered regions, such as Antarctica, or on deserts. By counting all the debris under the surface using metal detectors and other technologies, the actual number of impacts of each type of meteorite is determined.

9-10 The Allende meteorite and Tunguska mystery provide evidence of catastrophic collisions

A chance to study debris immediately after impact came shortly after midnight on February 8, 1969, when a brilliant, blue-white light shot across the night sky around Chihuahua, Mexico. Hundreds of people witnessed the dazzling display. The light disappeared in a spectacular and deafening explosion that dropped thousands of rocks and pebbles over the terrified onlookers. Within hours, teams of scientists were on their way to collect specimens of carbonaceous chondrites, collectively named the *Allende meteorite,* after the locality in which they fell (Figure 9-33).

One of the most significant discoveries to come from the Allende meteorite was evidence of the detonation of a nearby supernova 4.6 billion years ago. Among nature's most violent and spectacular phenomena, a *supernova explosion* occurs when a massive star dies. A massive star blows apart in a cataclysm that hurls matter outward at tremendous speeds, as we will see in Chapter 13. During this detonation, violent collisions between atomic nuclei produce a host of radioactive elements, including a short-lived radioactive isotope of aluminum. Based on its decay products, scientists found unmistakable evidence that this isotope once lay within the Allende meteorite. Some astronomers interpret this as evidence for a supernova in our vicinity at about the time the Sun was born. By compressing interstellar gas and dust, the supernova's shock wave may have helped stimulate the birth of our solar system.

At 7:14 A.M. local time on June 30, 1908, another spectacular explosion occurred, this one over the Tunguska region of Siberia. The blast, comparable to a nuclear detonation of several megatons, knocked a man off his porch some 60 km away and was audible more than 1000 km away. Millions of tons of dust were injected into the atmosphere, darkening the air as far away as California.

Preoccupied with wars, along with political and economic upheaval, neither Russia nor its successor, the former Soviet Union, sent a scientific expedition to the site until 1927. At that time, Soviet researchers found that trees had been seared and felled radially outward in an area about 30 km in diameter (Figure 9-34). There was no clear evidence of a crater. In fact, the trees at "ground zero" were left standing upright, although they were completely stripped of branches and leaves. Because no significant meteorite samples were found, for many years scientists assumed that a small comet had struck the Earth.

Recently, however, several teams of astronomers have argued that a small comet, composed primarily of light elements and ice, breaks up too high in the atmosphere to cause significant damage on the ground. They argue that the Tunguska explosion was actually caused by a small asteroid or large meteoroid traveling at supersonic speed. The Tunguska event is consistent with an explosion of an asteroid about 80 m (260 ft) in diameter entering the Earth's atmosphere at 79,000 km/h (50,000 mph) and exploding in the air as a result of becoming exceedingly hot.

a

R I V U X G

b

R I V U X G

FIGURE 9-33 **Pieces of the Allende Meteorite** (a) This carbonaceous chondrite fell near Chihuahua, Mexico, in February 1969. Note the meteorite's dark color, caused by a high abundance of carbon. Geologists believe that this meteorite is a specimen of primitive planetary material. The ruler is 15 cm long. (b) Sliced open, the Allende meteorite shows round, rocky *inclusions* called *chondrules* in a matrix of dark rock. (a: J. A. Wood; b: The R. N. Hartman Collection)

R I V U X G

FIGURE 9-34 **Aftermath of the Tunguska Event** In 1908, a stony asteroid traveling at supersonic speed struck the Earth's atmosphere and exploded over the Tunguska region of Siberia. Trees were blown down for many kilometers in all directions from the impact site. (SOVFOTO)

9-11 Asteroid impacts with Earth have caused mass extinctions

In the late 1970s, the geologist Walter Alvarez and his father, physicist Luis Alvarez, discovered a different sort of shock wave. Working at a site of exposed marine limestone in the Apennine Mountains in Italy that had been on the Earth's surface 65 million years ago, the Alvarez team discovered an exceptionally high abundance of iridium in a dark-colored layer of clay between limestone strata (Figure 9-35).

Since this discovery was announced in 1979, a comparable layer of iridium-rich material has been uncovered at numerous sites around the world. In every case, geologic dating reveals that this apparently worldwide iridium-rich layer was deposited about 65 million years ago. Paleontologists were quick to realize the significance of this date, because it was 65 million years ago when all the dinosaurs rather suddenly became extinct. In fact, two-thirds of all the species on Earth disappeared within a brief span of time back then.

The Alvarez discovery supported an astronomical explanation for the dramatic extinction of so much of the life that once inhabited our planet—an asteroid impact. There is no universal agreement on the disasters that befell the Earth during this episode. One scenario has an asteroid 10 km in diameter slamming into the Earth at high speed and throwing enough dust into the atmosphere to block out sunlight for several years. As the temperature dropped drastically and plants died for lack of sunshine, the dinosaurs would have perished, along with many other creatures in the food chain that were highly dependent on vegetation. The dust eventually settled, depositing an iridium-rich layer over the Earth. Another theory posits that the impact created enough heat to cause planetwide fires, followed by changes in the oceans that killed many species of life in them. Whatever the correct scenario turns out to be, tiny, rodentlike creatures capable of ferreting out seeds and nuts were among the animals that managed to survive this holocaust, setting the stage for the rise of mammals and, consequently, the evolution of humans.

In 1992, a team of geologists discovered that the hypothesized asteroid crashed into a site in Mexico. They based this conclusion on glassy debris and violently shocked grains of rock ejected from the multiringed, 195-km-diameter Chicxulub Crater buried under the Yucatán Peninsula in Mexico (Figure 9-36). From the known rate at which radioactive potassium decays, the scientists have pinpointed the date when the asteroid struck—64.98 million years ago. In 1998, geologists digging on the Pacific Ocean floor discovered a piece of meteoritic debris with precisely the same age, apparently a piece of the offending asteroid. While some geologists and paleontologists are not yet convinced that an asteroid impact led to the extinction of the dinosaurs, most agree that this hypothesis fits the available evidence better than any other explanation that has been offered so far.

R I V U X G

FIGURE 9-35 **Iridium-Rich Layer of Clay** This photograph of strata in the Apennine Mountains of Italy shows a dark-colored layer of iridium-rich clay sandwiched between white limestone (below) from the late Mesozoic era and grayish limestone (above) from the early Cenozoic era. The coin is the size of a U.S. quarter. (W. Alvarez)

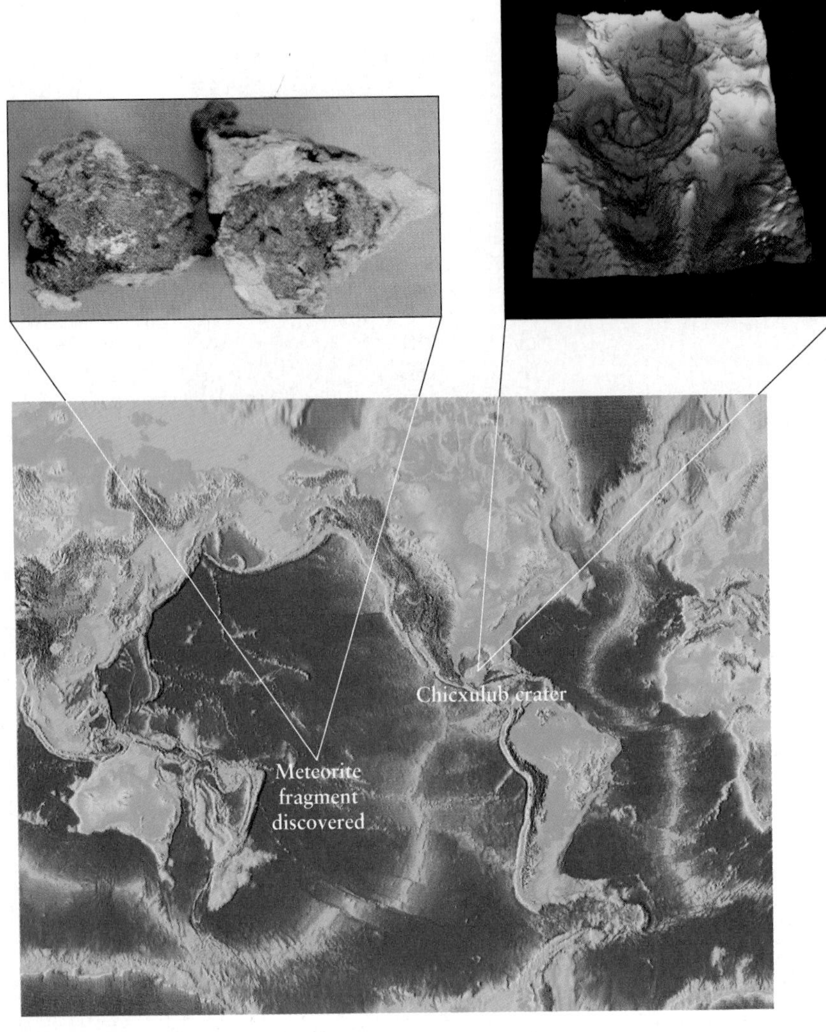

FIGURE 9-36 **Confirming an Extinction-Level Impact Site** **Right inset:** By measuring slight variations in the gravitational attraction of different materials under the Earth's surface, geologists create images of underground features. Concentric rings of the underground Chicxulub Crater, shown here, lie under a portion of the Yucatán Peninsula. This crater has been dated to 65 million years ago and is believed to be the site of the impact that led to the extinction of the dinosaurs. **Left inset:** A piece of 65-million-year-old meteorite discovered in the middle of the Pacific Ocean in 1998 and believed to be a fragment of the meteorite that struck the Yucatán Peninsula. The fragment, about a tenth of an inch long, was cut into two pieces, shown here. (Top left inset: Frank T. Kyte, UCLA; top right inset: Virgil L. Sharpton, Lunar and Planetary Institute; digital image by Peter W. Sloss, NOAA-NESDIS-NGDCD)

Another Impact Led to an Earlier, More Devastating Mass Extinction

The end of the dinosaurs' reign was not the only mass extinction caused by an impact. In 2001, evidence came to light from the Permian-Triassic boundary of 250 million years ago that suggests another devastating blow from space. That time was called the "Great Dying," a mass extinction during which some 80% of the species of life living on land and 90% of those living in the oceans perished. Rocks from that time discovered in places from Japan to Hungary show evidence, in the form of fullerenes, soccer ball–shaped molecules containing at least 60 carbon atoms, of an impact from space. Trapped inside these fullerenes were gases that could only have been forced into them from stars. (This signature of gas-filled fullerenes from space has now also been discovered in the layer of rock existing on the Earth's surface 65 million years ago.)

A 250-million-year-old crater was found in 2004 off the northwestern coast of modern-day Australia. Keep in mind that both this impact and the one that ended the reign of the dinosaurs occurred on a world whose surface is in continual motion. Therefore, their present locations are not the same as their initial impact sites. The Chicxulub impact actually occurred in the Pacific Ocean, while the Australian impact occurred before the continents were separated from Pangaea (see Section 6-2). The impact explanation of the Great Dying is not universally accepted. Another possible explanation is a massive amount of volcanic eruption that darkened the skies and changed the climate.

Could a catastrophic impact happen again? So many asteroids cross the Earth's path that scientists agree that it is a matter of "when" rather than "if." The good news is that studies of craters show that larger asteroids strike the Earth significantly less often than do smaller ones. While asteroids large enough to create Meteor Crater strike the Earth about once every 10,000 years, killer asteroids, like the one that killed off the dinosaurs, collide with Earth only once every 100 million years. The threat of a catastrophic impact by an asteroid or comet in our lifetimes, thankfully, is remote.

9-12 Frontiers yet to be discovered

The years of the *Pioneer* and *Voyager* spacecraft were the first golden era of solar system research. We are now in another such period, with spacecraft either in development or already going to planets, asteroids, comets, and even, as we will see shortly, out observing the weather in space. From the solar system debris astronomers hope to learn whether life on Earth was brought here from elsewhere by asteroid or meteoroid impacts. They also hope to answer such questions as the evolutionary history of the space debris, how much of Earth's water came here during the planet's formation and how much landed afterward from comet impacts, whether the asteroids have sufficiently valuable compositions to justify mining them, whether comets can be harvested to supply water and other materials for people colonizing the solar system, whether the Oort cloud really exists, and whether Pluto should be considered a Kuiper belt object.

Summary of Key Ideas

Asteroids

• Tens of thousands of belt asteroids with diameters larger than a kilometer are known to orbit the Sun between the orbits of Mars and Jupiter. The gravitational attraction of Jupiter depletes certain orbits within the asteroid belt. The resulting gaps, called Kirkwood gaps, occur at simple fractions of Jupiter's orbital period.

• Jupiter's and the Sun's gravity combine to capture Trojan asteroids in two locations, called stable Lagrange points, along Jupiter's orbit.

• The Apollo asteroids move in highly elliptical orbits that cross the orbits of Mars and Earth. Many of these asteroids will eventually strike the inner planets.

Comets

• Comets are fragments of ice and rock that generally move in highly elliptical orbits about the Sun often at a great inclination to the plane of the ecliptic.

• As a comet approaches the Sun, its icy nucleus develops a luminous coma surrounded by a vast hydrogen envelope. A gas (or ion) tail and a dust tail extend from the comet, pushed away from the Sun by the solar wind and radiation pressure.

• Many comets orbit the Sun in the Kuiper belt, a doughnut-shaped region beyond Pluto. Billions of cometary nuclei are also believed to exist in the spherical Oort cloud located far beyond Pluto.

Meteoroids, Meteors, and Meteorites

• Boulders and smaller rocks in space are called meteoroids. When a meteoroid enters the Earth's atmosphere, it produces a fiery trail, and it is then called a meteor. If part of the object survives the fall, the fragment that reaches the Earth's surface is called a meteorite.

• Meteorites are grouped in three major classes according to their composition: iron, stony-iron, and stony meteorites. Rare stony meteorites called carbonaceous chondrites may be relatively unmodified material from the primitive solar nebula. These meteorites often contain organic hydrocarbon compounds, including amino acids.

• Fragments of rock from "burned-out" comets produce meteor showers.

• An analysis of the Allende meteorite suggests that a nearby supernova explosion may have been involved in the formation of the solar system some 4.6 billion years ago.

• An asteroid that struck the Earth 65 million years ago probably contributed to the extinction of the dinosaurs and many other species. Another impact caused the "Great Dying" of life 250 million years ago. Such devastating impacts occur on average every 100 million years.

WHAT DID YOU THINK?

1 *Are the asteroids a planet that was somehow destroyed?* No. The gravitational pull from Jupiter prevented a planet from ever forming in the asteroid belt. Also, the total mass of the asteroids is much less than even the mass of tiny Pluto, the smallest planet.

2 *How far apart are the asteroids on average?* The distance between asteroids averages 10 million kilometers.

3 *Why do comets have tails?* Gas and dust that evaporate from a comet's nucleus are pushed away from the Sun by sunlight and the solar wind. We see the tails by the sunlight they scatter in our direction.

4 *In what direction does a comet tail point?* Comets' gas tails point directly away from the Sun; their dust tails make arcs pointing away from the Sun.

5 *What is a shooting star?* A shooting star is a piece of space debris plunging through the Earth's atmosphere—a meteor. It is not a star.

Key Words

amino acid, 259
Apollo asteroid, 244
asteroid (minor planet), 240
asteroid belt, 240
belt asteroid, 240
carbonaceous chondrite, 258
coma (of a comet), 249
comet, 246
dust tail (of a comet), 250
gas (ion) tail, 250
hydrogen envelope, 249
impact crater, 255
iron meteorite, 258
Kirkwood gaps, 242
Kuiper belt, 246
long-period comet, 252
meteor, 254
meteor shower, 255
meteorite, 257
meteoroid, 254
nucleus (of a comet), 247
Oort cloud, 247
radiation (photon) pressure, 251
short-period comet, 252
stable Lagrange points, 244
stony meteorite, 258
stony-iron meteorite, 258
tail (of a comet), 249
Trojan asteroid, 244
Widmanstätten patterns, 258

Review Questions

1. A piece of space debris that you pick up off the ground is called a(an): **a.** asteroid, **b.** meteoroid, **c.** meteor, **d.** meteorite, **e.** comet.

2. Space debris that is a roughly equal mix of rock and ice is called a(an): **a.** asteroid, **b.** comet, **c.** meteoroid, **d.** meteorite, **e.** meteor.

3. Of the following, which is the rarest type of meteorite? **a.** irons, **b.** stony-irons, **c.** stony meteorites.

4. Which part of a comet is solid? **a.** nucleus, **b.** halo, **c.** gas tail, **d.** dust tail.

5. Why are asteroids, meteoroids, and comets of special interest to astronomers who want to understand the early history of the solar system?

6. Describe the asteroid belt.

7. To test your understanding of the asteroid belt, do Interactive Exercise 9-1 on the Web. You can print out your results, if required.

8. Why are there many small asteroids but only a few very large ones?

9. Does each comet always have tails? Explain.

10. What are the Kirkwood gaps and what causes them?

11. What are the Trojan asteroids, and where are they located?

12. Describe the three main classifications of meteorites. How do astronomers believe these different types of meteorites originated?

13. Suppose you find a rock that you suspect to be a meteorite. Describe some of the things you could do to see if it is a meteorite or a "meteorwrong."

14. Why do astronomers believe that the debris that creates many isolated meteors comes from asteroids, whereas the debris that creates meteor showers is related to comets?

15. Make a drawing that describes the structure of a comet.

16. To test your understanding of comets, do Interactive Exercise 9-2 on the Web. You can print out your results, if required.

17. Why is the phrase "dirty iceberg" an appropriate characterization of a comet's nucleus?

18. What is the Kuiper belt, and how is it related to debris left over from the formation of the solar system?

19. Why do scientists think the Tunguska event was caused by a large meteoroid and not a comet?

20. What evidence in Figure 9-23 supports the labeling of the gas and dust tails?

Advanced Questions

The answers to computational problems, which are preceded by an asterisk (*), appear at the end of the book.

21. How did the regolith on Eros form?

22. Why are comets generally brighter after passing perihelion (closest approach to the Sun) than before reaching perihelion?

23. Can you think of another place in the solar system where a phenomenon similar to the Kirkwood gaps in the asteroid belt is likely to exist? Explain.

24. Where on Earth might you find large numbers of stony meteorites that have not been significantly changed by weathering?

***25.** Assuming a constant rate of meteor infall, how much mass has the Earth gained in the past 4.6 billion years?

Discussion Questions

26. Suppose it was discovered that the asteroid Hermes had been perturbed in such a way as to put it on a collision course with Earth. Describe what you would do to counter such a catastrophe using present technology.

27. From the abundance of craters on the Moon and Mercury, we know that numerous asteroids and meteoroids struck the inner planets during the very early history of the solar system. Is it reasonable to suppose that numerous comets also pelted the planets 4.0 to 4.5 billion years ago? What effects would such a cometary bombardment have had, especially with regard to the evolution of the primordial atmospheres of the terrestrial planets and oceans on Earth.

What If . . .

28. An ocean on Earth were struck by a comet nucleus a kilometer across? What physical effects would occur to the Earth?

29. We passed through the tail of a comet? What would happen to the Earth and life on it?

30. As some astronomers have recently argued, passage of the solar system through an interstellar cloud of gas could perturb the Oort cloud, causing many comets to deviate slightly from their original orbits? What might be the consequences for Earth?

31. All the space debris (asteroids, meteoroids, and comets) had been cleared out of the solar system 3.8 billion years ago? What would have been different in the history of the Earth and in the history of life on the Earth?

Web Questions

32. Search the Web to find out why some scientists disagree with the idea that a tremendous impact led to the demise of the dinosaurs. (They do not dispute that the impact occurred, only what its consequences were.) What are their arguments? From what you learn, what is your opinion?

33. Several scientific research programs are dedicated to the search for near-Earth objects (NEOs), especially those that might someday strike our planet. Search the Web for information about at least one of these programs. How does the program search for NEOs? How many NEOs are now known and how many has this program found? Will any of these NEOs pose a threat in your lifetime?

34. Search the Web to learn if there are any comets visible at present. List them. What constellations are they presently in? Are any visible to the naked eye? (Recall that the unaided human eye sees objects brighter than about sixth magnitude.)

Observing Projects

35. Make arrangements to view an asteroid. At opposition, some of the largest asteroids are bright enough to be seen through a modest telescope. You can use *Starry Night Enthusiast*™ to see which bright asteroids are visible. Update asteroids on-line using *LiveSky*. Set time to when you will observe. Press *Find* tab and the [+] by asteroids. The bright visible ones will appear in black. Double click on one and print the screen. Or, check the "Minor Planets" section of the current issue of the *Astronomical Almanac* to see if any bright asteroids are near opposition. If so, check the current issue as well as the most recent January issue of *Sky & Telescope* for a star chart showing the asteroid's path among the constellations. You will need such a chart to distinguish the asteroid from background stars. Observe the asteroid on at least two occasions separated by a few days. On each night, draw a star chart of the objects in your telescope's field of view. Has the position of one starlike object shifted between observing sessions? Does the position of the moving object agree with the path plotted on published star charts? Do you feel confident that you have in fact observed an asteroid? Explain.

36. Make arrangements to view a comet through a telescope. Because astronomers discover roughly a dozen comets each year, a comet is usually visible somewhere in the sky. Unfortunately, because most comets are quite dim, you will need access to a moderately large telescope. You can use *Starry Night Enthusiast*™ to see which comets are visible. First, update comets on-line using *LiveSky*. Set time to when you will observe, then find a comet that is up: *Find* tab, [+]comets. The visible ones will be in black. Find its apparent magnitude by pressing INSERT SYMBOL ▼ next to its name in *Find* tab and then choosing *Show Info/Other Data*. If it is too dim, try others. Double click on one and zoom in on it to see the direction of its tail. Print screen. Or consult the Web or recent issues of the *IAU Circular*, published by the International Astronomical Union's Central Bureau for Astronomical Telegrams, which contains predicted positions and the anticipated brightness of comets in the sky. Also, if there is an especially bright comet in the sky, the latest issue of *Sky & Telescope* might contain useful information. Observe the comet through a telescope. Can you distinguish it from background stars? Can you see its coma? How many tails do you see?

37. Make arrangements to view a meteor shower. The date of maximum intensity in Figure 9-27 is the best time to observe a particular shower, although good displays can often be seen a day or two before and after the maximum. In order to see a fine meteor display, you need a clear, moonless sky. The Moon's presence above the horizon can significantly detract from the number of faint meteors you will be able to see.

38. Use *Starry Night Enthusiast*™ software to observe several comets. (a) Set the program to Atlas mode (*Favourites/Guides/Atlas*). Set the date to 3/22/97. Go to either the North or South Pole (*Options/Viewing Location/North or South Pole*). Turn off the planets and Sun (*Options* tab/*SolarSystem/Planets-Moons*). Choose Comet Hale-Bopp (*Find* tab/[+] *Comets,* then scroll down and double-click on Hale-Bopp). Zoom in until you can see the comet and its tail. Predict in what direction the Sun is located relative to the comet and explain how you made your prediction. To verify that you are correct, zoom out to about 90°, turn on the planets and Sun (put check in the Planets-Moon box) to locate the Sun. If necessary, use the hand cursor to move the sky in the direction of your prediction toward the Sun. Were you correct? (b) Using the Comet List again, double-click on the name of another comet to center on it, but do not zoom in. Then move the sky with the hand cursor and locate the Sun. From your observations, predict the direction of the comet's tail on the sky, and explain how you made your prediction. Center again on the comet and zoom in until you can see its tail. (If this comet does not have a visible tail, repeat this part with yet another comet.) Was your prediction correct?

39. Halley's Comet In this exercise we will use *Deep Space Explorer*™ to locate Halley's comet. If you have not done so already, go to **Settings** and uncheck "Use magnitude cutoffs," slide the "Galaxy drawing/Brightness" slide to the middle of the range, and then click on the **Views** tab.

Now, go to the comet by clicking **Solar System/Halley's Comet** on the left side of the star field. The comet will be labeled, but the tail will not initially be visible. Explain why the tail is not on the screen. Before moving the comet to the left on the screen, write down the direction you expect the tail will point when you do move the comet to the left. Now grab the screen by holding down the left PC mouse button (the mouse button on a Mac), and move the image of Halley's comet to the left side far enough for the tail to appear. Write down the direction that the tail points. Explain why it points this way. From the information in front of you, can you predict the direction of Halley's motion when this image was taken? If so, give its direction and explain how you deduced it. If not, explain why not.

40. Asteroid Ceres In this exercise we will use the *Deep Space Explorer*™ program to locate the largest asteroid, Ceres. If you have not done so already, go to **Settings** and uncheck "Use magnitude cutoffs," slide the "Galaxy drawing/Brightness" slide to the middle of the range, and then click on the **Views** tab.

Now, go to the solar system by clicking **Solar System/Outer Solar System** on the left side of the star field. Ceres will not yet be labeled. Write down the general region of the solar system in which you expect to find it. Now turn on the label for Ceres by going to **Edit/Find/Ceres.** You won't be able to see the dot representing Ceres at this point, but you will be locked on it. Use the "zoom in" ▼ button to bring the dot of Ceres just barely into view. Make a sketch of its location. What is its size compared to the dots representing Earth and Venus, which should just be on your screen? Because the dots are to scale, what can you conclude about the size of Ceres compared to the inner planets? Zoom in until Ceres is clearly discernable. Move the screen around by grabbing it with the left button on your mouse (the mouse button on a Mac). Is Ceres in the plane of the ecliptic?

The Sun: Our Extraordinary Ordinary Star

CHAPTER 10

Magnetic Fields Carry Gases above the Sun's Surface in Loops That Can Extend Hundreds of Thousands of Kilometers (NASA)

R I V U X G

WHAT DO YOU THINK?

1. How does the mass of the Sun compare with that of the rest of the solar system?
2. Does the Sun have a solid and liquid interior like the Earth?
3. What is the surface of the Sun like?
4. Does the Sun rotate?
5. What makes the Sun shine?

Sunrise. Light and heat from the Sun begin to arouse our senses and signal the start of daily activity. Earliest societies, realizing that the Sun is essential to the existence and maintenance of life on Earth, revered it. The same respect for the Sun's awesome power motivated astronomers in the nineteenth and twentieth centuries to figure out how it shines.

1 A quick glance at the Sun (never more) shows a brilliant disk of light no larger than the Moon. Despite appearances, the Sun is a body of staggering size and mass. If the Sun were as close to the Earth as the Moon is, the Sun would spread two-thirds of the way across the sky, not to mention that it would be so hot here that the Earth would quickly vaporize. The Sun contains as much mass as 333,000 Earths. Indeed, all the planets and other bodies orbiting the Sun *combined* have only .15% as much mass as the Sun.

The Sun's tremendous mass not only holds the planets in orbit, but it also explains why the Sun shines. The Sun's mass creates a crushing force at its center that literally fuses atoms together—a process that generates the energy that eventually leaves the Sun and lights up our days. Some of the energy rushing into space (the ultraviolet radiation, visible light, and infrared radiation) is essential to maintaining the Earth's habitability.

The Sun is the closest star to Earth. By studying it astronomers have come to learn how most stars throughout the universe work. As powerful and majestic as it is in our sky, we are going to discover in the next few chapters that the Sun is an ordinary star, neither among the most massive nor among the least massive. Similarly, it is neither among the brightest nor among the dimmest of stars. Figure 10-1 lists the Sun's properties.

In this chapter you will discover

- why the Sun is a typical star
- how today's technology has led to new understanding of solar phenomena, from sunspots to the powerful ejections of matter that sometimes enter our atmosphere
- that some features of the Sun generated by its varying magnetic field occur in cycles
- how the Sun generates the energy that makes it shine
- new insights into the nature of matter from solar neutrinos

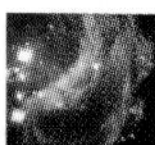

THE SUN'S ATMOSPHERE

Although astronomers often speak of the solar "surface," 2
the Sun is so hot that it has neither liquid nor solid matter anywhere inside it. Moving down through the Sun, one continually encounters ever denser and hotter gases.

10-1 The photosphere is the visible layer of the Sun

The Sun appears to have a surface only because most of its visible light comes from one specific gas layer (see Figure 10-1). This region, which is about 400 km thick, is appropriately called the **photosphere** ("sphere of light"). The density of the photosphere's gas is low by Earth standards, about 0.01% as thick as the air we breathe. The photosphere has a blackbody spectrum (recall Chapter 4) corresponding to an average temperature of 5800 K (review Figure 4-3).

The photosphere is the innermost of the three layers 3
comprising the Sun's atmosphere. Because the upper two layers (discussed in the following two sections) are transparent to most wavelengths of visible light, we see through them down to the photosphere. We cannot, however, see through the shimmering gases of the photosphere, so everything below the photosphere is called the Sun's interior.

As you can see in Figure 10-1, the photosphere appears darkest toward the edge, or **limb,** of the solar disk, a phenomenon called **limb darkening.** This occurs because wherever we look on the Sun, we see light through roughly equal amounts of the Sun's atmosphere. Since the Sun is spherical, the light we see leaving the Sun at different places actually comes from different levels of the photosphere (Figure 10-2). Looking from Earth at the center of the Sun's disk (bottom black line, Figure 10-2), we see farther into the Sun's atmosphere than when we look toward the limb (top black line, Figure 10-2). The photosphere is hottest and brightest at its base and so the center of the Sun's disk, where we see deepest into it, looks brighter than does its limb.

Under good observing conditions with a telescope and using special dark filters for protection, you can see a blotchy pattern called *granulation* on the photosphere

THE SUN: VITAL STATISTICS

Mass (1 $M_\odot$):	1.989×10^{30} kg = 3.33×10^5 Earth masses
Visual radius (1 $R_\odot$):	6.960×10^5 km = 109 Earth radii
Luminosity (1 $L_\odot$):	3.827×10^{26} W
Mean angular diameter:	32 arcmin
Rotation periods:	Equatorial: 25 days Polar: 35 days
Mean density:	1408 kg/m^3
Distances from Earth:	Mean (1 AU): 1.496×10^8 km Maximum: 1.520×10^8 km Minimum: 1.470×10^8 km
Mean light travel time to Earth:	8.32 min
Mean temperatures:	Surface: 5800 K Center: 1.55×10^7 K
Composition (by mass):	74% hydrogen, 25% helium, 1% other elements
Composition (by number of atoms):	92.1% hydrogen, 7.8% helium, 0.1% other elements
Distance from center of Galaxy:	26,000 ly = 8000 pc
Orbital period around center of Galaxy:	220 million years
Orbital speed around center of Galaxy:	220 km/s

R I V U X G

FIGURE 10-1 **Our Star, the Sun** The Sun emits most of its visible light from a thin layer of gas called the photosphere, shown here. Although the Sun has no solid or even liquid region, we see the photosphere as its "surface." Astronomers think of the photosphere as the lowest level of the Sun's atmosphere. Astronomers always take great care when viewing the Sun by using extremely dark filters or by projecting the Sun's image onto a screen. (Celestron International)

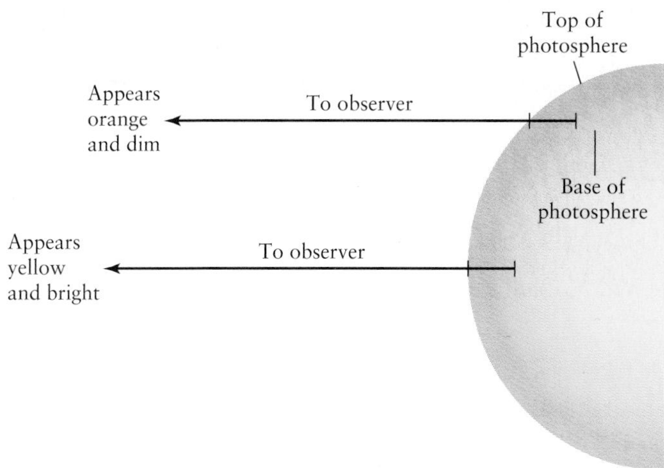

◀ FIGURE 10-2 **Limb Darkening** The Sun's edge or limb appears distinctly darker and more orange than does its center, as seen from Earth (see Figure 10-1). This occurs because we look through the same amount of solar atmosphere at all places. As a consequence, we see higher in the Sun's photosphere near its limb than when we look at its central regions. The higher photosphere is cooler and (since it is a blackbody) darker and more orange than the lower, hotter region of the photosphere.

(Figure 10-3). The lightly colored **granules** measure about 1000 km across and are surrounded by relatively dark boundaries. Time-lapse photography shows that granules form, disappear, and then re-form in cycles lasting several minutes. At any single moment, several million granules cover the solar surface.

Like a pot of simmering soup, the gases in granules rise and fall. In Section 3-2 we learned that radial motion of a light source affects the wavelengths of its spectral lines through the Doppler effect (see Figures 3-6 and 4-15 and An Astronomer's Toolbox 4-4). By carefully measuring the Doppler shifts of spectral lines in various parts of individual solar granules, astronomers have determined that hot gases move upward in the center of each granule and cooler gases cascade downward around its edges. This is caused by convection. In Section 6-3 we saw that convection moves the continents on Earth and in Section 8-1 that it also helps form the belts and zones on the giant planets. Once the Sun's gases arrive at the photosphere, they radiate energy out into space. We see this energy as visible light and other electromagnetic radiation. Upon radiating, the gases cool, spill over the edges of the granule, and plunge back down into the Sun along the boundaries between granules (see Figure 10-3 inset).

According to the Stefan-Boltzmann law (see Section 4-2), hotter regions emit more photons per square meter than do cooler regions. Note in Figure 10-3 that the centers of granules are brighter than their edges. Observations indicate that a granule's center is typically 100 K hotter than its edge, thus explaining the observed brightness difference.

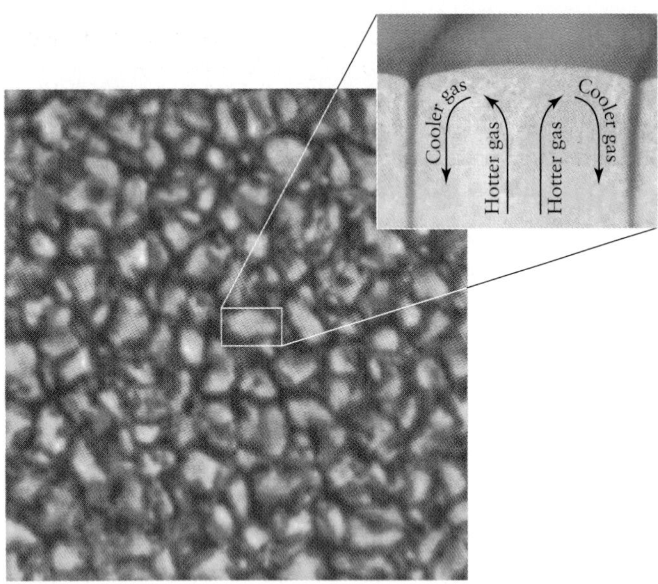

FIGURE 10-3 **Solar Granulation** High-resolution photographs of the Sun's surface reveal a blotchy pattern called granulation. Granules, which measure about 1000 km across, are convection cells in the Sun's photosphere. **Inset:** Gas rising upward produces bright granules. Cooler gas sinks downward along the darker, cooler boundaries between granules. This convective motion transports energy from the Sun's interior outward to the solar atmosphere. (MSFC/NASA; inset: Goran Scharmer, Lund Observatory)

10-2 The chromosphere is characterized by spikes of gas called spicules

Immediately above the photosphere is a dim layer of less dense stellar gas called the **chromosphere** ("sphere of color"). This unfortunate name suggests that it is the layer we normally see, but for centuries it was visible only when the photosphere was blocked during a total solar eclipse.

During an eclipse, the chromosphere is visible as a pinkish strip some 2000 km thick around the edge of the dark Moon (Figure 10-4). Today, astronomers can also study the chromosphere through filters that pass light with specific wavelengths strongly emitted by it—but not by the photosphere—or through telescopes sensitive to nonvisible wavelengths that the chromosphere emits intensely.

High-resolution images of the chromosphere reveal numerous spikes, which are jets of gas called **spicules** (Figure 10-5a). A typical spicule rises for several minutes at the rate of 72,000 km/h (45,000 mph) to a height of nearly 10,000 km (Figure 10-5b). Then it collapses and fades away. At any one time, roughly a third of a million spicules cover a few percent of the Sun's chromosphere.

Spicules are generally located on the boundaries of enormous regions of rising and falling chromospheric gas called **supergranules** (Figure 10-5a). A typical supergranule has a diameter slightly larger than the Earth's and contains about 900 granules.

FIGURE 10-4 **The Chromosphere** This photograph of the chromosphere was taken during an eclipse. It appears pinkish because the gas in the chromosphere emits only certain colors (wavelengths), among which the red one from hydrogen gas dominates. (NOAO)

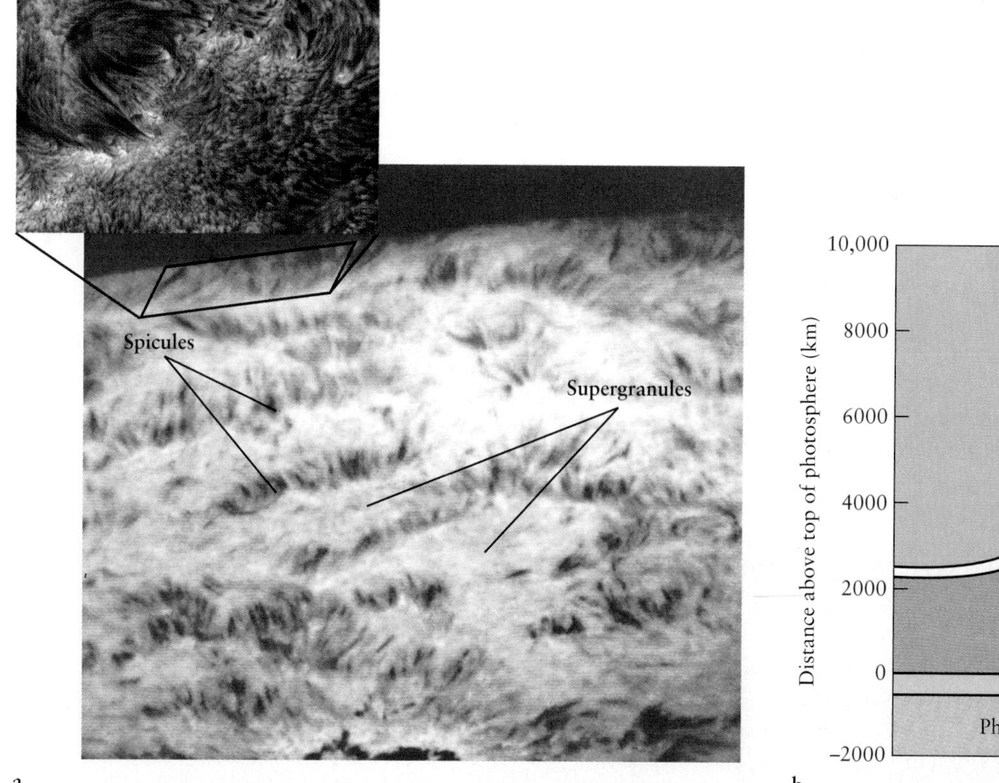

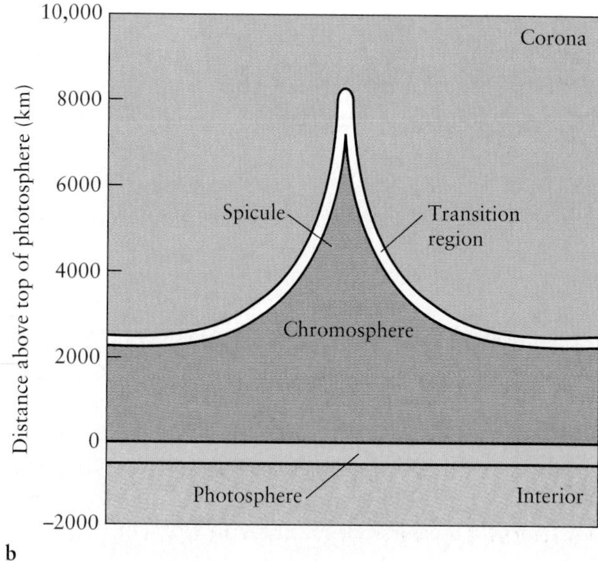

FIGURE 10-5 **Spicules and Supergranules** (a) Spicules are seen in this photograph of the Sun's chromosphere. The Sun appears rose-colored in this image because it was taken through an H_α filter that passes red light from hydrogen and effectively blocks most of the photosphere's light. Surrounded by spicules, supergranules are regions of rising and falling gas in the chromosphere. Each supergranule spans hundreds of granules in the photosphere below. **Inset:** A view of spicules from above. (b) The spicules are jets of gas that surge upward into the Sun's outer atmosphere. This schematic diagram shows a spicule and its relationship to the solar atmosphere's layers. The photosphere is about 400 km thick. The chromosphere above it extends to an altitude of about 2000 km, with spicules jutting up to nearly 10,000 km above the photosphere. The outermost layer, the corona (discussed in Section 10-3), extends millions of kilometers above the photosphere. (a: NASA; inset: Swedish Solar Telescope/Royal Swedish Academy of Sciences, La Palma, Spain, by Bert De Pontieu, Lockheed Martin Solar and Astrophysics Lab)

10-3 Temperatures increase higher in the Sun's atmosphere

It seems plausible that the temperature should fall as one rises through the Sun's atmosphere. After all, by moving upward, one moves farther from the sun's internal heat source. Indeed, starting in the photosphere at 5800 K, the temperature drops to around 4000 K in the lower chromosphere. Surprisingly, however, the temperature then begins to rise much higher as one ascends, reaching about 10,000 K at the top of the chromosphere. The outermost region of the Sun's atmosphere, the **corona,** extends several million kilometers from the top of the chromosphere (Figure 10-6). Between the chromosphere and corona is a **transition zone** in which the temperature skyrockets to about 1 million K (Figure 10-6c).

The unexpected increase in temperature was discovered around 1940 as a result of the high temperature's effect on the spectrum of the Sun's corona. The hotter a gas is, the more it is ionized (electrons are stripped off the atoms). Astronomers discovered that the corona contains the emission lines of a number of highly ionized (therefore very hot) elements. For example, a prominent green line caused by the presence of Fe XIV, an iron atom stripped of 13 electrons, appears in the coronal spectrum. (Recall that Fe I is a neutral iron atom.) It is now known that coronal temperatures are typically in the range of 1 to 2 million kelvins, while some regions are even hotter.

Although its temperature is extremely high, the density of gas in the corona is very low, about 10 trillion times less dense than the air at sea level on Earth. The low density partly accounts for the dimness of the corona, which otherwise would outshine the photosphere. Astronomers have mounting evidence that the corona is heated by energy carried aloft and released there by the Sun's complex magnetic fields, discussed in Sections 10-5 and 10-6.

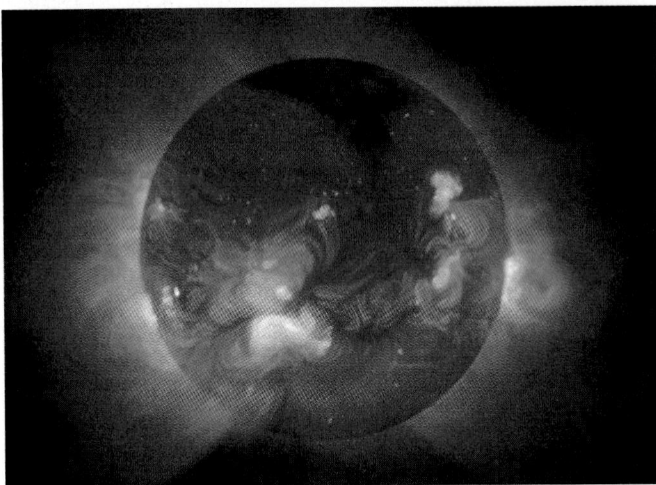

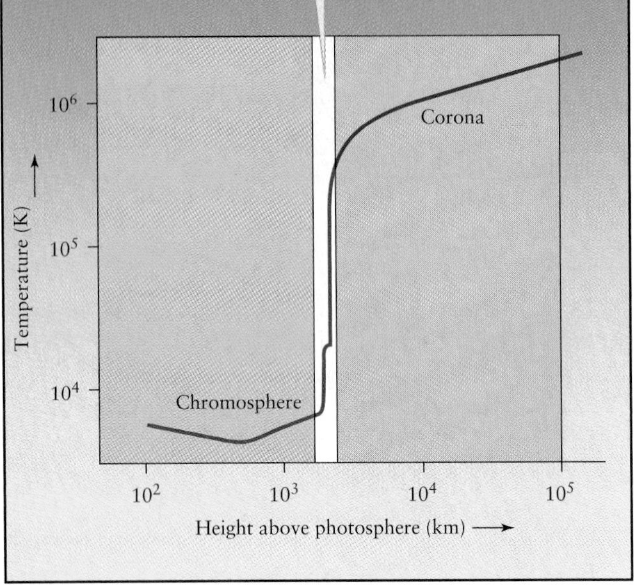

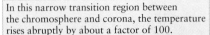

FIGURE 10-6 **The Solar Corona** (a) This visible-light photograph was taken during the total solar eclipse of July 11, 1991. Numerous streamers are visible, extending millions of kilometers above the solar surface. (b) This X-ray image of the Sun's corona, taken by *Yohkoh* in 1999, provides hints of the complex activity taking place on and in the Sun. The million-degree gases in the corona emit the X rays visible here. (c) This graph shows how temperature varies with altitude in the Sun's chromosphere and corona and in the transition region between them (in white). Note that both the height and temperature scales are nonlinear. (a: R. Christen and M. Christen, Astro-Physics Inc.; b: NASA; c: adapted from A. Gabriel)

The total amount of visible light that we receive from the solar corona is comparable to the brightness of the full Moon—or only about one-millionth as bright as the photosphere. As with the chromosphere, the corona can be seen only when the photosphere is blocked out or through special filters or at nonvisible wavelengths (such as ultraviolet and X-ray), at which the corona is especially bright compared to the photosphere. The photosphere is blocked naturally during a total eclipse or artificially with a specially designed telescope called a *coronagraph*. Figure 10-6a is a photograph of the corona taken during a total eclipse (see also Figure 1-29). Figure 10-6b shows a stunning X-ray image of the corona.

Just as the Earth's gravity prevents most of our atmosphere from escaping into space, so, too, does the Sun's gravity keep most of its outer layers from leaving. However, some of the gas in the corona is moving fast enough—around a million kilometers per hour—to escape the Sun's gravity forever and race into space. As we saw in Section 6-4 in discussing Earth's magnetosphere and in Section 9-5 in discussing comets, this outflow of particles is called the **solar wind.** The *heliosphere* is a bubble in space containing the Sun and planets created by the solar wind. This outflow of gases from the Sun prevents most of the gases flowing in space from other stars from entering our solar system.

The Sun ejects around a million tons of matter each second as the solar wind. Even at this rate of emission, the mass loss due to the solar wind will amount to only a few tenths of a percent of the Sun's total mass throughout its lifetime. While electrons and hydrogen and helium nuclei comprise 99.9% of the solar wind, silicon, sulfur, calcium, chromium, nickel, neon, and argon ions have also been detected in it. The solar wind particles reach speeds up to 2.9×10^6 km/h (1.8×10^6 mph). The wind achieves these high speeds in part by being accelerated by the Sun's magnetic field, discussed in Section 10-5. By the time the solar wind travels 1 AU to the vicinity of the Earth, it has spread out so much that there are typically just 5 particles per cubic centimeter and they are moving around 1.1–1.4 $\times$ 10^6 km/h (6.8–8.5 $\times$ 10^5 mi/h). For comparison, there are 6 $\times$ 10^{19} molecules per cubic centimeter in the air we breathe and the winds at the Earth's surface are moving less than 240 km/h (150 mi/h).

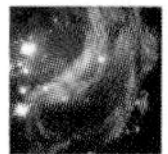

THE ACTIVE SUN

Granules, supergranules, spicules, and the solar wind occur continuously. But the Sun's atmosphere is periodically disrupted by magnetic fields that stir things up, creating the *active* Sun. The Sun's most obvious transient features are **sunspots,** regions of the photosphere that appear dark because they are cooler than the rest of the Sun's lower atmosphere. Sometimes sunspots occur in isolation (Figure 10-7a), but often they arise in clusters called *sunspot groups* (Figure 10-7b).

INSIGHT INTO SCIENCE

Perception versus Reality Our senses (and often our technology) are limited, and so we must be careful how we interpret what we perceive. For example, the sunspots in Figure 10-7 certainly look like black spots. However, they appear black only in contrast to the bright light around them. As we will discuss shortly, sunspots are actually red and orange.

10-4 Sunspots reveal the solar cycle and the Sun's rotation

Like other transient features of the active Sun, the average number and location of sunspots vary in fairly predictable cycles because of the changes in magnetic fields that produce them (as we will explore in Section 10-5). As shown in Figure 10-8a, the average sunspot cycle lasts approximately 11 years. Within an 11-year cycle, most sunspots appear at a **sunspot maximum** (Figure 10-8b). Sunspot maxima occurred most recently in 1979, 1989, and 2001. During a **sunspot minimum,** the Sun is almost devoid of sunspots, as it was in 1976, 1986, and 1996 (Figure 10-8c). The next trough in the cycle is projected to occur in 2007.

A typical sunspot is 10,000 km across and lasts between a few hours and a few months. Each sunspot has two parts: a dark, central region, called the *umbra,* and a brighter ring surrounding the umbra, called the *penumbra,* visible in Figures 10-7a and 10-7b. While these are the same names as the blocked areas of an eclipse (see Section 1-11), we will see in the next section that their causes are completely different—another good example of how certain astronomical terms can have different meanings in different contexts.

Seen without the surrounding brilliant granules that outshine it, a sunspot's umbra appears red and its penumbra orange. From Wien's law (see Section 4-2), these colors indicate that the umbra is typically 4300 K and the penumbra 5000 K, both cooler than the normal photosphere.

On rare occasions a sunspot group is so large that it can be seen with the unaided eye. Chinese astronomers recorded such sightings 2000 years ago, and a huge sunspot

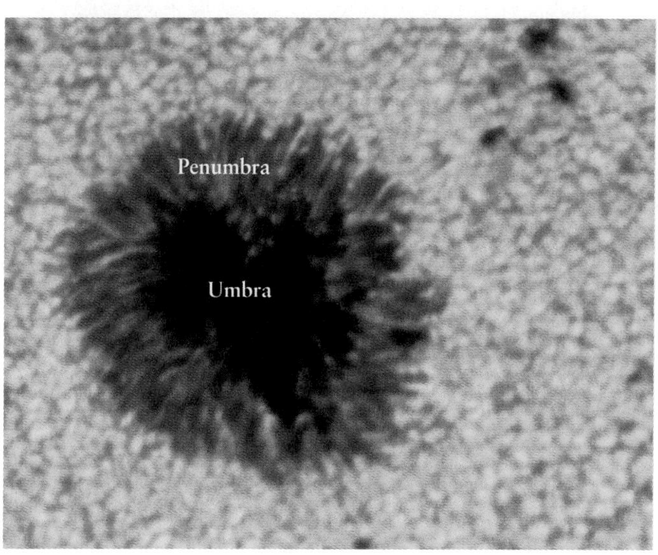

a

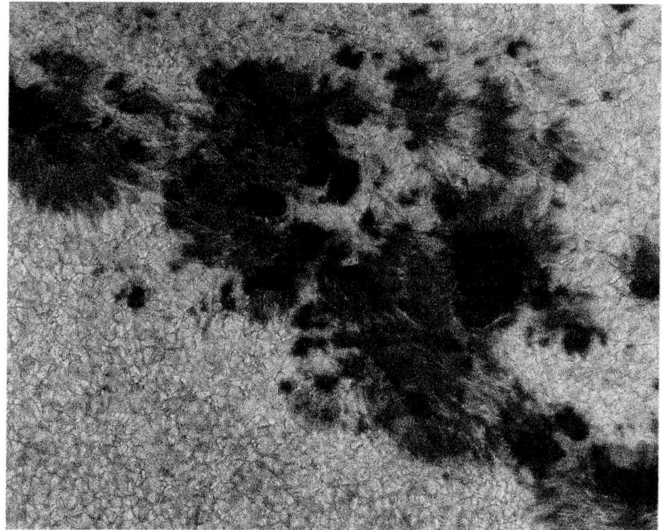

b

R I V U X G

FIGURE 10-7 **Sunspots** **(a)** This dark region on the Sun is a typical isolated sunspot. Granulation is visible in the surrounding, undisturbed photosphere. **(b)** This high-resolution photograph shows a sunspot group in which several sunspots overlap and others are nearby. (a: NOAO; b: National Solar Observatory)

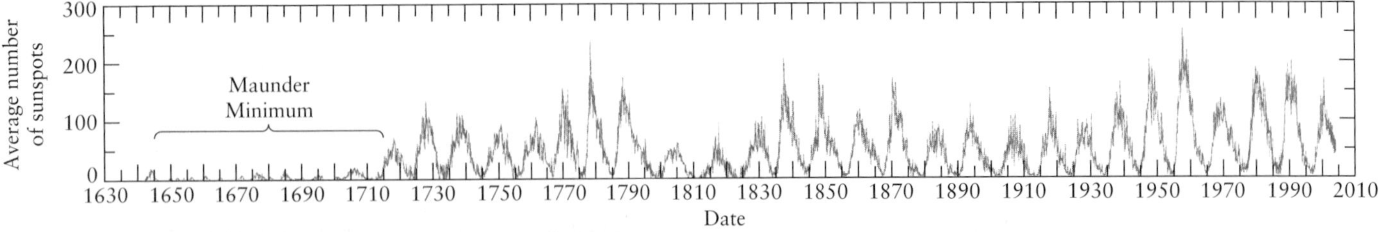

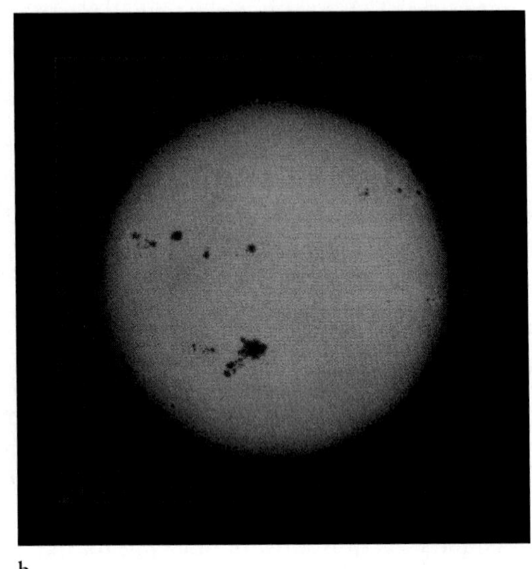

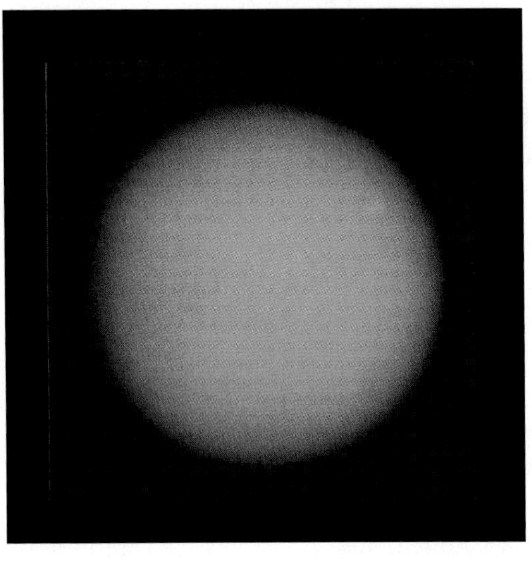

WEB LINK 10.3 **FIGURE 10-8 The Sunspot Cycle** **(a)** The number of sunspots on the Sun varies with a period of about 11 years. The most recent sunspot maximum occurred in 2001, and the most recent sunspot minimum occurred in 1996. **(b)** The active Sun has many sunspots (this photo was taken in 1979). **(c)** The Sun has many fewer sunspots when it is not active (this photo, showing a time with no sunspots, was taken in 1989). (b and c: NOAO)

group visible to the naked eye was seen in 2001. **Always use special dark filters or other means to protect your eyes when viewing the Sun. Looking directly at the Sun for more than a few moments can cause eye damage!** Of course, a telescope gives a much better view, so it was not until Galileo did so that anyone examined sunspots in detail.

4 By following sunspots as they moved across the solar disk (Figure 10-9), Galileo discovered that *the Sun rotates once in about 4 weeks.* Galileo initially observed the Sun directly, undoubtedly contributing to his eventual blindness. Eventually his protégé, Benedetto Castelli, developed the technique of projecting sunlight onto a screen to safely observe sunspots. Because a typical sunspot group lasts

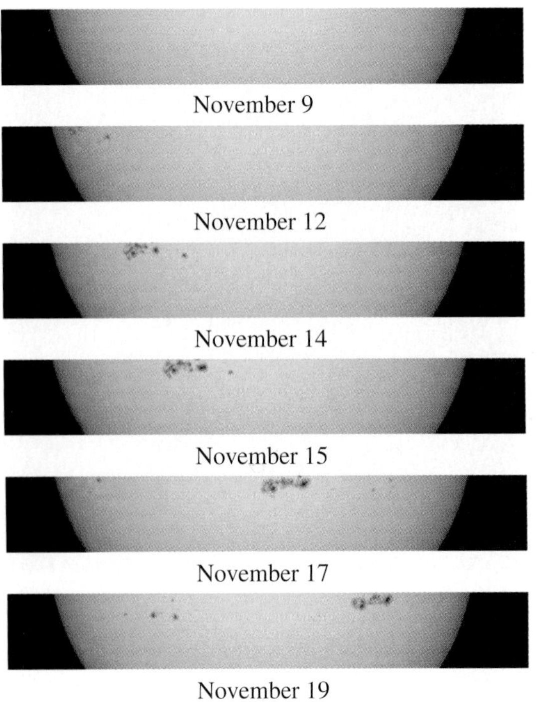

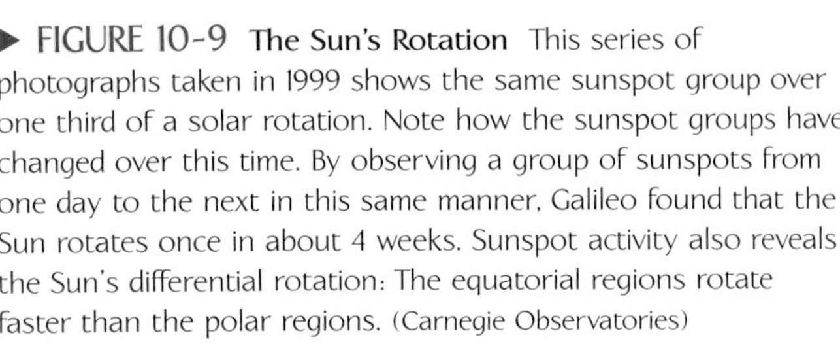

▶ **FIGURE 10-9 The Sun's Rotation** This series of photographs taken in 1999 shows the same sunspot group over one third of a solar rotation. Note how the sunspot groups have changed over this time. By observing a group of sunspots from one day to the next in this same manner, Galileo found that the Sun rotates once in about 4 weeks. Sunspot activity also reveals the Sun's differential rotation: The equatorial regions rotate faster than the polar regions. (Carnegie Observatories)

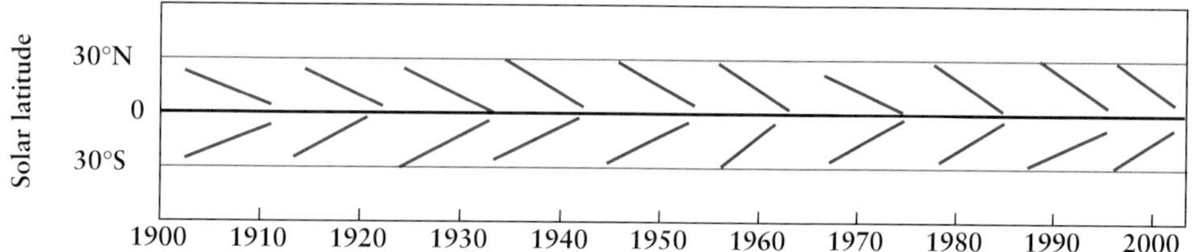
Solar latitude
30°N
0
30°S
1900
1910
1920
1930
1940
1950
1960
1970
1980
1990
2000

FIGURE 10-10 **Average Latitude of Sunspots throughout the Sunspot Cycle** This graph shows that at the beginning of each sunspot cycle, most sunspots are at moderate latitudes, around 30° north or south. Sunspots arising later in each cycle typically form closer and closer to the Sun's equator.

about 2 months, it can be followed for two solar rotations. Sunspot activity also lets us see the Sun's *differential rotation* (see Section 8-1): The equatorial regions rotate more rapidly than the polar regions. A sunspot near the solar equator takes 25 days to go once around the Sun and a sunspot at 30° north or south of the equator takes about 27 days. The rotation period at 75° north or south of the equator is about 33 days, and near the poles it is as long as 35 days.

The average latitude at which new sunspots appear changes throughout the sunspot cycle. At the beginning of each cycle, the sunspots appear mostly at about 30° north and south latitudes. Ones that form later in the cycle typically occur closer to the equator. Figure 10-10 shows that the average latitude at which sunspots are located varies at the same 11-year rate as does the number of sunspots. Satellite measurements reveal that the Sun emits about 0.1% more energy at the peak of the sunspot cycle than at its minimum.

10-5 The Sun's magnetic fields create sunspots

In 1908, the American astronomer George Ellery Hale discovered that sunspots are directly linked to intense magnetic fields on the Sun. When Hale focused a spectroscope on sunlight coming from a sunspot (Figure 10-11a), he

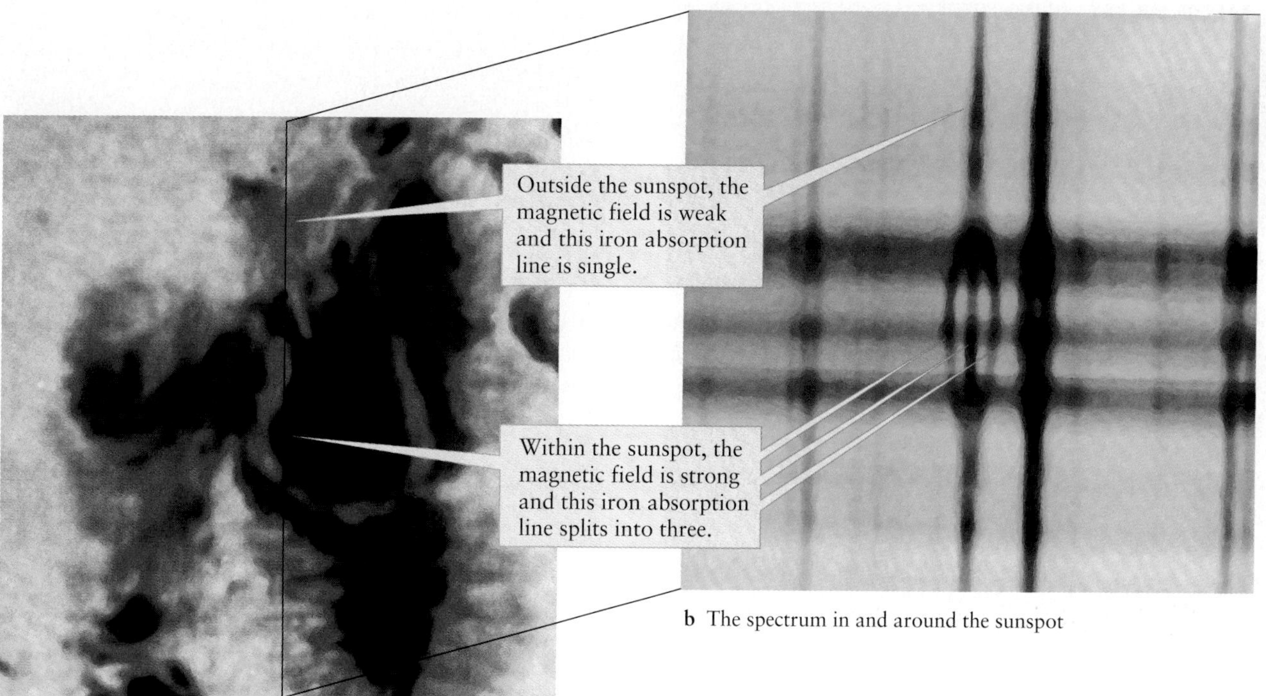
Outside the sunspot, the magnetic field is weak and this iron absorption line is single.
Within the sunspot, the magnetic field is strong and this iron absorption line splits into three.
R I V U X G

a A sunspot

b The spectrum in and around the sunspot

FIGURE 10-11 **Zeeman Splitting by a Sunspot's Magnetic Field** (a) The black line drawn across the sunspot indicates the location toward which the slit of the spectroscope was aimed. (b) In the resulting spectrogram, one line in the middle of the normal solar spectrum is split into three components by the Sun's magnetic field. The amount of splitting between the three lines is used to determine the magnetic field's strength. Typical sunspots have magnetic fields some 5000 times stronger than the Earth's magnetic field. (a and b: NOAO) found that each spectral line in the normal solar spectrum is flanked by additional, closely spaced spectral lines not usually observed (Figure 10-11b). This "splitting" of a single spectral line into two or more lines is called the **Zeeman effect.** The Dutch physicist Pieter Zeeman, who first observed it in the laboratory in 1896, showed that an intense magnetic field splits the spectral lines of a light source inside the field. The more intense the magnetic field, the more the split lines are separated.

Observationally, sunspots are areas where concentrated north or south magnetic fields project through the hot gases of the photosphere. But how do the fields create the spots? The answer lies in the interaction between the magnetic fields and the photosphere's gases. Because of the photosphere's high temperature, many atoms in it are *ionized:* One or more of their electrons have been stripped off by high-energy photons there. As a result, the photosphere is a mixture of electrically charged ions and electrons called a **plasma.**

Plasmas are extremely good conductors of electricity and are repelled from regions of high magnetic field. Therefore, the magnetic field protruding through the photosphere prevents hot, ionized gases inside the Sun from rising to the surface as they normally do. Thus, such regions of the photosphere are left relatively devoid of hot gas and are therefore cooler and darker than the surrounding solar surface. We observe these darker, cooler regions as sunspots.

The fact that magnetic fields create sunspots presented a problem for astronomers studying the Sun. If you have ever played with toy magnets, you may have discovered that two north poles or two south poles repel each other. How is it then that bundles of same-pole magnetic fields can stay together where they pierce the Sun's photosphere? The answer had to lay under the Sun's visible layer.

To learn more about the Sun's interior, astronomers record its vibrations—a study called **helioseismology**—just as geologists use earthquakes to study the Earth's interior structure. Although there are no true sunquakes, the Sun does vibrate at a variety of frequencies, somewhat like a ringing bell. These vibrations, first noted in 1960, can be detected with sensitive Doppler shift measurements. They have shown, for example, that portions of the Sun's surface move up and down by about 10 km every 5 minutes (Figure 10-12a).

Slower vibrations with periods ranging from 20 minutes to nearly an hour were discovered in the 1970s. Oscillations lasting several days have been detected, as have pulses 16 months long. One important discovery from helioseismology is that deep inside, the Sun rotates like a

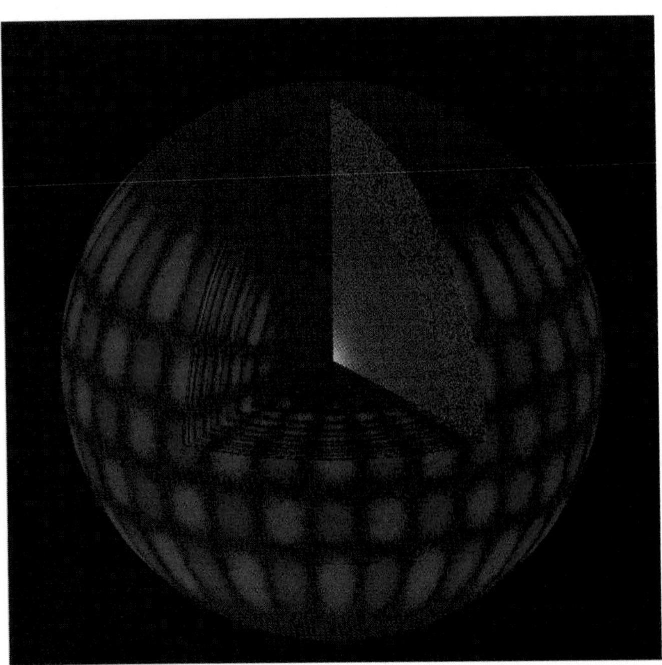

a

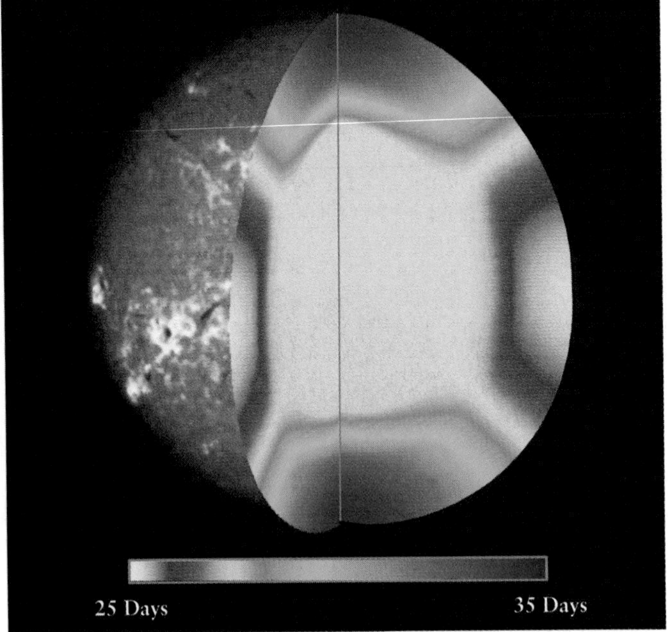

b

FIGURE 10-12 **Helioseismology** (a) This computer-generated image shows one of the millions of ways that the Sun vibrates because of sound waves resonating in its interior. The regions that are moving outward are colored blue; those moving inward are red. The cutaway shows how deep these oscillations are believed to extend. (b) This cutaway picture of the Sun shows how the rate of solar rotation varies with depth and latitude. Red and yellow denote faster-than-average motion; blue regions move more slowly than average. The pattern of differential surface rotation, which varies from 25 days at the equator to 35 days near the poles, persists at least 19,000 km down into the Sun's convective layer. Sunspots preferentially occur on the boundaries between different rotating regions. Earthlike jet streams and other wind patterns have also been discovered in the Sun's atmosphere.

rigid body, rather than with the differential rotation we see on its surface (Figure 10-12b).

Observing the vibrations of the Sun around sunspots, astronomers in 2001 apparently found the answer to the question of how sunspots persist for months despite the repulsion of the magnetic fields inside them. They discovered that below the photosphere, the gases surrounding each sunspot are whipping around like a hurricane as large across as the Earth's diameter. The circulation of charged gases around the magnetic fields hold them in place.

Recall from Section 6-4 that magnetic fields form complete loops, with each magnetic field having a north pole and a south pole. Likewise, when the Sun's magnetic field emerges through one sunspot or sunspot group, it forms a loop that reenters the Sun at another sunspot or sunspot group (Figure 10-13 insets). We associate the names "north pole" or "south pole" with each sunspot or sunspot group depending on whether the magnetic field there is pointing outward (a south pole) or pointing inward (a north pole). Sunspots and sunspot groups are connected in pairs, one where the magnetic field points out of the Sun, the other where it points into the Sun.

During his observations, Hale found that on one hemisphere of the Sun, sunspots with a magnetic north pole always come into view before the corresponding sunspots with a magnetic south pole. At the same time on the other hemisphere, the order is reversed (see Figure 10-13 insets). Hale also found that this pattern reverses itself about every 11 years. The hemisphere where north magnetic poles come first during one 11-year cycle has south magnetic poles coming first during the next. Astronomers therefore speak of the 22-year **solar cycle,** the time it takes the magnetic fields to return to the original orientation.

Based on these observations, we can now explain the Sun's magnetic field. This field is created as a result of the Sun's rotation and the resulting motion of the ionized particles found throughout it. This was proposed in 1960 by another American astronomer, Horace Babcock, as the **magnetic dynamo** model, in an effort to explain the 22-year solar cycle. The Sun's magnetic field normally lies just below the surface in the highly conducting plasma located there, unlike the Earth's field, which passes through our planet's center.

As shown in Figure 10-13, the Sun's differential rotation causes the field to become increasingly stretched. Like stretching a rubber band, stretching the magnetic field this way causes it to store energy. This magnetic field also becomes tangled as convection of the gases under the photosphere causes the fields to move vertically and horizontally. Unlike a rubber band, however, magnetic field lines cannot break to release the energy stored in them. Rather, the fields must untangle themselves.

This untangling process begins as the jumbled regions of magnetic field trap gases. This gas expands, thereby becoming buoyant and floating up through the solar surface, carrying the magnetic fields with them. Sunspots, with magnetic fields typically 5000 times stronger than the Earth's magnetic field, form where loops or tangles of solar

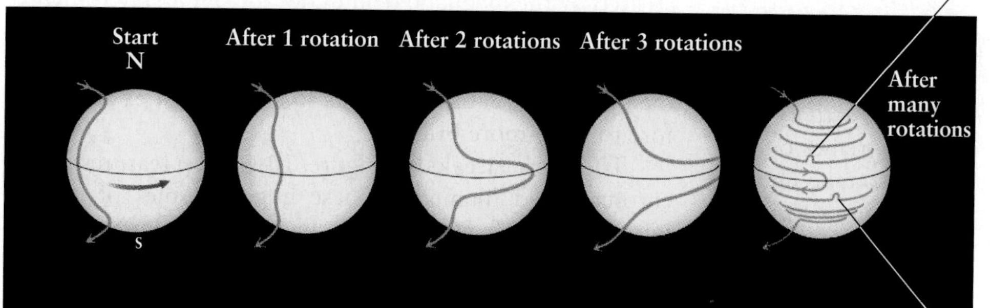

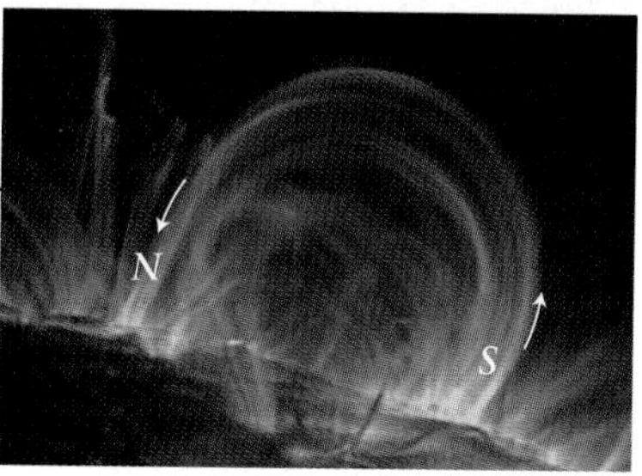

FIGURE 10-13 **Babcock's Magnetic Dynamo** In a plausible partial explanation for the sunspot cycle, differential rotation wraps a magnetic field around the Sun. Convection under the photosphere tangles the field, which becomes buoyant and rises through the photosphere, creating sunspots and sunspot groups. **Insets:** In each group, the sunspot that appears first has the same polarity as the Sun's magnetic pole in that hemisphere (in bottom inset, N in the upper hemisphere and S in the lower hemisphere). The Sun's magnetic fields are revealed by the radiation emitted from the gas they trap. These ultraviolet images show coronal loops up to 160,000 km (100,000 mi) high, with gases moving along the magnetic field lines at speeds of 100 k/s (60 mi/s). (top inset: TRACE, Standford-Lockheed Institute for Space Research, and NASA Small Explorer program; bottom inset: NASA)

magnetic field leave and reenter the Sun. The gas bottled up in the loops of magnetic field eventually leaks out, and the fields untangle, interact with other parts of the Sun's magnetic field, and gradually settle back under the photosphere, at which point the sunspots associated with them disappear. This process of magnetic fields piercing the Sun's surface begins at high latitudes, and over the next 11 years it occurs closer and closer to the Sun's equator (see Figure 10-10).

Because of how these fields interact and vanish, every 11 years the Sun's entire magnetic field is reversed—the Sun's north magnetic pole becomes its south magnetic pole and vice versa. After another 11-year cycle, the field is back to its original orientation. This is why the solar cycle is 22 years long. The most recent reversal of the Sun's magnetic field occurred in 2001.

Notable irregularities occur in the solar cycle. For example, the overall reversal of the Sun's magnetic field is often piecemeal and haphazard. More intriguing still is the strong historical evidence that all traces of sunspots and the sunspot cycle have vanished for decades at a time. For example, in 1893 the British astronomer E. Walter Maunder used historical observations to conclude that virtually no sunspots occurred from 1645 through 1715 (see Figure 10-8a). This period, called the *Maunder minimum*, coincides with a period of cold in Europe so extreme that it was called the Little Ice Age. At the same time, western North America was subject to severe drought.

Similar sunspot-free periods apparently occurred at irregular intervals in earlier times as well. Conversely, periods of increased sunspot activity in the eleventh and twelfth centuries coincided with periods of warmer-than-average temperatures. It remains to be seen, however, if scientists can find all the links between the number of sunspots and periods of extreme temperatures on Earth.

INSIGHT INTO SCIENCE

Cause and Effect While extremes of sunspot activity often occur at the same times as weather extremes on Earth, models of these phenomena must also take into account the possibility of other, terrestrial causes for these temperature changes. Remember, just because one event follows another does not mean that the first *causes* the second.

10-6 Solar magnetic fields also create other atmospheric phenomena

Figure 10-14 shows the active Sun's chromosphere and corona. The bright areas in this photograph are called **plages** (pronounced "plahzh," from the French word for "beaches"). Best seen in the light emitted by calcium or

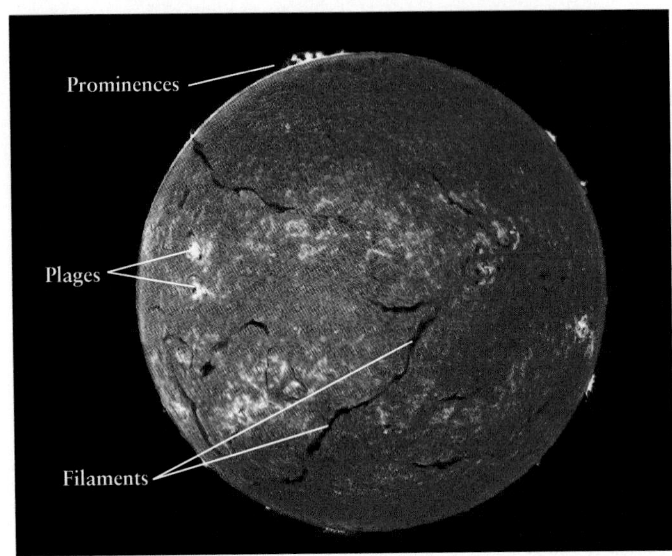

R I V U X G

FIGURE 10-14 **Active Sun in H_α** This photograph shows the chromosphere and corona during a solar maximum, when sunspots are abundant. The image was taken through a filter allowing only light from H_α emission to pass through. The hot upper layers of the Sun's atmosphere are strong emitters of H_α photons. (See Section 4-6 for details of the Balmer series, $H_\alpha - H_\infty$.) A few large sunspots are evident. Most notable are features that do not appear at the solar minimum, such as the snakelike features shown here called filaments, bright areas called plages, and prominences (filaments seen edge-on) observed at the solar limb. (NASA)

hydrogen atoms, they are hotter, and therefore brighter, than the surrounding chromosphere. Plages, which often appear just before nearby sunspots form, are believed to be created by the magnetic field under the photosphere crowding upward just before they emerge through the photosphere. In pushing upward, the fields compress the gases of the upper Sun, causing this gas to become hotter and therefore to glow more brightly.

The dark streaks in Figure 10-14 are features in the corona called **filaments.** These huge volumes of gas are lofted upward from the photosphere by the Sun's magnetic field. When viewed from the side rather than from above, filaments form gigantic loops or arches called **prominences** (see Figures 10-14 and 10-15a). The temperature of gas in prominences can reach 50,000 K. These features are almost always associated with sunspots. Some prominences last only for a few hours, while others persist for months. The most energetic prominences escape the magnetic fields that confine them and surge out into space (Figure 10-15b).

X-ray photographs also reveal numerous coronal bright and dark spots that are hotter or colder, respectively, than the surrounding corona (Figure 10-16). Temperatures in the bright regions occasionally reach 4 million K. Many

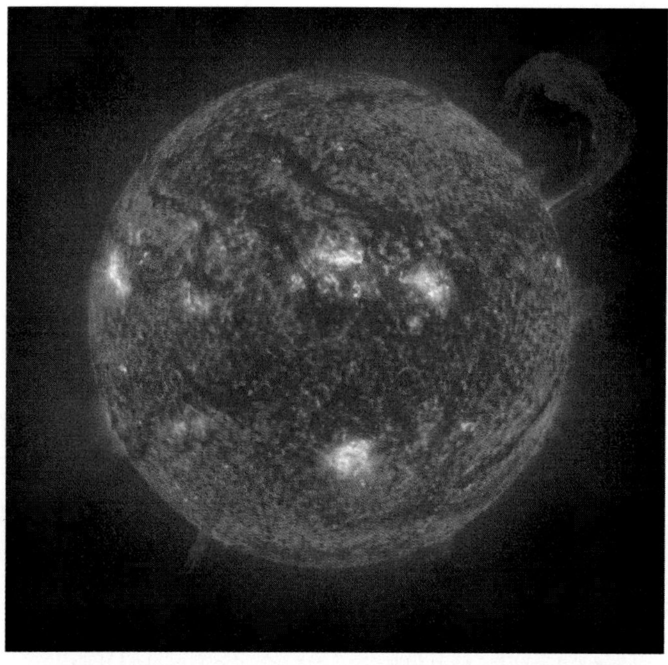

a

R I V U X G

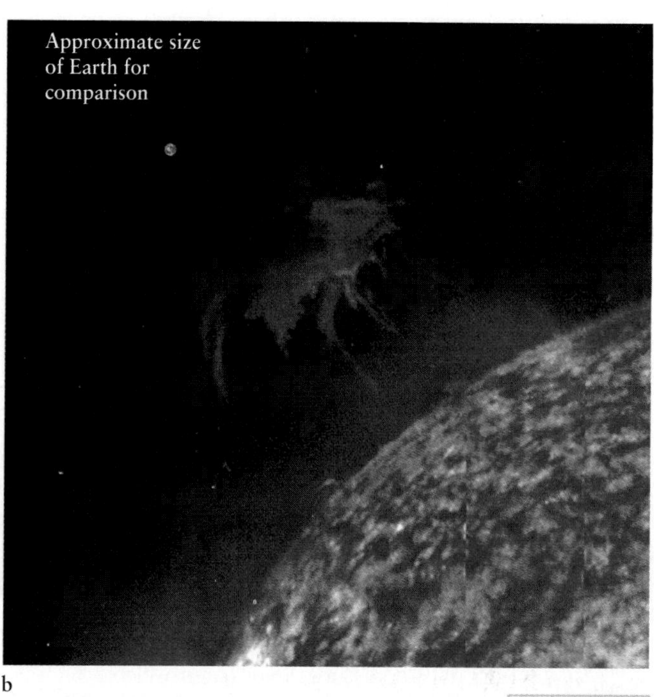

b

R I V U X G

FIGURE 10-15 **Prominences** (a) A huge prominence arches above the solar surface in this *SOHO* image taken in 2001. The radiation that exposed this picture is from singly ionized helium at a wavelength of 30.4 nm, corresponding to a temperature of about 50,000 K. (b) The gas in this prominence was so energetic that it broke free from the magnetic fields that shaped and confined it. This eruptive prominence occurred in 1999 and did not strike the Earth (shown for size). (a: ESA/NASA; b: Joseph B. Gurman, Solar Data Analysis Center, NASA)

of the bright coronal hot spots are located over sunspots. The darker, cooler **coronal holes** act as conduits for gases to flow out of the Sun. Therefore, when a coronal hole on the rotating Sun faces Earth, the solar wind in our direction increases dramatically.

Violent, eruptive events on the Sun, called **solar flares,** release vast quantities of high-energy particles, as well as X rays and ultraviolet radiation, from the Sun (Figure 10-17). Flares are such powerful events that they leave the region of the surface of the Sun in their vicinity quaking for an hour or more. At the maximum of the sunspot cycle, there are about 1100 flares per year. Most flares last for less than an hour, but during that time, temperatures soar to 5 million K. By following the paths of particles emitted by flares as they are guided outward by the Sun's magnetic field, astronomers have recently begun mapping that field as it extends out beyond the Earth (Figure 10-18).

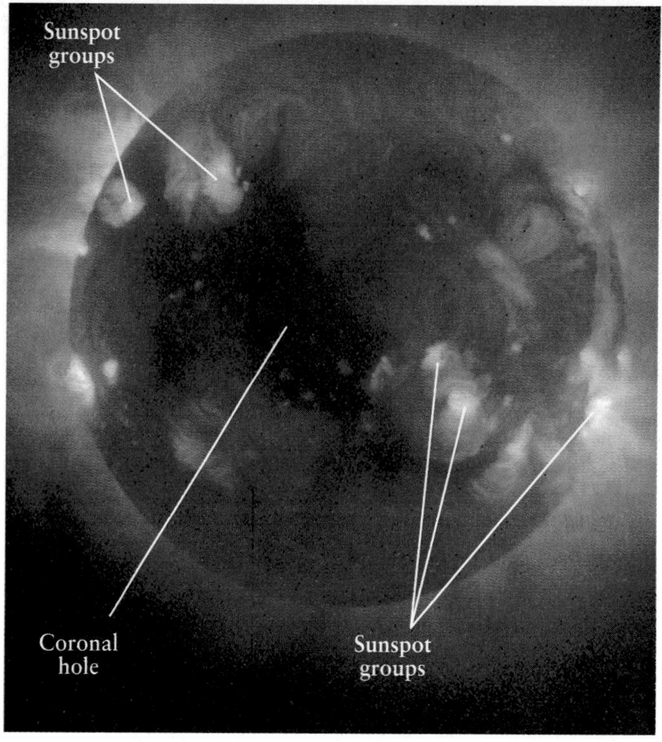

R I V U X G

WEB LINK 10.6

FIGURE 10-16 **A Coronal Hole** This X-ray picture of the Sun's corona was taken by the *Yohkoh* satellite on February 23, 2000. A huge coronal hole dominates the central region of the corona. Gases from this hole caused spectacular aurorae on Earth. The bright regions are emissions from sunspot groups. (NASA)

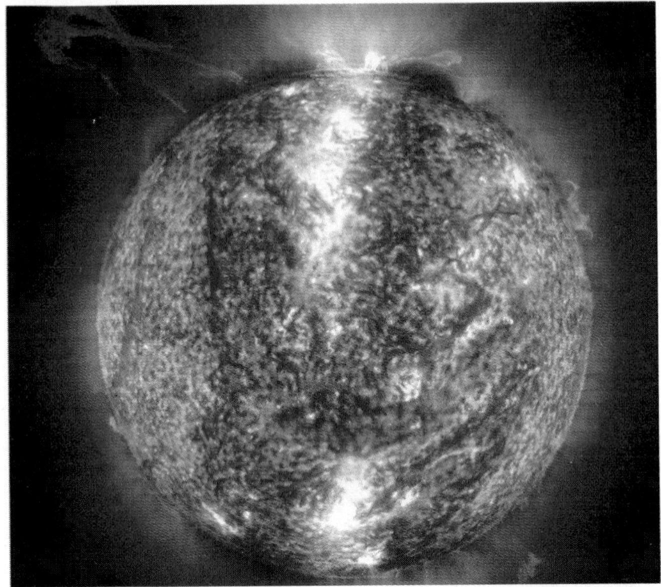

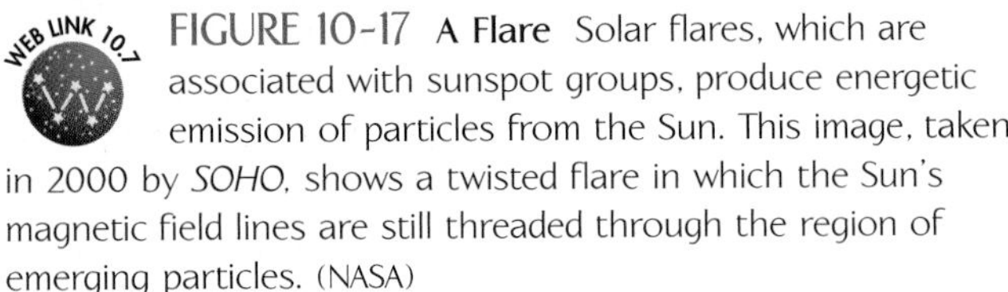

FIGURE 10-17 **A Flare** Solar flares, which are associated with sunspot groups, produce energetic emission of particles from the Sun. This image, taken in 2000 by *SOHO*, shows a twisted flare in which the Sun's magnetic field lines are still threaded through the region of emerging particles. (NASA)

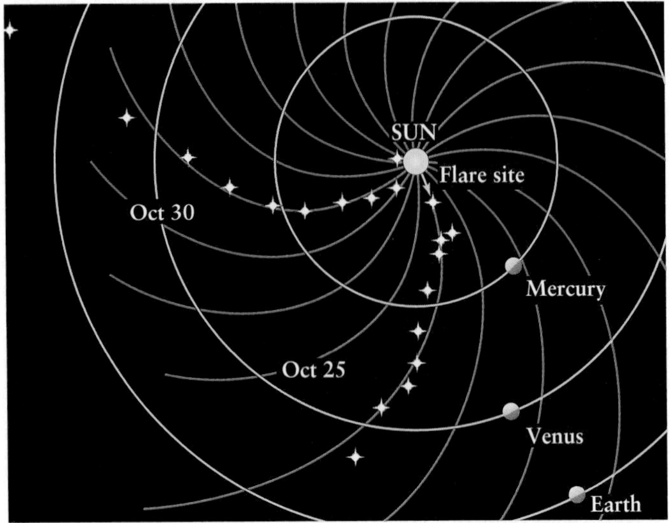

FIGURE 10-18 **A Snapshot of the Sun's Global Magnetic Field** By following the paths of particles emitted by a solar flare, astronomers have begun mapping the solar magnetic field outside the Sun. The field guides the outflowing particles, which in turn emit radio waves that indicate the position of the field. These data were collected by the *Ulysses* spacecraft in 1994. *Ulysses* was the first spacecraft to explore interplanetary space from high above the plane of the ecliptic.

Astronomers have also observed huge, balloon-shaped volumes of high-energy gas being ejected from the corona. These **coronal mass ejections** typically expel 2 trillion tons of matter at 400 km/s, and each one lasts for up to a few hours (Figure 10-19). Coronal mass ejections have enough energy to break through the Sun's magnetic fields that normally contain them or even to carry fields outward, altering the Sun's magnetic field. Flares have been observed to create shock waves that initiate some of the coronal mass ejections. The origins of others are still under investigation.

Some coronal mass ejections, solar flares, and filaments head toward Earth. It takes about 8 minutes for their electromagnetic radiation to get here; their high-energy particles arrive a few days later. At times when these surges of particles are *not* coming to us, the normal solar wind particles are trapped by the Earth's magnetic fields in the Van Allen belts (see Section 6-4). The additional particles from a coronal mass ejection or other solar event overwhelm the Van Allen belts, enabling matter in them to cascade earthward. One of the most spectacular such events in recent years occurred on and around October 31, 2003 (Halloween in the U.S.), filling the night skies with aurorae in many places around the world. Perhaps you saw them.

Using the technology described in Chapter 3, several spacecraft now monitor the Sun and the space between it and the Earth. These include the *Solar and Heliospheric Observatory* (*SOHO*) launched in 1995 by the European Space Agency (ESA) and NASA, the *Transition Region and Coronal*

R I V U X G

FIGURE 10-19 **A Coronal Mass Ejection Left:** X-ray image of a coronal mass ejection from the Sun taken by *SOHO*. **Right:** Two to four days later, the highest energy gases from the ejection reach 1 AU. If they come our way, most particles are deflected by the Earth's magnetic field (in blue). However, as shown, some particles leak earthward, causing aurorae, disrupting radio communications and electrical power transmission, damaging satellites, and ejecting some of Earth's atmosphere into interplanetary space. (NASA)

Explorer (*TRACE*) launched in 1998, the *Reuven Ramaty High Energy Solar Spectroscopic Imager (RHESSI)* launched in 2002, and the *Solar Radiation and Climate Experiment (SORCE)* launched in 2003, among others. These spacecraft continuously monitor the Sun in gamma-ray, X-ray, ultraviolet, and visible parts of the spectrum, and they provide both scientific and space weather information.

Space weather can profoundly affect some forms of technology. Strong surges of high-energy particles from the Sun can potentially damage satellites, disrupt radio communications, short out power grids, produce intense aurorae in the Earth's atmosphere, cause some of the Earth's atmosphere to gush into interplanetary space, and harm or kill people in space. The satellites mentioned above provide data, even from the side of the Sun facing away from the Earth, that allow astronomers to forecast several days in advance the arrival of dangerous levels of solar particles. This enables us to protect satellites and other assets. We can determine the state of the Sun's far side by measuring ripples on the Earth-facing side of the Sun's surface and then using computer models of how activity on the far side affects these ripples.

The numbers of plages, prominences, flares, and coronal mass ejections vary with the same 11-year cycle as the number of sunspots. Coronal mass ejections, the major source of hazardous particles from the Sun, occur with varying frequency throughout the sunspot cycle, but they never completely cease.

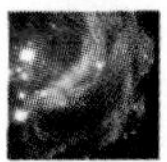

THE SUN'S INTERIOR

During the 1800s, geologists and biologists found convincing evidence that the Earth must have existed in more or less its present form for at least hundreds of millions of years. This fact posed severe problems for astrophysicists, because at that time it seemed impossible to explain how the Sun could continue to shine for so long, radiating immense amounts of energy into space. If the Sun were shining by burning coal or hydrogen gas as we could burn these substances on Earth, the Sun would be ablaze for only 5000 years before consuming all its fuel.

Everyday experience tells us that the Sun is the source of an enormous amount of energy. We have just seen that observations reveal hot gas, intense magnetic fields, and a variety of continuous and transient features on the Sun's surface. But the energy does not come from the surface gas or the magnetic fields—they have no mechanisms to create it. The origin of the Sun's electromagnetic radiation is deep within it. To understand why the Sun shines, we must understand where its energy originates and how that energy is transported to its surface.

10-7 Thermonuclear reactions in the core of the Sun produce its energy

In 1905, Albert Einstein provided an important clue to the source of the Sun's energy with his special theory of relativity. One of the implications of this theory is that matter and energy are related by the simple equation:

$$E = mc^2$$

In other words, a mass (m) can be converted into an amount of energy (E) equivalent to mc^2, where c is the speed of light. Because c^2 ($= c \times c$) is huge, namely 9×10^{10} km^2/s^2, a small amount of matter can be converted into an awesome amount of energy.

Inspired by Einstein's work, physicists discovered that the Sun's energy output comes from the conversion of matter into energy. In the 1920s, the British astrophysicist Arthur Eddington proposed that the temperatures at the center of the Sun, its **core**, are much greater than had ever been imagined. Subsequent calculations revealed that the temperature in the Sun's core is about 15.5×10^6 K. Physicists also showed that at temperatures above about 10×10^6 K hydrogen fuses into the element helium. In each such reaction, a tiny amount of mass is converted into energy in the form of gamma rays. Enough fusion occurs in its core to account for all the energy emitted by the Sun.

The process of fusing nuclei at such extreme temperatures is called **thermonuclear fusion.** For more details on thermonuclear fusion, see An Astronomer's Toolbox 10-1. In particular, conversion of hydrogen into helium is called **hydrogen fusion.** The same process provides the devastating energy released by a hydrogen bomb. (We will encounter the thermonuclear fusion of helium and other elements in later chapters.) Some of the energy generated by hydrogen fusion in the Sun's core eventually escapes through the photosphere into space. That energy makes the Sun shine.

You may have heard comments that matter is always conserved or that energy is always conserved. We know that both of these concepts are incorrect, because mass can be converted into energy and vice versa. What is true,

(© Sidney Harris)

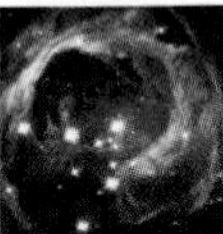

AN ASTRONOMER'S TOOLBOX 10-1
Thermonuclear Fusion

What drives the thermonuclear fusion that powers the Sun? For nuclei to fuse, they must be brought together at incredibly high temperatures and pressures. That is exactly what occurs in the Sun's core, where the entire mass of the Sun compresses inward. The core's temperature is 15.5 million K, its pressure is about 3.4×10^{11} atm, and its density is 160 times greater than that of water.

Normally, nuclei cannot penetrate each other because the positive electric charge on each proton prevents nearby protons from coming too close. (Remember that like charges repel each other.) But in the extreme heat and pressure of the Sun's center, the protons move so fast in such close proximity that they can stick, or fuse, together.

The nuclear transformations inside the Sun follow several routes, but each begins with the simplest atom, hydrogen (H). Most hydrogen nuclei consist of a single proton. The final outcome of fusion is creation of the nucleus of the next simplest atom, helium (He), consisting of two protons and two neutrons. The fusion of hydrogen into helium takes several steps.

The accompanying diagram shows the most common path for hydrogen fusion in the Sun. This particular sequence is called the *proton-proton*, or PP, *chain*.

Note that proton-proton fusion releases positively charged electrons, e^+, called **positrons,** and neutral, nearly massless particles called **neutrinos,** ν. When these positrons encounter regular electrons in the Sun's core, both particles are annihilated and their mass is converted into energy in the form of gamma-ray photons. The final fusion, in which the helium forms, also returns two protons to the core, which are then available to fuse again.

Because the PP chain produces neutrinos and the Sun's energy, we can summarize hydrogen fusion this way:

$$4\text{H} \rightarrow 1\text{He} + \text{neutrinos} + \text{energy}$$

Example: From our summary equation, we can easily calculate the energy released during a fusion reaction. We simply look at how much mass is converted into energy:

$$\begin{aligned} \text{Mass of 4 hydrogen atoms} &= 6.693 \times 10^{-27}\ \text{kg} \\ -\ \text{Mass of 1 helium atom} &= 6.645 \times 10^{-27}\ \text{kg} \\ \hline \text{Mass lost} &= 0.048 \times 10^{-27}\ \text{kg} \end{aligned}$$

Thus, a small fraction (0.7%) of the mass of the hydrogen going into the nuclear reactions does not show up in the mass of the helium. This last mass is converted into energy, as predicted by Einstein's famous equation,

$$\begin{aligned} E &= mc^2 \\ &= (0.048 \times 10^{-27}\ \text{kg}) \times (3 \times 10^8\ \text{m/s})^2 \\ &= 4.3 \times 10^{-12}\ \text{J} \end{aligned}$$

Compare! The energy released from the formation of a single helium atom would light a 10-watt lightbulb for almost one-half a trillionth of a second.

Now let us add up the energy emitted from the entire Sun. As noted earlier in this chapter, the Sun's mass, usually designated 1 $M_\odot$, is equal to 333,000 $M_\oplus$. Its total energy output per second, called the **solar luminosity** and

however, is that the total amount of mass plus energy is conserved. So, the destruction of mass and the creation of energy by the Sun do not violate any laws of nature.

Hydrogen fusion is also called *hydrogen burning,* even though nothing is burned in the conventional sense. The ordinary burning of wood, coal, or any flammable substance is a chemical process involving only the electrons orbiting the nuclei of the atoms. Thermonuclear fusion is a far more energetic process that involves violent collisions between the atomic nuclei themselves. An Astronomer's Toolbox 10-1 shows you just how energetic.

10-8 Solar models describe how energy escapes from the Sun's core

A scientific description of the Sun's interior, called a **solar model,** explains how the energy from nuclear fusion in the Sun's core gets to its photosphere. The model begins with the inward force of the Sun's gravity. This force raises the pressure and temperature in the Sun's core. Because the Sun is not shrinking today, there must be an outward force throughout its interior that counters the inward force of gravity. That outward force is created by the gamma-ray photons generated during fusion.

To begin their journey outward, newly created photons slam into nearby ions and electrons in the solar core. The ions and electrons absorb the energy of the photons and rebound at very high speeds. Because they are densely packed together, they do not travel far before striking other particles. In each collision, the ions and electrons exert forces on each other. They then rebound, emitting photons and colliding with other particles. The result of these frequent interactions is an outward force sufficient to counterbalance gravity. The balance between the inward force of gravity and the outward force from the motion of the hot gas is called **hydrostatic equilibrium** (Figure 10-20).

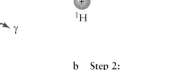

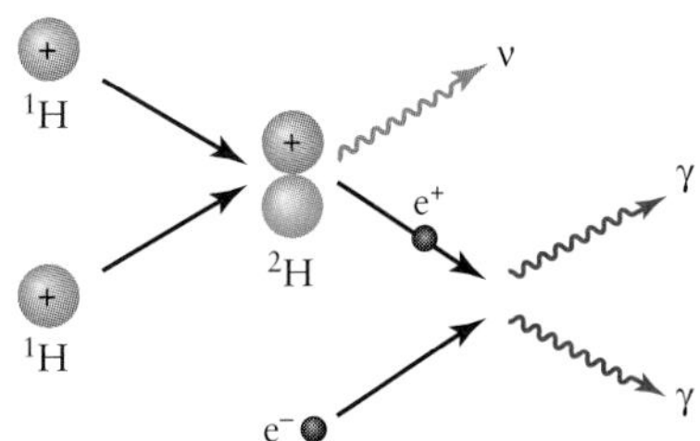

a Step 1:

- Two protons (hydrogen nuclei, ^{1}H) collide.
- One of the protons changes into a neutron (shown in blue), a neutral, nearly massless neutrino (ν), and a positron (e^+), a positively charged electron.
- The proton and neutron form a hydrogen isotope (^{2}H).
- The positron encounters an ordinary electron (e^-), annihilating both particles and converting them into gamma-ray photons (γ).

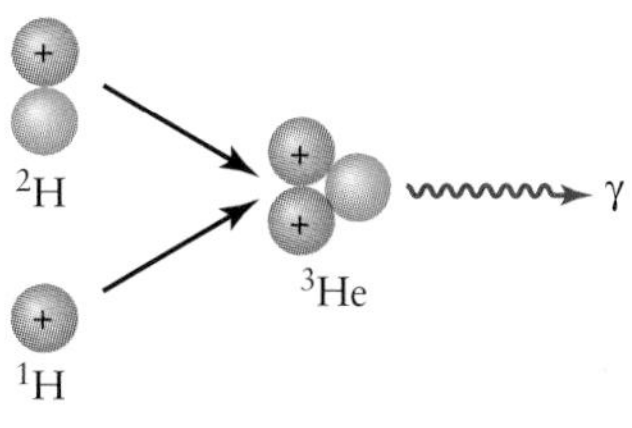

b Step 2:

- The ^{2}H nucleus from the first step collides with a third proton.
- A helium isotope (^{3}He) is formed and another gamma-ray photon is released.

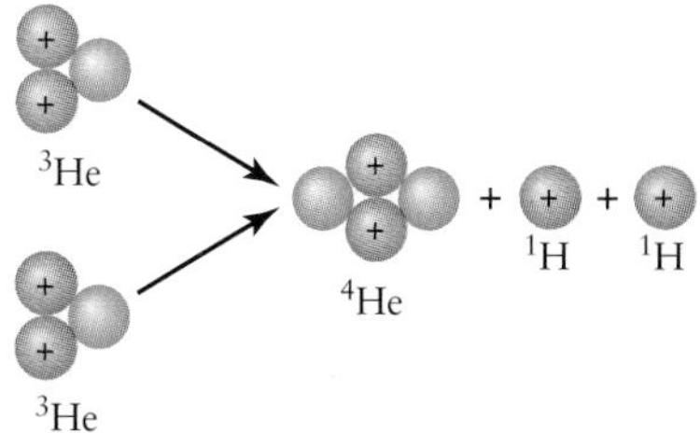

c Step 3:

- Two ^{3}He nuclei collide.
- A different helium isotope with two protons and two neutrons (^{4}He) is formed and two protons are released.

Steps to Fuse Hydrogen into Helium Note also that an electron, e^-, and a positron, or positively charged electron, e^+, annihilate each other and form gamma rays, γ. The energy in the gamma rays helps maintain hydrostatic equilibrium and to eventually provide sunlight and other electromagnetic emissions from the Sun.

denoted 1 $L_{\odot}$, is 3.9×10^{26} W. To produce this luminosity, the Sun converts 600 million metric tons of hydrogen into helium within its core each second. This is twice the weight of the Empire State Building in New York City. This enormous rate is possible because the Sun contains a vast supply of hydrogen—enough to continue the present rate of energy output for another five billion years.

Try these questions: How many helium atoms would have to be created from hydrogen to release enough energy to light a 60-watt bulb for 12 hours? How many joules of energy would be released if a U.S. penny (2.5×10^{-3} kg) were converted entirely into energy?

(Answers appear at the end of the book.)

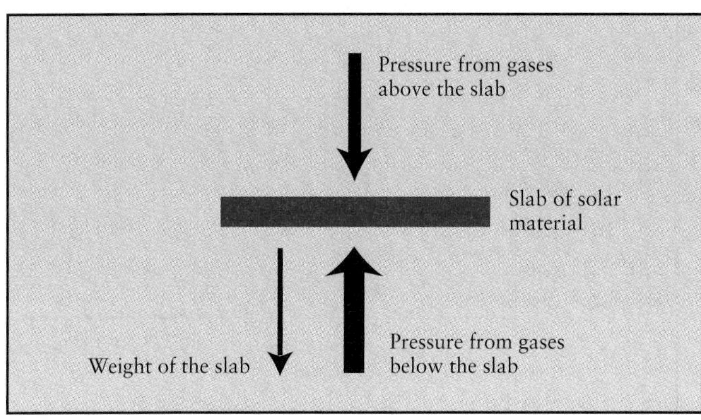

a Material inside the Sun is in hydrostatic equilibrium, so forces balance.

b A fish floating in water is in hydrostatic equilibrium, so forces balance.

FIGURE 10-20 **Hydrostatic Equilibrium** (a) Matter deep inside the Sun is in hydrostatic equilibrium, meaning that upward and downward forces on the gases there are balanced. (b) When the forces on a fish floating in water are in hydrostatic equilibrium, it neither sinks nor rises. (b: Ken Usami/PhotoDisc)

Maintaining the Sun's hydrostatic equilibrium requires that the photons created in its core give up some of their energy. Photons collide with particles, which collide with other particles and thereby emit new photons. The emitted photons have slightly less energy than the photons that these particles had previously absorbed. The energy the photons lose provides the outward force keeping the Sun in equilibrium. Photon energies slowly diminish as the photons travel outward from the Sun's core to the photosphere, an odyssey that typically takes 170,000 years.

The outward movement of energy by photons hitting particles, which then bounce off other particles and thereby reemit photons, is called *radiative transport,* because individual photons are responsible for carrying energy from collision to collision. Calculations show that radiative transport is the dominant means of outward energy flow in the **radiative zone,** extending from the core to 80% of the way out to the photosphere.

Near the photosphere, energy is carried the remaining distance to the surface by the bulk motion of the hot gas, rather than the flying, energetic photons. As we discussed earlier, this circulation of gas blobs is called *convection.* We thus say that the Sun has a **convective zone.** Once hot gas convects up to the photosphere, it emits photons into space, cools, and settles back into the Sun. As noted earlier, this convective flow is the origin of solar granules, and the departing energy is the visible light and other electromagnetic radiation that the Sun emits into space. Figure 10-21a sketches our model of the internal structure of the Sun.

Different gamma rays created in the Sun's core lose different amounts of energy as they and their successors travel upward through the Sun. Therefore, the photons emitted from the photosphere have a wide range of energies and, hence, wavelengths. The most intense emission is in the visible part of the electromagnetic spectrum. This process of energy loss by photons traveling up from the Sun's core is the origin of the blackbody nature of the photosphere's spectrum.

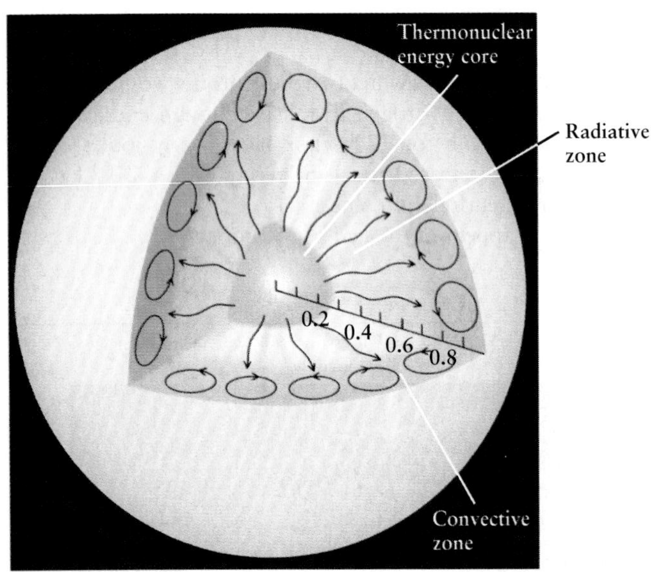

a

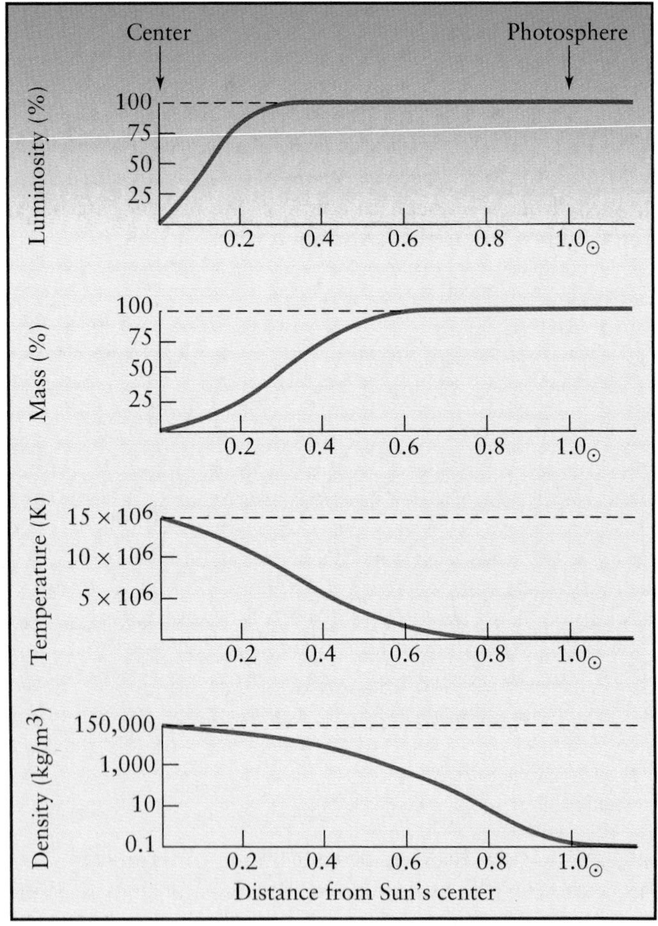

b

FIGURE 10-21 The Solar Model (a) Thermonuclear reactions occur in the Sun's core, which extends to a distance of 0.25 solar radius from the center. In this model, energy from the core radiates outward to a distance of 0.8 solar radius. Convection is responsible for energy transport in the Sun's outer layers. (b) The Sun's internal structure is displayed here with graphs that show how the luminosity, mass, temperature, and density vary with the distance from the Sun's center. A solar radius (the distance from the Sun's center to the photosphere) equals 696,000 km.

Our model of the Sun's interior is expressed as a set of mathematical equations called the *equations of stellar structure*. The model quantitatively determines the Sun's internal characteristics, such as its pressure, temperature, and density at various depths. These equations also describe the conditions inside other stars as they evolve. Because the equations are so complex, astrophysicists today use computers to solve them. Figure 10-21b presents their results graphically. The four graphs show how the Sun's luminosity, mass, temperature, and density vary from the Sun's center to its photosphere. For example, the upper graph gives the percentage of the Sun's luminosity created within that radius. The luminosity rises to 100% at about one-quarter of the way from the Sun's center to its photosphere. This tells us that all the Sun's energy is produced within a volume extending out to $0.25R_{\odot}$, where $R_{\odot}$ is the radius of the Sun.

The mass curve rises to nearly 100% at about 0.6 $R_{\odot}$ from the Sun's center. Almost all of the Sun's mass is therefore confined to a volume extending only 60% of the distance from the Sun's center to its visible surface. This helps explain why the density of the gas in the photosphere is 10^4 times lower than the density of the air we breathe.

10-9 The mystery of the missing neutrinos inspired research into the fundamental nature of matter

As explained in An Astronomer's Toolbox 10-1, for every proton that changes into a neutron during thermonuclear fusion, a *neutrino* is released. Neutrinos have no electric charge, and they are extraordinarily difficult to detect because they rarely interact with ordinary matter. In fact, neutrinos interact so infrequently with matter that they can easily pass through the entire Earth as if it were not there. The Sun is also largely transparent to neutrinos, allowing these particles to stream outward, unimpeded, from its core.

According to our model of fusion in the Sun, nearly 10^{38} *solar neutrinos* are produced at the Sun's center each second. This output is so huge that here on Earth roughly 100 billion solar neutrinos pass through every square centimeter of your body every second!

Occasionally, solar neutrinos strike neutrons and convert them into protons. If astronomers could detect even a few of these converted protons, it might be possible to build a "neutrino telescope" that could be used to "see" directly into the center of the Sun and the thermonuclear inferno there that is now hidden from ordinary view.

WEB LINK 10.12 Inspired by such possibilities, Raymond Davis of the Brookhaven National Laboratory designed and built a large neutrino detector. This device consisted of a huge tank containing 100,000 gallons of perchloroethylene (C_2Cl_4, the fluid your local dry cleaner uses) buried deep in a mine in South Dakota. All neutrino experiments are performed underground to help prevent them from being contaminated by other sources of energy, like high-energy protons from space. Because matter is virtually transparent to neutrinos, most of the solar neutrinos passed right through Davis's tank. On rare occasions, however, a solar neutrino hit the nucleus of one of the chlorine atoms in the cleaning fluid and converted one of its neutrons into a proton, creating a radioactive atom of argon. The rate at which argon was produced was therefore correlated with the number of solar neutrinos arriving at the Earth.

On average, solar neutrinos created one radioactive argon atom every 3 days in Davis's tank. To the consternation of astronomers, this rate corresponds to only one-third of the neutrinos predicted to be created by the fusion in the Sun's core. There were three possible explanations of this unexpected result: The experiment could be in error; the Sun might not be fusing at the expected rate; or our understanding of the properties of neutrinos could be in error. This experiment, which began in the mid-1960s, was repeated with extreme care by other researchers around the world who got the same results, suggesting that the experiment was not at fault. Calculations of how much energy the Sun must generate to shine as it does confirmed the number of neutrinos created per second. The problem had to lie in our understanding of the neutrino.

Instead of being massless, as scientists had originally believed, neutrinos have a small amount of mass. Massive neutrinos have a property that massless neutrinos lack, namely, massive neutrinos can spontaneously change into other particles over time. Experiments have shown that before reaching the Earth, two-thirds of the neutrinos emitted by the Sun change into either of two other types of neutrinos, neither of which could be observed by the early detectors. The two other types of neutrinos into which solar neutrinos transformed on their way to Earth were first observed directly in Japan in 1998 and then again in 2002 at the Sudbury Neutrino Observatory (Figure 10-22) in Ontario, Canada. At least a dozen other neutrino detectors have begun operations or are being built to further study solar neutrinos.

To understand how the present generation of neutrino detectors work, let us consider one of the several interactions that occur between neutrinos and other matter. The Sudbury Neutrino Observatory is filled with water that contains a rare type of hydrogen nucleus called *deuterium*, which consists of a proton and a neutron. When a solar neutrino strikes a deuterium nucleus, the neutrino is converted into an electron and the deuterium breaks apart and forms two protons. As these particles move, they emit flashes of light called **Cerenkov radiation**. As the Russian physicist Pavel A. Cerenkov (pronounced "Che-ren-kov") first observed, the flash occurs whenever a particle moves through water faster than light can. Such motion does not violate the tenet that the speed of light *in a vacuum* (3×10^5 km/s) is the ultimate speed limit in the universe. Light is

R I V U X G

FIGURE 10-22 **The Solar Neutrino Experiment** Located 6800 feet underground in the Creighton nickel mine in Sudbury, Canada, the Sudbury Neutrino Observatory is centered around a tank containing 1000 tons of water. Occasionally, a neutrino entering the tank interacts with one or another of the particles already there. Such interactions create flashes of light called Cerenkov radiation. Some 9600 light detectors sense this light. The numerous silver protrusions are the back sides of the light detectors prior to their being wired and connected to electronics in the lab, seen at the bottom of the photograph. (Ernest Orlando Lawrence/Berkeley National Laboratory)

slowed considerably as it passes through water, and high-energy particles can exceed this reduced speed without violating the laws of nature.

Thus, scientists detect neutrinos by observing Cerenkov radiation flashes with light-sensitive devices called *photomultipliers* mounted in the water (see Figure 10-22). The three types of neutrinos have different interactions with water, all of which have been detected. This evidence that neutrinos can change requires that they have mass and it explains the earlier low rate of solar neutrino observations.

10-10 Frontiers yet to be discovered

Despite decades of observing and modeling it, the Sun holds innumerable secrets to be uncovered. We have launched an armada of spacecraft specifically designed to observe the Sun in different parts of the spectrum. Astronomers hope to soon understand more about such questions as how the Sun's magnetic field is generated and how it changes throughout the solar cycle; how the Sun's atmosphere heats so dramatically in the transition zone; why the solar luminosity varies with time; just how the Sun's changing output affects the temperatures on the Earth; the details of how flares and coronal mass ejections occur; and why the Sun rotates differentially.

Summary of Key Ideas

The Sun's Atmosphere

- The visible surface of the Sun is a layer at the bottom of the solar atmosphere called the photosphere. The gases in this layer shine as a blackbody. Convection of gas from below the surface produces features there called granules.
- Above the photosphere is a layer of hotter but less dense gas called the chromosphere. Jets of gas called spicules rise up into the chromosphere along the boundaries of supergranules.
- The outermost layer of thin gases in the solar atmosphere is called the corona, which extends outward to become the solar wind at great distances from the Sun. The gases of the corona are very hot, but they have extremely low densities.

The Active Sun

- Some surface features on the Sun vary periodically in an 11-year cycle, with the magnetic fields that cause these changes actually varying over a 22-year cycle.
- Sunspots are relatively cool regions produced by local concentrations of the Sun's magnetic field protruding through the photosphere. The average number of sunspots and their average latitude increases and decreases in an 11-year cycle.
- A prominence is gas lifted into the Sun's corona by magnetic fields. A solar flare is a brief, but violent, eruption of hot, ionized gases from a sunspot group. Coronal mass ejections send out large quantities of gas from the Sun. Coronal mass ejections and flares that head our way affect satellites, communication and electrical power and cause aurorae.
- The magnetic dynamo model suggests that many transient features of the solar cycle are caused by the effects of differential rotation and convection on the Sun's magnetic field.

The Sun's Interior

- The Sun's energy is produced by the thermonuclear process called hydrogen fusion, in which four hydrogen nuclei release energy when they fuse to produce a single helium nucleus.
- The energy released in a thermonuclear reaction comes from the conversion of matter into energy according to Einstein's equation $E = mc^2$.

• A stellar model is a theoretical description of a star's interior derived from calculations based on the laws of physics. The solar model suggests that hydrogen fusion occurs in a core that extends from the Sun's center to about 0.25 solar radius.

• Throughout most of the Sun's interior, energy moves outward from the core by radiative diffusion. In the Sun's outer layers, energy is transported to the Sun's surface by convection.

• Neutrinos were originally believed to be massless. The neutrinos generated and emitted by the Sun were orginally detected at a lower rate than is predicted by our model of thermonuclear fusion. That occurred because neutrinos have mass; therefore, some of them change into other forms of neutrinos before they reach Earth. These alternative forms are now being detected and the fact that the neutrino has mass is now an accepted part of elementary particles physics.

WHAT DID YOU THINK?

1 *What percentage of the solar system's mass is in the Sun?* The Sun contains about 99.85% of the solar system's mass.

2 *Are there stars nearer the Earth than the Sun?* No, the Sun is our closest star.

3 *Does the Sun have a solid and liquid interior, like the Earth?* No. The entire Sun is composed of hot gases.

4 *What is the surface of the Sun like?* The Sun has no solid surface. Indeed, it has no solids or liquids anywhere. The level we see, the photosphere, is composed of hot, churning gases.

5 *Does the Sun rotate?* The Sun's surface rotates differentially, varying between once every 35 days near its poles and once every 25 days at its equator.

6 *What makes the Sun shine?* Thermonuclear fusion in the Sun's core is the source of the Sun's energy.

Key Words

Cerenkov radiation, 285
chromosphere, 270
convective zone, 284
core (of the Sun), 281
corona, 271
coronal hole, 279
coronal mass ejection, 280
filament, 278
granule, 270
helioseismology, 276
hydrogen fusion, 281
hydrostatic equilibrium, 282
limb (of the Sun), 268
limb darkening, 268
magnetic dynamo, 277
neutrino, 282
photosphere, 268
plage, 278
plasma, 276
positron, 282
prominence, 278
radiative zone, 284
solar cycle, 277
solar flare, 279
solar luminosity ($L_\odot$), 282
solar model, 282
solar wind, 272
spicule, 270
sunspot, 273
sunspot maximum, 273
sunspot minimum, 273
supergranule, 270
thermonuclear fusion, 281
transition zone, 271
Zeeman effect, 276

Review Questions

The answers to all computational problems, which are preceded by an asterisk (*), appear at the end of the book.

1. Describe the features of the Sun's atmosphere that are always present.

2. Describe the three main layers in the solar atmosphere and how you would best observe them.

3. Name and describe seven features of the active Sun. Which two are the same, seen from different angles? Which have a direct impact on the Earth? Explain.

4. Describe the three main layers of the Sun's interior.

***5.** When will the next sunspot minimum and sunspot maximum occur after the maximum in 2001 and the minimum in 2007? Explain your reasoning.

6. Why is the solar cycle said to have a period of 22 years, even though the sunspot cycle is only 11 years long?

7. How do astronomers detect the presence of a magnetic field in hot gases such as those in the solar photosphere?

8. Describe the dangers in attempting to observe the Sun. How have astronomers learned to circumvent these hazards?

9. Give an everyday example of hydrostatic equilibrium not presented in the book.

10. Give some everyday examples of heat transfer by convection and radiative transport.

11. What do astronomers mean by "a model of the Sun"?

12. Why do thermonuclear reactions in the Sun take place only in its core?

13. What is hydrogen fusion? This process is sometimes called "hydrogen burning." How is hydrogen burning fundamentally unlike the burning of a log in a fireplace?

14. Describe the Sun's interior, including the main physical processes that occur at various depths within the Sun.

15. What is a neutrino, and why are astronomers so interested in detecting neutrinos from the Sun?

Advanced Questions

***16.** Using the mass and size of the Sun, calculate the Sun's average density. Compare your answer to the average densities of the outer planets. *Hint:* The volume of a sphere of radius r is $^4/_3\pi r^3$.

*17. Assuming that the current rate of hydrogen fusion in the Sun remains constant, what fraction of the Sun's mass will be converted into helium over the next five billion years? How will this affect the chemical composition of the Sun?

*18. Calculate the wavelengths at which the photosphere, chromosphere, and corona emit the most radiation. Explain how the results of your calculations suggest the best way to observe these regions of the solar atmosphere. *Hint:* Use Wien's law and assume that the average temperatures of the photosphere, chromosphere, and corona are 5800 K, 50,000 K, and 1.5×10^6 K, respectively.

19. When we are near a sunspot maximum, the Hubble Space Telescope must be put in a higher orbit. Why? *Hint:* Think about how the increased solar energy affects the Earth's atmosphere.

Discussion Questions

20. Discuss the extent to which cultures around the world have worshiped the Sun as a deity throughout history. Why do you think our star inspires such widespread veneration?

21. Discuss some of the difficulties of correlating solar activity with changes in the terrestrial climate.

22. Describe some advantages and disadvantages of observing the Sun (a) from space and (b) from Earth's South Pole. What kinds of phenomena and issues do solar astronomers want to explore from both Earth-orbiting and Antarctic observatories?

What If . . .

23. The Sun was not rotating? What about it would be different?

24. The typical solar wind was much stronger (say 100 times stronger) than it is now? What differences would there be in the solar system?

25. The typical solar wind was much weaker (say 100 times weaker) than it is now? What differences would there be in the solar system?

26. The Sun's brightness and heat output periodically changed, as they do in many stars? What differences would there be in the solar system? Unless your teacher gives you another time frame, assume that the light and heat vary together in a cycle lasting 1 year.

Web Questions

27. Search the Web for the latest information from the Sudbury Neutrino Observatory. What are the most recent results from Sudbury? What is the current thinking about the solar neutrino problem? What is the status of a new detector called Borexino?

28. Search the Web for information about features of the solar atmosphere called *sigmoids*. What are they? What causes them? How do they provide a way to predict coronal mass ejections?

29. **Determining the Lifetime of a Solar Granule** Access and view the video "Granules on the Sun's Surface" in Chapter 10 of the *Discovering the Universe* Web site. You will use it to determine the approximate lifetime of a solar granule. Select an area on the Sun's image and slowly and rhythmically repeat "start, stop, start, stop" until you can consistently predict the appearance and disappearance of granules. While keeping your rhythm, move to a different area of the video and continue monitoring the appearance and disappearance of granules. When you are confident you have the timing right, move your eyes (or use a partner) to the clock shown in the video. Using your "start stop" cycle, determine the length of time between the appearance and disappearance of the granules and record your answer.

Observing Projects

30. Use a telescope to view the Sun by using a solar filter or by projecting the Sun's image onto a screen or sheet of white paper. **Do not look directly at the Sun! Looking at the Sun causes blindness.** Do you see any sunspots? If so, sketch their appearance. Can you distinguish between the umbras and penumbrae of the sunspots? Can you see limb darkening? Can you see granulation?

31. If you have access to an H_α filter attached to a telescope especially designed for viewing the Sun safely, use this instrument to examine the solar surface. How does the filtered appearance of the Sun differ from that in white light (see Observing Project 30)? What do sunspots look like in H_α? Can you see any prominences? Can you see any filaments? Are the filaments in the H_α image near any sunspots seen in white light?

*32. Use *Starry Night Enthusiast*™ software to determine the Sun's rotation rate. Go to Atlas mode (*Favourites/Guides/Atlas*) and find the Sun (*Find*tab/*Sun*). Zoom in until you can see surface features, and lock on the Sun. Note the sunspots. Set the time step to 1 sidereal day and use the single time step button (|▶) to observe the Sun's rotation. Step through enough time steps to determine the rotation rate for the Sun. Compare the number of days for one rotation to the rotation information for the Sun presented in the text and state which part of the Sun the program's rotation is portraying. (Note that the program does not show the Sun's differential rotation.)

UNDERSTANDING THE STARS

PART III

R I V U X G

An Infrared Image of Newly-Formed Stars and Brown Dwarfs in the Orion Nebula. (NASA; K. L. Luhman/Harvard-Smithsonian Center for Astrophysics and G. Schneider, E. Young, G. Rieke, A. Cotera, H. Chen, M. Rieke, R. Thomson/Steward Observatory, University of Arizona, Tucson, AZ)

An Astronomer's Almanac

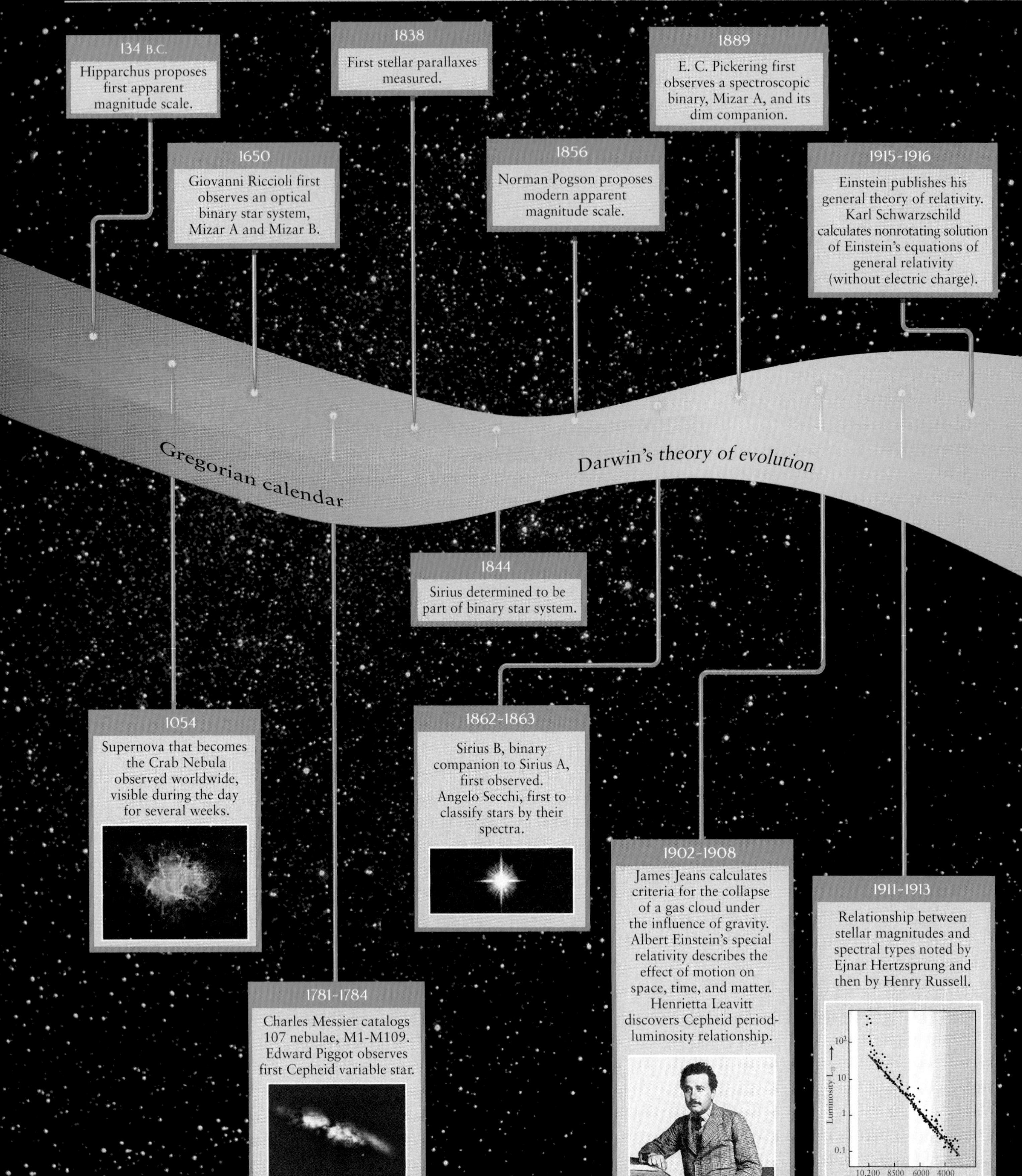

134 B.C.
Hipparchus proposes first apparent magnitude scale.
1054
Supernova that becomes the Crab Nebula observed worldwide, visible during the day for several weeks.
Gregorian calendar
1650
Giovanni Riccioli first observes an optical binary star system, Mizar A and Mizar B.
1781-1784
Charles Messier catalogs 107 nebulae, M1-M109. Edward Piggot observes first Cepheid variable star.
1838
First stellar parallaxes measured.
1844
Sirius determined to be part of binary star system.
1856
Norman Pogson proposes modern apparent magnitude scale.
Darwin's theory of evolution
1862-1863
Sirius B, binary companion to Sirius A, first observed. Angelo Secchi, first to classify stars by their spectra.
1889
E. C. Pickering first observes a spectroscopic binary, Mizar A, and its dim companion.
1902-1908
James Jeans calculates criteria for the collapse of a gas cloud under the influence of gravity. Albert Einstein's special relativity describes the effect of motion on space, time, and matter. Henrietta Leavitt discovers Cepheid period-luminosity relationship.
1911-1913
Relationship between stellar magnitudes and spectral types noted by Ejnar Hertzsprung and then by Henry Russell.
Luminosity $L_\odot$ →
10^2
10
1
0.1
10,200 8500 6000 4000
← Surface temperature (K)
1915-1916
Einstein publishes his general theory of relativity. Karl Schwarzschild calculates nonrotating solution of Einstein's equations of general relativity (without electric charge).

The Stars

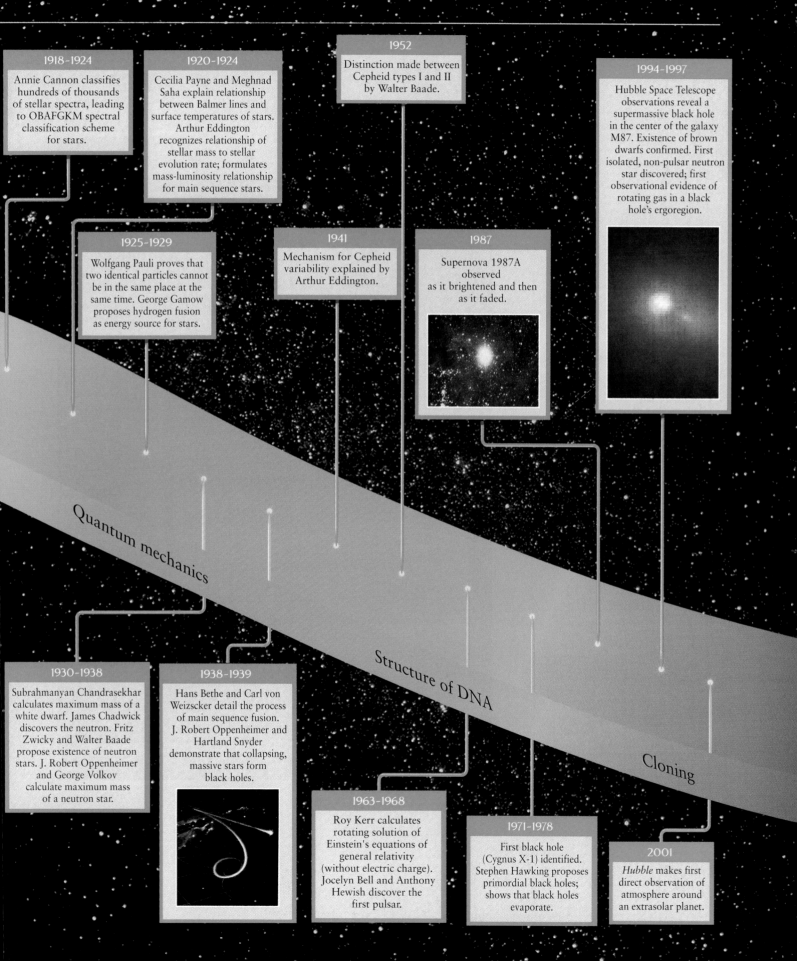

1918-1924
Annie Cannon classifies hundreds of thousands of stellar spectra, leading to OBAFGKM spectral classification scheme for stars.

1920-1924
Cecilia Payne and Meghnad Saha explain relationship between Balmer lines and surface temperatures of stars. Arthur Eddington recognizes relationship of stellar mass to stellar evolution rate; formulates mass-luminosity relationship for main sequence stars.

1925-1929
Wolfgang Pauli proves that two identical particles cannot be in the same place at the same time. George Gamow proposes hydrogen fusion as energy source for stars.

1930-1938
Subrahmanyan Chandrasekhar calculates maximum mass of a white dwarf. James Chadwick discovers the neutron. Fritz Zwicky and Walter Baade propose existence of neutron stars. J. Robert Oppenheimer and George Volkov calculate maximum mass of a neutron star.

1938-1939
Hans Bethe and Carl von Weizscker detail the process of main sequence fusion. J. Robert Oppenheimer and Hartland Snyder demonstrate that collapsing, massive stars form black holes.

1941
Mechanism for Cepheid variability explained by Arthur Eddington.

1952
Distinction made between Cepheid types I and II by Walter Baade.

1963-1968
Roy Kerr calculates rotating solution of Einstein's equations of general relativity (without electric charge). Jocelyn Bell and Anthony Hewish discover the first pulsar.

1971-1978
First black hole (Cygnus X-1) identified. Stephen Hawking proposes primordial black holes; shows that black holes evaporate.

1987
Supernova 1987A observed as it brightened and then as it faded.

1994-1997
Hubble Space Telescope observations reveal a supermassive black hole in the center of the galaxy M87. Existence of brown dwarfs confirmed. First isolated, non-pulsar neutron star discovered; first observational evidence of rotating gas in a black hole's ergoregion.

2001
Hubble makes first direct observation of atmosphere around an extrasolar planet.

Quantum mechanics

Structure of DNA

Cloning

CHAPTER 11

Characterizing Stars

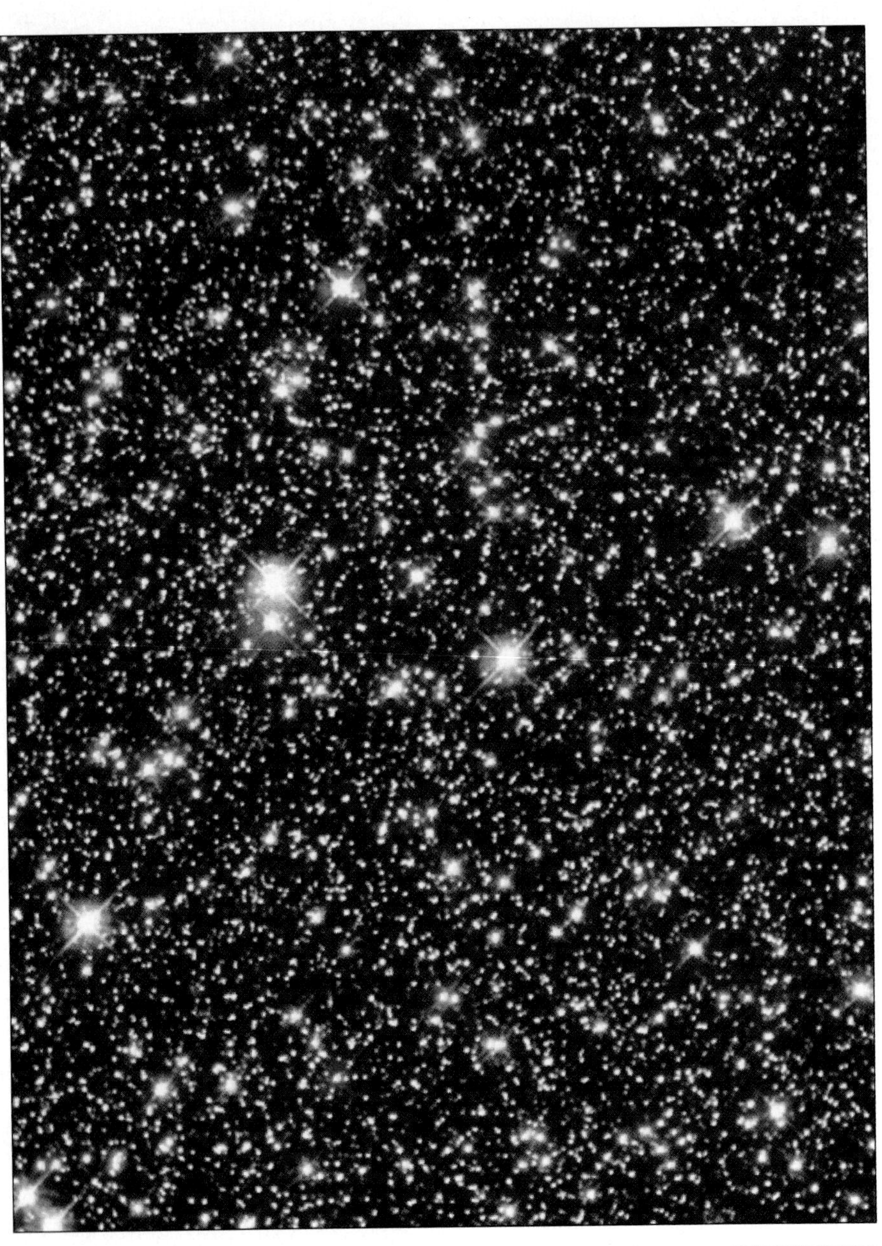

Stars Come in Many Colors (Hubble Heritage Team/AURA/STScI/NASA)

WHAT DO YOU THINK?

1. How near is the closest star other than the Sun?
2. Is the Sun brighter than other stars, or just closer?
3. What colors are stars?
4. Are brighter stars hotter?
5. What sizes are stars?
6. Are most stars isolated from other stars, as the Sun is?

Legend tells of a race of remarkable insects, the Ephemera, who inhabited a great forest. These noble creatures were blessed with great intelligence but were cursed with tragically short life spans. To the Ephemera, the forest seemed eternal and unchanging. Members of each generation lived out their brief lives without ever noticing any changes in their leafy world. Nevertheless, careful observations and reasoning led some Ephemera to postulate that the forest was not static. They began to suspect that small green shoots grew to become huge trees and that mature trees eventually died, toppled over, and littered the forest with rotting logs, enriching the soil for future trees. Although unable to witness the transformation personally, the Ephemera predicted the existence of life processes stretching over the mind-boggling periods of many years.

To us, the cosmos seems eternal and unchanging; the views that greet us every night are virtually indistinguishable from those seen by our ancestors. This permanence, however, is an illusion. We see the stars because they emit vast amounts of radiation, created as a result of converting one element into another in their interiors.

We do have one advantage over the Ephemera, who were so small that they could see only a few trees, leaves, shoots, and rocks. Astronomers are witnesses to literally billions of stars in various stages of evolution. Therefore, although we cannot see any one star go through more than a tiny fraction of the process, we see every stage of **stellar evolution.** By understanding the changes that stars undergo, we gain insight into our place in the cosmos. We begin our journey to the stars by learning the properties of stars that astronomers used to develop models of why stars shine and how they evolve.

In this chapter you will discover

- that the distances to many nearby stars can be measured directly, while the distances to farther ones are determined indirectly
- the observed properties of stars on which astronomers base their models of stellar evolution
- how astronomers analyze starlight to determine a star's temperature and chemical composition
- how the total energy emitted by stars and their surface temperatures are related
- the different classes of stars
- the variety and importance of binary star systems
- how astronomers calculate stellar masses

11-1 Distances to nearby stars are determined by stellar parallax

Nothing in the human experience prepares us for understanding the distances between the stars. On Earth, the greatest distance we encounter in our daily lives is perhaps halfway around the globe, 20,000 kilometers, which is about the distance from New York City to Perth, Australia. By observing the Sun, Moon, and planets, we have some comprehension of hundreds of thousands or even millions of kilometers.

How far away do you think the nearest stars are? Ask 10 people and you'll probably hear answers ranging from millions to billions of kilometers or miles. In fact, the closest star other than the Sun, Proxima Centauri, in the constellation Centaurus, is about 40 trillion kilometers (25 trillion miles) away. It takes light about 4 years to get here from there. Most of the stars you see in the night sky are many times farther away than Proxima Centauri. (Centaurus, by the way, is not visible from most of the northern hemisphere.) How do we measure the distance to a nearby star?

Because Proxima Centauri is closer to us than other stars, its position among the background stars changes as the Earth orbits the Sun. This is precisely the same effect that we see when, for example, Mars appears to have retrograde motion as we pass between it and the Sun (see Figures 2-2 and 2-3) or that Tycho Brahe sought for the supernova of 1572 (see Figure 2-7). The motion of nearby stars among the background of more distant stars due to the Earth's motion around the Sun is called **stellar parallax.**

Parallax is an everyday phenomenon. We experience it when nearby objects appear to shift their positions against a distant background as we move (review Figure 2-6). We also experience it continuously when we are awake, because our eyes change angle when looking at objects at different distances. As you look at a tree 10 m away, your

a

b

In January, the nearby star appears to be here.
In July, the nearby star appears to be here.
Nearby star
p
d
Earth (July)
Sun
Earth (January)
1 AU

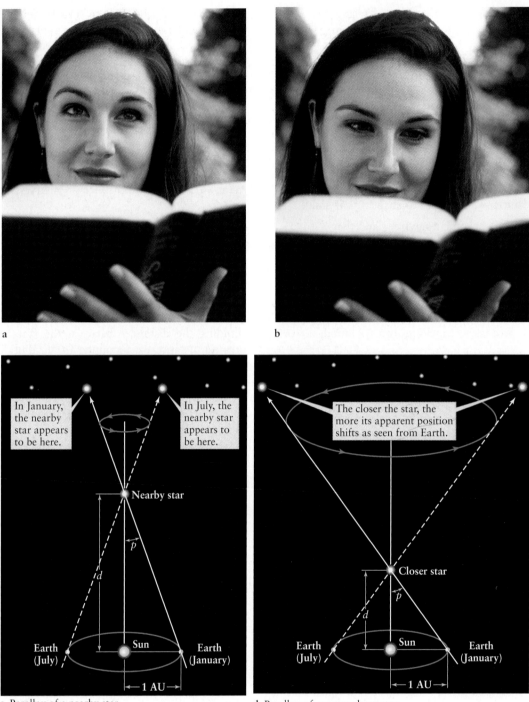

c Parallax of a nearby star

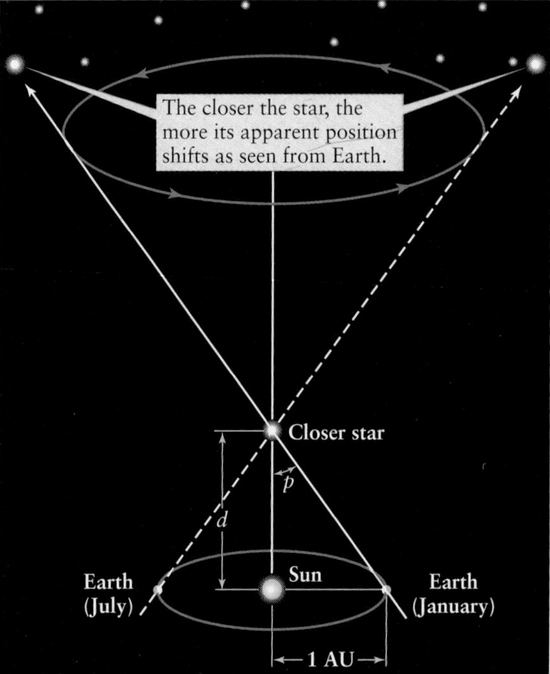

d Parallax of an even closer star

R I V U X G

FIGURE 11-1 **Using Parallax to Determine Distance** **(a, b)** Our eyes change angle as we look at things that are different distances away. Our eyes are adjusting for the parallax of the things we see. This change helps our brains determine the distances to objects and is analogous to how astronomers determine the distance to objects in space. **(c)** As the Earth orbits the Sun, a nearby star appears to shift its position against the background of distant stars. The star's parallax angle (p) is equal to the angle between the Sun and Earth as seen from the star. The stars on the scale of this drawing are shown much closer than they are in reality. If drawn to the correct scale, the closest star, other than the Sun, would be about 5 km (3.2 mi) away. **(d)** The closer the star is to us, the greater the parallax angle p. The distance to the star (in parsecs) is found by taking the inverse of the parallax angle p (in arcseconds), $d = 1/p$. (a and b: Richard Megna/Fundamental Photographs, NYC)

eyes cross only slightly (Figure 11-1a). Looking at something closer, say this book, your eyes cross much more in order for both of them to focus on the same word (Figure 11-1b). The parallax angle formed between your eyes and the tree or the book lets your brain judge just how close they are to you. Working out distances of nearby stars is done similarly, but it requires painstaking angle measurements and a little geometry.

As the Earth moves from one side of its orbit around the Sun to the other, a nearby star's apparent position shifts among the more distant stars. Referring to Figure 11-1c, the parallax angle, p, is half the angle by which the Earth shifts positions throught the year as seen from that star, measured in arcseconds. The difference in parallax angles for stars at different distances can be seen by comparing Figures 11-1c and d. An Astronomer's Toolbox 11-1 explores some details of how distances are determined from parallax angles.

The first stellar parallax measurement was made in 1838 by Friedrich Wilhelm Bessel, a German astronomer and mathematician. He found the parallax angle of the star 61 Cygni to be ⅓ arcsec, and so its distance is about 3 pc (see An Astronomer's Toolbox 2-1 to review astronomical distance units). The precision of stellar parallax measurements is limited by the angular resolution of the telescope, as discussed in Section 3-6. Because parallax angles smaller than about 0.01 arcsec are difficult to measure from Earth-based observatories, the stellar parallax method using Earth-based telescopes gives stellar distances only up to about 100 pc.

Telescopes in space are unhampered by our atmosphere and therefore have higher resolutions than Earth-based telescopes. Parallax measurements made in space thus enable astronomers to determine the distances to stars well beyond the reach of ground-based observations.

In 1989, the European Space Agency (ESA) launched a satellite called *Hipparcos* (an acronym for *h*igh *p*recision *par*allax *co*llecting *s*atellite, named for Hipparchus, an astronomer in ancient Greece who created an early classification system for stars). Although the satellite failed to achieve its proper orbit, astronomers used it to measure the distances to over 2.5 million of the nearest stars up to 500 ly (150 pc) away. The success of *Hipparcos* has led to plans for better satellites, scheduled for launch later this decade, to collect parallax data from stars even farther away.

Despite the information gained from stellar parallax, astronomers need to know the distances to more remote stars for which parallax cannot yet be measured. Several of

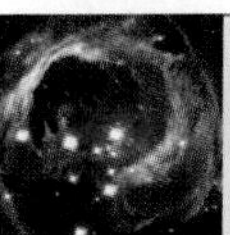

AN ASTRONOMER'S TOOLBOX 11-1
Distances to Nearby Stars

Recall from An Astronomer's Toolbox 2-1 that 1 parsec (1 pc) is the distance at which two objects 1 AU apart appear 1 arcsecond apart. This distance is 3.09×10^{13} km, or 206,265 AU. The word parsec (from *par*allax *sec*ond) originated in the use of parallax to measure distance. Using parsecs, we can write down an especially simple equation for the distances to stars:

$$\text{distance to a star in parsecs} = \frac{1}{\text{parallax angle of that star in arcseconds}}$$

or

$$d = 1/p$$

where d is the distance to the star in parsecs and p is the parallax angle of that star in arcseconds.

The equation is only this simple in these units, which is one of the main reasons that many astronomers discuss cosmic distances in parsecs rather than light-years. We will continue to primarily use light-years (ly) throughout this book, however, as they are more intuitive. In these latter units, the same equation becomes approximately

$$\text{distance to a star in light-years} \approx \frac{3.26}{\text{parallax angle of that star in arcseconds}}$$

or

$$d_{ly} \approx 3.26/p$$

where d_{ly} is the distance to a star in light-years.

Example: The nearest star, Proxima Centauri, has a parallax angle of 0.77 arcsec, and so its distance is 1/0.77, or approximately 1.3 pc. Equivalently, Proxima Centauri is 4.24 ly away. The parallax of Proxima Centauri is comparable to the angular diameter of a dime seen from a distance of 3 km. The parallax angles of the 25 nearest stars are listed in Appendix Table E (E-4).

Try these questions: What is the average distance from the Earth to the Sun in parsecs? What is the parallax angle of a star 100 ly away? How far away is a star with a parallax angle of 0.385″? Rigel is 773 ly away; how far away is it in parsecs? What is its parallax angle in arcseconds?

(Answers appear at the end of the book.)

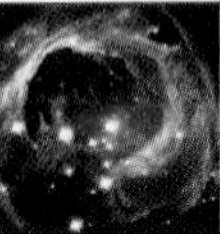

GUIDED DISCOVERY
Star Names

Most of the stars never received fanciful names such as Betelgeuse or Aldebaran. Indeed, most are so dim that they have been observed only through telescopes in the last two centuries. To study the stars yourself, you must be able to keep track of them. Astronomers have created a system of labels for all of them.

Up to the 24 most prominent stars in each constellation are assigned Greek lowercase letters:

α alpha	ι iota	ρ rho
β beta	κ kappa	σ sigma
γ gamma	λ lambda	τ tau
δ delta	μ mu	ϑ upsilon
ε epsilon	υ nu	ϕ phi
ζ zeta	ξ xi	χ chi
η eta	ο omicron	ψ psi
θ theta	π pi	ω omega

A bright star's name is a Greek letter together with its constellation. We use the Latin possessive form of the constellations. For example:

Constellation	Possessive
Aries	Arietis
Taurus	Tauri
Gemini	Geminorum
Cancer	Cancri
Leo	Leonis
Virgo	Virginis
Libra	Librae
Scorpius	Scorpii
Ophiuchus	Ophiuchi
Sagittarius	Sagittarii
Capricornus	Capricorni
Aquarius	Aquarii
Pisces	Piscium

In most cases the brightest star in the constellation is α, the second brightest is β, the third is γ, and so on. For example, the brightest star in the constellation Leo is called α Leonis. This name is more informative than its common name, Regulus.

For the millions of stars extending beyond the 24 brightest, a variety of catalogs list the stars numerically. For example, HDE 226868 is a bright, blue star, the 226,868th star in the *Henry Draper Extended Catalogue* of stars.

Caveat emptor: The naming of stars is the responsibility of the International Astronomical Union, which does not ever recognize the commercial sale of star names. Companies that offer to name a star for a price do not have any official standing or recognition in the astronomical community, nor do the names they promulgate. You can pay them to name a star for you, but the name is absolutely not official.

these methods of determining ever-greater distances will be introduced later in this chapter and in Chapters 12, 13, and 17. Having established the fact that different stars are at different distances from Earth, we now turn to considering the brightnesses that stars appear to have as seen from our planet. Combining the distances and the brightnesses we observe will enable us to calculate how much light stars actually emit and thereby explore their evolution.

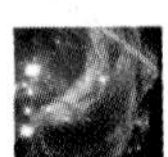

MAGNITUDE SCALES

Greek astronomers, from Hipparchus in the second century B.C. to Ptolemy in the second century A.D., undertook the classification of stars strictly by evaluating how bright they appear to be, compared to each other. This made sense because back then the stars were all assumed to be at the same distance from us, and, therefore, differences in brightness due to different stellar distances from Earth were not expected. Classification regardless of distance is still useful to help us navigate around the night sky and so we will study it. We will add the effects of different distances in the next section.

11-2 Apparent magnitude measures the brightness of stars as seen from Earth

The brightnesses of stars without regard to their distances are called **apparent magnitudes,** denoted with a lowercase m. The brightest stars were originally said to be of first magnitude, and their apparent magnitudes were designated $m = +1$. Those stars that appeared to be about half as bright as first-magnitude stars were said to be second-magnitude stars (designated $m = +2$), and so forth, down to sixth-magnitude stars, the dimmest ones visible to the unaided eye. (Greek astronomers did not try to classify the Sun's

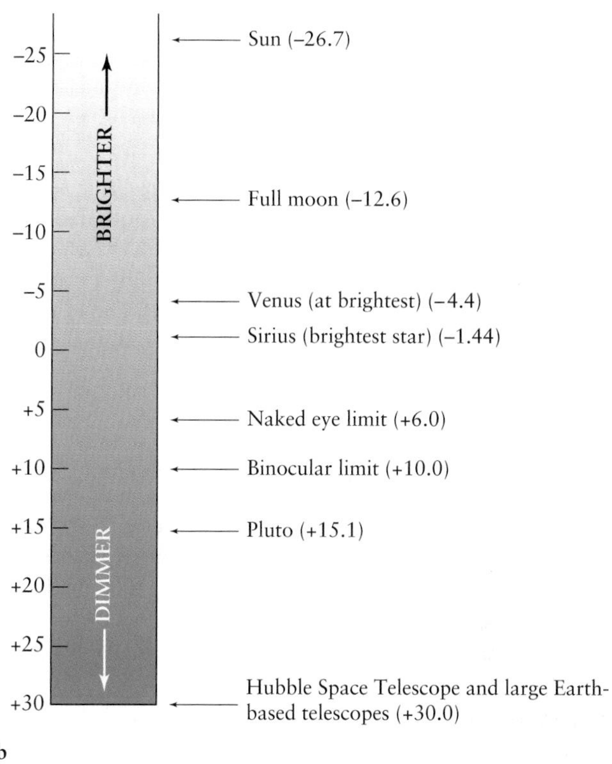

FIGURE 11-2 **Apparent Magnitude Scale** (a) Several stars in and around the constellation Orion are labeled with their names and apparent magnitudes. For a discussion of star names, see Guided Discovery: Star Names. (b) Astronomers denote the brightnesses of objects in the sky by their apparent magnitudes. Stars visible to the naked eye have magnitudes between $m = -1.44$ (Sirius) and about $m = +6$. CCD (charge-coupled device) photography through the Hubble Space Telescope or a large Earth-based telescope can reveal stars and other objects nearly as faint as magnitude $m = +30$. (a: Okiro Fujii, L'Astronomie)

dazzling brightness in this scheme.) Because stars do not appear with discrete levels of brightness, this system has noninteger magnitudes as well, such as +3.5 or +4.8. You can learn more about the relative brightnesses of stars with different magnitudes in An Astonomer's Toolbox 11-2.

The brightest stars and some other objects in the sky are actually brighter than the original apparent magnitude $m = +1.0$ (first magnitude) stars. In order to discuss these, astronomers resorted to *negative* numbers for the magnitudes of the very *brightest* objects. Sirius, for example, the brightest star in the night sky, has an apparent magnitude of $m = -1.44$. Figure 11-2a shows Sirius and the apparent magnitudes of some of the stars in Orion. With this convention we can describe other bright objects in the sky, such as the Sun, Moon, comets, and planets. At its brightest, Venus shines with an apparent magnitude of $m = -4.4$, the full Moon has an apparent magnitude of $m = -12.6$, and the Sun has an apparent magnitude of $m = -26.7$. Remember: Stars with negative apparent magnitudes appear brighter than stars with positive apparent magnitudes—the more negative, the brighter.

INSIGHT INTO SCIENCE

Bigger Isn't Necessarily Brighter Numbering as well as naming schemes may be counterintuitive in science. One would expect brighter stars to have larger, more positive numbers than dimmer stars, but the apparent magnitude scheme is just the opposite. Similarly, we will see shortly that on the standard plot of stars used by astronomers, the Hertzsprung-Russell diagram, the hottest stars fall on the *left* and the coolest stars on the *right*.

Astronomers also extended the magnitude scale so that dimmer stars, visible only through telescopes, could be included in the magnitude system. For example, the dimmest stars visible through a pair of powerful binoculars have an apparent magnitude of about $m = +10$. Time-exposure photographs through telescopes reveal even dimmer stars. The Keck telescopes and the Hubble Space Telescope,

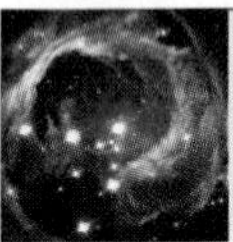

AN ASTRONOMER'S TOOLBOX 11-2
Details of the Magnitude Scales

The magnitude scales were created before accurate measurements of the relative brightnesses of stars could be made, and they have since been refined. Specifically, careful measurements reveal that the original first-magnitude stars were about 100 times brighter than the original sixth-magnitude stars. Astronomers chose the brightness factor of exactly 100 to define the range of brightness between modern first- and sixth-magnitude stars. In other words, it takes 100 stars of apparent magnitude $m = +6$ to provide as much light as we receive from a single star of apparent magnitude $m = +1$.

To find out how much brighter each magnitude is from the next dimmer one, we note that there are five integer magnitudes between first and sixth magnitude. Going from $m = +6$ to $m = +5$ increases (multiplies) the brightness we see by the same factor as going from $m = +5$ to $m = +4$, and so on. Going from $m = +6$ to $m = +1$ requires multiplying the brightness factor from one magnitude to the next 5 times. The number we must multiply 5 times to get the range of brightness of 100 is $100^{1/5} \approx 2.512$, or, put another way, $2.512 \times 2.512 \times 2.512 \times 2.512 \times 2.512 \approx 100$. This means that *each successively brighter magnitude is approximately 2.512 times brighter than the preceding magnitude.*

Example: An $m = +3$ star is approximately 2.512 times brighter than an $m = +4$ star. Equivalently, it takes 2.512 fourth-magnitude stars to provide as much light as we receive from a single third-magnitude star.

Try these questions after reading Section 11-3: How much brighter is an $m = 0$ star than an $m = +4$ star? How much brighter is an $M = -2$ star than an $M = +5$ star? If one star is 7.93 times brighter than another and the brighter star has an absolute magnitude of $M = +3$, what is the absolute magnitude of the dimmer star? (*Hint for last question:* Recall that magnitudes need not be integers.)

(Answers appear at the end of the book.)

among others, image stars nearly as dim as magnitude $m = +30$. Figure 11-2b illustrates the modern apparent magnitude scale. Similarly, apparent magnitudes from entire groups of stars, such as distant galaxies, can be measured.

Knowing, as we now do, that stars are at different distances from us, the apparent magnitudes do not directly reveal fundamental stellar properties. All other things being equal, the closer of two identical stars appears brighter to us (has a smaller apparent magnitude) than the farther star. We take the different distances into account with either of two measures, absolute magnitude and luminosity.

11-3 Absolute magnitudes and luminosities do not depend on distance

To determine the total energy emitted by stars, astronomers need to remove the effect of distance by calculating the brightnesses stars would have if they were all at the same distance from Earth. Knowing how far away they really are and how bright they appear (their apparent magnitudes), we can calculate how bright they all would be at any distance. **Absolute magnitude** is the brightness each star would have at a distance of 10 pc. Unfortunately, absolute brightnesses have the same counterintuitive numerical scale as apparent magnitudes.

To understand the relationship between the apparent magnitude and the absolute magnitude, we need to know how the brightness of an object changes with distance. Suppose that we observe two identical stars, one twice as far away as the other. How much dimmer will the farther one appear to be as seen from Earth? As light moves outward from a source, it spreads out over increasingly larger areas of space and its brightness decreases. Thus, the farther away a source of light is, the dimmer it appears. The **inverse-square law** provides the rule for just how quickly the brightnesses of objects change with distance.

Imagine light moving out from a star (Figure 11-3a). Start with a small square area of light that has moved out a distance $d = 1$. The light in that square has a certain brightness. When the same square of light has gone twice as far ($d = 2$), you can see on the figure that it has become 4 times larger. The light in each of the same-sized squares at $d = 2$ contains one-quarter of the photons that were in the single square at $d = 1$. Therefore, the small squares at $d = 2$ are one-quarter as bright as the same-sized square at $d = 1$. Similarly, when the square of light has moved to $d = 3$, there are now nine squares the same size as the original, each one-ninth as bright as the original square at $d = 1$. For example, the Sun emits 3.83×10^{26} W (for watts) of power (compared to 100 W for a bright home lightbulb). The Sun's power is spread over a wider and wider area as it travels through space. When it passes Mercury, sunlight provides 9140 W on every square meter of space. That same energy has spread out so much that by the time it passes us at 1 AU from the Sun, it provides only 1370 W per square meter. You see that effect everyday when you look at a car at different distances (Figure 11-3b).

The inverse-square law can be summarized mathematically as follows:

Apparent brightness decreases inversely with the square of the distance between the source and the observer.

Returning to our two identical stars, one twice as far away as the other, the farther one's brightness falls off to $(\frac{1}{2})^2 = \frac{1}{4}$ the brightness of its closer twin by the time the starlight gets to us. Absolute magnitudes are usually determined by calculations based on apparent magnitudes. Of course, we cannot just move a star to 10 pc distance and then remeasure its apparent magnitude. However, we can easily calculate the absolute magnitude of a nearby star: We first measure its apparent magnitude and then we find its distance by measuring its parallax angle. Combining these numbers, as shown in An Astronomer's Toolbox 11-3, gives the star's absolute magnitude.

Knowing the Sun's true distance and its apparent magnitude, we can use the inverse-square law to determine how bright it would appear at 10 pc. At that distance, it would have an apparent magnitude of $m = +4.8$. Therefore, the absolute magnitude of the Sun is $M = +4.8$.

Because absolute magnitudes tell astronomers how bright stars are compared with each other, this information is used to evaluate models of stellar evolution, which we discuss in the next few chapters. Absolute magnitudes range from roughly $M = -10$ for the brightest stars to $M = +17$ for the dimmest. While absolute magnitudes give us comparisons between

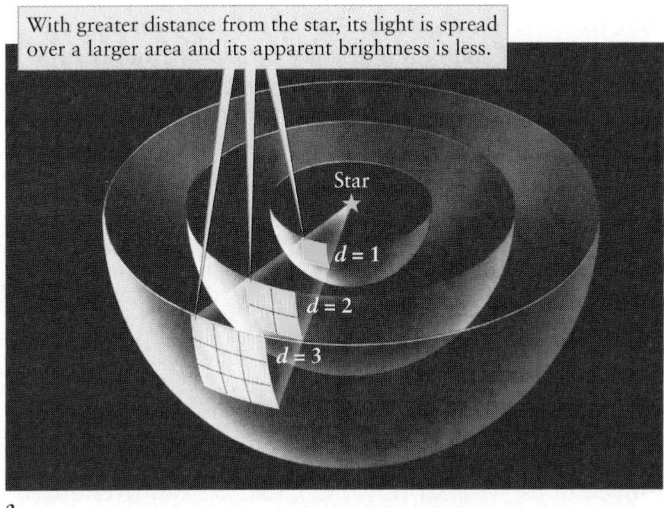
With greater distance from the star, its light is spread over a larger area and its apparent brightness is less.
Star
d = 1
d = 2
d = 3

a

10 meters away

20 meters away

R I V U X G

b

30 meters away

FIGURE 11-3 **The Inverse-Square Law** (a) This drawing shows how the same amount of radiation from a light source must illuminate an ever-increasing area as the distance from the light source increases. The decrease in brightness follows the inverse-square law, which means, for example, that tripling the distance decreases the brightness by a factor of 9. (b) Three cars with identical lights at distances of 10 m, 20 m, and 30 m showing the effect described in part (a). (b: Royalty Free/CORBIS)

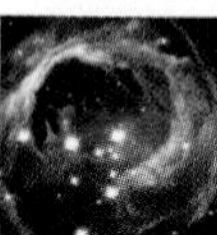

AN ASTRONOMER'S TOOLBOX 11-3
The Distance-Magnitude Relationship

The closer a star, the brighter it appears. The inverse-square law leads to a simple equation for absolute magnitude, M. Suppose a star's apparent magnitude is m, and its distance from the Earth is d (measured in parsecs). Then

$$M = m - 5 \log (d/10)$$

where log stands for the base-10 logarithm. This distance-magnitude relation can be rewritten as

$$m - M = 5 \log d - 5$$

Example: Consider Proxima Centauri, the nearest star in the night sky. By measuring its parallax angle, we know this star is at a distance from the Earth of $d = 1.3$ pc. Its apparent magnitude is $m = +11.1$. Therefore, its absolute magnitude is

$$M = 11.1 - 5 \log (1.3/10) = 11.1 - (-4.4) = 15.5$$

Compare! The Sun is an average star with $M = 4.8$, so Proxima Centauri is an intrinsically dim star. If you know any two of d, m, and M, you can calculate the third variable. For example, if we know a star's absolute and apparent magnitude, the equation can be used to determine its distance.

Try these questions: A star is observed to have an apparent magnitude $m = +0.268$ and an absolute magnitude $M = -0.01$. How far from Earth is the star in parsecs and light-years? A star is observed to have an apparent magnitude $m = +1.17$ and is at a distance of 25.1 ly from Earth. What is its absolute magnitude? (Remember to convert to parsecs first.) A star is at a distance from Earth of 8.61 ly and has an absolute magnitude of $m = +1.45$. What is its apparent magnitude? You can check your results and identify the stars by referring to Appendix Table E-4 and Table E-5.

(Answers appear at the end of the book.)

the energy outputs of stars, we also need to know the total energy they release. The total amount of electromagnetic power (energy emitted each second) is called a star's **luminosity**. We saw in Chapter 4 that the higher its luminosity, the brighter an object is. Therefore, the smaller or more negative a star's absolute magnitude, the greater its luminosity. For convenience, stellar luminosities are expressed in multiples of the Sun's luminosity, denoted $L_\odot$. As we saw earlier, this is about 3.83×10^{26} W (watts). The intrinsically brightest stars ($M = -10$) have luminosities of $10^6\ L_\odot$. In other words, each of these stars has the energy output of a million Suns. The dimmest stars ($M = +17$) have luminosities of $10^{-5}\ L_\odot$. We will provide both luminosity and absolute magnitude data about stars in the chapters that follow.

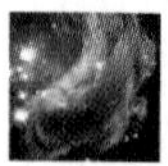

THE TEMPERATURES OF STARS

Armed with knowledge of stellar luminosities, astronomers early in the twentieth century searched for ways to use this information to determine why stars shine. A major step in this process was to find a graph on which physically similar stars are located near each other, because such groupings often provide insight into how objects work. They found this graph when they plotted either the luminosity or, equivalently, the absolute magnitude versus the stars' surface temperatures. The resulting graph led to useful models of both stellar activity and evolution. To find the surface temperatures, we start with a fact that is easily overlooked: Stars are not all the same color.

11-4 A star's color reveals its surface temperature

If you look carefully at the stars with your naked eyes, you 3
can see that they have different colors, most commonly red, orange, yellow, white, and blue-white, as Figure 11-2 and Figure 11-4a reveal. Because a star behaves very nearly like a perfect blackbody, its color is determined by its surface temperature. Recall from Figure 4-2 that different blackbody curves are associated with blackbodies of different temperatures. These curves differ in both their intensities and the locations of their peaks, as described by Wien's law.

Three typical blackbody curves are presented in Figure 11-4. The intensity of light from a relatively cool star peaks at long wavelengths, so the star looks red. The intensity of light from a very hot star peaks at shorter wavelengths, making it look blue. The maximum intensity of a star of intermediate temperature, such as the Sun, is found near the middle of the visible spectrum.

To accurately determine the peaks of stars' blackbody spectra and, hence, their surface temperatures, astronomers need to know which blackbody curve most accurately describes each star. **Photometry,** in which a telescope col-

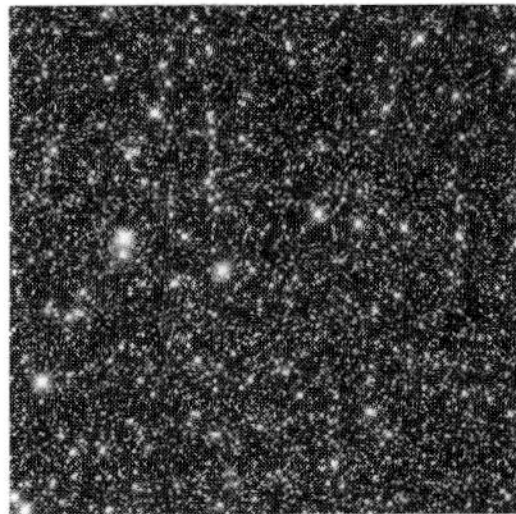

FIGURE 11-4 **Temperature and Color** (a) This beautiful Hubble Space Telescope image shows the variety of colors of stars. (b) This diagram shows the relationship between the color of a star and its surface temperature. The intensity of light emitted by three hypothetical stars is plotted against wavelength (compare with Figure 4-2). The range of visible wavelengths is indicated. Where the peak of a star's intensity curve lies relative to the visible light band determines the apparent color of its visible light. The insets show stars of about these surface temperatures. UV stands for ultraviolet, which extends to 10 nm. See Figure 3-4 for more on wavelengths of the spectrum. (a: Hubble Heritage Team/AURA/STScI/NASA; left inset: Andrea Dupree/Harvard-Smithsonian CFA, Ronald Gilliland/STScI, NASA and ESA; center inset: NSO/AURA/NSF; right inset: Till Credner, Allthesky.com)

a

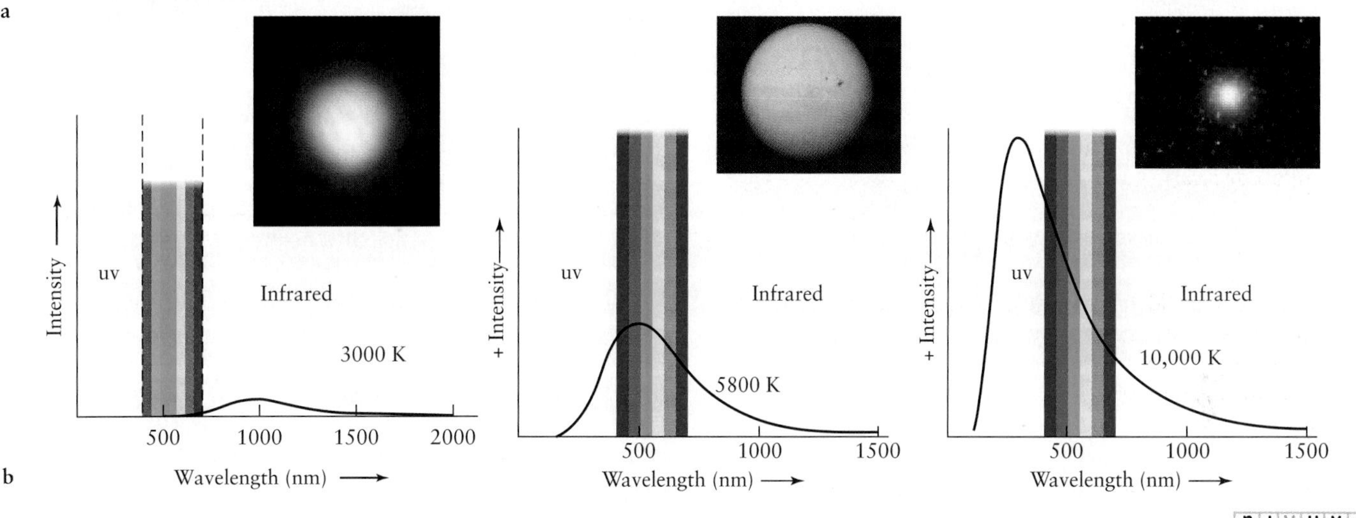

b

lects starlight that is then passed through one of a set of colored filters and recorded on a CCD (recall Figure 4-8), provides this information. At least three photometric images are taken of each star through filters passing different wavelengths of light. The intensity of a star's image is different at different wavelengths, and photometric data can thus be used to locate the peak of the star's blackbody radiation and, thereby, its surface temperature.

If a star's surface is very hot, for example 10,000 K, its radiation is skewed toward the ultraviolet, which makes the star bright as detected through an ultraviolet-passing filter, dimmer through a blue filter, and dimmer still through a yellow filter. Regulus, in Leo, is such a star. If a star is cool, say, 3000 K, its radiation peaks at long visible or even infrared wavelengths, making the star bright through a red filter, dimmer through a yellow filter, and dimmer still through a blue filter. Aldebaran and Betelgeuse are examples of such cool stars. Humans are blackbody radiators with peak temperatures around 300 K. Our peak is in the infrared.

11-5 A star's spectrum also reveals its surface temperature

Another way to determine a star's surface temperature is from a study of its spectrum, a technique called **stellar spectroscopy** (see Chapter 4). Recall that spectral lines result from the absorption and scattering of starlight by gases in the star's atmosphere, in interstellar space, and in the Earth's atmosphere. Astronomers begin by taking the spectrum of a star and then identifying and discarding spectral lines due to interstellar gas and Earth's atmosphere. What is left are spectral lines created in the star's atmosphere.

At first glance, there seems to be a bewildering variety of stellar spectra, some of which are shown in Figure 11-5. Some stellar spectra show prominent absorption lines of hydrogen. Some exhibit many absorption lines of calcium and iron. Still others are dominated by broad absorption lines created by molecules such as titanium oxide.

Fortunately, these spectra all hold clues about the nature of stars and, in particular, their surface temperatures. Consider the abundant hydrogen gas found in nearly every star's atmosphere. Although hydrogen accounts for about three-quarters of the mass of a typical star, strong (meaning dark) hydrogen absorption lines do not show up in every star's spectrum. The strength of the absorption lines depends on the star's temperature, and in the late 1920s, Harvard astronomer Cecilia Payne and physicist Meghnad Saha succeeded in explaining how a star's visible hydrogen spectrum is affected by its surface temperature.

Niels Bohr's model of the hydrogen atom (recall Figure 4-12) shows why. As Bohr proposed in the early 1900s, the visible hydrogen lines, called Balmer lines, are produced when photons excite electrons in the second, $n = 2$, energy

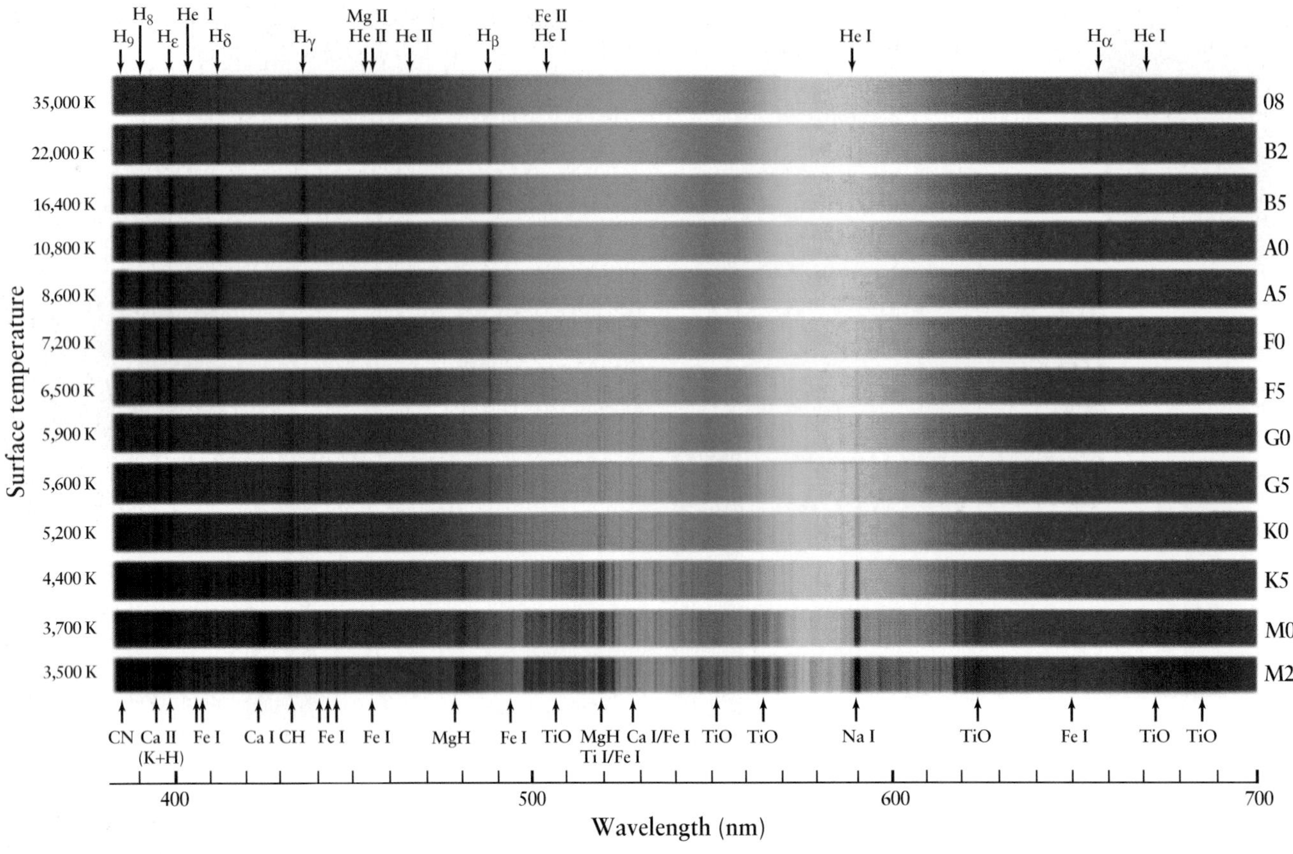

FIGURE 11-5 Principal Types of Stellar Spectra This figure shows the spectra for stars with different surface temperatures. The corresponding spectral types are indicated on the right side of each spectrum. (Note that stars of each spectral type have a range of temperature.) The hydrogen Balmer lines are strongest in stars with surface temperatures of about 10,000 K (called A-type stars). Cooler stars (G- and K-type stars) exhibit numerous atomic lines caused by various elements, indicating temperatures from 4000 to 6000 K. The broad, dark bands in the spectrum of the coolest stars (M-type stars) are caused by titanium oxide (TiO) molecules, which can exist only if the temperature is below about 3500 K. Recall from Section 4–5 that the roman numeral I after a chemical symbol means that the absorption line is caused by a neutral atom; a numeral II means that the absorption is caused by atoms that have each lost one electron. (R. Bell, University of Maryland, and M. Briley, University of Wisconsin at Oshkosh)

level of hydrogen to a higher energy level. At 10,000 K, most hydrogen electrons absorb enough energy from photons to normally reside in the $n = 2$ level. When these electrons gain further energy, they move to higher energy levels. Therefore, a star with that surface temperature produces the strongest Balmer lines.

Why do hotter or cooler stars produce weaker (less dark) Balmer absorption lines? Suppose first that the star is much hotter than 10,000 K. High-energy photons streaming through the photosphere completely strip away (ionize) electrons from most of the hydrogen atoms there. Because an ionized hydrogen atom cannot produce spectral lines, a very hot star has very dim hydrogen Balmer lines, even though it contains great quantities of hydrogen. Conversely, if a star is much cooler than 10,000 K, most of the photons escaping from it possess too little energy to keep many electrons in the $n = 2$ (first excited state) of the hydrogen atoms, or to boost electrons in the $n = 2$ state to higher energy states. Therefore, these stars also produce dim Balmer lines. To produce strong Balmer lines, a star must be hot enough to excite electrons out of the $n = 2$ state but not hot enough to ionize a significant fraction of the atoms.

As you can see in Figure 11-5, at temperatures much cooler or hotter than 10,000 K, the spectral lines of other elements dominate a star's spectrum. For example, the spectral lines of neutral helium are pronounced at around 25,000 K, because photons have enough energy to excite helium atoms without tearing away the electrons. Conversely, the spectral lines of neutral iron

a

b

R I V U X G

FIGURE 11-6 **Classifying the Spectra of Stars** The modern classification scheme for stars based on their spectra was developed at the Harvard College Observatory in the late nineteenth century. Women astronomers, initially led by Edward C. Pickering (not shown) and Williamina Fleming, standing in (a), and then by Annie Jump Cannon (b), analyzed hundreds of thousands of spectra. Social conventions of the time prevented most women astronomers from using research telescopes or receiving salaries comparable to those of men. (a: Harvard College Observatory; b: © Bettmann/CORBIS)

are especially strong at around 3500 K. By surveying the relative strengths of a variety of absorption lines, astronomers can determine a star's surface temperature to high precision.

11-6 Stars are classified by their spectra

WEB LINK 11.5

We have seen that astronomers can determine a star's surface temperature from either the peak of its blackbody or the strength of its various spectral lines. To make use of the diverse stellar spectra, astronomers since the middle of the nineteenth century have grouped similar spectra into classes, or **spectral types.** According to one classification scheme popular in the late 1800s, a star was assigned a letter from A through P, depending on the strength of the Balmer hydrogen lines in the star's spectrum. It was assumed that the strength of these lines was uniquely related to the star's surface temperature, but as we have seen in the preceding section, this belief is unjustified.

In the early 1900s, William Pickering and Williamina Fleming, followed by Annie Jump Cannon, and their colleagues at Harvard Observatory (Figure 11-6) set up the spectral classification scheme we use today. Many of the early A through P categories were dropped because the Balmer lines in a star's spectrum can be weak whether the star is very cool or very hot. The remaining Balmer-based spectral types were thus reordered by stellar surface temperature into the **OBAFGKM sequence.** This sequence is most easily learned with the aide of a mnemonic, such as: "**Oh, Be A Fine Guy, Kiss Me!**" or "**Oh, Be A Fine Girl, Kiss Me!**"

The hottest stars are the O type, with surface temperatures of 30,000 to 50,000 K; their spectra are dominated by He II and Si IV (triply ionized silicon). M stars are the coolest type, with surface temperatures of 2500 to 3000 K. Table 11-1 includes representative examples of each spectral type, which you can compare to the spectra in Figure 11-5.

The wide range of temperatures covered by each spectral type prompted astronomers to subdivide the OBAFGKM temperature sequence further. Each spectral type is now broken up into 10 temperature subranges. These 10 finer steps are indicated by adding an integer from 0 (hottest) through 9 (coolest). Thus, an A8 star is hotter than an A9 star, which is hotter than an F0 star, which is hotter than an F1 star, and so on. Test yourself on this: What class of star is just slightly cooler than a K9? (The caption for Figure 11-7 has the answer.) The Sun, whose spectrum is dominated by singly ionized metals (especially Fe II and Ca II), is a G2 star.

INSIGHT INTO SCIENCE

Tolerate Idiosyncrasies In astronomy, the word "metal" applies to all elements other than hydrogen or helium. In chemistry, sodium and iron are metals, while carbon and oxygen are not.

TABLE 11-1 The Spectral Sequence

Spectral class	Color	Temperature (K)	Spectral lines	Examples
O	Blue-violet	30,000–50,000	Ionized atoms, especially helium	Naos (ζ Puppis), Mintaka (δ Orionis)
B	Blue-white	11,000–30,000	Neutral helium, some hydrogen	Spica (α Virginis), Rigel (β Orionis)
A	White	7500–11,000	Strong hydrogen, some ionized metals	Sirius (α Canis Majoris), Vega (α Lyrae)
F	Yellow-white	5900–7500	Hydrogen and ionized metals such as calcium and iron	Canopus (α Carinae), Procyon (α Canis Minoris)
G	Yellow	5200–5900	Both neutral and ionized metals, especially ionized calcium	Sun, Capella (α Aurigae)
K	Orange	3900-5200	Neutral metals	Arcturus (α Boötis), Aldebaran (α Tauri)
M	Red-orange	2500-3900	Strong titanium oxide and some neutral calcium	Antares (α Scorpii), Betelgeuse (α Orionis)

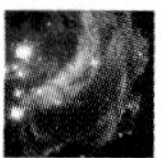

TYPES OF STARS

Keeping in mind that patterns of objects often reveal information about them, astronomers began searching for relationships between different properties of stars. They found the grail in the relationship between luminosities and surface temperatures. As often happens in such scientific quests, it was discovered nearly simultaneously by two independent researchers.

11-7 The Hertzsprung-Russell diagram identifies distinct groups of stars

Around 1911, the Danish astronomer Ejnar Hertzsprung noticed that patterns emerge when the luminosities of stars (or their equivalent absolute magnitudes) are plotted against their surface temperatures or spectral types. Luminosity and absolute magnitude indicate the total energy emitted by a star or other body. They refer to what we loosely call "brightness." Within two years, the American astronomer Henry Norris Russell independently discovered the same result. Graphs of stellar luminosity or absolute magnitude against surface temperature (or, equivalently, spectral type) are now known as **Hertzsprung-Russell diagrams,** or **H-R diagrams.**

The H-R diagram is valuable because it shows that stars do not have random surface temperatures and luminosities; the two factors are correlated. Figure 11-7 is a typical Hertzsprung-Russell diagram. Each dot represents a star whose luminosity and spectral type have been determined. The surface temperatures are plotted along the top of this figure and the absolute magnitudes along the right side. You can therefore see the equivalence of the temperature and spectral type and the equivalence of the luminosity and absolute magnitude.

Bright stars are near the top of the diagram; dim stars are near the bottom. Contrary to intuition, hot (O and B) stars are toward the left side and cool (M) stars are toward the right. Hertzsprung and Russell made this choice because of the standard sequence OBAFGKM.

INSIGHT INTO SCIENCE

From Patterns to Models Scientists look at patterns of behavior in related objects as valuable clues to underlying properties and their causes. Guided by this data, they then create theoretical models and make fresh predictions. For example, astronomers analyze the relationship between the luminosities and surface temperatures of stars as displayed on an H-R diagram to gain insight into the internal activities of stars.

The band of stars in Figure 11-7 stretching diagonally across the H-R diagram and on which a red curve is superimposed represents most of the stars we see in the nighttime sky. This band, called the **main sequence,** extends from the hot, bright, bluish stars in the upper left corner of the dia-

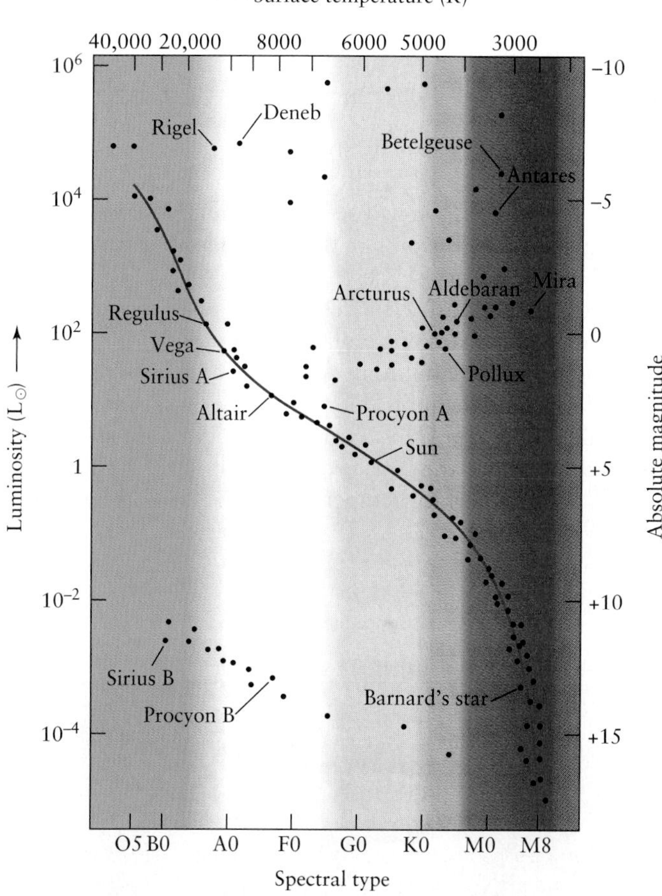

FIGURE 11-7 **A Hertzsprung-Russell Diagram** On an H-R diagram, the luminosities of stars are plotted against their spectral types. Each dot on this graph represents a star whose luminosity and spectral type have been determined. Some well-known stars are identified. The data points are grouped in just a few regions of the diagram, revealing that luminosity and spectral type are correlated: Main-sequence stars fall along the red curve, giants are to the right, supergiants are on the top, and white dwarfs are below the main sequence. The absolute magnitudes and surface temperatures are listed at the right and top of the graph, respectively. These are sometimes used on H-R diagrams instead of luminosities and spectral types. (Answer to text question: An M0 star is the next coolest after a K9.)

gram down to the cool, dim, reddish stars in the lower right corner. Each star on this band is called a **main-sequence star.** Just over 91% of the stars surrounding the solar system fall into this category.

Observations reveal that the number of main-sequence stars decreases with increasing surface temperature. Therefore, along the main sequence, the cooler M, K, and G stars are the most common ones and the hot O stars are the rarest. The Sun (spectral type G2, absolute magnitude +4.8) is a main-sequence star.

To the right of the main sequence on the H-R dia- 4
gram is a second major grouping of stars. These stars are bright but cool. From the Stefan-Boltzmann law (see An Astronomer's Toolbox 4-1), we know that a cool object radiates much less light from each unit of surface area than does a hot object. To be so luminous, these cool stars must therefore be huge compared to main-sequence stars of the same temperature, so they are called **giant stars.** By contrast, main-sequence stars are often called **dwarf stars.** The dimmest, coolest main sequence stars are called *red dwarfs*.

Giants are typically 10 to 100 times the radius of the Sun and have surface temperatures between 2000 and 20,000 K. The cooler members of this class of stars (those with surface temperatures between 2000 and 4500 K) are often called **red giants** because they appear reddish in the nighttime sky. Aldebaran in the constellation Taurus and Arcturus in Boötes are examples of red giants that you can easily see with the naked eye.

A few rare stars are considerably bigger and more luminous than typical giants. Located along the top of the H-R diagram, these superluminous stars are appropriately called **supergiants.** They can extend in radius up to about 1000 $R_\odot$. Betelgeuse in Orion and Antares in Scorpius are two examples that are visible in the nighttime sky. Together, giants and supergiants comprise less than 1% of the stars in our vicinity.

The remaining 8% of stars in our neighborhood of space fall in a final grouping toward the lower left and bottom of the Hertzsprung-Russell diagram. As their placement on the diagram shows, these stars are hot, dim, and tiny compared to the Sun. Called **white dwarfs,** we will see that they are actually remnants of stars. White dwarfs are roughly the same size as the Earth, and because of their great distances, they can be seen only with the aid of a telescope.

These results are summarized in Figure 11-8, where the 5
dashed lines indicate the radii of stars. Notice that most main-sequence stars are roughly the same size as the Sun. It is important to keep in mind that while the H-R diagram is invaluable in helping astronomers *organize* stars by their physical properties (temperature and absolute magnitude), the diagrams do not *explain* the physical mechanisms that produce these characteristics. As we will see shortly, the locations of stars on the H-R diagram provided clues that astrophysicists used in developing and testing their theories of how stars shine.

11-8 Luminosity classes set the stage for understanding stellar evolution

We have seen that a star's surface temperature largely determines which lines are prominent in its spectrum. Therefore, classifying stars by spectral type is the same as categorizing them according to surface temperature. However, as you

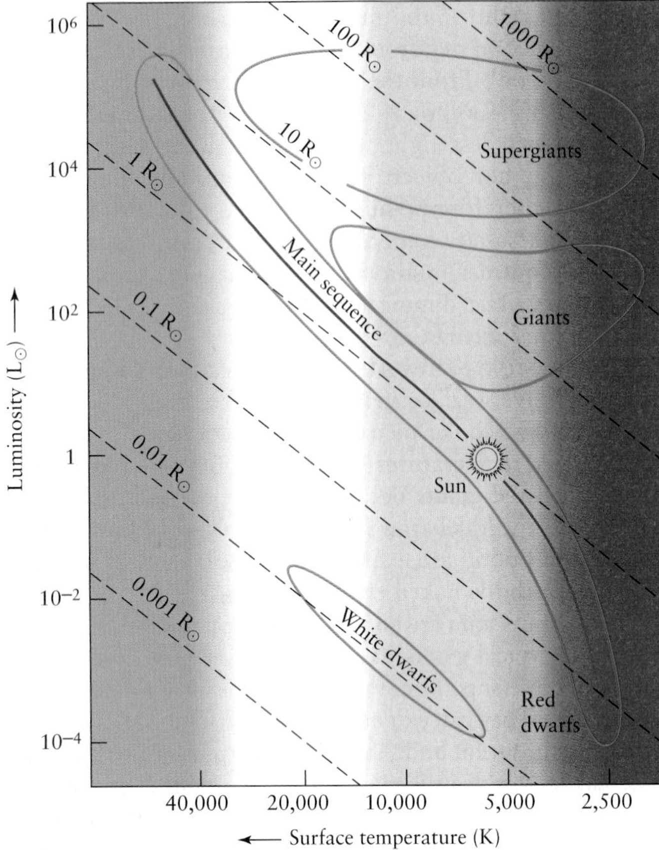

FIGURE 11-8 **The Types of Stars and Their Sizes** On this H-R diagram stellar luminosities are graphed against the surface temperatures of stars. The dashed diagonal lines indicate stellar radii. For stars of the same radius, hotter stars (corresponding to moving from right to left on the H-R diagram), glow more intensely and are more luminous (corresponding to moving upward on the diagram) than cooler stars. While individual stars are not plotted, we show the regions of the diagram in which main-sequence, giant, supergiant, and white dwarf stars are found. Note that the Sun is intermediate in luminosity, surface temperature, and radius; it is very much a middle-of-the-road star.

can see in Figure 11-8, stars of the same surface temperature can have different luminosities. As an example, a star with a surface temperature of 5800 K could be a white dwarf, a main-sequence star, a giant, or a supergiant.

By studying the absorption lines in detail, astronomers can categorize stars more accurately than just by determining their luminosities and temperatures. This refined classification is possible primarily because absorption lines are affected by the density and pressure of the gas in a star's atmosphere, both of which depend on whether the star is a white dwarf, main-sequence star, giant, or supergiant.

Based upon the differences in stellar spectra, W. W. Morgan and P. C. Keenan of the Yerkes Observatory developed a system of **luminosity classes** in the 1930s. Luminosity classes Ia and Ib include all the supergiants, giants of various luminosity are assigned classes II, III, and IV, and main-sequence stars are luminosity class V. Figure 11-9 summarizes these results. We will see in Chapter 12 that the different luminosity classes correspond to different stages of stellar evolution. White dwarfs are not given a luminosity class because, as we will see in Chapter 13, they are not creating energy by fusion like the Sun and other stars in the above five luminosity classes.

Plotting luminosity classes on the H-R diagram (see Figure 11-9) provides a useful subdivision of star types. In fact, astronomers commonly describe a star by both its spectral type and its luminosity class. The Sun, for example, is called a G2 V star. This notation supplies a great deal of information about the star, because its spectral type is correlated with its surface temperature. Thus, an astronomer

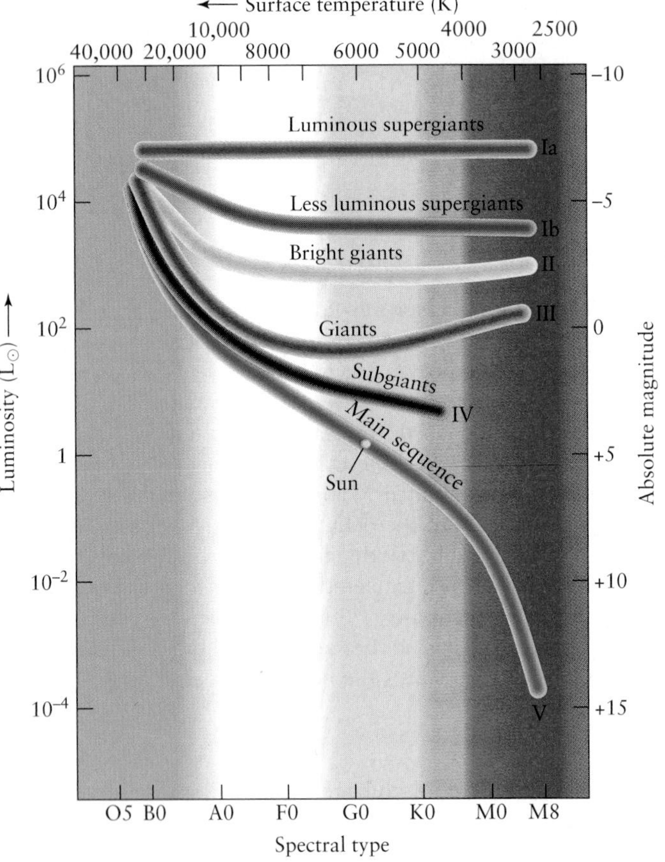

FIGURE 11-9 **Luminosity Classes** It is convenient to divide the H-R diagram into regions called luminosity classes. These subdivisions permit finer distinctions between giants and supergiants. Luminosity classes Ia and Ib encompass the supergiants. Luminosity classes II, III, and IV indicate giants of different brightness. Luminosity class V is the main-sequence stars. White dwarfs do not have their own luminosity class.

knows immediately that a G2 V star is a main-sequence star with a luminosity of 1 $L_\odot$ and a surface temperature of around 5800 K. Similarly, knowing that Aldebaran is a K5 III star tells an astronomer that it is a red giant with a luminosity of around 370 $L_\odot$ and a surface temperature of about 4400 K (see Figure 11-7).

11-9 A star's spectral type and luminosity class provide a second distance-measuring technique

A star's spectral type and luminosity class, combined with information from the H-R diagram, give astronomers information necessary for determining the distances to stars millions of light years away, far beyond the maximum distance that can be measured using stellar parallax. The process works like this:

1. Astronomers observe a distant star's apparent magnitude and spectral type.

2. From its spectrum, they determine what luminosity class the star belongs to.

3. The spectrum also provides the star's absolute magnitude, since stars of a given spectral type and classification (such as a main-sequence star or a giant) each have a fairly well-defined absolute magnitudes (see Figure 11-9).

4. Using the distance-magnitude relationship (An Astronomer's Toolbox 11-3), the star's distance can then be calculated.

Consider, for example, the star Regulus in the constellation Leo. Its spectrum reveals Regulus to be a B7 V star (a hot, blue, main-sequence star). Placing it on the H-R diagram (see Figure 11-7), we can read off its luminosity as 140 $L_\odot$ and an absolute magnitude of –0.52. Given the star's apparent magnitude, we can use the distance-magnitude relationship to determine its distance from Earth. Because both the spectral type and luminosity class are obtained spectroscopically, this method of determining distances is called **spectroscopic parallax.** The name is misleading, because no parallax angle is involved.

Spectroscopic parallax is limited in accuracy because of the spread of stars in each luminosity class—the stars in each class do not fall on a single, narrow line (hence the use of the words "fairly well-defined absolute magnitude" in the numbered list above). It is also limited in that spectra of distant stars become increasingly hard to determine. As a result, errors of 10% in distance are common using spectroscopic parallax. Accepting such errors, the power of this method is that it can be used for stars at much greater distances than those determined by stellar parallax. Indeed, it even provides distances to stars in other galaxies tens of millions of light-years away.

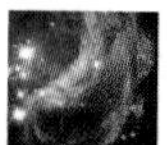

STELLAR MASSES

The final clue that astronomers needed in order to understand how stars work is their mass—how much matter stars have. If the masses were related to the stellar luminosities and locations on the H-R diagram (and they are), then astronomers would have enough information to start making scientific models of why stars shine. The mass, it turns out, determines the amount of energy each star has available to manufacture light and other electromagnetic radiation. Models of how stars generate that radiation, developed in the twentieth century and still being refined, provide us with further insights into how nature works. They also enable us to harness the power of stars.

The problem is that there is no way of determining stellar mass directly by examining isolated stars. The mass of a star must be determined by its gravitational effects on other bodies using Newton's law of gravity. When Newton derived the formula for Kepler's third law, he discovered that for any two objects in orbit, such as a planet around the Sun, a moon around a planet, or two stars around each other, the period of their orbits is related to the sum of their masses (see Section 2-5). You have seen that for objects orbiting the Sun, we can ignore the mass of the smaller body, thereby getting Kepler's original third law in Chapter 2. For stars orbiting each other, we use Newton's full equation, which includes the masses of both bodies.

Fortunately for astronomers, two-thirds of the stars near our solar system are members of star systems in which 6
two stars orbit each other. Put another way, *half the objects we see as single stars with our unaided eyes are actually pairs of stars in orbit around each other.* They are so distant that we cannot resolve the individual stars in the pairs and so they appear to us as single stars. Using telescopes to observe the periods of the orbits and the distances between stars, astronomers can determine the sum of the stellar masses of the pair.

11-10 Binary stars provide information about stellar masses

A pair of stars located at nearly the same position in the night sky is called a double star. Between 1782 and 1838, William Herschel, his sister, and his son John catalogued thousands of them. Some *double stars* are not really near each other in space and do not orbit each other. These **optical doubles** (sometimes called *apparent binaries*) just happen to lie in the same direction as seen from Earth. For example, δ Herculis, visible with the naked eye, is an optical double with a dim star. Through a telescope these stars appear close together, but that is an optical illusion.

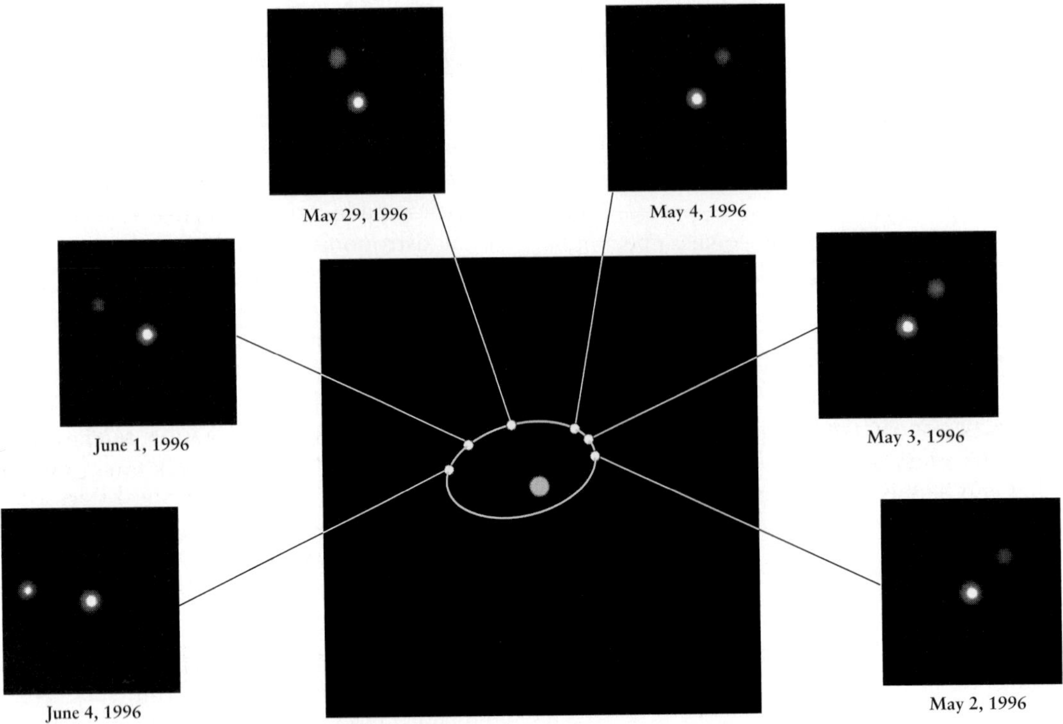

FIGURE 11-10 **A Binary Star System** About one-half of the visible "stars" are actually double stars. ζ (zeta) Ursae Majoris in Ursa Major is a binary system with stars separated by only about 0.01 arcseconds. The images surrounding this diagram show the relative positions of the two stars over half of their orbital period. The orbital motion of the two binary stars about each other is evident. Either star can be considered fixed in making such plots. (Navy Prototype Optical Interferometer, Flagstaff, Arizona. Courtesy of Dr. Christian A. Hummel)

Other double stars are true **binary stars**—pairs in which two stars orbit each other. In the case of **visual binaries,** both stars can be seen, using a telescope if necessary (Figure 11-10). Astronomers can plot the orbit of one star about the other in a visual binary.

To see how astronomers determine the masses of stars, consider a visual binary. Newton rewrote Kepler's third law as a relation between the masses of the stars (in solar masses), their orbital period around each other in Earth years (this period is the same for both so it does not matter which star we assume orbits the other), and the average separation between the stars in AU (which equals the semimajor axis of the elliptical orbit):

$$\text{The sum of the masses} = \frac{\text{the cube of the semimajor axis}}{\text{the square of the orbital period}}$$

So, by observing the separation between a pair of stars and how long one of them takes to complete its orbit, we can calculate the sum of the masses of the two stars. The details are presented in An Astronomer's Toolbox 11-4.

In many cases, we can also determine the individual masses of the stars. To do this, we need to know the distances of the stars from the center of mass of the pair. Both stars in a binary system actually move in elliptical orbits around a common point between them. This point, called the **center of mass,** is determined by the masses of the stars and can be understood by analogy to a whirling wrench sliding on a table (see Figure 6-24). There is a point on the wrench that moves in a straight line. This is its center of mass. Just as the ends of the wrench orbit around its center of mass, the two stars in a binary system orbit around their center of mass under the influence of their mutual gravitational attraction (Figure 11-11). The more massive star is closer to the center of mass, just as a heavier child is closer to the pivot point or fulcrum of a seesaw (Figure 11-11b).

The center of mass of a visual binary is located by plotting the separate orbits of the two stars, as in Figure 11-10, using the background stars as reference points. The center of mass lies at the common focus of the two elliptical orbits. Comparing the relative sizes of the two orbits around the center of mass yields the ratio of the two stars' masses, M_1/M_2. To get the relative sizes of the orbits, however, we must be able to determine the plane of orbit of the two stars. This is not always possible,

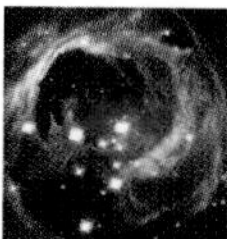

AN ASTRONOMER'S TOOLBOX 11-4
Kepler's Third Law and Stellar Masses

The same gravitational force that holds the Earth in orbit around the Sun also keeps pairs of stars in orbit about each other. For any pair of orbiting bodies, the orbits are ellipses, and Kepler's third law can be written as:

$$M_1 + M_2 = a^3/P^2$$

Here M_1 and M_2 are the two masses (expressed in solar masses), a is the length of the semimajor axis of the ellipse (in astronomical units), and P is the orbital period (in years). Note that a is also the average separation between the two bodies. Thus, the sum of the masses can be found once the orbital period and average separation are known.

Example: Suppose two stars make up a double system, and one star follows an ellipse with a semimajor axis of 4 AU. We find that it takes 2.5 years to complete one orbit. Then the sum of the stars' masses is

$$M_1 + M_2 = (4)^3/(2.5)^2 = 10.2\ M_\odot$$

The total mass of the system is 10.2 $M_\odot$.

Try these questions: Why can Kepler's third law for a planet orbiting the Sun be written as $P^2 = a^3$? *Hint:* For the Sun and a planet, the combined mass is extremely close to the Sun's mass alone. Two identical stars are observed to orbit each other with a period of 2 years at a separation of 2.32 AU. What are their masses? How long would it take a 1 $M_\odot$ and a ½ $M_\odot$ star to orbit each other if their separation were 20 AU?

(Answers appear at the end of the book.)

which is why we cannot always determine the individual masses of the stars. When we can determine M_1/M_2, we can find the individual masses, because we already know the sum $M_1 + M_2$ from Kepler's third law. The individual masses are determined by combining the ratio of the masses and the sum of the masses. The range of stellar masses thus determined extends from 0.08 $M_\odot$ to about 100 $M_\odot$.

11-11 There is a relationship between mass and luminosity for main-sequence stars

As binary star data accumulated, an important trend began to emerge. On the main sequence, the more luminous the star, the more massive it is. This mass-luminosity relation can be conveniently displayed graphically (Figure 11-12a). Noting that the axes of this figure are logarithmic, the Sun's

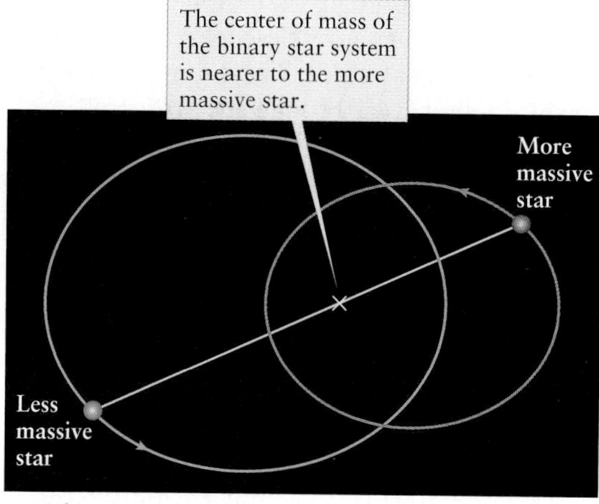

a A binary star system

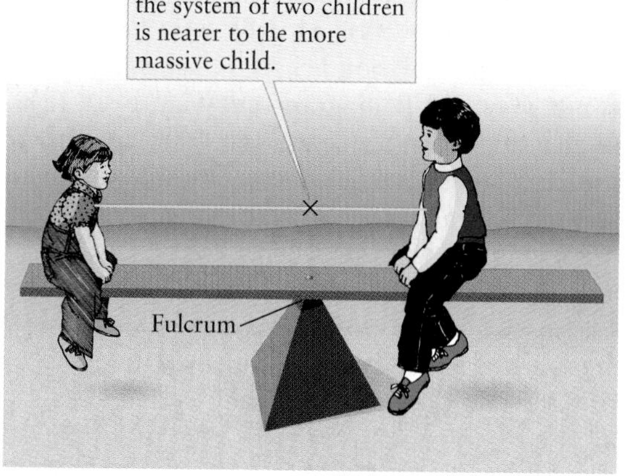

b A "binary system" of two children

FIGURE 11-11 **Center of Mass of a Binary Star System** (a) Two stars move in elliptical orbits around a common center of mass. Although the orbits cross each other, the two stars are always on opposite sides of the center of mass and thus never collide. (b) A seesaw balances if the fulcrum is at the center of mass of the two children. When balanced, the heavier child is always closer to the fulcrum.

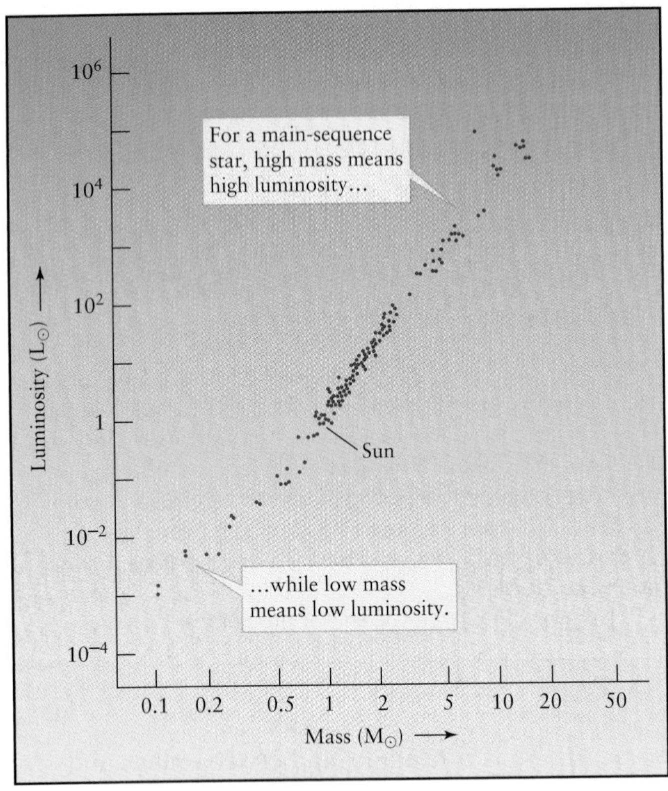

a

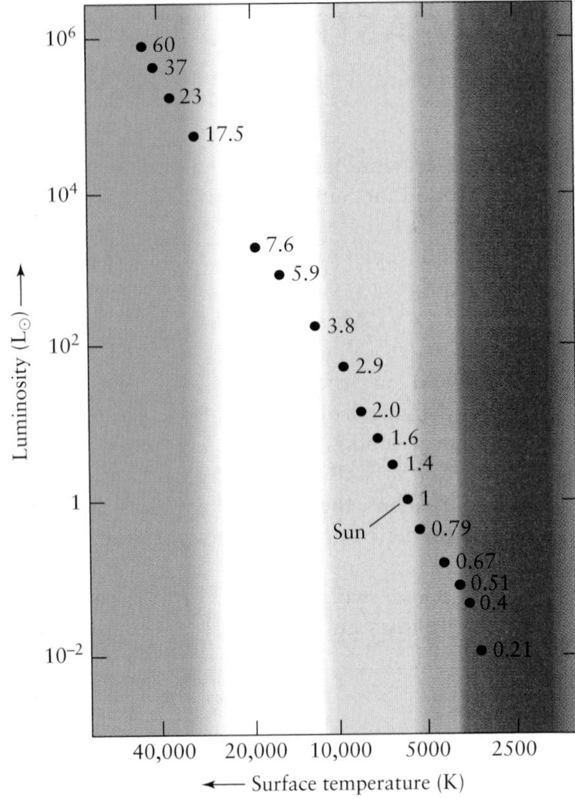

b

FIGURE 11-12 The Mass-Luminosity Relation (a) For main-sequence stars, mass and luminosity are directly correlated—the more massive a star, the more luminous it is. A main-sequence star of mass 10 $M_\odot$ has roughly 3000 times the Sun's luminosity (3000 $L_\odot$); one with 0.1 $M_\odot$ has a luminosity of only about 0.001 $L_\odot$. To fit them on the page, the luminosities and masses are plotted using logarithmic scales. (b) On this H-R diagram, each dot represents a main-sequence star. The number next to each dot is the mass of that star in solar masses ($M_\odot$). As you move up the main sequence from the lower right to the upper left, the mass, luminosity, and surface temperature of main-sequence stars all increase.

mass lies in the middle of this range. The **mass-luminosity relation** demonstrates that *the main sequence on the H-R diagram is a progression in mass as well as in luminosity and surface temperature.* The hot, bright, bluish stars in the upper left corner of the H-R diagram (see Figure 11-12b) are the most massive main-sequence stars. As we move down the sequence, stellar masses decrease until we reach the dim, cool, reddish stars in the lower right corner of the H-R diagram. The correlation of stellar luminosities with mass suggests (correctly as it turns out) that a star's energy production is closely linked to its mass. As we will see shortly, the mass is *the* crucial factor in explaining why stars are as bright as they are and how they evolve.

11-12 The orbital motion of binary stars affects the wavelengths of their spectral lines

Many binary stars are scattered throughout our Milky Way Galaxy, but only those that are nearby or are widely separated can be distinguished as visual binaries. A remote binary often presents the appearance of a single star because even our best telescopes cannot resolve the images of its individual stars. These binary systems can still be detected, however, through spectroscopy.

Spectral analysis yields incongruous spectral lines for some stars. For example, the spectrum of what appears at first to be a single star may include strong absorption lines for both hydrogen (indicating a hot, type A star) and also titanium oxide (indicating a cool, type M star). Because a single star cannot display both types of absorption lines prominently, such an observation must be revealing a binary system.

Spectroscopy can also detect the movements of stars orbiting each other because of the Doppler shift in spectral lines. As we saw in Section 4-7, an approaching source of light has shorter wavelengths than if the same source were stationary, and a receding source has longer wavelengths. The amount of the shift is proportional to the speed of the light source: The greater the speed, the greater the shift.

If the two stars in a binary are orbiting at more than a few kilometers per second, they will produce spectral lines

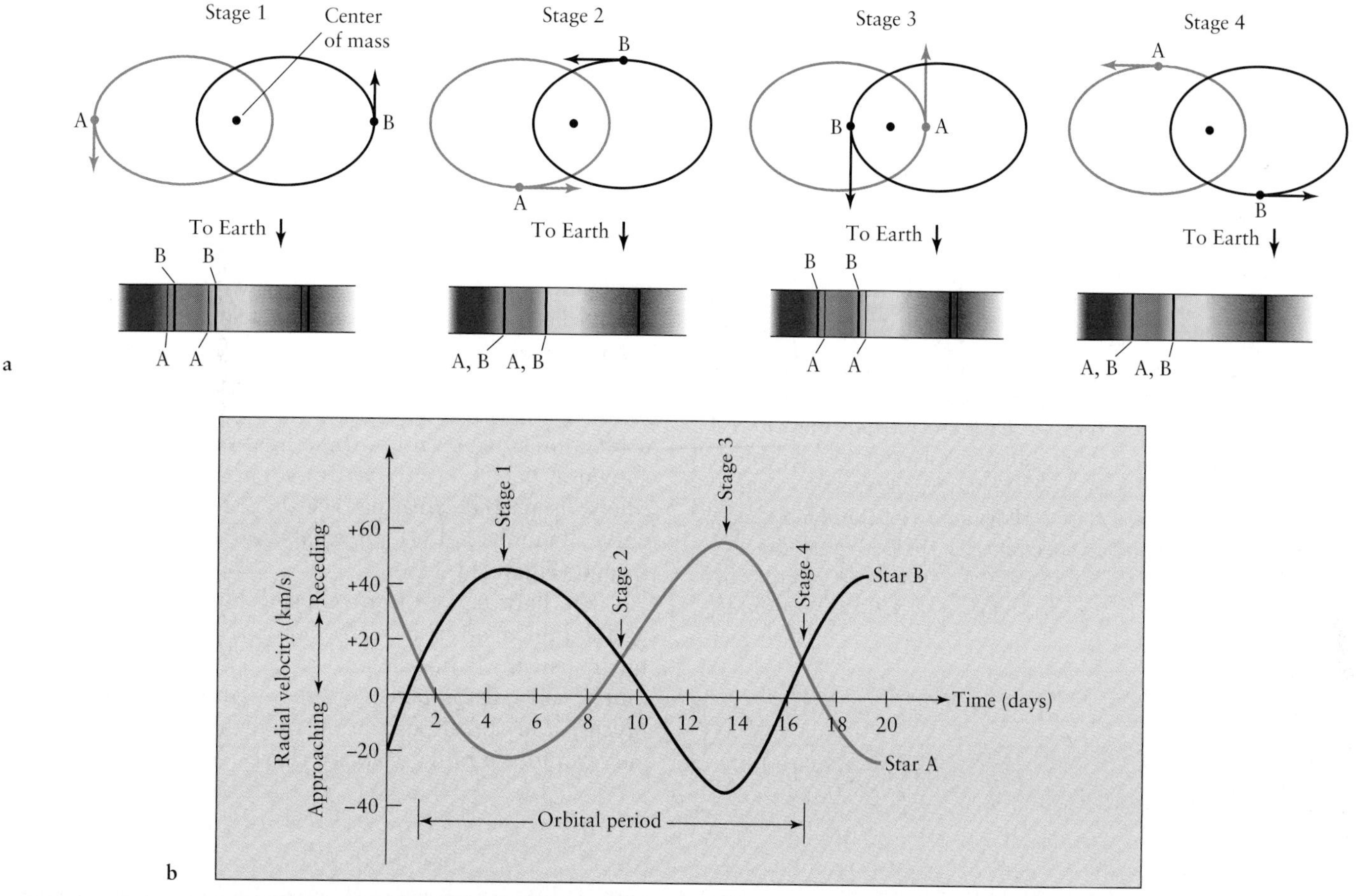

FIGURE 11-13 **Spectral Line Motion in Binary Star Systems** **(a)** The drawings at the top indicate the positions and motions of the stars, labeled A and B relative to the Earth (below the diagram), and their spectra at four selected moments (Stages 1, 2, 3, and 4) during an orbital period. **(b)** This graph displays the radial-velocity curves of the binary HD 171978. (The HD means that this is a star from the *Henry Draper Catalogue* of stars.) The entire binary is moving away from us at 12 km/s, which is why the pattern of radial velocity curves is displaced upward from the zero-velocity line.

that shift back and forth regularly. Such stars are called **spectroscopic binaries.** The motions of the stars revolving about their center of mass produce the periodic shifting of the spectral lines. In many spectroscopic binaries, one of the stars is so dim that its spectral lines cannot be detected. Instead, a single set of spectral lines from the star that we can see shifts regularly back and forth. Such a *single-line spectroscopic binary* yields less information about its two stars than does a *double-line spectroscopic binary,* in which the lines from both stars are visible. In particular, the single-line binary only gives the ratio of the stellar masses, whereas from a double-line binary, the individual star masses can be determined.

Figure 11-13a shows four spectra of a spectroscopic binary system. At Stage 1 star A is moving toward Earth; hence its spectrum is blueshifted. Conversely, star B is moving away, so its spectrum is redshifted. In the spectrum below the drawing you can see two sets of spectral lines, each slightly offset from each other. A few days later, at Stage 2, the stars have progressed along their orbits so that star A is moving toward the right and star B toward the left as seen from Earth. Because neither star is moving toward or away from us, their spectral lines return to their normal positions. Stage 3 shows the opposite spectra from Stage 1 (star A is now redshifted and star B blueshifted), while Stage 4 shows the same spectrum as Stage 2. Figure 11-14 shows spectra taken of the binary star system κ (kappa) Arietis.

The shifts in spectral lines can yield significant information about the orbital velocities of the stars in a spectroscopic binary. This information is best displayed as a **radial-velocity curve,** in which radial velocity is graphed over time (see Figure 11-13b). In Figure 11-13b, the wavy pattern repeats with a period of about 15 days, which is the orbital period of the binary. This pattern is displaced upward from the zero-velocity line by about 12 km/s, which is the overall

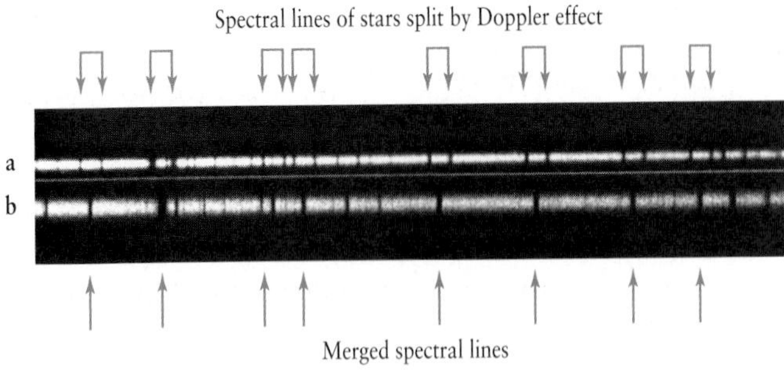

FIGURE 11-14 **A Double-Line Spectroscopic Binary** The spectrum of the double-line spectroscopic binary κ (kappa) Arietis has spectral lines that shift back and forth as the two stars revolve about each other. **(a)** The stars are moving parallel to the line of sight with one star approaching Earth, the other star receding as in Stages 1 or 3 of Figure 11-13a. These motions produce two sets of shifted spectral lines. **(b)** Both stars are moving perpendicular to our line of sight as in Stages 2 or 4 of Figure 11-13a. As a result, the spectral lines of the two stars have merged. (Lick Observatory)

motion of the binary system away from the Earth. Superimposed on this overall recessional motion are the periodic approaches and recessions of the two stars as they orbit around each other.

11-13 Some binary stars eclipse each other

Some binary systems are oriented so that the two stars periodically eclipse each other as seen from Earth. Such **eclipsing binaries** can be detected even when the stars cannot be resolved as two distinct images in a telescope. The apparent magnitude of the image of an eclipsing binary dims each time one star blocks out part of the other. An astronomer can measure light intensity from binaries very accurately as a function of time. From these **light curves,** such as those shown in Figure 11-15, we can see at a glance whether the eclipse is partial, creating a V-shaped trough in the light curve (Figure 11-15a), or total, creating a flat-bottomed trough (Figure 11-15b).

The light curve of an eclipsing binary can yield other useful information as well. For example, if an eclipsing binary is also a double-line spectroscopic binary, astronomers can calculate the masses, diameters, brightnesses, speeds, and stellar separation of each star from the light curves and the radial-velocity curves. These stars are rare,

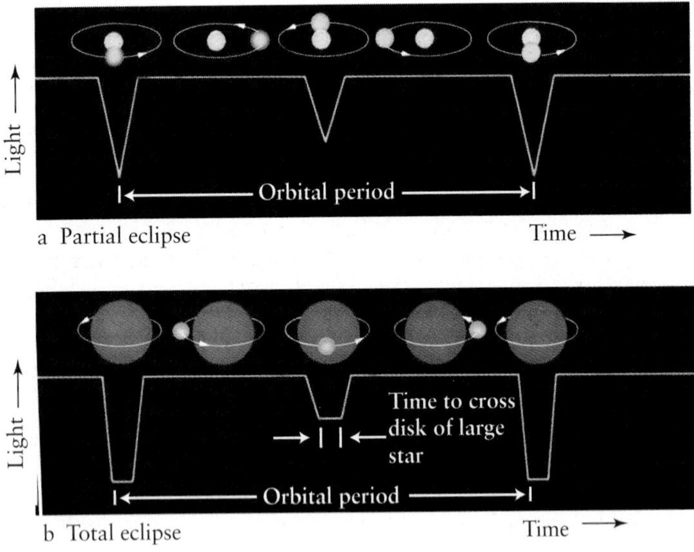

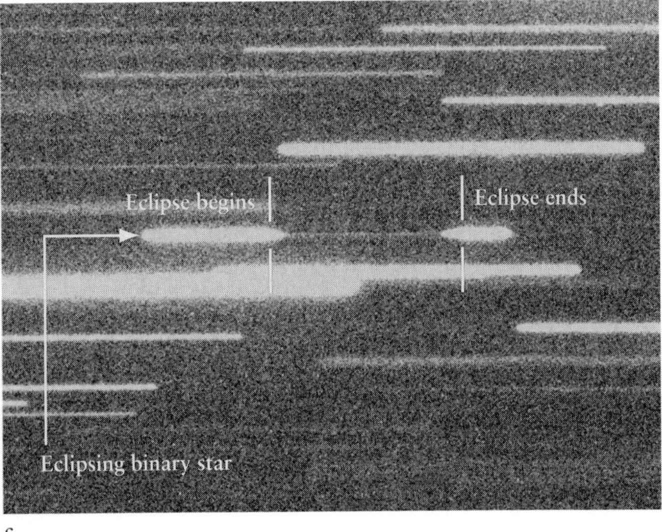

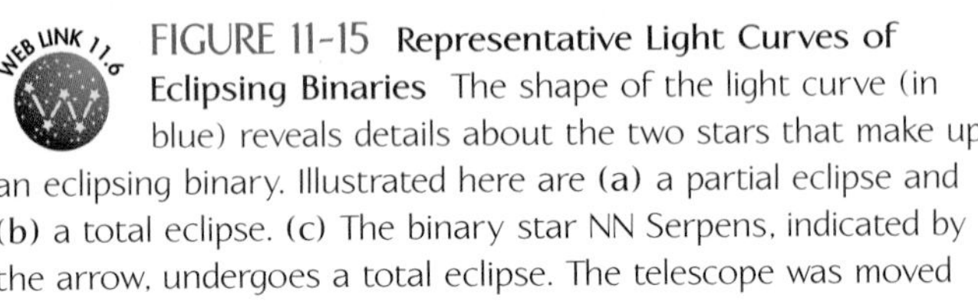

FIGURE 11-15 **Representative Light Curves of Eclipsing Binaries** The shape of the light curve (in blue) reveals details about the two stars that make up an eclipsing binary. Illustrated here are **(a)** a partial eclipse and **(b)** a total eclipse. **(c)** The binary star NN Serpens, indicated by the arrow, undergoes a total eclipse. The telescope was moved during the exposure so that the sky drifted slowly from left to right. During the 10.5-minute eclipse, the dimmer, but larger, star in the binary system (an M6 main-sequence star) passed in front of the more luminous, but smaller star (a white dwarf). The binary became so dim that it almost disappeared. (c: European Southern Observatory)

however, because the orbital planes of most spectroscopic binaries are tilted so that eclipses do not occur as seen from Earth.

Light curves can also reveal information about stellar atmospheres. Suppose that one star of a binary is a tiny white dwarf (a stellar remnant the size of the Earth) and the other is a giant (the late stage of stellar evolution in which the star has greatly expanded from its previous size). By observing exactly how the light from the white dwarf is gradually cut off as it begins to move behind the edge of the giant during an eclipse, astronomers can infer the pressure and density in the upper atmosphere of the giant. Such information is invaluable in testing models of stellar structure.

Many binary stars are separated by several AU or more. Other than orbiting each other, these stars behave as though they are isolated; that is, the models we develop in the coming chapters for the evolution of isolated stars apply to them as well. However, there are also **close binary** systems, with only a few stellar diameters separating the stars. Such stars are so close together that the gravity of each one dramatically affects the appearance and evolution of the other. If one member of a close binary is a giant, some of the gas of its outer layers is pulled onto its more compact companion—mass is transferred from one star to the other. We will explore more about such systems in Chapter 12.

This chapter has provided us with observational data that serve, in large measure, as the basis for the models of stellar activity and evolution presented in the next three chapters. The models make predictions, of course, that lead to new observations and refined models. We will take up some further observations thus driven as we proceed through the models.

11-14 Frontiers yet to be discovered

The legacy of *Hipparcos* is a better knowledge of the size of the universe. We will discover in the coming chapters that the more accurately we know the distance between stars and, indeed, between galaxies, the better we are able to understand the evolution of the universe. As noted earlier, stellar parallax is the most accurate means of measuring these distances. This technique requires the fewest assumptions about how stars work and the nature of the material between us and the stars. Therefore, a major goal of space-based observational astronomy is to get more and better parallax measurements. To improve them, satellites will be sent far from the Earth to increase the baseline used to make these measurements (see Figure 11-1) and hence the angular accuracy that is possible.

Classifying stars by luminosity and spectral type is an ongoing process providing an ever-growing body of data about the properties of stars. This information provides continual surprises. For example, there is a growing number of individual "stars" whose properties astronomers thought they understood but which turned out to be binary systems, each star having different properties from that of the single "star" they were originally believed to be.

Another significant piece of information astronomers need to better understand star formation is how many stars of each mass are formed. Observations strongly indicate that very high–mass stars form quite infrequently, but is the same true of very low–mass stars? Or do the numbers of stars formed increase with decreasing mass? Or perhaps there is a peak of star formation at some mass with decreasing numbers of stars with higher and lower mass. The number of stars formed with different masses is expressed as the **initial mass function.** While there are several theoretical initial mass functions, in the end, observations will determine which is correct.

Summary of Key Ideas

• Stars differ in size, luminosity, temperature, color, mass, and chemical composition—facts that help astronomers understand stellar structure and evolution.

Magnitude Scales

• Determining stellar distances from Earth is the first step to understanding the nature of the stars. Distances to the nearer stars can be determined by stellar parallax, the apparent shift of a star's location against the background stars while the Earth moves along its orbit around the Sun. The distances to more remote stars are determined using spectroscopic parallax.

• The apparent magnitude of a star, denoted *m,* is a measure of how bright the star appears to Earth-based observers. The absolute magnitude of a star, denoted M, is a measure of the star's true brightness and is directly related to the star's energy output, or luminosity.

• The absolute magnitude of a star is the apparent magnitude it would have if viewed from a distance of 10 pc. Absolute magnitudes can be calculated from the star's apparent magnitude and distance.

• The luminosity of a star is the amount of energy emitted by it each second.

The Temperatures of Stars

• Stellar temperatures can be determined from stars' colors or stellar spectra.

• Stars are classified into spectral types (O, B, A, F, G, K, and M) based on their spectra or, equivalently, their surface temperatures.

Types of Stars

• The Hertzsprung-Russell (H-R) diagram is a graph on which luminosities of stars are plotted against their spectral types (or, equivalently, their absolute magnitudes are plotted against surface temperatures). The H-R diagram reveals the existence of four major groupings of stars: main-sequence stars, giants, supergiants, and white dwarfs.

• The mass-luminosity relation expresses a direct correlation between a main-sequence star's mass and the total energy it emits.

• Distances to stars can be determined using their spectral type and luminosity class.

Stellar Masses

• Binary stars are surprisingly common. Those that can be resolved into two distinct star images (even if it takes a telescope to do this) are called visual binaries.

• The masses of the two stars in a binary system can be computed from measurements of the orbital period and orbital dimensions of the system.

• Some binaries can be detected and analyzed, even though the system may be so distant (or the two stars so close together) that the two star images cannot be resolved with a telescope.

• A spectroscopic binary is a system detected from the periodic shift of its spectral lines. This shift is caused by the Doppler effect as the orbits of the stars carry them alternately toward and away from the Earth.

• An eclipsing binary is a system whose orbits are viewed nearly edge-on from the Earth, so that one star periodically eclipses the other. Detailed information about the stars in an eclipsing binary can be obtained by studying its light curve.

• Mass transfer occurs between binary stars that are close together.

WHAT DID YOU THINK?

1 *How near is the closest star other than the Sun?* Proxima Centauri is about 40 trillion kilometers (25 trillion miles) away. It takes light about 4 years to reach the Earth from there.

2 *How luminous is the Sun compared with other stars?* The most luminous stars are about a million times brighter and the least luminous stars are about a hundred thousand times dimmer than the Sun.

3 *What colors are stars?* Stars are found in a wide range of colors, from red through violet as well as white.

4 *Are brighter stars hotter than dimmer stars?* Not necessarily. Many brighter stars, such as red giants, are cooler but larger, than hotter, dimmer stars, such as white dwarfs.

5 *What sizes are stars?* Stars range from more than 1000 times the Sun's diameter to less than 1/100 the Sun's diameter.

6 *Are most stars isolated from other stars, as the Sun is?* No. In the vicinity of the Sun, two-thirds of the stars are found in pairs or larger groups.

Key Words

absolute magnitude, 298
apparent magnitude, 296
binary star, 308
center of mass, 308
close binary, 313
dwarf star, 305
eclipsing binary, 312
giant star, 305
Hertzsprung-Russell (H-R) diagram, 304
initial mass function, 313
inverse-square law, 298
light curve, 312
luminosity, 300
luminosity class, 306
main sequence, 304
main-sequence star, 305
mass-luminosity relation, 310
OBAFGKM sequence, 303
optical double, 307
photometry, 300
radial-velocity curve, 311
red giant, 305
spectral types, 303
spectroscopic binary, 311
spectroscopic parallax, 307
stellar evolution, 293
stellar parallax, 293
stellar spectroscopy, 301
supergiant, 305
visual binary, 308
white dwarf, 305

Review Questions

The answers to all computational problems, which are preceded by an asterisk (*), appear at the end of the book.

1. Stellar parallax measurements are used in astronomy to measure which of the following properties of stars: **a.** speeds **b.** rotation rates **c.** distances **d.** colors **e.** temperatures.

2. The brightness a star would have if it were at 10 pc from the Earth is called its: **a.** absolute magnitude **b.** apparent magnitude **c.** luminosity **d.** spectral type **e.** center of mass.

3. Measurements of a binary star system are required to determine what property of stars? **a.** luminosity **b.** apparent magnitude **c.** distance **d.** mass **e.** temperature.

4. A star with which of the following apparent magnitudes appears brightest from Earth? **a.** 6.8 **b.** 3.2 **c.** 0.41 **d.** –0.44 **e.** –1.5.

5. A star of what spectral class has the strongest (darkest) H_α line? (*Hint:* See Figure 11-5): **a.** B2 **b.** A0 **c.** A5 **d.** G5 **e.** M0.

6. Describe how the parallax method of finding a star's distance is similar to the binocular (two-eye) vision of animals.

7. What is stellar parallax?

8. How do astronomers use stellar parallax to measure the distances to stars?

9. Why do stellar parallax measurements work only with relatively nearby stars?

10. What is the difference between apparent magnitude and absolute magnitude?

11. Briefly describe how you would determine the absolute magnitude of a nearby star.

12. What does a star's luminosity measure?

13. Why is the magnitude scale "backward" from what common sense dictates?

14. Does the star Betelgeuse, whose apparent magnitude is $m = +0.5$, look brighter or dimmer than the star Pollux, whose apparent magnitude is $m = +1.1$?

*15. Consider two identical stars, with one star 5 times farther away than the other. How much brighter will the closer star appear than the more distant one?

16. How and why is the spectrum of a star related to its surface temperature?

17. What is the primary chemical component of most stars?

18. A star of which spectral type has the strongest Na I absorption lines? At approximately what wavelength is this line normally found? *Hint:* See Figure 11-5.

19. Why does a G2 star have many more absorption lines than a B0 star?

20. Draw an H-R diagram and sketch the regions occupied by main-sequence stars, giants, supergiants, and white dwarfs. Briefly discuss the different ways you could have labeled the axes of your graph.

21. To test your understanding of the H-R diagram, do Interactive Exercise 11-1. You can print out your answers if required.

22. How can observations of a visual binary lead to information about the masses of its stars?

23. What is a radial-velocity curve? What kinds of stellar systems exhibit such curves?

24. What is the difference between a single-line and a double-line spectroscopic binary?

25. What is meant by the light curve of an eclipsing binary? What sorts of information can be determined from such a light curve?

26. What is the mass-luminosity relation? To what kind of stars does it apply?

27. Refer to Figure 11-7: **a.** What are the hottest and coolest named stars on the diagram? **b.** What are the brightest and dimmest named stars on the diagram? **c.** What are the hottest and coolest named main-sequence stars on the diagram? **d.** What named stars are white dwarfs? giants? supergiants?

Advanced Questions

28. What is the inverse-square law? Use it to explain why a headlight on a car can appear brighter than a star, even though the headlight emits far less light energy per second.

*29. Van Maanen's star, named after the Dutch astronomer who discovered it, has a parallax angle of 0.232 arcsec. How far away is the star?

30. Explain how sailors on a ship traveling parallel to a coastline at a known speed can use parallax angle measurements to determine the distance to the shore.

*31. Suppose that a dim star were located 2 million AU from the Sun. Find **a.** the distance to the star in parsecs and **b.** the parallax angle of the star.

*32. How many times brighter is a star of apparent magnitude $m = -1$ than a star of apparent magnitude $m = +7$?

33. Sketch the radial-velocity curve of a binary whose stars are moving in nearly circular orbits that are **a.** perpendicular and **b.** parallel to our line of sight.

34. Sketch the light curve of an eclipsing binary whose stars are moving along highly elongated orbits with the major axes of the orbits **a.** pointed toward the Earth and **b.** perpendicular to our line of sight.

*35. **a.** What is the approximate mass of a main-sequence star that is 10,000 times as luminous as the Sun? **b.** What is the approximate luminosity of a main-sequence star whose mass is one-tenth that of the Sun?

*36. What is the approximate surface temperature of a main-sequence star with a luminosity 100 times as bright as the Sun?

Discussion Questions

37. Discuss the advantages and disadvantages of measuring stellar parallax from a space telescope in a large solar orbit, say, at the distance of Jupiter from the Sun.

38. How does a star's rotation affect the appearance of its spectral lines? *Hint:* Assume we are looking at the star from above its equator. Then, at every instant, half of the spinning star is approaching the Earth, while the other half is receding. Consider the resulting Doppler shift in the spectral lines.

What If . . .

39. All the stars in our Milky Way Galaxy actually were the same distance from Earth, say, 10 pc? Describe what the night sky might look like. How about the daytime sky? *Hint:* If you are doing this quantitatively, assume for simplicity

that all the stars have the same absolute magnitudes as the Sun and that there are roughly 200 billion stars in the Galaxy.

40. All stellar parallax angles, p, were observed to be increasing? What would that imply about the motions of stars? Is there another observational technique that could be used to confirm this motion?

41. A star's parallax was observed to oscillate—regularly increase in angle and then decrease in angle—over a period of 100 years? What would that imply about the star?

42. The Sun was part of a binary star system of two main-sequence stars and the Earth orbited around one of the stars, while the other star just changed the Earth's distance of orbit slightly, compared to its orbit today? You might want to discuss such things as climate, tides, impacts from meteoroids and asteroids, and habitability.

43. The Sun was an M-type star, rather than a G-type star? Assuming that the Earth orbiting the M-type star had the same composition and orbit distance as it does today, what would be different here?

44. The Sun was a B-type star, rather than a G-type star? Assuming that the Earth orbiting the B-type star had the same composition and orbit distance as it does today, what would be different here? We will pick up this question again in Chapter 13 for further insights.

Web Questions

AIMM 11.1 *45. **Distances to Stars Using Parallax.** Access the Active Integrated Media Module "Using Parallax to Determine Distance" in Characterizing Stars of the *Discovering the Universe* Web site. Use this to determine the distance in parsecs and in light-years to each of the following stars: **a.** Betelgeuse (parallax $p = 0.00763''$); **b.** Vega ($p = 0.129''$); **c.** Antares ($p = 0.00540''$); and **d.** Sirius ($p = 0.379''$).

46. Search the Web for the periods of 10 binary star systems. Plot these on a graph of time (on the horizontal axis) versus number of systems. If possible, combine this information with similar data from your classmates. Do you see any patterns in the periods of these star systems?

47. To explore the range of periods of binary star systems locate on the Web binary star systems with periods of **a.** less than 1 week, **b.** between 1 day and 1 week, **c.** between 1 and 2 years, **d.** between 40 and 50 years, and **e.** more than 400 years.

Observing Projects

48. Use your *Starry Night Enthusiast*™ software to locate the bright stars listed in Appendix Table E-5. Set the time in the program to the time you will be observing tonight. Display the major features of the constellations (*options* tab/constellations/*Boundaries*, constellations/*Labels*, and constellations/*Stick Figures*). Locate the stars by selecting *Find* tab and then typing the name of the object from Appendix Table E-5. Ignore (choose *Cancel*) the stars that the computer says are not visible tonight. Make a list of stars that are up tonight, including their apparent magnitudes and the constellations they are in, and arrange to observe them. You may find it helpful to print out star charts showing them using the *Starry Night Enthusiast*™ program. By comparing stars of similar apparent magnitude, such as Antares and Spica, determine the difference in apparent magnitude for which you can tell stars have different apparent brightnesses. If you bring out a star chart, identify and observe other stars, estimating their apparent magnitudes based on the ones on your list.

49. Locate the stars Betelgeuse and Rigel in Orion. Observe them both by eye and through a small telescope. Are they the same color? To determine their color(s), it helps to compare them to their neighbors.

50. Locate one or more of the following double star systems: Regulus in Leo, Algieba in Leo, Irak (ε Boo) in Boötes, Ras Algethi in Hercules, Albireo in Cygnus, Vega in Lyra, Polaris in Ursa Minor, Rigel in Orion, Antares in Scorpius, Sirius in Canis Major, and Schedar in Cassiopeia. You can find their locations and print star charts using your *Starry Night Enthusiast*™ software. View them with and without a telescope. What are their colors?

51. Mizar and Alcor. In this exercise, we use *Deep Space Explorer*™ to explore two stars in the Big Dipper. If you have not done so already, go to **Settings** and uncheck "Use magnitude cutoffs," slide the "Galaxy drawing/Brightness" slide to the middle of the range, and then click on the **Views** tab.

Now, go to Earth by clicking **Solar System/Earth** on the left side of the star field. Locate Mizar by going to **Edit/Find/Mizar.** Although the Earth is off to one side, you will now be seeing Mizar from the vicinity of Earth. As seen from Earth, Mizar has a companion star, Alcor. Tap the "zoom in" button (downward-pointing triangle) until a second star is visible near Mizar. Stop about 30 ly from Mizar. Where is the smaller star Alcor located? Now continue zooming in until you essentially arrive at Mizar (any distance under an AU will do). Select Alcor and note the distance to Alcor from Mizar. This is in the "Distance" box with Alcor's name on it. Ignoring the slight distance you are away from Mizar, this represents the separation between Mizar and Alcor. The closest star to Earth is Proxima Centauri, about 4.3 ly away. Given all this data and what you already know, state whether you think that Mizar and Alcor are a visible binary, spectroscopic binary, or optical double, and explain your reasoning.

The Lives of Stars from Birth Through Middle Age

CHAPTER 12

Ionized Hydrogen in NGC 2264 (© David Malin/Anglo-Australian Observatory)

WHAT DO YOU THINK?

1. How do stars form?
2. Are stars still forming today?
3. Do more massive stars shine longer?

Time changes everything. We grow up, we grow older, and we die. We are most familiar with changes brought about by aging in our own lives and the lives of other people, but everything else in the universe alters with time, too. The changes are due to the forces in nature causing interactions between matter and energy. For example, exposure to the effects of weather and pollution causes even solid stone to crumble (Figure 12-1).

Gravity is the primary force causing stars to age. As we saw in studying the Sun (Chapter 10), gravity compresses stars, thereby heating them and causing fusion to occur. Fusion enables the stars to emit huge amounts of radiation. As they age, stars show changes in their chemical compositions, masses, and brightnesses. Stars are aging and changing all the time; they only seem unchanging to us because of the enormous spans of time over which these changes occur. Major stages in the life of a star can last for millions or even billions of years. We will explore these processes in the following two chapters.

You saw in Chapter 11 that by observing stars with different temperatures and brightnesses, astronomers have discovered groupings of stars with similar properties. We now begin to explore the scientific theories of **stellar evolution** based on the groupings on the H-R diagram. These theories are the foundation for our understanding of how stars change.

INSIGHT INTO SCIENCE

Beware Poetic License All too often, the names and descriptions we use for nonliving objects may appear to give them human qualities. Anthropomorphism—assigning human attributes to nonhuman creatures or even to nonliving objects like stars—is descriptive but not scientific. In fact, scientists avoid this practice where possible, because it creates unjustified expectations. Nevertheless, anthropomorphism does creep into astronomy. For lack of better words, astronomers describe stellar change with words such as "evolution," "birth," "aging," "maturing," "growing old," and "dying" when talking about stars. This is poetic license of a sort: It piques our interest but muddles our science. As you study stellar evolution, always be mindful of that.

In this chapter you will discover

- how stars form
- what a stellar "nursery" looks like
- how astronomers use the physical properties of stars to learn about stellar life cycles
- the remarkable transformations of older stars into giants and supergiants
- that some dying stars eject material that creates new generations of stars, while others act as beacons that enable astronomers to pinpoint distant galaxies
- that the H-R diagram is your guide to the stellar life cycle

a

b

FIGURE 12-1 **Everything Ages** In less than 60 years (1935 to 1994), corrosive gases in the air caused this statue of George Washington in New York City to erode even more rapidly than normal. Even without the significant pollution that humans add to the air, everything in contact with it decays. (a: © 1994 NYC Parks Photo Archive/Fundamental Photographs; b: © 1994 Kristen Brochmann/Fundamental Photographs)

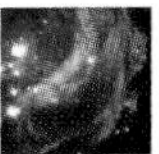

PROTOSTARS AND PRE-MAIN-SEQUENCE STARS

We saw in Chapter 5 that the solar system is believed to have formed from a collapsing, rotating cloud of gas and dust some 4.6 billion years ago. This scientific theory requires that interstellar matter existed before then so that the solar system had something from which to form. Taking this reasoning one step further, if such interstellar matter exists today, perhaps star formation continues.

12-1 Gas and dust exist between the stars

Matter does indeed exist between the stars. We can see a relatively small amount of it through visible light telescopes (see Figure 5-2); however, the bulk of it is too cold to be seen optically and requires the use of infrared and radio telescopes. We call this material between the stars the **interstellar medium,** and it contains at least 10% of all the known mass in our Galaxy. Observations of the spectra of the interstellar medium reveal that it is composed of atoms, molecules, and tiny pieces of dust.

Ninety percent of the particles comprising the interstellar medium are hydrogen atoms and molecules, 9% are helium atoms, and the remaining 1% are all the other naturally forming elements, often in the form of molecules or dust particles. Different atoms have different masses and, in order to understand the chemistry and evolution of planets and stars, astronomers often need to know the percentages of the interstellar medium's mass that is composed of the various elements. In that case, they present the composition information in terms of masses: About 74% of the interstellar medium's mass is hydrogen, 25% is helium, and all the other naturally forming elements make up the remaining 1%.

A variety of different molecules are found in interstellar space, including molecular hydrogen (H_2), carbon monoxide (CO), carbon dioxide (CO_2), water (H_2O), ammonia (NH_3), and formaldehyde (H_2CO), among many others. These molecules are identified by their unique spectral emissions, just as we can identify elements from atomic spectra.

The dust found in space has a variety of structures. Some carbon-based (that is, organic) particles out there are typically 0.005 micrometers across (10,000 times smaller than the diameter of a typical human hair). Among these are collections of molecules similar to those found in engine exhaust and burnt meat (Figure 12-2) called polycyclic aromatic hydrocarbons (PAHs). PAHs are composed only of carbon and hydrogen atoms. Much larger space dust has cores of carbon or silicon compounds surrounded with mantles of ice and other materials. These grow to more than .30 micrometers in diameter (nearly 200 times smaller than the diameter of a human hair).

FIGURE 12-2 **A Connection to Interstellar Space** The charred layer created by overcooking this chicken contains compounds called polycyclic aromatic hydrocarbons (molecules composed just of carbon and hydrogen). These molecules are also found in interstellar clouds. (Spike Mafford/Photodisc Green/Getty Images)

Working from the knowledge that new stars form from the gas and dust of the interstellar medium, astronomers map this matter to identify places to look for newly forming stars. When it can be seen in the visible part of the spectrum, the gas and dust in the interstellar medium glow as a result of scattering light from stars in its vicinity. For example, the interstellar medium is dramatically highlighted by stars in the Pleiades star cluster (Figure 12-3a) located in the constellation Taurus. The bluish haze, called a reflection nebula, is caused by fine grains of interstellar dust that scatter blue light from surrounding stars.

Finding more distant gas and dust is difficult because the nearby interstellar medium dims or even blocks light from behind it, in just the same way that thick clouds or several thin cloud layers in our sky prevent us from seeing the Sun. This darkening of light by intervening gas and dust in space is called **interstellar extinction.**

Even when we can see a star or other object through the interstellar medium, it appears redder than it actually is, a phenomenon called **interstellar reddening.** This occurs because short-wavelength starlight is scattered by dust grains in the cloud more than the longer-wavelength red light. (Violet is scattered most, followed by blue, green, yellow, orange, with red scattered least.) Therefore, when we observe a star through intervening gas and dust, we are

a

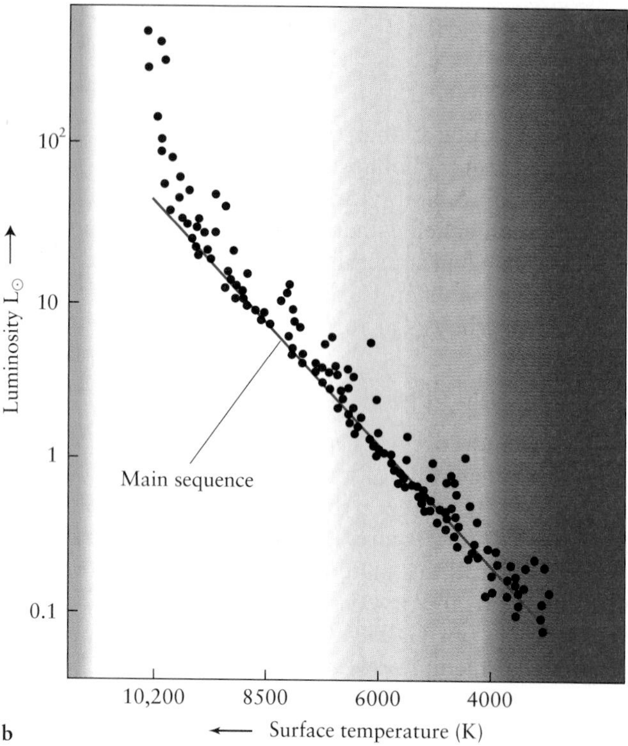

b

R I V U X G

FIGURE 12-3 The Pleiades (a) This open cluster called the Pleiades can easily be seen with the naked eye in the constellation Taurus (the Bull). It lies about 375 ly (116 pc) from Earth. The stars are not shedding mass, unlike the star in Figure 12-14a. The blue glow surrounding the stars of the Pleiades is a reflection nebula created as some of the stars' radiation scatters off preexisting dust grains in their vicinity. (b, used later in this chapter) Each dot plotted on this H-R diagram represents a star in the Pleiades whose luminosity and surface temperature have been determined. Note that most of the cool, low-mass stars have arrived at the main sequence, indicating that hydrogen fusion has begun in their cores. The cluster has a diameter of about 5 ly, is about 100 million years old, and contains about 500 stars. (a: Anglo-Australian Observatory)

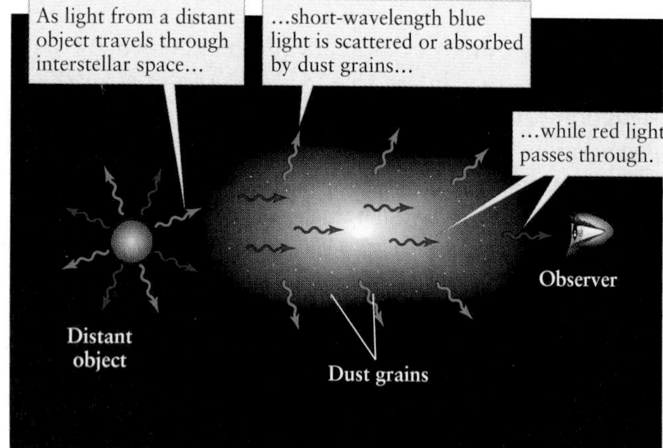

a How dust causes interstellar reddening

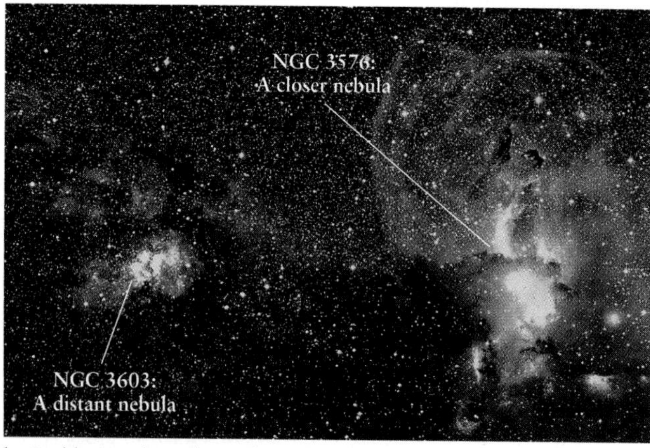

b Reddening depends on distance

R I V U X G

FIGURE 12-4 Interstellar Reddening (a) Dust in interstellar space scatters more short-wavelength (blue) light passing through it than longer-wavelength colors. Therefore, stars and other objects seen through interstellar clouds appear redder than they would otherwise. **(b)** Light from these two nebulae pass through different amounts of interstellar dust and therefore they appear to have different colors. Because NGC 3603 is farther away, it appears a ruddier shade of red than does NGC 3576. (Anglo-Australian Observatory)

seeing less blue short-wavelength light from it than we would if the cloud were not there (Figure 12-4a). Indeed, the more interstellar medium between them and us, the redder the objects appear (Figure 12-4b). Interstellar reddening is different from reddening due to the Doppler shift (see Section 3-2). The Doppler shift causes all wavelengths of electromagnetic radiation to lengthen equally, while interstellar reddening, due to the stronger scattering of shorter wavelengths, does not change the wavelengths of the starlight we receive—only their intensities.

We have been able to discover distant stars and interstellar clouds whose visible light is obscured by the interstellar medium because the distant objects emit radio or infrared photons, and these are scattered relatively little by intervening gas and dust. Therefore, we search for the interstellar medium (and otherwise invisible stars) primarily in the radio and infrared parts of the electromagnetic spectrum. Likewise, we use this technique to find nearby interstellar gas and dust that is too cold to emit much visible light.

Stars are formed from gas and dust that is sufficiently cool so that it can collapse together. The interstellar hydrogen that participates in this process is mostly in the form of molecular (rather than atomic) hydrogen. Molecular hydrogen is relatively hard to detect in space. Therefore, radio astronomers often search instead for carbon monoxide, which emits lots of photons at a wavelength of 2.6 mm. Calculations based on the known abundances of elements reveal that there are about 10,000 hydrogen molecules (H_2) for every CO molecule in a typical interstellar cloud. Consequently, wherever astronomers detect strong emission of CO, they deduce that an enormous amount of hydrogen gas must also be present.

In mapping the locations of CO emission (Figure 12-5a), astronomers came to realize that vast amounts of interstellar gas and dust are concentrated in **giant molecular clouds.** In some cases, these regions appear as dark areas, such as Orion's famous Horsehead Nebula (Figure 12-5c) silhouetted against a glowing background light. In other cases, the clouds appear as dark blobs that obscure the background stars

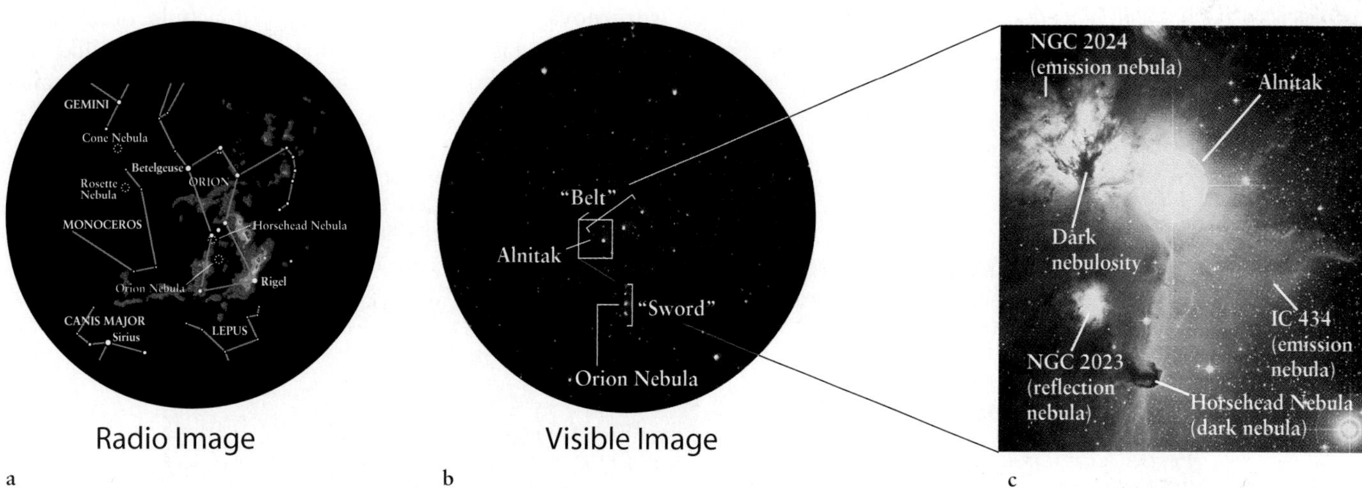

R I V U X G

FIGURE 12-5 **A Gas- and Dust-Rich Region of Orion** (a) This color-coded radio map of a large section of the sky shows the extent of giant molecular clouds in Orion and Monoceros. The intensity of carbon monoxide (CO) emission is displayed by colors in the order of the rainbow, from violet for the weakest to red for the strongest. Black indicates no detectable emission. The locations of four prominent star-forming nebulae are indicated on the star chart overlay. Note that the Orion and Horsehead nebulae are sites of intense CO emission, indicating that stars are forming in these regions. (b, c) A variety of nebulae appear in the sky around Alnitak, also called ζ (zeta) Orionis, the easternmost star in the belt of Orion. To the left of Alnitak is a bright, red emission nebula called NGC 2024. The glowing gases in emission nebulae are excited by ultraviolet radiation from young, massive stars.

R I V U X G

Dust grains obscure part of NGC 2024, giving the appearance of black streaks, while the distinctively shaped dust cloud called the Horsehead Nebula blocks the light from the background nebula IC 434. The Horsehead is part of a larger complex of dark interstellar matter, seen in the lower left of this image. Above and to the left of the Horsehead Nebula is the reflection nebula NGC 2023, whose dust grains scatter blue light from stars between us and it more effectively than any other color. All this nebulosity lies about 1600 ly from Earth, while the star Alnitak is only 815 ly away from us. NGC refers to the New General Catalog of stars and IC stands for Index Catalogs, two supplements to the NGC. (a: R. Maddalena, M. Morris, J. Moscowitz, and P. Thaddeus; b: Royal Observatory, Edinburgh; c: R. C. Mitchell, Central Washington University)

Barnard 86 (dark nebula)

NGC 6520 (star cluster)

R I V U X G

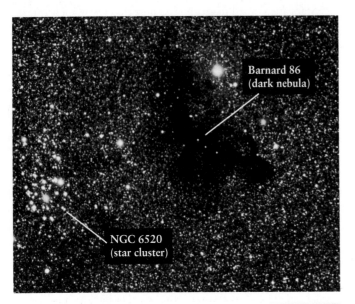

FIGURE 12-6 **A Dark Nebula** This dark nebula, Barnard 86, is located in Sagittarius. It is visible in this photograph simply because it blocks out light from the stars beyond it. The cluster of bluish stars to the left of the dark nebula is a star cluster called NGC 6520. (Anglo-Australian Observatory)

(Figure 12-6). Some 6000 of these molecular clouds are estimated to exist, with masses ranging from 10^5 to 2×10^6 $M_\odot$ and diameters ranging from 50 to 300 ly. The density inside one of these clouds ranges from 10^2 to 10^5 hydrogen molecules per cubic centimeter, several thousand times greater than the average density of the gas and dust dispersed throughout interstellar space, but some 10^{15} times less dense than the air you breathe. Having located interstellar matter, astronomers turned to explaining why some of this gas and dust collapses to form new stars (and planets).

12-2 Supernovae, collisions of interstellar clouds, and starlight trigger new star formation

We will see in detail in Chapter 13 that a *supernova* is a violent detonation that ends the life cycle of a massive star. In a matter of seconds the core of the doomed star collapses, releasing vast quantities of particles and energy that blow the star apart. The star's outer layers are blasted into space at speeds of several thousand kilometers per second.

Astronomers find the ashes (more properly, the gas and 1
dust) of many such dead stars scattered across the sky. These **supernova remnants** are one type of **nebula** or cloud of interstellar gas and dust. Supernova remnants, like the Cygnus Loop shown in Figure 12-7, have a distinctly arched

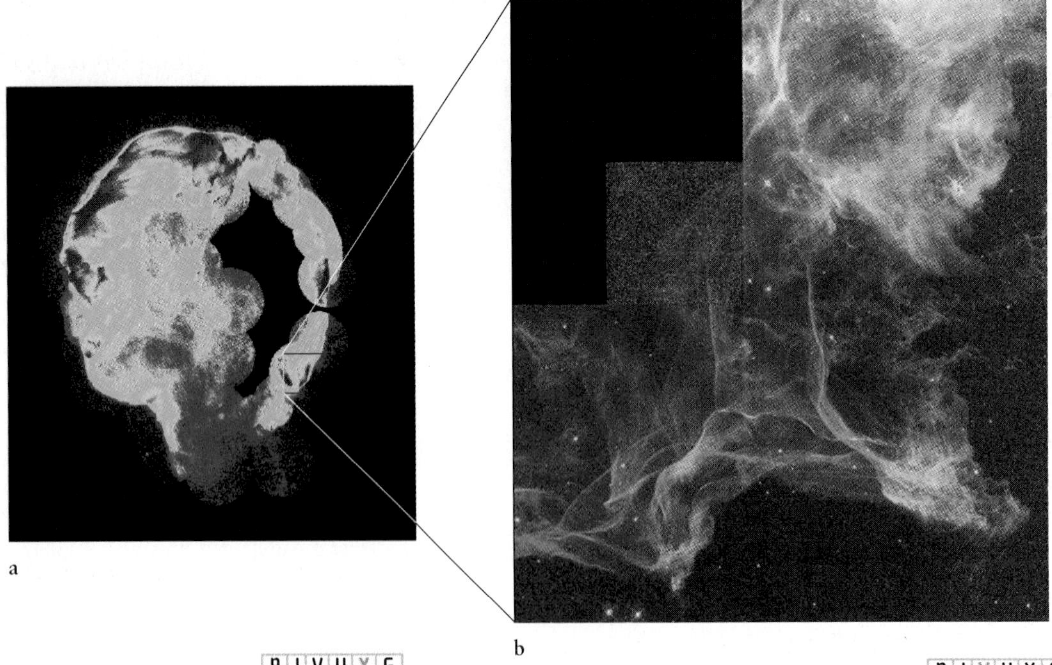

WEB LINK 12.3 FIGURE 12-7 **A Supernova Remnant** (a) X-ray image of the Cygnus Loop, the remnant of a supernova that occurred nearly 20,000 years ago. The expanding spherical shell of gas now has a diameter of about 120 ly. The entire Cygnus Loop has an angular diameter in our sky 6 times wider than the Moon. (b) This visible-light Hubble Space Telescope image of part of the Cygnus Loop shows emission from different atoms: blue from oxygen, red from sulfur, and green from hydrogen. (a: Nancy Levenson/NASA; b: Jeff Hester, Arizona State University and NASA)

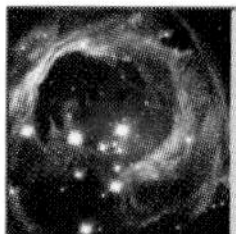

GUIDED DISCOVERY
Observing the Nebulae

Binoculars and the naked eye are enough to let you "get your hands dirty" exploring star dust. Distant nebulae—clusters of stars and glowing gases—are among the most impressive objects in the night sky.

You can observe the Great Nebula of Orion (M42) during winter in the northern hemisphere even with the naked eye. In all likelihood, you have seen it dozens of times without knowing it. To locate the Great Nebula, find the constellation Orion (using, for example, the star charts at the end of the book). Locate Orion's belt. Due south of the belt are three stars in a row making up Orion's sword. Examine the sword very carefully with your naked eye. Do any of the stars in it look at all odd? Now look at them through a pair of binoculars. Which one is different from the others? That one is the Great Nebula in Orion, not a star at all! How does what you see compare with Figure 12-16?

The North America Nebula and the Pelican Nebula in Cygnus are best spotted in autumn. Pick a dark, moonless night and use binoculars rather than a telescope. Higher magnification reveals too small a region of the sky for you to see the entirety of these vast, dim nebulae. To find them, first locate the bright star Deneb on the tail of Cygnus (using, for example, the star charts at the end of the book). The North America Nebula is located 3° east of Deneb, while the Pelican Nebula is located 2° southeast of it. These are both small angles, so sweep around the sky east of Deneb. If your binoculars are powerful enough, you should be able to see the outlines that give these nebulae their names.

The constellations of Orion and Monoceros encompass one of the most accessible regions of the sky for studying star formation and the interaction of young stars with the interstellar medium. Figure 12-5 shows a map of this region made with a radio telescope tuned to a wavelength of 2.6 mm, which is emitted by CO. Note the extensive areas covered by giant molecular clouds. Such comprehensive maps of CO emission help astronomers understand how the large-scale structure of the interstellar medium is related to the formation of stars.

appearance, as would be expected for a shell of gas expanding at supersonic speeds. As it passes through the surrounding interstellar medium, the supernova remnant slams into preexisting matter, exciting the electrons in the atoms and molecules there, causing the gases to glow. If the expanding shell of a supernova remnant rams into a giant molecular cloud, it can cause the cloud to contract, thus stimulating star birth in it. As we learned in Chapter 5, there is evidence that such an event happened around the time the solar system formed. For more about seeing nebulae on your own, see Guided Discovery: Observing the Nebulae.

A simple collision between two interstellar clouds can also create regions sufficiently dense and cool to collapse and form new stars. Likewise, radiation from O and B stars, which are especially bright and hot, will ionize (remove electrons from their orbit) the gas surrounding them, which then expands and compresses the nearby interstellar medium enough to give birth to stars (Figure 12-8).

FIGURE 12-8 The Core of the Rosette Nebula The large, circular Rosette Nebula (NGC 2237) is near one end of a sprawling giant molecular cloud in the constellation Monoceros (the Unicorn). Radiation from young, hot stars has blown gas away from the center of this nebula. Some of this gas has become clumped in dark globules that appear silhouetted against the glowing background gases. New star formation is taking place within these globules. The entire Rosette Nebula has an angular diameter on the sky nearly 3 times that of the Moon, and it lies some 3000 ly from Earth. (Anglo-Australian Observatory)

Once a giant molecular cloud contracts and cools enough, gravitational attraction causes small regions of gas and dust in it to collapse. As these regions become denser, they block light from behind and inside them, becoming dark **Bok globules,** named after astronomer Bart Bok, who studied them. Infrared observations show compact regions of gas and dust called **dense cores** inside Bok globules. These cores are regions of gas and dust destined to become stars.

In order to collapse and form stars, dark cores must be cool, because the hotter a gas, the more its particles collide with each other. The collisions between atoms and molecules create the pressure in the cloud. If the temperature is too high, this pressure overcomes the gravitational attraction between the particles, preventing them from drawing close enough together to form stars.

Dense cores have temperatures of around 10 K, which is sufficiently cold so that the gravitational attraction of the matter in this region overwhelms the pressure there and pulls the gas together to form new stars. This collapse is called a **Jeans instability,** after the British physicist Sir James Jeans, who in 1902 calculated the conditions for it to occur.

Often a giant molecular cloud has several hundred or even thousands of dense cores. In that case, hundreds or thousands of stars form together. Such stellar nurseries will become **open clusters** of stars like the Pleiades (see Figure 12-3). Open clusters are gravitationally unbound systems of stars, meaning that the stars in them eventually drift apart. It is likely that the solar system formed in an open cluster.

At first, a collapsing dense core is just a cool, dusty 2
region thousands of times larger than our solar system. The dense core actually collapses from the inside out. The inner region falls in rapidly, leaving the outer layers of the dense core to drift in at a more leisurely rate. This process of increasing mass in the central region is called *accretion,* and the newly forming object at the center is called a **protostar.** Although fusion has not begun, a protostar emits energy. Some of this energy comes from the compression of the gas it contains as it is pulled inward by the gravitational attraction of this growing mass of hot gas. However, most of the energy it releases comes from infalling gases colliding with the surface of the protostar. Figures 12-9a, b, and c illustrate the protostellar phase and Figure 12-10 is an infrared image showing the locations of myriad protostars in and around a nebula in Centaurus.

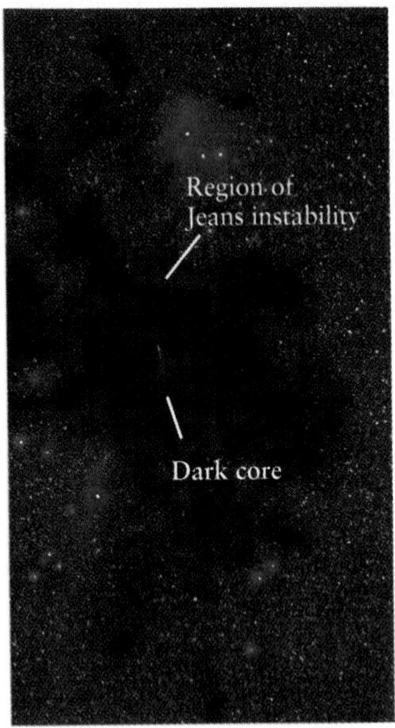

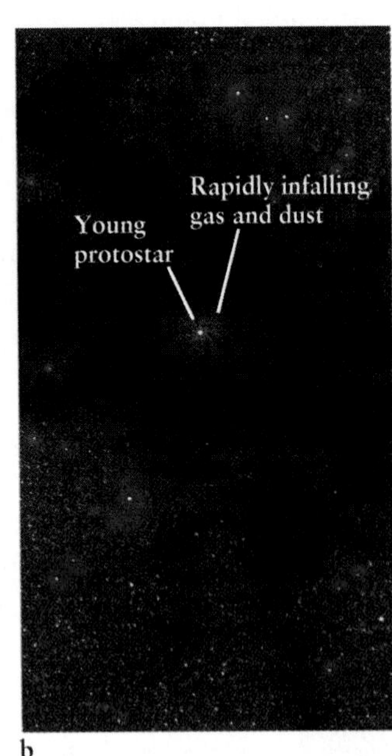

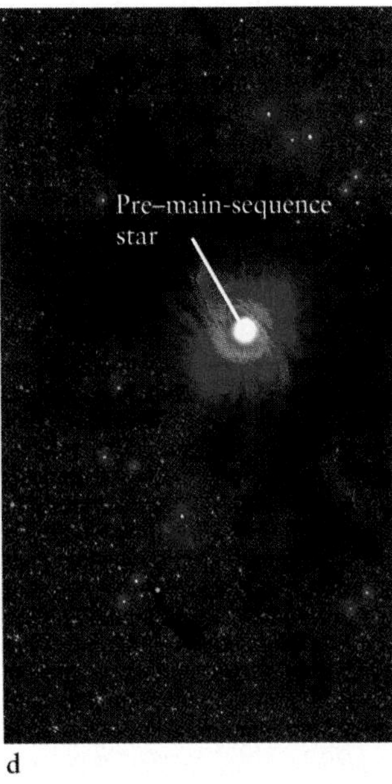

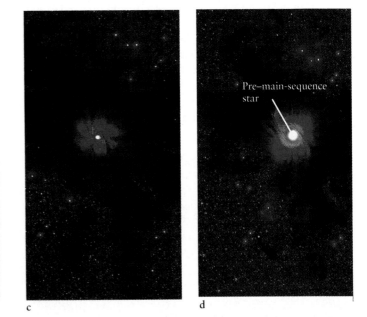

FIGURE 12-9 **Star Formation** (a) The dark region in this drawing is of a dark core a few tenths of a light-year across, the center of which has just developed a Jeans instability. (b) The central gases are heating as they fall freely and slam into the newly forming protostar. The surrounding dark core hides this activity from visible light observations. (c) As the protostar grows in mass, its surface gets brighter, while its core heats up. (d) When little gas is left in the center of the dark core to strike the protostar, its external growth and internal contraction slow significantly and the object becomes a pre–main-sequence star. (Paul DiMare)

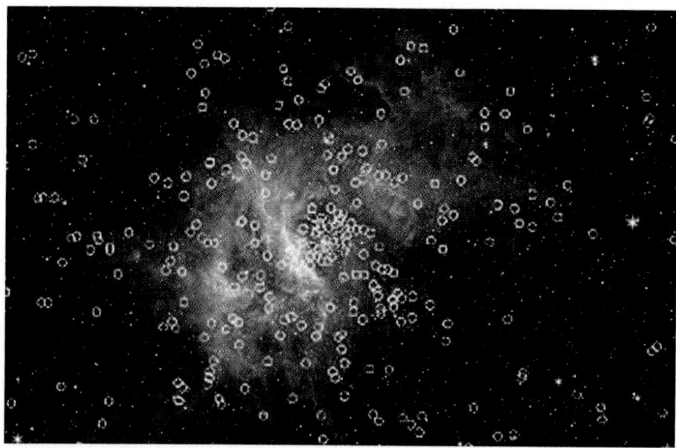

FIGURE 12-10 **A Cluster of Protostars** Protostars (indicated by yellow circles) were observed in the infrared by the Spitzer Space Telescope. This cluster of newly forming stars is 13,700 ly away in the constellation Centaurus. The nebula, some of whose gas is being converted into stars, is called RCW 49 and contains more than 2200 stars and protostars. Most of the interior of this nebula is hidden from our eyes by the dust it contains. The infrared telescope revealed over 300 protostars in it. (NASA/JPL-Caltech/E. Churchwell, University of Wisconsin)

If a dense core is not spinning, it collapses into a sphere, which ultimately becomes an isolated star. If it is spinning, it collapses into a disk, which may then condense into two or three stars. Or, if the disk has a low enough mass, it may become a single star with orbiting protoplanets, as we discussed in Chapter 5.

12-3 When a protostar ceases to accumulate mass, it becomes a pre–main-sequence star

Protostars are physically larger than the main-sequence stars into which they are evolving. A protostar of 1 $M_\odot$, for example, is about 5 times larger in diameter than the Sun. Because of their large sizes, protostars emit large quantities of radiation and gas (analogous to the solar wind), and they can be observed as sources of infrared radiation. At this point, they are not visible to the outside universe in visible light because they are enshrouded by their outer layer of gas and dust. Much matter is still slowly falling inward from the dense core's outer shell. However, the radiation and particles flowing off the protostar exert outward forces on this remaining gas and dust, preventing it from ever reaching the protostar. As a result, mass accretion stops, and the protostar becomes a **pre–main-sequence star** (see Figure 12-9d).

A pre–main-sequence star contracts slowly, unlike the rapid collapse of a protostar. When the temperature at its core reaches 10^7 K, hydrogen fusion begins there. As we saw in Chapter 10, this thermonuclear process releases enormous amounts of energy. The outpouring of energy from hydrogen fusion creates enough pressure inside the pre–main-sequence star to halt its contraction. In the final stages of pre–main-sequence evolution, the outer shell of gas and dust finally dissipates (Figure 12-11,

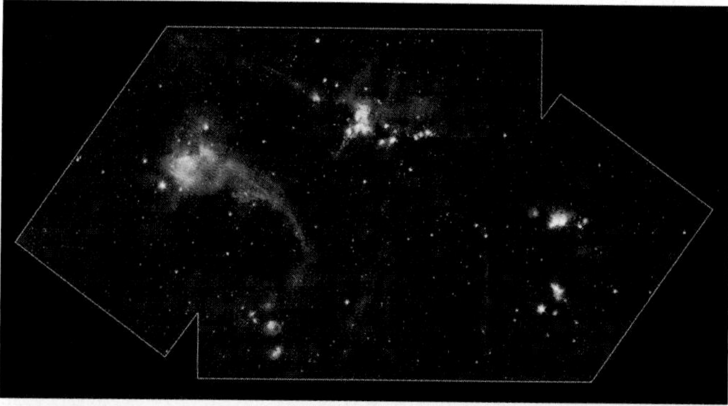
a

b

FIGURE 12-11 **Star-Forming Regions** (a) Surrounded by dust, and therefore virtually invisible to visual telescopes, this star-forming region of Cygnus, denoted DR21, is a strong radio and infrared emitter. The infrared reveals many protostars along with pre–main-sequence and main-sequence stars. The image is about the size of two full Moons. Green denotes areas of carbon monoxide and molecular hydrogen. The red filaments are polycyclic aromatic hydrocarbons. (b) Star formation in nearby galaxy M33, 2.7 million ly away. This region, called NGC 604, is about 1300 times larger than the nearby Orion star-forming region (see Figure 12-16). It is full of hundreds of bright young stars, as well as many dimmer ones. (a: NASA/JPL-Caltech/A. Marston/ESTEC/ESA, A. Noriega-Crespo/SSC/Caltech; b: NASA and the Hubble Heritage Team/AURA/STScI) see also Figure 12-8). For the first time, the star is directly revealed to the outside universe.

12-4 The evolutionary track of a pre–main-sequence star depends on its mass

The more massive a pre–main-sequence star is, the more rapidly it begins hydrogen fusion in its core. For example, calculations indicate that a 5-$M_{\odot}$ pre–main-sequence star starts fusing only 100,000 years after it first forms from a protostar, whereas a 1-$M_{\odot}$ pre–main-sequence star takes a few tens of millions of years to do the same. Stars more massive than about 7 $M_{\odot}$ start fusing so rapidly that they go directly from protostars to the main sequence.

Astrophysicists use computers and the equations of stellar structure (described in Section 10-8) to model the evolution of a pre–main-sequence star. By calculating changes in the energy that the contracting star emits, computer simulations can follow its changing position on a Hertzsprung-Russell diagram (Figure 12-12). Keep in mind that such an **evolutionary track** represents changes in a star's temperature and luminosity, not its motion in space.

Protostars transform into pre–main-sequence stars along a curve called the **birth line** (see blue line on Figure 12-12). A star's exact location on this curve depends primarily on its mass and, to a much smaller extent, on the amount of metal it contains. As a pre–main-sequence star of less than 2 $M_{\odot}$ contracts, its diminishing surface area causes its luminosity to drop significantly. On the H-R diagram, therefore, the track of the star drops below the birth line. Eventually, its surface temperature increases, and the star's track moves to the left in the diagram.

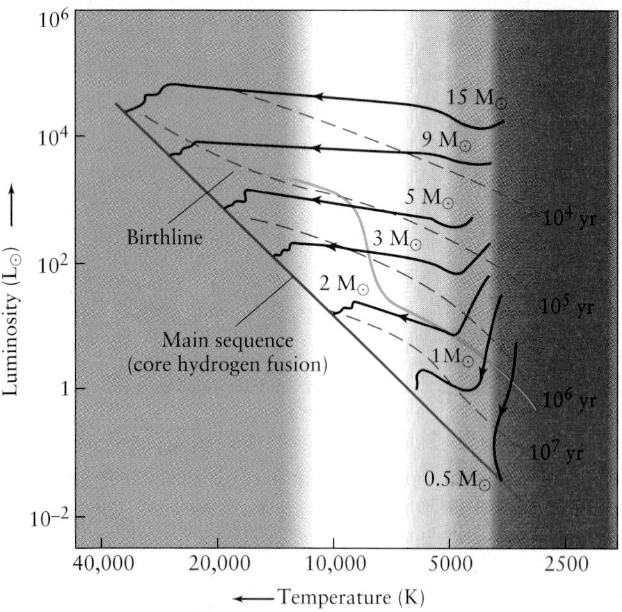

FIGURE 12-12 **Pre–Main-Sequence Evolutionary Tracks** The evolutionary tracks based on models of seven stars having different masses are shown in this H-R diagram. The dashed lines indicate the stage reached after the indicated number of years of evolution. The birth line, shown in blue, is the location where each protostar stops accreting matter and becomes a pre–main-sequence star. Note that all tracks terminate on the main sequence at points agreeing with the mass-luminosity relation.

Spectroscopic observations of pre–main-sequence stars show that many are vigorously ejecting gas just before they reach the main sequence. Gas-ejecting stars in spectral classes G and cooler (that is, G, K, and M) are called **T Tauri stars,** after the first example discovered in the constellation of Taurus. Some astronomers propose that the onset of hydrogen fusion is preceded by vigorous chromospheric activity marked by enormous spicules and flares that propel the star's outermost layers back into space. In fact, an infant star going through its T Tauri stage can lose as much as 0.4 $M_{\odot}$ of matter and also shed its cocoon while still a pre–main-sequence star. In light of all this activity, it is not surprising that observations reveal T Tauri stars to be variable.

Pre–main-sequence stars more massive than 2 $M_{\odot}$ become hotter without much change in overall luminosity. The evolutionary tracks of these pre–main-sequence stars thus traverse the H-R diagram horizontally, from right to left. A star more massive than about 7 $M_{\odot}$ has no pre–main-sequence phase at all. Its gravitational compression is so great that it begins to fuse hydrogen in its protostellar phase.

Calculations reveal that the minimum temperature required to start normal hydrogen fusion (see An Astronomer's Toolbox 10-1) in the core of a star is 10 million K and that pre–main-sequence stars less massive than 0.08 $M_{\odot}$ do not have enough gravitational force compressing and heating their cores to ever get this hot and thereby initiate fusion. Instead, these small bodies contract to become planetlike orbs of hydrogen and helium, called **brown dwarfs** (Figure 12-13). See Guided Discovery: Extrasolar Planets and Brown Dwarfs for further discussion of these important, albeit nonstellar, bodies.

Until 2004, the upper limit to main-sequence stellar mass was thought to be around 120 $M_{\odot}$. This number was based on calculations showing that above this mass, protostars rapidly develop extremely high fusion rates in their cores, which leads to extremely high surface temperatures—temperatures so great that their outer layers are superheated and thereby expelled into interstellar space. This, in turn, lowers their masses and their temperatures. An example of a very massive star in the process of shedding mass as it settles down onto the main sequence is the Pistol Star (Figure 12-14a). One of the most luminous stars in our Galaxy, the Pistol Star may have started its life with as much as 200 $M_{\odot}$. Observations reveal that every few thousand years it expels shells of gas, and it may have less than 10 $M_{\odot}$ left when the expulsion of matter stops. The entire process of mass loss from very massive stars takes only a few million years.

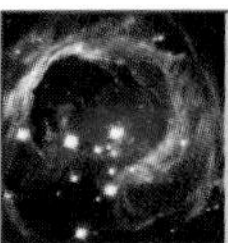

GUIDED DISCOVERY
Extrasolar Planets and Brown Dwarfs

The lowest mass that an object can have and still maintain the fusion of normal hydrogen into helium as occurs in the Sun (see An Astronomer's Toolbox 10-1) is 0.08 $M_{\odot}$ or about 75 times the mass of Jupiter. Astronomers have discovered hundreds of objects in our Galaxy with less than this mass. Like Jupiter, they are primarily composed of hydrogen and helium, with traces of other elements. Many of these objects are found in orbit around stars, while some are found as free-floating masses that apparently formed without ever orbiting a star. An intriguing question has arisen: What should they be called?

One school of thought is that all such low-mass objects orbiting stars should be called *extrasolar planets* and all the isolated ones should be called *brown dwarfs.* A second school of thought is that the distinction between extrasolar planets and brown dwarfs should be based on the fact that while normal hydrogen fusion does not occur in them, those bodies with more than 13 times Jupiter's mass do fuse deuterium (a rare form of hydrogen) into helium and those with more than 60 times Jupiter's mass also fuse lithium (three protons and four neutrons) into helium. Both of these types of fusion occur very briefly in cosmic terms because of the limited supplies of deuterium and lithium in any known object in space. Adherents of this second school of thought hold that all objects below 13 times Jupiter's mass should be considered as extrasolar planets regardless of their location, while all objects with more than this mass should be considered brown dwarfs.

An emerging third school of thought is that all objects above 13 times Jupiter's mass, no matter where they are located, are brown dwarfs. These astronomers suggest that objects orbiting stars with less than this mass be called extrasolar planets, while free-floating bodies with less than this mass be given a third name, perhaps *sub-brown dwarfs.* The debate rages in the literature and at professional meetings. We will hereafter use this third set of definitions.

Because they give off relatively little energy compared to stars, extrasolar planets and brown dwarfs are dim and therefore very challenging to observe. Those in orbit are detected by their gravitational or eclipsing effects on the stars they orbit. The first brown dwarf was discovered only in 1994. Named Gliese 229B, it is located in orbit around a star, Gliese 229A (see Figure 12-13), in the constellation Lepus about 18 ly (6 pc) from Earth. Since then, over 100 more brown dwarfs have been found, along with more than a dozen sub-brown dwarfs. Many of these are found in active star-forming regions, such as the Orion Nebula (see Figure 12-16) and the rho Ophiuchi cloud (see accompanying figure). Astronomers have also found more than 120 extrasolar planets. In 2002, astronomers observed clouds and storms on a brown dwarf similar to, but probably much larger than, the storms observed on the giant planets in our solar system.

Brown dwarfs have the interesting feature that when they fuse deuterium or lithium (in the higher mass ones), the helium they create moves upward, out of the core where it is formed. This helium is replaced with fresh deuterium or lithium fuel to fuse. The upward motion of the helium and downward motion of deuterium and lithium-rich hydrogen are due to convection, and as a result of this motion, eventually all the deuterium and lithium are consumed. Like the lowest-mass main-sequence stars, these brown dwarfs are *fully convective.* This is different behavior than we find in the Sun (see Chapter 10), which has a separate core, convective zone, and radiative zone that do not share atoms. Flares have been observed from brown dwarfs. By analogy to the Sun's flares caused by magnetic fields emerging from its surface, astronomers believe that some brown dwarfs rotate and have magnetic fields.

Based on the numbers and locations of the known brown dwarfs and sub-brown dwarfs, astronomers estimate that there may be as many of these bodies in our Milky Way Galaxy as there are stars. Even in these numbers, brown dwarfs do not contribute a substantial amount of mass or gravitational force in the Galaxy because they have such low individual masses.

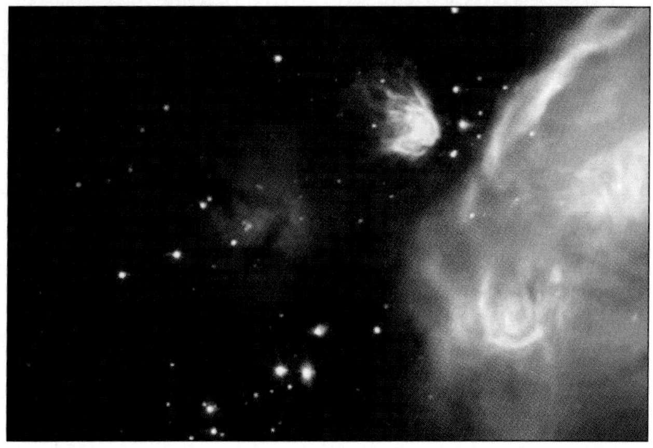

R I V U X G

A Stellar Nursery Full of Brown Dwarfs Besides containing more than 100 young stars, the rho Ophiuchi cloud, located 540 ly away in the constellation Ophiuchus, contains at least 30 brown dwarfs. By studying these objects, astronomers expect to learn more about early stellar evolution. This infrared image is color coded, with red indicating 7.7 micrometer radiation and blue indicating 14.5 micrometer radiation. (Infrared Space Observatory, NASA)

◄ FIGURE 12-13 **A Brown Dwarf** Located 18 ly (6 pc) from Earth in the constellation Lepus (the Hare), Gliese 229B (the smaller object in the image) was the first confirmed brown dwarf ever observed. With a surface temperature of about 1000 K, its spectrum is similar to that of Jupiter. Gliese 229B is in a binary star system. The overexposed image of part of its companion, Gliese 229A, is seen on the left. The two stars are separated by about 43 AU. Gliese 229B has from 20 to 50 times the mass of Jupiter, but the brown dwarf is compressed to the same size as our giant planet. The spike of light was produced when Gliese 229A overloaded part of the Hubble Space Telescope's electronics. (S. Kularni, California Institute of Technology; D. Golimowski, Johns Hopkins University; NASA)

In 2004, astronomers observed a star that apparently has 150 $M_\odot$ (Figure 12-14b). If this observation withstands the test of time, it will require a rethinking of the dynamics of very high mass stars.

12-5 H II regions harbor young star clusters

We can detect the formation of a cluster of stars as a magnificent glow in the nebula. Figure 12-5 and Figure 12-15 show these **emission nebulae.** Because these nebulae are predominantly ionized hydrogen, they are also called **H II**

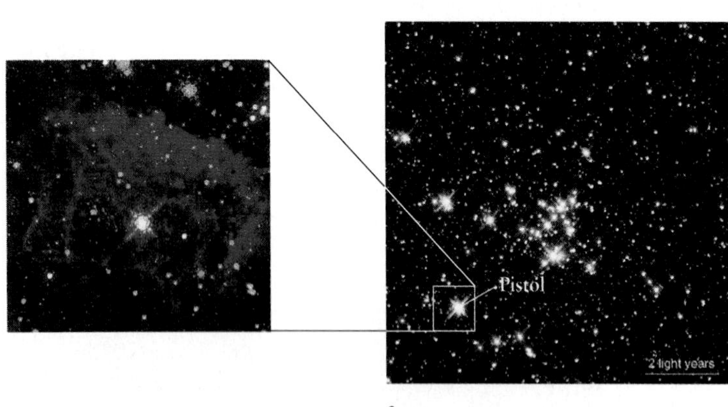

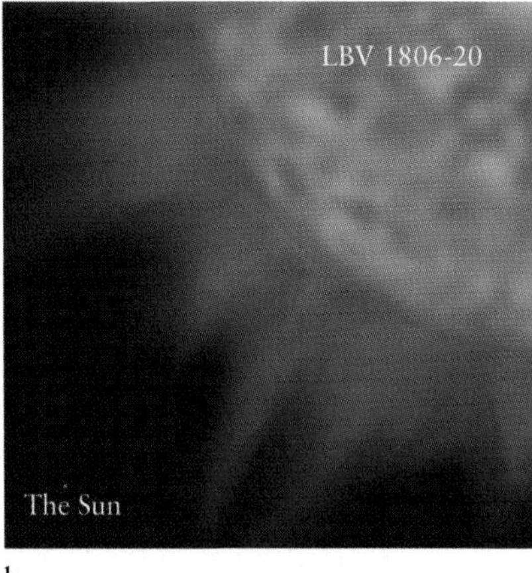

FIGURE 12-14 **Mass Loss from a Supermassive Star** **(a)** The Quintuplet Cluster is 25,000 ly from Earth. **Inset:** Within the cluster is one of the brightest known stars, called the Pistol. Astronomers calculate that the Pistol formed nearly 3 million years ago and originally had 100–200 $M_\odot$. The structure of the gas cloud suggests the star ejected the gas we see in two episodes 6000 and 4000 years ago. The gas from any previous ejections is so thinly spread now that we cannot see it. The nebula shown in the inset is more than 4 ly (1.25 pc) across—it would stretch from the Sun nearly to the closest star, Proxima Centauri. The image of the Quintuplet Cluster was taken in the infrared. The name Pistol was given to the star based on early, low-resolution radio images of its gas, which initially looked like an old-fashioned pistol aimed to the left near the top of the inset. **(b)** The largest, most massive known star, LBV 1806-20, is 5 million times brighter and apparently some 150 times more massive than the Sun. This drawing shows the star's color and its size compared to the Sun. (a: D. Filger, NASA; b: Image by Dr. Stephen Eikenberry, Meghan Kennedy/University of Florida)

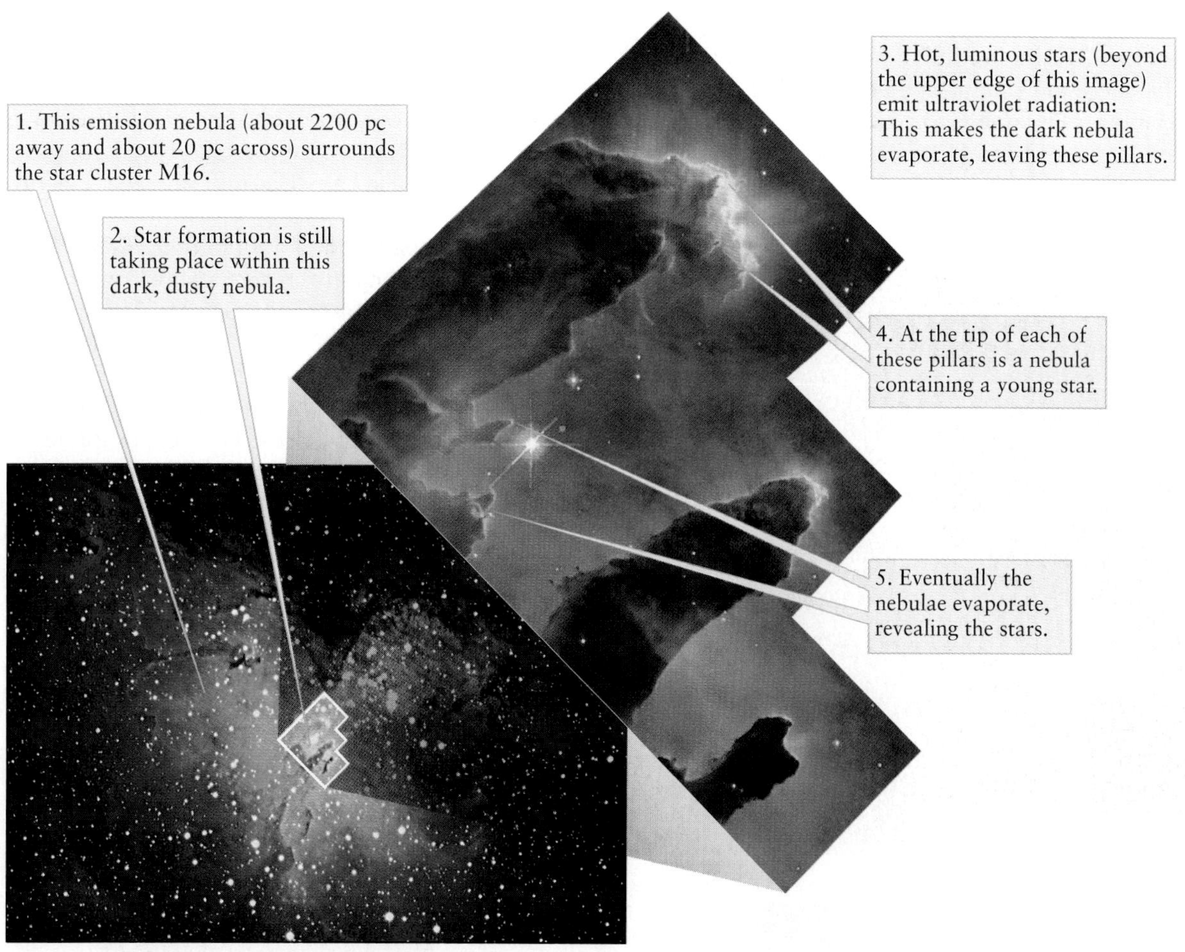

1. This emission nebula (about 2200 pc away and about 20 pc across) surrounds the star cluster M16.
2. Star formation is still taking place within this dark, dusty nebula.
3. Hot, luminous stars (beyond the upper edge of this image) emit ultraviolet radiation: This makes the dark nebula evaporate, leaving these pillars.
4. At the tip of each of these pillars is a nebula containing a young star.
5. Eventually the nebulae evaporate, revealing the stars.
R I V U X G

FIGURE 12-15 **An H II Region** This emission nebula, M16, called the Eagle Nebula because of its shape, surrounds a star cluster. Star formation is presently occurring in M16, which is located 7000 ly from Earth in the constellation of Serpens Cauda (the Serpent's Tail). Several bright, hot O and B stars are responsible for the ionizing radiation that causes the gases to glow. **Inset:** Star formation is occurring inside these dark pillars of gas and dust. Intense ultraviolet radiation from existing massive stars off to the right of this image is evaporating the dense cores in the pillars, thereby prematurely terminating star formation there. Newly revealed stars are visible at the tips of the columns. (Anglo-Australian Observatory; J. Hester and P. Scowen, Arizona State University; NASA)

regions. To see why H II regions exist, remember that most massive pre–main-sequence stars, those of spectral types O and B, are exceptionally hot. Because their surface temperatures are typically 15,000 to 35,000 K, they emit vast quantities of ultraviolet radiation. This energetic radiation easily ionizes any surrounding hydrogen gas, thereby creating an H II region. Photons from an O5 star can ionize hydrogen atoms up to 500 ly away.

But since H II denotes ionized hydrogen, how do we observe it? While some hydrogen atoms in the H II regions are being knocked apart by ultraviolet photons, some of the free protons and electrons manage to get back together. As these new hydrogen atoms assemble, their electrons return to their ground state ($n = 1$). This downward cascade through each atom's energy levels is what makes the nebula glow. Particularly prominent is the transition from $n = 3$ to $n = 2$, which produces H_α photons at 656 nm in the red portion of the visible spectrum (review the emission line spectrum in Figure 4-10). Thus, the nebula around a newborn star cluster often shines with a distinctive reddish hue (see Figure 12-15). Nebulae often look green when seen with the eye through a telescope because they contain oxygen gas, which has a green emission line at 501 nm. Because the eye is more sensitive to green light than red, the dimmer oxygen emission line appears brighter in our brains than does the H_α line, which shows up well on CCD images.

An H II region is a small, bright "hot spot" in a giant molecular cloud. The collection of hot, bright O and B stars that produces the ionizing ultraviolet radiation is called an **OB association.** The famous Orion Nebula (Figure 12-16) is an example. Four O and B stars at the heart of the Orion Nebula are responsible for the ionizing radiation that causes the surrounding gases to glow. The Orion Nebula is embedded in a giant molecular cloud whose mass is estimated at 500,000 $M_\odot$.

The OB association creating an H II region also affects the rest of the giant molecular cloud (see Figure 12-16 insets). Detailed models indicate that vigorous stellar winds, along with ionizing ultraviolet radiation from the O and B stars, carve out a cavity in the cloud. Where this outflow is supersonic, it creates a shock wave, like the sonic boom created by fast-flying aircraft. The shock wave forms along the outer edge of the expanding H II region, compressing hydrogen gas as it passes and thereby stimulating a new round of star birth. As more O and B stars form, they power the expansion of the H II region still farther into the giant molecular cloud. Meanwhile, the older O and B stars left behind begin to disperse (Figure 12-17). In this way, an OB association "eats into" a giant molecular cloud, creating stars in its wake.

To test the theory of star birth just described, astronomers have looked for protostars adjacent to stars in an OB association. Figure 12-16 shows star formation in the core of the Orion Nebula. Visible light observations easily reveal four O and B stars amid the glowing gas and dust, but

FIGURE 12-16 The Orion Nebula The middle "star" in Orion's sword is actually the Orion Nebula, a part of a huge system of interstellar gas and dust in which new stars are now forming. The Orion Nebula is a region visible to the naked eye. It is 1600 ly (490 pc) from Earth and has a diameter of roughly 16 ly (5 pc). This nebula's mass is about 300 $M_\odot$. **Inset (left):** This view at visible wavelengths shows the inner regions of the Orion Nebula. At the lower left are four massive stars, called the Trapezium, which cause the nebula to glow. **Inset (right):** This view shows that infrared radiation penetrates interstellar dust that absorbs visible photons. Numerous infrared objects, many of which are stars in the early stages of formation, can be seen, along with shock waves caused by matter flowing out of protostars faster than the speed of sound waves in the nebula. Shock waves from the Trapezium may have helped trigger the formation of the protostars in this view. (ESO—European Southern Observatory; inset (left): C. R. O'Dell, S. K. Wong, and NASA; inset (right): R. Thompson, M. Rieke, G. Schneider, S. Stolovy, E. Erickson, D. Axon, and NASA)

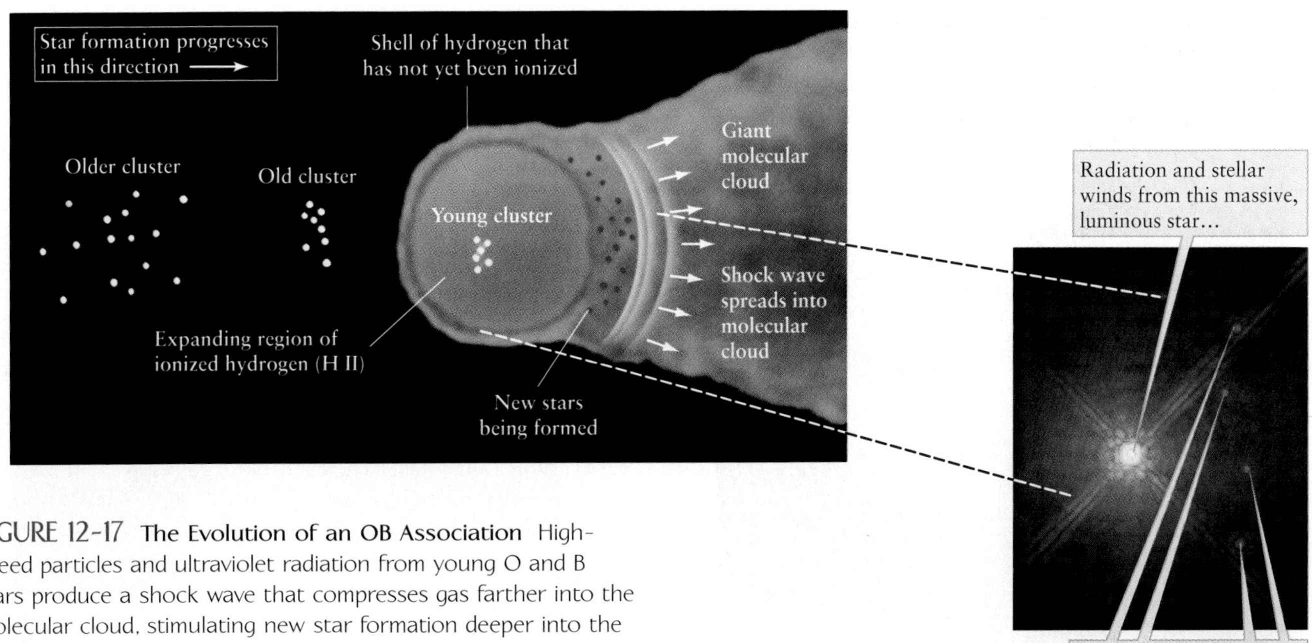

FIGURE 12-17 The Evolution of an OB Association High-speed particles and ultraviolet radiation from young O and B stars produce a shock wave that compresses gas farther into the molecular cloud, stimulating new star formation deeper into the cloud. Meanwhile, older stars are left behind. **Inset:** Stars forming around a massive star 2500 ly (770 pc) away in the constellation Monoceros's Cone Nebula. The six stars (small dots on the right side of the inset) arrayed around the bright, massive central star are believed to have formed as a result of the central star compressing surrounding gas with high-speed particles and radiation. The younger stars are just 0.04–0.08 ly from the central star. (Adapted from C. Lada, L. Blitz, and B. Elmegreen; inset: R. Thompson, M. Rieke, G. Schneider, and NASA)

infrared observations of the same region are showing us cocoons of warm dust still enveloping pre–main-sequence stars. The inset in Figure 12-17 shows star formation around a single O star.

12-6 Plotting a star cluster on an H-R diagram reveals its age

As noted in Section 12-2, stars are observed to form in open clusters, such as seen in the nearby Orion Nebula (see Figure 12-16). Numerous other star-forming regions have been identified and the young open clusters in them offer astronomers a rich source of information about stars in their infancy. By measuring each star's apparent magnitude, color, and distance, an astronomer can deduce its luminosity and surface temperature. The data for all the stars in the cluster can then be plotted on an H-R diagram, as shown for Pleiades in Figure 12-3 or for the cluster NGC 2264 in Figure 12-18.

Applying the facts that all the stars in a cluster begin forming at the same time and that stars with different masses arrive on the main sequence at different times, astronomers can use the H-R diagram to determine the age of a cluster. For example, note that the hottest stars in NGC 2264 lie on the main sequence. These hot stars, with surface temperatures around 20,000 K, are extremely bright and massive: Their radiation causes the surrounding gases to glow. Most of the stars cooler than about 10,000 K have not yet arrived at the main sequence. These less massive stars, which are in the final stages of pre–main-sequence contraction, are just now beginning to ignite thermonuclear reactions at their centers.

From the H-R diagram we can see which are the lowest-mass stars that have already entered the main sequence. Using this information with theories of stellar evolution reveals that the open cluster in Figure 12-18 is roughly 2 million years old.

In contrast to the H-R diagram for NGC 2264, nearly all the stars in the Pleiades (see Figure 12-3b) have completed their pre–main-sequence stage. The cluster's age is calculated to be about 100 million years, which is how long it takes for the least massive stars to finally begin hydrogen fusion in their cores.

As noted earlier, open clusters, such as the Pleiades and NGC 2264, possess barely enough mass to hold themselves together. A star moving faster than the average speed for the cluster occasionally escapes. This lowers the total gravitational force of the cluster, making it easier for other stars to leave. Astronomers predict that in a few hundred million years, after their stars have separated from each other and mixed with the rest of the stars in the Galaxy, most open clusters cease to exist.

a

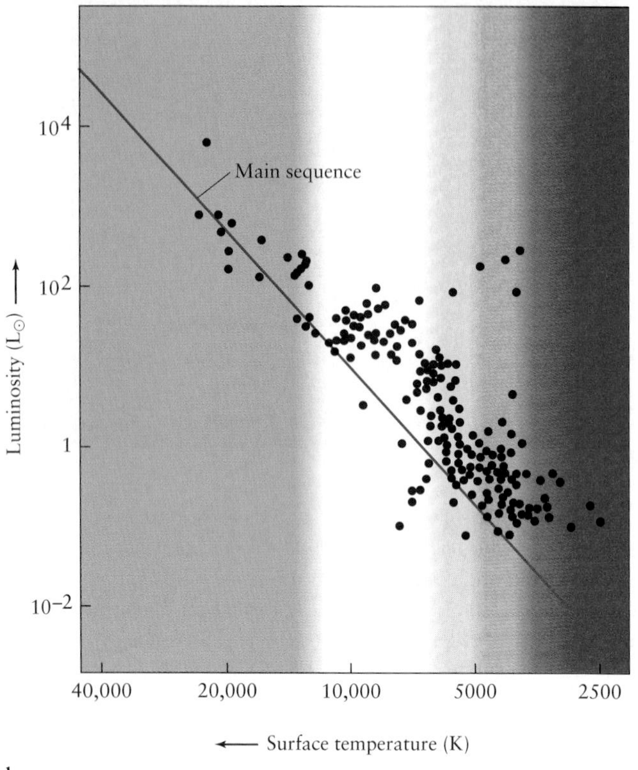

b

R I V U X G

FIGURE 12-18 **Plotting the Ages of Stars** (a) This photograph shows a region of ionized hydrogen and the young star cluster NGC 2264 in the constellation Monoceros. The red nebulosity is located about 2600 ly from Earth and contains numerous stars that are about to begin hydrogen fusion in their cores. (b) Each dot plotted on this H-R diagram represents a star in this cluster whose luminosity and surface temperature have been measured. Note that most of the cool, low-mass stars have not yet arrived at the main sequence. Calculations of stellar evolution indicate that this star cluster started forming about two million years ago. (a: David Malin/Anglo-Australian Observatory)

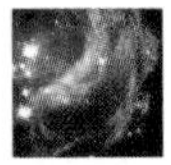

MAIN-SEQUENCE AND GIANT STARS

In the first part of this chapter, we saw how stars form from dense cores of gas and dust inside giant clouds. As these cores collapse, the gas and dust in their centers quickly form protostars. Eventually, protostars have pulled in (accreted) most of the mass available to them and they become pre-main-sequence stars that slowly contract. As a pre–main-sequence star evolves, fusion begins in its core.

Recall from Chapter 10 that in, our Sun, thermal pressure created by fusion in the core pushes outward, balancing the inward force of gravity. Indeed, the thermal pressure outward from the gas everywhere in our Sun is equal to the gravitational force pulling that gas inward and so the Sun neither expands nor collapses. The Sun is said to be in *hydrostatic equilibrium* (see Figure 10-20). When each layer of a pre–main-sequence star can finally support all the layers above it (meaning that collapse ceases), that star also comes into hydrostatic equilibrium and a main-sequence star is born (see Figure 12-12). *Main-sequence stars are those stars in hydrostatic equilibrium with nuclear reactions fusing hydrogen into helium in their cores at a nearly constant rate.*

12-7 Stars spend most of their life cycles on the main sequence

The **zero-age main sequence (ZAMS)** is the set of locations on the H-R diagram where pre–main-sequence stars of different masses first become stable objects, neither shrinking nor expanding. This is the solid red line that appears on several of the H-R diagrams. Note that the evolutionary tracks in Figure 12-12 end at locations along the main sequence that agree with the mass-luminosity relation (recall Figure 11-12): The most massive main-sequence stars are the most luminous, while the least massive stars are the least luminous.

The equations of nuclear fusion predict that the more 3
massive a star is, the faster it evolves and the less time it spends on the main sequence. This occurs because gravity presses down with greater pressure (force per area) on a more massive star's core than on a less massive star's core.

TABLE 12-1 Main-Sequence Lifetimes

Mass ($M_\odot$)	Surface temperature (K)	Luminosity ($L_\odot$)	Time on main sequence (10^6 yrs)	Spectral class
25	35,000	80,000	3	O
15	30,000	10,000	15	B
3	11,000	60	500	A
1.5	7000	5	3,000	F
1.0 (Sun)	6000	1	10,000	G
0.75	5000	0.5	15,000	K
0.50	4000	0.03	200,000	M

This tremendous pressure acting on O and B stars forces them to consume all their core hydrogen in only a few million years, as shown in Table 12-1. Conversely, stars of very low mass take hundreds of billions of years to convert their cores from hydrogen into helium.

The equations reveal that the conversion of hydrogen into helium in every star's core takes a long time compared to any other stage of its stellar evolution. (These stages are presented in the following sections and the next two chapters.) That is why the vast majority of stars represented on an H-R diagram are located on the main sequence. Our theory of solar evolution predicts that the Sun's total lifetime on the main sequence will be about 10 billion years, based on the time it will take to convert all the hydrogen in its core to helium.

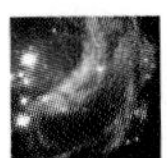

EVOLUTION OF STARS WITH MASSES BETWEEN 0.08 AND 0.4 $M_\odot$

Because they evolve beyond the ZAMS so differently, we explore separately the evolution of stars with less than 0.4 $M_\odot$ and stars with more than 0.4 $M_\odot$.

12-8 Red dwarfs convert essentially their entire mass into helium

The lowest-mass main-sequence stars, called **red dwarfs,** have masses between 0.08 $M_\odot$ and 0.4$M_\odot$. They have the lowest core temperatures and pressures of all stars. Fusion rates depend critically on temperature, and so low temperatures imply slow fusion of hydrogen into helium. As a result, these stars produce the least amount of energy and therefore emit the least energy. They are the coolest and dimmest of all main-sequence stars.

These stars are also different from higher-mass main-sequence stars in that as red dwarfs create helium in their cores, this helium is convected upward and out of the core, while hydrogen from the outer layers moves downward (Figure 12-19). As a result, these stars eventually convert their entire mass into helium.

Fusion in their cores is so slow that red dwarfs remain on the main sequence for hundreds of billions of years. None of them have yet left the main sequence. Our models predict their fates. When such a star has converted its entire mass into helium, fusion will cease inside it because it does not have enough pressure in its core to heat and fuse that helium into anything else. These helium bodies will then

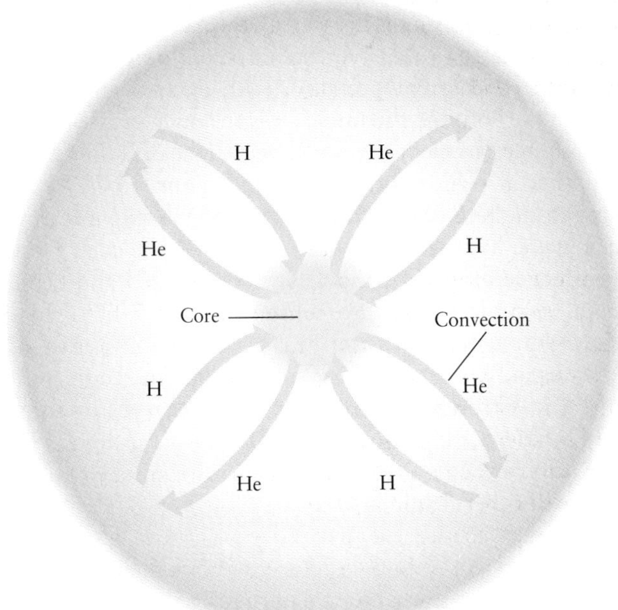

FIGURE 12-19 **Fully Convective Star** This drawing shows how the helium created in the cores of red dwarfs convects into the outer layers of the star, while the hydrogen from the outer layers descend into the core. This process continues until the entire star is helium.

radiate the heat they generated while on the main sequence and thereby become cooler, moving down and to the right from the main sequence.

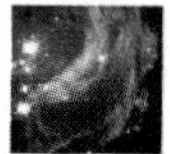

EARLY AND MIDLIFE EVOLUTION OF STARS WITH MORE THAN 0.4 $M_\odot$

The changes in stars with more than $0.4M_\odot$ are much more complicated. In this chapter, we consider what happens to them on the main sequence and into the giant stage following it. In the next two chapters, we explore the ends of their lives.

12-9 When core hydrogen fusion slows down, a main-sequence star with $M > 0.4\ M_\odot$ becomes a giant

Recall that the main sequence is defined by the fusion of hydrogen into helium in a star's core. Unlike lower mass stars, stars with more than $0.4M_\odot$ do not convect helium out of their cores. Therefore, the chemical compositions of their cores and outer layers remain different throughout their lives.

When the hydrogen in the core of a main-sequence star that is more massive than a red dwarf is mostly converted into helium, fusion in the core drastically slows down. The temperature of the core at that time is not high enough to enable its helium to fuse into other elements. However, the star continues to evolve. Let us see how the end of core fusion actually causes these stars to *expand* in size to become *giants.*

Recall from our discussion of the Sun in Chapter 10 that while on the main sequence a star's outer layers are supported by thermal pressure, with energy supplied by fusion in the core. As the rate of this fusion decreases, a star can no longer support the crushing weight of its outer layers, and so it begins compressing its helium core. We will return to the fate of the core shortly, but in the meantime, consider the hydrogen-rich gas just outside the core. Under the influence of gravity, this gas is compressed and heated enough to begin fusing into helium (Figure 12-20). This is called **hydrogen shell fusion** because it occurs in a shell a few thousand kilometers thick surrounding the core. Recall from An Astronomer's Toolbox 10-1 that the proton-proton chain is the major source of energy inside the Sun. This fusion process also occurs in hydrogen shell fusion, along with other fusion activity.

Why then does the star expand to become a giant? Essentially, it begins generating more energy in its shell than it did when it was in hydrostatic equilibrium as a main-sequence star. The extra energy is absorbed by the gases in the outer layers of the star, which are thereby heated more than they had been by energy from hydrogen core fusion. Since heated gas expands, this increase in internal temperature causes the entire star to swell. All the details involved in this transition to the giant phase are not yet understood. When our equations of stellar activity are used to create computer models of stars

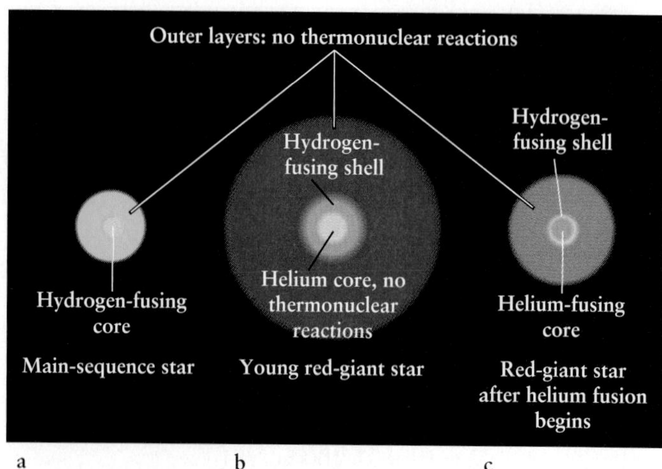

FIGURE 12-20 Evolution of Stars from the Main Sequence (a) Hydrogen fusion occurs in the cores of main-sequence stars. (b) When the core is converted into helium, fusion ceases there and then begins in a shell surrounding the core. The star expands into the giant phase. This newly formed helium sinks into the core, which heats up. (c) Eventually, the core reaches 10^8 K, whereupon core helium fusion begins. This causes the core to expand, slowing the hydrogen shell fusion and thereby forcing the outer layers of the star to contract.

leaving the main sequence, the simulated stars swell up, just like real red giants. However, the equations are so complex that astrophysicists have been unable to isolate all the effects that go into creating red giants.

INSIGHT INTO SCIENCE

More Than the Sum of Its Parts We have gotten to the point in science where some computer models that numerically solve complex sets of equations reproduce phenomena without fully explaining it. For example, computer models of the equations that describe how stars evolve show them expanding onto the giant phase. This gives us some confidence that the equations are meaningful representations of reality. However, the models are so complex that astrophysicists are unable to identify the specific physical effects taking place that cause the stars to expand.

Far from the fusing shell, the surface gases cool and soon the temperature of the star's bloated surface falls to between 3000 and 6000 K, depending on the star's total mass. In about 5 billion years, our Sun will have a helium core and will swell into a giant with a radius of about ½ AU, vaporizing Mercury and perhaps causing Venus to spiral into the Sun. The Earth will be scorched to a cinder.

Giant stars are so enormous that their bloated outer layers constantly leak gases into space. At times, this *mass loss* is

FIGURE 12-21 **A Mass-Loss Star** This unusual star in the constellation Norma (the Carpenter's Square) is the brightest of a system of three stars orbiting each other. This extremely hot, massive star, named HD 148937, is continuously shedding its outer layers. Other vigorous outbursts in the past gave rise to the two symmetric shells of material, called NGC 6164 and NGC 6165, on either side of the star. These shells absorb ultraviolet radiation from the star, causing them to glow with the characteristic red color of excited hydrogen gas. (David Malin/Anglo-Australian Observatory)

significant (Figure 12-21), and it can be detected spectroscopically. The escaping gases exhibit narrow absorption lines, and the lines from gases coming toward us are slightly blueshifted, owing to the Doppler effect. This small shift corresponds to a speed of 10 km/s, typical of the expansion velocities with which gases leave the tenuous outer layers of giants. A typical rate of mass loss for a giant is roughly 10^{-7} $M_{\odot}$ per year. For comparison, in a main-sequence star such as the Sun, mass loss rates are only around 10^{-14} $M_{\odot}$ per year.

Although the surface of a giant is cooler than that of the main-sequence star from which it evolved, the giant is more luminous: It can emit more photons each second because it has so much more surface area. As a full-fledged giant (Figure 12-22), our Sun will shine 2000 times more brightly than it does today.

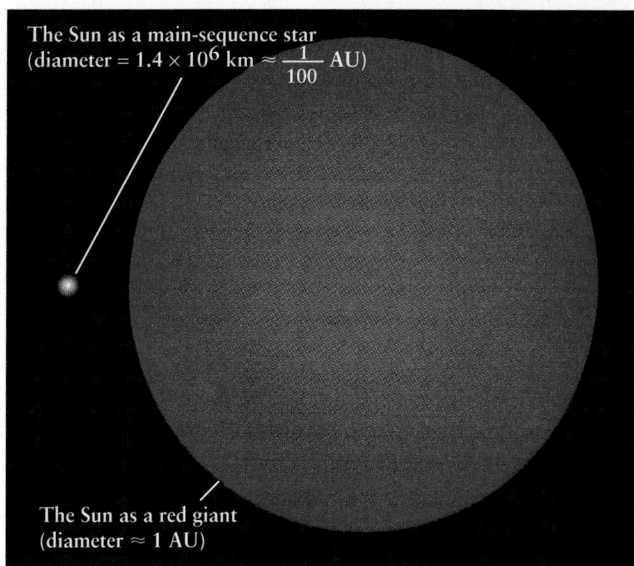

a The Sun today and as a red giant

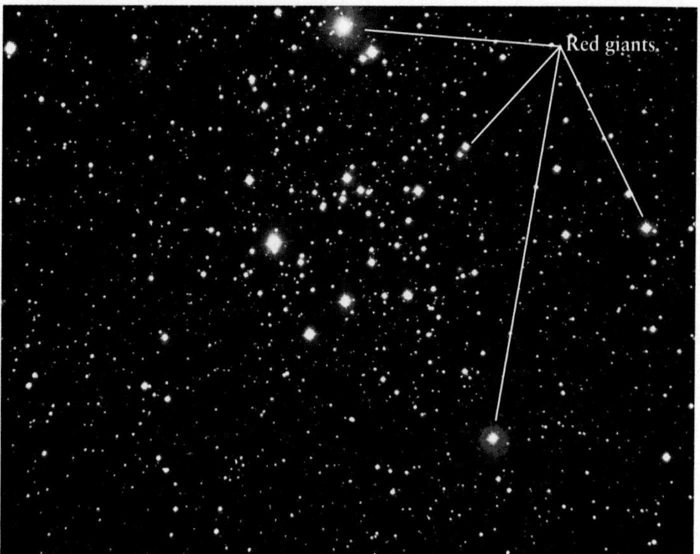

b Red giant stars in the star cluster M50

R I V U X G

FIGURE 12-22 **The Sun Today and as a Giant** (a) In about five billion years, when the Sun expands to become a giant, its diameter will increase a hundredfold from what it is now, while its core becomes more compact. Today, the Sun's energy is produced in a hydrogen-fusing core whose diameter is about 200,000 km. When the Sun becomes a giant, it will draw its energy from a hydrogen-fusing shell surrounding a compact helium-rich core. The helium core will have a diameter of only 30,000 km. The Sun's diameter will be about 100 times bigger, and it will be about 2000 times more luminous as a giant than it is today. (b) This composite of visible and infrared images shows red giant stars in the open cluster M60 in the constellation of Monoceros (the Unicorn). (T. Credner and S. Kohle, Astronomical Institutes of the University of Bonn)

12-10 Helium fusion begins at the center of a giant

When a star first becomes a giant, its hydrogen-fusing shell surrounds a small, compact core of almost pure helium. In a moderately low-mass giant, the dense helium core is about twice the size of the Earth.

At first, no thermonuclear reactions occur in the helium-rich core of a giant because the temperature there is too low to fuse helium nuclei. The hydrogen-fusing shell creates more helium, which becomes part of the core. The ever-more-massive helium core continues to contract and heat up, while the hydrogen-fusing shell surrounding it consumes more of the star's hydrogen.

When the central temperature reaches about 100 million K, **core helium fusion** begins at the giant's center (see Figure 12-20c). The aging star thus has a central energy source for the first time since it left the main sequence. The *triple-alpha process* provides the dominant path for helium in the core to fuse into carbon. (Early in the twentieth century, particles emitted by radioactive nuclei were called alpha particles. They were later identified as helium nuclei.) First, two helium nuclei combine to create beryllium. Then, within 10^{-8} seconds, a third helium nucleus is added to the short-lived beryllium to create carbon. This process also releases energy and can be summarized as follows:

$$^{4}\text{He} + {}^{4}\text{He} + {}^{4}\text{He} \rightarrow {}^{12}\text{C} + \gamma$$

where γ denotes energy emitted as photons. But the fusion in giants does not stop there. Some of the carbon created can then fuse with another helium nucleus to produce oxygen:

$$^{12}\text{C} + {}^{4}\text{He} \rightarrow {}^{16}\text{O} + \gamma$$

Again, energy is released as gamma rays.

A mature giant fuses helium in its core for about 10% of the time that it spent fusing hydrogen as a main-sequence star. For example, about 5 billion years from now, the Sun will start consuming the helium it is now creating in its core, and about a billion years later, that process will be complete. While helium fusion is occurring in a giant's core, further gravitational contraction of the star's core ceases.

In high-mass giant stars, helium fusion begins gradually as temperatures in the star's core approach 100 million K. In lower-mass stars, helium fusion begins explosively and suddenly in an event called the **helium flash.** Depending on whose calculations of stellar evolution you choose to believe, the upper limit to stars that undergo the helium flash is between 2 $M_\odot$ and 4 $M_\odot$ (Table 12-2). For convenience, we will adopt the limit of 2 $M_\odot$ in what follows.

The helium flash is the result of unusual conditions that develop in the core of a lower-mass star after it leaves the main sequence. To appreciate these conditions, we must first understand how an ordinary gas behaves, then explore how the densely packed electrons at the star's center alter this behavior.

When an ordinary gas is compressed, it heats up; when it expands, it cools down. The helium in the core of a young giant star with mass greater than about 2 $M_\odot$ is such a gas. The pressure on the core created by the gravitational force of the star's mass is enough to smoothly compress and heat the core until it reaches the fusion temperature of helium, at which point helium core fusion begins, as just described. If energy production overheats its core, the core expands, cooling the gases and slowing the rate of thermonuclear reactions. Conversely, if too little energy is being created to support the star's overlying layers, they move inward, compressing the core. The resulting increase in temperature speeds up the thermonuclear reactions and thus increases the energy output, which stops the contraction. Either way, the star has a "safety valve" to keep it from collapsing or exploding.

A young giant with less than 2 $M_\odot$ lacks that safety valve because its outer layers do not provide enough inward force (and therefore enough heat) to keep the core as a normal gas. Instead, the lower-mass star squeezes the nuclei in its nonfusing helium core into a crystallike solid. The densely packed electrons that come from these core atoms change their behavior before helium core fusion begins. At the extreme pressures deep inside a giant with less than 2 $M_\odot$, the atoms are completely ionized, separating into nuclei and electrons. The electrons, distributed between the nuclei, are so closely crowded together that another law of physics called the **Pauli exclusion principle** becomes important.

According to the Pauli exclusion principle, first formulated in 1925 by the Austrian physicist Wolfgang Pauli, two identical particles cannot exist in the same place at the same time. As the electrons are pressed closer and closer together by the star's gravitational force, the exclusion principle prevents them from freezing in place. Instead, many of them must vibrate faster and faster so that they do not become "identical," here meaning being in the same place and moving with the same speeds as adjacent electrons.

Astronomers say that electrons in this state are *degenerate* and that the helium-rich core of a low-mass giant is

TABLE 12-2 How Helium Core Fusion Begins in Different Red Giants

Mass of star	Onset of helium burning in core
Less than 2–4 solar masses	Explosive (helium flash)
More than 2–4 solar masses	Gradual

supported by **electron degeneracy pressure,** which provides a greater outward pressure than did the normal pressure in the star before it became degenerate. Therefore, electron degeneracy pressure prevents the core from collapsing further. The solid core just heats up under the gravitational influence of the star's mass and the photons generated by the hydrogen shell fusion occurring around it.

The equations describing degenerate matter such as the electrons in this star predict that *unlike the pressure of an ordinary gas, degeneracy pressure does not change with temperature.* This means that as the core's temperature grows, the pressure in the core does not increase. Without the "safety valve" of increasing pressure, the star's core cannot expand and cool.

Eventually, the core temperature reaches about 10^8 K, the point where helium begins to fuse. This fusion creates energy, so the core's temperature goes up, but the degenerate electrons still do not increase the pressure. Therefore, the core does not immediately expand, and for several hours both the core temperature and the fusion rate rise dramatically. This is the helium flash.

During the helium flash, temperatures become so high (around 3.5×10^8 K) that the helium again becomes an ordinary gas. Suddenly, the usual safety valve operates once again, namely that the high temperature raises the core pressure, causing the core to expand and cool. Within a matter of hours, the fusion rate drops. This decrease in the star's energy output causes its outer layers to again contract until the star gets into a new hydrostatic equilibrium. A low-mass, helium-fusing giant is left smaller, dimmer, and hotter than it was before it began fusing helium.

12-11 As stars evolve, their positions on the H-R diagram shift

Recalling that astrophysicists developed models of stellar evolution from the data on Hertzsprung-Russell diagrams, it is useful to plot on an H-R diagram the theoretical evolutionary tracks of post–main-sequence stars that these models predict, to see if the stars calculated from the models evolve in places on the diagram where stars are actually observed. If the luminosities and surface temperatures that the models predict are not consistent with the observed properties of giant stars, then the models need refinement. Indeed, such plots led to refined models that today do predict most of the observed properties of stars.

A Hertzsprung-Russell diagram (Figure 12-23) shows us how a star's brightness and temperature evolve from the time it arrives on the main sequence as a zero-age main-sequence star. As it ages, its core is being converted into helium. This and related internal changes cause its surface brightness and temperature to change, so its track slowly inches away from its ZAMS location. The dashed line in Figure 12-23 shows the locations of the stars when all their core hydrogen has been consumed. As we saw in Section 12-7, it takes a few million years for the most massive main-sequence stars to reach this point, while a 1-$M_\odot$ star takes 10 billion years. Calculations indicate that lower-mass stars take tens or even hundreds of billions of years to get there. Because, as we will see, the universe is roughly 14 billion years old, no main-sequence stars with masses less than about 0.75 $M_\odot$ have yet moved into the giant stage.

After core hydrogen fusion ceases, the points representing high-mass stars move rapidly from left to right across the H-R diagram. This is when each star's core begins to contract and its outer layers expand. Although the stars' surface temperatures are decreasing, their surface areas are increasing. Therefore, their overall luminosities remain roughly constant.

Just before core helium fusion begins, the evolutionary tracks of high-mass stars turn upward into the giant region of the H-R diagram. After the core helium fusion begins, however, the evolutionary tracks back away from these peak luminosities. The tracks then wander back and forth in the giant region while the stars readjust to their new energy sources.

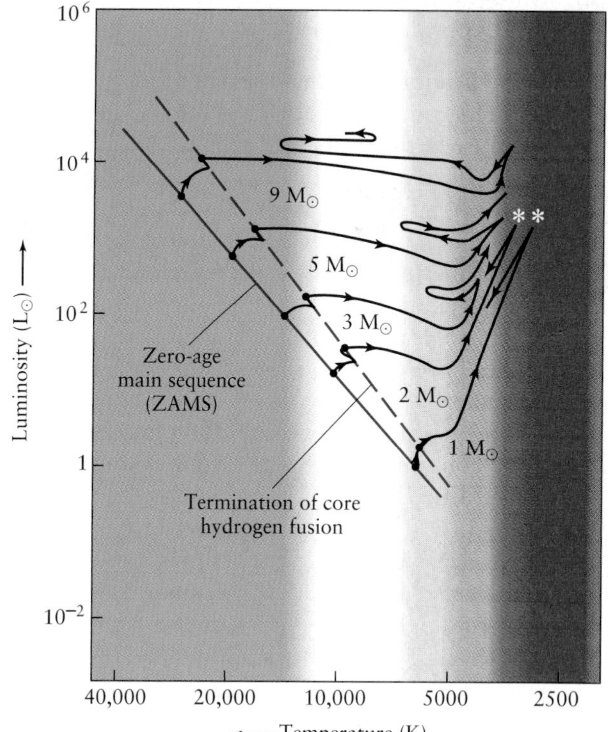

ANIMATION 12.2

FIGURE 12-23 Post-Main-Sequence Evolution Model-based evolutionary tracks of five stars are shown on this H-R diagram. In the high-mass stars, core helium fusion ignites smoothly where the evolutionary tracks make a sharp turn upward into the giant region of the diagram. The white asterisks on the 1-$M_\odot$ and 2-$M_\odot$ curves indicate where the helium flash in these stars occurs.

FIGURE 12-24 **A Globular Cluster** A globular cluster is a spherical cluster that typically contains up to a million stars within a region about 70 ly across. This cluster, called M10, is located in the constellation Ophiuchus (the Serpent Holder), roughly 16,000 ly from Earth. Most of the stars here are either red giants or blue, horizontal-branch stars with both core helium fusion and hydrogen shell fusion. (T. Credner and S. Kohle, Astronomical Institutes of the University of Bonn)

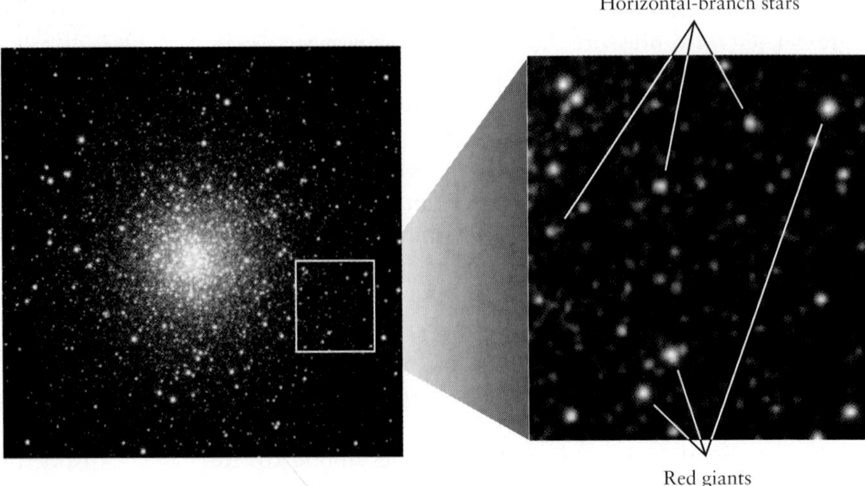

We saw that after the helium flash, lower-mass stars become dimmer but hotter. On the H-R diagram in Figure 12-23, note that the two lower-mass stars move down and to the left.

12-12 Globular clusters are bound groups of old stars

WEB LINK 12.8

Post–helium-flash stars are often found in old star clusters, called **globular clusters,** so named because of their spherical shapes. A typical globular cluster, like the one shown in Figure 12-24, may contain up to a million stars in a volume 300 ly across. Like open clusters, the stars in globular clusters all form at about the same time. Unlike open clusters of young stars, globular clusters are gravitationally bound groups of stars from which few stars escape. Globular clusters in our Galaxy (other galaxies have them, too) are distributed in a sphere around the center of the Milky Way.

Astronomers know that globular clusters are old because they contain no high-mass main-sequence stars. If you measure the luminosity and surface temperature of many stars in a globular cluster and plot the data on a color-magnitude diagram, as shown in Figure 12-25, you will find that the upper half of the main sequence is missing. All the high-mass main-sequence stars evolved long ago into giants, leaving behind only lower-mass, slowly evolving stars still undergoing core hydrogen fusion.

The H-R diagram of a globular cluster typically shows a horizontal grouping of stars to the left of the center portion of the diagram (see Figure 12-25). These hot stars, called **horizontal branch stars,** are post–helium-flash stars, and they have luminosities of about 50 $L_\odot$ (see Figure 12-25). Eventually, these stars will move back toward the giant region as core helium fusion and hydrogen shell fusion devour their fuel.

As noted earlier, an H-R diagram of a cluster can also be used to determine its age, assuming that all of the stars in the cluster formed at the same time. In the H-R diagram for a young open cluster (review Figure 12-18b), almost all the stars are on the main sequence. As a cluster gets older, like the Pleiades, stars begin to leave the main sequence (see Figure 12-3). The high-mass, high-luminosity stars are the first to become giants.

Over time, the main sequence in a cluster gets shorter and shorter. The top of the surviving portion of the main

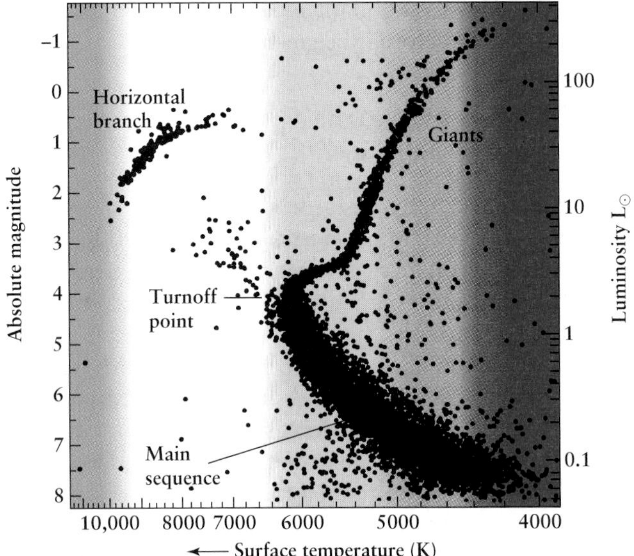

FIGURE 12-25 **An H-R Diagram of a Globular Cluster** Each dot on this graph represents the absolute magnitude and surface temperature of a star in the globular cluster called M55. Note that the upper half of the main sequence is missing. The horizontal branch stars are stars that recently experienced the helium flash in their cores and now exhibit core helium fusion and hydrogen shell fusion.

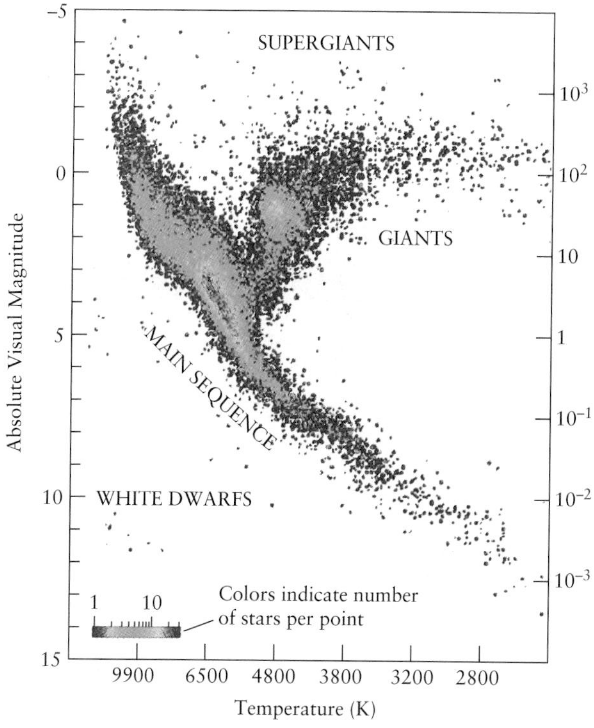

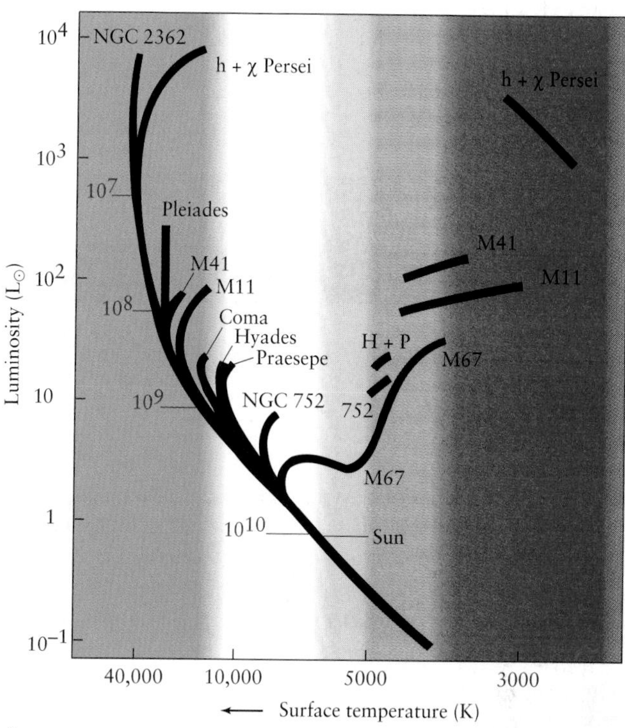

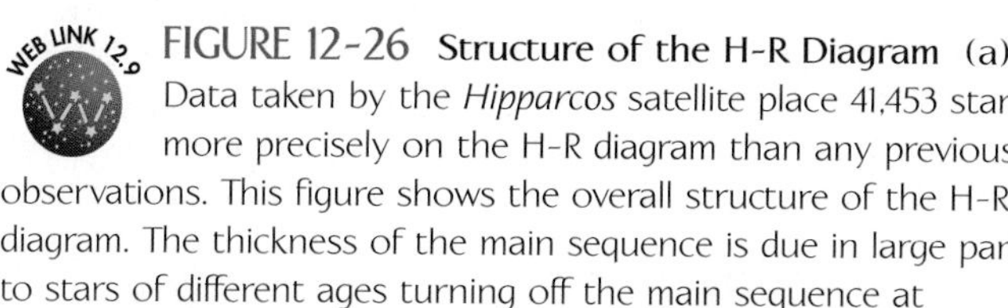

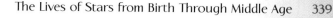

FIGURE 12-26 **Structure of the H-R Diagram (a)** Data taken by the *Hipparcos* satellite place 41,453 stars more precisely on the H-R diagram than any previous observations. This figure shows the overall structure of the H-R diagram. The thickness of the main sequence is due in large part to stars of different ages turning off the main sequence at different places, as shown in (b). **(b)** The black bands indicate where data from various star clusters fall on the H-R diagram. The ages of turnoff points (in years) are listed in red alongside the main sequence. The age of a cluster can be estimated from the location of the turnoff point, where the cluster's most massive stars are just now leaving the main sequence.

sequence is called the *turnoff point* (see Figure 12-25). Stars at the turnoff point are just beginning to exhaust the hydrogen in their cores. The theory of stellar evolution enables us to predict how long stars with different masses last on the main sequence before moving toward the giant stage. Therefore, the mass of the stars currently at the turnoff point is used to determine the age of the cluster (recall Table 12-1).

Consider the plot for the globular cluster M55 shown in Figure 12-25 (so named because it was fifty-fifth in the *Messier Catalogue* of astronomical objects). In M55, 0.8-$M_\odot$ stars are just leaving the main sequence, so, according to Table 12-1, the cluster's age is approximately 13.5-billion years.

Using data collected from the *Hipparcos* satellite, Figure 12-26a is a composite diagram of isolated and cluster stars, reflecting the overall structure of the H-R diagram. Using data from stellar evolution theory as presented in Table 12-1, the ages of star clusters can be estimated from the turnoff points, such as those depicted in Figure 12-26b.

While globular clusters contain many of the oldest stars, the youngest star clusters in the Milky Way (those with their main sequences still intact) are found in open clusters. Unlike globular clusters, which are distributed in a sphere around the center of the Milky Way, our Galaxy's open clusters exist in the plane of the Galaxy, that is, along the band of light sweeping majestically across the night sky.

Stars in open clusters are said to be *metal-rich,* because their spectra contain many prominent spectral lines of heavy elements. (Recall from Section 11-6 that all elements other than hydrogen and helium are considered metals by astronomers.) This material originally came from stars that exploded long ago, enriching the interstellar gases with the heavy elements formed in their cores. The young clusters are therefore formed from the debris of older generations of stars. The Sun is an example of a young, metal-rich star that was likely formed in an open cluster. Such stars are also called **Population I stars.**

Globular clusters, containing the oldest stars, are generally located above or below the plane of our Galaxy. Because their spectra show only weak lines of heavy elements, these ancient stars are said to be *metal-poor.* They were created long ago from gases that had not yet been substantially enriched with heavy elements. They are also called **Population II stars.** Figure 12-27 compares the spectra from a Population II star and the Sun.

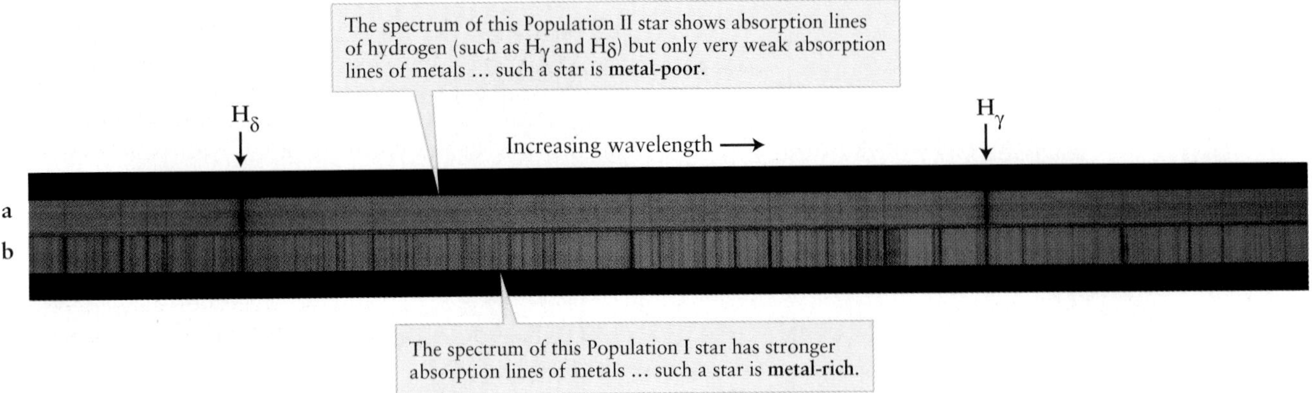

FIGURE 12-27 **Spectra of a Metal-Poor and a Metal-Rich Star** These spectra compare (a) a metal-poor (Population II) and (b) a metal-rich (Population I) star (the Sun) of the same surface temperature. Numerous spectral lines prominent in the solar spectrum are caused by elements heavier than hydrogen and helium. Note that corresponding lines in the metal-poor star's spectrum are weak or absent. Both spectra cover a wavelength range that includes two strong hydrogen absorption lines labeled H_γ (410 nm) and H_δ (434 nm). (Lick Observatory)

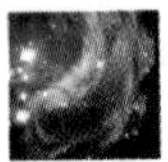

VARIABLE STARS

After core helium fusion begins, the evolutionary tracks of mature stars move across the middle of the H-R diagram. In Figure 12-23, we saw the evolutionary tracks of post–main-sequence stars. During these excursions across the H-R diagram, a star can become unstable and pulsate. The region on the H-R diagram between the main sequence and the giant branch is therefore called the **instability strip** (Figure 12-28). When a star's evolutionary track carries it through this region, the star slowly pulsates. As it does so, its brightness varies periodically. These so-called **variable stars** can be easily identified by their changes in brightness amid a field of stars of constant luminosity.

Lower-mass, post–helium-flash stars pass through the lower end of the instability strip as they move in the horizontal branch along their evolutionary tracks. These stars become **RR Lyrae variables,** named after the prototype in the constellation of Lyra (the Lyre). RR Lyrae variables all have periods shorter than one day, and all have roughly the same average brightness as stars on the horizontal branch. Higher-mass stars pass back and forth through the upper end of the instability strip on the H-R diagram. These stars become **Cepheid variables,** often simply called Cepheids.

12-13 A Cepheid pulsates because it is alternately expanding and contracting

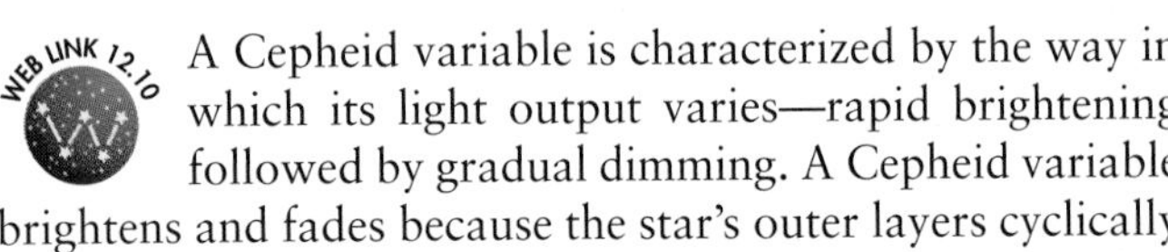

A Cepheid variable is characterized by the way in which its light output varies—rapid brightening followed by gradual dimming. A Cepheid variable brightens and fades because the star's outer layers cyclically expand and contract. Lines in the spectrum of δ (delta) Cephei shift back and forth with the same 5.4-day period that characterizes its variations in magnitude. According to the Doppler effect, these shifts mean that the star's surface is alternately approaching and receding from us.

When a Cepheid variable pulsates, the star's surface oscillates up and down like a spring. Consequently, the

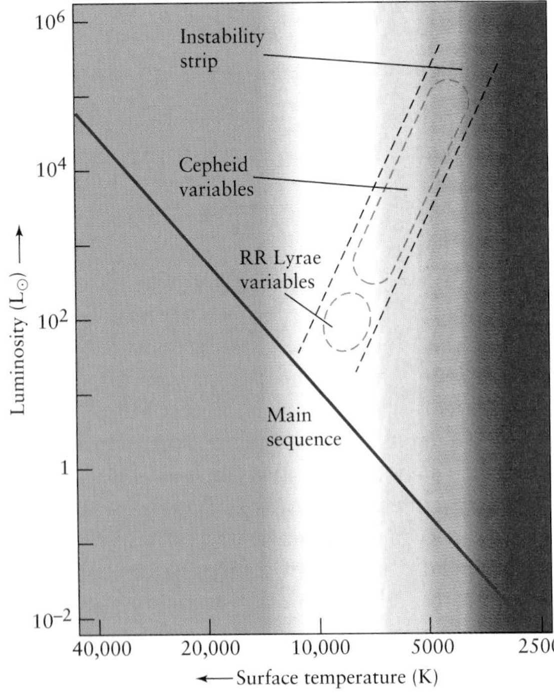

FIGURE 12-28 **The Instability Strip** The instability strip occupies a region between the main sequence and the giant branch on the H-R diagram. A star passing through this region along its evolutionary track becomes unstable and pulsates.

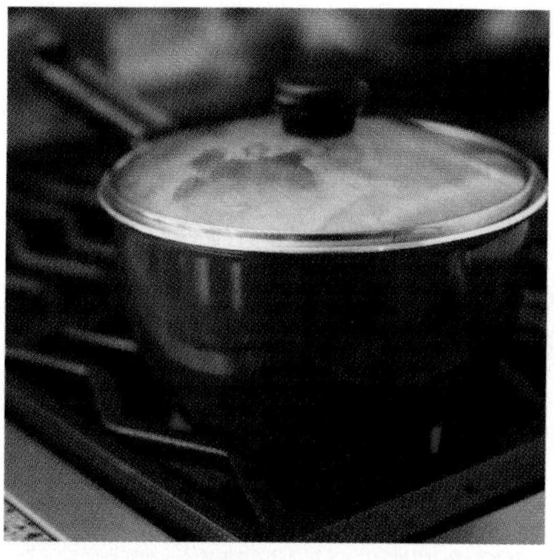
a

b

FIGURE 12-29 **Analogy for Cepheid Variability** **(a)** As pressure builds up in this pot, the force on the lid (Cepheid's outer layers) increases. **(b)** When the pressure inside the pot is sufficient, it lifts the lid off (expands the star's outer layers) and thereby allows some of the energy inside to escape. This process cycles, as do the luminosity and temperature of Cepheid stars. (a: Getty Images/FoodPix; b: Getty Images/Stone)

star's gases alternately heat up and cool down; the surface temperature changes from about 6300 K to about 5000 K and back. Thus, the characteristic light curve (luminosity versus time curve, as in Figure 11-15) of a Cepheid variable results from changes in both size and surface temperature.

Just as a bouncing ball eventually comes to rest, a pulsating star would soon stop pulsating without some mechanism to keep its oscillations going. The means by which it keeps going is to push from inside the star. These stars have a layer of gas rich in partially ionized helium (helium with an electron stripped off). This gas, absorbing a lot of photons leaving the core, heats, expands, and pushes the outer layers outward, just as steam raises the lid of a pot (Figure 12-29). Eventually, the ionized helium layer is spread so thin that photons pass through it. It cools, contracts, and the star's outer layers collapse, compressing the interior gases until the process repeats itself. In our analogy with the pot of boiling water, when enough of the steam inside escapes, the lid falls back and the steam inside builds up again. Variability only occurs when the conditions of temperature and pressure inside the star are appropriate. We see this in the fact that giants are variable only when the stars are in the instability strip.

12-14 Cepheids enable astronomers to estimate vast distances

Cepheids are important to astronomers because there is a direct relationship between a Cepheid's period of pulsation and its average luminosity. This relationship is called, appropriately enough, the **period-luminosity relation.** Dim Cepheid variables pulsate rapidly with periods of one to two days and have average brightnesses of a few hundred Suns. The most luminous Cepheids have the longest periods of all Cepheids, with variations occurring over 100 days and average brightnesses equaling 10,000 $L_{\odot}$. Because the changes in brightness of Cepheids can be seen even in distant galaxies where many other techniques for measuring distance fail, the period-luminosity relation plays an important role in determining the overall size and structure of the universe, as we will see in Chapter 16.

The details of a Cepheid's pulsation depend on the abundance of heavy elements in its atmosphere. The average luminosity of metal-rich Cepheids is roughly 4 times greater than the average luminosity of metal-poor Cepheids having the same period. Thus, there are two classes: **Type I Cepheids** (also called δ Cephei stars), which are the brighter, metal-rich stars, and **Type II Cepheids** (also called W Virginis stars), which are the dimmer, metal-poor stars. The period-luminosity relation for both types of variables is shown in Figure 12-30.

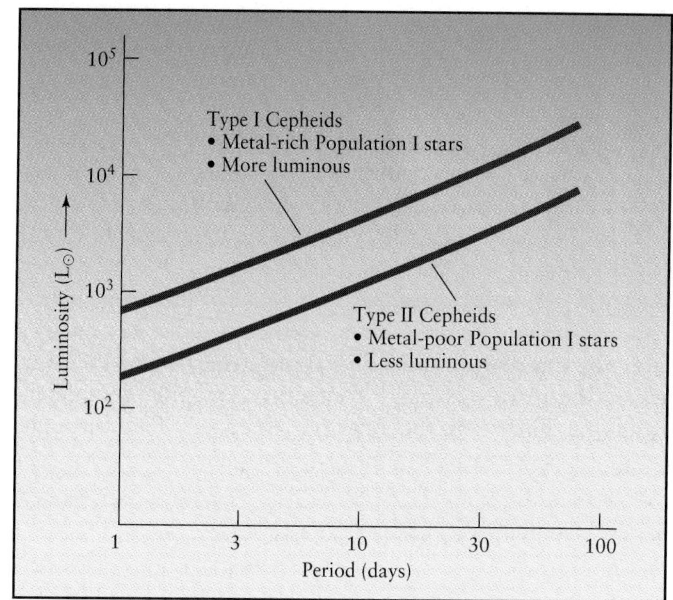

WEB LINK 12.11

FIGURE 12-30 **The Period-Luminosity Relation** The period of a Cepheid variable is directly related to its average luminosity: The more luminous the Cepheid, the longer its period and the slower its pulsations. Type I Cepheids (δ Cephei stars) are metal-rich Population I stars. They are brighter than the Type II Cepheids (W Virginis stars), which are metal-poor Population II stars. (Adapted from H. C. Arp)

In rare cases, stellar pulsations can give the gas of the star's outer layer enough speed so that it is completely ejected. Beyond the obvious mass loss in an isolated star, there are other implications of stars shedding matter. We end this chapter by relating the effect of mass loss in a close binary system, and in the next chapter we explore the mass shedding that occurs at the ends of stellar life cycles.

INSIGHT INTO SCIENCE

Firm Foundations? To determine things far from our everyday size and time scales, scientists often must base their results on the predictions of complex theories, even before all the testing and refinement of the theories are complete. An error anywhere along the chain of ideas leads to incorrect results. For example, using the Cepheid variable stars to measure distances requires combining several intermediate concepts, including the belief that the period-luminosity relationship applies well to all Cepheids, believing that there are just two different types of Cepheids, and assuming that the peak luminosity depends only on the Cepheid's distance from us (not correct if there is gas and dust between us and the star, as there often is). If any of these or myriad other assumptions is wrong, the calculated distances will be incorrect.

12-15 Mass transfer in close binary systems can produce unusual double stars

Some stars change their masses over time and thereby change the patterns of their evolution. This occurs for stars in binary systems that are sufficiently close to each other (see Chapter 11). In the mid-1800s, the French mathematician Édouard Roche pointed out that the atmospheres of two stars in a binary system must remain within a pair of teardrop-shaped regions surrounding the stars. Otherwise, the gas escapes from the star of its origin, either transferring to the other star, where the teardrops make contact, or escaping completely from the binary in the opposite direction. Cut through,

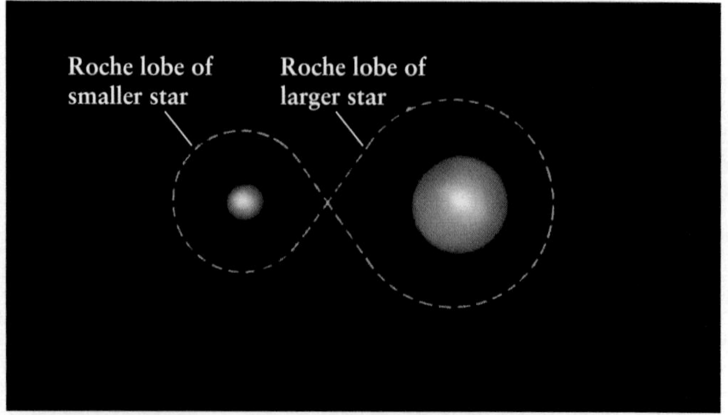

a Detached binary: Neither star fills its Roche lobe.

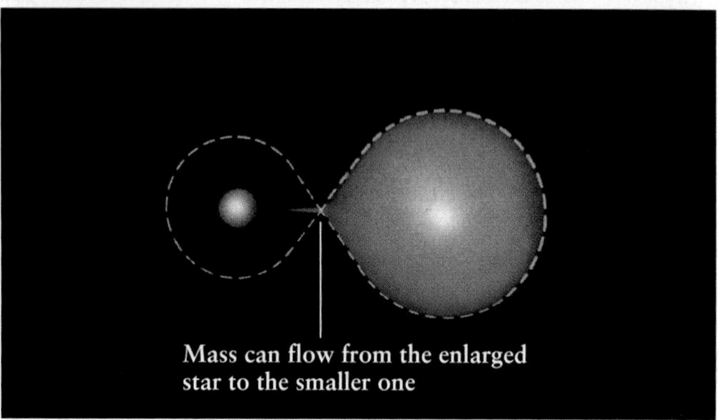

b Semi-detached binary: One star fills its Roche lobe.

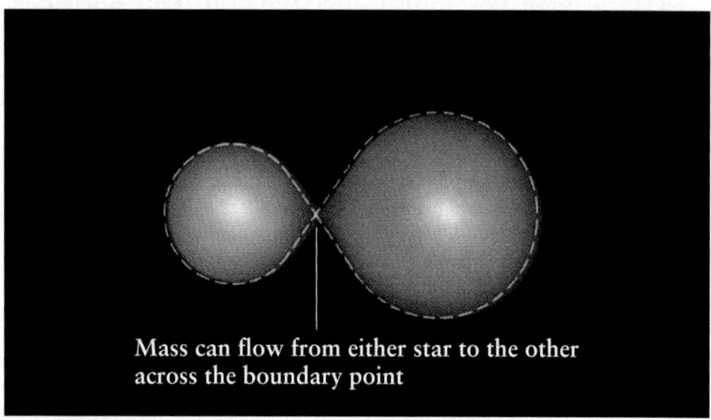

c Contact binary: Both stars fill their Roche lobes.

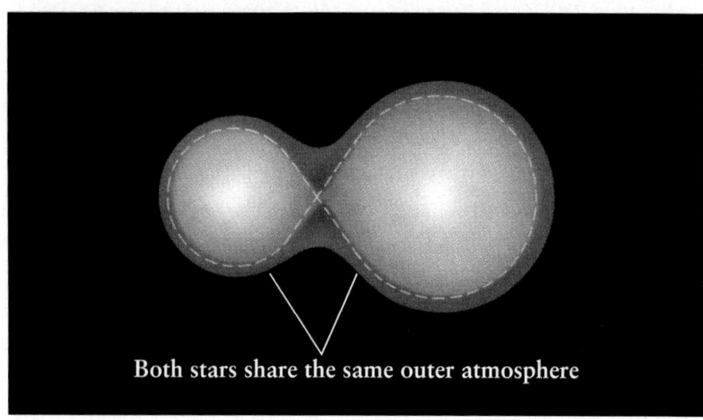

d Overcontact binary: Both stars overfill their Roche lobes.

▶ FIGURE 12-31 **Detached, Semidetached, Contact, and Overcontact Binaries** These figures show the various types of binary star systems. (a) In a detached binary, neither star fills its Roche lobe. (b) If one star fills its Roche lobe, the binary is semidetached. Mass transfer is often observed in semidetached binaries. (c) In a contact binary, both stars fill their Roche lobes. (d) The two stars in an overcontact binary both overfill their Roche lobes. The two stars actually share the same outer atmosphere.

these **Roche lobes** take on a figure-eight shape (dashed lines in Figure 12-31). The more massive star is always located inside the larger Roche lobe.

The stars in many binaries are so far apart that even during their giant stages the stars' surfaces remain well inside their Roche lobes. Other than orbiting one another, each star in such systems lives out its life cycle as if it were single and isolated, and the system is referred to as a **detached binary** (see Figure 12-31a). If two stars are relatively close together, however, one star may fill or overflow its Roche lobe as it expands into the giant phase. This system is called a **semidetached binary,** and gases then flow across the point where the two Roche lobes touch and fall onto the companion star (Figure 12-31b). When both stars completely fill their Roche lobes, the system is called a **contact binary,** because the two stars actually touch and exchange gas (Figure 12-31c). It is quite unlikely, however, that both stars will exactly fill their Roche lobes at the same time. It is more likely that they overflow their lobes, giving rise to a common atmospheric envelope. Such a system is called an **overcontact binary** (Figure 12-31d).

Semidetached and contact binaries are easiest to detect if they are also eclipsing binaries (recall Figure 11-15). Their light curves have a distinctly rounded appearance caused by these tidally distorted egg-shaped stars. The eclipsing binary called β (beta) Persei, or Algol (from an Arabic term for "demon"), is a semidetached binary that can easily be seen with the naked eye in the constellation Perseus. From β Persei's light curve (Figure 12-32a), astronomers have determined that the binary contains a star that fills its Roche lobe. Sometime in the past, as it expanded and became a giant, this star dumped a significant amount of gas onto its companion.

Mass transfer is still occurring in a semidetached eclipsing binary called β Lyrae in the constellation Lyra. Like β Persei, β Lyrae contains a giant that fills its Roche lobe (Figure 12-32b). For many years astronomers were puzzled by the fact that the detached companion star in β Lyrae is

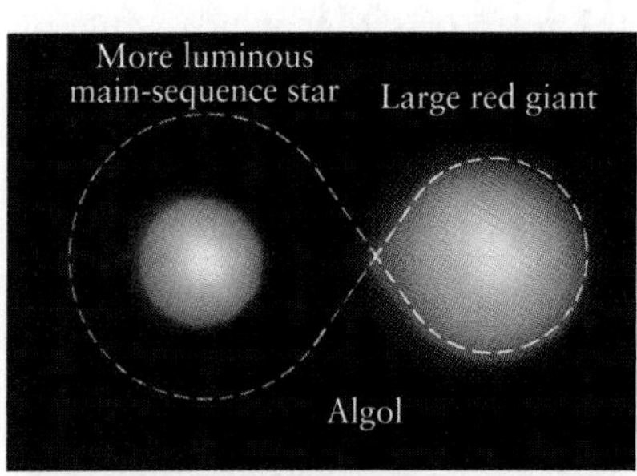

a A semidetached binary

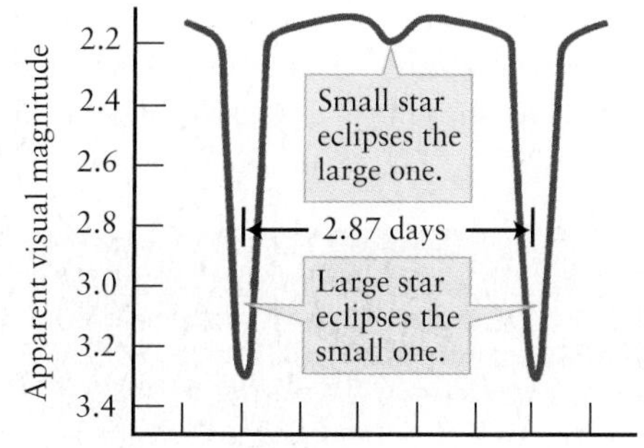

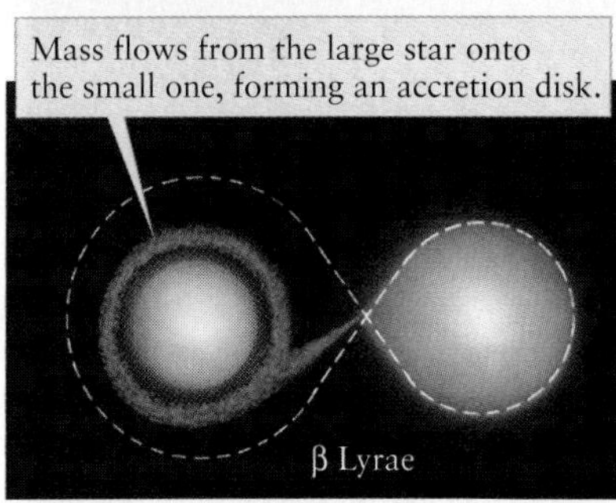

b A semidetached binary with mass transfer

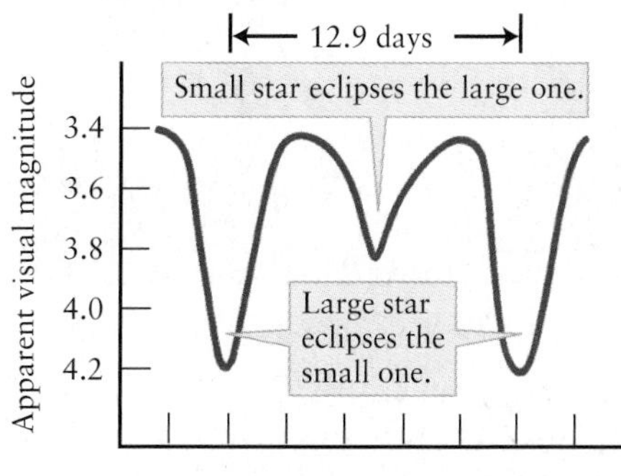

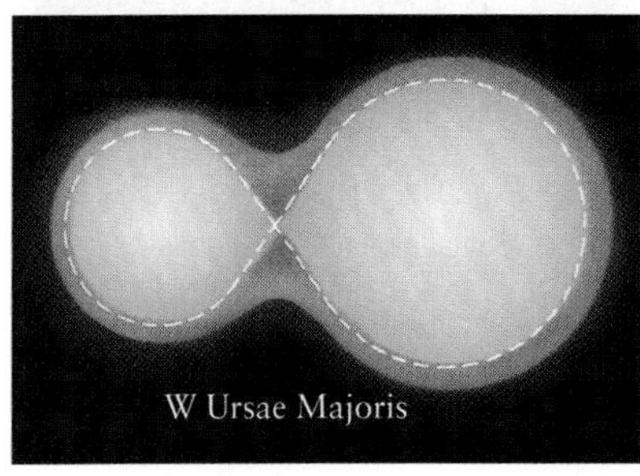

c An overcontact binary

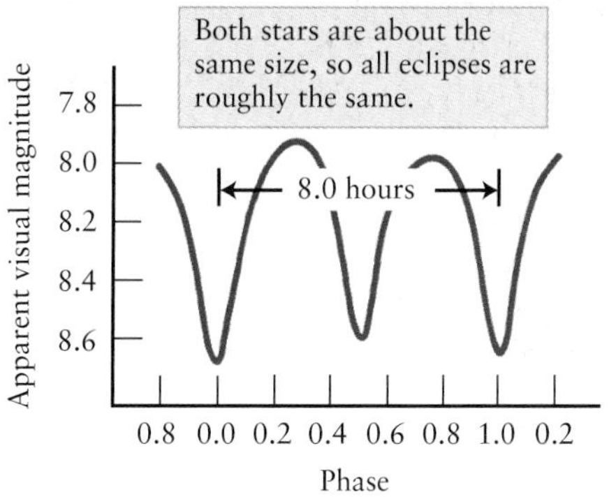

FIGURE 12-32 **Three Close Binaries** Sketches of and light curves for three eclipsing binaries are shown. The phase denotes the fraction of the orbital period from one primary minimum to the next. **(a)** Algol, also known as β Persei, is a semidetached binary. The deep eclipse occurs when the giant star (on the right) blocks the light from the smaller, but more luminous, main-sequence star. **(b)** β Lyrae is a semidetached binary in which mass transfer has produced an accretion disk surrounding the detached star. This disk is so thick and opaque that it renders the secondary star almost invisible. **(c)** W Ursae Majoris is an overcontact binary. Both stars therefore share their outer atmospheres. The short, 8-hour period of this binary indicates that the stars are very close to each other.

severely underluminous, contributing virtually no light at all to the visible radiation coming from the system. Furthermore, the spectrum of β Lyrae contains unusual features, some of which are consistent with gas flowing between the stars and around the system as a whole.

The β Lyrae system was explained in 1963, when Su-Shu Huang proposed that the underluminous star in β Lyrae is enveloped in a huge **accretion disk** of gas captured from its bloated companion. The disk is so large and thick that it completely shrouds the secondary star, making it impossible to observe at visible wavelengths. The primary star is overflowing its Roche lobe, with gases streaming onto the disk at the rate of 1 $M_\odot$ per hundred thousand years.

The fate of a semidetached system like β Persei or β Lyrae depends primarily on how fast its stars evolve. If the detached star expands to fill its Roche lobe while the companion star fills its own Roche lobe, then the result is an overcontact binary. An example is W Ursae Majoris, in which two stars share the same photosphere (Figure 12-32c).

Theories of the evolution of binary systems that transfer mass reveal that some remarkable transformations are possible. This is seen in the case of ϕ (phi) Persei. As shown in Figure 12-33, the more massive star in a binary can actually be transformed into the lower-mass star and vice versa! When stars in such systems gain mass, their increased gravitational force increases their rate of fusion, which increases the rate at which they evolve. Conversely, the stars that lose mass begin evolving more slowly than they were before their mass loss. In Chapters 13 and 14 we will see that mass transfer produces some of the most extraordinary objects in the sky.

12-16 Frontiers yet to be discovered

Stars, whether individually, in pairs, or in much larger groups, are our primary source of information about the universe. Details of each step of stellar evolution remain to be explored: the pre–main-sequence evolution of stars is being revealed in sources like the Eagle Nebula (see Figure 12-15) as the pre–main-sequence stars there are being seen through our telescopes. These observations provide data with which to compare the theories of early star formation. Over the past two decades, observations have revealed that more stars lose mass in the form of stellar winds while on the main sequence than previously expected. Consequently, we have a lot to learn about why such stars shed mass.

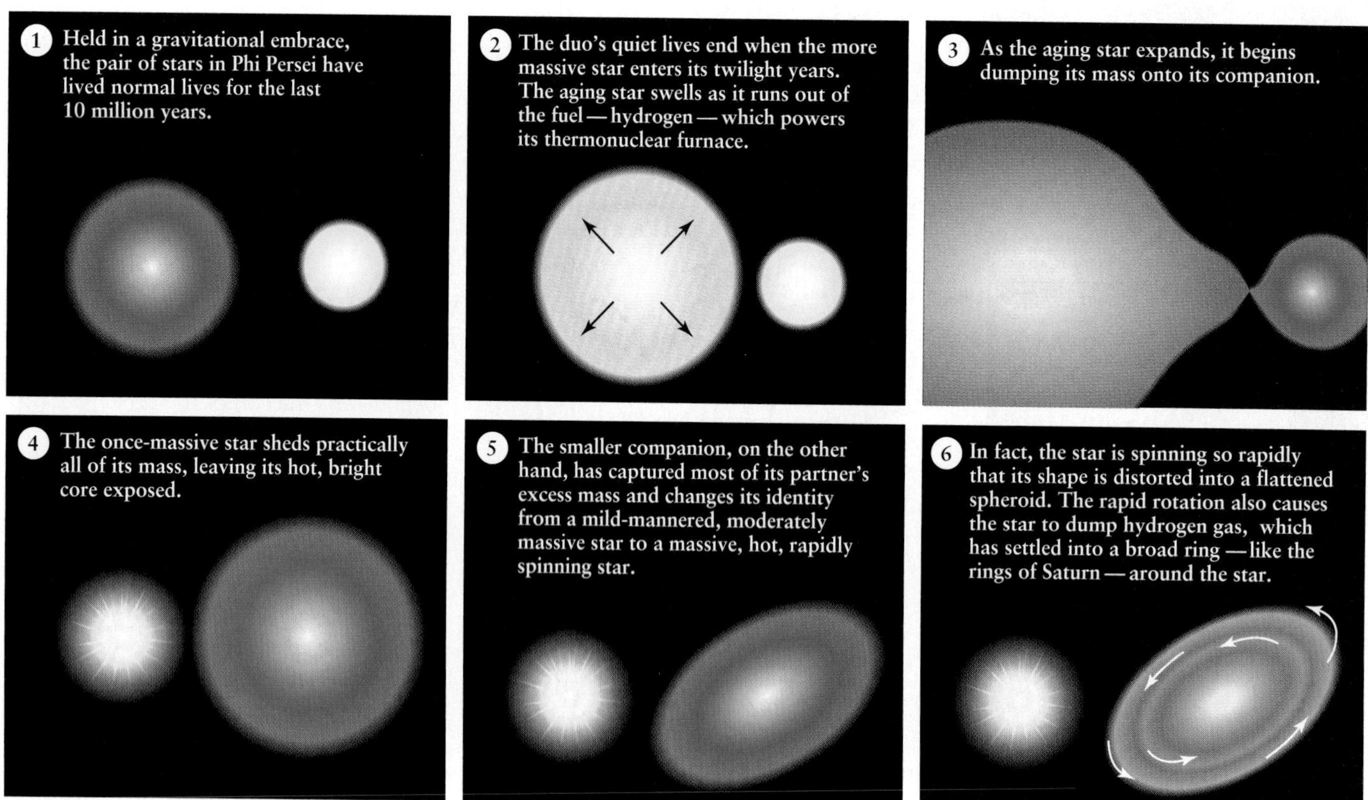

FIGURE 12-33 **Mass Exchange between Close Binary Stars** This sequence of drawings shows how close binary stars can initially be isolated but, as they age, grow and exchange mass. Such mass exchange leads to different fates than if the same stars had evolved in isolation.

As noted, we still need to understand the details of why stars swell into the red giant phase. Also, the mechanisms that cause stars to be variable are still under investigation.

Summary of Key Ideas

Protostars and Pre–Main-Sequence Stars

• Enormous, cold clouds of gas and dust, called giant molecular clouds, are scattered about the disk of the Galaxy.

• Star formation begins when gravitational attraction causes clumps of gas and dust called protostars to coalesce within a giant molecular cloud. As a protostar contracts, its matter begins to heat and glow. When the contraction slows down, the protostar becomes a pre–main-sequence star. When the pre–main-sequence star's core temperature becomes high enough to begin hydrogen fusion and stop contracting, it becomes a main-sequence star.

• The most massive pre–main-sequence stars take the shortest time to become main-sequence stars (O and B stars).

• In the final stages of pre–main-sequence contraction, when hydrogen fusion is about to begin in the core, the pre–main-sequence star may undergo vigorous chromospheric activity that ejects large amounts of matter into space. Such gas-ejecting stars are called T Tauri stars.

• A collection of a few hundred or a few thousand newborn stars is called an open cluster. Stars escape from open clusters, most of which eventually dissipate.

Main-Sequence and Giant Stars

• The Sun has been a main-sequence star for 4.6 billion years and should remain so for about another 5 billion years. Less massive stars than the Sun evolve more slowly and have longer main-sequence lifetimes. More massive stars than the Sun evolve more rapidly and have shorter main-sequence lifetimes.

• Main-sequence stars with less than $0.4M_\odot$ convert all of their mass into helium and then stop fusing. Their lifetimes are hundreds of billions of years and so none of these stars has yet left the main sequence.

• Core hydrogen fusion ceases when hydrogen is exhausted in the core of a main-sequence star with $M > 0.4M_\odot$, leaving a core of nearly pure helium surrounded by a shell where hydrogen fusion continues. Hydrogen shell fusion adds more helium to the star's core, which contracts and becomes hotter. The outer atmosphere expands considerably, and the star becomes a giant.

• When the central temperature of a giant reaches about 100 million K, the thermonuclear process of helium fusion begins. This process converts helium to carbon, then to oxygen. In a massive giant, helium fusion begins gradually. In a less massive giant, it begins suddenly in a process called the helium flash.

• The age of a stellar cluster can be estimated by plotting its stars on an H-R diagram. The upper portion of the main sequence disappears first, because more massive main-sequence stars become giants before low-mass stars do. Relatively young stars are metal-rich; ancient stars are metal-poor.

• Giants undergo extensive mass loss, sometimes producing shells of ejected material that surround the entire star.

Variable Stars

• When a star's evolutionary track carries it through a region called the instability strip in the H-R diagram, the star becomes unstable and begins to pulsate.

• RR Lyrae variables are low-mass, pulsating variables with short periods. Cepheid variables are high-mass, pulsating variables exhibiting a regular relationship between the period of pulsation and luminosity.

• Mass can be transferred from one star to another in close binary systems. When this occurs the evolution of the two stars changes.

WHAT DID YOU THINK?

1 *How do stars form?* Stars form from the mutual gravitational attraction between gas and dust inside giant molecular clouds.

2 *Are stars forming today?* Yes. Astronomers have seen stars that have just arrived on the main sequence, as well as infrared images of gas and dust clouds in the process of forming stars.

3 *Do stars with greater mass shine longer?* Lower-mass stars last longer because the lower gravitational force inside them causes fusion to take place at slower rates compared to the fusion inside higher-mass stars.

Key Words

accretion disk, 344
birth line, 326
Bok globule, 324
brown dwarf, 326
Cepheid variable, 340
contact binary, 343
core helium fusion, 336
dense core, 324
detached binary, 343
electron degeneracy pressure, 337
emission nebula, 328
evolutionary track, 326
giant molecular cloud, 321
globular cluster, 338
H II regions, 328
helium flash, 336
horizontal branch star, 338
hydrogen shell fusion, 334
instability strip, 340
interstellar extinction, 319
interstellar medium, 319
interstellar reddening, 319
Jeans instability, 324
nebulae (*singular* nebula), 322
OB association, 330
open cluster, 324
overcontact binary, 343
Pauli exclusion principle, 336
period-luminosity relation, 341

Review Questions

1. Consider a star behind a cloud of interstellar gas and dust as seen from our perspective. Which of the following would you see? **a.** The star appears brighter than it would if the cloud were not present. **b.** The star appears to be moving toward us. **c.** The star appears redder than it would if the cloud were not present. **d.** The star would always be invisible to us. **e.** The star would always appear green.

2. What is the lowest mass that a star can have on the main sequence? **a.** There is no lower limit. **b.** 0.003 $M_{\odot}$. **c.** 0.08 $M_{\odot}$. **d.** 0.4 $M_{\odot}$. **e.** 2.0 $M_{\odot}$.

3. What is the source of energy that enables a main-sequence star to shine? **a.** Friction between its atoms. **b.** Fusion of hydrogen in a shell surrounding the core. **c.** Fusion of helium in its core. **d.** Fusion of hydrogen in its core. **e.** Burning of gases on its surface.

4. What are giant molecular clouds, and what role do these clouds play in star formation?

5. Why are low temperatures necessary for dense cores to form and contract into protostars?

6. Why don't thermonuclear reactions occur on the surface of a main-sequence star?

7. What is an evolutionary track, and how can such tracks help us interpret the H-R diagram?

8. Why are most stars we see in the sky main-sequence stars?

9. Draw the pre–main-sequence evolutionary track of the Sun on an H-R diagram. Briefly describe what was occurring throughout the solar system at various stages along this track. (You may find reviewing Chapter 11 useful.)

10. On what grounds are astronomers able to say that the Sun has about 5 billion years remaining in its main-sequence stage?

11. What will happen inside the Sun 5 billion years from now when it begins to evolve into a giant?

12. How is the evolution of a main-sequence star with less than 0.4 $M_{\odot}$ fundamentally different from that of a main-sequence star with more than 0.4$M_{\odot}$?

13. Draw the post–main-sequence evolutionary track of the Sun on an H-R diagram up to the point when the Sun becomes a helium-fusing giant. Briefly describe what might occur throughout the solar system as the Sun undergoes this transition.

14. What does it mean when an astronomer says that a star "moves" from one place to another on an H-R diagram?

15. What is the helium flash and what causes it?

16. Explain how and why the turnoff point on the H-R diagram of a cluster is related to the cluster's age.

17. Why do astronomers believe that globular clusters are made of old stars?

18. What are Cepheid variables and how are they related to the instability strip?

19. What occurs in Cepheid stars that is analogous to the vapor leaking out of a pot of boiling water when the lid is raised?

20. Which are older stars: Type I or Type II Cepheids? Justify your answer.

21. What are RR Lyrae variables and how are they related to the instability strip?

22. What is a Roche lobe and what is its significance in close binary systems?

23. What are the differences between detached, semidetached, contact, and overcontact binaries?

Advanced Questions

The answers to all computational problems, which are preceded by an asterisk (*), appear at the end of the book.

24. How is a degenerate gas different from an ordinary gas?

25. If you took a spectrum of a reflection nebula, would you see absorption lines, emission lines, or no lines? Explain your answer.

26. Why is it useful to plot the apparent magnitudes of stars in a single cluster on an H-R diagram?

27. What observations would you make of a star to determine whether its primary source of energy was fusing hydrogen or helium?

28. What might happen to the massive outer planets when the Sun becomes a giant?

*29. How many 1.5-$M_{\odot}$ main-sequence stars would it take to equal the luminosity of one 15-$M_{\odot}$ star?

*30. How many times longer does a 1.5-$M_{\odot}$ star fuse hydrogen in its core than does a 15-$M_{\odot}$ star?

31. Why does a shock wave from a supernova produce relatively few high-mass O and B stars compared to the number of lower-mass A, F, G, K, and M stars produced?

32. How would you distinguish a newly formed protostar from a giant, given that they occupy the same location on the H-R diagram?

33. What observational consequences would we find in H-R diagrams for star clusters as a result of the universe having a finite age? Could we use these consequences to establish constraints on the possible age of the universe? Explain.

Discussion Questions

34. Discuss the possibility of life-forms and biological processes occurring in giant molecular clouds. In what ways might conditions favor or hinder biological evolution?

35. Is there any evidence that the Earth has ever passed through a star-forming region in space?

What If . . .

36. The solar system passed through a giant molecular cloud? How would this encounter affect the Earth and life on it?

37. The Sun had already entered its giant phase? What would happen to the Earth and to life on it?

38. The Earth were orbiting a 0.5-$M_{\odot}$ star at a distance of 1 AU? What would be different for the Earth and life on it? What effects would moving the Earth closer to the lower-mass Sun have?

39. The Sun were a variable star? How would this change life on Earth and, assuming we could live in orbit around such a star, how might it change our perspective of the cosmos?

Web Questions

40. To test your understanding of where stars are formed, do Interactive Exercise 12-1 on the Web. You can print out your results, if required.

41. To test your understanding of variable stars, do Interactive Exercise 12-2 on the Web. You can print out your results, if required.

42. To test your understanding of close binary star systems, do Interactive Exercise 12-3 on the Web. You can print out your results, if required.

Observing Projects

43. Use a telescope to observe at least two of the following interstellar gas clouds: M42 (Orion), M43, M20 (Trifid), M8 (Lagoon), M17 (Omega). You can easily locate them with the aid of your *Starry Night Enthusiast*™ program, star charts published in such magazines as *Astronomy* and *Sky & Telescope,* or the list of coordinates below. In each case, can you identify the stars responsible for the ionizing radiation that causes the nebula to glow? Draw a picture of what you see through the telescope and compare it with a photograph of the object. Which portions of the nebula are not visible through your telescope? Why are they not visible?

Nebula	Right ascension	Declination
M42 (Orion)	$5^h\ 35.4^m$	–5° 27′
M43	$5^h\ 35.6^m$	–5° 16′
M20 (Trifid)	$18^h\ 02.6^m$	–23° 02′
M8 (Lagoon)	$18^h\ 03.8^m$	–24° 23′
M17 (Omega)	$18^h\ 20.8^m$	–16° 11′

44. Several star clusters can be seen quite well with a good pair of binoculars. You can easily locate many of them with the aid of your *Starry Night Enthusiast*™ program, star charts published during the summer months in such magazines as *Astronomy* and *Sky & Telescope,* or the list of coordinates below. Observe as many of these clusters as you can. Look at them through a telescope, if available. Note the overall distribution of stars in each cluster. Can you see any of these clusters with the naked eye? What difference do you note between binocular and telescopic images of individual clusters?

Star cluster	Right ascension	Declination
M45 (Pleiades)	$3^h\ 47.0^m$	+24° 07′
Hyades	$4^h\ 27.0^m$	+16° 00′
Praesepe	$8^h\ 40.1^m$	+19° 59′
Coma	$12^h\ 25.0^m$	+26° 00′
M11	$18^h\ 51.1^m$	–06° 16′

45. Use the *Starry Night Enthusiast*™ program to examine a star-forming region. First put the program into Atlas mode (*Favourites/Guides/Atlas*). Center the field on M20 (tab *Find/Trifid*). Zoom out to the maximum field of view. Set the timestep to 1 lunar month and single step through the months of the year. **a.** In which month is M20 highest in the sky at noon? Explain how you determined this. **b.** In which month is M20 highest in the sky at midnight, so that it is best placed for observing with a telescope? Explain how you determined this. **c.** Zoom in on the Trifid nebula and explain why different areas of the nebula have the colors they have.

WHAT IF . . . THE EARTH ORBITED A 1.5 $M_\odot$ SUN?

The Earth is at a perfect distance from a wonderful star. The Sun provides just enough heat so that liquid water, necessary to sustain life, can exist here. Would our planet still be suitable for the evolution of life if the Sun were 1.5 $M_\odot$ rather than 1.0 $M_\odot$? To begin with, we need to know how far from the new Sun, let's call it Sol II, to place the Earth.

A Very Sunny Day Sol II's surface temperature would be 8400 K and it would appear blue-white in our sky. Sol II would give off 7 times as much energy per second as our present Sun, due to the combination of a higher temperature and a 20% larger radius than the Sun. The effect of Sol II's increased energy emission would require that Earth be located much farther away from Sol II than our present distance from the Sun. To understand why, consider the increase in infrared (heat) output from Sol II. That extra heat would have the initial effect of raising the average global temperature on Earth at our present distance by about 10 K (about 20°F). This does not seem like a lot, but the impact of that slight increase would actually boost the atmospheric temperature much higher.

The extra heat from Sol II would cause more ocean water to evaporate into the atmosphere. Because water is a greenhouse gas (that is, it traps infrared radiation), the air temperature would rise, causing even more water to evaporate from the oceans, which, in turn, would cause the air to heat even more. This vicious cycle, called a runaway greenhouse effect, would make the Earth's surface so hot and dry that it would be uninhabitable.

By moving the Earth about 2.6 times farther away from Sol II, the temperature would become suitable for life. At that distance, the year would be 1249 days long. While moving away from the heat is one thing, moving away from the ultraviolet radiation emitted by Sol II is something else altogether.

Ultraviolet Excess Just by increasing the Sun's mass by 50%, the ultraviolet radiation emitted would be several thousand times stronger. This is because the energy output of stars with different surface temperatures varies with wavelength. While the output of visible light would change slightly, the output of ultraviolet radiation would be vastly greater. Therefore, even though the Earth's surface temperature would be suitable farther from Sol II, the flood of ultraviolet radiation would be so strong that the ozone layer would be overwhelmed, and the level of ultraviolet radiation at the Earth's surface would be much higher than it is today. This would be so even with the greater concentration of ozone created by the increased ultraviolet from Sol II.

Life would have to evolve greater protection from ultraviolet radiation than it has today. And if this were not a great enough challenge, suppose intelligent life-forms were evolving on the Earth orbiting Sol II 4.6 billion years after the solar system formed. They would discover that their star had evolved so rapidly that it was just about to expand into the giant phase!

R I V U X G

(John Elk III/Bruce Coleman)

The Deaths of Stars

CHAPTER 13

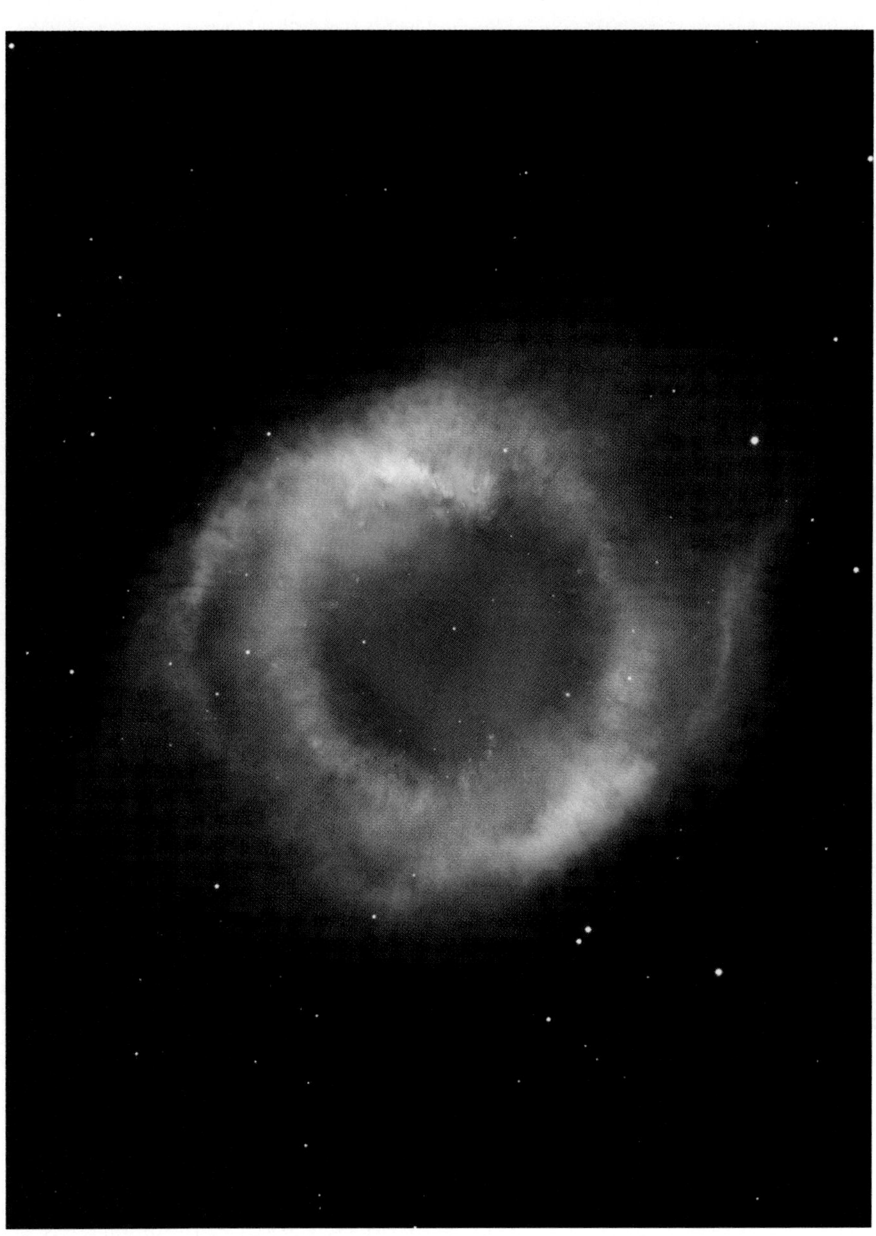

The Helix Nebula (NASA, NOAO, ESA, The Hubble Helix Nebula Team, M. Meixner/STScI, and T. A. Rector/NRAO)

WHAT DO YOU THINK?

1. Will the Sun someday cease to shine brightly? If so, how?
2. What is a nova?
3. Where do heavy elements on the Earth like carbon, silicon, oxygen, iron, and uranium come from?
4. What are cosmic rays?
5. What is a pulsar?

Most video and computer games have one thing in common: Players move through different levels, each progressively more difficult than the last. The biggest challenge comes at the final level, when you have to face the most powerful or evil entity. Stellar evolution is similar, with stars passing through different stages of stellar activity before they can move on to the next stage. There is one big difference—a game player can win the final level, but a star at that stage ceases to shine with the vigor that it had previously. At that stage, stars eject vast quantities of gas and dust into interstellar space. In human terms, they die. In this chapter and Chapter 14, we learn about the later stages of stellar evolution and their often spectacular finales.

In this chapter you will discover

- what happens to stars when core helium fusion ceases
- how heavy elements are created
- the characteristics of the end of stellar evolution
- why some stars go out relatively gently, while others go with a bang
- the incredible densities of neutron stars and how these stars are observed

LOW-MASS STARS AND PLANETARY NEBULAE

We have seen that red dwarf stars, with masses less than 0.4$_{\odot}$, never get to the giant phase. We will now explore the fates of stars greater than 0.4$M_{\odot}$ as they proceed along through the giant phase and later. The destinies of stars depend on their masses, and we have three mass ranges to explore: 0.4 to 8 $M_{\odot}$ (hereafter, *low-mass* stars), 8 to 25 $M_{\odot}$ (hereafter *intermediate-mass* stars), and greater than 25 $M_{\odot}$ (hereafter *high-mass* stars).

In Chapter 12, we discovered that when hydrogen shell fusion first begins, the energy from it causes a star to expand and become a giant. In this process, low-mass stars move over to, and ascend, the giant branch on the H-R diagram for the first time (Figure 13-1a). As this happens, mass is expelled into space in the form of stellar winds, which reduce the masses of these stars. Then comes core helium fusion (Figure 13-1b), with stars of less than 2 $M_{\odot}$ undergoing a core helium flash in the giant stage (Figure 13-1a). Stars with more than 2 $M_{\odot}$ begin helium fusion more gradually. After fusion begins, all these stars shrink

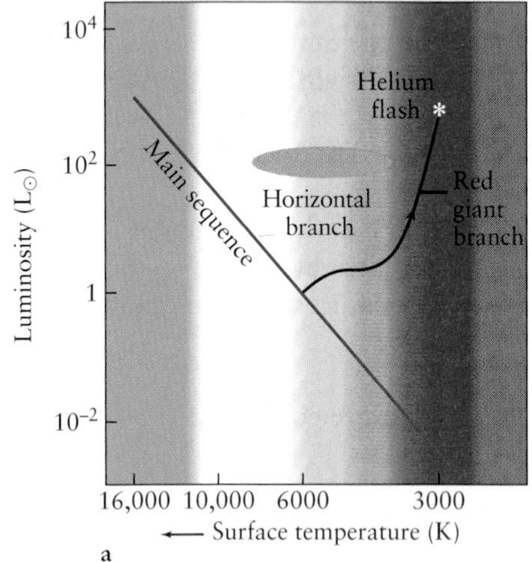

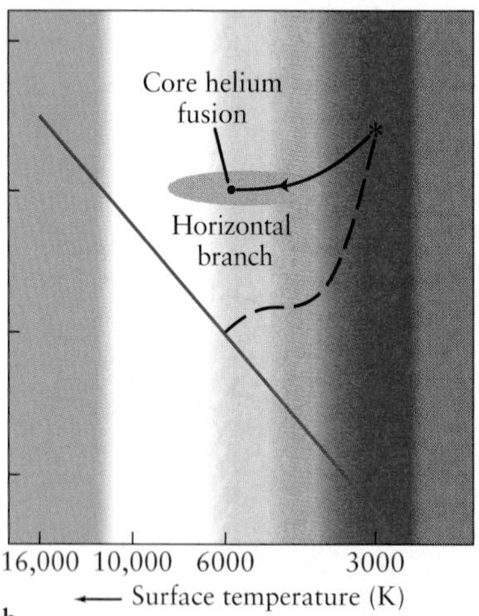

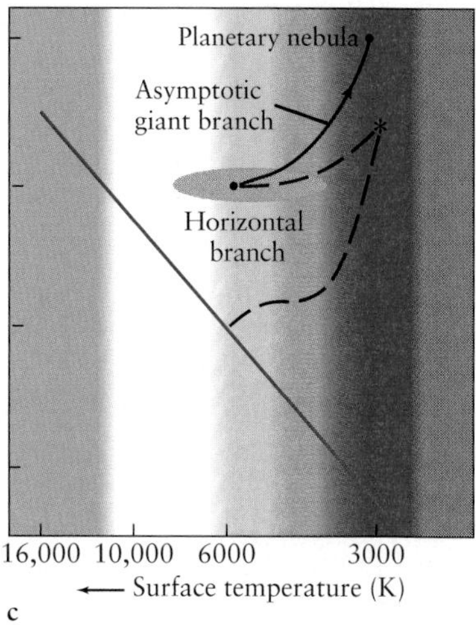

FIGURE 13-1 **Post-Main-Sequence Evolution of Low-Mass Stars** (a) The evolutionary track on the H-R diagram as a star makes the transition from the main sequence to the giant phase. The asterisk (*) shows the helium flash occurring in a low-mass star. (b) After the helium flash, the star converts its helium core into carbon and oxygen. While doing so, its core reexpands, decreasing shell fusion. As a result, the star's outer layers recontract. (c) After the helium core is completely transformed into carbon and oxygen, the core recollapses, and the outer layers reexpand, powered up the asymptotic giant branch by hydrogen shell fusion and helium shell fusion.

and move onto the *horizontal branch* (Figure 13-1b). As we will now see, their cores are eventually converted into carbon and oxygen, helium fusion ceases, and these stars undergo another stage that closely parallels the end of core hydrogen fusion (Figure 13-1c).

13-1 Lower-mass stars become supergiants before expanding into planetary nebulae

Calculations reveal that carbon and oxygen require a temperature of at least 600 million K to fuse. Because the core of a low-mass giant on the horizontal branch only reaches about 200 million K, fusion of these elements in the core does not occur. Photon production therefore drops off, and the inner regions of the star again contract, compressing and heating the shell of helium-rich gas just outside the core. As a result, **helium shell fusion** begins outside the core; this shell is itself surrounded by a hydrogen-fusing shell (Figure 13-2). All this takes place within a volume roughly the size of the Earth.

Once helium shell fusion commences, the new outpouring of energy pushes the outer envelope of the star out again. A low-mass star thus leaves the horizontal branch and ascends the giant branch for a second, and final, time. Powered by the fusion from two shells, it becomes brighter than ever before. Such an **asymptotic giant branch star,** or **AGB star,** of 8 $M_{\odot}$ will have a diameter as big as the orbit of Mars and shine with the luminosity of 10^4 $L_{\odot}$. As the Sun moves along the asymptotic giant branch, it will expand to a radius of about 1 AU, enveloping the Earth in its tenuous outer layers (Figure 13-2). At the top of the asymptotic giant branch, these stars are now so bright that they are classified as low-temperature, red supergiants (see Figure 13-1c).

AGB stars are destined to self-destruct. Giant stars of all spectral types have strong stellar winds. Low-mass stars expel their outer layers more slowly than higher-mass stars, but they all reduce their masses significantly from what they were on the main sequence. During their initial ascent up the giant branch, stars can lose as much as 30% of their masses. On the asymptotic giant branch they lose even more, often surrounding themselves with thickening cocoons of gas and dust. A star of the Sun's mass loses about 10^{-5} $M_{\odot}$ per year at this stage, eventually dumping half of its mass back into space.

This mass loss limits the amount of gravity available to compress the star's core and regions of shell fusion. A fine balance is struck: The core of our low-mass AGB star is compressed until its core of carbon and oxygen becomes degenerate, meaning that the electrons in the core provide a growing repulsive force that stops its contraction (as discussed in Section 12-10). However, even for the most massive of these low-mass stars (now much less than their original 8 $M_{\odot}$ due to their stellar winds), the core temperature cannot reach the 600 million K necessary to fuse carbon or oxygen. Therefore, no further core fusion occurs.

The final stage through which a low-mass star passes begins with a "thermal runaway" (meaning a rapid rise in temperature) in the helium shell, like the helium flash in its core earlier in its life. This occurs because the triple alpha process (see Section 12-10) is extremely sensitive to temperature, and as the temperature goes up slightly, the fusion rate skyrockets. The increase in energy output from the thin helium-fusing shell, called a **helium shell flash,** expands the star, thereby briefly decreasing the temperature in the shell and slowing the rate of fusion.

A star undergoes a number of helium shell flashes as its helium shell thickens. During each flash, the helium shell's energy output jumps a thousandfold. These brief outbursts are separated by relatively quiet intervals lasting about 100,000 years, during which the helium shell gradually becomes thicker. The energy from the increased fusion

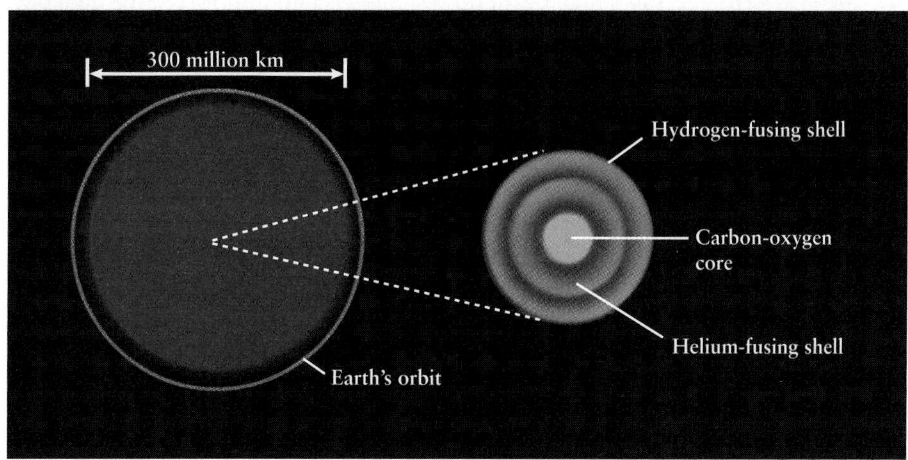

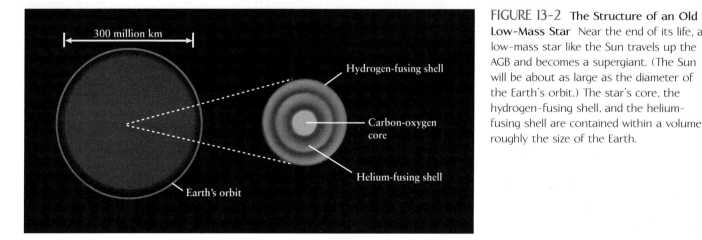

FIGURE 13-2 **The Structure of an Old Low-Mass Star** Near the end of its life, a low-mass star like the Sun travels up the AGB and becomes a supergiant. (The Sun will be about as large as the diameter of the Earth's orbit.) The star's core, the hydrogen-fusing shell, and the helium-fusing shell are contained within a volume roughly the size of the Earth.

during the helium shell flashes causes the star's outer layers to expand and, therefore, cool further. Eventually, the outer gases are sufficiently cool so that electrons and ions there can recombine. This is exactly the opposite of ionization (discussed in Section 4-5). Whereas ionization requires an electron to absorb a photon, recombination forces an electron to emit a photon.

Just as the collisions of particles create pressure (force acting over an area), the collisions of photons with matter do, too. The impacts of photons emitted during recombination, along with the photons from the helium flashes and the photons normally flowing outward from fusion, generate enough pressure to eject more and more of the star's outer layers into space. As the ejected material expands and cools, some of it condenses to form dust grains that are propelled outward by radiation pressure from the star's hot, burned-out core. When enough of the gas has left the star for the core to be visible, the expanding dust and gases are considered to be a **planetary nebula** (see Figure 13-1c). As the outer layers of the star are shed, an increasingly hot interior is revealed, and the star moves to the left across the H-R diagram (Figure 13-3). Low-mass stars lose as much as 80% of their masses by shedding their outer layers, including the mass lost prior to the planetary nebula phase.

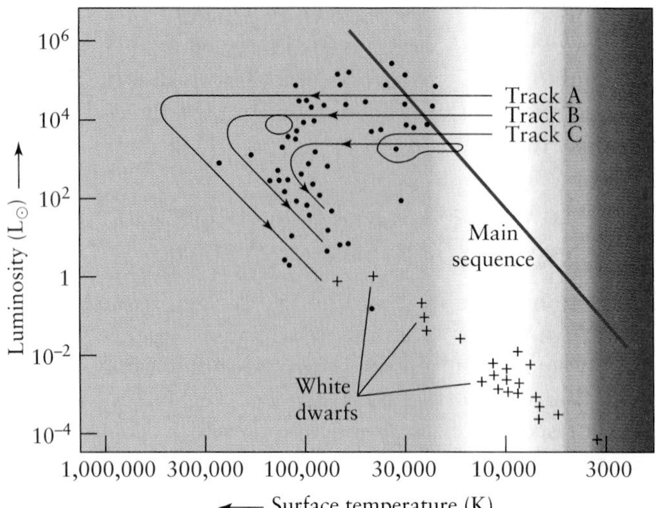

Evolutionary track	Mass ($M_\odot$)		
	Red supergiant	Ejected nebula	White dwarf
A	3.0	1.8	1.2
B	1.5	0.7	0.8
C	0.8	0.2	0.6

FIGURE 13-3 Evolution from Supergiants to White Dwarfs The evolutionary tracks of three low-mass, supergiants are shown as they eject planetary nebulae. The table gives their masses as supergiants, the amount of mass they lose as planetary nebulae, and their remaining (white dwarf) masses. The dots on this graph represent the central stars of planetary nebulae whose surface temperatures and luminosities have been determined. The crosses are white dwarfs for which similar data exist.

INSIGHT INTO SCIENCE

Names Don't Change as Understanding Improves Planetary nebulae have *nothing* to do with planets. This term was coined by the astronomer William Herschel in the eighteenth century. When viewed through the small telescopes of the day, these glowing objects, often green in appearance, looked like planets, such as Uranus.

1 Planetary nebulae are quite common in our Galaxy. More than 1800 have been identified, and astronomers estimate that 20,000 to 100,000 exist in the Milky Way. Indeed, the fate of the Sun is to become a planetary nebula. The rate at which the gases are shed in a planetary nebula is so slow that they really are not explosive compared either to our common conception of an explosion or to the supernova explosions we will examine shortly.

Because stellar winds from red giants and subsequent planetary nebulae are so plentiful, astronomers estimate they return a total of about 5 $M_\odot$ to the disk of our Galaxy. This amounts to about 85% of all matter expelled by all types of stars. This material goes into the formation of new, metal-rich Population I stars and associated planets. Thus, planetary nebulae play an important role in the chemical and physical evolution of the Galaxy.

The outflowing gases ejected in a planetary nebula are turned into a breathtaking variety of shapes when they interact with gases surrounding their stars, with companion stars, and with the stars' magnetic fields, which are often 10 to 100 times stronger than the Sun's field (Figures 13-4 and Figure 13-5). The Hourglass Nebula (Figure 13-5c) appears initially to have shed mass in a doughnut shape around itself. In the star's final death throes, this gas and dust forced the final outflow to go in two directions perpendicular to the plane of the doughnut, creating what is called a *bipolar planetary nebula*.

Spectroscopic observations of planetary nebulae show bright emission lines of hydrogen, carbon, neon, magnesium, oxygen, and nitrogen. From the Doppler shifts of these lines, we conclude that the expanding shell of gas is moving outward from the dying low-mass star at speeds of 10 to 30 km/s. A typical planetary nebula moving outward for 10,000 years has a diameter of a few light-years.

By astronomical standards, a planetary nebula is short-lived. After about 50,000 years, the nebula spreads over distances so far from the cooling central star that its nebulosity simply fades from view. The gases then mingle and mix with the surrounding interstellar medium, enriching it with moderately heavy elements, such as carbon.

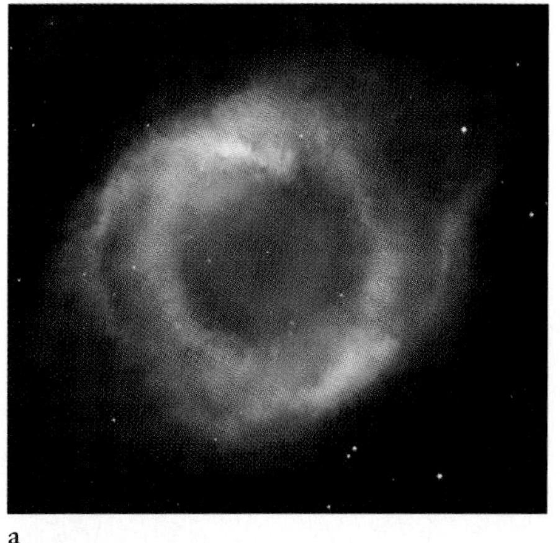
a

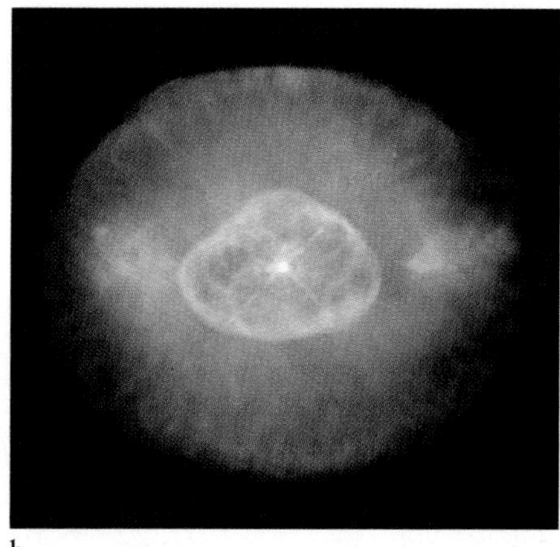
b

c

R I V U X G

WEB LINK 13.3 **FIGURE 13-4 Some Shapes of Planetary Nebulae** The outer shells of dying low-mass stars are ejected in a wonderful variety of patterns. **(a)** NGC 7293, the Helix Nebula, is located in the constellation Aquarius. The star that ejected these gases is seen at the center of the glowing shell. This nebula, located about 700 ly (215 pc) from Earth, has an angular diameter equal to about half that of the full Moon. **(b)** NGC 6826 shows jets of gas (in red) whose origin is as yet unknown. **(c)** Mz 3 (Menzel 3), in the constellation Norma (the Carpenter's Level), is 3000 ly (900 pc) from Earth. The dying star, creating these bubbles of gas, may be part of a binary system. (a: NASA, NOAO, ESA, The Hubble Helix Nebula Team, M. Meixner/STScI, and T. A. Rector/NRAO; b: Howard Bons, STScI/Robin Ciardullo, Pennsylvania State University/NASA; c: AURA/STScI/NASA)

13-2 The burned-out core of a low-mass star becomes a white dwarf

Our models of stellar evolution predict that low-mass stars that form a planetary nebula never develop the central pressures or temperatures necessary to ignite thermonuclear reactions of the carbon or oxygen in their cores. Instead, as we have just seen, the process of mass ejection strips away the stars' outer layers and exposes the carbon-oxygen cores, which simply cool off. Such burned-out hulks are called **white dwarfs.** This means that the electrons in them are compressed so close together that the Pauli exclusion

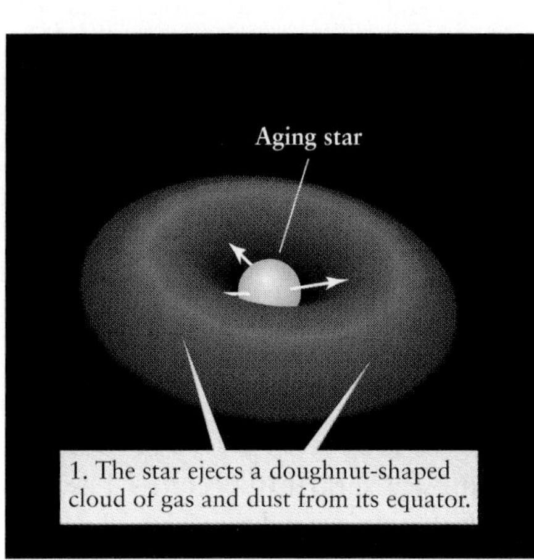
Aging star
1. The star ejects a doughnut-shaped cloud of gas and dust from its equator.
a

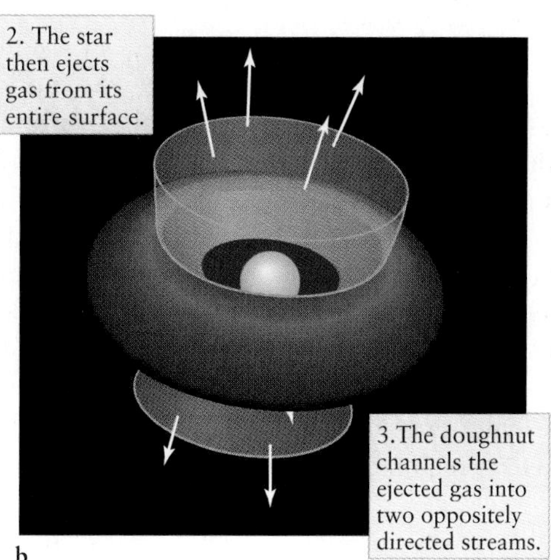
2. The star then ejects gas from its entire surface.
3.The doughnut channels the ejected gas into two oppositely directed streams.
b

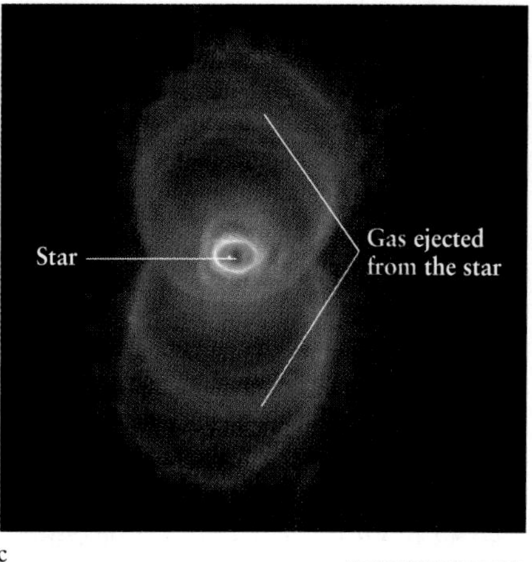
Star
Gas ejected from the star
R I V U X G
c

FIGURE 13-5 Formation of a Bipolar Planetary Nebula Bipolar planetary nebulae may form in two steps. Astronomers hypothesize that **(a)** first, a doughnut-shaped cloud of gas and dust is emitted from the star's equator, **(b)** followed by outflow that is channeled by the original gas to squirt out perpendicular to the plane of the doughnut. **(c)** The Hourglass Nebula appears to be a textbook example of such a system. The bright ring is believed to be the doughnut-shaped region of gas lit by energy from the planetary nebula. The Hourglass is located about 8000 ly (2500 pc) from Earth. (c: R. Sahai and J. Trauer, JPL; WFPC-2 Science Team; and NASA)

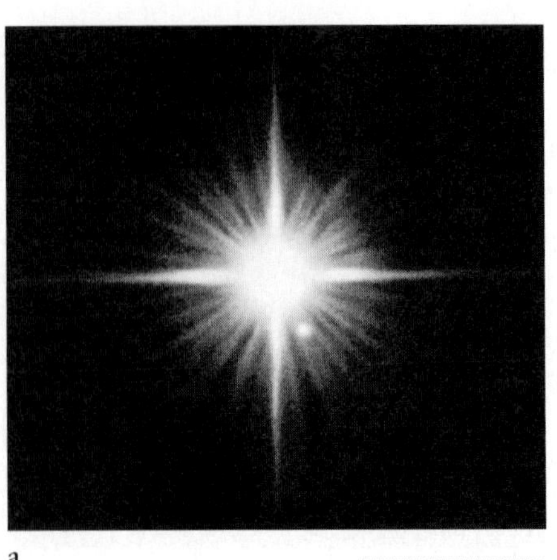

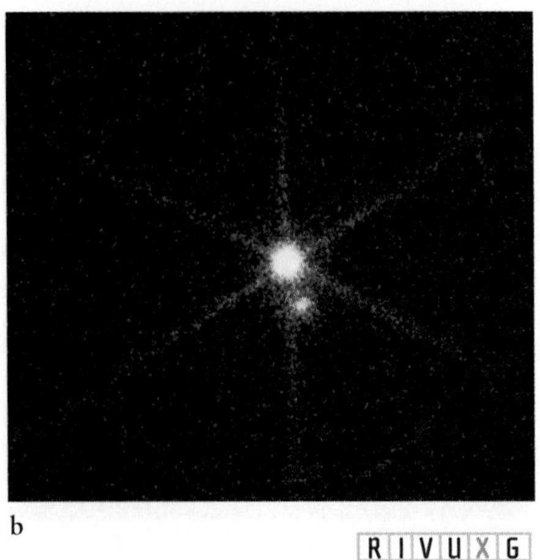

FIGURE 13-6 Sirius and Its White Dwarf Companion (a) Sirius, the brightest-appearing star in the night sky, is actually a double star. The smaller star, Sirius B, is a white dwarf, seen here at the five o'clock position in the glare of Sirius. The spikes and rays around the bright star, Sirius A, are created by optical effects within the telescope. (b) Since Sirius A (11,000 K) and Sirius B (30,000 K) are hot blackbodies, they are strong emitters of X rays. (a: R. B. Minton; b: NASA/SAO/CXC)

principle comes into effect. This principle says that identical particles cannot be in the same place at the same time. The word "identical" means two particles of the same kind that both have the same amount of energy. To avoid being identical to their neighbors, many electrons vibrate much faster than their neighbors and therefore have more energy (see Section 12-10). The energy of the electrons provides extra outward pressure that prevents further collapse. The Sun's core will eventually become a white dwarf.

Theoretical models predict that degenerate matter, such as the electrons in a white dwarf, can withstand only a finite amount of pressure, above which that matter cannot exert enough outward force to prevent the core from collapsing further. This limits the mass of a white dwarf to less than 1.4 $M_{\odot}$. This mass is called the **Chandrasekhar limit,** after the astrophysicist Subrahmanyan Chandrasekhar, who received a Nobel prize in 1990 for his pioneering theoretical studies of white dwarfs a half-century earlier.

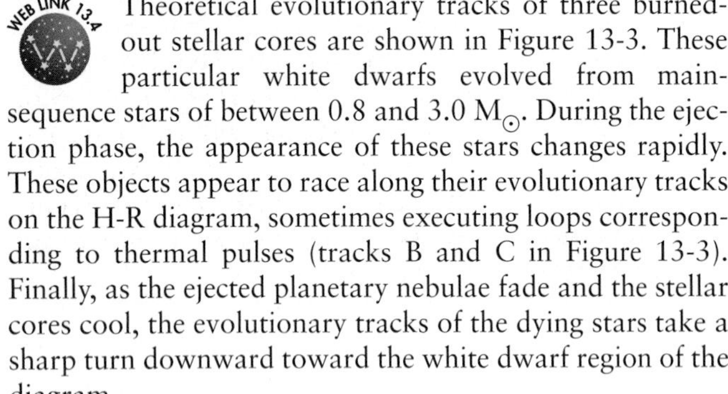

Theoretical evolutionary tracks of three burned-out stellar cores are shown in Figure 13-3. These particular white dwarfs evolved from main-sequence stars of between 0.8 and 3.0 $M_{\odot}$. During the ejection phase, the appearance of these stars changes rapidly. These objects appear to race along their evolutionary tracks on the H-R diagram, sometimes executing loops corresponding to thermal pulses (tracks B and C in Figure 13-3). Finally, as the ejected planetary nebulae fade and the stellar cores cool, the evolutionary tracks of the dying stars take a sharp turn downward toward the white dwarf region of the diagram.

As noted earlier, the temperature inside a white dwarf is too low to ignite the carbon or oxygen. Electron degeneracy pressure (discussed in Section 12-10) prevents the crushing weight of the carbon and oxygen from contracting the remnant core further, thereby creating a stable white dwarf roughly the size of the Earth and typically covered with a thin coating of hydrogen and helium. The densities of the carbon-oxygen interiors of these stellar "corpses" are typically 10^9 kg/m^3. In other words, a teaspoonful of white dwarf matter brought to Earth would weigh 5 tons.

As billions of years pass, an isolated white dwarf radiates its stored energy into space, cooling and decreasing in luminosity. It moves down and to the right on the H-R diagram. The Sun will become a cold, dark, dense sphere of degenerate gases rich in oxygen and carbon, about the size of the Earth. We can actually follow the theory a little further. It predicts that after billions of years of radiating away their heat, the interior temperatures of white dwarfs will decrease to about 4000 K, and the carbon and oxygen in them will solidify, transforming them into giant crystals.

Many white dwarfs are found in our solar neighborhood, but all are too faint to be seen with the naked eye. The first white dwarf to be discovered is a companion to the bright star Sirius. The binary nature of Sirius was first deduced in 1844 by the German astronomer Friedrich Bessel, who noticed that the star was moving back and forth, as if orbited by an unseen object. This companion, called Sirius B, was first glimpsed in 1862 (Figure 13-6). Recent satellite observations at ultraviolet wavelengths—where white dwarfs emit most of their light—demonstrate that the surface temperature of Sirius B is about 30,000 K.

13-3 White dwarfs in close binary systems can create powerful explosions

Occasionally, a star in the sky suddenly becomes between ten thousand and a million times brighter than it is normally. This phenomenon is called a **nova.** Its abrupt rise in brightness is followed by a gradual decline that may stretch over several months or more (Figure 13-7 and Figure 13-8). Astronomers typically observe two or three novae in our Galaxy each year. Novae are fairly common, with about 20 estimated to occur in the Milky Way Galaxy each year. We do not see all of them because of the gas and dust that obscures many parts of the Galaxy.

FIGURE 13-7 **Nova Herculis 1934** These two pictures show a nova (a) shortly after peak brightness as a magnitude −3 star and (b) 2 months later, when it had faded to magnitude +12. Novae are named after the constellation and year in which they appear. (UCO/Lick Observatory)

2 Painstaking observations strongly suggest that novae occur in close binary systems containing a white dwarf. The ordinary companion star fills its Roche lobe, so it gradually deposits fresh hydrogen onto the white dwarf. This new mass becomes a dense layer covering the hot surface of the white dwarf. As more gas is deposited and compressed, the temperature in the hydrogen layer continues to increase. Finally, at about 10^7 K, hydrogen fusion ignites throughout the layer, blowing it into interstellar space. This explosion is the nova. After a nova, fusion ceases on the white dwarf. The companion star, however, may retain enough mass to supply a new layer of surface hydrogen, enabling some novae to reoccur. White dwarfs are not damaged by novae initiated on their surfaces.

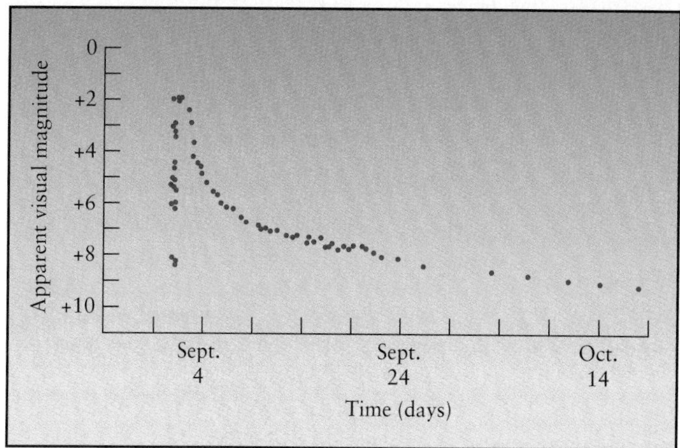

FIGURE 13-8 **The Light Curve of a Nova** This graph shows the history of Nova Cygni 1975, a nova that was observed to blaze forth in the constellation of Cygnus in September 1975. The rapid rise in magnitude followed by a gradual decline is characteristic of many novae, although some oscillate in intensity as they become dimmer.

Recent visible observations by the Hubble Space Telescope and X-ray observations by *Chandra* reveal that novae are much more complex than astronomers had thought. Rather than smooth shells of expanding gas, they are composed of thousands of clumps of gas. These observations are stimulating astrophysicists to reexamine the details of the nova mechanism described earlier to try to explain these newly observed effects.

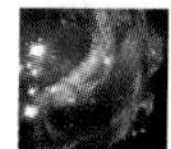

INTERMEDIATE-MASS STARS, HIGH-MASS STARS, AND SUPERNOVAE

In intermediate-mass (between 8 and 25 $M_\odot$) and high-mass (greater than 25 $M_\odot$) main-sequence stars, sufficient gravitational force compresses the core to heat it enough to enable the carbon and oxygen there to fuse. This leads to the creation of a host of other elements that are eventually ejected into interstellar space.

13-4 A series of fusion reactions in intermediate-mass and high-mass stars leads to luminous supergiants

When helium fusion ends in the core of a star more massive than 8 $M_{\odot}$, gravitational compression collapses the carbon-oxygen core, driving the star's central temperature above 600 million K. Now, carbon fusion begins, producing such elements as neon and magnesium. When a star compresses itself enough to drive its central temperature to 1.2 billion K, neon fusion occurs. When the central temperature of the star then reaches 1.5 billion K, oxygen fusion begins. As the star consumes increasingly heavier nuclei, it produces sulfur and isotopes of silicon and phosphorus, among other elements.

As the core temperature increases, its fusion rate soars. Also, each succeeding core stage has fewer and fewer nuclei than the previous one. For example, because each helium atom is made up of four hydrogen atoms, there are only one quarter as many helium atoms in a star's core than there were hydrogen atoms. Combining these two effects reveals that each cycle of core thermonuclear fusion occurs more rapidly than the last. For example, detailed calculations for a star that was initially 25 $M_{\odot}$ on the main sequence demonstrate that carbon fusion occurs for 600 years, neon fusion for 1 year, and oxygen fusion for only 6 months. After half a year of core oxygen fusion, gravitational compression forces the central temperature up to 2.7 billion K and silicon fusion begins. This thermonuclear process proceeds so furiously that the entire core supply of silicon in a 25-$M_{\odot}$ star is used up in 1 day.

Each stage of fusion adds a new shell of matter outside the core (Figure 13-9), creating a structure resembling the layers of an enormous onion. Together, the tremendous numbers of fusion-generated photons created in all these shells, as well as in the core, push the outer layers of the star outward. It expands to become a luminous supergiant almost as wide across as Jupiter's orbit around the Sun.

Intermediate-mass and high-mass main-sequence stars evolve into luminous supergiant stars. These latter stars are brighter than 10^5 $L_{\odot}$, emit winds throughout most of their existence, and have mass-loss rates exceeding those of giants (see Section 12-9). Figure 12-14a shows a supergiant star losing mass. Betelgeuse (see Figure 11-2a), 470 ly away in the constellation of Orion, is another good example of a supergiant experiencing mass loss. Spectroscopic observations show that Betelgeuse is losing mass at the rate of 1.7×10^{-7} $M_{\odot}$ per year and is surrounded by ejected gas, its *circumstellar shell*. This huge shell is expanding at 10 km/s, and escaping gases have been detected at distances of 10,000 AU from the star. The expanding circumstellar shell has an overall diameter of ⅓ ly.

Silicon fusion in intermediate- and high-mass stars involves many types of nuclear reactions, but its major final product is iron. Eventually, the core is converted entirely into iron, which is surrounded by layers of shell fusion that consume the star's remaining reserves of fuel (see Figure 13-9). Whereas the star's enormously bloated atmosphere is nearly as big as the orbit of Jupiter, its entire energy-producing region is again contained in a volume the size of the Earth. The buildup of an inert core of iron nuclei and electrons ionized from them signals the impending violent death of a massive star.

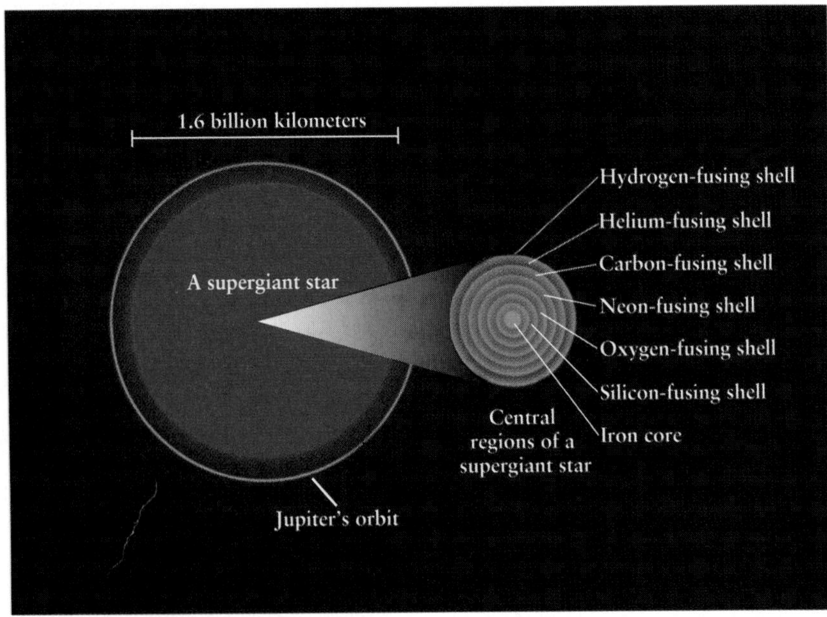

FIGURE 13-9 **The Structure of an Old High-Mass Star** Near the end of its life, a high-mass star becomes a supergiant with a diameter almost as wide as the orbit of Jupiter. The star's energy comes from six concentric fusing shells, all contained within a volume roughly the same size as the Earth.

13-5 Both intermediate-mass and high-mass stars blow apart in violent supernova explosions

Unlike lighter elements, iron cannot fuel further thermonuclear reactions. The protons and neutrons inside iron nuclei are already so tightly bound together that no further energy can be extracted by fusing still more nuclei together. The sequence of fusion stages in the cores of high-mass stars therefore ends.

Because iron atoms do not fuse and emit energy, the electrons in the core must now support the star's outer layers by the strength of electron degeneracy pressure alone. Soon, however, the continued deposition of fresh iron from the silicon-fusing shell causes the core's mass to exceed the Chandrasekhar limit. Electron degeneracy suddenly fails to support the star's enormous weight, and the core collapses.

Any isolated main-sequence star of more than 8 $M_\odot$ develops an iron core as a luminous supergiant. When the core collapses, a rapid series of cataclysms is triggered that tears the star apart in a few seconds. Let us see how this happens in the death of a 25-$M_\odot$ star according to computer simulations (Table 13-1).

In a 25-$M_\odot$ star, electron degeneracy pressure fails when the density is sufficiently high inside the iron core. The core, only some 3000 km in diameter, then collapses immediately. In roughly 1⁄10 second, the central temperature exceeds 5 billion K. Gamma-ray photons associated with this intense heat have so much energy that they begin to break apart the iron nuclei, a process called **photodisintegration.** Although millions of years passed from the time the star arrived on the main sequence with a hydrogen- and helium-filled core until its core became iron, it takes less than a second to convert the core back into elemental protons, neutrons, and electrons!

Within another 1⁄10 second, as the density continues to climb, the electrons in the core are forced to combine with protons there to produce neutrons, and the process releases a flood of neutrinos. As we saw in Section 10-9, neutrinos have no electric charge and very little mass, and most pass through the Earth or Sun without interaction. Nevertheless, the matter deep inside a collapsing high-mass star is so fantastically dense that the newly created neutrinos there cannot immediately escape from the star's core.

According to the calculations, about 1⁄4 second after the collapse begins, the density of the entire core reaches 4×10^{17} kg/m^3, which is *nuclear density,* the density at which neutrons and protons are normally packed together inside atomic nuclei. (Compare this with the density of water, 10^3 kg/m^3.) When the neutron-rich material of the core reaches nuclear density, it suddenly stiffens. This means that it resists the collapse much more than when it was less dense. The collapse of the core halts so abruptly that it rebounds and begins rushing back out in a process called *core bounce.*

The model predicts that during this critical stage, the star's unsupported layers of shell-fusing matter are plunging inward at up to 15% of the speed of light. The outward-flowing neutrinos and rebounding core slam into this matter (the red regions in Figure 13-10a). The impact stops the core's rebound while causing the infalling matter to reverse course. In just a fraction of a second, a tremendous volume of matter begins to move back up toward the star's surface. This matter accelerates rapidly as it encounters less and less resistance, and soon it forms an outgoing shock wave. After a few hours, this shock wave reaches the star's surface, lifting the star's outer layers away from the core in a mighty blast. The star becomes a **supernova.**

TABLE 13-1 Evolutionary Stages of a 25-$M_\odot$ Star

Stage	Central temperature (K)	Central density (kg/m^3)	Duration of stage
Hydrogen fusion	4×10^7	5×10^3	7×10^6 yr
Helium fusion	2×10^8	7×10^5	5×10^5 yr
Carbon fusion	6×10^8	2×10^8	600 yr
Neon fusion	1.2×10^9	4×10^9	1 yr
Oxygen fusion	1.5×10^9	1×10^{10}	6 mo
Silicon fusion	2.7×10^9	3×10^{10}	1 d
Core collapse	5.4×10^9	3×10^{12}	0.2 s
Core bounce	2.3×10^{10}	4×10^{17}	milliseconds
Supernova explosion	about 10^9	varies	hours

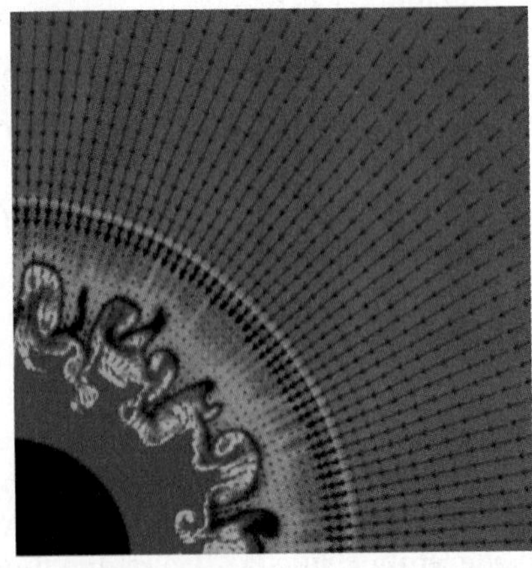

a 10 milliseconds

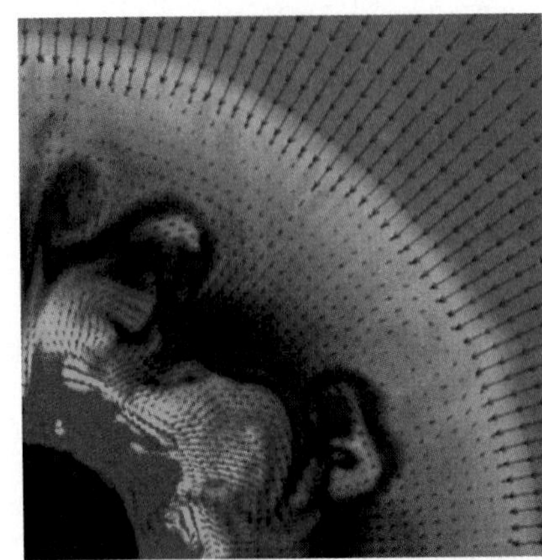

b 20 milliseconds

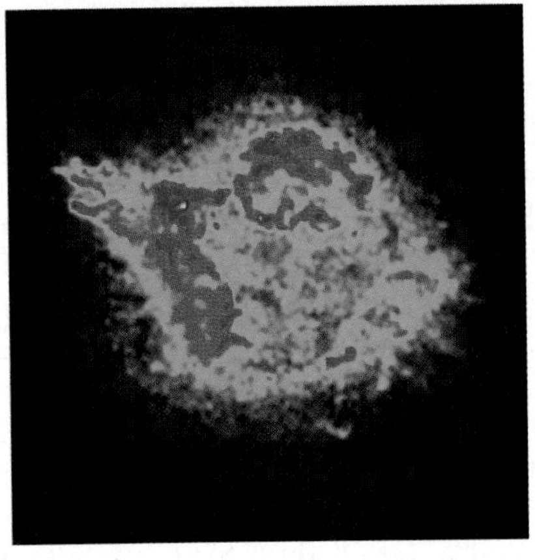

c Silicon

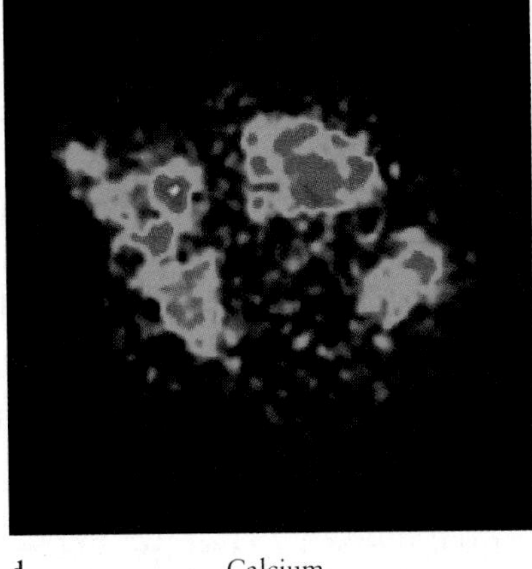

d Calcium

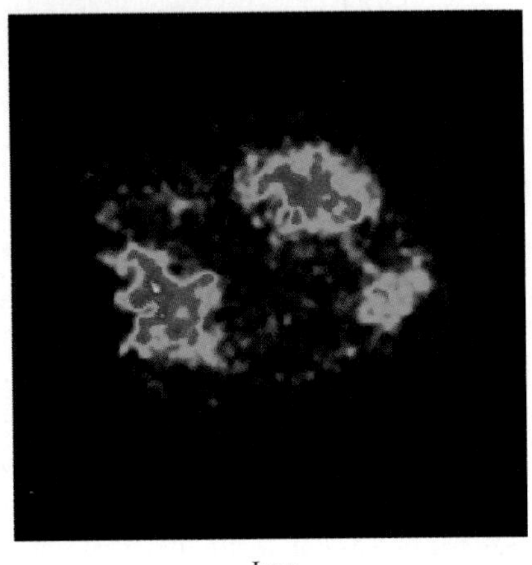

e Iron

R I V U X G

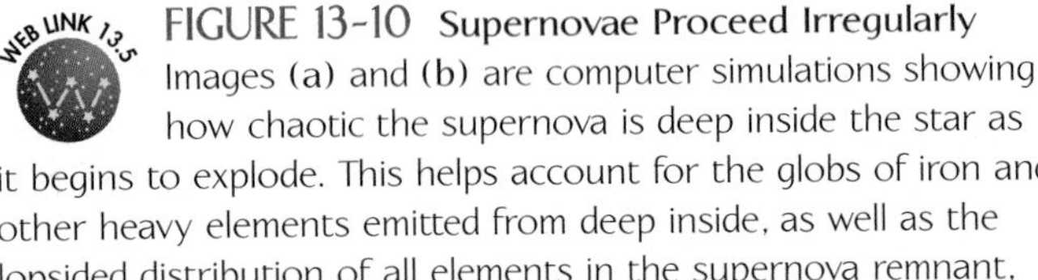

WEB LINK 13.5

FIGURE 13-10 Supernovae Proceed Irregularly Images (a) and (b) are computer simulations showing how chaotic the supernova is deep inside the star as it begins to explode. This helps account for the globs of iron and other heavy elements emitted from deep inside, as well as the lopsided distribution of all elements in the supernova remnant, as shown in (c), (d), and (e). The latter three are X-ray images of a supernova remnant taken by *Chandra* at different wavelengths. (a and b: Courtesy of Adam Burrows, University of Arizona and Bruce Fryxell, NASA/GSFC; c, d, and e: U. Hwang et al., NASA/GSFC)

As the supernova expands, the star's luminosity, which was already some 10^4 and 10^5 $L_{\odot}$, suddenly increases by another factor of between 10^3 and 10^6. The brightest supernova can therefore emit nearly as much energy as all 200 billion stars in the Milky Way Galaxy combined! In other words, for a few days following the explosion, a supernova may shine as brightly as an entire galaxy.

3 As the layers of a massive dying star are blasted into space, they are compressed so much by the neutrinos and the shock wave that fusion actually occurs during the explosion, creating a broad assortment of elements. Indeed, the products of shell fusion and this final burst of fusion during supernovae are where many of the elements on the Earth and other terrestrial planets come from, including many of the atoms in our bodies. We really are made of stardust.

Computer simulations indicate that the onion skins of different elements do not just move out of a star going supernova in lockstep (Figures 13-10a, b). Therefore, the ejected gases are not expected to be uniform shells of matter. Indeed, observations reveal that the material emerging in a supernova comes out quite irregularly. Figures 13-10c, d, and e show the distributions of some of the elements in the supernova Cassiopeia A (Cas A). Other observations show that some of the dense elements, like iron, ejected in supernovae are actually created deep inside their innermost silicon and oxygen layers and then shot out like bullets or geysers. Other elements are created throughout the expanding star. However, calculations and observations reveal that supernovae do not create enough of the heaviest elements, like gold, silver, and platinum, to account for the amounts of these elements found on Earth. We will explore their possible origin in Section 13-12.

Over its lifetime, our model 25-$M_\odot$ star ejects more than 20 $M_\odot$ of its mass back into space via stellar winds, gas ejected during unstable periods, and during its supernova phase. Less massive stars return proportionately less mass to the interstellar medium.

INSIGHT INTO SCIENCE

The Process of Science The best scientific theories and models also provide explanations about matters that they were not initially designed to study. In modeling stellar evolution and supernovae, the processes that create and eject metals (that is, elements heavier than hydrogen and helium) from stars also explain how the elements necessary for life are both created and made available.

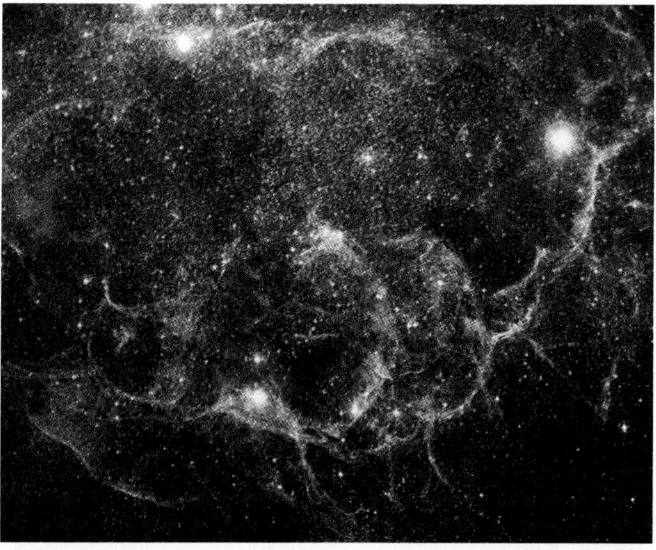

R I V U X G

FIGURE 13-11 **The Gum Nebula** The Gum Nebula is the largest known supernova remnant. It spans 60° across the sky and is centered roughly on the southern constellation of Vela. The nearest portions of this expanding nebula are only 300 ly from the Earth. The supernova explosion occurred about 11,000 years ago, and its remnant now has a diameter of about 2300 ly. Only the central regions of the nebula are shown here. (Royal Observatory, Edinburgh)

13-6 Supernovae occur in our Galaxy, among many others

Vigorous stellar evolution in our Milky Way Galaxy occurs primarily in the Galaxy's disk, because stars form in the giant molecular clouds located there. The disk (which we are in and which we see edge-on as the milky-white band of stars and glowing gas and dust across the sky) is also where supernovae explode. Astronomers find the debris of local supernova explosions in the Milky Way scattered across the Galaxy's disk. A beautiful example is the Cygnus Loop (see Figure 12-7). The star's outer layers were blasted into space 20,000 years ago so violently that they are still traveling at supersonic speeds. As this expanding shell of gas plows through the interstellar medium, it collides with atoms and molecules, making the gases glow.

Many supernova remnants cover sizable fractions of the sky. The largest is the Gum Nebula, with an angular diameter of 60° (the central region shown in Figure 13-11). This nebula looks so big because it is so close: Its near side is only about 300 ly from Earth. Studies of the Gum Nebula's expansion rate suggest that the supernova exploded about 11,000 years ago. At maximum brilliance, the exploding star probably was as bright as the Moon at first quarter.

The Milky Way's disk is so filled with interstellar gas and dust that we cannot see very far into it. Supernovae are believed to erupt every few decades in remote parts of our Galaxy, but their detonations are hidden from optical telescopes by the interstellar medium. Many supernova remnants in our Galaxy can be detected only at nonvisible wavelengths, ranging from X rays through radio waves. For example, Figure 13-12 shows images of the supernova remnant Cassiopeia A as seen in X ray (Figure 13-12a) and radio (Figure 13-12b). Visible-light photographs of this part of the sky reveal only a few small, faint wisps. Thus, radio searches for supernova remnants are more fruitful than visual searches. Only two dozen supernova remnants have been found in visible-light photographs, but more than 3100 remnants in our Galaxy and in others have been discovered by radio astronomers.

From the expansion rate of the nebulosity in Cassiopeia A, astronomers calculate that the light from that supernova explosion first reached Earth about 300 years ago. Although telescopes were in wide use by the late 1600s, no record of the event is known. In fact, the last supernova of a star near its maximum brightness in our Galaxy seen by the naked eye was observed by Johannes Kepler in 1604. In 1572, Tycho Brahe also recorded the sudden appearance of an exceptionally bright star in the sky. In 2002, astronomers discovered that 2 million years ago a group of bright O and B stars called the Scorpius-Centaurus OB association passed within 150 light-years of Earth and that one or more supernovae in this group may have occurred at that time. Such a close explosion could have damaged the Earth's ozone layer and caused the extinction of some ocean life at the end of the Pliocene epoch.

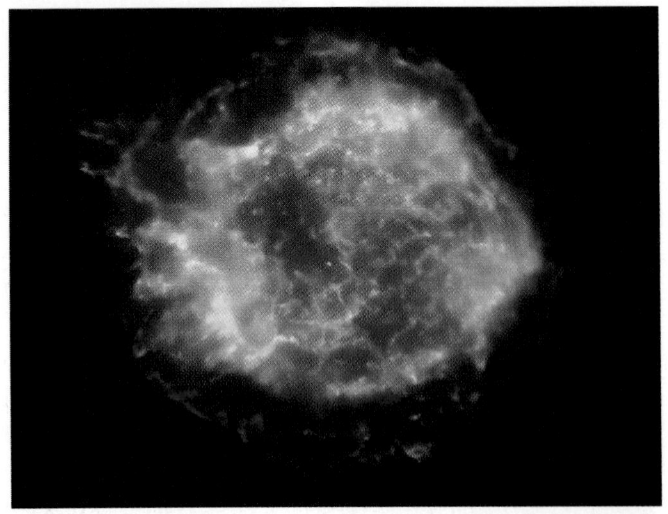

a

R I V U X G

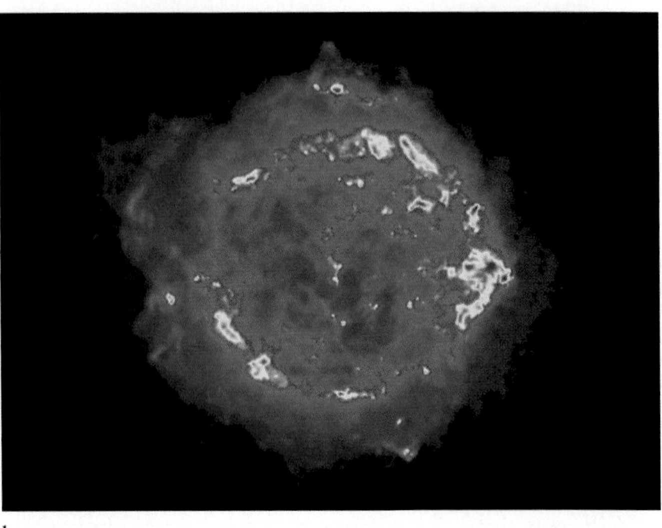

b

R I V U X G

WEB LINK 13.7

ANIMATION 13.2

FIGURE 13-12 **Cassiopeia A** Supernova remnants such as Cassiopeia A are typically strong sources of X rays and radio waves. (a) An X-ray picture of Cassiopeia A taken by *Chandra*. (b) A corresponding radio image produced by the Very Large Array (VLA). Radiation from the supernova that produced this nebula first reached Earth 300 years ago. The explosion occurred about 10,000 ly from here. (a: NASA/CSC/SAO; b: Very Large Array)

Besides the gas and dust enriched with heavy elements, supernovae leave a heritage long after their remnants have dissipated. Much of this gas and dust is recycled into new stars and planets, such as our solar system.

13-7 Cosmic rays are not rays at all

4 In observations connected with supernovae, astronomers have detected particles flying through space at speeds exceeding 90% the speed of light. These high-speed particles are called **cosmic rays.** They were named "rays" before their true identity as particles was known and, as often happens, this misleading name has "stuck." Cosmic rays are primarily hydrogen nuclei, along with nuclei of more massive elements like helium and carbon, and a few percent of electrons and positrons (positively charged electrons).

The origins of cosmic rays are just beginning to become clear. At first, many astronomers thought that moderate-energy cosmic rays were emitted directly by supernovae. This made sense because these explosions are the best-known source of the energy necessary to give such cosmic rays their tremendous speeds. However, observations by the

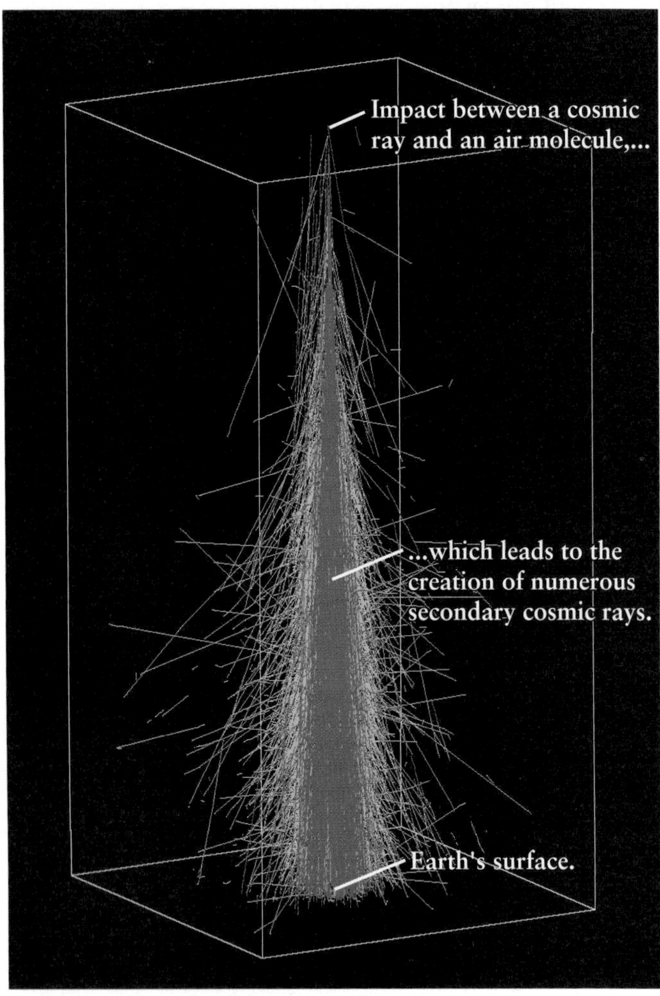
Impact between a cosmic ray and an air molecule,...
...which leads to the creation of numerous secondary cosmic rays.
Earth's surface.

▶ FIGURE 13-13 **Cosmic Ray Shower** Cosmic rays from space slam into particles in the atmosphere, breaking up the latter and sending them earthward. These debris are called secondary cosmic rays. This process of impact and breaking up continues as secondary cosmic rays travel downward, creating a cosmic ray shower, as depicted in this computer simulation. (Clem Pryke/University of Chicago)

Advanced Composition Explorer satellite reveal that the isotopes of nickel coming to Earth in the form of cosmic rays were not the isotopes emitted by supernovae. Instead, evidence mounts that most cosmic rays are created when supernova debris slams into gas already in space, causing some of this preexisting matter to accelerate and become cosmic rays.

Much less common than cosmic rays from supernova remnants are ultrahigh-energy cosmic rays that come from outside of our Galaxy. Each of these particles has as much energy as a baseball thrown by a major league pitcher. Observations of their paths suggest that many of them come from galaxies containing supermassive black holes, which we will discuss in Section 14-8. Whether these black holes generate the ultrahigh-energy cosmic rays is not yet known.

Most cosmic rays coming in our direction collide with gas in the atmosphere some 15 km above the Earth's surface. This is fortunate because each cosmic ray packs an enormous amount of energy that potentially could harm living tissue. The collision in the air divides a cosmic ray's energy among several gas particles, which are shoved earthward. These, in turn, often hit other gas atoms, creating a cascade of lower-energy particles that eventually reach the Earth's surface as a **cosmic ray shower** (Figure 13-13). These cosmic rays created from atoms in our atmosphere are called **secondary cosmic rays.**

13-8 Supernova 1987A offered a detailed look at a massive star's death

Supernovae are often seen in remote galaxies with the aid of a telescope. In 1885, a supernova in the Andromeda Galaxy was just barely visible to the unaided eye. On February 23, 1987, a supernova was observed in the Large Magellanic Cloud, a galaxy near our own Milky Way. The supernova, designated SN 1987A, was the first to be observed that year (Figure 13-14), and it was so bright that it could easily be seen with the naked eye. What made SN 1987A such an exciting discovery was that it gave astronomers a rare opportunity to study the death of a nearby massive star using modern equipment and thereby provide data with which to verify the theory of supernovae described earlier.

At first, SN 1987A reached only a tenth of the luminosity predicted for an exploding massive star. For the next 85 days, it gradually brightened before slowly dimming, as is characteristic of an ordinary supernova. Fortunately, the star had been observed before it became a supernova (Figure 13-14 insets). The Large Magellanic Cloud is about 160,000 ly from Earth—near enough to us that many of its stars have been individually observed and catalogued. The star had been identified as a B3 I supergiant. The theory of the evolution of such stars provided an explanation for SN 1987A's unusually slow brightening.

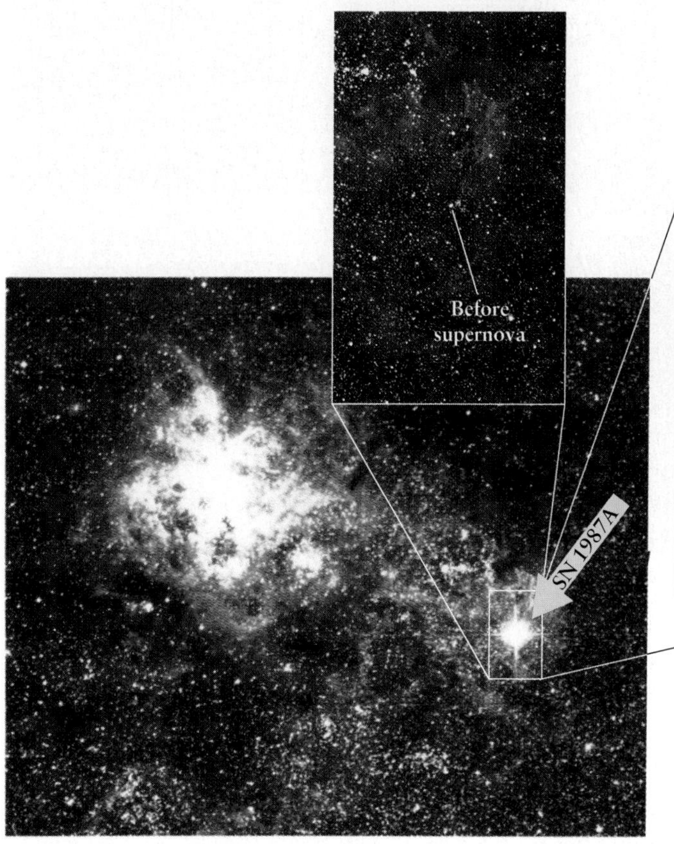

WEB LINK 13.8

FIGURE 13-14 **Supernova 1987A** A supernova was discovered in a nearby galaxy called the Large Magellanic Cloud (LMC) in 1987. This photograph shows a portion of the LMC that includes the supernova and a huge H II region called the Tarantula Nebula. At its maximum brightness, observers at southern latitudes saw the supernova without a telescope. **Insets:** The star before and after it exploded. (European Southern Observatory; insets: Anglo-Australian Observatory/David Malin Images)

When this star was on the main sequence, its mass was about 20 $M_\odot$, although by the time it exploded it had shed many solar masses. The evolutionary track for an aging 20-$M_\odot$ star wanders back and forth across the top of the H-R diagram, so the star alternates between being a hot (blue) supergiant and cool (red) supergiant. The star's size changes significantly as its surface temperature changes. A blue supergiant is only 10 times larger in diameter than the Sun, but a red supergiant of the same luminosity is 1000 times larger. Because the doomed star was a relatively small blue supergiant when it exploded, it reached only a tenth of the brightness that it would have attained had it been a red supergiant at that time.

WEB LINK 13.9 Ordinary telescopic observations of a supernova explosion can only show us the expanding outer layers of the dying star. Even X-ray or radio observations cannot see through the hot gases being blasted into space. Thus, using ordinary techniques, we cannot observe the extraordinary events occurring in and around the doomed star's core. However, most of the energy of a supernova explosion is carried away in a characteristic form: Recall that neutrinos are produced in great profusion as the star's core collapses. By detecting these neutrinos and measuring their properties, astronomers can learn many details about the star's collapsing core, especially about core bounce. Several neutrino detectors were operating when the neutrinos from SN 1987A reached Earth.

Nearly a day before SN 1987A was first observed in the sky, teams of scientists at neutrino detectors in Japan and the United States reported finding Cerenkov flashes (see Section 10-9) from a burst of neutrinos. The Kamiokande II detector in Japan detected 12 neutrinos at about the same time that 8 were found by the IMB (Irvine-Michigan-Brookhaven) detector in a salt mine under Lake Erie. Neutrinos preceded the visible supernova outburst because they escaped from the dying star before the shock wave from the collapsing core reached the star's surface. They were detected in the Earth's northern hemisphere, where the supernova is always below the horizon, after having passed through the Earth. The discovery of these neutrinos provided strong support for the theory that supernovae are caused by the collapse of the star's core, its subsequent bounce, and the flow of neutrinos that help cause the supernova, as described in Section 13-5.

About three and a half years after SN 1987A's detonation was first seen from Earth, astronomers using the Hubble

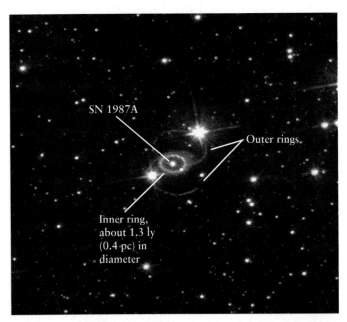

a Supernova 1987A seen in 1996

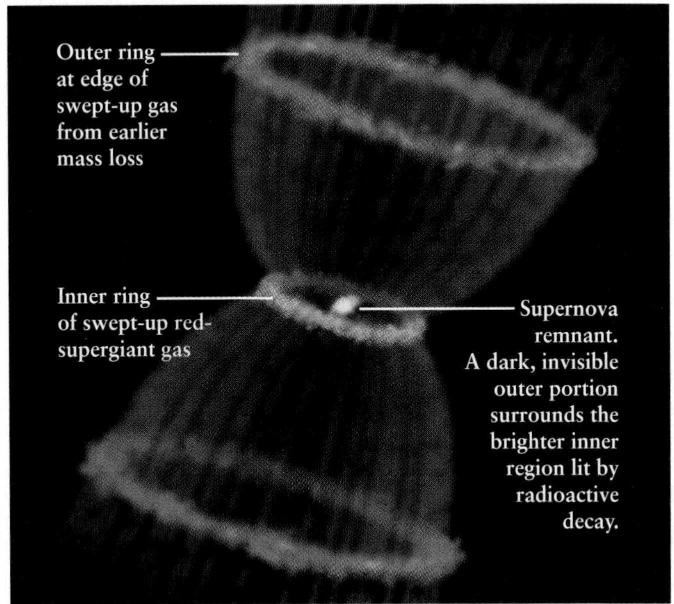

b An explanation of the rings

R I V U X G

FIGURE 13-15 Shells of Gas Around SN 1987A (a) Intense radiation from the supernova explosion caused three rings of gas surrounding SN 1987A to glow in this Hubble Space Telescope image. This gas was ejected from the star 20,000 years before it detonated. All three rings lie in parallel planes. The inner ring is about 1.3 ly across. The white and colored spots are unrelated stars. (b) When the progenitor star of SN 1987A was still a red supergiant, a slowly moving wind from the star filled the surrounding space with a thin gas. When the star contracted into a blue supergiant, it produced a faster-moving stellar wind. The interaction between the fast and slow winds somehow caused gases to pile up along an hourglass-shaped shell surrounding the star. The burst of ultraviolet radiation from the supernova ionized the gas in the rings, causing the rings to glow. The supernova itself, at the center of the hourglass, glows because of energy released from radioactive decay. (Robert P. Kirshner and Peter Challis, Harvard-Smithsonian Center for Astrophysics; STScI)

Space Telescope obtained a picture showing several rings of glowing gas around the exploded star (Figure 13-15). This gas had been emitted by the star 20,000 years before the supernova, when a hydrogen-rich stellar envelope was ejected by stellar winds from the doomed star, as discussed earlier in this chapter. This gas expanded in an hourglass shape, because it was blocked from expanding around the star's equator either by a preexisting ring of gas there or by the orbit of an as-yet-unseen companion star.

While emitting these gases, the doomed star was a red supergiant. In its final blue giant phase, the same star emitted higher-speed gas that compressed the hourglass gas into three rings—a narrow one around the equator and wider ones at the top and bottom of the hourglass. This relic gas is now being illuminated by photons from the supernova. In 1998, astronomers began to observe gas ejected from the supernova striking the preexisting circumstellar gas. This collision has caused the ring of gas to brighten, and it should eventually illuminate more of the earlier gas that had been ejected.

SN 1987A provided astronomers with invaluable information, in part because the doomed star had been studied before it exploded and its distance from Earth was known. The supernova was also located in an unobscured part of the sky, and neutrino detectors happened to be operating at the time of the outburst. Astronomers will be monitoring the progress of this supernova for years to come.

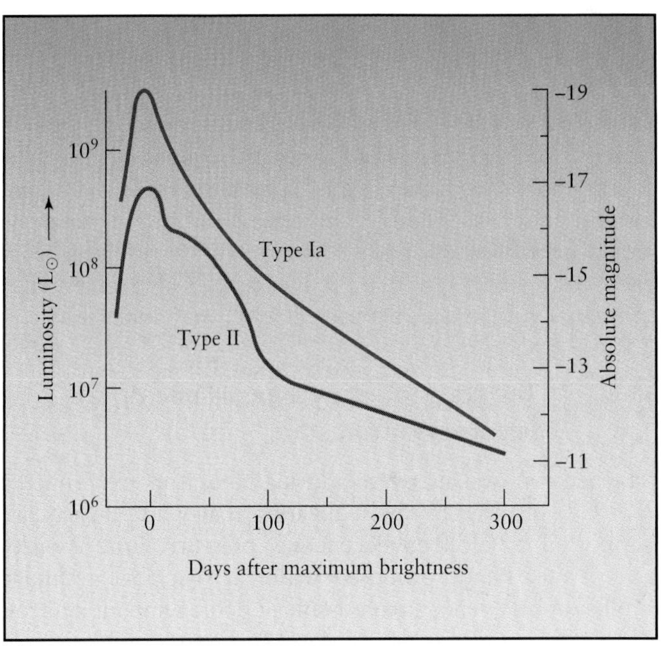

FIGURE 13-16 **Supernova Light Curves** A Type Ia supernova, which gradually declines in brightness, is caused by an exploding white dwarf in a close binary system. A Type II supernova is caused by the explosive death of a massive star and usually has alternating intervals of steep and gradual declines in brightness.

13-9 Accreting white dwarfs in close binary systems can also explode as supernovae

ANIMATION 13.3

Supernovae such as SN 1987A, which are the death throes of massive stars, are known as **Type II supernovae,** but not every supernova originated as a high-mass star. Observations reveal that others that we will encounter later in the book, known as **Type Ia supernovae,** begin as white dwarfs.

A supernova's spectrum indicates its type: Hydrogen lines are prominent in Type II supernovae but absent in Type Ia. Both types begin with a sudden rise in brightness (Figure 13-16). A Type Ia supernova typically reaches an absolute magnitude of –19 at peak brightness, while a Type II supernova usually peaks at –17. Type Ia supernovae then decline gradually for more than a year, whereas Type II supernovae alternate between periods of steep and gradual declines in brightness. Type II light curves therefore have a steplike appearance.

ANIMATION 13.4

A Type Ia supernova begins with a high-mass carbon-oxygen–rich white dwarf in a close, semi-detached binary system. As we saw in Chapters 11 and 12, mass transfer can occur in a close binary if one star overflows its Roche lobe (recall Figures 12-31, 12-32, and 12-33). To trigger a Type Ia supernova, a swollen giant companion star dumps gas onto a white dwarf. When the white dwarf's mass gets close to the Chandrasekhar limit, the increased pressure created by the additional mass enables carbon fusion to begin in the core. In a catastrophic runaway process reminiscent of the helium flash, the rate of carbon fusion skyrockets, and the star blows up.

A Type Ia supernova is powered by nuclear energy. What we see is simply the fallout from a gigantic thermonuclear explosion, which produces a wide array of radioactive isotopes. Especially abundant is an unstable isotope of nickel that decays into a radioactive isotope of cobalt. Most of the electromagnetic display of a Type Ia supernova, including the smooth decline of its light curve, results directly from the radioactive decay of this nickel and cobalt. The spectrum lacks hydrogen because the star that explodes is mostly carbon and oxygen to begin with.

Astronomers have seen more than 1000 supernovae in distant galaxies. These observations suggest that in a typical galaxy like the Milky Way, Type Ia supernovae occur approximately once every 36 years, while Type II supernovae occur about once every 44 years. Thus, there should be about five supernovae exploding in our Milky Way Galaxy each century. We now return to the cores of stars that explode as supernovae. Recall that the cores' expansions are stopped when they slam into the outer layers of their stars and help force them to explode.

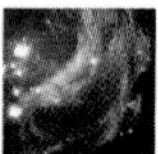

NEUTRON STARS AND PULSARS

When intermediate-mass main-sequence stars explode as supernovae, their expanding cores are stopped by the outer layers into which they slam. The cores recontract and remain intact as highly compressed clumps of neutrons called **neutron stars.** Isolated neutron stars resulting from these 8-$M_\odot$ to 25-$M_\odot$ main-sequence stars are stable stellar remnants with masses between 1.4 $M_\odot$ and about 3 $M_\odot$.

13-10 The cores of many Type II supernovae become neutron stars

The neutron was discovered during laboratory experiments in 1932. Inspired by the realization that white dwarfs are supported by electron degeneracy pressure, Fritz Zwicky and Walter Baade proposed within a year that a similar repulsion between neutrons could support remnant neutron cores. They used the fact that the Pauli exclusion principle also prevents neutrons with identical properties from packing too closely together. When neutrons are pressed together, many of them have to move rapidly (so as not to be identical to their neighbors). This motion provides a pressure, called **neutron degeneracy pressure,** that equations predict is even greater than electron degeneracy pressure (see Section 12-10) and thus allows stellar remnants with masses beyond the Chandrasekhar limit to be stable.

Most scientists ignored Zwicky and Baade's theory for years. After all, a neutron star seemed to be a rather unlikely object. To transform protons and electrons into neutrons, the density in the star would have to be equal to nuclear density, about 10^{17} kg/m^3. Thus, a teaspoon full of neutron star matter brought back to Earth would weigh about 1 billion tons. Furthermore, an object compacted to nuclear density would be very small. A 2-$M_\odot$ neutron star would have a diameter of only 20 km (12 mi) and would fit inside any major city on Earth. The surface gravity on one of these neutron stars would be so strong that the escape velocity would equal one-half the speed of light. All these conditions appeared to be so outrageous that few astronomers paid any serious attention to the subject of neutron stars—until 1968.

In that year, Jocelyn Bell, then a graduate student in radio astronomy at Cambridge University, noticed that the radio telescope she was using was detecting regular pulses from one particular location in the sky. Careful repetition of the observations demonstrated that the radio pulses were arriving with a regular period of 1.3373011 seconds. The regularity of this pulsating radio source was so striking that the Cambridge team suspected that they might be detecting signals from an advanced alien civilization, which accounts for the name first assigned to it: LGM1 (short for Little Green Men 1). This possibility was soon discarded as several more of these pulsating radio sources, which soon came to be known as **pulsars,** were discovered across the sky. The broad distribution of pulsars would require incredibly widespread civilizations, inconsistent with what we have found in our searches for alien intelligence (Chapter 19). In all cases, the periods were extremely regular, ranging between 0.2 second and 1.5 seconds (Figure 13-17).

When the discovery of pulsars was officially announced in early 1968, astronomers around the world began proposing all sorts of explanations. Many of these theories were bizarre, and arguments raged for months. We have already ruled out alien civilizations. Astronomers therefore looked for some recurring event in a star's life. We know that a star sheds matter as it swells to a giant. Perhaps some stars alternately expand and contract, emitting energy as they pulsate. However, this explanation also fails because any star expanding and contracting as fast as a pulsar pulses would explode.

Late in 1968, all controversy was laid to rest with the discovery of a pulsar in the middle of the Crab Nebula. In A.D. 1054, American Indian, Chinese, and possibly other peoples saw and recorded the appearance of a "guest star" in the constellation of Taurus. When we turn a telescope toward this location, we find the Crab Nebula, shown in Figure 13-18. It looks like the residue of an explosion and is, in fact, a supernova remnant. At the center of the Crab Nebula is the Crab pulsar (Figure 13-18b).

With expansion and contraction ruled out, the likely scenario is that pulsars are spinning objects. If the Crab pulsar is spinning, it is spinning too fast to be a white dwarf. Its period is 0.033 seconds, which means that it rotates 30 times each second. At that speed, something as wide as a white dwarf would immediately fly apart. To strengthen the case against pulsars being white dwarfs, astronomers discovered numerous other pulsars rotating

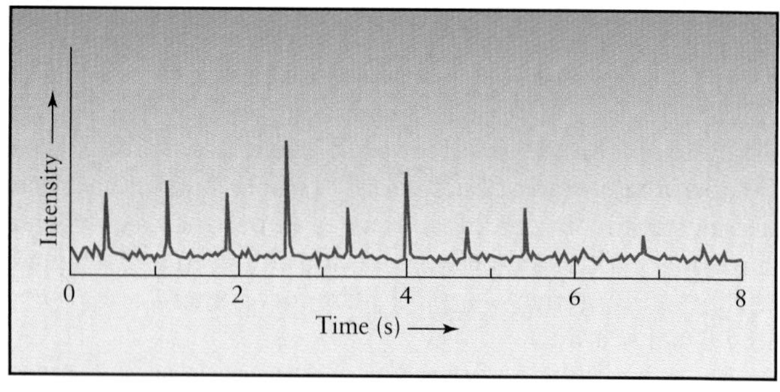

FIGURE 13-17 **A Recording of a Pulsar** This chart recording shows the intensity of radio emissions from one of the first pulsars to be discovered, PSR 0329+54. Note that some of the pulses are weak and others are strong. Nevertheless, the spacing between pulses is so regular (0.714 seconds) that it is more precise than most clocks on Earth.

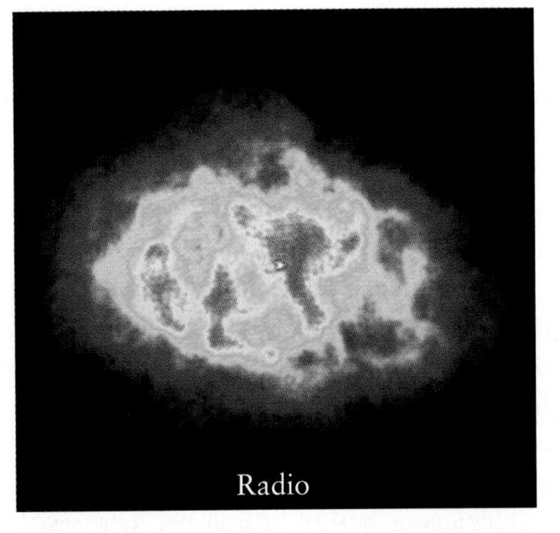

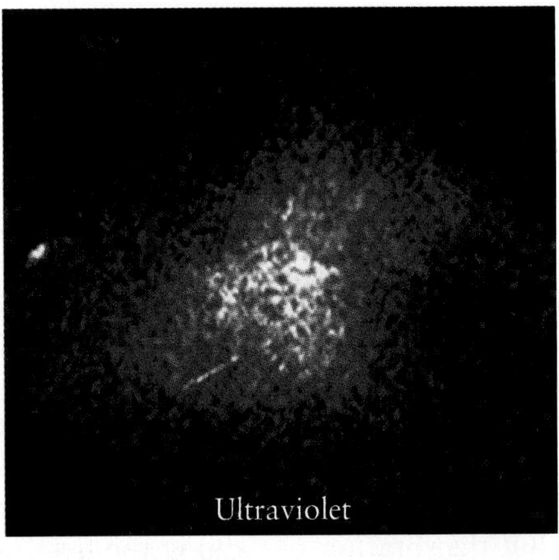

a

b

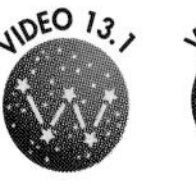

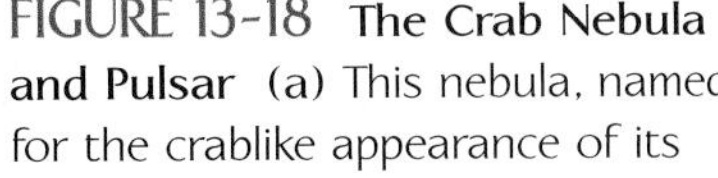

FIGURE 13-18 **The Crab Nebula and Pulsar** **(a)** This nebula, named for the crablike appearance of its filamentary structure in early visible-light telescope images, is the remnant of a supernova seen in A.D. 1054. The distance to the nebula is about 6000 ly, and its present angular size (4 by 6 arcmin) corresponds to linear dimensions of about 7 by 10 ly. Observations at different wavelengths give astronomers information about the nebula's chemistry, motion, history, and interactions with preexisting gas and dust. The gamma-ray image is only of the central region of the nebula. **(b)** The insets show the Crab pulsar in its "on" and "off" states. Both its visible flashes and X-ray pulses have identical periods of 0.033 second. (a: NASA; b: The FORS Team, VLT, European Southern Observatory; left insets: Lick Observatory; right insets: Einstein Observatory, Harvard-Smithsonian Center for Astrophysics)

even faster, including PSR 1937+21 (PSR for pulsar and 1937+21 for the object's right ascension and declination), a stellar remnant that pulses 640 times each second. Staying with the belief that they are spinning, astronomers concluded that pulsars must be incredibly compact. Calculations revealed that if neutron stars exist, they have a sufficiently small diameter to remain intact while rotating as fast as pulsars pulsate. Indeed, this was the only explanation that withstood the process of scientific scrutiny and, as a result, the theory of neutron stars described earlier was quickly developed.

13-11 A rotating magnetic field explains the pulses from a neutron star

Why are neutron stars rotating so fast and how do they emit pulses of electromagnetic energy? Based on observations of the rotation of the Sun, Betelgeuse, and other nearby stars, astronomers theorize that many, perhaps most, stars on the main sequence have some angular momentum (rotation). The Sun, for example, takes nearly a full month to rotate once about its axis. But the rotation rate of collapsing stars increases, just as pirouetting ice skaters speed up when they pull in their arms, as shown in Figure 2-12. (Recall from Section 2-7 that the total amount of angular momentum in an isolated system always remains constant.) The core of a high-mass main-sequence star rotating once a month spins faster than once a second when it collapses to the size of a neutron star.

In addition to rapid rotation, most neutron stars are believed to have intense magnetic fields. In an average star like our Sun, the magnetic field is spread out over millions upon millions of square kilometers just under the star's surface (see Chapter 10). However, when a star of solar dimensions collapses down to a neutron star, its magnetic field becomes very concentrated, in the same way as stalks of wheat in a field become concentrated when you gather them in a bundle (Figure 13-19). The strength of a collapsing star's magnetic field increases to as much as 10^{15} times the Earth's magnetic field. A magnet that strong located halfway to the Moon would lift metal rods off the Earth.

The axis of rotation of a typical neutron star is not the 5
same as the axis connecting its north and south magnetic poles (Figure 13-20), just as the magnetic and rotation axes of the Sun and planets (except Saturn) are different from each other. As the neutron star rotates, its powerful magnetic field therefore rapidly changes direction. Like a giant electric generator, the star creates intense electric fields, which act on protons and electrons near its surface. The powerful electric fields channel these charged particles, causing them to flow out from the neutron star's polar regions, as sketched in Figure 13-20. As the particles stream along the field, they accelerate and emit energy. The result is two very thin beams of radiation pouring out of the neutron star's north and south magnetic polar regions and sweeping through space—a pulsar.

This explanation for pulsars is often called the **lighthouse model.** A rotating, magnetized neutron star is somewhat like a lighthouse beacon. The magnetic field causes nearby gases to accelerate and radiate. Depending on the conditions, this radiation can range from radio waves to gamma rays, and many pulsars emit in more than one range. Calculations indicate that this radiation is beamed and as the star rotates, it sweeps around the sky. If the Earth happens to be located in the right direction, a brief flash can be observed each time a beam whips past our line of sight. As a

a

b

FIGURE 13-19 **Analogy for How Magnetic Field Strengths Increase** When planted, these wheat stalks cover a much larger area than when they are harvested and bound together. A star's magnetic field behaves similarly, as the collapsing star carries the field inward, thereby increasing its strength. (a: Corbis; b: Oscar Burriel/Photo Researchers, Inc.)

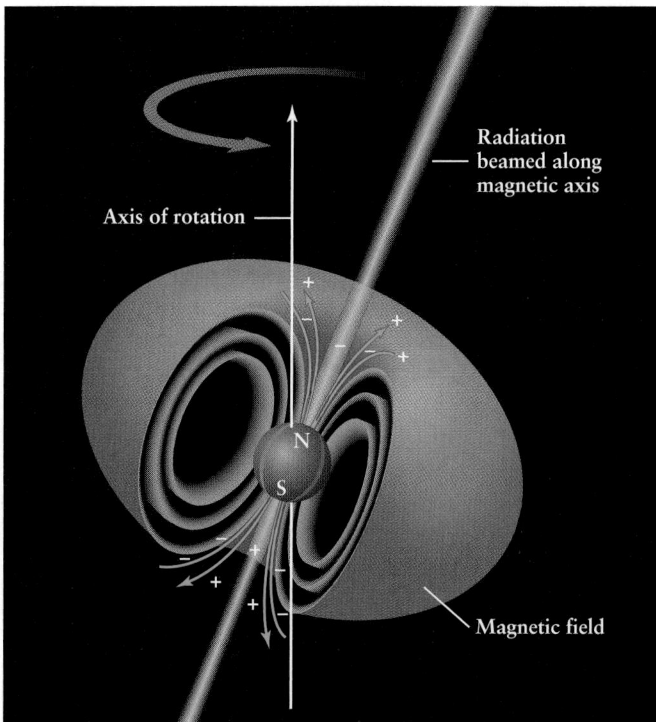

FIGURE 13-20 **A Rotating, Magnetized Neutron Star** Calculations reveal that many neutron stars rotate rapidly and possess powerful magnetic fields. Charged particles are accelerated near a neutron star's magnetic poles and produce two oppositely directed beams of radiation. As the star rotates, the beams sweep around the sky. If the Earth happens to lie in the path of a beam, we see the neutron star as a pulsar.

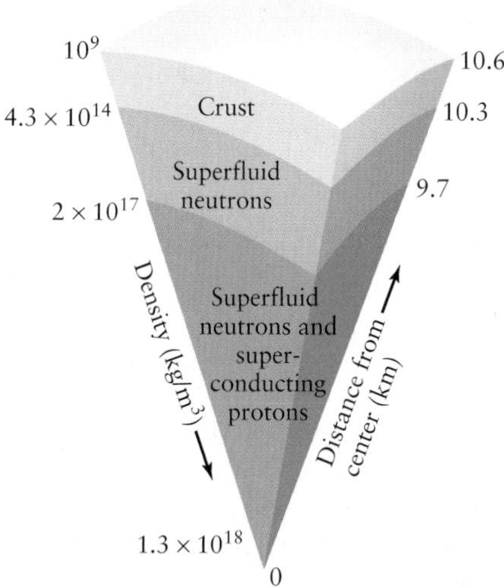

FIGURE 13-21 **Model of a Neutron Star's Interior** This drawing shows the theoretical model of a 1.4-$M_\odot$ neutron star. The neutron star has a superconducting, superfluid core 9.7 km in radius, surrounded by a 0.6-km-thick mantle of superfluid neutrons. The neutron star's crust is only 0.3 km thick (the length of four football fields) and is composed of heavy nuclei (such as neutrons) and free electrons. The thicknesses of the layers are not shown to scale.

result of its neutron star's rotation, the Crab Nebula is flashing on and off 30 times each second (see Figure 13-18).

The Crab pulsar is one of the youngest known pulsars, its creation having been observed some 950 years ago. Also visibly flashing is the Vela pulsar at the core of the Gum Nebula (see Figure 13-11). The Vela pulsar, with a period of 0.089 second, is the slowest pulsar ever detected at visual wavelengths. Because they use some of their energy of rotation to create the pulses they emit, *isolated pulsars slow down as they get older.* Only the very youngest pulsars are energetic enough to emit visible flashes along with their radio pulses, and the Vela pulsar was created about 11,000 years ago.

Neutron stars are not composed solely of neutrons. Astronomers have developed a theoretical model for the surface and internal structure of a neutron star (Figure 13-21). Its interior has a radius of about 10 km, with a core of superconducting protons and superfluid neutrons. A *superconductor* is a material in which electricity and heat flow without losing energy, while a *superfluid* has the strange property that it flows without any friction. Both superconductors and superfluids have been created in the laboratory. Surrounding a neutron star's core is a layer of superfluid neutrons. The surface of the neutron star is a solid, brittle crust of dense nuclei and electrons about ⅓ km thick. The gravitational force of the neutron star is so great at its surface that climbing a bump just a millimeter high there would take more energy than it takes to climb Mount Everest. Neutron stars may also have atmospheres, as indicated by absorption lines in the spectrum of at least one neutron star. The absorption lines are believed to be due to scattering of radiation from the neutron star by gases in its atmosphere, just as scattering in the Sun's atmosphere causes absorption lines in its spectrum.

Just as the Earth's crust has earthquakes, calculations show that the crusts of rotating neutron stars have equivalent events. When a rotating neutron star (pulsar) undergoes such a quake, its rotation rate suddenly changes to conserve its angular momentum. This creates a **glitch** in the rate at which the pulsar is slowing down. Such events have been observed in a number of pulsars (Figure 13-22).

While most pulsars appear to be isolated bodies, more than 90 have been discovered with companions. These pulsars are called *binary pulsars* and they are of interest for several reasons. First, many binary pulsars pull mass onto themselves from their companions. This infall causes the pulsar to speed up (like the skater in Figure 2-12 pulling in her arms). As a result of this effect, there are pulsars that spin nearly a thousand times a second and are called

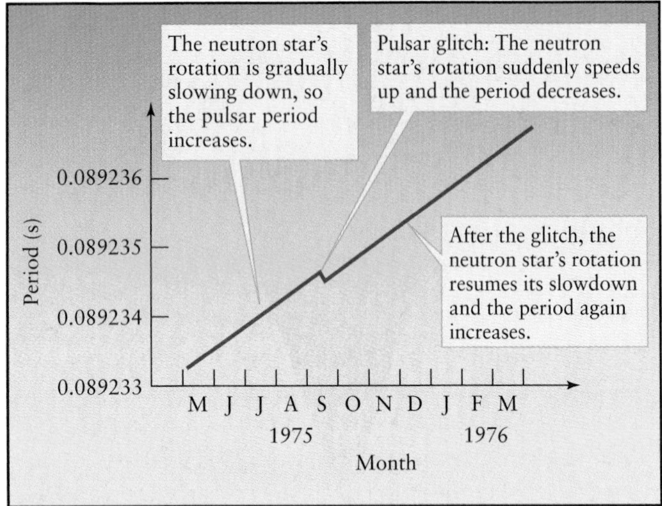
The neutron star's rotation is gradually slowing down, so the pulsar period increases.
Pulsar glitch: The neutron star's rotation suddenly speeds up and the period decreases.
After the glitch, the neutron star's rotation resumes its slowdown and the period again increases.
Period (s)
0.089236
0.089235
0.089234
0.089233
M J J A S O N D J F M
1975
1976
Month

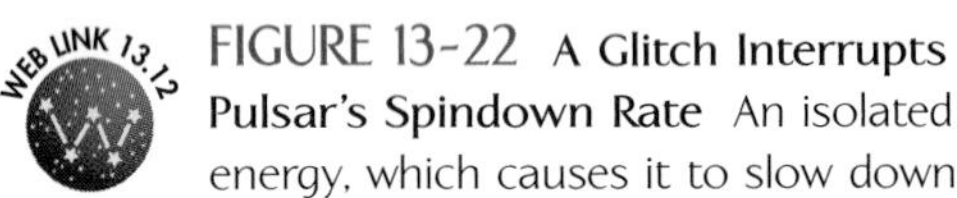

FIGURE 13-22 **A Glitch Interrupts the Vela Pulsar's Spindown Rate** An isolated pulsar radiates energy, which causes it to slow down. This "spin down" is not always smooth. As it slows down, it becomes more circular, and so its spinning, solid surface must readjust its shape. Since the surface is brittle, this readjustment is often sudden, like the cracking of glass, which causes the angular momentum of the pulsar to suddenly jump. Such an event, shown here for the Vela pulsar in 1975, changes the pulsar's rotation period and is called a glitch.

FIGURE 13-23 **Double Pulsar** This artist's conception shows two pulsars orbiting their center of mass. The double pulsar they represent is called PSR J0737-3039, which is about 1500 ly from Earth in the constellation Puppis. One pulsar has a 23-millisecond period and the other has a 2.8-second period. The two orbit once every 2.4 hours. (Michael Kramer/Jodrell Bank Observatory, University of Manchester)

millisecond pulsars. Second, some pulsars, like the one labeled PSR 1916+13, have compact massive companions, which are often also neutron stars. Observations of binary pulsars with neutron star companions reveal that the two bodies are spiraling toward each other, as predicted by Einstein's gravitational theory called general relativity (see Sections 2-1 and 14-5 for details).

In 2004, astronomers observed for the first time a pair of pulsars in orbit around each other (Figure 13-23). This discovery provides an exciting opportunity to further test the predictions of general relativity. For example, this theory predicts that these two pulsars should be precessing (the directions of their rotation axes should be changing; see Section 1-8) and that like the binary pulsars discussed directly above, they should be spiraling in toward each other. Observations of this system in the next few years will test these predictions.

13-12 Colliding neutron stars may provide some of the heavy elements in the universe

Nucleosynthesis, the process of creating the heavier elements, was discussed in Sections 13-4 and 13-5. We saw how a variety of elements up to iron are created inside massive stars before the supernova and then ejected into space during supernovae, while even heavier elements are created and ejected during the supernovae themselves. However, models indicate that supernovae do not create enough of the elements heavier than iron to account for the amounts of these elements found in the universe.

Computer simulations of what happens during collisions of two neutron stars predict that in these impacts, some of the neutrons are splashed into space, where many of them decay back into protons and form various elements. The interactions that follow create a variety of heavy elements in amounts consistent with observations.

13-13 Binary neutron stars create pulsating X-ray sources

Neutron stars in binary systems may also hold the key to another regular pulse from the sky—pulses from X-ray sources. During the 1960s, astronomers obtained tantalizing X-ray views of the sky during short rocket and balloon flights that briefly lifted X-ray detectors above the Earth's atmosphere. Several strong X-ray sources were discovered, and each was named after the constellation in which it was located.

Astronomers were so intrigued by these preliminary discoveries that NASA built and launched *Explorer 42*, an X-ray–detecting satellite that could make observations 24 hours a day. The satellite was launched from Kenya (to place it in an orbit above the Earth's equator) in 1970, on the seventh anniversary of Kenyan independence. To com-

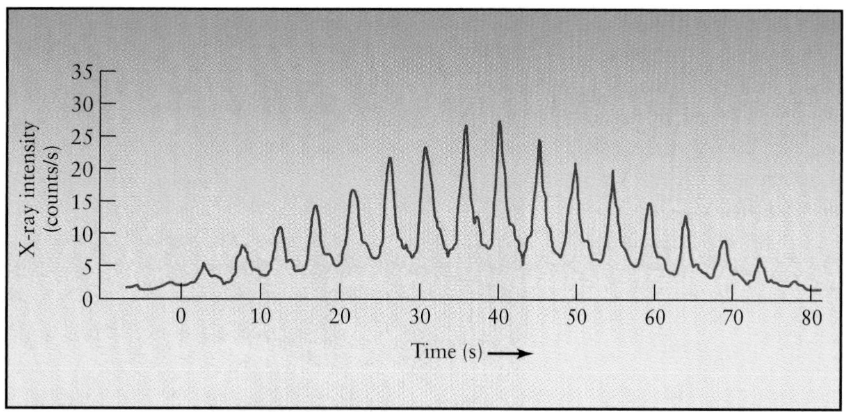

FIGURE 13-24 **X-Ray Pulses from Centaurus X-3** This graph shows the intensity of X rays detected by *Uhuru* as Centaurus X-3 moved across the satellite's field of view. The successive pulses are separated by 4.84 seconds. The variation in the height of the pulses was a result of the changing orientation of *Uhuru*'s X-ray detectors toward the source as the satellite rotated.

memorate the occasion, *Explorer 42* was renamed *Uhuru*, which means "freedom" in Swahili.

Uhuru gave us our first comprehensive look at the X-ray sky. As the satellite slowly rotated, its X-ray detectors swept across the heavens. Each time an X-ray source came into view, signals were transmitted to receiving stations on the ground. Before its battery and transmitter failed in early 1973, *Uhuru* had succeeded in locating 339 X-ray sources.

The discovery of pulsars was still fresh in everyone's mind when the *Uhuru* team discovered X-ray pulses coming from Centaurus X-3 in early 1971. Figure 13-24 shows data from one sweep of *Uhuru*'s detectors across Centaurus X-3. The pulses have a regular period of 4.84 seconds. A few months later, similar pulses were discovered coming from a source called Hercules X-1, which has a period of 1.24 seconds. Because the periods of these two X-ray sources are so short, astronomers suspected that they had found rapidly rotating neutron stars.

It soon became clear, however, that systems such as Centaurus X-3 and Hercules X-1 are not ordinary pulsars like the Crab or Vela pulsars. Every 2087 days, Centaurus X-3 completely turns off for almost 12 hours. This fact suggests that Centaurus X-3 is an eclipsing binary (see Figure 11-15) and that the X-ray source takes nearly 12 hours to pass behind its companion star.

The case for the binary nature of Hercules X-1 is even more compelling. It has an off state corresponding to a 6-hour eclipse every 1.7 days, and careful timing of the X-ray pulses shows a periodic Doppler shift every 1.7 days. This information provides direct evidence of orbital motion about a companion star. When the X-ray source is approaching us, its pulses are separated by slightly less than 1.24 seconds. When the source is receding from us, slightly more than 1.24 seconds elapse between the pulses.

Astronomers now realize that systems such as Centaurus X-3 and Hercules X-1 are examples of binary stars in which one is a neutron star. Because all these binaries have very short orbital periods, the distance between the ordinary star and the neutron star must be small. This proximity enables the neutron star to capture gas escaping from the companion star.

To explain the pulsations of X-ray sources such as Centaurus X-3 or Hercules X-1, astronomers propose that the ordinary star either fills or nearly fills its Roche lobe. Either way, matter escapes from the star. If the star fills its lobe, as in the case of Hercules X-1, mass loss results from direct overflow through the Roche lobe. If the star's surface lies just inside its lobe, as with Centaurus X-3 (Figure 13-25), a stellar wind carries off the mass. A typical rate of mass loss for the ordinary star in such a pair is approximately 10^{-9} $M_{\odot}$ per year.

The neutron star in a pulsating X-ray source, like an ordinary pulsar, rotates rapidly and has a powerful magnetic field inclined to the axis of rotation (recall Figure 13-20). As the gas from the companion star falls toward the neutron star, the magnetic field funnels the incoming matter down onto its magnetic polar regions. The neutron star's gravity is so strong that the gas is traveling at nearly half the speed of light by the time it crashes onto the star's surface. This violent impact creates hot spots at both poles, with temperatures of about 10^8 K, that emit abundant X rays with a luminosity nearly 100,000 times brighter than the Sun. As the neutron star rotates, two beams of X rays from the polar caps sweep around the sky. If the Earth happens to be in the path of one of the beams, we can observe this pulsating X-ray source. The pulse period is thus equal to the neutron star's rotation period. For example, the neutron star in Hercules X-1 completes one rotation every 1.24 seconds.

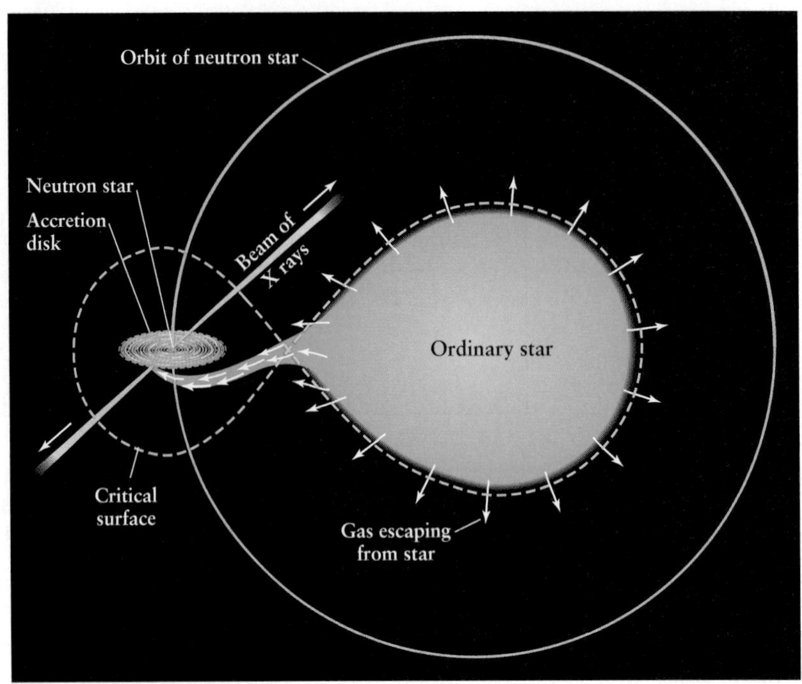

FIGURE 13-25 **A Model of a Pulsating X-Ray Source** Gas transfers from an ordinary star to the neutron star. The infalling gas is funneled down onto the neutron star's magnetic poles, where it strikes the star with enough energy to create two X-ray–emitting hot spots. As the neutron star spins, beams of X rays from the hot spots sweep around the sky.

13-14 Neutron stars in binary systems can also emit powerful isolated bursts of X rays

Neutron stars can acquire additional mass from a companion star, which leads to the flaring up of the neutron stars. Beginning in late 1975, astronomers analyzing data from X-ray satellites detected sudden, powerful bursts of radiation. The record of a typical burst is shown in Figure 13-26. The source, called an **X-ray burster,** emits X rays at a constant low level; then, suddenly and without warning, there is an abrupt increase followed by a gradual decline. A typical burst lasts for 20 seconds. Several dozen X-ray bursters have been located, most lying in the plane of the Milky Way Galaxy.

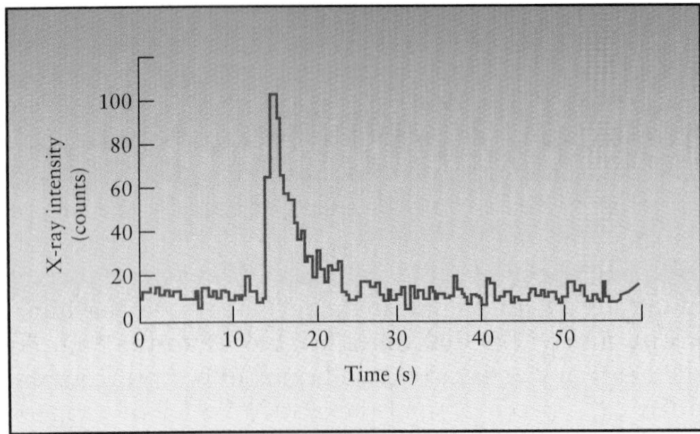

WEB LINK 13.13 FIGURE 13-26 **X Rays from an X-Ray Burster** A burster emits X rays with a constant low intensity interspersed with occasional powerful bursts. This burst was recorded in September 1975 by an X-ray telescope that was pointed toward the globular cluster NGC 6624.

X-ray bursters, like novae, are believed to arise from mass transfer in binary star systems. With a burster, however, the stellar corpse is a neutron star rather than a white dwarf. Gases escaping from the ordinary companion star fall onto the neutron star. The energy released as this gas crashes down onto the neutron star's surface produces the low-level X rays that are continuously emitted by the burster.

Most of the gas falling onto the neutron star is hydrogen, which becomes compressed against the hot surface of the star by the star's powerful surface gravity. In fact, temperatures and pressures in this accreting layer are so high that the arriving hydrogen is promptly converted by fusion into helium. Constant hydrogen fusion soon produces a layer of helium that covers the entire neutron star. When the helium layer is about 1 m thick, helium fusion ignites explosively, and we observe a sudden burst of X rays. In other words, whereas explosive hydrogen fusion on a white dwarf produces a nova, explosive helium fusion on a neutron star produces an X-ray burster.

13-15 There may exist smaller, more exotic stellar remnants composed of quarks

In death as well as in life, the mass of a star determines its fate. Just as there is an upper limit to the mass of a white dwarf (the Chandrasekhar limit), there is also an upper limit to the mass of a neutron star, called the *Oppenheimer–Volkov limit.* Above this limit, neutron degeneracy pressure cannot support the overbearing weight of the star's matter pressing in from all sides. The Chandrasekhar limit for a white dwarf

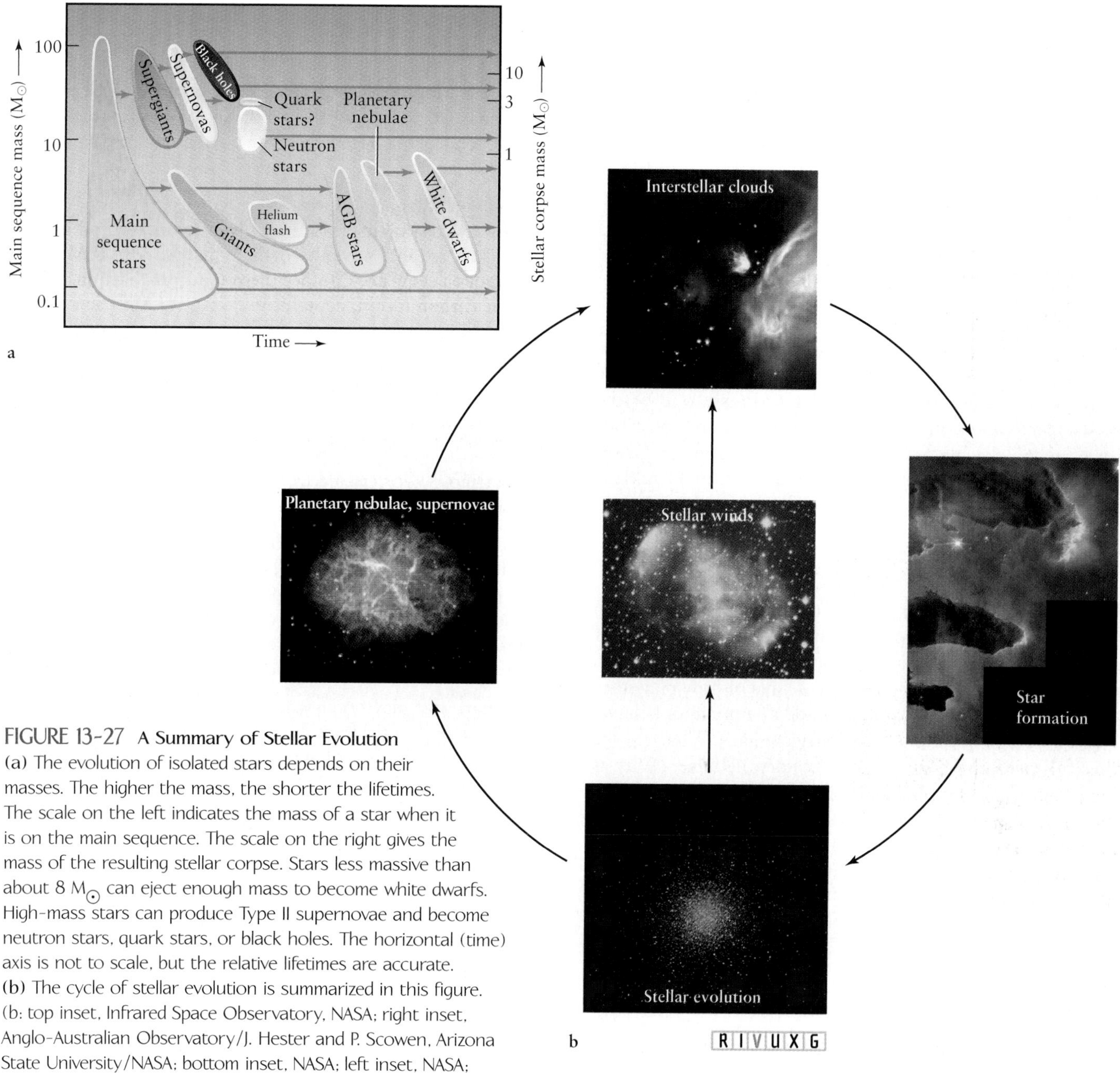

FIGURE 13-27 **A Summary of Stellar Evolution**
(a) The evolution of isolated stars depends on their masses. The higher the mass, the shorter the lifetimes. The scale on the left indicates the mass of a star when it is on the main sequence. The scale on the right gives the mass of the resulting stellar corpse. Stars less massive than about 8 $M_\odot$ can eject enough mass to become white dwarfs. High-mass stars can produce Type II supernovae and become neutron stars, quark stars, or black holes. The horizontal (time) axis is not to scale, but the relative lifetimes are accurate.
(b) The cycle of stellar evolution is summarized in this figure.
(b: top inset, Infrared Space Observatory, NASA; right inset, Anglo-Australian Observatory/J. Hester and P. Scowen, Arizona State University/NASA; bottom inset, NASA; left inset, NASA; middle inset, Anglo-Australian Observatory)

is 1.4 $M_\odot$, and the Oppenheimer–Volkov limit for a neutron star is about 3 $M_\odot$.

In 2002, astronomers observed X rays from two very dense objects, 3C58 and RXJ1856.5-3754, that are too small and too cool to fit the accepted model of neutron stars. These may be the first examples of stellar remnants with slightly more mass than the Oppenheimer–Volkov limit called *quark stars*. According to calculations first done in 1964 and experiments verifying the theory done in 1968, many elementary particles such as neutrons (and protons, but not electrons) are actually composed of smaller particles called **quarks.**

Normally, quarks cannot exist as separate objects—in the cases of neutrons and protons they clump together in groups of 3. However, when the pressure on them is great enough, neutrons can become so compressed that they dissolve into their constituent quarks. As free particles, quarks obey the Pauli exclusion principle, meaning that identical ones cannot occupy the same place at the same time. As with electrons and neutrons, quarks therefore create a pressure that resists collapse, thereby creating quark stars. The upper limit for the mass of quark stars is unknown, but it will not be much greater than the Oppenheimer–Volkov limit.

If the remnant core from a supernova is above the mass limit of neutron stars (and quark stars, if their existence is verified), then there is another fate in store for it. The gravitational attraction near a neutron star or quark star is so strong that the particles escaping from its surface must be moving at roughly half the speed of light. If the stellar corpse has slightly greater than 3 $M_{\odot}$, there is so much matter crushed into such a small volume that the repulsion created by the Pauli exclusion principle on its neutrons and quarks is quickly overcome. The remnant therefore collapses further and the escape velocity exceeds the speed of light. Because nothing can travel faster than light, nothing—not even light—can leave such a collapsed object. The discovery of neutron stars and the calculations regarding the limit to the Pauli exclusion principle inspired astrophysicists to consider seriously one of the most bizarre and fantastic objects ever predicted by modern science—the black hole, which we examine in Chapter 14. Figure 13-27 summarizes the evolutionary history of stars.

13-16 Frontiers yet to be discovered

The incredible telescopes now in existence or scheduled to go on-line soon are revolutionizing our understanding of the final stages of star formation. In coming years, astronomers hope to pin down the details of how planetary nebulae, novae, supernovae, and stellar remnants behave. Among specific questions to be answered are: What are the physical processes that cause planetary nebulae? Why is there such a variety of shapes of planetary nebulae? What causes the hourglass shape of so many stellar remnants? Why are novae composed of many blobs of gas? Do quark stars really exist, and, if so, what are their properties?

Summary of Key Ideas

- Stars with different masses fuse different elements.
- Stars lose mass via stellar winds throughout their lives.

Low-Mass Stars and Planetary Nebulae

- A low-mass (below 8 $M_{\odot}$) main-sequence star becomes a giant when hydrogen shell fusion begins. It becomes a horizontal-branch star when core helium fusion begins. It enters the asymptotic giant branch and becomes a supergiant when helium shell fusion starts.
- Stellar winds during the thermal pulse phase eject mass from the star's outer layers.
- The burned-out core of a low-mass star becomes a dense carbon-oxygen body, called a white dwarf, with about the same diameter as the Earth. The maximum mass of a white dwarf (the Chandrasekhar limit) is 1.4 $M_{\odot}$.
- Explosive hydrogen fusion may occur in the surface layer of a white dwarf in some close binary systems, producing sudden increases in luminosity that we call novae.
- An accreting white dwarf in a close binary system can also become a supernova when carbon fusion ignites explosively throughout such a degenerate star. Such a detonation is called a Type Ia supernova.

High-Mass Stars and Supernovae

- After exhausting its central supply of hydrogen and helium, the core of an intermediate- or high-mass (above 8 $M_{\odot}$) star undergoes a sequence of other thermonuclear reactions. These include carbon fusion, neon fusion, oxygen fusion, and silicon fusion. This last fusion eventually produces an iron core.
- An intermediate- or high-mass star dies in a supernova explosion that ejects most of the star's matter into space at very high speeds. This Type II supernova is triggered by the gravitational collapse of the doomed star's core.
- Neutrinos were detected from Supernova 1987A, which was visible to the naked eye.

Neutron Stars, Pulsars, and Quark Stars

- A high-mass star's dead core becomes a neutron star, a quark star, or even a black hole (Chapter 14). A neutron star is a very dense stellar corpse consisting of closely packed neutrons in a sphere roughly 20 km in diameter. The maximum mass of a neutron star, called the Oppenheimer–Volkov limit, is about 3 $M_{\odot}$.
- A quark star may occur when neutrons dissolve into quarks, which can oppose the force of gravity slightly more strongly than can neutrons. The upper limit to quark star masses is believed to be only slightly above the Oppenheimer–Volkov limit.
- A pulsar is a rapidly rotating neutron star with a powerful magnetic field that makes it a source of periodic radio and other electromagnetic pulses. Energy pours out of the polar regions of the neutron star in intense beams that sweep across the sky.
- Some X-ray sources exhibit regular pulses. These objects are thought to be neutron stars in close binary systems with ordinary stars.
- Explosive helium fusion may occur in the surface layer of a companion neutron star, producing the sudden increase in X-ray radiation called an X-ray burster.

WHAT DID YOU THINK?

1 *Will the Sun someday cease to shine brightly? If so, how?* The Sun will shed matter as a planetary nebula in about 6 billion years and then cease nuclear fusion. Its remnant white dwarf will dim over the succeeding billions of years.

2 *What is a nova?* A nova is a relatively gentle explosion of hydrogen gas on the surface of a white dwarf in a binary star system.

3 *What are the origins of the carbon, silicon, oxygen, iron, uranium, and other heavy elements on Earth?* These elements are created during stellar evolution, by supernovae, and by colliding neutron stars.

4 *What are cosmic rays?* Cosmic rays are high-speed particles (mostly hydrogen and other atomic nuclei) in space. Many of them are believed to have been created as a result of supernovae.

5 *What is a pulsar?* A pulsar is a rotating neutron star in which the magnetic field's axis does not coincide with the rotation axis. The beam of radiation it emits sweeps across our region of space.

Key Words

asymptotic giant branch (AGB) star, 351
Chandrasekhar limit, 354
cosmic ray, 360
cosmic ray shower, 361
glitch, 367
helium shell flash, 351
helium shell fusion, 351
lighthouse model, 366
neutron degeneracy pressure, 364
neutron star, 364
nova (*plural* novae), 354
photodisintegration, 357
planetary nebula, 352
pulsar, 364
quark, 371
secondary cosmic ray, 361
supernova, 357
Type Ia supernova, 363
Type II supernova, 363
white dwarf, 353
X-ray burster, 370

Review Questions

1. Will the Sun shed most of its mass and, if so, what is that event called? **a.** yes, as a planetary nebula. **b.** yes, as a supernova. **c.** yes, as a white dwarf. **d.** yes, as a neutron blast. **e.** no.

2. A white dwarf is primarily composed of **a.** neutrons. **b.** hydrogen and helium. **c.** iron. **d.** cosmic rays. **e.** carbon and oxygen.

3. What prevents a neutron star from collapsing? **a.** hydrogen fusion. **b.** friction. **c.** electron degeneracy pressure. **d.** neutron degeneracy pressure. **e.** helium fusion.

4. What is the difference between a giant and a supergiant?

5. Why is knowing the temperature in a star's core so important in determining which nuclear reactions can occur there?

6. What determines the temperature in the core of a star?

7. What is a planetary nebula and how does it form?

8. What is the Chandrasekhar limit?

9. What is a neutron star?

10. Compare a white dwarf and a neutron star. Which of these stellar corpses is more common? Why?

11. To test your understanding of the stages of stellar evolution, do Interactive Exercise 13-1. You can print out your results, if required.

12. What is the Oppenheimer–Volkov limit?

13. On an H-R diagram, sketch the evolutionary track that the Sun will follow between the time it leaves the main sequence and when it becomes a white dwarf. Approximately how much mass will the Sun have when it becomes a white dwarf? Where will the rest of the mass have gone?

14. Why have searches for supernova remnants at visible wavelengths been less fruitful than searches at other wavelengths?

15. To test your understanding of how stars "die," do Interactive Exercise 13-2. You can print out your results, if required.

16. To test your understanding of neutron stars, do Interactive Exercise 13-3. You can print out your results, if required.

17. Why do astronomers believe that pulsars are rapidly rotating neutron stars?

18. To test your understanding of rotating neutron stars, pulsars, and novae, do Interactive Exercise 13-4. You can print out your results, if required.

19. What is the difference between Type Ia and Type II supernovae?

20. Compare a nova with a Type Ia supernova. What do they have in common? How are they different?

21. Compare a nova and an X-ray burster. What do they have in common? How are they different?

22. Describe what X-ray pulsars, pulsating X-ray sources, and X-ray bursters have in common. How are they different manifestations of the same type of astronomical object?

Advanced Questions

The answers to all computational problems, which are preceded by an asterisk (*), appear at the end of the book.

23. What prevents thermonuclear reactions from occurring at the center of a white dwarf? If no thermonuclear reactions take place in its core, why doesn't the star collapse?

24. Suppose you wanted to determine the age of a planetary nebula. What observations would you make, and how would you use the resulting data?

25. Why is the rate of expansion of the gas shell in a planetary nebula often not uniform in all directions?

26. What kinds of stars would you monitor if you wished to observe a supernova explosion from its very beginning? Look up the tabulated lists of the nearest and brightness stars in Appendix Tables E-4 and E-5. Which, if any, of these stars are possible supernova candidates? Explain.

27. To determine the period of a pulsar accurately, astronomers must take the Earth's orbital motion about the Sun into account. Explain why.

*28. The distance to the Crab Nebula is about 2000 pc. When did it actually explode?

29. To test your overall understanding of stellar evolution, do Interactive Exercise 13-5. You can print out your results, if required.

Discussion Questions

30. Suppose that you discover a small glowing disk of light while searching the sky with a telescope. How would you determine if this object was a planetary nebula? What else could this object be?

31. Immediately after the first pulsar was discovered, one explanation offered was that the pulses were signals from an extraterrestrial civilization. Why did astronomers discard this idea?

32. Describe how astronomers can determine whether a supernova at a known distance is Type Ia or Type II, assuming that they can see the supernova from the time it begins to brighten. There are at least two valid answers to this question.

33. Describe how astronomers can determine whether a supernova at an unknown distance is Type Ia or Type II.

What If . . .

34. The Earth passed through an old supernova remnant? What would happen to the Earth and life on it?

35. The Sun were a B-type star, rather than a G-type star? Assuming that the Earth orbiting the B-type star had the same composition and orbital distance that it has today, what would be different here? If you answered this question in Chapter 11, you might want to see how the material in this chapter enhanced your previous answer.

36. The Sun were an M-type star, rather than a G-type star? Assuming that the Earth orbiting the M-type star had the same composition and orbital distance that it has today, what would be different here?

37. The Sun were expanding into a giant star today? How might we cope with this change?

Web Questions

38. It has been claimed that the Dogon tribe in western Africa has known for thousands of years that Sirius is a binary star. Search the Web for information about these claims. What is their basis? Why are scientists skeptical, and how do they refute these claims?

39. Search the Web for recent information about SN 1987A. Sketch the shape of the supernova remnant. Has a pulsar yet been detected in the center of this supernova remnant? If so, how fast is it spinning? Has the supernova debris thinned out enough to give a clear view of the neutron star?

40. Search the Web for information about SN 1994I, a supernova that occurred in the galaxy M51 (NGC 5194). Why was this supernova unusual? Was it bright enough to have been seen by amateur astronomers?

Observing Projects

41. Planetary nebulae are found all across the sky. Some of the brightest are listed in Chapter 13 of the *Discovering the Universe* Web site. With the help of star maps or your *Starry Night Enthusiast*™ software, locate and observe as many planetary nebulae as possible on clear, moonless nights, using the largest telescope at your disposal. Some of the more notable ones include: Little Dumbbell (M76), NGC 1535, Eskimo, Ghost of Jupiter, Owl (M97), Ring (M57), Blinking Planetary, Dumbbell, Saturn Nebula, and NGC 7662. Note and compare the various shapes of the different nebulae. In which cases can you see the central star?

42. Two supernova remnants can be seen through modest telescopes, one in the winter sky and the other in the summer sky. Both are quite faint, however, so you should schedule your observations for a moonless night. The winter sky contains the Crab Nebula, which is discussed in detail in this chapter. It is located near the star marking the eastern horn of Taurus (the Bull) and can be found using *Starry Night Enthusiast*™ software. Its coordinates are R.A. 5^h 34.5^m and Dec. = +22° 00′. Whereas the entire Crab Nebula easily fits into the field of view of an eyepiece, the Veil (or Cirrus) Nebula in the summer sky is so vast that you can see only a small fraction of it at a time. The easiest way to find the Veil Nebula is to aim the telescope at the star 52 Cygni (R.A. = 20^h 45.7^m and Dec. = +30° 43′, which lies on one of the brightest portions of the nebula. This star is also accessible in your *Starry Night Enthusiast*™. If you then move the telescope slightly north or south until 52 Cygni is just out of the field of view, you should see giant wisps of glowing gas.

43. Use a telescope to observe the triple star system 40 Eridani, whose coordinates are R.A. = 4^h 15.3^m and Dec. = −7° 39′. The primary, a 4.4 magnitude yellowish star like the Sun, has a 9.6 magnitude white dwarf companion, the most easily seen white dwarf in the sky. On a clear, dark night with a moderately large telescope, you should also see that the white dwarf has an 11th-magnitude companion, which completes this interesting trio.

WHAT IF . . . A SUPERNOVA EXPLODED NEAR EARTH?

The Flash What would happen here if a supernova occurred only 50 ly away? (Considering the titanic forces supernovae release, it should come as no surprise that the high-energy electromagnetic radiation from such an explosion detonating much closer than this distance would immediately kill virtually all life on Earth.) Without warning, the doomed star would grow tremendously luminous, 50 times brighter than the Moon and only 8000 times less bright than the Sun. It would be brighter than the light from all the other stars in the night sky combined.

Neutrinos would foreshadow by a few hours the visible flash and pending flood of X rays and gamma rays from the supernova. These lethal radiations would be the first causes of death on Earth. Within days individuals from virtually all species of plants and animals would begin dying of radiation poisoning. Entire radiation-sensitive species might be annihilated immediately. The radiation would also cause many cancers and other internal diseases over succeeding years, leading to many more plant and animal deaths. At the same time, ultraviolet radiation reaching the Earth's surface would cause an astronomical jump in the rates of skin cancer and cataracts.

The first blast of ultraviolet radiation from the supernova would destroy the Earth's ozone layer within a matter of days, transforming the ozone primarily into atomic oxygen. With the removal of this protective barrier, ultraviolet radiation from both the supernova and the Sun would saturate the Earth's surface. After the intensity of ultraviolet radiation from the supernova diminished, sunlight would begin repairing the ozone layer, eventually returning it to normal levels.

The brightness and emission of all the electromagnetic radiation from the supernova would peak after a month. It would then fade, but the supernova remnant would be visible for millennia as an expanding cloud of gas and dust in the night sky. Fortunately for life that survived the onslaught of high-energy radiation from the supernova, the remnant cloud would primarily emit harmless visible light.

The Aftermath The most energetic particles from the supernova and its environs, the so-called cosmic rays, would also affect the Earth's surface. Moving at 90% of the speed of light, cosmic rays would arrive here only 5 to 10 years after the photons. Cosmic ray energies are much higher than those of any particles normally existing on Earth. For example, a typical cosmic ray has enough energy to light a small lightbulb for 1 second; it takes over 600,000 trillion electrons flowing through wiring in your house to do the same thing.

Some cosmic rays would slam into nitrogen or oxygen molecules in the air, shattering the air molecules and creating cosmic ray showers. The highest energy cosmic rays from the supernova would reach the Earth intact, as do the highest energy cosmic rays from other sources today. These would break up atoms of the objects they strike on the Earth's surface, enhancing the earlier biological damage caused by the supernova's electromagnetic radiation.

The bulk of the matter ejected into space by the supernova would travel at speeds of 16,000 km/sec (60 million km/hr), nearly 20 times slower than the emitted photons. Thus, the bulk of the supernova remnant would take at least a thousand years longer to reach us than did its initial radiation. By the time the bulk of the blast wave reaches the solar system, it would have become so thin and diffuse that it would probably do little damage to the Earth's atmosphere. However, we could expect this material to deposit a thin but exotic mix of elements into the upper atmosphere. All this material would eventually fall to Earth and alter the chemistry of both the ocean and the soil.

The Outcome The damage done to life by both the electromagnetic radiation and cosmic ray particles would disrupt the global food chain. In the oceans large quantities of plankton and other microscopic organisms at the foundation of the chain would die off. As a result, many of the larger aquatic species that feed on these smaller organisms would starve. Similarly, on the surface, most plant life would wither and many herbivorous animals would starve as a result. This would, of course, lead to the death and dislocation of animals throughout the food chain. Surviving plants and animals would undergo genetic mutations due to the supernova's radiation. Most of these genetic changes would lead to immediate death of the altered plant or animal, or their offspring, but a few changes would be beneficial and enable the mutated organisms to thrive.

How would the human race fare after the supernova? Our long-term survival would depend, in part, on whether surviving humans could coexist with surviving plants and animals, and with each other.

CHAPTER 14

Black Holes: Matters of Gravity

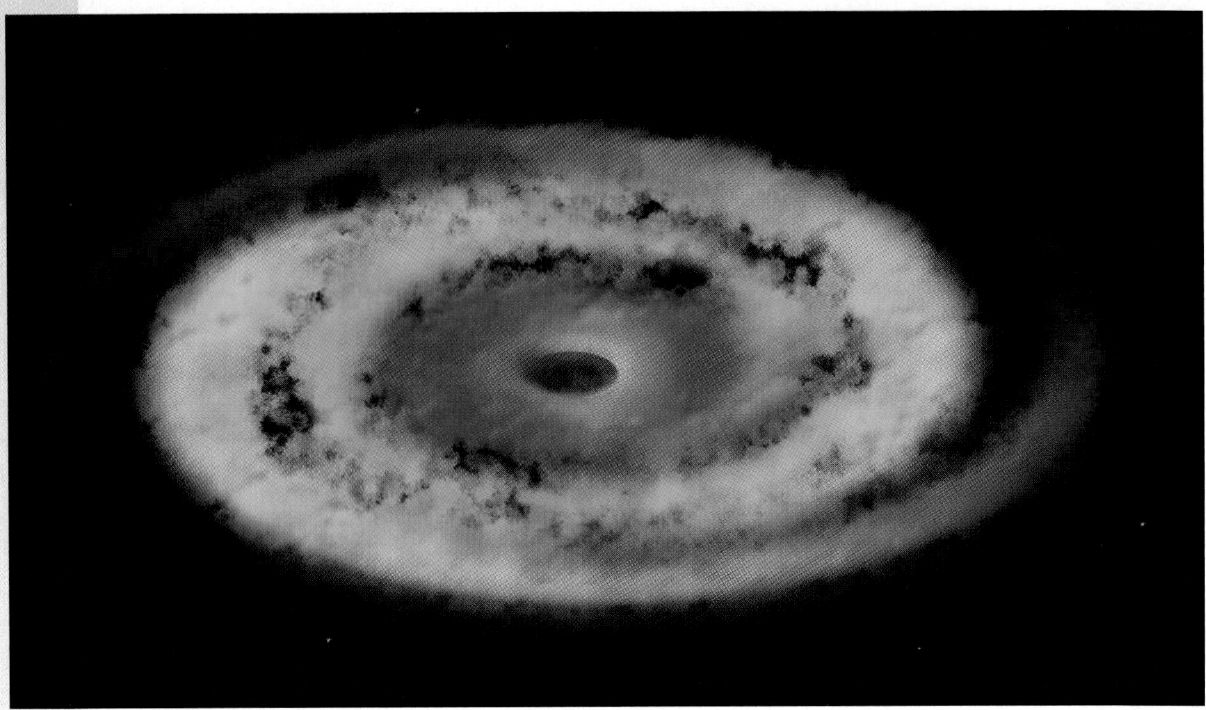

Accretion Disk Around a Black Hole (NASA/CXC/SAO)

R I V U X G

WHAT DO YOU THINK?

1 Are black holes empty holes in space?

2 Does a black hole have a solid surface?

3 What power or force enables black holes to draw things into them?

4 Can you use black holes to travel to different places in the universe?

5 Do black holes last forever?

Stars are formed and held together by the competition between the attractive force of gravitation and the outward forces generated by fusion and electromagnetism. In this chapter we explore the extreme situation where gravitation "wins" the battle, but eventually loses the war.

When the gravitational force of an object is so great that it overcomes all the opposing repulsive forces or pressures it has (like neutron degeneracy pressure), the object collapses in on itself and its gravitational attraction becomes so strong that nothing—not even light—can escape from it. When this happens, the matter and the space around it become a **black hole**.

Black holes inspire awe, fear, and uncertainty. Many people harbor the mistaken belief that black holes are giant vacuum cleaners destined to "suck up" all the matter in the universe. Happily, the equations describing them reveal that black holes are more benign than that, but they are still truly strange.

The idea that gravity could prevent matter from escaping was first put forward in the eighteenth century, but it was not until the twentieth century, when Albert Einstein presented his ideas of *relativity*—how matter and motion affect mass, distance, and time—that the modern concept of a black hole developed. We must turn to the relativity theories to understand these exotic objects.

In this chapter you will discover

- that Einstein's general theory of relativity predicts regions of space and time that are severely distorted by the extremely dense mass they contain
- that space and time are not separate entities
- how black holes arise
- the surprisingly simple theoretical properties of black holes
- that X rays and jets of gas are created near many black holes
- the apparent bizarre fate of black holes
- the unsurpassed energy emitted by gamma-ray bursts

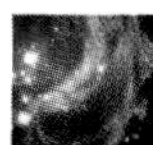

THE RELATIVITY THEORIES

At the beginning of the twentieth century, Einstein began a revolution in science by disregarding his common sense and applying physical assumptions directly to mathematics. The resulting equations revealed profound, albeit highly counterintuitive, insights into how nature works. Armed with the equations that Einstein derived, scientists have been exploring and using the consequences of his work to develop new technology and deeper understanding of how matter, energy, space, and time interact.

14-1 Special relativity changes our conception of space and time

In 1905, Albert Einstein began a revolution in physics with his **special theory of relativity,** a description of how motion affects our measurements of distance, time, and mass. He was guided by two innovative ideas. While the implications of the theory proved revolutionary, the first notion seems simple:

Your description of physical reality is the same regardless of the (constant) velocity at which you move.

In other words, as long as you are moving in a straight line at a constant speed, you experience the same laws of physics as anyone else moving at any other constant speed and in any other direction. To illustrate this first idea, suppose you were inside a closed boxcar moving smoothly in a straight line at 100 km/h and you dropped a pen from a height of 2 meters. You time how long it takes the pen to reach the floor (about 0.78 seconds). Then the train is stopped and you repeat the same experiment in the station. The time it would take the pen to fall would be identical.

The second idea seems more bizarre:

Regardless of your speed or direction, you always measure the speed of light to be the same.

Suppose that you are in a (very powerful) car moving toward a distant street lamp at 150,000 km/sec (93,000 mi/sec). In simpler terms, this is 50% the speed of light, denoted $0.5c$ (Figure 14-1a). A friend of yours leaning on the lamp sees you coming and also sees light from the lamp heading toward you at c. How fast do *you* see the photons of light coming at you? It would seem reasonable that you clock them at $c + 0.5c$, the speed of the photons toward you plus your speed toward them, but this is incorrect.

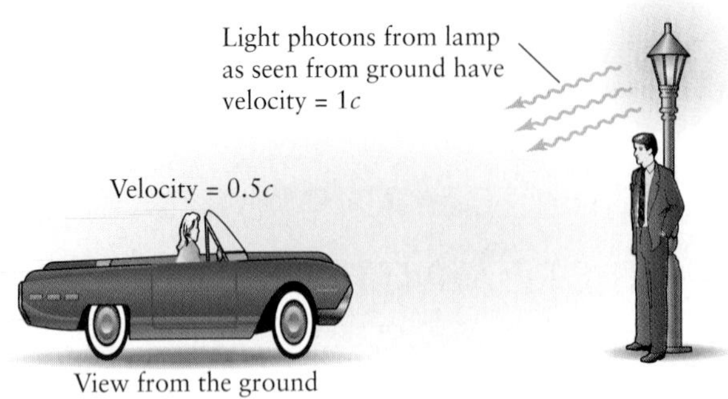

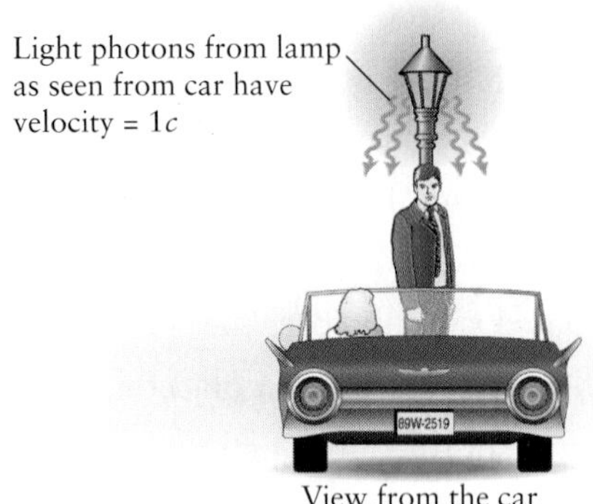

FIGURE 14-1 **The Speed of Light Is Constant** (a) As seen from the ground, the photons of light from the lamp are traveling toward the car with a speed of $v = c$ (ignoring the slight decrease in speed due to the presence of the air). (b) As seen from the car, moving toward the lamp at $v = 0.5c$, the photons are also traveling at the speed $v = c$. The difference in color between the light in the two figures is due to the Doppler shift, as explained in An Astronomer's Toolbox 4-4.

You will, in fact, see the photons coming toward you with the same speed, c, that your friend sees them leaving the lamp (Figure 14-1b). You would, however, see them have a different color than your friend would see. That difference is due to the Doppler shift, as explained in An Astronomer's Toolbox 4-4.

Einstein's special theory of relativity expresses these two assumptions mathematically, and the results of these assumptions have been confirmed in innumerable experiments. Here are three fundamental results of special relativity that also defy common sense:

1. The length of an object decreases as its speed increases. In other words, if a train moves past you, you would measure its length to be shorter than you would measure its length to be when it is stopped at the station (Figure 14-2a also why the car in Figure 14-1 appears too short). However, if you measure the length of the moving train while you are inside it and moving with it, your measurement of its length will be the same as you measured on the ground when it was at rest. The word *relativity* emphasizes the importance of the relative speed between the observer and the measured object.

2. Clocks passing by you run more slowly than do clocks at rest. This is called *time dilation*. Indeed, the faster a clock passes by, the slower it appears to tick from your perspective. For example, air travelers actually age more slowly than they would if they did not fly (although because of the relatively low speeds of aircraft travel compared to the speed of light, this difference is imperceptibly small). People flying in aircraft do not feel time moving more slowly, however, because their biological activities slow down at the same rate as the clocks around them. Only an observer moving more slowly sees that the clock and the activities of the high-speed travelers have slowed. These connections between motion and clocks mean that space and time cannot be considered as two separate concepts. Relativity requires us to combine them in a single entity, thus creating the concept of **spacetime.**

3. The mass of an object increases as it moves faster. The concept of mass is discussed in Section 2-7. The equations of special relativity reveal that a body moving at the speed of light would have an infinite amount of mass. It is impossible for any object with mass to move at this speed, much less go faster than the speed of light, because an infinite

This train is at rest relative to you.

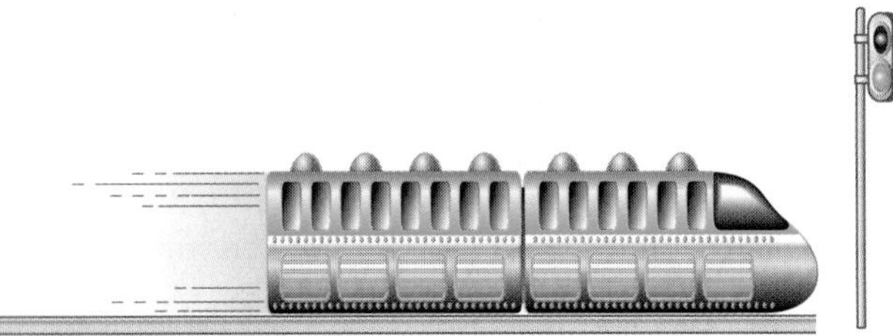
The same train is now moving at very high speed relative to you.

FIGURE 14-2 **Movement and Space** According to the special theory of relativity, the faster an object moves, the shorter it becomes in its direction of motion as observed by someone not moving with the object. It becomes infinitesimally short as its speed approaches the speed of light. The dimensions perpendicular to the object's motion are unchanged.

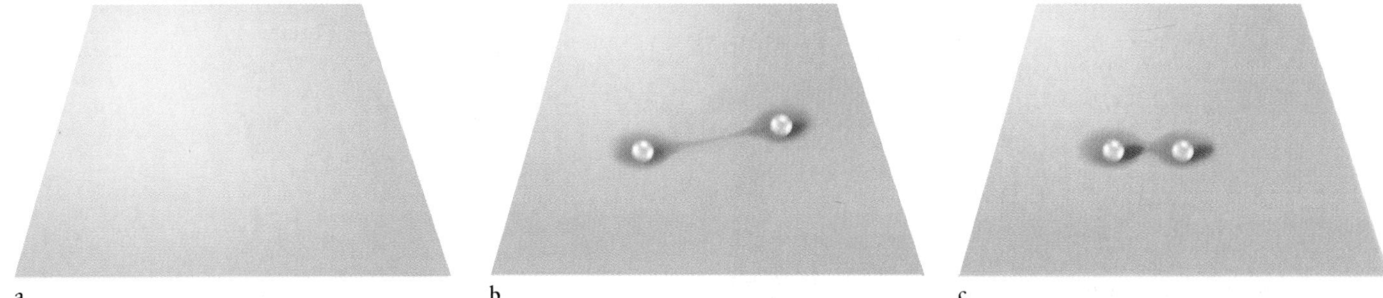

FIGURE 14-3 **Curved Spacetime** (a) This flat surface represents two dimensions in our spacetime. In the absence of any matter in the spacetime, straight lines are straight in our intuitive sense (the sheet is flat). (b) In the presence of matter, spacetime curves, as shown by the curvature of the sheet when mass is laid on it. Straight lines, defined by the paths that light rays take, are no longer straight in the "usual" sense. Besides changing the path of photons, this curvature also creates gravity, which (c) pulls the two masses toward each other.

amount of mass is more than the total mass that exists in the universe. Furthermore, the equation $F = ma$ (F is the force exerted on an object, m is its mass, and a is the acceleration it undergoes; see Section 2-7) reveals that if an object's mass is infinite, the force necessary to accelerate this object must also be infinite. Since the total force available to move matter in the universe is finite, no massive object can be sped up to the speed of light: The speed of light is the universal speed limit.

14-2 General relativity explains how matter warps spacetime, creating gravitational attraction

WEB LINK 14.2

First published in 1915, Einstein's **general theory of relativity** (or, more simply, general relativity) extends his special theory of relativity to include the effects of gravity. It describes how spacetime actually changes shape in the presence of matter (Figures 14-3a, b). The greater the mass, the more the distortion or *curvature*. Furthermore, the curvature of spacetime create attraction between all the pieces of matter in the universe (Figure 14-3c)—this attraction is what we call the *gravitational force*.

Another result of general relativity is that time slows down in the presence of matter. The greater the concentration of matter, the slower clocks tick (Figure 14-4). For example, time passes more slowly for us here on the Earth's surface than it would for astronauts on the Moon, which has less mass than the Earth.

Newton's laws of motion and his universal law of gravitation are accurate only for objects with relatively small masses, slow velocities, and low densities. For example, we use Newton's laws to determine the force necessary to put the Space Shuttle in orbit around the Earth. However, Newton's laws fail to correctly predict Mercury's orbit around the Sun or the paths of objects near a collapsing neutron star. In such cases, general relativity correctly predicts how objects move and how time passes.

FIGURE 14-4 **Time Slows Down Near Matter** (a) Two clocks in space set at exactly the same time are (b) brought to Earth and the Moon. As you see from your vantage point far from the Earth and Moon, the clock on the Earth is ticking more slowly than the clock on the Moon. This occurs because mass slows down the flow of time and the Earth has more mass (and a higher density, which adds to the effect) than the Moon.

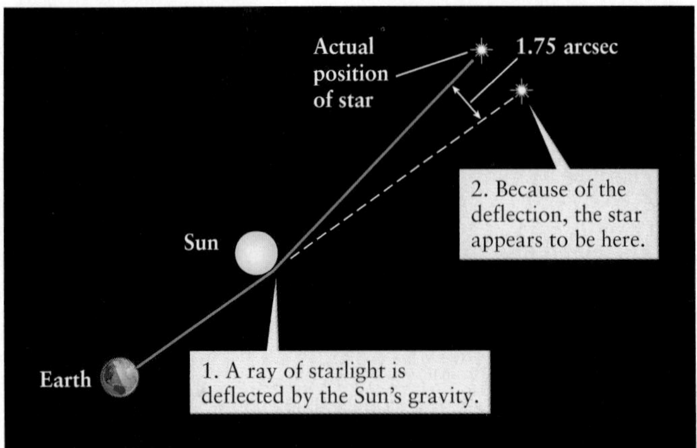

FIGURE 14-5 **Curved Spacetime and the Path of Light** The warping of space by matter causes light to be deflected. This was the first prediction of general relativity to be confirmed, in 1919. This confirmation came when stars behind the Sun were observed during an eclipse. The star in this drawing was not observed where it was supposed to be, as a result of the Sun's gravity changing the path of its light.

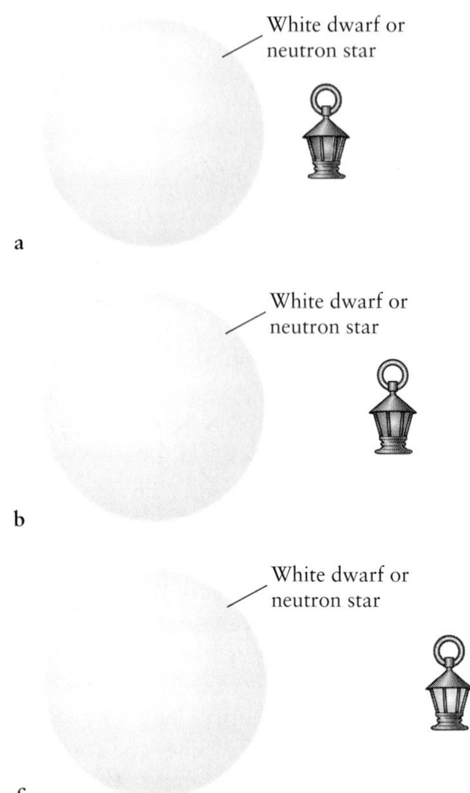

FIGURE 14-6 **Gravitational Redshift** As this sequence of drawings shows, the color of light from the same object located at different distances from a mass appears different as seen from far away. The photons leaving the vicinity of the massive object lose energy and therefore redshift. The closer it is to the mass, the redder the light source appears, and hence the name gravitational redshift.

14-3 Spacetime affects the behavior of light

Besides affecting the behavior of objects, the curvature of spacetime changes the path and wavelength of light that passes near any matter. (These behaviors are not predicted to occur by Newton's laws.) Light travels along paths in space called *geodesics*. You can get an idea of how geodesics work by noting that we do not walk in straight lines because the Earth is curved and its gravity determines the paths we travel in our daily lives. Similarly, the paths of geodesics (and hence light) are determined by the curvature of spacetime created by matter. The first experimental verification of general relativity was its prediction that the geodesics around the Sun are sufficiently curved to cause light from stars behind it to arc around the Sun (Figure 14-5).

Photons leaving the vicinity of a star also lose energy in climbing out of the star's gravitational field. They show this change by shifting to longer wavelengths (Figure 14-6), an effect we see in the spectra of some white dwarfs, whose light appears redder than it would if this effect did not occur. This shift of wavelengths leaving the vicinity of a massive object is called **gravitational redshift.**

General relativity has been confirmed again and again, as seen in these observations:

• Light is measurably deflected by the Sun's gravitational curving of space (see Figure 14-5).

• The perihelion position of Mercury as seen from the Sun shifts, or precesses, by 43 arcsec per year more than is predicted by Newtonian gravitational theory (Figure 14-7).

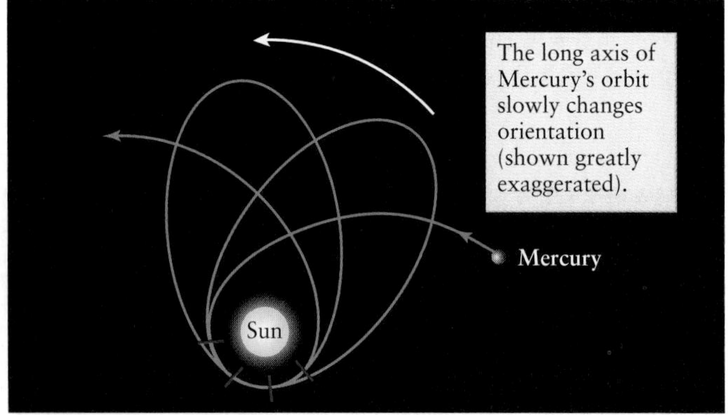

FIGURE 14-7 **Mercury's Orbit Explained by General Relativity** The location of Mercury's perihelion (its position closest to the Sun, indicated by the red lines) and its long axis of orbit change, or precess, with each orbit. This occurs because of the gravitational influences of the other planets plus the curvature of space as predicted by Einstein's general theory of relativity. The amount of precession is inconsistent with the prediction of the orbit made by Newton's law of gravity alone.

• The orbits of stars in binary systems follow the paths predicted by general relativity, rather than those predicted by Newtonian gravitation.

• The spectra of stars are observed to have the gravitational redshifts predicted by general relativity (see Figure 14-6).

14-4 General relativity predicts the fate of massive star cores—black holes

Let us now return to stellar evolution and consider what happens when high-mass main-sequence stars (more than 25 $M_\odot$) explode as supernovae. As with the intermediate-mass main-sequence stars (between 8 $M_\odot$ and 25 $M_\odot$), high-mass stars create neutron (and possibly quark) stars. Remnant bodies of high-mass stars, however, have more than 3 $M_\odot$. This is problematic because above this mass, the gravitational force of the neutron or quark star overcomes the repulsion created by the interactions between its particles. This causes such remnant stars to collapse and there is no force in nature that can stop their infall.

1 We can use the equations of general relativity to understand the fate of collapsing neutron or quark stars. We just saw that all matter warps the space around itself (see Figure 14-3 and Figure 14-5). When matter gets sufficiently dense, as in the case of the high-mass remnant bodies of certain stars, it actually causes nearby space to curve so much that it closes in on itself (Figure 14-8). Photons flying outward at an angle from such a collapsing star arc back inward. Photons flying straight outward lose all their energy, and they cease to exist.

If light, the fastest moving of all known things, cannot escape from the vicinity of such dense matter, then nothing can! Such regions out of which no matter or any form of electromagnetic radiation can escape are called black holes. Plummeting in on itself, a collapsing neutron or quark star becomes so dense that it ceases to consist of neutrons or quarks. General relativity predicts that in creating a black hole, matter compresses to infinite density (and zero value), a state called a **singularity.** However, general relativity and our other theories of physics are known to be invalid in such a situation. A more comprehensive theory of nature must be developed to explain the state of matter in a black hole's singularity. Efforts are under way to develop such a theory. The best-known nascent theory is called *superstrings,* but its mathematics are daunting and much remains to be done. All we can say with confidence is that the matter in black holes becomes incredibly dense and compact.

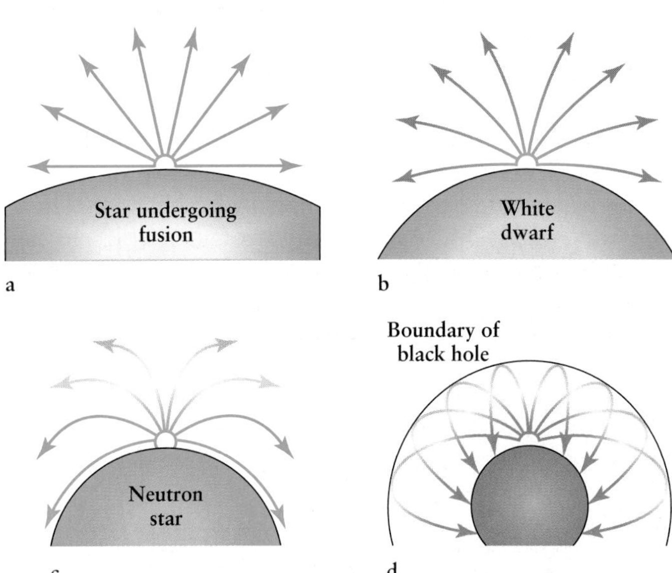

FIGURE 14-8 **Trapping of Light by a Black Hole** (a) The paths and color of light rays departing from a main-sequence, giant, or supergiant star are affected very little by the star's gravitational force. (b) Light leaving the vicinity of a white dwarf curves and redshifts more, while (c) near a neutron star, some of the photons actually return to the star's surface. (d) Inside a black hole, all light remains trapped. Most photons curve back in, except those flying straight upward, which become infinitely redshifted, thereby disappearing.

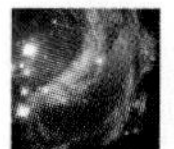

INSIDE A BLACK HOLE

The formation of a black hole is complicated, but its nature is surprisingly simple. It has a boundary shaped like a sphere, and it either rotates or does not rotate.

14-5 Matter in a black hole becomes much simpler than elsewhere in the universe

A black hole is separated from the rest of the universe by a 2
boundary called the **event horizon.** The event horizon is not like the surface of a solid or liquid body. No matter exists at this location except for the instant it takes infalling mass to cross the event horizon and enter the black hole.

We cannot look inside a black hole because no electromagnetic radiation escapes from it. Our understanding of its structure comes from the equations of general relativity. According to Einstein's theory, the event horizon is a sphere. The distance from the center of the black hole to its boundary is called the **Schwarzschild radius** (abbreviated R_{Sch}), after the German physicist Karl Schwarzschild, who first determined its properties. An Astronomer's Toolbox 14-1 shows how to calculate this distance, which depends only on the black hole's mass. The more massive the black hole, the larger its event horizon.

According to the equations of general relativity, when a stellar remnant collapses to a black hole, it loses its magnetic field. The field's energy radiates away in the form of **gravitational waves,** which are ripples in the very fabric of spacetime. Gravitational waves are also

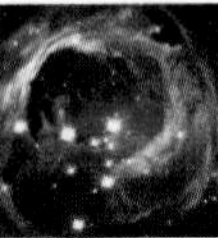

AN ASTRONOMER'S TOOLBOX 14-1
The Sizes of Black Holes

In 1918, Karl Schwarzschild, a German physicist, discovered the first solution of Einstein's equations. His solution describes the nature of nonrotating black holes. According to Einstein's general theory of relativity, the Schwarzschild radius R_{Sch} of any black hole can be found from its mass M:

$$R_{Sch} = \frac{2GM}{c^2}$$

where R_{Sch} is measured in meters; M is the black hole's mass in kilograms; c is the speed of light, 3×10^8 m/s^2; and G is the gravitational constant, 6.67×10^{-11} m^3/(kg•s^2). This equation can be approximated by:

$$R_{Sch} \approx 3\ M_{\odot}$$

where $M_{\odot}$ is the black hole's mass in solar masses and R_{Sch} is in kilometers.

Example: A 5-$M_{\odot}$ main-sequence star has a radius of 3×10^6 km, while the above equation reveals that a black hole with the same mass has a 15-km Schwarzschild radius.

Try these questions: A primordial black hole with the mass of Mount Everest would have a Schwarzschild radius of just 1.5×10^{-15} m! What is Mount Everest's mass? What is the Schwarzschild radius of a galactic black hole containing 3 billion solar masses? A black hole with a Schwarzschild radius of the Earth has how much mass?

(Answers appear at the end of the book.)

created when neutron stars or black holes collide or when stars and stellar remnants are in close orbits around each other. (Actually, gravitational radiation is emitted whenever any two things move around each other, such as a pair of dancers. When the moving bodies are less massive than stars or stellar remnants, however, we have no hope of detecting their gravitational radiation with either present or projected technology.) The binary and double pulsars mentioned in Section 13-11 are a pair of bodies orbiting each other and therefore emitting gravitational waves. The energy in these waves comes from the orbital energy of the stars and remnants, so they are spiraling toward each other.

Gravitational radiation has not yet been directly detected, as we can directly detect light by its effects on our eyes or a CCD, but the effects of gravitational radiation on the orbits of the neutron stars in the binary pulsar are in complete agreement with the predictions of general relativity for such a system. This agreement earned a Nobel prize in 1993 for Joseph Taylor and Russell Hulse, who discovered the binary pulsar in 1974 and measured the changes in its neutron stars' orbits.

The ripples in spacetime created by gravitational waves from stars or stellar remnants are incredibly tiny—each meter-wide volume of space changes by less than 10^{-20} m as the waves pass by. Astronomers around the world are building *gravitational wave detectors* to measure these small changes (Figure 14-9). They have begun operation, and astronomers expect that they will eventually provide a

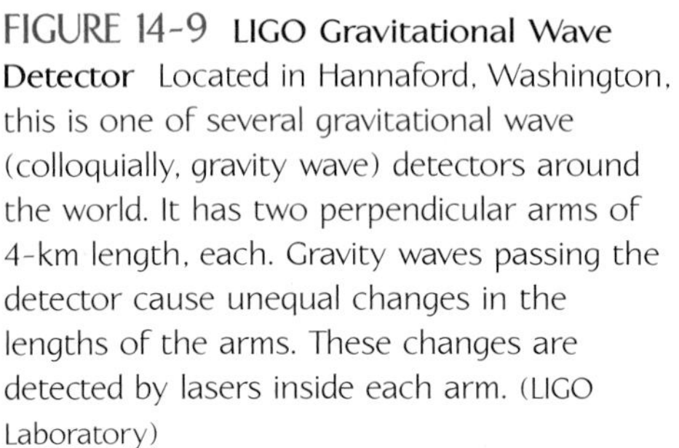

FIGURE 14-9 **LIGO Gravitational Wave Detector** Located in Hannaford, Washington, this is one of several gravitational wave (colloquially, gravity wave) detectors around the world. It has two perpendicular arms of 4-km length, each. Gravity waves passing the detector cause unequal changes in the lengths of the arms. These changes are detected by lasers inside each arm. (LIGO Laboratory)

unique way of observing high energy gravitational activity in the cosmos.

Characteristics of Black Holes

Besides losing its magnetic field, matter within a black hole loses almost all traces of its composition and origin. It retains only three properties that it had before entering the black hole: its *mass*, its *angular momentum*, and its *electrical charge*. Familiar concepts, such as proton, neutron, electron, atom, and molecule, no longer apply. In addition, because few large bodies appear to have a net charge, it is also doubtful that black holes do. We therefore predict that there are only two different types of black holes: those that rotate and those that do not.

Types of Black Holes

If the mass creating a black hole is not rotating, the black hole it forms does not rotate either. We call nonrotating black holes **Schwarzschild black holes** (Figure 14-10). General relativity predicts that all the mass in such a black hole collapses to a point of infinite density at its center, the singularity mentioned earlier. The rest of the volume from the event horizon to the singularity of a Schwarzschild black hole is empty space.

When the matter creating a black hole possesses angular momentum, that matter collapses to a ring-shaped singularity located inside the black hole between its center and the event horizon (Figure 14-11). Such rotating black holes are called **Kerr black holes** in honor of the New Zealand mathematician Roy Kerr, who first calculated their structure in 1963. Once again, the black hole is empty except for the singularity. Most Kerr black holes should be spinning thousands of times every second, even faster than the pulsars we studied in Section 13-10.

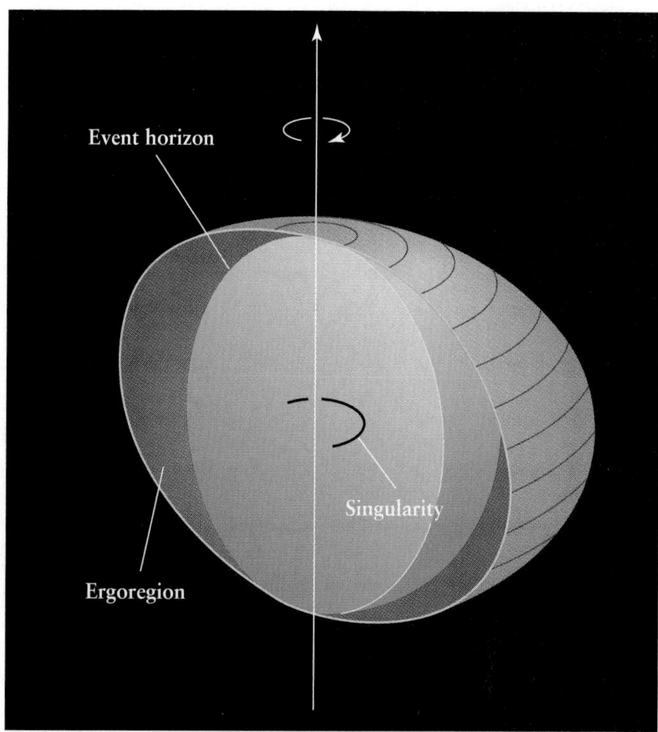

FIGURE 14-11 Structure of a Kerr (Rotating) Black Hole Rotating black holes are only slightly more complex than nonrotating ones. The singularity of a Kerr black hole is located in an infinitely thin ring around the center of the hole. It appears as an arc in this cutaway drawing. The event horizon is again a spherical surface. There is also a doughnut-shaped region, called the ergoregion, just outside the event horizon, in which nothing can remain at rest. Space in the ergoregion is being curved or pulled around by the rotating black hole.

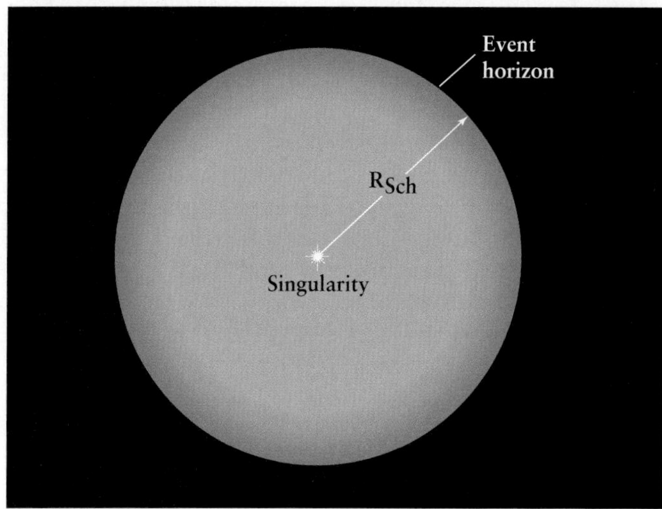

FIGURE 14-10 Structure of a Schwarzschild (Nonrotating) Black Hole A nonrotating black hole has a simple structure. In fact, it has only two notable parts: its singularity and its boundary. Its mass, called a singularity because it is so dense, collects at its center. The spherical boundary between the black hole and the outside universe is called the event horizon. The distance from the center to the event horizon is the Schwarzschild radius, R_{Sch}. There is no solid, liquid, or gas surface at the event horizon. In fact, except for its location at the boundary of the black hole, an event horizon lacks any features at all.

Unlike Schwarzschild black holes, the equations indicate that Kerr black holes possess a doughnut-shaped region directly *outside* their event horizons in which objects cannot remain at rest without falling into the black hole. Called **ergoregions,** they are regions of spacetime that the rotating black hole drags around, like so much batter in a blender (Figure 14-12). In 1997, matter orbiting a black hole was observed to behave in a fashion consistent with the existence of an ergoregion surrounding the hole. If it is moving fast enough, an object entering the ergoregion can fly out of it again; if it stops in the ergoregion, however, it must fall into the black hole.

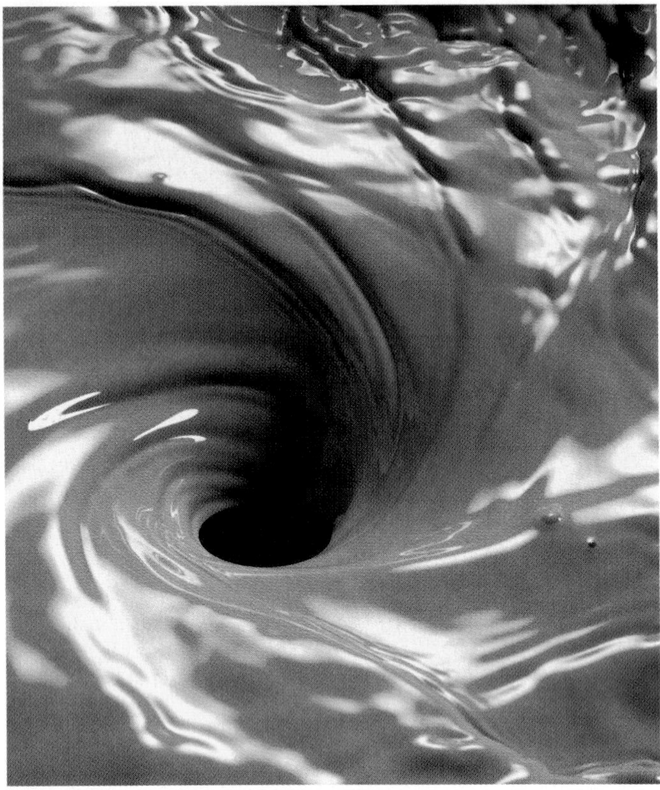

FIGURE 14-12 **Swirling Space in an Ergoregion** Just as the chocolate in this blender is being dragged by the spinning blade, so too is space dragged by a spinning (Kerr) black hole. (Jack Andersen/Food Pix)

14-6 Falling into a black hole is an infinite voyage

3 Imagine being in a spacecraft orbiting only 1000 Schwarzschild radii (15,000 km) from an isolated 5-$M_\odot$ black hole (see An Astronomer's Toolbox 14-1 for the equation of the Schwarzschild radius). You are held in orbit by the black hole's gravitational force. Even at that short distance, the only effect the black hole has on you is its gravitational attraction. It is only when you get very, very close to the event horizon that bizarre things begin to happen. To investigate these changes, you send a cube-shaped probe toward the black hole, with the same side of the cube always facing "downward" toward the black hole. The probe emits a blue glow so that you can follow its progress. What happens to the cube as it approaches the black hole?

From the time you eject it until it reaches about 100 Schwarzschild radii (1500 km), you see the probe descend as if it were falling toward a planet or moon (Figure 14-13). At 100 Schwarzschild radii, however, the probe begins to respond to a severe tidal effect from the black hole. The face of the probe closest to the event horizon receives more gravitational pull from the black hole than its farther-away parts, and it begins to stretch apart.

By the time the probe comes within a few Schwarzschild radii of the event horizon, the tidal forces on it are so great that it violently elongates. The part of the probe closest to the black hole accelerates downward and away from the rest of the probe. Furthermore, the sides of the probe are drawn together: They are falling in straight lines toward a common center. The net gravitational effect of moving close to the event horizon is for the probe to be pulled long and thin. From a practical perspective, this means that the probe would be violently torn apart, because it is not composed of perfectly elastic material.

As the probe nears the black hole, the blue photons leaving it must give up more and more energy to escape the increasing gravitational force. However, unlike a projectile fired upward, photons cannot slow down. Rather, they lose energy by increasing their wavelengths (see An Astronomer's Toolbox 4-4). This is another example of the gravitational redshift predicted by general relativity. The closer the probe gets to the event horizon, the more its light is redshifted—first to green, then yellow, then orange, then red, then infrared, and finally radio waves (see Figures 14-13b, c, d).

Stranger still is the black hole's effect on time. General relativity predicts that when the probe approaches within a

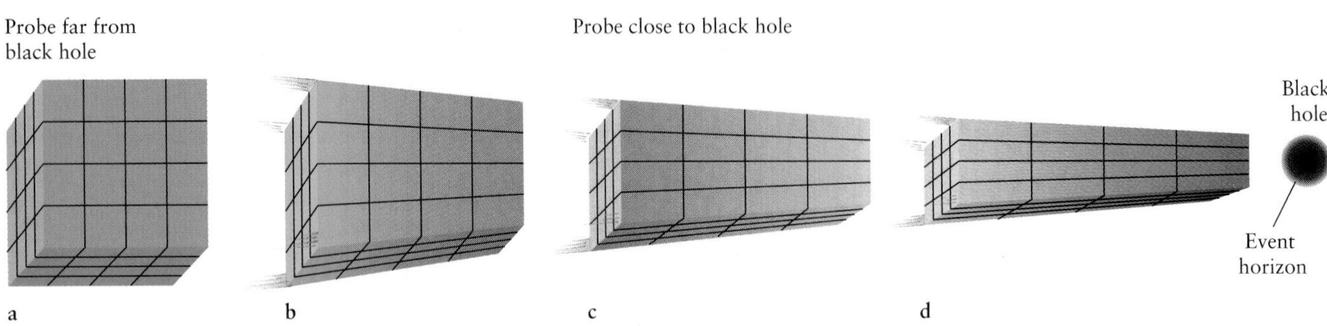

WEB LINK 14.4

FIGURE 14-13 **Effect of a Black Hole's Tidal Force on Infalling Matter** (a) A cube-shaped probe intact at 1500 km from a 5-$M_\odot$ black hole. (b, c, d) Near the Schwarzschild radius, the probe is pulled long and thin by the difference in the gravitational forces felt by its different sides. This tidal effect is a greatly magnified version of the Moon's gravitational force on the Earth. The probe changes color as its photons undergo extreme gravitational redshift.

few Schwarzschild radii of the black hole, its infall rate will slow down as seen from far away. Also, signals from the probe show you that its clocks are running more slowly than they did when it left your spacecraft. Time dilation is so great near the event horizon that the probe will appear to hover above it and its clocks will stop. Someone in the probe observes something else altogether: They see the probe actually cross the event horizon and continue falling toward the black hole's singularity in a normal period of time, according to their own watch. Pulled apart by tidal effects, the probe disintegrates as it falls inward. Contrary to the science fiction concept of traveling great distances quickly by passing through a black hole, calculations indicate that objects entering them could not survive passage through, even if there were a way to come out somewhere else.

4 Could a black hole be connected to another part of spacetime or even some other universe? General relativity predicts such connections, called **wormholes,** for Kerr black holes, but astrophysicists are skeptical that the equations are correct in this regard. Their conviction is called **cosmic censorship:** Nothing can leave a local region of space containing a singularity (that is, a black hole).

14-7 Several binary star systems contain black holes

Black holes are more than fine points of relativity theory: They are real, their presence has been observed by their effects on the orbits of other stars and on gas and dust near them, and more of them are being located all the time. To find evidence for black holes, we look first to binary star systems.

As we saw in Chapter 11, most stars occur in binary star systems. The technique for detecting black holes formed from collapsing stars is based on the interaction between the black hole and its binary companion. When one star in a close binary becomes a black hole, its gravitational attraction pulls off some of its companion's atmosphere. However, such black holes have diameters of only a few kilometers, so there is not enough room for all this gas to fall straight in. Rather, the infalling gas swirls into the black hole like water going down a bathtub drain (Figure 14-14). The gas waiting to fall in forms an **accretion disk**, a disk of gas and dust spiraling in toward the black hole. Calculations reveal that this gas is compressed and heats so much from the collisions of its particles that it gives off X rays (Figure 14-15). Thus, if a

FIGURE 14-14 **Formation of an Accretion Disk** Just as the water in this photograph swirls around waiting to get down the drain, the matter pulled toward a black hole spirals inward. Angular momentum of the infalling gas and dust causes it to form an accretion disk around the hole. (Chris Collins/Corbis)

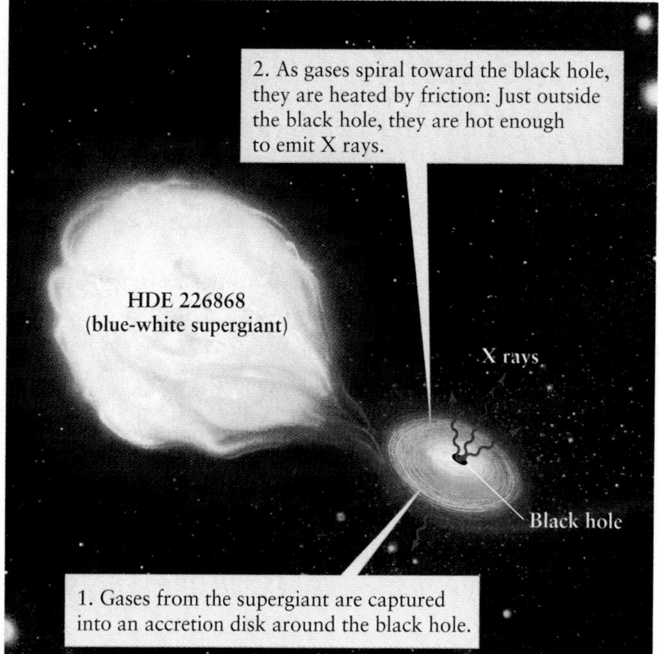
2. As gases spiral toward the black hole, they are heated by friction: Just outside the black hole, they are hot enough to emit X rays.
HDE 226868
(blue-white supergiant)
X rays
Black hole
1. Gases from the supergiant are captured into an accretion disk around the black hole.

WEB LINK 14.5

FIGURE 14-15 **X Rays Generated by Accretion of Matter Near a Black Hole** Stellar-remnant black holes, such as Cygnus X-1, LMC X-3, V404 Cygni, and probably A0620-00, are detected in close binary star systems. This drawing (of the Cygnus X-1 system) shows how gas from the 30 $M_\odot$ companion star, HDE 226868, transfers to the black hole, which has at least 11 $M_\odot$. This process creates an accretion disk. As the gas spirals inward, friction and compression heat it so much that the gas emits X rays, which astronomers can detect. (Courtesy of D. Norton, Science Graphics)

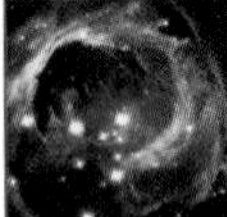

GUIDED DISCOVERY
Identifying Stellar-Remnant Black Holes

Shortly after the *Uhuru* X-ray satellite was launched in the early 1970s, astronomers found a promising black-hole candidate—an X-ray source called Cygnus X-1. This source is highly variable and irregular. Its strong X-ray emission flickers on time scales as short as a hundredth of a second. If different parts of an X-ray source grew bright at different times, its emission would be a continuous stream. In order for Cygnus X-1 to flicker, the entire star must brighten and dim as a unit. Therefore, light must have time to travel across Cygnus X-1 between pulses. Because light travels 3×10^8 km per second, it travels 3000 km in the "flicker time" of a hundredth of a second. This means that for all of Cygnus X-1 to brighten and darken simultaneously, it must be about 3000 km across, smaller in diameter than the Earth.

Cygnus X-1 occasionally emits radio radiation, and in 1971 radio astronomers succeeded in associating Cygnus X-1 with the visible star HDE 226868 (see accompanying figure), a B0 supergiant with a surface temperature of about 31,000 K. Because such stars do not emit significant X rays, HDE 226868 alone cannot be the Cygnus X-1 X-ray source. Spectroscopic observations soon showed that the lines in the spectrum of HDE 226868 shift back and forth within a period of 5.6 days. This behavior is characteristic of a single-line spectroscopic binary (see Section 11-12), and HDE 226868's companion is too dim to produce its own set of spectral lines. The clear implication is that HDE 226868 and Cygnus X-1 are the two components of a binary star system.

The B0 supergiant HDE 226868's mass is estimated at about 30 $M_\odot$, like other B0 supergiants. As a result, Cygnus X-1 must have at least 11 $M_\odot$; otherwise, it would not exert enough gravitational pull to make the B0 star wobble by the amount deduced from the periodic Doppler shift of its spectral lines. Cygnus X-1 cannot be a white dwarf or a neutron star, because its mass is too large for either of these objects. The only remaining possibility is that it must be a fully collapsed star—a black hole.

In the early 1980s, a similar binary system was identified in the nearby galaxy called the Large Magellanic Cloud. The X-ray source, called LMC X-3, exhibits rapid fluctuations, just like those of Cygnus X-1. LMC X-3 orbits a B3 main-sequence star every 1.7 days. From its orbital data, astronomers conclude that the mass of LMC X-3 is about 10 $M_\odot$, which would make it a black hole.

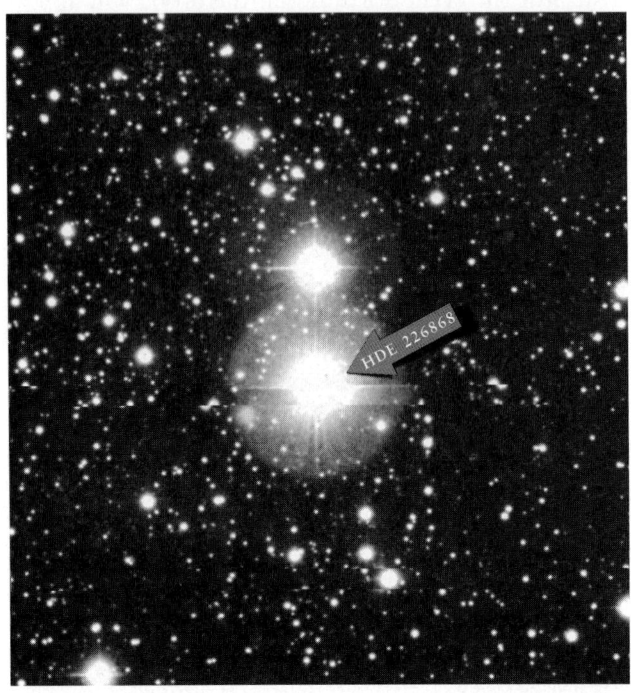

HDE 226868 This star is the visual companion of the X-ray source Cygnus X-1. This binary system is located about 8000 ly from Earth and contains a black hole of at least 11 $M_\odot$ in orbit with HDE 226868, a B0 blue supergiant star. The photograph was taken with the 200-in. telescope at Palomar Observatory on Palomar Mountain, north of San Diego. The slightly dimmer star above is an optical double that is not part of the binary system. (J. Kristian, Carnegie Observatories)

WEB LINK 14.6 Another black-hole candidate is a spectroscopic binary in the constellation Monoceros that contains the flickering X-ray source A0620-00. The visible companion of A0620-00 is an orange dwarf star. It is a low-mass, main-sequence star of spectral type K that orbits the X-ray source every 7.75 hours. From orbital data, astronomers conclude that the mass of A0620-00 must be greater than 5 $M_\odot$, more probably about 9 $M_\odot$.

visible star has a sufficiently tiny, sufficiently massive X-ray–emitting companion, we have located a black hole. See Guided Discovery: Identifying Stellar-Remnant Black Holes for details on how this is done.

To date, in the Milky Way Galaxy alone, at least nine stellar-remnant black holes in binary star systems have been identified (Table 14-1). There is also growing evidence that black holes can collide with each

TABLE 14-1 Stellar Black Holes in the Milky Way

X-ray source name	Mass* of companion	Mass* of black hole
Cygnus X-1	24–42	11–21
V404 Cygni	~ 0.6	10–15
GS 2000+25	~ 0.7	6–14
H 1705-250	0.3–0.6	6.4–6.9
GRO J1655-40	2.34	7.02
A 0620-00	0.2–0.7	5–10
GS 1124-T68	0.5–0.8	4.2–6.5
GRO J042+32	~ 0.3	6–14
4U 1543-47	~ 2.5	2.7–7.5

*All masses in solar masses.

"Revisiting the Black Hole," R. Blandford and N. Gehrels, *Physics Today*, June 1999.

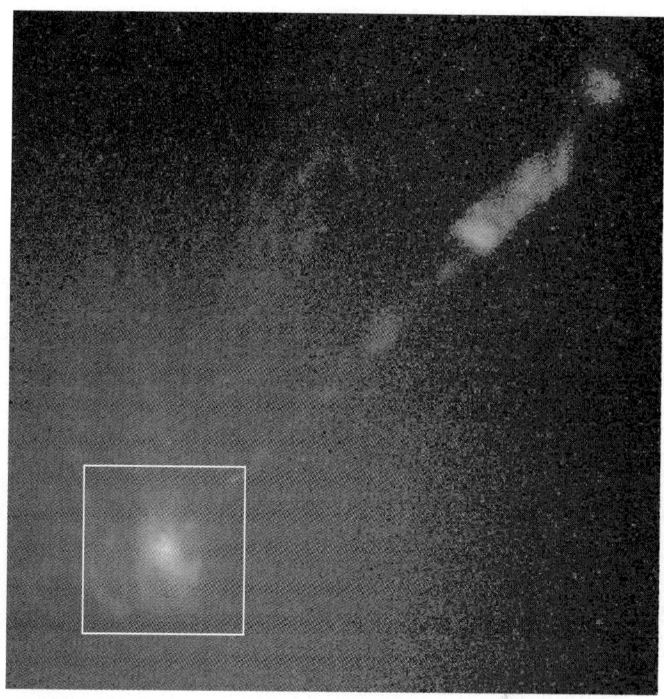

FIGURE 14-16 **Supermassive Black Hole** The bright region in the center of this galaxy, M87, has stars and gas held in tight orbits by a black hole. M87's bright nucleus (center of the region in the white box) is only about the size of the solar system and pulls on the nearby stars with so much force that astronomers calculate that it is a 3-billion-$M_\odot$ black hole. (Holland Ford, STScI/Johns Hopkins University; Richard Harms, Applied Research Corp.; Zlatan Tsvetanov, Arthur Davidsen, and Gerard Kriss at Johns Hopkins University; Ralph Bohlin and George Hartig at STScI; Linda Dressel and Ajay K. Kochhar at Applied Research Corp. in Landover, Md; and Bruce Margon, University of Washington, Seattle)

other or with different types of objects, such as neutron stars.

14-8 Other black holes range in mass up to billions of solar masses

Based on the equations of general relativity, the fate of massive neutron stars led to the idea of black holes as early as 1939. Since then, calculations have supported at least three other types of black holes:

1. Early in the life of the universe, black holes could have formed from the condensing of vast amounts of gas, as well as the collisions of stars, during the process of galaxy formation, thereby creating **supermassive black holes,** each with millions or billions of solar masses.

2. We saw in Section 12-2 that stars form in clusters. If a cluster forms with enough stars concentrated in a sufficiently small volume of space, collisions between stars in the center of the cluster can lead to the formation of a black hole with between a few hundred and a few thousand solar masses. These are called **intermediate-mass black holes.**

3. Black holes could have formed during the explosive beginning of the universe as tiny amounts of matter were compressed sufficiently to form **primordial black holes.** These bodies would have masses from grams to the mass of a planet.

Supermassive Black Holes

Supermassive black holes, with millions or billions of solar masses, have been observed at the centers of many galaxies. In May 1994, the Hubble Space Telescope obtained compelling evidence for a black hole at the very center of the galaxy M87. In the nucleus of M87 is a tiny, bright source of light. Spectra showed that nearby gas and stars are orbiting it extremely rapidly. They can be held in place only if the bright object contains some three *billion* solar masses (Figure 14-16). Given that its size is only slightly larger than the solar system, the source can only be a black hole.

Since 1994, numerous black holes in the centers of galaxies have been identified by their X-ray emissions and gravitational effects on surrounding gas and stars. For example, a distinct, frisbee-shaped accretion disk around the supermassive black hole in NGC 7052 was seen by the Hubble Space Telescope in 1998 (Figure 14-17). As we will see in the next chapter, a black hole containing several million solar masses has even been found at the center of our Milky Way, only 25,000 ly from the Earth. Indeed,

a

R I V U X G

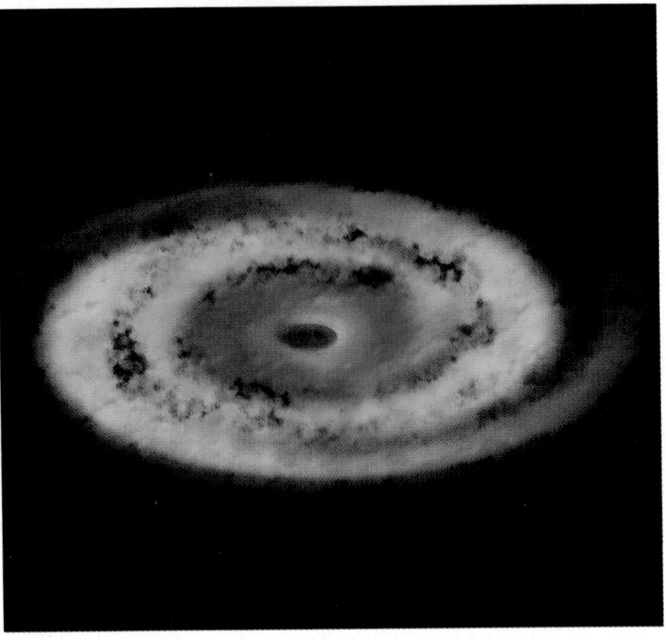

b

FIGURE 14-17 **Accretion Disk Around a Supermassive Black Hole** (a) Swirling around a 300-million-$M_\odot$ black hole in the center of the galaxy NGC 7052, this disk of gas and dust is 3700 ly across. The gas is cascading into the black hole, which will consume it over the next few billion years. The black hole appears bright because of light emitted by the hot, accreting gas outside its event horizon. NGC 7052 is 191 million ly from Earth in the constellation of Vulpecula. (b) This drawing shows how the gases spiralling inward in an accretion disk heat up as they approach the black hole. Color coding follows Wien's displacement law: red (coolest), followed by orange, yellow, green, blue, and violet (hottest). (R. P. van der Marel, STScI/F. C. van den Bosch, University of Washington/NASA; b: NASA/CXC/SAO)

evidence for central, supermassive black holes has been found in most of several dozen nearby galaxies.

We have studied early in the book the tidal effects of one body on another in several situations, including the Earth-Moon system and the heating of Io by Jupiter. In 2004, astronomers observed several supermassive black holes in other galaxies absorbing parts of passing stars. These black holes created such high tides on those nearby stars that the stars were pulled apart. Some of each star's mass was then absorbed by the black hole. These discoveries were made because infalling gas from the stars was rapidly heated to millions of degrees K, causing the gas to emit a burst of X rays that was observed by orbiting telescopes.

As we will explore in more detail in Chapter 18, galaxies were created from condensing gas and stars in the early universe. Our technology is now providing us with observations of that epoch of the universe, and the formation process of supermassive black holes in the centers of galaxies is now being studied. These black holes apparently formed in part from the gas that was condensing to form galaxies and in part from the collisions of stars and black holes formed shortly after the beginning of time. There is also observational evidence that the formation of supermassive black holes is still ongoing. We will explore this further in Chapter 16.

Intermediate-Mass Black Holes

At the end of the twentieth century, astronomers began discovering objects that appear to be black holes with masses between 10^2 and 10^5 solar masses. Again, these objects were identified as potential black holes by the intensity and spectra of the X rays that they emit. For example, near the center of the galaxy M82, astronomers found what appears to be a 500–1000 $M_\odot$ black hole (Figure 14-18), while in the galaxy NGC 1313, astronomers have observed what appear to be two intermediate-mass black holes, each with 200–500 $M_\odot$.

In support of the belief that these objects are black holes, computer simulations of stars in the crowded regions where such objects are found show that black holes in this mass range can form as a result of collisions and close tidal interactions between stars. Not to fear—such a black hole will not occur near Earth. There have to be at least a million times more stars per volume than there are in our stellar neighborhood for such frequent stellar collisions or near misses to occur.

Primordial Black Holes

Even more exotic black holes may have formed along with the universe itself. The British

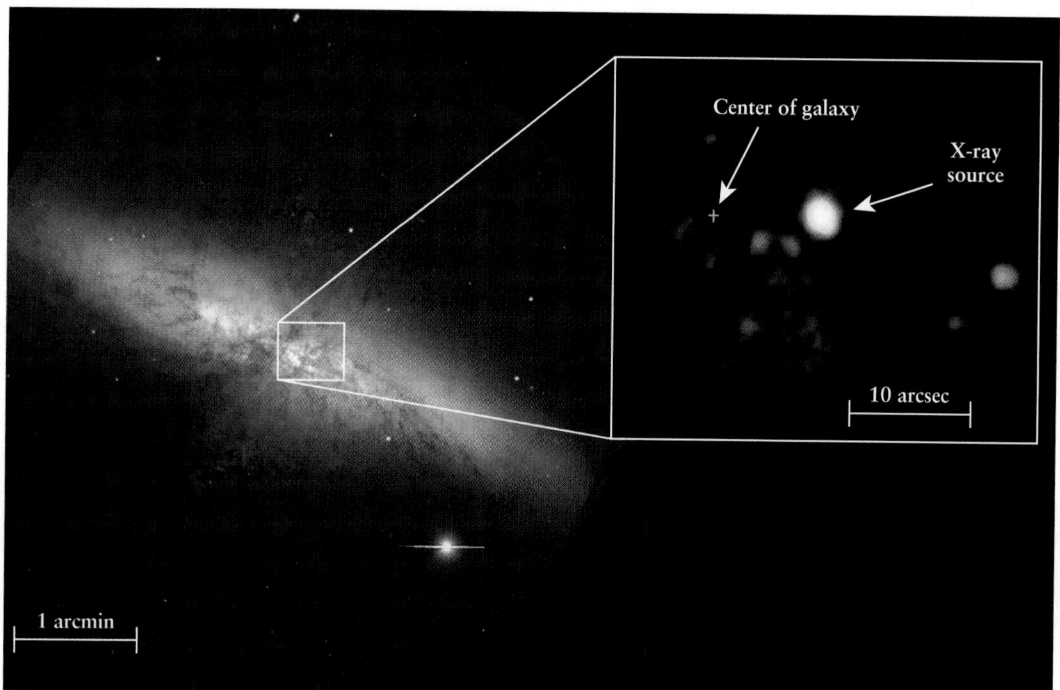

FIGURE 14-18 An "Intermediate-Mass" Black Hole M82 is an unusual galaxy in the constellation Ursa Major. The inset shows an image of the central region of M82 from the *Chandra* X-ray Observatory. The bright, compact X-ray source shown varies in its light output over a period of months. The properties of this source suggest that it is a black hole of roughly 500 $M_\odot$. (Subaru Telescope, National Astronomical Observatory of Japan; inset: NASA/SAO/CXC)

astrophysicist Stephen Hawking has proposed that the Big Bang explosion from which most astronomers believe the universe emerged may have been chaotic and powerful enough to have compressed tiny knots of matter into primordial black holes. Their masses may have ranged from a few grams to greater than the mass of the Earth. Astronomers have not yet observed evidence of primordial black holes, though that does not mean they do not necessarily exist—just that we cannot yet detect them.

14-9 Black holes and neutron stars in binary systems often create jets of gas

We saw in Section 13-11 how beams of radiation are emitted by neutron stars. Many neutron stars and black holes also frequently emit pairs of jets of gas shooting out in opposite directions. These jets result from the presence of an accretion disk around the black hole or neutron star. Consider a black hole (or neutron star) in a binary system. If the companion is still fusing, then its outer layers can be pulled off, as discussed in Section 12-15. As discussed earlier, the infall of this matter toward the compact companion is too rapid for all the mass to enter the event horizon immediately. The resulting accretion disk around the black hole is the key to the jets.

As the disk mass spirals down toward the event horizon, this gas is being compressed into a smaller and smaller volume. Such compression heats the gas, which, in turn, causes its pressure to increase. As a result of the increased pressure, the gas expands, forming a doughnut-shaped region around the black hole (Figure 14-19). As the gas starts the final plunge toward the black hole, its temperature skyrockets to tens or even hundreds of millions of degrees K. At such temperatures, the pressure is so great that much of the infalling gas expands and, finding little

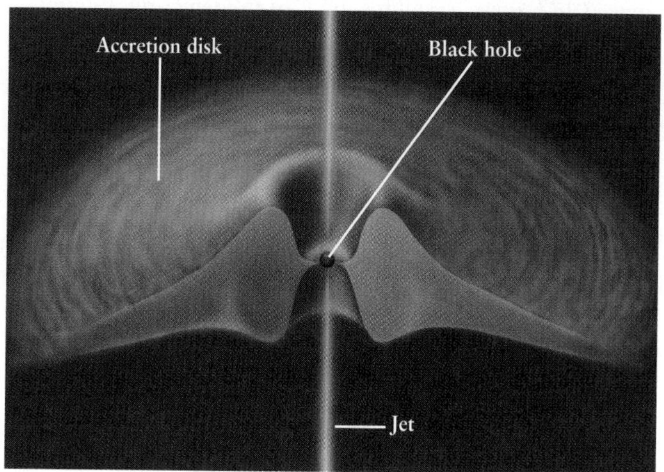

FIGURE 14-19 Jets Created by a Black Hole in a Binary System Some of the matter spiraling inward in the accretion disk around a black hole is superheated and redirected outward to produce two powerful jets of particles traveling at close to the speed of light. The companion star is off to one side of this drawing.

resistance perpendicular to the plane of the accretion disk, it squirts out as two jets (Figure 14-19). These jets are prevented from spreading out by the black hole's magnetic field and by pressure from surrounding gas.

Whereas neutron stars and stellar mass black holes create jets as a result of being in binary star systems, supermassive black holes have so much gravitational attraction that they pull huge quantities of nearby interstellar gas and dust into orbit around them without needing a binary companion. Figure 14-16 shows one of the jets emitted by the supermassive black hole in M87. We will explore more about jets created by black holes in Chapter 17.

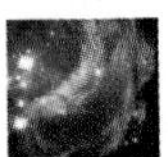

GAMMA-RAY BURSTS

Some black holes, neutron stars, and the supernovae that accompany their formation are apparently the sources of the most powerful and enigmatic energy emissions known in the universe. Called **gamma-ray bursts**, these pulses of gamma rays (the shortest wavelength electromagnetic radiation) emit more energy than supernovae and from even smaller volumes of space.

14-10 Gamma-ray bursts are the most powerful explosions in the known universe

Because atomic bomb blasts emit pulses of gamma rays, the U.S. military put the gamma-ray-detecting Vela satellites in orbit around the Earth to monitor illegal nuclear explosions in the 1960s. In 1973, astronomers were told that since 1967 these satellites had been detecting bursts of gamma rays from objects in space. This news intrigued the astronomical community because there was no known mechanism to emit such gamma rays.

Gamma-ray bursts each last between a few milliseconds and about 1000 seconds (just over 16 minutes). There appear to be at least two types of gamma-ray bursts. One group typically lasts for a few tenths of a second, while those in the other group typically last for about 40 seconds. Unlike X-ray bursts (see Section 13-14), each gamma-ray burst occurs only once.

More than 3000 gamma-ray bursts have been observed, and more are being discovered at a rate of about one per day. Plotting their locations on the celestial sphere (Figure 14-20), astronomers have discovered that the shorter-lived bursts occur slightly more commonly in the plane of the Milky Way Galaxy than elsewhere, while the longer-lived ones occur randomly over the sky. In 2004, astronomers detected the first remnant of a gamma-ray burst that occurred in the Milky Way. While some bursts are local; most occur outside our Galaxy.

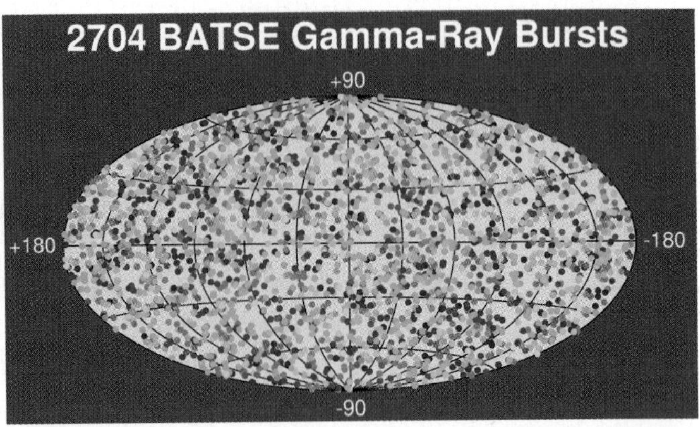

FIGURE 14-20 **The Most Powerful Known Bursts** Gamma-ray bursts have been observed everywhere in the sky, indicating that, unlike X-ray bursters, most do not originate in the disk of the Milky Way Galaxy. This map of the entire sky "unfolded" onto the page shows 2704 bursts detected by the Burst and Transient Source Experiment (BATSE) aboard the Compton Gamma-Ray Observatory. The colors indicate the brightness of the bursts; they are coded brightest in red, dimmest in violet. Gray dots indicate incomplete information about the burst strength. (NASA)

To help determine where else they occur and what causes them, astronomers have searched for optical and infrared objects located in the same place as ongoing bursts (Figure 14-21). Since 1997, optical and infrared "afterglows" of many gamma-ray bursts have been detected. These often persist for days, which allows astronomers to study their spectra in detail. Such observations reveal that the light has passed through intergalactic gas clouds. Combining this with observations of the bursts' spectra and redshifts reveals that the longer-lasting bursts typically originate more than 6 billion light-years away. Many of the shorter bursts occur in more nearby galaxies. The most distant gamma-ray burst was more than 12 billion light-years away. Taking their distance into account, a typical long-lived gamma-ray burst emits as much energy in 100 seconds as the Sun will emit over the 10 billion years that it will be on the main sequence.

Deep sky observations at the locations of some gamma-ray bursts have begun to show galaxies in the same places. As Figure 14-22 shows, the bursts do not appear in the centers of their host galaxies. Likewise, the bursts in the plane of the Milky Way are not located near the galactic nucleus. These observations indicate that bursts are not associated with the intermediate-mass or supermassive black holes in the galactic cores. This suggests that the gamma-ray bursts are instead associated with stellar cataclysms.

In 2003, observations by NASA's *High-Energy Transient Explorer* observed a 30-second gamma-ray burst that was seen to lie at the same place as a particularly

R I V U X G

FIGURE 14-21 **Gamma-Ray Bursts** In 1999, astronomers photographed the visible-light counterpart of a 100-second gamma-ray burst some 9 billion light-years away in the constellation Boötes. The times indicated are after the burst began. (Carl Akerlof, University of Michigan, Los Alamos National Laboratory, Lawrence Livermore National Laboratory)

powerful supernova called a *hypernova*. It appears that some gamma-ray bursts occur when the exploding star is spinning quickly and is surrounded by lots of gas and a strong magnetic field. The details of the explosions and why they generate gamma-ray bursts are still under investigation.

Other gamma-ray bursts are believed to occur when black holes collide, when a black hole swallows a neutron star, or when a pair of neutron stars collide and form a black hole. Calculations confirm that the collision of such dense objects should indeed lead to a rapid, intense burst of gamma rays.

In order to better understand gamma-ray bursts, the astronomy community has constructed telescopes both on Earth and in orbit dedicated to rapidly shifting to look in the direction in which these bursts occur in space. These telescopes, such as the *High Energy Transient Explorer,* have begun seeing them and providing new insights into their origins and dynamics.

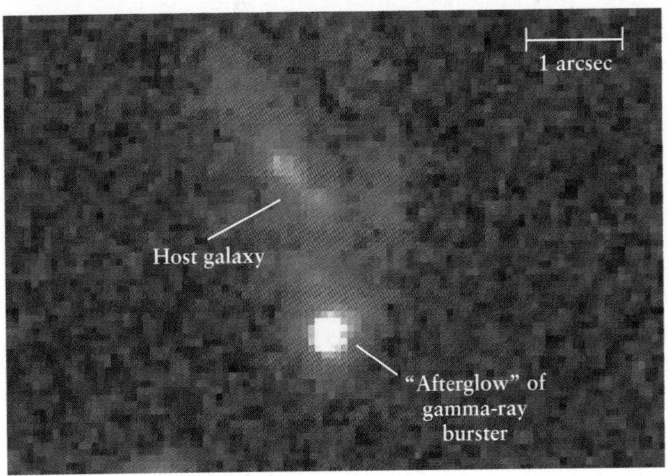

R I V U X G

FIGURE 14-22 **The Host Galaxy of a Gamma-Ray Burst** The Hubble Space Telescope recorded this visible-light image in 1999, 16 days after a gamma-ray burst was observed at this location. The image shows a faint galaxy that is presumed to be the home of the gamma-ray burst. The galaxy has a bright blue color, indicating the presence of many recently formed stars. (Andrew Fruchter, STScI; and NASA)

14-11 Black holes evaporate

In exploring the fate of black holes, astrophysicists find that, once again, these objects confound common sense—they evaporate! With a black hole cloaked behind the event horizon and its mass collapsed into a singularity, it must seem that there is no way of getting mass from the black hole back out into the universe again. No way, that is, until one recalls that mass and energy are two sides of the same coin. What if there was a way of converting the mass into a form of energy that *could* get out of the black hole, such as gravitational energy? The key to the conversion is Einstein's equation $E = mc^2$, and the conversion does occur, according to an idea first proposed by Stephen Hawking.

WEB LINK 14.9

Black holes convert their mass into energy by a process called **virtual particle** production. As we discussed in Section 4-6, physicists have discovered that nature allows pairs of particles to form spontaneously. These are called virtual particles. Each pair always consists of a particle and its antiparticle, such as a proton and an antiproton, an electron and a positron, or a pair of photons (because a photon is its own antiparticle). Normally, virtual particles form, come back together, and annihilate each other without a trace, all within the incredibly short time of about 10^{-21} seconds. For example, an electron and its antiparticle, a positron, might form and destroy each other in this process.

Here is how black holes evaporate: The virtual particles that form *just outside* the event horizon of a black hole feel an exceptionally powerful gravitational field. If one particle is created slightly farther from the hole than its companion, the two virtual particles feel a tidal force from the black hole's tremendous gravitational pull. When the gravitational tidal force is strong enough, it can pull the two virtual particles apart before they can annihilate each other. This separation makes them real (Figure 14-23).

5 To conserve momentum, at least one of the newly formed real particles always falls into the black hole. But the other particle will sometimes have enough energy to escape from the vicinity of the black hole. This latter particle flies free into space, and the black hole has effectively emitted energy equal to $E = mc^2$, where m is the mass of the freed particle or $E = hc/\lambda$, where λ is the wavelength of the photon, if a pair of photons is created. Where the virtual particles become real is a temporary void in space outside the event horizon. The black hole fills the void with energy that comes from its mass. Some of the black hole's mass is converted into gravitational energy. This energy is then transmitted outside the event horizon as gravitational radiation to replace the energy taken away by the escaped particle. This is called the **Hawking process,** and the particles leaving the vicinity of the black hole are called *Hawking radiation* (even though many of the particles are not electromagnetic radiation).

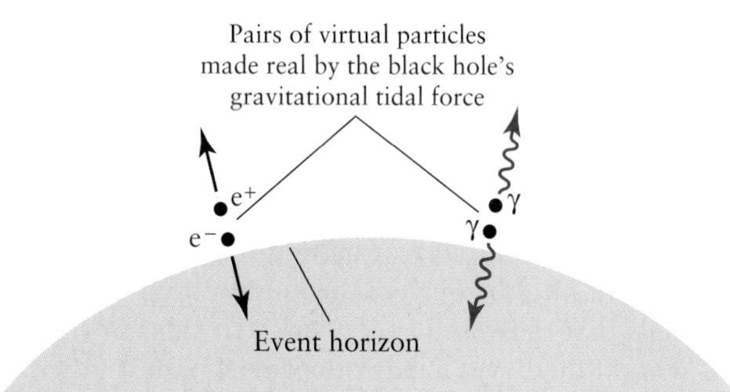

FIGURE 14-23 **Evaporation of a Black Hole** Throughout the universe, pairs of virtual particles spontaneously appear and disappear so quickly that they do not violate any laws of nature. The tidal force just outside the event horizon of a black hole is strong enough to tear apart two virtual particles that appear there before they destroy each other. The gravitational energy that goes into separating them makes them real and, therefore, permanent. At least one of each pair of newly created particles falls into the black hole. Sometimes one of the particles escapes into the universe. Because the gravitational energy used to create the particles came from the black hole, the hole loses mass and shrinks, eventually evaporating completely. Here we see just a few particles in the making: an electron (e^-) and positron (e^+) and a pair of photons (γ).

FIGURE 14-24 **Mount Everest** Primordial black holes, formed at the beginning of time, may have had masses similar to that of Mount Everest, shown here. Calculations indicate that the universe is old enough for such black holes to have evaporated. Their final particle production rate is so high, they should look as though they are exploding. (© Galen Rowell/Corbis)

The time it takes a black hole to completely evaporate by the Hawking process increases with the mass of the hole. More massive black holes have lower tidal forces at their event horizons than do less massive and, therefore, smaller-diameter, black holes. Therefore, the rate at which virtual particles are converted into real particles around bigger black holes is actually slower than it is around smaller ones. The faster the real particles are produced, the faster they wear down the mass of the black hole; thus, lower-mass black holes evaporate more rapidly. Indeed, calculations indicate that in its final moments of evaporation, a black hole should create particles so quickly that it appears to explode.

A stellar black hole of 5 $M_\odot$ would take more than 10^{62} years to evaporate (much longer than the present age of the universe), while a 10^{10}-kg primordial black hole (equivalent to the mass of Mount Everest, Figure 14-24) would take only about 15 billion years. Depending on the precise age of the universe, a black hole of this size and age should be in the final throes of evaporating. Astronomers are trying to identify events in space corresponding to the violent, particle-generating deaths of primordial black holes.

14-12 Frontiers yet to be discovered

Black holes take us to the limits of science. They represent a state of matter that we are not yet fully capable of explaining. This state of affairs makes black holes an especially intriguing field of study. Understanding the nature of black hole singularities is a major research goal. Numerous other frontiers exist in this realm. By directly observing gravitational radiation, we will have a new tool for exploring violent activity in the universe. We still need to understand the details of black hole formation, both from stars and by other means. We have yet to understand in detail why midmass and high-mass black holes have the masses they have.

If, as is now expected, black holes merge, the details of these events and their effects on the rest of the universe have still to be understood. It also remains to be shown whether or not the cosmic censorship theorem is correct. If not, the implications of the existence of wormholes need to be examined in more depth. Among the more intriguing issues surrounding black holes is whether primordial black holes exist. While the evaporation process has not yet been seen, we should soon have the technology to discover it, if it is occurring.

Gamma-ray bursts still require much explanation. How does so much energy get transformed so quickly into electromagnetic radiation in these events? Why do some supernovas emit most of their energy as gamma rays, while others emit most as less energetic electromagnetic photons?

Summary of Key Ideas

• If a stellar corpse is more massive than about 3 $M_{\odot}$, gravitational compression overcomes neutron degeneracy (or the theorized quark degeneracy pressure) and forces it to collapse further and become a black hole.

• A black hole is an object so dense that the escape velocity from it exceeds the speed of light.

The Relativity Theories

• According to general relativity, mass causes space to curve and time to slow down. These effects are significant only near large masses or compact objects.

• Observations indicate that some binary star systems harbor black holes. In such systems, gases captured by the black hole from the companion star heat up and emit detectable X rays and jets of gas.

• Supermassive black holes exist in the centers of some galaxies. Intermediate-mass black holes exist in clusters of stars. Very low mass (primordial) black holes may have formed at the beginning of the universe.

Inside a Black Hole

• The event horizon of a black hole is a spherical boundary where the escape velocity equals the speed of light. No matter or electromagnetic radiation can escape from inside the event horizon. The distance from the center of the black hole to the event horizon is called the Schwarzschild radius.

• The matter inside a black hole collapses to a singularity. The singularity for nonrotating matter is a point at the center of the black hole. For rotating matter, the singularity is a ring inside the event horizon.

• Matter inside a black hole has only three physical properties: mass, angular momentum, and electrical charge.

• Nonrotating black holes are called Schwarzschild black holes. Rotating black holes are called Kerr black holes. The event horizon of a Kerr black hole is surrounded by an ergoregion in which all matter must constantly move to avoid being pulled into the black hole.

• Matter approaching a black hole's event horizon is stretched and torn by the extreme tidal forces generated by the black hole, light from the matter is redshifted, and time slows down.

• Black holes can evaporate by the Hawking process, in which virtual particles near the black hole become real. This transition decreases the mass of a black hole until eventually it disappears.

Gamma-Ray Bursts

• Gamma-ray bursts are events believed to be caused by some supernovae and by the collisions of dense objects, such as neutron stars or black holes. Some occur in the Milky Way and nearby galaxies, while many occur billions of light-years away from Earth.

• Typical gamma-ray bursts occur for a few tens of seconds and emit more energy than the Sun will radiate over its entire 10-billion-year lifetime.

WHAT DID YOU THINK?

1 *Are black holes empty holes in space?* No, black holes contain highly compressed matter—they are not empty.

2 *Does a black hole have a solid surface?* The surface of a black hole, called the event horizon, is empty space. No stationary matter exists there.

3 *What power or force enables black holes to draw things into them?* The only force that pulls things in is the gravitational attraction of the matter in the black hole.

4 *Can you travel through black holes to get to different places in the universe?* No, most astronomers believe that the wormholes predicted by general relativity do not exist.

5 *Do black holes last forever?* No, black holes evaporate.

Key Words

Review Questions

1. Which property, if any, of normal matter ceases to exist in a black hole? **a.** mass. **b.** chemical composition. **c.** angular momentum. **d.** charge. **e.** all these properties exist in a black hole.

2. Supermassive black holes are found in which of the following locations? **a.** in the centers of galaxies. **b.** in globular clusters. **c.** in open (or galactic) clusters. **d.** between galaxies. **e.** in orbit with a single star.

3. Which feature is found with Kerr black holes but not Schwarzschild black holes? **a.** a singularity. **b.** an event horizon. **c.** gravitational redshift of photons outside the black hole. **d.** an ergoregion. **e.** warping of nearby space time.

4. Under what conditions do all outward pressures on a collapsing star fail to stop its inward motion?

5. In what way is a black hole blacker than black ink or a black piece of paper?

6. If the Sun suddenly became a black hole, how would the Earth's orbit be affected?

7. What is cosmic censorship?

8. What are the differences between rotating and nonrotating black holes?

9. Why are all the observed stellar-remnant black-hole candidates members of close binary systems?

10. If light cannot escape from a black hole, how can we detect X rays from such objects?

Advanced Questions

The answers to all computational problems, which are preceded by an asterisk (*), appear at the end of the book.

***11.** What is the Schwarzschild radius of a black hole, measured in kilometers, containing 3 $M_\odot$? 30 $M_\odot$?

12. If more massive stars evolve and die before less massive ones, why do some black-hole candidates have lower masses than their stellar companions?

13. Under what circumstances might a neutron star in a binary star system become a black hole?

14. Which type of black hole, nonrotating or rotating, do science fiction writers (implicitly) use in sending spaceships from one place to another through the hole? Why would the other type not be suitable?

***15.** You are standing in a train car at rest in the station as in Figure 14-2. You measure the car's length to be 10 meters long. The train speeds up to the considerable speed of 0.95*c* with you still inside the car. How long will you measure the car to be now? Explain how you got your answer.

What If . . .

16. A black hole of 5 $M_\odot$ passed by the Earth at, say, Pluto's average distance from the Sun? What would happen to our planet's orbit and to life here?

17. A primordial black hole with the mass of our Moon approached, passed through, and exited the Earth? What might happen to our planet and to life here? *Hints:* You may want to calculate the black hole's Schwartzschild radius. Also, specify if this is a "high speed" or "low speed" event.

18. A primordial black hole exploded nearby? What would astronomers observe?

Web Questions

19. To test your understanding of black hole structure, do Interactive Exercise 14-1 on the Web. You can print out your answers if required.

20. Search the Web for information about the stellar-mass black-hole candidate named V4641 Sgr. In what ways does it resemble other black-hole candidates such as Cygnus X-1 and V404 Cygni? In what ways is it different and more dramatic? How do astronomers currently explain why V4641 Sgr is different?

Understanding the Universe

Part IV

The Sombrero is a Disk Galaxy 50 Million-Light Years from Earth That We View Nearly Edge-on.
(NASA and the Hubble Heritage Team/STScI/AURA)

1755

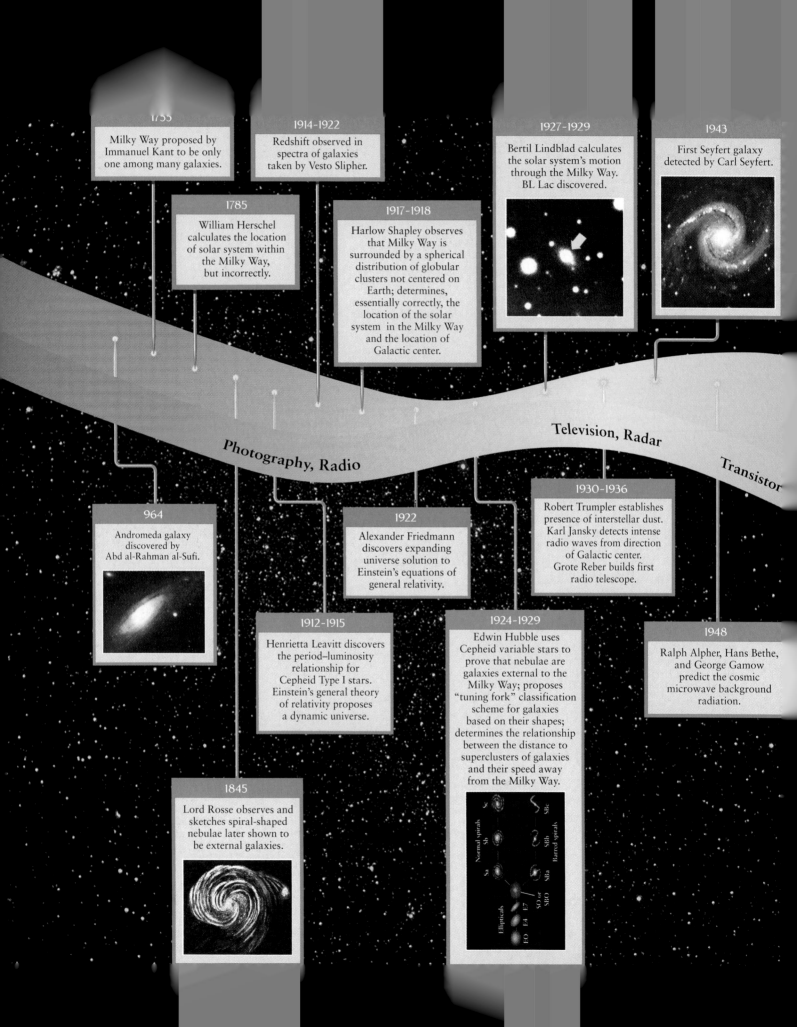

Milky Way proposed by Immanuel Kant to be only one among many galaxies.

1785

William Herschel calculates the location of solar system within the Milky Way, but incorrectly.

1914–1922

Redshift observed in spectra of galaxies taken by Vesto Slipher.

1917–1918

Harlow Shapley observes that Milky Way is surrounded by a spherical distribution of globular clusters not centered on Earth; determines, essentially correctly, the location of the solar system in the Milky Way and the location of Galactic center.

1927–1929

Bertil Lindblad calculates the solar system's motion through the Milky Way. BL Lac discovered.

1943

First Seyfert galaxy detected by Carl Seyfert.

Photography, Radio

Television, Radar

Transistor

964

Andromeda galaxy discovered by Abd al-Rahman al-Sufi.

1845

Lord Rosse observes and sketches spiral-shaped nebulae later shown to be external galaxies.

1912–1915

Henrietta Leavitt discovers the period–luminosity relationship for Cepheid Type I stars. Einstein's general theory of relativity proposes a dynamic universe.

1922

Alexander Friedmann discovers expanding universe solution to Einstein's equations of general relativity.

1924–1929

Edwin Hubble uses Cepheid variable stars to prove that nebulae are galaxies external to the Milky Way; proposes "tuning fork" classification scheme for galaxies based on their shapes; determines the relationship between the distance to superclusters of galaxies and their speed away from the Milky Way.

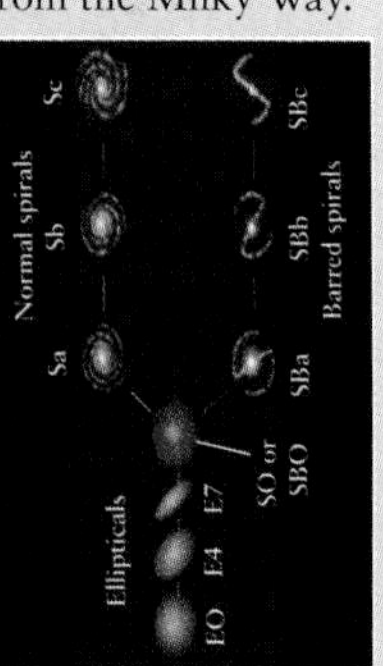

1930–1936

Robert Trumpler establishes presence of interstellar dust. Karl Jansky detects intense radio waves from direction of Galactic center. Grote Reber builds first radio telescope.

1948

Ralph Alpher, Hans Bethe, and George Gamow predict the cosmic microwave background radiation.

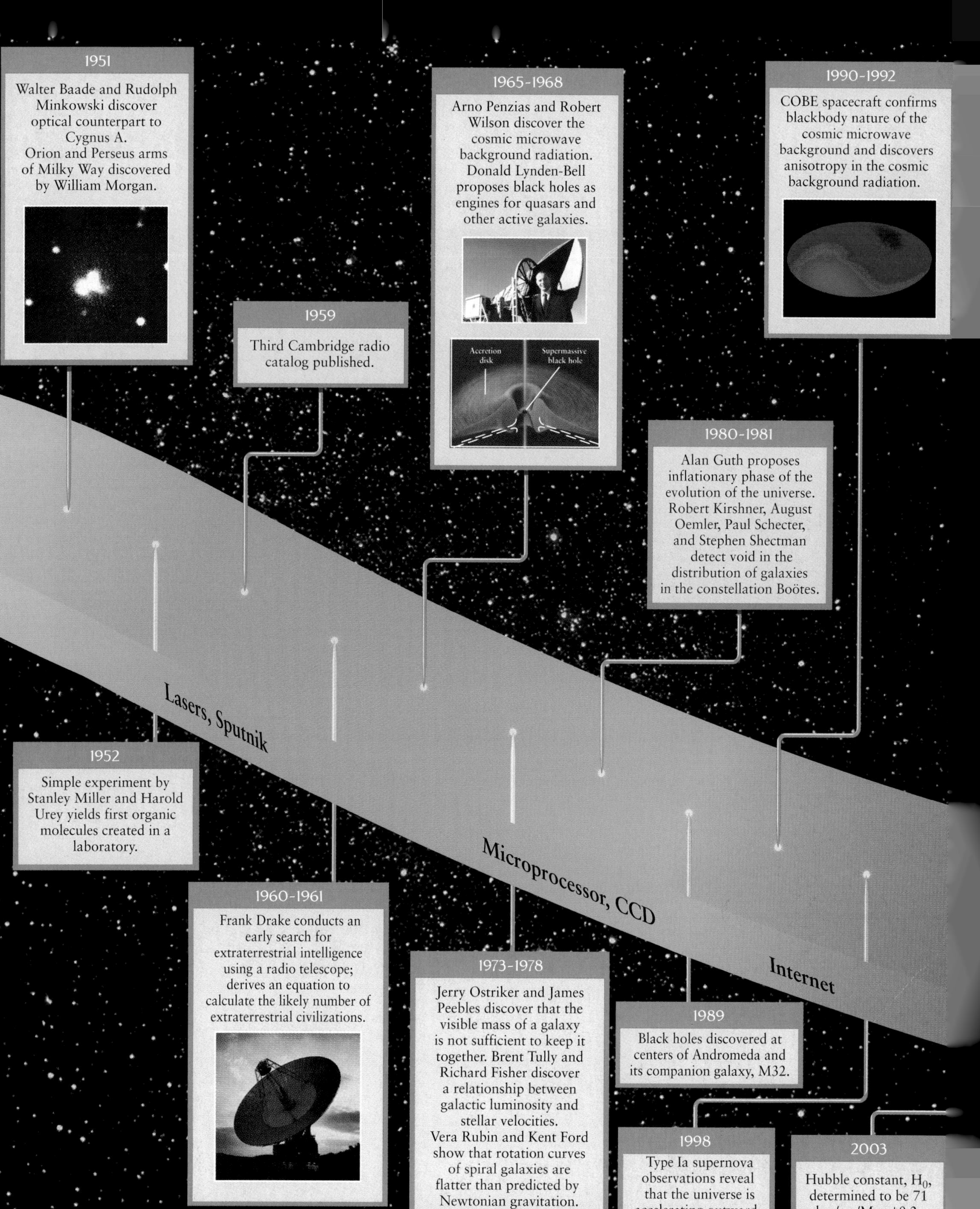

1951
Walter Baade and Rudolph Minkowski discover optical counterpart to Cygnus A.
Orion and Perseus arms of Milky Way discovered by William Morgan.
1959
Third Cambridge radio catalog published.
1965-1968
Arno Penzias and Robert Wilson discover the cosmic microwave background radiation.
Donald Lynden-Bell proposes black holes as engines for quasars and other active galaxies.
Accretion disk
Supermassive black hole
1990-1992
COBE spacecraft confirms blackbody nature of the cosmic microwave background and discovers anisotropy in the cosmic background radiation.
1980-1981
Alan Guth proposes inflationary phase of the evolution of the universe.
Robert Kirshner, August Oemler, Paul Schecter, and Stephen Shectman detect void in the distribution of galaxies in the constellation Boötes.
Lasers, Sputnik
1952
Simple experiment by Stanley Miller and Harold Urey yields first organic molecules created in a laboratory.
Microprocessor, CCD
1960-1961
Frank Drake conducts an early search for extraterrestrial intelligence using a radio telescope; derives an equation to calculate the likely number of extraterrestrial civilizations.
Internet
1973-1978
Jerry Ostriker and James Peebles discover that the visible mass of a galaxy is not sufficient to keep it together. Brent Tully and Richard Fisher discover a relationship between galactic luminosity and stellar velocities.
Vera Rubin and Kent Ford show that rotation curves of spiral galaxies are flatter than predicted by Newtonian gravitation.
1989
Black holes discovered at centers of Andromeda and its companion galaxy, M32.
1998
Type Ia supernova observations reveal that the universe is accelerating outward.
2003
Hubble constant, H_0, determined to be 71 km/sec/Mpc ±0.2.

CHAPTER 15

The Milky Way Galaxy

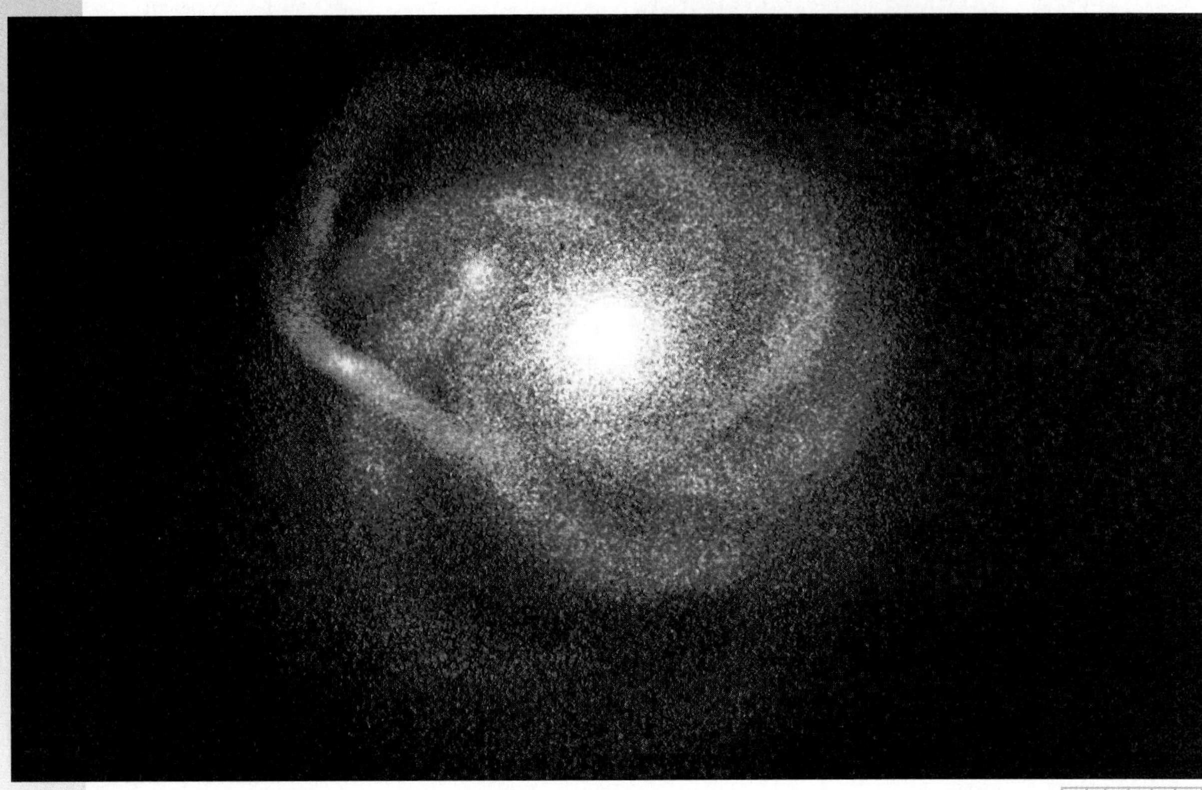

The Canis Major Dwarf galaxy (in red) being consumed by the Milky Way (in blue and white)
(R. Ibata/Strasbourgh Observatory, ULP/et al, 2MASS, NASA)

WHAT DO YOU THINK?

1 How many stars does the Milky Way Galaxy contain?

2 Where is our Solar System located in the Milky Way Galaxy?

3 Is the Sun moving through the Milky Way Galaxy and, if so, about how fast?

In this part of the book we explore how matter is distributed throughout the universe. We examine the huge accumulations of matter found in galaxies and groups of galaxies. Our understanding is at present hampered by the fact that what we can see of the universe (with any present technology) is only a very small percentage of all that it contains. The nature of the matter we have not yet observed is a mystery. Then we will examine both the earliest times of the universe and its future. We begin by learning more about the **Milky Way Galaxy,** in which the solar system resides, and how astronomers discovered that the Milky Way is just a very, very tiny piece of the universe.

For those fortunate enough to live away from bright outdoor lights, the Milky Way appears as a veiled band overlaid with the glow of individual stars. Galileo, the first person to view the Milky Way through a telescope, discovered that it contains countless dim stars. Centuries of observations since that time have established that our solar system is part of a galaxy, an enormous assemblage of hundreds of billions of stars, along with gas, dust, and other matter, all held together by their mutual gravitational attraction. Most of the stars in our Milky Way Galaxy are located in a disk that resembles a flying saucer in an old science fiction movie (Figure 15-1a). The inner part of the disk is uniformly filled with stars. Spiral

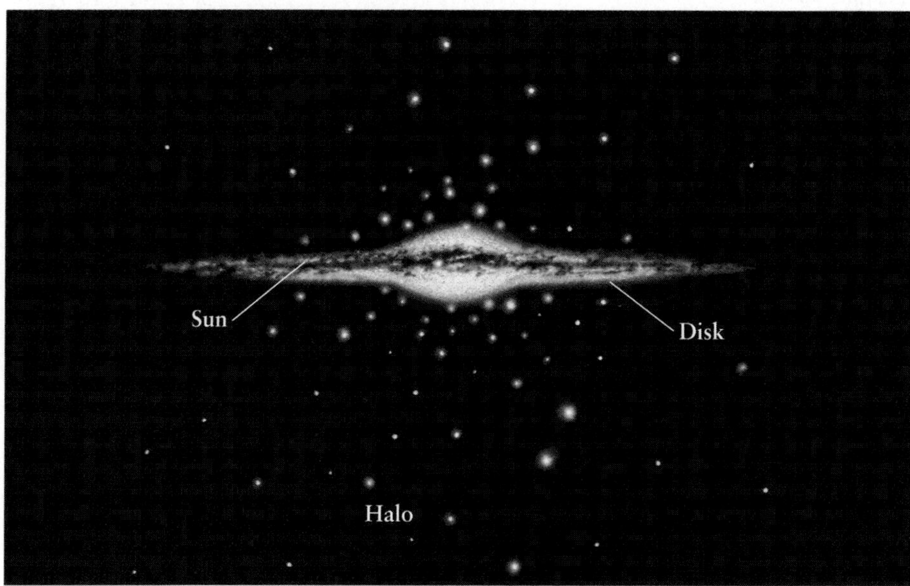

b

c

FIGURE 15-1 **Schematic Diagrams of the Milky Way** (a) This edge-on view shows the Milky Way's disk, containing most of the stars, gas, and dust, and its halo, containing many old stars. Individual stars in the halo are too dim to show, so the bright regions in the halo are clusters of stars. (b, c) Two possible distributions of the spiral arms of our Galaxy. Our Galaxy has at least four major spiral arms and several shorter arm segments. Because we are in the disk, surrounded by gas and dust, we cannot see much of the spiral structure directly. Therefore, we are not yet sure of its exact shape. (b, c: Illustration by Dennis Davidson, courtesy of AMNH/Hayden Planetarium) arms swirl out from this inner region and within these arms (Figures 15-1b and c), new stars form from the debris of earlier generations of stars. The Galaxy's remaining stars are located in a spherical halo surrounding the disk (note only our own Galaxy is capitalized).

In this chapter you will discover

- the Milky Way Galaxy—billions of stars along with gas and dust bound together by mutual gravitational attraction
- the properties of our Milky Way Galaxy
- Earth's location in the Milky Way
- how interstellar gas and dust enable star formation to continue in our Galaxy
- that observations reveal the presence of significant mass in the Milky Way that astronomers have yet to identify
- that there is a black hole at the center of our Galaxy

DISCOVERING THE MILKY WAY

Prior to the twentieth century, astronomers did not know the large-scale distribution of stars and other matter in the universe. Throughout history, most people, including many astronomers, believed that the Milky Way contains all the stars in the cosmos. In other words, they thought that the "Galaxy" and the "universe" were the same thing. We begin studying galaxies by learning how that belief changed.

15-1 Studies of Cepheid variable stars revealed that the Milky Way is only one of many galaxies

The belief that our Galaxy is but one of many was put forth in 1755, when the German philosopher Immanuel Kant suggested that vast collections of stars lie far beyond the confines of the Milky Way. Less than a century later, the Irish astronomer William Parsons observed the structure of some of those "island universes" proposed by Kant. Parsons was the third Earl of Rosse in Ireland. He was rich, he liked machines, and he was fascinated with astronomy. Accordingly, he set about building gigantic telescopes. In February 1845, his pièce de résistance was finished. This telescope's massive mirror measured 1.8 m (6 ft) in diameter and was mounted at one end of an 18-m (60-ft) tube controlled by cables, straps, pulleys, and cranes (Figure 15-2a). For many years, this triumph of nineteenth-century engineering enjoyed distinction as the largest telescope in the world.

With this new telescope, Lord Rosse examined many of the glowing interstellar clouds discovered and catalogued by William Herschel. Herschel, his sister Caroline, and his son John, among others, discovered and recorded details of many fuzzy-looking astronomical objects called **nebulae** (*singular* **nebula**). With the high resolution provided by his telescope, Lord Rosse observed that some of these nebulae have a distinct spiral structure. A particularly good example is M51, also called the Whirlpool or NGC 5194.

Lord Rosse had no photographic equipment in 1845, so he made drawings of what he observed. Figure 15-2b is his drawing of M51. Views like this inspired Lord Rosse to echo Kant's proposal of island universes. Figure 15-2c shows a modern photograph of M51. It is interesting to note the differences between the perception and interpretation of astronomical objects on the one hand and the camera's recording of them on the other.

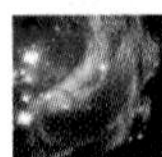

Most astronomers of Rosse's day did not agree with the notion of island universes outside our Galaxy. They thought that the Milky Way contained all the stars in the universe—the Milky Way was the universe. In April 1920, a formal discussion, now known as the **Shapley–Curtis debate**, was held at the National Academy of Sciences in Washington, D.C. Harlow Shapley argued that the spiral nebulae are relatively small, nearby objects scattered around our Galaxy. Heber D. Curtis championed the island universe theory, arguing that each of these spiral nebulae is a separate rotating system of stars much like our own Galaxy. While the Shapley–Curtis debate focused scientific attention on the size of the universe, nothing was decided, because no one had any firm evidence to demonstrate exactly how far away the spiral nebulae were. Astronomers desperately needed to devise a way to measure the distances to them. It took a young teacher and former basketball player from Kentucky who moved to Chicago to study astronomy to finally achieve this goal. His name was Edwin Hubble.

In 1923, Hubble took an historic photograph of M31, then called the Andromeda Nebula. It was one of the spiral nebulae around which controversy raged. On the photographic plate he discovered what first appeared to be a nova. Referring to previous plates of that region, he soon realized that the object was actually a Cepheid variable star. As we saw in Section 12-13, these pulsating stars vary in

a

b

c

RIVUXG

FIGURE 15-2 **High-Tech Telescope of the Mid-Nineteenth Century** **(a)** Built in 1845, this structure housed a 1.8-m-diameter telescope, the largest of its day. The improved resolution it provided over other telescopes is like the improvement that the Hubble Space Telescope provided over Earthbound optical instruments when it was launched. The telescope, as shown here, was restored to its original state during 1996–1998. **(b)** Using his telescope, Lord Rosse made this sketch of the spiral structure of the galaxy M51. **(c)** A modern photograph of M51 (also called NGC 5194). This spiral galaxy in the constellation of Canes Venatici is known as the Whirlpool Galaxy because of its distinctive appearance. Its distance from Earth is about 20 million light-years. The blob on the right, at the end of one of its spiral arms, is a companion galaxy. (a: Birr Castle Demesne; b: Lund Humphries; c: NOAO)

brightness periodically. Further scrutiny over the next several months revealed many other Cepheids. Figure 15-3 shows a Cepheid in the galaxy M 100 at different stages of brightness.

Only a decade before, in 1912, the American astronomer Henrietta Leavitt had published an important study of Cepheid variables. Leavitt studied numerous Cepheids in the

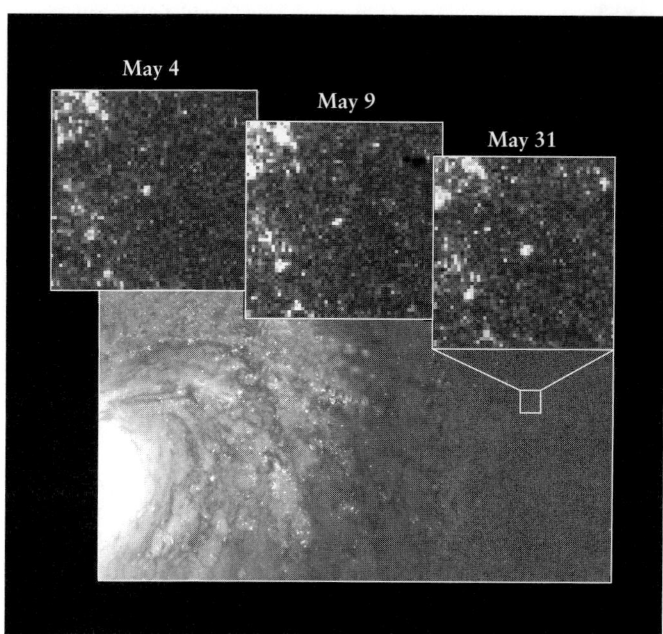

RIVUXG

◀ FIGURE 15-3 **M 100** One of the most reliable ways to determine the distance to moderately remote galaxies is to locate Cepheid variable stars in them, as discussed in the text. At 17 Mpc (56 Mly), the galaxy M 100 in the constellation Coma Berenices is among the most remote objects whose distances have been determined using Cepheids. **Insets:** The Cepheid in this view, one of 20 located to date in M 100, is shown at different stages in its brightness cycle, which recurs over several weeks. (Dr. Wendy L. Freedman, Observatories of the Carnegie Institution of Washington; NASA)

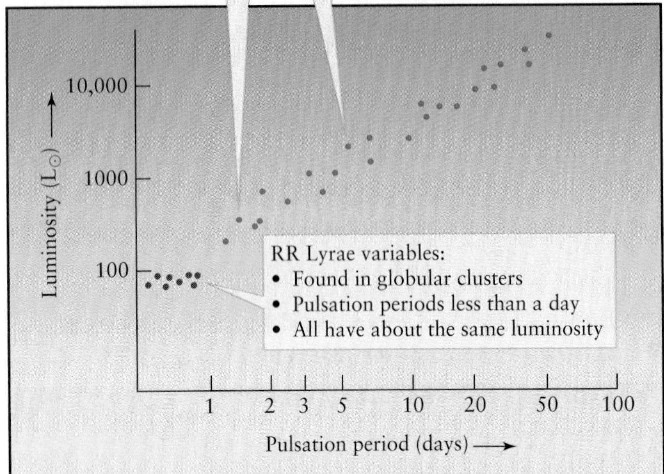

Cepheid variables:
• Found throughout the Galaxy
• Pulsation periods of 1 to 50 days
• Luminosity related to pulsation period
10,000
1000
100
Luminosity ($L_\odot$) ⟶
RR Lyrae variables:
• Found in globular clusters
• Pulsation periods less than a day
• All have about the same luminosity
1 2 3 5 10 20 50 100
Pulsation period (days) ⟶

FIGURE 15-4 **The Period-Luminosity Relation** This graph shows the relationship between the periods and average luminosities of classical (Type I) Cepheid variables and the closely related RR Lyrae stars (also discussed in Chapter 12). Each dot represents a Cepheid or RR Lyrae whose luminosity and period have been measured.

Small Magellanic Cloud, then also believed to be a nebula, but now known to be a galaxy orbiting the Milky Way. Leavitt's study led her to the period-luminosity relation for Type I Cepheids (see Section 12-14). As we saw in Figure 12-30 (presented in more detail in Figure 15-4), she established that a direct relationship exists between a Cepheid's luminosity (or absolute magnitude) and its period of oscillation. By observing the star's period and apparent magnitude, its distance can be calculated, as described in An Astronomer's Toolbox 15-1.

Thanks to this work, Hubble knew that he could use the characteristics of the fluctuating light of Cepheid variable stars to help him calculate the distance and, he figured, put the Shapley–Curtis debate to rest once and for all. His observations of Type I Cepheids led him to determine that M31 is some 2.2 million light-years *beyond* the Milky Way. This proves that M31 is not an open or globular cluster in our Galaxy, but rather an enormous separate stellar system—a separate galaxy. M31, now called the *Andromeda Galaxy,* is the most distant object in the universe that can be seen with the naked eye. Similar calculations have been done for all galaxies in which Cepheids can be observed. Observations of Type Ia supernovae also provide distance measurements to even more distant galaxies (see An Astronomer's Toolbox 15-1).

INSIGHT INTO SCIENCE

Room for Debate Lacking definitive data, competent scientists can develop and believe strikingly different explanations for the same observations. At the time of the Shapley–Curtis debate, the observations allowed both points of view. It was only with the advent of the distance measurement technique used by Hubble that the spiral nebulae were definitely shown to be outside the Galaxy.

Hubble's results, which he presented at the end of 1924, settled the Shapley–Curtis debate once and for all. The universe was recognized to be far larger and populated with far bigger objects than most astronomers had imagined. Hubble had discovered the realm of the galaxies. Today, we know that the universe contains myriad galaxies, of which the Milky Way is just one. Like the Milky Way, each **galaxy** is a grouping of millions, billions, or even trillions of stars, gas, dust, and matter in other forms, all gravitationally bound together.

We now apply Hubble's method to find Earth's place in the Milky Way Galaxy. Recall that the Sun's proximity to us makes it our best-understood star. It might seem that the nearness of the stars and clouds in the Milky Way would make it the best-understood galaxy. However, the clouds of gas and dust surrounding the solar system make it very challenging for astronomers to survey the distant parts of the Galaxy completely.

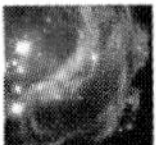

THE STRUCTURE OF OUR GALAXY

Because the band of the Milky Way completely encircles us, astronomers long ago suspected that the Sun and all the stars that we see are part of it. In the 1780s, William Herschel took the first steps toward mapping its structure. He attempted to deduce the Sun's location in the Galaxy by counting the number of stars in 683 regions of the sky. He reasoned that the greatest density of stars should be seen toward the Galaxy's center and a lesser density seen toward the edge. However, Herschel found roughly the same density of stars all along the Milky Way. He therefore concluded that we are at the center of the Galaxy.

Herschel was wrong: The Earth has no privileged place in the Milky Way. The Sun is about 26,000 ly (8000 pc) from the Galaxy's center, the **galactic nucleus.** Herschel's physical understanding of the cosmos was incomplete, so he misinterpreted his observations and thus came to an incorrect conclusion.

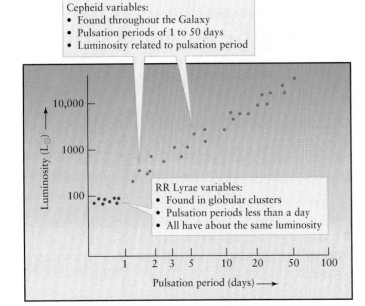

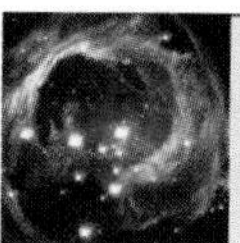

AN ASTRONOMER'S TOOLBOX 15-1

Cepheids and Supernovae as Indicators of Distance

Because their periods are directly linked to their luminosities, Cepheid variables are one of the most reliable tools astronomers have for determining the distances to galaxies. To this day, astronomers use this link—much as Hubble did back in the 1920s—to measure intergalactic distances. More recently, they have begun to use Type Ia supernovae, which are far more luminous and thus can be seen much farther away, to determine the distances to very remote galaxies.

Example: In 1992, a team of astronomers used Cepheid variables in a galaxy called IC 4182 to deduce that galaxy's distance from the Earth. The team of astronomers used the Hubble Space Telescope on 20 separate occasions to record images of the stars in IC 4182. By comparing these images, the astronomers could pick out which stars vary in brightness. In this way, they discovered 27 Cepheids in IC 4182. Using their observations, the astronomers estimated the brightnesses of the Cepheid variables and plotted their light curves. The Hubble Space Telescope is particularly well suited for studies of this kind, because its extraordinary angular resolution makes it possible to pick out individual stars at great distances. One such Cepheid has a period of 42.0 days and an average apparent magnitude (m) of +22.0. (See Section 11-2 for an explanation of the apparent magnitude scale.) By comparison, the dimmest star you can see with the naked eye has $m = +6$; this Cepheid in IC 4182 appears less than one one-millionth as bright. The star's spectrum shows that it is a metal-rich Type I Cepheid variable.

According to the period-luminosity relation shown in Figure 15-4, such a Type I Cepheid with a period of 42.0 days has an average luminosity of 33,000 $L_\odot$. This can be expressed by saying that this Cepheid has an average absolute magnitude (M) of –6.5. (This compares to $M = +4.8$ for the Sun). Hence, the difference between the Cepheid's apparent and absolute magnitudes, called its **distance modulus,** is

$$m - M = (+22.0) - (-6.5) = 22.0 + 6.5 = 28.5$$

From An Astronomer's Toolbox 11-3, we see that the distance modulus of a star is related to its distance in parsecs (d) by

$$m - M = 5 \log d - 5$$

This can be rewritten as

$$d = 10^{(m - M + 5)/5} \text{ parsecs}$$

Inserting the value for the distance modulus in this equation, we get the distance to the Cepheid variable and, hence, the distance to the galaxy of which the star is part:

$$d = 10^{(28.5 + 5)/5} \text{ parsecs} = 10^{6.7} \text{ parsecs} = 5.0 \times 10^6 \text{ parsecs}$$

The galaxy is 5 Mpc (1 Mpc = 1 megaparsec = 10^6 parsecs), or 16 million (1.6×10^7) light-years, from Earth.

Astronomers are interested in IC 4182 because a Type Ia supernova was observed there in 1937. All Type Ia supernovae are exploding white dwarfs that reach nearly the same maximum brightness at the peak of their outburst (see Section 13-9). Once astronomers knew the peak absolute magnitudes of Type Ia supernovae, they could use these supernovae as distance indicators. Because the distance to IC 4182 is known from its Cepheids, the 1937 observations of the supernova in that galaxy allow us to calibrate Type Ia supernovae as distance indicators. At maximum brightness, the 1937 supernova reached an apparent magnitude of $m = +8.6$. Since the distance modulus of the galaxy ($m - M$) is 28.5, we see that when a Type Ia supernova is at maximum brightness, its absolute magnitude is

$$M = m - (m - M) = 8.6 - 28.5 = -19.9$$

Whenever astronomers find a Type Ia supernova in a remote galaxy, they can combine this absolute magnitude with the observed maximum apparent magnitude to get the galaxy's distance modulus, from which the galaxy's distance can be easily calculated (just as we did above for the Cepheids in IC 4182). This technique has been used to determine the distances to galaxies hundreds of millions of parsecs away.

Try these questions: At what distance in parsecs is a star with distance modulus 20? Epsilon (ϵ) Indi has an apparent magnitude of +4.7 and distance of 3.6 pc. What is its absolute magnitude? What would the Sun's apparent magnitude and distance modulus be as seen from 100 pc away? Its absolute magnitude is +4.83.

(Answers appear at the end of the book.)

FIGURE 15-5 **Our Galaxy** This wide-angle photograph spans half the Milky Way as seen from the equatorial latitudes. The Northern Cross is at the left, and the Southern Cross is at the right. The center of the Galaxy is in the constellation Sagittarius, in the middle of this photograph. The dark lines and blotches are caused by hundreds of interstellar clouds of gas and dust that obscure the light from background stars, rather than by a lack of stars. (Dirk Hoppe)

15-2 Cepheid variables help us locate our Galaxy's center

While studying star clusters in the 1930s, R. J. Trumpler discovered the reason for Herschel's mistake. Herschel did not know about interstellar gas and dust, which affected his counts of the stars. Trumpler noticed that remote star clusters appear dimmer than would be expected just from their distance alone. Something must be blocking starlight on its way toward the Earth. He correctly concluded that interstellar space is not a perfect vacuum. Instead, it contains dust that absorbs light from distant stars. Great patches of this dust are clearly visible in wide-angle photographs (Figure 15-5). Like the stars, this dust is concentrated in the plane of the Galaxy.

This interstellar dust almost completely obscures from view visible light emanating from the center of our Galaxy. Visual photons from there are mostly absorbed or scattered before they reach us. Therefore, Herschel was seeing only nearby stars, and he measured apparent magnitudes that were dimmer than they would have been had there been no interstellar dust. Without adjusting for the effects of the dust, he concluded the stars were farther away than they really are. He also had no idea of the true size of the Galaxy, nor could he see the vast number of stars located in the general direction of the galactic center that are hidden by the dust.

INSIGHT INTO SCIENCE

A Little Knowledge Incomplete information often leads to incorrect interpretation of data and, therefore, to incorrect conclusions. Herschel's lack of knowledge about the matter in interstellar space prevented him from correctly interpreting the distribution of stars surrounding the Earth and, thus, led to an inaccurate position for the Sun within the Galaxy.

Because interstellar dust is concentrated in the plane of the Galaxy, the absorption of starlight is strongest in those parts of the sky covered by the Milky Way. Above or below the plane of the Galaxy, our view is relatively unobscured. Knowledge of our true position in the Galaxy eventually came from observations of globular clusters (see Section 12-12). Shapley used the period-luminosity relationship for variable stars to determine the distances to the then-known 93 globular clusters in the sky. (More than 150 are known today.) From their directions and distances, he mapped out the distribution of these clusters in three-dimensional space. By 1917, Shapley had discovered that the globular clusters

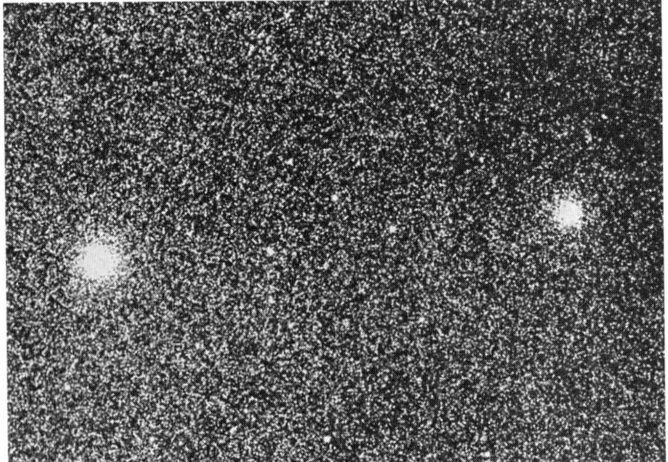

FIGURE 15-6 **A View Toward the Galactic Center** More than a million stars in the disk of our Galaxy fill this view, which covers a relatively clear window just 4° south of the galactic nucleus in Sagittarius. Beyond the disk stars you can see two prominent globular clusters. Although most regions of the sky toward Sagittarius are thick with dust, there is very little obscuring matter in this tiny section of the sky. (Harvard Observatory)

are located in a spherical distribution centered not on the Earth, but on a point in the Milky Way toward the constellation Sagittarius. Figure 15-6 shows two globular clusters in a relatively clear part of the sky in that direction. Shapley then made a bold conjecture: *The globular clusters orbit the center of the Milky Way located in Sagittarius.* His pioneering research has since been observationally verified. The Earth is not at the center of the Galaxy.

15-3 Nonvisible observations help map the galactic disk

In order to see into the dust-filled plane of the Milky Way, astronomers mainly use radio wave, infrared, X-ray, and gamma-ray telescopes. These wavelengths are scattered much less by the interstellar gas and dust located throughout the Galaxy's disk than are visible or ultraviolet wavelengths. Observations of the distant parts of the Galaxy were first made using radio telescopes. Radio waves penetrate the Earth's atmosphere, so we can observe them anywhere we can build a radio telescope. (Recall that infrared observations must be made at high altitudes or in space, and that X-ray and gamma-ray observations are almost always made from space.)

Detecting the radio emission directly from interstellar hydrogen, by far the most abundant element in the universe, is a primary means of mapping the Galaxy. Unfortunately, the major transitions of electrons in the hydrogen atom (see Figure 4-11) produce photons at ultraviolet and

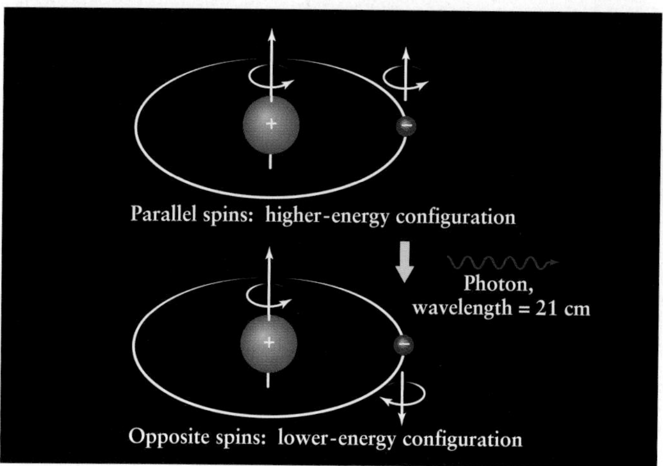

FIGURE 15-7 **Electron Spin and the Hydrogen Atom** Due to their spin, electrons and protons are both tiny magnets. When an electron and the proton it orbits are spinning in the same direction, their energy is higher than when they are spinning in opposite directions. When the electron flips from the higher-energy to the lower-energy configuration, the atom loses a tiny amount of energy that is radiated as a radio photon with a wavelength of 21 cm.

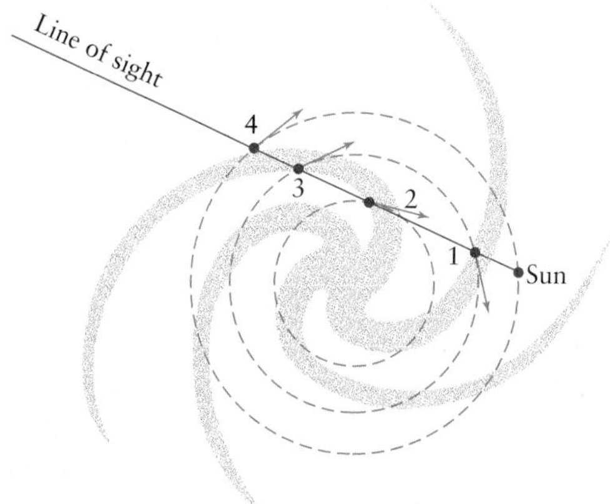

FIGURE 15-8 **A Technique for Mapping the Galaxy** Hydrogen clouds at different locations along our line of sight are moving around the center of the Galaxy at different speeds. Radio waves from the various gas clouds therefore exhibit slightly different Doppler shifts, permitting astronomers to sort out the gas clouds and map the Galaxy.

visible wavelengths that do not penetrate the interstellar medium. How, then, can radio telescopes detect all this hydrogen? The answer lies in atomic physics.

In addition to mass and charge, particles such as protons and electrons possess a tiny amount of angular momentum commonly called **spin.** According to the laws of quantum mechanics, the electron and proton in a hydrogen atom can only spin in either parallel or opposite directions (Figure 15-7); they can have no other spin orientations. If the electron in a hydrogen atom flips from one orientation to the other, the atom must gain or lose a tiny amount of energy. In particular, when flipping from parallel to opposite spins, the atom simultaneously emits a low-energy radio photon whose wavelength is 21 cm. This flip happens rarely in each atom, so it is only because the Galaxy has vast quantities of interstellar hydrogen gas that it can be detected at all. In 1951, a team of astronomers first succeeded in detecting the faint hiss of 21-cm radio static from spin flips.

The detection of **21-cm radio radiation** was a major breakthrough in mapping the **disk** of the Galaxy. To see why, suppose that you aim your radio telescope across the Galaxy as sketched in Figure 15-8. Your radio receiver picks up 21-cm emission from hydrogen clouds at points 1, 2, 3, and 4. However, the radio waves from these various clouds are Doppler shifted (see An Astronomer's Toolbox 4-4) by slightly different amounts, because they are moving at different speeds as they orbit around the center of the Galaxy. Because these radio waves from gas clouds in different parts of the Galaxy arrive at our radio telescopes with slightly different wavelengths as a result of being Doppler shifted, it is possible to identify which radio signals come from which

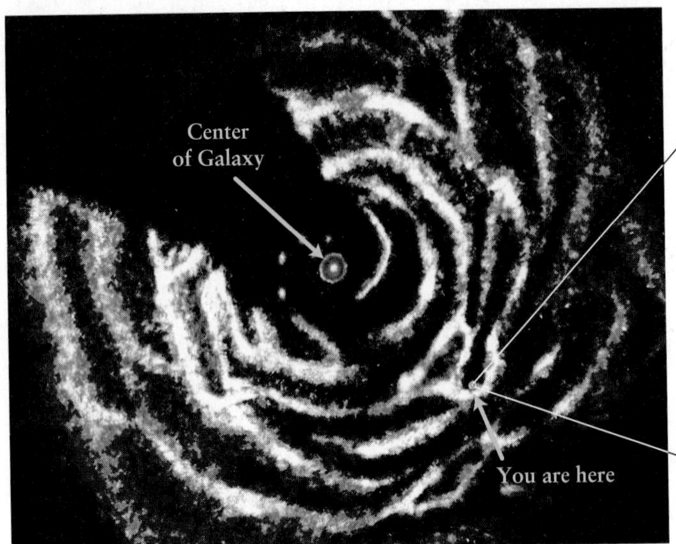

R I V U X G

WEB LINK 15.4

FIGURE 15-9 A Map of the Galaxy This map, based on radio telescope surveys of 21-cm radiation, shows the distribution of hydrogen gas in a face-on view of the Galaxy. This view just hints at spiral structure. The galactic nucleus is marked with a dot surrounded by a circle. Details in the large, blank, wedge-shaped region toward the upper left of the map are unknown, because gas in this part of the sky is moving perpendicular to our line of sight and thus does not exhibit a detectable Doppler shift. **Inset:** This drawing, based on visible-light data, shows that our solar system lies just inside the Orion arm of the Milky Way Galaxy. (Courtesy of G. Westerhout; inset: National Geographic)

gas clouds and thus to produce an initial map of the Galaxy, such as that shown in Figure 15-9.

Our radio map reveals numerous arched lanes of neutral hydrogen gas. If this were the overall structure of the Galaxy, then the Milky Way would appear unlike any other observed galaxy (see Chapter 16 for typical images of other galaxies). Indeed, the other disk-shaped galaxies we observe have spiral arms. We need more information from space to improve our understanding of the Galaxy's disk features. Note that photographs of a spiral galaxy (Figure 15-10)

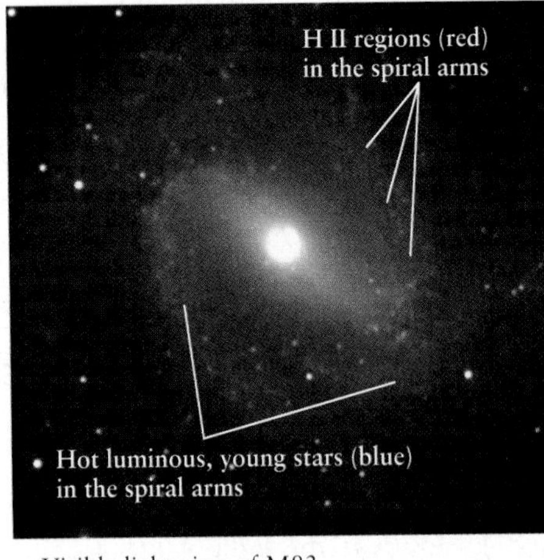

a Visible-light view of M83

R I V U X G

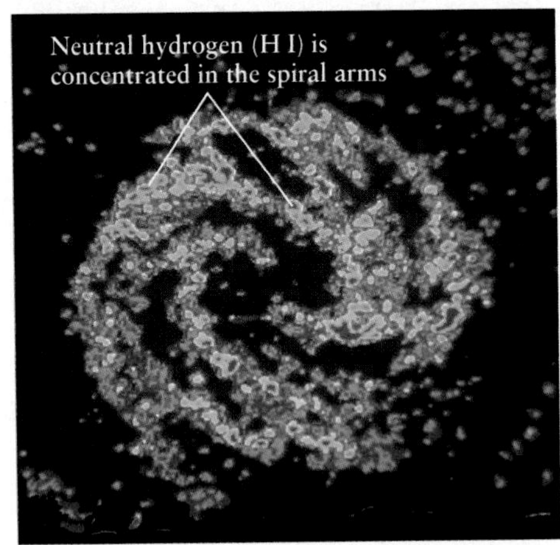

b 21-cm radio view of M83

R I V U X G

FIGURE 15-10 Two Views of a Spiral Galaxy The galaxy M83 is in the southern constellation of Centaurus, about 12 million light-years from Earth. (a) At visible wavelengths, spiral arms are clearly illuminated by young stars and glowing H II regions. (b) A radio view at 21-cm wavelength shows the emission from neutral hydrogen gas. Note that the spiral arms are more clearly demarcated by visible stars and H II regions than by 21-cm radio emission. (a: S. Van Dyk/IPAC; b: VLA, NRAO)

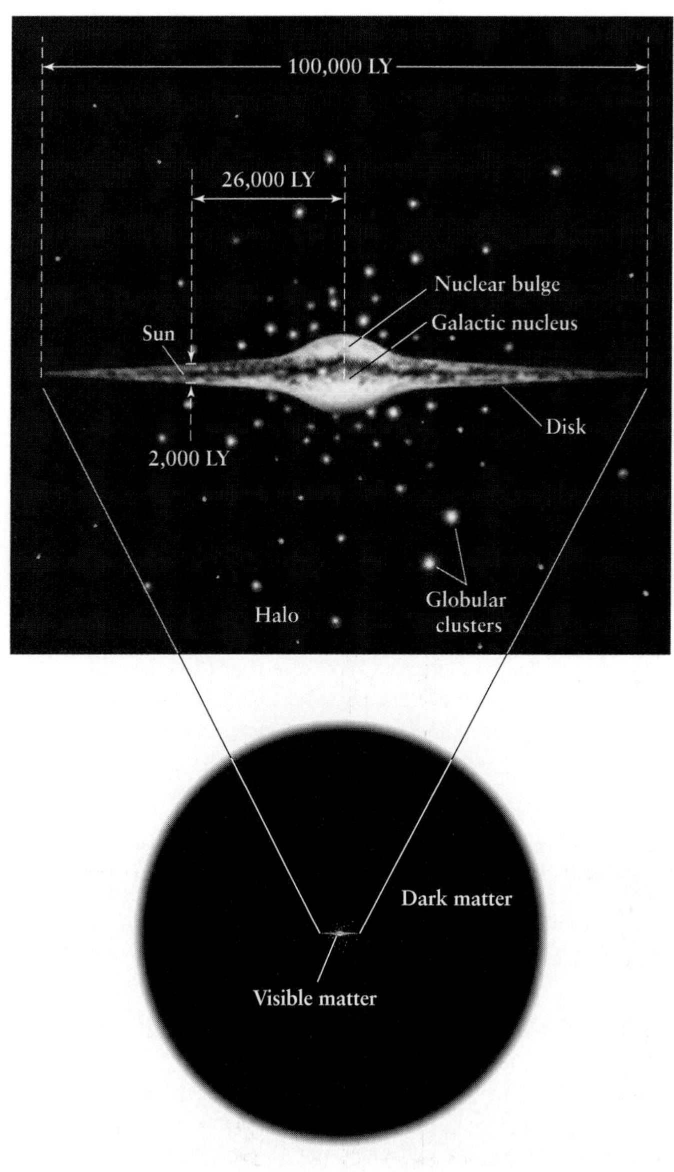

◀ **FIGURE 15-11** **Our Galaxy:** As seen from the side, the three major visible components of our Galaxy are a thin disk, a nuclear bulge, and a halo. The visible Galaxy's diameter is about 100,000 ly, and the Sun is about 26,000 ly from the galactic center. The disk contains gas and dust along with Population I (young, metal-rich) stars. The halo is composed almost exclusively of Population II (old, metal-poor) stars. **Inset:** The visible matter in our Galaxy fills only a small volume compared to the distribution of dark matter, whose composition is presently unknown. Its presence is felt by its gravitational effect on visible matter.

show arms outlined by "spiral tracers"—bright, Population I stars and emission nebulae. As we saw in Chapter 12, these features indicate active star formation. If the Milky Way is spiral, then another useful way to further chart its structure and show that it has **spiral arms** is to map the locations of star-forming complexes marked by H II regions, giant molecular clouds, and massive, hot, young stars in groups called *OB associations*.

1 Dust absorption limits the range of visual observations in the plane of the Galaxy to less than 10,000 ly from the Earth. Astronomers use visible observations of nearby, bright OB associations and associated H II regions to plot the spiral arms near the Sun. Radio observations of hydrogen and carbon monoxide molecules (discussed in Section 12-1) have been used to chart more remote star-forming regions of the Galaxy. Taken together, all these observations indicate that our Galaxy has about 200 billion stars located in and between at least four major spiral arms and several short arm segments (see Figure 15-1b). Recent observations also suggest that a bar of stars and gas may cross the center of the Galaxy. If so, the Milky Way resembles the galaxy in Figure 15-1c more than it does Figure 15-1b. We will explore the origin of the spiral arms in Chapter 16, when we present more evidence about the cause of spiral structure from observations of other galaxies.

The observable disk of our Galaxy is about 100,000 ly 2
in diameter and about 2000 ly thick (Figure 15-11a). The Sun is located near a relatively short arm segment called the Orion arm, which includes the Orion Nebula (see Figure 12-16) and neighboring sites of vigorous star formation in that constellation. Two major spiral arms border either side of the Sun's position. On the side toward the galactic center is the Sagittarius arm, which stargazers in the northern hemisphere see in the summer when they look at the portion of the Milky Way stretching across Scorpius and Sagittarius (see Figures 15-5 and 15-9). Directed away from the galactic center is the Perseus arm, which is visible in the northern hemisphere in the winter. The remaining two major spiral arms are usually referred to as the Centaurus arm and the Cygnus arm, neither of which can be seen at visible wavelengths because of obscuring dust in the interstellar medium.

Observations reveal that the Galaxy's arms spiral out from a flattened sphere of stars, called the **nuclear bulge**, that is about 20,000 ly in diameter. This feature is also seen in Figure 15-12, a wide-angle infrared image of the Galaxy taken by the *COBE* spacecraft. The nuclear bulge is centered on the galactic nucleus 26,000 ly away from us.

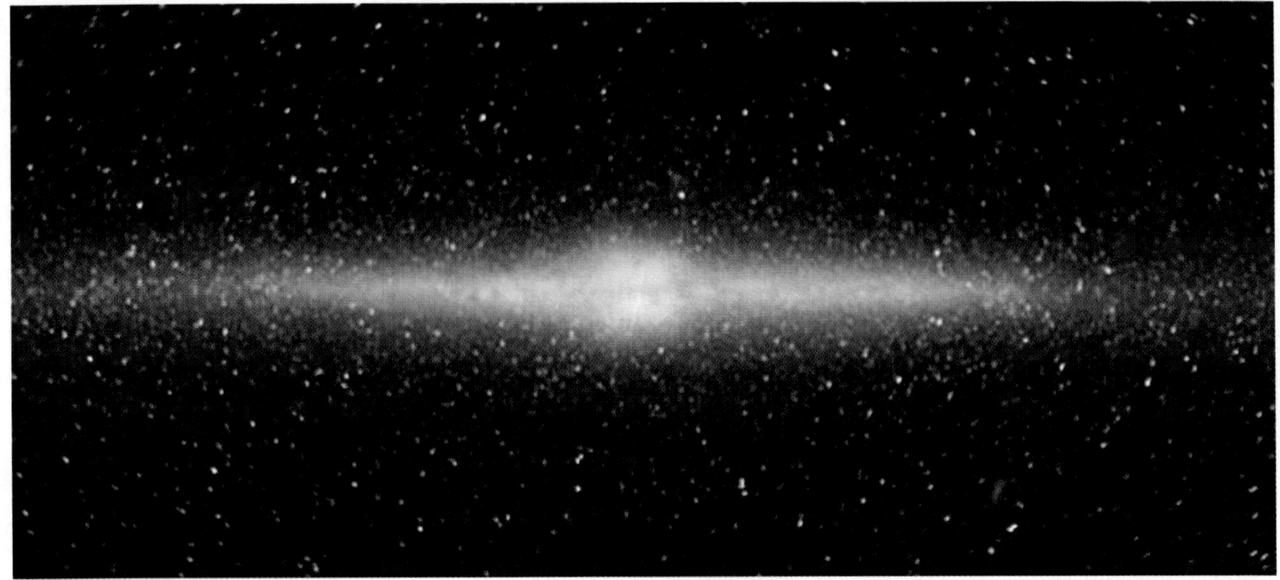

FIGURE 15-12 **Infrared View of the Milky Way** Taken by the *COBE* satellite in 1997, this infrared image shows the disk and nuclear bulge of our Galaxy. Most of the sources scattered above and below the disk are nearby stars. Stars appear white, while interstellar dust appears red. Note that the dust that obscures light from more distant stars in Figure 15-5 is quite bright in this infrared image. (The COBE Project, DIRBE, NASA)

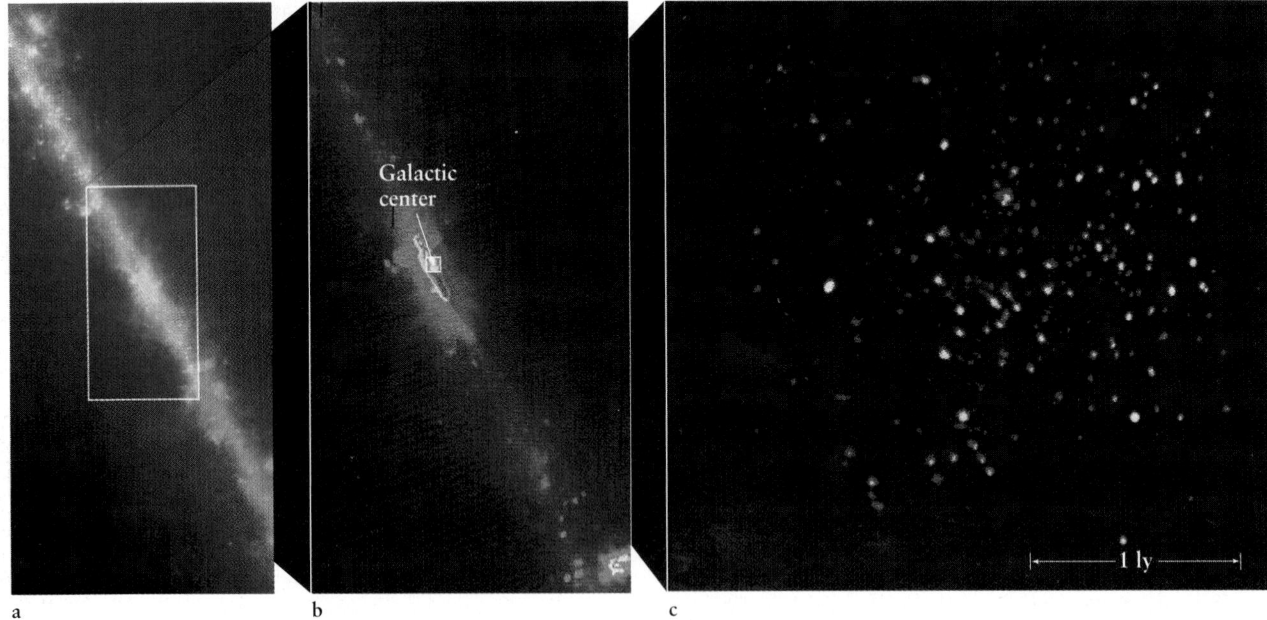

WEB LINK 15.5

FIGURE 15-13 **The Galactic Center** (a) This wide-angle view at infrared wavelengths shows a 50° segment of the Milky Way centered on the nucleus of the Galaxy. Black represents the dimmest regions of infrared emission, with blue the next strongest, followed by yellow and red; white represents the strongest emission. The prominent band diagonally across this photograph is a layer of dust in the plane of the Galaxy. Numerous knots and blobs along the plane of the Galaxy are interstellar clouds of gas and dust heated by nearby stars. (b) This close-up infrared view of the galactic center covers the area outlined by the white rectangle in (a). (c) This infrared image shows about 300 of the brightest stars less than 1 ly from Sagittarius A*, which is at the center of the picture. The distribution of stars and their observed motions around the galactic center imply a very high density (about a million solar masses per cubic light-year) of less luminous stars. (a, b: NASA; c: A. Eckart, R. Genzel, R. Hofman, B.J. Sams, and C.E. Tocconi-Garman, ESO)

15-4 The galactic nucleus is an active, crowded place

If you lived on a planet near the very center of the Galaxy, called the **galactic nucleus,** you would see a million stars as bright as Betelgeuse. The total intensity of starlight from all those nearby stars would be equivalent to 200 of our full Moons. Night would never really fall. Stranger still, the Galaxy around you would be filled with intense activity.

Figure 15-13 shows three infrared views looking toward the nucleus of the Galaxy. Figure 15-13a is a wide-angle view covering a 50° segment of the Milky Way through Sagittarius and Scorpius. The prominent band across this image is a thin layer of dust in the plane of the Galaxy. The numerous knots and blobs along the dust layer are interstellar clouds heated by young O and B stars. Figure 15-13b is an *IRAS* view of the galactic center. Numerous streamers of dust (in blue) surround it. The strongest infrared emission (in white) comes from **Sagittarius A,** which is also a grouping of several powerful sources of radio waves. One of these sources, called Sagittarius A* (pronounced A-star), is believed to be the galactic nucleus. Figure 15-13c shows stars within 1 ly of Sagittarius A*, with resolution of 0.02 ly.

Radio observations give a different picture of the center of our Galaxy. In 1960, Doppler shift measurements of 21-cm radiation revealed two enormous arms of hydrogen. One arm, which is located between Earth and the galactic nucleus, is approaching us at a speed of 53 km/s. The other arm, on the far side of the galactic nucleus, is receding from us at a rate of 135 km/s. The total amount of hydrogen in these expanding arms is at least several million solar masses. Given their speed and how far they have traveled, astronomers conclude that something quite extraordinary must have happened about 10 million years ago to expel such an enormous amount of gas from the central region of the Galaxy.

In addition to 21-cm radiation from neutral hydrogen gas, astronomers have also detected radio noise coming from the galactic center. This radio emission, which is produced by high-speed electrons spiraling around a magnetic field, is called **synchrotron radiation.** Despite its small size, Sagittarius A is one of the brightest sources of synchrotron radiation in the entire sky.

Some of the most detailed radio images of the galactic nucleus come from the Very Large Array (VLA). Figure 15-14a is a wide-angle view of Sagittarius A and surrounding features, including arcs of gas, at least three supernova remnants, and localized radio sources. Huge filaments, labeled "Arc," lie perpendicular to the plane of the Galaxy and stretch 200 ly northward of the galactic disk, then abruptly arch southward toward Sagittarius A. The orderly arrangement of these filaments suggests that a

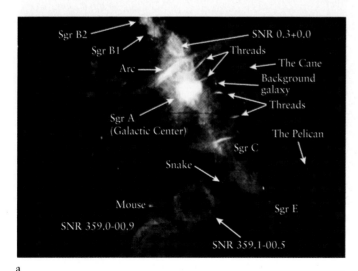

a

R I V U X G

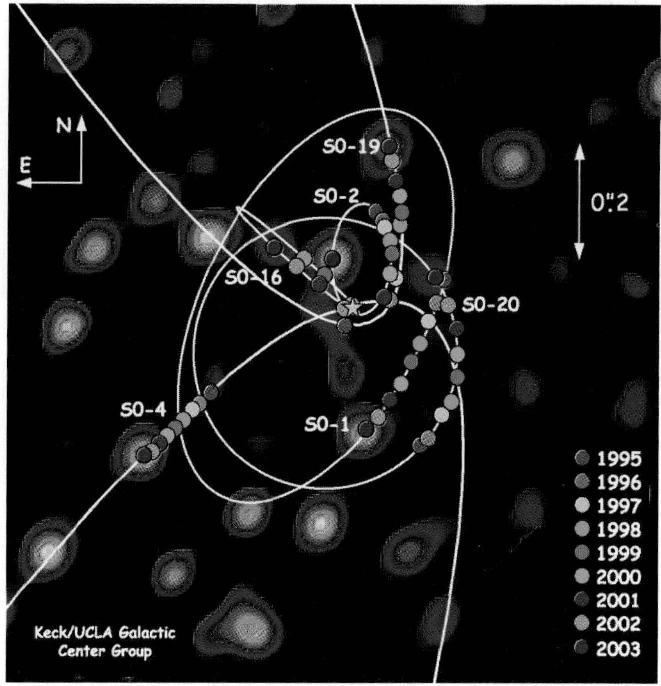

b

R I V U X G

FIGURE 15-14 **Two Views of the Galactic Nucleus** (a) A radio image taken at the VLA of the galactic nucleus and environs. This image covers an area of the sky 8 times wider than the Moon. SNR means supernova remnant. The numbers following each SNR are its right ascension and declination. The Sgr (Sagittarius) features are radio-bright objects. (b) The colored dots superimposed on this infrared image show the motion of six stars in the vicinity of the unseen massive object (denoted by the star) at the position of the radio source Sagittarius A*, part of Sgr A in (a). The orbits were measured over an 8-year period. This plot indicates that the stars are held in orbit by a $4 \times 10^6\ M_\odot$ black hole. (a: Naval Research Laboratory produced by N. E. Kassim, D. S. Briggs, T. J. W. Lazio, T. N. LaRosa, J. Imamura & S. D. Hyman. Originally from the NRAO Very Large Array. Courtesy of A. Pedlar, K. Anantharamiah, M. Gross & R. Ekers; b: UCLA Galactic Center Group)

magnetic field may be controlling the distribution and flow of ionized gas, just as magnetic fields on the Sun funnel such gas to create solar prominences.

X-ray observations taken by the *Chandra* telescope in 2004 reveal that the galactic nucleus is also bathed in ultrahot gas with a temperature of 100 million K. If this gas is more than a single outburst from an as-yet-unknown source, it must continually be replenished to remain as hot as it is. Astronomers are still trying to account for the source of this energetic gas.

If the center of our Galaxy is not active and bizarre enough, recent gamma-ray observations reveal positrons being ejected from that region. (Recall that positrons have positive electrical charges but are otherwise identical to electrons.) The source of these positrons is still unknown. Positrons can be detected because when a positron and an electron collide, they annihilate each other and release their energy as gamma rays with well-defined wavelengths. These special gamma rays have been observed emanating from the galactic center.

Infrared observations reveal stars and gas in very rapid orbits around Sagittarius A* (Figure 15-14b). Something must be holding this high-speed matter in such tight orbits about the galactic nucleus. Using Kepler's third law, astronomers calculate that 4×10^6 $M_{\odot}$ is needed to prevent the stars and gas from flying off into interstellar space. The observed broadening of spectral lines suggests that an object with the mass of 4 million Suns in a volume only the size of the solar system is concentrated at Sagittarius A*. Astronomers believe that the object is a supermassive black hole. In 2001, the *Chandra* X-ray telescope observed a burst of X rays from this black hole. The X rays were generated by gas heating up as it fell into the black hole. The bigger a black hole, the longer a burst of X ray can last. The duration of the observed blast indicated that the black hole is indeed no longer than 1 AU across, consistent with theoretical calculations.

As we saw in Chapter 14, extraordinary activity is also occurring in the nuclei of many other galaxies, implying the presence of supermassive black holes at their centers as well. Astronomers are actively studying these regions in an effort to understand the complex, intriguing activities that are happening there.

15-5 Our Galaxy's disk is surrounded by a spherical halo of stars and other matter

As mentioned at the beginning of this chapter, stars have been observed in a spherical distribution called the **halo** centered on the galactic nucleus and extending out far beyond the disk (see Figure 15-11a). The halo was originally discovered because of the globular clusters it contains. Looking out of the plane of the Milky Way's disk and between globular clusters, we see apparently unobstructed views of distant galaxies such as the Andromeda Galaxy observed by Lord Rosse (see Section 15-1). Appearances can be deceiving. While the brightness of the globular clusters suggests that they contain most of the stars in the halo, it turns out that about 99% of the halo's stars are isolated *halo field stars* spread all through the halo. The localized concentrations of stars in globular clusters account for only 1% of the halo's stars. It is worth noting, however, that some globular clusters are observed to contain intermediate-mass black holes, which contribute significantly to their total mass.

The various components of the Milky Way Galaxy overlap and interpenetrate each other. For example, the disk slices through the nuclear bulge, while globular clusters and halo field stars periodically pass through the plane of the disk. Figure 15-15 shows the shapes of the orbits of typical nuclear bulge, disk, and halo stars and clusters.

Since 1994, astronomers have observed several galaxies orbiting in the halo. The Sagittarius Dwarf and the Canis Major Dwarf (Figure 15-16) are named after the constellations in which their central regions lie. They have both spread widely through the halo and periodically pass through the disk of the Milky Way. These two galaxies are much smaller in size and mass than the Milky Way, and they are losing stars due to the disrupting influence of our Galaxy's strong gravitational attraction. Within the next 100 million years, they will cease to exist, their stars becom-

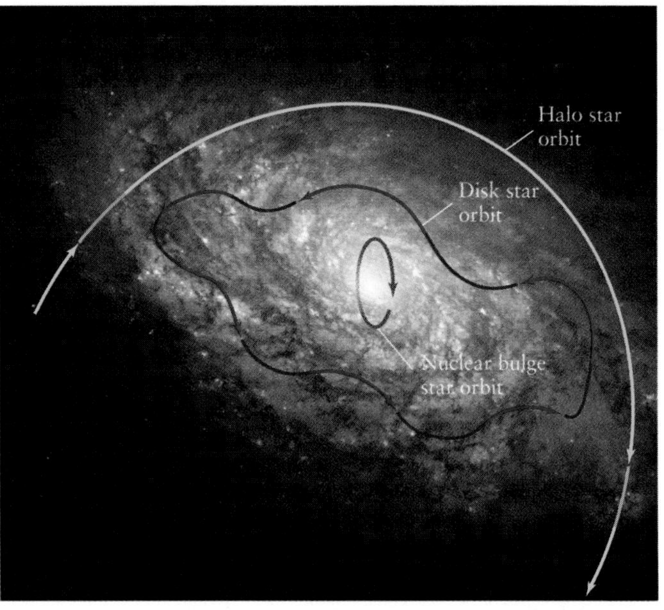

FIGURE 15-15 Orbits of Stars in Our Galaxy This disk galaxy, NGC 4144, looks very similar to what the Milky Way Galaxy would look like from far away. The colored arrows show typical orbits of stars in the nuclear bulge (blue), disk (red), and halo (yellow). Interstellar clouds, clusters, and other objects in the various components have similar orbits. (Hubble Heritage Team, AURA/STScI/NASA)

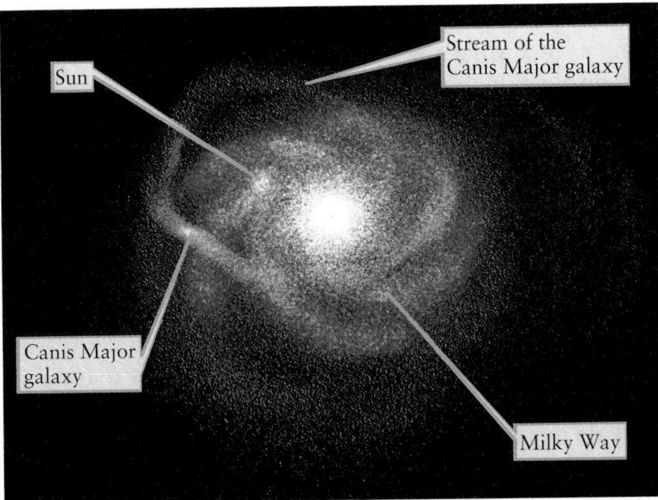

FIGURE 15-16 **The Nearest Galaxy** The Canis Major Dwarf is a dwarf elliptical galaxy that lies some 25,000 ly from the Milky Way. This infrared radiation-based image shows the Milky Way's spiral arms, as well as the distribution of stars being stripped from the Canis Major Dwarf by our Galaxy's gravitational tidal force. Containing only about 1 billion stars, the Canis Major Dwarf will be completely pulled apart within the next 100 million years or so by the Milky Way. (R. Ibata/Strasbourg Observatory, ULP/et al., 2MASS, NASA)

ing part of the Milky Way. This process of a bigger galaxy consuming a smaller one is called **galactic cannibalism.** The most recently discovered dwarf, called the Ursa Major dwarf spheroidal galaxy, was found in 2005.

15-6 The Galaxy is rotating

3 Our solar system is moving at 828,000 km (half a million miles) per hour around the center of our Galaxy. Just as the orbital motion of the planets keeps them from falling into the Sun, the motion of the stars and interstellar clouds around the galactic center keeps these bodies apart. If the stars and clouds in our Galaxy were not in orbit, their mutual gravitational forces would have caused them to fall together and form one massive black hole billions of years ago. (Indeed, such infall is believed to be the origin of the supermassive black holes located throughout the universe.) Just as detecting the positions of the stars and clouds has been difficult, so, too, has been measuring the orbital motion of the stars and gas.

Radio observations of 21-cm radiation from hydrogen gas provide important clues about our Galaxy's rotation. By measuring Doppler shifts, astronomers can determine the speed of objects toward or away from us across the Galaxy. These observations clearly indicate that our Galaxy does not rotate like a rigid body such as Earth or Mars, but rather exhibits *differential rotation,* meaning that stars at different distances from the galactic center orbit the Galaxy at different rates.

Observations show that the Sun has a nearly circular orbit around the center of the Galaxy. Because of the stars' differential rotation in the Galaxy, the Sun is like a car on a circular freeway with the fast lane on one side and the slow lane on the other. As sketched in Figure 15-17b, stars in the fast lane (closer to the center of the Galaxy) are

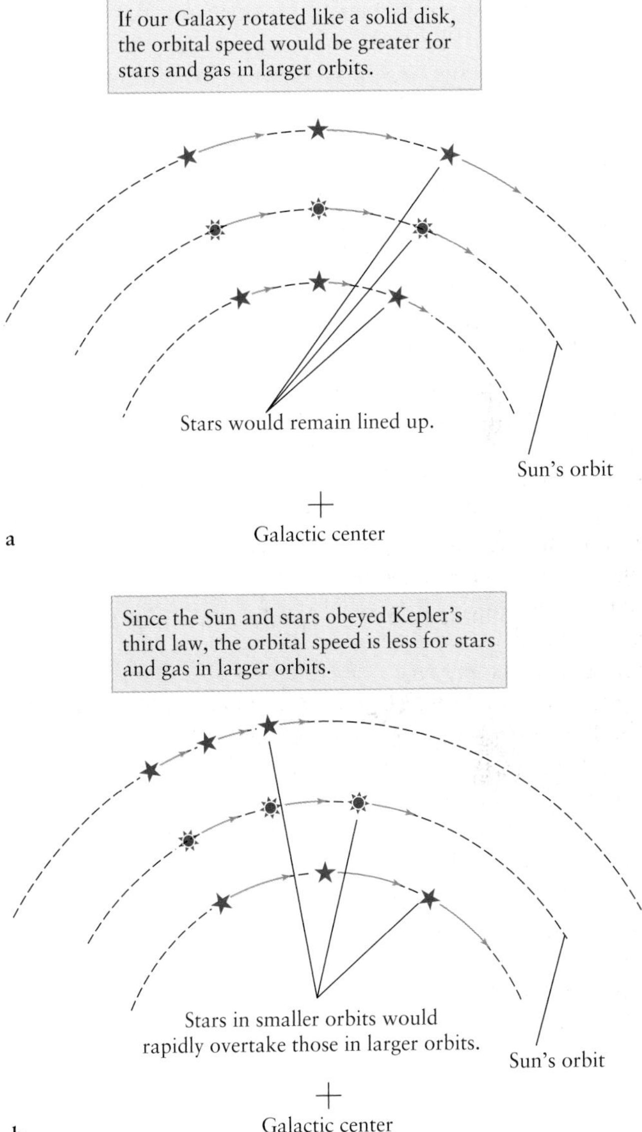

FIGURE 15-17 **Differential Rotation of the Galaxy** (a) If all the stars in the Galaxy had the same angular speed, they would orbit in lockstep. (b) However, stars at different distances from the galactic center have different angular speeds. It takes stars and clouds farther from the center longer to go around the Galaxy than it does stars closer to the center. As a result, stars closer to the Galaxy's center than the Sun are overtaking the solar system, while stars farther from the center are lagging behind us. By convention, galactic rotation is illustrated as proceeding clockwise.

passing the Sun and thus appear from our vantage point to be moving in one direction, while stars in the slow lane (farther from the center of the Galaxy than our solar system) are being overtaken by the Sun and therefore appear to be moving in the opposite direction. This is like the retrograde motion of the planets discussed in Section 2-2.

Unfortunately, like the 21-cm observations, studying the motion of nearby stars and gas reveals only how fast they are moving relative to the Sun. To get a complete picture of the Galaxy's rotation, we must find out how fast the Sun itself is orbiting the center of the Galaxy. The Swedish astronomer Bertil Lindblad proposed a method of computing this speed. He noted that not all the stars in the sky move in the orderly pattern shown in Figure 15-17. Different globular clusters in the halo of our Galaxy orbit in different planes and so they do not participate in the organized rotation of the objects in the Galaxy's disk. The combined velocities of the globular clusters around the center of the Galaxy must average to zero or else they would drift, en masse, relative to the rest of the Galaxy. Using the motions of globular clusters as a reference, astronomers have calculated that the speed of the Sun's orbit around the galactic center given earlier.

Knowing the Sun's speed and its distance from the galactic center, astronomers can calculate the Sun's orbital period. Traveling at 828,000 km per hour, our Sun takes about 230 million years to complete one trip around the Galaxy. When the solar system last passed our present location in the Milky Way, early dinosaurs of the Triassic period ruled the Earth.

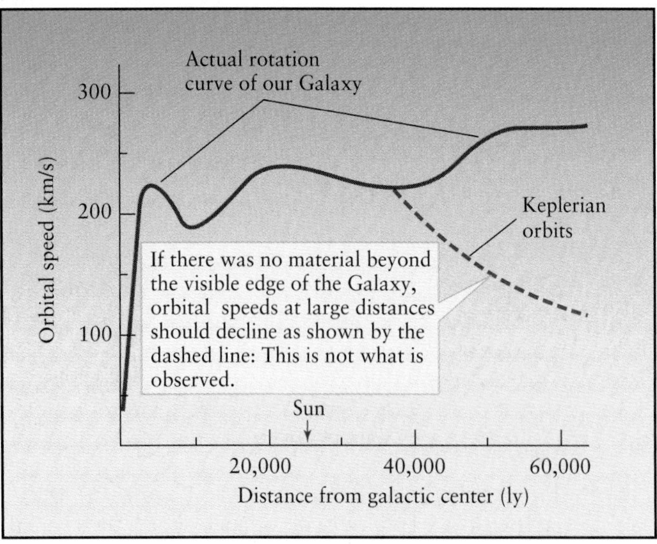

FIGURE 15-18 **The Galaxy's Rotation Curve** The blue curve shows the orbital speeds of stars and gas in the Galaxy, while the dashed red curve shows *Keplerian orbits* due to the gravitational force from known objects. Because the data (blue curve) do not show any such decline, there apparently is an abundance of dark matter that extends to great distances from the galactic center. This additional mass gives the outer stars higher speeds than they would have otherwise.

By combining the true speed of the Sun with the relative speed of the stars around us, as measured by radio astronomers, we can determine the actual orbital speeds of the stars. This computation gives us the **rotation curve** of the Galaxy, a graph showing the orbital speeds of stars and interstellar clouds at various distances from the center of the Milky Way (Figure 15-18).

Knowing the Sun's velocity around the Galaxy from the rotation curve, we can use Kepler's third law to estimate that the mass of the Galaxy that lies between us and the galactic nucleus about 1.1×10^{11} $M_{\odot}$.

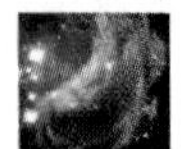

MYSTERIES AT THE GALACTIC FRINGES

Looking at the Milky Way extending all around the Earth, we know that stars and other matter exist beyond the Sun's orbit. In recent years, astronomers have been astonished to discover just how much matter lies beyond the orbit of the solar system. The nature of most of it is presently unknown.

15-7 Most of the matter in the Galaxy has not yet been identified

According to Kepler's third law, the farther a star or cloud is from the center of the Milky Way, the slower it should be moving (see the dashed line in Figure 15-18), just as the orbital speeds of the planets decrease with increasing distance from the Sun. Observations reveal, however, that galactic orbital speeds continue to *climb* well beyond the visible edge of the galactic disk. This means there must be more gravitational force from the Galaxy acting on the distant stars and clouds than we can see or have taken into account.

These observations led astronomers to calculate that a surprising amount of matter must therefore lie beyond the Sun's orbit in the Galaxy. Nearly 90% of the mass of our Galaxy has yet to be located. If we include the mass of the as-yet-unidentified matter, the Milky Way's total mass exceeds 1×10^{12} $M_{\odot}$.

To make matters even more mysterious, much of this mass is in the form of **dark matter:** Whatever it is, very little of it shows up on images at any wavelength. Given its effects on the stars and gas in the Galaxy that we can detect, astronomers deduce that this dark matter is spherically distributed all around the Galaxy in a massive halo extending at least 12 times farther from the galactic nucleus than we are, far beyond the halo defined by the visible globular clusters and halo field stars (see inset in Figure 15-11). Small fractions of the dark matter are composed of neutrinos and

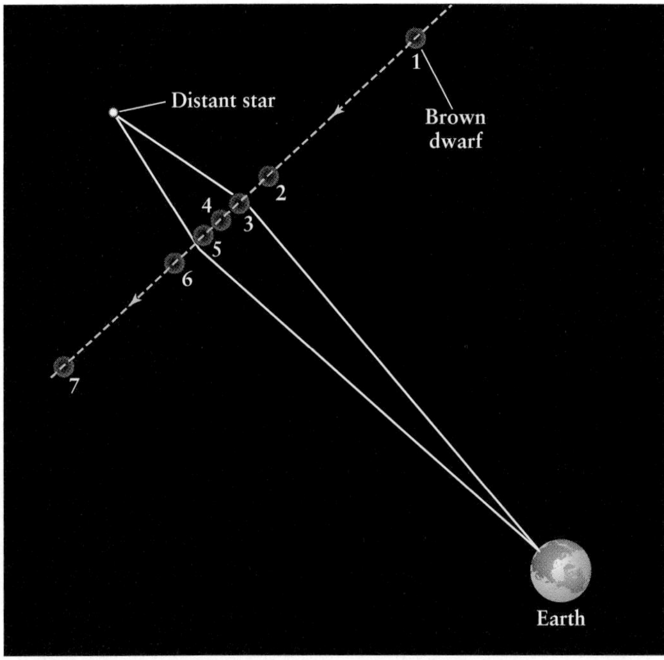

a

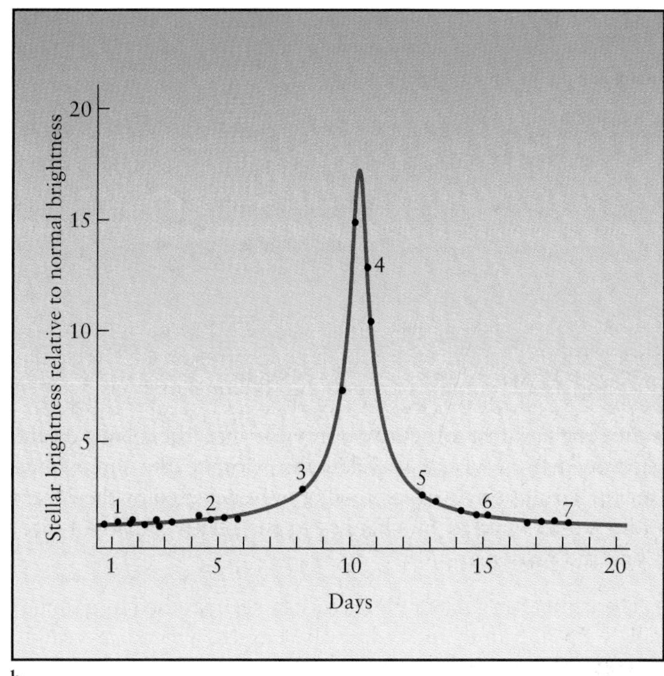

b

FIGURE 15-19 **Microlensing by Dark Matter in the Galactic Halo** (a) Gravitational fields cause light to change direction. A brown dwarf or black hole in the Galaxy's halo passing between the Earth and a more distant star will focus the starlight in our direction, making distant objects appear brighter than they are normally. (b) The light curve of the gravitational microlensing of light from a star in the Galaxy's nuclear bulge by an intervening object. Observational evidence indicates that some of the intervening objects are white dwarfs.

white dwarfs, but the nature of the rest of the dark matter—whether black holes, gas, Jupiterlike bodies, brown dwarfs, or something far more exotic—has yet to be determined. The unidentified matter in our Galaxy is sometimes called the **missing mass.** This is a poor name because the matter is not missing; we just do not yet know its location or nature.

Some searches for the dark matter rely on the fact from general relativity that matter causes passing light to change direction (see Figure 14-5). The Galaxy's dark matter may contain relatively massive objects, like brown dwarfs or black holes. If so, this dark matter will cause the light from more distant stars to change direction as the dark objects pass between those stars and the Earth (Figure 15-19a). Such objects are called *MACHOs,* for *Ma*ssive *C*ompact *H*alo *O*bjects. Indeed, as a dark object moves directly between a more distant star and the Earth, the star's light is focused toward us, causing the star to brighten. Called **microlensing,** this brightening of distant stars for several weeks has been observed (Figure 15-19b). Microlensing observations indicate that only a small fraction of the Galaxy's dark matter is in the form of these MACHOs.

15-8 Frontiers yet to be discovered

Despite our being part of it, our knowledge of the Milky Way is limited by the difficulty in making observations through its gas and dust. We have yet to locate all the globular clusters. Even more challenging will be locating and categorizing all the stars and clouds of gas and dust in the Galaxy. Are there any more small galaxies that the Milky Way is consuming? We also have to verify whether or not there is a bar-shaped distribution of stars in the nuclear bulge and to determine fully and accurately the structure and extent of the spiral arms.

Our observations of the Galaxy's nucleus hint at remarkable activity there that has yet to be explained. What is heating the 100 million K gas? What is creating the positrons there, whose resulting gamma rays we observe? In its farthest reaches, what else is there in the Galaxy's halo creating all the gravitational force we detect, but whose source is not yet observed? In other words, what is the dark matter? Another major unsolved mystery is the formation process of our Galaxy. Did stars form first? Did the supermassive black hole form and then pull gas and dust into orbit, which then formed stars? Did smaller galaxies form and then coalesce into the Milky Way as it is today?

Summary of Key Ideas

Discovering the Milky Way

• A century ago, astronomers were divided on whether all stars and nebulae are part of the Milky Way Galaxy.

• The Shapley–Curtis debate was the first major public discussion between astronomers as to whether the Milky Way contains all the stars in the universe.

• Cepheid variable stars are important in determining the distance to other galaxies.

• Edwin Hubble first determined that there are other galaxies far outside the Milky Way.

The Structure of Our Galaxy

• Our Galaxy has a disk about 100,000 ly in diameter and about 2000 ly thick, with a high concentration of interstellar dust and gas. It contains around 200 billion stars.

• Interstellar dust obscures our view into the plane of the galactic disk at visual wavelengths. Despite the intervening interstellar dust, hydrogen clouds can be detected by the 21-cm radio waves emitted by changes in the relative spins of electrons and protons.

• The center, or galactic nucleus, has been studied at gamma-ray, X-ray, infrared, and radio wavelengths, which pass readily through intervening interstellar dust and H II regions that illuminate the spiral arms. These observations have revealed the dynamic nature of the galactic nucleus, but much about it remains unexplained.

• A supermassive black hole of about 4×10^6 $M_\odot$ exists in the galactic nucleus.

• The galactic nucleus of the Milky Way is surrounded by a flattened sphere of stars called the nuclear bulge. The entire Galaxy is surrounded by a halo of matter that includes a spherical distribution of globular clusters, as well as large amounts of dark matter.

• Young OB associations, H II regions, and molecular clouds in the galactic disk outline huge spiral arms where stars are forming.

• The Sun is located about 26,000 ly from the galactic nucleus, between two major spiral arms. The Sun moves in its orbit at a speed of about 828,000 km per hour and takes about 230 million years to complete one orbit about the center of the Galaxy.

Mysteries at the Galactic Fringe

• From studies of the rotation of the Galaxy, astronomers estimate that its total mass is about 1×10^{12} $M_\odot$. Much of this mass is still undetectable.

WHAT DID YOU THINK?

1 *How many stars does the Milky Way Galaxy contain?* The Milky Way has about 200 billion stars.

2 *Where is our solar system located in the Milky Way Galaxy?* The solar system is between the Sagittarius and Perseus spiral arms about 26,000 ly from the center of the Galaxy.

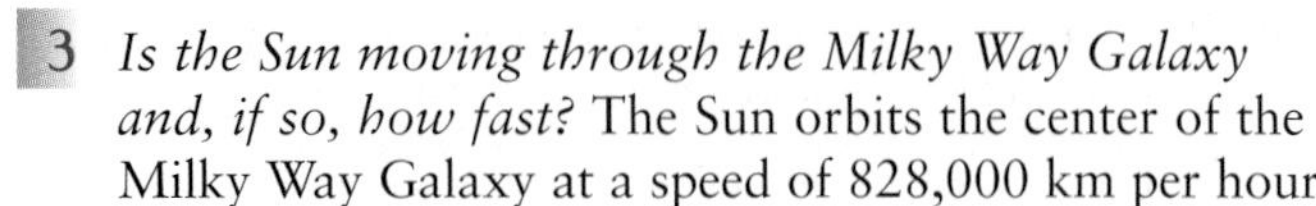

3 *Is the Sun moving through the Milky Way Galaxy and, if so, how fast?* The Sun orbits the center of the Milky Way Galaxy at a speed of 828,000 km per hour.

Key Words

dark matter (missing mass), 412
disk (of a galaxy), 405
distance modulus, 403
galactic cannibalism, 411
galactic nucleus, 409
galaxy, 402
halo (of a galaxy), 410
microlensing, 413
Milky Way Galaxy, 399
missing mass, 413
nebula (*plural* nebulae), 400
nuclear bulge, 407
rotation curve (of a galaxy), 412
Sagittarius A, 409
Shapley–Curtis debate, 400
spin (of an electron or proton), 405
spiral arm, 407
synchrotron radiation, 409
21-cm radio radiation, 405

Review Questions

The answers to computational problems, which are preceded by an asterisk (*), appear at the back of the book.

1. Where in the Galaxy is the solar system located? **a.** in the nucleus. **b.** in the halo. **c.** in a spiral arm. **d.** between the two spiral arms. **e.** in the nuclear bulge.

2. What is located in the nucleus of the Galaxy? **a.** a globular cluster. **b.** a spiral arm **c.** a black hole. **d.** the solar system. **e.** a MACHO.

3. Which statement about the Milky Way Galaxy is correct? **a.** Our Galaxy is but one of many galaxies. **b.** Our Galaxy contains all the stars in the universe. **c.** All the stars in our Galaxy take the same time to complete one orbit. **d.** Most of the stars in our Galaxy are in the nuclear bulge. **e.** None of the stars in our Galaxy move.

4. What was the Shapley–Curtis debate all about? Was a winner declared at the end of the debate? Whose ideas turned out to be correct?

5. How did Edwin Hubble prove that M31 is not a nebula in our Milky Way Galaxy?

6. As seen in the northern hemisphere, why is the Milky Way far more prominent in July than in December? Your *Starry Night Enthusiast*™ software may be useful in answering this question.

7. What observations led Harlow Shapley to conclude that we are not at the center of the Galaxy?

8. Explain why globular clusters spend more time in the galactic halo than in the plane of the disk, even though their eccentric orbits take them across the disk of the Galaxy.

9. How do hydrogen atoms generate 21-cm radiation? What do astronomers learn about our Galaxy from observations of that radiation?

10. Why do astronomers believe that vast quantities of dark matter surround our Galaxy?

11. What is synchrotron radiation?

12. Why are there no massive O and B stars in globular clusters?

13. What evidence indicates that a supermassive black hole is located at the center of our Galaxy?

***14.** Approximately how many times has the solar system orbited the center of the Galaxy since the Sun and planets were formed?

Advanced Questions

15. Why don't astronomers detect 21-cm radiation from the hydrogen in giant molecular clouds?

16. Determine the order of blueshifts in Figure 15-8 from highest blueshift to lowest.

17. Describe the rotation curve you would get if the Galaxy rotated like a rigid body.

18. Compare the apparent distribution of open clusters, which contain young stars, with the distribution of globular clusters in the Milky Way. Why are open clusters also referred to as galactic clusters?

***19.** The visible disk of the Galaxy is about 100,000 ly in diameter and 2000 ly thick. If about five supernovae explode in the Galaxy each century, how often on the average would you expect to see a supernova within 1000 ly of the Sun?

20. Give reasons for the rapid rise in the Galaxy's rotation curve (see Figure 15-18) at distances close to the galactic center.

What If . . .

21. The solar system were located in a globular cluster 26,000 ly above the plane of the Galaxy's disk? Make a drawing of what the Milky Way might look like from that location.

22. The solar system were located at the edge of the galactic disk? How would the Milky Way appear to us?

23. The solar system were located at the center of the Milky Way galaxy? What would we see and how would things be different on Earth?

24. A globular cluster were crossing the disk of the Milky Way in our vicinity right now? What would we see and how might things be different here?

25. The billions of what we now call galaxies, with their myriad stars, were actually as close as the stars in the Milky Way? What would be different about the night sky? Can you think of any other things that would be different under such conditions?

Web Questions

26. To test your understanding of our Galaxy's rotation, do Interactive Exercise 15-1 on the Web. You can print out your results, if required.

27. Search the Web for information on the Shapley–Curtis debate. Who proposed the debate in the first place? Briefly outline the relevant scientific beliefs held by the two men. What points did they agree upon and what did they disagree on? Why was the debate considered inconclusive? Support your argument with examples, if possible.

28. To test your understanding of the Milky Way's structure, do Interactive Exercise 15-2 on the Web. You can print out your results, if required.

29. Clouds of positively charged electrons, called positrons (see An Astronomer's Toolbox 10-1), have been discovered near the center of the Galaxy. Search the Web for information about this discovery. When was it made? How were they discovered? What is thought to be the origin of these particles? What happens when ordinary electrons and positrons collide?

Observing Projects

30. To determine the angle between the equatorial coordinate system and the Milky Way, turn on your *Starry Night Enthusiast*™ software and under View, choose *Hide Sky* and *Hide Horizon.* Turn on the Atlas mode (*Favourites/Guides/Altas*). Make sure that the Milky Way is visible. (Check on tab *Options/Stars/ Milky Way.* You can change its brightness by clicking on "Milky Way," which will turn to "Milky Way Options.") Next, drag the screen around until you see the Milky Way crossing the celestial equator. Move the celestial equator until it is horizontal. Being careful not to damage the screen, directly measure the angle between the celestial equator and the middle of the Milky Way with a protractor. What is the right ascension of the crossing between the Milky Way and the celestial equator? Now rotate the celestial sphere by 12 h around the celestial equator. Is the Milky Way again crossing the celestial equator? Measure the angle between the two. How does it compare with the previous measurement you made? Explain.

31. Solar System's Location in Milky Way In this exercise, we use *Deep Space Explorer*™ to find the solar system's location in the Milky Way. You will need a ruler. If you have not done so already, go to **Settings** and uncheck "Use magnitude cutoffs," slide the "Galaxy drawing/Brightness" slide to the middle of the range, and then click on the **Views** tab.

Click **Milky Way/Sun in Milky Way** on the left side of the star field. Make a sketch to scale of the image of the Milky Way on the screen, labeling the solar system and indicating the distance from the center of the Milky Way to the solar system. (*Hint:* See Figure 15-11.) On your paper, measure the distance from the edge of the galaxy to its center and then determine what fraction of the distance the solar system is from the center to the visible edge. Be careful not to damage the screen if you trace from it or measure from it.

32. Nuclear Bulge of the Milky Way Galaxy In this exercise we use *Deep Space Explorer*™ to determine the width of the galaxy's nuclear bulge. You will need a ruler. If you have not done so already, go to **Settings** and uncheck "Use magnitude cutoffs," slide the "Galaxy drawing/Brightness" slide to the middle of the range, and then click on the **Views** tab.

Click on **Milky Way/Sun in Milky Way** on the left side of the star field. The nuclear bulge of the galaxy is the bright central region that extends above and below the plane of the galaxy. Grab the galaxy by holding the left PC mouse button down (the mouse button on a Mac) and move the mouse until you see our galaxy edge-on. The white region is the Sun's neighborhood. With the galaxy edge-on, move the galaxy around until the Sun's neighborhood is as far to the right or left of the screen as you can make it go. Make a drawing to scale of the *edge view* of the galaxy. Be careful not to harm your computer screen if you trace from it or measure from it. Mark especially where the bulge meets the disk and the location of the solar system. Now determine the distance from the center of the galaxy to the solar system (see Figure 15-11 or the previous *Deep Space Explorer*™, exercise above). With this distance as a reference, determine the radius of the nuclear bulge (center of the Galaxy to the edge of the nuclear bulge) and to what fraction of the galaxy's visible diameter the nuclear bulge extends. Measure how many light years the solar system is from the edge of the nuclear bulge.

Galaxies

CHAPTER

16

R I V U X G

The Starburst Galaxy NGC 1512 (NASA/ESA/D. Maoz, Tel-Aviv University and Columbia University)

WHAT DO YOU THINK?

1 Do all galaxies have spiral arms?

2 Are most of the stars in spiral galaxies located in their spiral arms?

3 Are galaxies isolated objects?

4 Do other galaxies move relative to the Milky Way?

We live in a galactic suburb, far from the center of the action. The Sun is just one of about 200 billion stars in the Milky Way, and our Galaxy is but one of between 50 billion and 1 trillion galaxies in the visible universe. In studying galaxies, we have entered the realm of truly cosmic systems of stars, gas, and dust. While nearby stars are separated from us by light-years (trillions of miles), neighboring galaxies are hundreds of thousands of *times* farther away, and the most distant galaxies are at least 13.2 billion light-years from Earth.

Astronomers classify galaxies by various physical properties. This classification scheme is where we begin our study of the large-scale structure of the universe.

In this chapter you will discover

- how galaxies are categorized by their shapes
- the processes that produce galaxies of different shapes
- that galaxies are found in clusters that contain huge amounts of dark matter
- why clusters of galaxies form in superclusters
- how some galaxies merge while others devour their neighbors
- that the universe is expanding

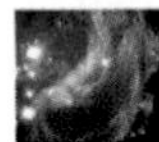

TYPES OF GALAXIES

Despite the incredible number of galaxies, there is surprising consistency in their overall shapes. Edwin Hubble began cataloging their appearance in the 1920s. The **Hubble classification** of galaxies—spirals, barred spirals, ellipticals, irregulars, and their subclasses—is still used today. 1

16-1 The winding of a spiral galaxy's arms is correlated to the size of its nuclear bulge

As we saw in Chapter 15, **spiral galaxies** like the Milky Way are characterized by a nuclear bulge and by arched lanes of stars and glowing interstellar clouds, which appear as spiral arms. Observations of many spiral galaxies reveal that the more tightly wound the spiral, the larger the nuclear bulge.

As shown in Figure 16-1, spirals with tightly wound spiral arms (and fat nuclear bulges) are called *Sa* (for spiral type *a*) *galaxies*. Those with moderately wound spiral arms (and a moderate nuclear bulge) are *Sb galaxies*. Loosely wound spirals (with tiny nuclear bulges) are *Sc galaxies*. A typical spiral galaxy contains an estimated 100 billion stars and measures nearly 10^5 ly in diameter.

a NGC 1357: an Sa galaxy

b M81: an Sb galaxy

c NGC 4321: an Sc galaxy

R I V U X G

FIGURE 16-1 Spiral Galaxies (Nearly Face-on Views) Edwin Hubble classified spiral galaxies according to the tightness of the spiral arms and the size of the nuclear bulge. Sa galaxies have the largest nuclear bulges and the most tightly wound spiral arms, while Sc galaxies have the smallest nuclear bulges and the least tightly wound arms. The images are different colors because they were taken through filters passing different colors. (left: NASA/Hubble Space Institute; middle: Robert Gendler; right: Anglo-Australian Observatory)

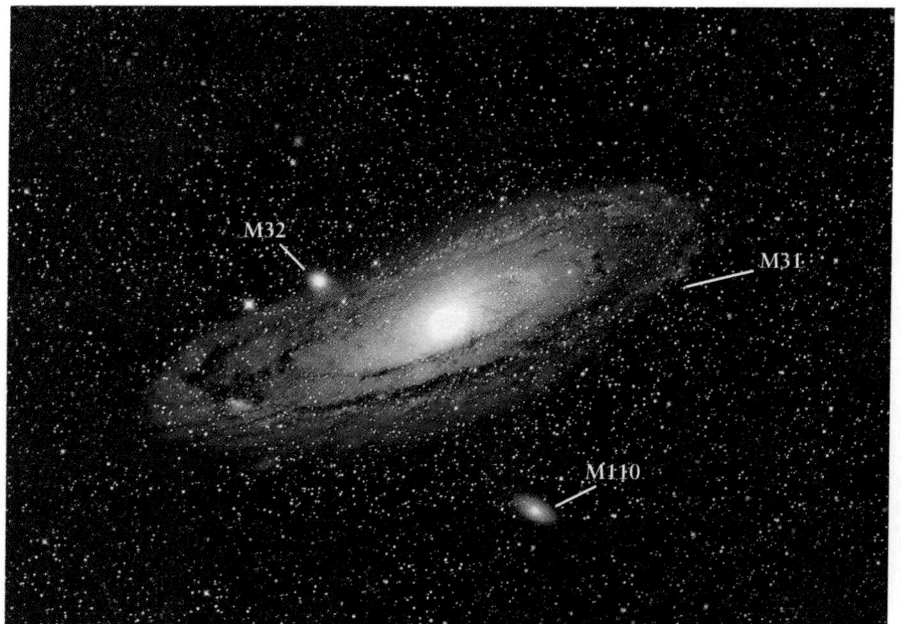

FIGURE 16-2 **Andromeda (M32)** Andromeda is a beautiful spiral galaxy and is the only galaxy visible to the naked eye from the Earth's northern hemisphere. Without a telescope, it appears to be a fuzzy blob in the constellation of Andromeda. Located only 0.9 Mpc (2.9 Mly) from us, Andromeda is gravitationally bound to the Milky Way, and it covers an area in the sky roughly 5 times as large as the full Moon. Two other galaxies, M32 and M110, are also labeled on this photograph. The points of light peppering the image are stars in our Galaxy. (Bill and Sally Fletcher/Tom Stack and Associates)

The closest spiral galaxy to the Milky Way is the Andromeda Galaxy, M31 (Figure 16-2). M31 is visible to the naked eye as a fuzzy blob in the constellation of Andromeda. You can see that the disk of this galaxy is tilted as seen from Earth. This is typical of spiral galaxies. Many are tilted so much that their spiral arms are not evident. We can still classify these latter galaxies as specific spiral types by observing the size of their central bulges. For example, M104 (Figure 16-3a) has a huge central bulge. It must therefore be an Sa galaxy with tightly wound arms. An Sb galaxy (Figure 16-3b) has a smaller central bulge. The tiny central bulge of an Sc galaxy (Figure 16-3c) is hardly noticeable at all in this galaxy, which is steeply tilted toward our view.

a M104: an Sa galaxy

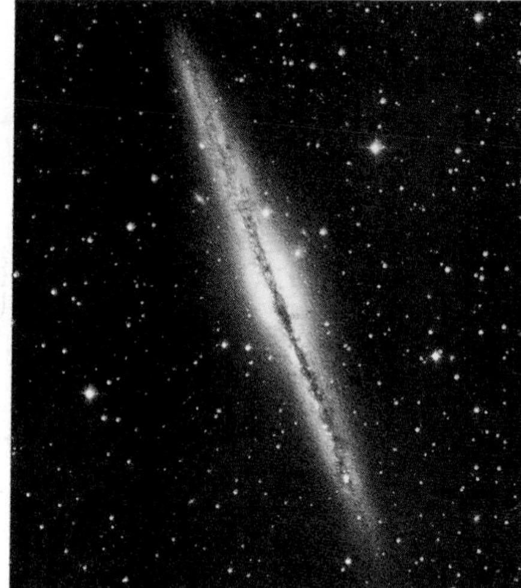
b NGC 891: an Sb galaxy

c NGC 4631: an Sc galaxy

R I V U X G

FIGURE 16-3 **Spiral Galaxies Seen Nearly Edge-on from the Milky Way** (a) Because of its large nuclear bulge, this galaxy is classified as an Sa. If we could see it face-on, the spiral arms would be tightly wound around a voluminous bulge. (b) Note the smaller nuclear bulge in this Sb galaxy. (c) At visible wavelengths, interstellar dust obscures the relatively insignificant nuclear bulge of this Sc galaxy. (a: European Southern Observatory; b: © Malin/IAC/RGO; c: Dr. Rudy Schild, Smithsonian Astrophysical Observatory)

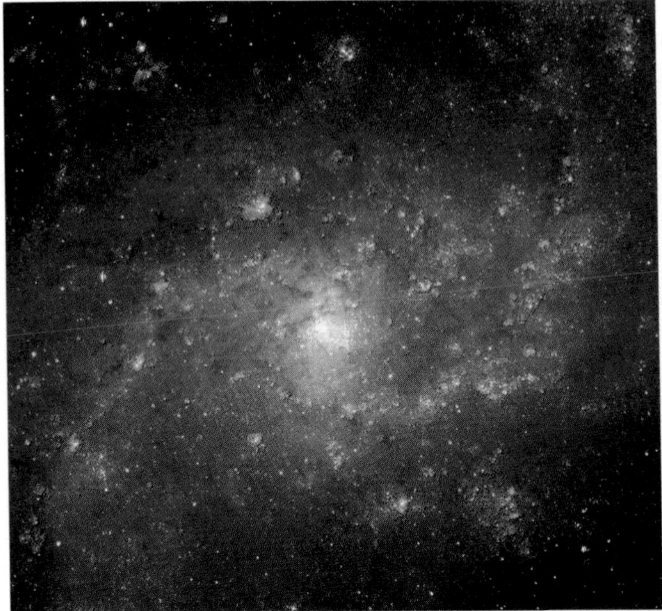

a M33: a Spiral Galaxy with Flocculent Spiral Arms

b M74: a Grand Design Spiral Galaxy

R I V U X G

FIGURE 16-4 **Variety in Spiral Arms** The differences in spiral galaxies suggest that at least two mechanisms create spiral arms. (a) This flocculent spiral galaxy has fuzzy, poorly defined spiral arms. (b) This grand design spiral galaxy has well-defined spiral arms. (a: NASA; b: Gemini Observatory/AURA)

Besides the degree of winding, the overall appearance of individual spiral arms varies from galaxy to galaxy. In some galaxies, called *flocculent spirals* (from the word meaning "fleecy"), the spiral arms are broad, fuzzy, chaotic, and poorly defined (Figure 16-4a). Other galaxies, called *grand-design spirals*, exhibit beautiful arching arms outlined by brilliant H II regions and OB associations. In these latter galaxies, the spiral arms are thin, delicate, graceful, and well-defined (Figure 16-4b).

At first glance, it may seem that the shapes of spiral galaxies are created by the orbital motions of strings of stars, with the ones closer to the center of the galaxy orbit-

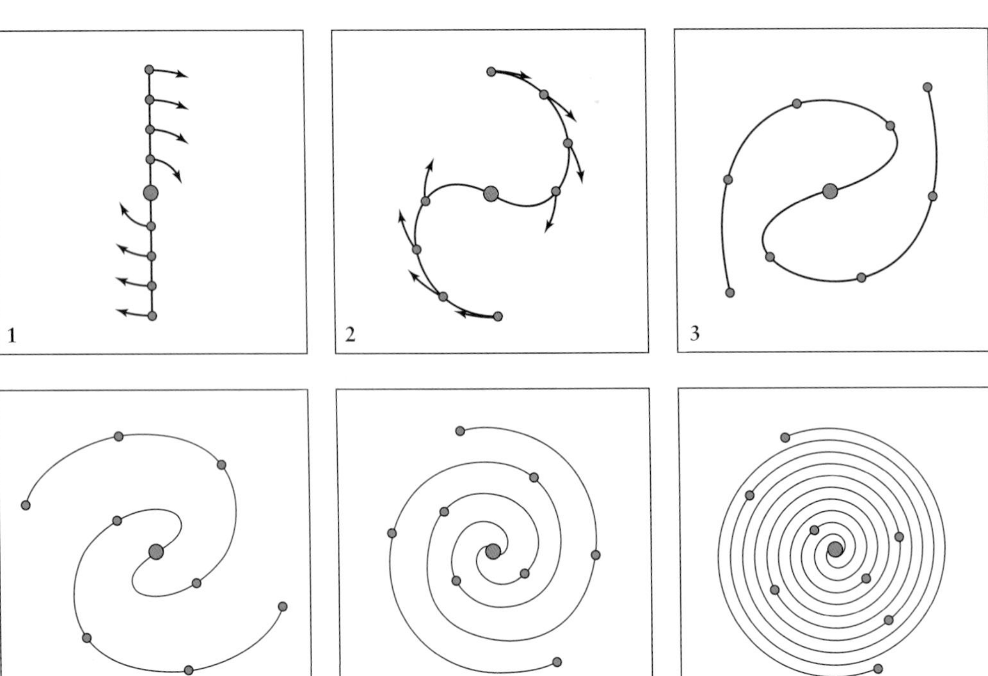

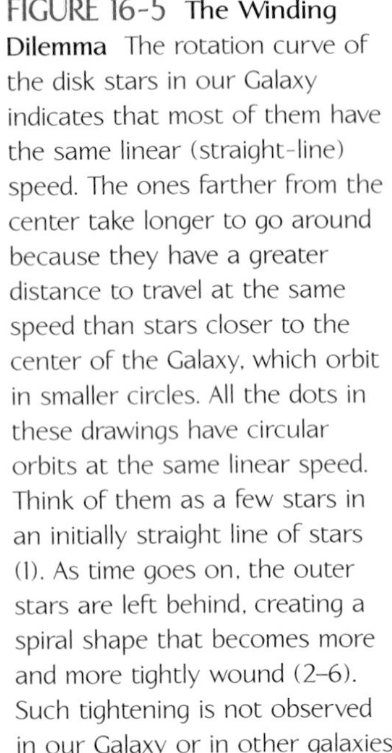

FIGURE 16-5 **The Winding Dilemma** The rotation curve of the disk stars in our Galaxy indicates that most of them have the same linear (straight-line) speed. The ones farther from the center take longer to go around because they have a greater distance to travel at the same speed than stars closer to the center of the Galaxy, which orbit in smaller circles. All the dots in these drawings have circular orbits at the same linear speed. Think of them as a few stars in an initially straight line of stars (1). As time goes on, the outer stars are left behind, creating a spiral shape that becomes more and more tightly wound (2–6). Such tightening is not observed in our Galaxy or in other galaxies.

ing more rapidly and thereby leading the outer edges of the spirals. But in this case, common sense leads us astray. While the stars closer to the center of the galaxy do orbit faster than those farther away, spiral arms cannot be orbiting strings of stars. If they were, the spiral arms should eventually "wind up," from the inside out (Figure 16-5), wrapping themselves tightly around the nucleus. This is not seen.

If star motions caused spirals, then after a few galactic rotations, the spiral structure should disappear altogether. Why, then, do we observe spiral arms in our Galaxy and many others? This *winding dilemma* led astronomers to two explanations, one for flocculent spirals and the other for grand design spirals.

16-2 Explosions create flocculent spirals and waves create grand design spirals

Flocculent Spiral Galaxies

Consider a disk galaxy so young that it has not yet formed spiral arms. Stars begin forming in a giant molecular cloud. Their radiation and stellar winds compress the nearby interstellar gas, triggering the formation of additional stars. Moreover, massive stars quickly explode as supernovae and produce shock waves, which further compress surrounding gases. Shock waves are very powerful compressions of the gas, like the sonic booms created by lightning, the snap of a whip, or a supersonic aircraft. As new stars thus trigger the birth of still other stars, the star-forming region grows, a process called *self-propagating star formation.*

The continuing birth of stars accounts for spiral arms in flocculent spiral galaxies. The galaxy's differential rotation (inner regions orbit faster than outer) drags the inner edge of each young region ahead of its outer edge, spreading the star-forming region into a spiral arm. This region of the galaxy is bright because it is highlighted by its bright O and B stars, and by the nebulae that these stars cause to glow. However, high-mass O and B stars explode relatively quickly and each spiral arm dims before it winds up (that is, no winding dilemma here). As old arms fade, new star-forming regions appear and create new spiral arms.

If self-propagating star formation were the only process leading to spiral arms, bits and pieces of spiral arms would appear only to disappear as the bright, massive stars in them die off. This process creates chaotic spiral arms, such as those observed in flocculent galaxies (see Figure 16-4a), but not the smooth spirals of grand-design galaxies. So, self-propagating star formation cannot be the whole story.

Grand Design Spiral Galaxies

Our search for the mechanism that generates grand design spirals begins in a pond. If you throw a rock into the water, you create ripples. As you know from experience, the ripples move outward in concentric rings from where the rock struck (Figure 16-6a). But suppose the pond water is rotating when you throw in a rock. *Now* what shape do the ripples have? According to calculations done by astronomer Bertil Lindblad in the 1920s, ripples created in a rotating disk-shaped system of liquid (the pond) or gas and dust

a

R I V U X G

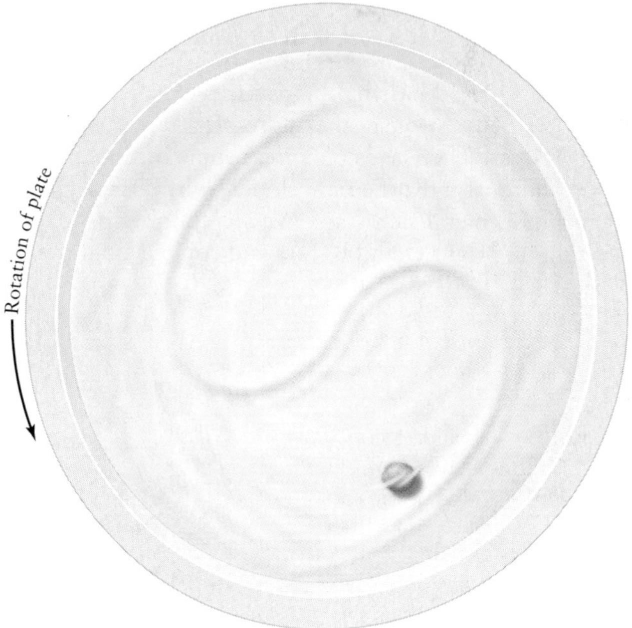

b

FIGURE 16-6 **Ripples in Water** (a) This photograph shows the usual circular ripples expanding from the place a rock was thrown into the water. (b) This illustration shows how ripples in rotating water create spiral arms, as do ripples in the gas and dust of a disk galaxy. (a: © Royalty Free/Corbis) (a disk galaxy) will be spiral. These ripples are called **spiral density waves.**

Russian scientists in the 1970s tested Lindblad's model. Lacking the computer power to create simulations of galaxies, they threw pebbles into pie pans of water rotating on phonograph turntables. The resulting water wave patterns were indeed spirals, as drawn in Figure 16-6b. Furthermore, spiral density waves never wind up. In the case of the gas and dust in a galaxy, these waves orbit in a rigid spiral pattern at about half the average speed of the interstellar medium through which they move.

In the mid-1960s, two American astronomers looked carefully at how galactic spiral density waves traveling through the disk of a galaxy lead to spiral arms. C. C. Lin and Frank Shu argued that density waves cause interstellar gas and dust to pile up temporarily. This occurs because, unlike water waves, which move up and down as they travel along, spiral density waves in the gas and dust of a galaxy are compressional waves, just like sound. As a result, they cause interstellar gas and dust to become denser as this material passes through the waves.

Lin and Shu demonstrated that a spiral arm is a galactic traffic jam. Imagine workers painting a line down a busy freeway. The cars normally cruise at 65 mph, but the crew of painters causes a bottleneck. The cars slow down temporarily to avoid hitting other cars and the slowly moving paint truck. As seen from the air, there is a noticeable congestion of cars around the painters (Figure 16-7). An individual car spends only a few moments in the moving traffic jam before resuming its usual speed, but the traffic jam itself lasts all day long.

The vehicles in the traffic jam represent the interstellar gas and dust entering a spiral arm. The paint truck represents a small part of the spiral density wave. Like the cars catching up to and passing the paint truck, the interstellar medium sweeps through the more slowly moving spiral density waves. The waves compress some of this interstellar gas and dust, creating new clouds, which then collapse to form new, metal-rich stars. We see spiral arms because they contain bright O and B stars and copious quantities of dust and gas that these stars illuminate. The sprawling dust lanes in Figure 16-4b attest to the recent passage of that material through a shock wave. While all this is going on, the density wave maintains its shape, and, overall, the spiral arms in these galaxies are better defined than the flocculent ones discussed earlier.

A few high-mass stars form in most open clusters, such 2
as those created by spiral density waves. So why aren't spiral galaxies eventually filled with enough of these bright stars to wash out the spiral arms and make the disk uniformly bright? The answer is that the massive stars highlighting each density wave are short-lived and explode before they finish passing through it. The remaining, longer-lived, lower-mass stars, like our Sun, fill the space between the spiral arms without emitting as much light (see Figure 16-4b). As pronounced as spiral arms may appear, they contain only 5% more stars than are found between them.

It takes an enormous amount of energy to compress interstellar gas and dust. After a billion years or so, even spiral density waves would begin to fade away. Some driving mechanism must keep them going, like throwing more rocks in the pond. We are not completely certain what that mechanism is yet, but the most likely explanation for reinvigorating spiral structure is the passage of a nearby companion galaxy. As this latter galaxy periodically passes close by, its gravitational attraction pulls on the gas, stars, and dust of the spiral galaxy to generate new density waves. Indeed, grand-design galaxies are usually found in the presence of a companion galaxy.

Although most spiral galaxies have two arms, a sizable minority have more. The Milky Way, for example, has at least four spiral arms. Astronomers have not yet established why the numbers of arms vary.

Since new stars are formed from interstellar gas and dust, it is plausible that galaxies with different amounts of this material will have different star formation rates and therefore different overall structures. Infrared and radio observations reveal that Sa galaxies typically contain about 4% gas and dust, while Sb galaxies contain 8% and Sc galaxies contain

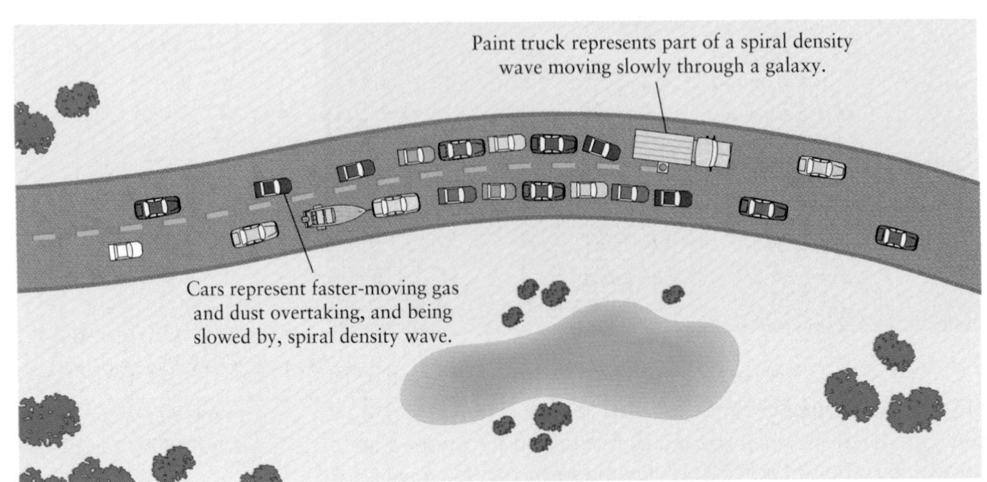
Paint truck represents part of a spiral density wave moving slowly through a galaxy.
Cars represent faster-moving gas and dust overtaking, and being slowed by, spiral density wave.

FIGURE 16-7 **Compression Wave in Traffic Flow** When normal traffic flow is slowed down, cars bunch together. In a grand design galaxy, a density wave moves through the stars and gas. The wave is merely a region of slightly denser matter, which, in turn, creates more gravitational force. This force compresses the gas and enhances star formation, which highlights the spiral density wave.

a M58: an SBa galaxy

b M83: an SBb galaxy

c NGC 1365: an SBc galaxy

FIGURE 16-8 **Barred Spiral Galaxies** As with spiral galaxies, Edwin Hubble classified barred spirals according to the tightness of their spiral arms (which correlates with the sizes of their nuclear bulges). SBa galaxies have the most tightly wound spirals and largest nuclear bulges, SBb have moderately tight spirals and medium-sized nuclear bulges, while SBc galaxies have the least tightly wound spirals and the smallest nuclear bulges. (**a:** Johan H. Knapen and Nik Szymanek, University of Hertfordshire; **b:** ESO, European Southern Observatory; **c:** Jean-Charles Cuillandre/CFHT/ Photo Researchers, Inc.)

25%. Continually improving computer simulations of galaxies reveal more and more of the details of how various concentrations of gas and dust affect galactic structure.

16-3 Bars of stars run through the nuclear bulges of barred spiral galaxies

Most astronomers believe that the Milky Way is a spiral galaxy. As we noted in Chapter 15, however, some evidence suggests that it is a **barred spiral,** a spiral galaxy with a bar of stars crossing through the nuclear bulge. Such bars are composed of stars and gas that basically flow first one way down the bar and then turn around and flow the other way. Bars form in spiral galaxies that have less total mass than the normal or "unbarred" spirals described earlier. Indeed, the higher total mass of normal spirals prevents bars from forming. About one-third of all spiral galaxies are barred spirals, the remainder being the normal spirals. The arms in barred spirals typically extend from the ends of the bar rather than from the nuclear bulge itself.

Edwin Hubble found that the winding of the spiral arms in barred spirals correlates with the size of the nuclear bulge (Figure 16-8), just as for normal spirals. An *SBa* (for spiral, barred, type *a*) galaxy has a large central bulge (and tightly wound spiral arms). Likewise, a barred spiral with moderately wound spiral arms (and a moderate central bulge) is an *SBb* galaxy, and an *SBc* galaxy has loosely wound spiral arms (and a tiny central bulge).

The motions of the stars in many nearby galaxies have been measured from Doppler shifts of their spectra. In all spiral and barred spiral galaxies discovered to date except one, the arms trail around behind as the galaxies rotate. They are thus called **trailing-arm spirals.** The exception is the galaxy NGC 4622. This galaxy has the points of its spirals leading the motion of the arms. It is a *leading-arm spiral.* Calculations predict that the leading arms are unstable and that it will eventually switch and become a trailing-arm spiral.

16-4 Elliptical galaxies display a wide variety of sizes and masses

Elliptical galaxies, named for their distinctive shapes, have no spiral arms. They range tremendously in size and mass—from the biggest to the smallest galaxies in the universe. Figure 16-9 includes two *giant elliptical galaxies,* M84 and M86, that form part of a cluster of galaxies in the

R I V U X G

WEB LINK 16.3

FIGURE 16-9 **Giant Elliptical Galaxies** The Virgo cluster is a rich, sprawling collection of more than 2000 galaxies about 50 million light-years from Earth. Only the center of this huge cluster appears in this photograph. The two largest galaxies in the cluster are the giant elliptical galaxies M84 and M86. (Royal Observatory, Edinburgh)

Leo I: an E4 galaxy

R I V U X G

FIGURE 16-10 **A Dwarf Elliptical Galaxy** This nearby E4 galaxy, called Leo I, is about 600,000 light-years from Earth. It is only 3000 ly in diameter and is so sparsely populated with stars that you can see right through its center. It is a satellite of the Milky Way. (NASA)

constellation Virgo. These enormous giant elliptical galaxies are each about 2 million light-years in diameter, 20 times the diameter of the Milky Way Galaxy.

Giant ellipticals, containing some 10 trillion solar masses, are rare compared to other types of galaxies, while *dwarf elliptical galaxies* are extremely common. Dwarf ellipticals are only a fraction of the size of an average elliptical galaxy and contain so few stars—only a few million—that we see these galaxies as nearly transparent. Because you can actually see straight through the center of a dwarf elliptical galaxy and out the other side (Figure 16-10), distant ones are hard to detect. Hence, many more dwarf ellipticals undoubtedly exist than have been identified, and the uncertainty in their number is a major reason we do not know the total number of galaxies in the visible universe.

Hubble subdivided elliptical galaxies according to how round or oval they look. The roundest elliptical galaxies are called *E0 galaxies,* while the most elongated are *E7 galaxies.* Elliptical galaxies of intermediate elongation are numbered E1 to E6 (Figure 16-11).

The Hubble scheme classifies galaxies solely by their appearance from Earth. But whenever we observe anything in the sky, we are seeing a two-dimensional view of a three-dimensional object. In the case of elliptical galaxies, we observe length and width, but we have no way of knowing anything about the third dimension of depth. What looks like an E0 galaxy (basically circular) might actually be egg-shaped when viewed from another angle. Conversely, an elongated E7 galaxy might look circular when viewed face-on.

Elliptical galaxies look far less dramatic than their spiral and barred spiral cousins because they contain relatively little interstellar gas and dust. Because stars form in interstellar clouds, relatively few stars should be forming in ellipticals compared to the numbers forming in spirals. Observations of spectra confirm the hypothesis that these galaxies contain primarily Population II, low-mass, long-lived stars. However, billions of years ago, many ellipticals contained bright, massive stars that have since exploded as supernovae. We know this because the *Chandra* X-ray telescope has discovered large numbers of remnant black holes and neutron stars in several of these galaxies.

16-5 Hubble presented spiral and elliptical galaxies in a tuning fork-shaped diagram

Edwin Hubble connected the three regularly shaped types of galaxies—spirals, barred spirals, and ellipticals—in a diagram shaped like a tuning fork (Figure 16-12). According to his scheme, S0 or SB0 galaxies, called **lenticulars** (lens-shaped), are an intermediate type between ellipticals and the two kinds of spirals. Although they often look somewhat

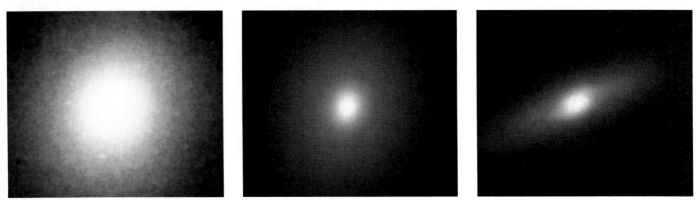

a M105: an E0 galaxy

b M49: an E4 galaxy

c NGC 4526: an E7 galaxy

R I V U X G

FIGURE 16-11 **Elliptical Galaxies** Hubble classified elliptical galaxies according to how round or elongated they appear. An E0 galaxy is round; a very elongated elliptical galaxy is an E7. Three examples are shown here. (**a:** J. D. Wray, McDonald Observatory; **b** & **c:** © 1999 Princeton University Press/Zsolt Frei and James E. Gunn)

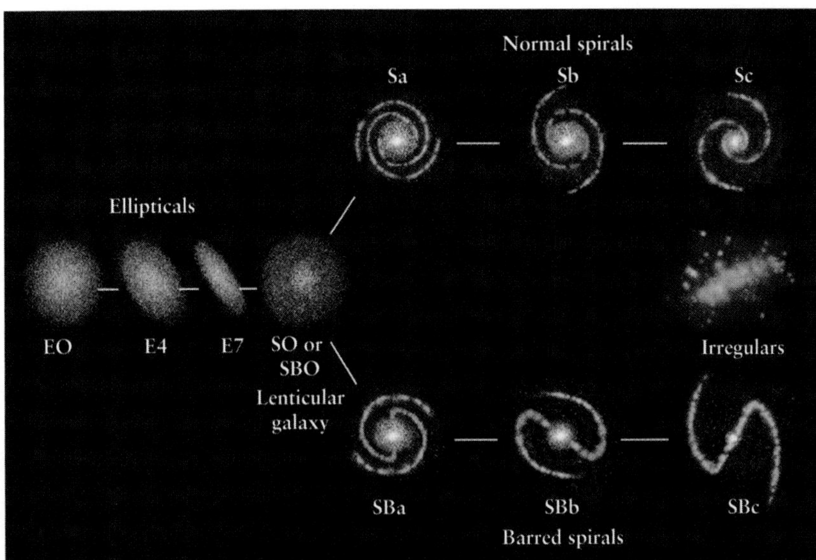

FIGURE 16-12 **Hubble's Tuning Fork Diagram** Hubble summarized his classification scheme for galaxies with this tuning fork diagram. Elliptical galaxies are classified by how oval they appear, while spirals and barred spirals are classified by the sizes of their central bulges and the correlated winding of their spiral arms. An S0 or SB0 galaxy, also called a lenticular galaxy, is an intermediate type between ellipticals and spirals. It has a disk but no spiral arms.

like ellipticals, lenticular galaxies have both a central bulge and a disk, like spiral galaxies, but they lack spiral arms.

Hubble found some galaxies that cannot be classified as spirals, barred spirals, or ellipticals. He called these **irregular galaxies.** They are generally rich in interstellar gas, dust, and both young and old stars. Irregular galaxies with only hints of organized structure and numerous OB associations are denoted Irr I. The nearby Large Magellanic Cloud (LMC) (Figure 16-13a) and the Small Magellanic Cloud (SMC), which orbit the Milky Way, are examples of Irr I. Both can be seen with the naked eye from southern latitudes.

Occasionally, irregulars are observed that appear highly distorted and completely asymmetrical, as though created by collisions between galaxies or by violent activity in their nuclei. These are denoted Irr II, as shown by NGC 4485 and NGC 4490 (Figure 16-13b). Irregular galaxies are typically smaller and less massive than spirals, containing between 10^8 and about 3×10^{10} $M_\odot$. For want of any better scheme, the irregular galaxies are sometimes placed between the ends of the tuning fork tines of the Hubble diagram.

The idea that one type of galaxy may change into another has been in and out of favor with astronomers for decades. Indeed, this was part of Hubble's motivation in relating the different types of galaxies as he did. As we will see later in this chapter and in Chapter 18, extensive observations by the Hubble Space Telescope and by

a Large Magellanic Cloud, an Irr 1 galaxy

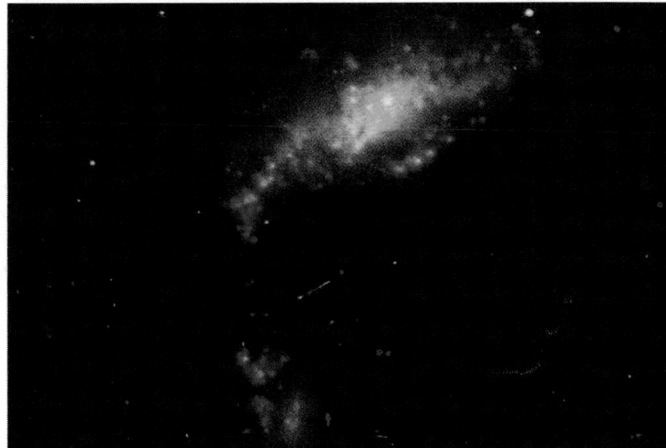

b NGC 4485 (Irr 2) and NGC 4490 (Sc) galaxies

R I V U X G

WEB LINK 16.4

FIGURE 16-13 **Irregular Galaxies** (a) At a distance of only 179,000 ly, the Large Magellanic Cloud (LMC), an Irr I irregular galaxy, is the third closest known companion of our Milky Way Galaxy. (The Milky Way's closest known companion, the Canis Major Dwarf, is shown in Figure 15-16.) About 62,000 ly across, the LMC spans 22° across the sky, about 44 times the angular size of the full Moon. Note the huge H II region (called the Tarantula Nebula or 30 Doradus) toward the left side of this image. Its diameter of 800 ly and mass of 5 million Suns makes it the largest known H II region. (b) The small irregular (Irr II) galaxy NGC 4485 (bottom galaxy) interacts with the highly distorted Sc galaxy NGC 4490, also called the Cocoon Galaxy. This pair is located in the constellation Canes Venatici. (a: Anglo-Australian Observatory; b: M. Altmann/Hoher Observatory)

TABLE 16-1 Some Properties of Galaxies

	Spiral (S) and barred spiral (SB) galaxies	Elliptical galaxies (E)	Irregular galaxies (Irr)
Mass ($M_\odot$)	10^9 to 4×10^{11}	10^5 to 10^{13}	10^8 to 3×10^{10}
Luminosity ($L_\odot$)	10^8 to 2×10^{10}	3×10^5 to 10^{11}	10^7 to 10^9
Diameter (ly)	1.6×10^5 to 8×10^5	3×10^3 to 6.5×10^5	3×10^3 to 3×10^4
Stellar populations	disk: young Population I central bulge and halo: Population II and old Population I	Population II and old Population I	mostly Population I
Percentage of observed galaxies	77%	*20%	3%

*This percentage does not include dwarf elliptical galaxies that are as yet too dim and distant to detect. Hence, the actual percentage of galaxies that are ellipticals is likely to be higher than shown here.

ground-based telescopes reveal that interactions between galaxies sometimes do lead to changes in their structures. For example, when two disk galaxies merge they are converted into a giant elliptical galaxy. However, there is no evidence that most galaxies change overall shape. The properties of the different types of galaxies are summarized in Table 16-1.

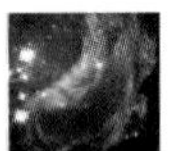

CLUSTERS AND SUPERCLUSTERS

As you consider the vastness of galaxies, it is hard to imagine that structures this huge orbit each other in groups. But they do.

16-6 Galaxies occur in clumps called clusters, which occur in clumps called superclusters

3 Galaxies are not scattered randomly throughout the universe but rather group together in **clusters.** The Fornax cluster (Figure 16-14), so called because it is located in the constellation of Fornax (the Furnace), is typical. Observations of galactic motion indicate that members of a cluster of galaxies are gravitationally bound together: The galaxies in a cluster orbit each other and occasionally even collide.

Clusters of galaxies are themselves grouped together in huge associations called **superclusters.** A typical supercluster contains dozens of individual clusters spread over a volume 150 million light-years across. Figure 16-15 shows the distribution of clusters in our vicinity of the universe. The nearer ones, out to the Virgo cluster, are members of our *Local Supercluster.* The others in this figure are members of other superclusters. Observations indicate that most superclusters are not gravitationally bound units—most clusters in each supercluster are drifting away from most of the other clusters in that same supercluster. Furthermore, the superclusters are all moving away from each other.

How superclusters are arranged in space became clearer beginning in the early 1980s, when astronomers first discovered enormous *voids* between superclusters where exceptionally few galaxies are observed. These voids are roughly spherical and measure between 100

FIGURE 16-14 **A Cluster of Galaxies** This group of galaxies, called the Fornax cluster, is about 60 million light-years from Earth. Both elliptical and spiral galaxies are easily identified. The barred spiral galaxy at the lower left is NGC 1365, the largest and most impressive member of the cluster. For a closer view of NGC 1365, see Figure 16-8. (Anglo-Australian Observatory/Royal Observatory, Edinburgh)

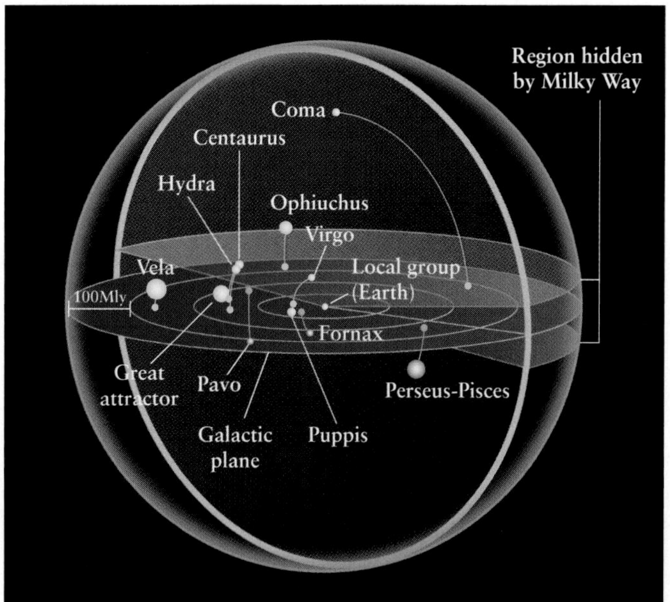

◀ **FIGURE 16-15 Clusters of Galaxies in Our Neighborhood** This is a drawing of a sphere of space 250 Mpc (800 Mly) across centered on the Earth in the Local Cluster. The spherical dots represent the locations of the nearby clusters of galaxies, while the flat circles represent the projection of the cluster locations onto the plane of the Milky Way. To better see the three-dimensionality of this figure, yellow arcs are drawn from each cluster down to the green projection of the Milky Way's plane extended out through the universe. The Great Attractor is an as-yet-unseen mass toward which the Local Group and other nearby clusters of galaxies are flowing.

million and 400 million light-years in diameter. They are not completely empty, however. Observations reveal hydrogen clouds in some of them, while others may be subdivided by strings of dim galaxies. Recent surveys show that most galaxies are distributed on the surfaces between these voids (Figure 16-16).

a

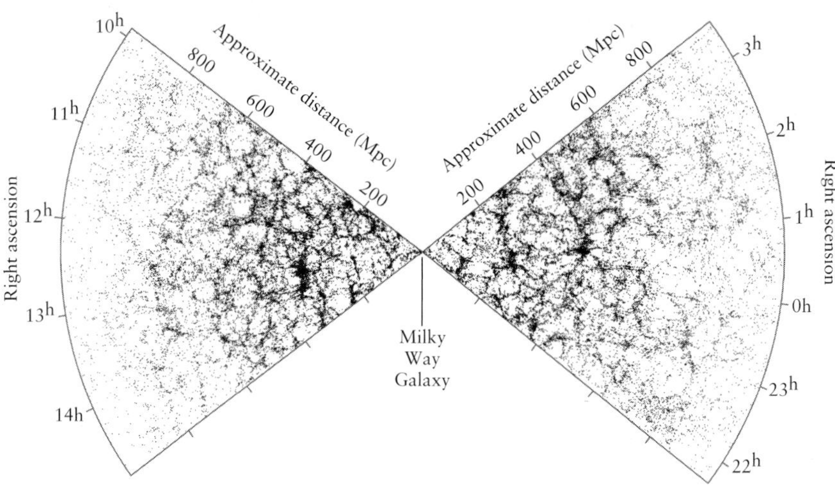

b The 2dF galaxy survey

◀ **FIGURE 16-16 Structure in the Universe** **(a)** This infrared map called 2MASS, for 2-Micron All Sky Survey, shows the light from 1.6 million galaxies. The entire sky is projected onto an oval; the blue band running vertically across the center of the image is light from the plane of the Milky Way. Note the filamentary structure with regions almost devoid of galaxies, surrounded by thin regions full of them. **(b)** This map shows the distribution of 62,559 galaxies in two wedges extending out in opposite directions from the Earth. For an explanation of right ascension, see Section 1-3. Note the prominent voids surrounded by thin areas full of galaxies. (a: 2MASS; IPAC/Caltech; and the University of Massachusetts; b: Courtesy of the 2dF Galaxy Redshift Survey Team)

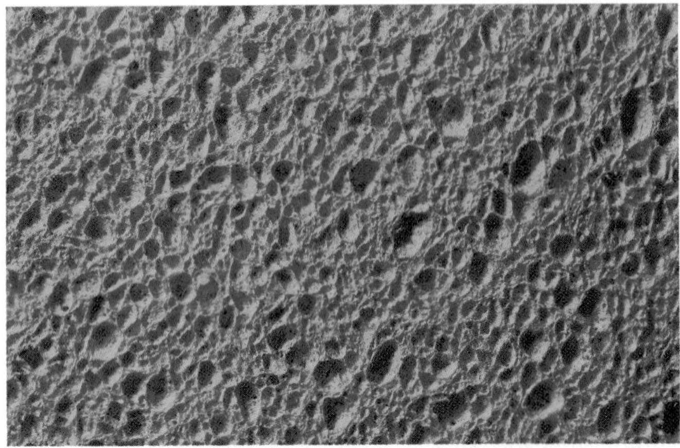

FIGURE 16-17 **Foamy Structure of the Universe** This photograph of a sponge recreates the distribution of bright clusters of galaxies throughout the universe. The empty spaces in the foam are analogous to the voids found throughout the universe. The spongy regions are analogous to the locations of most of the galaxies. (Image Source/Super Stock)

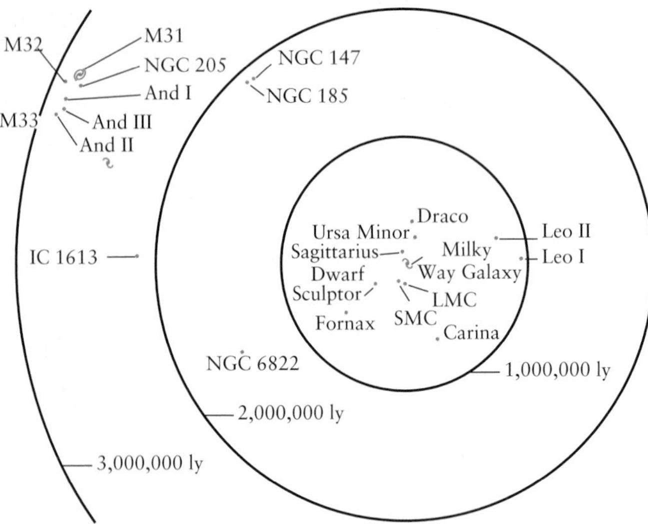

FIGURE 16-18 **The Local Group** Our Galaxy belongs to a poor, irregular cluster consisting of about 40 galaxies called the Local Group. This map shows the distribution of about three-quarters of the galaxies. The Andromeda Galaxy (M31) is the largest and most massive galaxy in the Local Group. The second largest is the Milky Way itself. M31 and the Milky Way are each surrounded by a dozen satellite galaxies. The recently discovered Canis Major Dwarf Galaxy is the Milky Way's nearest known neighbor.

WEB LINK 16.5

To better understand the distribution of galaxies throughout the universe, several surveys are underway to map the three-dimensional locations of many galaxies. The Sloan Digital Sky Survey is observing 100 million galaxies, stars, and other celestial objects and measuring redshifts of at least the nearest 1 million galaxies. This survey, made in the northern hemisphere, will cover about one-quarter of the sky and give us a three-dimensional map through 100 times the volume of space as we have mapped today. The Two Degree Field (2dF) Survey (Figure 16-16b) is observing regions of both the northern and southern hemispheres.

The distribution of clusters of galaxies throughout the universe is sometimes said to be "spongy" or "sudsy," with the galaxies located where the sponge or soapy material is, surrounded by voids (Figure 16-17). Astronomers believe that this spongy pattern contains important clues about conditions in the early universe that have not yet been fully fathomed.

16-7 Clusters of galaxies may appear densely or sparsely populated and regular or irregular in shape

A cluster of galaxies is said to be either a **poor cluster** or a **rich cluster,** depending on how many galaxies it contains. For example, the Milky Way Galaxy, the Andromeda Galaxy, and the Large and Small Magellanic Clouds belong to a poor cluster called the **Local Group.** The Local Group contains some 40 galaxies, over a third of which are dwarf ellipticals. Figure 16-18 shows a map covering most of the Local Group.

New galaxies in the Local Group continue to be found. This is happening primarily because astronomers are developing techniques to observe objects that lie in the same plane as, but beyond, the Milky Way. Besides the 1994 discovery of the Sagittarius Dwarf Galaxy, the dwarf galaxy Antlia (Figure 16-19) was first observed in 1997 and the Canis Major Dwarf (see Figure 15-16) was found in 2003. The Canis Major Dwarf is only about 25,000 light-years from the Milky Way.

Astronomers also categorize clusters of galaxies according to their shapes. A **regular cluster** is distinctly spherical, with a marked concentration of galaxies at its center. Numerous gravitational interactions over billions of years have spread the galaxies into this distribution. In contrast, galaxies in an **irregular cluster** are more randomly scattered about a sprawling region of the sky.

The nearest example of a rich, regular cluster is the Coma cluster, located about 300 million light-years from us in the constellation of Coma Berenices (Figure 16-20; see also Figure 16-15). Despite its great distance, more than 1000 bright galaxies within it are easily visible from Earth. It is likely that the Coma cluster contains many thousands of dwarf ellipticals soon to be detected—maybe as many as 10,000 galaxies overall.

Rich, regular clusters like the Coma cluster contain mostly elliptical and lenticular galaxies. Only 15% of the Coma cluster's galaxies are spirals and irregulars. Irregular clusters, such as the Virgo cluster and the Hercules cluster

FIGURE 16-19 **A Recently Discovered Member of the Local Group** The galaxy Antlia was first detected in 1997. It lies about 3 million ly away, outside the region depicted in Figure 16-18. This galaxy contains only about a million stars. (M. J. Irwin, Royal Greenwich Observatory, and A. B. Whiting and G. K. T. Hau, Institute of Astronomy, Cambridge University)

(Figure 16-21), display a more even mixture of galaxy types. Two-thirds of the 200 brightest galaxies in the Hercules cluster are spirals, nearly a fifth are ellipticals, and the rest are irregular.

16-8 Galaxies in a cluster can collide and combine

Occasionally, two or more galaxies in a cluster or two or more galaxies from different, nearby clusters collide. Sometimes they pass through one another (Figure 16-22) and sometimes they merge. In 2002, the Hubble Space Telescope observed four galaxies colliding simultaneously. There is so much space between stars that the probability of two stars crashing into each other during galactic collisions is extremely small. On the other hand, the galaxies' huge clouds of interstellar gas and dust are so large that they do collide, slamming into each other and producing strong shock waves. The colliding interstellar clouds are stopped in their tracks.

A violent collision can strip both galaxies of their interstellar gas and dust. The energy associated with the collision heats this gas and dust to extremely high temperatures. This process is a major source of the hot **intergalactic gas** often observed in rich, regular clusters. The *Chandra* X-ray Observatory has observed intergalactic gas with temperatures between 300,000 and 100 million K.

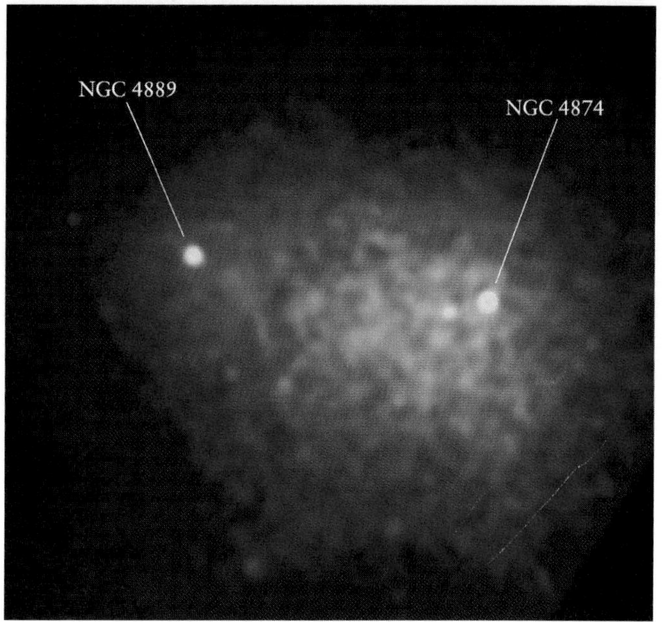

FIGURE 16-20 **The Coma Cluster** **(a)** This rich, regular cluster containing thousands of galaxies is about 300 million light-years from Earth. **(b)** Regular clusters are composed mostly of elliptical and lenticular galaxies and are sources of X rays. This *Chandra* image of Coma's central region is 1.5 million light-years across. The gas cloud emitting most of these X rays is 100 million K. (a: O. Lopez-Cruz and I. Shelton, NOAO/AURA/NSF; b: NASA/CXC/SAO/A. Vikhlinin et al.)

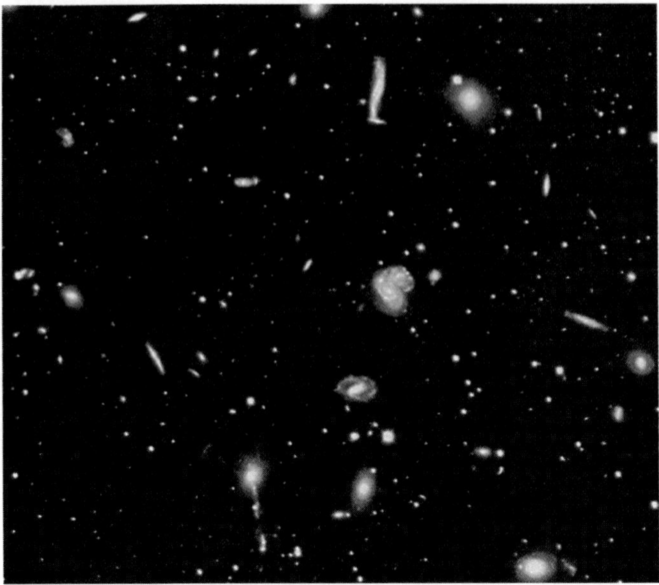

FIGURE 16-21 **The Hercules Cluster** This irregular cluster, which is about 700 million light-years from Earth, contains a high proportion of spiral galaxies, often associated in pairs and small groups. (NOAO)

Another source of intergalactic gas is gas escaping from dwarf galaxies. This gas, ejected by supernovae of stars created in a burst of star formation when the galaxies formed, escaped from the relatively weak gravitational pull of dwarf galaxies into intergalactic space.

Less violent collisions and near misses between galaxies allow the compressed interstellar gas and dust to cool sufficiently to have new stars form from it. These collisions can also stimulate galaxywide star formation, which may account for the **starburst galaxies** that blaze with the light of numerous newborn stars. These galaxies are characterized by bright centers surrounded by clouds of warm interstellar dust, indicating a recent, vigorous episode of star birth (Figure 16-23).

There is observational evidence that globular clusters are formed during collisions of interstellar gas and dust. Unlike the globular clusters in our Galaxy, some galaxies contain very young globular clusters. Whether the globular clusters in our Galaxy formed as a result of an early collision of the Milky Way has yet to be determined. Their warm dust is so abundant that starburst galaxies are among the most luminous objects in the universe at infrared wavelengths.

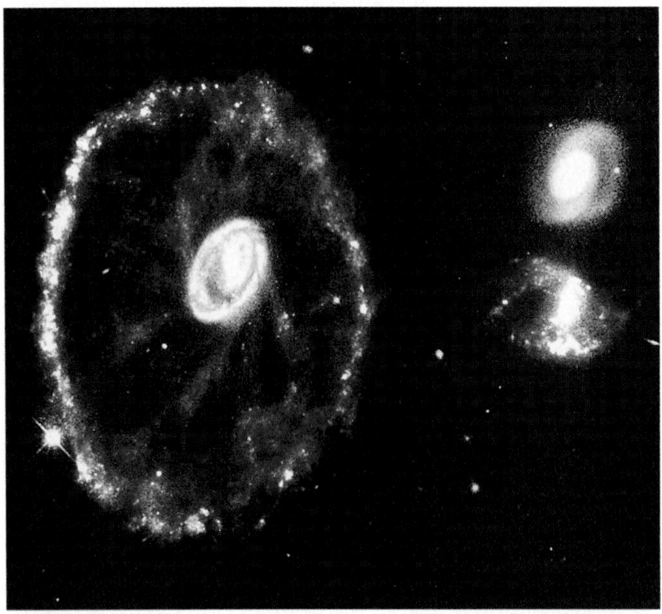

FIGURE 16-22 **The Cartwheel Galaxy** This ring-shaped assemblage 500 million light-years from Earth is the likely result of one galaxy, probably the blue-white one on the right, having passed through the middle of the larger one. Astronomers suspect that the passage created a circular density wave in the Cartwheel that stimulated a burst of star formation, creating many bright blue and white stars. (Kirk Borne, STScI; and NASA)

FIGURE 16-23 **A Starburst Galaxy** A ring of vigorous star formation 2400 ly wide highlights this stunning ultraviolet, visible light, and infrared composite image of the barred spiral galaxy NGC 1512 taken by the Hubble Space Telescope. Located 30 Mly away in the constellation Horologium, this galaxy is 70,000 ly across. Such rings of star formation are common in starburst galaxies. (NASA/ESA/D. Maoz, Tel-Aviv University and Columbia University)

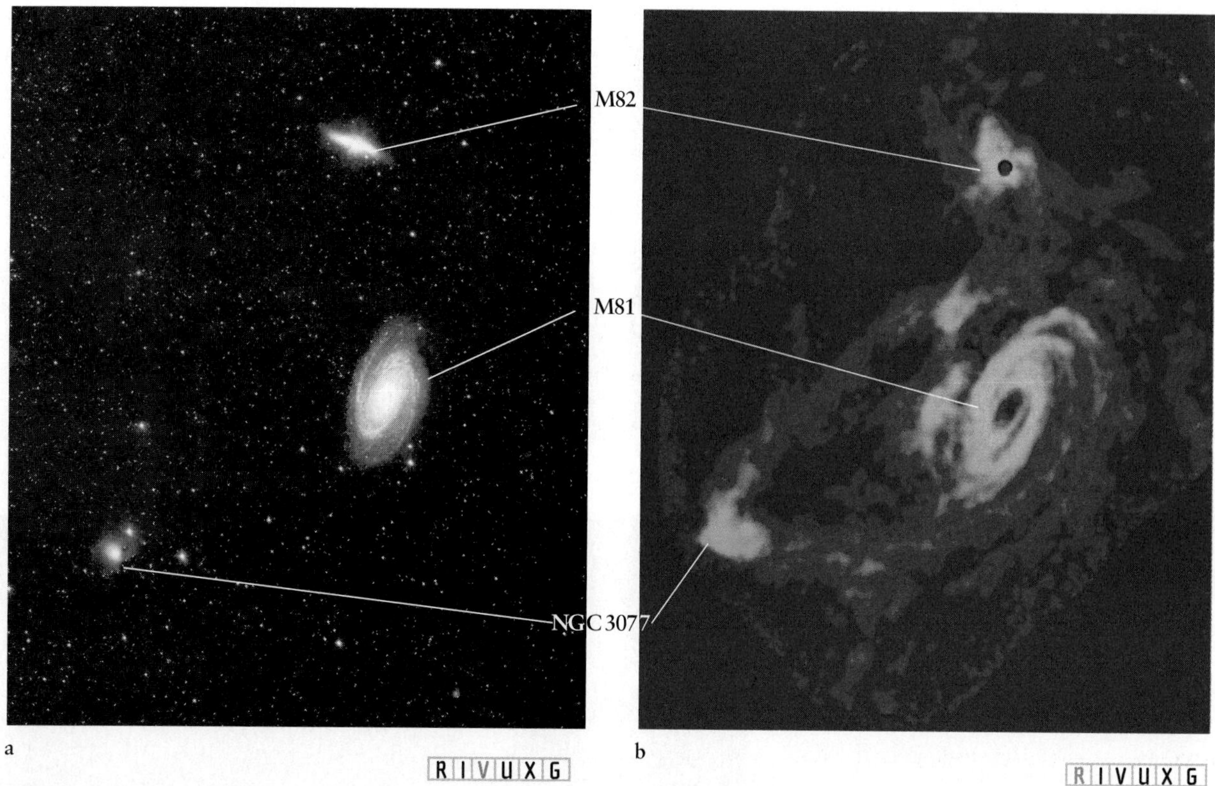

FIGURE 16-24 **The M81 Group** The Irr II, starburst galaxy M82 is in a nearby cluster of about a dozen galaxies, including the spectacular spiral M81. Several of the galaxies in this cluster are connected by streamers of hydrogen gas. **(a)** This photograph shows the three brightest galaxies at visual wavelengths. **(b)** This radio image, created from data taken by the Very Large Array, shows the streamers of hydrogen gas that connect the bright galaxies and also several dim ones, seen as regions of bright orange here. (a: Palomar Sky Survey; b: M. S. Yun, VLA, and Harvard)

The Irr II, starburst galaxy M82 is one member of a nearby cluster of about a dozen galaxies that include the beautiful spiral galaxy M81 and a fainter companion called NGC 3077 (Figure 16-24a). A radio survey of that region of the sky revealed enormous streams of hydrogen gas connecting these and other, smaller galaxies in that cluster (Figure 16-24b). The loops and twists in these streamers suggest that the three galaxies have had several close encounters over the ages, most recently about 600 million years ago. The effects of the gravitational interactions between colliding galaxies can also hurl thousands of stars out into intergalactic space along huge, arching streams as seen in Figure 16-25a.

In rich clusters, astronomers observe near misses between galaxies (Figure 16-25b). If galaxies are surrounded by extended halos of dim stars, these near misses strip the galaxies of these outlying stars. In this way, a loosely dispersed sea of dim stars populates the space between galaxies in a cluster.

WEB LINK 16.9

Similarly, after galaxies collide, some interior stars are flung far and wide, scattering material into intergalactic space. Such isolated stars have been observed by the Hubble Space Telescope. However, other stars slow down, often causing the remnants of the galaxies to merge. Several dramatic examples of **galactic mergers** have been discovered, such as the galaxy known as NGC 6240 (Figure 16-26a). The supermassive black holes in the nuclei of the original two galaxies (see Section 14-8) have not yet merged (Figure 16-26b). They are expected to do so within a few hundred million years.

As introduced in Chapter 15, with the Milky Way tearing apart the Canis Major Dwarf Galaxy, astronomers also speak of galactic cannibalism, which occurs when a large galaxy captures and "devours" a smaller one. Cannibalism differs from mergers in that the dining galaxy is significantly bigger than its dinner, whereas merging galaxies are about the same size.

Many astronomers suspect that giant ellipticals are the product of galactic mergers and cannibalism, including spiral-spiral collisions. Observations reveal that giant galaxies typically occupy the centers of rich clusters. Once formed, giant ellipticals can continue to grow because smaller galaxies are often located around them (see Figure 16-9). As they pass through the extended halo of a giant elliptical, these smaller galaxies slow down and are eventually consumed by the larger galaxy.

FIGURE 16-25 Interacting and Colliding Galaxies **(a)** Pairs of colliding galaxies often exhibit long "antennae" of stars ejected by the collision. This particular system is known as NGC 4676 or "the Mice" (because of its tails of stars and gas). It is 300 million ly from Earth in the constellation Coma Berenices. The collision has stimulated a firestorm of new star formation, as can be seen in the bright blue regions. Mass can also be seen flowing between the two galaxies, which will eventually merge. **(b)** These two galaxies, NGC 2207 (right) and IC 2163, are orbiting and tidally distorting each other. Their most recent close encounter occurred 40 Myr ago when the two were perpendicular to each other and about one galactic diameter apart. Computer simulations indicate that they should eventually coalesce. (a: NASA, H. Ford/JHU, G. Illingworth/UCSC/Lick, M. Clampin/STScI, G. Hartig/STScI, The ACS Science Team, and ESA; b: NASA)

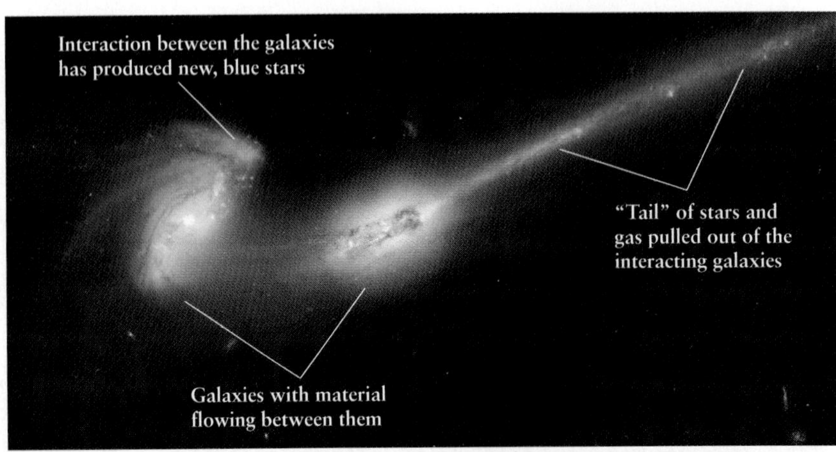

a NGC 4676

R I V U X G

b I2163 (left) and NHC 2207

R I V U X G

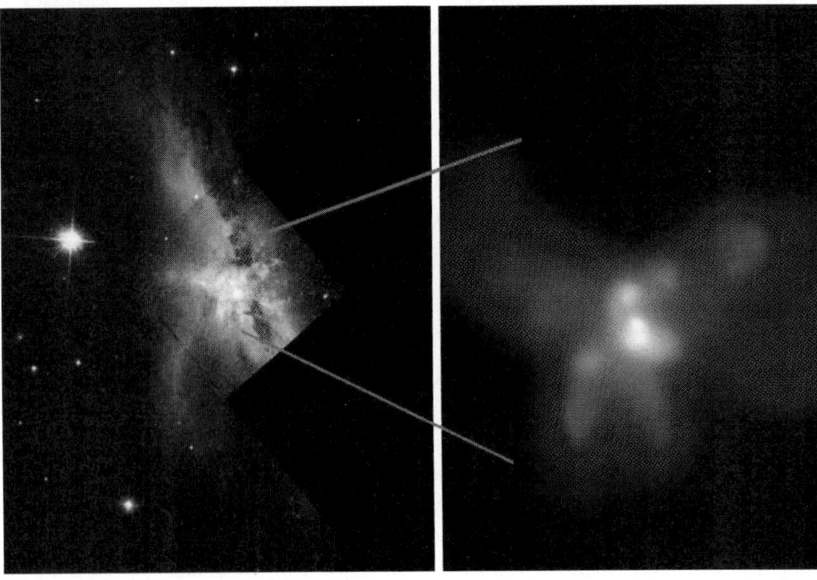

Hubble Optical

Chandra X-ray

NGC 6240

R I V U X G

R I V U X G

FIGURE 16-26 Merging Galaxies This contorted object, NGC 6240, in the constellation Ophiuchus is the result of two spiral galaxies in the process of merging. The widespread blue area reveals that the collision between the two galaxies has triggered an immense burst of star formation. **Inset:** The *Chandra* X-ray telescope has revealed that at the heart of this system are two supermassive black holes. They are in the bright areas of this image. Within a few hundred million years, these black holes are expected to merge into one more massive black hole. (left: R. P. van der Marel/J. Gersgen/STScI/NASA; right: S. Komossa/G. Hasinger/MPE, et al., CXC/NASA)

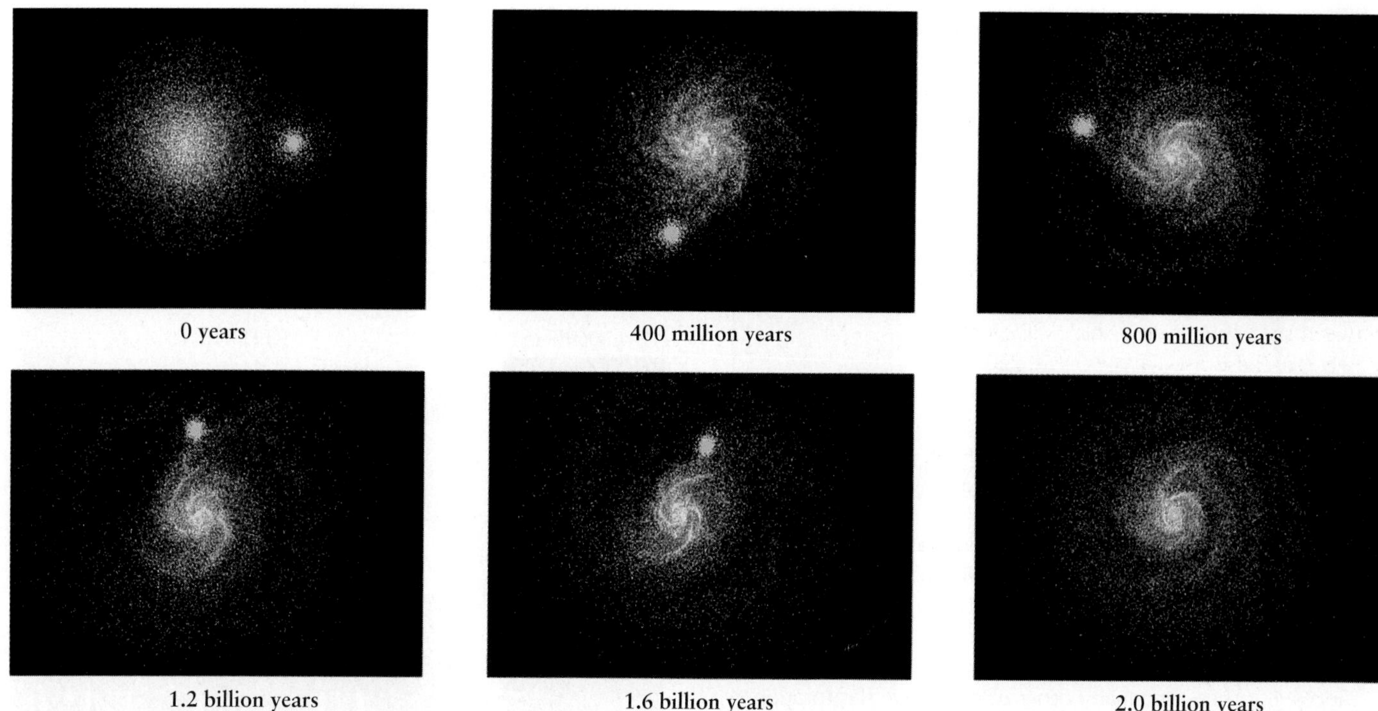

ANIMATION 16.2

FIGURE 16-27 Simulated Galactic Cannibalism This computer simulation shows a small galaxy (stars in yellow) being devoured by a larger, disk-shaped galaxy (stars in blue, gas in white). Note how spiral arms are generated in the disk galaxy by its interaction with the satellite galaxy. (Lars Hernquist, Institute for Advanced Study with simulations performed at the Pittsburgh Supercomputing Center)

Figure 16-27, a computer simulation, shows a large, disk-shaped galaxy devouring a small satellite galaxy. The large galaxy consists of 90% stars (in blue) and 10% gas (in white) by mass. It is surrounded by a halo of dark matter having a mass about 3.3 times that of the disk. The satellite galaxy, which has a tenth of the mass of the large galaxy, contains only stars (in yellow). Initially, the satellite is in circular orbit about the large galaxy. Note that spiral arms appear in the large galaxy as the collision proceeds. Two billion years elapse as the satellite spirals in toward the core of the large galaxy. Although much material is stripped from the satellite, most of its stars plunge into the nucleus of the large galaxy.

16-9 Galactic halos may account for some dark matter in the universe

What keeps the rapidly moving galaxies in clusters together? There must be sufficient matter to provide the necessary gravitational force to bind galaxies together in clusters and in some cases clusters into superclusters. However, *no cluster of galaxies contains enough visible matter to stay bound together.* Most of the matter binding clusters of galaxies is dark matter (matter that we cannot yet directly detect). This must be true because otherwise the galaxies would have long ago wandered apart in random directions, and the clusters would no longer exist. Recall that a similar problem arose in Chapter 15 regarding dark matter in individual galaxies. Analyses of the total visible mass demonstrate that the total mass needed to bind a typical rich cluster is 10 times greater than the mass of material that shows up on visible-light images. Furthermore, observations of the distribution of galaxies in clusters have determined that the dark matter has the same distribution as the galaxies in the cluster and that each galaxy corresponds to a concentration of dark matter.

There is quite a bit of evidence to support the idea of extended halos surrounding galaxies. Many galaxies have rotation curves similar to that of our Milky Way (recall Figure 15-18). These rotation curves remain remarkably flat out to surprisingly great distances from the galaxies' centers. For example, Figure 16-28 shows the rotation curves of four spiral galaxies. In all cases, the orbital speed is fairly constant. According to Kepler's third law, we should see a decline in orbital speed toward the outer portions of a galaxy. The fact that in most cases we do not implies that we still have not detected the true edge of these and many similar galaxies. As with the Milky Way Galaxy, astronomers conclude that a considerable amount of dark matter must extend well beyond the visible portions of most galaxies.

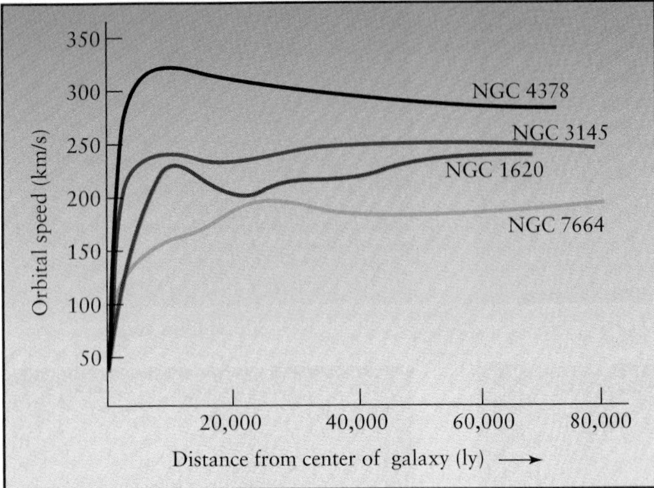

FIGURE 16-28 The Rotation Curves of Four Spiral Galaxies This graph shows how the orbital speed of material in the disks of four spiral galaxies varies with the distance from the center of each galaxy. If most of each galaxy's mass was concentrated near the center of the galaxy, these curves would fall off at large distances. But these and many other galaxies have flat rotation curves that do not fall off. This indicates the presence of extended halos of dark matter. See Figure 15-18 to compare these to the Milky Way's rotation curve.

Astronomers using X-ray telescopes have discovered the nature of some of the dark matter (which is therefore no longer considered dark). Satellite observations have revealed X rays pouring from the space between galaxies in rich clusters. This radiation is emitted by substantial amounts of hot intergalactic gas at temperatures between 10 million and 100 million K. Figure 16-20b shows this hot gas in the center of the Coma cluster. Observations by the *Chandra* X-ray Observatory indicate that this intergalactic gas contains twice as much mass as exists in all the galaxies.

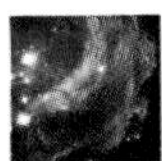

SUPERCLUSTERS IN MOTION

Whenever an astronomer finds an object in the sky, one of the first tasks is to determine its composition. As we saw in Chapters 3 and 4, that means attaching a spectrograph to a telescope and recording the object's spectrum. As long ago as 1914, V. M. Slipher, working at the Lowell Observatory in Arizona, took spectra of "spiral nebulae," now known to be spiral galaxies. He was surprised to discover that the spectral lines of 11 of the 15 spiral nebulae he studied were substantially redshifted. These redshifts indicate that they are moving away from the Milky Way at significant speeds. This left scientists with a major puzzle: Is the entire universe expanding?

16-10 The redshifts of superclusters indicate that the universe is expanding

WEB LINK 16.10

During the 1920s, Edwin Hubble and Milton Humason recorded the spectra of many galaxies with the 100-in. telescope on Mount Wilson, confirming that most galaxies are rapidly receding from the Milky Way. Using the Doppler effect (recall Figure 3-6 and

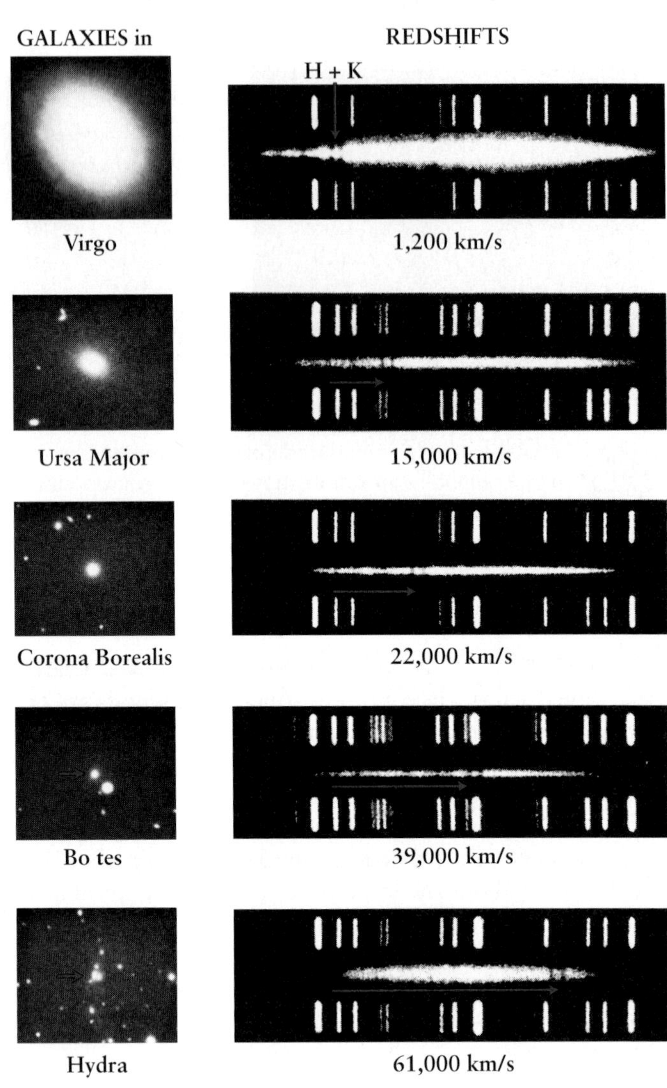

FIGURE 16-29 Five Galaxies and Their Spectra The photographs of these five elliptical galaxies were all taken at the same magnification. They are labeled according to the constellation in which each galaxy is located. The spectrum of each galaxy is the hazy band between the comparison spectra at the top and bottom of each plate. In all five cases, the so-called H and K lines of calcium are seen. The recessional velocity (calculated from the Doppler shifts of the H and K lines) appears below each spectrum. Note that the fainter—and thus more distant—a galaxy is, the greater is its redshift. (Carnegie Observatories)

An Astronomer's Toolbox 4-4), Hubble calculated the speed at which each galaxy is moving away from us.

4 Using techniques such as the brightness of Cepheid variables (see Section 15-1), Hubble also estimated the distances to a number of these galaxies. He found a direct correlation between the distance to a galaxy and the size of its redshift: *Galaxies in distant clusters and superclusters are moving away from us more rapidly than galaxies in nearby clusters and superclusters.* Figure 16-29 shows this correlation for five elliptical galaxies. This recessional motion pervades the universe and is now called the **Hubble flow.**

The separation of clusters in the same supercluster is slower than the general Hubble flow of the universe because these clusters are relatively close to each other, so their gravitational attractions slow each other down. Keep in mind that the Hubble flow does not occur for galaxies in any given cluster, since all those galaxies are gravitationally bound to each other. Specifically, the Hubble flow applies to the separation of superclusters from each other and many clusters of galaxies from each other.

When Hubble plotted the redshift data on a graph of distance versus speed, he found that the points lie nearly along a straight line. Figure 16-30 shows the relationship between the distances to galaxies and their recessional motion, including many whose distances were measured using the apparent magnitude of Type Ia supernovae. The information in this figure is easily stated as a formula called the **Hubble law:**

$$\text{Recessional velocity} = H_0 \times \text{distance}$$

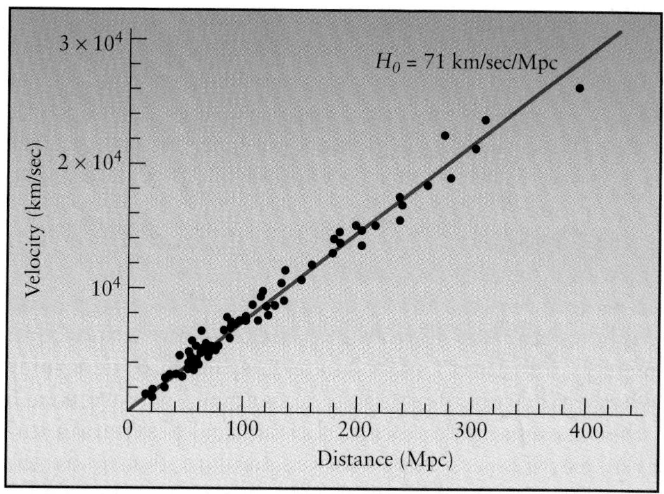

ANIMATION 16.3

FIGURE 16-30 **The Hubble Law** The distances and recessional velocities of distant galaxies are plotted on this graph. The straight line is the "best fit" for the data. This linear relationship between distance and speed is called the Hubble law.

This is simply the equation of a straight line, with the slope of the line denoted by the constant H_0, commonly called the **Hubble constant.** An Astronomer's Toolbox 16-1 shows how the Hubble law works.

The relationship between the distances to galaxies and their redshifts was one of the most important

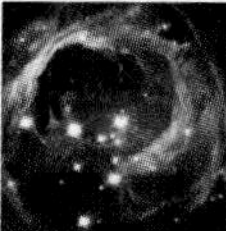

AN ASTRONOMER'S TOOLBOX 16-1
The Hubble Law

The Hubble law describes our expanding universe. It is a simple formula:

$$v = H_0 \times d$$

where v is the recessional velocity, d is the distance, and H_0 is the Hubble constant. It is the formula for the straight line displayed in Figure 16-30, and the Hubble constant is the slope of this line.

As Figure 16-30 shows, we usually measure a galaxy's velocity in kilometers per second and its distance in megaparsecs. You will see in section 16-11 that astronomers have determined Hubble's constant $H_0 \approx 71$ km/s/Mpc. If you have distances in millions of light years (Mly), you can convert H_0 using 1.00 Mpc = 3.26 Mly.

Example: A galaxy 1 Mpc from us is moving away at a rate of

$$v = (71\ \text{km/s/Mpc}) \times (1\ \text{Mpc}) = 71\ \text{km/s}$$

A galaxy 2 Mpc away is therefore receding at 142 km/s, and so on. A galaxy located 100 million parsecs from Earth rushes away from us with a speed of 7100 km/s or 16 million mph!

Compare! While a supercluster of galaxies 10 Mpc away recedes at 710 km/s, the solar system orbits the center of our Galaxy at a comparable 230 km/s.

Try these questions: At what speed is a galaxy observed to be 8 billion light-years away moving from us? How far away from us is a galaxy observed to be moving away at 216 km/s? If H_0 were 50 km/s/Mpc, would a galaxy 4 billion light-years away be moving away from us faster or slower than the same galaxy with H_0 = 71 km/s/Mpc?

(Answers appear at the end of the book.)

astronomical discoveries of the twentieth century. It tells us that we are living in an expanding universe, and the Hubble law reveals the speed of that expansion. This discovery raised the question of whether the universe will expand forever or whether the gravitational attraction between all its parts is enough to cause it someday to stop and recollapse. This is like trying to escape forever from the Earth. A normal jump might raise you up a third of a meter or so, but if you had a sufficiently powerful cannon to shoot you upward fast enough, you could escape the Earth's gravitational attraction completely and enter interplanetary space. To determine whether the superclusters are expanding away from each other fast enough to escape their mutual gravitation forever, astronomers have worked hard to determine the exact value of H_0. We explore that now and then pick up the question of the fate of the universe in Chapter 18.

INSIGHT INTO SCIENCE

Power in Numbers Much of the power of science comes from expressing systematic data in mathematical form. By themselves, the spectra of galaxies give a qualitative picture of galaxies' motion, and the pattern of recessional velocities suggests that a large-scale property of the universe is in effect here. When the data are converted into the equation called the Hubble law, it becomes clear that the universe is expanding, and we can begin to look for the reason behind this expansion.

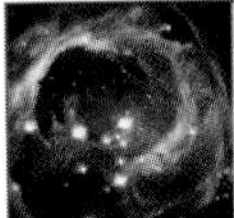

GUIDED DISCOVERY
The Tully-Fisher Relation and Other Distance-Measuring Techniques

Spectroscopic parallax (see Section 11-9) provides distances to stars up to 10 kpc (33 kly, where kly denotes 1000 light-years). This distance is still within the Milky Way. To determine distances to other galaxies, we need sources in them that become very bright compared to the luminosity of normal stars, and whose absolute magnitudes are well known. Using the brightnesses of RR Lyrae variable stars (see Sections 12-13 and 12-15), astronomers can measure distances to the Large and Small Magellanic Clouds or about 100 kpc (330 kly). To measure the Hubble flow, we must determine distances to even farther galaxies.

Cepheid variable stars (see Section 15-2 and Figure 15-4) can now be seen out to 30 Mpc (100 Mly, where Mly denotes a million light-years) from Earth. The distances to galaxies in this nearby volume of space can thus be determined from the Cepheid period-luminosity law. Plotting the distances to galaxies versus their recessional velocity for the small volume of space extending out 30 Mpc from Earth (see left side of Figure 16-30) enabled astronomers to get a first estimate for H_0 from the slope of the line through these data. This gave $H_0 \approx 75$ km/sec/Mpc. Errors in both distance and recessional velocities led to inaccuracies in the slope of this curve.

Beyond 30 Mpc, even the brightest Cepheid variables, which have absolute magnitudes of about –6, are not visible with current technology. In the 1970s, the astronomers Brent Tully and Richard Fisher developed another method for determining distances. They discovered that the width of the hydrogen 21-cm emission line of a spiral galaxy (discussed in Chapter 15) is related to the galaxy's absolute magnitude: *The broader the line, the brighter the galaxy.* The connection goes like this: Interstellar gas orbits in galaxies, as we have seen. Different hydrogen gas clouds in different places in the same galaxy have different speeds toward or away from us and so we observe a variety of slightly different wavelengths from this gas, centered around 21 centimeters. This combination of emissions makes the 21-cm emission line appear "broad."

The **Tully-Fisher relation** says that the greater the galaxy's mass, the greater its luminosity and the more rapidly the stars and gas in it orbit. The faster it rotates, the greater the range of speeds and hence Doppler shifts we see from it. Measuring the width of the 21-cm line, therefore, tells us the galaxy's mass and hence its *absolute* magnitude. Combining this with observations of the galaxy's *apparent* magnitude allows us to calculate its distance from us using the distance-magnitude relationship (see An Astronomer's Toolbox 11-3).

Because line widths can be measured quite accurately, astronomers can use the Tully-Fisher relation to determine the luminosities (absolute magnitudes) of spiral galaxies and thus their distances. Measurements of 21-cm-line widths for distant, receding galaxies yield a Hubble constant of 71 km/s/Mpc.

A key method for determining the distances to the most remote galaxies is to measure the brightness of type Ia supernova explosions within them. These supernovae reach an absolute magnitude of –19 at the peak of their outbursts (see accompanying figure). Using this method, astronomers in 2001 measured a supernova nearly 3 billion parsecs (10 billion light-years) away, and more distant type Ia supernovae have been measured since then.

16-11 Different techniques determine the expansion of the universe at different distances from Earth

To determine the Hubble constant, astronomers must measure the redshifts and distances to many galaxies. Although redshift measurements from spectra can be quite precise, it is very difficult to measure the distances to remote galaxies accurately. We cannot use the method of stellar parallax at the distances of galaxies (recall from Section 11-1 that the distance to the stars in our Galaxy within about 150 pc (500 ly) can be determined very precisely this way).

Distances to other galaxies are determined from observations of the apparent magnitudes of objects in them or of the entire galaxy. These observations are compared to known absolute magnitudes for similar objects. Having the apparent magnitude and the absolute magnitude enables astronomers to calculate distances, as discussed in An Astronomer's Toolbox 11-3. Astronomers use the term **standard candle** to denote any object whose absolute magnitude is known. The various techniques used to determine large-scale distances are presented in Guided Discovery: The Tully-Fisher Relation and Other Distance-Measuring Techniques. Figure 16-31 summarizes the distances for which all the measuring techniques described here are valid. Including data from all the methods yields a distance-velocity graph as shown in Figure 16-30. Using this diagram, astronomers have calculated that the Hubble constant is

$$H_0 = 71 \pm 4 \text{ km/s/Mpc}$$

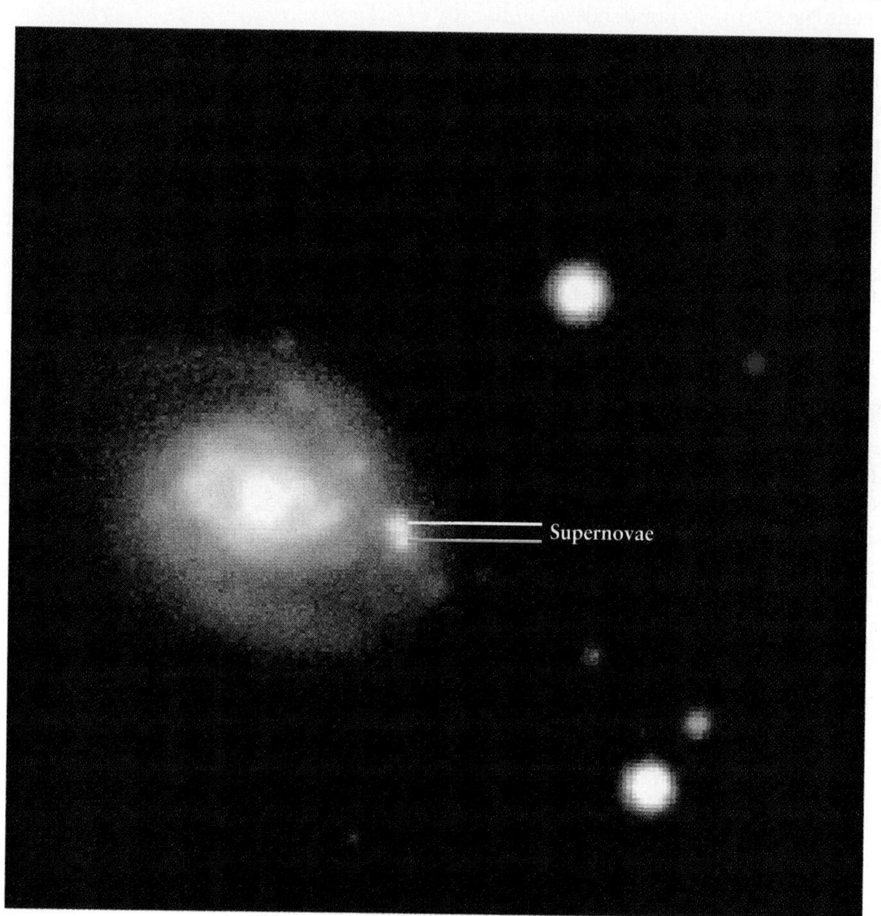

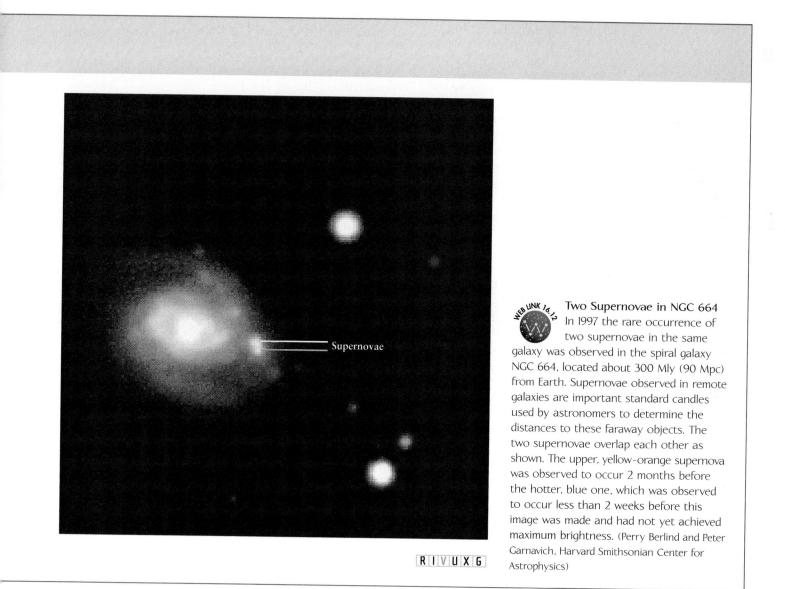

Two Supernovae in NGC 664 In 1997 the rare occurrence of two supernovae in the same galaxy was observed in the spiral galaxy NGC 664, located about 300 Mly (90 Mpc) from Earth. Supernovae observed in remote galaxies are important standard candles used by astronomers to determine the distances to these faraway objects. The two supernovae overlap each other as shown. The upper, yellow-orange supernova was observed to occur 2 months before the hotter, blue one, which was observed to occur less than 2 weeks before this image was made and had not yet achieved maximum brightness. (Perry Berlind and Peter Garnavich, Harvard Smithsonian Center for Astrophysics)

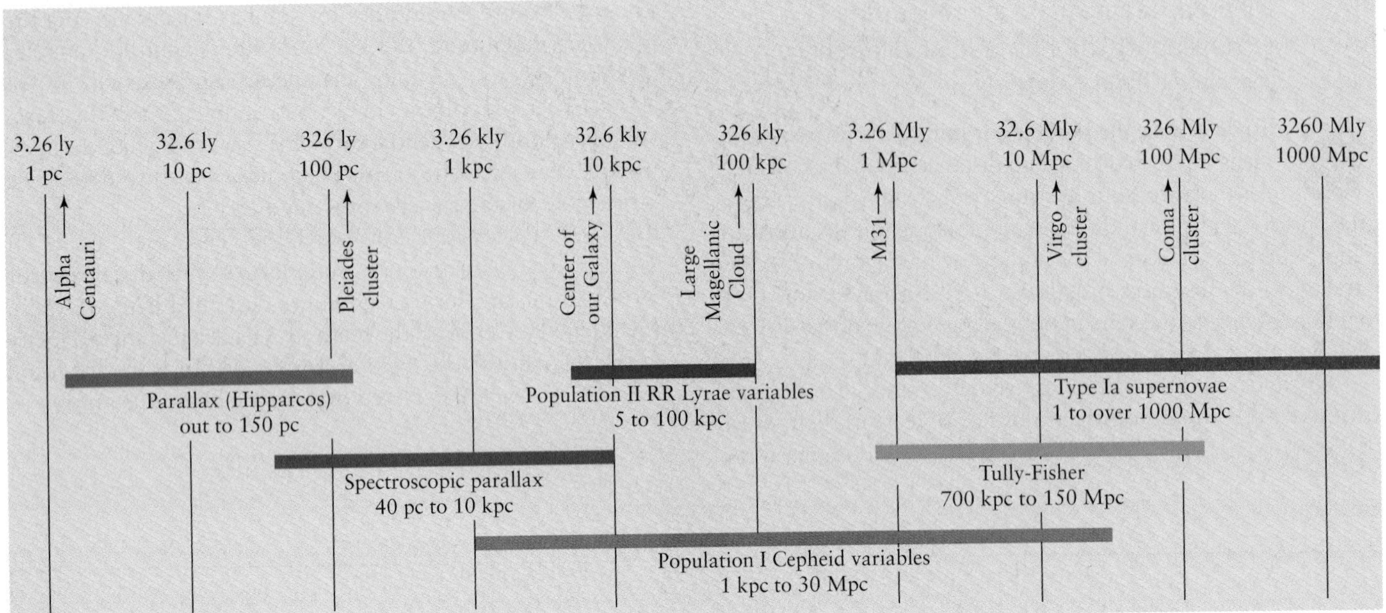

FIGURE 16-31 Techniques for Measuring Cosmological Distances Astronomers use different methods to determine different distances in the universe. All the methods shown here are discussed in the text.

The plus or minus 4 km/sec/Mpc indicates the possible range for H_0, when taking into account all the errors that may still remain in the observations and calculations. A significant source of error is the effect of intergalactic gas causing the standard candles to appear dimmer than they would if the gas were not present. Astronomers determine the amount of interstellar gas and take its effects into account, something that is often hard to do.

16-12 Astronomers are looking back to a time when galaxies were first forming

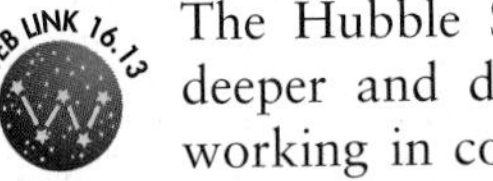

The Hubble Space Telescope has been looking deeper and deeper into the universe. In 1998, working in concert with the Keck I telescope, it observed galaxies 8 Bly (where Bly denotes a billion light-years) away, many of them in the process of merging

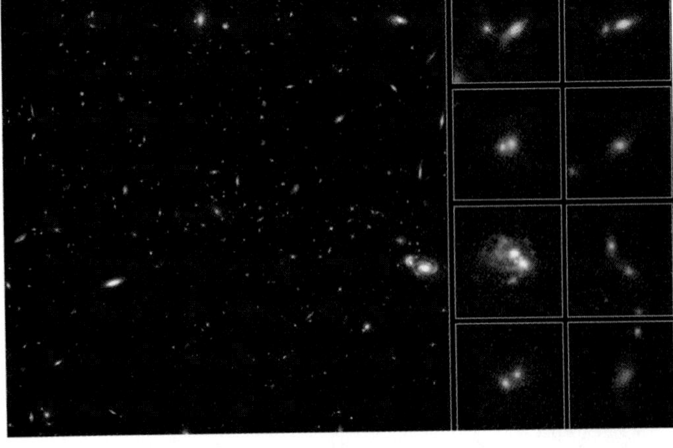

a

b

R I V U X G

FIGURE 16-32 Distant Galaxies (a) The young cluster of galaxies MS1054-03, shown on the left, contains many orbiting pairs of galaxies, as well as remnants of recent galaxy collisions. Several of these systems are shown at the right. This cluster is located 8 billion light-years away from Earth. (b) This image of more than 300 spiral, elliptical, and irregular galaxies contains several that are an estimated 12 billion light-years from Earth. Two of the most distant galaxies are shown in the images on the right, colored in red at the centers of the pictures. (a, b: P. Van Dokkum, Uner of Granengen, ESA and NASA)

(Figure 16-32). The same year it saw galaxies nearly 12 Bly away (Figure 16-32b). In 2004, galaxies up to 13.2 Bly away were observed. We are seeing these latter objects as they were less than a billion years after the universe came into existence. While many of the early galaxies are not as well-formed as nearby galaxies, the fact that there are galaxy-sized collections of stars so far back in time is an important clue to the processes of star and galaxy formation that occurred in the early universe.

Nature also provides astronomers with assistance in visualizing distant galaxies through the gravitational warping of space, as described by Einstein's general theory of relativity (see Section 14-2). We saw in Chapter 15 how the same effect occurs on small scales here in the Milky Way. While the local distortion is called microlensing, when entire galaxies or clusters of galaxies focus light from more distant galaxies, the effect is called **gravitational lensing.** Figure 16-33 shows the gravitationally lensed images of distant galaxies. Indeed, the most distant galaxies, mentioned in the preceding paragraph, were detected as a result of gravitational lensing.

The amount of gravitational lensing by a galaxy or cluster of galaxies provides astronomers with more than just enlarged images of distant objects. It tells how much mass the galaxy or cluster that does the lensing contains. The greater the mass of the lensing galaxy or cluster, the greater the effect on the distant object. The visible matter in the galaxies could only account for about 10% of the gravitational lensing that occurs. In other words, the amount of lensing confirms that at least 90% of the matter in the universe is still unaccounted for.

WEB LINK 16.14 The observational data now being accumulated from the period of time when the universe was less than a few billion years old are helping astronomers develop and test theories of the evolution of the early universe. We will explore this point further in Chapter 18.

16-13 Frontiers yet to be discovered

The farther out in the universe we explore, the more there is to discover. In the realm of the galaxies, we have yet to understand fully how the different types of galaxies form. Are nuclear bulges formed as a result of collisions between galaxies? Are globular clusters the remnants of previous galaxies that have been consumed by bigger galaxies? How long does it take interacting galaxies to combine? Do supermassive black holes form before, simultaneously with, or

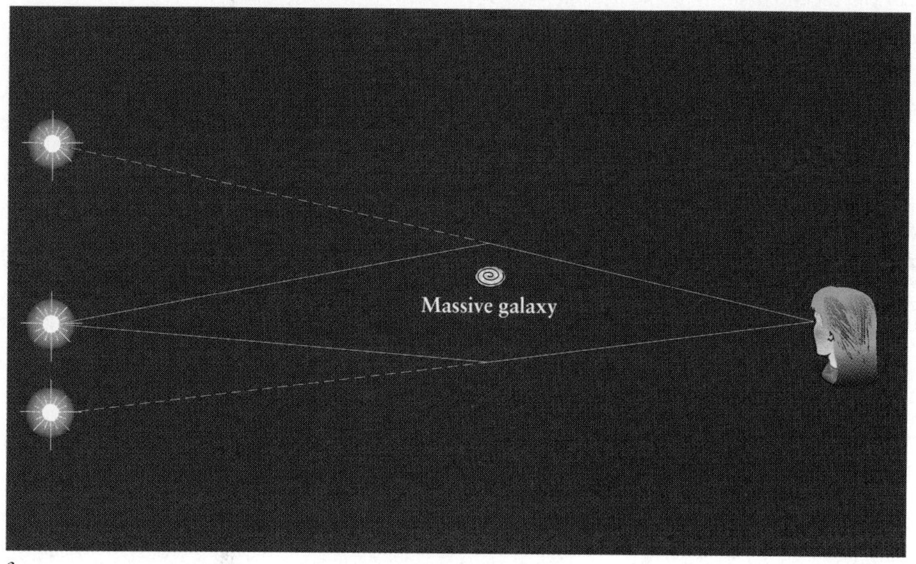

a

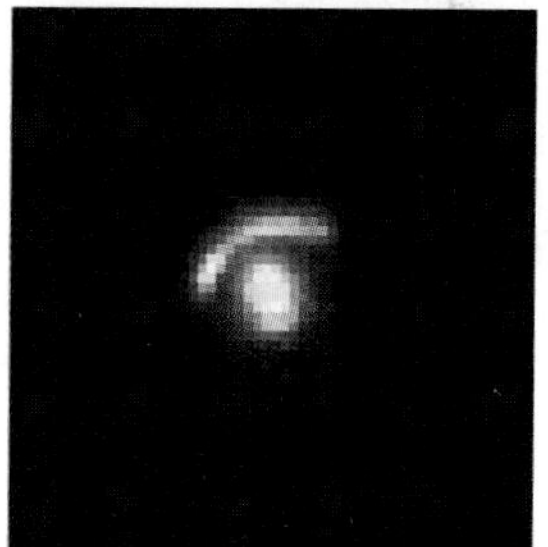

1

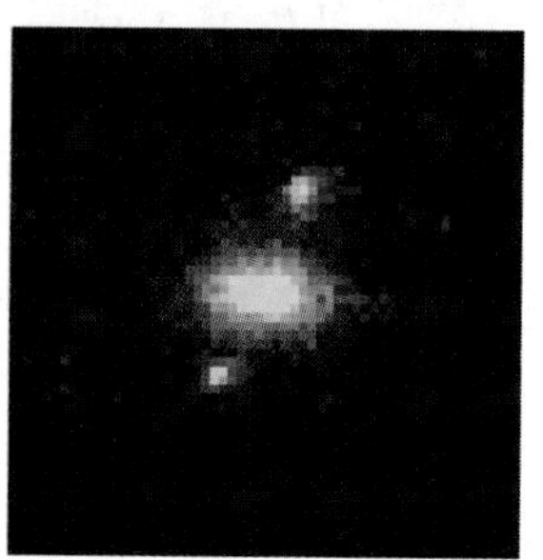

2

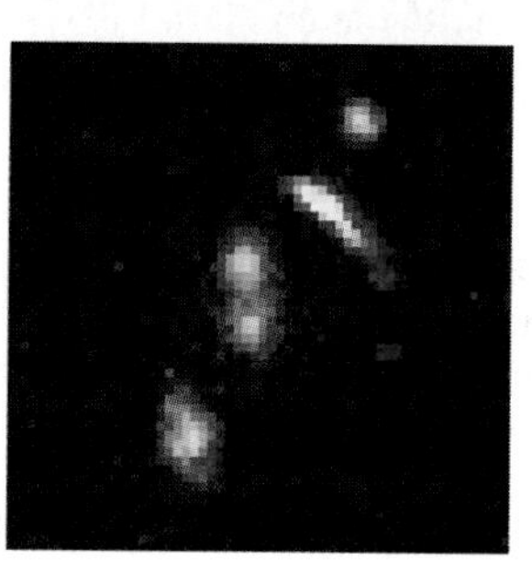

3

b

FIGURE 16-33 **Gravitational Lensing of Extremely Distant Galaxies** (a) Schematic of how a gravitational lens works. Light from the distant object changes direction due to the gravitational attraction of the intervening galaxy. The more distant galaxy appears in two places to the observer on the right. (b) Three examples of gravitational lensing: (1) The bluer arc has been lensed by the redder elliptical galaxy. (2) A pair of bluish images of the same object lensed symmetrically by the brighter, redder galaxy between them. (3) The lensed object appears as a blue arc under the gravitational influence of the group of four galaxies. (b: Kavan Ratnatunga, Carnegie Mellon University, ESA, and NASA) after galaxies come into existence? Moving into the realm of the global structure of the universe, what is the overall distribution of clusters and superclusters of galaxies in the universe? Are most of them really found on spongelike or soap-bubble–like surfaces and, if so, what is that structure telling us about the underlying distribution of dark matter?

Summary of Key Ideas

Types of Galaxies

• The Hubble classification system groups galaxies into four major types: spiral, barred spiral, elliptical, and irregular.

• The arms of spiral and barred spiral galaxies are sites of active star formation.

• According to the theory of self-propagating star formation, spiral arms of flocculent galaxies are caused by the births and deaths of stars over extended regions of a galaxy. Differential rotation of a galaxy stretches the star-forming regions into elongated arches of stars and nebulae that we see as spiral arms.

• According to the spiral density wave theory, spiral arms of grand-design galaxies are caused by density waves. The gravitational field of a spiral density wave compresses the interstellar clouds that pass through it, thereby triggering the formation of stars, including OB associations, which highlight the arms.

• Elliptical galaxies contain much less interstellar gas and dust than do spiral galaxies; little star formation is occurring in elliptical galaxies.

Clusters and Superclusters

• Galaxies group into clusters rather than being randomly scattered through the universe.

• A rich cluster contains hundreds or even thousands of galaxies; a poor cluster may contain only a few dozen. A regular cluster has a nearly spherical shape with a central concentration of galaxies; in an irregular cluster, the distribution of galaxies is asymmetrical.

• Our Galaxy is a member of a poor, irregular cluster called the Local Group.

• Rich, regular clusters contain mostly elliptical and lenticular galaxies; irregular clusters contain more spiral and irregular galaxies. Giant elliptical galaxies are often found near the centers of rich clusters.

• No cluster of galaxies has an observable mass large enough to account for the observed motions of its galaxies; a large amount of unobserved mass must be present between the galaxies.

• Hot intergalactic gases emit X rays in rich clusters.

• When two galaxies collide, their stars initially pass each other but their interstellar gas and dust collide violently, either stripping the gas and dust from the galaxies or triggering prolific star formation. The gravitational effects of a galactic collision can cast stars out of their galaxies into intergalactic space.

• Galactic mergers occur; a large galaxy in a rich cluster may grow steadily through galactic cannibalism, sometimes producing a giant elliptical galaxy.

Superclusters in Motion

• A simple linear relationship exists between the distance from the Earth to galaxies in other superclusters and the redshifts of those galaxies (a measure of the speed at which they are receding from us). This relationship is the Hubble law, recessional velocity = $H_0 \times$ distance, where H_0 is the Hubble constant.

• Astronomers use standard candles—Cepheid variables, the brightest supergiants, globular clusters, H II regions, supernovae in a galaxy and the Tully-Fisher relation—to calculate intergalactic distances. Because of difficulties in measuring the distances to remote galaxies, the value of the Hubble constant, H_0, is not known with complete certainty.

WHAT DID YOU THINK?

1 *Do all galaxies have spiral arms?* No. Galaxies may be either spiral, barred spiral, elliptical, or irregular. Only spirals and barred spirals have arms.

2 *Are most of the stars in spiral galaxies located in their spiral arms?* No. The spiral arms contain only 5% more stars than the regions between the arms.

3 *Are galaxies isolated objects?* No. Galaxies are grouped in clusters, and clusters are grouped in superclusters.

4 *Are all other galaxies moving away from the Milky Way?* All galaxies except those in our Local Cluster are receding from us. Some local ones are moving toward us.

Key Words

irregular galaxy, 425
lenticular galaxy, 424
Local Group, 428
poor cluster (of galaxies), 428
regular cluster (of galaxies), 428
rich cluster (of galaxies), 428
spiral density wave, 422
spiral galaxy, 418
standard candle, 437
starburst galaxy, 430
supercluster (of galaxies), 426
trailing-arm spiral galaxy, 423
Tully-Fisher relation, 436

Review Questions

1. What are the most massive galaxies in the universe? **a.** normal spirals. **b.** barred spirals. **c.** giant ellipticals. **d.** dwarf ellipticals. **e.** irregulars.

2. Spiral density waves are directly responsible for which of the following? **a.** flocculent spirals. **b.** grand design spirals. **c.** supernovae. **d.** the collisions between galaxies. **e.** galactic cannibalism.

3. Which of the following statements about the motion of galaxies is correct? **a.** All galaxies are moving apart. **b.** Superclusters of galaxies are all moving apart. **c.** Superclusters of galaxies are all moving toward each other. **d.** The Milky Way Galaxy is at the center of the universe. **e.** All the clusters of galaxies in each supercluster are moving toward each other.

4. What is the common name for the sonic boom created by lightning?

5. What 5% of stars are not found between the arms of spiral galaxies?

6. What is the Hubble classification scheme? Which category includes the biggest galaxies? Into which category do the smallest galaxies fall? Which type of galaxy is the most common?

7. In which Hubble types of galaxies are new stars most commonly forming? Describe the observational evidence that supports your answer.

8. What is the difference between a flocculent spiral galaxy and a grand-design spiral galaxy?

9. Briefly describe how the theory of self-propagating star formation accounts for the existence of spiral arms in some spiral galaxies.

10. Briefly describe how the spiral density wave theory accounts for the existence of spiral arms in some spiral galaxies.

11. How is it possible that galaxies in our Local Group still remain to be discovered? In what part of the sky are these galaxies located? What sorts of observations might reveal these galaxies?

12. Can any galaxies besides our own be seen with the naked eye? If so, which?

13. What is the difference between a rich cluster of galaxies and a poor one? What is the difference between a regular cluster of galaxies and an irregular one?

14. How can a collision between galaxies produce a starburst galaxy?

15. Why do astronomers believe that considerable quantities of dark matter must exist in clusters of galaxies?

16. Explain why the dark matter in galaxy clusters cannot be neutral hydrogen.

17. What is the Hubble law?

18. Some galaxies in the Local Group exhibit blueshifted spectral lines. Why aren't these blueshifts violations of the Hubble law?

19. What is a standard candle? Why are standard candles important to astronomers trying to measure the Hubble constant?

20. What kinds of stars would you expect to find populating space between galaxies in a cluster?

Advanced Questions

The answers to all computational problems, which are preceded by an asterisk (*), appear at the end of the book.

***21.** Suppose a spectrum of a distant galaxy showed that its redshift corresponds to a speed of 22,000 km/s. How far away is the galaxy in Mpc?

***22.** A cluster of galaxies in the southern constellation of Pavo (the Peacock) is located 100 Mpc from Earth. How fast, on average, are galaxies in this cluster receding from us? Why do different galaxies in the cluster show different velocities?

23. How would you determine what fraction of a distant galaxy's redshift is caused by the galaxy's orbital motion about the center of mass of its cluster?

Discussion Questions

24. Discuss the advantages and disadvantages of using the various standard candles to determine extragalactic distances.

25. Discuss whether the various Hubble types of galaxies actually represent some sort of evolutionary sequence.

26. Discuss the sorts of phenomena that can occur when galaxies collide. Do you think that such collisions can change the Hubble type of a galaxy? Explain.

27. From what you know about stellar evolution, the interstellar medium, and the spiral density wave theory, explain the appearance and structure of the spiral arms of grand-design spiral galaxies.

What If . . .

28. The solar system was located in an active star-forming region in a spiral arm, rather than on the edge of the Orion arm? How would the solar system be different?

29. The Milky Way collided with the Andromeda Galaxy? What might we experience here on Earth? (This merger will occur in the distant future.)

30. No distant galaxies showed any redshift or blueshift? Discuss what that would imply about the universe.

31. All distant galaxies showed a blueshift, rather than a redshift? Discuss what that would imply about the universe.

Web Questions

32. To test your understanding of the global properties of galaxies, do Interactive Exercise 16-1 on the Web. You can print out your answer, if required.

33. To test your understanding of the local group of galaxies, do Interactive Exercise 16-2 on the Web. You can print out your answer, if required.

34. To test your understanding of the various types of galaxies, do Interactive Exercise 16-3 on the Web. You can print out your answer, if required.

35. To test your understanding of star formation in the spiral density wave model, do Interactive Exercise 16-4 on the Web. You can print out your results, if required.

Observing Projects

36. Using a telescope with an aperture of at least 30 cm (12 in.) in a dark location, observe as many of the spiral galaxies listed in Table E-10, in the back of the book, as you can. You can locate them and print out their locations on a star chart using your *Starry Night Enthusiast*™ software. Many of these galaxies are members of the Virgo cluster, which can best be seen from the northern hemisphere during the Spring. Because all galaxies are quite faint, be sure to schedule your observations for a moonless night. The best view is obtained when a galaxy is high in the sky. While at the eyepiece, make a sketch of what you see. Can you distinguish any spiral structure? After completing your observations, compare your sketches with photographs found in such popular books as *Galaxies* by Timothy Ferris (Sierra Club Books, 1980) and *The Color Atlas of the Galaxies* by James Wray (Cambridge University Press, 1988).

37. Using a telescope with an aperture of at least 30 cm (12 in.), observe as many of the interacting galaxies listed in Table E-10, in the back of the book, as you can. You can locate them and print out their locations on a star chart using your *Starry Night Enthusiast*™ software. As in Observing Project 36, be sure to schedule your observations for a moonless night when the galaxies you wish to observe will be near the meridian. While at the eyepiece, make a sketch of what you see. Can you distinguish hints of interplay among the galaxies? After completing your observations, compare your sketches with photographs found in popular books such as *Galaxies* by Timothy Ferris (Sierra Club Books, 1980).

38. **Optical Double Galaxies** The galaxies M81 and M82 (Figure 16-24) appear to be near each other as seen on photographs taken from Sun. In this exercise, we use *Deep Space Explorer*™ to determine just how close they are. If you have not done so already, go to **Settings** and uncheck "Use magnitude cutoffs," slide the "Galaxy drawing/Brightness" slide to the middle of the range, and then click on the **Views** tab.

Now go to the "Home" setting to see the Milky Way. Then click on **Edit/Find/M81** on the left side of the star field. The program will take you to 0.119 Mly from M81. Write down the total distance from the Sun to M81 from the information on the bottom of the screen. Grab the screen (hold down left PC mouse button or the mouse button on a Mac) and move the mouse around until M82 is visible. It is red-orange. You can identify it by releasing the button and putting the mouse over it. It will read "Cigar Galaxy" with M82 below it. Right click on the mouse and choose "Go there." You will arrive 0.12 Mly from M82. Write down the total distance from M82 to the Sun from the data on the bottom of the screen. From the data you now have available, calculate the distance between M81 and M82. Are they as close to each other as the Milky Way and the Large Magallenic Cloud, which are about 0.195 Mly apart?

39. **Galaxy Orientation** Here is an opportunity to compare an image in the text with one in *Deep Space Explorer*™. If you have not done so already, go to **Settings** and uncheck "Use magnitude cutoffs," slide the "Galaxy drawing/Brightness" slide to the middle of the range, and then click on the **Views** tab.

Now press the **Home** button so that you see the Milky Way Galaxy on the screen. Then select **Edit/Find/M74.** This will take you to an image of the galaxy shown in Figure 16-4b in the text. Do not yet move the screen image. Describe the appearance of the galaxy in the text and then describe the appearance of the galaxy on the screen. Write down how these two Views of the same galaxy (book and program) are similar and how they differ. Write an explanation for the difference in appearance. Now grab the screen by holding down the left PC mouse button (the mouse button on a Mac) and move your mouse around until you see the galaxy on the screen face on. Determine whether or not the galaxy on the screen is the same as the galaxy shown in the text. Explain how you make this confirmation.

40. Galaxy Types To help you get a feel for interpreting the images of galaxies that you see, we will use *Deep Space Explorer*™ to visit a variety of galaxies and determine whether they are spiral, barred spiral, or irregular. If you have not done so already, go to **Settings** and uncheck "Use magnitude cutoffs," slide the "Galaxy drawing/Brightness" slide to the middle of the range, and then click on the **Views** tab. Go to the **Home** image. If you have information displayed in the upper left part of your screen, right click (ctrl-click for Mac) on the galaxy in the center of the screen and select **Hide Info.**

Now, use **Edit/Find/"Object Name"** to visit each of the galaxies listed below in turn. Be careful—include spaces, where indicated, when typing in object names. If it is seen at any angle other than face on, grab the screen by pressing the left PC mouse button (mouse button on a Mac) and moving the mouse until it is face on. Then write down whether you think it is a spiral (S), barred spiral (SB), or irregular (Irr). We will not worry about the subclasses in this question.

IC 4182	IC 5152	M33	M58	M74
M81	M82	M83	M94	M109
NGC 1232	NGC 1313	NGC 1365	LMC	SMC

41. Clusters of Galaxies Clusters of galaxies contain different numbers and distributions of galaxies. In this exercise we use *Deep Space Explorer*™ to compare a few of these groupings. If you have not done so already, go to **Settings** and uncheck "Use magnitude cutoffs," slide the "Galaxy drawing/Brightness" slide to the middle of the range, and then click on the **Views** tab.

Press the **Home** button. Now press the **Highlight** tab on the left of the star field. A long list of galaxy clusters and other features will appear. Double click on **GA Virgo Cluster** under "Supergalactic Eq Plane." You will see a check mark next to the name and the Virgo Cluster will appear in yellow. Grab the screen by holding the left PC mouse button (the mouse button on a Mac) and move the mouse around. What is the general shape of the Virgo cluster? If you see other groupings of galaxies (as white dots) as you move the screen, make a sketch of the screen and circle what you believe are other groups on your sketch. To see how astronomers have grouped these other galaxies, single click on one of the other clusters (and clouds and extensions) listed near Virgo in the "Highlight" list to the left of the star field. It will light up in yellow in turn. Unclick this cluster and click another. Repeat until you have identified the clusters around Virgo. Sketch them on your drawing and compare with your original groupings. Choose three of these clusters, study them by double clicking on each in the list, move space around, and describe their distributions of galaxies compared to Virgo. For example, what are their shapes and relative sizes compared to Virgo and to each other? When you are done, click on **Deselect all** from above the list of clusters.

42. Great Wall of Galaxies In this exercise we will use *Deep Space Explorer*™ to view and interpret the Great Wall of galaxies. If you have not done so already, go to **Settings** and uncheck "Use magnitude cutoffs," slide the "Galaxy drawing/Brightness" slide to the middle of the range, and then click on the **Views** tab.

Press the **Home** button. Now press **Local Universe/Great Wall.** The galaxies in the Great Wall will be displayed in yellow. Keep in mind that the image is centered on the Milky Way. Grab the screen by pressing the left PC mouse button (the mouse button on a Mac) and moving the mouse. You will see the Great Wall galaxies move around the Milky Way. Study the distribution of galaxies in the Great Wall as you move it and write down an explanation of where the name "Great Wall" originates. Write down any other appropriate names that come to mind from the shape. Now click on the **Highlight** tab on the left of the star field and then **Deselect all** at the top of the list that appears. This will turn off the yellow highlight of the Great Wall galaxies.

WHAT IF . . . THE SOLAR SYSTEM WERE LOCATED CLOSER TO THE CENTER OF THE GALAXY?

It is a clear, moonless night. Stars by the thousands twinkle serenely against the ebony darkness of space. Overhead the soft white span of the Milky Way catches your attention, and you try to see individual stars in its glowing haze.

Most of the 6000 stars visible to the naked eye throughout the year are within 300 ly of Earth. The rest of the Galaxy's 200 billion or so stars are too dim or too obscured by interstellar gas and dust to be easily observed. What if the solar system were one-third of its present distance from the center of our Galaxy? At that distance, we would still be in the realm of the spiral arms, extremely close to the nuclear bulge, and none of the stars we see now would be visible.

Population Explosion Out in the galactic suburbs, where we live today, there is about one star per 300 cubic light-years. The displaced Earth would be surrounded by nearly 5 times as many stars, or one per 60 cubic light-years. And the solar system would pass much more frequently through the dust-rich spiral arms of the Galaxy. Scattering starlight, this interstellar matter would glow as diaphanous wisps throughout our nighttime sky. Furthermore, whenever the solar system was actually in a spiral arm, several nearby stars would be of the high-mass, high-luminosity variety. The combined light from these stars and the shimmering clouds would be so great that for millions of years at a time night would never fall.

Evolution Interruption Recall from Chapter 13 that high-mass stars evolve much more rapidly than average-mass stars like the Sun. If we were closer to the center of the Milky Way and frequently passing through the spiral arms of the Galaxy, massive stars would explode near us much more frequently than they do today. Those nearby supernovae, occurring within 50 ly of Earth, would deposit lethal radiation, damage life, and cause mass extinctions. As a result, the direction of evolution would change more frequently.

At Earth's current location, the closest star, Proxima Centauri, is more than 4 ly away. Would Earth be in danger of colliding with a star if the solar system were closer to the center of the Galaxy? Several stars would certainly be much closer than Proxima Centauri is now. However, stars are so small compared to the vastness of a galaxy that the likelihood of a collision would still approach nil.

Much more likely would be the passage of a star so close to the Sun that the Earth's orbit would be disturbed. If the Earth's orbit became, say, more elliptical, the change of seasons would be noticeably affected. With the seasonal effect of the Earth's tilt compounded by greater changes in distance between the Earth and Sun, one hemisphere of the Earth would suffer much more extreme temperatures than it does today, while the other would see less variation. This would affect the evolution of life, of course, and the distribution of life-forms on the planet.

Close Encounters Earth-evolved life at our new location in the Galaxy would be more likely to encounter sentient beings on planets orbiting nearby stars. Today, after a century of broadcasting radio and television signals, a 200-ly-diameter sphere of space centered on the Earth is filled with such signals. There are about 6300 stars in that sphere. At our new location there would be 31,500 stars in the same volume and the probability of many more stars with life-supporting planets orbiting them. Perhaps one of the reasons that life on Earth has been able to evolve so long unaffected by other intelligent life is that the solar system exists near the fringe of the Galaxy.

Quasars, Active Galaxies, and Other Ultrahigh Energy Sources

CHAPTER

17

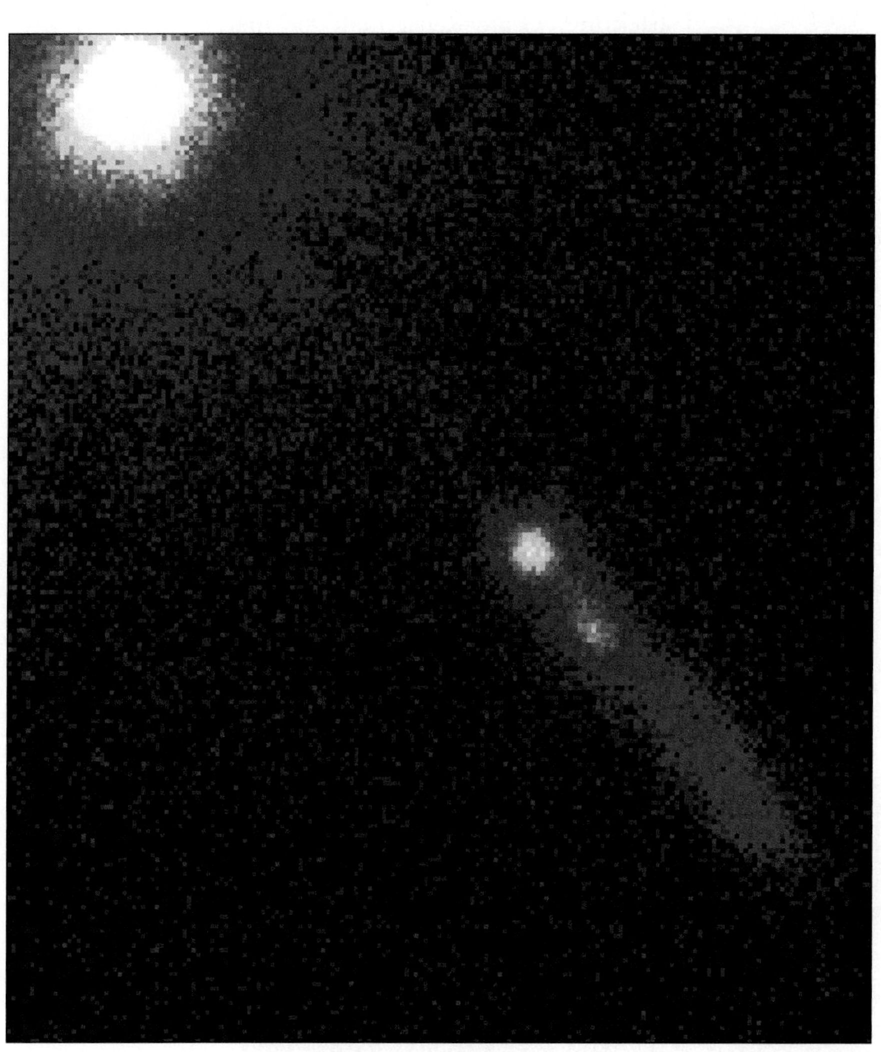

The Quasar 3C 273 (NASA/CXC/SAO/H. Marshall et al.)

WHAT DO YOU THINK?

1 What does "quasar" stand for?

2 What do quasars look like?

3 Where do quasars get their energy?

In this chapter we look at energy sources of almost unimaginable power. The short-lived outputs of supernovae boggle the mind, but they represent miniscule amounts of energy compared to quasars, active galaxies, and BL Lac objects. Quasars, for example, emit more energy each second than the Sun does in 200 years. And they continue to do so for millions, even billions, of years. Some truly remarkable activity must occur deep within quasars and other ultrahigh energy sources in order to make them so luminous.

In this chapter you will discover

- distant, luminous quasars
- the unusual spectra and small volumes of quasars
- bright and unusual objects called active galaxies
- the extremely powerful BL Lac objects
- supermassive black holes that serve as central engines for quasars, active galaxies, and BL Lac objects
- how observations from the central regions of quasars, active galaxies, and BL Lac objects have helped astronomers devise theories about their energy sources

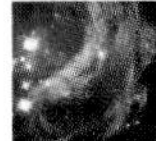

QUASARS

Our knowledge of quasars began in an amateur astronomer's backyard. Grote Reber built the first radio telescope in 1936 behind his home in Illinois, opening the realm of nonvisual astronomy. By 1944, Reber had detected strong radio emissions from sources in the constellations of Sagittarius, Cassiopeia, and Cygnus. Two of these sources, Sagittarius A (Sgr A) and Cassiopeia A (Cas A), are in our Galaxy. The first is the galactic nucleus (see Section 15-4) and the second is a supernova remnant (see Chapter 13). However, Reber's third source, called Cygnus A (Cyg A), proved hard to categorize (Figure 17-1). Others quickly refined Reber's observations, but the mystery only deepened in 1954, when Walter Baade and Rudolph Minkowski, using the 200-in. optical telescope on Mount Palomar, discovered a strange-looking galaxy at the position of Cygnus A (Figure 17-1 inset).

The galaxy associated with Cyg A is very dim. Nevertheless, Baade and Minkowski managed to photograph its

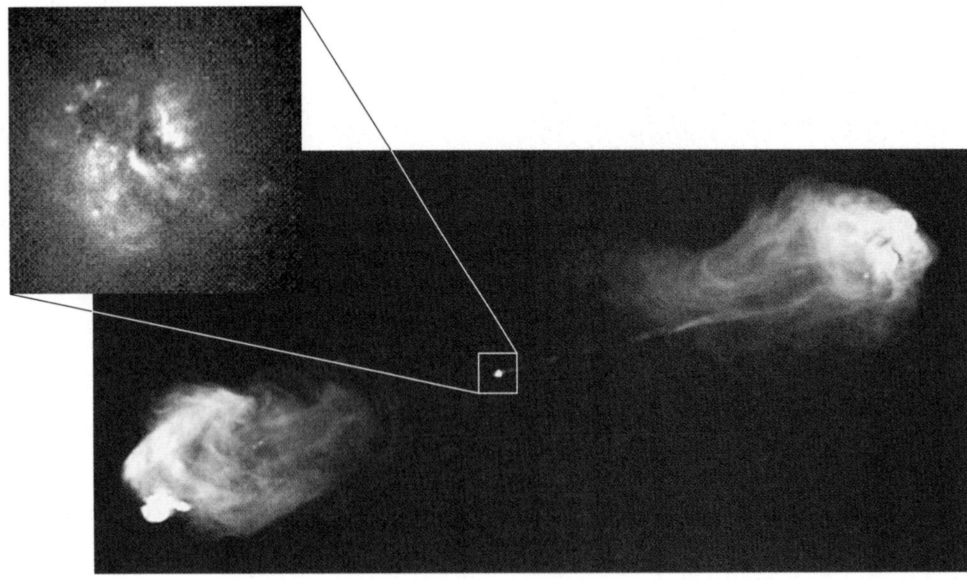

R I V U X G

FIGURE 17-1 **Cygnus A (3C 405)** This radio image was produced from observations made at the Very Large Array. Most of the radio emission from Cygnus A comes from the radio lobes located on either side of the peculiar galaxy seen in the inset. The two radio lobes each extend about 160,000 ly from the optical galaxy, and each contains a brilliant, condensed region of radio emission. **Inset:** At the heart of this system of gas lies a strange-looking galaxy that has a redshift corresponding to a recessional speed of 5% of the speed of light. According to the Hubble law, Cygnus A is therefore 635 million light-years from Earth. Because Cygnus A is one of the brightest radio sources in the sky, this remote galaxy's energy output must be enormous. (R. A. Perley, J. W. Dreher, J. J. Cowan, NRAO; inset: William C. Keel) spectrum. They detected a redshift corresponding to a speed of 14,000 km/s. According to the Hubble law, this speed indicates that Cyg A lies 635 million light-years (194 Mpc) from Earth.

Because it is one of the brightest radio sources in the sky, the enormous distance to Cyg A intrigued astronomers. Although barely visible through the giant optical telescope at Palomar, Cyg A's radio waves can be picked up by amateur astronomers with backyard equipment. Its energy output must therefore be colossal. In fact, Cyg A shines with a radio luminosity 10^7 times as bright as that of an ordinary galaxy, such as Andromeda, yet it is far more distant. The object creating the Cyg A radio emissions has to be something extraordinary.

17-1 Quasars look like stars but have huge redshifts

Cygnus A is not the only powerful radio source in the far-distant sky. Starting in the 1950s, radio astronomers created long lists of radio sources. One of the most famous lists, the *Third Cambridge Catalogue*, was published in 1959. (The first two catalogs were filled with inaccuracies.) Even today, astronomers often refer to its 471 radio sources by their "3C numbers." Cyg A, for example, is designated 3C 405, because it is the 405th source in the Cambridge list. Because of the extraordinary luminosity of Cyg A, astronomers were eager to learn whether any other sources in the 3C catalog had similar properties.

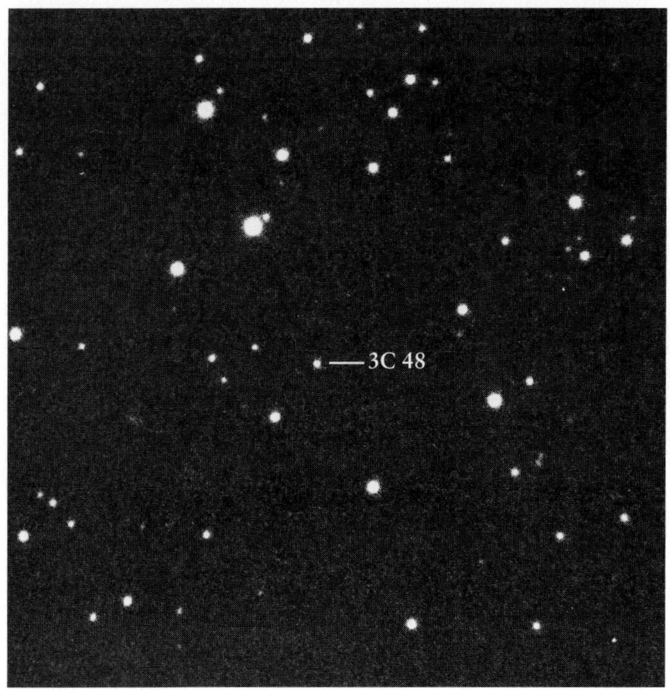

FIGURE 17-2 **The Quasar 3C 48** For several years astronomers erroneously believed that this object is simply a peculiar, nearby star that happens to emit radio waves. Actually, the redshift of this starlike object is so great that, according to the Hubble law, it must be roughly 4 billion light-years away. (Alex G. Smith, Rosemary Hill Observatory, University of Florida)

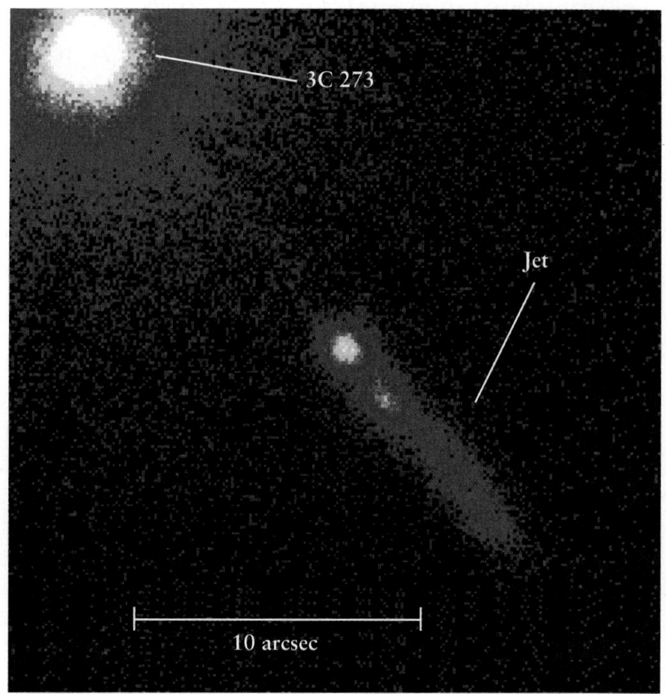

FIGURE 17-3 **The Quasar 3C 273** This greatly enlarged X-ray view shows the starlike object associated with the radio source 3C 273. Note the luminous jet that it has created. The jet is also visible in the radio and visible parts of the spectrum. By 1963, astronomers determined that the redshift of this quasar is so great that, according to the Hubble law, it is nearly 2 billion light-years from Earth. (NASA/CXC/SAO/H. Marshall et al.)

One interesting case was 3C 48. In 1960, Allan Sandage used the Palomar telescope to discover a "star" at the location of this radio source (Figure 17-2). Recall that stars are blackbodies whose peak intensity typically is visible light and whose radio emission is much less intense. Because ordinary stars are not strong sources of radio emission, 3C 48 had to be something unusual. Indeed, its spectrum showed a series of emission lines that, initially, no one could identify. Although 3C 48 was clearly an oddball, many astronomers thought it was just another strange star in our Galaxy.

Another such "star," called 3C 273, was discovered in 1962. Like 3C 48, this object and the luminous "jet" of bright gas found protruding from one side of it (Figure 17-3) emit a series of bright spectral emission lines that no one could then identify. These spectral lines are brighter than the background radiation at other wavelengths, called the continuum (recall Kirchhoff's laws in Section 4-4).

A breakthrough finally came in 1963, when Maarten Schmidt at the California Institute of Technology found that four of the brightest spectral lines of 3C 273 are four spectral lines of hydrogen. However, these emission lines from 3C 273 are found at much longer wavelengths than the usual wavelengths of hydrogen lines. Schmidt concluded that the hydrogen lines are subjected to a substantial redshift. Furthermore, the intense emission lines mean that something unusual is heating the gas.

Spectra for stars in our Galaxy exhibit comparatively small Doppler shifts, because these stars cannot move extremely fast relative to the Sun without soon escaping from the Galaxy. Schmidt thus concluded that 3C 273 is not a nearby star after all. Pursuing this conclusion, he promptly found that its redshift corresponds to a speed of almost 16% the speed of light. According to the Hubble law, this huge redshift implies an incredible distance to 3C 273 of roughly 2 billion light-years (Bly).

Figure 17-4 shows the spectrum of 3C 273. Remember from Chapter 4 that a spectrograph records energy intensity at different wavelengths. The emission lines appear as peaks, and absorption lines appear as valleys. The emission lines are caused by excited gas atoms emitting radiation at specific wavelengths.

Inspired by Schmidt's success, astronomers looked again at the spectral lines of 3C 48. The redshift corresponds to a velocity of nearly one-third the speed of light. Therefore, 3C 48 must be nearly twice as far away as 3C 273, or about 4 Bly from Earth, assuming that the Hubble constant, H_0, is 71 km/s/Mpc.

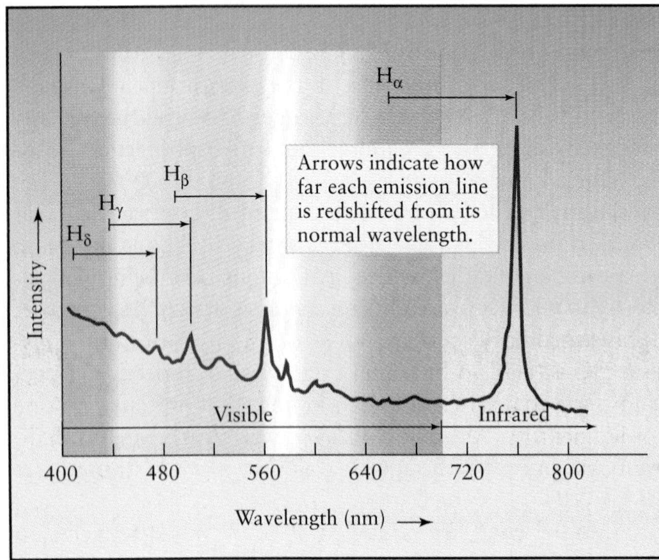

FIGURE 17-4 **The Spectrums of 3C 273** The visible and infrared spectrums of 3C 273 are dominated by four bright emission lines caused by hydrogen. This radiation is redshifted nearly 16% from its rest wavelengths.

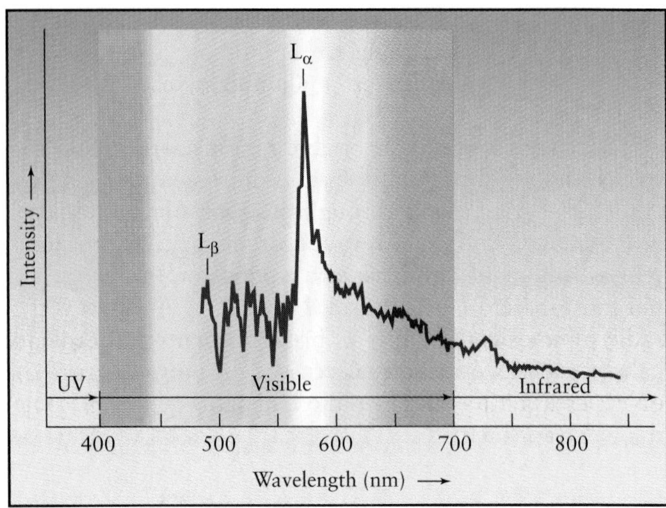

FIGURE 17-5 **The Spectrum of a High-Redshift Quasar** The light from this quasar, known as PKS 2000-330, is so highly redshifted that spectral emission lines normally in the far-ultraviolet (L_α and L_β) are seen at visible wavelengths. Note the many deep absorption lines on the short-wavelength side of L_α. These lines, collectively called the "Lyman-alpha forest," are believed to be created by remote clouds of gas along our line of sight to the quasar. Hydrogen in these clouds absorbs photons from the quasar at wavelengths less redshifted than the quasar's L_α emission line.

Because of their starlike appearances and strong emissions, 1 2 3C 48 and 3C 273 were dubbed **quasi-stellar radio sources,** a term soon shortened to **quasars.** Quasars, also called **quasi-stellar objects (QSOs),** look like stars but have energy outputs that are incredibly higher. Furthermore, quasars need not specifically be radio sources; in fact, only 10% are strong radio emitters. Others have strong emissions in all different parts of the electromagnetic spectrum, with most being strongest in the infrared. Nevertheless, the name quasar has stuck, and more than 10,000 have been discovered since Grote Reber discovered the first one, Cyg A.

All quasars have redshifts greater than about 0.06 or 6% the speed of light. The upper limit corresponds to speeds away from Earth of greater than 90% the speed of light. Figure 17-5 shows the spectrum of a quasar whose redshift corresponds to 92% of the speed of light. From the Hubble law, it follows that the distances to these high-redshift quasars are typically in the range of 10 billion to more than 13 billion light-years.

17-2 A quasar emits a huge amount of energy from a small volume

Galaxies are big and bright. A typical large galaxy, like our own Milky Way, contains hundreds of billions of stars and shines with the luminosity of 10 billion Suns. The most gigantic and most luminous galaxies, the giant ellipticals, are only 10 times brighter. But beyond 8 billion light-years from Earth, even the brightest galaxies are too faint to be easily detected. Most ordinary galaxies are too dim to be detected at half that distance. If quasars can easily be seen more than 13 billion light-years away, they must be far more luminous than galaxies. Indeed, a typical quasar is 100 times brighter than our Milky Way.

In the mid-1960s, several astronomers discovered that quasars fluctuate in brightness. Going back over old images, they found that some newly identified quasars had actually been photographed in the past but they were not considered as anything special because they looked like stars. One photograph of 3C 273, for example, dates back to 1887. By carefully examining these old images, as well as recent ones, astronomers could see that *quasars occasionally flare up*. As Figure 17-6 shows for another quasar, 3C 279, energy from prominent outbursts reached Earth around 1937 and 1943. During these outbursts, the luminosity of 3C 279 increased by a factor of at least 25. Because of the enormous distance to this quasar, during those peak periods it must have been emitting at least 10,000 times as much energy as the entire Milky Way.

Suppose that somewhere on a star or other object an event occurred that began making the object more luminous. The shortest time it would take that fluctuation in luminosity to travel across the object is the time it takes light to cross it. After its entire surface has gotten more luminous, it might begin decreasing in luminosity, but again, no faster than the time it takes light to cross it. Therefore, the timescale of fluctuation tells astronomers about the maximum size an object can have. For example, an object with a fluctuation period of 1 day cannot be bigger than 1 light-day across. (On the other hand, that same object could take longer to vary in output if the mechanism causing the change travels more slowly than the speed of light.) Because quasars fluctuate in luminosity, astronomers can place upper limits on their sizes.

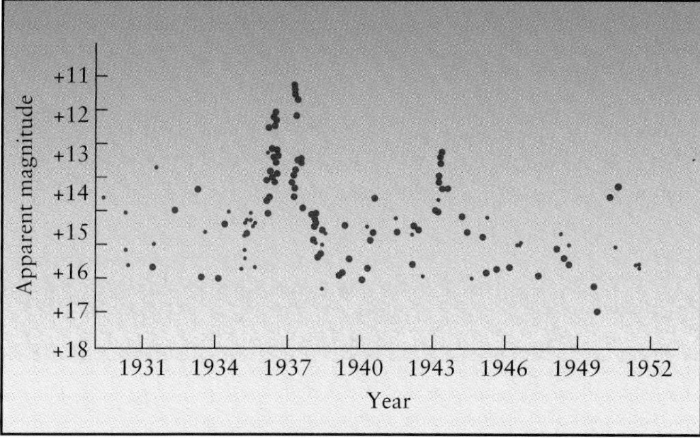

FIGURE 17-6 **The Brightness of 3C 279** This graph shows variations in the brightness of the quasar 3C 279. Note the especially large outburst observed in 1937. The data were obtained by carefully examining old photographic plates in the files of the Harvard College Observatory.

Many quasars vary in brightness over only a few
months, weeks, or days. In fact, X-ray observations reveal 2
large variations in as little as 3 hours. This rapid flickering means that the source of the quasar's energy must be quite small by galactic standards. The energy-emitting region of a typical quasar—the "powerhouse" that blazes with the luminosity of 100 galaxies—is less than 1 light-day in diameter. Something must be producing the luminosity of 100 galaxies from a volume with approximately the same diameter as our solar system. Before considering their energy source, let us examine objects that will turn out to be powered by the same "engine," as they are called by astronomers.

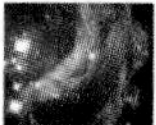

ACTIVE GALAXIES

Quasars are but one kind of very powerful energy sources in space. Astronomers have discovered objects called **active galaxies** (also called **active galactic nuclei,** or **AGN**), with luminosities between those of ordinary galaxies and quasars. Some active galaxies have unusually bright, starlike (that is, a pinpoint as seen from Earth) nuclei; others have strong emission lines in their spectra; still others are highly variable, meaning that their brightnesses change in different parts of the electromagnetic spectrum. Some have jets and beams of radiation emanating from their cores like the emissions observed for 3C 273, and most of these objects are more luminous than ordinary galaxies. We consider the two major classes of active galaxies: radio galaxies and Seyfert galaxies.

17-3 Active galaxies bridge the energy gap between ordinary galaxies and quasars

Some elliptical galaxies, called **radio galaxies** because of their strong radio emission, are like lower energy quasars. While their energy source is very small, like that of a quasar, the radio emissions of these galaxies are spread over very large areas, often hundreds of thousands of light-years across.

Radio galaxies are accompanied by unusual optical features; namely, the galaxy looks as if it is exploding, which it is not. These are called **peculiar galaxies** (denoted pec). "Peculiar" is not the classification of a galaxy shape, like "spiral," "barred spiral," or "elliptical." Any of the Hubble classes of galaxies (see Chapter 16) can be peculiar.

Figure 17-7 is a combined X-ray, optical, and radio image of the peculiar elliptical galaxy NGC 5128 (Centaurus A) in the southern constellation of Centaurus. The dark

FIGURE 17-7 **The Peculiar Galaxy NGC 5128 (Centaurus A)** This extraordinary radio galaxy is located in the constellation Centaurus, 11 million light-years from Earth. At visible wavelengths a dust lane crosses the face of the galaxy. Superimposed on this visible image are a false-color radio image (green) showing that vast quantities of radio radiation pour from matter ejected from the galaxy perpendicular to the dust lane, along with radio emission (rose-colored) along the dust lane, and X-ray emission detected by NASA's *Chandra* X-ray Observatory (blue). The X rays may be from material ejected by the black hole or from the collision of Centaurus A with a smaller galaxy. **Inset:** This X-ray image from the Einstein Observatory shows that NGC 5128 has a bright X-ray nucleus. An X-ray jet protrudes from the nucleus along a direction perpendicular to the galaxy's dust lane. (X-ray: NASA/CXC/M. Karovska et al.; Radio 21-cm: NRAO/VLA/J. Van Gorkom/Schminovich et al.; Radio continuum: NRAO/VLA/J. Conden et al.; Optical: Digitized Sky Survey U.K. Schmidt Image/STScI; inset: X-ray — NASA/CXC/Bristol U./M. Hardcastle; Radio — NRAO/VLA/Bristol U./M. Hardcastle)

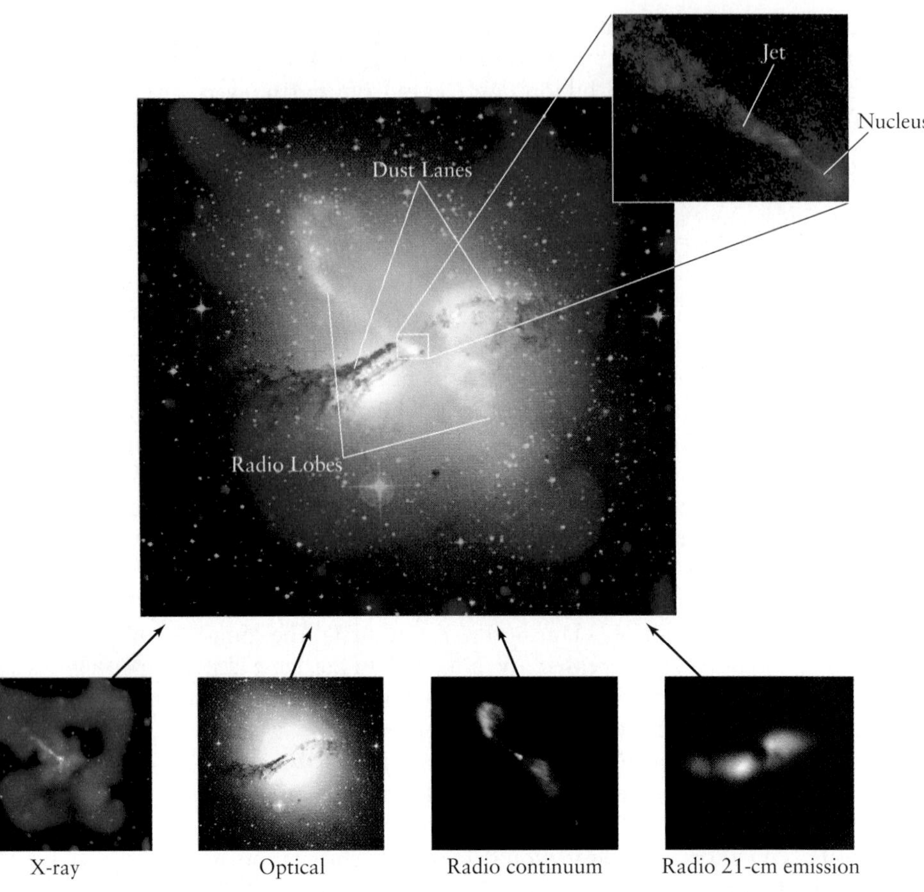

region cutting the elliptical galaxy in two is actually a region of very thick dust through which little light from the galaxy behind it can penetrate. The galaxy is ejecting large volumes of hot X-ray–emitting gas, as well as jets of gas in two **radio lobes** shown on this figure. NGC 5128 is classified as an "E0 (pec)." The E0 denotes a circular elliptical galaxy, while (pec) means that this galaxy has the features of a peculiar galaxy.

Carl Seyfert at the Mount Wilson Observatory discovered the first active galaxies in 1943 while surveying spiral galaxies. Now called **Seyfert galaxies,** these spirals reveal exceptionally bright, starlike nuclei and strong emission lines in their spectra. For example, the rich spectrum of NGC 4151 has many prominent emission lines. Some of these lines are produced by iron atoms with a dozen or more electrons stripped away, indicating that NGC 4151 contains some extremely hot gas. Seyfert galaxies also vary in brightness. For example, at times the magnitude of NGC 4151 changes over the course of a few days. Unlike radio galaxies, they do not have extended radio-emitting regions.

Another example of a Seyfert galaxy is NGC 1566, shown in Figure 17-8. At infrared wavelengths this galaxy shines with the brilliance of 10^{11} Suns. This extraordinary luminosity has

FIGURE 17-8 **The Seyfert Galaxy NGC 1566** This Sc galaxy is a Seyfert galaxy some 50 Mly (16 Mpc) from Earth in the southern constellation Dorado (the Goldfish). The nucleus of this galaxy is a strong source of radiation whose spectrum shows emission lines of highly ionized atoms. (Anglo-Australian Observatory)

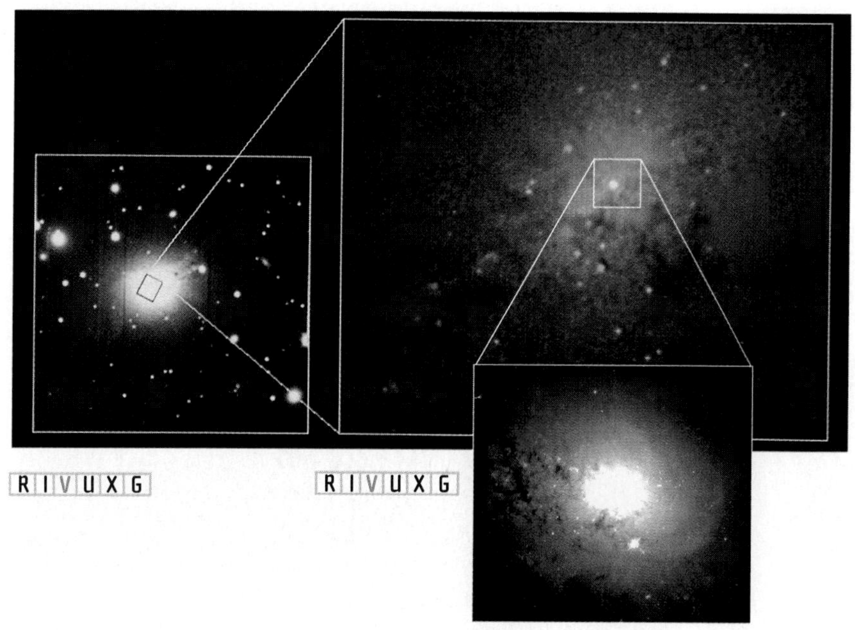

FIGURE 17-9 **The Active Galaxy NGC 1275 (3C 84)** This ground-based image (far left) is of the Seyfert galaxy NGC 1275, the largest and brightest member of the Perseus cluster of galaxies located about 250 million light-years from Earth. Spectroscopic studies indicate that NGC 1275 is actually two galaxies in collision. **Inset (right):** This Hubble Space Telescope image at visible wavelengths shows about 50 small, bright, blue objects that are part of NGC 1275. They are believed to be young globular clusters formed as a result of the collision. The small white spot is the nucleus of the galaxy. **Inset (bottom):** NGC 1275 is also a strong source of X rays and radio waves. This image from the *ROSAT* X-ray Observatory shows that most of the galaxy's X-ray emission comes from its nucleus. (J. Holtzman, NASA; inset bottom: Swinburne University of Technology)

been observed to vary by as much as $7 \times 10^9 L_\odot$ over only a few weeks. In other words, the infrared power output of the nucleus of NGC 1566 rises and falls by an amount nearly equal to the total luminosity of our entire Galaxy.

Many more Seyfert galaxies have been discovered in recent years. Approximately 10% of the most luminous galaxies in the sky are Seyferts. Some of the brightest Seyfert galaxies shine as brightly as faint quasars, which led most astronomers to suspect that the nuclei of Seyfert galaxies are, in fact, low-luminosity quasars.

Some Seyfert galaxies show vestiges of violent explosions in their nuclei. Filaments of gas tens of thousands of light-years long protrude from the nucleus of NGC 1275 in all directions (Figure 17-9). Spectroscopic studies indicate that this gas is being blasted away from the galaxy's nucleus at 3000 km/s. In 1977, Vera Rubin and her colleagues reported observations demonstrating that NGC 1275 actually consists of two colliding galaxies, a spiral and an elliptical. As we saw in Chapter 16, a collision or close encounter between two galaxies can eject matter into intergalactic space. More and more of this intergalactic medium is being discovered. Table 17-1 summarizes the relative power emitted by the Sun, the Milky Way Galaxy, quasars, Seyfert galaxies, and radio galaxies.

17-4 BL Lacertae objects and blazars are brighter than either quasars or active galaxies

While active galaxies emit less energy from the same small regions than do quasars, there are also objects of the same size that emit even more energy than quasars. These are the **BL Lacertae objects.** The name comes from their prototype, BL Lacertae (also called **BL Lac**) in the constellation Lacerta (the Lizard). BL Lac (Figure 17-10) was discovered in 1929, when it was mistaken for a variable star, largely because its brightness varies by a factor of 15 in only a few months. A BL Lac's most intriguing characteristic is an unusually weak set of spectral lines. In other words, its spectrum is not as easily identified as those of stars or interstellar clouds. Nevertheless, the spectra enable astronomers to determine their redshifts and hence their differences.

BL Lac objects are thought to be elliptical galaxies with bright quasars at their centers, much as Seyfert galaxies are spiral galaxies with quasarlike centers. BL Lac objects change intensities—they are variable. Those that have periods of a day or less are also called **blazars.** In 2004, a blazar with 10 billion solar masses located 12.5 billion light-years away was discovered. It is intriguing that this incredibly massive object formed less than a billion years after the universe formed.

TABLE 17-1 Galaxy and Quasar Luminosities

Object	Luminosity (watts)
Sun	4×10^{26}
Milky Way Galaxy	10^{37}
Seyfert galaxies	$10^{36} - 10^{38}$
Radio galaxies	$10^{36} - 10^{38}$
Quasars	$10^{38} - 10^{42}$

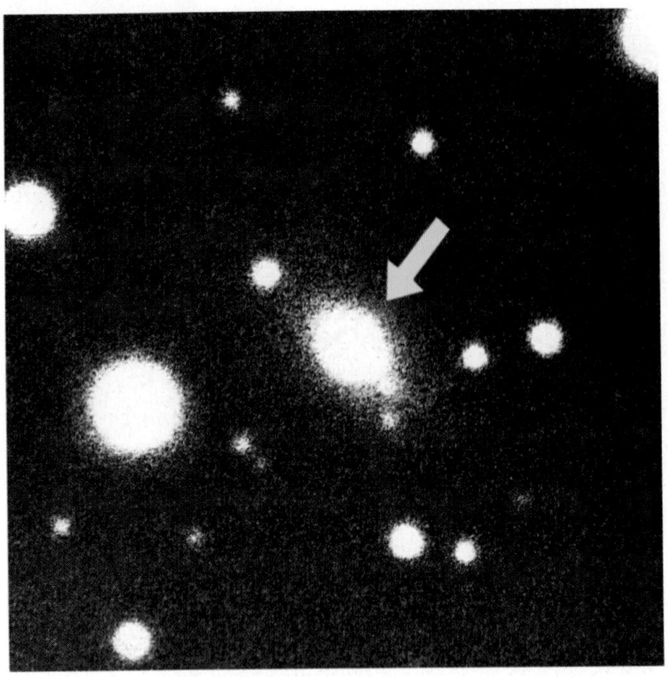

R I V U X G

FIGURE 17-10 **BL Lacertae** This photograph shows fuzz around BL Lacertae (arrow). The redshift of this fuzz indicates that BL Lacertae is about 900 Mly (280 Mpc) from the Earth. BL Lac objects appear to be giant elliptical galaxies with bright quasarlike nuclei, much as Seyfert galaxies are spiral galaxies with quasarlike nuclei. BL Lac objects contain much less gas and dust than Seyfert galaxies. (T. D. Kinman, NOAO/AURA)

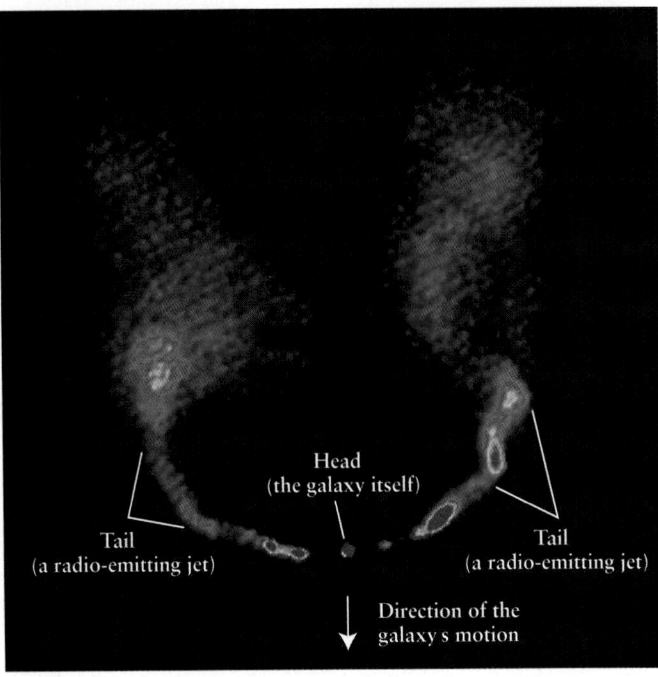

R I V U X G

FIGURE 17-11 **The Head-Tail Source NGC 1265** This active elliptical galaxy is moving at a high speed through the intergalactic medium. Because of this motion, the two tail jets trail the galaxy at its head, giving this radio source a distinctly windswept appearance. (NRAO)

17-5 Active galaxies emit twin jets of gas that span galaxies

Detailed observations of Centaurus A reveal both radio and X-ray jets (see Figure 17-7 inset) originating in the galaxy's nucleus. Particles and energy stream out of the galaxy's nucleus toward the radio lobes at nearly the speed of light. Such galaxies, with two radio lobes, are now called **double radio sources.**

By 1970, radio astronomers had discovered dozens of other double radio sources. An active galaxy, usually resembling a giant elliptical, is often found between the two radio lobes. Double radio sources are among the brightest radio objects in the universe.

All double radio sources appear to have a central engine that ejects charged particles and magnetic fields outward along two oppositely directed jets. After traveling many thousands or even millions of light-years, this material slows down, allowing ejected electrons and magnetic fields to produce the radio radiation that we detect. This type of radio emission, called *synchrotron radiation* (see Section 15-4), occurs whenever energetic electrons move in a spiral path within a magnetic field.

The idea that a double radio source emits jets of particles is supported by the existence of **head-tail sources,** so named because each such source appears to have a region of concentrated radio emission (the head) with two tails of gas streaming from it in opposite directions and then sweeping back. NGC 1265, an active elliptical galaxy in the Perseus cluster of galaxies, shows these properties. NGC 1265 is known to be moving at a high speed (2500 km/s) relative to the Perseus cluster as a whole. In a radio map (Figure 17-11), its radio emission has a distinctly windswept appearance.

Particles ejected in the two jets from NGC 1265 are deflected by the galaxy's passage through the sparse intergalactic medium. The head-tail source leaves behind a trail of particles, like the trail of smoke pouring from a rapidly moving steam train. But what is an active galaxy's engine? What gives rise to its enormous jets?

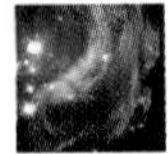

SUPERMASSIVE ENGINES

Active galaxies, quasars, BL Lac objects, and double radio sources are excellent examples of how astronomers use one mechanism to explain an apparently wide variety of objects in space. In 1968, the British astronomer Donald Lynden-Bell suggested how quasars and active galaxies could produce such enormous amounts of energy from such small volumes. He argued that the gravitational field of a supermassive black hole (see Chapter 14) could force this enormous release of energy.

17-6 Supermassive black holes exist at the centers of most galaxies

As we saw in Chapter 14, black holes with a wide range of masses are being discovered throughout the universe. Supermassive black holes (those exceeding a million solar masses or so) have been detected in a rapidly growing number of galaxies, including M31, M32, M87, M104, and the Milky Way.

The Andromeda Galaxy (M31) is the largest, most massive galaxy in the Local Group (see Figure 16-18). In the mid-1980s, several astronomers made careful spectroscopic observations of the core of M31. Using measurements of Doppler shifts, they determined that stars within 50 ly of the galactic core are orbiting the nucleus at exceptionally high speeds, which suggests that a massive object is located at the galaxy's center. Without the gravity of such an object to keep the stars in their high-speed orbits, they would have escaped from the core region long ago. From such observations, astronomers estimate the mass of the central object in M31 to be about 50 million solar masses. That much matter confined to such a small volume strongly suggests the existence of a supermassive black hole.

Located near M31 is a small elliptical galaxy called M32 (Figure 17-12). High-resolution spectroscopy indicates that stars close to the center of M32 are also orbiting this galaxy's nucleus at unusually high speeds, which is explained by the presence of a supermassive black hole there. Furthermore, a picture taken by the Hubble Space Telescope (Figure 17-12 inset) shows that the concentration of stars at the core of M32 is truly remarkable. The number density of stars there is more than 100 million times greater than the density of stars in the Sun's neighborhood. The concentration of stars and their high speeds strongly support the belief that a supermassive black hole exists at the center of M32.

Astronomers have uncovered evidence of supermassive black holes in other, more remote galaxies (eg., M104, Figure 17-13). John Kormendy used a 3.6-m telescope on Mauna Kea to examine the core of M104 spectroscopically. Once again, high-speed gas orbiting the galaxy's nucleus was found. These observations suggest that the center of this galaxy is dominated by a supermassive black hole containing a billion solar masses. Similar spectroscopic observations of the edge-on disk galaxy NGC 3151 also reveal evidence for a billion-solar-mass black hole at its core.

One early candidate galaxy for study by the Hubble Space Telescope was the giant elliptical galaxy M87. Located some 50 million light-years from Earth, M87 is an

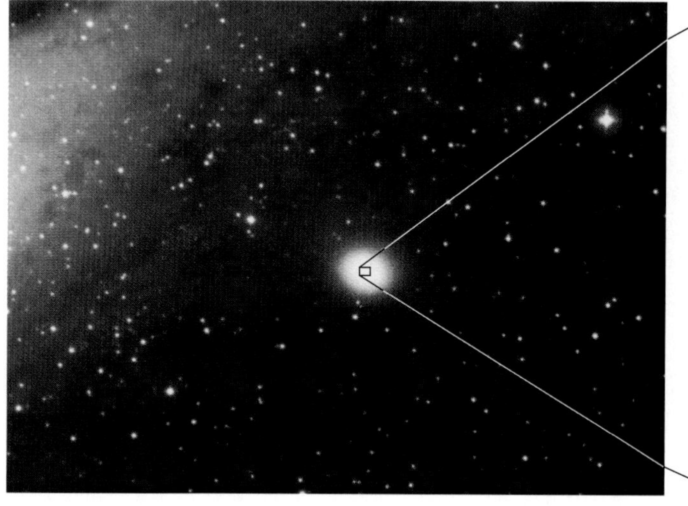

FIGURE 17-12 **The Elliptical Galaxy M32** This small galaxy is a satellite of M31, a portion of which is seen at the left of this wide-angle photograph. Both galaxies are roughly 2.2 million light-years from Earth. **Inset:** This high-resolution image from the Hubble Space Telescope shows the center of M32. Note the concentration of stars at the nucleus of the galaxy. The nucleus is only 175 ly across. (Palomar Observatory; inset: NASA, ESA)

FIGURE 17-13 **The Sombrero Galaxy (M104)** This spiral galaxy in Virgo is nearly edge-on to our Earth-based view. Spectroscopic observations indicate that a billion-solar-mass black hole is located at the galaxy's center. (ESO)

active galaxy that has long been recognized as unusual. In 1918, Heber Curtis at the Lick Observatory reported that the center of M87 has "a curious straight ray . . . apparently connected with the nucleus by a thin line of matter." Figure 17-14a shows M87 and several neighboring galaxies; M87's nucleus and jet are buried in the galaxy's glare. Figure 17-14b is a radio image showing the gas emission from M87, a radio galaxy that is also a powerful source of X rays. X rays emitted by such small sources are a signature of black holes (as discussed in Section 14-7).

The Hubble Space Telescope image of M87 in Figure 17-14c shows an exceptionally bright, starlike nucleus and a surrounding disk of gas with trailing spiral arms. To produce this fiery glow, stars must be packed so tightly at the center of M87 that their density is at least 300 times greater than that normally found at the centers of giant ellipticals. The motions of the stars in this central clustering support the hypothesis that a black hole with a mass of nearly 3 billion Suns resides at the center of M87. Many astronomers now believe that supermassive black holes lie at the hearts of most galaxies.

17-7 Jets of matter ejected from around black holes explain BL Lac objects, quasars, active galaxies, and double radio sources

The gravitational energy associated with supermassive black holes at the centers of galaxies creates huge jets of 3
gas. To see how this works, consider a black hole at the center of a young galaxy filled with gas and dust. Because the centers of galaxies are congested places, the black hole there will capture a massive accretion disk of this interstellar debris (Figure 17-15a). According to Kepler's third law, the material in the inner regions of such a disk orbits the most rapidly. The inner matter therefore constantly interacts with the more slowly moving gases in the outer regions, and the friction between them heats all the gas. As energized gases spiral toward the black hole, they are heated to millions of degrees. This temperature creates great pressure, causing the hot gas to expand and eventually squirt out where the resistance to its expansion is lowest, namely, perpendicular to the accretion disk.

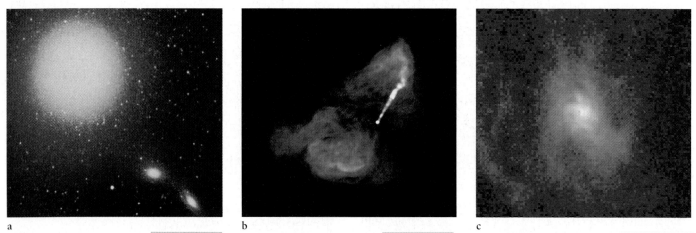

a

b

c

WEB LINK 17.6

FIGURE 17-14 **The Giant Elliptical Galaxy M87** M87 is located near the center of the sprawling, rich Virgo cluster, which is about 50 million light-years from Earth. **(a)** This long exposure shows the extent of M87 and some smaller, neighboring galaxies. The numerous fuzzy spots that surround M87 are globular clusters; each contains about a million stars. **(b)** This radio image shows the intergalactic gas emitted by M87. These gas clouds dwarf the giant galaxy, the bright spot near the center of the picture. **(c)** This Hubble Space Telescope image of M87 shows the gas disk in its nucleus. M87's extraordinarily bright nucleus and the gas jets result from a 3 billion-solar-mass black hole, whose gravity causes huge amounts of gas and an enormous number of stars to crowd around it. (a: © 1987 Anglo-Australian Observatory; b: NASA; c: STScI)

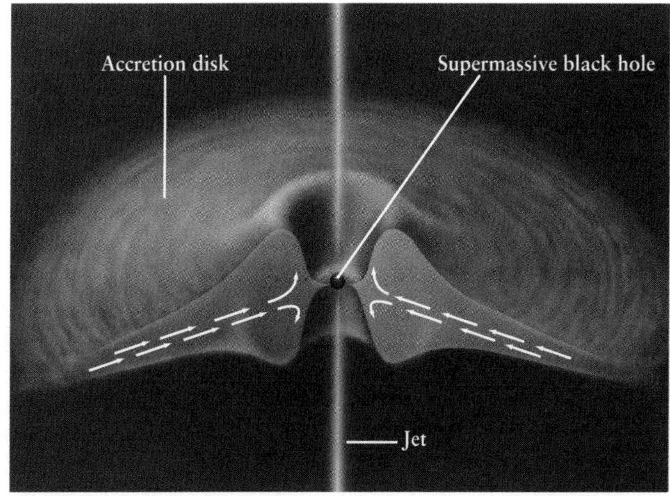

a

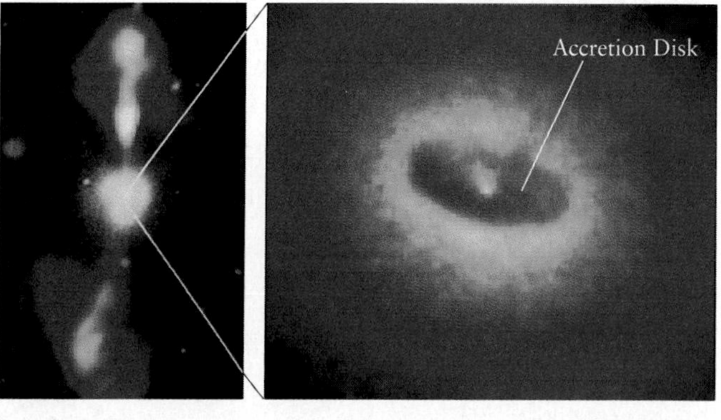

b

R I V U X G

FIGURE 17-15 Supermassive Black Holes as Engines for Galactic Activity **(a)** This drawing shows the accretion disk around a supermassive black hole. The inswirling gas heats and expands. Some of it is eventually expelled perpendicular to the disk in two jets. **(b)** The giant elliptical galaxy NGC 4261 is a double radio source located in the Virgo cluster, about 100 million light-years from Earth. An optical photograph of the galaxy (white) is combined with a radio image (orange and yellow) to show both the visible galaxy, which does not emit much radio energy, and its jets, which do. **Inset:** This Hubble Space Telescope image of the nucleus of NGC 4261 shows a disk of gas and dust about 800 ly (250 pc) in diameter orbiting a supermassive black hole. (b: NASA; inset: ESA)

In 1995, the Hubble Space Telescope took a picture of the giant elliptical galaxy NGC 4261 (Figure 17-15b) that astronomers interpret as showing this set of events. An accretion disk about 800 ly in diameter is seen orbiting a supermassive black hole. The speed of the gas and dust in the disk indicates that it is held in orbit by a 1.2-billion-solar-mass object. The inset in Figure 17-15b is a Hubble image of the galaxy's nucleus. The dark, horizontal oval in the picture is our oblique view of the accretion disk and ground-based radio and optical observations show a double-lobed structure (Figure 17-15b) that is bisected by the oval.

What confines the ejected gas to narrow jets? At first, the gas still falling toward the black hole prevents the jets from spreading. To see why, consider air spraying out of the nozzle of a hose. The stream of air broadens and fans out through a wide angle in the air. This would be the case around these black holes if the infalling gas was not there. If the nozzle is placed in a swimming pool, however, the stream of air does not fan out as much (Figure 17-16).

a

b

FIGURE 17-16 Focusing Jets by Pressure **(a)** If high-speed air emerging from a hose encounters little pressure (from the surrounding air), then the jet will spread out. **(b)** If the fluid encounters high pressure, such as surrounding water, then the jet will maintain its shape as a column longer. (a: Comstock Images/Getty Images; b: Fundamental Photographs, New York)

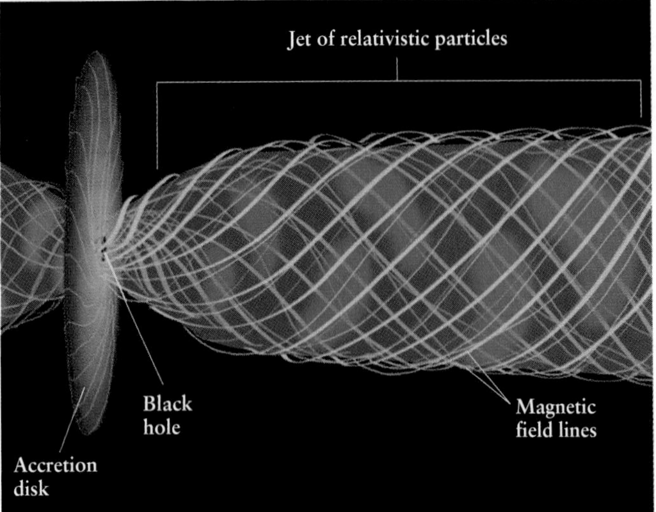

WEB LINK 17.8

◀ FIGURE 17-17 **Focusing Jets from a Supermassive Black Hole** The accretion disk (in red-yellow) around the black hole rotates and creates a magnetic field that is twisted into spring-shaped spirals above and below the disk. Some of the accretion disk's gas falling toward the black hole is overheated and squirted at high speeds into the two tubes created by the magnetic fields. The fields keep the gas traveling directly outward from above and below the disk, thus creating the two jets.

Similarly, as the two jets of hot gas leave the vicinity of the black hole, they must blast their way through the gas that is still crowding inward. Passage through this material causes the jets to become narrow, concentrated beams. Furthermore, any magnetic fields spiraling out from the region of the black hole help keep the ejecta in columns (Figure 17-17). These magnetic fields are generated by the gases orbiting around the black hole.

In 2003, astronomers using the *Chandra* X-ray Observatory observed that gas jetting out from a supermassive black hole in the center of the Perseus cluster of galaxies strikes gas residing in the cluster between galaxies. The *intracluster* gas becomes compressed by the jet and generates concentric rings or shells of sound waves 35,000 ly in wavelength. Put another way, the sound created by the jet is a B-flat that is 57 octaves below middle C.

This "rocket nozzle" explains not only double radio sources but also the jets and beams we see protruding from active galaxies, quasars, and BL Lac objects. We can now tie together all the various objects discussed in this chapter. *The difference between double radio sources, active galaxies, quasars, and BL Lac objects is the angle at which the central engine is viewed.* As Figure 17-18 shows, an observer sees a double radio source when the accretion disk is viewed nearly edge-on, because the jets are nearly in the plane of the sky. At a steeper angle, the observer sees an active galaxy; more steeply still, a quasar is seen. If one of the jets is aimed almost directly at Earth, the galaxy is a BL Lac object.

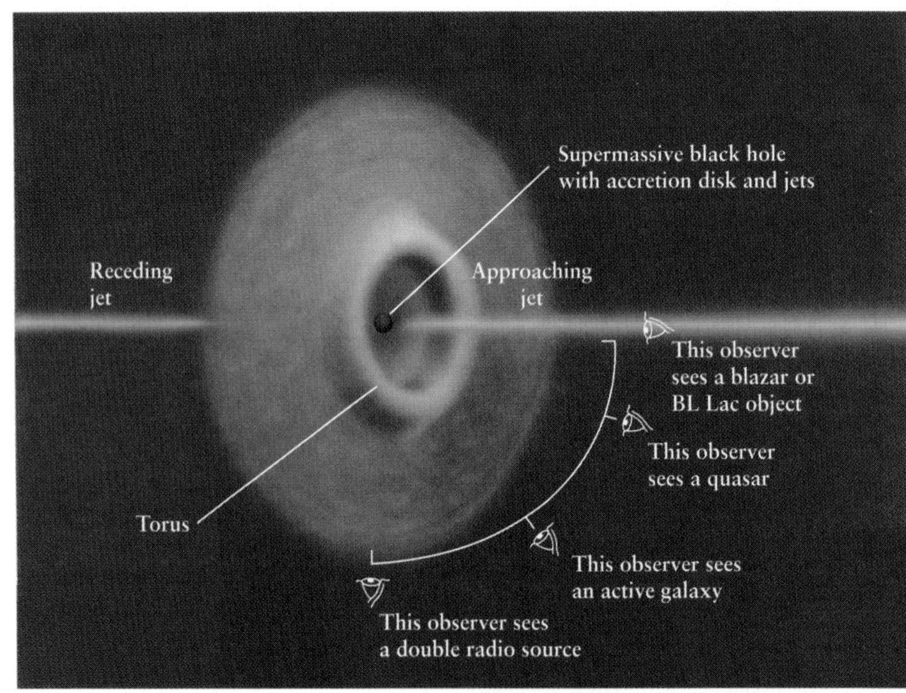

FIGURE 17-18 **The Orientation of the Central Engine and Its Jets** BL Lacertae objects, quasars, double radio sources, and active galaxies appear to be the same type of object viewed from different directions. If one of the jets is aimed almost directly at the Earth, we see a BL Lac object. If the jet is somewhat tilted to our line of sight, we see a quasar. Tilted further and we see an active galaxy. If the jets are nearly perpendicular to our line of sight, we see a double radio source. The central region of the system is shown in Figure 17-17.

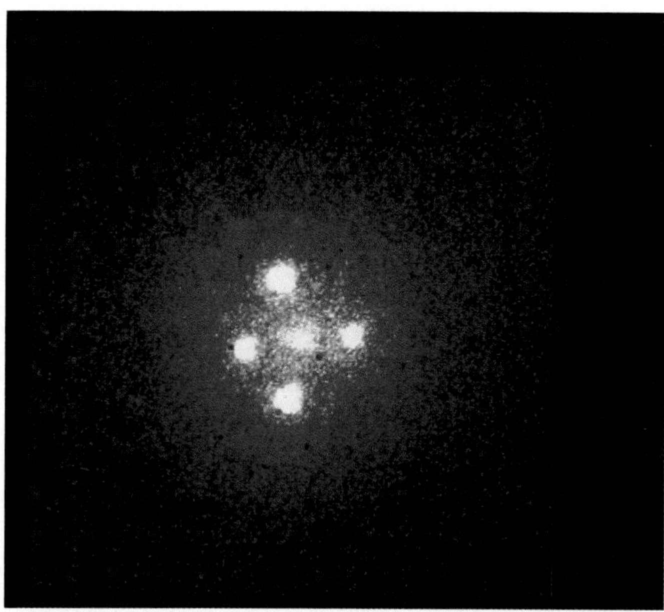

FIGURE 17-19 **Gravitational Lensing of Quasars** This image from the Hubble Space Telescope shows the gravitational lensing of a quasar in the constellation of Pegasus. The quasar, about 8 billion light-years from Earth, is seen as four separate images surrounding a galaxy that is only 400 million light-years away. This pattern is called an Einstein cross. The diffuse image at the center of this Einstein cross is the core of the intervening galaxy. The physical effect that creates these multiple images is the same as seen for galaxies as depicted in Figure 16-34. (NASA/ESO)

INSIGHT INTO SCIENCE

Occam's Razor Revisited It is possible to create *separate* theories for quasars, BL Lac objects, active galaxies, and other double radio sources. Scientists prefer a comprehensive single theory that explains all three of them, because it incorporates common properties where only unconnected phenomena existed before. This search for a simple, more powerful explanation is the central engine of progress in science.

17-8 Gravity focuses light from quasars

The radiation emitted by all quasars is subject to the gravitational lensing described by Einstein's general theory of relativity (see Sections 14-2 and 16-12). Like the light from stars behind the Sun or behind the dark matter in our Galaxy, the light from quasars is deflected as it travels past other galaxies toward us. As we saw in Section 16-12, this gravitational lensing can lead to our receiving several images of objects. Figure 17-19 shows four images of a quasar in a configuration called an **Einstein cross.** If the background object is exactly behind an intervening galaxy, the light is actually focused as a ring, called an **Einstein ring,** rather than as several separate images.

As you know, light travels at a finite speed. For example, light from the Sun takes about 8 minutes to reach the Earth. We see the Sun as it was 8 minutes ago. The farther an object is from us, the longer it takes light and other radiation from it to reach Earth. We see a galaxy one billion light-years away as it was a billion years ago. When we look at quasars, we are seeing objects as they appeared when the universe was younger still. In 2002, using a cluster of galaxies as a gravitational lens (see Section 16-12) to brighten its light, astronomers discovered a quasar at a distance corresponding to when the universe was only 800 million years old.

17-9 Frontiers yet to be discovered

Quasars, active galaxies, BL Lac objects, and double radio sources are rich areas for observational and theoretical research. Why do different quasars emit most strongly in different parts of the spectrum? Are quasars an early phase of most galaxies, including the Milky Way, as many astronomers are coming to believe? What causes quasar variability? Do we have the proper explanations that blazars, quasars, and radio galaxies are caused by supermassive black holes seen from different orientations? How did such supermassive black holes form so quickly in the life of the universe? New discoveries in this realm are being made nearly every week.

Summary of Key Ideas

• The development of radio astronomy in the late 1940s opened the realm of nonvisual astronomy.

Quasars

• A quasar, or quasi-stellar radio source, is an object that looks like a star but has a huge redshift. This redshift corresponds to a distance of billions of light-years from Earth, according to the Hubble law.

• To be seen from Earth, a quasar must be very luminous, typically about 100 times brighter than an ordinary galaxy. Relatively rapid fluctuations in the brightnesses of some quasars indicate that they cannot be much larger than the diameter of our solar system.

Active Galaxies

• An active galaxy is an extremely luminous galaxy that has one or more unusual features: an unusually bright, starlike nucleus; strong emission lines in its spectrum; rapid variations in luminosity; or jets or beams of radiation emanating from its core.

• An active spiral galaxy with a bright, starlike nucleus and strong emission lines in its spectrum is categorized as a Seyfert galaxy.

• BL Lacertae (BL Lac) objects (some of which are called blazars) have bright nuclei whose cores show relatively rapid variations in luminosity.

• Most double radio sources contain an active galaxy located between two characteristic radio lobes. A head-tail radio source shows evidence of jets of high-speed particles emerging from an active galaxy.

• Most quasars are probably very distant active galaxies.

Supermassive Central Engines

• There are huge concentrations of matter at the centers of many galaxies.

• The strong energy emission from quasars, active galaxies, BL Lac objects, and double radio sources is probably produced as matter falls toward a supermassive black hole at the center of each.

• Some matter spiraling in toward a supermassive black hole is squeezed into two oppositely directed beams that carry particles and energy into intergalactic space.

WHAT DID YOU THINK?

1 *What does "quasar" stand for?* Quasi-stellar radio source.

2 *What do quasars look like?* They look like stars, but they emit much more energy than any star.

3 *What powers a quasar?* A quasar is believed to be powered by a supermassive black hole with millions or billions of solar masses at the center of a galaxy.

Key Words

active galaxy (or active galactic nuclei or AGN), 449
blazar, 451
BL Lacertae (BL Lac) object, 451
double radio source, 452
Einstein cross, 457
Einstein ring, 457
head-tail source, 452
peculiar galaxy (pec), 449
quasar (quasi-stellar radio source), 448
quasi-stellar object (QSO), 448
radio galaxy, 449
radio lobe, 450
Seyfert galaxy, 450

Review Questions

1. A double radio source seen along one axis of a jet is called a? **a.** BL Lac object. **b.** quasar. **c.** active galaxy. **d.** double radio source. **e.** pulsar.

2. What two things does the engine of a quasar contain? **a.** a supermassive black hole and a binary companion. **b.** a stellar-mass black hole and a binary companion. **c.** a supermassive black hole and an accretion disk. **d.** a stellar-mass black hole and an accretion disk. **e.** a supergiant star and an accretion disk.

3. Suppose you suspected a certain object in the sky to be a quasar. What sort of observations would you perform to confirm your hypothesis?

4. Explain why astronomers do not use any of the standard candles described in Chapter 16 to determine the distances to quasars.

5. Explain how the rate of variability of a source of light can be used to place an upper limit on the size of the source.

6. What is an active galaxy? List the different kinds of active galaxies. How do they differ from one another?

7. Why do astronomers believe that the energy-producing region of a quasar is very small?

8. How is synchrotron radiation produced?

9. What evidence indicates that quasars are extremely distant active galaxies?

10. What is a double radio source?

11. What is a supermassive black hole? What observational evidence suggests that supermassive black holes might be located at the centers of many galaxies?

12. Why do many astronomers believe that the engine at the center of a quasar is a supermassive black hole surrounded by an accretion disk?

13. How does the orientation of the jets emanating from the center of a galaxy relative to our line of sight relate to the type of active galaxy that we observe?

Advanced Questions

14. In the 1960s, some astronomers suggested that quasars might be compact objects ejected at high speeds from the centers of nearby ordinary galaxies. Why does the absence of blueshifted quasars disprove this hypothesis?

15. When quasars were first discovered, many astronomers were optimistic that these extremely luminous objects could be used to probe distant regions of the universe. For example, it was hoped that quasars would provide high-redshift data from which the Hubble constant could be accurately determined. Why have these hopes not been realized?

Discussion Questions

16. Explore the belief that quasars, double radio sources, and giant elliptical galaxies represent an evolutionary sequence.

17. Some quasars show several sets of absorption lines whose redshifts are less than the redshift of the quasars' emission lines. For example, the quasar PKS 0237-23 has five sets of absorption lines, all with redshifts somewhat less than the redshift of the quasar's emission lines. Propose an explanation for these sets of absorption lines.

What If . . .

18. Earth passed through a jet emitted by a radio galaxy? What would we see and what might happen to the Earth?

19. A jet of gas suddenly appeared all across the night sky? What might that indicate has recently happened?

20. Sirius suddenly became ring-shaped for a few hours and then returned to normal? What might cause such an event?

Web Question

21. To test your understanding of active galaxies, do Interactive Exercise 17-1 on the Web. You can print out your answer if required.

CHAPTER 18

Cosmology

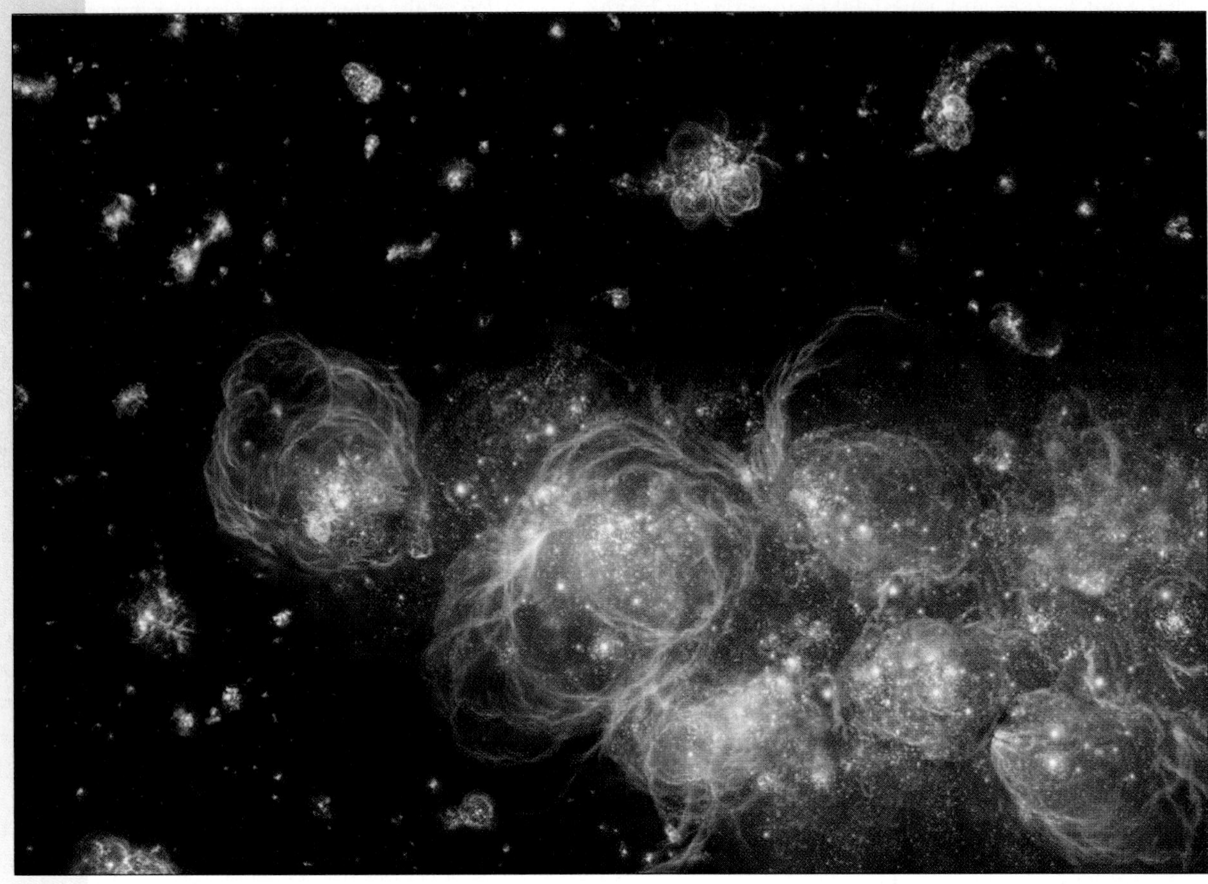

Galaxies Forming by Combining Smaller Units (Adolf Schaller, STScI/NASA/K. Lanzetta, SUNY) R I V U X G

WHAT DO YOU THINK?

1 What does the universe encompass?

2 Is the universe expanding, fixed in size, or contracting?

3 Will the universe last forever?

For millennia, our ancestors speculated about whether the universe has existed forever or whether it began some finite time in the past. Did God create the universe or was the creation of the universe an event that required no divine intervention? This question of the actual beginnings of the universe has yet to be determined. But astronomers and other scientists *have* accumulated a great deal of knowledge about what has happened to the universe *since* it began and what its fate will be. Current scientific theories provide us with insights into how the universe has evolved since a split second after it came into existence, how it is changing today, and what it will do in the distant future. While the overall theory of cosmic evolution is not complete (or astronomers would be out of their jobs), its major elements are consistent with observations.

In this chapter we will examine the distribution of matter on the largest scales in the universe and how this matter is moving. Using this information, we will explore the Big Bang model of how the universe has evolved from its earliest moments, when it was smaller than an atom. We will see how it grew and created matter as we know it, forming and evolving large-scale systems like galaxies and groupings of galaxies. Finally, we will consider the fate of the universe.

In this chapter you will discover

- cosmology, which seeks to explain how the universe began, how it evolves, and its fate
- the best theory we have for the evolution of the universe—the Big Bang
- how astronomers trace the emergence of matter and the formation of galaxies
- how astronomers explain the overall structure of the universe
- our understanding of the fate of the universe

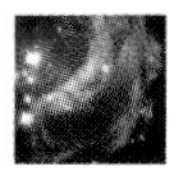

THE BIG BANG

1 2 The **universe** consists of all the matter, energy, and spacetime that we can ever detect or that will ever be able to affect us. (We use this definition because there may be matter and energy in other dimensions or matter and energy that are moving away from us so quickly that their influence will never reach us. These ideas are still very preliminary and theoretical, and for this book we will continue to deal just with concepts for which science has experimental evidence.) So far in this text we have explored matter on size scales from atoms to superclusters of galaxies. We also learned in Section 16-10 that the superclusters are all moving away from each other, implying that the universe is expanding. In this chapter we take the observational evidence from the rest of the book and use it to explore the Big Bang theory of cosmology. **Cosmology** is the study of the large-scale structure and evolution of the universe.

18-1 General relativity predicts an expanding (or contracting) universe

Modern cosmology almost began in 1915, when Einstein published his general theory of relativity. To his surprise and dismay, the equations predicted that the universe is not static: They predicted that it should be either expanding (which it is) or contracting. But Einstein was not ready for what the equations were telling him.

The prediction of a changing universe flew in the face of the widely accepted belief in an infinite, static universe, a concept promoted by Isaac Newton more than two centuries earlier. Newton believed that each star is fixed in place and held there under the influence of a uniform gravitational pull from every part of the cosmos. If the stars were not uniformly distributed, he argued, one region would have more mass than another. The denser region's gravity would then attract other stars, causing them to clump together further. Because he did not observe this clumping, Newton concluded that the mass of the universe must be distributed uniformly over an infinite space. At the time the general theory of relativity was published, the existence of galaxies and of clusters and superclusters of galaxies had not yet been established and the relative motion of the superclusters of galaxies had not been observed.

The apparently static universe and the prevailing philosophy that the universe had existed forever made Einstein doubt the implications of his own theory, so he missed the opportunity to propose that we live in a changing universe. Instead, he adjusted his elegant equations to yield a static cosmos. He did this by adding a repulsive (outward-pushing) term called the **cosmological constant** to his equations so that gravity's normal attractive force would be counterbalanced and the universe would be static. After observations revealed that the universe is expanding, Einstein said that adding the cosmological constant was the biggest blunder of his career.

Although the value of the cosmological constant Einstein inserted was wrong, the concept of such a constant may be correct. Observations since 1997 indicate that the universe is not just expanding, but actually *accelerating* outward.

This means that there must be an outward pressure that more than counteracts the effects of normal gravitation, which is trying to slow the universe's expansion. One of the two current theories that can explain that acceleration is the presence of a cosmological constant that creates the outward pressure. We discuss these two theories further in Section 18-15.

18-2 The expansion of the universe creates a Dopplerlike redshift

Edwin Hubble is credited with discovering that we live in an **expanding universe** (see Section 16-10). The redshift of clusters and superclusters of galaxies that Hubble found moving away from us appears to come from the Doppler effect, but it actually does not. Recall (see Section 4-7) that the normal Doppler shift is caused by an object moving toward or away from us through *fixed* spacetime. However, using Einstein's general theory of relativity, one finds that spacetime, the fabric of the universe, is *not fixed,* but is actually expanding. This expansion is what is carrying the superclusters away from each other, and in many cases it carries clusters in a given supercluster away from each other. (The gravitational force holding objects like planets and stars and entire galaxies together is so strong that these bodies and systems are not being pulled apart. The expansion of space just acts to increase the separation between large ensembles of galaxies.)

The redshift that Hubble observed, caused by the expansion of the universe, is properly called the **cosmological redshift.** In other words, the photons we observe from galaxies in other superclusters are all redshifted because space is expanding. To understand why, consider a wave drawn on a rubber band (Figure 18-1). The wave has an unstretched wavelength (see Section 3-1) of λ_0. As the rubber band stretches ($\lambda > \lambda_0$), the wavelength increases. Imagine a photon coming toward us from a distant galaxy. As the photon travels through space, space is expanding and, like stretching the rubber band, this expansion stretches the photon's wavelength. When the photon reaches our eyes, we see a drawn-out wavelength—the photon is redshifted. The longer the photon's journey, the more its wavelength is stretched by the expansion of the universe. Therefore, astronomers observe larger redshifts in photons from relatively distant galaxies than in photons from relatively nearby galaxies.

As we saw in Section 16-10, Hubble discovered the linear relationship between the distances to galaxies in other superclusters and the redshifts of those galaxies' spectral lines. For example, a galaxy twice as far from Earth as another galaxy has twice the cosmological redshift of the closer one. The normal Doppler shift has the same relationship. As a result of the two effects being described by the same equations, the normal Doppler shift and the cosmological redshift predict the same relationship between redshift and motion, except for the most distant galaxies and quasars where effects of relativity must be taken into account. Working with relatively nearby galaxies, Hubble was fully justified in using the Doppler equation to calculate the recession of galaxies.

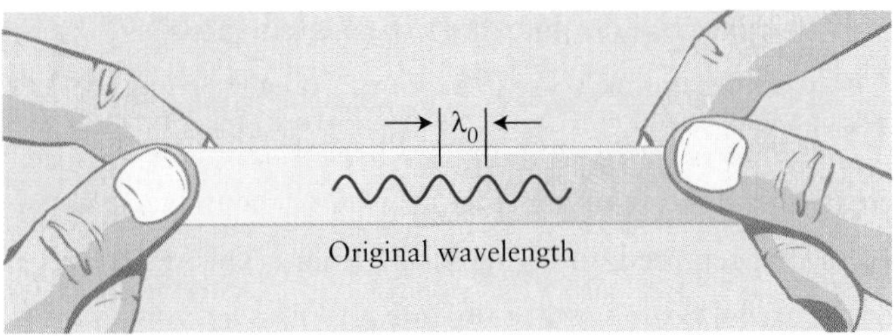

a A wave drawn on a rubber band ...

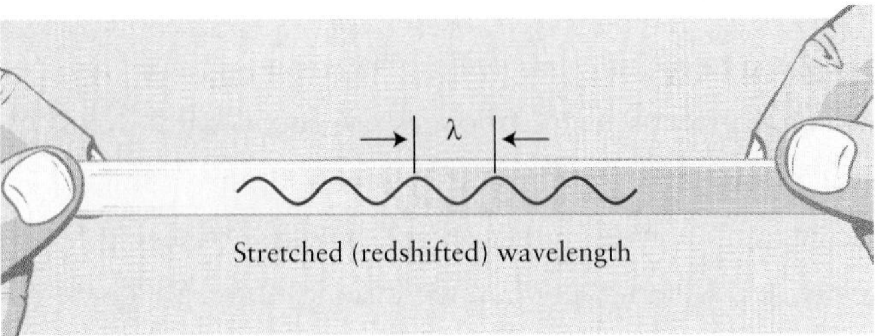

b ... increases in wavelength as the rubber band is stretched.

FIGURE 18-1 **Cosmological Redshift** Just as the waves drawn on this rubber band are stretched along with the rubber band, so too are the wavelengths of photons stretched as the universe expands.

18-3 The Hubble constant is related to the age of the universe

Hubble's law gives us a way to estimate the age of the universe. Imagine watching a movie of two superclusters receding from each other. If we then run the film backward, time runs in reverse, and we observe the superclusters approaching each other. The time it would take for them to collide is the time since the Big Bang. *Assuming the universe has always expanded at a constant rate,* we can use a simple equation to estimate the age of the universe:

$$\text{Time since the Big Bang} = \frac{\text{separation distance}}{\text{recessional velocity}}$$

Recall that Hubble found the relationship

$$\text{Recessional velocity} = H_0 \times \text{separation distance}$$

which we can rewrite as

$$H_0 = \frac{\text{recessional velocity}}{\text{separation distance}}$$

Comparing the first and last equations (see An Astronomer's Toolbox 18-1), we see that Hubble's constant is the reciprocal of the time since the universe began. Using a Hubble constant of 71 km/s/Mpc, we find:

$$1/H_0 = 1/71 \text{ km/s/Mpc} = 13.8 \text{ billion years}$$

The universe has *not* been expanding at a constant rate, and we do not yet know the complete history of its motion. Nevertheless, calculations reveal that the age of the universe given by $1/H_0$ is within 2 billion years of our best determination of the age of the universe, and we use 13.8 billion years as the age of the universe in what follows.

18-4 Remnants of the initial expansion, the Big Bang, have been detected

In 1927, the Belgian astrophysicist and Catholic priest Georges Lemaître proposed that, as a consequence of the equations of Einstein's general theory of relativity, the universe is expanding. Using Hubble's subsequent observations of an expanding universe, Lemaître logically concluded that the superclusters must have expanded from a much smaller volume. He proposed that the universe began as an extraordinarily dense *primordial atom* of energy. This idea led to the concept of the **Big Bang,** when all of spacetime, matter, and energy were created. Indeed, Lemaître is considered the "father of the Big Bang."

Shortly after World War II, the astrophysicists Ralph Alpher and George Gamow proposed that seconds after the Big Bang, the universe was an incredibly hot blackbody—far in excess of 10^{12} K. It must therefore have been filled with high-energy, electromagnetic radiation. As the universe expanded, the cosmological redshift stretched the wavelengths of this radiation so that most of the photons left over from this event are now radio waves. Recall from Section 3-3 that photon energies decrease with increasing wavelength. Because the early photons are being redshifted, the energy they have is decreasing. This energy is measured as heat, so the universe is cooling off as it expands.

WEB LINK 18.4

Calculations indicated that if the universe began as the Big Bang, the remnants of that energy should still fill all space today, giving the universe a background temperature only a few Kelvin above absolute zero. This radiation is called the **cosmic microwave background** because the peak of its blackbody spectrum should lie in the microwave part of the radio spectrum. In the early 1960s, Robert Dicke, P. J. E. Peebles, David Wilkinson, and their colleagues at Princeton University began designing a microwave radio telescope to detect it.

Meanwhile, just a few miles from Princeton, two physicists had already detected the cosmic background radiation. Arno Penzias and Robert Wilson of Bell Laboratories were working on a new horn antenna designed to relay telephone calls to Earth-orbiting communications satellites

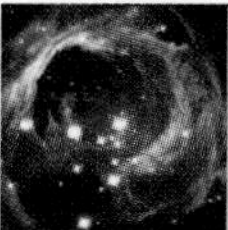

AN ASTRONOMER'S TOOLBOX 18-1
H_0 and the Age of the Universe

WEB LINK 18.5

The Hubble constant may not look like a measure of time, but it is. All we need to see this is a little practice with unit conversion. The trick is to remove all distance units from H_0 and then invert it. Recall that we are using H_0 = 71 km/s/Mpc, which has different units of distance in the numerator and denominator. As we saw in Chapter 16, this mix of units makes it easy to determine how fast a galaxy is receding once its distance is known.

Example: A galaxy 10 Mpc away is receding at 71 km/s/Mpc × 10 Mpc = 710 km/s. We know that 1 pc = 3.09×10^{13} km, and so, 1 Mpc = 3.09×10^{19} km. Because there are 3.156×10^7 seconds in a year, we get the conversion factors we need:

$$H_0 = \frac{1}{71 \text{ km/s/Mpc}} \times (3.09 \times 10^{19} \text{ km/Mpc}) \times \frac{1}{3.156 \times 10^7 \text{ s/yr}} = 13.8 \times 10^9 \text{ yr}$$

If the universe had always been expanding at the same speed, then $1/H_0$ would be the age of the universe. However, for the first several billion years of its existence, the expansion of the universe was slowing down, and since then it has been speeding up.

Decades of observations by many astronomers have confirmed that all distant galaxies are moving away from the Earth. Not only that, *the Hubble law relating velocities and distances is the same in all directions.* The Hubble constant, H_0, is the same no matter where you aim your telescope. The fact that the recession rates are the same in all directions is the condition called isotropy. The universe is isotropic.

The general expansion away from us suggests that the Earth is at the center of the universe. After all, where else could we be if all the distant galaxies are moving away from us? In fact, the answer is that we could be practically anywhere in the universe! To understand why, see Guided Discovery: The Expanding Universe.

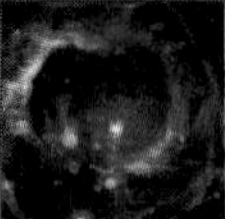

GUIDED DISCOVERY
The Expanding Universe

To understand the universe, scientists build models, mathematical representations of the world about us. To develop such pictures, it often helps to begin with a simple analogy. Because the expansion of the universe is hard to visualize, imagine for a moment the batter for a chocolate chip cake floating in an oven in the International Space Station. As the cake bakes, it expands and the chocolate chips move farther apart. They do not get larger, nor do they move through the batter. Each chocolate chip remains at rest in its own little bit of cake and is carried along as the cake spreads out.

Now, think of each chocolate chip as a supercluster of galaxies and the batter between the chocolate chips as the rest of spacetime. As the universe expands, the distance between widely separated superclusters of galaxies grows larger and larger. The expansion of the universe is the expansion of spacetime.

Suppose you were on one particular chocolate chip, say the chocolate chip labeled 3 at the left in the figure. You would see chocolate chips 2 and 4 moving away from you with equal velocities. Every 10 minutes chocolate chips 2 and 4 get 1 centimeter farther away as the cake expands, as shown on the right. In that same time interval, chocolate chips 1 and 5 move twice as far from you as chocolate chips 2 and 4. To go twice as far away in the same time, chocolate chips 1 and 5 must be moving twice as fast as the closer chocolate chips 2 and 4. This is exactly what Hubble's law of cosmological expansion (recessional velocity = $H_0 \times$ distance) says: *Double the distance and you double the velocity.*

To see why we need not occupy the center of the universe, put yourself on chocolate chip 4. You see chocolate chips 3 and 5 moving away at the same rate—in fact, they recede at the same rate as you saw chocolate chips 2 and 4 move away when you were on chocolate chip 3. From chocolate chip 4, chocolate chips 2 and 6 are moving away twice as fast as chocolate chips 3 and 5. From any chocolate chip, you will see all the other chocolate chips moving away. Similarly, no matter where we are in the universe, we see all the other superclusters receding from us.

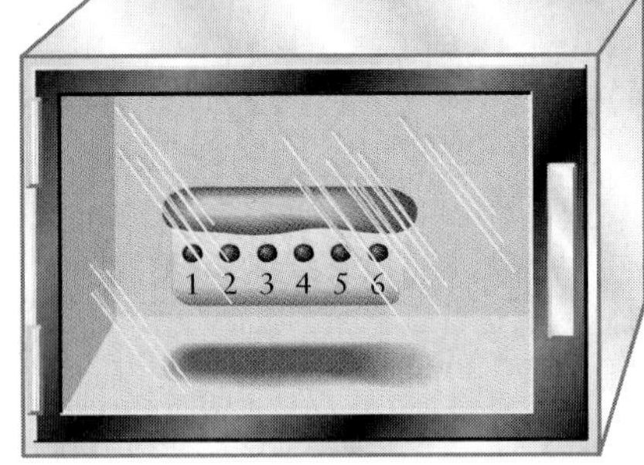

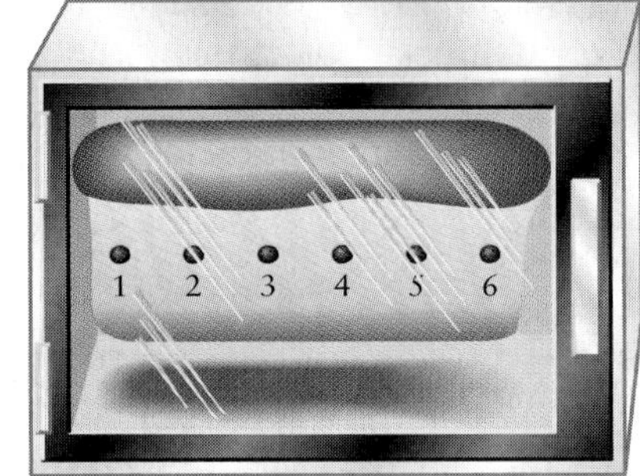

The Expanding Chocolate Chip Cake Analogy The expanding universe can be compared to a chocolate chip cake baking and expanding in the International Space Station. Just as all the chocolate chips move apart as the cake rises, all the superclusters of galaxies recede from each other as the universe expands.

(Figure 18-2). Penzias and Wilson were mystified. No matter where in the sky they pointed their antenna, they detected a faint background noise. All efforts to eliminate this background noise, even the careful removal of static noise–generating pigeon droppings from the antenna, failed.

Thanks to a colleague, they learned of the then-theoretical cosmic microwave background and the work of Dicke, Peebles, and Wilkinson in trying to locate it. Communicating with the Princeton astronomers, Penzias and Wilson presented their finding and thus were able to claim

▶ FIGURE 18-2 **The Bell Labs Horn Antenna** This Bell Laboratories horn antenna at Holmdel, New Jersey, was used by Arno Penzias (on the right) and Robert Wilson in 1965 to detect the cosmic microwave background. (Lucent Technologies, Bell Laboratories)

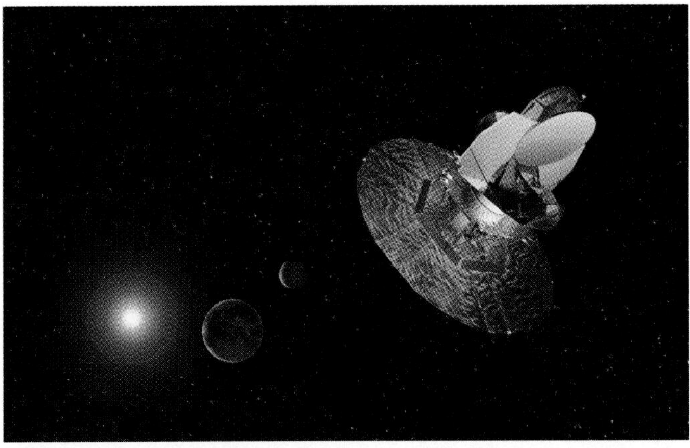

a

b

FIGURE 18-3 **In Search of Primordial Photons** (a) The *Wilkinson Microwave Anisotropy Probe* (*WMAP*) satellite, launched in 2001, improved upon the measurements of the spectrum and angular distribution of the cosmic microwave background taken by the *COBE* satellite. (b) The balloon-carried telescope BOOMERANG orbited above Antarctica for 10 days collecting data used to resolve the cosmic microwave background with 10 times higher resolution than that of *COBE*. All these experiments found local temperature variations across the sky, but no overall deviation from a perfect blackbody spectrum. (a: NASA/WMAP Science Team b: The BOOMERANG Group, University of California, Santa Barbara)

the first detection—their annoying noise was, in fact, the remnant energy of the Big Bang. The cosmic microwave background, also commonly called the 3-degree background radiation, is required by the Big Bang theory but is neither required nor predicted by the steady-state theory, nor any other competing cosmological theory. Detection of the cosmic microwave background is a principal reason why the Big Bang is accepted by astronomers as the correct cosmological theory.

INSIGHT INTO SCIENCE

Consider Plausible Alternatives Until They Are Shown to Be Inconsistent with Observations Sir Fred Hoyle, who aptly named the Big Bang in 1949, was one of the very few astrophysicists who never believed that it occurred. He, Herman Bondi, and Thomas Gold proposed the *steady-state theory* in which the universe has existed forever, is continuously expanding, and, as it expands, new matter is created that takes the place of the receding matter. Where this new matter might come from is unknown, but the laws of physics do not necessarily exclude its creation. Today, astronomers accept the Big Bang theory rather than the steady-state theory because cosmic microwave background observations are consistent with the former but not with the latter.

Precise measurements of the cosmic microwave background were first made by the *Cosmic Background Explorer* (*COBE*) satellite, which operated between 1989 and 1994, and has since been measured more precisely by other equipment, such as the *Wilkinson Microwave Anisotropy Probe* (*WMAP*) and the balloon-carried BOOMERANG (Figure 18-3). The data taken by *COBE* and shown in Figure 18-4 reveal that, as predicted, this ancient radiation has the spectrum of a blackbody with a temperature of approximately 2.73 K.

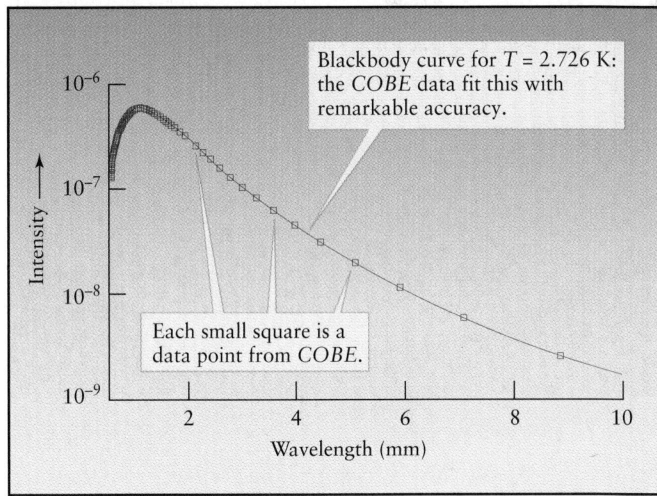

FIGURE 18-4 **The Spectrum of the Cosmic Microwave Background** The little squares on this graph are *COBE*'s measurements of the brightness of the cosmic microwave background plotted against wavelength. To a remarkably high degree of accuracy, the data fall along a blackbody curve for 2.73 K. The peak of the curve is at a wavelength of 1.1 mm, in accordance with Wien's law. (Courtesy of E. Cheng; NASA COBE Science Team)

18-5 The universe has two symmetries—isotropy and homogeneity

Observations show that the cosmic microwave background is almost perfectly isotropic: Its intensity is nearly the same in every direction in the sky. Figure 18-5 shows a map of the microwave sky. It is very, very slightly warmer than average in the direction of the constellation Leo and slightly cooler in the opposite direction, toward Aquarius. The tiny temperature variation depicted in Figure 18-5 results from the Earth's overall motion through the cosmos. Here is how that works: As the Milky Way moves toward Leo, our motion through the cosmic microwave background radiation creates a normal Doppler shift that causes photons from that direction to appear to have shorter wavelengths—radiation from the direction of Leo is blueshifted, while radiation from the opposite direction, Aquarius, is redshifted since we are moving away from that region (Figure 18-6). A photon's wavelength determines its energy (see Section 3-3). The energy of the photons then determines the temperature we measure for that part of the sky. Seeing slightly shorter-wavelength photons in the direction of Leo means that space in that direction appears slightly warmer than space in other directions.

The observed temperature differences shown in Figure 18-5 mean that the solar system is moving toward Leo at a speed of 390 km/s. Taking into account the known velocity of the Sun around the center of our Galaxy, astronomers calculate that the entire Milky Way Galaxy is moving relative to the cosmic microwave background at 600 km/s—some 1.3 million miles per hour—in the general direction of the Centaurus cluster. The gravitational attraction of four nearby clusters of galaxies, most significantly the *Great Attractor*, and an enormous supercluster called the *Shapley concentration* are believed to be pulling us in that direction. Subtracting the motion of the Earth, Sun, Milky Way, and the Local Cluster with respect to the microwave background, the average intensity of radiation is found to be the same in all directions to about 1 part in 100,000.

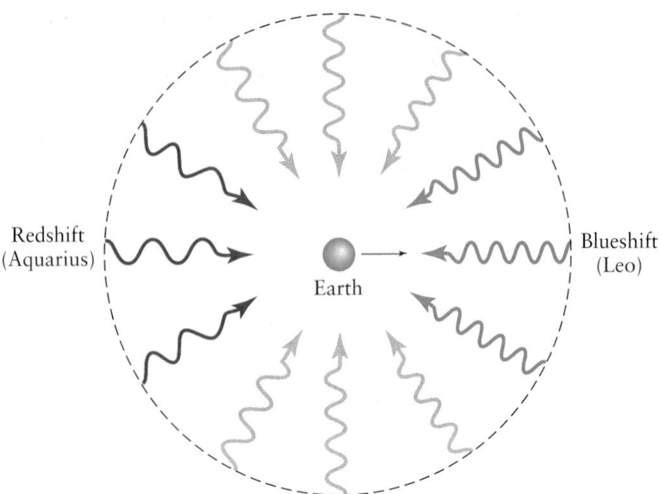

FIGURE 18-6 **Our Motion Through the Microwave Background** Because of the Doppler effect, we detect shorter wavelengths in the microwave background and a higher temperature of radiation in that part of the sky toward which we are moving. This part of the sky is the area shown in blue in Figure 18-5. In the opposite part of the sky, shown in red in Figure 18-5, the microwave radiation has longer wavelengths and a cooler temperature.

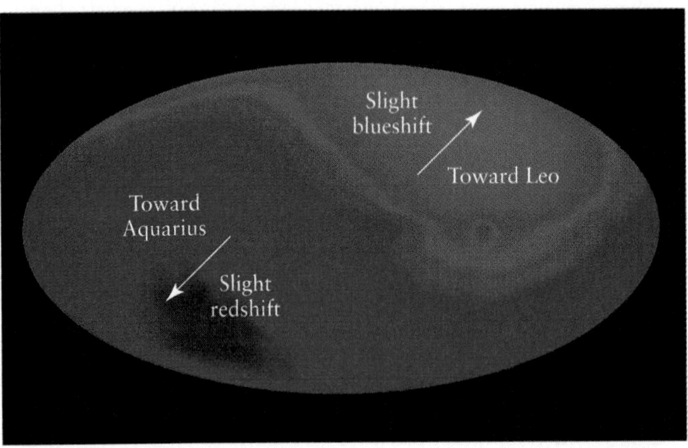

FIGURE 18-5 **The Microwave Sky** This map of the microwave sky was produced from data taken by instruments on board *COBE*. The galactic center is in the middle of the map, and the plane of the Milky Way runs horizontally across the map. Color indicates Doppler shift and temperature: Blue is blueshift and warmer, red is redshift and cooler. The Doppler shift across the sky is caused by the Earth's motion through the microwave background. The resulting variation in background temperature is quite small, only 0.0033 K above the average radiation temperature of 2.726 K (which we have rounded to 2.73 K in the text). (NASA)

While the existence of the cosmic background radiation provides compelling support for the Big Bang theory, its near-isotropy was, until recently, a major problem. The problem was that the original Big Bang theory did not require space to be isotropic—after accounting for the Earth's motion through the cosmos, the temperature in one direction *could* have been many degrees different than the temperature in any other direction, but it is not.

WEB LINK 18.7 **Isotropy** is not limited to the blackbody radiation observed throughout the universe. It is also found on sufficiently large scales when exploring the numbers of galaxies in different directions. Astronomers have counted the numbers of galaxies at different distances from us and in a variety of directions. The number of galaxies stays roughly constant, allowing, as we saw in Section 16-6, for the distribution of most galaxies on the boundaries of sponge-shaped regions or bubbles throughout the universe. Because of this *uniformity with distance,* we say that the universe is also homogeneous. Isotropy and

homogeneity must be explained by any viable theory of cosmology such as the Big Bang.

The need to explain isotropy and homogeneity, among other things, has led to numerous refinements in the Big Bang theory. As a result, this theory now provides an accurate scenario for the evolution of the universe from a tiny fraction of a second after it formed and onward.

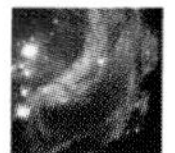

A BRIEF HISTORY OF SPACETIME, MATTER, ENERGY, AND EVERYTHING

The composition of the universe during its first few minutes of existence was profoundly different from what it is today. To understand why, we need to expand briefly on the nature of the four known physical forces: gravity, electromagnetism, and the strong and weak nuclear forces.

18-6 All physical forces in nature were unified at first

Only the effects of electromagnetism and gravity can extend over infinite distances. The electromagnetic force holds electrons in orbit about the nuclei in atoms. But over large volumes of space, the net effects of electromagnetism cancel out because a negative electric charge exists for every positive charge and a south magnetic pole exists for every north magnetic pole.

While gravity is the *weakest* of all four forces, it is the only force whose effect is both infinite *and* attractive for all ordinary matter. This is why gravity determines the evolutionary behavior of stars, the orbits of planets, the existence of galaxies, and many other large-scale phenomena. Indeed, the attractive force of gravity dominates the universe at astronomical distances. (Despite all this, we will learn shortly that under certain conditions, gravity can be repulsive.)

In contrast to gravity and electromagnetism, the strong and the weak nuclear forces both act over extremely short ranges. Their influences extend only over atomic nuclei, distances less than about 10^{-15} m. The **strong nuclear force** holds protons and neutrons together. Without this force, nuclei would disintegrate because of the electromagnetic repulsion of their positively charged protons. Thus, the strong nuclear force overpowers the electromagnetic force inside nuclei.

The **weak nuclear force** is at work in certain kinds of radioactive decay, such as the transformation of a neutron into a proton. Protons and neutrons are composed of more basic particles called **quarks.** A proton is composed of two "up" quarks and one "down" quark, whereas a neutron is made of two "down" quarks and one "up" quark. The weak nuclear force is at work whenever a quark changes from one variety to another. For example, in one kind of radioactive decay, a neutron transforms into a proton as a result of one of the neutron's "down" quarks changing into an "up" quark.

To examine details of the physical forces, scientists use particle accelerators that hurl high-speed particles at targets. High speed is equivalent to high energy and therefore high temperature. In such experiments, physicists find that *at sufficiently high temperatures, the different forces begin to behave the same way.* For example, at extremely high temperatures the electromagnetic force, which works over all distances under "normal" circumstances, and the weak force, which only works over very short distances under the same "normal" circumstances, become identical. They are no longer separate forces but one that is called the *electroweak force.* Experiments at the CERN accelerator in Europe in the 1980s slammed particles together with such violence (that is, created such high temperatures) that the electromagnetic force and the weak nuclear force were indistinguishable. Thus was confirmed the electroweak force.

The equations describing the electromagnetic, weak, and strong forces indicate that they all have the same strength when particles have energies or, equivalently, temperatures much greater than will ever be possible to create in the laboratory on Earth. But recall from the earlier discussion that when the universe was young, it was very hot—hot enough for this unifying of the forces to occur. Detailed calculations reveal that in the first 10^{-35} seconds of existence, the temperature of the universe was more than high enough for the electromagnetic as well as the weak and strong nuclear forces to have the same strength.

The relationship between gravitation and the other three forces is still uncertain. Scientists have not yet been able to include its effects, although it is generally expected that the next generation of theories that explain elementary particles will include it. Efforts to create such a theory are under way under the general name of **superstring** theories. Assuming gravity will be joined to the other forces, we include its expected effects here. Figure 18-7 shows that during the beginning moments of the universe (from 0 to 10^{-43} s, called the **Planck era**), the four forces are believed to have been unified—only one physical force existed in nature. However, this remains to be proven because the equations we use to describe nature break down when applied to this era.

INSIGHT INTO SCIENCE

Different Realms of Science Are Working Together Because physicists who study elementary particles have no hope of building accelerators powerful enough to test the unified theory of the fundamental forces described here, they have joined with astrophysicists and astronomers to use the entire universe as a large laboratory. If there is a way to see what happened during the earliest moments of the universe, they could use those observations to test the theory. Amazingly, such observations can be, and are being, made.

FIGURE 18-7 **Unification of the Four Forces** The strength of the four physical forces depends on the speed with which particles interact or, equivalently, their temperature. As shown in this figure, the higher the temperature of the universe, the more the forces resemble each other. Also included here are the ages of the universe at which the various forces were equal.

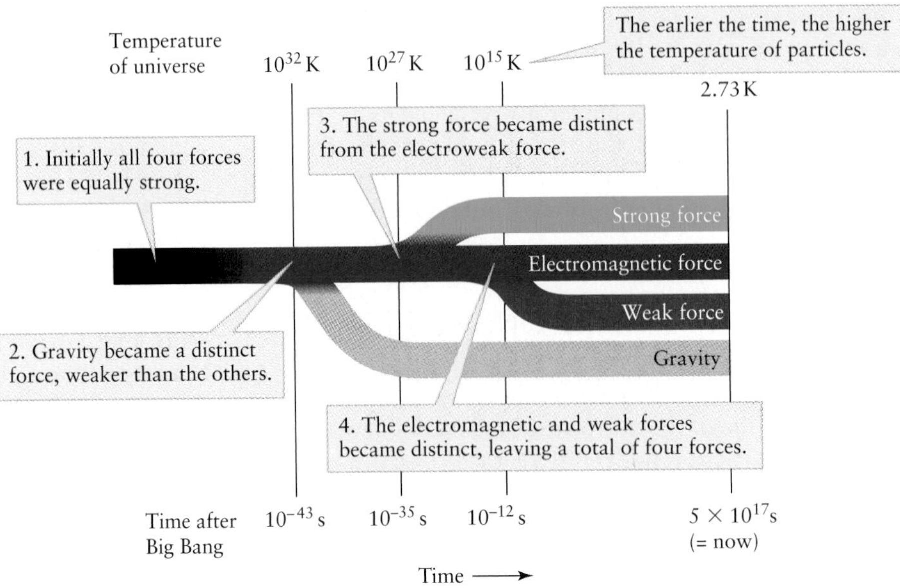

18-7 Equations explain the evolution of the universe even before matter and energy as we know it existed

The earliest time at which our current equations can explain the behavior of the universe is at the end of the Planck era, about 10^{-43} seconds. This is called the **Planck time** in honor of Max Planck, who helped derive some of the fundamental quantum physics that was used to calculate this time. It was at the Planck time that gravity is believed to have become a separate force, leaving electromagnetism and the weak and strong nuclear forces still united as one, called the GUT (for **grand unified theory**) force. The Planck time corresponded to when the currently observable universe was much smaller than the size of an atom today. Keep in mind that the universe was not (and is not) expanding into preexisting spacetime. Rather, the expansion was (and is still) creating spacetime. Therefore, you could not stick a thermometer into the early universe from "outside" to take its temperature. However, if you could measure the total amount of energy in the universe at that early time, it would have had a temperature of about 10^{32} K. There were no separate particles and electromagnetic radiation then, just gravitation and the GUT force, so matter and energy as we know them did not exist.

Since disorder is more physically likely to occur than order, astronomers theorize that the matter in the early universe was distributed chaotically. Some places had higher temperatures than other places. *If* the universe had continued expanding only at the rate calculated for the Planck time, today we should be able to see different parts of the universe with significantly different temperatures. As we saw in Figure 18-5, the temperatures of every region of the universe are the same to within a tiny fraction of 1° K. This is called the **isotropy problem** or **horizon problem.** Realizing how uniform the universe's temperature is, astronomers have hypothesized that very early in its existence, the universe underwent a short period of dramatically higher expansion, called the **inflationary epoch,** that smoothed out the universe we can see.

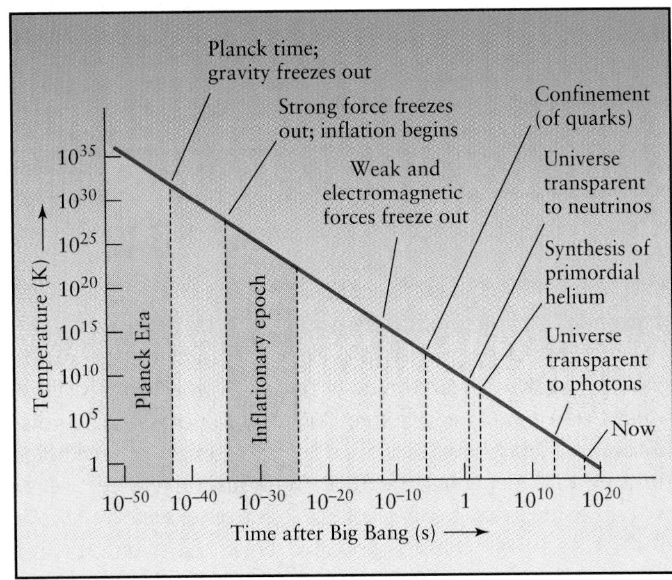

FIGURE 18-8 **The Early History of the Universe** Current theory holds that, as the universe cooled, the four forces separated from their initial unified state. The inflationary epoch lasted from 10^{-35} s to 10^{-24} s after the Big Bang. Quarks became confined together, thereby creating neutrons and protons 10^{-6} seconds after the Big Bang. The universe became transparent to light (that is, photons decoupled from matter) when the universe was about 1.6×10^{13} s (379,000 years) old. The physics of the Planck era is presently unknown.

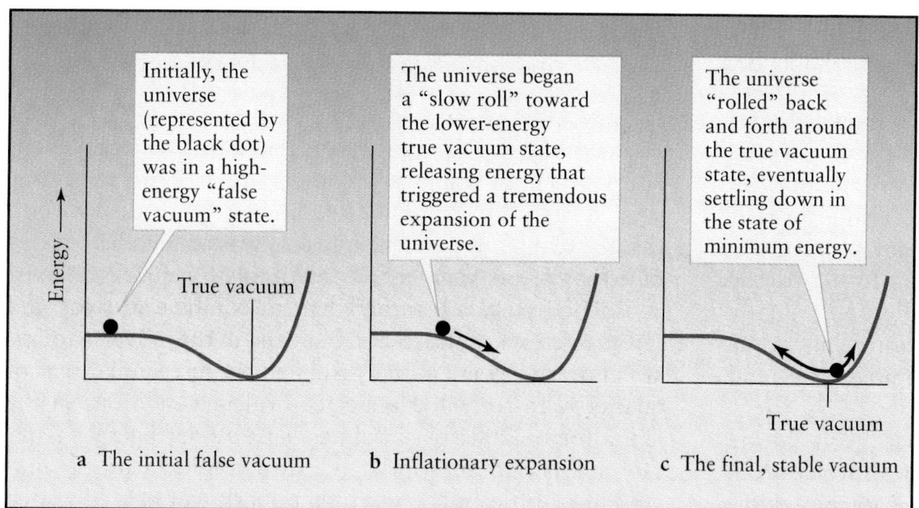

FIGURE 18-9 **The Cause of Inflation** (a) The universe formed in an unstable energy state that (b) began to transition to a stable configuration. (c) This transition provided the energy that caused inflation.

For the first 10^{-35} seconds of the universe's existence (the blink of an eye takes about one-tenth of a second), calculations indicate that the universe expanded at the initial rate and that its temperature dropped to 10^{27} K (Figure 18-8). (For comparison, room temperature is about 300 K.) By then, the energy in each volume of the universe had decreased to the point at which the strong nuclear force could no longer remain unified with the electromagnetic and weak nuclear forces. At that instant, the strong nuclear force emerged (see Figure 18-7), distinct from the other two forces, which remained together as the electroweak force, introduced earlier.

Many astronomers hypothesize that before the strong force decoupled from the electroweak force, the universe was in an unstable state called a *false vacuum*. It was unstable in the same sense as you would be if you woke up one day to find yourself on a greased, 10-centimeter shelf on the face of a cliff thousands of meters above the cliff bottom. The slightest move and you fall. Analogous to the transition you make by falling off the shelf, the young universe is believed to suddenly have made a transition (Figure 18-9) into a stable *true vacuum* state. The universe's transition to the true vacuum state caused it to momentarily accelerate.

As the universe began moving toward the true vacuum state, it was already expanding. The transition caused the expansion rate to increase dramatically. From about 10^{-35} until about 10^{-33} seconds after the Big Bang, the universe ballooned outward about 10^{50} times larger (Figure 18-10). The currently observable universe expanded to about the size of a soccer ball. This huge rate of expansion of the universe has

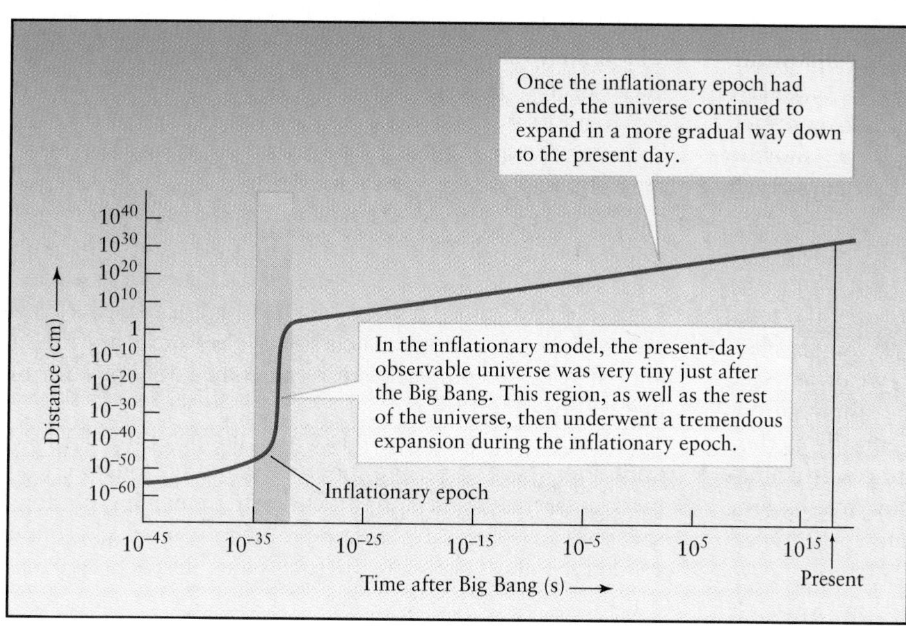

FIGURE 18-10 **The Observable Universe Before and After Inflation** Shortly after the Big Bang, the universe expanded by a factor of about 10^{50} due to inflation. This growth in the size of the presently observable universe occurred in a very brief time (shaded interval).

come to be called **inflation.** The end of the inflationary epoch heralded the first formation of elementary particles: quarks, electrons, photons, and their antiparticles.

Every part of the Big Bang became so much larger during inflation that only a tiny fraction of it (we do not know what that fraction is) became the universe that we can observe. The energy and matter distributed in that initially very small patch of spacetime that became our universe was very, but not perfectly, uniform, so the entire visible universe created from it became equally uniform as a result of undergoing inflation. This is the process by which the universe became homogeneous and isotropic.

However, quantum theory played a vital role during the inflationary epoch that prevented the universe from becoming perfectly homogeneous and isotropic. Normally, quantum fluctuations (the spontaneous formation of pairs of virtual particles; see Section 4-6) occur on the size scales of pairs of particles, as we have seen in Section 14-11 with the creation and annihilation of virtual particles throughout all space. The quantum fluctuations on microscopic scales that were occurring at the beginning of the inflationary period were stretched by inflation to scales that spanned clusters of galaxies and beyond! Imagine a sponge compressed very tightly so that the spaces and fiber are initially pressed very close together. Released, the sponge expands, with the relationship between the spaces and the fibers remaining intact, while each part of the sponge grows to a much larger scale. The quantum fluctuations created a spongelike universe, with some regions nearly empty and others relatively full of matter. Because of this internal structure, the universe can only be considered homogeneous on size scales greater than about 1 billion light-years, so that the voids and supercluster regions average out.

Indeed, the theory predicts that tiny quantum fluctuations ultimately became the seeds for the formation of the large-scale structures of the universe that exist today. Since inflation is part of a scientific theory, its occurrence leads to predictable observations. As we will see shortly, telescopes have indeed seen its predicted distributions of matter.

When inflation ended, the universe resumed the relatively leisurely expansion initiated by the Big Bang (see Figure 18-10). Arriving at the time 10^{-12}s, when the temperature of the universe had dropped to 10^{15} K, the electromagnetism force separated from the weak nuclear force (see Figures 18-7 and 18-8). From that moment on, all four forces interacted with particles essentially as they do today. At that time, the universe's matter was primarily the building blocks of protons and neutrons, namely quarks, antiquarks, and the other elementary particles: neutrinos, antineutrinos, electrons, and positrons.

18-8 During the first second, most of the matter and antimatter in the universe annihilated each other

The next significant event is calculated to have occurred at 10^{-6} s, when the universe's temperature was 10^{13} K. Just as it is energetically favorable for boiling water to turn into gas, the temperature of the universe was so high before this time that it was energetically favorable for quarks to exist as isolated particles, rather than to combine in twos and threes to create the particles that exist in the universe today. We know they were initially isolated because similar seas of quarks were recreated under equivalent conditions in the laboratory in 2001 by smashing dense atoms together. After 10^{-6} s, a period appropriately called **confinement,** the universe was sufficiently cool so that quarks could finally stick together to form individual protons, neutrons, and their antiparticles. Figure 18-8 summarizes the connections between particle physics and cosmology.

Equations show that early in the first second, the universe was so hot that virtually all photons were gamma rays possessing incredibly high energies. These energies were high enough so that photons could create matter and antimatter, according to Einstein's equation $E = mc^2$. This is the reverse process to what we saw in Section 10-7, where we discussed how matter is converted into energy in nuclear fusion. To make a particle of mass m, you need an amount of energy E at least as great as mc^2, where c is the speed of light.

The creation of matter from energy is routinely observed in laboratory experiments involving high-energy gamma rays. A highly energetic photon colliding with an atomic nucleus can create a pair of particles, as sketched in Figure 18-11a. This process, called **pair production,** always creates one ordinary particle and its so-called antiparticle. For example, if one of the particles is an electron, the other is an antielectron or positron (recall from Section 14-11 that similar pairs of particles are also believed to be produced near a black hole).

A positron has the same mass as an electron but the opposite charge. When an electron and a positron meet, they annihilate each other and their energy is converted into photons (Figure 18-11b). Theory predicts that at times earlier than 1 second in the evolution of the universe, photon collisions created vast numbers of pairs of particles. Consequently, shortly after the Big Bang, all space was chock full of protons, neutrons, electrons, and their antiparticles, all immersed in a phenomenally hot bath of high-energy photons.

As the universe expanded, the temperature declined until after the first second, photons could no longer create pairs of particles and antiparticles. The remaining photons still exist and they are the cosmic microwave background we detect today. Thereafter, particles and antiparticles often

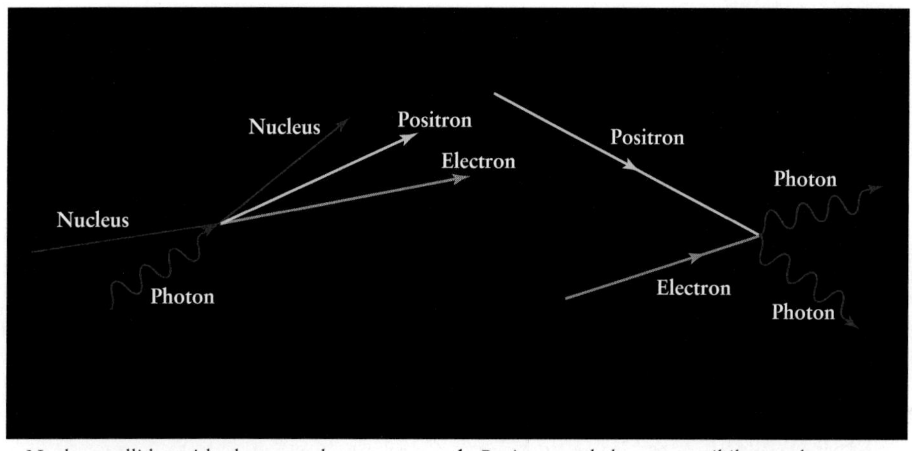

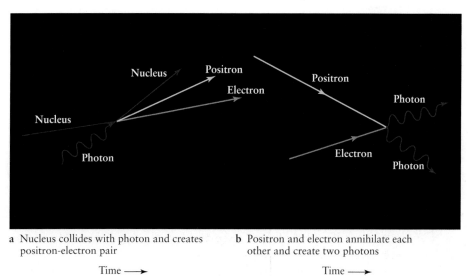

a Nucleus collides with photon and creates positron-electron pair

Time ⟶

b Positron and electron annihilate each other and create two photons

Time ⟶

FIGURE 18-11 **Pair Production and Annihilation** (a) A particle and an antiparticle can be created when a high-energy photon collides with a nucleus. (b) Conversely, a particle and an antiparticle can annihilate each other and emit energy in the form of gamma rays. (The processes are more complex and are only summarized in these drawings.)

collided, annihilating each other and converting their masses back into high-energy photons, and nothing could replenish the dwindling supply of particles and antiparticles.

If all particles had annihilated their antiparticles in the early universe, no matter would be left at all and we would not be here. And yet, here we are and astronomers observe very few antiparticles in the universe. Since particles and antiparticles are always created or destroyed in pairs, why aren't they found in equal numbers in the universe today?

Physicists theorize that the symmetry or equality between the number of particles and the number of antiparticles was broken—a *symmetry breaking*. This means that the number of particles was actually slightly greater than the number of antiparticles. For every billion antiprotons, perhaps a billion plus one protons formed; for every billion antielectrons, a billion plus one electrons formed. As far as theorists can tell, virtually all the antiparticles created in the first second of time annihilated normal particles shortly thereafter. The remaining particles in the universe were the slight excess of normal matter created as a result of this symmetry breaking.

18-9 The universe changed from being controlled by radiation to being controlled by matter

As the universe expanded and cooled further, the remaining protons and neutrons began colliding and fusing together. For the most part, they were quickly separated again by the gamma rays in which they were bathed. However, during the first 3 minutes, enough fusions occurred that were not subsequently split apart to create most of the helium and at least 10% of the trace element lithium that exists today. Lithium is the element with three protons. Hydrogen, helium, and lithium are the three lowest-mass elements. After a few minutes, the universe was too cool to allow fusion to create more massive elements, such as carbon and oxygen. All elements other than the three lowest-mass ones formed later as a result of stellar evolution, as discussed in Chapters 10, 12, and 13.

INSIGHT INTO SCIENCE

Share Information While the evolution of the universe is a truly large-scale event, the activity in the first 3 minutes was nevertheless intimately connected to the creation and properties of elementary particles, the smallest objects in the universe. Physicists who study elementary particles and astrophysicists who study cosmology now regularly hold meetings to compare notes and connect their fields.

Calculations indicate that for tens of thousands of years, photons dominated the behavior of the universe in two important ways. One effect of the photons was to contribute to the gravitational attraction in the universe. This contradicts intuition, because photons are massless and mass is what we normally associate with causing gravitational attraction. Nevertheless, the gravitational effect of photons is a tested prediction of general relativity. For about 30,000 years after the Big Bang, the energy density, and hence the gravitational effect, of photons was greater than the gravitational effect of matter. This was called the **radiation-dominated universe.**

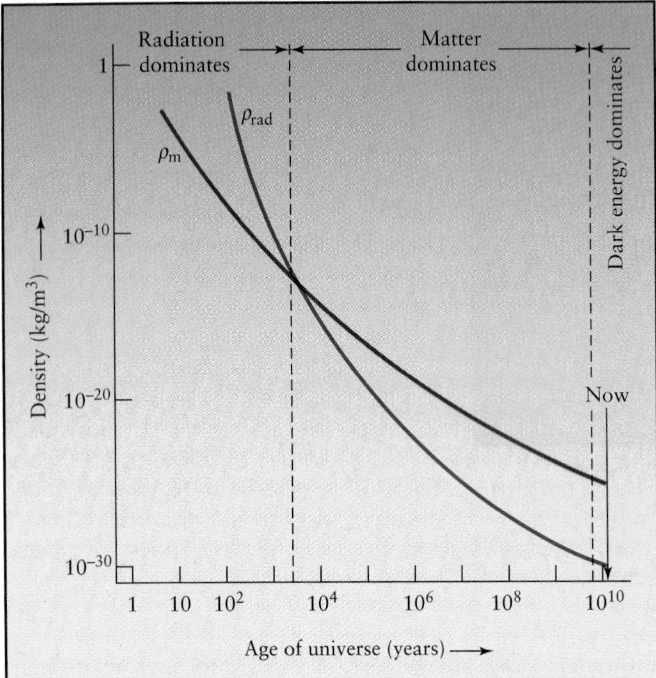

FIGURE 18-12 **The Evolution of Density** For approximately 30,000 years after the Big Bang, the gravitational effects from photons (ρ_{rad}, shown in red) exceeded the effects of the matter density (ρ_m, shown in blue). This early period is said to have been radiation-dominated. Later, however, continued expansion of the universe caused ρ_{rad} to become less than ρ_m, at which time the universe became matter-dominated.

The major effect of the radiation-dominated universe was that photons prevented matter from collapsing together and forming stars and galaxies. As the universe expanded, the gravitational effects of both photons and matter decreased as they became more spread out. The effect of photons fell faster than that of matter (Figure 18-12), however, because as the universe expanded, the photons lost energy due to the cosmological redshift, but particles with mass did not lose energy similarly. After about 30,000 years, matter came to dominate the gravitational behavior of the universe, a situation that has persisted ever since. Thus developed the **matter-dominated universe** in which we live today. It was only after the universe became matter-dominated that gas could respond to gravitational interactions and begin clumping where the force of gravity was sufficiently strong.

The second way that photons dominated the early universe occurred for about the first 379,000 years. During this time, photons prevented ions and electrons from connecting to form neutral atoms. During that epoch, the universe was completely filled with a shimmering expanse of high-energy photons colliding vigorously with protons and electrons. This state of matter, called a *plasma*, prevents any photons from traveling through it. In other words, in a plasma you would see a glow from local photons, but you would not be able to see photons from even a few meters or yards away (Figure 18-13a). The term **primordial fireball** describes the universe during this time.

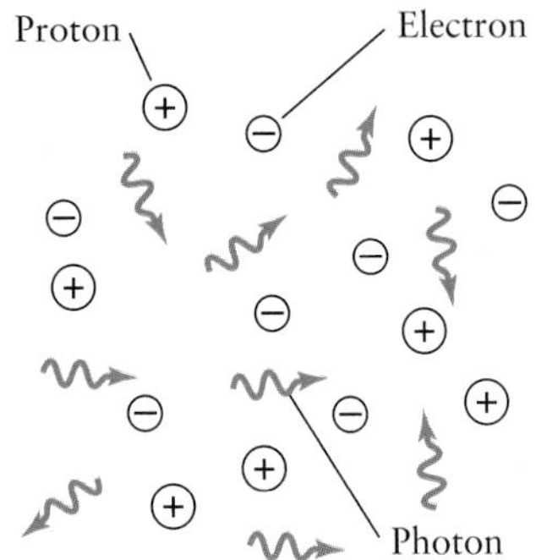

a Before recombination:
- Temperatures were so high that electrons and protons could not combine to form hydrogen atoms.
- The universe was opaque: Photons underwent frequent collisions with electrons.
- Matter and radiation were at the same temperature.

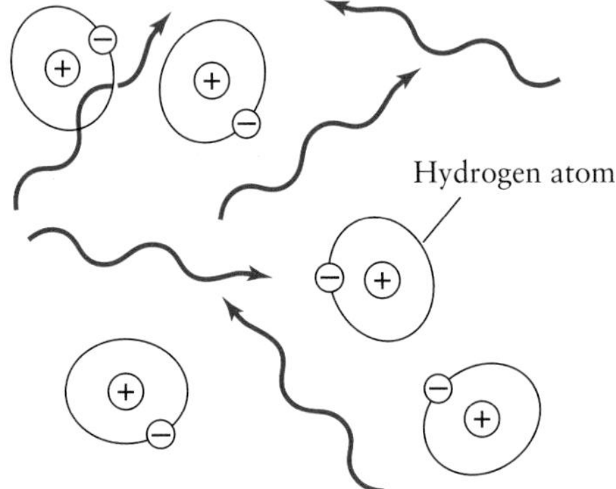

b After recombination:
- Temperatures became low enough for hydrogen atoms to form.
- The universe became transparent: Collisions between photons and atoms became infrequent.
- Matter and radiation were no longer at the same temperature.

FIGURE 18-13 **The Era of Recombination** (a) Before recombination, the energies of photons in the cosmic background were high enough to prevent protons and electrons from forming hydrogen atoms. (b) As soon as the energy of the background radiation became too low to ionize hydrogen, neutral atoms came into existence.

By around 379,000 years, the universe was so large that the cosmological redshift had stretched photon wavelengths (see Figure 18-1) and therefore decreased photons' energies to the point where they could no longer ionize hydrogen. This is called **decoupling,** meaning that the massive particles and the photons were effectively disconnected. Electrons began orbiting nuclei, thereby creating neutral atoms (Figure 18-13b). The space between nuclei became so large that most primordial photons now passed by these particles, rather than colliding with them. Without these collisions to scatter photons, the universe ceased being hazy. It became clear, as it is today, enabling electromagnetic radiation to travel long distances unimpeded and, important from our point of view, enabling us to see far into the universe.

This dramatic period, when the universe transformed from being opaque to transparent, took about 100,000 years to complete and is referred to as the **era of recombination,** referring to nuclei and electrons combining. There are many instances, both in space and in the laboratory, where electrons are stripped from atoms. This is the process of ionization mentioned in Section 4-5. After ionization occurs, electrons often recombine with nuclei to recreate neutral atoms. However, the gases created early in the evolution of the universe were never neutral, so the "re" in recombination is misleading. Because the universe was opaque in its first 379,000 years, we cannot see any further into the past than the era of recombination with telescopes that detect electromagnetic radiation of any wavelength.

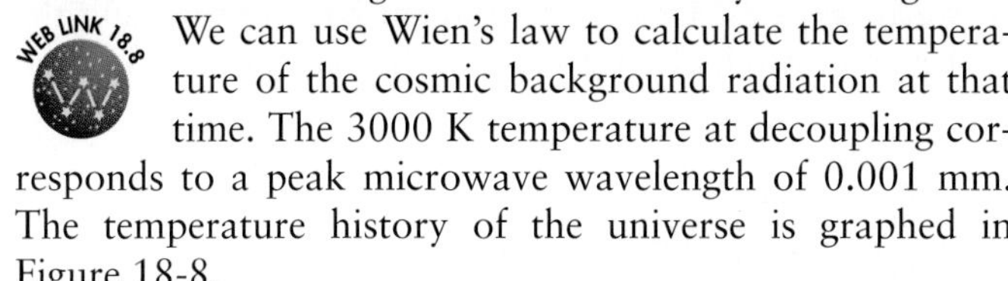

We can use Wien's law to calculate the temperature of the cosmic background radiation at that time. The 3000 K temperature at decoupling corresponds to a peak microwave wavelength of 0.001 mm. The temperature history of the universe is graphed in Figure 18-8.

Ever since the primordial photons formed, the expansion of the universe has been increasing their wavelengths, so today we see these same photons as the cosmic microwave background. The universe is about 1000 times larger than it was at decoupling. Although the universe is dominated by matter today, the number of photons left over from the Big Bang is still immense. From the physics of blackbody radiation, astronomers calculate that there are now 400 million cosmic background microwave photons in every cubic meter of space. In contrast, if all the visible matter in the universe were uniformly spread throughout space, there would be roughly one hydrogen atom in every 3 cubic meters of space. In other words, photons continue to outnumber atoms by more than a billion to one. The microwave background contains the most ancient photons we expect ever to be able to observe. This microwave background is a ghostly relic of the former dazzling splendor of the universe.

18-10 Inflation explains why the universe is isotropic and homogeneous

We can now summarize the issues raised earlier in the chapter of how the Big Bang cosmology explains the universe's isotropy and homogeneity. Recall that there is no reason to expect that the entire universe created at the Big Bang was everywhere the same. It plausibly had different temperatures, pressures, and densities in different places. Recall that light takes time to travel to Earth, and the farther we look into space, the further back in time we are seeing. We can only see a finite distance into it, namely back to the time of decoupling, about 13.8 billion light-years away. Farther than this (that is, earlier in the life of the universe), the universe was opaque and so we cannot see into it.

At a distance of 13.8 billion light-years, we see objects as they were just being formed. Therefore, all that we can see from the Earth is contained in an enormous sphere (Figure 18-14). The boundary of this sphere is called the **cosmic light horizon.** This sphere, with a radius of 13.8 billion light-years (assuming a 13.8-billion-year-old universe), is our entire observable universe.

Light from all the luminous objects in this volume has had time to reach the Earth, but light from objects beyond the cosmic light horizon has not yet arrived here, so we are unaware of objects in these regions. The key to homogeneity and isotropy is that inflation expanded the universe so much and so swiftly that our cosmic light horizon only encompasses what was initially a tiny volume of the whole universe, a volume so small that it was initially isotropic and homogeneous.

18-11 Galaxies formed from huge clouds of primordial gas

The Big Bang theory explains the existence of spacetime, matter, and energy. It accounts for the hydrogen, most of the helium, and some of the lithium that exist today. But how do we explain the existence of superclusters of galaxies, clusters of galaxies, and individual galaxies? As noted earlier, these must have formed after the beginning of the matter-dominated era of the universe.

To explain large-scale structures in the early matter-dominated universe, recall the quantum fluctuations that were stretched by inflation. These clumps throughout the inflated universe were initially extremely small differences in density with a wide range of sizes in the microwave background. We know this because the *COBE* and *WMAP* satellites observed the cosmic microwave background and discovered these variations in density back when the universe was only 379,000 years old. Their data show that they were only about 1 part in 100,000 denser than the average density of matter in the early universe

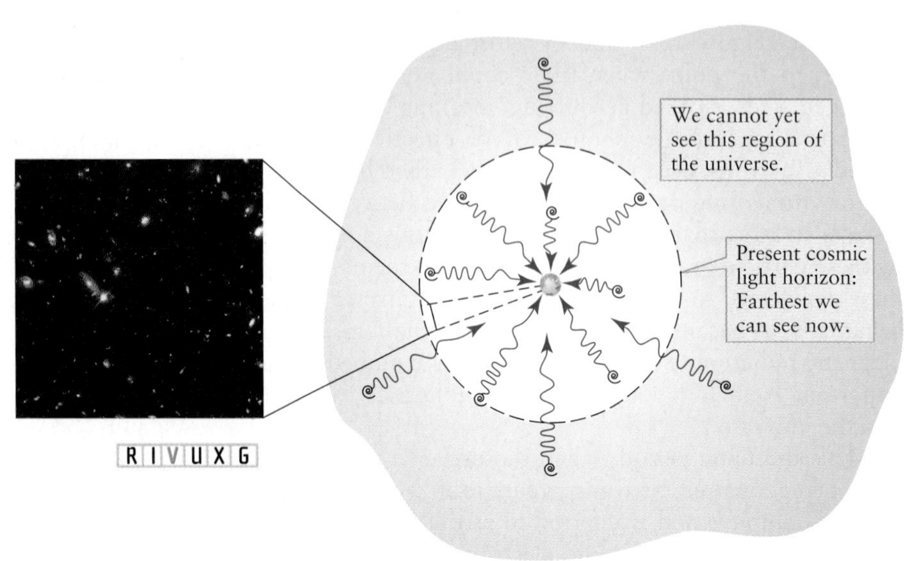

FIGURE 18-14 **The Observable Universe** This diagram shows why we only see part of the entire universe. As time passes, this volume grows, meaning that light from more distant galaxies reaches us. The galaxies we see at the farthest reaches of our telescopes' resolving power are as they were within a few hundred million years after the Big Bang (see **inset**). These galaxies, formed at the same time as the Milky Way, appear young because the light from their beginnings is just now reaching us. The radius of the cosmic light horizon is equal to the distance that light has traveled since the Big Bang. Because the Big Bang occurred about 13.8 billion years ago, the cosmic light horizon today is about 13.8 billion light-years away in all directions. **Inset:** This image of the Hubble Deep Field shows some of the most distant galaxies we have seen. (inset: Robert Williams and the Hubble Deep Field Team, STScI and NASA)

(Figure 18-15). But these small variations on various size scales are all that is needed to explain the present large-scale structure of the universe.

When the universe was radiation-dominated, the particles in it were moving too fast to clump together and form structure. When the universe became matter-dominated and particles had cooled down sufficiently, the extra gravitational attraction in regions where the expanded quantum fluctuations created concentrations of matter slightly greater than average caused the particles to draw together and ultimately form galaxies, clusters, and superclusters.

WEB LINK 18.9 In the first stage of the clustering process, the contracting matter collapsed inward, heated up, and rebounded outward due to the increased pressure between particles. Gravity drew them back inward and the process repeated itself over many cycles. This contracting and

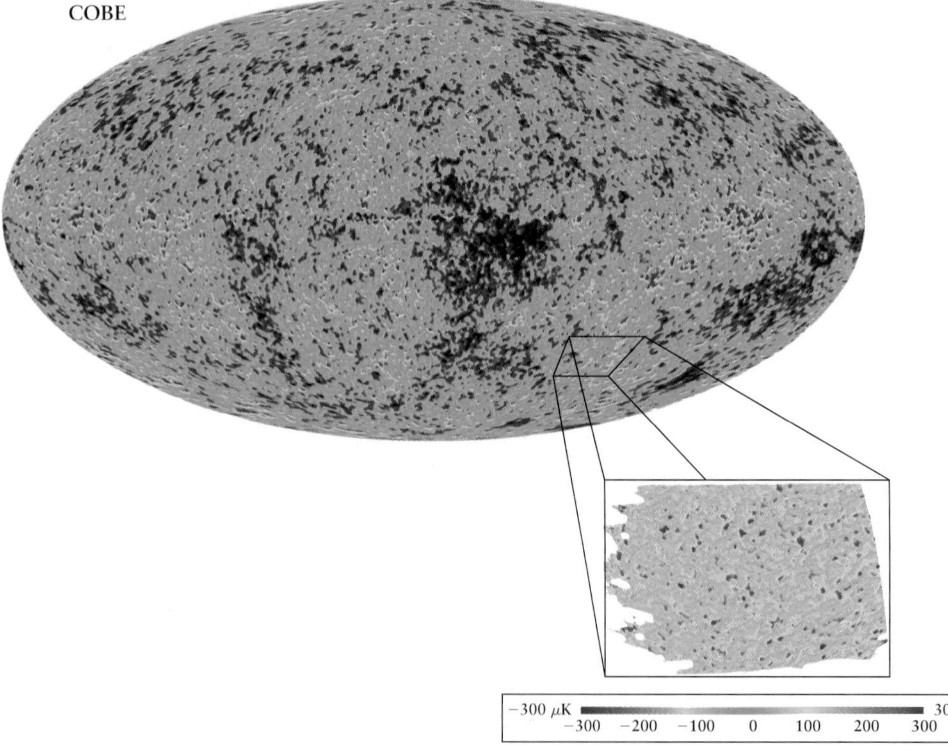

FIGURE 18-15 **Structure of the Early Universe** This microwave map of the entire sky, produced from data taken by the *Wilkinson Microwave Anisotropy Probe* (*WMAP*), shows temperature variations in the cosmic microwave background. Red regions are about 0.00003 K warmer than the average temperature of 2.73 K; blue regions are about 0.00003 K cooler than the average. **Inset:** These tiny temperature fluctuations, observed by BOOMERANG, are related to the large-scale structure of the universe today, indicating where superclusters and voids grew. The radiation detected to make this map is from a time 379,000 years after the Big Bang. (NASA/WMAP Science Team; inset: NSF/NASA)

reexpanding of vast quantities of gas produced oscillation analogous to sound waves in our air. The gas transferred some of the oscillating energy to the photons that are today the cosmic microwave background, creating regions of slightly hotter and slightly cooler gas. These oscillations in the microwave background have been observed. As we will see, these observations have an important bearing on our understanding of the overall shape of the universe.

Virtually all the matter in the early universe consisted of clouds of hydrogen and helium. Between the time of decoupling and the first burst of star formation was a period of little starlight, called the **dark ages.** Note, however, that the dark ages were actually bright due to light from energy created in the Big Bang. There is growing observational evidence that Population II star formation (see Section 12-12) began within 200 million years of the Big Bang (Figure 18-16a) and that by 600 million years after the Big Bang, galaxies were beginning to coalesce (Figure 18-16b). We also have observational evidence that galaxies were in clusters within 2 billion years of the Big Bang (Figure 18-16c).

Computer simulations indicate that the first stars were typically 100–500 $M_{\odot}$, a million times or more brighter than the Sun, and lived for no more than a few million years before exploding. During the first billion years, it appears that clumps of gas containing millions or billions of solar masses also collapsed to create supermassive black holes. These would have attracted other matter into orbit around them and thereby served as the seeds for the growth of some galaxies. By observing remote galaxies, astronomers have discovered that most galaxies initially emitted energy we associate with quasars and other active galaxies. This implies that most young galaxies have supermassive black holes at their centers. Recent observations reveal that massive and supermassive black holes typically grow until they have about 0.2% of the mass of a galaxy's nuclear bulge.

From about 600 million years to about 6 billion years after the universe formed, it underwent heavy star formation. Galaxies grew in size continuously during this period, after which star formation fell off as the amount of hydrogen available for star formation diminished. Observations also

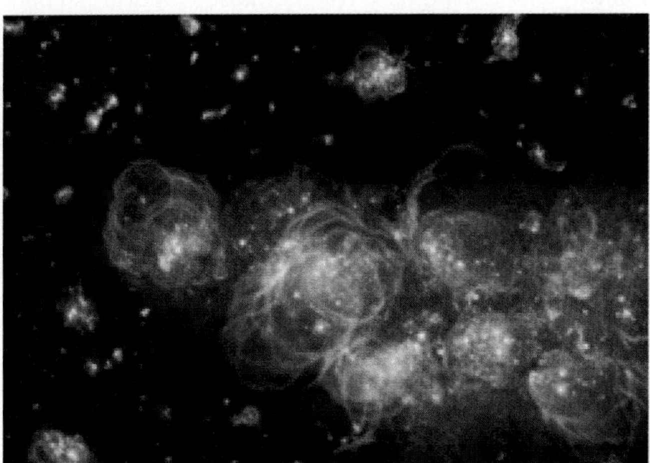

a Early bursts of star formation

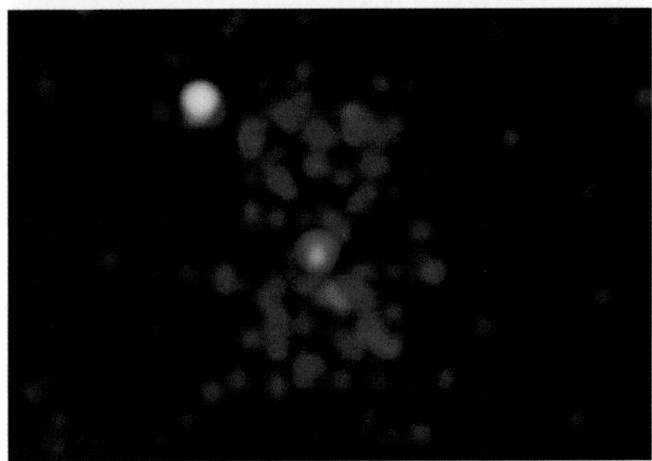

c Massive cluster of galaxies formed 11.2 years ago R I V U X G

R I V U X G

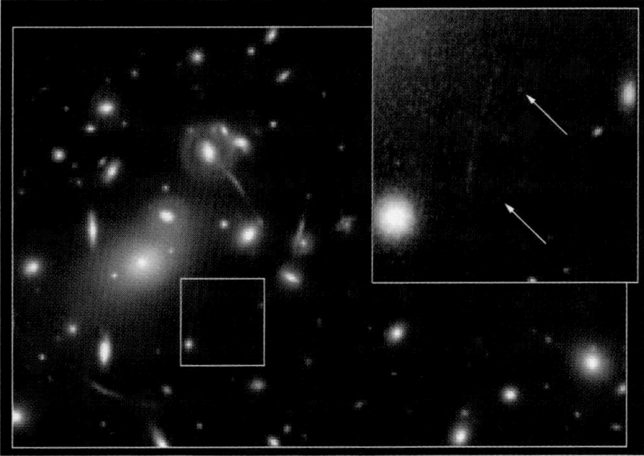

b Arrows indicate galaxies beginning to form 13.4 billion years ago

FIGURE 18-16 **Galaxies Forming by Combining Smaller Units** **(a)** This painting indicates how astronomers visualize the burst of star formation that occurred within a few hundred million years after the Big Bang. The arcs and irregular circles represent interstellar gas that is illuminated by supernovae. **(b)** Using the Hubble and Keck telescopes, astronomers discovered two groups of stars (arrows) 13.4 Bly away that are believed to be protogalaxies, from which bigger galaxies grew. These protogalaxies were discovered because they were enlarged by the gravitational lensing of an intervening cluster of galaxies. **(c)** The *Chandra* X-ray telescope imaged gravitationally bound gas around the distant galaxy 3C 294. The X-ray emission from this gas is the signature of an extremely massive cluster of galaxies, in this case at a distance of about 11.2 Bly from us. (a: Adolf Schaller, STScI/NASA/K. Lanzetta, SUNY; b: Richard Ellis (Caltech) and Jean-Paul Kneib (Observatorie Midi-Pyrenees, France), NASA, ESA; c: NASA)

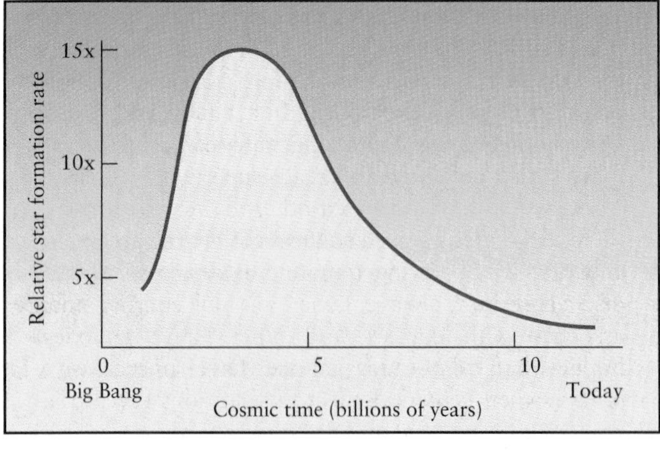

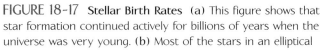

a

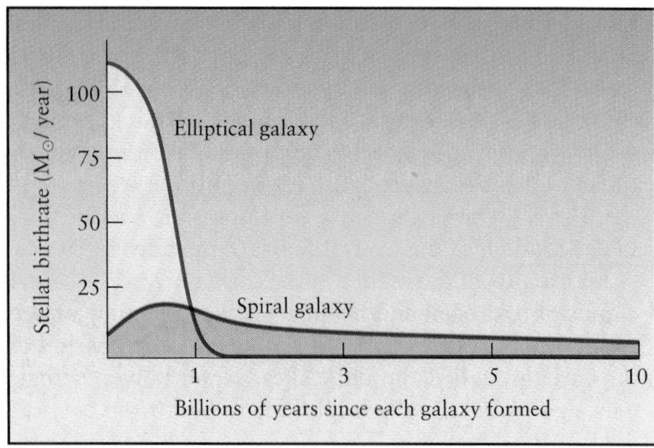

b

FIGURE 18-17 **Stellar Birth Rates** (a) This figure shows that star formation continued actively for billions of years when the universe was very young. (b) Most of the stars in an elliptical galaxy are created in a brief burst of star formation when the galaxy is very young. In spiral galaxies, stars form at a more leisurely pace that extends over billions of years.

reveal that galaxies were bluer and brighter in the past than they are today. These changes in color and brightness suggest a high abundance of young, bright, hot, massive stars in newly formed galaxies (Figure 18-17a). As galaxies age, these blue O and B stars become supergiants and eventually die off. Therefore, galaxies grow somewhat redder and dimmer. This is especially true of those elliptical galaxies that formed early on. They appear to form nearly all their stars in one vigorous burst of activity that lasts for about a billion years (Figure 18-17b).

Sequence of events ⟶

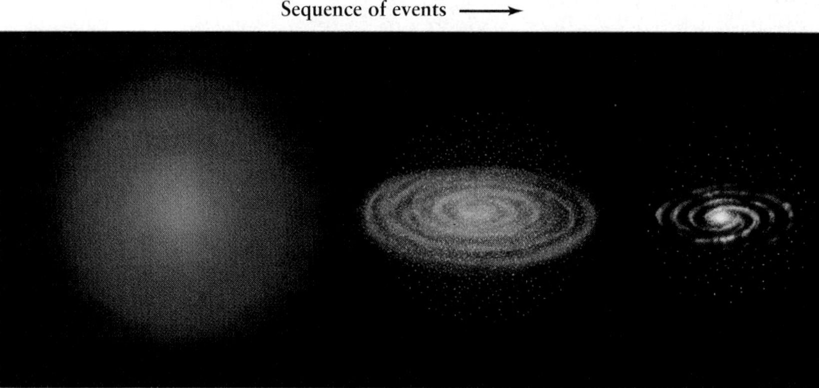

a Formation of a disk galaxy

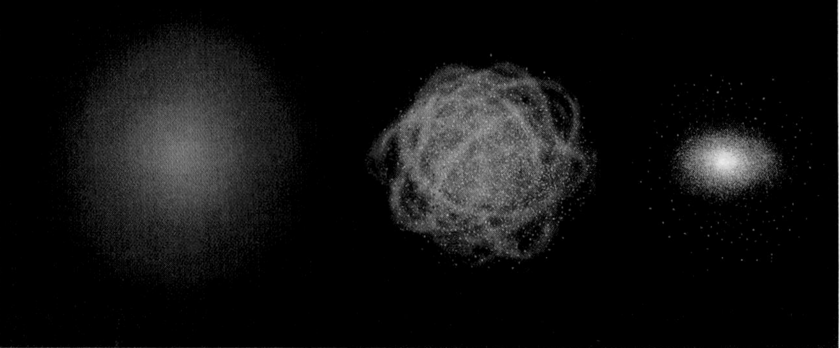

b Formation of an elliptical galaxy

FIGURE 18-18 **The Creation of Spiral and Elliptical Galaxies** A galaxy begins as a huge cloud of primordial gas that collapses gravitationally. (a) If the rate of star birth is low, then much of the gas collapses to form a disk, and a spiral galaxy is created. (b) If the rate of star birth is high, then the gas is converted into stars before a disk can form, resulting in an elliptical galaxy.

In contrast, spiral galaxies have been forming stars for at least the past 10 billion years, although at a gradually decreasing rate. There is still plenty of interstellar hydrogen in the disks of spiral galaxies like our Milky Way to fuel star formation today. That is why O and B stars still highlight their spiral arms. Figure 18-17b compares the rates at which spiral and elliptical galaxies form stars. Evidence suggests that, overall, star formation peaked at about 15 times its present rate when the universe was about one-third its present age.

18-12 Star formation activity determines a galaxy's initial structure

Imagine a developing galaxy, called a *protogalaxy*, forming from a cloud of gas. Theory proposes that the rate of star formation determines whether this protogalaxy becomes a spiral or an elliptical. If stars form slowly enough, then the gas surrounding them has plenty of time to settle by collision with other infalling gas into a flattened disk, just like the early solar system. Star formation continues because the protogalactic disk contains an ample supply of hydrogen, and a spiral or lenticular (disk-shaped but without spiral arms) galaxy is created. If, however, the initial stellar birthrate is high in the protogalaxy, the theory predicts that virtually all pregalactic gas is used up in the creation of stars before a disk can form. In this case, an elliptical galaxy is created. Figure 18-18 depicts these contrasting sequences of events. Figure 18-19 summarizes the major evolutionary activities in the universe through the present day.

Not all galaxies maintain their initial structure over the evolution of the universe. As we discussed in Section 16-8, some galaxies collide, and these collisions can change a galaxy's structure from spiral, for example, to elliptical. Indeed, astronomers using the Hubble Deep Field images have observed elliptical galaxies during the first few billion years of the universe's existence that are far less uniform in color than are closer ellipticals. They observed blue stars in the young ellipticals consistent with the merger of spirals and a resulting burst of star formation, as well as lots of lower-mass (yellow and red) stars. Our understanding of galactic formation and evolution is far from complete. For instance, why wasn't gas in the early universe eventually transformed into systems of stars a million times bigger or smaller than galaxies?

Determining the location and nature of the dark matter in the universe (see Sections 15-7 and 16-9) is still of paramount importance in understanding the cosmos. The observable stars, gas, and dust in a galaxy or cluster of galaxies account for only about 10% to 20% of each object's mass. (Recall that this observable matter does not have enough mass to hold galaxies or clusters of galaxies together.) Other than massive neutrinos, we have very little idea of what the remaining 80% to 90% of each galaxy or cluster of galaxies is composed of. However, some progress on the distribution of this dark matter is being made. In 2002, astronomers mapping large numbers of galaxies and using computer simulations of the effects of dark matter determined that on the scales of clusters of galaxies, the locations of the galaxies coincide with the concentrations of the dark matter and voids between clusters coincide with voids of dark matter.

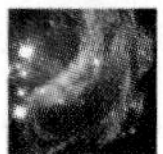

THE FATE OF THE UNIVERSE

We now turn to the future. Will the universe last forever? Or will it someday stop expanding and collapse?

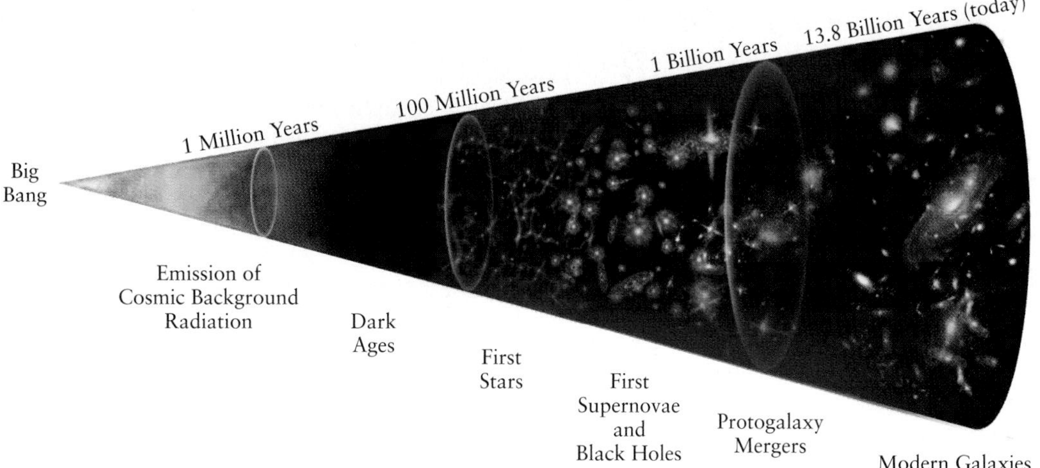

FIGURE 18-19 **A Cosmic Timeline** This figure shows our current thinking about the evolution of star and galaxy formation in the universe. (Don Dixon)

18-13 The average density of matter is one factor that determines the future of the universe

As different superclusters move apart in the expanding universe, their mutual gravitational attractions act on each other and thereby slow the expansion rate. If that were all that mattered in determining the fate of the universe, one would simply have to locate and add up all the visible and (presently) dark matter in existence and see if its gravitational force is enough to eventually stop the expansion.

This is analogous to what engineers do in calculating, for example, how high a cannonball shot upward from the surface of the Earth will travel. The Earth's gravitational force slows the ball's ascent. If the cannonball's speed upward is less than the escape velocity from the Earth's gravity (about 11 km/s straight up), it will fall back to Earth. If the ball's speed exceeds the escape velocity, it will leave the Earth and continue outward forever, despite the relentless pull of the Earth's gravity. On the boundary between these two scenarios is the situation when the cannonball's speed equals the escape velocity. In that case, the cannonball will just barely escape falling back to Earth, slowing forever, and come to rest an infinite distance away.

By analogy with the cannonball fired from Earth, it would seem that if the universe is moving outward too slowly to overcome the mutual gravitational attraction of its parts, it should stop expanding and someday collapse. If it is moving outward at exactly its escape velocity, it should expand until coming to a stop an infinite time in the future. Or, if it is moving outward fast enough, it should continue to expand forever. None of these options is correct!

The laws of physics pertaining to the evolution of the universe, as spelled out in the general theory of relativity and in recent observations of distant objects, reveal that reality is more complex than this simple analogy. Just when astronomers were getting comfortable with the idea that the gravitational force is always attractive, general relativity once again showed that reality ignores our common sense by demonstrating that gravity can be repulsive.

There are two additional effects we now know must be considered: The effect of the matter on the shape of the universe and the presence of the repulsive gravitational force called **dark energy** (*not to be confused with dark matter*). We begin by considering the effect that the universe's mass has on the shape of the universe.

During the 1920s, Alexandre Friedmann in Russia, Georges Lemaître in Belgium, Willem de Sitter in the Netherlands, and, of course, Einstein himself applied the general theory of relativity to the expanding universe. General relativity predicts that the presence of matter curves the fabric of spacetime, as we saw in Chapters 14, 15, and 17 in the form of gravitational lensing and the distortion of spacetime around black holes. Similarly, the presence of energy also curves spacetime—recall that matter and energy are related by $E = mc^2$.

18-14 The overall shape of spacetime affects the future of the universe

There are three possibilities for the overall shape of the universe. They are determined by the amount of mass and energy it contains and how fast it is expanding. Imagine shining two powerful laser beams out into space. Suppose that we can align these two beams so that they are perfectly parallel as they leave the Earth. Suppose, further, that nothing gets in the way of these two beams. We follow the beams across the spacetime whose shape we wish to determine. The light beams will begin to diverge due to the expanding universe carrying them apart. This effect will happen regardless of any properties of the spacetime, and since it is not what we are interested in we compensate for it (that is, we ignore it) in what follows. The three possibilities for the paths of the laser beams due to the actual curvature of the universe, are:

1. Two beams of light, starting out parallel, gradually get closer and closer together as they move across the universe, eventually intersecting at some enormous distance from Earth (Figure 18-20a). In this case, we say that space is *positively curved.* Analogously, the lines of longitude on the Earth's surface are parallel at the equator but intersect at the poles. Thus, if the universe has this effect on light, the shape of the universe is like the surface of a sphere: Space is spherical and the universe has positive curvature. This is what would happen if the universe had such a high mass density that the mass literally curved space back in on itself. In the absence of any outward-pushing force, such as would be supplied by a sufficiently large cosmological constant (see Section 18-1) or other form of dark energy, such a universe does not have enough energy to keep expanding forever. It would someday recollapse.

It is easiest to understand such a universe by visualizing a two-dimensional version of the universe as being the surface of a balloon on whose surface you must remain (Figure 18-20a). In such a universe, you could, in principle, keep going in a straight line (which is a curve on the balloon's surface) and eventually end up back where you started. Like the surface of a balloon, a three-dimensional universe with positive curvature has no outside or center and is said to be a **closed universe.**

2. Two beams of light remain parallel regardless of how far they travel (Figure 18-20b). In this case, space is not curved. That is, space is flat and it extends without limit. This is the structure that meets our commonsense belief.

3. Two initially parallel beams of light gradually diverge farther and farther apart as they move across the universe (Figure 18-20c). This is what would happen if the energy of the Big Bang was sufficiently great to assure that the universe is going to expand forever. In this case, the universe has negative curvature. A horse's saddle is a good example of a negatively curved or *hyperbolic* surface. Parallel lines drawn on a saddle always diverge. Thus, in a negatively curved universe, we would describe space as hyperbolic. A hyperbolic universe extends without limit and is called an **open universe.** Both the flat and the hyperbolic universes are open and will therefore grow larger forever.

Most astronomers have a distinct preference for the flat universe. The reason for this expectation that the universe is flat stems from the observations that the universe is homogeneous and isotropic. The only mechanism we have at present to explain these properties of the matter distribution in space is inflation, as discussed earlier. However, the equations predict that if inflation occurred, the universe must be very nearly flat. The period of inflation stretched the volume of the universe and the matter in it so much that it became virtually flat.

Telescopic observations strongly support the belief that the universe is flat. This conclusion comes from examining the sizes of the regions of slightly higher and lower temperature in the cosmic microwave background (see Figure 18-15) and comparing them to the sizes predicted for a flat universe. The observations are consistent with the theoretical variations. Furthermore, if the universe had positive curvature, light from the hot regions of the early universe would be curved and thereby focused, creating bigger, brighter images than are observed. Figure 18-21 summarizes the effects of space curvature on observations of the cosmic microwave background.

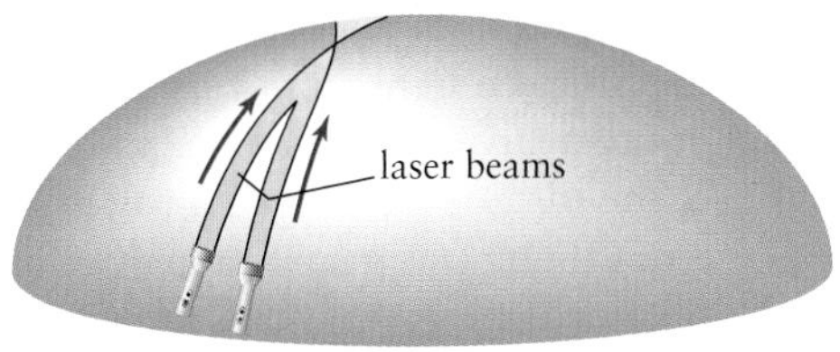
laser beams

a Parallel light beams converge

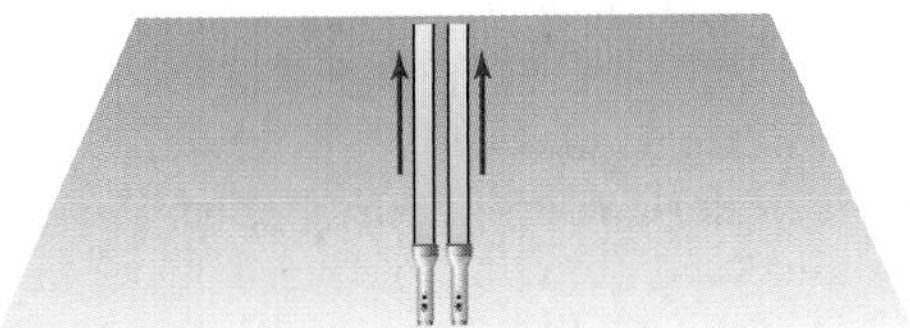

b Parallel light beams remain parallel

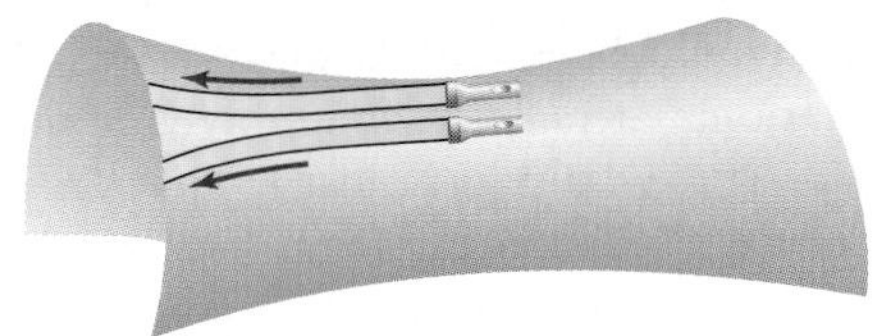

c Parallel light beams diverge

FIGURE 18-20 **The Possible Geometries of the Universe** The shape of space (represented here as two-dimensional for ease of visualization) is determined by the matter and energy contained in the universe. The curvature is either **(a)** positive, **(b)** zero, or **(c)** negative, depending on whether the average matter and energy density throughout space is greater than, equal to, or less than a critical value. The lines on each curve are initially parallel. They converge, remain parallel, or diverge depending on the curvature of space.

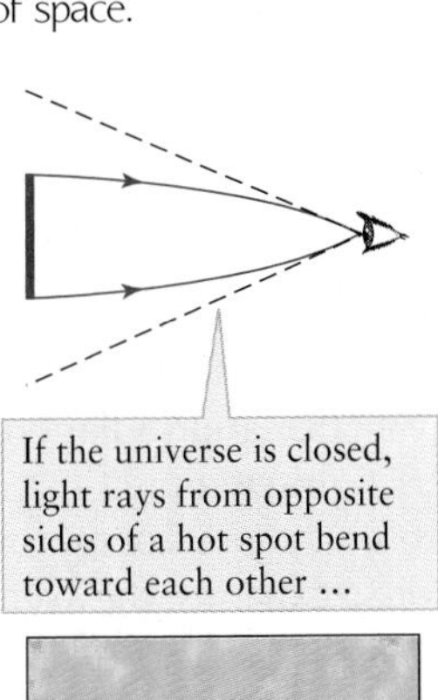
If the universe is closed, light rays from opposite sides of a hot spot bend toward each other ...

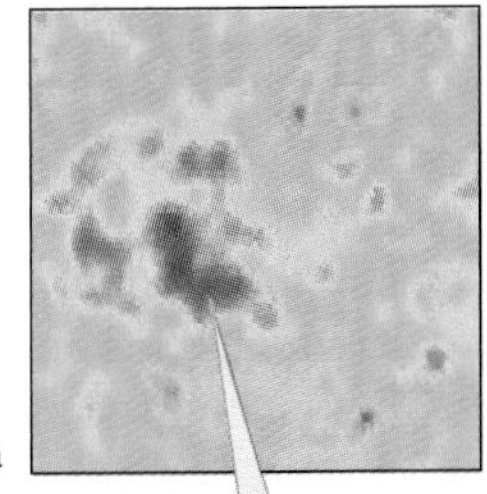

a

... and as a result, the hot spot appears to us to be larger than it actually is.

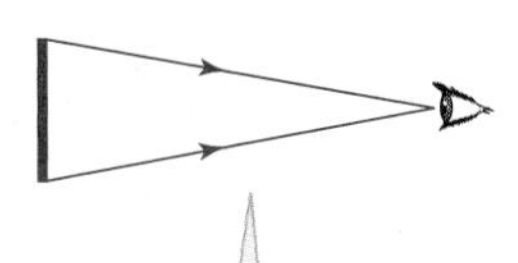

If the universe is flat, light rays from opposite sides of a hot spot do not bend at all ...

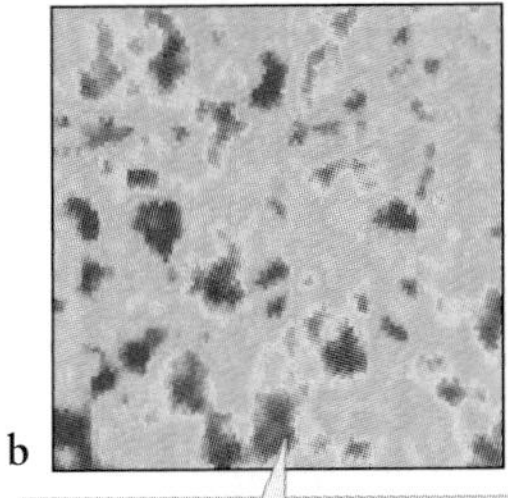

b

... and so the hot spot appears to us with its true size.

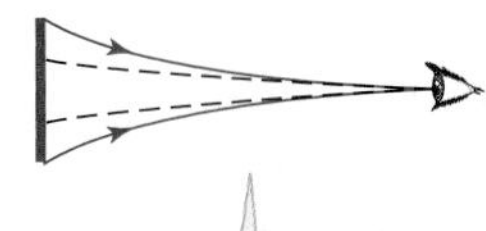

If the universe is open, light rays from opposite sides of a hot spot bend away from each other ...

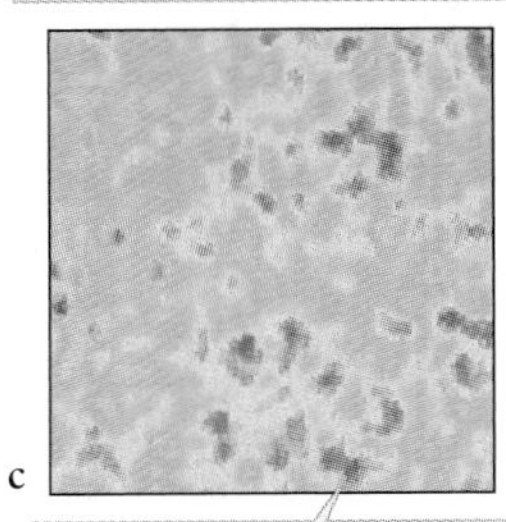

c

... and as a result, the hot spot appears to us to be smaller than it actually is.

R I V U X G

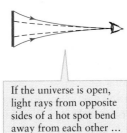

FIGURE 18-21 **The Cosmic Microwave Background and the Curvature of Space** Temperature variations in the early universe appear as "hot spots" in the cosmic microwave background. The apparent sizes of these spots depend on the curvature of space. **(a)** In a closed universe with positive curvature, light rays from opposite sides of a hot spot bend toward each other. Hence, the hot spot appears larger than it actually is, as shown by the dashed lines. **(b)** The light rays do not bend in a flat universe. **(c)** In an open universe, light rays bend apart. The dashed lines show that a hot spot would appear smaller than its actual size. (The BOOMERANG Group, University of California, Santa Barbara)

18-15 Dark energy is causing the universe to accelerate outward

Until the past few years, there was a major inconsistency between the distribution of observed matter and energy in the universe and the flatness of space. Just as there is not enough visible matter to account for galaxies and clusters of galaxies remaining as bound systems, there is not enough observed matter or energy to account for a flat universe. Visible matter accounts for only 4% of the required mass. The cosmic microwave background photons only add 0.005% of the required gravitational effects needed for flatness. And calculations of the mass necessary to keep galaxies and clusters bound, combined with gravitational lensing by dark matter of distant galaxies and quasars, reveal that the dark matter known to exist accounts for only about 23% of the required mass. Therefore, the universe has only about 27% of the required mass and energy necessary to make it flat. Allowing for the errors that still exist in all these observations, the possible range of mass and energy from all matter and photons in the universe is still only between 20% and 40% of that required for flatness. Yet flat it certainly appears to be.

By the mid-1990s, combined evidence of the microwave background and large-scale structure had forced astronomers to the conclusion that the remaining energy required to make the universe flat must be a form of dark energy with the remarkable feature that it causes the universe to accelerate outward. Corroborating evidence for the existence of such dark energy was found in the recent observations of Type Ia supernovae in extremely distant galaxies (Figure 18-22).

The light curves (see Section 13-9) for Type Ia supernovae (explosions of white dwarfs in binary star systems) are very well known. Studies show that these supernovae in nearby galaxies behave similarly to those observed in the Milky Way, regardless of their environment. Because they are so bright, such supernovae are excellent standard candles for determining distances to objects billions of light-years away. Assuming that supernovae in distant galaxies also behave similarly to those in our Galaxy, observations first made in 1998 revealed that the supernovae in distant galaxies appear dimmer than they would if the universe had been continually decelerating or even expanding at a constant speed. In other words, the universe is now expanding faster than it was, so the distant galaxies are farther away than they would otherwise have been. Because they are farther away, the supernovae in them appear dimmer than in a continually slowing universe. The universe is *accelerating* outward.

There must be some kind of repulsive force acting to increase the rate at which superclusters separate today. It is hypothesized that this is due to some kind of dark energy that has a repulsive gravitational effect, as introduced earlier. Confirmation that dark energy exists came in 2003, when astronomers observed light from very distant galaxies passing through more nearby clusters of galaxies on its way toward Earth. As the distant light moved toward a cluster on its way to us, that light was gravitationally blueshifted, meaning the light gained energy as it was pulled toward the cluster. If there was no dark energy in a flat universe, then as that light left the vicinity of the cluster it would have been redshifted (lost energy) by exactly the same amount and come to us with the same wavelengths it would have had if it had never passed through the cluster.

However, if dark energy exists, then the universe is accelerating outward (moving outward faster and faster) and so the universe would have expanded more during the time the light was leaving the cluster than when that light was first moving toward the cluster. Under those conditions, the light would have gained more energy from the cluster's gravitational attraction while traveling toward the cluster than the light would have lost to that gravitational attraction as it was moving away from the cluster. In other words, the amount of blue shift the light underwent falling into the cluster would be greater than the amount of red shift it underwent leaving the vicinity of the cluster. This was precisely what astronomers discovered: The light from distant galaxies undergoes a net blueshift when it passes through clusters of galaxies on its way to us—the universe is accelerating outward.

WEB LINK 18.10

There are two theoretical explanations for dark energy. Let us consider first what would happen if the cosmological constant that Einstein introduced and then rejected actually did exist. As Einstein had intended, it would provide a repulsive force to nature. The energy associated with that force would appear in the vacuum of space. Recall from Sections 4-6 and 14-11 that even in a vacuum, particles are constantly appearing and disappearing. They add energy to the cosmos, so even empty space has energy. The cosmological constant form of energy has the property of contributing a repulsive gravitational force that competes with the attractive gravitational force from matter and radiation, which slows down the expansion. Which gravitational force wins depends on whether there is more vacuum energy or more matter and radiation. As the universe expands, the average density of matter and energy decreases. In other words, on average there is less matter and energy in each cubic meter of space every second than there was the second before. However, the density of energy created by the cosmological constant is unchanged—as the universe expands the amount of vacuum energy in each cubic meter of space remains constant. If the universe is still expanding when the vacuum energy exceeds the matter and energy per unit volume, then the repulsive gravitational force would dominate and force the universe to start accelerating outward, as is observed.

The concern some astronomers express about the cosmological constant is that the repulsive force it creates has

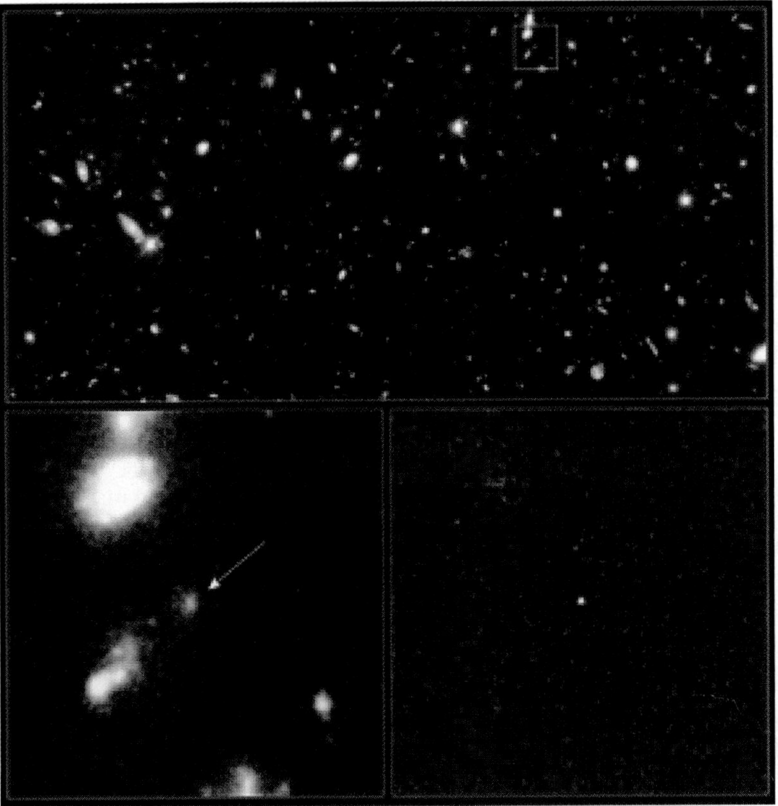

◀ FIGURE 18-22 **Dimmer Distant Supernova** (a) These Hubble Space Telescope images show the galaxy in which the supernova SN 1997ff occurred. This supernova, more than 10 Bly away, was dimmer than expected, indicating that the distance to it is greater than the distance it would have if the universe had been continually slowing down since the Big Bang. This supports the notion that an outward (cosmological) force is acting over vast distances in the universe. The arrow on the first inset shows the galaxy in which the supernova was discovered. The bright spot on the second inset shows the supernova by subtracting the constant light emitted by all the other nearby objects. (b) The distances and brightnesses of many very distant supernovae are plotted on this diagram. The location of the most distant supernovae in the upper region strongly indicates that the universe has been accelerating outward for the past 6 billion years. (a: Adam Riess, Space Telescope Science Institute, NASA)

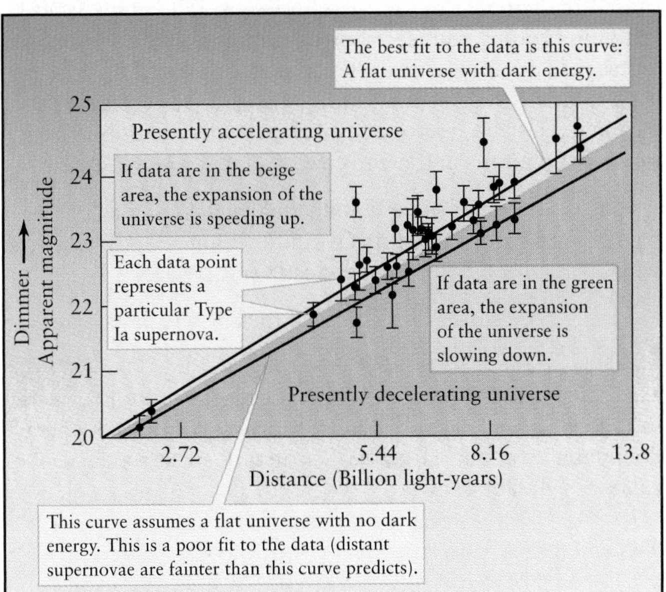

just the right strength to have allowed the universe to slow down for several billion years and then to slowly cause it to accelerate outward, as seen today. To explain this apparent coincidence would require fine tuning the ratio of the vacuum energy to the matter and radiation to a remarkably high degree of precision. Scientists do not yet have a good reason why the repulsion created by a cosmological constant is not, say, a hundred times greater, in which case structure throughout the universe such as stars and galaxies would not have formed yet, or a billion times less, in which case the repulsive effects of the cosmological constant would be negligible today.

Despite these concerns, observations of distant Type Ia supernovae suggest that the outward force acting on the universe is nearly constant, consistent with the dark energy being created by the physics we characterized with the cosmological constant. However, these observations do not rule out the major competing theory of dark energy, called **quintessence.** Scientists are exploring a variety of mathematical descriptions of quintessence. It differs from the vacuum energy associated with the cosmological constant in that the energy of quintessence is not constant, but changes slowly with the expansion of the universe and can also be changed by the flow of energy and matter through the universe. Because it can change, quintessence need not have been fine-tuned to a strength that makes the universe flat and creates just the slight acceleration we are seeing. Instead, it could have started out with some arbitrary strength and then been adjusted by physical properties of the universe to match the growth of the universe.

As the universe grew, the energy in quintessence would eventually come to dominate the gravitational energy from matter. If this took place a few billion years ago, quintessence would have begun exerting a negative gravitational force and so the universe would have begun accelerating.

The unexpected discovery that the universe is accelerat-
3 ing outward completely alters astronomers' perspectives about the fate of the cosmos. *The universe will expand forever.* At the same time, it neatly solves the mystery of why the universe is flat, as required by the occurrence of inflation in the early universe. Dark energy provides the remaining 73% of the energy needed to account for the universe being flat. The contributions of all its major components to the mass and energy in the universe are listed in Appendix E-11.

18-16 Frontiers yet to be discovered

Some of the unanswered questions about the cosmos follow from the discoveries presented in this chapter. What caused the Big Bang? What happened during the Planck time before 10^{-43} seconds? Is our universe the only one, or are there others, perhaps with profoundly different physical properties than ours, that were created at the same time but that are inaccessible to us? What caused the universe to come into existence in a false vacuum? What are the details of the formation processes of galaxies? Did black holes grow with the evolution of the galaxies, as is currently believed? What are the origins of the dark energy? Which, if either, of the current candidates for dark energy, the cosmological constant or quintessence, is correct? One of the things that makes this time in human existence so fascinating is that it is likely we will have answers to most of these questions within your lifetime.

Summary of Key Ideas

The Big Bang

• Astronomers believe that the universe began as an exceedingly dense cosmic singularity that expanded explosively in an event called the Big Bang. The Hubble law describes the ongoing expansion of the universe and the rate at which superclusters of galaxies move apart.

• The observable universe extends about 13.8 billion light-years in every direction from the Earth to what is called the cosmic light horizon. We cannot see any objects that may exist beyond the cosmic light horizon because light from these objects has not had enough time to reach us.

• According to the theory of inflation, early in its existence, the universe expanded super rapidly for a short period, spreading matter that was originally near our location throughout a volume of the universe so large that we cannot yet observe much of it. The observable universe today is thus a growing volume of space containing matter and radiation that was in close contact with our matter and radiation during the first instant after the Big Bang. This explains the isotropic and homogeneous appearance of the universe.

A Brief History of Spacetime, Matter, Energy, and Everything

• Four basic forces—gravity, electromagnetism, the strong nuclear force, and the weak nuclear force—explain the interactions observed in the universe.

• According to current theory, all four forces were identical just after the Big Bang. At the end of the Planck time (about 10^{-43} s after the Big Bang), gravity became a separate force. A short time later, the strong nuclear force became a distinct force. A final separation created the electromagnetic force and the weak nuclear force.

• Before the Planck time, the universe was so dense that known laws of physics do not describe the behavior of spacetime, matter, and energy.

• In its first 30,000 years, the universe was radiation-dominated, during which time photons prevented matter from forming clumps. Then it was matter-dominated, during which time superclusters and smaller clumps of matter formed. Today it is dark-energy-dominated. Dark energy of some sort supplies a repulsive gravitational force that causes superclusters to accelerate away from each other.

• During the first 379,000 years of the universe, astronomers believe that matter and energy formed an opaque plasma called the primordial fireball. Cosmic microwave background radiation is the greatly redshifted remnant of the universe as it existed about 379,000 years after the Big Bang.

• By 379,000 years after the Big Bang, spacetime expansion caused the temperature of the universe to fall below 3000 K, enabling protons and electrons to combine to form hydrogen atoms. This event is called the era of recombination. The universe became transparent during the era of recombination, meaning that the microwave background radiation contains the oldest photons in the universe.

• Clusters of galaxies and individual galaxies formed from pieces of enormous hydrogen and helium clouds, each of which became a separate supercluster of galaxies.

• All the superclusters and most of the clusters of galaxies are moving away from each other.

• During the matter-dominated era, structure formed in the universe. As the universe goes farther into the dark-energy-dominated era, the large-scale structure of superclusters of galaxies will fade away.

The Fate of the Universe

• The average density of matter and dark energy in the universe determines the curvature of space and the ultimate fate of the universe.

• Observations show that the universe is flat and that the cosmic microwave background is almost perfectly isotropic, resulting from a brief period of very rapid expansion (the inflationary epoch) in the very early universe.

• The universe is accelerating outward and it will expand forever.

WHAT DID YOU THINK?

1 *What is the universe?* It is all the matter, energy, and spacetime that will ever be detectable from the Earth or that will ever affect us.

2 *Did the universe have a beginning?* Yes, it occurred about 13.8 billion years ago in an event called the Big Bang.

3 *Will the universe last forever?* Current observations support the belief that the universe will last forever.

Key Words

Review Questions

1. Which force in nature is believed to have formed second? **a.** gravity. **b.** electromagnetic force. **c.** weak force. **d.** strong force. **e.** all formed at the same time.

2. Inflation directly explains which of the following? **a.** Why the universe is homogeneous. **b.** Why the universe is expanding. **c.** Why the universe is going to last forever. **d.** Why the universe has a background temperature. **e.** Why the universe has particles.

3. What does it mean when astronomers say that we live in an expanding universe?

4. Explain the difference between a Doppler redshift and a cosmological redshift.

5. In what ways are the fate of the universe, the shape of the universe, and the average density of the universe related?

6. Assuming that the universe will expand forever, what will eventually become of the microwave background radiation?

7. What does it mean to say that the universe is dark-energy-dominated? When was the universe radiation-dominated? When was it matter-dominated? How did radiation domination show itself?

8. Explain the difference between an electron and a positron.

9. Where do astronomers believe most of the photons in the cosmic microwave background originated?

10. Give examples of the actions or roles of each of the four basic physical forces in the universe.

11. What is the observational evidence for **a.** the Big Bang, **b.** the inflationary epoch, and **c.** the confinement of quarks?

Advanced Questions

12. Explain why the detection of cosmic microwave background radiation was a major blow to the steady-state theory.

13. The separation of nuclei by energetic photons in the early universe was caused by the same mechanism as what other transformation covered earlier in this book? Explain.

Discussion Questions

14. Discuss the implications of the fact that science cannot yet tell us what caused the Big Bang or what, if anything, existed before the Big Bang occurred.

15. Explain why gravitational attraction has dominated the behavior of the universe until recently and why the dark energy determines the fate of the universe.

What If . . .

16. The universe were destined to collapse and the collapse were under way? What would be different in space and on Earth under those conditions?

17. Our solar system formed very early in the evolution of our Galaxy, when the universe was just 2 billion years old? What would be different in space and on Earth?

18. Our solar system formed much later in the evolution of the universe than it actually did? How would observations of stars and galaxies be different than they are now?

19. Our solar system formed with the first generation of stars? What would be different about the solar system? Would the Earth exist as an inhabitable world? Why or why not?

Web Questions

20. **Temperatures in the Early Universe.** Access the Active Integrated Media Module "Black Body Curves"

in the *Discovering the Universe* Web site. **a.** Use the module to determine by trial and error the temperature at which a blackbody spectrum has its peak at a wavelength of 1 μm. **b.** At the time when the temperature of the cosmic background radiation was equal to the value you found in **a.**, was the universe matter-dominated or radiation-dominated? Explain your answer. *Hint:* Figure 18-8 might help.

Observing Projects

21. Evolution of the Universe In this exercise we will use *Deep Space Explorer*™ to watch the major epochs of evolution in the history of the universe. Start by clicking on **Home.** Now, click and hold the "zoom out" (upward-pointing triangle) near the "Spaceship" button until the image on the screen stops with a "Big Bang" essay in the upper-left corner of the screen. Keep in mind that the Milky Way is located in the center of the sphere you see. Does that mean that we are at the center of the universe? Why or why not? Read the essay. What does the blue haze inside the outermost spherical shell physically represent? What is and is not occurring in the region between the outermost shell and the orange shell inside it? Now "zoom in" until the "CMB" essay is visible in the upper-left corner. Read the essay and state what the orange-yellow represents. Now "zoom in" until the "young galaxies" essay is visible. Read it and write down what the wavy blue lines represent. Move in until the orange-yellow blobs are visible. What do these represent?

The Search for Extraterrestrial Life

CHAPTER 19

The Region of the Search for Extraterrestrial Intelligence
(NASA/Space Telescope Science Institute)

WHAT DO YOU THINK?

1. How do astronomers search for extraterrestrial life?
2. Have astronomers located any extraterrestrial civilizations?
3. If advanced alien civilizations exist, is there any way they might know of our existence?

If, as we said in the first chapter of this book, the starry night sky inspires us to look beyond ourselves, the study of astronomy motivates us to contemplate the formation of the Earth, the nature of the stars, and even the creation of the universe. Perhaps no question is as compelling, however, as whether there is other intelligent life in the universe. We can begin looking for intelligent alien life here at home. From a scientist's perspective, the search is not hard. Despite all the alleged UFO sightings and personal encounters, no one has produced a single piece of evidence convincing to scientists that an intelligent extraterrestrial life-form has *ever* visited Earth.

Are there other advanced civilizations elsewhere in the cosmos? Some people find comfort in the belief that Earth is the only home for life, while others look at the vastness of the universe and size and diversity of the objects in the cosmos and find reason to expect that there are other worlds suitable for sustaining life. Some of these scientists are now undertaking high-tech searches for extraterrestrial life to try to answer this question of whether or not we are alone in the universe or whether others are searching it just as we do, looking for company in the cosmos.

In this chapter you will discover

- what qualities scientists believe are necessary for life to exist
- how scientists estimate the number of planets orbiting other stars that could support complex life
- how scientists search for life beyond our own solar system—and the results of those searches
- how we are trying to communicate with extraterrestrial life

19-1 The existence of life depends on chemical and physical properties of matter

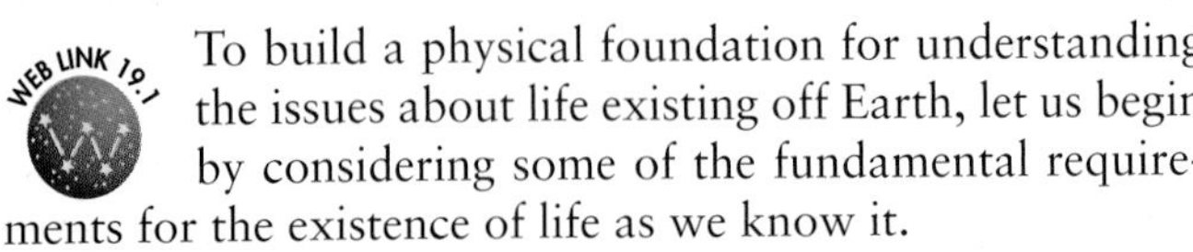

To build a physical foundation for understanding the issues about life existing off Earth, let us begin by considering some of the fundamental requirements for the existence of life as we know it.

First is the presence of liquid water. Water enables many atoms and molecules to bond and separate in large numbers. Such activity is essential for the creation and evolution of the complex molecules and systems of molecules that exist in living organisms. Biologists believe that life on Earth began developing in primordial bodies of water.

Second, some chemical element must bond strongly with at least three other atoms in order to allow for complex molecules necessary for life. Atoms able to make only two bonds can form strings or loops of atoms, but cannot create complex structures containing atoms of other elements, as is necessary for making the variety of molecules necessary for life (Figure 19-1). Five elements can bond with three or more elements: boron (B), carbon (C), nitrogen (N), silicon (Si), and phosphorus (P). Carbon is unique among these elements in that its bonds are flexible yet strong. As counterexamples, silicon-silicon bonds come apart under the slightest disturbance; silicon-oxygen bonds create gels and liquids that are very hard to alter (Figure 19-2a); and silicon-oxygen-oxygen bonds are so strong they create rocks (Figure 19-2b). The other three elements that might serve as the backbone of life—boron, nitrogen, and phosphorus—have similar bonding problems. Despite the enormous range of conditions under which terrestrial life exists, all of it is based on the unique properties of carbon.

Carbon atoms form chemical bonds that can combine to create especially long, complex molecules. These molecules can be further linked in elaborate chains, lattices, and

a

Z − X − X − X − X − Y

Linear molecule

b

```
O     OH  H  OH  OH
 \\    |   |   |   |
   C − C − C − C − C − CH2OH
 /     |   |   |   |
H      H  OH   H   H
```

Glucose

FIGURE 19-1 **Creating Complex Molecules** **(a)** Atoms, denoted by X, that can only bond strongly to two other atoms can make linear chains as depicted here. However, when such atoms bond to atoms that can only make one bond, denoted Y and Z, the chain stops. In no case can atoms X, Y, and Z combine to create nonlinear chains other than loops. **(b)** When an atom, like carbon (C), can combine with more than two other atoms (in carbon's case, with four atoms), then the complex, nonlinear chains essential for life can form. Chains of carbon atoms form the *backbone* of organic molecules. Glucose, with carbon, oxygen (O), and hydrogen (H), is a simple example of such a nonlinear molecule. The lines indicate bonds between atoms.

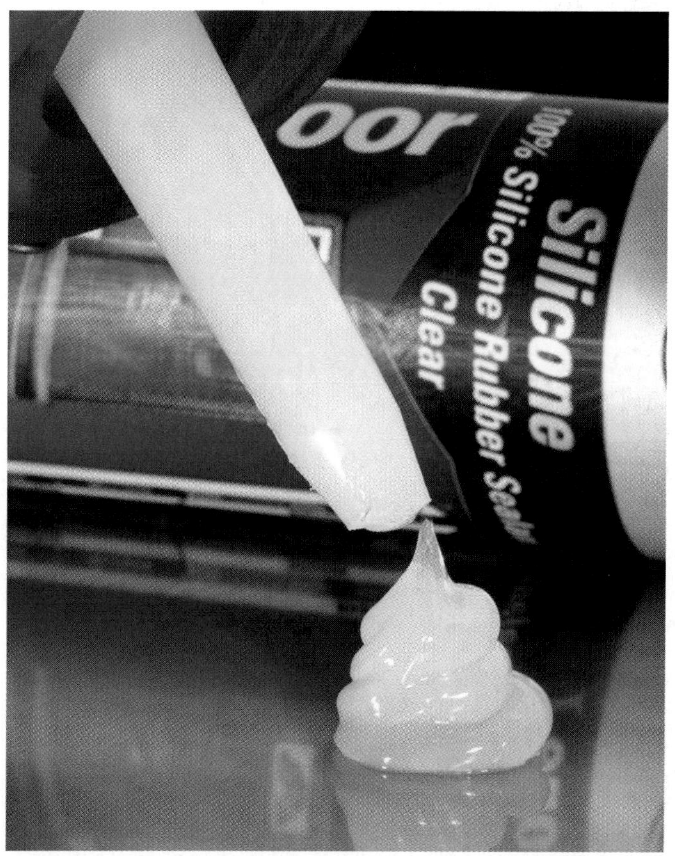

a

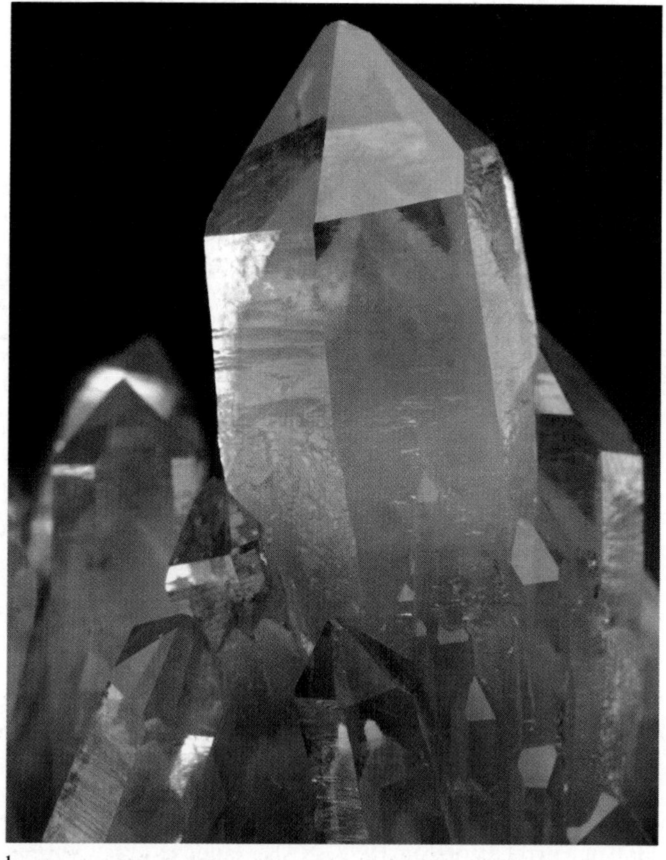

b

FIGURE 19-2 **Non-Carbon Organic Molecules?** When any molecule that can make three or more strong bonds (other than carbon) combines, it makes compounds that are either too soft or too hard or too reactive or too inert to be useful in supporting life. Consider silicon. **(a)** The silicon-oxygen pair that creates the backbone of silicone is too inert to allow such molecules to change rapidly and thereby serve as organic molecules. Furthermore, these bonds are gel or liquid, as shown. **(b)** Conversely, when the backbone is silicon-oxygen-oxygen, the bonds are rigid, as in this quartz rock. (a: Richard Megna/Fundamental Photographs, New York; b: © Mark A. Schneider/Visuals Unlimited)

fibers. It is for these reasons that carbon-based compounds, called **organic molecules,** are the stuff of life. Indeed, biologists have already identified more than six million carbon compounds.

The variety and stability of living organisms depend on the complex, self-regulating, chemical reactions of organic molecules. Carbon is also among the most abundant elements in the universe. The unique versatility and abundance of carbon imply that extraterrestrial biology would also be based on organic chemistry.

Third, a life-supporting environment must also have a variety of elements essential for developing life. Other constituents of organic molecules, such as hydrogen, oxygen, nitrogen, and sulfur, are plentiful in our Galaxy. The chemicals needed to form life may not just combine on planets as these bodies form. Some astronomers have proposed that after planets come into existence, organic material from space may also rain down on these worlds, thereby expediting the evolution of life. This belief is supported by the fact that spectra taken of comets in our solar system reveal that they contain an assortment of organic compounds.

In interstellar clouds, carbon atoms have combined with other elements to produce an impressive variety of organic compounds. Since the 1960s, radio astronomers have detected telltale microwave emission lines from interstellar clouds that help to identify dozens of these carbon-based compounds. Examples include polycyclic aromatic hydrocarbons (PAHs—molecules composed just of carbon and hydrogen atoms), as well as ethyl alcohol (CH_3CH_2OH), formaldehyde (H_2CO), methyl cyanoacetylene (CH_3C_3N), and acetaldehyde (CH_3CHO); in 2003 scientists detected the amino acid glycine. Since stars and planets condense from interstellar gas and dust enriched in these elements, these latter materials can help provide the building blocks of life elsewhere in the Galaxy. Indeed, in 2004, astronomers using the Spitzer Space Telescope detected some of the ingredients for life in orbit around

a

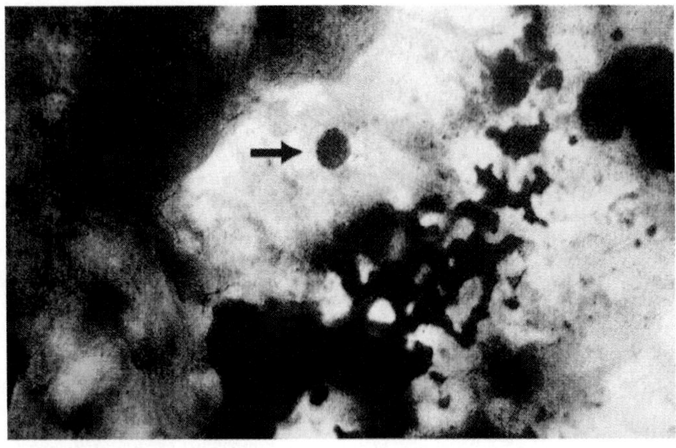

b

FIGURE 19-3 **A Carbonaceous Chondrite** (a) Carbonaceous chondrites are meteorites that date back to the formation of the solar system. This sample is a piece of the Allende meteorite, a large carbonaceous chondrite that fell in Mexico in 1969. (b) Chemical analysis of newly fallen carbonaceous chondrites discloses that they are rich in organic compounds (arrow). (a: From the collection of Ronald A. Oriti; b: Harvard Smithsonian Center for Astrophysics)

newly forming stars. These materials included methanol, water, and carbon dioxide.

Evidence for the existence of organic molecules in interplanetary space was collected by, among others, the spacecraft *Giotto* as it passed close to the nucleus of Comet Halley. Furthermore, in 1997, NASA launched the *Stardust* mission, which rendezvoused with Comet Wild 2 in January 2004, collecting dust and other debris from the comet. The spacecraft will return to Earth in 2006. Scientists hope to study this material for organic compounds uncontaminated by exposure to the Earth's atmosphere and surface.

Further evidence of extraterrestrial organic molecules comes from meteorites called carbonaceous chondrites (Figure 19-3). Many of the carbonaceous chondrites that have fallen to Earth contain a variety of organic substances. As noted in Section 9-9, carbonaceous chondrites are ancient meteorites dating from the formation of the solar system. If they contained organic molecules while in space, it seems reasonable to conclude that even from their earliest days, the planets have been continually bombarded with organic compounds.

Interstellar space is not the only source of organic material. In a classic experiment performed in 1952, the American chemists Stanley Miller and Harold Urey demonstrated that simple chemicals can combine on liquid water-bearing planets to form organic compounds. Hoping to capture conditions on the primitive Earth and in its atmosphere, as then understood, Miller and Urey mixed hydrogen, ammonia, methane, and water vapor in a closed container. They then subjected their mixture to an electric arc to simulate lightning bolts. At the end of a week, the inside of the container had become coated with a reddish-brown substance rich in compounds essential to life (Figure 19-4).

Since that time, geochemists have come to realize that Earth's early atmosphere was more likely to have been a mixture of carbon dioxide, nitrogen, and hydrogen, along with water vapor "outgassed" from volcanoes and deposited by comet impacts. Modern versions of the Miller-Urey experiment (Figure 19-4) have therefore used these common gases and, once again, organic compounds were produced. Could these compounds be forming elsewhere in the universe today?

Keep in mind that scientists have not yet created life in a test tube. Biologists have yet to figure out, among other things, how simple organic molecules gathered into cells and developed systems for self-replication.

The fourth issue surrounding the existence of life elsewhere relates to physical properties of planets and life—the environments supporting life-forms must have suitable temperature, radiation, and other factors. If a planet is too close to a star, the temperature and deadly ultraviolet radiation levels will be too high for life to evolve. If the planet is too far away, it will be too cold to have liquid water on its surface. If a planet's environment is otherwise hostile, such as too windy or too seismically active, life may either not be able to spread or may quickly become extinct. If the star around which a life-giving world orbits is too massive, the star will explode before advanced life on any of its planets has a chance to evolve. If the star is too small, the planet would have to be very close to it to be warm enough for life, but then tidal forces from the star would lock the planet in synchronous rotation, making most of its surface either too hot (daytime side) or too cold (nighttime side) for life to flourish there.

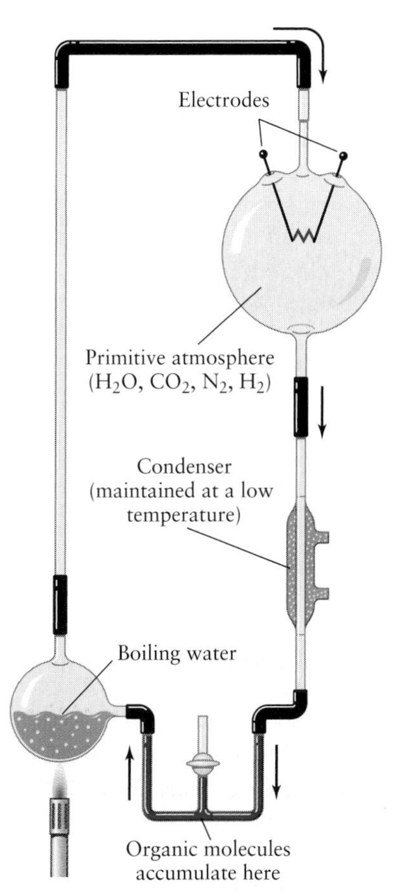

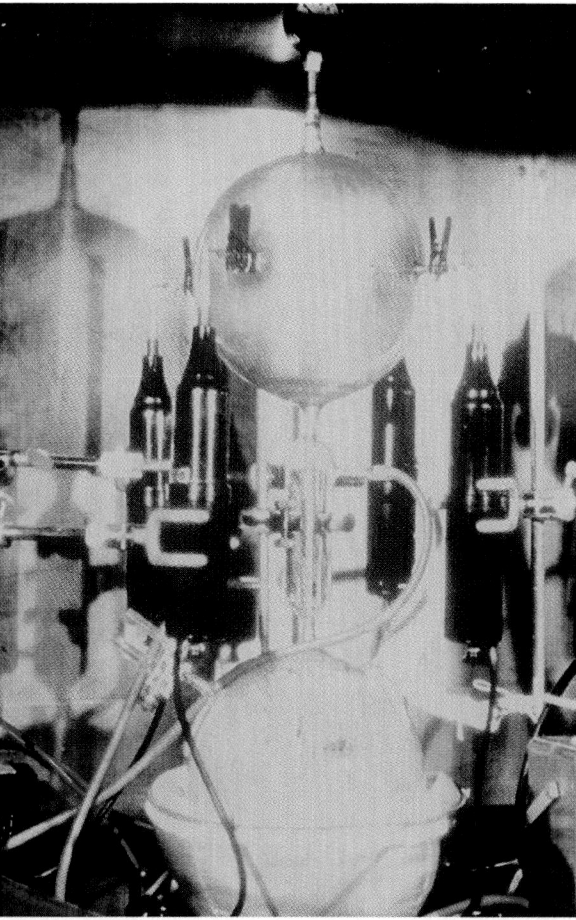

FIGURE 19-4 The Miller-Urey Experiment Updated **(a)** Modern versions of this classic experiment prove that numerous organic compounds important to life can be synthesized from gases that were present in Earth's primordial atmosphere. This experiment supports the hypothesis that life on Earth arose as a result of ordinary chemical reactions. **(b)** This photograph shows a slime of organic material created in the original Miller-Urey experiment. (b: NASA)

19-2 Evidence is mounting that life might exist elsewhere in our solar system

Despite all these potential limitations, scientists have strong evidence that everything does not have to be optimal in order for life to form. Geologists and biologists have discovered life on Earth in some incredibly challenging environments, such as on the ocean bottom, in a lake far under the Antarctic ice pack, deep inside the planet's crust, and even in hot geothermal vents. It therefore seems reasonable to believe that life could have originated off Earth under similarly challenging conditions. There are at least four places here in our solar system that scientists are considering as possible habitats for life past or present. They are Mars and Jupiter's moons Europa, Ganymede, and Callisto.

Astronomers have discovered (see Chapter 7) that Mars once had surface water and that it may still have underground bodies of water. We saw in Sections 8-5 through 8-7 that Europa, Ganymede, and Callisto are also likely to have liquid water under their frozen surfaces. These four bodies, then, are candidates for having evolved simple life-forms. Indeed, we saw in Chapter 7 what may be fossilized bacterial life from Mars (see Figure 7-36).

Confirmation that life evolved on any of these worlds will demonstrate that the formation of complex molecules and structures are possible off Earth. However, these locations are unsuitable for the evolution of sufficiently complex life, like ourselves, that is self-aware and able to create advanced civilizations.

19-3 Searches under way for advanced civilizations try to detect their radio signals

We have not yet discovered Earthlike planets orbiting Sunlike stars. Furthermore, we saw in Section 5-6 that many stars have Jupiter-mass planets orbiting very close to them. The presence of massive planets near stars would make it very hard for those stars to support habitable planets. The gravitational tugs from the massive planets would put Earthlike planets in highly elliptical orbits. During parts of such elliptical orbits, terrestrial worlds would be overheated and overradiated by their stars, and during other parts of their orbits, the planets

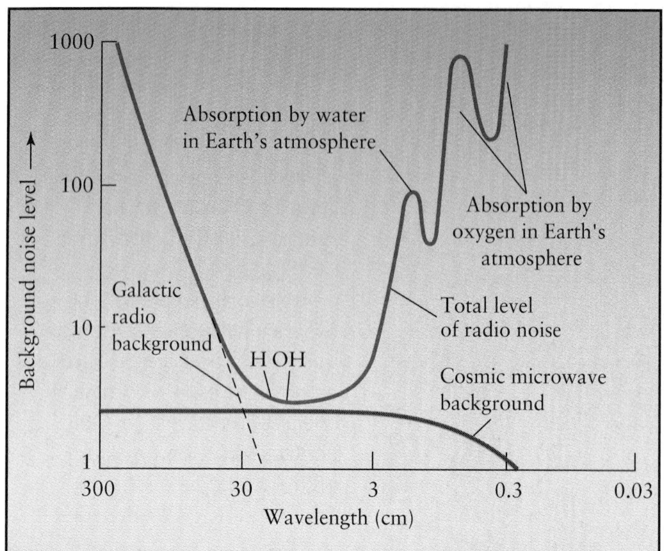

FIGURE 19-5 **The Water Hole** The so-called water hole is a range of radio wavelengths from about 3 to 30 centimeters (cm) that happens to have relatively little cosmic noise. Some scientists suggest that this noise-free region would be well-suited for interstellar communication.

would freeze. However, not all Sunlike stars have close-orbiting massive planets.

There are billions of Sunlike stars in our Galaxy to be studied, and most astronomers and many other people expect that we will eventually discover habitable planets around some of them. This belief helps motivate the search for extraterrestrial intelligence, or **SETI.** Unlike the fossil evidence for life on Mars, the effort to locate life outside the solar system is based on searching for high-tech evidence of advanced life.

1 How might we ascertain whether extraterrestrial civilizations exist, given the tremendous distances that separate us from other stars and the huge number of stars with the potential to support life-bearing planets? Looking for individual, habitable, extrasolar planets is technically demanding and time consuming. As we saw in Section 5-6, it requires careful observation of data available only in the past 25 years. Furthermore, the discovery of a planet does not imply that it harbors life. SETI astronomers therefore need a way to bypass the daunting search for habitable planets, and starting in the 1950s and 1960s, they identified a promising approach to searching directly for extraterrestrial intelligence. They look for radio transmissions from distant civilizations.

As we have seen, radio waves can travel immense distances without being significantly altered by the interstellar medium. Because they penetrate gas and dust, radio waves are a logical choice for interstellar communication. Even if alien civilizations are not trying to communicate with us, they probably use radio waves in their own technology. But at what frequencies should we look to maximize the odds of detecting alien signals?

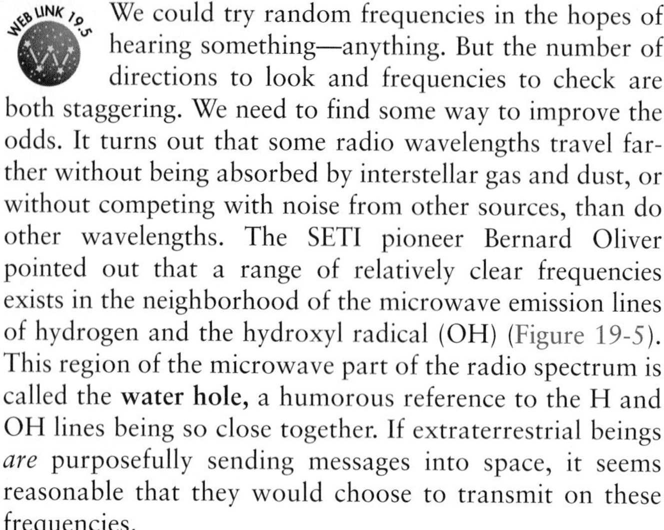

WEB LINK 19.5 We could try random frequencies in the hopes of hearing something—anything. But the number of directions to look and frequencies to check are both staggering. We need to find some way to improve the odds. It turns out that some radio wavelengths travel farther without being absorbed by interstellar gas and dust, or without competing with noise from other sources, than do other wavelengths. The SETI pioneer Bernard Oliver pointed out that a range of relatively clear frequencies exists in the neighborhood of the microwave emission lines of hydrogen and the hydroxyl radical (OH) (Figure 19-5). This region of the microwave part of the radio spectrum is called the **water hole,** a humorous reference to the H and OH lines being so close together. If extraterrestrial beings *are* purposefully sending messages into space, it seems reasonable that they would choose to transmit on these frequencies.

Even if only a few alien civilizations are scattered across the Galaxy, we have the technology to detect radio transmissions from them. One of the first searches occurred in 1960, when the American astronomer Frank Drake carried out Project Ozma (named after the land in L. Frank Baum's Oz series of books). In it he used a radio telescope at the National Radio Astronomy Observatory in West Virginia to "listen" to two Sunlike stars, τ (tau) Ceti and ε (epsilon) Eridani, but without success.

WEB LINK 19.6 SETI searches use radio telescopes (Figure 19-6) along with sophisticated electronic equipment and powerful computers to conduct both a targeted

FIGURE 19-6 **A Radio Telescope Used for SETI** This radio telescope, located in the Mojave Desert in California, was originally built by NASA to track interplanetary spacecraft. In the past decade, it was used in an all-sky survey, along with an antenna at the Arecibo Observatory in Puerto Rico, to search for extraterrestrial intelligence. In 1996, an antenna in Canberra, Australia, joined the network. (JPL)

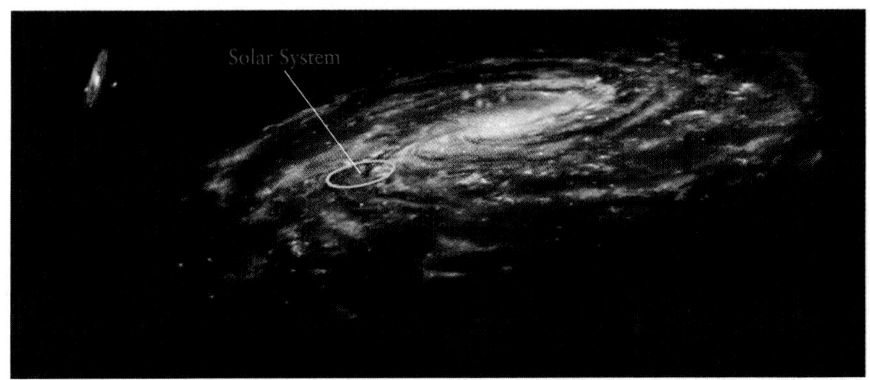

FIGURE 19-7 **The Region of the Search for Extraterrestrial Intelligence** The yellow circle, centered on the solar system, shows the region of the Milky Way in which SETI searches can reasonably expect to detect radio emissions from alien civilizations. This assumes that the signals are similar in nature to the kinds of radio emissions we generate here on Earth. (NASA/Space Telescope Science Institute)

search and an all-sky survey. While NASA has created astrobiology centers to study issues regarding life in the solar system and Galaxy, most of the SETI programs in operation today are privately funded. Several are actively listening and analyzing enormous volumes of data.

Indeed, there is so much data to analyze that one SETI project, called SETI@home, has enlisted the help of home computer users. These SETI researchers provide actual data from radio telescopes to be analyzed, along with a data-analysis program. While the computer's screensaver is on, the program runs, the data are analyzed, and the results are sent back to the researchers to be combined with data from other home-computer users. The processed data is replaced with fresh data to be analyzed. Another SETI program uses an array of small, inexpensive TV satellite antennas to look for signals from space.

2 Figure 19-7 shows the region of the Galaxy from which radio signals similar to the kinds of transmissions we make could be detected. Occasionally, a SETI project does detect powerful or unusual signals from space. However, because none of these signals is ever repeated, SETI researchers do not believe they have yet discovered any extraterrestrial civilization. Despite more than 40 unsuccessful searches to date using radio telescopes around the world, the search for extraterrestrial intelligence continues.

The detection of a message from an alien civilization would be one of the greatest events in human history. Such a message could dramatically change the course of civilization through the sharing of scientific information or an awakening of social or humanistic enlightenment. In only a few years, our industry and social structure could advance centuries into the future. The effects of such change would touch every person on Earth.

19-4 The Drake equation: How many civilizations are likely to exist in the Milky Way?

Just how many planets are likely to harbor complex life throughout our Galaxy? The first person to tackle this question quantitatively was Frank Drake. Drake proposed that the number of technologically advanced civilizations in the Galaxy (designated by the letter N) could be estimated by what is now called the **Drake equation:**

$$N = R^* f_p n_e f_l f_i f_c L$$

in which

R^* = rate at which solar-type stars form in the Galaxy

f_p = fraction of stars that have planets

n_e = number of planets per star system suitable for life

f_l = fraction of those habitable planets on which life actually arises

f_i = fraction of those life-forms that evolve into intelligent species

f_c = fraction of those species that develop adequate technology and then choose to send messages out into space

L = lifetime of that technologically advanced civilization

The Drake equation expresses the number of extraterrestrial civilizations quantitatively as a product of terms, some of which can be estimated from what we know about stars and stellar evolution. For instance, thanks to new evidence for extrasolar planets, astronomers can now hope to determine the first two terms, R^* and f_p, by observation. We should probably exclude stars larger than about 1.5 $M_\odot$, because they have main-sequence lifetimes shorter than the time it took for intelligent life to develop here on Earth—some 3.8 to 4.0 billion years. If that is typical of the time needed to evolve higher life-forms, then a massive star becomes a giant or even explodes as a supernova before self-aware creatures can evolve on any of its planets.

Although low-mass stars have much longer lifetimes, they are less suited for supporting life on their planets because they are so cool. As noted earlier, only planets very near a low-mass star would be sufficiently warm for water to be a liquid. But, as also mentioned earlier, a planet that close would become tidally coupled to the star, developing synchronous rotation. One side would have continuous daylight, leading to the evaporation of oceans, while the other side would be in perpetual, frigid darkness. The only place that life could survive on such planets is in the narrow ring at the boundary between day and night. Such a small fraction of the land available greatly reduces the likelihood that life there would be able to evolve to the complexity necessary for technological civilizations to develop.

As "ideal" life-supporting stars, this leaves main-sequence stars with masses near that of the Sun. These are stars in spectral types between F5 and M0. Based on statistical studies of star formation in the Milky Way, astronomers calculate that roughly one of these Sunlike stars forms in the Galaxy each year, yielding a value of $R^* = 1$ per year.

We learned in Sections 5-2 and 5-3 that the planets in our solar system formed in conjunction with the birth of the Sun. We have also seen evidence that similar processes of planetary formation may be commonplace around isolated stars. Many astronomers therefore give f_p a value of 1, meaning they believe it likely that most Sunlike stars have planets.

Unfortunately, the rest of the factors in the Drake equation are very hypothetical. The chances that a planetary system has an Earthlike world are not known. Were we to consider our own solar system as representative, we could put n_e at 1. Let us be more conservative, however, and suppose that 1 in 10 solar-type stars is orbited by a habitable planet, making $n_e = 0.1$.

From what we know about the evolution of life on Earth, we might assume that, given appropriate conditions, the development of life is a certainty, which would make $f_l = 1$. This is, of course, an area of intense interest to biologists. For the sake of argument, we might also assume that evolution naturally leads to the development of intelligence (a conjecture that is hotly debated) and also make $f_i = 1$. It is anyone's guess as to whether these intelligent extraterrestrial beings would attempt communication with other civilizations in the Galaxy, but if we assume they all would, f_c would also be put at 1.

The last variable, L, the longevity of technological civilization, is the most uncertain of all—it cannot be tested! Looking at our own example, we see a planet whose atmosphere and oceans are increasingly polluted, potentially destroying the food chain. When we add in how close we have come to destroying ourselves with weapons of mass destruction, it may be that we humans are among the lucky few technological civilizations to squeak through its first years. In other words, L may be as short as 100 years. Putting all these numbers together, we arrive at

$$N = 1 \times 1 \times 0.1 \times 1 \times 1 \times 1 \times 100 = 10$$

Therefore, out of the hundreds of billions of stars in the Galaxy, there may be only 10 civilizations technologically advanced enough to communicate with us.

Of course, this is just an estimate. A wide range of values has been proposed for the terms in the Drake equation, and these various numbers produce vastly different estimates of N. Some scientists argue that there is exactly one advanced civilization in the Galaxy and that we are it. Others speculate that there may be tens of millions of planets inhabited by intelligent creatures. Although we do not know yet, science enables us to home in on the number.

19-5 Humans have been sending signals into space for more than a century

We have been doing more than just listening passively for 3
voices from the cosmos. Astronomers have intentionally broadcast well-focused signals through radio telescopes toward star systems likely to harbor advanced life, including a 1974 "hello" message aimed at the globular cluster M13 (Figure 19-8a). Unintentional broadcasts into space have occurred for longer than that. Since 1895, with the first radio transmission by the Italian engineer Guglielmo Marconi, we humans have been sending messages about ourselves into space. The spherical region within some 110 ly of the solar system is now filled with signals from radio and television shows. If other advanced civilizations fall within that sphere, they could very well be listening to

"AS I READ IT, WE'RE RECEIVING A MESSAGE FROM OUTER SPACE TELLING US TO STOP BOMBARDING THEM WITH UNINTELLIGIBLE MESSAGES."

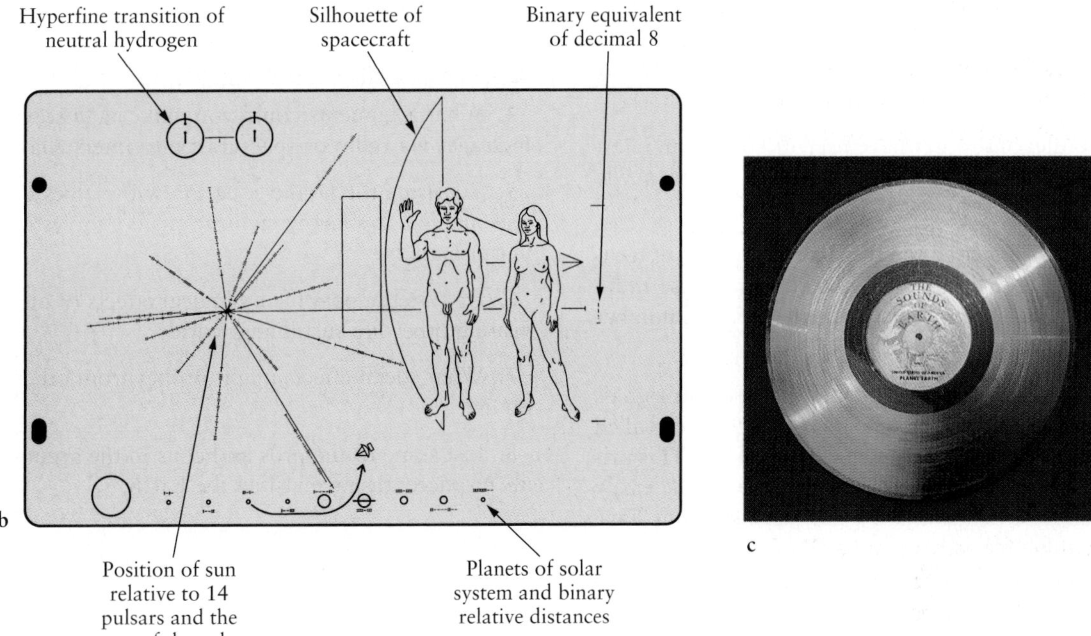

FIGURE 19-8 Human Memorabilia in Space (a) Humans have beamed radio signals into space, hoping that the message will someday be intercepted by an alien civilization. This is a visual version of the signal sent in 1974 from the Arecibo radio telescope toward the globular cluster M13. The *Pioneer* and *Voyager* spacecraft, now in interstellar space, also carry messages from Earth. (b) The plaques on *Pioneer 1* and *Pioneer 2* provide information about where we are, what we look like, and some of the science we know. (c) Images and sounds sent on *Voyager 1* and *Voyager 2* were stored on phonograph records, long before DVDs were even a twinkle in an engineer's eye. There are also instructions for playing the record, which contains information about our biology, our technology, and our knowledge base. Each record also contains the sounds of children's voices. It is remotely possible that another race might someday discover the spacecraft. (a: Frank Drake/UCSC, et al., Arecibo Observatory/Cornell/NAIC; b and c: NASA)

radio programs or watching television shows from decades past, trying to make sense of our species.

Four spacecraft also carry information about the human race. Launched in 1972 and 1973, respectively, the *Pioneer 10* and *Pioneer 11* spacecraft carry plaques about us (Figure 19-8b). They are now making their way into interstellar space. In 1977, the two *Voyager* spacecraft were sent toward the outer solar system. They carry old-fashioned phonograph records encoded with human images and voices. Because space is so vast and the spacecraft are so small, the likelihood of their being discovered is truly remote. And yet millions, perhaps billions, of years from now, one of the spacecraft just might reach another race of intelligent creatures. From its path and the information on board, those creatures might be able to determine where the *Pioneer* and *Voyager* spacecraft came from. If their travel budgets are sufficiently large, they might even decide to come and visit us. At the least, they could send us radio messages. Will our descendants be here to receive them?

19-6 Frontiers yet to be discovered

We still have much to learn about the formation and evolution of life both here and elsewhere in the cosmos. We need to confirm whether life existed on Mars, Europa, Ganymede, or Callisto. Astronomers are still working to refine their ideas of how organic compounds formed in space, and biologists still need to understand crucial steps taken by life as it began and evolved here on Earth. A paramount frontier is detecting life elsewhere in the universe.

Summary of Key Ideas

- The chemical building blocks of life exist throughout the Milky Way Galaxy.
- Organic molecules and water have been discovered in interstellar clouds, in some meteorites, in comets, and in newly forming star and planet systems.
- The Drake equation is used to estimate the number of technologically advanced civilizations in the Galaxy whose radio transmissions we might discover. Estimates of this number vary from 1 to millions.
- Astronomers are using radio telescopes to search for signals from other self-aware life in the Galaxy. This effort is called the search for extraterrestrial intelligence, or SETI. SETI is primarily done at frequencies where radio waves pass most easily through the interstellar medium. So far, these searches have not detected any life off Earth.
- Everyday radio and television transmissions from Earth, along with intentional broadcasts into space, may be detected by other life-forms.

WHAT DID YOU THINK?

1 *How do astronomers search for extraterrestrial intelligence?* They search for radio signals from other advanced civilizations.

2 *Have astronomers located any extraterrestrial civilizations?* No extraterrestrial civilizations have yet been discovered.

3 *If advanced civilizations exist, is there any way they might know of our existence?* Yes, they might detect our intentional radio broadcasts into space, our everyday radio and television broadcasts, or possibly one of our spacecraft traveling in interstellar space.

Key Words

Drake equation, 491
organic molecule, 487
SETI, 490
water hole, 490

Discussion Questions

1. A star of what spectral type is most likely to have a planet on which advanced life exists? **a.** O. **b.** B. **c.** A. **d.** G. **e.** M.

2. Which element serves as the foundation or "backbone" of life? **a.** oxygen. **b.** carbon. **c.** silicon. **d.** hydrogen. **e.** nitrogen.

3. What information would you have put on the record sent out on the *Voyager* spacecraft? Compare your answer to the results listed in the Web link associated with that record.

4. What arguments could you make against sending messages via radio or spacecraft into interstellar space?

5. Try using the Drake equation with values that you find reasonable. How many civilizations do you estimate there are in our Galaxy?

6. Discuss the possible biological effects of our probes visiting other life-sustaining worlds.

7. What social effects might probes from other worlds have on us?

8. List some of the pros and cons in the argument that alien spacecraft have visited the Earth.

What If . . .

9. Our planet were the moon of a much larger, Jupiterlike planet orbiting the Sun at 1 AU? What might the Earth and life inhabiting it be like?

10. Our Moon were Earth-sized and also had life on it?

11. We discover living organisms in Europa's subsurface oceans? What precautions should we take in the search for such life, and why?

12. We develop life in a Miller-Urey–like experiment? Explore some of the implications this event would have.

Web Questions

13. To test your understanding of the Miller-Urey experiment, do Interactive Exercise 19-1 on the Web. You can print out your answer if required.

14. To test your understanding of the Drake equation, do Interactive Exercise 19-2 on the Web. You can print out your answer if required.

WHAT IF . . . LIFE HAD BEGUN ON AN OLDER EARTH?

Our view of the universe is hardly more than a snapshot in time. The universe evolves, and Earth's surface, the eclipses, and the stars in the sky appear the way they do only because we experience them at this unique time. What would the universe look like if evolutionary processes on Earth had taken twice as long to produce creatures who were aware of their own existence? What if humans had evolved 9 billion years after the solar system formed, rather than after only 4.5 billion years?

A Different Earth As you have learned, our planet's crust is composed of tectonic plates floating on Earth's mantle. The continents are still drifting, and 4.5 billion years from now their distribution will be completely different. No pattern of land that we recognize today will remain intact. The Mediterranean Sea will be gone, and the Atlantic Ocean may be as wide as the Pacific Ocean is today, while the Pacific would probably be much narrower. The islands of the Pacific, from Japan south to Australia, may be a single connected unit. New islands will grow over hot spots in the mantle, just as the Hawaiian Islands do today. The Andes and the Himalayas will have tried to rise higher, only to come tumbling down due to weathering.

Another major change is the length of a day on an older Earth. It will be almost twice as long as our day, and the Moon will be about 1.5 times farther away, appearing correspondingly smaller in the sky and taking 1.8 times as long to orbit the planet. Both the longer day and more distant Moon will result from the gravitational interplay between ocean tides and the Moon. As tidal waters rub against the solid Earth, this friction slows the Earth's rotation and currently causes the day to grow 0.002 seconds longer each century. The energy lost as Earth spins down must be retained in the Earth-Moon system, so the Moon must gain energy in its orbit around Earth. This increase in energy currently forces the Moon to move farther away from Earth at a rate of 3.8 centimeters per year.

Because the Moon will appear smaller in the sky, "humans" on the older Earth will never see a total eclipse of the Sun. The best they will see is an annular eclipse, with a broad ring of the Sun showing around the edges of the Moon. Total lunar eclipses will still occur because the older Earth's shadow will remain wider than the Moon.

Changes in the Night Sky Currently, Polaris appears almost exactly over Earth's north rotation pole. But citizens of the older Earth will not see a pole star for two reasons. First, Earth's rotation axis precesses, or wobbles, by 23.5° over a period of 25,800 years, making the pole wander around the sky. It is just by chance that in recent centuries there has been a bright star to mark the north pole. It is unlikely that such a bright star would mark the pole when humans evolve on the older Earth. Second, many bright stars that could be near the pole in the future may no longer be visible, having used up their nuclear fuels in the next 4.5 billion years and exploded.

Perhaps new bright stars will form that will take turns as future polar guide stars as Earth's rotation axis precesses, much as Polaris does now and Thuban in Draco did in the past. Asterisms (patterns of stars), such as the Big Dipper or Orion, will all be different, too, because many stars that exist in our sky now will have exploded within the next 4.5 billion years, while the others will all move relative to each other.

Older Earth's astronomers will, like Edwin Hubble, discover that distant galaxies appear to move away from us because of the expansion of the universe. However, the Hubble constant is not at all constant. When those astronomers measure the speed of distant galaxies moving away from us, they will get a very different number for the Hubble constant.

A Brighter Sun The Sun in the time of the older Earth will still have nearly the same surface temperature and yellowish hue that it has now, but it will appear about twice as bright because it will be about 60% bigger. The extra 4.5 billion years will have begun to take their toll on the Sun's nuclear fuel supply.

The Sun has fused hydrogen into helium throughout its present lifetime of 4.5 billion years, using up less than half of the available hydrogen in its core. In another 4.5 billion years, a total of 80% to 90% of the available hydrogen in the core will have been converted into helium. While the Sun's giant phase would still be a billion years off, the humans of the older Earth would face changes to the planet and the solar system as the Sun's output increased. These changes, such as a runaway greenhouse effect in which much of Earth's surface water becomes vaporized, could potentially force those humans to devise a way to quickly leave the planet and colonize elsewhere in the Galaxy.

APPENDIX A: Powers-of-Ten Notation

Astronomy is a science of extremes. As we examine various cosmic environments, we find an astonishing range of conditions, from the incredibly hot, dense centers of stars to the frigid, near-perfect vacuum of interstellar space. To describe such divergent conditions accurately, we need a wide range of both large and small numbers. Astronomers avoid such confusing terms as "a million billion billion" (1,000,000,000,000,000,000,000,000) by using a standard shorthand system. All the cumbersome zeros that accompany such a large number are consolidated into one term consisting of 10 followed by an *exponent,* which is written as a superscript and called the **power of ten.** The exponent merely indicates how many zeros you would need to write out the long form of the number. Thus,

$$10^0 = 1$$

$$10^1 = 10$$

$$10^2 = 100$$

$$10^3 = 1000$$

$$10^4 = 10{,}000$$

and so forth. Equivalently, the exponent tells you how many tens must be multiplied together to yield the desired number. For example, ten thousand can be written as 10^4 ("ten to the fourth") because $10^4 = 10 \times 10 \times 10 \times 10 = 10{,}000$. Similarly, 273,000,000 can be written as 2.73×10^8.

In scientific notation, numbers are written as a figure between 1 and 10 multiplied by the appropriate power of 10. The distance between the Earth and the Sun, for example, can be written as 1.5×10^8 km. Once you get used to it, you will find this notation more convenient than writing "150,000,000 kilometers" or "one hundred and fifty million kilometers."

This powers-of-ten system can also be applied to numbers that are less than 1 by using a minus sign in front of the exponent. A negative exponent tells you that the location of the decimal point is as follows:

$$10^0 = 1.0$$

$$10^{-1} = 0.1$$

$$10^{-2} = 0.01$$

$$10^{-3} = 0.001$$

$$10^{-4} = 0.0001$$

and so forth. For example, the diameter of a hydrogen atom is 1.1×10^{-8} cm. That is more convenient than saying "0.000000011 centimeter" or "11 billionths of a centimeter." Similarly, .000728 equals 7.28×10^{-4}.

Using the powers-of-ten shorthand, one can write large or small numbers like these compactly:

$$3{,}416{,}000 = 3.416 \times 10^6$$

$$0.000000807 = 8.07 \times 10^{-7}$$

Because powers-of-ten notation bypasses all the cumbersome zeros, a wide range of circumstances can be numerically described conveniently:

$$\text{one thousand} = 10^3$$

$$\text{one million} = 10^6$$

$$\text{one billion} = 10^9$$

$$\text{one trillion} = 10^{12}$$

and also

$$\text{one thousandth} = 10^{-3} = 0.001$$

$$\text{one millionth} = 10^{-6} = 0.000001$$

$$\text{one billionth} = 10^{-9} = 0.000000001$$

$$\text{one trillionth} = 10^{-12} = 0.000000000001$$

Try these questions: Write 3,141,000,000 and .0000000031831 in scientific notation. Write 2.718282×10^{10} and 3.67879×10^{-11} in standard notation.

(Answers appear at the end of the book.)

APPENDIX B: Guidelines for Solving Math Problems and Reading Graphs

Astronomy relies on mathematics. While we have distilled the concepts into words, you can gain further insights into the distances, intensities, and other astronomical quantities from the equations presented in this book. In addition, much scientific information is presented in the form of graphs, a technique that presents data very compactly and often provides a good way to see the trends that the data reveal. This appendix thus provides guidance on three things: setting up problems to be solved analytically (using equations), solving algebra equations, and reading graphs.

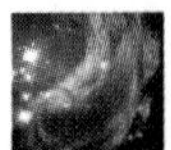

SETTING UP AND SOLVING ANALYTICAL PROBLEMS

At the end of most chapters, there are questions, indicated by an asterisk (*), that require mathematical solution. You will need to translate the question into an equation or two in order to solve for some value. This is best done systematically. In what follows, we will work a question, appropriate to Chapter 2: How fast does a meteoroid (small piece of rocky space debris) of mass 5 kg (kilograms) move if it has a kinetic energy of 10 joules (1 joule = 1 kg $\times$ m^2/sec^2)? This approach should work for most of the questions in the book.

Step 1. ***Write down all the information you are given in terms of the variable names used in the book.*** For example, denoting mass by m and kinetic energy by KE, write: mass m = 5 kg, kinetic energy KE = 10 joules.

Step 2. ***Identify and write down the thing you are trying to find in terms of its variable name.*** In this case: velocity, v = ?.

Step 3. ***Find (usually) one or (sometimes) two equations that are needed to solve for the unknown.*** All the equations presented in the book are listed in Appendix C. For our problem, use the kinetic energy equation $KE = \frac{1}{2}mv^2$, because this equation has the unknown, v, along with only variables and constants that are given, namely KE and m. (As in this case, you will often need only one equation, but sometimes you will have to solve first one equation and then another to get the answer. As an example, you might be given the information above but asked for the momentum, p, of the meteoroid. The equation for momentum, also in Chapter 2, is $p = mv$. First you would solve for the velocity, as we are now doing, and then use that v in this last equation.)

Step 4. ***Manipulate the equation(s) until you have the unknown variable on one side and all the other variables and constants on the other side.*** Variables can be added, subtracted, multiplied, and divided in the same way that numbers are combined. For our example, we want to find v, so we start with $KE = \frac{1}{2}mv^2$, multiply by 2, and divide by m. This gives $v^2 = 2\ KE/m$. Since we want v, we now take the square root of both sides. $v = \sqrt{2\ KE/m}$.

Step 5. ***Make sure that all the units match up.*** Units on both sides of the equation must be identical. Period. If you have kilometers (km) in one place and meters (m) on another, then the result you will get by combining numbers is not meaningful. If you end up with inconsistent units on opposite sides, convert one to the other. For example, change kilometers into meters or vice versa. In this case, use the fact that 1 km = 10^3 m.

Step 6: ***Plug in all the numbers and solve for the unknown.*** In our case, $v = \sqrt{2 \times 10 \text{ joules}/5 \text{ kg}}$ or v = 2 m/sec.

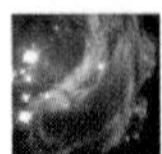

READING GRAPHS

The graphs you will encounter in this book are compact ways of displaying patterns of information relating two variables, like the temperature and luminosity of stars or the temperature at different altitudes in an atmosphere. The relevant values of one of the variables are presented along the horizontal or x axis and the relevant values of the other variable are presented along the vertical or y axis. The word "relevant" here indicates that often graphs do not start at 0. Consider three examples: First is data presented in Chapter 7 concerning the temperature of Venus's atmosphere at different altitudes (Figure 1).

The horizontal axis of the graph indicates the temperature in degrees Kelvin (K). The vertical axis denotes the altitude above Venus's surface in kilometers (km). Note that both axes are always labeled with a name (temperature or altitude, here) and units (K or km, respectively). Bear in mind that some variables in graphs you will see in this book increase to the right or upward (these are more common), but some variables will increase in the opposite directions.

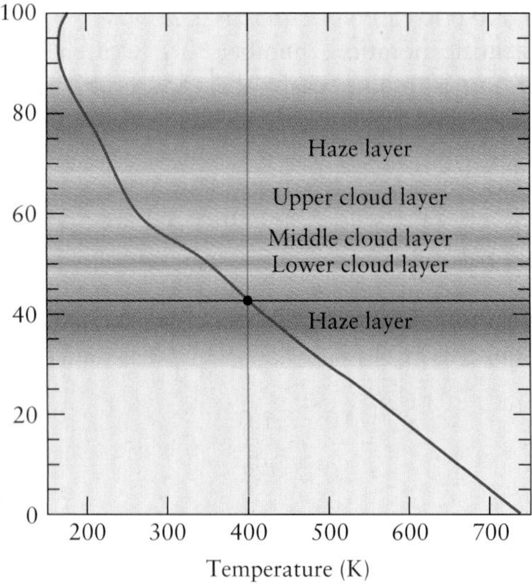

FIGURE 1 Linear Graph

To read a graph, note that a value on the horizontal axis is then transferred directly upward through the graph. For example, all points on the blue line in Figure 1 (which extends upward from 400 K) have a temperature of 400 K. Equivalently, the value given on the vertical axis is transferred to all points horizontally across from this value. All the points on the black line in Figure 1 are at an altitude of about 43 km above Venus's surface. We have interpolated between 40 and 45 to get this value (Figure 2).

A curve or a set of points on the graph presents the *relationship* between the variable represented on the horizontal axis and the variable on the vertical axis. Each point on a curve or each separate point relates the two variables. Choose a point, say the dot on Figure 1. It represents the temperature at a certain altitude above Venus's surface. To find the temperature at that point, you slide directly down (along the blue line in this example) from the point and read the value of the horizontal variable under it. To find the altitude for that point, you slide directly over to the side (along the black line) and read the value of the vertical variable there. In our example, sliding along the blue line leads to 400 K on the temperature line. Therefore, this point has a temperature of 400 K. Moving horizontally from the point, you encounter, by interpolation, the 43 km indicator. Combining the data, you conclude that *the temperature of Venus's atmosphere 43 km above its surface is 400 K.*

The red curve in Figure 1 provides the relationship between altitude and temperature for a wide range of locations above Venus in a representation that is much more informative than a table of heights and temperatures. Specifically, this curve shows you the trend of temperature with altitude.

Try these questions: What is the temperature at 20 km? What is the altitude at which the temperature is 300 K? For most of this graph, what is the general *trend* of the temperature with height? Answers are in Answers to Computational Questions on page Ans-2.

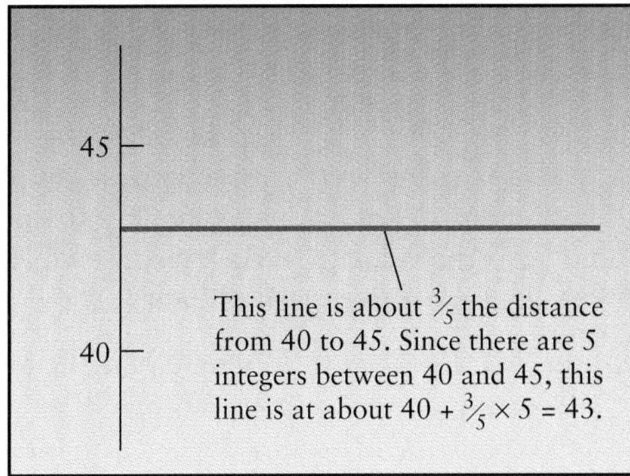

FIGURE 2 Interpolation

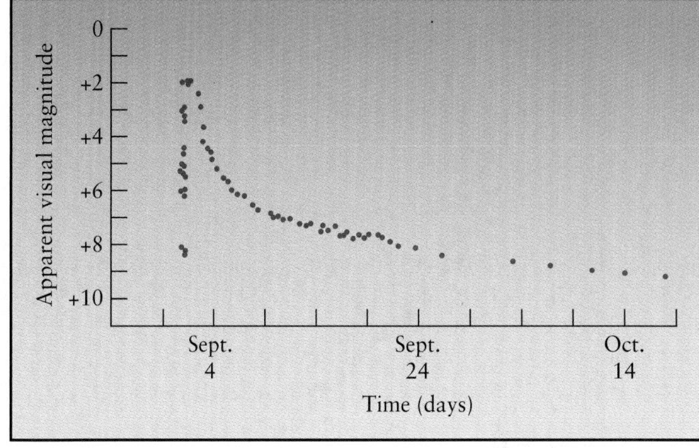

FIGURE 3 The Brightness of a Nova

Sometimes, the known information is not a curve, but rather a set of points, as in our second example (Figure 3), from Chapter 13. In this case, the graph connects the apparent magnitude of a nova (how bright it appears to be as seen from Earth regardless of its distance or other factors) and time. Each dot indicates how bright the nova (an explosion on the surface of certain stars) was at different times. For example, the peak brightness of the nova was an apparent magnitude of about +2 and it occurred on September 2. Noting that time passes to the right, you can immediately see that the trend of the nova's brightness is to rise rapidly and fall more slowly.

Try these questions: What is the apparent magnitude on September 24? October 9? On what two days was the apparent magnitude +6? Answers are in Answers to Computational Questions on page Ans-2.

The graphs so far have shown variables that change uniformly along the axes. For example, the distance on Figure 1 from 300 K to 400 K is the same as the distance from 400 K to 500 K, and so on. Many graphs you will encounter have variables that do not change uniformly (that is, linearly) along the axes. That means that the change in value going along each axis varies—there is not the same amount of change per centimeter along the axis. In Figure 4, from Chapter 12 and typical of graphs in the second half of the book, the temperature decreases going from left to right nonuniformly and the luminosity (total energy emitted per second) increases upward nonuniformly. These are called logarithmic scales.

The purpose of logarithmic and other nonlinear axes is to present in compact form data that varies very widely in values. For example, the dimmest star represented in Figure 4 is just less than 0.1 times as luminous as the Sun, while the brightest star there is nearly 1000 times as luminous. The process of

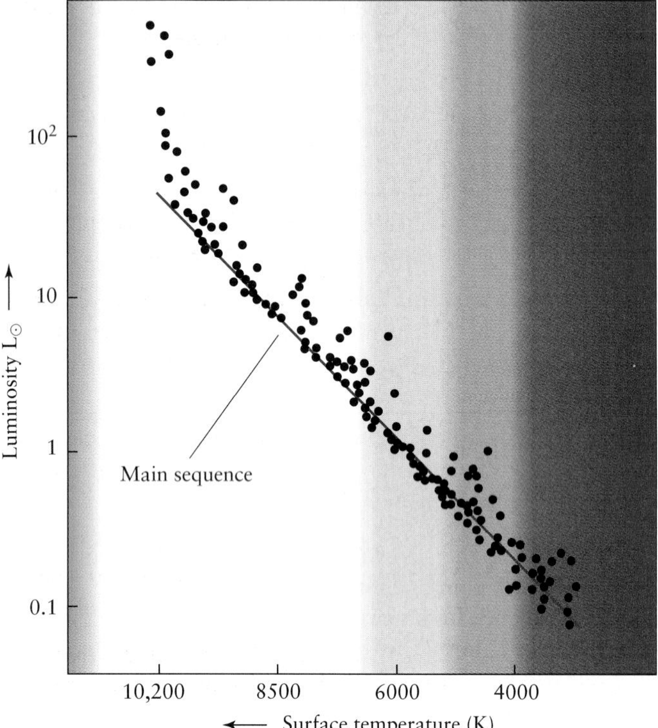

FIGURE 4 Logarithmic Graph

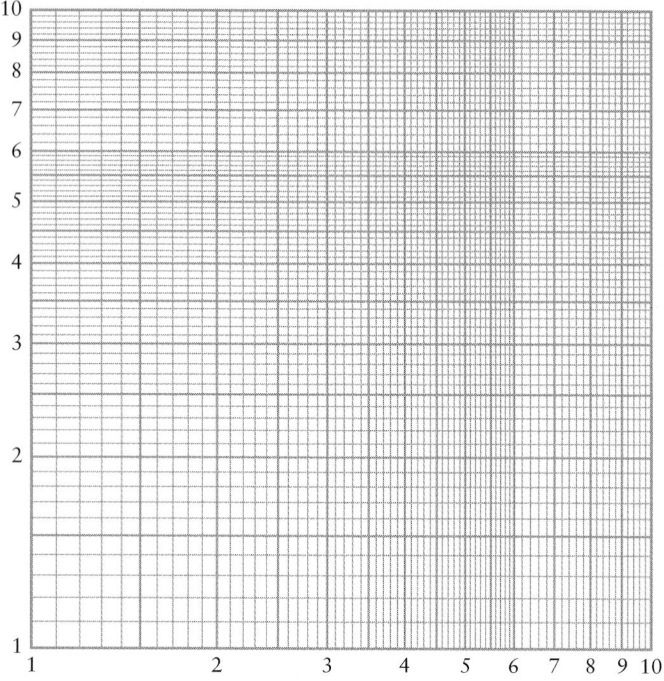

FIGURE 5 Logarithmic Scale

getting information off logarithmic graphs is the same as linear graphs. You must just be careful not to think of values as doubling or tripling when you go over or up two or three intervals. Figure 5 shows how a logarithmic scale varies over one decade of values. The same numbering intervals apply for any decade of values, for example, 1 to 10 or 10^5 to 10^6. As you can see, the numbers bunch up near the highest value, so you need to interpolate these graphs more carefully than linear graphs.

Referring to Figure 4, the luminosity of a star with surface temperature of 4000 K is about 0.1 $L_\odot$. Note, also, that some graphs are linear on one axis and logarithmic on the other axis.

Try these questions: Approximately what is the luminosity of a star with surface temperature 8500 K? What is the surface temperature of a star with the same luminosity as the Sun (1 $L_\odot$)? Answers are in Answers to Computational Questions on page Ans-2.

APPENDIX C: Key Formulas

Angular Momentum $L = I\omega$, where L is the angular momentum, I is the object's moment of inertia, and ω is the object's angular velocity (Chapter 2).

Area of a Circle $A = \pi d^2/4$, where π is approximately 3.14, and d is the diameter of the circle (Chapter 3).

Average Density $\rho = m/V$, where ρ is the average density, m is the total mass, and V is the volume of the object (Chapter 5).

Distance from Parallax $d = 1/p$ where d is the distance in parsecs and p is the parallax angle in arcseconds, or $d_{ly} = 3.26/p$, where d_{ly} is the distance in light-years (Chapter 11).

Distance-Magnitude Relationship $M = m - 5 \log(d/10)$, where M is the star's absolute magnitude, m is its apparent magnitude, and d is its distance in parsecs (Chapter 11).

Doppler Shift $\Delta\lambda/\lambda_o = v/c$, where $\Delta\lambda$ is the change in wavelength $(\lambda - \lambda_0)$, λ is the observed wavelength, λ_0 is the wavelength the object emits as seen by someone not moving toward or away from it, v is the velocity of the object toward or away from the observer, and c is the speed of light (Chapter 4).

Drake Equation $N = R f_p n_e f_l f_i f_c L$, where N is the number of advanced civilizations in our Galaxy estimated by the equation, R is the rate at which solar-type stars form in the Galaxy, f_p is the fraction of stars that have planets, n_e is the number of planets per star system suitable for life, f_l is the fraction of habitable planets on which life arises, f_i is the fraction of lifeforms that develope advanced intelligence, f_c is the fraction of species that develop technology and send signals into space, and L is the lifetime of technological civilizations (Chapter 19).

Energy Flux $F = \sigma T^4$, where σ is Stefan's constant and T is the blackbody's temperature in K (Chapter 4).

Energy-Mass Equation $E = mc^2$, where c is the speed of light and E is the energy released when mass m is converted to energy (Chapter 10).

Gravitational Force $F = G(m_1 m_2/r^2)$, where G is the gravitational constant, m_1 and m_2 are the masses of the two interacting objects, and r is the distance between them (Chapter 2).

Gravitational Potential Energy (always true) $PE = GmM/r$, where PE is the potential energy, G is the gravitational constant, m is the mass of the object you are measuring, M is the mass of the object creating the gravitational potential, and r is the distance between the centers of the two (Chapter 2).

Gravitational Potential Energy (near Earth) $PE = mgh$, where m is the mass of the object you are measuring, $g = 9.8$ m/s^2, and h is the height of the object above the Earth's surface (Chapter 2).

Kepler's Third Law $P^2 = a^3$, where P is the sidereal period in Earth years and a is its semimajor axis in AU (Chapter 2).

Kinetic Energy $KE = \frac{1}{2} mv^2 = p^2/2m = L^2/2I$, where KE is its kinetic energy, m is its mass, v is its velocity (or speed), p is the momentum, L is the angular momentum, and I is the object's moment of inertia. The first two equalities apply if the object is moving in a straight line. The last equality applies if the object is revolving or rotating (Chapter 2).

Luminosity $L = F \cdot 4\pi r^2$, where L is the luminosity (total energy per second) emitted, F is the energy flux, and r is the radius of the object (Chapter 4).

Magnification $M = f_o/f_e$, where f_o is the focal length of the objective and f_e is the focal length of the eyepiece (Chapter 3).

Momentum $\mathbf{p} = m\mathbf{v}$, where $\mathbf{p}$ is the momentum, m is the mass, and $\mathbf{v}$ is the velocity (Chapter 2).

Newton's Force Law $\mathbf{F} = m\mathbf{a}$, where F is the force acting on an object, m is its mass, and a is its acceleration ($\mathbf{F}$ and $\mathbf{a}$ are in boldface to denote that they act in some direction or other) (Chapter 2).

Newton's Version of Kepler's Third Law for Binary Star Systems $M_1 + M_2 = a^3/P^2$, where M_1 and M_2 are the masses of the stars, a is the average distance between them, and P is the period of their orbit in years (Chapter 11).

Photon Energy $E = hc/\lambda$, where E is the photon's energy, h is Planck's constant, c is the speed of light, and λ is the wavelength of the light (Chapter 3).

Pressure $P = F/A$, where P is the pressure, F is the force, and A is the area over which the force acts (Chapter 6).

Recessional Velocity of Galaxies $v = H_0 \times d$, where v is the recessional velocity in km/s, H_0 is the Hubble constant, and d is the distance to the galaxy in Mpc (Chapter 16).

Schwarzschild Radius of a Black Hole $R_{Sch} = 2\,GM/c^2$, where R_{Sch} is the Schwarzschild radius in meters, G is the gravitational constant, M is the mass of the black hole in kg, and c is the speed of light (Chapter 14).

Size of a Distant Object $D = R \times \tan\theta$, where D is the physical diameter of an object, R is the distance to it, and $\tan\theta$ is the tangent of the angle it makes in the sky (Chapter 1).

Wien's law $\lambda_{max} = 2.9 \times 10^3/T$, where λ_{max} is the peak wavelength of the blackbody and T is the blackbody's temperature in K (Chapter 4).

Work $W = Fd$, where W is the work, F is the force acting, and d is the distance moved in the direction that the force acts.

APPENDIX D: Temperature Scales

Three temperature scales are in common use. Throughout most of the world, temperatures are expressed in degrees Celsius (°C), named in honor of the Swedish astronomer Anders Celsius, who proposed it in 1742. The **Celsius temperature scale** (also known as the "centigrade scale") is based on the behavior of water, which freezes at 0°C and boils at 100°C at sea level on Earth.

Scientists usually prefer the **Kelvin scale,** named after the British physicist Lord Kelvin (William Thomson), who made many important contributions to our knowledge about heat and temperature. On the Kelvin temperature scale, water freezes at 273 K and boils at 373 K. Note that we do not use the degree symbol with the Kelvin temperature scale.

Because water must be heated by 100 K or 100°C to go from its freezing point to its boiling point, you can see that the size of a Kelvin is the same as the size of a degree Celsius. When considering temperature *changes,* measurements in Kelvins and in degrees Celsius lead to the same number.

A temperature expressed in Kelvins is always equal to the temperature in degrees Celsius plus 273. Scientists prefer the Kelvin scale because it is closely related to the physical meaning of temperature. All substances are made of atoms, which are very tiny (a typical atom has a diameter of about 10^{-10} m) and constantly in motion. The temperature of a substance is directly related to the average speed of its atoms. If something is hot, it atoms are moving at high speeds. If a substance is cold, its atoms are moving much more slowly.

The coldest possible temperature is the temperature at which atoms move as slowly as possible (they can never quite stop completely). This minimum possible temperature, called absolute zero, is the starting point for the Kelvin scale. Absolute zero is 0 K, or –273°C. Because it is impossible for anything to be colder than 0 K, there are no negative temperatures on the Kelvin scale.

In the United States, many people still use the now archaic Fahrenheit scale, which expresses temperature in degrees Fahrenheit (°F). When the German physicist Gabriel Fahrenheit introduced this scale in the early 1700s, he intended 0°F to represent the coldest temperature then achievable (with a mixture of ice and saltwater) and 100°F to represent the temperature of a healthy human body. On the Fahrenheit scale, water freezes at 32°F and boils at 212°F. Because there are 180 degrees Fahrenheit between the freezing and boiling points of water, a degree Fahrenheit is only 100/180 (= 5/9) the size of the other scales.

The following equation converts from degrees Fahrenheit to degrees Celsius:

$$T_C = 5/9\ (T_F - 32)$$

To convert from Celsius to Fahrenheit, a simple rearrangement of terms gives the relationship

$$T_F = 9/5\ T_C + 32$$

where T_F is the temperature in degrees Fahrenheit and T_C is the temperature in degrees Celsius.

For example, consider a typical room temperature of 68°F. Using the first equation, we can convert this measurement to the Celsius scale as follows:

$$T_C = 5/9\ (68 - 32) = 20°\text{C}$$

To arrive at the Kelvin scale, we simply add 273 degrees to the value in degrees Celsius. Thus, 68°F = 20°C = 293 K. The accompanying figure displays the relationships among these three temperature scales.

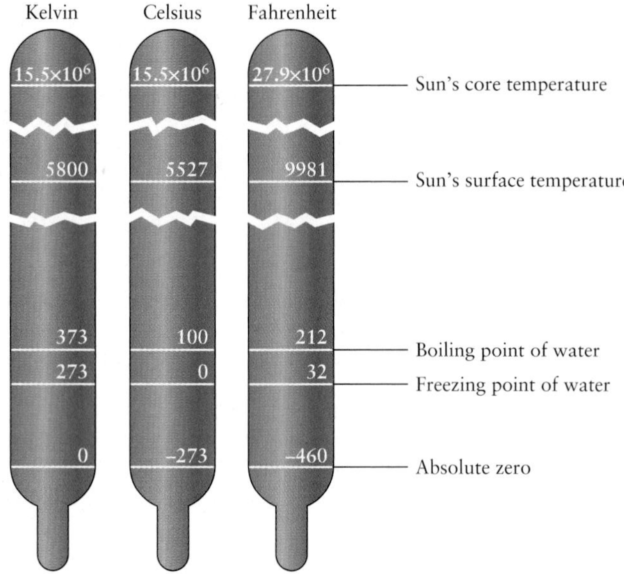

Try these questions: The Sun's surface temperature is about 5800 K. What is its temperature in Celsius and Fahrenheit? The temperature of empty space is about 3 K. What is its temperature in Celsius and Fahrenheit?

(Answers appear at the end of the book.)

APPENDIX E: Data Tables

TABLE E-1 The Planets: Orbital Data

Planet	Semimajor axis (AU)	Semimajor axis (10^6 km)	Sidereal period (yr)	Sidereal period (d)	Synodic period (d)	Mean orbital speed (km/s)	Orbital eccentricity	Inclination of orbit to ecliptic (°)
Mercury	0.3871	57.9	0.2408	87.97	115.88	47.9	0.206	7.00
Venus	0.7233	108.2	0.6152	224.70	583.92	35.0	0.007	3.39
Earth	1.0000	149.6	1.0000	365.26	—	29.8	0.017	0.00
Mars	1.5237	227.9	1.8809	686.98	779.94	24.1	0.093	1.85
(Ceres)	2.7656	413.7	4.603	1,681.3	466.6	17.9	0.097	10.61
Jupiter	5.2034	778.3	11.862	4,332.7	398.9	13.1	0.048	1.31
Saturn	9.5371	1426.7	29.458	10,760	378.1	9.6	0.054	2.49
Uranus	19.1913	2871.0	84.01	30,685	369.7	6.8	0.047	0.77
Neptune	30.0690	4498.3	164.79	60,082	367.5	5.4	0.009	1.77
Pluto	39.4817	5906.4	247.9	90,767	366.7	4.7	0.249	17.15

TABLE E-2 The Planets: Physical Data

Planet	Equatorial diameter (km)	Equatorial diameter (Earth = 1)	Mass (kg)	Mass (Earth = 1)	Average density (kg/m^3)	Rotation period* (days)	Inclination of equator to orbit (°)	Surface gravity (Earth = 1)	Albedo	Escape speed (km/s)
Mercury	4,879	0.383	3.302×10^{23}	0.055	5430	58.646	0.0	0.39	0.106	4.3
Venus	12,104	0.949	4.869×10^{24}	0.815	5240	243.01^{R}	177.3	0.91	0.65	10.4
Earth	12,756	1.000	5.974×10^{24}	1.000	5515	0.997	23.45	1.000	0.37	11.2
Mars	6,794	0.533	6.419×10^{23}	0.107	3940	1.026	25.19	0.38	0.15	5.0
Jupiter	142,984	11.209	1.899×10^{27}	317.83	1330	0.414	3.12	2.5	0.52	59.5
Saturn	120,536	9.449	5.685×10^{26}	95.16	700	0.444	26.73	1.1	0.47	35.5
Uranus	51,118	4.007	8.685×10^{25}	14.54	1300	0.720^{R}	97.86	0.90	0.50	21.3
Neptune	49,528	3.883	1.024×10^{26}	17.204	1760	0.671	29.56	1.1	0.5	23.5
Pluto	2,300	0.18	1.3×10^{22}	0.002	2000	6.387^{R}	119.6	0.07	0.5	1.3

* *For Jupiter, Saturn, Uranus, and Neptune, the internal rotation period is given. A superscript R means that the rotation is retrograde (opposite the planet's orbital motion).*

TABLE E-3 Satellites of the Planets

Planet	Satellite	Discoverers of planet	Average distance from center (km)	Orbital (sidereal) period* (days)	Orbital eccentricity	Diameter of satellite (km)	Mass (kg)
EARTH	Moon	—	384,400	27.322	0.0549	3476	7.348×10^{22}
MARS	Phobos	Hall (1877)	9,380	0.319	0.01	28 × 23 × 2	1.1×10^{16}
	Deimos	Hall (1877)	23,460	1.263	0.00	16 × 12 × 10	1.8×10^{15}
JUPITER	Metis	Synott (1979)	127,960	0.2948	0.00	40	1×10^{17}
	Adrastea	Jewitt et al. (1979)	128,980	0.2983	0 (?)	24 × 16 × 20	1.9×10^{16}
	Amalthea	Barnard (1892)	181,300	0.4981	0.00	270 × 200 × 155	7.2×10^{18}
	Thebe	Synott (1979)	221,900	0.6745	0.01	100	8×10^{17}
	Io	Galileo (1610)	421,600	1.769	0.00	3630	8.94×10^{22}
	Europa	Galileo (1610)	670,900	3.551	0.01	3138	4.92×10^{22}
	Ganymede	Galileo (1610)	1,070,000	7.155	0.00	5262	1.48×10^{23}
	Callisto	Galileo (1610)	1,883,000	16.689	0.01	4800	1.08×10^{23}
	Themisto	Kowal (1975)	7,435,000	131	0.20	5	(?)
	Leda	Kowal (1974)	11,094,000	238.72	0.15	16	6×10^{15}
	Himalia	Perrine (1904)	11,480,000	250.57	0.16	180	1×10^{19}
	Lysithea	Nicholson (1938)	11,720,000	259.22	0.11	40	8×10^{16}
	Elara	Perrine (1905)	11,737,000	259.65	0.21	80	8×10^{17}
	S/2000/J11	Sheppard et al. (2000)	12,654,000	290	0.22	4	(?)
	Euporie	Sheppard et al. (2001)	19,017,000	534^{R}	0.16	2	(?)
	Chaldene	Sheppard et al. (2000)	20,375,000	591^{R}	0.16	4	(?)
	Iocaste	Sheppard et al. (2000)	20,733,000	605^{R}	0.27	5	(?)
	Kale	Sheppard et al. (2001)	20,804,000	609^{R}	0.48	2	(?)
	Orthosie	Sheppard et al. (2001)	20,876,000	617^{R}	0.27	2	(?)
	Thyone	Sheppard et al. (2001)	20,876,000	615^{R}	0.30	4	(?)
	Euanthe	Sheppard et al. (2001)	20,947,000	622^{R}	0.18	3	(?)
	Harpalyke	Sheppard et al. (2000)	21,019,000	618^{R}	0.20	4	(?)
	Praxidike	Sheppard et al. (2000)	21,162,000	626^{R}	0.15	7	(?)
	Ananke	Nicholson (1951)	21,200,000	631^{R}	0.17	30	4×10^{16}
	Hermippe	Sheppard et al. (2001)	21,376,000	630^{R}	0.25	4	(?)
	Taygete	Sheppard et al. (2000)	21,734,000	652^{R}	0.25	5	(?)
	Erinome	Sheppard et al. (2000)	21,948,000	661^{R}	0.35	3	(?)
	Carme	Nicholson (1938)	22,600,000	692^{R}	0.21	44	1×10^{17}
	Isonoe	Sheppard et al. (2000)	22,806,000	704^{R}	0.28	4	(?)
	S/2001 J6	Sheppard et al. (2001)	22,949,000	715^{R}	0.29	2	(?)

TABLE E-3 Satellites of the Planets (continued)

Planet	Satellite	Discoverers of planet	Average distance from center (km)	Orbital (sidereal) period* (days)	Orbital eccentricity	Diameter of satellite (km)	Mass (kg)
	Eurydome	Sheppard et al. (2001)	23,378,000	713^R	0.35	3	(?)
	Aitne	Sheppard et al. (2001)	23,449,000	736^R	0.29	3	(?)
	Pasiphae	Melotte (1908)	23,500,000	735^R	0.38	2	(?)
	Megaclite	Sheppard et al. (2000)	23,521,000	733^R	0.53	5	(?)
	Sponde	Sheppard et al. (2001)	23,592,000	732^R	0.45	2	(?)
	Sinope	Nicholson (1914)	23,700,000	758^R	0.28	40	8×10^{16}
	Autonoe	Sheppard et al. (2001)	23,979,000	753^R	0.42	4	(?)
	Kalyke	Sheppard et al. (2000)	24,164,000	766^R	0.32	5	(?)
	Callirrhoe	Spacewatch (1999)	24,200,000	(?)	0.13	5	(?)
SATURN	Pan	Showalter (1990)	133,570	0.573	0.00	20	(?)
	Atlas	Terrile (1980)	137,640	0.602	0 (?)	$40 \times 30 \times 30$	(?)
	Prometheus	Collins et al. (1980)	139,350	0.613	0.00	$140 \times 80 \times 100$	3×10^{17}
	Pandora	Collins et al. (1980)	141,700	0.629	0.00	$110 \times 70 \times 100$	2×10^{17}
	Epimetheus	Walker (1966)	151,422	0.694	0.01	$140 \times 100 \times 100$	6×10^{17}
	Janus	Dolfuss (1966)	151,472	0.695	0.01	$220 \times 160 \times 200$	2×10^{18}
	Mimas	Herschel (1789)	185,520	0.942	0.02	390	3.8×10^{19}
	Enceladus	Herschel (1789)	238,020	1.370	0.00	500	7.3×10^{19}
	Tethys	Cassini (1684)	294,660	1.888	0.00	1050	6.2×10^{20}
	Calypso	Smith et al. (1980)	294,660	1.888	0 (?)	$30 \times 20 \times 25$	(?)
	Telesto	Smith et al. (1980)	294,660	1.888	0 (?)	24	(?)
	Dione	Cassini (1684)	377,400	2.737	0.00	1120	1.1×10^{21}
	Helene	Laques et al. (1980)	377,400	2.737	0.01	$40 \times 30 \times 30$	(?)
	Rhea	Cassini (1672)	527,040	4.518	0.00	1530	2.3×10^{21}
	Titan	Huygens (1655)	1,221,850	15.945	0.03	5150	1.4×10^{23}
	Hyperion	Bond (1848)	1,481,000	21.277	0.10	$410 \times 260 \times 220$	2×10^{19}
	Iapetus	Cassini (1671)	3,561,300	79.331	0.03	1440	1.6×10^{21}
	Phoebe	Pickering (1898)	12,952,000	9.274^R	0.16	220	4×10^{18}
URANUS	Cordelia	*Voyager 2* (1986)	49,750	0.335	0 (?)	40	(?)
	Ophelia	*Voyager 2* (1986)	53,760	0.376	0 (?)	30	(?)
	Bianca	*Voyager 2* (1986)	59,160	0.435	0 (?)	40	(?)
	Cressida	*Voyager 2* (1986)	61,770	0.464	0 (?)	70	(?)
	Desdemona	*Voyager 2* (1986)	62,660	0.474	0 (?)	60	(?)
	Juliet	*Voyager 2* (1986)	64,360	0.493	0 (?)	80	(?)

TABLE E-3 Satellites of the Planets (continued)

Planet	Satellite	Discoverers of planet	Average distance from center (km)	Orbital (sidereal) period* (days)	Orbital eccentricity	Diameter of satellite (km)	Mass (kg)
	Portia	*Voyager 2* (1986)	66,100	0.513	0 (?)	110	(?)
	Rosalind	*Voyager 2* (1986)	69,930	0.558	0 (?)	60	(?)
	S/2003 U1	2003	97,734	(?)	0 (?)	26	(?)
	Belinda	*Voyager 2* (1986)	75,260	0.624	0 (?)	70	(?)
	S/1986 U10	*Voyager 2* (1986)	76,420	0.638	0 (?)	20	(?)
	Puck	*Voyager 2* (1986)	86,010	0.762	0 (?)	150	(?)
	S/2003 U2	2003	74,800	(?)	0 (?)	10	(?)
	Miranda	Kuiper (1948)	129,780	1.414	0.00	470	6.6×10^{19}
	Ariel	Lassell (1851)	191,240	2.520	0.00	1160	1.4×10^{21}
	Umbriel	Lassell (1851)	265,790	4.144	0.00	1170	1.2×10^{21}
	Titania	Herschel (1787)	435,840	8.706	0.00	1580	3.5×10^{21}
	Oberon	Herschel (1787)	582,600	13.463	0.00	1520	3.0×10^{21}
	Caliban	Gladman et al. (1997)	5,753,400	415^{R}	0.16	80	(?)
	Sycorax	Gladman et al. (1997)	8,000,000	415^{R}	0.52	160	(?)
	Stephano	Kavelaars et al. (1999)	>10,000,000	675	0.15	0 (?)	(?)
	Prospero	Kavelaars et al. (1999)	>10,000,000	1997	0.32	0 (?)	(?)
	Setebos	Kavelaars et al. (1999)	>10,000,000	2194	0.56	0 (?)	(?)
NEPTUNE	Naiad	*Voyager 2* (1989)	48,230	0.296	0 (?)	60	(?)
	Thalassa	*Voyager 2* (1989)	50,070	0.312	0 (?)	80	(?)
	Despina	*Voyager 2* (1989)	52,530	0.333	0 (?)	150	(?)
	Galatea	*Voyager 2* (1989)	61,950	0.429	0 (?)	180	(?)
	Larissa	*Voyager 2* (1989)	73,550	0.554	0 (?)	190	(?)
	Proteus	*Voyager 2* (1989)	117,640	1.121	0 (?)	415	(?)
	Triton	Lassell (1846)	354,800	5.877^{R}	0.00	2700	2.15×10^{22}
	Nereid	Kuiper (1949)	5,513,400	359.2	0.75	340	(?)
PLUTO	Charon	Christy (1978)	19,640	6.387	0.00	1190	1.9×10^{21}

* *A superscript R means that the satellite orbits in a retrograde direction (opposite to the planet's rotation).*

TABLE E-4 The Nearest Stars

Name*	Parallax (arcsec)	Distance (light-years)	Spectral type	Radial velocity** (km/s)	Proper motion (arcsec/year)	Apparent visual magnitude	Absolute visual magnitude	Luminosity (Sun = 1)
Sun			G2 V			–26.7	+4.83	1.00
Proxima Centauri	0.772	4.22	M5.5 V	–22	3.853	+11.01	+15.45	8.2×10^{-4}
Alpha Centauri A	0.742	4.40	G2 V	–25	3.710	–0.01	+4.34	1.77
Alpha Centauri B	0.742	4.40	K0 V	–21	3.724	+1.35	+5.70	0.55
Barnard's Star	0.549	5.94	M4 V	–111	10.358	+9.54	+13.24	3.6×10^{-3}
Wolf 359	0.418	7.80	M6 V	+13	4.689	+13.45	+16.6	3.5×10^{-4}
Lalande 21185	0.392	8.32	M2 V	–84	4.802	+7.49	+10.46	0.023
L 726-8 A	0.381	8.56	M5.5 V	+29	3.366	+12.41	+15.3	9.4×10^{-4}
L 726-8 B	0.381	8.56	M6 V	+32	3.366	+13.2	+16.1	5.6×10^{-4}
Sirius A	0.379	8.61	A1 V	–9	1.33	–1.44	+1.45	26.1
Sirius B	0.379	8.61	white dwarf	–9	1.339	+8.44	+11.3	2.4×10^{-3}
Ross 154	0.336	9.71	M3.5 V	–12	0.666	+10.37	+13.00	4.1×10^{-3}
Ross 248	0.316	10.32	M5.5 V	–78	1.626	+12.29	+14.8	1.5×10^{-3}
Epsilon Eridani	0.311	10.49	K2 V	+17	0.977	+3.72	+6.18	0.40
Lacaille 9352	0.304	10.73	M1.5 V	+10	6.896	+7.35	+9.76	0.051
Ross 128	0.300	10.87	M4 V	–31	1.361	+11.12	+13.50	2.9×10^{-3}
L 789-6	0.294	11.09	M5 V	–60	3.259	+12.33	+14.7	1.3×10^{-3}
61 Cygni A	0.287	11.36	K5 V	–65	5.281	+5.20	+7.49	0.16
61 Cygni B	0.285	11.44	K7 V	–64	5.172	+6.05	+8.33	0.095
Procyon A	0.286	11.40	F5 IV–V	–4	1.259	+0.40	+2.68	7.73
Procyon B	0.286	11.40	white dwarf	–4	1.259	+10.7	+13.0	5.5×10^{-4}
BD +59° 1915 A	0.281	11.61	M3 V	–1	2.238	+8.94	+11.18	0.020
BD +59° 1915 B	0.281	11.61	M3.5 V	+1	2.313	+9.70	+11.97	0.010
Groombridge 34 A	0.280	11.65	M1.5 V	+12	2.918	+8.09	+10.33	0.030
Groombridge 34 B	0.280	11.65	M3.5 V	+11	2.918	+11.07	+13.3	3.1×10^{-3}
Epsilon Indi	0.276	11.82	K5 V	–40	4.704	+4.69	+6.89	0.27
GJ 1111	0.276	11.82	M6.5 V	–5	1.288	+14.79	+17.0	2.7×10^{-4}
Tau Ceti	0.274	11.90	G8 V	–17	1.922	+3.49	+5.68	0.62
GJ 1061	0.270	12.08	M5.5 V	–20	0.836	+13.03	+15.2	1.0×10^{-3}
L 725-32	0.269	12.12	M4.5 V	+28	1.372	+12.10	+14.25	1.7×10^{-3}
BD +05° 1668	0.263	12.40	M3.5 V	+18	3.738	+9.84	+11.94	0.011
Kapteyn's star	0.255	12.79	M0 V	+246	8.670	+8.86	+10.89	0.013
Lacaille 8760	0.253	12.89	M0 V	+28	3.455	+6.69	+8.71	0.094
Krüger 60 A	0.250	13.05	M3 V	–33	0.990	+9.85	+11.9	0.010
Krüger 60 B	0.250	13.05	M4 V	–32	0.990	+11.3	+13.3	3.4×10^{-3}

* *Stars that are components of binary systems are labeled A and B.*

** *A positive radial velocity means the star is receding; a negative radial velocity means the star is approaching.*

Compiled from the Hipparcos General Catalogue *and from data reported by the Research Consortium on Nearby Stars. The table lists all known stars within 4.00 parsecs (13.05 light-years).*

TABLE E-5 The Visually Brightest Stars

Name	Designation	Distance (light-years)	Spectral type	Radial velocity* (km/s)	Proper motion (arcsec/year)	Apparent visual magnitude	Apparent visual brightness** (Sirius = 1)	Absolute visual magnitude	Luminosity (Sun = 1)
Sirius A	α CMa A	8.61	A1 V	−9	1.339	−1.44	1.000	+1.45	26.1
Canopus	α Car	313	F0 I	+21	0.031	−0.62	0.470	−5.53	1.4×10^4
Arcturus	α Boo	36.7	K2 III	−5	2.279	−0.05	0.278	−0.31	190
Rigil Kentaurus	α Cen A	4.4	G2 V	−25	3.71	−0.01	0.268	+4.34	1.77
Vega	α Lyr	25.3	A0 V	−14	0.035	+0.03	0.258	+0.58	61.9
Capella	α Aur	42.2	G8 III	+30	0.434	+0.08	0.247	−0.48	180
Rigel	β Ori A	773	B8 Ia	+21	0.002	+0.18	0.225	−6.69	7.0×10^5
Procyon	α CMi A	11.4	F5 IV–V	−4	1.259	+0.4	0.184	+2.68	7.73
Achernar	α Eri	144	B3 IV	+19	0.097	+0.45	0.175	−2.77	5250
Betelgeuse	α Ori	427	M2 Iab	+21	0.029	+0.45	0.175	−5.14	4.1×10^4
Hadar	β Cen	525	B1 II	−12	0.042	+0.61	0.151	−5.42	8.6×10^4
Altair	α Aql	16.8	A7 IV–V	−26	0.661	+0.76	0.132	+2.2	11.8
Aldebaran	α Tau A	65.1	K5 III	+54	0.199	+0.87	0.119	−0.63	370
Spica	α Vir	262	B1 V	+1	0.053	+0.98	0.108	−3.55	2.5×10^4
Antares	α Sco A	604	M1 Ib	−3	0.025	+1.06	0.100	−5.28	3.7×10^4
Pollux	β Gem	33.7	K0 III	+3	0.627	+1.16	0.091	+1.09	46.6
Fomalhaut	α PsA	25.1	A3 V	+7	0.368	+1.17	0.090	+1.74	18.9
Deneb	α Cyg	3230	A2 Ia	−5	0.002	+1.25	0.084	−8.73	3.2×10^5
Mimosa	β Cru	353	B0.5 III	+20	0.05	+1.25	0.084	−3.92	3.4×10^4
Regulus	α Leo A	77.5	B7 V	+4	0.249	+1.36	0.076	−0.52	331

Data in this table was compiled from the Hipparcos General Catalogue.

* *A positive radial velocity means the star is receding; a negative radial velocity means the star is approaching.*

** *This is the ratio of the star's apparent brightness to that of Sirius, the brightest star in the night sky*

Note: *Acrux, or α Cru (the brightest star in Crux, the Southern Cross), appears to the naked eye as a star of apparent magnitude +0.87, the same as Aldebaran, but it does not appear in this table because Acrux is actually a binary star system. The blue-white component stars of this binary system have apparent magnitudes of +1.4 and +1.9, and thus they are dimmer than any of the stars listed here.*

TABLE E-6 The Constellations

Name	Meaning	R.A.	Dec.	Genitive*	Abbreviation
Andromeda	proper name; princess	1	+40	Andromedae	And
Antlia	air pump	10	−35	Antliae	Ant
Apus	bee	16	−75	Apodis	Aps
Aquarius[1,2]	waterman	22	−10	Aquarii	Aqr
Aquila	eagle	20	+15	Aquilae	Aql
Ara	altar	17	−55	Arae	Ara
Aries[2]	ram	3	+20	Arietis	Ari
Auriga	charioteer	6	+40	Aurigae	Aur
Boötes	proper name; herdsman, wagoner	15	+30	Boötis	Boo
Caelum	engraving tool	5	−40	Caeli	Cae
Camelopardalis	giraffe	6	+70	Camelopardalis	Cam
Cancer[2]	crab	8.5	+15	Cancri	Cnc
Canes Venatici	hunting dogs	13	+40	Canum Venaticorum	CVn
Canis Major	larger dog	7	−20	Canis Majoris	CMa
Canis Minor	smaller dog	8	+5	Canis Minoris	CMi
Capricornus[1,2]	water-goat	21	−20	Capricornii	Cap
Carina	keel	9	−60	Carinae	Car
Cassiopeia	proper name; queen	1	+60	Cassiopeiae	Cas
Centaurus	centaur	13	−45	Centauri	Cen
Cepheus	proper name; king	22	+65	Cephei	Cep
Cetus	whale	2	−10	Ceti	Cet
Chamaeleon	chameleon	10	−80	Chamaeleontis	Cha
Circinus	compasses	15	−65	Circini	Cir
Columba	dove	6	−35	Columbae	Col
Coma Berenices	Berenice's hair	13	+20	Comae Berenices	Com
Corona Australis[3]	southern crown	19	+40	Coronae Australis	CrA
Corona Borealis[4]	northern crown	16	+30	Coronae Borealis	CrB
Corvus[5]	crow, raven	12	−20	Corvi	Crv
Crater	cup	11	−15	Crateris	Crt
Crux[6]	southern cross	12	−60	Crucis	Cru
Cygnus	swan	21	+40	Cygni	Cyg
Delphinus[1]	Dolphin	21	+10	Delphini	Del
Dorado[7]	swordfish	6	−55	Doradus	Dor
Draco[8]	dragon	15	+60	Draconis	Dra
Equuleus	little horse	21	+10	Equulei	Equ

TABLE E-6 The Constellations (continued)

Name	Meaning	R.A.	Dec.	Genitive*	Abbreviation
Eridanus	proper name; river	4	–30	Eridani	Eri
Fornax	furnace	3	–30	Fornacis	For
Gemini[2]	twins	7	+20	Geminorum	Gem
Grus	crane	22	–45	Gruis	Gru
Hercules[9]	proper name; hero	17	+30	Herculis	Her
Horologium	clock	3	–55	Horologii	Hor
Hydra	water serpant	12	–25	Hydrae	Hya
Hydrus	water snake	2	–70	Hydri	Hyi
Indus	Indian	22	–70	Indi	Ind
Lacerta	lizard	22	+45	Lacertae	Lac
Leo[2]	lion	11	+15	Leonis	Leo
Leo Minor	smaller lion	10	+35	Leonis Minoris	LMi
Lepus	hare	6	–20	Leporis	Lep
Libra[2,10]	scales	15	–15	Librae	Lib
Lupus	wolf	15	–45	Lupi	Lup
Lynx	lynx	8	+45	Lyncis	Lyn
Lyra[5]	lyre	19	+35	Lyrae	Lyr
Microscopium	microscope	21	–40	Microscopii	Mic
Monoceros	unicorn	7	0	Monocerotis	Mon
Mensa	table	6	–75	Mensae	Men
Musca (Australis)[11]	(southern) fly	12	–70	Muscae	Mus
Norma	square	16	–50	Normae	Nor
Octans[12]	octant	—	–90	Octantis	Oct
Ophiuchus[13]	serpent-bearer	17	0	Ophiuchi	Oph
Orion	proper name; hunter, giant	6	0	Orionis	Ori
Pavo	peacock	20	–70	Pavonis	Pav
Pegasus	proper name; winged horse	23	+20	Pegasi	Peg
Perseus	proper name; hero	3	+45	Persei	Per
Phoenix	phoenix	1	–50	Phoenicis	Phe
Pictor	easel	6	–55	Pictoris	Pic
Pices[1,2]	fishes	1	+10	Piscium	Psc
Piscis Austrinus[1]	southern fish	22	–30	Piscis Austrini	PsA
Puppis	stern	8	–30	Puppis	Pup
Pyxis	compass	9	–30	Pyxidis	Pyx
Reticulum	net	4	–60	Reticuli	Ret
Sagitta	arrow	20	+20	Sagittae	Sge

TABLE E-6 The Constellations (continued)

Name	Meaning	R.A.	Dec.	Genitive*	Abbreviation
Sagittarius[2,14]	archer	19	−25	Sagittarii	Sgr
Scorpius[2]	scorpion	17	−30	Scorpii	Sco
Sculptor[15]	sculptor's workshop	1	−30	Sculptoris	Scl
Scutum[16]	shield	19	−10	Scuti	Sct
Serpens[13]	serpent	17	0	Serpentis	Ser
Sextans	sextant	10	0	Sextantis	Sex
Taurus[2]	bull	5	+20	Tauri	Tau
Telescopium	telescope	19	−50	Telescopii	Tel
Triangulum	triangle	2	+30	Trianguli	Tri
Triangulum Australe	southern triangle	16	−65	Trianguli Australis	TrA
Tucana[17]	toucan	0	−65	Tucanae	Tuc
Ursa Major	larger bear	11	+60	Ursae Majoris	UMa
Ursa Minor[18]	smaller bear	16	+80	Ursae Minoris	UMi
Vela	sails	10	−45	Velorum	Vel
Virgo[2]	Virgin	13	0	Virginis	Vir
Volans	flying fish	8	−70	Volantis	Vol
Vulpecula	fox	20	+25	Vulpeculae	Vul

** Genitive is the grammatical case denoting possession. For example, astronomers denote the brightest or α (alpha) star in Orion (Betelgeuse) as α Orionis.*

[1]Constellations of the area of the sky known as the wet quarter for its many watery images.

[2]A zodiac constellation.

[3]Sometimes considered as Sagittarius's crown.

[4]Ariadne's crown.

[5]Corvus was Orpheus's companion, Lyra his harp.

[6]Originally a part of Centaurus.

[7]Contains the Large Magellanic Cloud and the south ecliptic pole.

[8]Contains the north ecliptic pole.

[9]One of the oldest constellations known.

[10]Originally the claws of Scorpius.

[11]Originally named Musca Australis to distinguish it from Musca Borealis, the northern fly, which is now defunct; "Australis" is now dropped.

[12]Contains the south celestial pole.

[13]Ophiucus is identified with the physician Aesculapius, and Serpens with the caduceus.

[14]Contains the galactic center.

[15]Originally named by Lacaille l'Atelier du Sculpteur (in Latin, Apparatus Sculptoris); now known simply as Sculptor. Contains the south galactic pole.

[16]Shield of the Polish hero John Sobieski.

[17]Contains the Small Magellanic Cloud.

[18]Contains the north celestial pole.

TABLE E-7 Some Useful Astronomical Quantities

Astronomical unit:	1 AU = 1.496×10^{11} m
Light-year:	1 ly = 9.461×10^{15} m = 63,240 AU
Parsec:	1 pc = 3.086×10^{16} m = 3.262 ly
Solar mass:	$1\ M_{\odot} = 1.989 \times 10^{30}$ kg
Solar radius:	$1\ R_{\odot} = 6.960 \times 10^{8}$ m
Solar luminosity:	$1\ L_{\odot} = 3.827 \times 10^{26}$ W
Earth's mass:	$1\ M_{\oplus} = 5.974 \times 10^{24}$ kg
Earth's equatorial radius:	$1\ R_{\oplus} = 6.378 \times 10^{6}$ m
Moon's mass:	$1\ M_{Moon} = 7.348 \times 10^{22}$ kg
Moon's equatorial radius:	$1\ R_{Moon} = 1.738 \times 10^{6}$ m

TABLE E-8 Some Useful Physical Constants

Speed of light:	$c = 2.998 \times 10^{8}$ m/s
Gravitational constant:	$G = 6.668 \times 10^{-11}$ N m^2 kg^{-2}
Planck constant:	$h = 6.625 \times 10^{-34}$ J s $= 4.136 \times 10^{-15}$ eV s
Boltzmann constant:	$k = 1.380 \times 10^{-23}$ J K^{-1} $= 8.617 \times 10^{-5}$ eV K^{-1}
Stefan–Boltzmann constant:	$\sigma = 5.669 \times 10^{-8}$ W m^{-2} K^{-4}
Mass of electron:	$m_e = 9.108 \times 10^{-31}$ kg
Mass of 1H atom:	$m_H = 1.673 \times 10^{-27}$ kg

TABLE E-9 Common Conversions between British and Metric Units

1 inch = 2.54 centimeters (cm)
1 cm = 0.394 inch
1 yard = 0.914 meter (m)
1 meter = 1.09 yards = 39.37 inches
1 mile = 1.61 kilometers (km)
1 km = 0.621 mile

TABLE E-10 Spiral Galaxies and Interacting Galaxies

Spiral galaxy	R.A.	Decl.	Hubble type
M31 (NGC 224)	$0^h\ 42.7^m$	+41° 16′	Sb
M58 (NGC 4579)	12 37.7	+11 49	Sb
M61 (NGC 4303)	12 21.9	+ 4 28	Sc
M63 (NGC 5055)	13 15.8	+42 02	Sb
M64 (NGC 4826)	12 56.7	+21 41	Sb
M74 (NGC 628)	1 36.7	+15 47	Sc
M83 (NGC 5236)	13 37.0	−29 52	Sc
M88 (NGC 4501)	12 32.0	+14 25	Sb
M90 (NGC 4569)	12 36.8	+13 10	Sb
M94 (NGC 4736)	12 50.9	+41 07	Sb
M98 (NGC 4192)	12 13.8	+14 54	Sb
M99 (NGC 4254)	12 18.8	+14 25	Sc
M100 (NGC 4321)	12 22.9	+15 49	Sc
M101 (NGC 5457)	14 03.2	+54 21	Sc
M104 (NGC 4594)	12 40.0	−11 37	Sa
M108 (NGC 3556)	11 11.5	+55 40	Sc

Interacting galaxies	R.A.	Decl.
M51 (NGC 5194)	$13^h\ 29.9^m$	+47°12′
NGC 5195	13 30.0	+47 16
M65 (NGC 3623)	11 18.9	+13 05
M66 (NGC 3627)	11 20.2	+12 59
M81 (NGC 3031)	9 55.6	+69 04
M82 (NGC 3034)	9 55.8	+69 41
M95 (NGC 3351)	10 44.0	+11 42
M96 (NGC 3368)	10 46.8	+11 49
M105 (NGC 3379)	10 47.8	+12 35

TABLE E-11 Mass and Energy Inventory for the Universe

	Individual Contribution	Section Total
Dark matter and dark energy contributions		0.954±0.003*
Dark Energy	0.72±0.03	
Dark Matter	0.23±0.03	
Primeval Graviational Radiation	$\leq 10^{-10}$	
Contributions from Big Bang era		0.0010±0.0005
Electromagnetic Radiation	$10^{-4.3\pm0.000001}$	
Neutrinos	$10^{-2.9\pm0.1}$	
Normal Particle (Baryon) Rest Mass		0.045±0.003
Charged particles (plasma) between stars and galaxies	0.0418±0.003	
Main sequence stars in elliptical galaxies and nuclear bulges	0.0015±0.0004	
Neutral hydrogen & helium	0.00062±0.00010	
Main sequence stars in galactic disks and in irregular galaxies	0.00055±0.00014	
White dwarfs	0.00036±0.00008	
Molecular gas	0.00016±0.00006	
Substellar objects	0.00014±0.00007	
Black holes	0.00007±0.00002	
Neutron stars	0.00005±0.00002	
Planets	$10^{-6\pm0.1}$	

** The numbers after the plus or minus (±) symbol indicate the possible errors in the given numbers.*
This table lists the major contributions to the mass and energy of the universe.

APPENDIX F: More to Know

For expanded coverage of these key concepts in the text, visit our More to Know resource at www.whfreeman.com/dtu7e.

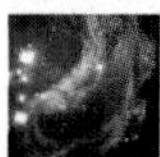

UNDERSTANDING ASTRONOMY

Chapter 1

- Earth Satellites
- Stars' Technical Names
- Height of the Sun
- Maximum Number of Eclipses Annually

Chapter 2

- The Conservation of Angular Momentum

Chapter 3

- Seeing Nonvisible Radiation

Chapter 4

- Radioactive Age-Dating

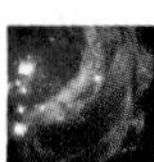

UNDERSTANDING THE SOLAR SYSTEM

Chapter 5

- Computer Simulations of Planet Formation
- Spectra

Chapter 6

- Waves in and on the Earth
- The Origin of the Earth's Magnetic Field
- Radioactive Age-Dating

Chapter 7

- *Mariner 10* and the Exploration of Mercury
- Radar Doppler Measurements

Chapter 8

- Roche Limit

Chapter 9

- Orbital Resonances
- Discovering Comets
- Meteors

Chapter 10

- Zeeman Effect
- Fusion

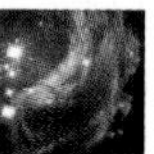

UNDERSTANDING THE STARS

Chapter 11

- Exact Spectra
- Stellar Sizes

Chapter 13

- Origins of Cosmic Rays
- Brightest Planetary Nebulae

Chapter 14

- Gravitational Waves

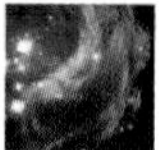

UNDERSTANDING THE UNIVERSE

Chapter 16

- Computers in Astronomical Research

Chapter 17

- Black Hole Masses or How Plausible Are Extremely Massive Black Holes?

APPENDIX G: Using the *Deep Space Explorer*™ and *Starry Night Enthusiast*™ CD-ROM

The *Deep Space Explorer Manual,* which can be found in the *Deep Space Explorer*™ program on the CD, contains detailed directions on how install the *Deep Space Explorer*™ software and how to use its various features. Once installed, this user's guide is accessible from the **Help** button on the program's main menu. Likewise, the *Starry Night Enthusiast Manual,* which can be found in the *Starry Night Enthusiast*™ program on the CD, contains detailed directions on how install the *Starry Night Enthusiast*™ software and how to use its various features. Once installed, this user's guide is also accessible from the **Help** button on the program's main menu.

It might be useful, and it is certainly good scientific practice, to keep a notebook specifically for these exercises in which you can write down the date and time of each observation as well as notes on set-up procedures as you become familiar with them.

What follows are some major concepts concerning these two programs to help you jumpstart your use of them.

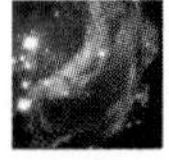

DEEP SPACE EXPLORER™: Getting Started

Open *Deep Space Explorer*™. Your screen consists of two major sections separated by tab buttons:

1. **Tabbed lists** contain all of the objects and regions that you can view with *Deep Space Explorer*™.

2. **Main view window** comprises the majority of the screen and shows the region of the sky that you select or to which you travel.

You will mostly be using the **View** tab, which has a list of most of the objects or regions you can view. Go to the **Settings** tab, uncheck "Use magnitude cutoffs" and slide the "Galaxy drawing/Brightness" slide to the middle of the range, and then press the **Views** tab. The **Highlight** tab opens a list of large-scale regions of the nearby universe that will be useful in the later chapters of the book. The check marks next to these regions will cause them to appear in yellow if you observe a region of space containing them. You can uncheck them individually or use the **Deselect all** button to return all the galaxies to white.

You can pivot around the object locked in the center of your screen in order to see the surrounding star fields by grabbing the screen and moving the mouse. You grab by holding down the left PC mouse button (hold the mouse on a Mac).

You can quickly visit different places in the solar system and many different galaxies by pressing **Edit,** then **Find,** and then typing in the name of the object of interest. In the *Deep Space Explorer*™ questions, these instructions will be presented as, for example, **Edit/Find/M82** when you are asked to find the galaxy M82.

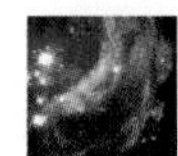

STARRY NIGHT ENTHUSIAST™: Getting Started

Open *Starry Night Enthusiast*™. Your screen will now display what is called a *Starry Night Enthusiast*™ document. This consists of four sections:

1. The **menu** line contains all of the menu commands available in *Starry Night Enthusiast*.

2. The **control panel**, with the date and time displays, allows you to change many of the important parameters, like the time, location, and area of the sky you view.

3. The **Tabs** on the left edge, when pressed, open a set of vertical windows on the left side of the main window that expedite navigating the sky and setting the parameters.

4. The **main view window** comprises the majority of the document window and shows a virtual view of the sky consistent with the parameters shown in the control panel and in the tabs on the left.

The first time you open the program, you will be prompted to set your home location. You can change "home" at any time by clicking **File/Set Home Location.** The home location window gives you several ways of setting home. You can scroll through the cities listed, click points on a map, or type in latitude and longitude.

The program will normally open with a display of the sky at your location for the time and date set up on your computer system clock and with time running at a normal clock rate. This time and date are displayed on the control panel. Also shown are **time controls** similar to those on a VCR. These allow single time-steps backward and forward, continuous running backward and forward, and stop and real-time running forward. The length of each **time-step** for backward and forward motion can be chosen from a list displayed when you click on the time-step box located next to the time. The normal time rate is denoted 1x. You should experiment with these controls.

The **Field of View** slide bar on the right side of the *control panel* permits zooming in and out.

Click on the **Find** tab on the left side of the main view window to see a list of planets and planetary system objects. Press on the plus signs to see more objects. Double clicking on a name will slew the sky to that object.

If you want the daylight off, you can toggle it using **View/Hide Daylight** or remove the check by **Daylight** in the *Options* tab. If you want the horizon off, remove the check by **Local Horizon** in the *Options* tab or chose **Favourites/ Guides/Atlas.**

Click on the check marks for the labels you want to see using the **Labels** button on the control panel.

Here are some practice exercises:

1. Grab the sky and move it around until you have the eastern horizon in the center of the screen. Set the time to 11 P.M. tonight. Do you see any planets on the screen? If you are using a notebook, record date and time and the names of the visible planets.

To answer this and other questions, you can use a powerful function built into the **Hand Tool**—the hand-shaped icon that shows the position of the mouse on the screen. Whenever this icon is moved close to an object, it will automatically display the object's name and other details.

2. What is the brightest star (other than the Sun) that appears in the southern sky from your home location? (Again, enter this name in your notebook.)

3. Find the Andromeda Galaxy by opening the **Find** tab and beginning to type the name into the box at the top. You will see the name "Andromeda Galaxy," among others, appear below. Double click on "Andromeda Galaxy." *Starry Night Enthusiast*™ will automatically slew to this nearby galaxy. You can adjust the field of view using the zoom slide in the Control Panel. (If this object is below your horizon, *Starry Night Enthusiast*™ will ask whether you want time adjusted to bring this object into view. If this occurs, click on **Yes.**) Zooming in, you will see a bright circular object below Andromeda. What is this object?

You should experiment with the various controls and choices provided by *Starry Night Enthusiast*™. For example, you should use the **Hand Tool** to move your direction of view. Clicking on the mouse changes the hand icon to a closed fist. Movement of this mouse will now result in the sky moving to provide you with the view toward different directions.

Take some time to just "play" with the program. Try different views of the sky. Locate an object in which you are interested (e.g., a planet) and zoom in on it to see its surface and/or moons. Change the viewing date, time, or location by clicking on the relevant boxes and resetting values. Change the Time Step to 1 Day and watch the movement of the planets across the sky. (These are just suggestions; use your imagination!) Take notes on your observations; your instructor may ask you to share them with the class.

See the *Starry Night Enthusiast*™ ***Manual*** on the CD for detailed directions on how to use the *Starry Night Enthusiast*™ software.

A ring The outermost ring of Saturn visible from Earth; it is located just beyond Cassini's division.

absolute magnitude The apparent magnitude that a star would have if it were 10 parsecs from Earth.

absorption line A dark line in a continuous spectrum created when photons are removed from the continuum.

absorption line spectrum Dark lines superimposed on a continuous spectrum.

acceleration A change in the direction or magnitude of a velocity.

accretion The gradual accumulation of matter by an astronomical body, usually caused by gravity.

accretion disk An orbiting disk of matter spiraling in toward a star or black hole.

achromatic lens A compound lens designed to minimize the effect of chromatic aberration.

active galaxy A very luminous galaxy, often containing an active galactic nucleus.

active optics A system that adjusts a reflecting telescope in response to changes in temperature and shape of the mount; it helps optimize an image.

adaptive optics Primary telescope mirrors that are continuously and automatically adjusted to compensate for the distortion of starlight due to the motion of the Earth's atmosphere.

AGB star *See* **asymptotic giant branch (AGB) star.**

albedo The fraction of sunlight that a planet, asteroid, or satellite scatters directly back into space.

amino acids A class of chemical compounds that are the building blocks of proteins.

angle The opening between two straight lines that meet at a point.

angular diameter (angular size) The arc angle across an object.

angular momentum A measure of how much energy an object has stored in its rotation and/or revolution.

angular resolution The angular size of the smallest detail of an astronomical object that can be distinguished with a telescope.

annular eclipse An eclipse of the Sun in which the Moon is too distant to cover the Sun completely so that a ring of sunlight is seen around the Moon at mideclipse.

anorthosite A light-colored rock found throughout the lunar highlands and in some very old mountains on Earth.

aphelion The point in its orbit where a planet or other solar system body is farthest from the Sun.

Apollo asteroid An asteroid that is sometimes closer to the Sun than the Earth is.

apparent magnitude A measure of the brightness of light from a star or other object as seen from Earth.

arc angle The measurement of the angle between two objects or two parts of the same object.

arcminute (arcmin) One-sixtieth of a degree of arc.

arcsecond (arcsec) One-sixtieth of a minute of arc.

asteroid (minor planet) Any of the rocky objects larger than a few hundred meters in diameter (and not classified as a planet or moon) that orbits the Sun.

asteroid belt A 1½-astronomical-unit-wide region between the orbits of Mars and Jupiter in which most of the asteroids are found.

astronomical unit (AU) The average distance between the Earth and the Sun: 1.5×10^8 kilometers = 93 million miles.

astronomy The branch of science dealing with objects and phenomena that lie beyond the Earth's atmosphere.

astrophysics That part of astronomy dealing with the physics of astronomical objects and phenomena.

asymptotic giant branch (AGB) star A red giant star that has completed core helium fusion and has re-expanded for a second time.

atom The smallest particle of an element that has the properties characterizing that element.

atomic number The number of protons in the nucleus of an atom.

aurora (*plural* **aurorae**) Light radiated by atoms and ions formed by the solar wind in the Earth's upper atmosphere; seen most commonly in the polar regions.

autumnal equinox The intersection of the ecliptic and the celestial equator where the Sun crosses the equator moving from north to south. The beginning of autumn (around September 23).

average density The mass of an object divided by its volume.

B ring The brightest of the three rings of Saturn visible from Earth; it lies just inside the Cassini division.

barred spiral galaxy A spiral galaxy in which the spiral arms begin from the ends of a bar running through the nuclear bulge.

belt asteroid An asteroid whose orbit lies in the asteroid belt.

belts (of Jupiter) Dark, reddish bands in Jupiter's cloud cover.

Big Bang An explosion that took place roughly 15 billion years ago, creating all space, time, matter, and energy in which the universe emerged.

binary star Two stars revolving about each other; a double star.

birth line A line on the Hertzsprung-Russell diagram corresponding to where stars with different masses transform from protostars to pre–main-sequence stars.

BL Lacertae (BL Lac) object A type of active galaxy; a blazar.

black hole An object whose gravity is so strong that the escape velocity from it exceeds the speed of light.

blackbody A hypothetical perfect radiator that absorbs and reemits all radiation falling upon it.

blackbody curve The curve obtained when the intensity of radiation from a blackbody at a particular temperature is plotted against wavelength.

blackbody radiation Electromagnetic radiation emitted by a blackbody.

blazar A BL Lacertae object.

blueshift A shift of all spectral features toward shorter wavelengths; the Doppler shift of light from an approaching source.

Bohr atom A model of the atom, described by Niels Bohr, in which electrons revolve about the nucleus in various circular orbits.

Bok globule A small, roundish, dark nebula in which stars are forming.

brown dwarf Any of the planetlike bodies with less than 0.08 $M_\odot$ and more than about 13 M_{Jupiter}; such bodies do not have enough mass to sustain fusion in their cores.

C ring The faint, inner portion of Saturn's main ring system.

caldera The crater at the summit of a volcano.

capture theory The idea that the Moon was created at a different location in the solar system and subsequently captured by Earth's gravity.

carbon fusion The thermonuclear fusion of carbon nuclei to produce oxygen and neon.

carbonaceous chondrites A class of extremely ancient, carbon-rich meteorites.

Cassegrain focus An optical arrangement in a reflecting telescope in which light rays are reflected by a secondary mirror through a hole in the primary mirror.

Cassini division A prominent gap between Saturn's A and B rings discovered in 1675 by J. D. Cassini.

celestial equator A great circle on the celestial sphere 90° from the celestial poles.

celestial poles Points about which the celestial sphere appears to rotate.

celestial sphere A hypothetical sphere of very large radius centered on the observer; the apparent sphere of the night sky.

Celsius temperature scale *See* **temperature (Celsius).**

center of mass The point around which a rigid system is perfectly balanced in a gravitational field; also, the point in space around which mutually orbiting bodies have elliptical orbits.

Cepheid variable star One of two types of yellow, supergiant, pulsating stars.

Cerenkov radiation Radiation produced by particles traveling through a substance faster than light can.

Ceres The largest known asteroid and the first to be discovered.

Chandrasekhar limit The maximum mass of a white dwarf, about 1.4 $M_\odot$.

charge-coupled device (CCD) A type of solid-state silicon wafer designed to detect photons.

chromatic aberration An optical property whereby different colors of light passing through a lens are focused at different distances from it.

chromosphere The layer in the solar atmosphere between the photosphere and the corona.

circumpolar stars All the stars that never set at a given latitude; all the stars between Polaris and the northern horizon.

close binary A binary star whose members are separated by a few stellar diameters.

closed universe A universe that contains enough matter to cause it to recollapse. It is finite in extent and has no "outside."

cluster (of galaxies) A collection of a few hundred to a few thousand galaxies bound by gravity.

cocreation theory The theory that the Moon formed simultaneously with the Earth and in orbit around it.

collision-ejection theory The theory that the Moon was created by the impact of a planet-sized object with the Earth; presently considered the most plausible theory of the Moon's formation.

color-magnitude diagram A plot of the surface temperatures (colors) versus the absolute magnitudes of stars.

coma (of a comet) The nearly spherical, diffuse gas surrounding the nucleus of a comet near the Sun.

comet A small body of ice and dust in orbit about the Sun. While passing near the Sun, a comet's vaporized ices give rise to a coma, tails, and a hydrogen envelope.

conduction (thermal) The transfer of heat by passing energy directly from atom to atom.

configuration (of a planet) A particular geometric arrangement of the Earth, a planet, and the Sun.

confinement The moment shortly after the Big Bang when quarks bound together to form particles like protons and neutrons.

conic section The curve of intersection between a circular cone and a plane. This curve can be a circle, ellipse, parabola, or hyperbola.

conjunction The alignment of two bodies in the solar system so that they appear in the same part of the sky as seen from Earth.

conservation of angular momentum The law of physics stating that the total amount of angular momentum in an isolated system remains constant.

constellation Any of the 88 contiguous regions that cover the entire celestial sphere, including all the objects in each region; also, a configuration of stars often named after an object, a person, or an animal.

contact binary A close binary system in which both stars fill or overflow their Roche lobes.

continental drift The gradual movement of the continents over the surface of the Earth due to plate tectonics.

continuous spectrum (continuum) A spectrum of light over a range of wavelengths without any spectral lines.

convection The transfer of energy by moving currents of fluid or gas containing that energy.

convective zone A layer in a star where energy is transported outward by means of convection; also known as the *convective envelope* or *convection zone*.

core The central portion of any astronomical object.

core helium fusion The fusion of helium to form carbon and oxygen at the center of a star.

core hydrogen fusion The fusion of hydrogen to form helium at the center of a star.

corona The Sun's outer atmosphere.

coronagraph A specially designed telescope with a baffle that blocks out the solar disk so that the corona can be photographed.

coronal hole A dark region of the Sun's inner corona as seen at X-ray wavelengths.

coronal mass ejection Large volumes of high energy gas released from the Sun's corona.

cosmic censorship The belief that the only connection between a black hole and the universe is the black hole's event horizon.

cosmic light horizon A sphere, centered on the Earth, whose radius equals the distance traveled by light since the Big Bang.

cosmic microwave background Photons from every part of the sky with a blackbody spectrum at 2.73 K; the cooled-off radiation from the primordial fireball that originally filled all space.

cosmic ray High speed particles traveling through space.

cosmic ray shower Groups of particles from Earth's atmosphere propelled Earthward by the impact of a cosmic ray.

cosmological constant A number sometimes inserted in the equations of general relativity that represents a pressure that opposes gravity throughout the universe.

cosmological redshift An increase in wavelength from distant galaxies and quasars caused by the expansion of the universe.

cosmology The study of the formation, organization, and evolution of the universe.

coudé focus A reflecting telescope in which a series of mirrors direct light to a remote focus away from the moving parts of the telescope.

crater A circular depression on a celestial body caused by the impact of a meteoroid, asteroid, or comet or by a volcano.

crescent Moon A lunar phase during which the Moon appears less than half full.

crust The solid surface layer of some astronomical bodies, including the terrestrial planets, the moons, the asteroids, and some stellar remnants.

dark ages The age of the universe between the time of decoupling and the first burst of star formation.

dark energy A repulsive gravitational effect that is causing the universe to accelerate outward.

dark matter (missing mass) The as-yet-undetected matter in the universe that is underluminous and probably quite different from ordinary matter.

dark nebula A cloud of interstellar gas and dust that obscures the light of more distant stars.

declination The coordinate on the celestial sphere exactly analogous to latitude on Earth; measured north and south of the celestial equator.

decoupling The epoch in the early universe when electrons and ions first combined to create stable atoms; the time when electromagnetic radiation ceased to dominate over matter.

deferent A fixed circle in the Earth-centered universe along which a smaller circle (an epicycle) moves carrying a planet, the Sun, or the Moon.

degree (°) A unit of angular measure or a temperature measure.

dense core Any of the regions of interstellar gas clouds that are slightly denser than normal and destined to collapse to form one or a few stars.

density The ratio of the mass of an object to its volume.

density wave A spiral-shaped compression of the gas and dust in a spiral galaxy.

density-wave theory An explanation of spiral arms in galaxies elaborated by C. C. Lin and F. Shu. (*See* **spiral density wave.**)

detached binary A binary system in which the surfaces of both stars are inside their Roche lobes.

differential rotation The rotation of a nonrigid object in which parts at different latitudes or different radial distances move at different speeds.

differentiation *See* **planetary differentiation.**

diffraction grating An optical device consisting of closely spaced lines ruled on a piece of glass that is used like a prism to disperse light into a spectrum.

direct motion The gradual, eastward apparent motion of a planet against the background stars as seen from Earth.

disk (of a galaxy) A flattened assemblage of stars, gas, and dust in a spiral galaxy like the Milky Way.

distance modulus The difference between the apparent and absolute magnitudes of an object.

diurnal motion Cyclic motion with a 1-day period.

Doppler effect (or **Doppler shift**) The change in wavelength of radiation due to relative motion between the source and the observer along the line of sight.

double-line spectroscopic binary A spectroscopic binary whose spectrum exhibits spectral lines of both stars.

double radio source An extragalactic radio source characterized by two large lobes of radio emission, often located on either side of an active galaxy.

Drake equation A mathematical equation used to estimate the number of extraterrestrial civilizations that may exist in our Galaxy.

dust devil Whirlwind found in dry or desert areas on both Earth and Mars.

dust tail A comet tail caused by dust particles escaping from the comet's nucleus.

dwarf elliptical galaxy A small elliptical galaxy with far fewer stars than a typical galaxy.

dwarf star Any star smaller than a giant, such as a main-sequence star or a white dwarf.

dynamo theory The generation of a magnetic field by circulating electric charges.

eccentricity *See* **orbital eccentricity.**

eclipse The blocking of part or all of the light from the Moon by the Earth (lunar eclipse) or from the Sun by the Moon (solar eclipse).

eclipse path The track of the tip of the Moon's shadow along the Earth's surface during a total or annular solar eclipse.

eclipsing binary A double star system in which stars periodically pass in front of each other as seen from Earth.

ecliptic The annual path of the Sun on the celestial sphere; the plane of the Earth's orbit around the Sun.

Einstein cross The appearance of four images of the same galaxy or quasar due to gravitational lensing by an intervening galaxy.

Einstein ring The circular or arc-shaped image of a distant galaxy or quasar created by gravitational lensing by an intervening galaxy.

ejecta blanket The ring of material surrounding a crater that was ejected during the crater-forming impact.

electromagnetic radiation Radiation consisting of oscillating electric and magnetic fields, namely gamma rays, X rays, visible light, ultraviolet and infrared radiation, and radio waves.

electromagnetic spectrum The entire array or family of electromagnetic radiation.

electron A negatively charged subatomic particle usually found in orbit about the nucleus of an atom.

electron degeneracy pressure A powerful pressure produced by repulsion of closely packed (degenerate) electrons.

element A substance that cannot be decomposed by chemical means into simpler substances. Every atom of the same element contains the same number of protons.

ellipse A closed curve obtained by cutting completely through a circular cone with a plane; the shape of planetary orbits.

elliptical galaxy A galaxy with an elliptical shape, little interstellar matter, and no spiral arms.

elongation The angle between a planet and the Sun as seen from Earth.

emission line A bright line of electromagnetic radiation.

emission line spectrum A spectrum that contains only bright emission lines.

emission nebula A glowing gaseous nebula whose light comes from fluorescence caused by a nearby star.

Encke division A thin gap in Saturn's A ring, possibly first seen by J. F. Encke in 1838.

energy The ability to do work.

energy flux The amount of energy emitted from each square meter of an object's surface per second.

energy level (in an atom) A particular amount of energy possessed by an electron in orbit around a nucleus.

epicycle In the Earth-centered universe, a moving circle about which planets revolve.

equations of stellar structure A set of relationships that describe the interactions of matter, energy, and gravity inside a star.

equinox Either of the two days of the year when the Sun crosses the celestial equator and is therefore directly over the Earth's equator; *see also* **autumnal equinox** *and* **vernal equinox.**

era of recombination The time, roughly 500,000 years after the Big Bang, when the universe became transparent.

ergoregion The region of space immediately outside the event horizon of a rotating black hole where it is impossible to remain at rest.

event horizon The location around a black hole where the escape velocity equals the speed of light; the boundary of a black hole.

evolutionary track On the Hertzsprung-Russell diagram, the path followed by a point representing an evolving star.

excited state Orbit of an electron with energy greater than the lowest energy orbit (or state) available to that election.

expanding universe The motion of the superclusters of galaxies away from each other.

eyepiece lens A magnifying lens used to view the image produced at the focus of a telescope.

F ring A thin ring just beyond the outer edge of Saturn's main ring system.

Fahrenheit scale *See* **temperature (Fahrenheit).**

filament A dark curve seen above the Sun's photosphere that is the top view of a solar prominence.

first quarter Moon A phase of the waxing Moon when Earth-based observers see half of the Moon's illuminated hemisphere.

fission theory The theory that the Moon formed from matter flung off the Earth because the planet was rotating extremely fast.

flare *See* **solar flare.**

flocculent spiral galaxy A spiral galaxy whose spiral arms are broad, fuzzy, and poorly demarcated.

focal length The distance from a lens or concave mirror to where converging light rays meet.

focal plane The plane at the focal length of a lens or concave mirror on which an extended object is focused.

focal point *See* **focus.**

focus (of a lens or concave mirror) The place at the focal length where light rays from a point object (that is, one that is too distant or tiny to resolve) are converged by a lens or concave mirror.

focus (*plural* **foci**) (of an ellipse) The two points inside an ellipse, the sum of whose distances from any point on the ellipse is constant.

force That which can change the momentum of an object.

full Moon A phase of the Moon during which its full daylight hemisphere can be seen from Earth.

galactic cannibalism A collision between two galaxies of unequal mass and size in which the smaller galaxy is absorbed by the larger galaxy.

galactic merger A collision and subsequent merger of two roughly equal-sized galaxies.

galactic nucleus The center of a galaxy; the center of the Milky Way Galaxy.

galaxy A large assemblage of stars, gas, and dust bound together by their mutual gravitational attraction.

Galilean satellite (Galilean moon) Any one of the four large moons of Jupiter (Callisto, Europa, Ganymede, Io) that is visible from Earth through a small telescope.

gamma ray The most energetic form of electromagnetic radiation.

gamma-ray burst A short burst of gamma rays; the sources of the bursts are outside our Galaxy.

gas (ion) tail The relatively straight tail of a comet produced by the solar wind acting on ions in a comet's coma.

general theory of relativity A description of spacetime formulated by Einstein explaining how gravity affects the geometry of space and the flow of time.

geocentric cosmology The belief that the Earth is at the center of the universe.

giant elliptical galaxy A very large, extremely massive elliptical galaxy, usually located near the center of a rich cluster of galaxies.

giant molecular cloud A large interstellar cloud of cool gas and dust in a galaxy.

giant star A star whose diameter is roughly 10 to 100 times that of the Sun.

gibbous Moon A phase of the Moon in which more than half, but not all, of the Moon's daylight hemisphere is visible from Earth.

glitch A sudden speedup in the period of a pulsar.

globular cluster A large spherical cluster of gravitationally bound stars usually found in the outlying regions of a galaxy.

grand-design spiral galaxy A spiral galaxy whose spiral arms are thin, graceful, and well defined.

grand unified theory (GUT) A theory that describes and explains the four physical forces.

granulation The rice-grain–like structure of the solar photosphere due to convection of solar gases.

granules Lightly colored convection features about 1000 kilometers in diameter seen constantly in the solar photosphere.

gravitation (gravity) The tendency of all matter to attract all other matter.

gravitational lensing The distortion of the appearance of an object by a source of gravity between it and the observer.

gravitational redshift The redshift of photons leaving the gravitational field of any massive object, such as a star or black hole.

gravitational waves (gravitational radiation) Ripples in the overall geometry of space produced by nonspherical moving objects.

gravity *See* **gravitation.**

Great Dark Spot A large, dark, oval-shaped storm that used to be in Neptune's southern hemisphere.

Great Red Spot A large, red-orange, oval-shaped storm in Jupiter's southern hemisphere.

Great Wall A huge arc of galaxies between two voids in the cosmos.

greatest elongation The largest possible angle between the Sun and an inferior planet.

greenhouse effect The trapping of infrared radiation near a planet's surface by the planet's atmosphere.

ground state The lowest energy level of an atom.

H II region A region of ionized hydrogen in interstellar space.

halo (of a galaxy) A spherical distribution of globular clusters, isolated stars, and possibly dark matter that surrounds a galaxy.

Hawking process The formation of real particles from virtual ones just outside a black hole's event horizon; the means by which black holes evaporate.

head-tail source A radio galaxy whose radio emission is deflected from the galaxy.

heliocentric cosmology A theory of the formation and evolution of the solar system with the Sun at the center.

helioseismology The study of vibrations of the solar surface.

helium flash The explosive ignition of helium fusion in the core of a low-mass, giant star.

helium fusion The thermonuclear fusion of helium to produce carbon and oxygen.

helium shell flash The explosive ignition of helium fusion in a thin shell surrounding the core of a low-mass star.

helium shell fusion Helium fusion that occurs in a thin shell surrounding the core of a star.

Hertzsprung-Russell (H-R) diagram A plot of the absolute magnitude or luminosity of stars versus their surface temperatures or spectral classes.

highlands Heavily cratered, mountainous regions of the lunar surface.

homogeneity The property of the universe being smooth or uniform as measured over suitably large distance intervals.

horizon problem (isotropy problem) The difficulty in explaining why seemingly disconnected regions of the universe have the same temperature.

horizontal-branch stars A group of post–helium-flash stars near the main sequence on the Hertzsprung-Russell diagram of a typical globular cluster.

hot-spot volcanism The creation of volcanoes on a planet's surface caused by a reservoir of hot magma in the planet's mantle under a thin part of the crust.

Hubble classification A system of classifying galaxies according to their appearance into one of four broad categories: spirals, barred spirals, ellipticals, and irregulars.

Hubble constant (H_0) The constant of proportionality in the relation between the recessional velocities of remote galaxies and their distances; the correct value will determine the age of the universe.

Hubble flow The recession of the galaxies caused by the expansion of the universe.

Hubble law The relationship that states that the redshifts of remote galaxies are directly proportional to their distances from Earth.

hydrocarbon A molecule based on hydrogen and carbon.

hydrogen envelope An extremely large, tenuous sphere of hydrogen gas surrounding the head of a comet.

hydrogen fusion (hydrogen burning) The thermonuclear fusion of hydrogen to produce helium.

hydrogen shell fusion Hydrogen fusion that occurs in a thin shell surrounding the core of a star.

hydrostatic equilibrium A balance between the weight of a layer in a star and the pressure that supports it.

hyperbola An open curve obtained by cutting a cone with a plane.

impact breccia A rock consisting of various fragments cemented together by the impact of a meteoroid.

impact crater A crater on the surface of a planet or moon produced by the impact of an asteroid, meteoroid, or comet.

inferior conjunction The configuration when Mercury or Venus is directly between the Sun and the Earth.

inflation A sudden expansion of space.

inflationary epoch A brief period shortly after the Big Bang during which the scale of the universe increased very rapidly.

infrared radiation Electromagnetic radiation of a wavelength longer than visible light but shorter than radio waves.

initial mass function The numbers of stars on the main sequence at all different masses.

instability strip A region on the Hertzsprung-Russell diagram occupied by pulsating stars.

interferometry A method of increasing resolving power by combining electromagnetic radiation obtained by two or more telescopes.

intergalactic gas Gas located between the galaxies within a cluster of galaxies.

intermediate-mass black hole A black hole with a mass between a few hundred and few thousand solar masses.

interstellar dust Microscopic solid grains of various compounds in interstellar space usually encased in ice.

interstellar extinction The dimming of starlight as it passes through the interstellar medium.

interstellar gas Sparse gas in interstellar space.

interstellar medium Interstellar gas and dust.

interstellar reddening The reddening of starlight passing through the interstellar medium resulting from the scattering of short wavelength light more than long wavelength light.

inverse-square law The gravitational attraction between two objects and the apparent brightness of a light source are both inversely proportional to the square of its distance.

ion An atom that has become electrically charged due to the loss or addition of one or more electrons.

ionization The process by which an atom loses or gains electrons.

ionosphere Region of Earth's atmosphere, above the mesosphere, in which sunlight ionizes many atoms.

iron meteorite A meteorite composed primarily of iron with an admixture of nickel; also called an *iron.*

irregular cluster (of galaxies) An unevenly distributed group of galaxies bound together by their mutual gravitational attraction.

irregular galaxy An asymmetrical galaxy having neither spiral arms nor an elliptical shape.

isotope Any of several forms of the same chemical element whose nuclei all have the same number of protons but different numbers of neutrons. Their nuclear properties often differ greatly.

isotropy The fact that the average number of galaxies at different distances from Earth is the same in all directions; also, the fact that the temperature of the cosmic microwave background is essentially the same in all directions.

isotropy problem (horizon problem) The difficulty in explaining why seemingly disconnected regions of the universe have the same temperature.

Jeans instability The condition under which gravitational forces overcome thermal forces to cause part of an interstellar cloud to collapse and form stars and planets.

Kelvin *See* **temperature (Kelvin)**.

Kepler's laws Three statements, formulated by Johannes Kepler, that describe the motions of the planets.

Kerr black hole Any rotating, uncharged black hole.

kiloparsec (kpc) One thousand parsecs; about 3260 light-years.

kinetic energy The energy an object has as a result of its motion.

Kirchhoff's laws Three statements formulated by Gustav Kirchhoff describing what physical conditions produce each type of spectra.

Kirkwood gaps Gaps in the spacing of asteroid orbits discovered by Daniel Kirkwood caused by gravitational attractions of planets.

Kuiper belt A doughnut-shaped ring of space around the Sun beyond Pluto that contains many frozen comet bodies, some of which are occasionally deflected toward the inner solar system.

Large Magellanic Cloud (LMC) An irregular galaxy, companion to the Milky Way about 179,000 ly away.

last quarter Moon A phase of the waning Moon when Earth-based observers see half of the Moon's illuminated hemisphere.

law of equal areas Kepler's second law.

law of inertia (Newton's first law of motion) The physical law that an object will stay at rest or move at a constant speed in a fixed direction unless acted upon by an outside force.

law of universal gravitation Newton's law of gravitation, which describes how the gravitational force between two bodies depends on their masses and separation.

lenticular galaxy A disk-shaped galaxy without spiral arms.

light Electromagnetic radiation, which travels in packets called photons.

light curve A graph that displays variations in the brightness of a star or other astronomical object over time.

light-gathering power A measure of how much light a telescope intercepts and brings to a focus.

light-year (ly) The distance that light travels in a vacuum in 1 year.

lighthouse model The explanation that a pulsar pulses by rotating and funneling energy outward via magnetic fields that are not aligned with the rotation axis.

limb (of the Sun) The apparent edge of the Sun as seen in the sky.

limb darkening The phenomenon whereby the Sun is darker near its limb than near the center of its disk.

line of nodes The line along which the plane of the Moon's orbit intersects the plane of the ecliptic.

liquid metallic hydrogen A metallike form of hydrogen that is produced under extreme pressure.

Local Group The cluster of about 40 galaxies, of which our own Galaxy is a member.

long-period comet A comet that takes tens of thousands of years or more to orbit the Sun once.

luminosity The rate at which electromagnetic radiation is emitted from a star or other object.

luminosity class The classification of a star of a given spectral type according to its luminosity and density; the classes are supergiant, bright giant, giant, subgiant, and main sequence.

lunar Referring to the Moon.

lunar eclipse An eclipse during which the Earth blocks light that would have struck the Moon.

lunar month *See* **synodic month.**

lunar phases The names given to the apparent shapes of the Moon as seen from Earth.

Lyman series A series of spectral lines of hydrogen produced by electron transitions to and from the lowest energy state of the hydrogen atom.

magnetic dynamo A theory that explains phenomena of the solar cycle as a result of periodic winding and unwinding of the Sun's magnetic field in the solar atmosphere.

magnetic field A region of space near a magnetized body within which magnetic forces can be detected.

magnetosphere The region around a planet occupied by its magnetic field.

magnification (magnifying power) The number of times larger in angular diameter an object appears through a telescope than when it is seen by the naked eye.

magnitude A measure of brightness. *See* **absolute magnitude; apparent magnitude.**

magnitude scale The system of denoting magnitudes.

main sequence A grouping of stars on the Hertzsprung-Russell diagram extending diagonally across the graph from the hottest, brightest stars to the dimmest, coolest stars.

main-sequence star A star, fusing hydrogen to helium in its core, whose surface temperature and luminosity place it on the main sequence on the Hertzsprung-Russell diagram.

mantle (of a planet) That portion of a terrestrial planet located between its crust and core.

mare (*plural* **maria**) Latin for "sea," a large, relatively crater-free plain on the Moon.

mare basalt Dark, solidified lava that covers the lunar maria.

mascons Regions of high density matter near the surface of the Moon.

mass A measure of the total amount of material in an object.

mass-luminosity relation The direct relationship between the masses and luminosities of main-sequence stars.

matter-dominated universe A universe in which the radiation field that fills all space is unable to prevent the existence of neutral atoms.

megaparsec (Mpc) One million parsecs.

mesosphere The layer in the Earth's atmosphere above the stratosphere.

metal All elements except hydrogen and helium.

metal-poor star *See* **Population II star.**

metal-rich star *See* **Population I star.**

meteor The streak of light seen when any space debris vaporizes in the Earth's atmosphere; a "shooting star."

meteor shower Frequent meteors that seem to originate from a common point in the sky.

meteorite A fragment of space debris that has survived passage through the Earth's atmosphere.

meteoroid A small rock in interplanetary space.

microlensing The gravitational focusing of light from a distant star by a closer object to give a brighter image of the star.

Milky Way Galaxy The galaxy in which our solar system resides.

minor planet *See* **asteroid.**

missing mass *See* **dark matter.**

model A hypothesis that has withstood observational or experimental tests.

molecule A bound combination of two or more atoms.

momentum A measure of the inertia of an object; an object's mass multiplied by its velocity.

neap tide The least change from high to low tide during a day; it occurs during the first and third quarter phases of the Moon.

nebula (*plural* **nebulae**) A cloud of interstellar gas and dust.

neon fusion The thermonuclear fusion of neon nuclei.

neutrino A subatomic particle, with no electric charge and little mass, that is important in many nuclear reactions and in supernovae.

neutron A nuclear particle with no electric charge and with a mass nearly equal to that of the proton.

neutron degeneracy pressure A powerful pressure produced by degenerate neutrons.

neutron star A very compact, dense stellar remnant composed almost entirely of neutrons.

New General Catalogue **(NGC)** A catalog of star clusters, nebulae, and galaxies, first published in 1888.

new Moon The phase of the Moon when it is nearest the Sun in the sky.

Newton's laws of motion Newton's equations that describe the motion of matter as a result of forces acting on it.

Newtonian reflector An optical arrangement in a reflecting telescope in which a small, flat mirror reflects converging light rays to a focus on one side of the telescope tube.

nonthermal radiation *See* **synchrotron radiation.**

north celestial pole The location on the celestial sphere directly above the Earth's northern rotation pole.

northern lights (aurora borealis) Light radiated by atoms and ions in the Earth's upper atmosphere due to high-energy particles from the Sun and seen mostly in the northern polar regions.

northern vastness (northern lowlands) Relatively young, crater-free terrain in the northern hemisphere of Mars.

nova (*plural* **novae**) A star in a binary system that experiences a sudden outburst of radiant energy, temporarily increasing its luminosity by a factor of between 10^4 and 10^6.

nuclear Referring to the nucleus of an atom.

nuclear bulge A distribution of stars in the shape of a flattened sphere that surrounds the nucleus of a spiral galaxy like the Milky Way.

nuclear density The density of matter in the nucleus of an atom; about 10^{17} kilograms per cubic meter (kg/m^3).

nucleus (of an atom) The massive part of an atom, composed of protons and neutrons, about which electrons revolve.

nucleus (of a comet) A collection of ices and dust that constitute the solid part of a comet.

nucleus (of a galaxy) *See* **galactic nucleus.**

OB association An unbound group of very young, massive stars predominantly of spectral types O and B.

OBAFGKM sequence The sequence of stellar spectral classifications from hottest to coolest stars.

objective lens The principal lens of a refracting telescope.

observable universe All space that is nearer to us than the distance traveled by light since the time of the Big Bang.

Occam's razor The principle of choosing the simplest scientific theory that correctly explains any phenomenon.

occultation The eclipsing of an astronomical object other than the Moon or Sun by another astronomical body.

Oort cloud A hollow spherical region of the solar system beyond the Kuiper belt where most comets are believed to spend most of their time.

open cluster A loosely bound group of young stars in the disk of the galaxy; a galactic cluster.

open universe A universe with a hyperbolic shape; lacks the mass necessary to someday stop expanding and recollapse. It will expand forever.

opposition The configuration of a planet when it is at an elongation of 180° and thus appears opposite the Sun in the sky.

optical double A pair of stars that appear to be near each other but are unbound and at very different distances from Earth.

optics The branch of physics dealing with the behavior and properties of light.

orbit The path of an object that is moving about a second object.

orbital eccentricity A measure between 0 and 1 indicating how close to circular a planet's orbit is (the eccentricity of a circular orbit is 0).

orbital inclination The tilt or angle of an object's orbital plane around the Sun compared to the ecliptic.

organic molecule A carbon-based compound.

overcontact binary A close binary system in which the two stars share a common atmosphere.

oxygen fusion The thermonuclear fusion of oxygen nuclei.

ozone layer The lower stratosphere, where most of the ozone in the air exists.

pair production The creation of a particle and an antiparticle from energetic photons.

parabola An open curve formed by cutting a circular cone at an angle parallel to the sides of the cone.

parallax The apparent displacement of an object relative to more distant objects caused by viewing it from different locations.

parsec (pc) A unit of distance equal to 3.26 light-years.

partial eclipse A lunar or solar eclipse in which the eclipsed object does not appear completely covered.

Pauli exclusion principle A principle of quantum mechanics that says that two identical particles cannot simultaneously have the same position and momentum.

peculiar galaxy Any Hubble class of galaxy that appears to be blowing apart.

penumbra The portion of a shadow in which only part of the light source is covered by the shadow-making body.

penumbral eclipse A lunar eclipse in which the Moon passes only through the Earth's penumbra.

perihelion The point in its orbit where a planet is nearest the Sun.

period The interval of time between successive repetitions of a periodic phenomenon.

period-luminosity relation A relationship between the period and average luminosity of a pulsating star.

periodic table A listing of the chemical elements according to their properties; created by D. Mendeleev.

phase (of the Moon) The appearance of the Moon at different points in its orbit of the Earth.

photodisintegration The breakup of nuclei in the core of a massive star due to the effects of energetic gamma rays.

photometry The measurement of light intensities.

photon A discrete unit of electromagnetic energy.

photon pressure The force per unit area exerted by photons on stellar or interstellar gas.

photosphere The region in the solar atmosphere from which most of the visible light escapes into space.

physics Basic principles that govern the behavior of physical reality.

pixel A contraction of the term "picture element"; usually refers to one square of a grid into which the light-sensitive component of a charge-coupled device is divided.

plage A bright spot on the Sun believed to be associated with an emerging magnetic field.

Planck era Time from the Big Bang until the Planck time (10^{-43} sec).

Planck time The brief interval of time, about 10^{-43} second, immediately after the Big Bang, when all four forces (gravity, electromagnetism, weak, strong) had the same strength.

Planck's law A relationship between the energy carried by a photon and its wavelength.

planetary differentiation The process early in the life of each planet whereby denser elements sank inward and lighter ones rose.

planetary nebula A luminous shell of gas ejected from an old, low-mass star.

planetesimal Primordial asteroidlike object from which the planets accreted.

plasma A hot, ionized gas.

plate tectonics The motions of large segments (plates) of the Earth's surface caused by convective motions in the underlying mantle.

polymer A long molecule composed of many smaller molecules.

poor cluster (of galaxies) A cluster of galaxies with only a few members.

Population I star A star, such as the Sun, whose spectrum exhibits spectral lines of many elements heavier than helium; a metal-rich star.

Population II star A star whose spectrum exhibits comparatively few spectral lines of elements heavier than helium; a metal-poor star.

positron An electron with a positive rather than negative electric charge; an antielectron.

potential energy The energy stored in an object as a result of its location in space.

powers of ten A convenient method of writing large and small numbers that uses a number between 1 and 10 multiplied by a power of 10.

pre–main-sequence star The stage of star formation just before the main sequence; it involves slow contraction of the young star.

precession (of the Earth) A slow, conical motion of the Earth's axis of rotation caused by the gravitational pull of the Moon and Sun on the Earth's equatorial bulge.

precession of the equinoxes The slow westward motion of the equinoxes along the ecliptic because of the Earth's precession.

primary mirror The large, concave, light-gathering mirror in a reflecting telescope, analogous to the objective lens on a refracting telescope.

prime focus The point in a reflecting telescope where the primary mirror focuses light.

primordial black hole A relatively low-mass black hole hypothetically formed at the beginning of the universe.

primordial fireball The extremely hot gas that filled the universe immediately following the Big Bang.

prism A wedge-shaped piece of glass used to disperse white light into a spectrum.

prograde orbit An orbit of a moon or satellite around a planet that is in the same direction as the planet's rotation.

prominence Flamelike protrusion seen near the limb of the Sun and extending into the solar corona. The side view of a filament.

proper motion The change in the location of a star on the celestial sphere.

proton A heavy, positively charged nuclear particle.

protoplanet The embryonic stage of a planet when it is growing because of collisions with planetesimals.

protoplanetary disk (proplyd) A disk of material encircling a protostar or a newborn star.

protostar The earliest stage of a star's life before fusion commences and when gas is rapidly falling onto it.

protosun The Sun prior to the time when hydrogen fusion began in its core.

pulsar A pulsating source associated with a rapidly rotating neutron star with an off-axis magnetic field.

quantum mechanics The branch of physics dealing with the structure and behavior of atoms and their interactions with each other and with light.

quark A particle that is a building block of the heavy nuclear particles such as protons and neutrons.

quarter Moon A phase of the Moon when it is located 90° from the Sun in the sky; halfway between the new and full phases.

quasar (quasi-stellar radio source) A starlike object with a very large redshift.

quasi-stellar object (QSO) A quasar.

quintessence One of the explanations of the dark energy causing the universe to accelerate outward.

radial velocity That portion of an object's velocity parallel to the line of sight.

radial-velocity curve A plot showing the variation of radial velocity with time for a binary star or variable star.

radiation Electromagnetic energy; photons.

radiation-dominated universe The time at the beginning of the universe when the electromagnetic radiation prevented ions and electrons from combining to make neutral atoms.

radiation (photon) pressure The transfer of momentum carried by radiation to an object on which the radiation falls.

radiative zone A region inside a star where energy is transported outward by the movement of photons through a gas from a hot location to a cooler one.

radio astronomy The branch of astronomy dealing with observations at radio wavelengths.

radio galaxy A galaxy that emits an unusually large amount of radio waves.

radio lobes Vast regions of radio emission on opposite sides of a radio galaxy.

radio telescope A telescope designed to detect radio waves.

radio wave Long-wavelength electromagnetic radiation.

radioactive The property whereby certain atomic nuclei naturally decompose by spontaneously emitting particles.

red dwarf A low-mass main-sequence star.

red giant A large, cool star of high luminosity.

red supergiant An extremely large, cool star of high luminosity; a star in the upper right corner of the Hertzsprung-Russell diagram.

redshift The shifting to longer wavelengths of the light from remote galaxies and quasars; the Doppler shift of light from any receding source.

reflecting telescope (reflector) A telescope in which the principal light-gathering component is a concave mirror.

reflection The rebounding of light rays off a smooth surface.

reflection nebula A comparatively dense cloud of gas and dust in interstellar space that is illuminated by a star between it and the Earth.

refracting telescope (refractor) A telescope in which the principal light-gathering component is a lens.

refraction The bending of light rays when they pass from one transparent medium to another.

regolith The powdery, lifeless material on the surface of a moon or planet.

regular cluster (of galaxies) An evenly distributed group of galaxies bound together by mutual gravitational attraction.

resolution The degree to which fine details in an optical image can be distinguished.

resolving power A measure of the ability of an optical system to distinguish fine details in the image it produces.

resonance The large response of an object to a small periodic gravitational tug from another object.

retrograde motion The occasional backward (that is, westward) apparent motion of a planet against the background stars as seen from Earth. Retrograde motion is an optical illusion.

retrograde orbit The orbit of a moon or satellite around a planet that is in the direction opposite to the planet's rotation.

retrograde rotation The rotation of a planet opposite to its direction of revolution around the Sun. Only Pluto, Uranus, and Venus have retrograde rotation.

revolution The orbit of one body about another.

rich cluster (of galaxies) A cluster of galaxies with many members.

right ascension The celestial coordinate analogous to longitude on Earth and measured around the celestial equator from the vernal equinox.

rille A winding crack or depression in the lunar surface caused by the collapse of a solidified lava tube.

ringlet Any one of numerous, closely spaced, thin bands of particles in Saturn's ring system.

Roche limit The shortest distance from a planet or other object at which a second object can be held together by its own gravitational forces.

Roche lobe The teardrop-shaped regions around each star in a binary star system inside of which gas is gravitationally bound to that star.

rotation The spinning of a body about an axis passing through it.

rotation curve (of a galaxy) A graph showing how the orbital speed of material in a galaxy depends on the distance from the galaxy's center.

RR Lyrae variable star A type of pulsating star with a period less than 1 day.

Sa, Sb, Sc Categories of spiral galaxies determined by the sizes of their nuclear bulges or how tightly wound their spiral arms are; an Sa galaxy is the most tightly wound. Spiral arms are directly attached to the nuclear bulge.

Sagittarius A The strong radio source associated with the nucleus of the Milky Way galaxy.

satellite A body that revolves about a larger one.

SBa, SBb, SBc Categories of barred spiral galaxies determined by how tightly wound their spiral arms are; SBa galaxies are the most tightly wound.

scarp A cliff on Mercury believed to have formed when the planet cooled and shrank.

Schmidt corrector plate A specially shaped lens used with spherical mirrors that corrects for spherical aberration and provides an especially wide field of view.

Schwarzschild black hole Any nonrotating, uncharged black hole.

Schwarzschild radius The distance from the center to the event horizon in any black hole.

scientific method The method of doing science based on observation, experimentation, and the formulation of hypotheses (theories) that can be tested.

scientific notation The style of writing large and small numbers using powers of ten.

scientific theory (hypothesis) An idea about the natural world that is subject to verification and refinement.

seafloor spreading The process whereby magma upwelling along rifts in the ocean floor causes adjacent segments of the Earth's crust to separate.

secondary cosmic rays Particles from the Earth's atmosphere given high speeds Earthward by cosmic rays from space.

secondary mirror A relatively small mirror used in reflecting telescopes to guide the light out the side or bottom of the telescope.

seeing disk The size that a star appears to have on a photographic or charge-coupled-device image as a result of the changing refraction of the starlight passing through the Earth's atmosphere.

seismic waves Vibrations traveling through or around an astronomical body usually associated with earthquakelike phenomena.

seismograph A device used to record and measure seismic waves, such as those produced by earthquakes.

seismology The study of earthquakes and related phenomena.

self-propagating star formation The process whereby the deaths of stars in one part of a galaxy stimulates star formation in a neighboring region of that galaxy.

semidetached binary A close binary system in which one star fills or is overflowing its Roche lobe.

semimajor axis (of an ellipse) Half of the longest dimension of an ellipse.

SETI The search for extraterrestrial intelligence.

Seyfert galaxy A spiral galaxy with a bright nucleus whose spectrum exhibits emission lines.

Shapley–Curtis debate An inconclusive debate between Harlow Shapley and Heber Curtis in 1920 about whether certain nebulae were beyond the Milky Way.

shepherd satellite (moon) A small satellite whose gravitational tug is responsible for maintaining a sharply defined ring of matter around a planet such as Saturn or Uranus.

shock wave An abrupt, localized region of compressed gas caused by an object traveling through the gas at a speed greater than the speed of sound.

shooting star *See* **meteor.**

short-period comet A comet that orbits the Sun in the vicinity of the planets, thereby reappearing with tails every 200 years or less.

sidereal month The period of the Moon's revolution about the Earth measured with respect to the Moon's location among the stars; 27⅓ Earth days.

sidereal period The orbital period of one object about another measured with respect to the stars.

single-line spectroscopic binary A spectroscopic binary whose periodically varying spectrum exhibits the spectral lines of only one of its two stars.

singularity A place of infinite curvature of spacetime in a black hole.

Small Magellanic Cloud (SMC) An irregular galaxy that is a companion to the Milky Way.

solar corona The Sun's outer atmosphere.

solar cycle A 22-year cycle during which the Sun's magnetic field reverses its polarity twice.

solar day From noontime to the next noontime; for Earth it is 24 hours.

solar eclipse An eclipse during which the Moon blocks the Sun.

solar flare A violent eruption on the Sun's surface.

solar luminosity ($L_\odot$) The total energy emitted by the Sun each second.

solar model A set of equations that describe the internal structure and energy generation of the Sun.

solar nebula The cloud of gas and dust from which the Sun and the rest of the solar system formed.

solar seismology The study of the Sun's interior from observations of vibrations of its surface.

solar system The Sun, planets, their satellites, asteroids, comets, and related objects that orbit the Sun.

solar wind An outward flow of particles (mostly electrons and protons) from the Sun.

solstice Either of two points along the ecliptic at which the Sun reaches its maximum distance north or south of the celestial equator.

south celestial pole The location on the celestial sphere directly above the Earth's south rotation pole.

southern highlands Older, cratered terrain in the Martian southern hemisphere.

southern lights (aurora australis) Light radiated by atoms and ions in the Earth's upper atmosphere due to high-energy particles from the Sun; seen mostly in the southern polar regions.

Southern Wall A huge sheet of galaxies between two voids in the cosmos.

spacetime The concept from special relativity that space and time are both essential in describing the position, motion, and action of any object or event.

special theory of relativity A description of mechanics and electromagnetic theory formulated by Einstein according to which measurements of distance, time, and mass are affected by the observer's motion.

spectral analysis The identification of chemicals by the appearance of their spectra.

spectral lines Dark or bright lines at specific wavelengths in a spectrum.

spectral type A classification of stars according to the appearance of their spectra.

spectrogram The photograph of a spectrum.

spectrograph A device for photographing a spectrum.

spectroscope A device for directly viewing a spectrum.

spectroscopic binary A double star whose binary nature can be deduced from the periodic Doppler shifting of lines in its spectrum.

spectroscopic parallax A method of determining a star's distance from the Earth by measuring its surface temperature, luminosity, and apparent magnitude.

spectroscopy The study of spectra.

spectrum (*plural* **spectra**) The result of electromagnetic radiation passing through a prism or grating so that different wavelengths are separated.

speed The rate at which an object moves.

spherical aberration An optical property whereby different portions of a spherical lens or spherical, concave mirror have slightly different focal lengths, thereby producing a fuzzy image.

spicule A narrow jet of rising gas in the solar chromosphere.

spin (of an electron or proton) A small, well-defined amount of angular momentum possessed by electrons, protons, and other particles.

spiral arms Lanes of interstellar gas, dust, and young stars that wind outward in a plane from the central regions of some galaxies.

spiral density wave A spiral-shaped pressure wave that orbits the disk of a spiral galaxy and induces new star formation.

spiral galaxy A flattened, rotating galaxy with pinwheel-like spiral arms winding outward from the galaxy's nuclear bulge.

spoke A moving dark region of Saturn's rings.

spring tide The greatest daily difference between high tide and low tide, occurring when the Moon is new or full.

stable Lagrange points Locations throughout the solar system where the gravitational forces from the Sun and a planet keep space debris trapped.

standard candle An object whose known luminosity can be used to deduce the distance to a galaxy.

star A self-luminous sphere of gas.

starburst galaxy A galaxy where there is an exceptionally high rate of star formation.

Stefan-Boltzmann law The relationship stating that an object emits energy at a rate proportional to the fourth power of its temperature, in Kelvins.

stellar evolution The changes in size, luminosity, temperature, and chemical composition that occur as a star ages.

stellar model The result of theoretical calculations that give details of physical conditions inside a star.

stellar parallax The apparent shift in a nearby star's position on the celestial sphere resulting from the Earth's orbit around the Sun.

stellar spectroscopy The study of the properties of stars encoded in their spectra.

stony meteorite A meteorite composed of rock with very little iron; also called a *stone.*

stony-iron meteorite A meteorite composed of roughly equal amounts of rock and iron.

stratosphere The second layer in the Earth's atmosphere, directly above the troposphere.

strong nuclear force The force that binds protons and neutrons together in nuclei.

subgiant A star whose luminosity is between that of main-sequence stars and normal giants of the same spectral type.

summer solstice The point on the ecliptic where the Sun is farthest north of the celestial equator; the day with the largest number of daylight hours in the northern hemisphere, around June 21.

Sun The star about which the Earth and other planets revolve.

sunspot A temporary cool region in the solar photosphere created by protruding magnetic fields.

sunspot cycle The semiregular 11-year period with which the number and location of sunspots fluctuate.

sunspot maximum The time during the solar cycle when the number of sunspots is greatest.

sunspot minimum The time during the solar cycle when the number of sunspots is minimum.

supercluster (of galaxies) A gravitationally bound collection of many clusters of galaxies.

supergiant A star of very high luminosity.

supergranule A large convective cell in the Sun's chromosphere containing many granules.

superior conjunction The configuration when a planet is behind the Sun as seen from Earth.

supermassive black hole A black hole whose mass exceeds 1000 solar masses.

supernova (*plural* **supernovae**) A stellar explosion during which a star suddenly increases its brightness roughly a millionfold.

supernova remnant A nebula left over after a supernova detonates.

superstring A set of theories that hope to describe the nature of spacetime and matter at a more fundamental level than is presently possible.

synchronous rotation The condition when a moon's rotation rate and revolution rate are equal or when a planet's rotation rate equals its moon's revolution rate.

synchrotron radiation The radiation emitted by charged particles moving through a magnetic field; nonthermal radiation.

synodic month (lunar month) The period of revolution of the Moon with respect to the Sun; the length of one cycle of lunar phases; 29½ Earth days.

synodic period The interval between successive occurrences of the same configuration of a planet as seen from Earth.

T Tauri stars Young, variable pre–main-sequence stars associated with interstellar matter that show erratic changes in luminosity.

tail (of a comet) Gas and dust particles from a comet's nucleus that have been swept away from the comet's nucleus by the radiation pressure of sunlight and impact of the solar wind.

telescope An instrument for viewing remote objects.

temperature (Celsius) Temperature measured on a scale where water freezes at 0° and boils at 100°.

temperature (Fahrenheit) Temperature measured on a scale where water freezes at 32° and boils at 212°.

temperature (Kelvin) Absolute temperature measured in Celsius degree intervals. Water freezes at 273 K and boils at 373 K.

terminator The line dividing day and night on the surface of any body orbiting the Sun; the line of sunset or sunrise.

terrestrial planet Any of the planets Mercury, Venus, Earth, or Mars; a planet with a composition and density similar to that of Earth.

theory *See* **scientific theory**.

thermal energy The energy associated with the motions of atoms or molecules in a substance.

thermal equilibrium A balance between the input and outflow of heat in a system.

thermonuclear fusion A reaction in which the nuclei of atoms are fused together at a high temperature.

3-to-2 spin-orbit coupling The rotation of Mercury, which makes three complete rotations on its axis for every two complete orbits around the Sun.

tidal force A gravitational force whose strength and/or direction varies over a body and thus tends to deform the body.

time zone One of 24 divisions of the Earth's surface separated by 15° along lines of constant longitude (with allowances for some political boundaries).

total eclipse A solar eclipse during which the Sun is completely hidden by the Moon, or a lunar eclipse during which the Moon is completely immersed in the Earth's umbra.

trailing-arm spiral A spiral arm pointing away from the direction of rotation, characteristic of all spiral galaxies.

transition (electronic) The change in energy and orbit of an electron around an atom or molecule.

transition zone Region between the Sun's chromosphere and corona where the temperature skyrockets to about 1 million K.

Trojan asteroid One of several asteroids at stable Lagrange points that share Jupiter's orbit about the Sun.

troposphere The lowest level of the Earth's atmosphere.

Tully-Fisher relation A correlation between the width of the 21-centimeter line of a spiral galaxy and its absolute magnitude.

turnoff point The location of the brightest main-sequence stars on the Hertzsprung-Russell diagram of a globular cluster.

21-cm radiation Radio emission from a hydrogen atom caused by the flip of the electron's spin orientation.

twinkling The apparent change in a star's brightness, position, or color due to the motion of gases in the Earth's atmosphere.

Type I Cepheid A Population I Cepheid variable star found in the disks of spiral galaxies.

Type Ia supernova A supernova occurring after a white dwarf accretes enough mass from a companion star to exceed the Chandrasekhar limit.

Type II Cepheid A Population II Cepheid variable star found in elliptical galaxies and in the halos of disk galaxies that is 1.5 magnitudes dimmer than a Type I Cepheid.

Type II supernova A supernova occurring after a massive star's core is converted to iron.

ultraviolet (UV) radiation Electromagnetic radiation of wavelengths shorter than those of visible light but longer than those of X rays.

umbra The central, completely dark portion of a shadow.

universal constant of gravitation The constant of proportionality in Newton's law of gravitation, usually denoted G.

universe All space along with all the matter and radiation in space.

Van Allen radiation belts Two flattened, doughnut-shaped regions around the Earth where many charged particles (mostly protons and electrons) are trapped by the Earth's magnetic field.

variable star A star whose luminosity varies.

velocity A quantity that specifies both direction and speed of an object.

vernal equinox The point on the ecliptic where the Sun crosses the celestial equator from south to north; the beginning of spring, around March 21.

very-long-baseline interferometry (VLBI) A method of connecting widely separated radio telescopes to make observations of very high resolution.

virtual particle A particle and its antiparticle, created simultaneously in pairs and which quickly disappear without a trace.

visual binary star A double star in which the two components can be resolved through a telescope.

void A huge, roughly spherical region of the universe where exceptionally few galaxies are found.

water hole The part of the electromagnetic spectrum at a few thousand megahertz where there is very little background noise from space.

waning An adjective that means "decreasing," as in the "waning crescent Moon" or the "waning gibbous Moon."

wavelength The distance between two successive peaks in a wave.

waxing An adjective that means "increasing," as in the "waxing crescent Moon" or the "waxing gibbous Moon."

weak nuclear force A nuclear interaction involved in certain kinds of radioactive decay.

weight The force with which a body presses down on the surface of the Earth.

white dwarf A low-mass stellar remnant that has exhausted all its thermonuclear fuel and contracted to a size roughly equal to the size of the Earth.

Widmanstätten patterns Crystalline structure seen inside iron meteorites.

Wien's law The relationship that the dominant wavelength of radiation emitted by a blackbody varies inversely with its temperature.

winter solstice The point on the ecliptic where the Sun is farthest south of the celestial equator; fewest hours of daylight in the northern hemisphere, around December 22.

work Change in an object's energy as a result of a force being applied to it.

wormhole A hypothetical connecting passage between black holes and other places in the universe.

X rays Electromagnetic radiation whose wavelength is between that of ultraviolet light and gamma rays.

X-ray burster A neutron star in a binary star system that accretes mass, undergoes thermonuclear fusion on its surface, and therefore emits short bursts of X rays.

year The sidereal period of revolution of a planet around the Sun.

Zeeman effect A splitting of spectral lines in the presence of a magnetic field.

zenith The point on the celestial sphere directly overhead.

zero-age main sequence (ZAMS) The positions of stars on the Hertzsprung-Russell diagram that have just begun to fuse hydrogen in their cores.

zodiac A band of 13 constellations around the sky through which the Sun appears to move throughout the year.

zones (on Jupiter) Light-colored bands in Jupiter's cloud cover.

Chapter 1

43. 1 more sidereal month than synodic month

53. $R_\odot \approx 6.53 \times 10^8$ m

64. Sun's diameter $\approx 13.9 \times 10^9$ km, Saturn's diameter $\approx 1.21 \times 10^5$ km, Pluto's diameter $\approx 2.30 \times 10^3$ km

Toolbox 1-1: ~0.53°. This is the same angle as the Moon makes in the sky; ~0.27°; ~1.1°

Chapter 2

17. **(a)** 8.3 pc = 27.1 ly;
(b) 6.52 ly = 2.0 pc;
(c) 8459 AU = 1.26×10^{12} km;
(d) 2.7×10^3 Mpc = 2.7×10^6 kiloparsec

20. 5.2 square AU in 1995; 26 square AU in 5 years

21. a = 100 AU; maximum distance is almost 200 AU

22. 2.8 yr

23. use $P^2 = a^3$

29. At 2 AU, 1 year = 2.8 present years; at ½ AU, 1 year = .35 present years

30. .01 × present gravitational attraction

31. same acceleration; same length of the year

Toolbox 2-1: ~2.48×10^{13} mi, ~3.99×10^{13} km, ~1.29 pc

Toolbox 2-2: Energy increases 9-fold; energy decreases to ¼; 2000 joules; 0 joules; change its moment of inertia, I, and its angular velocity, v.

Toolbox 2-3: Your weight in pounds; same; 9.8 m/s² *or* 32 ft/s²; 9×10^{21} N; it is one-quarter the present force.

Chapter 3

4. 9 times more

18. 1/25

19. Palomar gathers 10^6 times more light than the human eye.

21. **(a)** 222×; **(b)** 100×; **(c)** 36×

Chapter 4

11. 7½ times

12. 238 nm

13. 6520 K

14. approaching; 13 km/s

15. receding; 21 km/s

16. about 90,000 km/s

20. λ = 500 nm (same as now), 4× brighter than now

24. **(a)** Taking the visible range to be from 700 nm to 400 nm, visible peaks occur for stars between 4143 K and 7250K; **(b)** Rigel, no; Deneb, no; Arcturus, yes; Vega, no; Betelgeuse, no

Toolbox 4-1: ~5270K; ~9700 nm; infrared

Toolbox 4-2: ~408 nm; violet; 204 nm; ~3.64×10^{-19}J

Toolbox 4-3: ⅛ kg; ~.35 kg; ~17½

Toolbox 4-4: 3×10^4 km/s *or* 0.1c; ~3.3×10^{-4}; –0.1

Chapter 9

25. 5.0×10^{14} tons

Chapter 10

5. next maximum in 2005, next minimum in 2012

16. 1400 kg/m³

17. 4.8%

18. 500 nm = visible light; 58 nm = ultraviolet; 1.9 nm = X ray

32. $\approx$ 25d, equator

Toolbox 10-1: 6×10^{17} helium atoms; 2.25×10^{14} J

Chapter 11

15. 25 times brighter

29. 4.3 pc

31. **(a)** 9.7 pc;
(b) 0.10 arcsec

32. 1585 times brighter

35. **(a)** $\approx 10\ M_\odot$;
(b) $\approx 10^{-3}\ L_\odot$

36. $\approx$ 10,000–15,000K

45. **(a)** Betelgeuse $\approx$ 131.1 pc, 427.4 ly;
(b) Vega $\approx$ 7.752 pc, 25.27 ly;
(c) Antares $\approx$ 185.2 pc, 603.8 ly;
(d) Sirius $\approx$ 2.639 pc, 8.603 ly

Toolbox 11-1: ~4.85×10^{-6} pc; 3.26×10^{-2}″; ~2.6 pc *or* ~8.47 ly; ~237 pc; ~4.22×10^{-3}″

Toolbox 11-2: 40; 631; 5.5

Toolbox 11-3: ~11.4 pc, ~37.1 ly; +1.74 (Fomalhaut); –1.44 (Sirius A)

Toolbox 11-4: We can ignore M_2 (planet's mass) as tiny compared to the Sun's mass, M_1. Since the Sun's mass in these units is 1, the equation becomes $1 = a^3/P^2$, or $P^2 = a^3$; ~1.56 $M_\odot$; 73 years

Chapter 12

29. 2000

30. 200 times longer

Chapter 13

28. about 7470 years ago

Chapter 14

11. 8.9 km, 89 km

15. 10 m

Toolbox 13-1: ~1.0×10^{12} kg ; 9×10^{9} km; ~2130 $M_{\odot}$

Chapter 15

14. 20 times

19. about once every 3750 years

Toolbox 15-1: 1×10^{5} pc; ~6.92; m = 9.83; m – M = 5

Chapter 16

21. 306 Mpc

22. 7200 km/s

Toolbox 16-1: v = 1.74×105 km/s; d = 3.0 Mpc; slower

Appendix A

3.141×10^{9}, 3.1831×10^{-9};
27182820000, 0.0000000000367879

Appendix B

580 K, 55 km, temperature decreases with height; +8, +10, September 1, September 6; 8$L_{\odot}$, 6000 K

Appendix D

5527°C, 9980°F; –270°C, –454°F

Page numbers in *italics* indicate figures.

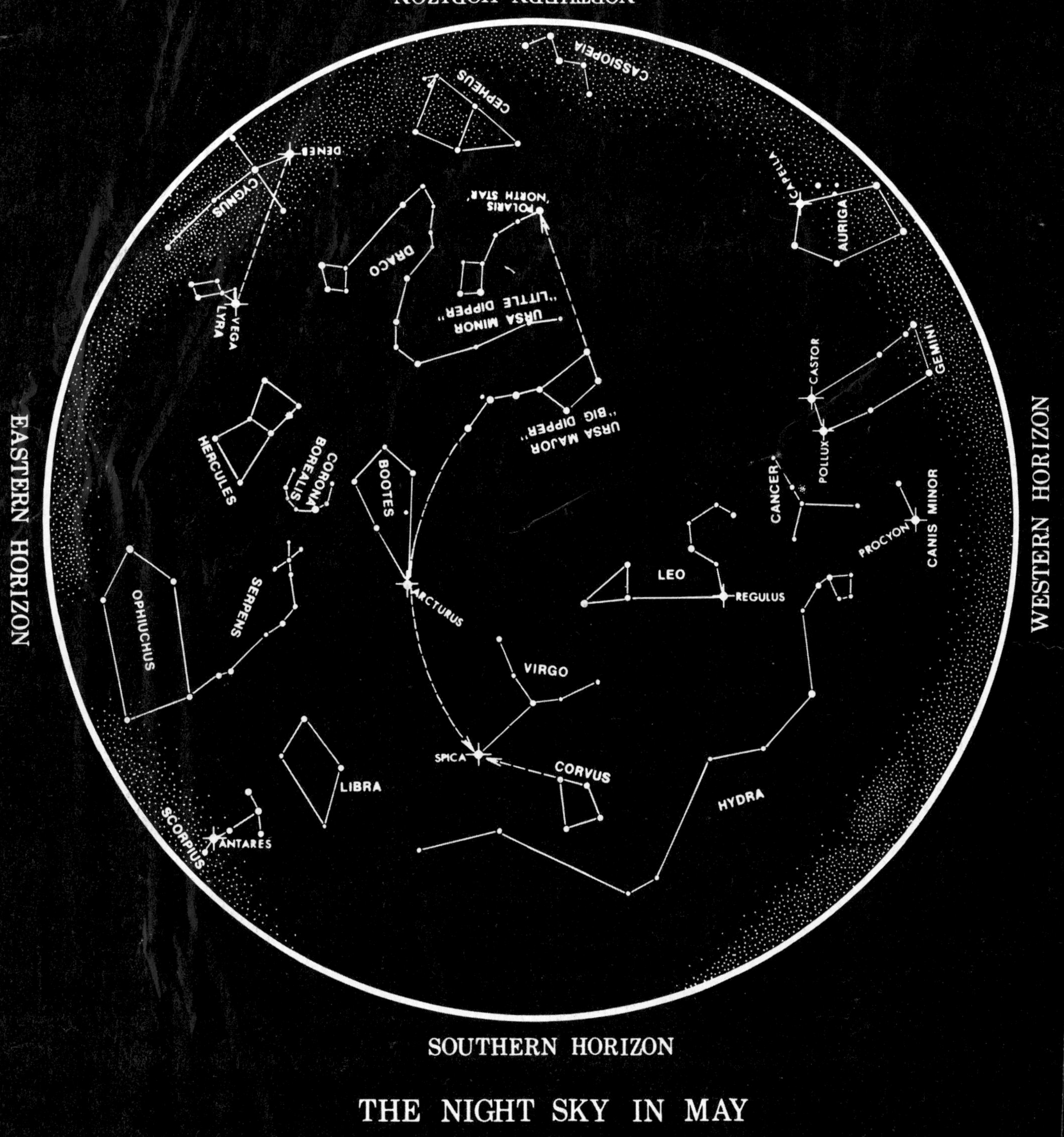

THE NIGHT SKY IN MAY

Chart time (Daylight Savings Time):

11 pm . . . First of May
10 pm . . . Middle of May
9 pm . . . Last of May

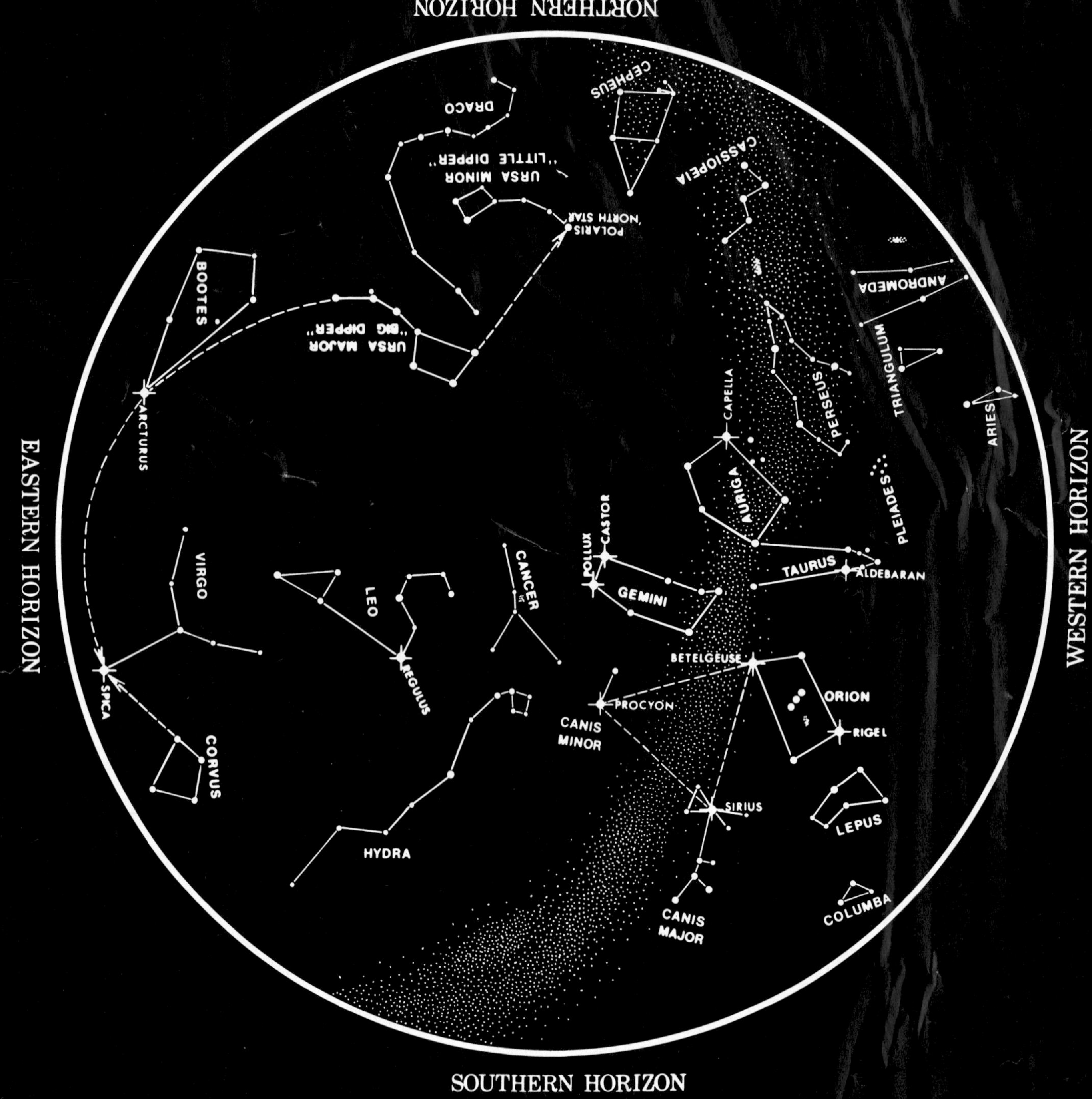

THE NIGHT SKY IN MARCH

Chart time (Local Standard Time):

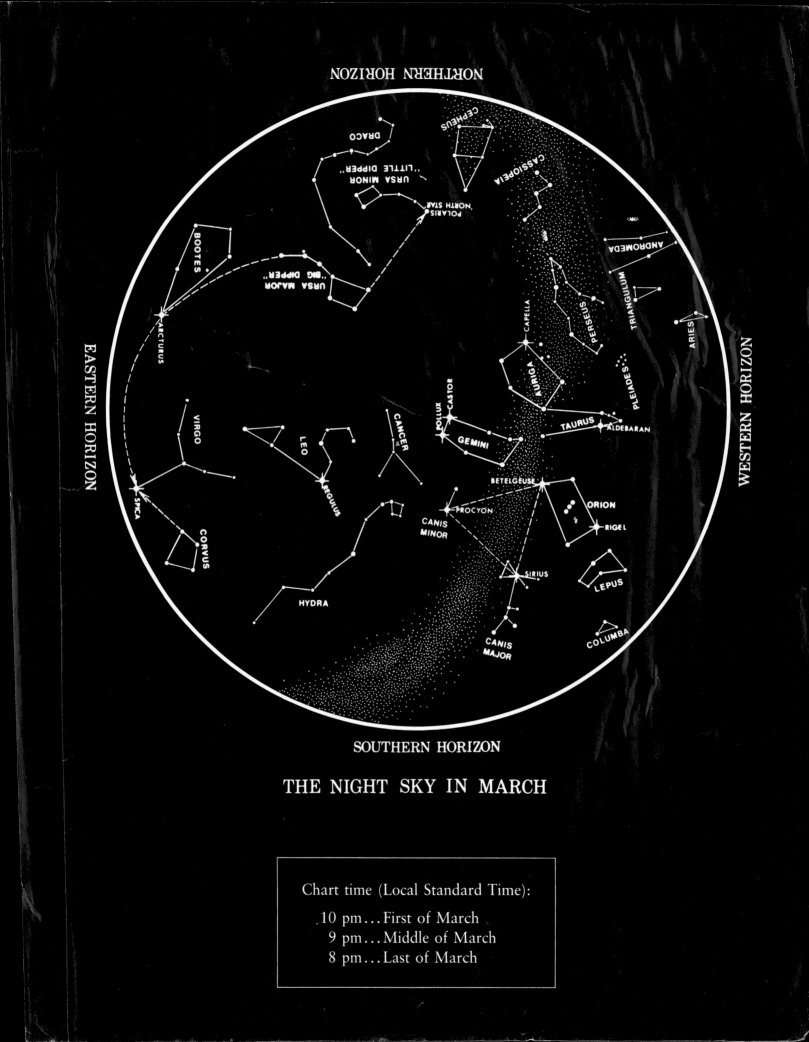

10 pm . . . First of March
9 pm . . . Middle of March
8 pm . . . Last of March